STUDY SMARTER

CHAPTER Test Prep VIDEO CD

Step-by-step solutions on video for all chapter test exercises from the text

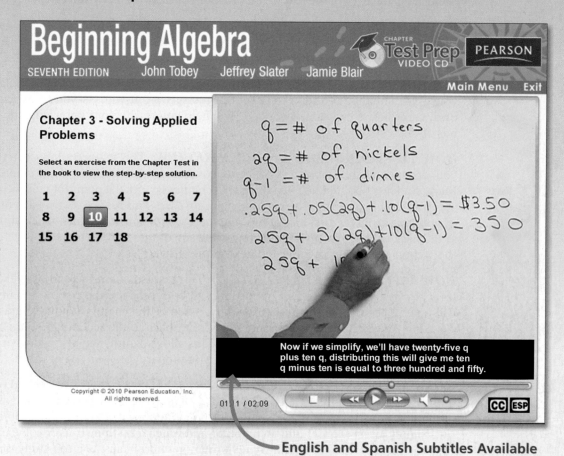

English and Spanish Subtitles Available

INCLUDED WITH EVERY NEW COPY OF THIS TEXTBOOK!

To the Student

With this edition of the Tobey/Slater/Blair Developmental Mathematics series, we are committed to helping you get the most out of your learning experience by showing you how important math can be in your daily lives. For example, the new *Use Math to Save Money* feature presents practical examples of how math can help you save money, cut costs, and spend less. One of the best ways you can save money is to pass your course the first time you take it. This text and its features and MyMathLab can help you do that.

The best place to start is the *How Am I Doing? Guide to Math Success.* This clear path for you to follow is based upon how our successful students have utilized the textbook in the past. Here is how it works:

EXAMPLES and PRACTICE PROBLEMS: When you study an Example, you should immediately do the Practice Problem that follows to make sure you understand each step in solving a particular problem. The worked-out solution to every Practice Problem can be found in the back of the text, starting at page SP-1, so you can check your work and receive immediate guidance in case you need to review.

EXERCISE SETS—Practice, Practice, Practice: You learn math by *doing* math. The best way to learn math is to *practice, practice, practice.* Be sure that you complete every exercise your instructor assigns as homework. In addition, check your answers to the odd-numbered exercises in the back of the text to see whether you have correctly solved each problem.

QUICK QUIZ: After every exercise set, be sure to do the problems in the Quick Quiz. This will tell you immediately if you have understood the key points of the homework exercises.

CONCEPT CHECK: At the end of the Quick Quiz is a Concept Check. This will test your understanding of the key concept of the section. It will ask you to explain in your own words how a procedure works. It will help you clarify your knowledge of the methods of the section.

HOW AM I DOING? MID-CHAPTER REVIEW: This feature allows you to check if you understand the important concepts covered to that point in a particular chapter. Many students find that halfway through a chapter is a crucial point for review because so many different types of problems have been covered. This review covers each of the types of problems from the first half of the chapter. Do these problems and check your answers at the back of the text. If you need to review any of these problems, simply refer back to the section and objective indicated next to the answer.

HOW AM I DOING? CHAPTER TEST: This test (found at the end of every chapter) provides you with an excellent opportunity to both practice and review for any test you will take in class. Take this test to see how much of the chapter you have mastered. By checking your answers, you can once again refer back to the section and the objective of any exercise you want to review further. This allows you to see at once what has been learned and what still needs more study as you prepare for your test or exam.

HOW AM I DOING? CHAPTER TEST PREP VIDEO CD: If you need to review any of the exercises from the *How Am I Doing? Chapter Test,* this video CD found at the front of the text provides a clear explanation of how to do each step of every problem on the test. Simply insert the CD into a computer and watch a math instructor solve each of the Chapter Test exercises in detail. By reviewing these problems, you can study through any points of difficulty and better prepare yourself for your upcoming test or exam.

These steps provide a clear path you can follow in order to successfully complete your math course. More importantly, the *How Am I Doing? Guide to Math Success* is a tool to help you achieve an understanding of mathematics. We encourage you to take advantage of this new feature.

John Tobey and Jeffrey Slater
North Shore Community College

Annotated Instructor's Edition

7th Edition

Beginning Algebra

John Tobey

North Shore Community College
Danvers, Massachusetts

Jeffrey Slater

North Shore Community College
Danvers, Massachusetts

With Contributions from Jennifer Crawford

Prentice Hall
is an imprint of

Upper Saddle River, NJ 07458

Editorial Director, Mathematics: *Christine Hoag*
Editor in Chief: *Paul Murphy*
Executive Project Manager: *Kari Heen*
Senior Project Editor: *Lauren Morse*
Assistant Editors: *Georgina Brown and Christine Whitlock*
Production Management: *Elm Street Publishing Services*
Senior Managing Editor: *Linda Mihatov Behrens*
Operations Specialist: *Ilene Kahn*
Senior Operations Supervisor: *Diane Peirano*
Marketing Manager: *Marlana Voerster*
Marketing Assistant: *Nathaniel Koven*
Art Director: *Heather Scott*
Interior/Cover Designer: *Tamara Newnam*
AV Project Manager: *Thomas Benfatti*
Executive Manager, Course Production: *Peter Silvia*
Media Producer: *Audra J. Walsh*
Associate Producer: *Emilia Yeh*
Manager, Content Development: *Rebecca Williams*
QA Manager: *Marty Wright*
Senior Content Developer: *Mary Durnwald*
Photo Research Development Manager: *Elaine Soares*
Image Permission Coordinator: *Kathy Gavilanes*
Photo Researcher: *Michael Downey*
Manager, Cover Visual Research and Permissions: *Karen Sanatar*
Cover Image: *Ryan McVay/Photodisc/Getty Images, Inc.*
Compositor: *Macmillan Publishing Solutions*
Art Studios: *Scientific Illustrators and Laserwords*

Photo credits appear on page P-1, which constitutes a continuation of the copyright page.

Prentice Hall
is an imprint of

© 2010, 2006, 2002, 1998, 1995, 1991, 1984 by Pearson Education, Inc.
Pearson Prentice Hall
Pearson Education, Inc.
Upper Saddle River, New Jersey 07458

Printed in the United States of America

10 9 8 7 6 5 4 3 2 1

ISBN 10: 0-321-57377-3 (Annotated Instructor's Edition)
ISBN 13: 978-0-321-57377-3 (Annotated Instructor's Edition)

ISBN 10: 0-321-57375-7 (Student Edition)
ISBN 13: 978-0-321-57375-9 (Student Edition)

Pearson Education Ltd., London
Pearson Education Singapore, Pte. Ltd.
Pearson Education Canada, Inc.
Pearson Education, Japan
Pearson Education Australia PTY, Limited

Pearson Education North Asia, Ltd., Hong Kong
Pearson Educación de Mexico, S.A. de C.V.
Pearson Education Malaysia, Pte. Ltd.
Pearson Education Upper Saddle River,
    New Jersey

This book is dedicated to the memory of Lexie Tobey and John Tobey, Sr.
They have left a legacy of love, a memory of four decades of faithful teaching,
and a sense of helping others that will influence generations to come.
For their grandchildren they have left an inspiring model of a loving family,
true character, and service to God and community.

Contents

CHAPTER 0

A Brief Review of Arithmetic Skills   1

CHAPTER 1

Real Numbers and Variables   67

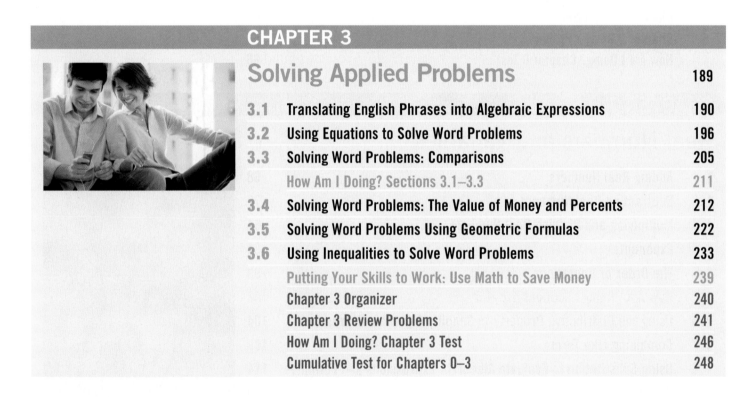

CHAPTER 4

Exponents and Polynomials   249

CHAPTER 5

Factoring   301

CHAPTER 8

Systems of Equations   470

CHAPTER 9

Radicals   515

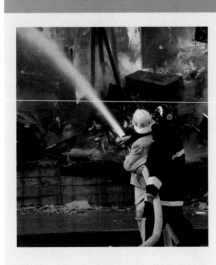

Preface

TO THE INSTRUCTOR

One of the hallmark characteristics of *Beginning Algebra* that makes the text easy to learn and teach from is the building-block organization. Each section is written to stand on its own, and every homework set is completely self-testing. Exercises are paired and graded and are of varying levels and types to ensure that all skills and concepts are covered. As a result, the text offers students an effective and proven learning program suitable for a variety of course formats—including lecture-based classes; discussion-oriented classes; distance learning classes; modular, self-paced courses; mathematics laboratories; and computer-supported centers. The book has been written to be especially helpful in online courses. The authors usually teach at least one course online each semester.

   Beginning Algebra is part of a series that includes the following:

Tobey/Slater/Blair, *Basic College Mathematics,* Sixth Edition

Tobey/Slater/Blair, *Essentials of Basic College Mathematics,* Second Edition

Blair/Tobey/Slater, *Prealgebra,* Fourth Edition

Tobey/Slater, *Beginning Algebra,* Seventh Edition

Tobey/Slater/Blair, *Beginning Algebra: Early Graphing,* Second Edition

Tobey/Slater, *Intermediate Algebra,* Sixth Edition

Tobey/Slater/Blair, *Beginning and Intermediate Algebra,* Third Edition

We have visited and listened to teachers across the country and have incorporated a number of suggestions into this edition to help you with the particular learning delivery system at your school. The following pages describe the key continuing features and changes in the seventh edition.

NEW! FEATURES IN THE SEVENTH EDITION

Quick Quiz

At the end of each problem section, there is a **Quick Quiz** for the student. The quiz contains three problems that cover all of the essential content of that entire section of the book. If a student can do those three problems, the student has mastered the mathematics skills of that section. If a student cannot do those three problems, the student is made aware that further study is needed to obtain mastery. At the end of the Quick Quiz is a **Concept Check** question. This question stresses a mastery of the concepts of each section of the book. The question asks the student to explain how and why a solution method actually works. It forces the student to analyze problems and reflect on the mathematical concepts that have been learned. The student is asked to explain in his or her own words the mathematical procedures that have been practiced in a given section.

Classroom Quiz

Adjacent to each Quick Quiz for the student, the Annotated Instructor's Edition of the book contains a **Classroom Quiz.** This quiz allows the instructor to give a short quiz in class that covers the essential content of each section of the book. Immediately, an instructor can find out if the students have mastered the material in this section or not. No more will an instructor have to rush to pick out the right balance of problems that test a student's knowledge of a section. The Classroom Quiz is instantly available to assist the instructor in assessing student knowledge.

Putting Your Skills to Work: Use Math to Save Money

Each chapter of the book presents a simple, down-to-earth, practical example of how to save money. Students are given a straightforward, realistic way to cut costs and spend less. They are shown logical ways to get out of debt. They are given motivating examples of how other students saved money. Students are very motivated to read these articles and soon begin to see how the course will actually help them in everyday life. Many of these activities were contributed by professors, and our thanks go out to Mary Pearce, Suellen Robinson, Mike Yarbrough, Betty Ludlum, Connie Buller, Armando Perez, and Maria Luisa Mendez for their insightful contributions.

Teaching Examples

Having in-class practice problems readily available is extremely helpful to both new and experienced instructors. These instructor examples, called **Teaching Examples,** are included in the margins of the Annotated Instructor's Edition for practice in class.

KEY FEATURES IN THE SEVENTH EDITION

The **How Am I Doing? Guide to Math Success** shows students how they can effectively use this textbook to succeed in their mathematics course. This clear path for them to follow is based upon how students have successfully utilized this textbook in the past.

The text has been designed so the **Examples and Practice Problems** are clearly connected in a cohesive unit. This encourages students to try the Practice Problem associated with each Example to ensure they understand each step in solving a particular problem. The worked-out solution to every Practice Problem can be found in the back of the text so students can check their work. **New!** to this edition, the answers are now included next to each Practice Problem in the Annotated Instructor's Edition.

Each **exercise set** progresses from easy to medium to challenging problems, with appropriate quantities of each, and has paired even and odd problems. All concepts are fully represented, with every Example from the section covered by a group of exercises. Exercise sets include **Mixed Practice** problems, which require students to identify the type of problem and the best method they should use to solve it, as well as **Verbal and Writing Skills** exercises, which allow students to explain new concepts fully in their own words. Throughout the text the application exercises have been updated. These **Applications** relate to everyday life, global issues beyond the borders of the United States, and other academic disciplines. Roughly 25 percent of the applications have been contributed by actual students based on scenarios they have encountered in their home or work lives.

Many students find that halfway through a chapter is a crucial point for review because so many different types of problems have been covered. The **How Am I Doing? Mid-Chapter Review** covers each of the types of problems from the first half of the chapter and allows instructors to check if students understand the important concepts covered to that point. Specific section and objective references are provided with each answer to indicate where a student should look for further review.

Developing Problem-Solving Abilities

As authors, we are committed to producing a textbook that emphasizes mathematical reasoning and problem-solving techniques as recommended by AMATYC, NCTM, AMS, NADE, MAA, and other bodies. To this end, the problem sets are built on a wealth of real-life and real-data applications. Unique problems have been developed and incorporated into the exercise sets that help train students in data interpretation, mental mathematics, estimation, geometry and graphing, number sense, critical thinking, and decision making.

The successful **Mathematics Blueprint for Problem Solving** strengthens problem-solving skills by providing a consistent and interactive outline to help students organize their approach to problem solving. Once students fill in the blueprint, they can refer back to their plan as they do what is needed to solve the problem. Because of its flexibility, this feature can be used with single-step problems, multi-step problems, applications, and nonroutine problems that require problem-solving strategies.

The **Developing Your Study Skills** boxes are integrated throughout the text to provide students with techniques for improving their study skills and succeeding in math courses.

Integration and Emphasis on Geometry Due to the emphasis on geometry on many statewide exams, geometry problems are integrated throughout the text. Examples and exercises that incorporate a principle of geometry are marked with a triangle icon for easy identification.

When students encounter mathematics in real-world publications, they often encounter data represented in a **graph, chart, or table** and are asked to make a reasonable conclusion based on the data presented. This emphasis on graphical interpretation is a continuing trend with today's expanding technology. In this text, students are asked to make simple interpretations, to solve medium-level problems, and to investigate challenging applied problems based on the data shown in a chart, graph, or table.

Mastering Mathematical Concepts

Text features that develop the mastery of concepts include the following:

Concise **Learning Objectives** listed at the beginning of each section allow students to preview the goals of that section.

To Think About questions extend the concept being taught, providing the opportunity for all students to stretch their minds, to look for patterns, and to make conclusions based on their previous experience. These critical-thinking questions may follow Examples in the text and appear in the exercise sets.

Almost every exercise set concludes with a section of **Cumulative Review** problems. These problems review topics previously covered, and are designed to assist students in retaining the material.

Calculator boxes are placed in the margin of the text to alert students to a scientific calculator application. In the exercise section a scientific calculator icon is used to indicate problems that are designed for solving with a calculator. There is also instruction on how to use a scientific calculator in an appendix.

Reviewing Mathematical Concepts

At the end of each chapter, we have included problems and tests to provide your students with several different formats to help them review and reinforce the ideas that they have learned. This assists them not only with that specific chapter, but reviews previously covered topics as well.

The concepts and mathematical procedures covered are reviewed at the end of each chapter in a unique **Chapter Organizer.** It lists concepts and methods, and provides a completely worked-out example for each type of problem.

Chapter Review Problems are grouped by section as a quick refresher at the end of the chapter. They can also be used by the student as a quiz of the chapter material.

Found at the end of the chapter, the **How Am I Doing? Chapter Test** is a representative review of the material from that particular chapter that simulates an actual testing format. This provides the students with a gauge of their preparedness for the actual examination.

At the end of each chapter is a **Cumulative Test.** One-half of the content of each cumulative test is based on the math skills learned in previous chapters. By completing these tests for each chapter, the students build confidence that they have mastered not only the contents of the chapter but those of previous chapters as well.

RESOURCES FOR THE STUDENT

Student Solutions Manual
(ISBNs: 0-321-57376-5, 978-0-321-57376-6)

- Solutions to all odd-numbered section exercises
- Solutions to every exercise (even and odd) in the Quick Quiz, mid-chapter reviews, chapter reviews, chapter tests, and cumulative reviews

Worksheets for Classroom or Lab Practice
(ISBNs: 0-321-58950-5, 978-0-321-58950-7)

- Extra practice exercises for every section of the text with ample space for students to show their work

Chapter Test Prep Video CD

Provides step-by-step video solutions to each problem in each How Am I Doing? Chapter Test in the textbook. Automatically included with every new copy of the text, inside the front cover.

Lecture Series on DVD
(ISBNs: 0-321-59385-5, 978-0-321-59385-6)

- Organized by section, contain problem-solving techniques and examples from the textbook
- Step-by-step solutions to selected exercises from each textbook section

MathXL® Tutorials on CD
(ISBNs: 0-321-58952-1, 978-0-321-58952-1)
This interactive tutorial CD-ROM provides:

- Algorithmically generated practice exercises correlated at the objective level
- Practice exercises accompanied by an example and a guided solution
- Tutorial video clips within the exercise to help students visualize concepts
- Easy-to-use tracking of student activity and scores and printed summaries of students' progress

RESOURCES FOR THE INSTRUCTOR

Annotated Instructor's Edition
(ISBNs: 0-321-57377-3, 978-0-321-57377-3)

- Complete student text with answers to all practice problems, section exercises, mid-chapter reviews, chapter reviews, chapter tests, cumulative tests, and practice final exam.
- Teaching Tips placed in the margin at key points where students historically need extra help
- New! Teaching Examples provide in-class practice problems and are placed in the margins accompanying each example.

Instructor's Solutions Manual
(ISBNs: 0-321-57812-0, 978-0-321-57812-9)

- Detailed step-by-step solutions to the even-numbered section exercises
- Solutions to every exercise (odd and even) in the Classroom Quiz, mid-chapter reviews, chapter reviews, chapter tests, cumulative tests, and practice final

Instructor's Resource Manual with Tests and Mini-Lectures
(ISBNs: 0-321-57813-9, 978-0-321-57813-6)

- For each section there is one Mini-Lecture with key learning objectives, classroom examples, and teaching notes.

- Two short group activities per chapter are provided in a convenient ready-to-use handout format.
- Three forms of additional practice exercises that help instructors support students of different ability and skill levels.
- Answers are included for all items.
- Alternate test forms with answers:
  - Two Chapter Pretests per chapter (1 free response, 1 multiple choice)
  - Six Chapter Tests per chapter (3 free response, 3 multiple choice)
  - Two Cumulative Tests per even-numbered chapter (1 free response, 1 multiple choice)
  - Two Final Exams (1 free response, 1 multiple choice)

TestGen®

- Enables instructors to build, edit, print, and administer tests.
- Features a computerized bank of questions developed to cover all text objectives.
- Creates multiple but equivalent versions of the same question or test with the click of a button.
- Instructors can modify questions or add new questions.
- Tests can be printed or administered online.

The software and testbank are available for download from Pearson Education's online catalog.

Pearson Adjunct Support Center

The Pearson Adjunct Support Center is staffed by qualified mathematics instructors with more than 50 years of combined experience at both the community college and university level. Assistance is provided for faculty in the following areas:

- Suggested syllabus consultation
- Tips on using materials packed with your book
- Book-specific content assistance
- Teaching suggestions including advice on classroom strategies

MEDIA RESOURCES

MathXL® www.mathxl.com

MathXL is a powerful online homework, tutorial, and assessment system that accompanies Pearson Education textbooks in mathematics and statistics. With MathXL, instructors can create, edit, and assign online homework and tests using algorithmically generated exercises correlated at the objective level to the textbook. They can also create and assign their own online exercises and import TestGen tests for added flexibility. All student work is tracked in MathXL's online gradebook. Students can take chapter tests in MathXL and receive personalized study plans based on their test results. The study plan diagnoses weaknesses and links students directly to tutorial exercises for the objectives they need to study and retest. Students can also access supplemental animations and video clips directly from selected exercises. MathXL is available to qualified adopters. For more information, visit our Web site at www.mathxl.com, or contact your sales representative.

MyMathLab® www.mymathlab.com

MyMathLab is a series of text-specific, easily customizable online courses for Pearson Education textbooks in mathematics and statistics. Powered by CourseCompass™ (our online teaching and learning environment) and MathXL® (our online homework, tutorial, and assessment system), MyMathLab gives instructors the tools they need to deliver all or a portion of their course online, whether students are in a lab or working at home or elsewhere. MyMathLab provides a rich and flexible set of course materials, featuring free-response exercises that are algorithmically generated for unlimited practice and mastery. Students can also use online tools, such as video lectures, animations, and a multimedia textbook, to independently improve their understanding and performance. Instructors can use MyMathLab's homework and test managers to select and assign online exercises correlated directly to the textbook, and they can create and assign their own online exercises and import TestGen tests for added flexibility. MyMathLab's online gradebook—designed specifically for mathematics and statistics—automatically tracks students' homework and test results and gives the instructor control over how to calculate final grades. Instructors can also add offline (paper-and-pencil) grades to the gradebook. MyMathLab also includes access to the Pearson Tutor Center, which provides students with tutoring via toll-free phone, fax, e-mail, and interactive Web sessions. MyMathLab is available to qualified adopters. For more information, visit our Web site at www.mymathlab.com, or contact your sales representative.

ACKNOWLEDGMENTS

This book is the product of many years of work and many contributions from faculty and students across the country. We would like to thank the many reviewers and participants in focus groups and special meetings with the authors in preparation of previous editions. Our deep appreciation to each of the following:

Carla Ainsworth, *Salt Lake Community College*

George J. Apostolopoulos, *DeVry Institute of Technology*

Mary Lou Baker, *Columbia State Community College*

Jana Barnard, *Angelo State University*

Katherine Barringer, *Central Virginia Community College*

Linda Beattie, *Western New Mexico University*

Mark Billiris, *St. Petersburg Junior College*

Jamie Blair, *Orange Coast College*

Larry Blevins, *Tyler Junior College*

Lydia Botsford, *Heald Business College, Milpitas*

Bonnie Brooks, *Sussex County Community College*

David Brown, *Heald Business College, Vacaville*

John Buckley, *Delaware Technical and Community College*

Robert Christie, *Miami-Dade Community College*

Gene Coco, *Heald Business College, Roseville*

Nelson Collins, *Joliet Junior College*

Mike Contino, *California State University at Heyward*

Kevin Cooper, *National American University*

Gregory Daubenmire, *Las Positas College*

Judy Dechene, *Fitchburg State University*

Floyd L. Downs, *Arizona State University*

Robert Dubuc, Jr., *New England Institute of Technology*

Barbara Edwards, *Portland State University*

Mahmoud El-Hashash, *Bridgewater State College*

Janice F. Gahan-Rech, *University of Nebraska at Omaha*

Naomi Gibbs, *Pitt Community College*

Mark Glines, *Salt Lake Community College*

Cynthia Gubitose, *Southern Connecticut State University*

Renu Gupta, *Louisiana State University, Alexandria*

Philip Hammersley, *Ivy Tech Community College*

Mary Beth Headlee, *Manatee Community College*

Autumn Hoover, *Angelo State University*

Laura Kaufmann, *Orange Coast College*

Carolyn Krause, *Delaware Technical and Community College*

Euniz Lochte, *Florida Metropolitan University, Orlando North*

Elizabeth Lucas, *North Shore Community College*

Jim Lynn, *Florida Metropolitan University, Orlando North*

Doug Mace, *Baker College*

Valerie Maley, *Cape Fear Community College*

Carl Mancuso, *William Paterson College*

James Matovina, *Community College of Southern Nevada*

Todd Mattson, *DeVry University, DuPage Campus*

Jim McKee, *Florida Metropolitan University, Tampa*

Janet McLaughlin, *Montclair State College*

Beverly Meyers, *Jefferson College*

Nancy Meyers, *University of Southern Indiana*

Wayne L. Miller, *Lee College*

Gloria Mills, *Tarrant County Junior College*

Norman Mittman, *Northeastern Illinois University*

Kathy Morgan, *Coastal Carolina Community College*

Christina Morian, *Lincoln University*

Sharon L. Morrison, *St. Petersburg Junior College*

Joe Nagengast, *Florida Career College*

Charlcie Neal, *Clayton College and State University*

Jim Osborn, *Baker College*

Linda Padilla, *Joliet Junior College*

Cathy Panik, *Manatee Community College*

Elizabeth A. Polen, *County College of Morris*

Larry Pontaski, *Pueblo Community College*

Joel Rappaport, *Miami-Dade Community College*

Juan Rivera, *Florida Metropolitan University, Orlando South*

Graciela Rodriguez, *Laredo Community College*

Ronald Ruemmler, *Middlesex County College*

Dennis Runde, *Manatee Community College*

Sally Search, *Tallahassee Community College*

Jeff Simmons, *Ivy Tech Community College*

Carolyn Gigi Smith, *Armstrong Atlantic State University*

Jed Soifer, *Atlantic Cape Community College*

Lee Ann Spahr, *Durham Technical Community College*

Richard Sturgeon, *University of Southern Maine*

Ara B. Sullenberger, *Tarrant County Community College*

Gwen Terwilliger, *University of Toledo*

Sharon Testone, *Onondaga Community College*

Margie Thrall, *Manatee Community College*

Michael Trappuzanno, *Arizona State University*

Georgina Vastola, *Middlesex Community College*

Jonathan Weissman, *Essex County College*

Alice Williamson, *Sussex County Community College*

Mike Wilson, *Florida Metropolitan University, Tampa*

Jerry Wisnieski, *Des Moines Community College*

Mary Jane Wolfe, *University of Rio Grande*

Michelle Younker, *Terra Community College*

Peter Zimmer, *West Chester University*

In addition, we want to thank the following individuals for providing splendid insight and suggestions for this new edition:

Suzanne Battista, *St. Petersburg Junior College, Clearwater*

Annette Benbow, *Tarrant County College, Northwest*

Karen Bingham, *Clarion College*

Nadine Branco, *Western Nevada Community College*

Connie Buller, *Metropolitan Community College*

Judy Carter, *North Shore Community College*

John Close, *Salt Lake Community College*

Cecil Ellard, *Ivy Tech Community College, Bloomington*

Colin Godfrey, *University of Massachusetts, Boston*

Shanna Goff, *Grand Rapids Community College*

Edna Greenwood, *Tarrant County College, Northwest*

Veronica Hupper, *Hesser College, Manchester*

Peter Kaslik, *Pierce College*

Joyce Keenan, *Horry-Georgetown Technical College*

Rajeesh Lal, *Pierce College*

Nam Lee, *Griffin Technical Institute*

Tanya Lee, *Career Technical College*

Betty Ludlum, *Austin Community College*

Phyllis Lurvey, *Hesser College, Manchester*

Mary Marlin, *West Virginia Northern Community College*

Carolyn T. McIntyre, *Horry-Georgetown Technical College*

Maria Luisa Mendez, *Laredo Community College*

Steven J. Meyer, *Erie Community College*

Marcia Mollé, *Metropolitan Community College*

Jay L. Novello, *Horry-Georgetown Technical College*

Sandra Lee Orr, *West Virginia State University*

Mary Pearce, *Wake Technical Community College*

Armando Perez, *Laredo Community College*

Regina Pierce, *Davenport University*

Anne Praderas, *Austin Community College*

Jose Rico, *Laredo Community College*

Suellen Robinson, *North Shore Community College*

Kathy Ruggieri, *Lansdale School of Business*

Randy Smith, *Des Moines Area Community College*

Dina Spain, *Horry-Georgetown Technical College*

Lori Welder, *PACE Institute*

Michael Yarbrough, *Cosumnes River College*

We have been greatly helped by a supportive group of colleagues who not only teach at North Shore Community College but have also provided a number of ideas as well as extensive help on all of our mathematics books. Our special best wishes to our colleague Bob Campbell, who recently retired. He has given us a friendly smile and encouraging ideas for 35 years! Also, a special word of thanks to Wally Hersey, Judy Carter, Rick Ponticelli, Lora Connelly, Sharyn Sharaf, Donna Stefano, Nancy Tufo, Elizabeth Lucas, Anne O'Shea, Marsha Pease, Walter Stone, Evangeline Cornwall, Rumiya Masagutova, Charles Peterson, and Neha Jain.

Jenny Crawford provided major contributions to this revision. She provided new problems, new ideas, and great mental energy. She greatly assisted us during the production process. She made helpful decisions. Her excellent help was much appreciated. She has become an essential part of our team as we work to provide the best possible textbook.

A special word of thanks goes to Lenore Parens, Cindy Trimble and Associates, and Beverly Fusfield for their excellent work in accuracy checking manuscript and reviewing page proofs.

Each textbook is a combination of ideas, writing, and revisions from the authors and wise editorial direction and assistance from the editors. We especially want to thank our editor at Pearson Education—Paul Murphy. He has a true vision of how authors and editors can work together as partners, and it has been a rewarding experience to create and revise textbooks together with him. We especially want to thank our Project Manager Lauren Morse for patiently answering questions and solving many daily problems. We also want to thank our entire team at Pearson Education—Marlana Voerster, Nathaniel Koven, Christine Whitlock, Georgina Brown, Linda Behrens, Tom Benfatti, Heather Scott, Ilene Kahn, Audra Walsh, and MiMi Yeh—as well as Allison Campbell and Karin Kipp at Elm Street Publishing Services for their assistance during the production process.

Nancy Tobey served as our administrative assistant. Daily she was involved with mailing, photocopying, collating, and taping. A special thanks goes to Nancy. We could not have finished the book without you.

Book writing is impossible for us without the loyal support of our families. Our deepest thanks and love to Nancy, Johnny, Melissa, Marcia, Shelley, Rusty, and Abby. Your understanding, your love and help, and your patience have been a source of great encouragement. Finally, we thank God for the strength and energy to write and the opportunity to help others through this textbook.

We have spent more than 37 years teaching mathematics. Each teaching day, we find that our greatest joy is helping students learn. We take a personal interest in ensuring that each student has a good learning experience in taking this course. If you have some personal comments, suggestions, or ideas for future editions of this textbook, please write to us at:

Prof. John Tobey and Prof. Jeffrey Slater
Pearson Education
Office of the College Mathematics Editor
75 Arlington Street, Suite 300
Boston, MA 02116

or e-mail us at

jtobey@northshore.edu

We wish you success in this course and in your future life!

John Tobey
Jeffrey Slater

Diagnostic Pretest: Beginning Algebra

Follow the directions for each problem. Simplify each answer.

Chapter 0

1. Add. $3\frac{1}{4} + 2\frac{3}{5}$

2. Multiply. $\left(1\frac{1}{6}\right)\left(2\frac{2}{3}\right)$

3. Divide. $\dfrac{15}{4} \div \dfrac{3}{8}$

4. Multiply. $(1.63)(3.05)$

5. Divide. $120 \div 0.0006$

6. Find 7% of 64,000.

Chapter 1

7. Add. $-3 + (-4) + (+12)$

8. Subtract. $-20 - (-23)$

9. Combine. $5x - 6xy - 12x - 8xy$

10. Evaluate. $2x^2 - 3x - 4$ when $x = -3$

11. Remove the grouping symbols. $2 - 3\{5 + 2[x - 4(3 - x)]\}$

12. Evaluate. $-3(2 - 6)^2 + (-12) \div (-4)$

Chapter 2

In questions 13–16, solve each equation for x.

13. $40 + 2x = 60 - 3x$

14. $7(3x - 1) = 5 + 4(x - 3)$

15. $\dfrac{2}{3}x - \dfrac{3}{4} = \dfrac{1}{6}x + \dfrac{21}{4}$

16. $\dfrac{4}{5}(3x + 4) = 20$

17. Solve for p. $A = \dfrac{1}{2}(3p - 4f)$

18. Solve for x and graph the result. $42 - 18x < 48x - 24$

$x > 1$

$$\begin{array}{cccccccc} & & & \circ & & & & \\ \hline -2 & -1 & 0 & 1 & 2 & 3 & \end{array}$$

Chapter 3

19. The length of a rectangle is 7 meters longer than the width. The perimeter is 46 meters. Find the dimensions.

20. One side of a triangle is triple the second side. The third side is 3 meters longer than double the second side. Find each side of the triangle if the perimeter of the triangle is 63 meters.

21. Hector has four test scores of 80, 90, 83, and 92. What does he need to score on the fifth test to have an average of 86 on the five tests?

22. Marcia invested $6000 in two accounts. One earned 5% interest, while the other earned 7% interest. After one year she earned $394 in interest. How much did she invest in each account?

---

1. $5\dfrac{17}{20}$

2. $3\dfrac{1}{9}$

3.   10

4.   4.9715

5.   200,000

6.   4480

7.   5

8.   3

9. $-7x - 14xy$

10.   23

11. $59 - 30x$

12. -45

13. $x = 4$

14. $x = 0$

15. $x = 12$

16. $x = 7$

17. $\dfrac{2A + 4f}{3} = p$

18. $x > 1$

19.   Width $= 5\dfrac{1}{3}$ meters; length $= 17\dfrac{2}{3}$ meters

20.   30 m; 10 m; 23 m

21.   85

22. $1300 at 5%; $4700 at 7%

23. 6 nickels; 9 dimes; 12 quarters

24. 2000 Friday, 1700 Thursday, 2510 Saturday

25. $8x^4y^6$

26. $-\dfrac{2x^2}{y^4}$

27. $-32x^{15}y^{20}$

28. $\dfrac{1}{81}$

29. $12x^3 + 5x^2 - 22x + 5$

30. $x^2 + 8x^2 + 15$

31. $5(x + 1)(x - 1)$

32. $(x - 4)(x - 8)$

33. $(2x + 1)(4x - 3)$

34. $(x - 2)(3a + 4b)$

35. $x = 0, x = -3, x = -4$

36. $x = \dfrac{3}{4}$

37. $\dfrac{x + 6}{2}$

38. $\dfrac{x + 3}{2(x + 2)}$

39. $\dfrac{x}{x + 2}$

40. $\dfrac{7x - 27}{(x - 3)(x - 4)(x - 5)}$

41. $x = -5$

42. $\dfrac{x}{3}$

---

23. Melissa has three more dimes than nickels. She has twice as many quarters as nickels. The value of the coins is $4.20. How many of each coin does she have?

24. The drama club put on a play for Thursday, Friday, and Saturday nights. The total attendance for the three nights was 6210. Thursday night had 300 fewer people than Friday night. Saturday night had 510 more people than Friday night. How many people came each night?

Chapter 4

25. Multiply. $(-2xy^2)(-4x^3y^4)$

26. Divide. $\dfrac{36x^5y^6}{-18x^3y^{10}}$

27. Raise to the indicated power. $(-2x^3y^4)^5$

28. Evaluate. $(-3)^{-4}$

29. Multiply. $(3x^2 + 2x - 5)(4x - 1)$

30. Divide. $(x^3 + 6x^2 - x - 30) \div (x - 2)$

Chapter 5

Factor completely.

31. $5x^2 - 5$

32. $x^2 - 12x + 32$

33. $8x^2 - 2x - 3$

34. $3ax - 8b - 6a + 4bx$

Solve for x.

35. $x^3 + 7x^2 + 12x = 0$

36. $16x^2 - 24x + 9 = 0$

Chapter 6

37. Simplify. $\dfrac{x^2 + 3x - 18}{2x - 6}$

38. Multiply. $\dfrac{6x^2 - 14x - 12}{6x + 4} \cdot \dfrac{x + 3}{2x^2 - 2x - 12}$

39. Divide and simplify. $\dfrac{x^2}{x^2 - 4} \div \dfrac{x^2 - 3x}{x^2 - 5x + 6}$

40. Add. $\dfrac{3}{x^2 - 7x + 12} + \dfrac{4}{x^2 - 9x + 20}$

41. Solve for x. $2 - \dfrac{5}{2x} = \dfrac{2x}{x + 1}$

42. Simplify. $\dfrac{3 + \dfrac{1}{x}}{\dfrac{9}{x} + \dfrac{3}{x^2}}$

Chapter 7

43. Graph $y = 2x - 4$.

44. Graph $3x + 4y = -12$.

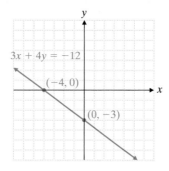

45. What is the slope of a line passing through $(6, -2)$ and $(-3, 4)$?

46. If $f(x) = 2x^2 - 3x + 1$, find $f(3)$.

47. Graph the region. $y \geq -\dfrac{1}{3}x + 2$

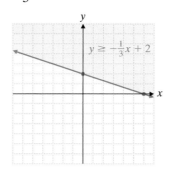

48. Find the equation of a line with a slope of $\dfrac{3}{5}$ that passes through the point $(-1, 3)$.

Chapter 8

Solve each system by the appropriate method.

49. Substitution method $x + y = 17$
$$2x - y = -5$$

50. Addition method $-5x + 4y = 8$
$$2x + 3y = 6$$

51. Any method $2(x - 2) = 3y$
$$6x = -3(4 + y)$$

52. Any method $x + \dfrac{1}{3}y = \dfrac{10}{3}$
$$\dfrac{3}{2}x + y = 8$$

53. Is $(2, -3)$ a solution for the system? $3x + 5y = -9$
$$2x - 3y = 13$$

---

43.   See graph

44.   See graph

45. $m = -\dfrac{2}{3}$

46.   10

47.   See graph

48. $3x - 5y = -18$

49. $x = 4; y = 13$

50. $x = 0; y = 2$

51. $x = -1$
      $y = -2$

52. $y = \dfrac{4}{3}$
      $y = 6$

53.   Yes

54.
Pair of gloves, $7.00
scarf, $8.00

54. A man bought three pairs of gloves and four scarves for $53. A woman bought two pairs of the same-priced gloves and three scarves of the same-priced scarves for $38. How much did each item cost?

55. 11

Chapter 9

55. Evaluate. $\sqrt{121}$

56. $5xy^2\sqrt{5xy}$

56. Simplify. $\sqrt{125x^3y^5}$

57. Multiply and simplify. $\left(\sqrt{2} + \sqrt{6}\right)\left(2\sqrt{2} - 3\sqrt{6}\right)$.

57. $-14 - 2\sqrt{3}$

58. Rationalize the denominator. $\dfrac{\sqrt{5} - \sqrt{3}}{\sqrt{6}}$

58. $\dfrac{\sqrt{30} - 3\sqrt{2}}{6}$

59. In the right triangle with sides a, b, and c, find side c if side $a = 4$ and side $b = 6$.

59. $2\sqrt{13}$

60. y varies directly with x. When $y = 56$, then $x = 8$. Find y when $x = 11$.

60. $y = 77$

Chapter 10

In questions 61–64, solve for x.

61. $x = 0; x = -\dfrac{3}{2}$

61. $14x^2 + 21x = 0$

62. $2x^2 + 1 = 19$

62. $x = \pm 3$

63. $2x^2 - 4x - 5 = 0$

64. $x^2 - x + 8 = 5 + 6x$

63. $x = \dfrac{2 \pm \sqrt{14}}{2}$

65. Graph the equation. $y = x^2 + 8x + 15$

64. $x = \dfrac{7 \pm \sqrt{37}}{2}$

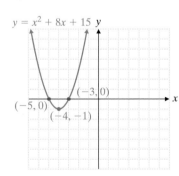

65. See graph

66.
Width 8 inches;
length 12 inches

66. A rectangular box is 4 inches longer in length than in width. The area of the box is 96 square inches. Find the length and the width of the box.

CHAPTER

0

What image comes to mind when you think of the Internal Revenue Service? A once-a-year challenge to get all your financial records together? Whether you file income taxes online or on paper it is a difficult task for most people. However, if you master the mathematics in this chapter you will be able to accurately complete your income tax returns. It is just one more way that mathematical knowledge will help you.

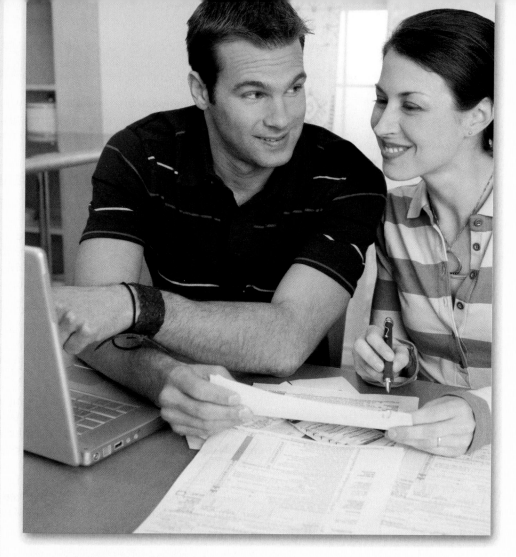

A Brief Review of Arithmetic Skills

Chapter 0 is designed to give you a mental "warm-up." In this chapter you'll be able to step back a bit and tone up your math skills. This brief review of arithmetic will increase your math flexibility and give you a good running start into algebra.

1 Understanding Basic Mathematical Definitions

Whole numbers are the set of numbers 0, 1, 2, 3, 4, 5, 6, 7, ... They are used to describe whole objects, or entire quantities.

Fractions are a set of numbers that are used to describe parts of whole quantities. In the object shown in the figure there are four equal parts. The *three of the four* parts that are shaded are represented by the fraction $\frac{3}{4}$. In the fraction $\frac{3}{4}$ the number 3 is called the **numerator** and the number 4, the **denominator.**

$$\frac{3}{4}$$

3 ← *Numerator* is on the top
4 ← *Denominator* is on the bottom

The *denominator* of a fraction shows the number of equal parts in the whole and the *numerator* shows the number of these parts being talked about or being used.

Numerals are symbols we use to name numbers. There are many different numerals that can be used to describe the same number. We know that $\frac{1}{2} = \frac{2}{4}$. The fractions $\frac{1}{2}$ and $\frac{2}{4}$ both describe the same number.

Usually, we find it more useful to use fractions that are simplified. A fraction is considered to be in **simplest form** or **reduced form** when the numerator (top) and the denominator (bottom) can both be divided exactly by no number other than 1, and the denominator is greater than 1.

$\frac{1}{2}$ is in simplest form.

$\frac{2}{4}$ is *not* in simplest form, since the numerator and the denominator can both be divided by 2.

If you get the answer $\frac{2}{4}$ to a problem, you should state it in simplest form, $\frac{1}{2}$. The process of changing $\frac{2}{4}$ to $\frac{1}{2}$ is called **simplifying** or **reducing** the fraction.

2 Simplifying Fractions to Lowest Terms Using Prime Numbers

Natural numbers or **counting numbers** are the set of whole numbers excluding 0. Thus the natural numbers are the numbers 1, 2, 3, 4, 5, 6, ...

When two or more numbers are multiplied, each number that is multiplied is called a **factor.** For example, when we write $3 \times 7 \times 5$, each of the numbers 3, 7, and 5 is called a factor.

Prime numbers are natural numbers greater than 1 whose only natural number factors are 1 and itself. The number 5 is prime. The only natural number factors of 5 are 5 and 1.

$$5 = 5 \times 1$$

The number 6 is not prime. The natural number factors of 6 are 3 and 2 or 6 and 1.

$$6 = 3 \times 2 \qquad 6 = 6 \times 1$$

The first 15 prime numbers are

$$2, 3, 5, 7, 11, 13, 17, 19, 23, 29, 31, 37, 41, 43, 47.$$

Any natural number greater than 1 either is prime or can be written as the product of prime numbers. For example, we can take each of the numbers 12, 30, 14,

19, and 29 and either indicate that they are prime or, if they are not prime, write them as the product of prime numbers. We write as follows:

$$12 = 2 \times 2 \times 3 \qquad 30 = 2 \times 3 \times 5 \qquad 14 = 2 \times 7$$

19 is a prime number. 29 is a prime number.

To reduce a fraction, we use prime numbers to factor the numerator and the denominator. Write each part of the fraction (numerator and denominator) as a product of prime numbers. Note any *factors* that appear in both the *numerator* (top) and *denominator* (bottom) of the fraction. If we divide numerator and denominator by these values we will obtain an equivalent fraction in *simplest form*. When the new fraction is simplified, it is said to be in **lowest terms.** Throughout this text, to *simplify* a fraction will always mean to write the fraction in lowest terms.

Teaching Tip Students will sometimes say, "I don't simplify the fraction $\frac{14}{21}$ that way. I just say 7 goes into 14 twice, and 7 goes into 21 three times; so the fraction simplifies to $\frac{2}{3}$. Can I do it that way?" Stress the importance of making sure they understand the concept. Explain that if they understand "removing the common factor of 7" it will make the simplifying of algebraic fractions easier later. They are welcome to simplify a fraction any way they wish, but the more they understand WHAT is going on and WHY it works, the more mathematics they will be able to learn.

EXAMPLE 1 Simplify each fraction.

(a) $\dfrac{14}{21}$ (b) $\dfrac{15}{35}$ (c) $\dfrac{20}{70}$

Solution

(a) $\dfrac{14}{21} = \dfrac{\cancel{7} \times 2}{\cancel{7} \times 3} = \dfrac{2}{3}$ We factor 14 and factor 21. Then we divide numerator and denominator by 7.

(b) $\dfrac{15}{35} = \dfrac{\cancel{5} \times 3}{\cancel{5} \times 7} = \dfrac{3}{7}$ We factor 15 and factor 35. Then we divide numerator and denominator by 5.

(c) $\dfrac{20}{70} = \dfrac{2 \times \cancel{2} \times \cancel{5}}{7 \times \cancel{2} \times \cancel{5}} = \dfrac{2}{7}$ We factor 20 and factor 70. Then we divide numerator and denominator by both 2 and 5.

Teaching Example 1 Simplify each fraction.

(a) $\dfrac{22}{33}$ (b) $\dfrac{36}{45}$ (c) $\dfrac{50}{65}$

Ans: (a) $\dfrac{2}{3}$ (b) $\dfrac{4}{5}$ (c) $\dfrac{10}{13}$

Practice Problem 1 Simplify.

(a) $\dfrac{10}{16}$ $\dfrac{5}{8}$ (b) $\dfrac{24}{36}$ $\dfrac{2}{3}$ (c) $\dfrac{36}{42}$ $\dfrac{6}{7}$

Sometimes when we simplify a fraction, all the prime factors in the top (numerator) are divided out. When this happens, we must remember that a 1 is left in the numerator.

EXAMPLE 2 Simplify each fraction.

(a) $\dfrac{7}{21}$ (b) $\dfrac{15}{105}$

Solution

(a) $\dfrac{7}{21} = \dfrac{\cancel{7} \times 1}{\cancel{7} \times 3} = \dfrac{1}{3}$ (b) $\dfrac{15}{105} = \dfrac{\cancel{5} \times \cancel{3} \times 1}{7 \times \cancel{5} \times \cancel{3}} = \dfrac{1}{7}$

Teaching Example 2 Simplify each fraction.

(a) $\dfrac{9}{54}$ (b) $\dfrac{12}{96}$

Ans: (a) $\dfrac{1}{6}$ (b) $\dfrac{1}{8}$

Practice Problem 2 Simplify.

(a) $\dfrac{4}{12}$ $\dfrac{1}{3}$ (b) $\dfrac{25}{125}$ $\dfrac{1}{5}$ (c) $\dfrac{73}{146}$ $\dfrac{1}{2}$

NOTE TO STUDENT: *Fully worked-out solutions to all of the Practice Problems can be found at the back of the text starting at page SP-1*

If all the prime numbers in the bottom (denominator) are divided out, we do not need to leave a 1 in the denominator, since we do not need to express the answer as a fraction. The answer is then a whole number and is not usually expressed as a fraction.

Teaching Example 3 Simplify each fraction.

(a) $\dfrac{24}{8}$ (b) $\dfrac{60}{15}$

Ans: (a) 3   (b) 4

EXAMPLE 3 Simplify each fraction.

(a) $\dfrac{35}{7}$

(b) $\dfrac{70}{10}$

Solution

(a) $\dfrac{35}{7} = \dfrac{5 \times \cancel{7}}{\cancel{7} \times 1} = 5$

(b) $\dfrac{70}{10} = \dfrac{7 \times \cancel{5} \times \cancel{2}}{\cancel{5} \times \cancel{2} \times 1} = 7$

Practice Problem 3 Simplify.

(a) $\dfrac{18}{6}$ 3   (b) $\dfrac{146}{73}$ 2   (c) $\dfrac{28}{7}$ 4

Sometimes the fraction we use represents how many of a certain thing are successful. For example, if a baseball player was at bat 30 times and achieved 12 hits, we could say that he had a hit $\frac{12}{30}$ of the time. If we reduce the fraction, we could say he had a hit $\frac{2}{5}$ of the time.

Teaching Example 4 A basketball player made 9 out of 12 free throws. Write this as a fraction in simplest form.

Ans: $\dfrac{3}{4}$

EXAMPLE 4 Cindy got 48 out of 56 questions correct on a test. Write this as a fraction in simplest form.

Solution Express as a fraction in simplest form the number of correct responses out of the total number of questions on the test.

$$48 \text{ out of } 56 \rightarrow \dfrac{48}{56} = \dfrac{6 \times \cancel{8}}{7 \times \cancel{8}} = \dfrac{6}{7}$$

Cindy answered the questions correctly $\frac{6}{7}$ of the time.

NOTE TO STUDENT: Fully worked-out solutions to all of the Practice Problems can be found at the back of the text starting at page SP-1

Practice Problem 4 The major league pennant winner in 1917 won 56 games out of 154 games played. Express as a fraction in simplest form the number of games won in relation to the number of games played.

The number *one* can be expressed as $1, \frac{1}{1}, \frac{2}{2}, \frac{6}{6}, \frac{8}{8}$, and so on, since

$$1 = \dfrac{1}{1} = \dfrac{2}{2} = \dfrac{6}{6} = \dfrac{8}{8}.$$

We say that these numerals are *equivalent ways* of writing the number *one* because they all express the same quantity even though they appear to be different.

SIDELIGHT: **The Multiplicative Identity**
When we simplify fractions, we are actually using the fact that we can multiply any number by 1 without changing the value of that number. (Mathematicians call the number 1 the **multiplicative identity** because it leaves any number it multiplies with the same identical value as before.)
Let's look again at one of the previous examples.

$$\dfrac{14}{21} = \dfrac{7 \times 2}{7 \times 3} = \dfrac{7}{7} \times \dfrac{2}{3} = 1 \times \dfrac{2}{3} = \dfrac{2}{3}$$

So we see that

$$\dfrac{14}{21} = \dfrac{2}{3}$$

When we simplify fractions, we are using this property of multiplying by 1.

❸ Converting Between Improper Fractions and Mixed Numbers

If the numerator is less than the denominator, the fraction is a **proper fraction.** A proper fraction is used to describe a quantity smaller than a whole.

Fractions can also be used to describe quantities larger than a whole. The following figure shows two bars that are equal in size. Each bar is divided into 5 equal pieces. The first bar is shaded in completely. The second bar has 2 of the 5 pieces shaded in.

The shaded-in region can be represented by $\frac{7}{5}$ since 7 of the pieces (each of which is $\frac{1}{5}$ of a whole box) are shaded. The fraction $\frac{7}{5}$ is called an improper fraction. An **improper fraction** is one in which the numerator is larger than or equal to the denominator.

The shaded-in region can also be represented by 1 whole added to $\frac{2}{5}$ of a whole, or $1 + \frac{2}{5}$. This is written as $1\frac{2}{5}$. The fraction $1\frac{2}{5}$ is called a mixed number. A **mixed number** consists of a whole number added to a proper fraction (the numerator is smaller than the denominator). The addition is understood but not written. When we write $1\frac{2}{5}$, it represents $1 + \frac{2}{5}$. The numbers $1\frac{7}{8}, 2\frac{3}{4}, 8\frac{1}{3}$, and $126\frac{1}{10}$ are all mixed numbers. From the preceding figure it seems clear that $\frac{7}{5} = 1\frac{2}{5}$. This suggests that we can change from one form to the other without changing the value of the fraction.

From a picture it is easy to see how to *change improper fractions to mixed numbers*. For example, suppose we start with the fraction $\frac{11}{3}$ and represent it by the following figure (where 11 of the pieces, each of which is $\frac{1}{3}$ of a box, are shaded). We see that $\frac{11}{3} = 3\frac{2}{3}$, since 3 whole boxes and $\frac{2}{3}$ of a box are shaded.

Changing Improper Fractions to Mixed Numbers
You can follow the same procedure without a picture. For example, to change $\frac{11}{3}$ to a mixed number, we can do the following:

$$\frac{11}{3} = \frac{3}{3} + \frac{3}{3} + \frac{3}{3} + \frac{2}{3} \quad \text{By the rule for adding fractions (which is discussed in detail in Section 0.2)}$$

$$= 1 + 1 + 1 + \frac{2}{3} \quad \text{Write 1 in place of } \frac{3}{3}, \text{ since } \frac{3}{3} = 1.$$

$$= 3 + \frac{2}{3} \quad \text{Write 3 in place of } 1 + 1 + 1.$$

$$= 3\frac{2}{3} \quad \text{Use the notation for mixed numbers.}$$

Now that you know how to change improper fractions to mixed numbers and why the procedure works, here is a shorter method.

> ## TO CHANGE AN IMPROPER FRACTION TO A MIXED NUMBER
> 1. Divide the denominator into the numerator.
> 2. The quotient is the whole-number part of the mixed number.
> 3. The remainder from the division will be the numerator of the fraction. The denominator of the fraction remains unchanged.

We can write the fraction as a division statement and divide. The arrows show how to write the mixed number.

$$\frac{7}{5} \qquad 5\overline{)7} \qquad \text{Whole-number part} \qquad 1\frac{2}{5} \qquad \text{Numerator of fraction}$$

$$\underline{5}$$
$$2 \qquad \text{Remainder}$$

Thus, $\frac{7}{5} = 1\frac{2}{5}$.

$$\frac{11}{3} \qquad 3\overline{)11}$$

$$\frac{9}{2}$$

Whole-number part

Numerator of fraction

$$3\frac{2}{3}$$

Remainder

Thus, $\frac{11}{3} = 3\frac{2}{3}$.

Sometimes the remainder is 0. In this case, the improper fraction changes to a whole number.

Teaching Example 5 Change to a mixed number or a whole number.

(a) $\frac{11}{4}$ (b) $\frac{26}{2}$

Ans: (a) $2\frac{3}{4}$ (b) 13

EXAMPLE 5   Change to a mixed number or to a whole number.

(a) $\frac{7}{4}$

(b) $\frac{15}{3}$

Solution

(a) $\frac{7}{4} = 7 \div 4$ $4\overline{)7}$

$$\frac{4}{3}$$ Remainder

Thus $\frac{7}{4} = 1\frac{3}{4}$

(b) $\frac{15}{3} = 15 \div 3$ $3\overline{)15}$

$$\frac{15}{0}$$ Remainder

Thus $\frac{15}{3} = 5$

NOTE TO STUDENT: Fully worked-out solutions to all of the Practice Problems can be found at the back of the text starting at page SP-1

Practice Problem 5   Change to a mixed number or to a whole number.

(a) $\frac{12}{7}$ $1\frac{5}{7}$

(b) $\frac{20}{5}$ 4

Changing Mixed Numbers to Improper Fractions   It is not difficult to see how to change mixed numbers to improper fractions. Suppose that you wanted to write $2\frac{2}{3}$ as an improper fraction.

$$2\frac{2}{3} = 2 + \frac{2}{3}$$ The meaning of mixed number notation

$$= 1 + 1 + \frac{2}{3}$$ Since $1 + 1 = 2$

$$= \frac{3}{3} + \frac{3}{3} + \frac{2}{3}$$ Since $1 = \frac{3}{3}$

When we draw a picture of $\frac{3}{3} + \frac{3}{3} + \frac{2}{3}$, we have this figure:

$$\frac{3}{3} \qquad\qquad \frac{3}{3} \qquad\qquad \frac{2}{3}$$

If we count the shaded parts, we see that

$$\frac{3}{3} + \frac{3}{3} + \frac{2}{3} = \frac{8}{3}.$$ Thus $2\frac{2}{3} = \frac{8}{3}.$

Now that you have seen how this change can be done, here is a shorter method.

TO CHANGE A MIXED NUMBER TO AN IMPROPER FRACTION

1. Multiply the whole number by the denominator.

2. Add this to the numerator. The result is the new numerator. The denominator does not change.

EXAMPLE 6 Change to an improper fraction.

(a) $3\frac{1}{7}$ **(b)** $5\frac{4}{5}$

Solution

(a) $3\frac{1}{7} = \frac{(3 \times 7) + 1}{7} = \frac{21 + 1}{7} = \frac{22}{7}$

(b) $5\frac{4}{5} = \frac{(5 \times 5) + 4}{5} = \frac{25 + 4}{5} = \frac{29}{5}$

Practice Problem 6 Change to an improper fraction.

(a) $3\frac{2}{5}$ $\frac{17}{5}$ **(b)** $1\frac{3}{7}$ $\frac{10}{7}$ **(c)** $2\frac{6}{11}$ $\frac{28}{11}$ **(d)** $4\frac{2}{3}$ $\frac{14}{3}$

Teaching Example 6 Change to an improper fraction.

(a) $4\frac{1}{3}$ (b) $2\frac{6}{7}$

Ans: (a) $\frac{13}{3}$ (b) $\frac{20}{7}$

④ Changing a Fraction to an Equivalent Fraction with a Given Denominator

Fractions can be changed to equivalent fractions with a different denominator by multiplying both numerator and denominator by the same number.

$$\frac{5}{6} = \frac{5 \times 2}{6 \times 2} = \frac{10}{12} \qquad\qquad \frac{3}{7} = \frac{3 \times 3}{7 \times 3} = \frac{9}{21}$$

So $\frac{5}{6}$ is equivalent to $\frac{10}{12}$. $\frac{3}{7}$ is equivalent to $\frac{9}{21}$.

We often multiply in this way to obtain an equivalent fraction with a *particular denominator*.

EXAMPLE 7 Find the missing number.

(a) $\frac{3}{5} = \frac{?}{25}$ **(b)** $\frac{4}{7} = \frac{?}{14}$ **(c)** $\frac{2}{9} = \frac{?}{36}$

Solution

(a) $\frac{3}{5} = \frac{?}{25}$ Observe that we need to multiply the denominator by 5 to obtain 25. So we multiply the numerator 3 by 5 also.

$\frac{3 \times 5}{5 \times 5} = \frac{15}{25}$ The desired numerator is 15.

(b) $\frac{4}{7} = \frac{?}{14}$ Observe that $7 \times 2 = 14$. We need to multiply the numerator by 2 to get the new numerator.

$\frac{4 \times 2}{7 \times 2} = \frac{8}{14}$ The desired numerator is 8.

(c) $\frac{2}{9} = \frac{?}{36}$ Observe that $9 \times 4 = 36$. We need to multiply the numerator by 4 to get the new numerator.

$\frac{2 \times 4}{9 \times 4} = \frac{8}{36}$ The desired numerator is 8.

Teaching Example 7 Find the missing number.

(a) $\frac{2}{3} = \frac{?}{15}$ (b) $\frac{5}{6} = \frac{?}{24}$ (c) $\frac{4}{11} = \frac{?}{33}$

Ans: (a) 10   (b) 20   (c) 12

Practice Problem 7 Find the missing number.

(a) $\frac{3}{8} = \frac{?}{24}$ 9   **(b)** $\frac{5}{6} = \frac{?}{30}$ 25   **(c)** $\frac{2}{7} = \frac{?}{56}$ 16

Verbal and Writing Skills

1. In the fraction $\frac{12}{13}$, what number is the numerator?
12

2. In the fraction $\frac{13}{17}$, what is the denominator?
17

3. What is a factor? Give an example.
When two or more numbers are multiplied, each number that is multiplied is called a factor. In 2×3, 2 and 3 are factors.

4. Give some examples of the number 1 written as a fraction.
$\frac{3}{3}$ or $\frac{17}{17}$

5. Draw a diagram to illustrate $2\frac{2}{3}$.

6. Draw a diagram to illustrate $3\frac{3}{4}$.

Simplify each fraction.

7. $\frac{18}{24}$ $\frac{3}{4}$

8. $\frac{20}{35}$ $\frac{4}{7}$

9. $\frac{12}{36}$ $\frac{1}{3}$

10. $\frac{12}{48}$ $\frac{1}{4}$

11. $\frac{60}{12}$ 5

12. $\frac{45}{15}$ 3

Change to a mixed number.

13. $\frac{17}{6}$ $2\frac{5}{6}$

14. $\frac{19}{5}$ $3\frac{4}{5}$

15. $\frac{111}{9}$ $12\frac{1}{3}$

16. $\frac{124}{8}$ $15\frac{1}{2}$

17. $\frac{38}{7}$ $5\frac{3}{7}$

18. $\frac{41}{6}$ $6\frac{5}{6}$

Change to an improper fraction or whole number.

19. $3\frac{1}{5}$ $\frac{16}{5}$

20. $4\frac{2}{5}$ $\frac{22}{5}$

21. $6\frac{3}{5}$ $\frac{33}{5}$

22. $5\frac{1}{12}$ $\frac{61}{12}$

23. $\frac{72}{9}$ 8

24. $\frac{78}{6}$ 13

Find the missing numerator.

25. $\frac{3}{8} = \frac{?}{64}$ 24

26. $\frac{5}{9} = \frac{?}{54}$ 30

27. $\frac{3}{5} = \frac{?}{35}$ 21

28. $\frac{5}{9} = \frac{?}{45}$ 25

29. $\frac{4}{13} = \frac{?}{39}$ 12

30. $\frac{13}{17} = \frac{?}{51}$ 39

Applications *Solve.*

31. ***Basketball*** During the regular 2006–07 WNBA basketball season, Asjha Jones of the Connecticut Sun scored 378 points in 24 games. Express as a mixed number in simplified form how many points she averaged per game.

$15\frac{3}{4}$

32. ***Kentucky Derby*** In 2006, 440 horses were nominated to compete in the Kentucky Derby. Only 20 horses were actually chosen to compete in the Derby. What simplified fraction shows what portion of the nominated horses actually competed?

$\frac{1}{22}$

33. *Income Tax* Last year, my parents had a combined income of $64,000. They paid $13,200 in federal income taxes. What simplified fraction shows how much my parents spent on their federal taxes?

$\dfrac{33}{160}$

34. *Transfer Students* The University of California system accepted 14,664 out of 18,330 California Community College students who applied as transfers in the fall of 2003. What simplified fraction shows what portion of applications were accepted?

$\dfrac{4}{5}$

Trail Mix *The following chart gives the recipe for a trail mix.*

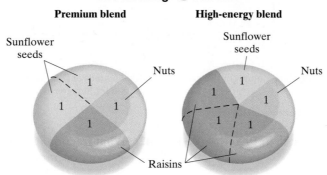

The Rocking ℛ trail mix

35. What part of the premium blend is nuts?

$\dfrac{1}{4}$

36. What part of the high-energy blend is raisins?

$\dfrac{3}{5}$

37. What part of the premium blend is not sunflower seeds? $\dfrac{1}{2}$

38. What part of the high-energy blend does not contain nuts? $\dfrac{4}{5}$

Theme Parks *The following chart provides some statistics about the number of visitors to the two most popular theme parks in the United States.*

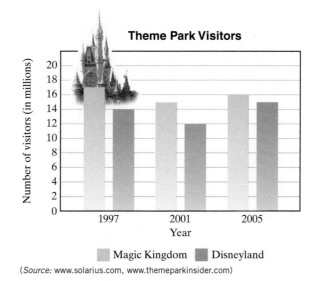

(*Source:* www.solarius.com, www.themeparkinsider.com)

39. What fractional part of the visitors to the two parks in 2001 went to Disneyland?

$\dfrac{4}{9}$

40. What fractional part of the visitors to the two parks in 2005 went to the Magic Kingdom?

$\dfrac{16}{31}$

Fish Catch *The following chart provides some statistics about the fish catch in the United States.*

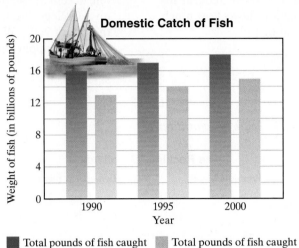

Domestic Catch of Fish

Source: National Marine Fisheries Service

41. What fractional part of the fish catch in 1990 was used as food for humans?

$\dfrac{13}{16}$

42. What fractional part of the fish catch in 2000 was used as food for humans?

$\dfrac{5}{6}$

Quick Quiz 0.1

1. Simplify. $\dfrac{84}{92}$ $\dfrac{21}{23}$

2. Write as an improper fraction. $6\dfrac{9}{11}$ $\dfrac{75}{11}$

3. Write as a mixed number. $\dfrac{103}{21}$ $4\dfrac{19}{21}$

4. Concept Check Explain in your own words how to change a mixed number to an improper fraction.
Answers may vary

Classroom Quiz 0.1 You may use these problems to quiz your students' mastery of Section 0.1.

1. Simplify. $\dfrac{56}{88}$
Ans: $\dfrac{7}{11}$

2. Write as an improper fraction. $5\dfrac{11}{14}$
Ans: $\dfrac{81}{14}$

3. Write as a mixed number. $\dfrac{109}{12}$
Ans: $9\dfrac{1}{12}$

0.2 ADDING AND SUBTRACTING FRACTIONS

① Adding or Subtracting Fractions with a Common Denominator

If fractions have the same denominator, the numerators may be added or subtracted. The denominator remains the same.

> **TO ADD OR SUBTRACT TWO FRACTIONS WITH A COMMON DENOMINATOR**
>
> 1. Add or subtract the numerators.
> 2. Keep the same (common) denominator.
> 3. Simplify the answer whenever possible.

Student Learning Objectives

After studying this section, you will be able to:

① Add or subtract fractions with a common denominator.

② Use prime factors to find the least common denominator of two or more fractions.

③ Add or subtract fractions with different denominators.

④ Add or subtract mixed numbers.

EXAMPLE 1 Add the fractions. Simplify your answer whenever possible.

(a) $\dfrac{5}{7} + \dfrac{1}{7}$ (b) $\dfrac{2}{3} + \dfrac{1}{3}$ (c) $\dfrac{1}{8} + \dfrac{3}{8} + \dfrac{2}{8}$ (d) $\dfrac{3}{5} + \dfrac{4}{5}$

Solution

(a) $\dfrac{5}{7} + \dfrac{1}{7} = \dfrac{5+1}{7} = \dfrac{6}{7}$ (b) $\dfrac{2}{3} + \dfrac{1}{3} = \dfrac{2+1}{3} = \dfrac{3}{3} = 1$

(c) $\dfrac{1}{8} + \dfrac{3}{8} + \dfrac{2}{8} = \dfrac{1+3+2}{8} = \dfrac{6}{8} = \dfrac{3}{4}$ (d) $\dfrac{3}{5} + \dfrac{4}{5} = \dfrac{3+4}{5} = \dfrac{7}{5}$ or $1\dfrac{2}{5}$

Practice Problem 1 Add.

(a) $\dfrac{3}{6} + \dfrac{2}{6}$ $\dfrac{5}{6}$ (b) $\dfrac{3}{11} + \dfrac{8}{11}$ 1

(c) $\dfrac{1}{8} + \dfrac{2}{8} + \dfrac{1}{8}$ $\dfrac{1}{2}$ (d) $\dfrac{5}{9} + \dfrac{8}{9}$ $\dfrac{13}{9}$ or $1\dfrac{4}{9}$

Teaching Example 1 Add the fractions. Simplify your answer whenever possible.

(a) $\dfrac{4}{15} + \dfrac{7}{15}$ (b) $\dfrac{1}{10} + \dfrac{9}{10}$

(c) $\dfrac{3}{20} + \dfrac{7}{20}$ (d) $\dfrac{1}{9} + \dfrac{2}{9} + \dfrac{3}{9}$

Ans: (a) $\dfrac{11}{15}$ (b) 1 (c) $\dfrac{1}{2}$ (d) $\dfrac{2}{3}$

Teaching Tip For some weaker students it is helpful to say the rule as follows:
WHEN ADDING OR SUBTRACTING FRACTIONS WITH THE SAME DENOMINATOR,
1. Add or subtract the numerators.
2. Keep the common (same) denominator.

EXAMPLE 2 Subtract the fractions. Simplify your answer whenever possible.

(a) $\dfrac{9}{11} - \dfrac{2}{11}$ (b) $\dfrac{5}{6} - \dfrac{1}{6}$

Solution

(a) $\dfrac{9}{11} - \dfrac{2}{11} = \dfrac{9-2}{11} = \dfrac{7}{11}$ (b) $\dfrac{5}{6} - \dfrac{1}{6} = \dfrac{5-1}{6} = \dfrac{4}{6} = \dfrac{2}{3}$

Practice Problem 2 Subtract.

(a) $\dfrac{11}{13} - \dfrac{6}{13}$ $\dfrac{5}{13}$ (b) $\dfrac{8}{9} - \dfrac{2}{9}$ $\dfrac{2}{3}$

Teaching Example 2 Subtract the fractions. Simplify your answer whenever possible.

(a) $\dfrac{5}{13} - \dfrac{2}{13}$ (b) $\dfrac{7}{9} - \dfrac{1}{9}$

Ans: (a) $\dfrac{3}{13}$ (b) $\dfrac{2}{3}$

Although adding and subtracting fractions with the same denominator is fairly simple, most problems involve fractions that do not have a common denominator. Fractions and mixed numbers such as halves, fourths, and eighths are often used. To add or subtract such fractions, we begin by finding a common denominator.

Using Prime Factors to Find the Least Common Denominator of Two or More Fractions

Before you can add or subtract fractions, they must have the same denominator. To save work, we select the smallest possible common denominator. This is called the **least common denominator** or LCD (also known as the *lowest common denominator*).

The LCD of two or more fractions is the smallest whole number that is exactly divisible by each denominator of the fractions.

EXAMPLE 3   Find the LCD. $\frac{2}{3}$ and $\frac{1}{4}$

Solution   The numbers are small enough to find the LCD by inspection. The LCD is 12, since 12 is exactly divisible by 4 and by 3. There is no smaller number that is exactly divisible by 4 and 3.

Practice Problem 3   Find the LCD. $\frac{1}{8}$ and $\frac{5}{12}$ 24

In some cases, the LCD cannot easily be determined by inspection. If we write each denominator as the product of prime factors, we will be able to find the LCD. We will use ($\cdot$) to indicate multiplication. For example, $30 = 2 \cdot 3 \cdot 5$. This means $30 = 2 \times 3 \times 5$.

PROCEDURE TO FIND THE LCD USING PRIME FACTORS

1. Write each denominator as the product of prime factors.
2. The LCD is a product containing each different factor.
3. If a factor occurs more than once in any one denominator, the LCD will contain that factor repeated the greatest number of times that it occurs in any one denominator.

EXAMPLE 4   Find the LCD of $\frac{5}{6}$ and $\frac{1}{15}$ by this new procedure.

Solution

$$6 = 2 \cdot 3$$
$$15 = \;\; 3 \cdot 5$$
$$LCD = 2 \cdot 3 \cdot 5$$
$$LCD = 2 \cdot 3 \cdot 5 = 30$$

Write each denominator as the product of prime factors.

The LCD is a product containing each different prime factor. The different factors are 2, 3, and 5, and each factor appears at most once in any one denominator.

Practice Problem 4   Use prime factors to find the LCD of $\frac{8}{35}$ and $\frac{6}{15}$.   105

Great care should be used to determine the LCD in the case of repeated factors.

EXAMPLE 5 Find the LCD of $\frac{4}{27}$ and $\frac{5}{18}$.

Solution

$$27 = 3 \cdot 3 \cdot 3$$

$$18 = \;\; 3 \cdot 3 \cdot 2$$

$$LCD = 3 \cdot 3 \cdot 3 \cdot 2$$

$$LCD = 3 \cdot 3 \cdot 3 \cdot 2 = 54$$

Write each denominator as the product of prime factors. We observe that the factor 3 occurs three times in the factorization of 27.

The LCD is a product containing each different factor. The factor 3 *occurred most* in the factorization of 27, where it occurred *three* times. Thus the LCD will be the product of *three* 3's and *one* 2.

Practice Problem 5 Find the LCD of $\frac{5}{12}$ and $\frac{7}{30}$.   60

Teaching Example 5 Find the LCD of $\frac{5}{8}$ and $\frac{7}{12}$ using the prime factor method.

Ans: 24

NOTE TO STUDENT: *Fully worked-out solutions to all of the Practice Problems can be found at the back of the text starting at page SP-1*

EXAMPLE 6 Find the LCD of $\frac{5}{12}$, $\frac{1}{15}$, and $\frac{7}{30}$.

Solution

$$12 = 2 \cdot 2 \cdot 3$$

$$15 = \;\;\;\;\; 3 \cdot 5$$

$$30 = \;\; 2 \cdot 3 \cdot 5$$

$$LCD = 2 \cdot 2 \cdot 3 \cdot 5$$

$$LCD = 2 \cdot 2 \cdot 3 \cdot 5 = 60$$

Write each denominator as the product of prime factors. Notice that the only repeated factor is 2, which occurs twice in the factorization of 12.

The LCD is the product of each different factor with the factor 2 appearing twice since it occurred twice in one denominator.

Practice Problem 6 Find the LCD of $\frac{2}{27}$, $\frac{1}{18}$, and $\frac{5}{12}$.   108

Teaching Example 6 Find the LCD of $\frac{1}{6}$, $\frac{5}{9}$, and $\frac{2}{45}$ using the prime factor method.

Ans: 90

3 Adding or Subtracting Fractions with Different Denominators

Before you can add or subtract them, fractions must have the same denominator. Using the LCD will make your work easier. First you must find the LCD. Then change each fraction to a fraction that has the LCD as the denominator. Sometimes one of the fractions will already have the LCD as the denominator. Once all the fractions have the same denominator, you can add or subtract. Be sure to simplify the fraction in your answer if this is possible.

TO ADD OR SUBTRACT FRACTIONS THAT DO NOT HAVE A COMMON DENOMINATOR

1. Find the LCD of the fractions.
2. Change each fraction to an equivalent fraction with the LCD for a denominator.
3. Add or subtract the fractions.
4. Simplify the answer whenever possible.

Let us return to the two fractions of Example 3. We have previously found that the LCD is 12.

EXAMPLE 7   Bob picked $\frac{2}{3}$ of a bushel of apples on Monday and $\frac{1}{4}$ of a bushel of apples on Tuesday. How much did he have in total?

Solution   To solve this problem we need to add $\frac{2}{3}$ and $\frac{1}{4}$, but before we can do so, we must change $\frac{2}{3}$ and $\frac{1}{4}$ to fractions with the same denominator. We change each fraction to an equivalent fraction with a common denominator of 12, the LCD.

$$\frac{2}{3} = \frac{?}{12} \qquad \frac{2 \times 4}{3 \times 4} = \frac{8}{12} \quad \text{so} \quad \frac{2}{3} = \frac{8}{12}$$

$$\frac{1}{4} = \frac{?}{12} \qquad \frac{1 \times 3}{4 \times 3} = \frac{3}{12} \quad \text{so} \quad \frac{1}{4} = \frac{3}{12}$$

Then we rewrite the problem with common denominators and add.

$$\frac{2}{3} + \frac{1}{4} = \frac{8}{12} + \frac{3}{12} = \frac{8 + 3}{12} = \frac{11}{12}$$

In total Bob picked $\frac{11}{12}$ of a bushel of apples.

Practice Problem 7   Carol planted corn in $\frac{5}{12}$ of the farm fields at the Old Robinson Farm. Connie planted soybeans in $\frac{1}{8}$ of the farm fields. What fractional part of the farm fields of the Old Robinson Farm was planted in corn or soybeans?

$\frac{13}{24}$ of the farm fields

Sometimes one of the denominators is the LCD. In such cases the fraction that has the LCD for the denominator will not need to be changed. If every other denominator divides into the largest denominator, the largest denominator is the LCD.

EXAMPLE 8   Find the LCD and then add. $\dfrac{3}{5} + \dfrac{7}{20} + \dfrac{1}{2}$

Solution   We can see by inspection that both 5 and 2 divide exactly into 20. Thus 20 is the LCD. Now add.

$$\frac{3}{5} + \frac{7}{20} + \frac{1}{2}$$

We change $\frac{3}{5}$ and $\frac{1}{2}$ to equivalent fractions with a common denominator of 20, the LCD.

$$\frac{3}{5} = \frac{?}{20} \qquad \frac{3 \times 4}{5 \times 4} = \frac{12}{20} \quad \text{so} \quad \frac{3}{5} = \frac{12}{20}$$

$$\frac{1}{2} = \frac{?}{20} \qquad \frac{1 \times 10}{2 \times 10} = \frac{10}{20} \quad \text{so} \quad \frac{1}{2} = \frac{10}{20}$$

Then we rewrite the problem with common denominators and add.

$$\frac{3}{5} + \frac{7}{20} + \frac{1}{2} = \frac{12}{20} + \frac{7}{20} + \frac{10}{20} = \frac{12 + 7 + 10}{20} = \frac{29}{20} \quad \text{or} \quad 1\frac{9}{20}$$

Practice Problem 8   Find the LCD and add.

$$\frac{3}{5} + \frac{4}{25} + \frac{1}{50} \quad \frac{39}{50}$$

Now we turn to examples where the selection of the LCD is not so obvious. In Examples 9 through 11 we will use the prime factorization method to find the LCD.

EXAMPLE 9   Add. $\dfrac{7}{18} + \dfrac{5}{12}$

Solution   First we find the LCD.

$$18 = 3 \cdot 3 \cdot 2$$
$$12 = \;\; 3 \cdot 2 \cdot 2$$
$$\downarrow \; \downarrow \; \downarrow \; \downarrow$$
$$\text{LCD} = 3 \cdot 3 \cdot 2 \cdot 2 = 9 \cdot 4 = 36$$

Now we change $\frac{7}{18}$ and $\frac{5}{12}$ to equivalent fractions that have the LCD.

$$\frac{7}{18} = \frac{?}{36} \qquad \frac{7 \times 2}{18 \times 2} = \frac{14}{36}$$

$$\frac{5}{12} = \frac{?}{36} \qquad \frac{5 \times 3}{12 \times 3} = \frac{15}{36}$$

Now we add the fractions.

$$\frac{7}{18} + \frac{5}{12} = \frac{14}{36} + \frac{15}{36} = \frac{29}{36} \quad \text{This fraction cannot be simplified.}$$

Practice Problem 9   Add.

$$\frac{1}{49} + \frac{3}{14} \quad \frac{23}{98}$$

NOTE TO STUDENT: Fully worked-out solutions to all of the Practice Problems can be found at the back of the text starting at page SP-1

Teaching Example 9 Add. $\dfrac{7}{15} + \dfrac{3}{10}$

Ans: $\dfrac{23}{30}$

EXAMPLE 10   Subtract. $\dfrac{25}{48} - \dfrac{5}{36}$

Solution   First we find the LCD.

$$48 = 2 \cdot 2 \cdot 2 \cdot 2 \cdot 3$$
$$36 = \;\; 2 \cdot 2 \cdot 3 \cdot 3$$
$$\downarrow \downarrow \; \downarrow \; \downarrow \; \downarrow \; \downarrow$$
$$\text{LCD} = 2 \cdot 2 \cdot 2 \cdot 2 \cdot 3 \cdot 3 = 16 \cdot 9 = 144$$

Now we change $\frac{25}{48}$ and $\frac{5}{36}$ to equivalent fractions that have the LCD.

$$\frac{25}{48} = \frac{?}{144} \qquad \frac{25 \times 3}{48 \times 3} = \frac{75}{144}$$

$$\frac{5}{36} = \frac{?}{144} \qquad \frac{5 \times 4}{36 \times 4} = \frac{20}{144}$$

Now we subtract the fractions.

$$\frac{25}{48} - \frac{5}{36} = \frac{75}{144} - \frac{20}{144} = \frac{55}{144} \quad \text{This fraction cannot be simplified.}$$

Practice Problem 10   Subtract.

$$\frac{1}{12} - \frac{1}{30} \quad \frac{1}{20}$$

Teaching Example 10 Subtract. $\dfrac{11}{18} - \dfrac{3}{10}$

Ans: $\dfrac{14}{45}$

Teaching Example 11 Combine. $\dfrac{3}{4} - \dfrac{1}{5} + \dfrac{3}{10}$

Ans: $\dfrac{17}{20}$

Teaching Tip Encourage students to line up the factors in the way that is shown in the text. In this way, factors that occur in more than one number will not be repeated.

EXAMPLE 11   Combine. $\dfrac{1}{5} + \dfrac{1}{6} - \dfrac{3}{10}$

Solution   First we find the LCD.

$$5 = 5$$
$$6 = \quad 2 \cdot 3$$
$$10 = 5 \cdot 2 \quad \Big|$$
$$\downarrow \quad \downarrow \quad \downarrow$$
$$\text{LCD} = 5 \cdot 2 \cdot 3 = 10 \cdot 3 = 30$$

Now we change $\frac{1}{5}, \frac{1}{6}$, and $\frac{3}{10}$ to equivalent fractions that have the LCD for a denominator.

$$\dfrac{1}{5} = \dfrac{?}{30} \qquad \dfrac{1 \times 6}{5 \times 6} = \dfrac{6}{30}$$

$$\dfrac{1}{6} = \dfrac{?}{30} \qquad \dfrac{1 \times 5}{6 \times 5} = \dfrac{5}{30}$$

$$\dfrac{3}{10} = \dfrac{?}{30} \qquad \dfrac{3 \times 3}{10 \times 3} = \dfrac{9}{30}$$

Now we combine the three fractions.

$$\dfrac{1}{5} + \dfrac{1}{6} - \dfrac{3}{10} = \dfrac{6}{30} + \dfrac{5}{30} - \dfrac{9}{30} = \dfrac{2}{30} = \dfrac{1}{15}$$

Note the important step of simplifying the fraction to obtain the final answer.

Practice Problem 11   Combine.

$$\dfrac{2}{3} + \dfrac{3}{4} - \dfrac{3}{8} \quad \dfrac{25}{24} \text{ or } 1\dfrac{1}{24}$$

4 Adding or Subtracting Mixed Numbers

If the problem you are adding or subtracting has mixed numbers, change them to improper fractions first and then combine (add or subtract). As a convention in this book, if the original problem contains mixed numbers, express the result as a mixed number rather than as an improper fraction.

Teaching Example 12 Combine. Simplify your answer whenever possible.

(a) $4\dfrac{2}{15} + 7\dfrac{1}{2}$ (b) $4\dfrac{1}{2} - 2\dfrac{2}{3}$

(c) $3\dfrac{3}{10} + \dfrac{3}{4}$

Ans: (a) $11\dfrac{19}{30}$ (b) $1\dfrac{5}{6}$ (c) $4\dfrac{1}{20}$

EXAMPLE 12   Combine. Simplify your answer whenever possible.

(a) $5\dfrac{1}{2} + 2\dfrac{1}{3}$ (b) $2\dfrac{1}{5} - 1\dfrac{3}{4}$ (c) $1\dfrac{5}{12} + \dfrac{7}{30}$

Solution

(a) First we change the mixed numbers to improper fractions.

$$5\dfrac{1}{2} = \dfrac{5 \times 2 + 1}{2} = \dfrac{11}{2} \qquad 2\dfrac{1}{3} = \dfrac{2 \times 3 + 1}{3} = \dfrac{7}{3}$$

Next we change each fraction to an equivalent form with the common denominator of 6.

$$\dfrac{11}{2} = \dfrac{?}{6} \qquad \dfrac{11 \times 3}{2 \times 3} = \dfrac{33}{6}$$

$$\dfrac{7}{3} = \dfrac{?}{6} \qquad \dfrac{7 \times 2}{3 \times 2} = \dfrac{14}{6}$$

Finally, we add the two fractions and change our answer to a mixed number.

$$\frac{33}{6} + \frac{14}{6} = \frac{47}{6} = 7\frac{5}{6}$$

Thus $5\frac{1}{2} + 2\frac{1}{3} = 7\frac{5}{6}$.

(b) First we change the mixed numbers to improper fractions.

$$2\frac{1}{5} = \frac{2 \times 5 + 1}{5} = \frac{11}{5} \qquad 1\frac{3}{4} = \frac{1 \times 4 + 3}{4} = \frac{7}{4}$$

Next we change each fraction to an equivalent form with the common denominator of 20.

$$\frac{11}{5} = \frac{?}{20} \qquad \frac{11 \times 4}{5 \times 4} = \frac{44}{20}$$

$$\frac{7}{4} = \frac{?}{20} \qquad \frac{7 \times 5}{4 \times 5} = \frac{35}{20}$$

Now we subtract the two fractions.

$$\frac{44}{20} - \frac{35}{20} = \frac{9}{20}$$

Thus $2\frac{1}{5} - 1\frac{3}{4} = \frac{9}{20}$.

Note: It is not necessary to use these exact steps to add and subtract mixed numbers. If you know another method and can use it to obtain the correct answers, it is all right to continue to use that method throughout this chapter.

(c) Now we add $1\frac{5}{12} + \frac{7}{30}$.

The LCD of 12 and 30 is 60. Why? Change the mixed number to an improper fraction. Then change each fraction to an equivalent form with a common denominator.

$$1\frac{5}{12} = \frac{17 \times 5}{12 \times 5} = \frac{85}{60} \qquad \frac{7 \times 2}{30 \times 2} = \frac{14}{60}$$

Then add the fractions, simplify, and write the answer as a mixed number.

$$\frac{85}{60} + \frac{14}{60} = \frac{99}{60} = \frac{33}{20} = 1\frac{13}{20}$$

Thus $1\frac{5}{12} + \frac{7}{30} = 1\frac{13}{20}$.

Practice Problem 12 Combine.

(a) $1\frac{2}{3} + 2\frac{4}{5} \quad 4\frac{7}{15}$

(b) $5\frac{1}{4} - 2\frac{2}{3} \quad 2\frac{7}{12}$

Teaching Tip After discussing Example 12, students may ask if they must change to improper fractions first before adding or subtracting. The logical answer is that it is not necessary, only suggested. By following this approach, it more logically leads to the methods of adding rational expressions in algebra. You might want to show them an additional example worked both ways:

Add. $2\frac{5}{8} + 3\frac{7}{12}$

METHOD OF CHANGING TO IMPROPER FRACTIONS

$$2\frac{5}{8} = \frac{21}{8} = \frac{63}{24}$$

$$3\frac{7}{12} = \frac{43}{12} = \frac{86}{24}$$

$$\frac{63}{24} + \frac{86}{24} = \frac{149}{24} = 6\frac{5}{24}$$

METHOD OF ADDING INTEGERS AND FRACTIONS SEPARATELY

$$2\frac{5}{8} = \qquad 2\frac{15}{24}$$

$$+3\frac{7}{12} = +3\frac{14}{24}$$

$$\overline{\qquad 5\frac{29}{24} = 6\frac{5}{24}}$$

When subtracting mixed numbers, students need not worry about borrowing from the whole number part if they use the technique illustrated in the text.

NOTE TO STUDENT: *Fully worked-out solutions to all of the Practice Problems can be found at the back of the text starting at page SP-1*

Teaching Example 13 Find the perimeter of a triangle with sides of $3\frac{1}{10}$ meters, $2\frac{2}{5}$ meters, and $4\frac{2}{15}$ meters.

Ans: $9\frac{19}{30}$ m

Teaching Tip Point out the fact that the answer includes the units of measure. The perimeter of the triangle is $54\frac{3}{8}$ feet. The answer to the word problem is information that Manuel needs to know. An answer of $54\frac{3}{8}$ would not be very useful, and any other unit of measure, for example inches, would be incorrect.

▲ **EXAMPLE 13** Manuel is enclosing a triangle-shaped exercise yard for his new dog. He wants to determine how many feet of fencing he will need. The sides of the yard measure $20\frac{3}{4}$ feet, $15\frac{1}{2}$ feet, and $18\frac{1}{8}$ feet. What is the perimeter of (total distance around) the triangle?

Solution

Understand the problem. Begin by drawing a picture.

We want to add up the lengths of all three sides of the triangle. This distance around the triangle is called the **perimeter.**

$$20\frac{3}{4} + 15\frac{1}{2} + 18\frac{1}{8} = \frac{83}{4} + \frac{31}{2} + \frac{145}{8}$$

$$= \frac{166}{8} + \frac{124}{8} + \frac{145}{8} = \frac{435}{8} = 54\frac{3}{8}$$

He will need $54\frac{3}{8}$ feet of fencing.

▲ **Practice Problem 13** Find the perimeter of a rectangle with sides of $4\frac{1}{5}$ cm and $6\frac{1}{2}$ cm. Begin by drawing a picture. Label the picture by including the measure of *each* side.

$21\frac{2}{5}$ cm

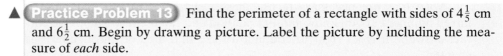

Developing Your Study Skills

Class Attendance

A student of mathematics needs to get started in the right direction by choosing to attend class every day, beginning with the first day of class. Statistics show that class attendance and good grades go together. Classroom activities are designed to enhance learning and you must be in class to benefit from them. Vital information and explanations that can help you in understanding concepts are given each day. Do not be deceived into thinking that you can just find out from a friend what went on in class. There is no good substitute for firsthand experience. Give yourself a push in the right direction by developing the habit of going to class every day.

Class Participation

People learn mathematics through active participation, not through observation from the sidelines. If you want to do well in this course, be involved in classroom activities. Sit near the front where you can see and hear well and where your focus is on the instruction process and not on the students around you. Ask questions, be ready to contribute toward solutions, and take part in all classroom activities. Your contributions are valuable to the class and to yourself. Class participation requires an investment of yourself in the learning process, which you will find pays huge dividends.

Verbal and Writing Skills

1. Explain why the denominator 8 is the least common denominator of $\frac{3}{4}$ and $\frac{5}{8}$.

Answers may vary. A sample answer is: 8 is exactly divisible by 4.

2. What must you do before you add or subtract fractions that do not have a common denominator?

You must change the fractions to equivalent fractions with a common denominator.

Find the LCD (least common denominator) of each set of fractions. Do not combine the fractions; only find the LCD.

3. $\frac{4}{5}$ and $\frac{11}{16}$
LCD = 80

4. $\frac{21}{25}$ and $\frac{8}{15}$
LCD = 75

5. $\frac{7}{10}$ and $\frac{1}{4}$
LCD = 20

6. $\frac{3}{16}$ and $\frac{1}{24}$
LCD = 48

7. $\frac{5}{18}$ and $\frac{7}{54}$
LCD = 54

8. $\frac{7}{24}$ and $\frac{5}{48}$
LCD = 48

9. $\frac{1}{2}, \frac{1}{18},$ and $\frac{13}{30}$
LCD = 90

10. $\frac{5}{8}, \frac{3}{14},$ and $\frac{11}{16}$
LCD = 112

Combine. Be sure to simplify your answer whenever possible.

11. $\frac{3}{8} + \frac{2}{8}$ $\frac{5}{8}$

12. $\frac{2}{9} + \frac{5}{9}$ $\frac{7}{9}$

13. $\frac{5}{14} - \frac{1}{14}$ $\frac{2}{7}$

14. $\frac{9}{20} - \frac{3}{20}$ $\frac{3}{10}$

15. $\frac{3}{8} + \frac{5}{6}$ $\frac{29}{24}$ or $1\frac{5}{24}$

16. $\frac{7}{10} + \frac{8}{15}$ $\frac{37}{30}$ or $1\frac{7}{30}$

17. $\frac{5}{7} - \frac{2}{9}$ $\frac{31}{63}$

18. $\frac{7}{8} - \frac{2}{3}$ $\frac{5}{24}$

19. $\frac{1}{3} + \frac{2}{5}$ $\frac{11}{15}$

20. $\frac{2}{7} + \frac{1}{3}$ $\frac{13}{21}$

21. $\frac{5}{9} + \frac{5}{12}$ $\frac{35}{36}$

22. $\frac{11}{15} + \frac{1}{10}$ $\frac{5}{6}$

23. $\frac{2}{3} + \frac{7}{12} + \frac{1}{4}$ $\frac{3}{2}$ or $1\frac{1}{2}$

24. $\frac{4}{7} + \frac{7}{9} + \frac{1}{3}$ $\frac{106}{63}$ or $1\frac{43}{63}$

25. $\frac{5}{36} + \frac{7}{9} - \frac{5}{12}$ $\frac{1}{2}$

26. $\frac{5}{24} + \frac{3}{8} - \frac{1}{3}$ $\frac{1}{4}$

27. $4\frac{1}{3} + 3\frac{2}{5}$ $7\frac{11}{15}$

28. $3\frac{1}{8} + 2\frac{1}{6}$ $5\frac{7}{24}$

29. $1\frac{5}{24} + \frac{5}{18}$ $1\frac{35}{72}$

30. $6\frac{2}{3} + \frac{3}{4}$ $7\frac{5}{12}$

31. $7\frac{1}{6} - 2\frac{1}{4}$ $4\frac{11}{12}$

32. $7\frac{2}{5} - 3\frac{3}{4}$ $3\frac{13}{20}$

33. $8\frac{5}{7} - 2\frac{1}{4}$ $6\frac{13}{28}$

34. $7\frac{8}{15} - 2\frac{3}{5}$ $4\frac{14}{15}$

35. $2\frac{1}{8} + 3\frac{2}{3}$ $5\frac{19}{24}$

36. $3\frac{1}{7} + 4\frac{1}{3}$ $7\frac{10}{21}$

37. $11\frac{1}{7} - 6\frac{5}{7}$ $4\frac{3}{7}$

38. $17\frac{1}{5} - 10\frac{3}{5}$ $6\frac{3}{5}$

39. $3\frac{5}{12} + 5\frac{7}{12}$ 9

40. $2\frac{13}{16} + 2\frac{3}{16}$ 5

Mixed Practice

41. $\frac{7}{8} + \frac{1}{12}$ $\frac{23}{24}$

42. $\frac{19}{30} + \frac{3}{10}$ $\frac{14}{15}$

43. $3\frac{3}{16} + 4\frac{3}{8}$ $7\frac{9}{16}$

44. $5\frac{2}{3} + 7\frac{2}{5}$ $13\frac{1}{15}$

45. $\frac{16}{21} - \frac{2}{7}$ $\frac{10}{21}$

46. $\frac{15}{24} - \frac{3}{8}$ $\frac{1}{4}$

47. $5\frac{1}{5} - 2\frac{1}{2}$ $2\frac{7}{10}$

48. $6\frac{1}{3} - 4\frac{1}{4}$ $2\frac{1}{12}$

Applications

49. *Inline Skating* Nancy and Sarah meet three mornings a week to skate. They skated $8\frac{1}{4}$ miles on Monday, $10\frac{2}{3}$ miles on Wednesday, and $5\frac{3}{4}$ miles on Friday. What was their total distance for those three days?

$24\frac{2}{3}$ mi

50. *Marathon Training* Paco and Eskinder are training for the Boston Marathon. Their coach gave them the following schedule: run $10\frac{1}{2}$ miles on Thursday, a short run of $5\frac{1}{4}$ miles on Friday, a rest day on Saturday, and a long run of $18\frac{2}{3}$ miles on Sunday. How many miles did they run over these four days?

$34\frac{5}{12}$ mi

51. *MTV Video* Sheryl has $8\frac{1}{2}$ hours this weekend to work on her new video. She estimates that it will take $2\frac{2}{3}$ hours to lip sync the new song and $1\frac{3}{4}$ hours to learn the new dance steps. How much time will she have left over for her MTV interview?

$4\frac{1}{12}$ hr

52. *Aquariums* Carl bought a 20-gallon aquarium. He put $17\frac{3}{4}$ gallons of water into the aquarium, but it looked too low, so he added $1\frac{1}{4}$ more gallons of water. He then put in the artificial plants and the gravel but now the water was too high, so he siphoned off $2\frac{2}{3}$ gallons of water. How many gallons of water are now in the aquarium?

$16\frac{1}{3}$ gal

To Think About

Carpentry Carpenters use fractions in their work. The picture below is a diagram of a spice cabinet. The symbol " means inches. Use the picture to answer exercises 53 and 54.

53. Before you can determine where the cabinet will fit, you need to calculate the height, A, and the width, B. Don't forget to include the $\frac{1}{2}$-inch thickness of the wood where needed.

$A = 12$ in.; $B = 15\frac{7}{8}$ in.

54. Look at the close-up of the drawer. The width is $4\frac{9}{16}''$. In the diagram, the width of the opening for the drawer is $4\frac{5}{8}''$. What is the difference? $\frac{1}{16}$ in.

Why do you think the drawer is smaller than the opening?   So that it can slide in and out

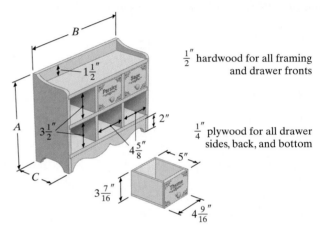

Cumulative Review

55. [0. 1. 2] Simplify. $\dfrac{36}{44}$ $\dfrac{9}{11}$

56. [0. 1. 3] Change to an improper fraction. $26\frac{3}{5}$ $\dfrac{133}{5}$

Quick Quiz 0.2 Perform the operations indicated. Simplify your answers whenever possible.

1. $\dfrac{3}{4} + \dfrac{1}{2} + \dfrac{5}{12}$ $\dfrac{5}{3}$ or $1\dfrac{2}{3}$

2. $2\dfrac{3}{5} + 4\dfrac{14}{15}$ $7\dfrac{8}{15}$

3. $6\dfrac{1}{9} - 3\dfrac{5}{6}$ $2\dfrac{5}{18}$

4. Concept Check Explain how you would find the LCD of the fractions $\frac{4}{21}$ and $\frac{5}{18}$.
Answers may vary

Classroom Quiz 0.2 You may use these problems to quiz your students' mastery of Section 0.2.

Perform the operations indicated. Simplify your answers whenever possible.

1. $\dfrac{1}{6} + \dfrac{2}{3} + \dfrac{3}{8}$ **Ans:** $\dfrac{29}{24}$ or $1\dfrac{5}{24}$

2. $2\dfrac{17}{18} + 4\dfrac{5}{9}$ **Ans:** $7\dfrac{1}{2}$

3. $7\dfrac{2}{5} - 3\dfrac{2}{3}$ **Ans:** $3\dfrac{11}{15}$

0.3 MULTIPLYING AND DIVIDING FRACTIONS

 Multiplying Fractions, Whole Numbers, and Mixed Numbers

Multiplying Fractions During a recent snowstorm, the runway at Beverly Airport was plowed. However, the plow cleared only $\frac{3}{5}$ of the width and $\frac{2}{7}$ of the length. What fraction of the total runway area was cleared? To answer this question, we need to multiply $\frac{3}{5} \times \frac{2}{7}$.

The answer is that $\frac{6}{35}$ of the total runway area was cleared.

The multiplication rule for fractions states that to multiply two fractions, we multiply the two numerators and multiply the two denominators.

> **TO MULTIPLY ANY TWO FRACTIONS**
>
> 1. Multiply the numerators.
> 2. Multiply the denominators.

Student Learning Objectives

After studying this section, you will be able to:

1 Multiply fractions, whole numbers, and mixed numbers.

2 Divide fractions, whole numbers, and mixed numbers.

EXAMPLE 1 Multiply.

(a) $\dfrac{3}{5} \times \dfrac{2}{7}$ **(b)** $\dfrac{1}{3} \times \dfrac{5}{4}$ **(c)** $\dfrac{7}{3} \times \dfrac{1}{5}$ **(d)** $\dfrac{6}{5} \times \dfrac{2}{3}$

Solution

(a) $\dfrac{3}{5} \times \dfrac{2}{7} = \dfrac{3 \cdot 2}{5 \cdot 7} = \dfrac{6}{35}$

(b) $\dfrac{1}{3} \times \dfrac{5}{4} = \dfrac{1 \cdot 5}{3 \cdot 4} = \dfrac{5}{12}$

(c) $\dfrac{7}{3} \times \dfrac{1}{5} = \dfrac{7 \cdot 1}{3 \cdot 5} = \dfrac{7}{15}$

(d) $\dfrac{6}{5} \times \dfrac{2}{3} = \dfrac{6 \cdot 2}{5 \cdot 3} = \dfrac{12}{15} = \dfrac{4}{5}$

Note that we must simplify this fraction.

Practice Problem 1 Multiply.

(a) $\dfrac{2}{7} \times \dfrac{5}{11}$ $\dfrac{10}{77}$ **(b)** $\dfrac{1}{5} \times \dfrac{7}{10}$ $\dfrac{7}{50}$ **(c)** $\dfrac{9}{5} \times \dfrac{1}{4}$ $\dfrac{9}{20}$ **(d)** $\dfrac{8}{9} \times \dfrac{3}{10}$ $\dfrac{4}{15}$

Teaching Example 1 Multiply.

(a) $\dfrac{2}{5} \times \dfrac{7}{11}$ **(b)** $\dfrac{1}{10} \times \dfrac{9}{8}$

(c) $\dfrac{5}{3} \times \dfrac{1}{2}$ **(d)** $\dfrac{8}{9} \times \dfrac{5}{6}$

Ans: (a) $\dfrac{14}{55}$ **(b)** $\dfrac{9}{80}$ **(c)** $\dfrac{5}{6}$ **(d)** $\dfrac{20}{27}$

NOTE TO STUDENT: *Fully worked-out solutions to all of the Practice Problems can be found at the back of the text starting at page SP-1*

It is possible to avoid having to simplify a fraction as the last step. In many cases we can divide by a value that appears as a factor in both a numerator and a denominator. Often it is helpful to write the numbers as products of prime factors in order to do this.

EXAMPLE 2 Multiply.

(a) $\dfrac{3}{5} \times \dfrac{5}{7}$ **(b)** $\dfrac{4}{11} \times \dfrac{5}{2}$ **(c)** $\dfrac{15}{8} \times \dfrac{10}{27}$

Teaching Example 2 Multiply.

(a) $\dfrac{4}{5} \times \dfrac{10}{13}$ **(b)** $\dfrac{1}{21} \times \dfrac{7}{8}$ **(c)** $\dfrac{6}{25} \times \dfrac{15}{16}$

Ans: (a) $\dfrac{8}{13}$ **(b)** $\dfrac{1}{24}$ **(c)** $\dfrac{9}{40}$

Solution

(a) $\dfrac{3}{5} \times \dfrac{5}{7} = \dfrac{3 \cdot 5}{5 \cdot 7} = \dfrac{3 \cdot \overset{1}{\cancel{5}}}{7 \cdot \cancel{5}} = \dfrac{3}{7}$ Note that here we divided numerator and denominator by 5.

If we factor each number, we can see the common factors.

(b) $\dfrac{4}{11} \times \dfrac{5}{2} = \dfrac{2 \cdot \overset{1}{\cancel{2}}}{11} \times \dfrac{5}{\underset{1}{\cancel{2}}} = \dfrac{10}{11}$ **(c)** $\dfrac{15}{8} \times \dfrac{10}{27} = \dfrac{\overset{1}{\cancel{3}} \cdot 5}{2 \cdot 2 \cdot \underset{1}{\cancel{2}}} \times \dfrac{5 \cdot \overset{1}{\cancel{2}}}{\underset{1}{\cancel{3}} \cdot 3 \cdot 3} = \dfrac{25}{36}$

After dividing out common factors, the resulting multiplication problem involves smaller numbers and the answers are in simplified form.

21

Practice Problem 2 Multiply.

(a) $\dfrac{3}{5} \times \dfrac{4}{3}$ $\dfrac{4}{5}$

(b) $\dfrac{9}{10} \times \dfrac{5}{12}$ $\dfrac{3}{8}$

SIDELIGHT: Dividing Out Common Factors

Why does this method of dividing out a value that appears as a factor in both numerator and denominator work? Let's reexamine one of the examples we have solved previously.

$$\frac{3}{5} \times \frac{5}{7} = \frac{3 \cdot 5}{5 \cdot 7} = \frac{3 \cdot \overset{1}{\cancel{5}}}{7 \cdot \underset{1}{\cancel{5}}} = \frac{3}{7}$$

Consider the following steps and reasons.

$\dfrac{3}{5} \times \dfrac{5}{7} = \dfrac{3 \cdot 5}{5 \cdot 7}$ Definition of multiplication of fractions.

$= \dfrac{5 \cdot 3}{5 \cdot 7}$ Change the order of the factors in the numerator, since $3 \cdot 5 = 5 \cdot 3$. This is called the commutative property of multiplication.

$= \dfrac{5}{5} \cdot \dfrac{3}{7}$ Definition of multiplication of fractions.

$= 1 \cdot \dfrac{3}{7}$ Write 1 in place of $\frac{5}{5}$, since 1 is another name for $\frac{5}{5}$.

$= \dfrac{3}{7}$ $1 \cdot \frac{3}{7} = \frac{3}{7}$, since any number can be multiplied by 1 without changing the value of the number.

Think about this concept. It is an important one that we will use again when we discuss rational expressions.

Multiplying a Fraction by a Whole Number   Whole numbers can be named using fractional notation. 3, $\frac{9}{3}$, $\frac{6}{2}$, and $\frac{3}{1}$ are ways of expressing the number *three*. Therefore,

$$3 = \frac{9}{3} = \frac{6}{2} = \frac{3}{1}.$$

When we multiply a fraction by a whole number, we merely express the whole number as a fraction whose denominator is 1 and follow the multiplication rule for fractions.

EXAMPLE 3 Multiply.

(a) $7 \times \dfrac{3}{5}$

(b) $\dfrac{3}{16} \times 4$

Solution

(a) $7 \times \dfrac{3}{5} = \dfrac{7}{1} \times \dfrac{3}{5} = \dfrac{21}{5}$ or $4\dfrac{1}{5}$

(b) $\dfrac{3}{16} \times 4 = \dfrac{3}{16} \times \dfrac{4}{1} = \dfrac{3}{4 \cdot \cancel{4}} \times \dfrac{\cancel{4}}{1} = \dfrac{3}{4}$

Notice that in (b) we did not use *prime* factors to factor 16. We recognized that $16 = 4 \cdot 4$. This is a more convenient factorization of 16 for this problem. Choose the factorization that works best for each problem. If you cannot decide what is best, factor into primes.

Practice Problem 3 Multiply.

(a) $4 \times \dfrac{2}{7}$ $\dfrac{8}{7}$ or $1\dfrac{1}{7}$

(b) $12 \times \dfrac{3}{4}$ 9

Multiplying Mixed Numbers   When multiplying mixed numbers, we first change them to improper fractions and then follow the multiplication rule for fractions.

EXAMPLE 4 How do we find the area of a rectangular field $3\frac{1}{3}$ miles long by $2\frac{1}{2}$ miles wide?

Solution To find the area, we multiply length times width.

$$3\frac{1}{3} \times 2\frac{1}{2} = \frac{10}{3} \times \frac{5}{2} = \frac{\cancel{2} \cdot 5}{3} \times \frac{5}{\cancel{2}} = \frac{25}{3} = 8\frac{1}{3}$$

The area is $8\frac{1}{3}$ square miles.

 Practice Problem 4 Delbert Robinson has a farm with a rectangular field that measures $5\frac{3}{5}$ miles long and $3\frac{3}{4}$ miles wide. What is the area of that field?

21 mi²

EXAMPLE 5 Multiply. $2\frac{2}{3} \times \frac{1}{4} \times 6$

Solution $2\frac{2}{3} \times \frac{1}{4} \times 6 = \frac{8}{3} \times \frac{1}{4} \times \frac{6}{1} = \frac{\cancel{4} \cdot 2}{\cancel{3}} \times \frac{1}{\cancel{4}} \times \frac{2 \cdot \cancel{3}}{1} = \frac{4}{1} = 4$

Practice Problem 5 Multiply.

$$3\frac{1}{2} \times \frac{1}{14} \times 4 \quad 1$$

② Dividing Fractions, Whole Numbers, and Mixed Numbers

Dividing Fractions To divide two fractions, we invert the second fraction (that is, the divisor) and then multiply the two fractions.

> **TO DIVIDE TWO FRACTIONS**
> 1. Invert the second fraction (that is, the divisor).
> 2. Now multiply the two fractions.

EXAMPLE 6 Divide.

(a) $\frac{1}{3} \div \frac{1}{2}$ **(b)** $\frac{2}{5} \div \frac{3}{10}$ **(c)** $\frac{2}{3} \div \frac{7}{5}$

Solution

(a) $\frac{1}{3} \div \frac{1}{2} = \frac{1}{3} \times \frac{2}{1} = \frac{2}{3}$ Note that we always invert the *second* fraction.

(b) $\frac{2}{5} \div \frac{3}{10} = \frac{2}{5} \times \frac{10}{3} = \frac{2}{\cancel{5}} \times \frac{\cancel{5} \cdot 2}{3} = \frac{4}{3}$ or $1\frac{1}{3}$ **(c)** $\frac{2}{3} \div \frac{7}{5} = \frac{2}{3} \times \frac{5}{7} = \frac{10}{21}$

Practice Problem 6 Divide.

(a) $\frac{2}{5} \div \frac{1}{3}$ $\frac{6}{5}$ or $1\frac{1}{5}$ **(b)** $\frac{12}{13} \div \frac{4}{3}$ $\frac{9}{13}$

Dividing a Fraction and a Whole Number The process of inverting the second fraction and then multiplying the two fractions should be done very carefully when one of the original values is a whole number. Remember, a whole number such as 2 is equivalent to $\frac{2}{1}$.

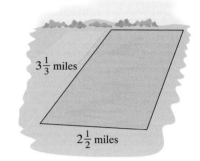

Teaching Example 7 Divide.

(a) $4 \div \dfrac{6}{7}$ (b) $\dfrac{5}{8} \div 3$

Ans: (a) $\dfrac{14}{3}$ or $4\dfrac{2}{3}$ (b) $\dfrac{5}{24}$

EXAMPLE 7 Divide.

(a) $\dfrac{1}{3} \div 2$

(b) $5 \div \dfrac{1}{3}$

Solution

(a) $\dfrac{1}{3} \div 2 = \dfrac{1}{3} \div \dfrac{2}{1} = \dfrac{1}{3} \times \dfrac{1}{2} = \dfrac{1}{6}$

(b) $5 \div \dfrac{1}{3} = \dfrac{5}{1} \div \dfrac{1}{3} = \dfrac{5}{1} \times \dfrac{3}{1} = \dfrac{15}{1} = 15$

Practice Problem 7 Divide.

(a) $\dfrac{3}{7} \div 6$ $\dfrac{1}{14}$

(b) $8 \div \dfrac{2}{3}$ 12

SIDELIGHT: Number Sense

Look at the answers to the problems in Example 7. In part (a), you will notice that $\frac{1}{6}$ is less than the original number $\frac{1}{3}$. Does this seem reasonable? Let's see. If $\frac{1}{3}$ is divided by 2, it means that $\frac{1}{3}$ will be divided into two equal parts. We would expect that each part would be less than $\frac{1}{3}$. $\frac{1}{6}$ is a reasonable answer to this division problem.

In part (b), 15 is greater than the original number 5. Does this seem reasonable? Think of what $5 \div \frac{1}{3}$ means. It means that 5 will be divided into thirds. Let's think of an easier problem. What happens when we divide 1 into thirds? We get *three* thirds. We would expect, therefore, that when we divide 5 into thirds, we would get 5×3 or 15 thirds. 15 is a reasonable answer to this division problem.

Complex Fractions Sometimes division is written in the form of a **complex fraction** with one fraction in the numerator and one fraction in the denominator. It is best to write this in standard division notation first; then complete the problem using the rule for division.

Teaching Example 8 Divide.

(a) $\dfrac{\frac{5}{9}}{\frac{2}{3}}$ (b) $\dfrac{\frac{3}{7}}{\frac{12}{21}}$

Ans: (a) $\dfrac{5}{6}$ (b) $\dfrac{3}{4}$

EXAMPLE 8 Divide.

(a) $\dfrac{\frac{3}{7}}{\frac{3}{5}}$

(b) $\dfrac{\frac{2}{9}}{\frac{5}{7}}$

Solution

(a) $\dfrac{\frac{3}{7}}{\frac{3}{5}} = \dfrac{3}{7} \div \dfrac{3}{5} = \dfrac{\cancel{3}}{7} \times \dfrac{5}{\cancel{3}} = \dfrac{5}{7}$

(b) $\dfrac{\frac{2}{9}}{\frac{5}{7}} = \dfrac{2}{9} \div \dfrac{5}{7} = \dfrac{2}{9} \times \dfrac{7}{5} = \dfrac{14}{45}$

Practice Problem 8 Divide.

(a) $\dfrac{\frac{3}{11}}{\frac{5}{7}}$ $\dfrac{21}{55}$

(b) $\dfrac{\frac{12}{5}}{\frac{8}{15}}$ $\dfrac{9}{2}$ or $4\dfrac{1}{2}$

SIDELIGHT: Invert and Multiply
Why does the method of "invert and multiply" work? The division rule really depends on the property that any number can be multiplied by 1 without changing the value of the number. Let's look carefully at an example of division of fractions:

$$\frac{2}{5} \div \frac{3}{7} = \frac{\frac{2}{5}}{\frac{3}{7}}$$ We can write the problem using a complex fraction.

$$= \frac{\frac{2}{5}}{\frac{3}{7}} \times 1$$ We can multiply by 1, since any number can be multiplied by 1 without changing the value of the number.

$$= \frac{\frac{2}{5}}{\frac{3}{7}} \times \frac{\frac{7}{3}}{\frac{7}{3}}$$ We write 1 in the form $\dfrac{\frac{7}{3}}{\frac{7}{3}}$, since any nonzero number divided by itself equals 1. We choose this value as a multiplier because it will help simplify the denominator.

$$= \frac{\frac{2}{5} \times \frac{7}{3}}{\frac{3}{7} \times \frac{7}{3}}$$ Definition of multiplication of fractions.

$$= \frac{\frac{2}{5} \times \frac{7}{3}}{1}$$ The product in the denominator equals 1.

$$= \frac{2}{5} \times \frac{7}{3}.$$

Thus we have shown that $\frac{2}{5} \div \frac{3}{7}$ is equivalent to $\frac{2}{5} \times \frac{7}{3}$ and have shown some justification for the "invert and multiply rule."

Dividing Mixed Numbers   This method for division of fractions can be used with mixed numbers. However, we first must change the mixed numbers to improper fractions and then use the rule for dividing fractions.

EXAMPLE 9   Divide.

(a) $2\dfrac{1}{3} \div 3\dfrac{2}{3}$ **(b)** $\dfrac{2}{3\frac{1}{2}}$

Solution

(a) $2\dfrac{1}{3} \div 3\dfrac{2}{3} = \dfrac{7}{3} \div \dfrac{11}{3} = \dfrac{7}{\cancel{3}} \times \dfrac{\cancel{3}}{11} = \dfrac{7}{11}$

(b) $\dfrac{2}{3\frac{1}{2}} = 2 \div 3\dfrac{1}{2} = \dfrac{2}{1} \div \dfrac{7}{2} = \dfrac{2}{1} \times \dfrac{2}{7} = \dfrac{4}{7}$

Practice Problem 9   Divide.

(a) $1\dfrac{2}{5} \div 2\dfrac{1}{3}$ $\dfrac{3}{5}$ **(b)** $4\dfrac{2}{3} \div 7$ $\dfrac{2}{3}$ **(c)** $\dfrac{1\frac{1}{5}}{1\frac{2}{7}}$ $\dfrac{14}{15}$

Teaching Example 9 Divide.

(a) $6\dfrac{2}{3} \div \dfrac{8}{15}$ **(b)** $\dfrac{4}{1\frac{1}{3}}$

Ans: **(a)** $12\dfrac{1}{2}$ **(b)** 3

NOTE TO STUDENT: Fully worked-out solutions to all of the Practice Problems can be found at the back of the text starting at page SP-1

Teaching Example 10 A grocer has 12 pounds of hard candy. He divides it into individual bags. Each bag holds $\frac{2}{3}$ pound. How many bags can he fill?

Ans: 18 bags

Teaching Tip After explaining Examples 10 and 11, you might want to show an applied example of division of mixed numbers. A land developer purchased $29\frac{3}{4}$ acres of land in northern Maine. He wants to divide the land into several equal-sized lots that each contain $1\frac{3}{4}$ acres of land. How many lots will he have? The number of lots is obtained by

$$\left(\begin{array}{c}\text{total acres}\\\text{of land}\end{array}\right) \div \left(\begin{array}{c}\text{number of acres}\\\text{in each lot}\end{array}\right)$$

$$29\frac{3}{4} \div 1\frac{3}{4} = \text{number of lots}$$

$$29\frac{3}{4} \div 1\frac{3}{4} = \frac{119}{4} \div \frac{7}{4}$$

$$= \frac{119}{4} \times \frac{4}{7}$$

$$= \frac{119}{7}$$

$$= 17 \text{ lots.}$$

Teaching Example 11 A car traveled 169 miles on $6\frac{1}{2}$ gallons of gas. How many miles per gallon did it get?

Ans: 26 mi/gal

EXAMPLE 10 A chemist has 96 fluid ounces of a solution. She pours the solution into test tubes. Each test tube holds $\frac{3}{4}$ fluid ounce. How many test tubes can she fill?

Solution We need to divide the total number of ounces, 96, by the number of ounces in each test tube, $\frac{3}{4}$.

$$96 \div \frac{3}{4} = \frac{96}{1} \div \frac{3}{4} = \frac{96}{1} \times \frac{4}{3} = \frac{\cancel{3} \cdot 32}{1} \times \frac{4}{\cancel{3}} = \frac{128}{1} = 128$$

She will be able to fill 128 test tubes.

Pause for a moment to think about the answer. Does 128 test tubes filled with solution seem like a reasonable answer? Did you perform the correct operation?

Practice Problem 10 A chemist has 64 fluid ounces of a solution. He wishes to fill several jars, each holding $5\frac{1}{3}$ fluid ounces. How many jars can he fill?

12 jars

Sometimes when solving word problems involving fractions or mixed numbers, it is helpful to solve the problem using simpler numbers first. Once you understand what operation is involved, you can go back and solve using the original numbers in the word problem.

EXAMPLE 11 A car traveled 301 miles on $10\frac{3}{4}$ gallons of gas. How many miles per gallon did it get?

Solution Use simpler numbers: 300 miles on 10 gallons of gas. We want to find out how many miles the car traveled on 1 gallon of gas. You may want to draw a picture.

10 gallons

Divide. $300 \div 10 = 30$.

300 miles

Now use the original numbers given in the problem.

$$301 \div 10\frac{3}{4} = \frac{301}{1} \div \frac{43}{4} = \frac{301}{1} \times \frac{4}{43} = \frac{1204}{43} = 28$$

This is 28 miles per gallon.

Practice Problem 11 A car traveled 126 miles on $5\frac{1}{4}$ gallons of gas. How many miles per gallon did it get? 24 mi/gal

---

Developing Your Study Skills

Why Is Homework Necessary?

Mathematics is a set of skills that you learn by doing, not by watching. Your instructor may make solving a mathematics problem look very easy, but for you to learn the necessary skills, you must practice them over and over again, just as your instructor once had to do. There is no other way. Learning mathematics is like learning to play a musical instrument, to type, or to play a sport. No matter how much you watch someone else do it, how many books you may read on "how to" do it, or how easy it may seem to be, the key to success is to practice on a regular basis. Homework provides this practice.

Verbal and Writing Skills

1. Explain in your own words how to multiply two mixed numbers.

First change each mixed number to an improper fraction. Look for a common factor in the numerator and the denominator to divide by; if one is found, perform the division. Multiply the numerators. Multiply the denominators.

2. Explain in your own words how to divide two proper fractions.

Invert the second fraction. Multiply the result by the first fraction.

Multiply. Simplify your answer whenever possible.

3. $\dfrac{36}{7} \times \dfrac{5}{9}$ $\quad \dfrac{20}{7}$ or $2\dfrac{6}{7}$

4. $\dfrac{3}{8} \times \dfrac{24}{5}$ $\quad \dfrac{9}{5}$ or $1\dfrac{4}{5}$

5. $\dfrac{17}{18} \times \dfrac{3}{5}$ $\quad \dfrac{17}{30}$

6. $\dfrac{21}{22} \times \dfrac{11}{6}$ $\quad \dfrac{7}{4}$ or $1\dfrac{3}{4}$

7. $\dfrac{4}{5} \times \dfrac{3}{10}$ $\quad \dfrac{6}{25}$

8. $\dfrac{5}{6} \times \dfrac{7}{11}$ $\quad \dfrac{35}{66}$

9. $\dfrac{24}{25} \times \dfrac{5}{2}$ $\quad \dfrac{12}{5}$ or $2\dfrac{2}{5}$

10. $\dfrac{15}{24} \times \dfrac{8}{9}$ $\quad \dfrac{5}{9}$

11. $9 \times \dfrac{2}{5}$ $\quad \dfrac{18}{5}$ or $3\dfrac{3}{5}$

12. $\dfrac{8}{11} \times 3$ $\quad \dfrac{24}{11}$ or $2\dfrac{2}{11}$

Divide. Simplify your answer whenever possible.

13. $\dfrac{8}{5} \div \dfrac{8}{3}$ $\quad \dfrac{3}{5}$

14. $\dfrac{13}{9} \div \dfrac{13}{7}$ $\quad \dfrac{7}{9}$

15. $\dfrac{3}{7} \div 3$ $\quad \dfrac{1}{7}$

16. $\dfrac{7}{8} \div 4$ $\quad \dfrac{7}{32}$

17. $10 \div \dfrac{5}{7}$ $\quad 14$

18. $18 \div \dfrac{2}{9}$ $\quad 81$

19. $\dfrac{\frac{7}{8}}{\frac{3}{4}}$ $\quad \dfrac{7}{6}$ or $1\dfrac{1}{6}$

20. $\dfrac{\frac{5}{6}}{\frac{10}{13}}$ $\quad \dfrac{13}{12}$ or $1\dfrac{1}{12}$

21. $\dfrac{\frac{5}{6}}{\frac{7}{9}}$ $\quad \dfrac{15}{14}$ or $1\dfrac{1}{14}$

22. $\dfrac{\frac{3}{4}}{\frac{11}{12}}$ $\quad \dfrac{9}{11}$

23. $1\dfrac{3}{7} \div 6\dfrac{1}{4}$ $\quad \dfrac{8}{35}$

24. $4\dfrac{1}{2} \div 3\dfrac{3}{8}$ $\quad 1\dfrac{1}{3}$

25. $3\dfrac{1}{3} \div 2\dfrac{1}{2}$ $\quad 1\dfrac{1}{3}$

26. $5\dfrac{1}{2} \div 3\dfrac{3}{4}$ $\quad 1\dfrac{7}{15}$

27. $\dfrac{15}{2\frac{2}{5}}$ $\quad 6\dfrac{1}{4}$

28. $\dfrac{18}{4\frac{1}{2}}$ $\quad 4$

29. $\dfrac{\frac{2}{3}}{1\frac{1}{4}}$ $\quad \dfrac{8}{15}$

30. $\dfrac{\frac{5}{6}}{2\frac{1}{2}}$ $\quad \dfrac{1}{3}$

Mixed Practice *Perform the proper calculations. Reduce your answer whenever possible.*

31. $\dfrac{4}{7} \times \dfrac{21}{2}$ $\quad 6$

32. $\dfrac{12}{18} \times \dfrac{9}{2}$ $\quad 3$

33. $\dfrac{5}{14} \div \dfrac{2}{7}$ $\quad \dfrac{5}{4}$ or $1\dfrac{1}{4}$

34. $\dfrac{5}{6} \div \dfrac{11}{18}$ $\quad \dfrac{15}{11}$ or $1\dfrac{4}{11}$

35. $10\dfrac{3}{7} \times 5\dfrac{1}{4}$ $\quad 54\dfrac{3}{4}$

36. $10\dfrac{2}{9} \div 2\dfrac{1}{3}$ $\quad 4\dfrac{8}{21}$

37. $30 \div \dfrac{3}{4}$ $\quad 40$

38. $8 \div 1\dfrac{1}{3}$ $\quad 6$

39. $6 \times 4\dfrac{2}{3}$ $\quad 28$

40. $5\dfrac{1}{2} \times 10$ $\quad 55$

41. $2\dfrac{1}{2} \times \dfrac{1}{10} \times \dfrac{3}{4}$ $\quad \dfrac{3}{16}$

42. $3\dfrac{1}{3} \times \dfrac{1}{5} \times \dfrac{2}{3}$ $\quad \dfrac{4}{9}$

Applications

43. *Shirt Manufacturing* A denim shirt at the Gap requires $2\frac{3}{4}$ yards of material. How many yards would be needed to make 26 shirts?

$71\frac{1}{2}$ yd

44. *Jeans Manufacturing* A pair of jeans requires $2\frac{1}{6}$ yards of denim. How many pairs of jeans can be made from 26 yards of denim?

12 pairs

45. *Mountain Biking* Jennifer rode her mountain bike for $4\frac{1}{5}$ miles after work. Two-thirds of the distance was over a mountain bike trail. How long is the mountain bike trail?

$2\frac{4}{5}$ mi

46. *Cross-Country Racing* Mens' college cross-country races are $6\frac{1}{5}$ km long. One-fourth of the conference race course has been described as uphill. How much of the course is uphill?

$1\frac{11}{20}$ km

Cumulative Review   *In exercises 47 and 48, find the missing numerator.*

47. **[0.1.4]** $\dfrac{11}{15} = \dfrac{?}{75}$ 55

48. **[0.1.4]** $\dfrac{7}{9} = \dfrac{?}{63}$ 49

Quick Quiz 0.3 Perform the operations indicated. Simplify answers whenever possible.

1. $\dfrac{7}{15} \times \dfrac{25}{14}$ $\dfrac{5}{6}$

2. $3\frac{1}{4} \times 4\frac{1}{2}$ $14\frac{5}{8}$

3. $3\frac{3}{10} \div 2\frac{1}{2}$ $1\frac{8}{25}$

4. **Concept Check** Explain the steps you would take to perform the calculation $3\frac{1}{4} \div 2\frac{1}{2}$.

Answers may vary

Classroom Quiz 0.3 You may use these problems to quiz your students' mastery of Section 0.3.

Perform the operations indicated. Simplify answers whenever possible.

1. $\dfrac{4}{13} \times \dfrac{26}{29}$ **Ans:** $\dfrac{8}{29}$

2. $5\frac{1}{2} \times 2\frac{5}{6}$ **Ans:** $15\frac{7}{12}$

3. $1\frac{3}{7} \div 2\frac{1}{14}$ **Ans:** $\dfrac{20}{29}$

1 Understanding the Meaning of Decimals

We can express a part of a whole as a fraction or as a decimal. A **decimal** is another way of writing a fraction whose denominator is 10, 100, 1000, and so on.

$$\frac{3}{10} = 0.3 \qquad \frac{5}{100} = 0.05 \qquad \frac{172}{1000} = 0.172 \qquad \frac{58}{10,000} = 0.0058$$

The period in decimal notation is known as the **decimal point.** The number of digits in a number to the right of the decimal point is known as the number of **decimal places** of the number. The place value of decimals is shown in the following chart.

Hundred thousands	Ten thousands	Thousands	Hundreds	Tens	Ones	← Decimal point	Tenths	Hundredths	Thousandths	Ten-thousandths	Hundred-thousandths
100,000	10,000	1000	100	10	1	•	$\frac{1}{10}$	$\frac{1}{100}$	$\frac{1}{1000}$	$\frac{1}{10,000}$	$\frac{1}{100,000}$

EXAMPLE 1 Write each of the following decimals as a fraction or mixed number. State the number of decimal places. Write out in words the way the number would be spoken.

(a) 0.6 **(b)** 0.29 **(c)** 0.527 **(d)** 1.38 **(e)** 0.00007

Solution

Decimal Form	Fraction Form	Number of Decimal Places	The Words Used to Describe the Number
(a) 0.6	$\frac{6}{10}$	one	six tenths
(b) 0.29	$\frac{29}{100}$	two	twenty-nine hundredths
(c) 0.527	$\frac{527}{1000}$	three	five hundred twenty-seven thousandths
(d) 1.38	$1\frac{38}{100}$	two	one and thirty-eight hundredths
(e) 0.00007	$\frac{7}{100,000}$	five	seven hundred-thousandths

Practice Problem 1 Write each decimal as a fraction or mixed number and in words.

(a) 0.9 **(b)** 0.09 **(c)** 0.731 **(d)** 1.371 **(e)** 0.0005

(a) $\frac{9}{10}$ = nine tenths **(b)** $\frac{9}{100}$ = nine hundredths **(c)** $\frac{731}{1000}$ = seven hundred thirty-one thousandths

(d) $1\frac{371}{1000}$ = one and three hundred seventy-one thousandths **(e)** $\frac{5}{10,000}$ = five ten-thousandths

Student Learning Objectives

After studying this section, you will be able to:

1 Understand the meaning of decimals.

2 Change a fraction to a decimal.

3 Change a decimal to a fraction.

4 Add and subtract decimals.

5 Multiply decimals.

6 Divide decimals.

7 Multiply or divide a decimal by a multiple of 10.

Teaching Example 1 Write each of the following decimals as a fraction or mixed number. State the number of decimal places. Write out in words the way the number would be spoken.

(a) 0.3 **(b)** 0.67 **(c)** 0.403
(d) 2.79 **(e)** 0.0017

Ans:

(a) $\frac{3}{10}$, one decimal place, three tenths

(b) $\frac{67}{100}$, two decimal places, sixty-seven hundredths

(c) $\frac{403}{1000}$, three decimal places, four hundred three thousandths

(d) $2\frac{79}{100}$, two decimal places, two and seventy-nine hundredths

(e) $\frac{17}{10,000}$, four decimal places, seventeen ten-thousandths

Teaching Tip Remind students that what we call *decimals* are really decimal fractions. They are a special kind of fraction with a denominator of 10, 100, 1000, 10,000, etc. Decimal fractions are usually referred to simply as decimals.

You have seen that a given fraction can be written in several different but equivalent ways. There are also several different equivalent ways of writing the decimal form of fractions. The decimal 0.18 can be written in the following equivalent ways:

$$\text{Fractional form:} \quad \frac{18}{100} = \frac{180}{1000} = \frac{1800}{10,000} = \frac{18,000}{100,000}$$

$$\text{Decimal form:} \quad 0.18 = 0.180 = 0.1800 = 0.18000.$$

Thus we see that *any number of terminal zeros may be added to the right-hand side of a decimal* without changing its value.

$$0.13 = 0.1300 \qquad 0.162 = 0.162000$$

Similarly, *any number of terminal zeros may be removed from the right-hand side of a decimal* without changing its value.

② Changing a Fraction to a Decimal

A fraction can be changed to a decimal by dividing the denominator into the numerator.

Teaching Example 2 Write each of the following fractions as a decimal.

(a) $\frac{1}{4}$ (b) $\frac{5}{8}$ (c) $\frac{13}{10}$ (d) $\frac{9}{400}$

Ans: (a) 0.25 **(b)** 0.625 **(c)** 1.3
      (d) 0.0225

EXAMPLE 2 Write each of the following fractions as a decimal.

(a) $\frac{3}{4}$ (b) $\frac{21}{20}$ (c) $\frac{1}{8}$ (d) $\frac{3}{200}$

Solution

(a) $\frac{3}{4} = 0.75$ since $\begin{array}{r} 0.75 \\ 4\overline{)3.00} \\ \underline{2\,8} \\ 20 \\ \underline{20} \\ 0 \end{array}$

(b) $\frac{21}{20} = 1.05$ since $\begin{array}{r} 1.05 \\ 20\overline{)21.00} \\ \underline{20} \\ 1\,00 \\ \underline{1\,00} \\ 0 \end{array}$

(c) $\frac{1}{8} = 0.125$ since $\begin{array}{r} 0.125 \\ 8\overline{)1.000} \\ \underline{8} \\ 20 \\ \underline{16} \\ 40 \\ \underline{40} \\ 0 \end{array}$

(d) $\frac{3}{200} = 0.015$ since $\begin{array}{r} 0.015 \\ 200\overline{)3.000} \\ \underline{2\,00} \\ 1\,000 \\ \underline{1\,000} \\ 0 \end{array}$

NOTE TO STUDENT: *Fully worked-out solutions to all of the Practice Problems can be found at the back of the text starting at page SP-1*

Practice Problem 2 Write as decimals.

(a) $\frac{3}{8}$ 0.375 (b) $\frac{7}{200}$ 0.035 (c) $\frac{33}{20}$ 1.65

Sometimes division yields an infinite repeating decimal. We use three dots to indicate that the pattern continues forever. For example,

$$\frac{1}{3} = 0.3333\ldots$$

$$\begin{array}{r} 0.333 \\ 3\overline{)1.000} \\ \underline{9} \\ 10 \\ \underline{9} \\ 10 \\ \underline{9} \\ 1 \end{array}$$

An alternative notation is to place a bar over the repeating digit(s):

$$0.3333\ldots = 0.\overline{3} \qquad 0.575757\ldots = 0.\overline{57}$$

EXAMPLE 3 Write each fraction as a decimal.

(a) $\dfrac{2}{11}$

(b) $\dfrac{5}{6}$

Solution

(a) $\dfrac{2}{11} = 0.181818\ldots$ or $0.\overline{18}$

$$\begin{array}{r} 0.1818 \\ 11\overline{)2.0000} \\ \underline{11} \\ 90 \\ \underline{88} \\ 20 \\ \underline{11} \\ 90 \\ \underline{88} \\ 2 \end{array}$$

(b) $\dfrac{5}{6} = 0.8333\ldots$ or $0.8\overline{3}$

$$\begin{array}{r} 0.8333 \\ 6\overline{)5.0000} \\ \underline{4\,8} \\ 20 \\ \underline{18} \\ 20 \\ \underline{18} \\ 20 \\ \underline{18} \\ 2 \end{array}$$

Note that the 8 does not repeat. Only the digit 3 is repeating.

Practice Problem 3 Write each fraction as a decimal.

(a) $\dfrac{1}{6}$ $0.1666\ldots$ or $0.1\overline{6}$

(b) $\dfrac{5}{11}$ $0.454545\ldots$ or $0.\overline{45}$

Sometimes division must be carried out to many places in order to observe the repeating pattern. This is true in the following example:

$$\frac{2}{7} = 0.285714285714285714\ldots \qquad \text{This can also be written as } \frac{2}{7} = 0.\overline{285714}.$$

It can be shown that the denominator determines the maximum number of decimal places that might repeat. So $\frac{2}{7}$ must repeat in the seventh decimal place or sooner.

3 Changing a Decimal to a Fraction

To convert from a decimal to a fraction, merely write the decimal as a fraction with a denominator of 10, 100, 1000, 10,000, and so on, and simplify the result when possible.

Teaching Example 3 Write each fraction as a decimal.

(a) $\dfrac{5}{9}$

(b) $\dfrac{7}{12}$

Ans: (a) $0.\overline{5}$ **(b)** $0.58\overline{3}$

Calculator

Fraction to Decimal

You can use a calculator to change $\frac{3}{5}$ to a decimal.
Enter:

$$\boxed{3} \;\; \boxed{\div} \;\; \boxed{5} \;\; \boxed{=}$$

The display should read

$$\boxed{0.6}$$

Try the following.

(a) $\dfrac{17}{25}$ **(b)** $\dfrac{2}{9}$

(c) $\dfrac{13}{10}$ **(d)** $\dfrac{15}{19}$

Answers:

(a) 0.68   **(b)** $0.\overline{2}$ **(c)** 1.3

(d) 0.7894737 (Rounded to seven decimal places.)

EXAMPLE 4   Write each decimal as a fraction.

(a) 0.2   (b) 0.35   (c) 0.516   (d) 0.74   (e) 0.138   (f) 0.008

Solution

(a) $0.2 = \dfrac{2}{10} = \dfrac{1}{5}$

(b) $0.35 = \dfrac{35}{100} = \dfrac{7}{20}$

(c) $0.516 = \dfrac{516}{1000} = \dfrac{129}{250}$

(d) $0.74 = \dfrac{74}{100} = \dfrac{37}{50}$

(e) $0.138 = \dfrac{138}{1000} = \dfrac{69}{500}$

(f) $0.008 = \dfrac{8}{1000} = \dfrac{1}{125}$

Practice Problem 4   Write each decimal as a fraction and simplify whenever possible.

(a) 0.8   (b) 0.88   (c) 0.45   (d) 0.148   (e) 0.612   (f) 0.016

$\dfrac{4}{5}$ $\dfrac{22}{25}$ $\dfrac{9}{20}$ $\dfrac{37}{250}$ $\dfrac{153}{250}$ $\dfrac{2}{125}$

All repeating decimals can also be converted to fractional form. In practice, however, repeating decimals are usually rounded to a few places. It will not be necessary, therefore, to learn how to convert $0.\overline{033}$ to $\frac{11}{333}$ for this course.

4 Adding and Subtracting Decimals

Last week Bob spent $19.83 on lunches purchased at the cafeteria at work. During this same period, Sally spent $24.76 on lunches. How much did the two of them spend on lunches last week?

Adding and subtracting decimals is similar to adding and subtracting whole numbers, except that it is necessary to line up decimal points. To perform the operation $19.83 + 24.76$, we line up the numbers in column form and add the digits:

$$\begin{array}{r} 19.83 \\ + 24.76 \\ \hline 44.59 \end{array}$$

Thus Bob and Sally spent $44.59 on lunches last week.

ADDITION AND SUBTRACTION OF DECIMALS

1. Write in column form and line up the decimal points.
2. Add or subtract the digits.

EXAMPLE 5   Perform the following operations.

(a) $3.6 + 2.3$ (b) $127.32 - 38.48$ (c) $3.1 + 42.36 + 9.034$ (d) $5.0006 - 3.1248$

Solution

(a) $\begin{array}{r} 3.6 \\ + 2.3 \\ \hline 5.9 \end{array}$

(b) $\begin{array}{r} 127.32 \\ - 38.48 \\ \hline 88.84 \end{array}$

(c) $\begin{array}{r} 3.1 \\ 42.36 \\ + 9.034 \\ \hline 54.494 \end{array}$

(d) $\begin{array}{r} 5.0006 \\ - 3.1248 \\ \hline 1.8758 \end{array}$

Practice Problem 5   Add or subtract.

(a) $3.12 + 5.08 + 1.42$ 9.62   (b) $152.003 - 136.118$ 15.885

(c) $1.1 + 3.16 + 5.123$ 9.383   (d) $1.0052 - 0.1234$ 0.8818

SIDELIGHT: Adding Zeros to the Right-Hand Side of the Decimal
When we added fractions, we had to have common denominators. Since decimals are really fractions, why can we add them without having common denominators? Actually, we have to have common denominators to add any fractions, whether they are in decimal form or fraction form. However, sometimes the notation does not show this. Let's examine Example 5(c).

Original Problem   We are adding the three numbers:

$$
\begin{array}{r}
3.1 \\
42.36 \\
+\ 9.034 \\
\hline
54.494
\end{array}
$$

$$3\tfrac{1}{10} + 42\tfrac{36}{100} + 9\tfrac{34}{1000}$$

$$3\tfrac{100}{1000} + 42\tfrac{360}{1000} + 9\tfrac{34}{1000}$$

$$3.100 + 42.360 + 9.034 \quad \text{This is the new problem.}$$

Original Problem   **New Problem**

$$
\begin{array}{r}
3.1 \\
42.36 \\
+\ 9.034 \\
\hline
54.494
\end{array}
\qquad
\begin{array}{r}
3.100 \\
42.360 \\
+\ 9.034 \\
\hline
54.494
\end{array}
$$

We notice that the results are the same. The only difference is the notation. We are using the property that any number of zeros may be added to the right-hand side of a decimal without changing its value.

This shows the convenience of adding and subtracting fractions in decimal form. Little work is needed to change the decimals so that they have a common denominator. All that is required is to add zeros to the right-hand side of the decimal (and we usually do not even write out that step except when subtracting).

As long as we line up the decimal points, we can add or subtract any decimal fractions.

In the following example we will find it useful to add zeros to the right-hand side of the decimal.

EXAMPLE 6   Perform the following operations.

(a) $1.0003 + 0.02 + 3.4$ **(b)** $12 - 0.057$

Solution   We will add zeros so that each number shows the same number of decimal places.

(a)
$$
\begin{array}{r}
1.0003 \\
0.0200 \\
+\ 3.4000 \\
\hline
4.4203
\end{array}
$$

(b)
$$
\begin{array}{r}
12.000 \\
-\ 0.057 \\
\hline
11.943
\end{array}
$$

Practice Problem 6   Perform the following operations.

(a) $0.061 + 5.0008 + 1.3$ 6.3618   **(b)** $18 - 0.126$ 17.874

Teaching Example 6 Perform the following operations.

(a) $0.04 + 5.3 + 2.157$

(b) $24 - 0.104$

Ans: (a) 7.497 **(b)** 23.896

NOTE TO STUDENT: *Fully worked-out solutions to all of the Practice Problems can be found at the back of the text starting at page SP-1*

Multiplying Decimals

MULTIPLICATION OF DECIMALS

To multiply decimals, you first multiply as with whole numbers. To determine the position of the decimal point, you count the total number of decimal places in the two numbers being multiplied. This will determine the number of decimal places that should appear in the answer.

Teaching Example 7 Multiply. 0.6×0.7

Ans: 0.42

EXAMPLE 7 Multiply. 0.8×0.4

Solution

$$
\begin{array}{rl}
0.8 & (\text{ one decimal place}) \\
\times 0.4 & (\text{ one decimal place}) \\
\hline
0.32 & (\text{ two decimal places})
\end{array}
$$

Practice Problem 7 Multiply. 0.5×0.3 0.15

Note that you will often have to add zeros to the left of the digits obtained in the product so that you obtain the necessary number of decimal places.

Teaching Example 8 Multiply. 0.17×0.3

Ans: 0.051

EXAMPLE 8 Multiply. 0.123×0.5

Solution

$$
\begin{array}{rl}
0.123 & (\text{ three decimal places}) \\
\times \quad 0.5 & (\text{ one decimal place}) \\
\hline
0.0615 & (\text{ four decimal places})
\end{array}
$$

Practice Problem 8 Multiply. 0.12×0.4 0.048

Here are some examples that involve more decimal places.

Teaching Example 9 Multiply.

(a) 1.14×0.003 **(b)** 0.004×0.00003

Ans: **(a)** 0.00342   **(b)** 0.00000012

EXAMPLE 9 Multiply.

(a) 2.56×0.003 **(b)** 0.0036×0.008

Solution

$$
\textbf{(a)} \quad
\begin{array}{rl}
2.56 & (\text{ two decimal places}) \\
\times \ 0.003 & (\text{ three decimal places}) \\
\hline
0.00768 & (\text{ five decimal places})
\end{array}
$$

$$
\textbf{(b)} \quad
\begin{array}{rl}
0.0036 & (\text{ four decimal places}) \\
\times \ 0.008 & (\text{ three decimal places}) \\
\hline
0.0000288 & (\text{ seven decimal places})
\end{array}
$$

Practice Problem 9 Multiply.

(a) 1.23×0.005 0.00615   **(b)** 0.003×0.00002 0.00000006

SIDELIGHT: Counting the Number of Decimal Places

Why do we count the number of decimal places? The rule really comes from the properties of fractions. If we write the problem in Example 8 in fraction form, we have

$$
0.123 \times 0.5 = \frac{123}{1000} \times \frac{5}{10} = \frac{615}{10{,}000} = 0.0615.
$$

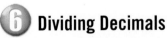 **Dividing Decimals**

When discussing division of decimals, we frequently refer to the three primary parts of a division problem. Be sure you know the meaning of each term.

The **divisor** is the number you divide into another.

The **dividend** is the number to be divided.

The **quotient** is the result of dividing one number by another.

In the problem $6 \div 2 = 3$ we represent each of these terms as follows:

When dividing two decimals, count *the number of decimal places* in the divisor. Then *move the decimal point to the right* that *same number of places* in both *the divisor* and *the dividend*. Mark that position with a caret ($_\wedge$). Finally, perform the division. Be sure to line up the decimal point in the quotient with the position indicated by the caret in the dividend.

EXAMPLE 10 Four friends went out for lunch. The total bill, including tax, was $32.68. How much did each person pay if they shared the cost equally?

Solution   To answer this question, we must calculate $32.68 \div 4$.

$$
\begin{array}{r}
8.17 \\
4\overline{)32.68} \\
\underline{32} \\
06 \\
\underline{4} \\
28 \\
\underline{28} \\
0
\end{array}
$$

Since there are no decimal places in the divisor, we do not need to move the decimal point. We must be careful, however, to place the decimal point in the quotient directly above the decimal point in the dividend.

Thus $32.68 \div 4 = 8.17$, and each friend paid $8.17.

Practice Problem 10 Sally Keyser purchased 6 boxes of paper for an inkjet printer. The cost was $31.56. There was no tax since she purchased the paper for a charitable organization. How much did she pay for each box of paper? $5.26

Teaching Example 10 Chris bought 3 identical shirts for his sons. The total cost was $28.47. How much did each shirt cost?

Ans: $9.49

NOTE TO STUDENT: Fully worked-out solutions to all of the Practice Problems can be found at the back of the text starting at page SP-1

Note that sometimes we will need to place extra zeros in the dividend in order to move the decimal point the required number of places.

EXAMPLE 11 Divide. $16.2 \div 0.027$

Solution

$$0.027\overline{)16.200}$$

There are **three** decimal places in the divisor, so we move the decimal point **three places** to **the right** in the **divisor** and **dividend** and mark the new position by a caret. Note that we must add two zeros to 16.2 in order to do this.

three decimal places

$$
\begin{array}{r}
600. \\
0.027\overline{)16.200} \\
\underline{16\,2} \\
000
\end{array}
$$

Now perform the division as with whole numbers. The decimal point in the answer is directly above the caret.

Thus $16.2 \div 0.027 = 600$.

Practice Problem 11 Divide. $1800 \div 0.06$ 30,000

Teaching Example 11 Divide. $35.1 \div 0.078$

Ans: 450

Special care must be taken to line up the digits in the quotient. Note that sometimes we will need to place zeros in the quotient after the decimal point.

Teaching Example 12 Divide. $0.03496 \div 2.3$

Ans: 0.0152

EXAMPLE 12   Divide. $0.04288 \div 3.2$

Solution

$$3.\overset{\wedge}{2}\,)\,\overline{0.0\overset{\wedge}{4}288}$$

There is **one** decimal place in the divisor, so we move the decimal point **one place** to **the right** in the **divisor** and **dividend** and mark the new position by a caret.

one decimal place

$$
\begin{array}{r}
0.0134 \\
3.\overset{\wedge}{2}\,)\,\overline{0.0\overset{\wedge}{4}288} \\
\underline{3\,2} \\
108 \\
\underline{96} \\
128 \\
\underline{128} \\
0
\end{array}
$$

Now perform the division as for whole numbers. The decimal point in the answer is directly above the caret. Note the need for the initial zero after the decimal point in the answer.

Thus $0.04288 \div 3.2 = 0.0134$.

Practice Problem 12   Divide. $0.01764 \div 4.9$ 0.0036

SIDELIGHT: Dividing Decimals by Another Method

Why does this method of dividing decimals work? Essentially, we are using the steps we used in Section 0.1 to change a fraction to an equivalent fraction by multiplying both the numerator and denominator by the same number. Let's reexamine Example 12.

$$0.04288 \div 3.2 = \frac{0.04288}{3.2}$$

Write the original problem using fraction notation.

$$= \frac{0.04288 \times 10}{3.2 \quad \times 10}$$

Multiply the numerator and denominator by 10. Since this is the same as multiplying by 1, we are not changing the fraction.

$$= \frac{0.4288}{32}$$

Write the result of multiplication by 10.

$$= 0.4288 \div 32$$

Rewrite the fraction as an equivalent problem with division notation.

Notice that we have obtained a new problem that is the same as the problem in Example 12 when we moved the decimal one place to the right in the divisor and dividend. We see that the reason we can move the decimal point as many places as necessary to the right in divisor and dividend is that this is the same as multiplying the numerator and denominator of a fraction by a power of 10 to obtain an equivalent fraction.

 Multiplying and Dividing a Decimal by a Multiple of 10

When multiplying by 10, 100, 1000, and so on, a simple rule may be used to obtain the answer. For every zero in the multiplier, move the decimal point one place to the right.

EXAMPLE 13 Multiply.

(a) 3.24×10 **(b)** 15.6×100 **(c)** 0.0026×1000

Solution

(a) $3.24 \times 10 = 32.4$ One zero—move decimal point one place to the right.
(b) $15.6 \times 100 = 1560$ Two zeros—move decimal point two places to the right.
(c) $0.0026 \times 1000 = 2.6$ Three zeros—move decimal point three places to the right.

Practice Problem 13 Multiply.

(a) 0.0016×100 0.16   **(b)** 2.34×1000 2340   **(c)** $56.75 \times 10,000$ 567,500

Teaching Example 13 Multiply.

(a) 5.142×100 **(b)** 0.04×1000

(c) 23.4×10

Ans: (a) 514.2 **(b)** 40 **(c)** 234

Teaching Tip Point out to students that once you can multiply by 100 and by 1000 using this simple rule, you can expand it to other situations.

The reverse rule is true for division. When dividing by 10, 100, 1000, 10,000, and so on, move the decimal point one place to the left for every zero in the divisor.

EXAMPLE 14 Divide.

(a) $52.6 \div 10$ **(b)** $0.0038 \div 100$ **(c)** $5936.2 \div 1000$

Solution

(a) $\dfrac{52.6}{10} = 5.26$ Move one place to the left.

(b) $\dfrac{0.0038}{100} = 0.000038$ Move two places to the left.

(c) $\dfrac{5936.2}{1000} = 5.9362$ Move three places to the left.

Teaching Example 14 Divide.

(a) $5.67 \div 10$ **(b)** $0.034 \div 1000$

(c) $265.9 \div 100$

Ans: (a) 0.567 **(b)** 0.000034 **(c)** 2.659

Practice Problem 14 Divide.

(a) $\dfrac{5.82}{10}$ 0.582   **(b)** $123.4 \div 1000$ 0.1234   **(c)** $\dfrac{0.00614}{10,000}$ 0.000000614

Developing Your Study Skills

Making a Friend in the Class

Attempt to make a friend in your class. You may find that you enjoy sitting together and drawing support and encouragement from one another. Exchange phone numbers so you can call each other whenever you get stuck in your study. Set up convenient times to study together on a regular basis, to do homework, and to review for exams.

You must not depend on a friend or fellow student to tutor you, do your work for you, or in any way be responsible for your learning. However, you will learn from one another as you seek to master the course. Studying with a friend and comparing notes, methods, and solutions can be very helpful. And it can make learning mathematics a lot more fun!

Verbal and Writing Skills

1. A decimal is another way of writing a fraction whose denominator is _____10, 100, 1000, 10,000, and so on_____.

2. We write 0.42 in words as _____forty-two hundredths_____.

3. When dividing 7432.9 by 1000 we move the decimal point ___3___ places to the ___left___.

4. When dividing 96.3 by 10,000 we move the decimal point ___4___ places to the ___left___.

Write each fraction as a decimal.

5. $\frac{5}{8}$ 0.625   **6.** $\frac{6}{25}$ 0.24   **7.** $\frac{3}{15}$ 0.2   **8.** $\frac{12}{15}$ 0.8   **9.** $\frac{7}{11}$ $0.\overline{63}$ **10.** $\frac{2}{3}$ $0.\overline{6}$

Write each decimal as a fraction in simplified form.

11. 0.8   **12.** 0.15   **13.** 0.625   **14.** 0.08   **15.** 3.4   **16.** 6.25
$\frac{4}{5}$ $\frac{3}{20}$ $\frac{5}{8}$ $\frac{2}{25}$ $\frac{17}{5}$ or $3\frac{2}{5}$ $\frac{25}{4}$ or $6\frac{1}{4}$

Add or subtract.

17. 1.71 + 0.38   **18.** 3.42 + 0.38   **19.** 2.5 + 3.42 + 4.9   **20.** 6.31 + 4.2 + 8.5
2.09   3.8   10.82   19.01

21. 46.03 + 215.1 + 0.078   **22.** 33.01 + 0.38 + 175.401   **23.** 147.18 − 15.39   **24.** 131.43 − 86.95
261.208   208.791   131.79   44.48

Multiply or divide.

25. 7.21 × 4.2 30.282   **26.** 7.12 × 2.6 18.512   **27.** 4.23 × 0.025 0.10575   **28.** 3.84 × 0.0017 0.006528

29. 58,200 × 0.0015 87.3   **30.** 23,000 × 0.0042 96.6   **31.** 7.9728 ÷ 3.02 2.64   **32.** 6.519 ÷ 2.05 3.18

33. 0.5230 ÷ 0.002 261.5   **34.** 0.031 ÷ 0.005 6.2   **35.** 0.03048 ÷ 0.06 0.508   **36.** 0.00855 ÷ 0.09 0.095

Multiply or divide by moving the decimal point.

37. 3.45 × 1000 3450   **38.** 1.36 × 1000 1360   **39.** 0.76 ÷ 100 0.0076   **40.** 175,318 ÷ 1000 175.318

41. 7.36 × 10,000   **42.** 0.00243 × 100,000   **43.** 73,892 ÷ 100,000   **44.** 3.52 ÷ 1000
73,600   243   0.73892   0.00352

Mixed Practice *Perform the indicated calculations.*

45. 54.8 × 0.15   **46.** 8.252 × 0.005   **47.** 13.75 + 2.55 + 0.078   **48.** 1.109 + 0.088 + 16.4
8.22   0.04126   16.378   17.597

49. 0.05724 ÷ 0.027   **50.** 77.136 ÷ 0.003   **51.** 0.7683 × 1000   **52.** 34.72 × 10,000
2.12   25,712   768.3   347,200

Applications

53. ***Measurement*** While mixing solutions in her chemistry lab, Mia needed to change the measured data from pints to liters. There are 0.4732 liters in one pint and the original measurement was 5.5 pints. What is the measured data in liters?

$5.5 \times 0.4732 = 2.6026$ L

54. ***Hybrid Cars*** Chris Smith bought a Toyota Prius, a hybrid electric-gas-powered car, because of its great gas mileage. The car averages 51 miles per gallon in the city, and the tank holds 11.9 gallons of gas. How many miles can Chris drive in the city on a full tank of gas?

$51 \times 11.9 = 606.9$ mi

55. ***Drinking Water*** The EPA standard for safe drinking water is a maximum of 1.3 milligrams of copper per liter of water. A water testing firm found 6.8 milligrams of copper in a 5-liter sample drawn from Jim and Sharon LeBlanc's house. Is the water safe or not? By how much does the amount of copper exceed or fall short of the maximum allowed?

No; there were 0.3 milligrams of copper in excess of the standard for 5 liters of water.

56. ***Wages*** Harry has a part-time job at Stop and Shop. He earns $8.50 an hour. Last week he worked 19 hours. He had hoped to earn at least $150. Did he reach his goal or not? By how much did he exceed or fall short of his goal?

Yes; he exceeded his goal by $11.50.

Cumulative Review *Perform each operation. Simplify all answers.*

57. [0.3.2] $3\dfrac{1}{2} \div 5\dfrac{1}{4}$

$\dfrac{2}{3}$

58. [0.3.1] $\dfrac{3}{8} \cdot \dfrac{12}{27}$

$\dfrac{1}{6}$

59. [0.2.3] $\dfrac{12}{25} + \dfrac{9}{20}$

$\dfrac{48 + 45}{100} = \dfrac{93}{100}$

60. [0.2.4] $1\dfrac{3}{5} - \dfrac{1}{2}$

$\dfrac{16 - 5}{10} = \dfrac{11}{10}$ or $1\dfrac{1}{10}$

Quick Quiz 0.4 Perform the calculations indicated.

1. $8.0567 - 2.3489$ 5.7078

2. 58.7×0.06 3.522

3. $4.608 \div 0.16$ 28.8

4. **Concept Check** Explain how you would place the decimal points when performing the calculation $0.252 \div 0.0035$. Answers may vary

Classroom Quiz 0.4 You may use these problems to quiz your students' mastery of Section 0.4.

Perform the calculations indicated.

1. $9.0038 - 4.6329$ **Ans:** 4.3709

2. 47.9×0.07 **Ans:** 3.353

3. $5.957 \div 0.23$ **Ans:** 25.9

How are you doing with your homework assignments in Sections 0.1 to 0.4? Do you feel you have mastered the material so far? Do you understand the concepts you have covered? Before you go further in the textbook, take some time to do each of the following problems.

0.1

In exercises 1 and 2, simplify each fraction.

1. $\dfrac{15}{55}$

2. $\dfrac{46}{115}$

3. Write $\dfrac{15}{4}$ as a mixed number.

4. Change $4\dfrac{5}{7}$ to an improper fraction.

Find the missing number.

5. $\dfrac{4}{5} = \dfrac{?}{30}$

6. $\dfrac{5}{9} = \dfrac{?}{81}$

0.2

7. Find the LCD, but do not add. $\dfrac{3}{8}, \dfrac{5}{6},$ and $\dfrac{7}{15}$

Perform the calculation indicated. Write the answer in simplest form.

8. $\dfrac{3}{7} + \dfrac{2}{7}$

9. $\dfrac{5}{14} + \dfrac{2}{21}$

10. $2\dfrac{3}{4} + 5\dfrac{2}{3}$

11. $\dfrac{17}{18} - \dfrac{5}{9}$

12. $\dfrac{6}{7} - \dfrac{2}{3}$

13. $3\dfrac{1}{5} - 1\dfrac{3}{8}$

0.3

Perform the calculations indicated. Write the answer in simplest form.

14. $\dfrac{25}{7} \times \dfrac{14}{45}$

15. $12 \times 3\dfrac{1}{2}$

16. $4 \div \dfrac{8}{7}$

17. $2\dfrac{1}{3} \div 3\dfrac{1}{4}$

0.4

Change to a decimal. Use bar notation for repeating decimals.

18. $\dfrac{3}{8}$

19. $\dfrac{5}{6}$

20. $\dfrac{3}{200}$

Perform the calculations indicated.

21. $15.23 + 3.6 + 0.821$

22. 3.28×0.63

23. $3.015 \div 6.7$

24. $12.13 - 9.884$

Now turn to page SA-1 for the answer to each of these problems. Each answer also includes a reference to the objective in which the problem is first taught. If you missed any of these problems, you should stop and review the Examples and Practice Problems in the referenced objective. A little review now will help you master the material in the upcoming sections of the text.

**Answers (left margin):**

1. $\dfrac{3}{11}$
2. $\dfrac{2}{5}$
3. $3\dfrac{3}{4}$
4. $\dfrac{33}{7}$
5. 24
6. 45
7. LCD $= 120$
8. $\dfrac{5}{7}$
9. $\dfrac{19}{42}$
10. $8\dfrac{5}{12}$
11. $\dfrac{7}{18}$
12. $\dfrac{4}{21}$
13. $1\dfrac{33}{40}$
14. $\dfrac{10}{9}$ or $1\dfrac{1}{9}$
15. 42
16. $\dfrac{7}{2}$ or $3\dfrac{1}{2}$
17. $\dfrac{28}{39}$
18. 0.375
19. $0.8\overline{3}$
20. 0.015
21. 19.651
22. 2.0664
23. 0.45
24. 2.246

Understanding the Meaning of Percent

A **percent** is a fraction that has a denominator of 100. When you say "sixty-seven percent" or write 67%, you are just expressing the fraction $\frac{67}{100}$ in another way. The word *percent* is a shortened form of the Latin words *per centum,* which means "by the hundred." In everyday use, percent means per one hundred.

Russell Camp owns 100 acres of land in Montana. 49 of the acres are covered with trees. The rest of the land is open fields. We say that 49% of his land is covered with trees.

It is important to see that 49% means 49 parts out of 100 parts. It can also be written as a fraction, $\frac{49}{100}$, or as a decimal, 0.49. Understanding the meaning of the notation allows you to change from one form to another. For example,

$$49\% = 49 \text{ out of } 100 \text{ parts} = \frac{49}{100} = 0.49.$$

Similarly, you can express a fraction with denominator 100 as a percent or a decimal.

$\frac{11}{100}$ means 11 parts out of 100 or 11%. So $\frac{11}{100} = 11\% = 0.11.$

Student Learning Objectives

After studying this section, you will be able to:

1 Understand the meaning of percent.

2 Change a decimal to a percent.

3 Change a percent to a decimal.

4 Find the percent of a given number.

5 Find the missing percent when given two numbers.

 Changing a Decimal to a Percent

Now that we understand the concept, we can use some quick procedures to change from decimals to percent, and vice versa.

CHANGING A DECIMAL TO A PERCENT

1. Move the decimal point two places to the right.

2. Add the % symbol.

EXAMPLE 1 Change to a percent.

(a) 0.23   **(b)** 0.461   **(c)** 0.4

Solution We move the decimal point two places to the right and add the % symbol.

(a) 0.23 = 23%   **(b)** 0.461 = 46.1%   **(c)** 0.4 = 0.40 = 40%

Practice Problem 1 Change to a percent.

(a) 0.92 92%   **(b)** 0.418 41.8%   **(c)** 0.7 70%

Teaching Example 1 Change to a percent.

(a) 0.46   **(b)** 0.3   **(c)** 0.523

Ans: (a) 46%   **(b)** 30%   **(c)** 52.3%

NOTE TO STUDENT: Fully worked-out solutions to all of the Practice Problems can be found at the back of the text starting at page SP-1

Be sure to follow the same procedure for percents that are less than 1%. Remember, 0.01 is 1%. Thus we would expect 0.001 to be less than 1%. 0.001 = 0.1% or one-tenth (0.1) of a percent.

EXAMPLE 2 Change to a percent.

(a) 0.0364   **(b)** 0.0026   **(c)** 0.0008

Solution   We move the decimal point two places to the right and add the % symbol.

(a) 0.0364 = 3.64%   **(b)** 0.0026 = 0.26%   **(c)** 0.0008 = 0.08%

Practice Problem 2 Change to a percent.

(a) 0.0019 0.19%   **(b)** 0.0736 7.36%   **(c)** 0.0003 0.03%

Be sure to follow the same procedure for percents that are greater than 100%. Remember that 1 is 100%. Thus we would expect 1.5 to be greater than 100%. In fact, 1.5 = 150%.

EXAMPLE 3 Change to a percent.

(a) 1.48   **(b)** 2.938   **(c)** 4.5

Solution   We move the decimal point two places to the right and add the % symbol.

(a) 1.48 = 148%   **(b)** 2.938 = 293.8%   **(c)** 4.5 = 4.50 = 450%

Practice Problem 3 Change to a percent.

(a) 3.04 304%   **(b)** 5.186 518.6%   **(c)** 2.1 210%

Calculator

 Percent to Decimal

You can use a calculator to change 52% to a decimal. If your calculator has a % key, do the following:

Enter: 52 %

The display should read:

    0.52

If your calculator does not have a % key, divide the number by 100.
Enter:

    52 ÷ 100 =

Try the following:

(a) 46%   **(b)** 137%
(c) 9.3%   **(d)** 6%

Note: The calculator divides by 100 when the percent key is pressed. If you do not have a % key, then you can divide by 100 instead.

③ Changing a Percent to a Decimal

In this procedure we move the decimal point to the left and remove the % symbol.

CHANGING A PERCENT TO A DECIMAL
1. Move the decimal point two places to the left.
2. Remove the % symbol.

EXAMPLE 4 Change to a decimal.

(a) 4%   **(b)** 3.2%   **(c)** 0.6%

Solution   First we move the decimal point two places to the left. Then we remove the % symbol.

(a) 4% = 4.% = 0.04
            ↑

The unwritten decimal point is understood to be here.

(b) 3.2% = 0.032
(c) 0.6% = 0.006

Practice Problem 4 Change to a decimal.

(a) 7% 0.07   **(b)** 9.3% 0.093   **(c)** 0.2% 0.002

EXAMPLE 5 Change to a decimal.

(a) 192%   **(b)** 254.8%   **(c)** 0.027%

Solution   First we move the decimal point two places to the left. Then we remove the % symbol.

(a) 192% = 192.% = 1.92

↑

The unwritten decimal point is understood to be here.

(b) 254.8% = 2.548

(c) 0.027% = 0.00027

Practice Problem 5 Change to a decimal.

(a) 131% 1.31   **(b)** 301.6% 3.016   **(c)** 0.04% 0.0004

Teaching Example 5 Change to a decimal.

(a) 231%   **(b)** 634.9%   **(c)** 0.018%

Ans: (a) 2.31 **(b)** 6.349 **(c)** 0.00018%

NOTE TO STUDENT: Fully worked-out solutions to all of the Practice Problems can be found at the back of the text starting at page SP-1

4 Finding the Percent of a Given Number

How do we find 60% of 20? Let us relate it to the work we did in Section 0.3. Consider the following problem.

What	is	$\frac{3}{5}$	of	20?
↓	↓	↓	↓	↓
?	=	$\frac{3}{5}$	×	20

$$? = \frac{3}{\cancel{5}} \times \cancel{20}^{4} = 12 \quad \text{The answer is 12.}$$

Since a percent is really a fraction, a percent problem is solved similarly to the way a fraction problem is solved. Since $\frac{3}{5} = \frac{6}{10} = 60\%$, we could write the problem as

What   is   60%   of   20?
↓   ↓   ↓   ↓   ↓

? = 60% × 20

? = 0.60 × 20

? = 12.0   The answer is 12.

Thus we have developed the following rule.

FINDING THE PERCENT OF A NUMBER

To find the percent of a number, change the percent to a decimal and multiply the number by the decimal.

EXAMPLE 6 Find.

(a) 10% of 36   **(b)** 2% of 350   **(c)** 182% of 12   **(d)** 0.3% of 42

Solution

(a) 10% of 36 = 0.10 × 36 = 3.6   **(b)** 2% of 350 = 0.02 × 350 = 7

(c) 182% of 12 = 1.82 × 12 = 21.84   **(d)** 0.3% of 42 = 0.003 × 42 = 0.126

Practice Problem 6 Find.

(a) 18% of 50 9   **(b)** 4% of 64 2.56   **(c)** 156% of 35 54.6

Teaching Tip Some students may find it easier to change 60% to a fraction and multiply. Allow those students who can change easily from a percent to a fraction to use fractions in those situations where fractions work best. Point out, however, that a decimal is always safe to use, and some percents such as 1.3% are best expressed as a decimal for calculation.

Calculator

 Finding the Percent of a Number

You can use a calculator to find 12% of 48. Enter:

12 [%] [×] 48 [=]

The display should read

5.76

If your calculator does not have a [%] key, do the following:

Enter: 0.12 [×] 48 [=]

What is 54% of 450?

Teaching Example 6 Find.

(a) 6% of 83   **(b)** 11% of 26

(c) 0.2% of 34   **(d)** 315% of 13

Ans:

(a) 4.98   **(b)** 2.86

(c) 0.068   **(d)** 40.95

There are many real-life applications for finding the percent of a number. When you go shopping in a store, you may find sale merchandise marked 35% off. This means that the sale price is 35% off the regular price. That is, 35% of the regular price is subtracted from the regular price to get the sale price.

EXAMPLE 7 A store is having a sale of 35% off the retail price of all sofas. Melissa wants to buy a particular sofa that normally sells for $595.

(a) How much will Melissa save if she buys the sofa on sale?

(b) What will the purchase price be if Melissa buys the sofa on sale?

Solution

(a) To find 35% of $595 we will need to multiply 0.35×595.

$$
\begin{array}{r}
595 \\
\times 0.35 \\
\hline
2975 \\
1785 \\
\hline
208.25
\end{array}
$$

Thus Melissa will save $208.25 if she buys the sofa on sale.

(b) The purchase price is the difference between the original price and the amount saved.

$$
\begin{array}{r}
595.00 \\
- 208.25 \\
\hline
386.75
\end{array}
$$

If Melissa buys the sofa on sale, she will pay $386.75.

Practice Problem 7 John received a 4.2% pay raise at work this year. He had previously earned $38,000 per year.

(a) What was the amount of his pay raise in dollars? $1596

(b) What is his new salary? $39,596

⑤ Finding the Missing Percent When Given Two Numbers

Recall that we can write $\frac{3}{4}$ as $\frac{75}{100}$ or 75%. If we were asked the question, "What percent is 3 of 4?" we would say 75%. This gives us a procedure for finding what percent one number is of a second number.

> **FINDING THE MISSING PERCENT**
>
> 1. Write a fraction with the two numbers. The number *after* the word "of" is always the denominator, and the other number is the numerator.
> 2. Simplify the fraction (if possible).
> 3. Change the fraction to a decimal.
> 4. Express the decimal as a percent.

EXAMPLE 8 What percent of 24 is 15?

Solution This can be solved quickly as follows.

Step 1 $\dfrac{15}{24}$ Write the relationship as a fraction. The number after "of" is 24, so 24 is the denominator.

Step 2 $\dfrac{15}{24} = \dfrac{5}{8}$ Simplify the fraction (when possible).

Step 3 $= 0.625$ Change the fraction to a decimal.

Step 4 $= 62.5\%$ Change the decimal to percent.

Practice Problem 8 What percent of 148 is 37? 25%

The question in Example 8 can also be written as "15 is what percent of 24?" To answer the question, we begin by writing the relationship as $\frac{15}{24}$. Remember that "of 24" means 24 will be the denominator.

EXAMPLE 9

(a) 82 is what percent of 200? **(b)** What percent of 16 is 3.8?

(c) $150 is what percent of $120?

Solution

(a) 82 is what percent of 200?

$$\frac{82}{200} \quad \text{Write the relationship as a fraction with 200 in the denominator.}$$

$$\frac{82}{200} = \frac{41}{100} = 0.41 = 41\%$$

(b) What percent of 16 is 3.8?

$$\frac{3.8}{16} \quad \text{Write the relationship as a fraction.}$$

You can divide to change the fraction to a decimal and then change the decimal to a percent.

$$\begin{array}{r} 0.2375 \rightarrow 23.75\% \\ 16\overline{)3.8000} \end{array}$$

(c) $150 is what percent of $120?

$$\frac{150}{120} = \frac{5}{4} \quad \text{Reduce the fraction whenever possible to make the division easier.}$$

$$\begin{array}{r} 1.25 \rightarrow 125\% \\ 4\overline{)5.00} \end{array}$$

Practice Problem 9

(a) What percent of 48 is 24? 50%   **(b)** 4 is what percent of 25? 16%

EXAMPLE 10

Marcia made 29 shots on goal during the last high school field hockey season. She actually scored a goal eight times. What percent of her total shots were goals? Round your answer to the nearest whole percent.

Solution   Marcia scored a goal eight times out of 29 tries. We want to know what percent of 29 is 8.

Step 1 $\dfrac{8}{29}$ Express the relationship as a fraction. The number after the word *of* is 29, so 29 appears in the denominator.

Step 2 $\dfrac{8}{29} = \dfrac{8}{29}$ Note that this fraction cannot be reduced.

Step 3 $= 0.2758\ldots$ The decimal equivalent of the fraction has many digits.

Step 4 $= 27.58\ldots\%$ Change the decimal to a percent, rounded to the nearest whole percent. (We review rounding in detail in Section 0.6.)

$\approx 28\%$ (The $\approx$ symbol means approximately equal to.)

Therefore, Marcia scored a goal approximately 28% of the time she made a shot on goal.

Practice Problem 10

Roberto scored a basket 430 times out of 1256 attempts during his high school basketball career. What percent of the time did he score a basket? Round your answer to the nearest whole percent. 34%

Calculator

Finding the Percent Given Two Numbers

You can use a calculator to find a missing percent.
What percent of 95 is 19?

1. Enter as a fraction.

19 ÷ 95

2. Change to a percent.

19 ÷ 95 × 100 =

The display should read

20

This means 20%.
What percent of 625 is 250?

Verbal and Writing Skills

1. When you write 19%, what do you really mean? Describe the meaning in your own words.

Answers may vary. Sample answers follow. 19% means 19 out of 100 parts. Percent means per 100. 19% is really a fraction with a denominator of 100. In this case it would be $\frac{19}{100}$.

2. When you try to solve a problem like "What percent of 80 is 30?" how do you know if you should write the fraction as $\frac{80}{30}$ or as $\frac{30}{80}$?

The number directly following the word *of* should be the denominator. In this case you should use $\frac{30}{80}$ since 80 follows the word *of.*

Change to a percent.

3. 0.28   28%

4. 0.61   61%

5. 0.568   56.8%

6. 0.089   8.9%

7. 0.076   7.6%

8. 0.325   32.5%

9. 2.39   239%

10. 7.4   740%

Change to a decimal.

11. 3%   0.03

12. 66%   0.66

13. 0.4%   0.004

14. 0.62%   0.0062

15. 250%   2.5

16. 175%   1.75

17. 3.75%   0.0375

18. 62.3%   0.623

Find the following.

19. What is 8% of 65?   5.2

20. What is 7% of 69?   4.83

21. What is 10% of 130?   13

22. What is 25% of 600?   150

23. What is 112% of 65?   72.8

24. What is 154% of 270?   415.8

25. 36 is what percent of 24?   150%

26. 49 is what percent of 28?   175%

27. What percent of 340 is 17?   5%

28. 12 is what percent of 48?   25%

29. 10 is what percent of 50?   20%

30. What percent of 35 is 28?   80%

Applications

31. *Exam Grades* Alana took a biology exam with 75 questions. She answered 60 correctly. What was her grade on the exam? Write the grade as a percent.
80%

32. *Hockey* For the 2006–07 season, Jordan Staal of the Pittsburgh Penguins led the National Hockey League in shooting percentage. Of his 131 shots on goal, 29 went in the net. What percent of the time did one of Staal's shots on goal result in a goal? (Round to the nearest whole percent.)   22%

33. *Tipping* Diana and Russ ate a meal costing $32.80 when they went out to dinner. If they want to leave the standard 15% tip for their server, how much will they tip and what will their total bill be?
$4.92 tip; $37.72 total bill

34. *CD Failure* Music CDs have a failure rate of 1.8%. (They skip or get stuck on a song.) If 36,000 CDs were manufactured last week, how many of them were defective?
648 CDs

35. *Food Budget* The Gonzalez family has a combined monthly income of $1850. Their food budget is $380/month. What percentage of their monthly income is budgeted for food? (Round your answer to the nearest whole percent.)   21%

36. *Nonprofit Associations* In 2006, there were 23,772 nonprofit associations in the United States. Of these, 2218 were social welfare agencies. What percent of the total nonprofit associations were social welfare agencies? Round your answer to the nearest hundredth of a percent. (*Source: Statistical Abstract of the U.S. 2007*)   9.33%

37. *Gift Returns* Last Christmas season, the Jones Mill Outlet Store chain calculated that they sold 36,000 gift items. If they assume that there will be a 1.5% return rate after the holidays, how many gifts can they expect to be exchanged?
540 gifts

38. *Rent* The total cost of a downtown Boston apartment, including rent, heat, electricity, and water, is $3690 per month. Sarah is renting this apartment with four friends. Based on a complicated formula, which takes into account the size of her bedroom and the furniture she brought to the apartment, she must pay 17% of the apartment cost each month. What is her monthly cost? $627.30

39. *Wages* Jim is earning $12.50 per hour as a cook. He has earned an 8% pay raise this year. What will his raise be, and what will his new hourly rate be?
$1 an hr raise; $13.50/hr

40. *On-Time Arrivals* Eastern Shores charter airline had 2390 flights last year. Of these flights, 560 did not land on time. What percent of the flights landed on time? Round your answer to the nearest whole percent.
77%

41. *Salary* Abdul sells medical devices for a local pharmaceutical company. He gets paid $600 per month plus a commission of 4% on all the devices he sells. Last year he sold $920,000 worth of medical devices. (a) What was his sales commission on the $920,000 worth of devices he sold? (b) What was his annual salary (that is, his monthly pay and commission combined)?
(a) $36,800
(b) $7200 + $36,800 = $44,000

42. *Business Travel* Bruce sells medical supplies on the road. He logged 18,600 miles in his car last year. He declared 65% of his mileage as business travel. (a) How many miles did he travel on business last year? (b) If his company reimburses him 31 cents for each mile traveled, how much should he be paid for travel expenses?
(a) 12,090 mi
(b) $3747.90

Cumulative Review

43. **[0.4.4]** *Checking Account Balance* Samuel had an opening balance of $132.80 in his checking account this month. He deposited two paychecks of $850 each, and withdrew $160 from the ATM. Then he wrote checks for $42.60, $775, $82.76, and $159.45. What will his balance be at the end of the month? $612.99

44. **[0.4.5]** *Car Loan* Tally is buying a car and has a bank loan. He has to make 33 more monthly payments of $188.50. How much money will he pay to the bank over the next 33 months? $6220.50

45. **[0.4.4]** *Gas Mileage* Linda took a trip in her Chevrolet Malibu. At the start of the trip her odometer read 42,443.8 miles. At the end of the trip her odometer read 43,556.8 miles. She used 42 gallons of gas during the trip. How many miles per gallon did her car achieve? 26.5 mi/gal

46. **[0.4.6]** *Rainfall* In Hilo, Hawaii, Brad spent three months working in a restaurant for a summer job. In June he observed that 4.6 inches of rain fell. In July the rainfall was 4.5 inches and in August it was 2.9 inches. What was the average monthly rainfall that summer for those three months in Hilo? 4.0 in.

Quick Quiz 0.5

1. What is 4.5% of 260? 11.7

2. 63 is what percent of 420? 15%

3. What is 114% of 85? 96.9

4. **Concept Check** Explain how you would change 0.0078 to a percent. Answers may vary

Classroom Quiz 0.5 You may use these problems to quiz your students' mastery of Section 0.5.

1. What is 6.4% of 380? **Ans:** 24.32

2. 247 is what percent of 380? **Ans:** 65%

3. What is 124% of 62? **Ans:** 76.88

Using Rounding to Estimate

As we begin this section, we will take some time to be sure you understand the idea of rounding a number. You will probably recall the following simple rule from your previous mathematics courses.

ROUNDING A NUMBER

If the first digit to the right of the round-off place is

1. less than 5, we make no change to the digit in the round-off place.

2. 5 or more, we increase the digit in the round-off place by 1.

To illustrate, 4689 rounded to the nearest hundred is 4700. Rounding 233,987 to the nearest ten thousand, we obtain 230,000. We will now use our experience in rounding as we discuss the general area of estimation.

Estimation is the process of finding an approximate answer. It is not designed to provide an exact answer. Estimation will give you a rough idea of what the answer might be. For any given problem, you may choose to estimate in many different ways.

ESTIMATION BY ROUNDING

1. Round each number so that there is one nonzero digit.

2. Perform the calculation with the rounded numbers.

Teaching Example 1 Find an estimate of the product 685×4123.

Ans: 2,800,000

EXAMPLE 1 Find an estimate of the product 5368×2864.

Solution

Step 1 Round 5368 to 5000.

Round 2864 to 3000.

Step 2 Multiply.

$$5000 \times 3000 = 15,000,000$$

An estimate of the product is 15,000,000.

Practice Problem 1 Find an estimate of the product $128,621 \times 378$.

NOTE TO STUDENT: Fully worked-out solutions to all of the Practice Problems can be found at the back of the text starting at page SP-1

40,000,000

Teaching Example 2 Enrico wants to buy wallpaper for two walls of his living room. One wall is $7\frac{1}{4}$ feet high and $15\frac{3}{4}$ feet long. The other wall is $7\frac{1}{4}$ feet high and $11\frac{1}{2}$ feet long. Estimate the number of square feet of wallpaper that Enrico will need.

Ans: 210 ft²

▲ **EXAMPLE 2** The four walls of a college classroom are $22\frac{1}{4}$ feet long and $8\frac{3}{4}$ feet high. A painter needs to know the area of these four walls in square feet. Since paint is sold in gallons, an estimate will do. Estimate the area of the four walls.

Solution

Step 1 Round $22\frac{1}{4}$ feet to 20 feet.

Round $8\frac{3}{4}$ feet to 9 feet.

Step 2 Multiply 20×9 to obtain an estimate of the area of one wall.

Multiply $20 \times 9 \times 4$ to obtain an estimate of the area of all four walls.

$$20 \times 9 \times 4 = 720$$

Our estimate for the painter is 720 square feet of wall space.

▲ **Practice Problem 2** Mr. and Mrs. Ramirez need to carpet two rooms of their house. One room measures $12\frac{1}{2}$ feet by $9\frac{3}{4}$ feet. The other room measures $11\frac{1}{4}$ feet by $18\frac{1}{2}$ feet. Estimate the number of square feet (square footage) in these two rooms. 300 ft^2

> *NOTE TO STUDENT: Fully worked-out solutions to all of the Practice Problems can be found at the back of the text starting at page SP-1*

EXAMPLE 3 Won Lin has a small compact car. He drove 396.8 miles in his car and used 8.4 gallons of gas.

(a) Estimate the number of miles he gets per gallon.

(b) Estimate how much it will cost him for fuel to drive on a cross-country trip of 2764 miles if gasoline usually costs $\$3.59\frac{9}{10}$ per gallon.

Solution

(a) Round 396.8 miles to 400 miles. Round 8.4 gallons to 8 gallons. Now divide.

$$8\overline{)400} \quad \begin{array}{c} 50 \\ \hline \end{array}$$

Won Lin's car gets about 50 miles per gallon.

(b) We will need to use the information we found in part (a) to determine how many gallons of gasoline Won Lin will use on his trip. Round 2764 miles to 3000 miles and divide 3000 miles by 50 miles per gallon.

$$50\overline{)3000} \quad \begin{array}{c} 60 \\ \hline \end{array}$$

Won Lin will use about 60 gallons of gas for his cross-country trip.

To estimate the cost, we need to ask ourselves, "What kind of an estimate are we looking for?" It may be sufficient to round $\$3.59\frac{9}{10}$ to $\$4.00$ and multiply.

$$60 \times \$4.00 = \$240.00$$

Keep in mind that this is a broad estimate. You may want an estimate that will be closer to the exact answer. In that case round $\$3.59\frac{9}{10}$ to $\$3.60$ and multiply.

$$60 \times \$3.60 = \$216.00$$

CAUTION: An estimate is only a rough guess. If we are estimating the cost of something, we may want to round the unit cost to a value that is closer to the actual amount so that our estimate is more accurate. When we estimate costs, it is a good idea to round to a value above the actual unit cost to make sure that we will have enough money for the expenditure. The actual cost to the nearest penny of the cross-country trip in Example 3 is $210.59.

> **Teaching Example 3** The Sheffield family's car was driven 295 miles last week and used 11.3 gallons of gas.
> **(a)** Estimate the number of miles the car gets per gallon.
> **(b)** Estimate the cost of driving on the family's vacation of 858 miles if gasoline usually costs $\$3.87\frac{9}{10}$ per gallon.
> **Ans:** **(a)** 30 mi/gal   **(b)** $120

Practice Problem 3 Roberta drove 422.8 miles in her truck and used 19.3 gallons of gas. Assume that gasoline costs $\$3.69\frac{9}{10}$ per gallon.

(a) Estimate the number of miles she gets per gallon.   20 mi/gal

(b) Estimate how much it will cost her to drive to Chicago and back, a distance of 3862 miles. $800

Other words that indicate estimation in word problems are *about* and *approximate*.

EXAMPLE 4   A local manufacturing company releases 1684 pounds of sulfur dioxide per day from its plant. The company operates 294 days per year and has been operating at this level for 32 years. Approximately how many pounds of sulfur dioxide have been released into the air by this company over the last 32 years?

Solution   The word *approximately* means we are looking for an estimate. We begin by rounding the given data so that there is only one nonzero digit in each number.

Round 1684 pounds to 2000 pounds.

Round 294 days to 300 days.

Round 32 years to 30 years.

Now determine the calculation you would perform to find the answer and calculate using the rounded numbers. We multiply the amount released in one day × the number of days in an operating year × years.

$$2000 \times 300 \times 30 = 18,000,000$$

Thus 18,000,000 pounds is an approximation of the sulfur dioxide released by the company during its 32 years of operation.

Practice Problem 4   A space probe is sent from Earth at 43,300 miles per hour toward the planet Pluto, which has an average distance from Earth of 3,580,000,000 miles.

(a) About how many hours will it take for the space probe to travel from Earth to Pluto? 100,000 hr

(b) About how many days will it take for the space probe to travel from Earth to Pluto? 5000 days

EXAMPLE 5   Find an estimate for 2.68% of $54,361.92.

Solution   Begin by writing 2.68% as a decimal.

$$2.68\% = 0.0268$$

Next round 0.0268 to the nearest hundredth.

0.0268 is 0.03 rounded to the nearest hundredth.

Finally, round $54,361.92 to $50,000.
    Now multiply.

$0.03 \times 50,000 = 1500$ Remember that to find a percent of a number you multiply.

Thus we estimate the answer to be $1500.
    To get a better estimate of 2.68% of $54,361.92, you may wish to round $54,361.92 to the nearest thousand. Try it and compare this estimate to the previous one. Remember, an estimate is only an approximation. How close you should try to get to the exact answer depends on what you need the estimate for.

Practice Problem 5   Find an estimate for 56.93% of $293,567.12. $180,000

In exercises 1–10, follow the principles of estimation to find an approximate value. Round each number so that there is one nonzero digit. Do not find the exact value.

1. 693×307
210,000

2. 437×892
360,000

3. 2862×5986
18,000,000

4. 4893×6174
30,000,000

5. $14 + 73 + 80 + 21 + 56$
240

6. $318 + 494 + 613 + 243$
1600

7. $41\overline{)829,346}$
20,000

8. $16\overline{)5,846,213}$
300,000

9. $\dfrac{2714}{31,500}$
0.1

10. $\dfrac{53,610}{786}$
62.5

11. Find 17% of $21,365.85. $4000.00

12. Find 4.9% of $9321.88. $450

Applications *In exercises 13–22, determine an estimate of the exact answer. Use estimation by rounding. Do not find the exact value.*

▲ **13.** *Carpeting* Estimate the cost of new carpeting that costs $22.50 per square yard, if you want to carpet a room that is 18.5 square yards in size. $400

▲ **14.** *Garden Design* Estimate the size of Mr. Brown's garden if it measures $15\frac{2}{3}$ feet by $11\frac{2}{3}$ feet.
200 ft²

15. *Checkout Sales* A typical customer at the local Piggly Wiggly supermarket spends approximately $82 at the checkout register. The store keeps four registers open, and each handles 22 customers per hour. Estimate the amount of money the store receives in one hour.
$6400

16. *Weekend Spending Money* The Westerly Community Credit Union in Rhode Island has found that on Fridays, the average customer withdraws $85 from the ATM for weekend spending money. Each ATM averages 19 customers per hour. There are five machines. Estimate the amount of money withdrawn by customers in one hour. $9000

17. *Gas Mileage* Rod's trip from Salt Lake City to his cabin in the mountains was 117.7 miles. If his car used 3.8 gallons of gas for the trip, estimate the number of miles his car gets to the gallon.   25 mi/gal

18. *Book Transport* The Paltrows are transferring their stock of books from one store to another. To save money, they are moving the books themselves. If their car can transport 430 books per trip, and they have 11,900 books to move, estimate the number of trips it will take.   25 trips

19. *Revenue* Total revenue for May, June, and July of 2007 for Krispy Kreme stores was $104,100,000. If there were 411 stores systemwide, estimate the average revenue per store. $250,000

20. *Happy Meals Sales* The local McDonald's registered $437,690 in sales last month. They know from experience that 18% of their sales are Happy Meals. Estimate the amount of money spent on Happy Meals last month. $80,000

21. *Education Budget* In the year 2000, it was estimated that 26% of the California state budget of $217,970,000,000 was spent for education. Estimate the amount of money spent on education in California in 2000. (*Source:* U.S. Census Bureau)
$60,000,000,000

22. *Operations* A recent survey of local hospitals by a state medical board estimates that 6.23% of all operations performed in hospitals are unnecessary. If 28,364,122 operations were performed last year, estimate the number of unnecessary operations.
1,800,000 operations

Income and Expenditures The data for exercises 23–26 were obtained at www.bls.gov.

23. In 2005, the average consumer aged 25–34 years had annual expenditures of $45,068 and spent 34% of this amount on housing. Estimate the annual expenditure for housing for a consumer in this age group.
$15,000

24. In 2005, the average consumer aged 35–44 years had annual expenditures of $55,190 and spent 13% of this amount on food. Estimate the annual expenditure for food for a consumer in this age group.
$6000

25. In 2005, the average consumer aged 65–74 years had annual expenditures of $38,573. Of this amount, $4176 was spent on health care. Estimate the percent of total expenditures spent on health care for consumers in this age group. 10%

26. In 2005, the average consumer aged 45–54 years had annual expenditures of $55,854. Of this amount, $3034 was spent on entertainment. Estimate the percent of total expenditures spent on entertainment for consumers in this age group. 5%

Cumulative Review

27. [0.5.4] What is 0.6% of 350? 2.1

28. [0.5.5] *Basketball* If Nicole plays basketball and makes 12 baskets out of every 20 attempts, what percent of the time does she score a basket? 60%

29. [0.5.4] *Tipping* Michael had breakfast at the Agawam Diner. The cost of his breakfast listed on the menu was $5.85. If he left a 15% tip, what was the total amount he paid for his meal? $6.73

30. [0.5.4] *Tire Inspection* In a recent tire inspection for Ford Explorers, local dealers found that 12% of the treads were defective. For each Explorer that was defective, the dealer offered immediate replacement of all five tires (including the spare). If the local dealers inspected the tires on 450 Explorers, how many new tires did they need? 270 tires

Quick Quiz 0.6 Estimate the answer for each indicated problem. First round each number to one nonzero digit. Then calculate.

1. 5687.98 + 8945.34 + 3429.57   18,000

2. 786 × 543   400,000

3. 34,987)‾567,238‾   20

4. **Concept Check** Explain how you would round 567,993,276 to the nearest hundred thousand.
Answers may vary

Classroom Quiz 0.6 You may use these problems to quiz your students' mastery of Section 0.6.
Estimate the answer for each indicated problem. First round each number to one nonzero digit. Then calculate.

1. 873.56 + 329.47 + 258.15 **Ans:** 1500

2. 4498.56 × 6721.18 **Ans:** 28,000,000

3. 19,045)‾7,956,000‾ **Ans:** 400

Using the Mathematics Blueprint to Solve Real-Life Problems

When a builder constructs a new home or office building, he or she often has a blueprint. This accurate drawing shows the basic form of the building. It also shows the dimensions of the structure to be built. This blueprint serves as a useful reference throughout the construction process.

Similarly, when solving real-life problems, it is helpful to have a "mathematics blueprint." This is a simple way to organize the information provided in the word problem in a chart or in a graph. You can record the facts you need to use. You can determine what it is you are trying to find and how you can go about actually finding it. You can record other information that you think will be helpful as you work through the problem.

As we solve real-life problems, we will use three steps.

Step 1 *Understand the problem.* Here we will read through the problem. We draw a picture if it will help, and use the Mathematics Blueprint as a guide to assist us in thinking through the steps needed to solve the problem.

Step 2 *Solve and state the answer.* We will use arithmetic or algebraic procedures along with problem-solving strategies to find a solution.

Step 3 *Check.* We will use a variety of techniques to see if the answer in step 2 is the solution to the word problem. This will include estimating to see if the answer is reasonable, repeating our calculation, and working backward from the answer to see if we arrive at the original conditions of the problem.

Student Learning Objective

After studying this section, you will be able to:

 Use the Mathematics Blueprint to solve real-life problems.

▲ **EXAMPLE 1** Nancy and John want to install wall-to-wall carpeting in their living room. The floor of the rectangular living room is $11\frac{2}{3}$ feet wide and $19\frac{1}{2}$ feet long. How much will it cost if the carpet is $18.00 per square yard?

Solution

1. Understand the problem. First, read the problem carefully. Drawing a sketch of the living room may help you see what is required. The carpet will cover the floor of the living room, so we need to find the area. Now we fill in the Mathematics Blueprint.

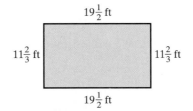

Teaching Example 1 Nate wants to lay ceramic tile in his entry hall and in his kitchen. The entry hall is $4\frac{1}{2}$ feet wide and 7 feet long. The kitchen is 15 feet wide and $12\frac{1}{3}$ feet long. The tile costs $3.50 per square foot. How much will the tile for this project cost?

Ans: $757.75

Mathematics Blueprint for Problem Solving

Gather the Facts	What Am I Solving For?	What Must I Calculate?	Key Points to Remember
The living room measures $11\frac{2}{3}$ ft by $19\frac{1}{2}$ ft. The carpet costs $18.00 per square yard.	**(a)** the area of the room in square feet **(b)** the area of the room in square yards **(c)** the cost of the carpet	**(a)** Multiply $11\frac{2}{3}$ ft by $19\frac{1}{2}$ ft to get area in square feet. **(b)** Divide the number of square feet by 9 to get the number of square yards. **(c)** Multiply the number of square yards by $18.00.	There are 9 square feet, 3 feet × 3 feet, in 1 square yard; therefore, we must divide the number of square feet by 9 to obtain square yards.

2. *Solve and state the answer.*

(a) To find the area of a rectangle, we multiply the length times the width.

$$11\frac{2}{3} \times 19\frac{1}{2} = \frac{35}{3} \times \frac{39}{2}$$

$$= \frac{455}{2} = 227\frac{1}{2}$$

A minimum of $227\frac{1}{2}$ square feet of carpet will be needed. We say a minimum because some carpet may be wasted in cutting. Carpet is sold by the square yard. We will want to know the amount of carpet needed in square yards.

(b) To determine the area in square yards, we divide $227\frac{1}{2}$ by 9. (9 sq ft = 1 sq yd)

$$227\frac{1}{2} \div 9 = \frac{455}{2} \div \frac{9}{1}$$

$$= \frac{455}{2} \times \frac{1}{9} = \frac{455}{18} = 25\frac{5}{18}$$

A minimum of $25\frac{5}{18}$ square yards of carpet will be needed.

(c) Since the carpet costs $18.00 per square yard, we will multiply the number of square yards needed by $18.00.

$$25\frac{5}{18} \times 18 = \frac{455}{18} \times \frac{18}{1} = 455$$

The carpet will cost a minimum of $455.00 for this room.

3. *Check.* We will estimate to see if our answers are reasonable.

(a) We will estimate by rounding each number to the nearest 10.

$$11\frac{2}{3} \times 19\frac{1}{2} \longrightarrow 10 \times 20 = 200$$

This is close to our answer of $227\frac{1}{2}$ sq ft. Our answer is reasonable. ✓

(b) We will estimate by rounding to the nearest hundred and ten, respectively.

$$227\frac{1}{2} \div 9 \longrightarrow 200 \div 10 = 20$$

This is close to our answer of $25\frac{5}{18}$ sq yd. Our answer is reasonable. ✓

(c) We will estimate by rounding each number to the nearest 10.

$$25\frac{5}{18} \times 18 \longrightarrow 30 \times 20 = 600$$

This is close to our answer of $455. Our answer seems reasonable. ✓

"Remember to estimate. It will save you time and money!"

NOTE TO STUDENT: *Fully worked-out solutions to all of the Practice Problems can be found at the back of the text starting at page SP-1*

▲ **Practice Problem 1** Jeff went to help Abby pick out wall-to-wall carpet for her new house. Her rectangular living room measures $16\frac{1}{2}$ feet by $10\frac{1}{2}$ feet. How much will it cost to carpet the room if the carpet costs $20 per square yard? $385

Mathematics Blueprint for Problem Solving

Gather the Facts	What Am I Solving For?	What Must I Calculate?	Key Points to Remember

TO THINK ABOUT: Example 1 Follow-Up   Assume that the carpet in Example 1 comes in a standard width of 12 feet. How much carpet will be wasted if it is laid out on the living room floor in one strip that is $19\frac{1}{2}$ feet long? How much carpet will be wasted if it is laid in two sections side by side that are each $11\frac{2}{3}$ feet long? Assuming you have to pay for wasted carpet, what is the minimum cost to carpet the room?   6.5 ft²; 52.5 ft²; $468

EXAMPLE 2   The following chart shows the 2006 sales of Micropower Computer Software for each of the four regions of the United States. Use the chart to answer the following questions (round all answers to the nearest whole percent):

(a) What percent of the sales personnel are assigned to the Northeast?

(b) What percent of the volume of sales is attributed to the Northeast?

(c) What percent of the sales personnel are assigned to the Southeast?

(d) What percent of the volume of sales is attributed to the Southeast?

(e) Which of these two regions of the country has sales personnel that appear to be more effective in terms of the volume of sales?

Teaching Example 2 Use the chart from Example 2 to answer the following questions. Round all answers to the nearest whole percent.

(a) What percent of the sales personnel are assigned to the East (Northeast and Southeast combined)?

(b) What percent of the volume of sales is attributed to the East?

(c) What percent of the sales personnel are assigned to the West (Northwest and Southwest combined)?

(d) What percent of the volume of sales is attributed to the West?

(e) Which of these two regions has sales personnel that appear to be more effective?

Ans: **(a)** 55%   **(b)** 44%   **(c)** 45%   **(d)** 56%   **(e)** the West

Region of the U.S.	Number of Sales Personnel	Dollar Volume of Sales
Northeast	12	1,560,000
Southeast	18	4,300,000
Northwest	10	3,660,000
Southwest	15	3,720,000
Total	55	13,240,000

Solution

1. Understand the problem. We will only need to deal with figures from the Northeast region and the Southeast region.

Mathematics Blueprint for Problem Solving

Gather the Facts	What Am I Solving For?	What Must I Calculate?	Key Points to Remember
Personnel: 12 Northeast 18 Southeast 55 total Sales Volume: $1,560,000 NE $4,300,000 SE $13,240,000 Total	**(a)** the percent of the total personnel in the Northeast **(b)** the percent of the total sales in the Northeast **(c)** the percent of the total personnel in the Southeast **(d)** the percent of the total sales in the Southeast **(e)** compare the percentages from the two regions	**(a)** 12 is what percent of 55? Divide. $12 \div 55$ **(b)** 1,560,000 is what percent of 13,240,000? $1,560,000 \div 13,240,000$ **(c)** $18 \div 55$ **(d)** $4,300,000 \div 13,240,000$	We do not need to use the numbers for the Northwest or the Southwest.

2. *Solve and state the answer.*

(a) $\frac{12}{55} = 0.21818\ldots$

$\approx 22\%$

(b) $\frac{1,560,000}{13,240,000} = \frac{156}{1324} \approx 0.1178$

$\approx 12\%$

(c) $\frac{18}{55} = 0.32727\ldots$

$\approx 33\%$

(d) $\frac{4,300,000}{13,240,000} = \frac{430}{1324} \approx 0.3248$

$\approx 32\%$

(e) We notice that 22% of the sales force in the Northeast made 12% of the sales. The percent of the sales compared to the percent of the sales force is about half (12% of 24% would be half) or 50%. 33% of the sales force in the Southeast made 32% of the sales. The percent of sales compared to the percent of the sales force is close to 100%. We must be cautious here. *If there are no other significant factors,* it would appear that the Southeast sales force is more effective. (There may be other significant factors affecting sales, such as a recession in the Northeast, new and inexperienced sales personnel, or fewer competing companies in the Southeast.)

3. *Check.* You may want to use a calculator to check the division in step 2, or you may use estimation.

(a) $\frac{12}{55} \rightarrow \frac{10}{60} \approx 0.17$

$= 17\%$ ✓

(b) $\frac{1,560,000}{13,240,000} \rightarrow \frac{1,600,000}{13,000,000} \approx 0.12$

$= 12\%$ ✓

(c) $\frac{18}{55} \rightarrow \frac{20}{60} \approx 0.33$

$= 33\%$ ✓

(d) $\frac{4,300,000}{13,240,000} \rightarrow \frac{4,300,000}{13,000,000} \approx 0.33$

$= 33\%$ ✓

NOTE TO STUDENT: Fully worked-out solutions to all of the Practice Problems can be found at the back of the text starting at page SP-1

Practice Problem 2 Using the chart for Example 2, answer the following questions. (Round all answers to the nearest whole percent.)

(a) What percent of the sales personnel are assigned to the Northwest? 18%

(b) What percent of the sales volume is attributed to the Northwest? 28%

(c) What percent of the sales personnel are assigned to the Southwest? 27%

(d) What percent of the sales volume is attributed to the Southwest? 28%

(e) Which of these two regions of the country has sales personnel that appear to be more effective in terms of volume of sales? the Northwest

Mathematics Blueprint for Problem Solving

Gather the Facts	What Am I Solving For?	What Must I Calculate?	Key Points to Remember

TO THINK ABOUT: Example 2 Follow-Up Suppose in 2007 the number of sales personnel (55) increased by 60%. What would the new number of sales personnel be? Suppose in 2008 that the number of sales personnel decreased by 60% from the number of sales personnel in 2007. What would the new number be? Why is this number not 55, since we have increased the number by 60% and then decreased the result by 60%? Explain. 88; 35

Applications *Use the Mathematics Blueprint for Problem Solving to help you solve each of the following exercises.*

▲ **1.** *Patio Pavers* Shaun wants to place pavers to make a new backyard patio. The patio will measure $3\frac{1}{3}$ yards by $4\frac{1}{3}$ yards. If the pavers are each 1 square foot and cost $2.35 each, how much will the pavers cost?
$305.50

▲ **2.** *Decking Costs* The Carters need to replace their deck floor. The deck is $18\frac{1}{2}$ feet by $10\frac{1}{2}$ feet. If the new decking costs $5.25 per square foot, how much will it cost them to replace the deck? (Round your answer to the nearest cent.) $1019.81

▲ **3.** *Garden Design* Jenna White has room for a rectangular garden that can measure as much as $15\frac{1}{2}$ feet wide and $25\frac{2}{3}$ feet long.

(a) She decides the garden should be as large as possible, but wants to fence it off to protect the vegetables from the neighborhood raccoons. How much fencing does she need?

$82\frac{1}{3}$ ft

(b) She finds that she can buy a prepackaged 90 feet of fencing for $155.00 or she can get a cut-to-order length for $2.10 per foot. Which should she buy? How much money does she save?
She should purchase the prepackaged 90 feet.
She would save $172.90 − $155.00 = $17.90.

▲ **4.** *Pond Design* The Lee family is installing a backyard pond for their prize koi. (A koi is a kind of fish bred in Japan. They are noted for their large size and variety of color.)

(a) If the pond measures $11\frac{1}{2}$ feet by 7 feet, and is $3\frac{1}{2}$ feet deep, how much water will it take to fill the pond?

$\frac{1127}{4} = 281\frac{3}{4} = 281.75$ ft^3

(b) If each cubic foot is 7.5 gallons, how many gallons of water will the pond hold?

$2113\frac{1}{8}$ or 2113.125 gal

(c) If at least 200 gallons of water is recommended for each koi, how many koi can the Lee family place in their pond? $2113.125 \div 200 \approx 10.6$ or 10 koi

---

Exercise Training *The following directions are posted on the wall at the gym.*

Beginning exercise training schedule

On day 1, each athlete will begin the morning as follows:

Jog.................. $1\frac{1}{2}$ miles

Walk.............. $1\frac{3}{4}$ miles

Rest.............. $2\frac{1}{2}$ minutes

Walk.............. 1 mile

5. Betty's athletic trainer told her to follow the beginning exercise training schedule on day 1. On day 2, she is to increase all distances and times by $\frac{1}{3}$ that of day 1. On day 3, she is to increase all distances and times by $\frac{1}{3}$ that of day 2. What will be her training schedule on day 3?
Jog $2\frac{2}{3}$ mi; walk $3\frac{1}{9}$ mi; rest $4\frac{4}{9}$ min; walk $1\frac{7}{9}$ mi

6. Melinda's athletic trainer told her to follow the beginning exercise training schedule on day 1. On day 2, she is to increase all distances and times by $\frac{1}{3}$ that of day 1. On day 3, she is to once again increase all distances and times by $\frac{1}{3}$ that of day 1. What will be her training schedule on day 3?
Jog $2\frac{1}{2}$ mi; walk $2\frac{11}{12}$ mi; rest $4\frac{1}{6}$ min; walk $1\frac{2}{3}$ mi

To Think About

Refer to exercises 5 and 6 in working exercises 7–10.

7. Who will have a more demanding schedule on day 3, Betty or Melinda? Why?
Betty; Melinda increases each activity by $\frac{2}{3}$ by day 3 and Betty increases each activity by $\frac{7}{9}$ by day 3.

8. If Betty kept up the same type of increase day after day, how many miles would she be jogging on day 5?
$4\frac{20}{27}$ mi

9. If Melinda kept up the same type of increase day after day, how many miles would she be jogging on day 7?

$4\frac{1}{2}$ mi

11. **House Prices** In 1985, the median selling price of an existing single-family home in Atlanta, Georgia, was $66,200. Between 1985 and 1990, the average price increased by 30%. Between 1990 and 2005, the average price increased again, this time by 15%. What was the median house price in Atlanta in 2005?

$98,969.00

10. Which athletic trainer would appear to have the best plan for training athletes if they used this plan for 14 days? Why? Melinda's trainer has the best plan. Over 14 days Betty's schedule increases more. The amount of increase each day is larger. Eventually it is unrealistic.

12. **Egg Weight** Chicken eggs are classified by weight per dozen eggs. Large eggs weigh 24 ounces per dozen and medium eggs weigh 21 ounces per dozen.

(a) If you do not include the shell, which is 12% of the total weight of an egg, how many ounces of eggs do you get from a dozen large eggs? From a dozen medium eggs? 21.12 oz; 18.48 oz

(b) At a local market, large eggs sell for $1.79 a dozen, and medium eggs for $1.39 a dozen. If you do not include the shell, which is a better buy, large or medium eggs? Medium eggs are a better buy; large is 1.79/21.12 ≈ $0.085/oz, medium is 1.39/18.48 ≈ $0.075/oz

Family Budgets *For the following problems, use the chart below.*

The Whittier family has created the following budget:

Rent	20%	Clothing	10%
Food	31%	Medical	12%
Utilities	5%	Savings	10%
Entertainment	6%	Miscellaneous	6%

13. The income for this family comes from Mrs. Whittier's annual salary of $62,500.

(a) If 28% of her salary is withheld for various taxes, how much money does the family have available for their budget? $62,500 × 0.72 = $45,000

(b) How much of Mrs. Whittier's take-home pay is budgeted for food? $45,000 × 0.31 = $13,950

14. With one child ready to head off to college, Mrs. Whittier is looking for ways to save money.

(a) She figures out that if they plant a garden this year, the family can save 12% of their food costs. How much will she save in food costs? (This does not account for the cost of planting a garden.) What is the new food budget amount?

$13,950 × 0.12 = $1674

$13,950 − $1674 = $12,276

(b) What percentage of Mrs. Whittier's budget is the new food amount?

$\dfrac{12{,}276}{45{,}000} = 0.2728 = 27.28\%$

Paycheck Stub *Use the following information from a paycheck stub to solve exercises 15–18.*

TOBEY & SLATER INC. 5000 Stillwell Avenue Queens, NY 10001		Check Number	Payroll Period		Pay Date
			From Date	To Date	
		495885	10-30-07	11-30-07	12-01-07

Name	Social Security No.	I.D. Number	File Number	Rate/Salary	Department	MS	DEP	Res
Fred J. Gilliani	012-34-5678	01	1379	1150.00	0100	M	5	NY

	Current	Year to Date		Current	Year to Date
GROSS	1,150.00	6,670.00	STATE	67.76	388.45
FEDERAL	138.97	781.07	LOCAL	5.18	30.04
FICA	87.98	510.28	DIS-SUI	.00	.00
W-2 GROSS		6,670.00	NET	790.47	4,960.16

Earnings						Special Deductions		
No.	Type	Hours	Rate	Amount	Dept/Job No.	No.	Description	Amount
96	REGULAR			1,150.00	0100	82	Retirement	12.56
						75	Medical	36.28
						56	Union Dues	10.80

Gross pay is the pay an employee receives for his or her services before deductions. Net pay is the pay the employee actually gets to take home. You may round each amount to the nearest whole percent for exercises 15–18.

15. What percent of Fred's gross pay is deducted for federal, state, and local taxes? 18%

16. What percent of Fred's gross pay is deducted for retirement and medical? 4%

17. What percent of Fred's gross pay does he actually get to take home? 69%

18. What percent of Fred's deductions are special deductions? 5%

Quick Quiz 0.7

1. The square college courtyard has 525 square yards of space. It is being paved with squares made of Vermont granite stone. Each stone has an area of 1.5 square yards. How many stones will be required to pave to courtyard? 350 stones

2. Melinda is paid $16,200 per year. She is also paid a sales commission of 4% of the value of her sales. Last year she sold $345,000 worth of products. What percent of her total income was her commission? Round to the nearest whole percent. 46%

3. A carpenter is building a fence for a rectangular yard that measures 40 feet by 95 feet. The fence costs $4.50 per foot. If he makes a fence large enough to surround the yard, how much will the fence cost? $1215

4. Concept Check Hank knows that 1 kilometer ≈ 0.62 mile. Explain how Hank could find out how many miles he traveled on a 12-kilometer trip in Mexico.
Answers may vary

Classroom Quiz 0.7 You may use these problems to quiz your students' mastery of Section 0.7.

1. The Stonehill Mall has a square courtyard that contains 650 square yards of space. It is being paved with square marble tiles. Each square tile measures 2.5 square yards in area. How many tiles will be required to pave the courtyard? **Ans:** 260 tiles

2. Steven is paid $20,150 per year. He is also paid a sales commission of 3% of the value of his sales. Last year he sold $495,000 worth of products. What percent of his total income was his commission? Round to the nearest whole percent. **Ans:** 42%

3. Jon Swensen is fencing in his back field in Perham, Maine. The rectangular field measures 230 yards by 420 yards. The fencing will cost $2.50 per yard. If he makes the fence large enough to surround the field, how much will the fence cost? **Ans:** $3250

Developing Your Study Skills

Reading the Textbook

Your homework time each day should begin with the careful reading of the section(s) assigned in your textbook. Usually, much time and effort have gone into the selection of a particular text, and your instructor has decided that this is the book that will help you to become successful in this mathematics class. Textbooks are expensive, but they can be a wise investment if you take advantage of them by reading them.

Reading a mathematics textbook is unlike reading many other types of books that you may find in your literature, history, psychology, or sociology courses. Mathematics texts are technical books that provide you with exercises for practice. Reading a mathematics text requires slow and careful reading of each word, which takes time and effort.

Begin reading your textbook with a paper and pencil in hand. As you come across a new definition or concept, underline it in the text and/or write it down in your notebook. Whenever you encounter an unfamiliar term, look it up and make a note of it. When you come to an example, work through it step-by-step. Be sure to read each word and to follow directions carefully.

Notice the helpful hints that the author provides. They guide you to correct solutions and prevent you from making errors. Take advantage of these pieces of expert advice.

Be sure that you understand what you are reading. Make a note of any of those things that you do not understand and ask your instructor about them. Do not hurry through the material. Learning mathematics takes time.

Putting Your Skills to Work: Use Math to Save Money

MANAGING DEBTS AND PAYMENTS

Can you imagine the joy of taking a great vacation without going into debt for it? Can you think about how great it would be to have all your debts paid off? It is a wonderful feeling! Paying off your debts and then being able to take a vacation is an excellent goal. But how is that done? Consider the story of Tracy and Max.

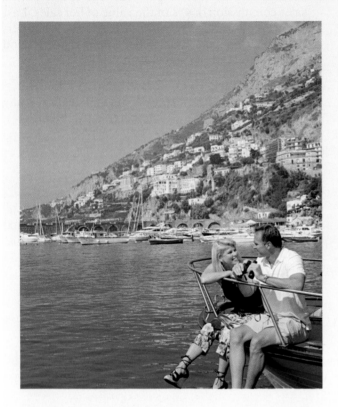

Facing Up to the Debt

Tracy and Max were overwhelmed with debt. Besides their ordinary living expenses, they had so much debt that they could barely make the money they earned last until the end of the month. They had little money left for extras and for having fun, and no money for a vacation. Each of their three credit cards was maxed out at $8000; they had a hospital debt of $12,000; they still owed $2000 on their car; and they had also borrowed money from friends in sums of $100 and $300.

Making a Plan

They decided to write down all of their debts, putting them in order from smallest to largest.

Then they made minimum payments on all the debts, but aimed at paying off the three smallest debts first.

1. Put the couple's debts in order from smallest to largest. Remember, there are three credit cards. $100, $300, $2000, $8000, $8000, $8000, $12,000

The minimum payment on each credit card averaged $25 per month. They had arranged with the hospital to pay the debt off at $50 per month. Their car payment was $200 per month, and they agreed to pay each of their friends $20 per month.

2. What is the total amount of their minimum monthly payment? $365

What Tracy and Max Accomplished

Tracy and Max decided they would eliminate any extras and not spend money on fun activities so they could use this money to pay off their three smallest debts. As a result they were able to pay off the first two smallest debts in just two months while making minimum payments on all other debts. Then each month they took the $40 they would have used to pay these two small debts and applied it towards the third smallest debt. Again, they made sure they made minimum payments on all other debts.

3. How many more months will it take Tracy and Max to pay off the third smallest debt if they follow the plan stated above? $2000 − $400 = $1600; $1600 ÷ $240 = 6.6... or 7 months

After eliminating the smaller debts, they took the money they would have spent on those debts and used it on the principal of the remaining debts. In other words, they paid more than the minimum payment on the remaining debts. Because they hated the way they felt when they were in debt, they stopped using credit cards for new purchases. Max also took a temporary part-time job so they could pay off their debts faster. Finally they paid off all the debts!

Applying It to Your Life

Many debt counselors have a simple, practical suggestion for people in debt. Arrange debts in order, pay off the smallest first, and then let the consequences of that action help pay off the rest of the debts more quickly. Many people have been able to get out of debt in about two years by using this approach, other strategies for budgeting, and wise choices for living.

Topic	Procedure	Examples
Simplifying fractions, p. 2.	1. Write the **numerator** and **denominator** as products of prime factors. 2. Use the basic rule of fractions that $$\frac{a \times c}{b \times c} = \frac{a}{b}$$ for any factor that appears in both the numerator and the denominator. 3. Multiply the remaining factors for the numerator and for the denominator separately.	$$\frac{15}{25} = \frac{\cancel{5} \cdot 3}{\cancel{5} \cdot 5} = \frac{3}{5}$$ $$\frac{36}{48} = \frac{\cancel{2} \cdot \cancel{2} \cdot 3 \cdot \cancel{3}}{\cancel{2} \cdot \cancel{2} \cdot 2 \cdot 2 \cdot \cancel{3}} = \frac{3}{4}$$ $$\frac{26}{39} = \frac{2 \cdot \cancel{13}}{3 \cdot \cancel{13}} = \frac{2}{3}$$
Changing improper fractions to mixed numbers, p. 5.	1. Divide the denominator into the numerator to obtain the whole-number part of the mixed number. 2. The remainder from the division will be the numerator of the fraction. 3. The denominator remains unchanged.	$\frac{14}{3} = 4\frac{2}{3}$ $\frac{19}{8} = 2\frac{3}{8}$ since $3\overline{)14}$ $\underline{12}$ $\,2$ since $8\overline{)19}$ $\underline{16}$ $\,3$
Changing mixed numbers to improper fractions, p. 6.	1. Multiply the whole number by the denominator and add the result to the numerator. This will yield the new numerator. 2. The denominator does not change.	$$4\frac{5}{6} = \frac{(4 \times 6) + 5}{6} = \frac{24 + 5}{6} = \frac{29}{6}$$ $$3\frac{1}{7} = \frac{(3 \times 7) + 1}{7} = \frac{21 + 1}{7} = \frac{22}{7}$$
Changing fractions to equivalent fractions with a given denominator, p. 7.	1. Divide the original denominator into the new denominator. This result is the value that we use for multiplication. 2. Multiply the numerator and the denominator of the original fraction by that value.	$$\frac{4}{7} = \frac{?}{21}$$ $\dfrac{3}{7\overline{)21}} \leftarrow$ Use this to multiply $\dfrac{4 \times 3}{7 \times 3} = \dfrac{12}{21}$.
Finding the LCD (least common denominator) of two or more fractions, p. 12.	1. Write each denominator as the product of prime factors. 2. The LCD is a product containing each different factor. 3. If a factor occurs more than once in any one denominator, the LCD will contain that factor repeated the greatest number of times that it occurs in any one denominator.	Find the LCD of $\dfrac{4}{15}$ and $\dfrac{3}{35}$. $$15 = 5 \cdot 3$$ $$35 = 5 \cdot 7$$ $$\text{LCD} = 3 \cdot 5 \cdot 7 = 105$$ Find the LCD of $\dfrac{11}{18}$ and $\dfrac{7}{45}$. $18 = 3 \cdot 3 \cdot 2 \quad$ (factor 3 appears twice) $45 = 3 \cdot 3 \cdot 5 \quad$ (factor 3 appears twice) $\text{LCD} = 2 \cdot 3 \cdot 3 \cdot 5 = 90$
Adding and subtracting fractions that do not have a common denominator, p. 13.	1. Find the LCD. 2. Change each fraction to an equivalent fraction with the LCD for a denominator. 3. Add or subtract the fractions and simplify the answer if possible.	$$\frac{3}{8} + \frac{1}{3} = \frac{3 \cdot 3}{8 \cdot 3} + \frac{1 \cdot 8}{3 \cdot 8} = \frac{9}{24} + \frac{8}{24} = \frac{17}{24}$$ $$\frac{11}{12} - \frac{1}{4} = \frac{11}{12} - \frac{1 \cdot 3}{4 \cdot 3} = \frac{11}{12} - \frac{3}{12} = \frac{8}{12} = \frac{2}{3}$$
Adding and subtracting mixed numbers, p. 16.	1. Change the mixed numbers to improper fractions. 2. Follow the rules for adding and subtracting fractions. 3. Simplify. If necessary, change your answer to a mixed number.	$$1\frac{2}{3} + 1\frac{3}{4} = \frac{5}{3} + \frac{7}{4} = \frac{5 \cdot 4}{3 \cdot 4} + \frac{7 \cdot 3}{4 \cdot 3}$$ $$= \frac{20}{12} + \frac{21}{12} = \frac{41}{12} = 3\frac{5}{12}$$ $$2\frac{1}{4} - 1\frac{3}{4} = \frac{9}{4} - \frac{7}{4} = \frac{2}{4} = \frac{1}{2}$$
Multiplying fractions, p. 21.	1. If there are no common factors, multiply the numerators. Then multiply the denominators. 2. If possible, write the numerators and denominators as the product of prime factors. Use the basic rule of fractions to divide out any value that appears in both a numerator and a denominator. Multiply the remaining factors in the numerator. Multiply the remaining factors in the denominator.	$$\frac{3}{7} \times \frac{2}{13} = \frac{6}{91}$$ $$\frac{6}{15} \times \frac{35}{91} = \frac{2 \cdot \cancel{3}}{3 \cdot \cancel{5}} \times \frac{\cancel{5} \cdot \cancel{7}}{\cancel{7} \cdot 13} = \frac{2}{13}$$ $$3 \times \frac{5}{8} = \frac{3}{1} \times \frac{5}{8} = \frac{15}{8} \text{ or } 1\frac{7}{8}$$

(Continued on next page)

Topic	Procedure	Examples
Dividing fractions, p. 23.	1. Invert the second fraction. 2. Multiply the fractions.	$\dfrac{4}{7} \div \dfrac{11}{3} = \dfrac{4}{7} \times \dfrac{3}{11} = \dfrac{12}{77}$ $\dfrac{5}{9} \div \dfrac{5}{7} = \dfrac{\cancel{5}}{9} \times \dfrac{7}{\cancel{5}} = \dfrac{7}{9}$
Multiplying and dividing mixed numbers, pp. 22, 25.	1. Change each mixed number to an improper fraction. 2. Use the rules for multiplying or dividing fractions. 3. If necessary, change your answer to a mixed number.	$2\dfrac{1}{4} \times 3\dfrac{3}{5} = \dfrac{9}{4} \times \dfrac{18}{5}$ $= \dfrac{3 \cdot 3}{2 \cdot \cancel{2}} \times \dfrac{\cancel{2} \cdot 3 \cdot 3}{5} = \dfrac{81}{10} = 8\dfrac{1}{10}$ $1\dfrac{1}{4} \div 1\dfrac{1}{2} = \dfrac{5}{4} \div \dfrac{3}{2} = \dfrac{5}{2 \cdot \cancel{2}} \times \dfrac{\cancel{2}}{3} = \dfrac{5}{6}$
Changing fractional form to decimal form, p. 30.	Divide the denominator into the numerator.	$\dfrac{5}{8} = 0.625$ since $8\overline{)5.000}$ with quotient 0.625
Changing decimal form to fractional form, p. 31.	1. Write the decimal as a fraction with a denominator of 10, 100, 1000, and so on. 2. Simplify the fraction, if possible.	$0.37 = \dfrac{37}{100}$ $0.375 = \dfrac{375}{1000} = \dfrac{3}{8}$
Adding and subtracting decimals, p. 32.	1. Carefully line up the decimal points as indicated for addition and subtraction. (Extra zeros may be added to the right-hand side of the decimals if desired.) 2. Add or subtract the appropriate digits.	Add. Subtract. $1.236 + 7.825$ $2 - 1.32$ $\begin{array}{r} 1.236 \\ + 7.825 \\ \hline 9.061 \end{array}$ $\begin{array}{r} 2.00 \\ - 1.32 \\ \hline 0.68 \end{array}$
Multiplying decimals, p. 33.	1. First multiply the digits. 2. Count the total number of decimal places in the numbers being multiplied. This number determines the number of decimal places in the answer.	$\begin{array}{r} 0.9 \text{ (one place)} \\ \times\ 0.7 \text{ (one place)} \\ \hline 0.63 \text{ (two places)} \end{array}$ $\begin{array}{r} 0.009 \text{ (three places)} \\ \times\ 0.07 \text{ (two places)} \\ \hline 0.00063 \text{ (five places)} \end{array}$
Dividing decimals, p. 34.	1. Count the number of decimal places in the divisor. 2. Move the decimal point to the right the same number of places in both the divisor and dividend. 3. Mark that position with a caret ($\wedge$). 4. Perform the division. Line up the decimal point in the quotient with the position indicated by the caret.	Divide. $7.5 \div 0.6$ Move decimal point one place to the right. $\begin{array}{r} 12.5 \\ 0.6_{\wedge})\overline{7.5_{\wedge}0} \\ \underline{6} \\ 15 \\ \underline{12} \\ 3\ 0 \\ \underline{3\ 0} \\ 0 \end{array}$ Therefore, $7.5 \div 0.6 = 12.5$
Changing a decimal to a percent, p. 41.	1. Move the decimal point two places to the right. 2. Add the % symbol.	$0.46 = 46\%$ $0.002 = 0.2\%$ $1.59 = 159\%$ $0.013 = 1.3\%$ $0.0007 = 0.07\%$
Changing a percent to a decimal, p. 42.	1. Remove the % symbol. 2. Move the decimal point two places to the left.	$49\% = 0.49$ $180\% = 1.8$ $59.8\% = 0.598$ $0.13\% = 0.0013$
Finding a percent of a number, p. 43.	1. Convert the percent to a decimal. 2. Multiply the decimal by the number.	Find 12% of 86. $12\% = 0.12$ $86 \times 0.12 = 10.32$ Therefore, 12% of 86 = 10.32.

Topic	Procedure	Examples
Finding what percent one number is of another number, p. 44.	1. Place the number after the word *of* in the denominator. 2. Place the other number in the numerator. 3. If possible, simplify the fraction. 4. Change the fraction to a decimal. 5. Express the decimal as a percent.	What percent of 8 is 7? $$\frac{7}{8} = 0.875 = 87.5\%$$ 42 is what percent of 12? $$\frac{42}{12} = \frac{7}{2} = 3.5 = 350\%$$
Estimation, p. 48.	1. Round each number so that there is one nonzero digit. 2. Perform the calculation with the rounded numbers.	Estimate the number of square feet in a room that is 22 feet long and 13 feet wide. Assume that the room is rectangular. 1. We round 22 to 20. We round 13 to 10. 2. To find the area of a rectangle, we multiply length times width. $$20 \times 10 = 200.$$ We estimate that there are 200 square feet in the room.

Problem solving, p. 53.	In solving a real-life problem, you may find it helpful to complete the following steps. You will not use all of the steps all of the time. Choose the steps that best fit the conditions of the problem.
	1. **Understand the problem.**
	  (a) Read the problem carefully.
	  (b) Draw a picture if this helps you.
	  (c) Use the Mathematics Blueprint for Problem Solving.
	2. **Solve and state the answer.**
	3. **Check.**
	  (a) Estimate to see if your answer is reasonable.
	  (b) Repeat your calculation.
	  (c) Work backward from your answer. Do you arrive at the original conditions of the problem?
		Susan is installing wall-to-wall carpeting in her $10\frac{1}{2}$-ft-by-12-ft bedroom. How much will it cost at \$20 a square yard?
		1. **Understand the problem.**
		  We need to find the area of the room in square yards. Then we can find the cost.
		2. **Solve and state the answer.**
		  Area: $10\frac{1}{2} \times 12 = \frac{21}{2} \times \frac{12}{1} = 126$ sq ft
		  $126 \div 9 = 14$ sq yd
		
		  Cost: $14 \times 20 = \$280$
		  The carpeting will cost \$280.
		3. **Check.**
		  Estimate: $10 \times 10 = 100$ sq ft
		             $100 \div 10 = 10$ sq yd
		             $10 \times 20 = \$200$
		  Our answer is reasonable. ✓

Chapter 0 Review Problems

Section 0.1

In exercises 1–4, simplify.

1. $\dfrac{36}{48}$ $\dfrac{3}{4}$

2. $\dfrac{15}{50}$ $\dfrac{3}{10}$

3. $\dfrac{36}{82}$ $\dfrac{18}{41}$

4. $\dfrac{18}{30}$ $\dfrac{3}{5}$

5. Write $7\dfrac{1}{8}$ as an improper fraction.

$\dfrac{57}{8}$

6. Write $\dfrac{34}{5}$ as a mixed number.

$6\dfrac{4}{5}$

7. Write $\dfrac{80}{3}$ as a mixed number.

$26\dfrac{2}{3}$

Find the missing numerator.

8. $\dfrac{5}{8} = \dfrac{?}{24}$ 15

9. $\dfrac{1}{7} = \dfrac{?}{35}$ 5

10. $\dfrac{3}{5} = \dfrac{?}{75}$ 45

11. $\dfrac{2}{5} = \dfrac{?}{55}$ 22

Section 0.2

Combine.

12. $\dfrac{3}{5} + \dfrac{1}{4}$ $\dfrac{17}{20}$

13. $\dfrac{7}{12} + \dfrac{5}{8}$ $\dfrac{29}{24}$ or $1\dfrac{5}{24}$

14. $\dfrac{7}{20} - \dfrac{1}{12}$ $\dfrac{4}{15}$

15. $\dfrac{7}{10} - \dfrac{4}{15}$ $\dfrac{13}{30}$

16. $3\frac{1}{6} + 2\frac{3}{5}$ $5\frac{23}{30}$ **17.** $2\frac{7}{10} + 3\frac{3}{4}$ $6\frac{9}{20}$ **18.** $6\frac{2}{9} - 3\frac{5}{12}$ $2\frac{29}{36}$ **19.** $3\frac{1}{15} - 1\frac{3}{20}$ $1\frac{11}{12}$

Section 0.3

Multiply.

20. $6 \times \frac{5}{11}$ $\frac{30}{11}$ or $2\frac{8}{11}$ **21.** $2\frac{1}{3} \times 4\frac{1}{2}$ $10\frac{1}{2}$ **22.** $16 \times 3\frac{1}{8}$ 50   **23.** $\frac{4}{7} \times 5$ $\frac{20}{7}$ or $2\frac{6}{7}$

Divide.

24. $\frac{3}{8} \div 6$ $\frac{1}{16}$ **25.** $\dfrac{\frac{8}{3}}{\frac{5}{9}}$ $\frac{24}{5}$ or $4\frac{4}{5}$ **26.** $\frac{15}{16} \div 6\frac{1}{4}$ $\frac{3}{20}$ **27.** $2\frac{6}{7} \div \frac{10}{21}$ 6

Section 0.4

Combine.

28. $1.634 + 3.007 + 2.560$
7.201

29. $24.831 - 17.094$
7.737

30. $47.251 - 17.69$
29.561

31. $1.9 + 2.53 + 0.006$
4.436

Multiply.

32. 0.007×5.35
0.03745

33. 362.341×1000
362,341

34. $2.6 \times 0.03 \times 1.02$
0.07956

35. $2.51 \times 100 \times 0.5$
125.5

Divide.

36. $0.186 \div 100$
0.00186

37. $71.32 \div 1000$
0.07132

38. $0.523 \div 0.4$
1.3075

39. $1.35 \div 0.015$
90

40. $4.186 \div 2.3$
1.82

41. $0.19 \div 0.38$
0.5

42. Write as a decimal: $\frac{3}{8}$.   0.375

43. Write as a fraction in simplified form: 0.36. $\frac{9}{25}$

Section 0.5

In exercises 44–47, change to a decimal.

44. 1.4%   0.014   **45.** 36.1%   0.361   **46.** 0.02%   0.0002

47. 125.3%   1.253   **48.** What is 30% of 400?   120   **49.** Find 7.2% of 55.   3.96

50. 76 is what percent of 80?
95%

51. What percent of 120 is 15?
12.5%

52. What percent of 1250 is 750?
60%

53. *Education Level* In 2005, 80% of California residents 25 years old or older reported that they had at least a 4-year high school education. If the population of California 25 years or older was 22,299,041, how many people reported having at least a 4-year high school education? (*Source*: www.census.gov)
17,839,233 people

54. *Math Deficiency* In a given university, 720 of the 960 freshmen had a math deficiency. What percentage of the class had a math deficiency?
75%

Section 0.6

In exercises 55–60, estimate. Do not find an exact value.

55. $234{,}897 \times 1{,}936{,}112$
400,000,000,000

56. $357 + 923 + 768 + 417$
2500

57. $634{,}318 - 284{,}000$
300,000

58. $21\frac{1}{5} - 8\frac{4}{5} - 1\frac{2}{3}$ 10

59. Find 18% of $56,297.
$12,000

60. $12{,}482 \div 389$
25

61. *Salary* Estimate Carmen's salary for the week if she earns $8.35 per hour, and worked 38.5 hours.
$320

62. *Vacation Cost* Estimate the weekly amount owed by each of five families who are renting a resort that costs $3875 per week. $800

Section 0.7

Solve. You may use the Mathematics Blueprint for Problem Solving.

▲ **63.** *Carpeting* Mr. and Mrs. Carr are installing wall-to-wall carpeting in a room that measures $12\frac{1}{2}$ ft by $9\frac{2}{3}$ ft. How much will it cost if the carpet is $26.00 per square yard?
$349.07

64. *Colonial Population* In 1620, the population of the U.S. colonies was 2300. By 1630, the population had doubled to 4600. What was the percent of increase in population?
100%

Gas Mileage   A six-passenger Piper Cub airplane has a gas tank that holds 240 gallons. Use this information to answer exercises 65 and 66.

65. When flying at cruising speed, the plane averages $7\frac{2}{3}$ miles per gallon. How far can the plane fly at cruising speed? If the pilot never plans to fly more than 80% of his maximum cruising distance, what is the longest trip he would plan to fly?
1840 mi; 1472 mi

66. When flying at maximum speed, the plane averages $6\frac{1}{4}$ miles per gallon. How far can the plane fly at maximum speed? If the pilot never plans to fly more than 70% of his maximum flying distance when flying at full speed, what is the longest trip he would plan to fly at full speed?   1500 mi; 1050 mi

▲ **67.** *Geometry* What is the volume of a matchbox toy box if it measures 4″ long $\times$ $2\frac{1}{2}$″ long $\times$ $\frac{3}{4}$″ high?
$7\frac{1}{2}$ in.^{3}

▲ **68.** *Geometry* Using your results from exercise 67, if a packing crate measures 16″ long $\times$ $12\frac{1}{2}$″ wide $\times$ $7\frac{1}{2}$″ high, how many matchbox toy boxes can be packed inside the crate?   200 boxes

69. *Sales Commission* Mike sells sporting goods on an 8% commission. During the first week in July, he sold goods worth $5785. What was his commission for the week?
$462.80

70. *Package Weight* Sam took to the post office several packages that weighed 0.75 pounds each. The total weight of all the packages was 34.5 pounds. How many packages did he take to the post office?
46 packages

71. *Car Loan* Dick and Ann Wright purchased a new car. They took out a loan of $9214.50 to help pay for the car and paid the rest in cash. They paid off the loan with payments of $225 per month for four years. How much more did they pay back than the amount of the car loan? (This is the amount of interest they were charged for the car loan.)
$1585.50

72. *Wages* Kevin works as an overnight stocker at Target. He is paid $7.50 an hour for a 40-hour week. For any additional time he gets paid 1.5 times the normal rate. Last week he worked 49 hours. How much did he get paid last week?
$401.25

How Am I Doing? Chapter 0 Test

Note to Instructor: The Chapter 0 Test file in the TestGen program provides algorithms specifically matched to these problems so you can easily replicate this test for additional practice or assessment purposes.

Remember to use your Chapter Test Prep Video CD to see the worked-out solutions to the test problems you want to review.

In exercises 1 and 2, simplify.

1. $\dfrac{16}{18}$

2. $\dfrac{48}{36}$

3. Write as an improper fraction. $6\dfrac{3}{7}$

4. Write as a mixed number: $\dfrac{105}{9}$.

In exercises 5–12, perform the operations indicated. Simplify answers whenever possible.

5. $\dfrac{2}{3} + \dfrac{5}{6} + \dfrac{3}{8}$

6. $1\dfrac{1}{8} + 3\dfrac{3}{4}$

7. $3\dfrac{2}{3} - 2\dfrac{5}{6}$

8. $\dfrac{5}{7} \times \dfrac{28}{15}$

9. $\dfrac{5}{18} \times \dfrac{3}{4}$

10. $\dfrac{7}{4} \div \dfrac{1}{2}$

11. $2\dfrac{1}{2} \times 3\dfrac{1}{4}$

12. $5\dfrac{3}{8} \div 2\dfrac{3}{4}$

In exercises 13–18, perform the calculations indicated.

13. $1.6 + 3.24 + 9.8$

14. $7.0046 - 3.0149$

15. 32.8×0.04

16. 0.07385×1000

17. $12.88 \div 0.056$

18. $26,325.9 \div 100$

19. Write as a percent. 0.073

20. Write as a decimal. 196.5%

21. What is 3.5% of 180?

22. What is 2% of 16.8?

23. 39 is what percent of 650?

24. What percent of 460 is 138?

25. A 4-inch stack of computer chips is on the table. Each computer chip is $\dfrac{2}{9}$ of an inch thick. How many computer chips are in the stack?

In exercises 26–27, estimate. Round each number to one nonzero digit. Then calculate.

26. $52,344\overline{)4,678,987}$

27. $285.36 + 311.85 + 113.6$

Solve. You may use the Mathematics Blueprint for Problem Solving.

28. Allison is paid \$14,000 per year plus a sales commission of 3% of the value of her sales. Last year she sold \$870,000 worth of products. What percent of her total income was her commission? Round to the nearest whole percent.

▲ **29.** Fred and Melinda are laying wall tile in the kitchen. Each tile covers $3\dfrac{1}{2}$ square inches of space. They plan to cover 210 square inches of wall space. How many tiles will they need?

1. $\dfrac{8}{9}$

2. $\dfrac{4}{3}$

3. $\dfrac{45}{7}$

4. $11\dfrac{2}{3}$

5. $\dfrac{15}{8}$ or $1\dfrac{7}{8}$

6. $4\dfrac{7}{8}$

7. $\dfrac{5}{6}$

8. $\dfrac{4}{3}$ or $1\dfrac{1}{3}$

9. $\dfrac{5}{24}$

10. $\dfrac{7}{2}$ or $3\dfrac{1}{2}$

11. $8\dfrac{1}{8}$

12. $1\dfrac{21}{22}$

13. 14.64

14. 3.9897

15. 1.312

16. 73.85

17. 230

18. 263.259

19. 7.3%

20. 1.965

21. 6.3

22. 0.336

23. 6%

24. 30%

25. 18 computer chips

26. 100

27. 700

28. 65%

29. 60 tiles

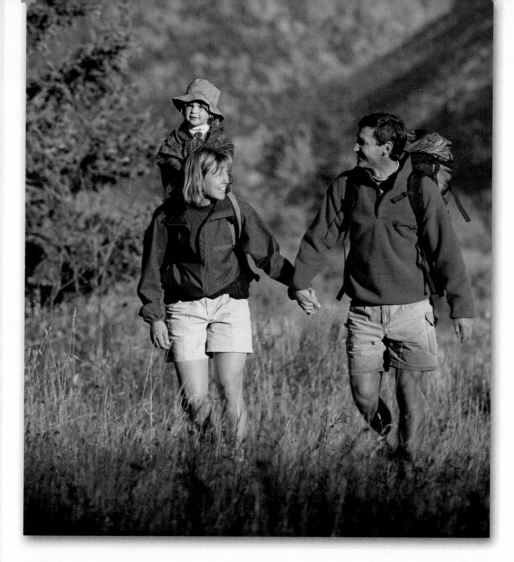

CHAPTER

1

Doctors and health fitness professionals are discovering that certain mathematical measurements of your body can be used in a formula to determine the level of fitness of your body. Suppose you would like to improve your fitness level. What can you do to change it? Doctors have found that through regular exercise, the level of fitness and overall health of almost any person can be significantly improved. Doctors and exercise physiologists use the mathematics that is covered in this chapter to study the progress that people can make through regular exercise sessions of walking or riding a bicycle.

Real Numbers and Variables

1.1 ADDING REAL NUMBERS

1 Identifying Different Types of Numbers

Let's review some of the basic terms we use to talk about numbers.

Whole numbers are numbers such as $0, 1, 2, 3, 4, \ldots$

Integers are numbers such as $\ldots, -3, -2, -1, 0, 1, 2, 3, \ldots$.

Rational numbers are numbers such as $\frac{3}{2}, \frac{5}{7}, -\frac{3}{8}, -\frac{4}{13}, \frac{6}{1},$ and $-\frac{8}{2}$.

Rational numbers can be written as one integer divided by another integer (as long as the denominator is not zero!). Integers can be written as fractions ($3 = \frac{3}{1}$, for example), so we can see that all integers are rational numbers. Rational numbers can be expressed in decimal form. For example, $\frac{3}{2} = 1.5, -\frac{3}{8} = -0.375$, and $\frac{1}{3} = 0.333\ldots$ or $0.\overline{3}$. It is important to note that rational numbers in decimal form are either terminating decimals or repeating decimals.

Irrational numbers are numbers that cannot be expressed as one integer divided by another integer. The numbers π, $\sqrt{2}$, and $\sqrt[3]{7}$ are irrational numbers.

Irrational numbers can be expressed in decimal form. The decimal form of an irrational number is a nonterminating, nonrepeating decimal. For example, $\sqrt{2} = 1.414213\ldots$ can be carried out to an infinite number of decimal places with no repeating pattern of digits.

Finally, **real numbers** are all the rational numbers and all the irrational numbers.

EXAMPLE 1 Classify as an integer, a rational number, an irrational number, and/or a real number.

(a) 5   **(b)** $-\frac{1}{3}$ **(c)** 2.85   **(d)** $\sqrt{2}$ **(e)** $0.777\ldots$

Solution Make a table. Check off the description of the number that applies.

	Number	Integer	Rational Number	Irrational Number	Real Number
(a)	5	✓	✓		✓
(b)	$-\frac{1}{3}$		✓		✓
(c)	2.85		✓		✓
(d)	$\sqrt{2}$			✓	✓
(e)	$0.777\ldots$		✓		✓

Practice Problem 1 Classify.

(a) $-\frac{2}{5}$ **(b)** $1.515151\ldots$ **(c)** -8 **(d)** π

(a) Rational, real   **(b)** Rational, real   **(c)** Integer, rational, real **(d)** Irrational, real

Any real number can be pictured on a **number line.**

Positive numbers are to the right of 0 on a number line.

Negative numbers are to the left of 0 on a number line.

The **real numbers** include the positive numbers, the negative numbers, and zero.

② Using Real Numbers in Real-Life Situations

We often encounter practical examples of number lines that include positive and negative rational numbers. For example, we can tell by reading the accompanying thermometer that the temperature is 20° below 0. From the stock market report, we see that the stock opened at 36 and closed at 34.5, so the net change for the day was −1.5.

Temperature in degrees Fahrenheit

The temperature is 20° below zero.

Net change of −1.5 for the day

The stock opened at 36.

The stock closed at 34.5.

A stock market report

In the following example we use real numbers to represent real-life situations.

EXAMPLE 2 Use a real number to represent each situation.

(a) A temperature of 128.6°F below zero is recorded at Vostok, Antarctica.
(b) The Himalayan peak K2 rises 29,064 feet above sea level.
(c) The Dow gains 10.24 points.
(d) An oil drilling platform extends 328 feet below sea level.

Solution A key word can help you to decide whether a number is positive or negative.

(a) 128.6°F *below* zero is −128.6.
(b) 29,064 feet *above* sea level is +29,064.
(c) A *gain* of 10.24 points is +10.24.
(d) 328 feet *below* sea level is −328.

Practice Problem 2 Use a real number to represent each situation.

(a) A population growth of 1259 +1259   **(b)** A depreciation of $763 −763
(c) A wind-chill factor of minus 10 −10

In everyday life we consider positive numbers the opposite of negative numbers. For example, a gain of 3 yards in a football game is the opposite of a loss of 3 yards; a check written for $2.16 on a checking account is the opposite of a deposit of $2.16.

Each positive number has an opposite negative number. Similarly, each negative number has an opposite positive number. **Opposite numbers,** also called **additive inverses,** have the same magnitude but different signs and can be represented on a number line.

Teaching Example 3 Find the additive inverse (the opposite).

(a) $-\dfrac{7}{8}$ (b) 328

Ans: (a) $\dfrac{7}{8}$ (b) -328

EXAMPLE 3   Find the additive inverse (that is, the opposite).

(a) -7 (b) $\dfrac{1}{4}$

Solution

(a) The opposite of -7 is $+7$.

(b) The opposite of $\dfrac{1}{4}$ is $-\dfrac{1}{4}$.

Practice Problem 3   Find the additive inverse (the opposite).

(a) $\dfrac{2}{5}$ $-\dfrac{2}{5}$ (b) -1.92 $+1.92$ (c) A loss of 12 yards on a football play
                                                                                 Gain of 12 yards

---

3 Adding Real Numbers with the Same Sign

To use a real number, we need to be clear about its sign. When we write the number three as $+3$, the sign indicates that it is a positive number. The positive sign can be omitted. If someone writes three (3), it is understood that it is a positive three ($+3$). To write a negative number such as negative three (-3), we must include the sign.

A concept that will help us add and subtract real numbers is the idea of absolute value. The **absolute value** of a number is the distance between that number and zero on a number line. The absolute value of 3 is written $|3|$.

$$|-3| = 3 \qquad\qquad |3| = 3$$

Distance is always a positive number regardless of the direction of travel. This means that the absolute value of any number will be a positive value or zero. We place the symbols $|$ and $|$ around a number to mean the absolute value of the number.

The distance from 0 to 3 is 3, so $|3| = 3$. This is read "the absolute value of 3 is 3."

The distance from 0 to -3 is 3, so $|-3| = 3$. This is read "the absolute value of -3 is 3."

Some other examples are

$$|-22| = 22, \qquad |5.6| = 5.6, \qquad \text{and} \qquad |0| = 0.$$

Thus, the absolute value of a number can be thought of as the magnitude of the number, without regard to its sign.

Teaching Example 4 Find the absolute value.

(a) $\left|-\dfrac{5}{9}\right|$ (b) $|2057|$ (c) $|-\sqrt{11}|$

Ans: (a) $\dfrac{5}{9}$ (b) 2057   (c) $\sqrt{11}$

EXAMPLE 4   Find the absolute value.

(a) $|-4.62|$ (b) $\left|\dfrac{3}{7}\right|$ (c) $|0|$

Solution

(a) $|-4.62| = 4.62$ (b) $\left|\dfrac{3}{7}\right| = \dfrac{3}{7}$ (c) $|0| = 0$

Practice Problem 4   Find the absolute value.

(a) $|-7.34|$ 7.34   (b) $\left|\dfrac{5}{8}\right|$ $\dfrac{5}{8}$ (c) $\left|\dfrac{0}{2}\right|$ 0

Now let's look at addition of real numbers when the two numbers have the same sign. Suppose that you are keeping track of your checking account at a local bank. When you make a deposit of 5 dollars, you record it as +5. When you write a check for 4 dollars, you record it as −4, as a debit. Consider two situations.

SITUATION 1: Total Deposit   You made a deposit of 20 dollars on one day and a deposit of 15 dollars the next day. You want to know the total value of your deposits.

Your record for situation 1.

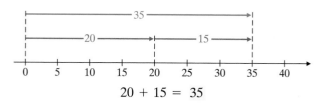

$$20 + 15 = 35$$

The amount of the deposit on the first day added to the amount of the deposit on the second day is the total of the deposits made over the two days.

SITUATION 2: Total Debit   You write a check for 25 dollars to pay one bill and two days later write a check for 5 dollars. You want to know the total value of the debits to your account for the two checks.

Your record for situation 2.

$$-25 + (-5) = -30$$

The value of the first check added to the value of the second check is the total debit to your account.

In each situation we found that we added the absolute value of each number. (That is, we added the numbers without regarding their sign.) The answer always contained the sign that was common to both numbers.

We will now state these results as a formal rule.

> **ADDITION RULE FOR TWO NUMBERS WITH THE SAME SIGN**
>
> To add two numbers with the same sign, add the absolute values of the numbers and use the common sign in the answer.

EXAMPLE 5   Add.

(a) $14 + 16$

(b) $-8 + (-7)$

Solution

(a) $14 + 16$

  $14 + 16 = 30$ Add the absolute values of the numbers.
  $14 + 16 = +30$ Use the common sign in the answer. Here the common sign is the + sign.

(b) $-8 + (-7)$

  $8 + 7 = 15$ Add the absolute values of the numbers.
  $-8 + (-7) = -15$ Use the common sign in the answer. Here the common sign is the − sign.

Practice Problem 5   Add.

(a) $37 + 19$ +56

(b) $-23 + (-35)$ −58

Teaching Tip The idea behind these situations is to help students see logical places in the real world where real number operations are needed, and the outcome is logical. Feel free to use examples other than banking.

Teaching Example 5 Add.

(a) $13 + 41$ **(b)** $-11 + (-5)$

Ans: (a) 54   **(b)** −16

Teaching Example 6 Add. $-\dfrac{4}{5} + \left(-\dfrac{2}{15}\right)$

Ans: $-\dfrac{14}{15}$

EXAMPLE 6 Add. $\dfrac{2}{3} + \dfrac{1}{7}$

Solution

$$\dfrac{2}{3} + \dfrac{1}{7}$$

$$\dfrac{14}{21} + \dfrac{3}{21} \qquad \text{Change each fraction to an equivalent fraction with a common denominator of 21.}$$

$$\dfrac{14}{21} + \dfrac{3}{21} = +\dfrac{17}{21} \text{ or } \dfrac{17}{21} \qquad \text{Add the absolute values of the numbers. Use the common sign in the answer. Note that if no sign is written, the number is understood to be positive.}$$

Practice Problem 6 Add.

$$-\dfrac{3}{5} + \left(-\dfrac{4}{7}\right) \quad -\dfrac{41}{35} \text{ or } -1\dfrac{6}{35}$$

NOTE TO STUDENT: *Fully worked-out solutions to all of the Practice Problems can be found at the back of the text starting at page SP-1*

Teaching Example 7 Add. $8.7 + 6.1$

Ans: 14.8

EXAMPLE 7 Add. $-4.2 + (-3.94)$

Solution

$$-4.2 \ + \ (-3.94)$$

$$4.20 + \quad 3.94 \ = 8.14 \qquad \text{Add the absolute values of the numbers.}$$

$$-4.20 + (-3.94) = -8.14 \qquad \text{Use the common sign in the answer.}$$

Practice Problem 7 Add. $-12.7 + (-9.38)$ -22.08

The rule for adding two numbers with the same sign can be extended to more than two numbers. If we add more than two numbers with the same sign, the answer will have the sign common to all.

Teaching Example 8 Add. $-7 + (-6) + (-11)$

Ans: -24

EXAMPLE 8 Add. $-7 + (-2) + (-5)$

Solution

$$-7 + (-2) + (-5) \qquad \text{We are adding three real numbers, all with the same sign. We begin by adding the first two numbers.}$$

$$= -9 + (-5) \qquad \text{Add } -7 + (-2) = -9.$$

$$= -14 \qquad \text{Add } -9 + (-5) = -14.$$

Of course, this can be shortened by adding the three numbers without regard to sign and then using the common sign for the answer.

Practice Problem 8 Add. $-7 + (-11) + (-33)$ -51

 Adding Real Numbers with Opposite Signs

What if the signs of the numbers you are adding are different? Let's consider our checking account again to see how such a situation might occur.

SITUATION 3: Net Increase You made a deposit of 30 dollars on one day. On the next day you write a check for 25 dollars. You want to know the result of your two transactions.

Your record for situation 3.

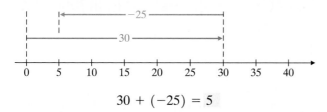

$$30 + (-25) = 5$$

A positive 30 for the deposit added to a negative 25 for the check, which is a debit, gives a net increase of 5 dollars in the account.

SITUATION 4: Net Decrease   You made a deposit of 10 dollars on one day. The next day you write a check for 40 dollars. You want to know the result of your two transactions.

Your record for situation 4.

$$10 + (-40) = -30$$

A positive 10 for the deposit added to a negative 40 for the check, which is a debit, gives a net decrease of 30 dollars in the account.

The result is a negative thirty (-30), because the check was larger than the deposit. If you did not have at least 30 dollars in your account at the start of situation 4, you have overdrawn your account.

What do we observe from situations 3 and 4? In each case, first we found the difference of the absolute values of the two numbers. Then the sign of the result was always the sign of the number with the greater absolute value. Thus, in situation 3, 30 is larger than 25. The sign of 30 is positive. The sign of the answer (5) is positive. In situation 4, 40 is larger than 10. The sign of 40 is negative. The sign of the answer (-30) is negative.

We will now state these results as a formal rule.

> **ADDITION RULE FOR TWO NUMBERS WITH DIFFERENT SIGNS**
>
> 1. Find the difference between the larger absolute value and the smaller one.
>
> 2. Give the answer the sign of the number having the larger absolute value.

EXAMPLE 9   Add. $8 + (-7)$

Solution

$8 + (-7)$	We are to add two numbers with opposite signs.
$8 - 7 = 1$	Find the difference between the two absolute values, which is 1.
$+8 + (-7) = +1$ or 1	The answer will have the sign of the number with the larger absolute value. That number is $+8$. Its sign is **positive,** so the answer will be $+1$.

Practice Problem 9   Add. $-9 + 15$ 6

Teaching Example 9 Add. $-3 + 5$

Ans: 2

Using the Addition Properties for Real Numbers

It is useful to know the following three properties of real numbers.

---

1. *Addition is commutative.*
This property states that if two numbers are added, the result is the same no matter which number is written first. The order of the numbers does not affect the result.

$$3 + 6 = 6 + 3 = 9$$
$$-7 + (-8) = (-8) + (-7) = -15$$
$$-15 + 3 = 3 + (-15) = -12$$

2. *Addition of zero to any given number will result in that given number again.*

$$0 + 5 = 5$$
$$-8 + 0 = -8$$

3. *Addition is associative.*
This property states that if three numbers are added, it does not matter which two numbers are grouped by parentheses and added first.

$$3 + (5 + 7) = (3 + 5) + 7$$ First combine numbers inside parentheses; then
$$3 + (12) = (8) + 7$$ combine the remaining numbers. The results are
$$15 = 15$$ the same no matter which numbers are grouped
and added first.

We can use these properties along with the rules we have for adding real numbers to add three or more numbers. We go from left to right, adding two numbers at a time.

Teaching Example 10 Add.
$$-\frac{4}{11} + \frac{2}{11} + \left(-\frac{1}{11}\right)$$
Ans: $-\frac{3}{11}$

EXAMPLE 10   Add. $\dfrac{3}{17} + \left(-\dfrac{8}{17}\right) + \dfrac{4}{17}$

Solution

$$-\frac{5}{17} + \frac{4}{17}$$ Add $\frac{3}{17} + \left(-\frac{8}{17}\right) = -\frac{5}{17}$.
The answer is negative since the larger of the two absolute values is negative.

$$= -\frac{1}{17}$$ Add $-\frac{5}{17} + \frac{4}{17} = -\frac{1}{17}$.
The answer is negative since the larger of the two absolute values is negative.

Practice Problem 10   Add.

$$-\frac{5}{12} + \frac{7}{12} + \left(-\frac{11}{12}\right) - \frac{3}{4}$$

Sometimes the numbers being added have the same signs; sometimes the signs are different. When adding three or more numbers, you may encounter both situations.

EXAMPLE 11 Add. $-1.8 + 1.4 + (-2.6)$

Solution

$-0.4 + (-2.6)$	We take the difference of 1.8 and 1.4 and use the sign of the number with the larger absolute value.
$= -3.0$	Add $-0.4 + (-2.6) = -3.0$. The signs are the same; we add the absolute values of the numbers and use the common sign.

Practice Problem 11 Add. $-6.3 + (-8.0) + 3.5$ -10.8

Teaching Example 11 Add.
$-7.2 + (-4.0) + 8.1$

Ans: -3.1

If many real numbers are added, it is often easier to add numbers with like signs in a column format. Remember that addition is commutative and associative; therefore, real numbers can be added *in any order*. You do *not* need to combine the first two numbers as your first step.

EXAMPLE 12 Add. $-8 + 3 + (-5) + (-2) + 6 + 5$

Solution

-8		$+3$	
-5	All the signs are the same.	$+6$	All the signs are the same.
-2	Add the three negative	$+5$	Add the three positive
-15	numbers to obtain -15.	$+14$	numbers to obtain $+14$.

Add the two results.

$$-15 + 14 = -1$$

The answer is negative because the number with the larger absolute value is negative.

Practice Problem 12 Add. $-6 + 5 + (-7) + (-2) + 5 + 3$ -2

Teaching Example 12 Add.
$-9 + (-4) + 2 + (-3) + 18 + (-1)$

Ans: 3

A word about notation: The only time we really need to show the sign of a number is when the number is negative—for example, -3. The only time we need to show parentheses when we add real numbers is when we have two different signs preceding a number. For example, $-5 + (-6)$.

EXAMPLE 13 Add.

(a) $2.8 + (-1.3)$ **(b)** $-\dfrac{2}{5} + \left(-\dfrac{3}{4}\right)$

Solution

(a) $2.8 + (-1.3) = 1.5$

(b) $-\dfrac{2}{5} + \left(-\dfrac{3}{4}\right) = -\dfrac{8}{20} + \left(-\dfrac{15}{20}\right) = -\dfrac{23}{20}$ or $-1\dfrac{3}{20}$

Practice Problem 13 Add.

(a) $-2.9 + (-5.7)$ -8.6 **(b)** $\dfrac{2}{3} + \left(-\dfrac{1}{4}\right)$ $\dfrac{5}{12}$

Teaching Example 13 Add.

(a) $-1.8 + (-5)$ **(b)** $-\dfrac{2}{3} + \dfrac{3}{4}$

Ans: (a) -6.8 **(b)** $\dfrac{1}{12}$

Verbal and Writing Skills

Check off any description of the number that applies.

	Number	Integer	Rational Number	Irrational Number	Real Number
1.	23	✓	✓		✓
2.	$-\frac{4}{5}$		✓		✓
3.	π			✓	✓
4.	2.34		✓		✓
5.	$-6.666\ldots$		✓		✓

	Number	Integer	Rational Number	Irrational Number	Real Number
6.	$-\frac{7}{9}$		✓		✓
7.	$-2.3434\ldots$		✓		✓
8.	14	✓	✓		✓
9.	$\sqrt{2}$			✓	✓
10.	$3.232232223\ldots$			✓	✓

Use a real number to represent each situation.

11. Jules Verne wrote a book with the title *20,000 Leagues Under the Sea*. $-20{,}000$

12. The value of the dollar is up $0.07 with respect to the yen. $+\$0.07$

13. Ramona lost $37\frac{1}{2}$ pounds on Weight Watchers. $-37\frac{1}{2}$

14. The scouts hiked from sea level to the top of a 3642-foot-high mountain. $+3642$

15. The temperature rises 7°F. $+7$

16. The lowest point in Australia is Lake Eyre at 52 feet below sea level. -52

Find the additive inverse (opposite).

17. 8 -8

18. $-\frac{4}{5}$ $\frac{4}{5}$

19. -2.73 2.73

20. 85.4 -85.4

Find the absolute value.

21. $|-1.3|$ 1.3

22. $|-5.9|$ 5.9

23. $\left|\frac{5}{6}\right|$ $\frac{5}{6}$

24. $\left|\frac{7}{12}\right|$ $\frac{7}{12}$

Add.

25. $-6 + (-5)$ -11

26. $-13 + (-3)$ -16

27. $-20 + (-30)$ -50

28. $-15 + (-25)$ -40

29. $-\frac{7}{20} + \frac{13}{20}$ $\frac{3}{10}$

30. $-\frac{2}{9} + \left(\frac{-4}{9}\right)$ $-\frac{2}{3}$

31. $-\frac{2}{13} + \left(-\frac{5}{13}\right)$ $-\frac{7}{13}$

32. $-\frac{5}{18} + \frac{11}{18}$ $\frac{1}{3}$

33. $-\frac{2}{5} + \frac{3}{7}$ $\frac{1}{35}$

34. $-\frac{2}{7} + \frac{3}{14}$ $-\frac{1}{14}$

35. $-10.3 + (-8.9)$ -19.2

36. $-5.4 + (-12.8)$ -18.2

37. $0.6 + (-0.2)$ 0.4

38. $-0.8 + 0.5$ -0.3

39. $-5.26 + (-8.9)$ -14.16

40. $-6.48 + (-3.7)$ -10.18

41. $-8 + 5 + (-3)$ -6

42. $7 + (-8) + (-4)$ -5

43. $-2 + (-8) + 10$ 0

44. $-9 + 6 + (-12)$ -15

45. $-\dfrac{3}{10} + \dfrac{3}{4}$ $\dfrac{9}{20}$ **46.** $-\dfrac{3}{8} + \dfrac{11}{24}$ $\dfrac{1}{12}$ **47.** $-14 + 9 + (-3)$ -8 **48.** $-18 + 10 + (-5)$ -13

Mixed Practice *Add.*

49. $8 + (-11)$ -3 **50.** $16 + (-24)$ -8 **51.** $-83 + 142$ 59 **52.** $-114 + 186$ 72

53. $-\dfrac{4}{9} + \dfrac{5}{6}$ $\dfrac{7}{18}$ **54.** $-\dfrac{3}{5} + \dfrac{2}{3}$ $\dfrac{1}{15}$ **55.** $-\dfrac{1}{10} + \dfrac{1}{2}$ $\dfrac{2}{5}$ **56.** $-\dfrac{2}{3} + \left(-\dfrac{1}{4}\right)$ $-\dfrac{11}{12}$

57. $5.18 + (-7.39)$ -2.21 **58.** $8.33 + (-14.2)$ -5.87 **59.** $4 + (-8) + 16$ 12

60. $27 + (-11) + (-4)$ 12 **61.** $34 + (-18) + 11 + (-27)$ 0 **62.** $-23 + 4 + (-11) + 17$ -13

63. $17.85 + (-2.06) + 0.15$ 15.94 **64.** $23.17 + 5.03 + (-11.81)$ 16.39

Applications

65. *Profit/Loss* Holly paid $47 for a vase at an estate auction. She resold it to an antiques dealer for $214. What was her profit or loss? $167 profit

66. *Temperature* When we skied at Jackson Hole, Wyoming, yesterday, the temperature at the summit was $-12°$F. Today when we called the ski report, the temperature had risen $7°$F. What is the temperature at the summit today? $-5°$F

67. *Home Equity Line of Credit* Ramon borrowed $2300 from his home equity line of credit to pay off his car loan. He then borrowed another $1500 to pay to have his kitchen repainted. Represent how much Ramon owed on his home equity line of credit as a real number. $-$3800

68. *Time Change* During the winter, New York City is on Eastern Standard Time (EST). Melbourne, Australia, is 15 hours ahead of New York. If it is 11 P.M. in Melbourne, what time is it in New York? 8 A.M.

69. *Football* During the Homecoming football game, Quentin lost 15 yards, gained 3 yards, and gained 21 yards in three successive running plays. What was his total gain or loss? 9-yd gain

70. *School Fees* Wanda's financial aid account at school held $643.85. She withdrew $185.50 to buy books for the semester. Does she have enough left in her account to pay the $475.00 registration fee for the next semester? If so, how much extra money does she have? If not, how much is she short? No; $16.65 short

71. *Butterfly Population* The population of a particular butterfly species was 8000. Twenty years later there were 3000 fewer. Today, there are 1500 fewer. Study the graph to the right. What is the new population? 3500

72. *Credit Card Balance* Aaron owes $258 to a credit card company. He makes a purchase of $32 with the card and then makes a payment of $150 on the account. How much does he still owe? $140

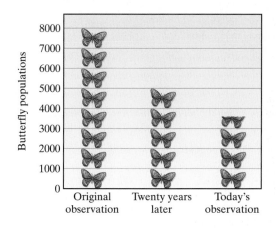

Profit/Loss *During the first five months of 2005 a regional Midwest airline posted profit and loss figures for each month of operation, as shown in the accompanying bar graph.*

73. For the first three months of 2005, what were the total earnings of the airline? $32,000,000

74. For the first five months of 2005, what were the total earnings for the airline? $22,000,000

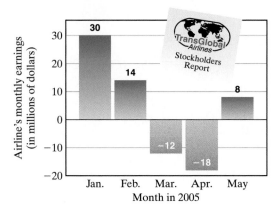

To Think About

75. What number must be added to -13 to get 5?   18

76. What number must be added to -18 to get 10?   28

Cumulative Review   *Perform the indicated calculations.*

77. [0.2.3] $\dfrac{15}{16} + \dfrac{1}{4}$ $\dfrac{19}{16}$ or $1\dfrac{3}{16}$

78. [0.3.1] $\left(\dfrac{3}{7}\right)\left(\dfrac{14}{9}\right)$ $\dfrac{2}{3}$

79. [0.2.3] $\dfrac{2}{15} - \dfrac{1}{20}$ $\dfrac{1}{12}$

80. [0.3.2] $2\dfrac{1}{2} \div 3\dfrac{2}{5}$ $\dfrac{25}{34}$

81. [0.4.4] $0.72 + 0.8$ 1.52

82. [0.4.4] $1.63 - 0.98$ 0.65

83. [0.4.5] $(1.63)(0.7)$ 1.141

84. [0.4.6] $0.208 \div 0.8$ 0.26

Quick Quiz 1.1   Add.

1. $-18 + (-16)$ -34

2. $-2.7 + 8.6 + (-5.4)$ 0.5

3. $-\dfrac{5}{6} + \dfrac{7}{24}$ $-\dfrac{13}{24}$

4. Concept Check Explain why when you add two negative numbers you always obtain a negative number but when you add one negative number and one positive number you may obtain zero, a positive number, or a negative number.

Answers may vary

Classroom Quiz 1.1   You may use these problems to quiz your students' mastery of Section 1.1.

Add.

1. $27 + (-34)$ **Ans:** -7

2. $-3.4 + 8.9 + (-4.8)$ **Ans:** 0.7

3. $-\dfrac{7}{8} + \dfrac{11}{24}$ **Ans:** $-\dfrac{5}{12}$

1.2 SUBTRACTING REAL NUMBERS

Subtracting Real Numbers with Like or Unlike Signs

So far we have developed the rules for adding real numbers. We can use these rules to subtract real numbers. Let's look at a checkbook situation to see how.

SITUATION 5: Subtract a Deposit and Add a Debit You have a balance of 20 dollars in your checking account. The bank calls you and says that a deposit of 5 dollars that belongs to another account was erroneously added to your account. They say they will correct the account balance to 15 dollars. The bank tells you that since they cannot take away the erroneous credit, they will add a debit to your account. You want to keep track of what's happening to your account.

Your record for situation 5.

$$20 - (+5) = 15$$

From your present balance subtract the deposit to give the new balance. This equation shows what needs to be done to your account. The bank tells you that because the error happened in the past they cannot "take it away." However, they can add to your account a debit of 5 dollars. Here is the equivalent addition.

$$20 + (-5) = 15$$

To your present balance add a debit to give the new balance. Subtracting a positive 5 has the same effect as adding a negative 5.

SUBTRACTION OF REAL NUMBERS

To subtract real numbers, add the opposite of the second number (that is, the number you are subtracting) to the first.

The rule tells us to do three things when we subtract real numbers. First, change subtraction to addition. Second, replace the second number by its opposite. Third, add the two numbers using the rules for addition of real numbers.

EXAMPLE 1 Subtract. $6 - (-2)$

Solution
$$
\begin{array}{ccc}
6 & - & (-2)
\end{array}
$$

Change subtraction to addition.

Write the opposite of the second number.

$$= \quad 6 \quad + \quad (+2)$$

Add the two real numbers with the same sign.

$$= \qquad 8$$

Practice Problem 1 Subtract. $9 - (-3)$ $\quad$ 12

EXAMPLE 2 Subtract. $-8 - (-6)$

Solution

-8 $-$ (-6)

| Change subtraction to addition. | | Write the opposite of the second number. |

$= \quad -8 \quad\quad + \quad\quad\quad\quad (+6)$

| Add the two real numbers with opposite signs. |

$= \quad\quad\quad -2$

Practice Problem 2 Subtract. $-12 - (-5)$ -7

Calculator

More with Negative Numbers

Subtract. Remember to use the $\boxed{+/-}$ key to change the sign of a number from $+$ to $-$ or from $-$ to $+$.

(a) $-18 - (-24)$

(b) $-6 + (-10) - (-15)$

Answers:

(a) 6 (b) -1

EXAMPLE 3 Subtract.

(a) $\dfrac{3}{7} - \dfrac{6}{7}$ (b) $-\dfrac{7}{18} - \left(-\dfrac{1}{9}\right)$

Solution

(a) $\dfrac{3}{7} - \dfrac{6}{7} = \dfrac{3}{7} + \left(-\dfrac{6}{7}\right)$ Change the subtraction problem to one of adding the opposite of the second number. We note that the problem has two fractions with the same denominator.

$= -\dfrac{3}{7}$ Add two numbers with different signs.

(b) $-\dfrac{7}{18} - \left(-\dfrac{1}{9}\right) = -\dfrac{7}{18} + \dfrac{1}{9}$ Change subtracting to adding the opposite.

$= -\dfrac{7}{18} + \dfrac{2}{18}$ Change $\dfrac{1}{9}$ to $\dfrac{2}{18}$ since LCD $= 18$.

$= -\dfrac{5}{18}$ Add two numbers with different signs.

Practice Problem 3 Subtract.

(a) $\dfrac{5}{9} - \dfrac{7}{9}$ $-\dfrac{2}{9}$ (b) $-\dfrac{5}{21} - \left(-\dfrac{3}{7}\right)$ $\dfrac{4}{21}$

EXAMPLE 4 Subtract. $-5.2 - (-5.2)$

Solution

$-5.2 - (-5.2) = -5.2 + 5.2$ Change the subtraction problem to one of adding the opposite of the second number.

$= 0$ Add two numbers with different signs.

Practice Problem 4 Subtract. $-17.3 - (-17.3)$ 0

Example 4 illustrates what is sometimes called the **additive inverse property.** When you add two real numbers that are opposites of each other, you will obtain zero. Examples of this are the following:

$$5 + (-5) = 0 \qquad -186 + 186 = 0 \qquad -\frac{1}{8} + \frac{1}{8} = 0.$$

EXAMPLE 5 Subtract.

(a) $-8 - 2$ **(b)** $23 - 28$ **(c)** $5 - (-3)$ **(d)** $\dfrac{1}{4} - 8$

Solution

(a) $-8 - 2 = -8 + (-2)$ Notice that we are subtracting a positive 2. Change to addition.

$\qquad\qquad = -10$ Add.

In a similar fashion we have

(b) $23 - 28 = 23 + (-28) = -5$

(c) $5 - (-3) = 5 + 3 = 8$

(d) $\dfrac{1}{4} - 8 = \dfrac{1}{4} + (-8) = \dfrac{1}{4} + \left(-\dfrac{32}{4}\right) = -\dfrac{31}{4}$

Practice Problem 5 Subtract.

(a) $-21 - 9$ **(b)** $17 - 36$ **(c)** $12 - (-15)$ **(d)** $\dfrac{3}{5} - 2$

(a) -30 **(b)** -19 **(c)** 27 **(d)** $-\dfrac{7}{5}$ or $-1\dfrac{2}{5}$

Teaching Example 5 Subtract.

(a) $12 - 18$ **(b)** $6 - (-5)$

(c) $\dfrac{5}{6} - 4$ **(d)** $-1 - 7$

Ans: (a) -6 **(b)** 11 **(c)** $-\dfrac{19}{6}$

(d) -8

Teaching Tip Emphasize that when you translate the expression "Subtract -3 from -20," the number -20 is written first. Thus we have

$$-20 - (-3).$$

EXAMPLE 6 A satellite is recording radioactive emissions from nuclear waste buried 3 miles below sea level. The satellite orbits Earth at 98 miles above sea level. How far is the satellite from the nuclear waste?

Solution We want to find the difference between $+98$ miles and -3 miles. This means we must subtract -3 from 98.

$$98 - (-3) = 98 + 3$$
$$= 101$$

The satellite is 101 miles from the nuclear waste.

Practice Problem 6 A helicopter is directly over a sunken vessel. The helicopter is 350 feet above sea level. The vessel lies 186 feet below sea level. How far is the helicopter from the sunken vessel? 536 ft

Teaching Example 6 The top of an oil rig is 120 feet above sea level. The base of the oil rig is 97 feet below sea level. What is the total height of the oil rig?

Ans: 217 ft

Developing Your Study Skills

Reading the Textbook

Begin reading your textbook with paper and pencil in hand. As you come across a new definition or concept, underline it in the text and/or write it down in your notebook. Whenever you encounter an unfamiliar term, look it up and make a note of it. When you come to an example, work through it step-by-step. Be sure to read each word and follow directions carefully.

Notice the helpful hints the author provides. They guide you to correct solutions and prevent you from making errors. Take advantage of these pieces of expert advice.

Be sure that you understand what you are reading. Make a note of any of those things that you do not understand and ask your instructor about them. Do not hurry through the material. Learning mathematics takes time.

Verbal and Writing Skills

1. Explain in your own words how you would perform the necessary steps to find $-8 - (-3)$.
First change subtracting -3 to adding a positive three. Then use the rules for addition of two real numbers with different signs. Thus $-8 - (-3) = -8 + 3 = -5$.

2. Explain in your own words how you would perform the necessary steps to find $-10 - (-15)$.
First change subtracting -15 to adding a positive fifteen. Then use the rules for addition of two real numbers with different signs. Thus $-10 - (-15) = -10 + 15 = 5$.

Subtract by adding the opposite.

3. $18 - 35$ -17
4. $16 - 48$ -32
5. $19 - 23$ -4
6. $10 - 20$ -10

7. $-14 - (-3)$ -11
8. $-24 - (-7)$ -17
9. $-52 - (-60)$ 8
10. $-48 - (-80)$ 32

11. $0 - (-5)$ 5
12. $0 - (-7)$ 7
13. $-18 - (-18)$ 0
14. $-24 - (-24)$ 0

15. $-17 - (-20)$ 3
16. $-11 - (-19)$ 8
17. $\frac{2}{5} - \frac{4}{5}$ $-\frac{2}{5}$
18. $\frac{2}{9} - \frac{7}{9}$ $-\frac{5}{9}$

19. $\frac{3}{4} - \left(-\frac{3}{5}\right)$ $\frac{27}{20}$ or $1\frac{7}{20}$
20. $-\frac{2}{3} - \frac{1}{4}$ $-\frac{11}{12}$
21. $-\frac{3}{4} - \frac{5}{6}$ $-\frac{19}{12}$ or $-1\frac{7}{12}$
22. $-\frac{7}{10} - \frac{10}{15}$ $-\frac{41}{30}$ or $-1\frac{11}{30}$

23. $-0.6 - 0.3$ -0.9
24. $-0.9 - 0.5$ -1.4
25. $2.64 - (-1.83)$ 4.47
26. $-0.03 - 0.06$ -0.09

Mixed Practice   *Calculate.*

27. $\frac{3}{5} - 4$ $-\frac{17}{5}$ or $-3\frac{2}{5}$
28. $\frac{5}{6} - 3$ $-\frac{13}{6}$ or $-2\frac{1}{6}$
29. $-\frac{2}{3} - 4$ $-\frac{14}{3}$ or $-4\frac{2}{3}$
30. $-\frac{1}{6} - 5$ $-\frac{31}{6}$ or $-5\frac{1}{6}$

31. $34 - 87$ -53
32. $19 - 76$ -57
33. $-25 - 48$ -73
34. $-74 - 11$ -85

35. $2.3 - (-4.8)$ 7.1
36. $8.4 - (-2.7)$ 11.1
37. $8 - \left(-\frac{3}{4}\right)$ $\frac{35}{4}$ or $8\frac{3}{4}$
38. $\frac{2}{3} - (-6)$ $\frac{20}{3}$ or $6\frac{2}{3}$

39. $\frac{5}{6} - 7$ $-\frac{37}{6}$ or $-6\frac{1}{6}$
40. $9 - \frac{2}{3}$ $\frac{25}{3}$ or $8\frac{1}{3}$
41. $-\frac{2}{7} - \frac{4}{5}$ $-\frac{38}{35}$ or $-1\frac{3}{35}$
42. $-\frac{5}{6} - \frac{1}{5}$ $-\frac{31}{30}$ or $-1\frac{1}{30}$

43. $-135 - (-126.5)$ -8.5
44. $-97.6 - (-146)$ 48.4
45. $\frac{1}{5} - 6$ $-\frac{29}{5}$ or $-5\frac{4}{5}$
46. $\frac{2}{7} - (-3)$ $\frac{23}{7}$ or $3\frac{2}{7}$

47. $4.5 - (-1.56)$ 6.06
48. $5.2 - (-3.88)$ 9.08
49. $-3 - 2.047$ -5.047
50. $-1.043 - 4$ -5.043

51. Subtract -9 from -2. 7
52. Subtract -12 from 20. 32
53. Subtract 13 from -35. -48

One Step Further   *Change each subtraction operation to "adding the opposite." Then combine the numbers.*

54. $9 + 6 - (-5)$ 20
55. $7 + (-6) - 3$ -2
56. $11 + (-3) - 8$ 0

57. $-13 + 12 - (-1)$ 0

58. $18 - (-15) - 3$ 30

59. $7 + (-42) - 27$ -62

60. $-4.2 - (-3.8) + 1.5$ 1.1

61. $-6.4 - (-2.7) + 5.3$ 1.6

Applications

62. *Sea Rescue* A rescue helicopter is 300 feet above sea level. The captain has located an ailing submarine directly below it that is 126 feet below sea level. How far is the helicopter from the submarine? 426 ft

63. *Checking Account Balance* Yesterday Jackie had $112 in her checking account. Today her account reads "balance $-$37." Find the difference in these two amounts. $149

64. *Federal Taxes* In 2007, Damian had $2435 withheld from his paychecks in federal taxes. When he filed his tax return, he received a $308 refund. How much did he actually pay in taxes? $2127

+300 feet

Sea level

−126 feet

Cumulative Review *In exercises 65–69, perform the indicated operations.*

65. **[1.1.4]** $-37 + 16$ -21

66. **[1.1.3]** $-37 + (-14)$ -51

67. **[1.1.3]** $-3 + (-6) + (-10)$ -19

68. **[1.1.4]** *Temperature* On a winter morning, Alisa noticed that the outside temperature was $-5°$F. By the afternoon the temperature had risen $20°$F. What was the afternoon temperature? 15°F

69. **[0.3.1]** *Hiking* Sean and Khalid went hiking in the Blue Ridge Mountains. During their $8\frac{1}{3}$-mile hike, $\frac{4}{5}$ of the distance was covered with snow. How many miles were snow covered? $6\frac{2}{3}$ mi

Quick Quiz 1.2 Subtract.

1. $-8 - (-15)$ 7

2. $-1.3 - 0.6$ -1.9

3. $\dfrac{5}{8} - \left(-\dfrac{2}{7}\right)$ $\dfrac{51}{56}$

4. Concept Check Explain the different results that are possible when you start with a negative number and then subtract a negative number.
Answers may vary

Classroom Quiz 1.2 You may use these problems to quiz your students' mastery of Section 1.2.
Subtract.

1. $-19 - (-13)$ **Ans:** -6

2. $-4.2 - 1.5$ **Ans:** -5.7

3. $\dfrac{2}{9} - \left(-\dfrac{3}{5}\right)$ **Ans:** $\dfrac{37}{45}$

Multiplying Real Numbers

We are familiar with the meaning of multiplication for positive numbers. For example, $5 \times 90 = 450$ might mean that you receive five weekly checks of 90 dollars each and you gain $450. Let's look at a situation that corresponds to $5 \times (-90)$. What might that mean?

SITUATION 6: Checking an Account Balance You write a check for five weeks in a row to pay your weekly room rent of 90 dollars. You want to know the total impact on your checking account balance.

Your record for situation 6.

$(+5)$	$\times$	(-90)	$=$	-450
The number of checks you have written	times	negative 90, the value of each check that was a debit to your account,	gives	negative 450 dollars, a net debit to your account.

Note that a multiplication symbol is not needed between the $(+5)$ and the (-90) because the two sets of parentheses indicate multiplication. The multiplication $(5)(-90)$ is the same as repeated addition of five (-90)'s. Note that 5 multiplied by -90 can be written as $5(-90)$ or $(5)(-90)$.

$$(-90) + (-90) + (-90) + (-90) + (-90) = -450$$

This example seems to show that a positive number multiplied by a negative number is negative.

What if the negative number is the one that is written first? If $(5)(-90) = -450$, then $(-90)(5) = -450$ by the commutative property of multiplication. This is an example showing that *when two numbers with opposite signs (one positive, one negative) are multiplied, the result is negative.*

But what if both numbers are negative? Consider the following situation.

SITUATION 7: Renting a Room Last year at college you rented a room at 90 dollars per week for 36 weeks, which included two semesters and summer school. This year you will not attend the summer session, so you will be renting the room for only 30 weeks. Thus the number of weekly rental checks will be six less than last year. You are making out your budget for this year. You want to know the financial impact of renting the room for six fewer weeks.

Your record for situation 7.

(-6)	$\times$	(-90)	$=$	540
The difference in the number of checks this year compared to last is −6, which is negative to show a decrease,	times	−90, the value of each check paid out,	gives	+540 dollars. The product is positive, because your financial situation will be 540 dollars better this year.

You could check that the answer is positive by calculating the total rental expenses.

	Dollars in rent last year	$(36)(90) =$	3240
(subtract)	Dollars in rent this year	$-(30)(90) =$	-2700
	Extra dollars available this year	$=$	$+540$

This agrees with our previous answer: $(-6)(-90) = +540$.

Note that negative 6 times -90 can be written as $-6(-90)$ or $(-6)(-90)$.

In this situation it seems reasonable that a negative number times a negative number yields a positive answer. We already know from arithmetic that a positive number times a positive number yields a positive answer. Thus we might see the general rule that *when two numbers with the same sign* (both positive or both negative) *are multiplied, the result is positive.*

We will now state our rule.

> ### MULTIPLICATION OF REAL NUMBERS
>
> To multiply two real numbers with **the same sign,** multiply the absolute values.
> The sign of the result is **positive.**
>
> To multiply two real numbers with **opposite signs,** multiply the absolute values.
> The sign of the result is **negative.**

Teaching Tip Sometimes students incorrectly memorize vague rules such as "A negative and a negative give a positive." This of course will confuse them later on. If they wish to memorize a short rule like that, it should be: "A negative **times** a negative gives a positive." Emphasize to students that they will need to keep all these rules straight when they encounter addition, subtraction, multiplication, and division all in one problem set.

EXAMPLE 1   Multiply.

(a) $(3)(6)$ **(b)** $\left(-\dfrac{5}{7}\right)\left(-\dfrac{2}{9}\right)$ **(c)** $-4(8)$ **(d)** $\left(\dfrac{2}{7}\right)(-3)$

Solution

(a) $(3)(6) = 18$

> When multiplying two numbers with the same sign, the result is a positive number.

(b) $\left(-\dfrac{5}{7}\right)\left(-\dfrac{2}{9}\right) = \dfrac{10}{63}$

(c) $-4(8) = -32$

> When multiplying two numbers with opposite signs, the result is a negative number.

(d) $\left(\dfrac{2}{7}\right)(-3) = \left(\dfrac{2}{7}\right)\left(-\dfrac{3}{1}\right) = -\dfrac{6}{7}$

Teaching Example 1 Multiply.

(a) $(5)(7)$ **(b)** $\left(-\dfrac{2}{3}\right)\left(\dfrac{4}{9}\right)$

(c) $6(-3)$ **(d)** $\left(-\dfrac{4}{13}\right)(-3)$

Ans: (a) 35 **(b)** $-\dfrac{8}{27}$

      (c) -18 **(d)** $\dfrac{12}{13}$

Practice Problem 1   Multiply.

(a) $(-6)(-2)$ 12 **(b)** $(7)(9)$ 63 **(c)** $\left(-\dfrac{3}{5}\right)\left(\dfrac{2}{7}\right)$ $-\dfrac{6}{35}$ **(d)** $\left(\dfrac{5}{6}\right)(-7)$

                                                       (d) $-\dfrac{35}{6}$ or $-5\dfrac{5}{6}$

NOTE TO STUDENT: *Fully worked-out solutions to all of the Practice Problems can be found at the back of the text starting at page SP-1*

To multiply more than two numbers, multiply two numbers at a time.

EXAMPLE 2   Multiply. $(-4)(-3)(-2)$

Solution

$(-4)(-3)(-2) = (+12)(-2)$ We begin by multiplying the first two numbers, (-4) and (-3). The signs are the same. The answer is positive 12.

$= -24$ Now we multiply $(+12)$ and (-2). The signs are different. The answer is negative 24.

Teaching Example 2 Multiply. $(-2)(4)(-6)$

Ans: 48

Practice Problem 2   Multiply. $(-5)(-2)(-6)$ -60

Teaching Example 3 Multiply.

(a) $-4(-8.1)$ **(b)** $(-3)(-1)\left(-\dfrac{1}{8}\right)$

(c) $-5(-2)(-1)(-4)$

Ans: (a) 32.4 **(b)** $-\dfrac{3}{8}$ **(c)** 40

EXAMPLE 3   Multiply.

(a) $-3(-1.5)$ **(b)** $\left(-\dfrac{1}{2}\right)(-1)(-4)$ **(c)** $-2(-2)(-2)(-2)$

Solution   Multiply two numbers at a time. See if you find a pattern.

(a) $-3(-1.5) = 4.5$ Be sure to place the decimal point in your answer.

(b) $\left(-\dfrac{1}{2}\right)(-1)(-4) = +\dfrac{1}{2}(-4) = -2$

(c) $-2(-2)(-2)(-2) = +4(-2)(-2) = -8(-2) = +16$ or 16

What kind of answer would we obtain if we multiplied five negative numbers? If you guessed "negative," you probably see the pattern.

Practice Problem 3   Determine the sign of the product. Then multiply to check.

(a) $-2(-3)$ Positive, 6   **(b)** $(-1)(-3)(-2)$ Negative, -6

(c) $-4\left(-\dfrac{1}{4}\right)(-2)(-6)$ Positive, 12

NOTE TO STUDENT: *Fully worked-out solutions to all of the Practice Problems can be found at the back of the text starting at page SP-1*

When you multiply two or more nonzero real numbers:

1. The result is always **positive** if there is an **even** number of negative signs.

2. The result is always **negative** if there is an **odd** number of negative signs.

② Using the Multiplication Properties for Real Numbers

For convenience, we will list the properties of multiplication.

1. *Multiplication is commutative.*

This property states that if two real numbers are multiplied, the order of the numbers does not affect the result. The result is the same no matter which number is written first.

$$(5)(7) = (7)(5) = 35, \quad \left(\dfrac{1}{3}\right)\left(\dfrac{2}{7}\right) = \left(\dfrac{2}{7}\right)\left(\dfrac{1}{3}\right) = \dfrac{2}{21}$$

2. *Multiplication of any real number by zero will result in zero.*

$$(5)(0) = 0, \quad (-5)(0) = 0, \quad (0)\left(\dfrac{3}{8}\right) = 0, \quad (0)(0) = 0$$

3. *Multiplication of any real number by 1 will result in that same number.*

$$(5)(1) = 5, \quad (1)(-7) = -7, \quad (1)\left(-\dfrac{5}{3}\right) = -\dfrac{5}{3}$$

4. *Multiplication is associative.*

This property states that if three real numbers are multiplied, it does not matter which two numbers are grouped by parentheses and multiplied first.

$$2 \times (3 \times 4) = (2 \times 3) \times 4$$ First multiply the numbers in parentheses. Then multiply the remaining numbers.

$$2 \times (12) = (6) \times 4$$ The results are the same no matter which

$$24 = 24$$ numbers are grouped and multiplied first.

Teaching Tip Point out that multiplication in algebra can be written in four different ways.

3×5 Using the symbol $\times$

$(3)(5)$
$3(5)$ Using parentheses

$3 \cdot 5$ Using a centered dot

Each of these expressions indicates "three times five." Most of these forms of notation will be used in the exercises for this section. The most common notation will be parentheses.

Dividing Real Numbers

What about division? Any division problem can be rewritten as a multiplication problem.

We know that $20 \div 4 = 5$ because $4(5) = 20$.

Similarly, $-20 \div (-4) = 5$ because $-4(5) = -20$.

In both division problems the answer is positive 5. Thus we see that *when you divide two numbers with the same sign* (both positive or both negative), *the answer is positive.* What if the signs are different?

We know that $-20 \div 4 = -5$ because $4(-5) = -20$.

Similarly, $20 \div (-4) = -5$ because $-4(-5) = 20$.

In these two problems the answer is negative 5. So we have reasonable evidence to see that *when you divide two numbers with different signs* (one positive and one negative), *the answer is negative.*

We will now state our rule for division.

DIVISION OF REAL NUMBERS

To divide two real numbers with **the same sign,** divide the absolute values. The sign of the result is **positive.**

To divide two real numbers with **different signs,** divide the absolute values. The sign of the result is **negative.**

EXAMPLE 4   Divide.

(a) $12 \div 4$ **(b)** $(-25) \div (-5)$ **(c)** $\dfrac{-36}{18}$ **(d)** $\dfrac{42}{-7}$

Solution

(a) $12 \div 4 = 3$

> When dividing two numbers with the same sign, the result is a positive number.

(b) $(-25) \div (-5) = 5$

(c) $\dfrac{-36}{18} = -2$

> When dividing two numbers with different signs, the result is a negative number.

(d) $\dfrac{42}{-7} = -6$

Teaching Example 4 Divide.

(a) $-15 \div 3$ **(b)** $\dfrac{-24}{-3}$

(c) $\dfrac{-55}{11}$ **(d)** $38 \div 19$

Ans: (a) -5 **(b)** 8 **(c)** -5 **(d)** 2

Practice Problem 4   Divide.

(a) $-36 \div (-2)$ 18   **(b)** $-49 \div 7$ -7 **(c)** $\dfrac{50}{-10}$ -5 **(d)** $\dfrac{-39}{13}$ -3

EXAMPLE 5   Divide.   **(a)** $-36 \div 0.12$ **(b)** $-2.4 \div (-0.6)$

Teaching Example 5 Divide.

(a) $25 \div (-0.5)$ **(b)** $-3.2 \div (-0.02)$

Ans: (a) -50 **(b)** 160

Solution

(a) $-36 \div 0.12$ Look at the problem to determine the sign. When dividing two numbers with different signs, the result will be a negative number.

We then divide the absolute values.

$$0.12_\wedge \overline{)36.00_\wedge} \quad \begin{array}{r} 3\,00. \\ \hline \underline{36} \\ 00 \end{array}$$

Thus $-36 \div 0.12 = -300$. The answer is a negative number.

(b) $-2.4 \div (-0.6)$ Look at the problem to determine the sign. When dividing two numbers with the same sign, the result will be positive.

We then divide the absolute values.

$$0.6_\wedge \overline{)2.4_\wedge} \overset{4.}{}$$
$$\underline{2\ 4}$$

Thus $-2.4 \div (-0.6) = 4$. The answer is a positive number.

NOTE TO STUDENT: *Fully worked-out solutions to all of the Practice Problems can be found at the back of the text starting at page SP-1*

Practice Problem 5   Divide.

(a) $-12.6 \div (-1.8)$ 7   **(b)** $0.45 \div (-0.9)$ -0.5

Note that the rules for multiplication and division are the same. When you **multiply** or **divide** two numbers with the **same** sign, you obtain a **positive** number. When you **multiply** or **divide** two numbers with **different** signs, you obtain a **negative** number.

Teaching Example 6 Divide. $4 \div -\dfrac{3}{7}$

Ans: $-\dfrac{28}{3}$ or $-9\dfrac{1}{3}$

EXAMPLE 6   Divide. $-\dfrac{12}{5} \div \dfrac{2}{3}$

Solution

$$= \left(-\frac{12}{5}\right)\left(\frac{3}{2}\right) \qquad \text{Divide two fractions. We invert the second fraction and multiply by the first fraction.}$$

$$= \left(-\frac{\overset{6}{\cancel{12}}}{5}\right)\left(\frac{3}{\underset{1}{\cancel{2}}}\right)$$

$$= -\frac{18}{5} \quad \text{or} \quad -3\frac{3}{5} \qquad \text{The answer is negative since the two numbers divided have different signs.}$$

Practice Problem 6   Divide.

$$-\frac{5}{16} \div \left(-\frac{10}{13}\right) \quad \frac{13}{32}$$

Note that division can be indicated by the symbol $\div$ or by the fraction bar $-$. $\frac{2}{3}$ means $2 \div 3$.

Teaching Example 7 Divide.

(a) $\dfrac{-\dfrac{5}{9}}{\dfrac{2}{3}}$ **(b)** $\dfrac{-\dfrac{3}{7}}{-9}$

Ans: **(a)** $-\dfrac{5}{6}$ **(b)** $\dfrac{1}{21}$

EXAMPLE 7   Divide.   **(a)** $\dfrac{\dfrac{7}{8}}{-21}$ **(b)** $\dfrac{-\dfrac{2}{3}}{-\dfrac{7}{13}}$

Solution

(a) $\dfrac{\dfrac{7}{8}}{-21}$

$$= \frac{7}{8} \div \left(-\frac{21}{1}\right) \qquad \text{Change } -21 \text{ to a fraction. } -21 = -\frac{21}{1}$$

$$= \frac{\overset{1}{\cancel{7}}}{8}\left(-\frac{1}{\underset{3}{\cancel{21}}}\right) \qquad \text{Change the division to multiplication. Cancel where possible.}$$

$$= -\frac{1}{24} \qquad \text{Simplify.}$$

(b) $\dfrac{-\dfrac{2}{3}}{-\dfrac{7}{13}} = -\dfrac{2}{3} \div \left(-\dfrac{7}{13}\right) = -\dfrac{2}{3}\left(-\dfrac{13}{7}\right) = \dfrac{26}{21}$　or　$1\dfrac{5}{21}$

Practice Problem 7 Divide.

(a) $\dfrac{-12}{-\dfrac{4}{5}}$　15　**(b)** $\dfrac{-\dfrac{2}{9}}{\dfrac{8}{13}}$　$-\dfrac{13}{36}$

---

1. *Division of 0 by any nonzero real number gives 0 as a result.*

$$0 \div 5 = 0, \qquad 0 \div \frac{2}{3} = 0, \qquad \frac{0}{5.6} = 0, \qquad \frac{0}{1000} = 0$$

   You can divide zero by 5, $\frac{2}{3}$, 5.6, 1000, or any number (except 0).

2. *Division of any real number by 0 is* **undefined.**

$$7 \div 0 \qquad\qquad \frac{64}{0}$$
$$\uparrow \qquad\qquad\quad \uparrow$$

   Neither of these operations is possible. **Division by zero is undefined.**

Teaching Tip Students often confuse 0 divided by 3 and 3 divided by 0. Take the time to make sure they realize that division by 0 is never allowed. They must remember this in future problem sets. 0 divided by a nonzero number is always zero.

You may be wondering why division by zero is undefined. Let us think about it for a minute. We said that $7 \div 0$ is undefined. Suppose there were an answer. Let us call the answer a. So we assume for a minute that $7 \div 0 = a$. Then it would have to follow that $7 = 0(a)$. But this is impossible. Zero times any number is zero. So we see that if there were such a number, it would contradict known mathematical facts. Therefore there is no number a such that $7 \div 0 = a$. Thus we conclude that division by zero is undefined.

When combining two numbers, it is important to be sure you know which rule applies. Think about the concepts in the following chart. See if you agree with each example.

Operation	Two Real Numbers with the Same Sign	Two Real Numbers with Different Signs
Addition	Result may be positive or negative. $9 + 2 = 11$ $-5 + (-6) = -11$	Result may be positive or negative. $-3 + 7 = 4$ $4 + (-12) = -8$
Subtraction	Result may be positive or negative. $15 - 6 = 15 + (-6) = 9$ $-12 - (-3) = -12 + 3 = -9$	Result may be positive or negative. $-12 - 3 = -12 + (-3) = -15$ $5 - (-6) = 5 + 6 = 11$
Multiplication	Result is always positive. $9(3) = 27$ $-8(-5) = 40$	Result is always negative. $-6(12) = -72$ $8(-3) = -24$
Division	Result is always positive. $150 \div 6 = 25$ $-72 \div (-2) = 36$	Result is always negative. $-60 \div 10 = -6$ $30 \div (-6) = -5$

Teaching Example 8 Use the information provided in Example 8 to answer the following:

(a) How many yards were lost by the Generals in the plays that were considered large losses?

(b) How many yards were gained by the Generals in the plays that were considered large gains?

(c) If the total yards gained in large gains were combined with the total yards lost in large losses, what would be the result?

Ans:

(a) Approximately 30 yd

(b) Approximately 50 yd

(c) A total of 20 yards was gained.

EXAMPLE 8   The Hamilton-Wenham Generals recently analyzed the 48 plays their team made while in possession of the football during their last game. The following bar graph illustrates the number of plays made in each category. The team statistician prepared the following chart indicating the average number of yards gained or lost during each type of play.

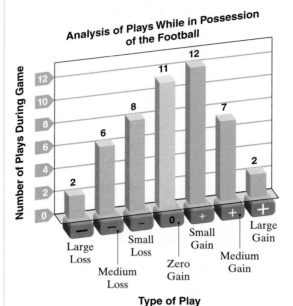

Type of Play	Average Yards Gained or Lost for Play
Large gain	+25
Medium gain	+15
Small gain	+5
Zero gain	0
Small loss	−5
Medium loss	−10
Large loss	−15

(a) How many yards were lost by the Generals in the plays that were considered small losses?

(b) How many yards were gained by the Generals in the plays that were considered small gains?

(c) If the total yards gained in small gains were combined with the total yards lost in small losses, what would be the result?

Solution

(a) We multiply the number of small losses by the average number of total yards lost on each small loss:

$$8(-5) = -40.$$

The team lost approximately 40 yards with plays that were considered small losses.

(b) We multiply the number of small gains by the average number of yards gained on each small gain:

$$12(5) = 60.$$

The team gained approximately 60 yards with plays that were considered small gains.

(c) We combine the results for **(a)** and **(b)**:

$$-40 + 60 = 20.$$

A total of 20 yards was gained during the plays that were small losses and small gains.

NOTE TO STUDENT: Fully worked-out solutions to all of the Practice Problems can be found at the back of the text starting at page SP-1

Practice Problem 8   Using the information provided in Example 8, answer the following:

(a) How many yards were lost by the Generals in the plays that were considered medium losses? Approximately 60 yd

(b) How many yards were gained by the Generals in the plays that were considered medium gains? Approximately 105 yd

(c) If the total yards gained in medium gains were combined with the total yards lost in medium losses, what would be the result? Gain of 45 yd

Verbal and Writing Skills

1. Explain in your own words the rule for determining the correct sign when multiplying two real numbers.
To multiply two real numbers, multiply the absolute values. The sign of the result is positive if both numbers have the same sign, but negative if the two numbers have opposite signs.

2. Explain in your own words the rule for determining the correct sign when multiplying three or more real numbers.
To multiply three or more real numbers, multiply the absolute values. The sign of the result is positive if there is an even number of negative signs. It is negative if there is an odd number of negative signs.

Multiply. Be sure to write your answer in the simplest form.

3. $5(-4)$ -20

4. $8(-2)$ -16

5. $0(-12)$ 0

6. $0(-150)$ 0

7. $14(3.5)$ 49

8. $8.5(6)$ 51

9. $(-1.32)(-0.2)$ 0.264

10. $(-2.3)(-0.11)$ 0.253

11. $1.8(-2.5)$ -4.5

12. $2.3(-4.4)$ -10.12

13. $\left(\dfrac{3}{8}\right)(-4)$ $-\dfrac{3}{2}$ or $-1\dfrac{1}{2}$

14. $(5)\left(-\dfrac{7}{10}\right)$ $-\dfrac{7}{2}$ or $-3\dfrac{1}{2}$

15. $\left(-\dfrac{3}{5}\right)\left(-\dfrac{15}{11}\right)$ $\dfrac{9}{11}$

16. $\left(-\dfrac{4}{9}\right)\left(-\dfrac{3}{5}\right)$ $\dfrac{4}{15}$

17. $\left(\dfrac{12}{13}\right)\left(\dfrac{-5}{24}\right)$ $-\dfrac{5}{26}$

18. $\left(\dfrac{14}{17}\right)\left(-\dfrac{3}{28}\right)$ $-\dfrac{3}{34}$

Divide.

19. $-36 \div (-9)$ 4

20. $0 \div (-15)$ 0

21. $-48 \div (-8)$ 6

22. $-45 \div (9)$ -5

23. $-120 \div (-8)$ 15

24. $-180 \div (-4)$ 45

25. $156 \div (-13)$ -12

26. $-0.6 \div 0.3$ -2

27. $-9.1 \div (0.07)$ -130

28. $8.1 \div (-0.03)$ -270

29. $0.54 \div (-0.9)$ -0.6

30. $-7.2 \div 8$ -0.9

31. $-6.3 \div 7$ -0.9

32. $\dfrac{2}{7} \div \left(-\dfrac{3}{5}\right)$ $-\dfrac{10}{21}$

33. $\left(-\dfrac{1}{5}\right) \div \left(\dfrac{2}{3}\right)$ $-\dfrac{3}{10}$

34. $\left(-\dfrac{5}{6}\right) \div \left(-\dfrac{7}{18}\right)$ $\dfrac{15}{7}$ or $2\dfrac{1}{7}$

35. $-\dfrac{5}{7} \div \left(-\dfrac{3}{28}\right)$ $\dfrac{20}{3}$ or $6\dfrac{2}{3}$

36. $\left(-\dfrac{4}{9}\right) \div \left(-\dfrac{8}{15}\right)$ $\dfrac{5}{6}$

37. $\left(-\dfrac{7}{12}\right) \div \left(-\dfrac{5}{6}\right)$ $\dfrac{7}{10}$

38. $\dfrac{12}{-\dfrac{2}{5}}$ -30

39. $\dfrac{-6}{-\dfrac{3}{7}}$ 14

40. $\dfrac{-\dfrac{3}{8}}{-\dfrac{2}{3}}$ $\dfrac{9}{16}$

41. $\dfrac{-\dfrac{2}{3}}{\dfrac{8}{15}}$ $-\dfrac{5}{4}$ or $-1\dfrac{1}{4}$

42. $\dfrac{\dfrac{9}{2}}{-3}$ $-\dfrac{3}{2}$ or $-1\dfrac{1}{2}$

43. $\dfrac{\dfrac{8}{3}}{-4}$ $-\dfrac{2}{3}$

Multiply. You may want to determine the sign of the product before you multiply.

44. $-6(2)(-3)(4)$ 144

45. $-1(-2)(-3)(4)$ -24

46. $-3(2)(-1)(-2)(5)$ -60

47. $-2(4)(3)(-1)(-3)$ -72

48. $-3(2)(-4)(0)(-2)$ 0

49. $-3(-2)\left(\dfrac{1}{3}\right)(-4)(2)$ -16

50. $-4(0.5)(-0.02)(2)$ 0.08

51. $25(-0.04)(-0.3)(-1)$ -0.3

52. $\left(\dfrac{3}{8}\right)\left(\dfrac{1}{2}\right)\left(-\dfrac{5}{6}\right)$ $-\dfrac{5}{32}$

53. $\left(-\dfrac{4}{5}\right)\left(-\dfrac{6}{7}\right)\left(-\dfrac{1}{3}\right)$ $-\dfrac{8}{35}$

54. $\left(-\dfrac{1}{2}\right)\left(\dfrac{4}{5}\right)\left(-\dfrac{7}{8}\right)\left(-\dfrac{2}{3}\right)$ $-\dfrac{7}{30}$

55. $\left(-\dfrac{3}{4}\right)\left(-\dfrac{7}{15}\right)\left(-\dfrac{8}{21}\right)\left(-\dfrac{5}{9}\right)$ $\dfrac{2}{27}$

Mixed Practice *Take a minute to review the chart before Example 8. Be sure that you can remember the sign rules for each operation. Then do exercises 56–65. Perform the indicated calculations.*

56. $-5 - (-2)$ -3

57. $-36 \div (-4)$ 9

58. $-3(-9)$ 27

59. $12 + (-8)$ 4

60. $(-30) \div 5$ -6

61. $8 - (-9)$ 17

62. $-6 + (-3)$ -9

63. $6(-12)$ -72

64. $18 \div (-18)$ -1

65. $-37 \div 37$ -1

Applications

66. *Stock Trading* In September 2007, the value of one share of NutriSystem stock opened at $56.95. At the end of the day, it closed at $50.08. If you owned 80 shares, what was your profit or loss that day?
A loss of $549.60

67. *Equal Contributions* Ed, Ned, Ted, and Fred went camping. They each contributed an equal share of money toward food. Fred did the shopping. When he returned from the store, he had $17.60 left. How much money did Fred give back to each person?
He gave $4.40 to each person and to himself.

68. *Student Loans* Ramon will pay $6480 on his student loan over the next 3 years. If $180 is automatically deducted from his bank account each month to pay the loan off, how much does he still owe after one year?
$4320

69. *Car Payments* Muriel will pay the Volkswagen dealer a total of $14,136 to be paid in 60 equal monthly installments. What is her monthly bill? $235.60

Football *The Beverly Panthers recently analyzed the 37 plays their team made while in possession of the football during their last game. The team statistician prepared the following chart indicating the number of plays in each category and the average number of yards gained or lost during each type of play. Use this chart to answer exercises 70–75.*

Type of Play	Number of Plays	Average Yards Gained or Lost per Play
Large gain	1	+25
Medium gain	6	+15
Small gain	4	+5
Zero gain	5	0
Small loss	10	-5
Medium loss	7	-10
Large loss	4	-15

70. How many yards were lost by the Panthers in the plays that were considered small losses?
Approximately 50 yd

71. How many yards were gained by the Panthers in the plays that were considered small gains?
Approximately 20 yd

72. If the total yards gained in small gains were combined with the total yards lost in small losses, what would be the result? A loss of 30 yd

73. How many yards were lost by the Panthers in the plays that were considered medium losses? Approximately 70 yd

74. How many yards were gained by the Panthers in the plays that were considered medium gains? Approximately 90 yd

75. If the total yards gained in medium gains were combined with the total yards lost in medium losses, what would be the result? A gain of 20 yd

Cumulative Review

76. **[1.1.5]** $-17.4 + 8.31 + 2.40$ -6.69

77. **[1.1.5]** $-\dfrac{3}{4} + \left(-\dfrac{2}{3}\right) + \left(-\dfrac{5}{12}\right)$ $-\dfrac{11}{6}$ or $-1\dfrac{5}{6}$

78. **[1.2.1]** $-47 - (-32)$ -15

79. **[1.2.1]** $-37 - 51$ -88

Quick Quiz 1.3 Perform the indicated operations.

1. $\left(-\dfrac{3}{8}\right)(5)$ $-\dfrac{15}{8}$ or $-1\dfrac{7}{8}$

2. $-4(3)(-5)(-2)$ -120

3. $-2.4 \div (-0.6)$ 4

4. **Concept Check** Explain how you can determine the sign of the answer if you multiply several negative numbers. Answers may vary

Classroom Quiz 1.3 You may use these problems to quiz your students' mastery of Section 1.3.

Perform the indicated operation.

1. $\left(-\dfrac{4}{7}\right)(6)$ **Ans:** $-\dfrac{24}{7}$ or $-3\dfrac{3}{7}$

2. $-2(-5)(3)(-6)$ **Ans:** -180

3. $-4.5 \div (-0.9)$ **Ans:** 5

Developing Your Study Skills

Steps Toward Success in Mathematics

Mathematics is a building process, mastered one step at a time. The foundation of this process consists of a few basic requirements. Those who are successful in mathematics realize the absolute necessity for building a study of mathematics on the firm foundation of these six minimum requirements.

1. Attend class every day.
2. Read the textbook.
3. Take notes in class.
4. Do assigned homework every day.
5. Get help immediately when needed.
6. Review regularly.

Student Learning Objectives

After studying this section, you will be able to:

 Write numbers in exponent form.

 Evaluate numerical expressions that contain exponents.

 Writing Numbers in Exponent Form

In mathematics, we use exponents as a way to abbreviate repeated multiplication.

Long Notation		Exponent Form
$2 \cdot 2 \cdot 2 \cdot 2 \cdot 2 \cdot 2$	$=$	2^6

There are two parts to exponent notation: (1) the **base** and (2) the **exponent.** The **base** tells you what number is being multiplied and the **exponent** tells you how many times this number is used as a factor. (A *factor,* you recall, is a number being multiplied.)

$$2 \cdot 2 \cdot 2 \cdot 2 \cdot 2 \cdot 2 = 2^6$$

The *base* is 2
(the number being multiplied)

The *exponent* is 6
(the number of times 2 is used as a factor)

If the base is a *positive* real number, the exponent appears to the right and slightly above the level of the number as in, for example, 5^6 and 8^3. If the base is a *negative* real number, then parentheses are used around the number and the exponent appears outside the parentheses. For example, $(-2)(-2)(-2) = (-2)^3$.

In algebra, if we do not know the value of a number, we use a letter to represent the unknown number. We call the letter a **variable.** This is quite useful in the case of exponents. Suppose we do not know the value of a number, but we know the number is multiplied by itself several times. We can represent this with a variable base and a whole-number exponent. For example, when we have an unknown number, represented by the variable x, and this number occurs as a factor four times, we have

$$(x)(x)(x)(x) = x^4.$$

Likewise, if an unknown number, represented by the variable w, occurs as a factor five times, we have

$$(w)(w)(w)(w)(w) = w^5.$$

Teaching Example 1 Write in exponent form.

(a) $(5)(5)$ (b) $-8(-8)(-8)(-8)(-8)$
(c) $-1(-1)(-1)(-1)(-1)(-1)(-1)$
(d) $(20)(20)(20)$ (e) $(a)(a)(a)(a)$
(f) $(m)(m)(m)(m)(m)(m)$

Ans: (a) 5^2 (b) $(-8)^5$ (c) $(-1)^7$
(d) 20^3 (e) a^4 (f) m^6

EXAMPLE 1 Write in exponent form.

(a) $9(9)(9)$ (b) $13(13)(13)(13)$ (c) $-7(-7)(-7)(-7)(-7)$
(d) $-4(-4)(-4)(-4)(-4)(-4)$ (e) $(x)(x)$ (f) $(y)(y)(y)$

Solution

(a) $9(9)(9) = 9^3$ (b) $13(13)(13)(13) = 13^4$
(c) The -7 is used as a factor five times. The answer must contain parentheses. Thus $-7(-7)(-7)(-7)(-7) = (-7)^5$.
(d) $-4(-4)(-4)(-4)(-4)(-4) = (-4)^6$
(e) $(x)(x) = x^2$ (f) $(y)(y)(y) = y^3$

Practice Problem 1 Write in exponent form.

(a) $6(6)(6)(6)$ 6^4 (b) $-2(-2)(-2)(-2)(-2)$ $(-2)^5$
(c) $108(108)(108)$ 108^3 (d) $-11(-11)(-11)(-11)(-11)(-11)$ $(-11)^6$
(e) $(w)(w)(w)$ w^3 (f) $(z)(z)(z)(z)$ z^4

NOTE TO STUDENT: Fully worked-out solutions to all of the Practice Problems can be found at the back of the text starting at page SP-1

If the base has an exponent of 2, we say the base is **squared.**

If the base has an exponent of 3, we say the base is **cubed.**

If the base has an exponent greater than 3, we say the base is raised **to the (exponent)-th power.**

x^2 is read "x squared."

y^3 is read "y cubed."

3^6 is read "three to the sixth power" or simply "three to the sixth."

Evaluating Numerical Expressions That Contain Exponents

EXAMPLE 2   Evaluate.

(a) 2^5

(b) $2^3 + 4^4$

Solution

(a) $2^5 = (2)(2)(2)(2)(2) = 32$

(b) First we evaluate each power.

$2^3 = 8$ $4^4 = 256$

Then we add. $8 + 256 = 264$

Practice Problem 2   Evaluate.

(a) 3^5 243

(b) $2^2 + 3^3$ 31

If the base is negative, be especially careful in determining the sign. Notice the following:

$$(-3)^2 = (-3)(-3) = +9 \qquad (-3)^3 = (-3)(-3)(-3) = -27.$$

From Section 1.3 we know that when you multiply two or more real numbers, first you multiply their absolute values.

- The result is positive if there is an even number of negative signs.
- The result is negative if there is an odd number of negative signs.

SIGN RULE FOR EXPONENTS

Suppose that a number is written in exponent form and the base is negative. The result is **positive** if the exponent is **even.** The result is **negative** if the exponent is **odd.**

Be careful how you read expressions with exponents and negative signs.

$$(-3)^4 \text{ means } (-3)(-3)(-3)(-3) \text{ or } +81.$$

$$-3^4 \text{ means } -(3)(3)(3)(3) \text{ or } -81.$$

EXAMPLE 3   Evaluate.

(a) $(-2)^3$ **(b)** $(-4)^6$ **(c)** -3^6 **(d)** $-(5^4)$

Solution

(a) $(-2)^3 = -8$ The answer is negative since the base is negative and the exponent 3 is odd.

(b) $(-4)^6 = +4096$ The answer is positive since the exponent 6 is even.

(c) $-3^6 = -729$ The negative sign is not contained in parentheses. Thus we find 3 raised to the sixth power and then take the opposite of that value.

(d) $-(5^4) = -625$ The negative sign is outside the parentheses.

Practice Problem 3   Evaluate.

(a) $(-3)^3$ −27   **(b)** $(-2)^6$ 64   **(c)** -2^4 −16   **(d)** $-(3^6)$ −729

 Calculator

Exponents

You can use a calculator to evaluate 3^5. Press the following keys:

$\boxed{3}\ \boxed{y^x}\ \boxed{5}\ \boxed{=}$

The display should read

$\boxed{243}$

Try the following.

(a) 4^6 **(b)** $(0.2)^5$

(c) 18^6 **(d)** 3^{12}

Answers:

(a) 4096   **(b)** 0.00032

(c) 34,012,224   **(d)** 531,441

The steps needed to raise a number to a power are slightly different on some calculators.

Teaching Example 4 Evaluate.

(a) $\left(\dfrac{1}{4}\right)^2$ **(b)** $(0.1)^4$

(c) $\left(\dfrac{3}{4}\right)^3$ **(d)** $(5)^2(2)^3$

(e) $8^2 - 3^3$

Ans: (a) $\dfrac{1}{16}$ **(b)** 0.0001   **(c)** $\dfrac{27}{64}$

(d) 200   **(e)** 37

EXAMPLE 4 Evaluate.

(a) $\left(\dfrac{1}{2}\right)^4$ **(b)** $(0.2)^4$ **(c)** $\left(\dfrac{2}{5}\right)^3$

(d) $(3)^3(2)^5$ **(e)** $2^3 - 3^4$

Solution

(a) $\left(\dfrac{1}{2}\right)^4 = \left(\dfrac{1}{2}\right)\left(\dfrac{1}{2}\right)\left(\dfrac{1}{2}\right)\left(\dfrac{1}{2}\right) = \dfrac{1}{16}$

(b) $(0.2)^4 = (0.2)(0.2)(0.2)(0.2) = 0.0016$

(c) $\left(\dfrac{2}{5}\right)^3 = \left(\dfrac{2}{5}\right)\left(\dfrac{2}{5}\right)\left(\dfrac{2}{5}\right) = \dfrac{8}{125}$

(d) First we evaluate each power.
$3^3 = 27 \qquad 2^5 = 32$
Then we multiply. $(27)(32) = 864$

(e) $2^3 - 3^4 = 8 - 81 = -73$

Practice Problem 4 Evaluate.

(a) $\left(\dfrac{1}{3}\right)^3$ $\dfrac{1}{27}$ **(b)** $(0.3)^4$ 0.0081   **(c)** $\left(\dfrac{3}{2}\right)^4$ $\dfrac{81}{16}$

(d) $(3)^4(4)^2$ 1296   **(e)** $4^2 - 2^4$ 0

NOTE TO STUDENT: *Fully worked-out solutions to all of the Practice Problems can be found at the back of the text starting at page SP-1*

Verbal and Writing Skills

1. Explain in your own words how to evaluate 4^4.
The base is 4 and the exponent is 4. Thus you multiply $(4)(4)(4)(4) = 256$.

2. Explain in your own words how to evaluate 9^2.
The base is 9 and the exponent is 2. Thus you multiply $(9)(9) = 81$.

3. Explain how you would determine whether $(-5)^3$ is negative or positive.
The answer is negative. When you raise a negative number to an odd power the result is always negative.

4. Explain how you would determine whether $(-2)^5$ is negative or positive.
The answer is negative. When you raise a negative number to an odd power the result is always negative.

5. Explain the difference between $(-2)^4$ and -2^4. What answers do you obtain when you evaluate the expressions?
If you have parentheses surrounding the -2, then the base is -2 and the exponent is 4. The result is 16. If you do not have parentheses, then the base is 2. You evaluate to obtain 16 and then take the opposite of 16, which is -16. Thus $(-2)^4 = 16$, but $-2^4 = -16$.

6. Explain the difference between $(-3)^4$ and -3^4. What answers do you obtain when you evaluate the expressions?
If you have parentheses surrounding the -3, then the base is -3 and the exponent is 4. The result is 81. If you do not have parentheses, then the base is 3. You evaluate to obtain 81 and then take the opposite of 81, which is -81. Thus $(-3)^4 = 81$, but $-3^4 = -81$.

Write in exponent form.

7. $(5)(5)(5)(5)(5)(5)(5)$ 5^7

8. $(7)(7)(7)(7)(7)$ 7^5

9. $(w)(w)$ w^2

10. $(z)(z)(z)$ z^3

11. $(p)(p)(p)(p)$ p^4

12. $(y)(y)(y)(y)(y)(y)$ y^6

13. $(3q)(3q)(3q)$ $(3q)^3$ or 3^3q^3

14. $(2w)(2w)(2w)(2w)(2w)$ $(2w)^5$ or 2^5w^5

Evaluate.

15. 3^3 27

16. 4^2 16

17. 3^4 81

18. 8^3 512

19. 6^3 216

20. 15^2 225

21. $(-3)^3$ -27

22. $(-2)^3$ -8

23. $(-4)^2$ 16

24. $(-5)^4$ 625

25. -5^2 -25

26. -4^2 -16

27. $\left(\dfrac{1}{4}\right)^2$ $\dfrac{1}{16}$

28. $\left(\dfrac{3}{4}\right)^2$ $\dfrac{9}{16}$

29. $\left(\dfrac{2}{5}\right)^3$ $\dfrac{8}{125}$

30. $\left(\dfrac{3}{2}\right)^3$ $\dfrac{27}{8}$ or $3\dfrac{3}{8}$

31. $(2.1)^2$ 4.41

32. $(1.5)^2$ 2.25

33. $(0.2)^4$ 0.0016

34. $(0.7)^3$ 0.343

35. $(-16)^2$ 256

36. $(-7)^4$ 2401

37. -16^2 -256

38. -7^4 -2401

Evaluate.

39. $5^3 + 6^2$ 161

40. $7^2 + 6^3$ 265

41. $10^2 - 11^2$ −21

42. $3^3 - 2^4$ 11

43. $(-4)^2 - (12)^2$ −128

44. $13^2 - (-7)^2$ 120

45. $2^5 - (-3)^2$ 23

46. $8^2 - (-2)^3$ 72

47. $(-4)^3(-3)^2$ −576

Cumulative Review *Evaluate.*

48. **[1.1.5]** $(-11) + (-13) + 6 + (-9) + 8$ −19

49. **[1.3.3]** $\dfrac{3}{4} \div \left(-\dfrac{9}{20}\right)$ $-\dfrac{5}{3}$ or $-1\dfrac{2}{3}$

50. **[1.2.1]** $-17 - (-9)$ −8

51. **[1.3.1]** $(-2.1)(-1.2)$ 2.52

Quick Quiz 1.4 Evaluate.

1. $(-4)^4$ 256

2. $(1.8)^2$ 3.24

3. $\left(\dfrac{3}{4}\right)^3$ $\dfrac{27}{64}$

4. **Concept Check** Explain the difference between $(-2)^6$ and -2^6. How do you decide if the answers are positive or negative?
Answers may vary

Classroom Quiz 1.4 You may use these problems to quiz your students' mastery of Section 1.4.
Evaluate.

1. $(-6)^3$ **Ans:** −216

2. $(2.3)^2$ **Ans:** 5.29

3. $\left(\dfrac{2}{5}\right)^4$ **Ans:** $\dfrac{16}{625}$

1.5 THE ORDER OF OPERATIONS

Using the Order of Operations to Simplify Numerical Expressions

It is important to know *when* to do certain operations as well as how to do them. For example, to simplify the expression $2 - 4 \cdot 3$, should we subtract first or multiply first?

Also remember that multiplication can be written several ways. Thus $4 \cdot 3$, 4×3, $4(3)$, and $(4)(3)$ all indicate that we are multiplying 4 times 3.

The following list will assist you. It tells which operations to do first: the correct **order of operations.** You might think of it as a *list of priorities*.

Student Learning Objective

After studying this section, you will be able to:

1 **Use the order of operations to simplify numerical expressions.**

ORDER OF OPERATIONS FOR NUMBERS

Follow this order of operations:

Do first   **1.** Do all operations inside parentheses.

            2. Raise numbers to a power.

            3. Multiply and divide numbers from left to right.

Do last   **4.** Add and subtract numbers from left to right.

Let's return to the problem $2 - 4 \cdot 3$. There are no parentheses or numbers raised to a power, so the first thing we do is multiply. Then we subtract since this comes last on our list.

$$2 - 4 \cdot 3 = 2 - 12 \quad \text{Follow the order of operations by first multiplying } 4 \cdot 3 = 12.$$
$$= -10 \quad \text{Combine } 2 - 12 = -10.$$

Teaching Tip Students sometimes ask why the order of operations is so important. Remind them that scientific calculators and computers follow the same order of operations. If we want to know how to use either of them, then we must understand which operations these devices will perform in what order. Tell them that Section 1.6 will be easier if they master this order of operations now.

EXAMPLE 1   Evaluate. $8 \div 2 \cdot 3 + 4^2$

Solution

$$8 \div 2 \cdot 3 + 4^2 = 8 \div 2 \cdot 3 + 16 \quad \text{Evaluate } 4^2 = 16 \text{ because the highest priority in this problem is raising to a power.}$$
$$= 4 \cdot 3 + 16 \quad \text{Next multiply and divide from left to right. So } 8 \div 2 = 4 \text{ and } 4 \cdot 3 = 12.$$
$$= 12 + 16$$
$$= 28 \quad \text{Finally, add.}$$

Teaching Example 1 Evaluate.
$12 \div 2 \cdot 3 + 5^2$

Ans: 43

Practice Problem 1   Evaluate. $25 \div 5 \cdot 6 + 2^3$ 38

Note: Multiplication and division have equal priority. We do not do multiplication first. Rather, we work from left to right, doing any multiplication or division that we encounter. Similarly, addition and subtraction have equal priority.

EXAMPLE 2   Evaluate. $(-3)^3 - 2^4$

Solution   The highest priority is to raise the expressions to the appropriate powers.

$$(-3)^3 - 2^4 = -27 - 16 \quad \text{In } (-3)^3 \text{ we are cubing the number } -3 \text{ to obtain } -27.$$
$$\text{Be careful; } -2^4 \text{ is not } (-2)^4!$$
$$\text{Raise 2 to the fourth power and subtract the result.}$$
$$= -43 \quad \text{The last step is to add and subtract from left to right.}$$

Teaching Example 2 Evaluate. $(-6)^2 - 1^8$

Ans: 35

NOTE TO STUDENT: *Fully worked-out solutions to all of the Practice Problems can be found at the back of the text starting at page SP-1*

Practice Problem 2   Evaluate. $(-4)^3 - 2^6$ −128

Teaching Example 3 Evaluate.
$5 + 2(7 - 9)^2 - 18 \div 2 + (5 - 6)^3$

Ans: 3

EXAMPLE 3 Evaluate. $2 \cdot (2 - 3)^3 + 6 \div 3 + (8 - 5)^2$

Solution

$$2 \cdot (2 - 3)^3 + 6 \div 3 + (8 - 5)^2$$
$$= 2 \cdot (-1)^3 + 6 \div 3 + 3^2$$

Combine the numbers inside the parentheses. Note that we need parentheses for -1 because of the negative sign, but they are not needed for 3.

$$= 2 \cdot (-1) + 6 \div 3 + 9$$

Next, raise to a power.

$$= -2 + 2 + 9$$

Next, multiply and divide from left to right.

$$= 9$$

Finally, add and subtract from left to right.

Practice Problem 3 Evaluate. $6 - (8 - 12)^2 + 8 \div 2$ -6

Teaching Example 4 Evaluate.
$$\left(-\frac{2}{3}\right)^3 + \frac{10}{21}\left(\frac{7}{15}\right)$$

Ans: $-\frac{2}{27}$

Calculator

Order of Operations

Use your calculator to evaluate $3 + 4 \cdot 5$. Enter

$$\boxed{3}\ \boxed{+}\ \boxed{4}\ \boxed{\times}\ \boxed{5}\ \boxed{=}$$

If the display is $\boxed{\ \ 23\ }$, the correct order of operations is built in. If the display is not 23, you will need to modify the way you enter the problem. You should use

$$\boxed{4}\ \boxed{\times}\ \boxed{5}\ \boxed{+}\ \boxed{3}\ \boxed{=}$$

Try $6 + 3 \cdot 4 - 8 \div 2$.

Answer:
14

EXAMPLE 4 Evaluate. $\left(-\frac{1}{5}\right)\left(\frac{1}{2}\right) - \left(\frac{3}{2}\right)^2$

Solution The highest priority is to raise $\frac{3}{2}$ to the second power.

$$\left(\frac{3}{2}\right)^2 = \left(\frac{3}{2}\right)\left(\frac{3}{2}\right) = \frac{9}{4}$$

$$\left(-\frac{1}{5}\right)\left(\frac{1}{2}\right) - \left(\frac{3}{2}\right)^2 = \left(-\frac{1}{5}\right)\left(\frac{1}{2}\right) - \frac{9}{4}$$

$$= -\frac{1}{10} - \frac{9}{4}$$

Next we multiply.

$$= -\frac{1 \cdot 2}{10 \cdot 2} - \frac{9 \cdot 5}{4 \cdot 5}$$

We need to write each fraction as an equivalent fraction with the LCD of 20.

$$= -\frac{2}{20} - \frac{45}{20}$$

$$= -\frac{47}{20} \text{ or } -2\frac{7}{20}$$

Add.

Practice Problem 4 Evaluate.

$$\left(-\frac{1}{7}\right)\left(-\frac{14}{5}\right) + \left(-\frac{1}{2}\right) \div \frac{3}{4} - \frac{4}{15}$$

Developing Your Study Skills

Previewing New Material

Part of your study time each day should consist of looking over the sections in your text that are to be covered the following day. You do not necessarily need to study and learn the material on your own, but a survey of the concepts, terminology, diagrams, and examples will help the new ideas seem more familiar as the instructor presents them.

You can look for concepts that appear confusing or difficult and be ready to listen carefully for your instructor's explanations. You can be prepared to ask questions that will increase your understanding. Previewing new material enables you to see what is coming and prepares you to be ready to absorb it.

Verbal and Writing Skills

Game Points *You have lost a game of UNO and are counting the points left in your hand. You announce that you have three fours and six fives.*

1. Write this as a number expression. $3(4) + 6(5)$

2. How many points have you in your hand? 42

3. What answer would you get for the number expression if you simplified it by
   (a) performing the operations from left to right? 90
   (b) following the order of operations? 42

4. Which procedure in exercise 3 gives the correct number of total points? **(b)**

Evaluate.

5. $(2 - 5)^2 \div 3 \times 4$
   12

6. $(3 - 7)^2 \div 2 \times 5$
   40

7. $9 + 4(5 + 2 - 8)$
   5

8. $12 + 5(3 - 9 + 4)$
   2

9. $8 - 2^3 \cdot 5 + 3$
   -29

10. $6 - 3^2 \cdot 6 + 4$
   -44

11. $4 + 42 \div 3 \cdot 2 - 8$
   24

12. $7 + 36 \div 12 \cdot 3 - 14$
   2

13. $3 \cdot 5 + 7 \cdot 3 - 5 \cdot 3$
   21

14. $2 \cdot 6 + 5 \cdot 3 - 7 \cdot 4$
   -1

15. $8 - 5(2)^3 \div (-8)$
   13

16. $11 - 3(4)^2 \div (-6)$
   19

17. $3(5 - 7)^2 - 6(3)$
   -6

18. $-2(3 - 6)^2 - (-2)$
   -16

19. $5 \cdot 6 - (3 - 5)^2 + 8 \cdot 2$
   42

20. $(-3)^2 \cdot 6 \div 9 + 4 \cdot 2$
   14

21. $\dfrac{1}{2} \div \dfrac{2}{3} + 6 \cdot \dfrac{1}{4}$ $\dfrac{9}{4}$ or $2\dfrac{1}{4}$

22. $\dfrac{5}{6} \div \dfrac{2}{3} - 6 \cdot \left(\dfrac{1}{2}\right)^2$ $-\dfrac{1}{4}$

23. $0.8 + 0.3(0.6 - 0.2)^2$ 0.848

24. $0.05 + 1.4 - (0.5 - 0.7)^3$ 1.458

25. $\dfrac{3}{8}\left(-\dfrac{1}{6}\right) - \dfrac{7}{8} + \dfrac{1}{2}$ $-\dfrac{7}{16}$

26. $\dfrac{1}{2} \div \dfrac{4}{5} - \dfrac{3}{4}\left(\dfrac{5}{6}\right)$ 0

Mixed Practice

27. $(3 - 7)^2 \div 8 + 3$ 5

28. $\left(\dfrac{3}{5}\right)\left(\dfrac{5}{6}\right) - \dfrac{3}{4} \div 6$ $\dfrac{3}{8}$

29. $\left(\dfrac{3}{4}\right)^2(-16) + \dfrac{4}{5} \div \dfrac{-8}{25}$ $\dfrac{-23}{2}$ or $-11\dfrac{1}{2}$

30. $\left(2\dfrac{4}{7}\right) \div \left(-1\dfrac{1}{5}\right)$ $-\dfrac{15}{7}$ or $-2\dfrac{1}{7}$

31. $-6.3 - (-2.7)(1.1) + (3.3)^2$ 7.56

32. $4.35 + 8.06 \div (-2.6) - (2.1)^2$ −3.16

33. $\left(\dfrac{1}{2}\right)^3 + \dfrac{1}{4} - \left(\dfrac{1}{6} - \dfrac{1}{12}\right) - \dfrac{2}{3} \cdot \left(\dfrac{1}{4}\right)^2$ $\dfrac{1}{4}$

34. $(2.4 \cdot 1.2)^2 - 1.6 \cdot 2.2 \div 4.0 - 3.6$ 3.8144

Applications *In January 2007, Tiger Woods won the Buick Invitational. The following scorecard shows the results of Round 1. There are 18 holes per round and the table shows the number of times Woods got each type of score.*

Score on a Hole	Number of Times the Score Occurred
Eagle (−2)	2
Birdie (−1)	4
Par (0)	10
Bogey (+1)	2

35. Write his score as the sum of eagles, birdies, pars, and bogeys.
$2(-2) + 4(-1) + 10(0) + 2(+1)$

36. What was his final score for the round, when compared to par?
−6 or 6 under par

37. What answer do you get if you do the arithmetic left to right?
2 above par

38. Explain why the answers in exercises 36 and 37 do not match.
Exercise 37 did not use the order of operations.

Cumulative Review *Simplify.*

39. **[1.4.2]** $(0.5)^3$ 0.125

40. **[1.2.1]** $-\dfrac{3}{4} - \dfrac{5}{6}$
$-\dfrac{19}{12}$ or $-1\dfrac{7}{12}$

41. **[1.4.2]** -1^{20} −1

42. **[0.3.2]** $3\dfrac{3}{5} \div 6\dfrac{1}{4}$ $\dfrac{72}{125}$

Quick Quiz 1.5 Evaluate.

1. $7 - 3^4 + 2 - 5$ −77

2. $(0.3)^2 - (4.2)(-4) + 0.07$ 16.96

3. $2(7 - 9)^4 + 22 \div (-2) + 6$ 27

4. **Concept Check** Explain in what order you would perform the calculations to evaluate the expression $4 - (-3 + 4)^3 + 12 \div (-3)$.
Answers may vary

Classroom Quiz 1.5 You may use these problems to quiz your students' mastery of Section 1.5.
Evaluate.

1. $6 - 4^3 + 7 - 12$ **Ans:** −63

2. $(0.6)^2 - (-2)(-3.5) + 0.23$ **Ans:** −6.41

3. $-3(4 - 5)^5 + 36 \div (9) + 8$ **Ans:** 7

How are you doing with your homework assignments in Sections 1.1 to 1.5? Do you feel you have mastered the material so far? Do you understand the concepts you have covered? Before you go further in the textbook, take some time to do each of the following problems.

Simplify each of the following. If the answer is a fraction, be sure to leave it in reduced form.

1.1

1. $3 + (-12)$

2. $-\dfrac{5}{6} + \left(-\dfrac{7}{8}\right)$

3. $0.34 + 0.9$

4. $-3.5 + 9 + 2.3 + (-3)$

1.2

5. $-23 - (-34)$

6. $-\dfrac{1}{6} - \dfrac{4}{5}$

7. $4.5 - (-7.8)$

8. $-4 - (-5) + 9$

1.3

9. $(-3)(-8)(2)(-2)$

10. $\left(-\dfrac{6}{11}\right)\left(-\dfrac{5}{3}\right)$

11. $-0.072 \div 0.08$

12. $\dfrac{5}{8} \div \left(-\dfrac{17}{16}\right)$

1.4

Evaluate.

13. $(0.7)^3$

14. $(-4)^4$

15. -2^8

16. $\left(\dfrac{2}{3}\right)^3$

17. $-3^3 + 3^4$

1.5

18. $20 - 12 \div 3 - 8(-1)$

19. $15 + 3 - 2 + (-6)$

20. $(9 - 13)^2 + 15 \div (-3)$

21. $-0.12 \div 0.6 + (-3)(1.2) - (-0.5)$

22. $\left(\dfrac{3}{4}\right)\left(-\dfrac{2}{5}\right) + \left(-\dfrac{1}{2}\right)\left(\dfrac{4}{5}\right) + \left(\dfrac{1}{2}\right)^2$

1.	-9
2.	$-\dfrac{41}{24}$ or $-1\dfrac{17}{24}$
3.	1.24
4.	4.8
5.	11
6.	$-\dfrac{29}{30}$
7.	12.3
8.	10
9.	-96
10.	$\dfrac{10}{11}$
11.	-0.9
12.	$-\dfrac{10}{17}$
13.	0.343
14.	256
15.	-256
16.	$\dfrac{8}{27}$
17.	54
18.	24
19.	10
20.	11
21.	-3.3
22.	$-\dfrac{9}{20}$

Now turn to page SA-3 for the answer to each of these problems. Each answer also includes a reference to the objective in which the problem is first taught. If you missed any of these problems, you should stop and review the Examples and Practice Problems in the referenced objective. A little review now will help you master the material in the upcoming sections of the text.

Student Learning Objective

After studying this section, you will be able to:

 Use the distributive property to simplify algebraic expressions.

Teaching Tip You may want to mention, as you are discussing the distributive property, that it is called the distributive property no matter how many terms appear inside the parentheses. For example,

$$a(b + c + d) = ab + ac + ad$$

and

$$a(b + c + d + e) = ab + ac + ad + ae.$$

① Using the Distributive Property to Simplify Algebraic Expressions

As we learned previously, we use letters called *variables* to represent unknown numbers. If a number is multiplied by a variable, we do not need any symbol between the number and variable. Thus, to indicate $(2)(x)$, we write $2x$. To indicate $3 \cdot y$, we write $3y$. If one variable is multiplied by another variable, we place the variables next to each other. Thus, $(a)(b)$ is written ab. We use exponent form if an unknown number (a variable) is used several times as a factor. Thus, $x \cdot x \cdot x = x^3$. Similarly, $(y)(y)(y)(y) = y^4$.

In algebra, we need to be familiar with several definitions. We will use them throughout the remainder of this book. Take some time to think through how each of these definitions is used.

An **algebraic expression** is a quantity that contains numbers and variables, such as $a + b, 2x - 3$, and $5ab^2$. In this chapter we will be learning rules about adding and multiplying algebraic expressions. A **term** is a number, a variable, or a product of numbers and variables. 17, x, $5xy$, and $22xy^3$ are all examples of terms. We will refer to terms when we discuss the distributive property.

An important property of algebra is the **distributive property.** We can state it in an equation as follows:

DISTRIBUTIVE PROPERTY

For all real numbers a, b, and c,

$$a(b + c) = ab + ac.$$

A numerical example shows that it does seem reasonable.

$$5(3 + 6) = 5(3) + 5(6)$$
$$5(9) = 15 + 30$$
$$45 = 45$$

We can use the distributive property to multiply any term by the sum of two or more terms. In Section 0.1, we defined the word *factor*. Two or more algebraic expressions that are multiplied are called **factors.** Consider the following examples of multiplying algebraic expressions.

Teaching Example 1 Multiply.

(a) $4(2m + n)$ **(b)** $-3(5x - 2y)$

Ans: (a) $8m + 4n$ **(b)** $-15x + 6y$

EXAMPLE 1 Multiply.

(a) $5(a + b)$ **(b)** $-3(3x + 2y)$

Solution

(a) $5(a + b) = 5a + 5b$ Multiply the factor $(a + b)$ by the factor 5.

(b) $-3(3x + 2y) = -3(3x) + (-3)(2y)$ Multiply the factor $(3x + 2y)$ by the factor (-3).

$$= -9x - 6y$$

Practice Problem 1 Multiply.

(a) $3(x + 2y)$ $3x + 6y$ **(b)** $-2(a - 3b)$ $-2a + 6b$

NOTE TO STUDENT: Fully worked-out solutions to all of the Practice Problems can be found at the back of the text starting at page SP-1

If the parentheses are preceded by a negative sign, we consider this to be the product of (-1) and the expression inside the parentheses.

EXAMPLE 2 Multiply. $-(a - 2b)$

Teaching Example 2 Multiply. $-(-7a + b)$

Ans: $7a - b$

Solution

$$-(a - 2b) = (-1)(a - 2b) = (-1)(a) + (-1)(-2b) = -a + 2b$$

Practice Problem 2 Multiply. $-(-3x + y)$ $3x - y$

In general, we see that in all these examples we have multiplied each term of the expression in the parentheses by the expression in front of the parentheses.

EXAMPLE 3 Multiply.

(a) $\frac{2}{3}(x^2 - 6x + 8)$ **(b)** $1.4(a^2 + 2.5a + 1.8)$

Teaching Example 3 Multiply.

(a) $\frac{7}{2}(y^2 + 6y - 2)$

(b) $2.1(m^2 - 3m + 0.4)$

Ans:

(a) $\frac{7}{2}y^2 + 21y - 7$

(b) $2.1m^2 - 6.3m + 0.84$

Solution

(a) $\frac{2}{3}(x^2 - 6x + 8) = \left(\frac{2}{3}\right)(1x^2) + \left(\frac{2}{3}\right)(-6x) + \left(\frac{2}{3}\right)(8)$

$$= \frac{2}{3}x^2 + (-4x) + \frac{16}{3}$$

$$= \frac{2}{3}x^2 - 4x + \frac{16}{3}$$

(b) $1.4(a^2 + 2.5a + 1.8) = 1.4(1a^2) + (1.4)(2.5a) + (1.4)(1.8)$

$$= 1.4a^2 + 3.5a + 2.52$$

Practice Problem 3 Multiply.

(a) $\frac{3}{5}(a^2 - 5a + 25)$ $\frac{3}{5}a^2 - 3a + 15$ **(b)** $2.5(x^2 - 3.5x + 1.2)$

(b) $2.5x^2 - 8.75x + 3$

There are times we multiply a variable by itself and use exponent notation. For example, $(x)(x) = x^2$ and $(x)(x)(x) = x^3$. In other cases there will be numbers and variables multiplied at the same time.

We will see problems like $(2x)(x) = (2)(x)(x) = 2x^2$. Some expressions will involve multiplication of more than one variable. We will see problems like $(3x)(xy) = (3)(x)(x)(y) = 3x^2y$. There will be times when we use the distributive property and all of these methods will be used. For example,

$$2x(x - 3y + 2) = 2x(x) + (2x)(-3y) + (2x)(2)$$
$$= 2x^2 + (-6)(xy) + 4(x)$$
$$= 2x^2 - 6xy + 4x.$$

We will discuss this type of multiplication of variables with exponents in more detail in Section 4.1. At that point we will expand these examples and other similar examples to develop the general rule for multiplication $(x^a)(x^b) = x^{a+b}$.

EXAMPLE 4 Multiply. $-2x(3x + y - 4)$

Teaching Example 4 Multiply.
$-3x(4x - 5y + 10)$

Ans: $-12x^2 + 15xy - 30x$

Solution

$$-2x(3x + y - 4) = -2(x)(3)(x) + (-2)(x)(y) + (-2)(x)(-4)$$
$$= -2(3)(x)(x) + (-2)(xy) + (-2)(-4)(x)$$
$$= -6x^2 - 2xy + 8x$$

Practice Problem 4 Multiply. $-4x(x - 2y + 3)$ $-4x^2 + 8xy - 12x$

The distributive property can also be presented with the a on the right.

$$(b + c)a = ba + ca$$

The a is "distributed" over the b and c inside the parentheses.

Teaching Example 5 Multiply.
$(6a - a^2)(-5)$

Ans: $-30a + 5a^2$

EXAMPLE 5 Multiply. $(2x^2 - x)(-3)$

Solution

$$(2x^2 - x)(-3) = 2x^2(-3) + (-x)(-3)$$
$$= -6x^2 + 3x$$

Practice Problem 5 Multiply.

$$(3x^2 - 2x)(-4) \quad -12x^2 + 8x$$

Teaching Example 6 A rectangular athletic field is 200 feet wide. One portion of the field is $5x$ feet long. The other portion of the field is $4y$ feet long. Use the distributive property to find an expression for the area of this field.

Ans: $1000x + 800y$ ft^2

▲ **EXAMPLE 6**   A farmer has a rectangular field that is 300 feet wide. One portion of the field is $2x$ feet long. The other portion of the field is $3y$ feet long. Use the distributive property to find an expression for the area of this field.

Solution   First we draw a picture of a field that is 300 feet wide and $2x + 3y$ feet long.

To find the area of the field, we multiply the width times the length.

$$300(2x + 3y) = 300(2x) + 300(3y) = 600x + 900y$$

Thus the area of the field in square feet is $600x + 900y$.

▲ **Practice Problem 6**   A farmer has a rectangular field that is 400 feet wide. One portion of the field is $6x$ feet long. The other portion of the field is $9y$ feet long. Use the distributive property to find an expression for the area of this field.

$$2400x + 3600y \text{ ft}^2$$

Developing Your Study Skills

How to Do Homework

As you begin your homework assignments, read the directions carefully. You need to understand what is being asked for. Concentrate on each exercise, taking time to solve it accurately. Rushing through your work usually causes errors. Check your answers with those given in the back of the textbook. If your answer is incorrect, check to see that you are doing the right exercise. Redo the exercise, watching for little errors. If it is still wrong, check with a friend. Perhaps the two of you can figure it out.

Verbal and Writing Skills

In exercises 1 and 2, complete each sentence by filling in the blank.

1. A ___variable___ is a symbol used to represent an unknown number.

2. When we write an expression with numbers and variables such as $7x$, it indicates that we are ___multiplying___ 7 by x.

3. Explain in your own words how we multiply a problem like $(4x)(x)$.
Here we are multiplying 4 by x by x. Since we know from the definition of exponents that x multiplied by x is x^2, this gives us an answer of $4x^2$.

4. Explain why you think the property $a(b + c) = ab + ac$ is called the distributive property. What does distribute mean?
Distribute means to disperse. We distribute the factor to each term in the parentheses.

5. Does the following distributive property work?
$$a(b - c) = ab - ac$$
Why or why not? Give an example.
Yes, $a(b - c)$ can be written as $a[b + (-c)]$.
$3(10 - 2) = (3 \times 10) - (3 \times 2)$
$\quad 3 \times 8 = 30 - 6$
$\quad\quad 24 = 24$

6. Susan tried to use the distributive property and wrote
$$-5(x + 3y - 2) = -5x - 15y - 10.$$
What did she do wrong?
She made an error with the sign rules. She should multiply $(-5)(-2)$ to get $+10$. The answer to the problem is $-5x - 15y + 10$.

Multiply. Use the distributive property.

7. $3(x - 2y)$
$3x - 6y$

8. $4(3x - y)$
$12x - 4y$

9. $-2(4a - 3b)$
$-8a + 6b$

10. $-3(2a - 5b)$
$-6a + 15b$

11. $3(3x + y)$
$9x + 3y$

12. $2(4x + y)$
$8x + 2y$

13. $8(-m - 3n)$
$-8m - 24n$

14. $10(-2m - n)$
$-20m - 10n$

15. $-(x - 3y)$
$-x + 3y$

16. $-(-4x + y)$
$4x - y$

17. $-9(9x - 5y + 8)$
$-81x + 45y - 72$

18. $-5(3x + 9 - 7y)$
$-15x - 45 + 35y$

19. $2(-5x + y - 6)$
$-10x + 2y - 12$

20. $3(2x - 6y - 5)$
$6x - 12y - 15$

21. $\dfrac{5}{6}(12x^2 - 24x + 18)$
$10x^2 - 20x + 15$

22. $\dfrac{2}{3}(-27a^4 + 9a^2 - 21)$
$-18a^4 + 6a^2 - 14$

23. $\dfrac{x}{5}(x + 10y - 4)$ $\left(Hint: \dfrac{x}{5} = \dfrac{1}{5}x \right)$
$\dfrac{x^2}{5} + 2xy - \dfrac{4x}{5}$

24. $\dfrac{y}{3}(3y - 4x - 6)$ $\left(Hint: \dfrac{y}{3} = \dfrac{1}{3}y \right)$
$y^2 - \dfrac{4xy}{3} - 2y$

25. $5x(x + 2y + z)$
$5x^2 + 10xy + 5xz$

26. $3a(2a + b - c)$
$6a^2 + 3ab - 3ac$

27. $(-4.5x + 5)(-3)$
$13.5x - 15$

28. $(-2.3x + 4)(-5)$
$11.5x - 20$

29. $(6x + y - 1)(3x)$
$18x^2 + 3xy - 3x$

30. $(3x - 3y + 4)(2x)$
$6x^2 - 6xy + 8x$

Mixed Practice *Multiply. Use the distributive property.*

31. $(3x + 2y - 1)(-xy)$
$-3x^2y - 2xy^2 + xy$

32. $(4a - 2b - 1)(-ab)$
$-4a^2b + 2ab^2 + ab$

33. $(-a - 2b + 4)5ab$
$-5a^2b - 10ab^2 + 20ab$

34. $(3a - b - 2)2ab$
$6a^2b - 2ab^2 - 4ab$

35. $\dfrac{1}{3}(6a^2 - 12a + 8)$
$2a^2 - 4a + \dfrac{8}{3}$

36. $\dfrac{1}{2}(-4a^2 + 10a - 5)$
$-2a^2 + 5a - \dfrac{5}{2}$

37. $-0.3x(-1.2x^2 - 0.3x + 0.5)$
Hint: $x(x^2) = x(x)(x) = x^3$
$0.36x^3 + 0.09x^2 - 0.15x$

38. $-0.6q(1.2q^2 + 2.5r - 0.7s)$
Hint: $q(q^2) = q(q)(q) = q^3$
$-0.72q^3 - 1.5qr + 0.42qs$

39. $0.4q(-3.3q^2 - 0.7r - 10)$
Hint: $q(q^2) = q(q)(q) = q^3$
$-1.32q^3 - 0.28qr - 4q$

Applications

▲ **40.** *Geometry* Gary Roswell owns a large rectangular field where he grows corn and wheat. The width of the field is 850 feet. The portion of the field where corn grows is $10x$ feet long and the portion where wheat grows is $7y$ feet long. Use the distributive property to find an expression for the area of this field.
$850(10x + 7y) = 8500x + 5950y$ ft^2

▲ **41.** *Geometry* Kathy DesMaris has a rectangular field that is 800 feet wide. One portion of the field is $5x$ feet long. The other portion of the field is $14y$ feet long. Use the distributive property to find an expression for the area of this field.
$800(5x + 14y) = 4000x + 11,200y$ ft^2

To Think About

▲ **42.** *Athletic Field* The athletic field at Gordon College is $2x$ feet wide. It used to be 1800 feet long. An old, run-down building was torn down, making the field $3y$ feet longer. Use the distributive property to find an expression for the area of the new field.
$2x(1800 + 3y) = 3600x + 6xy$ ft^2

▲ **43.** *Airport Runway* The Beverly Airport runway is $4x$ feet wide. The airport was supposed to have a 3000-foot-long runway. However, some of the land was wetland, so a runway could not be built on all of it. Therefore, the length of the runway was decreased by $2y$ feet. Use the distributive property to find an expression for the area of the final runway.
$4x(3000 - 2y) = 12,000x - 8xy$ ft^2

Cumulative Review *In exercises 44–48, evaluate.*

44. **[1.1.4]** $-18 + (-20) + 36 + (-14)$ $\quad -16$

45. **[1.4.2]** $(-2)^6$ $\quad 64$

46. **[1.2.1]** $-27 - (-41)$ $\quad 14$

47. **[1.5.1]** $25 \div 5(2) + (-6)$ $\quad 4$

48. **[1.5.1]** $(12 - 10)^2 + (-3)(-2)$ $\quad 10$

Quick Quiz 1.6 *Multiply. Use the distributive property.*

1. $5(-3a - 7b)$ $\quad -15a - 35b$

2. $-2x(x - 4y + 8)$ $\quad -2x^2 + 8xy - 16x$

3. $-3ab(4a - 5b - 9)$ $\quad -12a^2b + 15ab^2 + 27ab$

4. **Concept Check** Explain how you would multiply to obtain the answer for $\left(-\frac{3}{7}\right)(21x^2 - 14x + 3)$.
Answers may vary

Classroom Quiz 1.6 *You may use these problems to quiz your students' mastery of Section 1.6.*
Multiply. Use the distributive property.

1. $6(-4x - 9y)$ **Ans:** $-24x - 54y$

2. $-3x(2x + 4y - 7)$ **Ans:** $-6x^2 - 12xy + 21x$

3. $-4ab(-5a + 7b + 8)$ **Ans:** $20a^2b - 28ab^2 - 32ab$

1.7 COMBINING LIKE TERMS

Identifying Like Terms

We can add or subtract quantities that are *like quantities*. This is called **combining like quantities.**

$$5 \text{ inches} + 6 \text{ inches} = 11 \text{ inches}$$
$$20 \text{ square inches} - 16 \text{ square inches} = 4 \text{ square inches}$$

However, we cannot combine things that are not the same.

$$16 \text{ square inches} - 4 \text{ inches (Cannot be done!)}$$

Similarly, in algebra we can **combine like terms.** This means to add or subtract like terms. Remember, we cannot combine terms that are not the same. Recall that a *term* is a number, a variable, or a product of numbers and variables. **Like terms** are terms that have identical variables and exponents. In other words, like terms must have exactly the same letter parts.

EXAMPLE 1 List the like terms of each expression.

(a) $5x - 2y + 6x$

(b) $2x^2 - 3x - 5x^2 - 8x$

Solution

(a) $5x$ and $6x$ are like terms. These are the only like terms.
(b) $2x^2$ and $-5x^2$ are like terms.
    $-3x$ and $-8x$ are like terms.
    Note that x^2 and x are not like terms.

Practice Problem 1 List the like terms of each expression.

(a) $5a + 2b + 8a - 4b$

(b) $x^2 + y^2 + 3x - 7y^2$

(a) $5a$ and $8a$; $2b$ and $-4b$

(b) y^2 and $-7y^2$

Do you really understand what a term is? A term is a number, a variable, or a product of numbers and variables. Terms are the parts of an algebraic expression separated by plus or minus signs. The sign in front of the term is considered part of the term.

Combining Like Terms

It is important to know how to combine like terms. Since

$$4 \text{ inches} + 5 \text{ inches} = 9 \text{ inches},$$

we would expect in algebra that $4x + 5x = 9x$.

Why is this true? Let's take a look at the distributive property.

> Like terms may be added or subtracted by using the distributive property:
> $$ab + ac = a(b + c) \quad \text{and} \quad ba + ca = (b + c)a.$$

For example,

$$-7x + 9x = (-7 + 9)x = 2x$$
$$5x^2 + 12x^2 = (5 + 12)x^2 = 17x^2.$$

EXAMPLE 2   Combine like terms.

(a) $-4x^2 + 8x^2$ **(b)** $5x + 3x + 2x$

Solution

(a) Notice that each term contains the factor x^2. Using the distributive property, we have

$$-4x^2 + 8x^2 = (-4 + 8)x^2 = 4x^2.$$

(b) Note that each term contains the factor x. Using the distributive property, we have

$$5x + 3x + 2x = (5 + 3 + 2)x = 10x.$$

Practice Problem 2   Combine like terms.

(a) $16y^3 + 9y^3$ $25y^3$ **(b)** $5a + 7a + 4a$ $16a$

In this section, the direction *simplify* means to remove parentheses and/or combine like terms.

EXAMPLE 3   Simplify. $5a^2 - 2a^2 + 6a^2$

Solution

$$5a^2 - 2a^2 + 6a^2 = (5 - 2 + 6)a^2 = 9a^2$$

Practice Problem 3   Simplify. $-8y^2 - 9y^2 + 4y^2$ $-13y^2$

After doing a few problems, you will find that it is not necessary to write out the step of using the distributive property. We will omit this step for the remaining examples in this section.

EXAMPLE 4   Simplify.

(a) $5.6a + 2b + 7.3a - 6b$

(b) $3x^2y - 2xy^2 + 6x^2y$

(c) $2a^2b + 3ab^2 - 6a^2b^2 - 8ab$

Solution

(a) $5.6a + 2b + 7.3a - 6b = 12.9a - 4b$ We combine the a terms and the b terms separately.

(b) $3x^2y - 2xy^2 + 6x^2y = 9x^2y - 2xy^2$ **Note:** x^2y and xy^2 are not like terms because of different powers.

(c) $2a^2b + 3ab^2 - 6a^2b^2 - 8ab$ These terms cannot be combined; there are no like terms in this expression.

Practice Problem 4   Simplify.

(a) $1.3x + 3a - 9.6x + 2a$ $-8.3x + 5a$

(b) $5ab - 2ab^2 - 3a^2b + 6ab$ $11ab - 2ab^2 - 3a^2b$

(c) $7x^2y - 2xy^2 - 3x^2y^2 - 4xy$ Cannot be simplified

The two skills in this section that a student must practice are identifying like terms and correctly adding and subtracting like terms. If a problem involves many terms, you may find it helpful to rearrange the terms so that like terms are together.

EXAMPLE 5 Simplify. $3a - 2b + 5a^2 + 6a - 8b - 12a^2$

Solution   There are three pairs of like terms.

$$3a + 6a - 2b - 8b + 5a^2 - 12a^2$$
  a terms b terms a^2 terms

You can rearrange the terms so that like terms are together, making it easier to combine them.
Combine like terms.

$$= 9a - 10b - 7a^2$$

Because of the commutative property, the order of terms in an answer to this problem is not significant. These three terms can be rearranged in a different order. $-10b + 9a - 7a^2$ and $-7a^2 + 9a - 10b$ are also correct. Later, we will learn the preferred way to write the answer.

Practice Problem 5 Simplify. $5xy - 2x^2y + 6xy^2 - xy - 3xy^2 - 7x^2y$

$$4xy - 9x^2y + 3xy^2$$

Use extra care with fractional values.

EXAMPLE 6 Simplify. $\frac{3}{4}x^2 - 5y - \frac{1}{8}x^2 + \frac{1}{3}y$

Solution   We need a least common denominator for the x^2 terms, which is 8. Change $\frac{3}{4}$ to eighths by multiplying the numerator and denominator by 2.

$$\frac{3}{4}x^2 - \frac{1}{8}x^2 = \frac{3 \cdot 2}{4 \cdot 2}x^2 - \frac{1}{8}x^2 = \frac{6}{8}x^2 - \frac{1}{8}x^2 = \frac{5}{8}x^2$$

The least common denominator for the y terms is 3. Change -5 to thirds.

$$-\frac{5}{1}y + \frac{1}{3}y = \frac{-5 \cdot 3}{1 \cdot 3}y + \frac{1}{3}y = \frac{-15}{3}y + \frac{1}{3}y = -\frac{14}{3}y$$

Thus, our solution is $\frac{5}{8}x^2 - \frac{14}{3}y$.

Practice Problem 6 Simplify.

$$\frac{1}{7}a^2 - \frac{5}{12}b + 2a^2 - \frac{1}{3}b \qquad \frac{15}{7}a^2 - \frac{3}{4}b$$

EXAMPLE 7 Simplify. $6(2x + 3xy) - 8x(3 - 4y)$

Solution   First remove the parentheses; then combine like terms.

$$6(2x + 3xy) - 8x(3 - 4y) = 12x + 18xy - 24x + 32xy \qquad \text{Use the distributive property.}$$

$$= -12x + 50xy \qquad \text{Combine like terms.}$$

Practice Problem 7 Simplify. $5a(2 - 3b) - 4(6a + 2ab) \quad -14a - 23ab$

Verbal and Writing Skills

1. Explain in your own words the mathematical meaning of the word *term*.
A term is a number, a variable, or a product of numbers and variables.

2. Explain in your own words the mathematical meaning of the phrase *like terms*.
Like terms are terms that have identical variables and exponents.

3. Explain which terms are like terms in the expression $5x - 7y - 8x$.
The two terms $5x$ and $-8x$ are like terms because they both have the variable x with the exponent of one.

4. Explain which terms are like terms in the expression $12a - 3b - 9a$.
The two terms $12a$ and $-9a$ are like terms because they both have the variable a with the exponent of one.

5. Explain which terms are like terms in the expression $7xy - 9x^2y - 15xy^2 - 14xy$.
The only like terms are $7xy$ and $-14xy$ because the other two have different exponents even though they have the same variables.

6. Explain which terms are like terms in the expression $-3a^2b - 12ab + 5ab^2 + 9ab$.
The only like terms are $-12ab$ and $9ab$ because the other two have different exponents even though they have the same variables.

Combine like terms.

7. $-14b^2 - 11b^2$
$-25b^2$

8. $-17x^5 + 3x^5$
$-14x^5$

9. $5a^3 - 7a^2 + a^3$
$6a^3 - 7a^2$

10. $a^5 + 9a^3 - 3a^5$
$-2a^5 + 9a^3$

11. $3x + 2y - 8x - 7y$
$-5x - 5y$

12. $4a - 3b - 2a - 8b$
$2a - 11b$

13. $1.3x - 2.6y + 5.8x - 0.9y$
$7.1x - 3.5y$

14. $3.1a - 0.2b - 0.8a + 5.3b$
$2.3a + 5.1b$

15. $1.6x - 2.8y - 3.6x - 5.9y$
$-2x - 8.7y$

16. $1.9x - 2.4b - 3.8x - 8.2b$
$-1.9x - 10.6b$

17. $3p - 4q + 2p + 3 + 5q - 21$
$5p + q - 18$

18. $6x - 5y - 3y + 7 - 11x - 5$
$-5x - 8y + 2$

19. $2ab + 5bc - 6ac - 2ab$
$5bc - 6ac$

20. $7ab - 3bc - 12ac + 8ab$
$15ab - 3bc - 12ac$

21. $2x^2 - 3x - 5 - 7x + 8 - x^2$
$x^2 - 10x + 3$

22. $5x + 7 - 6x^2 + 6 - 11x + 4x^2$
$-2x^2 - 6x + 13$

23. $2y^2 - 8y + 9 - 12y^2 - 8y + 3$
$-10y^2 - 16y + 12$

24. $3y^2 + 9y - 12 - 4y^2 - 6y + 2$
$-y^2 + 3y - 10$

25. $\frac{1}{3}x - \frac{2}{3}y - \frac{2}{5}x + \frac{4}{7}y$ $-\frac{1}{15}x - \frac{2}{21}y$

26. $\frac{2}{5}s - \frac{3}{8}t - \frac{4}{15}s - \frac{5}{12}t$ $\frac{2}{15}s - \frac{19}{24}t$

27. $\frac{3}{4}a^2 - \frac{1}{3}b - \frac{1}{5}a^2 - \frac{1}{2}b$ $\frac{11}{20}a^2 - \frac{5}{6}b$

28. $\frac{2}{5}y - \frac{3}{4}x^2 - \frac{1}{3}y + \frac{7}{8}x^2$ $\frac{1}{15}y + \frac{1}{8}x^2$

29. $3rs - 8r + s - 5rs + 10r - s$
$-2rs + 2r$

30. $-rs + 10s + 5r - rs + 6s - 2r$
$-2rs + 16s + 3r$

Simplify. Use the distributive property to remove parentheses; then combine like terms.

31. $5(2a - b) - 3(5b - 6a)$
$28a - 20b$

32. $8(3x - 2y) + 4(3y - 5x)$
$4x - 4y$

33. $-3b(5a - 3b) + 4(-3ab - 5b^2)$
$-27ab - 11b^2$

34. $2x(x - 3y) - 4(-3x^2 - 2xy)$
$14x^2 + 2xy$

35. $6(c - 2d^2) - 2(4c - d^2)$
$-2c - 10d^2$

36. $-5(2cd + c^2) + 3c(3d - 2c)$
$-cd - 11c^2$

37. $4(2 - x) - 3(-5 - 12x)$
$32x + 23$

38. $7(3 - x) - 6(8 - 13x)$
$71x - 27$

To Think About

▲ **39.** *Fencing in a Pool* Mr. Jimenez has a pool behind his house that needs to be fenced in. The backyard is an odd quadrilateral shape. The four sides are $3a$, $2b$, $4a$, and $7b$ in length. How much fencing (the length of the perimeter) would he need to enclose the pool?
$7a + 9b$

▲ **40.** *Framing a Masterpiece* The new Degas masterpiece purchased by the Museum of Fine Arts in Boston needs to be reframed. If the rectangular picture measures $6x - 3$ wide by $8x - 7$ high, what is the perimeter of the painting?
$28x - 20$

▲ **41.** *Geometry* A rectangle is $5x - 10$ feet long and $2x + 6$ feet wide. What is the perimeter of the rectangle?
$14x - 8$ ft

▲ **42.** *Geometry* A triangle has sides of length $3a + 8$ meters, $5a - b$ meters, and $12 - 2b$ meters. What is the perimeter of the triangle?
$8a - 3b + 20$ m

Cumulative Review *Evaluate.*

43. **[1.2.1]** $-\dfrac{3}{4} - \dfrac{1}{3} - \dfrac{13}{12}$ or $-1\dfrac{1}{12}$

44. **[1.3.1]** $\left(\dfrac{2}{3}\right)\left(-\dfrac{9}{16}\right)$ $-\dfrac{3}{8}$

45. **[1.1.4]** $\dfrac{4}{5} + \left(-\dfrac{1}{25}\right) + \left(-\dfrac{3}{10}\right)$ $\dfrac{23}{50}$

46. **[1.3.3]** $\left(\dfrac{5}{7}\right) \div \left(-\dfrac{14}{3}\right)$ $-\dfrac{15}{98}$

Quick Quiz 1.7 *Combine like terms.*

1. $3xy - \dfrac{2}{3}x^2y - \dfrac{5}{6}xy + \dfrac{7}{3}x^2y$ $\dfrac{13}{6}xy + \dfrac{5}{3}x^2y$

2. $8.2a^2b + 5.5ab^2 - 7.6a^2b - 9.9ab^2$
$0.6a^2b - 4.4ab^2$

3. $2(3x - 5y) - 2(-7x - 4y)$
$20x - 2y$

4. **Concept Check** Explain how you would remove parentheses and then combine like terms to obtain the answer for $1.2(3.5x - 2.2y) - 4.5(2.0x + 1.5y)$.
Answers may vary

Classroom Quiz 1.7 *You may use these problems to quiz your students' mastery of Section 1.7.*
Combine like terms.

1. $5ab - \dfrac{3}{5}a^2b + \dfrac{1}{4}ab + \dfrac{4}{5}a^2b$ **Ans:** $\dfrac{21}{4}ab + \dfrac{1}{5}a^2b$

2. $-6.5x^2y + 3.8xy^2 - 7.6x^2y - 6.8xy^2$ **Ans:** $-14.1x^2y - 3xy^2$

3. $-3(5x - 2y) - 7(2x - 3y)$ **Ans:** $-29x + 27y$

1.8 USING SUBSTITUTION TO EVALUATE ALGEBRAIC EXPRESSIONS AND FORMULAS

Student Learning Objectives

After studying this section, you will be able to:

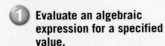 Evaluate an algebraic expression for a specified value.

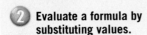 Evaluate a formula by substituting values.

1 Evaluating an Algebraic Expression for a Specified Value

You will use the order of operations to **evaluate** variable expressions. Suppose we are asked to evaluate

$$6 + 3x \text{ for } x = -4.$$

In general, x represents some unknown number. Here we are told x has the value -4. We can replace x with -4. Use parentheses around -4. Note that we always put replacement values in parentheses.

$$6 + 3(-4) = 6 + (-12) = -6$$

When we replace a variable by a particular value, we say we have **substituted** the value for the variable. We then evaluate the expression (that is, find a value for it).

Teaching Example 1 Evaluate $6 - \frac{4}{5}y$ for $y = -10$.

Ans: 14

EXAMPLE 1 Evaluate $\frac{2}{3}x - 5$ for $x = -6$.

Solution

$$\frac{2}{3}x - 5 = \frac{2}{3}(-6) - 5 \qquad \text{Substitute } -6 \text{ for } x. \text{ Be sure to enclose the } -6 \text{ in parentheses.}$$

$$= -4 - 5 \qquad \text{Multiply } \left(\frac{2}{3}\right)\left(-\frac{6}{1}\right) = -4.$$

$$= -9 \qquad \text{Combine.}$$

Practice Problem 1 Evaluate $4 - \frac{1}{2}x$ for $x = -8$. 8

NOTE TO STUDENT: Fully worked-out solutions to all of the Practice Problems can be found at the back of the text starting at page SP-1

Compare parts **(a)** and **(b)** in the next example. The two parts illustrate that you must be careful what value you raise to a power. *Note:* In part (b) we will need parentheses within parentheses. To avoid confusion, we use brackets [ ] to represent the outside parentheses.

Teaching Example 2 Evaluate for $y = -2$.
(a) $5y^2$ **(b)** $(5y)^2$

Ans: (a) 20 **(b)** 100

EXAMPLE 2

(a) Evaluate $2x^2$ for $x = -3$. **(b)** Evaluate $(2x)^2$ for $x = -3$.

Solution

(a) Here the value x is squared. **(b)** Here the value $(2x)$ is squared.

$$2x^2 = 2(-3)^2 \qquad\qquad (2x)^2 = [2(-3)]^2$$

$$= 2(9) \quad \text{First square } -3. \qquad = (-6)^2 \quad \text{First multiply the numbers inside the brackets.}$$

$$= 18 \quad \text{Then multiply.} \qquad = 36 \quad \text{Then square } -6.$$

Practice Problem 2 Evaluate for $x = -3$.

(a) $4x^2$ 36 **(b)** $(4x)^2$ 144

Carefully study the solutions to Example 2(a) and Example 2(b). You will find that taking the time to see *how* and *why* they are different is a good investment of time.

EXAMPLE 3 Evaluate $x^2 + 3x$ for $x = -4$.

Solution

$$
\begin{aligned}
x^2 + 3x &= (-4)^2 + 3(-4) & &\text{Replace each } x \text{ by } -4 \text{ in the original expression.}\\
&= 16 + (3)(-4) & &\text{Raise to a power.}\\
&= 16 - 12 & &\text{Multiply.}\\
&= 4 & &\text{Finally, add.}
\end{aligned}
$$

Practice Problem 3 Evaluate $2x^2 - 3x$ for $x = -2$. 14

② Evaluating a Formula by Substituting Values

We can *evaluate a formula* by substituting values for the variables. For example, the area of a triangle can be found using the formula $A = \frac{1}{2}ab$, where b is the length of the base of the triangle and a is the altitude of the triangle (see figure). If we know values for a and b, we can substitute those values into the formula to find the area. The units for area are *square units*.

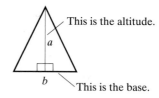

This is the altitude.

This is the base.

Because some of the examples and exercises in this section involve geometry, it may be helpful to review this topic.

The following information is very important. If you have forgotten some of this material (or if you have never learned it), please take the time to learn it completely now. Throughout the entire book we will be using this information in solving applied problems.

Perimeter is the distance around a plane figure. Perimeter is measured in linear units (inches (in.), feet (ft), centimeters (cm), miles (mi)). **Area** is a measure of the amount of surface in a region. Area is measured in square units (square inches (in.²), square feet (ft²), square centimeters (cm²)).

In our sketches we will show angles of 90° by using a small square (⌐). This indicates that the two lines are at right angles. All angles that measure 90° are called **right angles.** An **altitude** is perpendicular to the base of a figure. That is, the altitude forms right angles with the base. The small corner square in a sketch helps us identify the altitude of the figure.

The following box provides a handy guide to some facts and formulas you will need to know. Use it as a reference when solving word problems involving geometric figures.

GEOMETRIC FORMULAS: TWO-DIMENSIONAL FIGURES

A **parallelogram** is a four-sided figure with opposite sides parallel. In a parallelogram, opposite sides are equal and opposite angles are equal.

Perimeter = the sum of all four sides

Area = ab

A **rectangle** is a parallelogram with all interior angles measuring 90°.

Perimeter = $2l + 2w$

Area = lw

A **square** is a rectangle with all four sides equal.

Perimeter $= 4s$

Area $= s^2$

A **trapezoid** is a four-sided figure with two sides parallel. The parallel sides are called the *bases* of the trapezoid.

Perimeter $=$ the sum of all four sides

Area $= \dfrac{1}{2}a(b_1 + b_2)$

A **triangle** is a closed plane figure with three sides.

Perimeter $=$ the sum of the three sides

Area $= \dfrac{1}{2}ab$

A **circle** is a plane curve consisting of all points at an equal distance from a given point called the center.

Circumference is the distance around a circle.

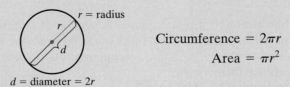

Circumference $= 2\pi r$

Area $= \pi r^2$

π (the number *pi*) is a constant associated with circles. It is an irrational number that is approximately 3.141592654. We usually use 3.14 as a sufficiently accurate approximation. Thus we write $\pi \approx 3.14$ for most of our calculations involving π.

Teaching Example 4 Find the area of a triangle with a base of 9 inches and a height of 20 inches.

Ans: 90 in.^{2}

▲ **EXAMPLE 4**   Find the area of a triangle with a base of 16 centimeters (cm) and a height of 12 centimeters (cm).

Solution

Use the formula $A = \dfrac{1}{2}ab$.

Substitute 12 centimeters for a and 16 centimeters for b.

$A = \dfrac{1}{2}(12\text{ centimeters})(16\text{ centimeters})$

$= \dfrac{1}{2}(12)(16)(\text{cm})(\text{cm})$ If you take $\frac{1}{2}$ of 12 first, it will make your calculation easier.

$= (6)(16)(\text{cm})^2 = 96$ square centimeters

The area of the triangle is 96 square centimeters or 96 cm^2.

NOTE TO STUDENT: Fully worked-out solutions to all of the Practice Problems can be found at the back of the text starting at page SP-1

▲ **Practice Problem 4**   Find the area of a triangle with an altitude of 3 meters and a base of 7 meters.   10.5 m^2

The area of a circle is given by

$$A = \pi r^2.$$

We will use 3.14 as an approximation for the *irrational number* π.

▲ **EXAMPLE 5**   Find the area of a circle if the radius is 2 inches.

Solution

$A = \pi r^2 \approx (3.14)(2 \text{ inches})^2$ Write the formula and substitute the given values for the letters.

$\quad = (3.14)(4)(\text{in.})^2$ Raise to a power. Then multiply.

$\quad = 12.56$ square inches or 12.56 in.^2

▲ **Practice Problem 5**   Find the area of a circle if the radius is 3 meters.

28.26 m²

The formula $C = \frac{5}{9}(F - 32)$ allows us to find the Celsius temperature if we know the Fahrenheit temperature. That is, we can substitute a value for F in degrees Fahrenheit into the formula to obtain a temperature C in degrees Celsius.

EXAMPLE 6   What is the Celsius temperature when the Fahrenheit temperature is $F = -22°$?

Solution   Use the formula.

$C = \dfrac{5}{9}(F - 32)$

$\quad = \dfrac{5}{9}[(-22) - 32]$ Substitute -22 for F in the formula.

$\quad = \dfrac{5}{9}(-54)$ Combine the numbers inside the brackets.

$\quad = (5)(-6)$ Simplify.

$\quad = -30$ Multiply.

The temperature is $-30°$ Celsius or $-30°$C.

Practice Problem 6   What is the Celsius temperature when the Fahrenheit temperature is $F = 68°$? Use the formula $C = \frac{5}{9}(F - 32)$.   20°C

When driving in Canada or Mexico, we must observe speed limits posted in kilometers per hour. A formula that converts r (miles per hour) to k (kilometers per hour) is $k \approx 1.61r$. Note that this is an approximation.

Teaching Example 7 The posted speed limit in a Canadian town is 50 kilometers per hour. If you are traveling at 30 miles per hour, are you exceeding the speed limit?

Ans: No; you are traveling at approximately 48.3 kilometers per hour.

EXAMPLE 7   You are driving on a highway in Mexico. It has a posted maximum speed of 100 kilometers per hour. You are driving at 61 miles per hour. Are you exceeding the speed limit?

Solution   Use the formula.

$$k \approx 1.61r$$
$$= (1.61)(61) \quad \text{Replace } r \text{ by 61.}$$
$$= 98.21 \quad \text{Multiply the numbers.}$$

You are driving at approximately 98 kilometers per hour. You are not exceeding the speed limit.

Practice Problem 7   You are driving behind a heavily loaded truck on a Canadian highway. The highway has a posted minimum speed of 65 kilometers per hour. When you travel at exactly the same speed as the truck ahead of you, you observe that the speedometer reads 35 miles per hour. Assuming that your speedometer is accurate, determine whether the truck is violating the minimum speed law.

The truck is traveling at approximately 56.35 kilometers per hour. It is violating the law.

Developing Your Study Skills

Problems with Accuracy

Strive for accuracy. Mistakes are often made as a result of human error rather than from lack of understanding. Such mistakes are frustrating. A simple arithmetic or sign error can lead to an incorrect answer.

These five steps will help you cut down on errors.

1. Work carefully and take your time. Do not rush through a problem just to get it done.
2. Concentrate on one problem at a time. Sometimes problems become mechanical, and your mind begins to wander. You can become careless and make a mistake.
3. Check your problem. Be sure that you copied it correctly from the book.
4. Check your computations from step to step. Check the solution in the problem. Does it work? Does it make sense?
5. Keep practicing new skills. Remember the old saying, "Practice makes perfect." An increase in practice results in an increase in accuracy. Many errors are due simply to lack of practice.

There is no magic formula for eliminating all errors, but these five steps will be a tremendous help in reducing them.

PRACTICE   WATCH   DOWNLOAD   READ   REVIEW

Evaluate.

1. $-2x + 1$ for $x = 3$
-5

2. $-4x - 2$ for $x = 5$
-22

3. $\frac{2}{5}y - 8$ for $y = -10$
-12

4. $\frac{3}{8}y - 9$ for $y = -8$
-12

5. $5x + 10$ for $x = \frac{1}{2}$
$\frac{25}{2}$ or $12\frac{1}{2}$

6. $7x + 20$ for $x = -\frac{1}{2}$
$\frac{33}{2}$ or $16\frac{1}{2}$

7. $2 - 4x$ for $x = 7$
-26

8. $3 - 5x$ for $x = 8$
-37

9. $3.5 - 2x$ for $x = 2.4$
-1.3

10. $6.3 - 3x$ for $x = 2.3$
-0.6

11. $9x + 13$ for $x = -\frac{3}{4}$
$\frac{25}{4}$ or $6\frac{1}{4}$

12. $5x + 7$ for $x = -\frac{2}{3}$
$\frac{11}{3}$ or $3\frac{2}{3}$

13. $x^2 - 3x$ for $x = -2$
10

14. $x^2 + 3x$ for $x = 4$
28

15. $5y^2$ for $y = -1$
5

16. $8y^2$ for $y = -1$
8

17. $-3x^3$ for $x = 2$
-24

18. $-7x^2$ for $x = 5$
-175

19. $-5x^2$ for $x = -2$
-20

20. $-2x^2$ for $x = -3$
-18

21. $2x^2 + 3x$ for $x = -3$
9

22. $18 + 3x^2$ for $x = -3$
45

23. $(2x)^2 + x$ for $x = 3$
39

24. $2 - x^2$ for $x = -2$
-2

25. $2 - (-x)^2$ for $x = -2$ -2

26. $2x - 3x^2$ for $x = -4$ -56

27. $10a + (4a)^2$ for $a = -2$ 44

28. $9a - (2a)^2$ for $a = -3$ -63

29. $4x^2 - 6x$ for $x = \frac{1}{2}$ -2

30. $5 - 9x^2$ for $x = \frac{1}{3}$ 4

31. $x^2 - 7x + 3$ for $x = 3$ -9

32. $4x^2 - 3x + 9$ for $x = 2$ 19

33. $\frac{1}{2}x^2 - 3x + 9$ for $x = -4$ 29

34. $\frac{1}{3}x^2 + 2x - 5$ for $x = -3$ -8

35. $2r^2 + 3s^2 - rs$ for $r = -1$ and $s = 3$ 32

36. $-r^2 + 5rs + 4s^2$ for $r = -2$ and $s = 3$ 2

37. $a^3 + 2abc - 3c^2$ for $a = 5, b = 9$, and $c = -1$ 32

38. $a^2 - 2ab + 2c^2$ for $a = 3, b = 2$, and $c = -4$ 29

39. $\frac{a^2 + ab}{3b}$ for $a = -1$ and $b = -2$ $-\frac{1}{2}$

40. $\frac{x^2 - 2xy}{2y}$ for $x = -2$ and $y = -3$ $\frac{4}{3}$ or $1\frac{1}{3}$

Applications

▲ **41. Geometry** A sign is made in the shape of a parallelogram. The base measures 22 feet. The altitude measures 16 feet. What is the area of the sign?
352 ft^2

▲ **42. Geometry** A field is shaped like a parallelogram. The base measures 92 feet. The altitude measures 54 feet. What is the area of the field?
4968 ft^2

▲ **43. TV Parts** A square support unit in a television is made with a side measuring 3 centimeters. A new model being designed for next year will have a larger square with a side measuring 3.2 centimeters. By how much will the area of the square be increased?
1.24 cm^2

▲ **44. Computer Chips** A square computer chip for last year's computer had a side measuring 23 millimeters. This year the computer chip has been reduced in size. The new square chip has a side of 20 millimeters. By how much has the area of the chip decreased?
129 mm^2

▲ **45. Carpentry** A carpenter cut out a small trapezoid as a wooden support for the front step. It has an altitude of 4 inches. One base of the trapezoid measures 9 inches and the other base measures 7 inches. What is the area of this support?
32 in.^{2}

▲ **46. Signal Tower** The Comcast signal tower has a small trapezoid frame on the top of the tower. The frame has an altitude of 9 inches. One base of the trapezoid is 20 inches and the other base measures 17 inches. What is the area of this small trapezoidal frame?
166.5 in.^{2}

▲ **47. Geometry** Bradley Palmer State Park has a triangular piece of land on the border. The altitude of the triangle is 400 feet. The base of the triangle is 280 feet. What is the area of this piece of land?
56,000 ft^2

▲ **48. Roofing** The ceiling in the Madisons' house has a leak. The roofer exposed a triangular region that needs to be sealed and then reroofed. The region has an altitude of 14 feet. The base of the region is 19 feet. What is the area of the region that needs to be reroofed?
133 ft^2

▲ **49. Geometry** The radius of a circular tablecloth is 3 feet. What is the area of the tablecloth?
28.26 ft^2

▲ **50. Geometry** A new shot put ring is being constructed. The ring is required to have a concrete surface and a diameter of 8 feet. What will the area of the concrete ring be?
50.24 ft^2

Temperature *For exercises 51 and 52, use the formula* $C = \dfrac{5}{9}(F - 32)$ *to find the Celsius temperature.*

51. Dry ice is solid carbon dioxide. Dry ice does not melt, it goes directly from the solid state to the gaseous state. Dry ice changes from a solid to a gas at $-109.3°F$. What is this temperature in Celsius?
−78.5°C

52. Ivana was an exchange student from Russia, attending school in Montgomery, Alabama. Her host family told her to take a jacket to school because the temperature was not supposed to rise above 50°. Since Ivana was used to temperatures in Celsius, this made no sense at all, for 50°C is incredibly warm. The host family realized they needed to change the temperature into Celsius for Ivana. What was the temperature in Celsius?
10°C

Solve.

▲ **53.** *Sail Dimensions* Find the total cost of making a triangular sail that has a base dimension of 12 feet and a height of 20 feet if the price for making the sail is $19.50 per square foot.
$2340.00

▲ **54.** *Window Coating* A semicircular window of radius 15 inches is to be laminated with a sunblock coating that costs $0.85 per square inch to apply. What is the total cost of coating the window, to the nearest cent? (Use $\pi \approx 3.14$.)
$300.26

55. *Temperature Extremes* The highest temperature ever recorded in Europe was 50° Celsius, and the lowest temperature was −55°C. What are the corresponding temperatures in degrees Fahrenheit? (Use the formula $F = \frac{9}{5}C + 32$.)
122°F; −67°F

56. *Tour de France* The 2007 Tour de France was 3550 kilometers long and was completed in 20 stages. What was the average length of each stage in miles? Use the formula $r = 0.62k$, where r is the number of miles and k is the number of kilometers.
110.05 mi

Cumulative Review *In exercises 57–58, simplify.*

57. **[1.5.1]** $(-2)^4 - 4 \div 2 - (-2)$
16

58. **[1.7.2]** $3(x - 2y) - (x^2 - y) - (x - y)$
$-x^2 + 2x - 4y$

Quick Quiz 1.8 Evaluate.

1. $2x^2 - 4x - 14$ for $x = -2$ 2

2. $5a - 6b$ for $a = \frac{1}{2}$ and $b = -\frac{1}{3}$ $\frac{9}{2}$ or $4\frac{1}{2}$

3. $x^3 + 2x^2y + 5y + 2$ for $x = -2$ and $y = 3$ 33

4. **Concept Check** Explain how you would find the area of a circle if you know its diameter is 12 meters.
Answers may vary

Classroom Quiz 1.8 You may use these problems to quiz your students' mastery of Section 1.8.
Evaluate.

1. $3x^2 + 5x - 15$ for $x = -3$ **Ans:** −3

2. $-7a + 4b$ when $a = \frac{2}{3}$ and $b = -\frac{1}{4}$ **Ans:** $-\frac{17}{3}$ or $-5\frac{2}{3}$

3. $x^3 - 2x^2y - 3y + 4$ for $x = 3$ and $y = -2$ **Ans:** 73

Student Learning Objective

After studying this section, you will be able to:

 Simplify algebraic expressions by removing grouping symbols.

 Simplifying Algebraic Expressions by Removing Grouping Symbols

Many expressions in algebra use **grouping symbols** such as parentheses, brackets, and braces. Sometimes expressions are inside other expressions. Because it can be confusing to have more than one set of parentheses, brackets and braces are also used. How do we know what to do first when we see an expression like $2[5 - 4(a + b)]$?

To simplify the expression, we start with the innermost grouping symbols. Here it is a set of parentheses. We first use the distributive law to multiply.

$$2[5 - 4(a + b)] = 2[5 - 4a - 4b]$$

We use the distributive law again.

$$= 10 - 8a - 8b$$

There are no like terms, so this is our final answer.

Notice that we started with two sets of grouping symbols, but our final answer has none. So we can say we *removed* the grouping symbols. Of course, we didn't just take them away; we used the distributive law and the rules for real numbers to simplify as much as possible. Although simplifying expressions like this involves many steps, we sometimes say "remove parentheses" as a shorthand direction. Sometimes we say "simplify."

Remember to remove the innermost grouping symbols first. Keep working from the inside out.

Teaching Example 1 Simplify.
$2[5y - 3(y - 1)]$

Ans: $4y + 6$

EXAMPLE 1 Simplify. $3[6 - 2(x + y)]$

Solution We want to remove the innermost parentheses first. Therefore, we first use the distributive property to simplify $-2(x + y)$.

$$3[6 - 2(x + y)] = 3[6 - 2x - 2y] \quad \text{Use the distributive property.}$$
$$= 18 - 6x - 6y \quad \text{Use the distributive property again.}$$

Practice Problem 1 Simplify. $5[4x - 3(y - 2)]$ $20x - 15y + 30$

NOTE TO STUDENT: *Fully worked-out solutions to all of the Practice Problems can be found at the back of the text starting at page SP-1*

You recall that a negative sign in front of parentheses is equivalent to having a coefficient of negative 1. You can write the -1 and then multiply by -1 using the distributive property.

$$-(x + 2y) = -1(x + 2y) = -x - 2y$$

Notice that this has the effect of removing the parentheses. Each term in the result now has its sign changed.

Similarly, a positive sign in front of parentheses can be viewed as multiplication by $+1$.

$$+(5x - 6y) = +1(5x - 6y) = 5x - 6y$$

If a grouping symbol has a positive or negative sign in front, we mentally multiply by $+1$ or -1, respectively.

Fraction bars are also considered grouping symbols. In exercises 39 and 40 of Section 1.8, your first step was to simplify expressions above and below the fraction bars. Later in this book we will encounter further examples requiring the same first step. This type of operation will have some similarities to the operation of removing parentheses.

EXAMPLE 2 Simplify. $-2[3a - (b + 2c) + (d - 3e)]$

Solution

$= -2[3a - b - 2c + d - 3e]$ Remove the two innermost sets of parentheses. Since one is not inside the other, we remove both sets at once.

$= -6a + 2b + 4c - 2d + 6e$ Now we remove the brackets by multiplying each term by -2.

Practice Problem 2 Simplify. $-3[2a - (3b - c) + 4d]$

$-6a + 9b - 3c - 12d$

EXAMPLE 3 Simplify. $2[3x - (y + w)] - 3[2x + 2(3y - 2w)]$

Solution

$= 2[3x - y - w] - 3[2x + 6y - 4w]$ In each set of brackets, remove the inner parentheses.

$= 6x - 2y - 2w - 6x - 18y + 12w$ Remove each set of brackets by multiplying by the appropriate number.

$= -20y + 10w$ or $10w - 20y$ Combine like terms. (Note that $6x - 6x = 0x = 0$.)

Practice Problem 3 Simplify. $3[4x - 2(1 - x)] - [3x + (x - 2)]$

$14x - 4$

You can always simplify problems with many sets of grouping symbols by the method shown. Essentially, you just keep removing one level of grouping symbols at each step. Finally, at the end you add up the like terms if possible.

Sometimes it is possible to combine like terms at each step.

EXAMPLE 4 Simplify. $-3\{7x - 2[x - (2x - 1)]\}$

Solution

$= -3\{7x - 2[x - 2x + 1]\}$ Remove the inner parentheses by multiplying each term within the parentheses by -1.

$= -3\{7x - 2[-x + 1]\}$ Combine like terms by combining $+x - 2x$.

$= -3\{7x + 2x - 2\}$ Remove the brackets by multiplying each term within them by -2.

$= -3\{9x - 2\}$ Combine the x terms.

$= -27x + 6$ Remove the braces by multiplying each term by -3.

Practice Problem 4 Simplify. $-2\{5x - 3[2x - (3 - 4x)]\}$ $26x - 18$

Teaching Example 2 Simplify.
$-3[6y - (4y - 3) + (2x - 7)]$

Ans: $-6y - 6x + 12$

Teaching Example 3 Simplify.
$3[2a - (b + 3c)] + (-2)[-8c - 3(b + 2a)]$

Ans: $18a + 3b + 7c$

Teaching Example 4 Simplify.
$-2\{2a^2 - 3[a + 4(a - 2)]\}$

Ans: $-4a^2 + 30a - 48$

Developing Your Study Skills

The Night Before an Exam

With adequate preparation, you can spend the night before an exam pulling together the final details.

1. Look over each section to be covered in the exam. Review the steps needed to solve each type of problem.

2. Review your list of terms, rules, and formulas that you are expected to know for the exam.

3. Take the Practice Test at the end of the chapter just as though you were taking the actual exam. Do not look in your text or get help in any way. Time yourself so that you know how long it takes you to complete the test.

4. Check the Practice Test. Redo the problems you missed.

5. Be sure you have ready the necessary supplies for taking your exam.

Verbal and Writing Skills

1. Rewrite the expression $-3x - 2y$ using a negative sign and parentheses.
$-(3x + 2y)$

2. Rewrite the expression $-x + 5y$ using a negative sign and parentheses.
$-(x - 5y)$

3. To simplify expressions with grouping symbols, we use the __distributive__ property.

4. When an expression contains many grouping symbols, remove the __innermost__ grouping symbol first.

Simplify. Remove grouping symbols and combine like terms.

5. $6x - 3(x - 2y)$
$3x + 6y$

6. $-4x - 2(y - 3x)$
$2x - 2y$

7. $5(c - 3d) - (3c + d)$
$2c - 16d$

8. $6(2d - c) - (9d + c)$
$3d - 7c$

9. $-3(x + 3y) + 2(2x + y)$
$x - 7y$

10. $-4(a + 2b) + 5(2b - a)$
$-9a + 2b$

11. $2[4x^2 - 2(x - 3)]$
$8x^2 - 4x + 12$

12. $4[-3y^2 + 2(4 - y)]$
$-12y^2 - 8y + 32$

13. $2[5(x + y) - 2(3x - 4y)]$
$-2x + 26y$

14. $-3[2(3a + b) - 5(a - 2b)]$
$-3a - 36b$

15. $[10 - 4(x - 2y)] + 3(2x + y)$
$2x + 11y + 10$

16. $[-5(-x + 3y) - 12] - 4(2x - 3)$
$-3x - 15y$

17. $5[3a - 2a(3a + 6b) + 6a^2]$
$15a - 60ab$

18. $3[x - y(3x + y) + y^2]$
$3x - 9xy$

19. $6(2a^2 - 3a - 4) - 3(a - 2)$
$12a^2 - 21a - 18$

20. $7(3b^2 - 2b - 5) - 2(4 - b)$
$21b^2 - 12b - 43$

21. $3a^2 - 4[2b - 3b(b + 2)]$
$3a^2 + 16b + 12b^2$

22. $2b^2 - 3[5b + 2b(2 - b)]$
$8b^2 - 27b$

23. $5b + \{-[3a + 2(5a - 2b)] - 1\}$
$-13a + 9b - 1$

24. $8a - \{-2[a - 3(2a + b)] - 5\}$
$-2a - 6b + 5$

25. $3\{3b^2 + 2[5b - (2 - b)]\}$
$9b^2 + 36b - 12$

26. $2\{3x^2 + 4[2x - (3 - x)]\}$
$6x^2 + 24x - 24$

27. $-4\{3a^2 - 2[4a^2 - (b + a^2)]\}$
$12a^2 - 8b$

28. $-2\{x^2 - 3[x - (x - 2x^2)]\}$
$10x^2$

Cumulative Review

29. **[1.8.2]** *Melting Point of Gold* The melting point of pure gold is $1064.18°C$. Use $F = 1.8C + 32$ to find the melting point of pure gold in degrees Fahrenheit. Round to the nearest hundredth of a degree.
$1947.52°F$

▲ **30.** **[1.8.2]** *Geometry* Use 3.14 as an approximation for π to compute the area covered by a circular irrigation system with radial arm of length 380 feet. Use $A = \pi r^2$.
$453,416 \text{ ft}^2$

31. **[1.8.2]** *Dog Weight* An average Great Dane weighs between 120 and 150 pounds. Express the range of weight for a Great Dane in kilograms. Use the formula $k = 2.205p$ (where $k =$ kilograms, $p =$ pounds).
264.6 to 330.75 kg

32. **[1.8.2]** *Dog Weight* An average Miniature Pinscher weighs between 9 and 14 pounds. Express the range of weight for a Miniature Pinscher in kilograms. Use the formula $k = 2.205p$ (where $k =$ kilograms, $p =$ pounds).
19.845 to 30.87 kg

Quick Quiz 1.9 Simplify.

1. $2[3x - 2(5x + y)]$ $-14x - 4y$

2. $3[x - 3(x + 4) + 5y]$ $-6x + 15y - 36$

3. $-4\{2a + 2[2ab - b(1 - a)]\}$
$-8a - 24ab + 86$

4. **Concept Check** Explain how you would simplify the following expression if you collect like terms whenever possible. $3\{2 - 3[4x - 2(x + 3) + 5x]\}$
Answers may vary

Classroom Quiz 1.9 You may use these problems to quiz your students' mastery of Section 1.9.
Simplify.

1. $3[4x - 6(2x + y)]$ **Ans:** $-24x - 18y$

2. $2[x - 4(x + 2y) - 4y]$ **Ans:** $-6x - 24y$

3. $-3\{a^2 - 4[a - (a - 2a^2)]\}$ **Ans:** $21a^2$

Putting Your Skills to Work: Use Math to Save Money

TIME TO BUDGET

One way to improve your financial situation is to learn to manage the money you have with a budget. A budget can maximize your efforts to ensure you have enough money to cover your fixed expenses (including housing costs, insurance payments, taxes, credit or loan payments, and savings) as well as your variable expenses (including food, education expenses, clothing expenses, and entertainment). Consider the story of Michael.

Going Back to School

Michael is a teacher in Revere, Massachusetts. Many states, including Massachusetts, require teachers to earn an advanced degree to maintain their teaching certificates. One of Michael's goals is to go back to school to earn his Master's Degree in Education. He knows that this will not only help him further his career, but also provide better financial stability in the long run. His net monthly income, after deductions for items such as taxes and insurance, is currently $2500. So, Michael knows he'll need to put himself on a budget for a period of time in order to save money to go back to school. Michael's research shows that consumer credit counseling services recommend allocating the following percentages for each category of the monthly budget:

Housing	25%
Transportation	10%
Savings	5%
Utilities	5%
Medical	5%
Debt Payments	20%
Food	15%
Clothing	5%
Misc.	10%

1. If Michael follows these recommendations, how much will he have saved at the end of one year? $1500

After investigating several schools in his area, Michael chooses to attend a state college that offers a one-year program for the degree he wishes to pursue. Michael will need $3000 for tuition and fees the first semester plus $450 for textbooks.

2. What is the total cost for each semester? $3450

3. How much will Michael need to save for two semesters of tuition and fees and textbooks? $6900

4. How many months will it take Michael to save the entire amount needed to complete his degree? About 55 months, or 4 years and 7 months

5. How much will Michael need to save per month to have the total tuition cost in two years? About $288

6. Michael decided to cut back on some of his variable monthly expenses such as entertainment so that he could increase his savings to 10% of his net monthly income instead of 5%. How much will he now save per month? How many months will it take Michael to save the entire amount needed for college? $250; about 28 months, or 2 years and 4 months

7. Michael's friend told him he could save about 50% on textbook costs if he bought eBooks. If he plans to buy eBooks each semester, how many fewer months would it take him to save the entire amount needed for college? 2 months

Once Michael earns his advanced degree, his salary will increase on the following schedule:

Year 1 Michael will earn an additional $4200
Year 2 Michael will earn an additional $4800
Year 3 Michael will earn an additional $5200
Year 4 Michael will earn an additional $5500
Year 5 Michael will earn an additional $5800

8. Over 5 years, how much more money will Michael earn because he went back to school to get an advanced degree? $25,500

Making It Personal for You

9. Do you have a budget? Answers may vary

10. How would you adjust Michael's budget to fit your needs? Answers may vary

11. What advice would you give Michael for achieving his goal? Answers may vary

Topic	Procedure	Examples
Absolute value, p. 70.	The absolute value of a number is the distance between that number and zero on the number line. The absolute value of any number will be positive or zero.	$\|3\| = 3$ $\|-2\| = 2$ $\|0\| = 0$ $\left\|-\dfrac{5}{6}\right\| = \dfrac{5}{6}$ $\|-1.38\| = 1.38$
Adding real numbers with the same sign, p. 71.	If the signs are the same, add the absolute values of the numbers. Use the common sign in the answer.	$-3 + (-7) = -10$
Adding real numbers with opposite signs, p. 73.	If the signs are different: 1. Find the difference between the larger and the smaller absolute values. 2. Give the answer the sign of the number having the larger absolute value.	$(-7) + 13 = 6$ $7 + (-13) = -6$
Adding several real numbers, p. 75.	When adding several real numbers, separate them into two groups by sign. Find the sum of all the positive numbers and the sum of all the negative numbers. Combine these two subtotals by the method described above.	$-7 + 6 + 8 + (-11) + (-13) + 22$ $\begin{array}{rr} -7 & +6 \\ -11 & +8 \\ \underline{-13} & \underline{+22} \\ -31 & +36 \end{array}$ $-31 + 36 = 5$ The answer is positive since 36 is positive.
Subtracting real numbers, p. 79.	Add the opposite of the second number.	$-3 - (-13) = -3 + (+13) = 10$
Multiplying and dividing real numbers, pp. 85, 87.	1. If the two numbers have the same sign, multiply (or divide) the absolute values. The result is positive. 2. If the two numbers have different signs, multiply (or divide) the absolute values. The result is negative.	$-5(-3) = +15$ $-36 \div (-4) = +9$ $28 \div (-7) = -4$ $-6(3) = -18$
Exponent form, p. 94.	The base tells you what number is being multiplied. The exponent tells you how many times this number is used as a factor.	$2^5 = 2 \cdot 2 \cdot 2 \cdot 2 \cdot 2 = 32$ $4^3 = 4 \cdot 4 \cdot 4 = 64$ $(-3)^4 = (-3)(-3)(-3)(-3) = 81$
Raising a negative number to a power, p. 95.	When the base is negative, the result is positive for even exponents and negative for odd exponents.	$(-3)^3 = -27$ but $(-2)^4 = 16$
Order of operations, p. 99.	Remember the proper order of operations: 1. Perform operations inside parentheses. 2. Raise to powers. 3. Multiply and divide from left to right. 4. Add and subtract from left to right.	$3(5 + 4)^2 - 2^2 \cdot 3 \div (9 - 2^3)$ $= 3 \cdot 9^2 - 4 \cdot 3 \div (9 - 8)$ $= 3 \cdot 81 - 12 \div 1$ $= 243 - 12 = 231$
Removing parentheses, p. 104.	Use the distributive property to remove parentheses. $a(b + c) = ab + ac$	$3(5x + 2) = 15x + 6$ $-4(x - 3y) = -4x + 12y$

(Continued on next page)

Topic	Procedure	Examples
Combining like terms, p. 109.	Combine terms that have identical variables and exponents.	$7x^2 - 3x + 4y + 2x^2 - 8x - 9y = 9x^2 - 11x - 5y$
Substituting into variable expressions, p. 114.	1. Replace each variable by the numerical value given for it. 2. Follow the order of operations in evaluating the expression.	Evaluate $2x^3 + 3xy + 4y^2$ for $x = -3$ and $y = 2$. $2(-3)^3 + 3(-3)(2) + 4(2)^2$ $\quad = 2(-27) + 3(-3)(2) + 4(4)$ $\quad = -54 - 18 + 16$ $\quad = -56$
Using formulas, p. 115.	1. Replace the variables in the formula by the given values. 2. Evaluate the expression. 3. Label units carefully.	Find the area of a circle with radius 4 feet. Use $A = \pi r^2$, with π as approximately 3.14. $A \approx (3.14)(4 \text{ ft})^2$ $\quad = (3.14)(16 \text{ ft}^2)$ $\quad = 50.24 \text{ ft}^2$ The area of the circle is approximately 50.24 square feet.
Removing grouping symbols, p. 122.	1. Remove innermost grouping symbols first. 2. Then remove remaining innermost grouping symbols. 3. Continue until all grouping symbols are removed. 4. Combine like terms.	$5\{3x - 2[4 + 3(x - 1)]\}$ $\quad = 5\{3x - 2[4 + 3x - 3]\}$ $\quad = 5\{3x - 8 - 6x + 6\}$ $\quad = 15x - 40 - 30x + 30$ $\quad = -15x - 10$

Chapter 1 Review Problems

Section 1.1

Add.

1. $-6 + (-2)$ -8

2. $-12 + 7.8$ -4.2

3. $5 + (-2) + (-12)$ -9

4. $3.7 + (-1.8)$ 1.9

5. $\dfrac{1}{2} + \left(-\dfrac{5}{6}\right)$
$-\dfrac{1}{3}$

6. $-\dfrac{3}{11} + \left(-\dfrac{1}{22}\right)$
$-\dfrac{7}{22}$

7. $\dfrac{3}{4} + \left(-\dfrac{1}{12}\right) + \left(-\dfrac{1}{2}\right)$
$\dfrac{1}{6}$

8. $\dfrac{2}{15} + \dfrac{1}{6} + \left(-\dfrac{4}{5}\right)$
$-\dfrac{1}{2}$

Section 1.2

Add or subtract.

9. $5 - (-3)$ 8

10. $-2 - (-15)$ 13

11. $-30 - (+3)$ -33

12. $8 - (-1.2)$ 9.2

13. $-\dfrac{7}{8} + \left(-\dfrac{3}{4}\right)$
$-\dfrac{13}{8}$ or $-1\dfrac{5}{8}$

14. $-\dfrac{3}{8} + \dfrac{5}{6}$
$\dfrac{11}{24}$

15. $-20.8 - 1.9$
-22.7

16. $-151 - (-63)$
-88

Section 1.3

17. $87 \div (-29)$ -3

18. $-10.4 \div (-0.8)$ 13

19. $\dfrac{-24}{-\dfrac{3}{4}}$ 32

20. $-\dfrac{2}{3} \div \left(-\dfrac{4}{5}\right)$ $\dfrac{5}{6}$

21. $\dfrac{5}{7} \div \left(-\dfrac{5}{25}\right)$
$-\dfrac{25}{7}$ or $-3\dfrac{4}{7}$

22. $-6(3)(4)$
-72

23. $-1(-4)(-3)(-5)$
60

24. $(-5)\left(-\dfrac{1}{2}\right)(4)(-3)$
-30

Mixed Practice

Sections 1.1–1.3

Perform the operations indicated. Simplify all answers.

25. $-5 + (-2) - (-3)$
-4

26. $6 - (-4) + (-2) + 8$
16

27. $-16 + (-13)$
-29

28. $-11 - (-12)$
1

29. $-\dfrac{4}{3} + \dfrac{2}{3} + \dfrac{1}{6}$ $-\dfrac{1}{2}$

30. $-\dfrac{6}{7} + \dfrac{1}{2} + \left(-\dfrac{3}{14}\right)$ $-\dfrac{4}{7}$

31. $-3(-2)(-5)$ -30

32. $-6 + (-2) - (-3)$
-5

33. $3.5(-2.6)$
-9.1

34. $-5.4 \div (-6)$
0.9

35. $5 - (-3.5) + 1.6$
10.1

36. $-8 + 2 - (-4.8)$
-1.2

37. $17 + 3.4 + (-16) + (-2.5)$
1.9

38. $37 + (-44) + 12.5 + (-6.8)$
-1.3

Solve.

39. *Football* The Dallas Cowboys football team had three plays in which they lost 8 yards each time. What was the total yardage lost?   24 yd

40. *Temperature Change* The low temperature in Anchorage, Alaska, last night was $-34°F$. During the day the temperature rose $12°F$. What was the temperature during the day? $-22°F$

41. *Elevation* The highest point in California is Mount Whitney, at 14,494 feet above sea level. The lowest point in California is in Death Valley, at 282 feet below sea level. What is the difference in height between these two elevations?   14,776 ft

42. *Stock Prices* One day in 2007, the value of Stock A rose $2.28 on Monday, rose $2.45 on Tuesday, and fell $6.58 on Wednesday. The value of Stock B fell $3.35 on Monday, rose $2.75 on Tuesday, and fell $1.84 on Wednesday. Which stock had the greatest loss over the three days?   Stock B

Section 1.4

Evaluate.

43. $(-3)^5$ -243

44. $(-2)^6$ 64

45. $(-5)^4$ 625

46. $\left(-\dfrac{2}{3}\right)^3$ $-\dfrac{8}{27}$

47. -9^2 -81

48. $(0.6)^2$ 0.36

49. $\left(\dfrac{5}{6}\right)^2$ $\dfrac{25}{36}$

50. $\left(\dfrac{3}{4}\right)^3$ $\dfrac{27}{64}$

Section 1.5

Simplify using the order of operations.

51. $5(-4) + 3(-2)^3$
-44

52. $20 - (-10) - (-6) + (-5) - 1$
30

53. $(3 - 6)^2 + (-12) + -3(-2)$
3

Section 1.6

Use the distributive property to multiply.

54. $7(-3x + y)$
$-21x + 7y$

55. $3x(6 - x + 3y)$
$18x - 3x^2 + 9xy$

56. $-(7x^2 - 3x + 11)$
$-7x^2 + 3x - 11$

57. $(2xy + x - y)(-3y^2)$
$-6xy^3 - 3xy^2 + 3y^3$

Section 1.7

Combine like terms.

58. $3a^2b - 2bc + 6bc^2 - 8a^2b - 6bc^2 + 5bc$
$-5a^2b + 3bc$

59. $9x + 11y - 12x - 15y$
$-3x - 4y$

60. $4x^2 - 13x + 7 - 9x^2 - 22x - 16$
$-5x^2 - 35x - 9$

61. $-x + \dfrac{1}{2} + 14x^2 - 7x - 1 - 4x^2$
$10x^2 - 8x - \dfrac{1}{2}$

Section 1.8

Evaluate for the given value of the variable.

62. $7x - 6$ for $x = -7$ -55

63. $7 - \dfrac{3}{4}x$ for $x = 8$ 1

64. $x^2 + 3x - 4$ for $x = -3$ -4

65. $-x^2 + 5x - 9$ for $x = 3$ -3

66. $2x^3 - x^2 + 6x + 9$ for $x = -1$ 0

67. $b^2 - 4ac$ for $a = -1, b = 5,$ and $c = -2$ 17

68. $\dfrac{mMG}{r^2}$ for $m = -4, M = 15, G = -1,$ and $r = -2$ 15

Solve.

69. Simple Interest Find the simple interest on a loan of $6000 at an annual interest rate of 18% per year for $\frac{3}{4}$ of a year. Use $I = prt$, where p = principal, r = rate per year, and t = time in years. $810

70. Medication The label of a medication warns the user that it must be stored at a temperature between 20°C and 25°C. What is this temperature range in degrees Fahrenheit? Use the formula $F = \dfrac{9C + 160}{5}$.
68°F to 77°F

▲ **71. Sign Painting** How much will it cost to paint a circular sign with a radius of 4 feet if the painter charges $1.50 per square foot? Use $A = \pi r^2$, where π is approximately 3.14. $75.36

72. Profit Find the daily profit P at a furniture factory if the initial cost of setting up the factory C = $1200, rent R = $300, and sales of furniture S = $56. Use the profit formula $P = 180S - R - C$. $8580.00

▲ **73. Parking Lot Sealer** A parking lot is in the shape of a trapezoid. The altitude of the trapezoid is 200 feet, and the bases of the trapezoid are 300 feet and 700 feet. What is the area of the parking lot? If the parking lot had a sealer applied that costs $2 per square foot, what was the cost of the amount of sealer needed for the entire parking lot?
100,000 ft^2; $200,000

▲ **74. Signal Paint** The Green Mountain Telephone Company has a triangular signal tester at the top of a communications tower. The altitude of the triangle is 3.8 feet and the base is 5.5 feet. What is the area of the triangular signal tester? If the signal tester is painted with a special metallic surface paint that costs $66 per square foot, what was the cost of the amount of paint needed to paint one side of the triangle?
10.45 ft^2; $689.70

Section 1.9

Simplify.

75. $5x - 7(x - 6)$
$-2x + 42$

76. $3(x - 2) - 4(5x + 3)$
$-17x - 18$

77. $2[3 - (4 - 5x)]$
$-2 + 10x$

78. $-3x[x + 3(x - 7)]$
$-12x^2 + 63x$

79. $2xy^3 - 6x^3y - 4x^2y^2 + 3(xy^3 - 2x^2y - 3x^2y^2)$
$5xy^3 - 6x^3y - 13x^2y^2 - 6x^2y$

80. $-5(x + 2y - 7) + 3x(2 - 5y)$
$x - 10y + 35 - 15xy$

81. $-(a + 3b) + 5[2a - b - 2(4a - b)]$
$-31a + 2b$

82. $-5\{2a - [5a - b(3 + 2a)]\}$
$15a - 15b - 10ab$

83. $-3\{2x - [x - 3y(x - 2y)]\}$
$-3x - 9xy + 18y^2$

84. $2\{3x + 2[x + 2y(x - 4)]\}$
$10x + 8xy - 32y$

Mixed Practice

Simplify the following.

85. $-6.3 + 4$
-2.3

86. $4 + (-8) + 12$
8

87. $-\dfrac{2}{3} - \dfrac{4}{5}$
$-\dfrac{22}{15}$ or $-1\dfrac{7}{15}$

88. $-\dfrac{7}{8} - \left(-\dfrac{3}{4}\right)$
$-\dfrac{1}{8}$

89. $3 - (-4) + (-8)$
-1

90. $-1.1 - (-0.2) + 0.4$
-0.5

91. $\left(-\dfrac{9}{10}\right)\left(-2\dfrac{1}{4}\right)$
$\dfrac{81}{40}$ or $2\dfrac{1}{40}$

92. $3.6 \div (-0.45)$
-8

93. $-14.4 \div (-0.06)$
240

94. $(-8.2)(3.1)$
-25.42

95. *Jeopardy* A Jeopardy quiz show contestant began the second round (Double Jeopardy) with $400. She buzzed in on the first two questions, answering a $1000 question correctly, but then giving the incorrect answer to a $800 question. What was her score? $600

Simplify the following.

96. $(-0.3)^4$
0.0081

97. -0.5^4
-0.0625

98. $9(5) - 5(2)^3 + 5$
10

99. $3.8x - 0.2y - 8.7x + 4.3y$
$-4.9x + 4.1y$

100. Evaluate $\dfrac{2p + q}{3q}$ for $p = -2$ and $q = 3$. $-\dfrac{1}{9}$

101. Evaluate $\dfrac{4s - 7t}{s}$ for $s = -3$ and $t = -2$.
$-\dfrac{2}{3}$

102. *Dog Body Temperature* The normal body temperature of a dog is 38.6°C. Your dog has a temperature of 101.1°F. Does your dog have a fever? Use the formula $F = \dfrac{9}{5}C + 32$ to convert normal temperature in Fahrenheit.
No; the dog's temperature is below the normal temperature of 101.48°F.

103. $-7(x - 3y^2 + 4) + 3y(4 - 6y)$
$3y^2 + 12y - 7x - 28$

104. $-2\{6x - 3[7y - 2y(3 - x)]\}$
$-12x + 6y + 12xy$

1.	-0.3
2.	2
3.	$-\dfrac{14}{3}$ or $-4\dfrac{2}{3}$
4.	-70
5.	4
6.	-3
7.	-64
8.	2.56
9.	$\dfrac{16}{81}$
10.	6.8
11.	-25
12.	$-5x^2 - 10xy + 35x$
13.	$6a^2b^2 + 4ab^3 - 14a^2b^3$
14.	$2a^2b + \dfrac{15}{2}ab$
15.	$-1.8x^2y - 4.7xy^2$
16.	$5a + 30$
17.	$14x - 16y$
18.	122
19.	37
20.	$\dfrac{13}{6}$ or $2\dfrac{1}{6}$
21.	96.6 km/hr
22.	$22,800$ ft^2
23.	$\$23.12$
24.	3 cans
25.	$3x - 6xy - 21y^2$
26.	$-3a - 9ab + 3b^2 - 3ab^2$

132

How Am I Doing? Chapter 1 Test

Remember to use your Chapter Test Prep Video CD to see the worked-out solutions to the test problems you want to review

Simplify.

1. $-2.5 + 6.3 + (-4.1)$

2. $-5 - (-7)$

3. $\left(-\dfrac{2}{3}\right)(7)$

4. $-5(-2)(7)(-1)$

5. $-12 \div (-3)$

6. $-1.8 \div (0.6)$

7. $(-4)^3$

8. $(1.6)^2$

9. $\left(\dfrac{2}{3}\right)^4$

10. $(0.2)^2 - (2.1)(-3) + 0.46$

11. $3(4 - 6)^3 + 12 \div (-4) + 2$

12. $-5x(x + 2y - 7)$

13. $-2ab^2(-3a - 2b + 7ab)$

14. $6ab - \dfrac{1}{2}a^2b + \dfrac{3}{2}ab + \dfrac{5}{2}a^2b$

15. $2.3x^2y - 8.1xy^2 + 3.4xy^2 - 4.1x^2y$

16. $3(2 - a) - 4(-6 - 2a)$

17. $5(3x - 2y) - (x + 6y)$

In questions 18–20, evaluate for the value of the variable indicated.

18. $x^3 - 3x^2y + 2y - 5$ for $x = 3$ and $y = -4$

19. $3x^2 - 7x - 11$ for $x = -3$

20. $2a - 3b$ for $a = \dfrac{1}{3}$ and $b = -\dfrac{1}{2}$

21. If you are traveling 60 miles per hour on a highway in Canada, how fast are you traveling in kilometers per hour? (Use $k = 1.61r$, where r = rate in miles per hour and k = rate in kilometers per hour.)

▲**22.** A field is in the shape of a trapezoid. The altitude of the trapezoid is 120 feet and the bases of the trapezoid are 180 feet and 200 feet. What is the area of the field?

▲**23.** Jeff Slater's garage has a triangular roof support beam. The support beam is covered with a sheet of plywood. The altitude of the triangular region is 6.8 feet and the base is 8.5 feet. If the triangular piece of plywood was painted with paint that cost $0.80 per square foot, what was the cost of the amount of paint needed to coat one side of the triangle?

▲**24.** You wish to apply blacktop sealer to your driveway, but do not know how much to buy. If your rectangular driveway measures 60 feet long by 10 feet wide, and a can of blacktop sealer claims to cover 200 square feet, how many cans should you buy?

Simplify.

25. $3[x - 2y(x + 2y) - 3y^2]$

26. $-3\{a + b[3a - b(1 - a)]\}$

CHAPTER

2

In states where there is significant snowfall in the winter, much of the snowplowing is done by four-wheel-drive pickup trucks with an adjustable snowplow attached to the front. With rising fuel prices, the owners of these trucks are concerned about gasoline consumption while operating the snowplows. Most of these trucks can operate in four-wheel drive or two-wheel drive. Those who operate snowplowing services can use the mathematics presented in this chapter to determine when it is more fuel efficient to operate snowplows using two-wheel drive and when it is actually more fuel efficient to operate them with four-wheel drive. In general, owners of small businesses will find the mathematics of Chapter 2 to be most useful in making wise business decisions.

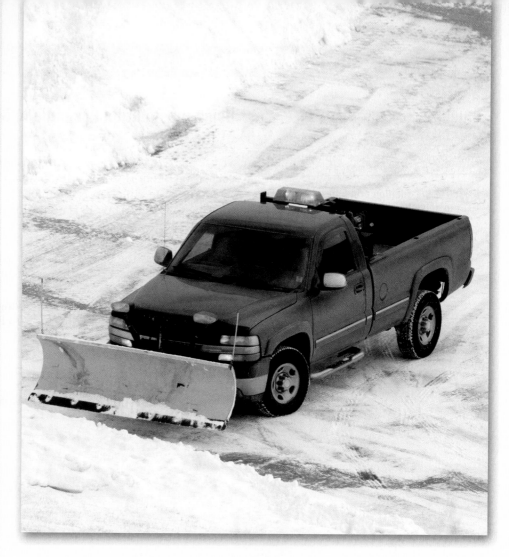

Equations and Inequalities

Student Learning Objective

After studying this section, you will be able to:

 Use the addition principle to solve equations of the form $x + b = c$.

 Using the Addition Principle to Solve Equations of the Form $x + b = c$

When we use an equals sign ($=$), we are indicating that two expressions are equal in value. Such a statement is called an **equation.** For example, $x + 5 = 23$ is an equation. A **solution** of an equation is a number that when substituted for the variable makes the equation true. Thus 18 is a solution of $x + 5 = 23$ because $18 + 5 = 23$. Equations that have exactly the same solutions are called **equivalent equations.** By following certain procedures, we can often transform an equation to a simpler, equivalent one that has the form $x =$ some number. Then this number is a solution of the equation. The process of finding all solutions of an equation is called **solving the equation.**

One of the first procedures used in solving equations has an application in our everyday world. Suppose that we place a 10-kilogram box on one side of a seesaw and a 10-kilogram stone on the other side. If the center of the box is the same distance from the balance point as the center of the stone, we would expect the seesaw to balance. The box and the stone do not look the same, but their weights are equal. If we add a 2-kilogram lead weight to the center of weight of each object at the same time, the seesaw should still balance. The weights are still equal.

There is a similar principle in mathematics. We can state it in words as follows.

THE ADDITION PRINCIPLE

If the same number is added to both sides of an equation, the results on both sides are equal in value.

We can restate it in symbols this way.

For real numbers a, b, and c,   if $a = b$,   then $a + c = b + c$.

Here is an example.

$$\text{If } 3 = \frac{6}{2}, \quad \text{then } 3 + 5 = \frac{6}{2} + 5.$$

Since we added the same amount, 5, to both sides, the sides remain equal to each other.

$$3 + 5 = \frac{6}{2} + 5$$

$$8 = \frac{6}{2} + \frac{10}{2}$$

$$8 = \frac{16}{2}$$

$$8 = 8$$

We can use the addition principle to solve certain equations.

EXAMPLE 1 Solve for x. $x + 16 = 20$

Solution

$$x + 16 + (-16) = 20 + (-16)$$ Use the addition principle to add -16 to both sides.

$$x + 0 = 4$$ Simplify.

$$x = 4$$ The value of x is 4.

Teaching Example 1 Solve for x.
$x + 11 = 28$

Ans: $x = 17$

We have just found a solution of the equation. A **solution** is a value for the variable that makes the equation true. We then say that the value 4 in our example **satisfies** the equation. We can easily verify that 4 is a solution by substituting this value into the original equation. This step is called **checking** the solution.

Check.
$$x + 16 = 20$$
$$4 + 16 \stackrel{?}{=} 20$$
$$20 = 20 \ \checkmark$$

Teaching Tip Remind students that when using the addition principle of equality, the number added to each side of the equation can be any real number. It may be positive or negative, an integer, a rational number in decimal form, a rational number in fractional form, or even an irrational number.

When the same value appears on both sides of the equals sign, we call the equation an **identity.** Because the two sides of the equation in our check have the same value, we know that the original equation has been solved correctly. We have found a solution, and since no other number makes the equation true, it is the only solution.

Practice Problem 1 Solve for x and check your solution. $x + 14 = 23$

$x = 9$

Notice that when you are trying to solve these types of equations, you must add a particular number to both sides of the equation. What is the number to choose? Look at the number that is on the same side of the equation with x, that is, the number added to x. Then think of the number that is **opposite in sign.** This is called the **additive inverse** of the number. The additive inverse of 16 is -16. The additive inverse of -3 is 3. The number to add to both sides of the equation is precisely this additive inverse.

It does not matter which side of the equation contains the variable. The x-term may be on the right or left. In the next example the x-term will be on the right.

EXAMPLE 2 Solve for x. $14 = x - 3$

Solution

$$14 + 3 = x - 3 + 3$$ Notice that -3 is being added to x in the original equation. Add 3 to both sides, since 3 is the additive inverse of -3. This will eliminate the -3 on the right and isolate x.

$$17 = x + 0$$ Simplify.

$$17 = x$$ The value of x is 17.

Check.
$$14 = x - 3$$
$$14 \stackrel{?}{=} 17 - 3$$ Replace x by 17.
$$14 = 14 \ \checkmark$$ Simplify. It checks. The solution is 17.

Teaching Example 2 Solve for x. $9 = x - 5$

Ans: $x = 14$

Teaching Tip Mention the definition of *additive inverse* a couple of times during the class session. It is an important term that will be used frequently in mathematics. Students should understand the concept that to isolate the variable, they should add the additive inverse of the term they want to remove to both sides of the equation.

Practice Problem 2 Solve for x and check your solution. $17 = x - 5$

$x = 22$

Before you add a number to both sides, you should always simplify the equation. The following example shows how combining numbers by addition—separately, on both sides of the equation—simplifies the equation.

EXAMPLE 3 Solve for x. $1.5 + 0.2 = 0.3 + x + 0.2$

Solution

$1.7 = x + 0.5$	Simplify by adding.
$1.7 + (-0.5) = x + 0.5 + (-0.5)$	Add the value -0.5 to both sides, since -0.5 is the additive inverse of 0.5.
$1.2 = x$	Simplify. The value of x is 1.2.

Check.

$$1.5 + 0.2 = 0.3 + x + 0.2$$
$$1.5 + 0.2 \overset{?}{=} 0.3 + 1.2 + 0.2 \quad \text{Replace } x \text{ by } 1.2 \text{ in the original equation.}$$
$$1.7 = 1.7 \checkmark \quad \text{It checks.}$$

Practice Problem 3 Solve for x and check your solution.

$$0.5 - 1.2 = x - 0.3 \quad x = -0.4$$

In Example 3 we added -0.5 to each side. You could subtract 0.5 from each side and get the same result. In Chapter 1 we discussed how subtracting a 0.5 is the same as adding a negative 0.5. Do you see why?

Just as it is possible to add the same number to both sides of an equation, it is also possible to subtract the same number from both sides of an equation. This is so because any subtraction problem can be rewritten as an addition problem. For example, $17 - 5 = 17 + (-5)$. Thus the addition principle tells us that we can subtract the same number from both sides of the equation.

We can determine whether a value is the solution to an equation by following the same steps used to check an answer. Substitute the value to be tested for the variable in the original equation. We will obtain an identity if the value is the solution.

EXAMPLE 4 Is 10 the solution to the equation $-15 + 2 = x - 3$? If it is not, find the solution.

Solution We substitute 10 for x in the equation and see if we obtain an identity.

$$-15 + 2 = x - 3$$
$$-15 + 2 \overset{?}{=} 10 - 3$$

$-13 \neq 7$	The values are not equal. The statement is not an identity.

Thus, 10 is not the solution. Now we take the original equation and solve to find the solution.

$-15 + 2 = x - 3$	
$-13 = x - 3$	Simplify by adding.
$-13 + 3 = x - 3 + 3$	Add 3 to both sides. 3 is the additive inverse of -3.
$-10 = x$	

Check to see if -10 is the solution. The value 10 was incorrect because of a sign error. We must be especially careful to write the correct sign for each number when solving equations.

Practice Problem 4 Is -2 the solution to the equation $x + 8 = -22 + 6$? If it is not, find the solution. No; $x = -24$

EXAMPLE 5 Find the value of x that satisfies the equation

$$\frac{1}{5} + x = -\frac{1}{10} + \frac{1}{2}.$$

Teaching Example 5 Find the value of x that satisfies the equation $\frac{2}{3} + \frac{1}{6} = \frac{1}{3} + x$.

Ans: $x = \frac{1}{2}$

Solution To be combined, the fractions must have common denominators. The least common denominator (LCD) of the fractions is 10.

$$\frac{1 \cdot 2}{5 \cdot 2} + x = -\frac{1}{10} + \frac{1 \cdot 5}{2 \cdot 5}$$ Change each fraction to an equivalent fraction with a denominator of 10.

$$\frac{2}{10} + x = -\frac{1}{10} + \frac{5}{10}$$ This is an equivalent equation.

$$\frac{2}{10} + x = \frac{4}{10}$$ Simplify by adding.

$$\frac{2}{10} + \left(-\frac{2}{10}\right) + x = \frac{4}{10} + \left(-\frac{2}{10}\right)$$ Add the additive inverse of $\frac{2}{10}$ to each side. You could also say that you are subtracting $\frac{2}{10}$ from each side.

$$x = \frac{2}{10}$$ Add the fractions.

$$x = \frac{1}{5}$$ Simplify the answer.

Check. We substitute $\frac{1}{5}$ for x in the original equation and see if we obtain an identity.

$$\frac{1}{5} + x = -\frac{1}{10} + \frac{1}{2}$$

$$\frac{1}{5} + \frac{1}{5} \overset{?}{=} -\frac{1}{10} + \frac{1}{2}$$ Substitute $\frac{1}{5}$ for x.

$$\frac{2}{5} \overset{?}{=} -\frac{1}{10} + \frac{5}{10}$$

$$\frac{2}{5} \overset{?}{=} \frac{4}{10}$$

$$\frac{2}{5} = \frac{2}{5} \checkmark$$ It checks.

Practice Problem 5 Find the value of x that satisfies the equation.

$$\frac{1}{20} - \frac{1}{2} = x + \frac{3}{5} \quad x = -\frac{21}{20} \text{ or } -1\frac{1}{20}$$

Developing Your Study Skills

Why Study Mathematics?

In our present-day, technological world, it is easy to see mathematics at work. Many vocational and professional areas—such as the fields of business, statistics, economics, psychology, finance, computer science, chemistry, physics, engineering, electronics, nuclear energy, banking, quality control, and teaching—require a certain level of expertise in mathematics. Those who want to work in these fields must be able to function at a given mathematical level. Those who cannot will not make it. So if your field of study requires you to take higher-level mathematics courses, be sure to master the topics of this course. Then you will be ready for the next one.

Verbal and Writing Skills *In exercises 1–3, fill in the blank with the appropriate word.*

1. When we use the ___equals___ sign, we indicate two expressions are ___equal___ in value.

2. If the ___same___ ___number___ is added to both sides of an equation, the results on each side are equal in value.

3. The ___solution___ of an equation is a value of the variable that makes the equation true.

4. What is the additive inverse of -20?
20

5. Why do we add the additive inverse of a to each side of $x + a = b$ to solve for x?
Answers may vary. A sample answer is to isolate the variable x.

6. What is the additive inverse of a?
$-a$

Solve for x. Check your answers.

7. $x + 11 = 15$
$x = 4$

8. $x + 12 = 18$
$x = 6$

9. $20 = 9 + x$
$x = 11$

10. $18 = 10 + x$
$x = 8$

11. $x - 3 = 14$
$x = 17$

12. $x - 11 = 5$
$x = 16$

13. $0 = x + 5$
$x = -5$

14. $0 = x - 7$
$x = 7$

15. $x - 6 = -19$
$x = -13$

16. $x - 11 = -13$
$x = -2$

17. $-12 + x = 50$
$x = 62$

18. $-18 + x = 48$
$x = 66$

19. $3 + 5 = x - 7$
$x = 15$

20. $8 - 2 = x + 5$
$x = 1$

21. $32 - 17 = x - 6$
$x = 21$

22. $27 - 12 = x - 9$
$x = 24$

23. $4 + 8 + x = 6 + 6$
$x = 0$

24. $19 - 3 + x = 10 + 6$
$x = 0$

25. $18 - 7 + x = 7 + 9 - 5$
$x = 0$

26. $3 - 17 + 8 = 8 + x - 3$
$x = -11$

27. $-12 + x - 3 = 15 - 18 + 9$
$x = 21$

28. $-19 + x - 7 = 20 - 42 + 10$
$x = 14$

In exercises 29–36, determine whether the given solution is correct. If it is not, find the solution.

29. Is $x = 5$ the solution to $-7 + x = 2$?
No; $x = 9$

30. Is $x = 7$ the solution to $-13 + x = 4$?
No; $x = 17$

31. Is -6 the solution to $-11 + 5 = x + 8$?
No; $x = -14$

32. Is -9 the solution to $-13 - 4 = x - 8$?
Yes

33. Is -33 the solution to $x - 23 = -56$?
Yes

34. Is -8 the solution to $-39 = x - 47$?
No; $x = 8$

35. Is 35 the solution to $15 - 3 + 20 = x - 3$?
Yes

36. Is -12 the solution to $x + 8 = 12 - 19 + 3$?
Yes

Find the value of x that satisfies each equation.

37. $2.5 + x = 0.7$ $x = -1.8$

38. $8.2 + x = 3.2$ $x = -5$

39. $12.5 + x - 8.2 = 4.9$ $x = 0.6$

40. $4.3 + x - 2.6 = 3.4$ $x = 1.7$

41. $x - \dfrac{1}{4} = \dfrac{3}{4}$ $x = 1$

42. $x + \dfrac{1}{3} = \dfrac{2}{3}$ $x = \dfrac{1}{3}$

43. $\dfrac{2}{3} + x = \dfrac{1}{6} + \dfrac{1}{4}$ $x = -\dfrac{1}{4}$

44. $\dfrac{2}{5} + x = \dfrac{1}{2} - \dfrac{3}{10}$ $x = -\dfrac{1}{5}$

Mixed Practice

45. $3 + x = -12 + 8$ $x = -7$

46. $12 + x = -7 + 20$ $x = 1$

47. $5\dfrac{1}{6} + x = 8$ $x = \dfrac{17}{6}$ or $2\dfrac{5}{6}$

48. $7\dfrac{1}{8} = -20 + x$ $x = \dfrac{217}{8}$ or $27\dfrac{1}{8}$

49. $\dfrac{5}{12} - \dfrac{5}{6} = x - \dfrac{3}{2}$ $x = \dfrac{13}{12}$ or $1\dfrac{1}{12}$

50. $\dfrac{4}{15} - \dfrac{3}{5} = x - \dfrac{4}{3}$ $x = 1$

51. $1.6 + x - 3.2 = -2 + 5.6$ $x = 5.2$

52. $0.7 - 4.2 = 3.6 + x$ $x = -7.1$

53. $x - 18.225 = 1.975$ $x = 20.2$

54. $x - 10.012 = -16.835$ $x = -6.823$

Cumulative Review *Simplify by adding like terms.*

55. **[1.7.2]** $x + 3y - 5x - 7y + 2x$
$-2x - 4y$

56. **[1.7.2]** $y^2 + y - 12 - 3y^2 - 5y + 16$
$-2y^2 - 4y + 4$

Quick Quiz 2.1 Solve for the variable.

1. $x - 4.7 = 9.6$ $x = 14.3$

2. $-8.6 + x = -12.1$ $x = -3.5$

3. $3 - 12 + 7 = 8 + x - 2$ $x = -8$

4. **Concept Check** Explain how you would check to verify whether $x = 3.8$ is the solution to $-1.3 + 1.6 + 3x = -6.7 + 4x + 3.2$.
Answers may vary

Classroom Quiz 2.1 You may use these problems to quiz your students' mastery of Section 2.1.
Solve for the variable.

1. $x - 8.3 = 12.8$ **Ans:** $x = 21.1$ **2.** $-7.8 + x = -14.2$ **Ans:** $x = -6.4$ **3.** $5 - 16 + 3 = -9 + x + 3$
Ans: $x = -2$

Student Learning Objectives

After studying this section, you will be able to:

 Solve equations of the form $\frac{1}{a}x = b$.

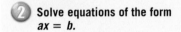 Solve equations of the form $ax = b$.

Solving Equations of the Form $\frac{1}{a}x = b$

The addition principle allows us to add the same number to both sides of an equation. What would happen if we multiplied each side of an equation by the same number? For example, what would happen if we multiplied each side of an equation by 3?

To answer this question, let's return to our simple example of the box and the stone on a balanced seesaw. If we triple the weight on each side (that is, multiply the weight on each side by 3), the seesaw should still balance. The weight values of both sides remain equal.

In words we can state this principle thus.

MULTIPLICATION PRINCIPLE

If both sides of an equation are multiplied by the same nonzero number, the results on both sides are equal in value.

In symbols we can restate the multiplication principle this way.

For real numbers a, b, and c with $c \neq 0$, if $a = b$, then $ca = cb$.

It is important that we say $c \neq 0$. We will explore this idea in later sections. Let us look at an equation where it would be helpful to multiply each side by 3.

Teaching Example 1 Solve for x.

$\frac{1}{6}x = -3$

Ans: $x = -18$

EXAMPLE 1 Solve for x. $\frac{1}{3}x = -15$

Solution We know that $(3)\left(\frac{1}{3}\right) = 1$. We will multiply each side of the equation by 3 because we want to isolate the variable x.

$$3\left(\frac{1}{3}x\right) = 3(-15) \qquad \text{Multiply each side of the equation by 3 since } (3)\left(\frac{1}{3}\right) = 1.$$

$$\left(\frac{3}{1}\right)\left(\frac{1}{3}\right)(x) = -45$$

$$1x = -45 \qquad \text{Simplify.}$$

$$x = -45 \qquad \text{The solution is } -45.$$

Check. $\qquad \frac{1}{3}(-45) \stackrel{?}{=} -15 \qquad$ Substitute -45 for x in the original equation.

$$-15 = -15 \quad \checkmark \quad \text{It checks.}$$

NOTE TO STUDENT: *Fully worked-out solutions to all of the Practice Problems can be found at the back of the text starting at page SP-1*

Practice Problem 1 Solve for x.

$$\frac{1}{8}x = -2 \quad x = -16$$

Note that $\frac{1}{5}x$ can be written as $\frac{x}{5}$. To solve the equation $\frac{x}{5} = 3$, we could multiply each side of the equation by 5. Try it. Then check your solution.

 Solving Equations of the Form $ax = b$

We can see that using the multiplication principle to multiply each side of an equation by $\frac{1}{2}$ is the same as dividing each side of the equation by 2. Thus, it would seem that the multiplication principle would allow us to divide each side of the equation by any nonzero real number. Is there a real-life example of this idea?

Let's return to our simple example of the box and the stone on a balanced seesaw. Suppose that we were to cut the two objects in half (so that the amount of weight of each was divided by 2). We then return the objects to the same places on the seesaw. The seesaw would still balance. The weight values of both sides remain equal.

In words we can state this principle thus.

Teaching Tip Emphasize the fact that the division principle of equality—if $a = b$, then $\frac{a}{c} = \frac{b}{c}$—is valid only if $c \neq 0$. We may never divide by zero. This concept requires continual emphasis.

DIVISION PRINCIPLE

If both sides of an equation are divided by the same nonzero number, the results on both sides are equal in value.

Note: We put a restriction on the number by which we are dividing. We cannot divide by zero. We say that expressions like $\frac{2}{0}$ are not defined. Thus we restrict our divisor to *nonzero* numbers. We can restate the division principle this way.

For real numbers a, b, and c where $c \neq 0$,   if $a = b$,   then $\dfrac{a}{c} = \dfrac{b}{c}$.

EXAMPLE 2   Solve for x. $5x = 125$

Solution
$$\frac{5x}{5} = \frac{125}{5} \qquad \text{Divide both sides by 5.}$$
$$x = 25 \qquad \text{Simplify. The solution is 25.}$$

Check.
$$5x = 125$$
$$5(25) \overset{?}{=} 125 \qquad \text{Replace } x \text{ by 25.}$$
$$125 = 125 \ \checkmark \qquad \text{It checks.}$$

Teaching Example 2 Solve for x. $7x = 84$

Ans: $x = 12$

Practice Problem 2   Solve for x. $9x = 72$ $x = 8$

For equations of the form $ax = b$ (a number multiplied by x equals another number), we solve the equation by choosing to divide both sides by a particular number. What is the number to choose? We look at the side of the equation that contains x. We notice the number that is multiplied by x. We divide by that number. The division principle tells us that we can still have a true equation provided that we divide by that number *on both sides* of the equation.

The solution to an equation may be a proper fraction or an improper fraction.

EXAMPLE 3   Solve for x. $4x = 38$

Solution $\dfrac{4x}{4} = \dfrac{38}{4}$ Divide both sides by 4.

$x = \dfrac{19}{2}$ Simplify. The solution is $\frac{19}{2}$.

If you leave the solution as a fraction, it will be easier to check that solution in the original equation.

Check. $4x = 38$

$\overset{2}{\cancel{4}}\left(\dfrac{19}{\cancel{2}}\right) \overset{?}{=} 38$ Replace x by $\frac{19}{2}$.

$38 = 38$ ✓   It checks.

Practice Problem 3   Solve for x.

$$6x = 50 \quad x = \dfrac{25}{3}$$

In Examples 2 and 3 we *divided by the number multiplied by x.* This procedure is followed regardless of whether the sign of that number is positive or negative. In equations of the form $ax = b$, a is a number multiplied by x. The **coefficient** of x is a. A coefficient is a multiplier.

EXAMPLE 4   Solve for x. $-3x = 48$

Solution $\dfrac{-3x}{-3} = \dfrac{48}{-3}$ Divide both sides by -3.

$x = -16$ The solution is -16.

Check.   Can you check this solution?

Practice Problem 4   Solve for x. $-27x = 54$ $x = -2$

The coefficient of x may be 1 or -1. You may have to rewrite the equation so that the coefficient of 1 or -1 is obvious. With practice you may be able to recognize the coefficient without actually rewriting the equation.

EXAMPLE 5   Solve for x. $-x = -24$.

Solution $-1x = -24$ Rewrite the equation. $-1x$ is the same as $-x$. Now the coefficient of -1 is obvious.

$\dfrac{-1x}{-1} = \dfrac{-24}{-1}$ Divide both sides by -1.

$x = 24$ The solution is 24.

Check.   Can you check this solution?

Practice Problem 5   Solve for x. $-x = 36$ $x = -36$

The variable can be on either side of the equation. The equation $-78 = -3x$ can be solved in exactly the same way as $-3x = -78$.

EXAMPLE 6 Solve for x. $-78 = -3x$

Teaching Example 6 Solve for x. $55 = -5x$

Ans: $x = -11$

Solution

$$\frac{-78}{-3} = \frac{-3x}{-3} \qquad \text{Divide both sides by } -3.$$

$$26 = x \qquad \text{The solution is 26.}$$

Check.

$$-78 = -3x$$

$$-78 \stackrel{?}{=} -3(26) \qquad \text{Replace } x \text{ by 26.}$$

$$-78 = -78 \;\checkmark \quad \text{It checks.}$$

Practice Problem 6 Solve for x.

$$-51 = -6x \qquad x = \frac{17}{2}$$

There is a mathematical concept that unites what we have learned in this section. The concept uses the idea of a multiplicative inverse. For any nonzero number a, the **multiplicative inverse** of a is $\frac{1}{a}$. Likewise, for any nonzero number a, the multiplicative inverse of $\frac{1}{a}$ is a. So to solve an equation of the form $ax = b$, we say that we need to multiply each side by the multiplicative inverse of a. Thus to solve $5x = 45$, we would multiply each side of the equation by the multiplicative inverse of 5, which is $\frac{1}{5}$. In similar fashion, if we wanted to solve the equation $\frac{1}{6}x = 4$, we would multiply each side of the equation by the multiplicative inverse of $\frac{1}{6}$, which is 6. In general, all the problems we have covered so far in this section can be solved by multiplying both sides of the equation by the multiplicative inverse of the coefficient of x.

EXAMPLE 7 Solve for x. $31.2 = 6.0x - 0.8x$

Teaching Example 7 Solve for x.
$12.3 = 6.7x - 2.6x$

Ans: $x = 3$

Solution

$$31.2 = 6.0x - 0.8x \qquad \text{There are like terms on the right side.}$$

$$31.2 = 5.2x \qquad \text{Combine like terms.}$$

$$\frac{31.2}{5.2} = \frac{5.2x}{5.2} \qquad \begin{array}{l}\text{Divide both sides by 5.2 (which is the same as multiplying} \\ \text{both sides by the multiplicative inverse of 5.2).}\end{array}$$

$$6 = x \qquad \text{The solution is 6.}$$

Note: Be sure to place the decimal point in the quotient directly above the caret ($\wedge$) when performing the division.

$$\begin{array}{r} 6. \\ 5.2_\wedge \overline{)31.2_\wedge} \\ \underline{31\,2} \\ 0 \end{array}$$

Check. The check is up to you.

Practice Problem 7 Solve for x. $16.2 = 5.2x - 3.4x$ $x = 9$

Developing Your Study Skills

Getting Help

Getting the right kind of help at the right time can be a key ingredient in being successful in mathematics. When you have gone to class on a regular basis, taken careful notes, methodically read your textbook, and diligently done your homework—all of which means making every effort possible to learn the mathematics—you may find that you are still having difficulty. If this is the case, then you need to seek help. Make an appointment with your instructor to find out what help is available to you. The instructor, tutoring services, a mathematics lab, videotapes, and computer software may be among the resources you can draw on.

Once you discover the resources available in your school, you need to take advantage of them. Do not put it off, or you will find yourself getting behind. You cannot afford that. When studying mathematics, you must keep up with your work.

Verbal and Writing Skills

1. To solve the equation $6x = -24$, divide each side of the equation by ___6___.

2. To solve the equation $-7x = 56$, divide each side of the equation by ___-7___.

3. To solve the equation $\frac{1}{7}x = -2$, multiply each side of the equation by ___7___.

4. To solve the equation $\frac{1}{9}x = 5$, multiply each side of the equation by ___9___.

Solve for x. Be sure to reduce your answer. Check your solution.

5. $\frac{1}{9}x = 4$ $x = 36$

6. $\frac{1}{7}x = 6$ $x = 42$

7. $\frac{1}{2}x = -15$ $x = -30$

8. $\frac{1}{8}x = -10$ $x = -80$

9. $\frac{x}{5} = 16$ $x = 80$

10. $\frac{x}{10} = 8$ $x = 80$

11. $-3 = \frac{x}{5}$ $x = -15$

12. $\frac{x}{3} = -12$ $x = -36$

13. $13x = 52$ $x = 4$

14. $15x = 60$ $x = 4$

15. $56 = 7x$ $x = 8$

16. $46 = 2x$ $x = 23$

17. $-16 = 6x$ $x = -\frac{8}{3}$

18. $-35 = 21x$ $x = -\frac{5}{3}$

19. $1.5x = 75$ $x = 50$

20. $2x = 0.36$ $x = 0.18$

21. $-15 = -x$ $x = 15$

22. $32 = -x$ $x = -32$

23. $-112 = 16x$ $x = -7$

24. $-108 = -18x$ $x = 6$

25. $0.4x = 0.08$ $x = 0.2$ or $\frac{1}{5}$

26. $2.1x = 0.3$ $x = \frac{1}{7}$

27. $-3.9x = -15.6$ $x = 4$

28. $-4.7x = -14.1$ $x = 3$

Determine whether the given solution is correct. If it is not, find the correct solution.

29. Is 7 the solution for $-3x = 21$?   No; $x = -7$

30. Is 8 the solution for $5x = -40$?   No; $x = -8$

31. Is -15 the solution for $-x = 15$?   Yes

32. Is -8 the solution for $-11x = 88$?   Yes

Mixed Practice   *Find the value of the variable that satisfies the equation.*

33. $7y = -0.21$ $y = -0.03$

34. $-3y = 0.42$ $y = -0.14$

35. $-56 = -21t$ $t = \frac{8}{3}$

36. $34 = -51q$ $q = -\frac{2}{3}$

37. $4.6y = -3.22$ $y = -0.7$

38. $-2.8y = -3.08$ $y = 1.1$

39. $4x + 3x = 21$ $x = 3$

40. $5x + 4x = 36$ $x = 4$

41. $2x - 7x = 20$ $x = -4$

42. $3x - 9x = 18$ $x = -3$

43. $-8x - 4x = -5$ $x = \frac{5}{12}$

44. $y - 10y = 6$ $y = -\frac{2}{3}$

45. $12 - 19 = -7x$ $x = 1$

46. $31 - 16 = -5m$ $m = -3$

47. $8m = -14 + 30$ $m = 2$

48. $11x = -20 + 42$ $x = 2$

49. $\dfrac{2}{3}x = 18$ $x = 27$

50. $\dfrac{3}{5}x = 39$ $x = 65$

51. $-2.5133x = 26.38965$ $x = -10.5$

52. $-5.42102x = -45.536568$ $x = 8.4$

Cumulative Review *Evaluate using the correct order of operations. (Be careful to avoid sign errors.)*

53. **[1.5.1]** $(-6)(-8) + (-3)(2)$
$48 - 6 = 42$

54. **[1.5.1]** $(-3)^3 + (-20) \div 2$
$-27 - 10 = -37$

55. **[1.5.1]** $5 + (2 - 6)^2$
$5 + 16 = 21$

56. **[1.1.4]** *Humpback Whales* During the summer months, a group of humpback whales gather at Stellwagen Bank, near Gloucester, Massachusetts, to feed. When they return to the Caribbean for winter, they will lose up to 25% of their body weight in blubber. If a humpback whale weighs 30 tons after feeding at Stellwagen Bank, how much will it weigh after losing 25% of its body weight? 22.5 tons

57. **[0.7.1]** *Earthquakes* In an average year, worldwide, there are 20 earthquakes of magnitude 7 on the Richter scale. If next year is predicted to be an exceptional year, and the number of earthquakes of magnitude 7 is expected to increase by 35%, about how many earthquakes of magnitude 7 can be expected? (Round off to the nearest whole number.)
27 earthquakes

Quick Quiz 2.2 Solve for the variable.

1. $2.5x = -95$ $x = -38$

2. $-3.9x = -54.6$ $x = 14$

3. $7x - 12x = 60$ $x = -12$

4. Concept Check Explain how you would check to verify whether $x = 36\frac{2}{3}$ is the solution to $-22 = -\frac{3}{5}x$.
Answers may vary

Classroom Quiz 2.2 You may use these problems to quiz your students' mastery of Section 2.2.

Solve for the variable.

1. $2.2x = -88$ **Ans:** $x = -40$

2. $-5.2x = -62.4$ **Ans:** $x = 12$

3. $15x - 18x = 21$ **Ans:** $x = -7$

Student Learning Objectives

After studying this section, you will be able to:

 Solve equations of the form $ax + b = c$.

② **Solve equations with the variable on both sides of the equation.**

③ **Solve equations with parentheses.**

① Solving Equations of the Form $ax + b = c$

Jenny Crawford scored several goals in field hockey during April. Her teammates scored three more than five times the number she scored. Together the team scored 18 goals in April. How many did Jenny score? To solve this problem, we need to solve the equation $5x + 3 = 18$.

To solve an equation of the form $ax + b = c$, we must use both the addition principle and the multiplication principle.

> **EXAMPLE 1** Solve for x to determine how many goals Jenny scored and check your solution.
>
> $$5x + 3 = 18$$
>
> **Solution** We first want to isolate the variable term.
>
> | | |
> |---|---|
> | $5x + 3 + (-3) = 18 + (-3)$ | Use the addition principle to add -3 to both sides. |
> | $5x = 15$ | Simplify. |
> | $\dfrac{5x}{5} = \dfrac{15}{5}$ | Use the division principle to divide both sides by 5. |
> | $x = 3$ | The solution is 3. Thus Jenny scored 3 goals. |
>
> *Check.*
> $$5(3) + 3 \stackrel{?}{=} 18$$
> $$15 + 3 \stackrel{?}{=} 18$$
> $$18 = 18 \checkmark \qquad \text{It checks.}$$

> **Practice Problem 1** Solve for x and check your solution. $9x + 2 = 38$
>
> $x = 4$

Teaching Example 1 Solve for x and check your solution. $3x - 7 = 8$

Ans: $x = 5$

② Solving Equations with the Variable on Both Sides of the Equation

In some cases the variable appears on both sides of the equation. We would like to rewrite the equation so that all the terms containing the variable appear on one side. To do this, we apply the addition principle to the variable term.

Teaching Example 2 Solve for x. $8x = 5x - 21$

Ans: $x = -7$

> **EXAMPLE 2** Solve for x. $9x = 6x + 15$
>
> **Solution**
>
> | | |
> |---|---|
> | $9x + (-6x) = 6x + (-6x) + 15$ | Add $-6x$ to both sides. Notice $6x + (-6x)$ eliminates the variable on the right side. |
> | $3x = 15$ | Combine like terms. |
> | $\dfrac{3x}{3} = \dfrac{15}{3}$ | Divide both sides by 3. |
> | $x = 5$ | The solution is 5. |
>
> *Check.* The check is left to the student.

> **Practice Problem 2** Solve for x. $13x = 2x - 66$ $x = -6$

In many problems the variable terms and constant terms appear on both sides of the equations. You will want to get all the variable terms on one side and all the constant terms on the other side.

EXAMPLE 3 Solve for x and check your solution. $9x + 3 = 7x - 2$

Solution First we want to isolate the variable term.

$$9x + (-7x) + 3 = 7x + (-7x) - 2 \quad \text{Add } -7x \text{ to both sides of the equation.}$$
$$2x + 3 = -2 \quad \text{Combine like terms.}$$
$$2x + 3 + (-3) = -2 + (-3) \quad \text{Add } -3 \text{ to both sides.}$$
$$2x = -5 \quad \text{Simplify.}$$
$$\frac{2x}{2} = \frac{-5}{2} \quad \text{Divide both sides by 2.}$$
$$x = -\frac{5}{2} \quad \text{The solution is } -\frac{5}{2}.$$

Check. $\quad 9x + 3 = 7x - 2$

$$9\left(-\frac{5}{2}\right) + 3 \overset{?}{=} 7\left(-\frac{5}{2}\right) - 2 \quad \text{Replace } x \text{ by } -\frac{5}{2}.$$
$$-\frac{45}{2} + 3 \overset{?}{=} -\frac{35}{2} - 2 \quad \text{Simplify.}$$
$$-\frac{45}{2} + \frac{6}{2} \overset{?}{=} -\frac{35}{2} - \frac{4}{2} \quad \text{Change to equivalent fractions with a common denominator.}$$
$$-\frac{39}{2} = -\frac{39}{2} \checkmark \quad \text{It checks. The solution is } -\frac{5}{2}.$$

Practice Problem 3 Solve for x and check your solution.

$$3x + 2 = 5x + 2 \quad x = 0$$

In our next example we will study equations that need simplifying before any other steps are taken. Where it is possible, you should first collect like terms on one or both sides of the equation. The variable terms can be collected on the right side or the left side. In this example we will collect all the x-terms on the right side.

EXAMPLE 4 Solve for x. $5x + 26 - 6 = 9x + 12x$

Solution

$$5x + 20 = 21x \quad \text{Combine like terms.}$$
$$5x + (-5x) + 20 = 21x + (-5x) \quad \text{Add } -5x \text{ to both sides.}$$
$$20 = 16x \quad \text{Combine like terms.}$$
$$\frac{20}{16} = \frac{16x}{16} \quad \text{Divide both sides by 16.}$$
$$\frac{5}{4} = x \quad \text{Don't forget to reduce the resulting fraction.}$$

Check. The check is left to the student.

Practice Problem 4 Solve for z.

$$-z + 8 - z = 3z + 10 - 3 \quad z = \frac{1}{5}$$

Do you really need all these steps? No. As you become more proficient you will be able to combine or eliminate some of these steps. However, it is best to write each step in its entirety until you are consistently obtaining the correct solution.

It is much better to show every step than to take a lot of shortcuts and possibly obtain a wrong answer. This is a section of the algebra course where working neatly and accurately will help you—both now and as you progress through the course.

③ Solving Equations with Parentheses

The equations that you just solved are simpler versions of equations that we will now discuss. These equations contain parentheses. If the parentheses are first removed, the problems then become just like those encountered previously. We use the distributive property to remove the parentheses.

Teaching Example 5 Solve for x and check your solution.
$11 - (6x + 1) = 5x - 2(x + 4)$

Ans: $x = 2$

EXAMPLE 5 Solve for x and check your solution.
$$4(x + 1) - 3(x - 3) = 25$$

Solution

$$4(x + 1) - 3(x - 3) = 25$$
$$4x + 4 - 3x + 9 = 25 \qquad \text{Multiply by 4 and } -3 \text{ to remove parentheses. Be careful of the signs. Remember that } (-3)(-3) = 9.$$

After removing the parentheses, it is important to combine like terms on each side of the equation. Do this before going on to isolate the variable.

$$x + 13 = 25 \qquad \text{Combine like terms.}$$
$$x + 13 - 13 = 25 - 13 \qquad \text{Subtract 13 from both sides to isolate the variable.}$$
$$x = 12 \qquad \text{The solution is 12.}$$

$$\text{\textit{Check.}} \quad 4(12 + 1) - 3(12 - 3) \overset{?}{=} 25 \qquad \text{Replace } x \text{ by 12.}$$
$$4(13) - 3(9) \overset{?}{=} 25 \qquad \text{Combine numbers inside parentheses.}$$
$$52 - 27 \overset{?}{=} 25 \qquad \text{Multiply.}$$
$$25 = 25 \ \checkmark \qquad \text{Simplify. It checks.}$$

Practice Problem 5 Solve for x and check your solution.
$$4x - (x + 3) = 12 - 3(x - 2) \quad x = \frac{7}{2}$$

NOTE TO STUDENT: *Fully worked-out solutions to all of the Practice Problems can be found at the back of the text starting at page SP-1*

Teaching Example 6 Solve for x.
$5(2 - x) = 8 - (3 + 6x)$

Ans: $x = -5$

EXAMPLE 6 Solve for x. $3(-x - 7) = -2(2x + 5)$

Solution $-3x - 21 = -4x - 10 \qquad \text{Remove parentheses. Watch the signs carefully.}$
$$-3x + 4x - 21 = -4x + 4x - 10 \qquad \text{Add } 4x \text{ to both sides.}$$
$$x - 21 = -10 \qquad \text{Simplify.}$$
$$x - 21 + 21 = -10 + 21 \qquad \text{Add 21 to both sides.}$$
$$x = 11 \qquad \text{The solution is 11.}$$

Check. The check is left to the student.

Practice Problem 6 Solve for x. $4(-2x - 3) = -5(x - 2) + 2 \quad x = -8$

In problems that involve decimals, great care should be taken. In some steps you will be multiplying decimal quantities, and in other steps you will be adding them.

EXAMPLE 7 Solve for x. $0.3(1.2x - 3.6) = 4.2x - 16.44$

Solution

$$0.36x - 1.08 = 4.2x - 16.44 \qquad \text{Remove parentheses.}$$
$$0.36x - 0.36x - 1.08 = 4.2x - 0.36x - 16.44 \qquad \text{Subtract } 0.36x \text{ from both sides.}$$
$$-1.08 = 3.84x - 16.44 \qquad \text{Combine like terms.}$$
$$-1.08 + 16.44 = 3.84x - 16.44 + 16.44 \qquad \text{Add } 16.44 \text{ to both sides.}$$
$$15.36 = 3.84x \qquad \text{Simplify.}$$
$$\frac{15.36}{3.84} = \frac{3.84x}{3.84} \qquad \text{Divide both sides by } 3.84.$$
$$4 = x \qquad \text{The solution is 4.}$$

Check. The check is left to the student.

Practice Problem 7 Solve for x.

$$0.3x - 2(x + 0.1) = 0.4(x - 3) - 1.1 \quad x = 1$$

Teaching Example 7 Solve for x.
$2.5(3x - 0.2) + 1.3 = 4x + 7.8$

Ans: $x = 2$

EXAMPLE 8 Solve for z and check. $2(3z - 5) + 2 = 4z - 3(2z + 8)$

Solution

$$6z - 10 + 2 = 4z - 6z - 24 \qquad \text{Remove parentheses.}$$
$$6z - 8 = -2z - 24 \qquad \text{Combine like terms.}$$
$$6z + 2z - 8 = -2z + 2z - 24 \qquad \text{Add } 2z \text{ to both sides.}$$
$$8z - 8 = -24 \qquad \text{Simplify.}$$
$$8z - 8 + 8 = -24 + 8 \qquad \text{Add 8 to both sides.}$$
$$8z = -16 \qquad \text{Simplify.}$$
$$\frac{8z}{8} = \frac{-16}{8} \qquad \text{Divide both sides by 8.}$$
$$z = -2 \qquad \text{Simplify. The solution is } -2.$$

Check.

$$2[3(-2) - 5] + 2 \overset{?}{=} 4(-2) - 3[2(-2) + 8] \qquad \text{Replace } z \text{ by } -2.$$
$$2[-6 - 5] + 2 \overset{?}{=} -8 - 3[-4 + 8] \qquad \text{Multiply.}$$
$$2[-11] + 2 \overset{?}{=} -8 - 3[4] \qquad \text{Simplify.}$$
$$-22 + 2 \overset{?}{=} -8 - 12$$
$$-20 = -20 \ \checkmark \qquad \text{It checks.}$$

Teaching Example 8 Solve and check.
$-4(1 - y) + 3(2y + 3) = 7y - 10$

Ans: $y = -5$

Practice Problem 8 Solve for z and check.

$$5(2z - 1) + 7 = 7z - 4(z + 3) \quad z = -2$$

Find the value of the variable that satisfies the equation in exercises 1–22. Check your solution. Answers that are not integers may be left in fractional form or decimal form.

1. $4x + 13 = 21$
$x = 2$

2. $7x + 4 = 53$
$x = 7$

3. $4x - 11 = 13$
$x = 6$

4. $5x - 11 = 39$
$x = 10$

5. $7x - 18 = -46$
$x = -4$

6. $6x - 23 = -71$
$x = -8$

7. $-4x + 17 = -35$
$x = 13$

8. $-6x + 25 = -83$
$x = 18$

9. $2x + 3.2 = 9.4$
$x = 3.1$

10. $4x + 4.6 = 9.2$
$x = 1.15$

11. $\frac{1}{4}x + 6 = 13$
$x = 28$

12. $\frac{1}{5}x + 2 = 9$
$x = 35$

13. $\frac{1}{3}x + 5 = -4$
$x = -27$

14. $\frac{1}{8}x - 3 = -9$
$x = -48$

15. $8x = 48 + 2x$
$x = 8$

16. $5x = 22 + 3x$
$x = 11$

17. $-6x = -27 + 3x$
$x = 3$

18. $-7x = -26 + 6x$
$x = 2$

19. $44 - 2x = 6x$ $x = \frac{11}{2}$

20. $21 - 5x = 7x$ $x = \frac{7}{4}$

21. $54 - 2x = -8x$
$-9 = x$

22. $72 - 4x = -12x$
$-9 = x$

In exercises 23–25, determine whether the given solution is correct. If it is not, find the solution.

23. Is 2 the solution for $2y + 3y = 12 - y$?
Yes

24. Is 4 the solution for $5y + 2 = 6y - 6 + y$?
Yes

25. Is 11 a solution for $7x + 6 - 3x = 2x - 5 + x$?
No; $x = -11$

26. Is -12 a solution for $9x + 2 - 5x = -8 + 5x - 2$?
No; $x = 12$

Solve for the variable. You may move the variable terms to the right or to the left.

27. $14 - 2x = -5x + 11$
$x = -1$

28. $8 - 3x = 7x + 8$
$x = 0$

29. $x - 6 = 8 - x$
$x = 7$

30. $-x + 12 = -4 + x$
$x = 8$

31. $0.6y + 0.8 = 0.1 - 0.1y$
$y = -1$

32. $1.1y + 0.3 = -1.3 + 0.3y$
$y = -2$

33. $5x - 9 = 3x + 23$
$x = 16$

34. $9x - 5 = 7x + 43$
$x = 24$

To Think About *First combine like terms on each side of the equation. Then solve for y by getting all the y-terms on the left. Then solve for y by getting all the y-terms on the right. Which approach is better?*

35. $-3 + 10y + 6 = 15 + 12y - 18$
$10y + 3 = 12y - 3$ $10y + 3 = 12y - 3$
$\quad -2y = -6$ $\qquad 6 = 2y$
$\qquad y = 3$ $\qquad 3 = y$

36. $7y + 21 - 5y = 5y - 7 + y$
$2y + 21 = 6y - 7$ $2y + 21 = 6y - 7$
$\quad -4y = -28$ $\qquad 28 = 4y$
$\qquad y = 7$ $\qquad 7 = y$

Remove the parentheses and solve for the variable. Check your solution. Answers that are not integers may be left in fractional form or decimal form.

37. $5(x + 3) = 35$
$5x + 15 = 35$
$5x = 20$
$x = 4$

38. $7(x + 3) = 28$
$7x + 21 = 28$
$7x = 7$
$x = 1$

39. $5(4x - 3) + 8 = -2$
$20x - 7 = -2$
$20x = 5$
$x = \dfrac{1}{4}$

40. $4(2x + 1) - 7 = 6 - 5$
$8x - 3 = 1$
$x = \dfrac{1}{2}$

41. $7x - 3(5 - x) = 10$
$7x - 15 + 3x = 10$
$10x = 25$
$x = 2.5 \text{ or } \dfrac{5}{2}$

42. $8x - 2(4 - x) = 14$
$8x - 8 + 2x = 14$
$10x = 22$
$x = 2.2 \text{ or } \dfrac{11}{5}$

43. $0.5x - 0.3(2 - x) = 4.6$
$0.5x - 0.6 + 0.3x = 4.6$
$x = 6.5 \text{ or } \dfrac{13}{2}$

44. $0.4x - 0.2(3 - x) = 1.8$
$0.4x - 0.6 + 0.2x = 1.8$
$x = 4$

45. $4(a - 3) + 2 = 2(a - 5)$
$4a - 12 + 2 = 2a - 10$
$a = 0$

46. $6(a + 3) - 2 = -4(a - 4)$
$6a + 18 - 2 = -4a + 16$
$a = 0$

47. $-2(x + 3) + 4 = 3(x + 4) + 2$
$-2x - 6 + 4 = 3x + 12 + 2$
$x = -\dfrac{16}{5}$

48. $-3(x + 5) + 2 = 4(x + 6) - 9$
$-3x - 15 + 2 = 4x + 24 - 9$
$x = -4$

49. $-3(y - 3y) + 4 = -4(3y - y) + 6 + 13y$
$6y + 4 = 5y + 6$
$y = 2$

50. $2(4x - x) + 6 = 2(2x + x) + 8 - x$
$6x + 6 = 5x + 8$
$x = 2$

Mixed Practice *Solve for the variable.*

51. $5.7x + 3 = 4.2x - 3$
$1.5x = -6$
$x = -4$

52. $4x - 3.1 = 5.3 - 3x$
$7x = 8.4$
$x = 1.2$

53. $5z + 7 - 2z = 32 - 2z$
$5z = 25$
$z = 5$

54. $8 - 7z + 2z = 20 + 5z$
$-12 = 10z$
$-1.2 = z$

55. $-0.3a + 1.4 = -1.2 - 0.7a$
$0.4a = -2.6$
$a = -6.5$

56. $-0.7b + 1.6 = -1.7 - 1.5b$
$0.8b = -3.3$
$b = -4.125$

57. $6x + 8 - 3x = 11 - 12x - 13$
$3x + 8 = -2 - 12x$
$15x = -10$
$x = -\dfrac{2}{3}$

58. $4 - 7x - 13 = 8x - 3 - 5x$
$-7x - 9 = 3x - 3$
$-10x = 6$
$x = -\dfrac{3}{5}$

59. $-3.5x + 1.3 = -2.7x + 1.5$
$-0.2 = 0.8x$
$-0.25 \text{ or } -\dfrac{1}{4} = x$

60. $2.8x - 0.9 = 5.2x - 3.3$
$2.4 = 2.4x$
$1 = x$

61. $5(4 + x) = 3(3x - 1) - 9$
$20 + 5x = 9x - 3 - 9$
$8 = x$

62. $6(3x - 1) = 4(2x + 5) - 6$
$18x - 6 = 8x + 20 - 6$
$x = 2$

63. $4x + 3.2 - 1.9x = 0.3x - 4.9$
$1.8x = -8.1$
$x = -4.5 \text{ or } -\dfrac{9}{2}$

64. $3x + 2 - 1.7x = 0.6x + 31.4$
$1.3x + 2 = 0.6x + 31.4$
$0.7x = 29.4$
$x = 42$

Cumulative Review *Simplify.*

65. **[1.7.2]** $-3y(2x + y) + 5(3xy - y^2)$
$9xy - 8y^2$

66. **[1.9.1]** $-\{2(x - 3) + 3[x - (2x - 5)]\}$
$x - 9$

67. **[0.7.1]** *Investments* On October 5, 2007, Marcella owned three different stocks: Motorola, General Mills, and CVS. Her portfolio contained the following:

18.5 shares of Motorola valued at $18.66 per share,
9.25 shares of General Mills valued at $57.56 per share, and
5.0 shares of CVS valued at $29.26 per share.

Find the market value of Marcella's stock holdings on October 5, 2007.
$1023.94

68. **[0.7.1]** *Employee Discount* Marvin works at Best Buy and gets a 10% discount on anything he buys from the store. A GPS navigation system that Marvin wishes to purchase costs $899 and is on sale at a 20% discount.

(a) What is the sale price if Marvin has a total discount of 30%? (Disregard sales tax.)
$629.30

(b) What is the price if Marvin gets a 10% discount on the 20% sale price? (Disregard sales tax.)
$647.28

Quick Quiz 2.3 Solve for the variable.

1. $7x - 6 = -4x - 10$
$x = -\dfrac{4}{11}$

2. $-3x + 6.2 = -5.8$
$x = 4$

3. $2(3x - 2) = 4(5x + 3)$ $\quad x = -\dfrac{8}{7}$

4. **Concept Check** Explain how you would solve the equation $3(x - 2) + 2 = 2(x - 4)$.
Answers may vary

Classroom Quiz 2.3 You may use these problems to quiz your students' mastery of Section 2.3.

Solve for the variable.

1. $8x + 3 = -12x - 7$ **Ans:** $x = -\dfrac{1}{2}$ **2.** $-7x + 3.5 = 16.8$ **Ans:** $x = -1.9$ **3.** $-3(4x - 2) = 2(6x + 1)$

Ans: $x = \dfrac{1}{6}$

Solving Equations with Fractions

Equations with fractions can be rather difficult to solve. This difficulty is simply due to the extra care we usually have to use when computing with fractions. The actual equation-solving procedures are the same, with fractions or without. To avoid unnecessary work, we transform the given equation with fractions to an equivalent equation that does not contain fractions. How do we do this? We multiply each side of the equation by the least common denominator of all the fractions contained in the equation. We then use the distributive property so that the LCD is multiplied by each term of the equation.

Student Learning Objective

After studying this section, you will be able to:

 Solve equations with fractions.

EXAMPLE 1 Solve for x. $\frac{1}{4}x - \frac{2}{3} = \frac{5}{12}x$

Solution First we find that the LCD = 12.

$$12\left(\frac{1}{4}x - \frac{2}{3}\right) = 12\left(\frac{5}{12}x\right) \qquad \text{Multiply both sides by 12.}$$

$$\left(\frac{12}{1}\right)\left(\frac{1}{4}\right)(x) - \left(\frac{12}{1}\right)\left(\frac{2}{3}\right) = \left(\frac{12}{1}\right)\left(\frac{5}{12}\right)(x) \qquad \text{Use the distributive property.}$$

$$3x - 8 = 5x \qquad \text{Simplify.}$$

$$3x + (-3x) - 8 = 5x + (-3x) \qquad \text{Add } -3x \text{ to both sides.}$$

$$-8 = 2x \qquad \text{Simplify.}$$

$$\frac{-8}{2} = \frac{2x}{2} \qquad \text{Divide both sides by 2.}$$

$$-4 = x \qquad \text{Simplify.}$$

Check.

$$\frac{1}{4}(-4) - \frac{2}{3} \stackrel{?}{=} \frac{5}{12}(-4)$$

$$-1 - \frac{2}{3} \stackrel{?}{=} -\frac{5}{3}$$

$$-\frac{3}{3} - \frac{2}{3} \stackrel{?}{=} -\frac{5}{3}$$

$$-\frac{5}{3} = -\frac{5}{3} \checkmark \qquad \text{It checks.}$$

Teaching Example 1 Solve for x.

$$\frac{1}{15}x - \frac{2}{3} = \frac{4}{5}x$$

Ans: $x = -\frac{10}{11}$

Practice Problem 1 Solve for x.

$$\frac{3}{8}x - \frac{3}{2} = \frac{1}{4}x \quad x = 12$$

NOTE TO STUDENT: Fully worked-out solutions to all of the Practice Problems can be found at the back of the text starting at page SP-1

In Example 1 we multiplied both sides of the equation by the LCD. However, most students prefer to go immediately to the second step and multiply each term by the LCD. This avoids having to write out a separate step using the distributive property.

Teaching Example 2 Solve for x and check your solution.

$$\frac{x}{4} + \frac{1}{2} = x - \frac{7}{4}$$

Ans: $x = 3$

Teaching Tip Point out that it is important to multiply each term on both sides of the equation by the least common denominator. A common student mistake is to multiply only the fractions by the LCD. Be sure students multiply every term by the LCD.

EXAMPLE 2   Solve for x and check your solution. $\dfrac{x}{3} + 3 = \dfrac{x}{5} - \dfrac{1}{3}$

Solution

$$15\left(\frac{x}{3}\right) + 15(3) = 15\left(\frac{x}{5}\right) - 15\left(\frac{1}{3}\right)$$ The LCD is 15. Use the multiplication principle to multiply each term by 15.

$$5x + 45 = 3x - 5$$ Simplify.

$$5x - 3x + 45 = 3x - 3x - 5$$ Subtract $3x$ from both sides.

$$2x + 45 = -5$$ Combine like terms.

$$2x + 45 - 45 = -5 - 45$$ Subtract 45 from both sides.

$$2x = -50$$ Simplify.

$$\frac{2x}{2} = \frac{-50}{2}$$ Divide both sides by 2.

$$x = -25$$ The solution is -25.

Check.

$$\frac{-25}{3} + 3 \overset{?}{=} \frac{-25}{5} - \frac{1}{3}$$

$$-\frac{25}{3} + \frac{9}{3} \overset{?}{=} -\frac{5}{1} - \frac{1}{3}$$

$$-\frac{16}{3} \overset{?}{=} -\frac{15}{3} - \frac{1}{3}$$

$$-\frac{16}{3} = -\frac{16}{3} \checkmark$$

Practice Problem 2   Solve for x and check your solution.

$$\frac{5x}{4} - 1 = \frac{3x}{4} + \frac{1}{2} \quad x = 3$$

Teaching Example 3 Solve for x.

$$\frac{5x + 2}{15} + \frac{1}{3} = -\frac{1}{5}$$

Ans: $x = -2$

EXAMPLE 3   Solve for x. $\dfrac{x + 5}{7} = \dfrac{x}{4} + \dfrac{1}{2}$

Solution

$$\frac{x}{7} + \frac{5}{7} = \frac{x}{4} + \frac{1}{2}$$ First we rewrite the left side as two fractions. This is actually multiplying $\frac{1}{7}(x + 5) = \frac{x}{7} + \frac{5}{7}$.

$$28\left(\frac{x}{7}\right) + 28\left(\frac{5}{7}\right) = 28\left(\frac{x}{4}\right) + 28\left(\frac{1}{2}\right)$$ We observe that the LCD is 28, so we multiply each term by 28.

$$4x + 20 = 7x + 14$$ Simplify.

$$4x - 4x + 20 = 7x - 4x + 14$$ Subtract $4x$ from both sides.

$$20 = 3x + 14$$ Combine like terms.

$$20 - 14 = 3x + 14 - 14$$ Subtract 14 from both sides.

$$6 = 3x$$ Combine like terms.

$$\frac{6}{3} = \frac{3x}{3}$$ Divide both sides by 3.

$$2 = x$$ The solution is 2.

Check.   The check is left to the student.

Practice Problem 3   Solve for x.

$$\frac{x + 6}{9} = \frac{x}{6} + \frac{1}{2} \quad x = 3$$

If a problem contains both parentheses and fractions, it is best to remove the parentheses first. Many students find it is helpful to have a written procedure to follow in solving these more involved equations.

PROCEDURE TO SOLVE EQUATIONS

1. Remove any parentheses.

2. If fractions exist, multiply all terms on both sides by the least common denominator of all the fractions.

3. Combine like terms if possible.

4. Add or subtract terms on both sides of the equation to get all terms with the variable on one side of the equation.

5. Add or subtract a constant value on both sides of the equation to get all terms not containing the variable on the other side of the equation.

6. Divide both sides of the equation by the coefficient of the variable.

7. Simplify the solution (if possible).

8. Check your solution.

Teaching Tip Fractional notation is encountered with this type of problem in two ways. Explain to students that both forms of notation are correct and that either may be used. In the first step of solving Example 4, we may write

$$\frac{x}{3} - \frac{2}{3} = \frac{x}{5} + \frac{4}{5} + 2.$$

An equivalent form for this equation is

$$\frac{1}{3}x - \frac{2}{3} = \frac{1}{5}x + \frac{4}{5} + 2.$$

Make sure students understand that these forms are equivalent.

Let's use each step in solving the next example.

EXAMPLE 4   Solve for x and check your solution.

$$\frac{1}{3}(x - 2) = \frac{1}{5}(x + 4) + 2$$

Teaching Example 4 Solve for x and check your solution.

$$\frac{1}{2}(2x - 27) + \frac{3x}{4} = \frac{1}{2}(x + 3)$$

Ans: $x = 12$

Solution

Step 1 $\dfrac{x}{3} - \dfrac{2}{3} = \dfrac{x}{5} + \dfrac{4}{5} + 2$ Remove parentheses.

Step 2 $15\left(\dfrac{x}{3}\right) - 15\left(\dfrac{2}{3}\right) = 15\left(\dfrac{x}{5}\right) + 15\left(\dfrac{4}{5}\right) + 15(2)$ Multiply by the LCD, 15.

$5x - 10 = 3x + 12 + 30$ Simplify.

Step 3 $5x - 10 = 3x + 42$ Combine like terms.

Step 4 $5x - 3x - 10 = 3x - 3x + 42$ Subtract $3x$ from both sides.

$2x - 10 = 42$ Simplify.

Step 5 $2x - 10 + 10 = 42 + 10$ Add 10 to both sides.

$2x = 52$ Simplify.

Step 6 $\dfrac{2x}{2} = \dfrac{52}{2}$ Divide both sides by 2.

Step 7 $x = 26$ Simplify the solution.

Step 8   *Check.* $\dfrac{1}{3}(26 - 2) \overset{?}{=} \dfrac{1}{5}(26 + 4) + 2$ Replace x by 26.

$\dfrac{1}{3}(24) \overset{?}{=} \dfrac{1}{5}(30) + 2$ Combine values within parentheses.

$8 \overset{?}{=} 6 + 2$ Simplify.

$8 = 8$ ✓   The solution is 26.

Teaching Tip With fractional equations that contain parentheses, some students prefer to remove parentheses first, while others prefer to clear the equation of fractions first. If the work is done carefully, either method is correct. It seems that for weaker students, it is advisable to always remove the parentheses first. It tends to minimize errors in solving the equation.

Practice Problem 4   Solve for x and check your solution.

$$\frac{1}{2}(x + 5) = \frac{1}{5}(x - 2) + \frac{1}{2} \quad x = -8$$

NOTE TO STUDENT: Fully worked-out solutions to all of the Practice Problems can be found at the back of the text starting at page SP-1

Remember that not every step will be needed in each problem. You can combine some steps as well, *as long as you are consistently obtaining the correct solution.* However, you are encouraged to write out every step as a way of helping you to avoid careless errors.

It is important to remember that when we write decimals, these numbers are really fractions written in a special way. Thus, $0.3 = \frac{3}{10}$ and $0.07 = \frac{7}{100}$. It is possible to take an equation containing decimals and to multiply each term by the appropriate value to obtain integer coefficients.

EXAMPLE 5 Solve for x. $0.2(1 - 8x) + 1.1 = -5(0.4x - 0.3)$

Solution

$0.2 - 1.6x + 1.1 = -2.0x + 1.5$	Remove parentheses.
$10(0.2) - 10(1.6x) + 10(1.1) = 10(-2.0x) + 10(1.5)$	Multiply each term by 10.
$2 - 16x + 11 = -20x + 15$	Multiplying by 10 moves the decimal point one place to the right.
$-16x + 13 = -20x + 15$	Simplify.
$-16x + 20x + 13 = -20x + 20x + 15$	Add $20x$ to both sides.
$4x + 13 = 15$	Simplify.
$4x + 13 - 13 = 15 - 13$	Subtract 13 from both sides.
$4x = 2$	Simplify.
$\dfrac{4x}{4} = \dfrac{2}{4}$	Divide both sides by 4.
$x = \dfrac{1}{2} \quad \text{or} \quad 0.5$	Simplify.

Check.

$$0.2[1 - 8(0.5)] + 1.1 \overset{?}{=} -5[0.4(0.5) - 0.3]$$
$$0.2[1 - 4] + 1.1 \overset{?}{=} -5[0.2 - 0.3]$$
$$0.2[-3] + 1.1 \overset{?}{=} -5[-0.1]$$
$$-0.6 + 1.1 \overset{?}{=} 0.5$$
$$0.5 = 0.5 \quad \checkmark$$

Practice Problem 5 Solve for x.

$$2.8 = 0.3(x - 2) + 2(0.1x - 0.3) \quad x = 8$$

TO THINK ABOUT: Does Every Equation Have One Solution? Actually, no. There are some rare cases where an equation has no solution at all. Suppose we try to solve the equation

$$5(x + 3) = 2x - 8 + 3x.$$

If we remove the parentheses and combine like terms, we have

$$5x + 15 = 5x - 8.$$

If we add $-5x$ to each side, we obtain

$$15 = -8.$$

Clearly this is impossible. There is no value of x for which these two numbers are equal. We would say this equation has **no solution.**

One additional surprise may happen. An equation may have an infinite number of solutions. Suppose we try to solve the equation

$$9x - 8x - 7 = 3 + x - 10.$$

If we combine like terms on each side, we have the equation

$$x - 7 = x - 7.$$

If we add $-x$ to each side, we obtain

$$-7 = -7.$$

Now this statement is always true, no matter what the value of x. We would say this equation has **an infinite number of solutions.**

In the To Think About exercises in this section, we will encounter some equations that have no solution or an infinite number of solutions.

Developing Your Study Skills

Taking Notes in Class

An important part of studying mathematics is taking notes. To take meaningful notes, you must be an active listener. Keep your mind on what the instructor is saying, and be ready with questions whenever you do not understand something.

If you have previewed the lesson material, you will be prepared to take good notes. The important concepts will seem somewhat familiar. If you frantically try to write all that the instructor says or copy all the examples done in class, you may find your notes nearly worthless when you are home alone. Write down *important* ideas and examples as the instructor lectures, making sure that you are listening and following the logic. Include any helpful hints or suggestions that your instructor gives you or refers to in your text.

In exercises 1–16, solve for the variable and check your answer. Noninteger answers may be left in fractional form or decimal form.

1. $\dfrac{1}{2}x + \dfrac{2}{3} = \dfrac{1}{6}$

$x = -1$

2. $\dfrac{1}{3} + \dfrac{5}{12}x = \dfrac{3}{4}$

$x = 1$

3. $\dfrac{2}{3}x = \dfrac{1}{15}x + \dfrac{3}{5}$

$x = 1$

4. $\dfrac{5}{21}x = \dfrac{2}{3}x - \dfrac{1}{7}$

$x = \dfrac{1}{3}$

5. $\dfrac{x}{2} + \dfrac{x}{5} = \dfrac{7}{10}$

$5x + 2x = 7$

$x = 1$

6. $\dfrac{x}{4} - \dfrac{x}{16} = \dfrac{3}{8}$

$4x - x = 6$

$x = 2$

7. $5 - \dfrac{1}{3}x = \dfrac{1}{12}x$

$60 - 4x = x$

$x = 12$

8. $15 - \dfrac{1}{2}x = \dfrac{1}{4}x$

$60 - 2x = x$

$x = 20$

9. $2 + \dfrac{y}{2} = \dfrac{3y}{4} - 3$

$8 + 2y = 3y - 12$

$20 = y$

10. $\dfrac{x}{3} - 1 = -\dfrac{1}{2} - x$

$2x - 6 = -3 - 6x$

$x = \dfrac{3}{8}$

11. $\dfrac{x - 3}{5} = 1 - \dfrac{x}{3}$

$3x - 9 = 15 - 5x$

$x = 3$

12. $\dfrac{y - 5}{4} = 1 - \dfrac{y}{5}$

$5y - 25 = 20 - 4y$

$y = 5$

13. $\dfrac{x + 3}{4} = \dfrac{x}{2} + \dfrac{1}{6}$

$x = \dfrac{7}{3}$

14. $\dfrac{x - 2}{3} = \dfrac{x}{12} + \dfrac{5}{4}$

$x = \dfrac{23}{3}$

15. $0.6x + 5.9 = 3.8$

$x = -\dfrac{7}{2}$ or -3.5

16. $-1.2x - 4.3 = 1.1$

$x = -\dfrac{9}{2}$ or -4.5

17. Is 4 a solution to $\dfrac{1}{2}(y - 2) + 2 = \dfrac{3}{8}(3y - 4)$?

Yes

18. Is 2 a solution to $\dfrac{1}{5}(y + 2) = \dfrac{1}{10}y + \dfrac{3}{5}$?

Yes

19. Is $\dfrac{5}{8}$ a solution to $\dfrac{1}{2}\left(y - \dfrac{1}{5}\right) = \dfrac{1}{5}(y + 2)$?

No

20. Is $\dfrac{13}{3}$ a solution to $\dfrac{y}{2} - \dfrac{7}{9} = \dfrac{y}{6} + \dfrac{2}{3}$?

Yes

Remove parentheses first. Then combine like terms. Solve for the variable. Noninteger answers may be left in fractional form or decimal form.

21. $\dfrac{3}{4}(3x + 1) = 2(3 - 2x) + 1$

$9x + 3 = 28 - 16x$

$x = 1$

22. $\dfrac{1}{4}(3x + 1) = 2(2x - 4) - 8$

$3x + 1 = 16x - 64$

$x = 5$

23. $2(x - 2) = \dfrac{2}{5}(3x + 1) + 2$

$10x - 20 = 6x + 2 + 10$

$x = 8$

24. $2(x - 4) = \dfrac{5}{6}(x + 6) - 6$

$12x - 48 = 5x - 6$

$x = 6$

25. $0.3x - 0.2(3 - 5x) = -0.5(x - 6)$

$0.3x - 0.6 + 1.0x = -0.5x + 3.0$

$1.8x = 3.6$

$x = 2$

26. $0.2(x + 1) + 0.5x = -0.3(x - 4)$

$2x + 2 + 5x = -3x + 12$

$10x = 10$

$x = 1$

27. $-8(0.1x + 0.4) - 0.9 = -0.1$

$-8x - 32 - 9 = -1$

$x = -5$

28. $0.6x + 1.5 = 0.3x - 0.6(2x + 5)$

$6x + 15 = 3x - 12x - 30$

$x = -3$

Mixed Practice *Solve. Noninteger answers may be left in fractional form or decimal form.*

29. $\frac{1}{3}(y + 2) = 3y - 5(y - 2)$

$y + 2 = 9y - 15y + 30$

$7y = 28$

$y = 4$

30. $\frac{1}{4}(y + 6) = 2y - 3(y - 3)$

$y + 6 = 8y - 12y + 36$

$y = 6$

31. $\frac{1 + 2x}{5} + \frac{4 - x}{3} = \frac{1}{15}$

$3 + 6x + 20 - 5x = 1$

$x = -22$

32. $\frac{1 + 3x}{2} + \frac{2 - x}{3} = \frac{5}{6}$

$3 + 9x + 4 - 2x = 5$

$x = -\frac{2}{7}$

33. $\frac{1}{5}(x + 3) = 2x - 3(2 - x) - 3$

$x + 3 = 10x - 30 + 15x - 15$

$48 = 24x$

$x = 2$

34. $\frac{2}{3}(x + 4) = 6 - \frac{1}{4}(3x - 2) - 1$

$8x + 32 = 72 - 9x + 6 - 12$

$17x = 34$

$x = 2$

35. $\frac{1}{3}(x - 2) = 3x - 2(x - 1) + \frac{16}{3}$

$x - 2 = 9x - 6x + 6 + 16$

$-24 = 2x$

$-12 = x$

36. $\frac{3}{4}(x - 2) + \frac{3}{5} = \frac{1}{5}(x + 1)$

$15x - 30 + 12 = 4x + 4$

$11x = 22$

$x = 2$

37. $\frac{4}{5}x - \frac{2}{3} = \frac{3x + 1}{2}$

$x = -\frac{5}{3}$

38. $\frac{4}{7}x + \frac{1}{3} = \frac{3x - 2}{14}$

$x = -\frac{4}{3}$

39. $0.4x - 0.5(2x + 3) = -0.7(x + 3)$

$x = -6$

40. $0.6(x + 0.1) = 2(0.4x - 0.2)$

$x = \frac{23}{10}$ or 2.3

To Think About *Solve. Be careful to examine your work to see if the equation may have no solution or an infinite number of solutions.*

41. $-1 + 5(x - 2) = 12x + 3 - 7x$

No solution

42. $x + 3x - 2 + 3x = -11 + 7(x + 2)$

No solution

43. $9(x + 3) - 6 = 24 - 2x - 3 + 11x$

Infinite number of solutions

44. $7(x + 4) - 10 = 3x + 20 + 4x - 2$

Infinite number of solutions

45. $-3(4x - 1) = 5(2x - 1) + 8$

$x = 0$

46. $11x - 8 = -4(x + 3) + 4$

$x = 0$

47. $3(4x + 1) - 2x = 2(5x - 3)$

No solution

48. $5(-3 + 4x) = 4(2x + 4) + 12x$

No solution

Cumulative Review

49. **[0.3.1]** Multiply. $\left(-3\frac{1}{4}\right)\left(5\frac{1}{3}\right)$

$-\dfrac{52}{3}$ or $-17\dfrac{1}{3}$

50. **[0.3.2]** Divide. $5\frac{1}{2} \div 1\frac{1}{4}$

$\dfrac{22}{5}$ or $4\dfrac{2}{5}$

51. **[0.7.5]** *Peregrine Falcons* Peregrine falcons are known for being the fastest birds on record, reaching horizontal speeds of 40–55 miles per hour. They also are one of the few animals in which the females are larger than the males. If female peregrine falcons are 30% larger than the males, and males measure 440–750 grams, what is the weight range for females?

572–975 g

52. **[1.8.2]** *Auditorium Seating* The seating area of an auditorium is shaped like a trapezoid, with front and back sides parallel. The front of the auditorium measures 88 feet across, the back of the auditorium measures 150 feet across, and the auditorium is 200 feet from front to back. If each seat requires a space that is 2.5 feet wide by 3 feet deep, how many seats will the auditorium hold? (This will only be an approximation because of the angled side walls. Round off to the nearest whole number.)

3173 seats

88 feet

200 feet

150 feet

Quick Quiz 2.4 Solve for the variable. When needed, leave your answer in fractional form. Reduce all fractions.

1. $\dfrac{3}{4}x + \dfrac{5}{12} = \dfrac{1}{3}x - \dfrac{1}{6}$ $x = -\dfrac{7}{5}$

2. $\dfrac{2}{3}x - \dfrac{3}{5} + \dfrac{7}{5}x + \dfrac{1}{3} = 1$ $x = \dfrac{19}{31}$

3. $\dfrac{2}{3}(x + 2) + \dfrac{1}{4} = \dfrac{1}{2}(5 - 3x)$ $x = \dfrac{11}{26}$

4. Concept Check Explain how you would solve the equation $\dfrac{x+5}{6} = \dfrac{x}{2} + \dfrac{3}{4}$. Answers may vary

Classroom Quiz 2.4 You may use these problems to quiz your students' mastery of Section 2.4.

Solve for the variable. You may leave your answers in fractional form. Reduce all fractions.

1. $\dfrac{3}{7}x + \dfrac{5}{14} = \dfrac{1}{2}x - \dfrac{2}{7}$ **Ans:** $x = 9$

2. $\dfrac{2}{5}x - \dfrac{3}{2} + \dfrac{4}{5} - \dfrac{7}{2}x = 2$ **Ans:** $x = -\dfrac{27}{31}$

3. $\dfrac{3}{4}(x + 3) - \dfrac{1}{2} = \dfrac{1}{8}(6 - 2x)$ **Ans:** $x = -1$

How are you doing with your homework assignments in Sections 2.1 to 2.4? Do you feel you have mastered the material so far? Do you understand the concepts you have covered? Before you go further in the textbook, take some time to do each of the following problems.

Solve for x. If the solution is not an integer, you may express your answer as a fraction or as a decimal.

2.1

1. $5 - 8 + x = -12$

2. $-2.8 + x = 4.7$

2.2

3. $-45 = -5x$

4. $12x - 6x = -48$

2.3

5. $-1.2x + 3.5 = 2.7$

6. $-14x + 9 = 2x + 7$

7. $14x + 2(7 - 2x) = 20$

8. $0.5(1.2x - 3.4) = -1.4x + 5.8$

9. $3(x + 6) = -2(4x - 1) + x$

2.4

10. $\dfrac{x}{3} + \dfrac{x}{4} = \dfrac{5}{6}$

11. $\dfrac{1}{4}(x + 3) = 4x - 2(x - 3)$

12. $\dfrac{1}{2}(x - 1) + 2 = 3(2x - 1)$

13. $\dfrac{1}{7}(7x - 14) - 2 = \dfrac{1}{3}(x - 2)$

14. $0.2(x - 3) = 4(0.2x - 0.1)$

1. $x = -9$

2. $x = \dfrac{15}{2}$ or 7.5

3. $x = 9$

4. $x = -8$

5. $x = \dfrac{2}{3}$

6. $x = \dfrac{1}{8}$

7. $x = \dfrac{3}{5}$

8. $x = \dfrac{15}{4}$ or 3.75

9. $x = -1.6$ or $-\dfrac{8}{5}$

10. $x = \dfrac{10}{7}$

11. $x = -3$

12. $x = \dfrac{9}{11}$

13. $x = 5$

14. $x = -\dfrac{1}{3}$

Now turn to page SA-5 for the answer to each of these problems. Each answer also includes a reference to the objective in which the problem is first taught. If you missed any of these problems, you should stop and review the Examples and Practice Problems in the referenced objective. A little review now will help you master the material in the upcoming sections of the text.

Student Learning Objective

After studying this section, you will be able to:

 Solve a formula for a specified variable.

Solving a Formula for a Specified Variable

Formulas are equations with one or more variables that are used to describe real-life situations. The formula describes the relationship that exists among the variables. For example, in the formula $d = rt$, distance (d) is equal to the rate of speed (r) multiplied by the time (t). We can use this formula to find distance if we know the rate and time. Sometimes, however, we are given the distance and the rate, and we are asked to find the time.

Teaching Example 1 A flight of 1550 miles from Boston to Dallas takes 4 hours. Find the airplane's average rate of speed for that flight.

Ans: 387.5 mi/hr

EXAMPLE 1 American Airlines recently scheduled a nonstop flight in a new Boeing 777 jet from Chicago to London. The approximate air distance traveled on the flight was 3975 miles. The average speed of the aircraft on the trip was 530 miles per hour. How many hours did it take the Boeing 777 to fly this trip? (*Source*: www.aa.com)

Solution

$d = rt$	Use the distance formula.
$3975 = 530t$	Substitute the known values for the variables.
$\dfrac{3975}{530} = \dfrac{530t}{530}$	Divide both sides of the equation by 530 to solve for t.
$7.5 = t$	Simplify.

It took the Boeing 777 about 7.5 hours to fly this trip from Chicago to London.

Practice Problem 1 The airlines are planning a nonstop flight from Chicago to Prague. This distance is approximately 4565 miles and the time of the flight is 8.3 hours. Find the average rate of speed for the flight. $r = 550$ mi/hr

NOTE TO STUDENT: Fully worked-out solutions to all of the Practice Problems can be found at the back of the text starting at page SP-1

If we have many problems that ask us to find the time given the distance and rate, it may be worthwhile to rewrite the formula in terms of time.

Teaching Example 2 Solve for w. $A = lw$

Ans: $\dfrac{A}{l} = w$

EXAMPLE 2 Solve for t. $d = rt$

Solution

$\dfrac{d}{r} = \dfrac{rt}{r}$	We want to isolate t. Therefore we divide both sides of the equation by the coefficient of t, which is r.
$\dfrac{d}{r} = t$	We have solved for the variable indicated.

Practice Problem 2 Einstein's equation relating energy E to mass m and the speed of light c is $E = mc^2$. Solve it for m.

$$m = \frac{E}{c^2}$$

A straight line can be described by an equation of the form $Ax + By = C$, where A, B, and C are real numbers and A and B are not both zero. We will study this in later chapters. Often it is useful to solve such an equation for the variable y in order to make graphing the equation easier.

EXAMPLE 3 Solve for y. $3x - 2y = 6$

Solution

$$-2y = 6 - 3x \qquad \text{We want to isolate the term containing } y,$$
$$\text{so we subtract } 3x \text{ from both sides.}$$

$$\frac{-2y}{-2} = \frac{6 - 3x}{-2} \qquad \text{Divide both sides by the coefficient of } y.$$

$$y = \frac{6}{-2} + \frac{-3x}{-2} \qquad \text{Rewrite the fraction on the right side as two fractions.}$$

$$y = \frac{3}{2}x - 3 \qquad \text{Simplify and reorder the terms on the right.}$$

This is known as the slope–intercept form of the equation of a line.

Practice Problem 3 Solve for y.

$$8 - 2y + 3x = 0 \qquad y = \frac{3}{2}x + 4$$

Teaching Example 3 Solve for y.
$5x - 2y = 12$

Ans: $y = \frac{5}{2}x - 6$

Our procedure for solving an equation can be rewritten to give us a procedure for solving a formula for a specified variable.

PROCEDURE TO SOLVE A FORMULA FOR A SPECIFIED VARIABLE

1. Remove any parentheses.

2. If fractions exist, multiply all terms on both sides by the LCD of all the fractions.

3. Combine like terms on each side if possible.

4. Add or subtract terms on both sides of the equation to get all terms with the desired variable on one side of the equation.

5. Add or subtract the appropriate quantities to get all terms that do *not* have the desired variable on the other side of the equation.

6. Divide both sides of the equation by the coefficient of the desired variable.

7. Simplify if possible.

▲ **EXAMPLE 4** A trapezoid is a four-sided figure with two parallel sides. If the parallel sides are a and b and the altitude is h, the area is given by

$$A = \frac{h}{2}(a + b).$$

In some houses, two of the sides of the roof are in the shape of a trapezoid. Solve this equation for a.

Teaching Example 4 Solve for x.

$$M = \frac{2}{5}(y + 4x)$$

Ans: $x = \dfrac{5M - 2y}{8}$ or $x = \dfrac{5M}{8} - \dfrac{y}{4}$

Solution
$$A = \frac{h}{2}(a + b)$$

$$A = \frac{ha}{2} + \frac{hb}{2} \qquad \text{Remove the parentheses.}$$

$$2(A) = 2\left(\frac{ha}{2}\right) + 2\left(\frac{hb}{2}\right) \qquad \text{Multiply all terms by the LCD of 2.}$$

$$2A = ha + hb \qquad \text{Simplify.}$$

$$2A - hb = ha \qquad \text{We want to isolate the term containing } a.$$
We want to isolate the term containing a.
Therefore, we subtract hb from both sides.

$$\frac{2A - hb}{h} = \frac{ha}{h} \qquad \text{Divide both sides by } h \text{ (the coefficient of } a\text{).}$$

$$\frac{2A - hb}{h} = a \qquad \text{The solution is obtained.}$$

Note: Although the solution is in simple form, it could be written in an alternative way. Since

$$\frac{2A - hb}{h} = \frac{2A}{h} - \frac{hb}{h} = \frac{2A}{h} - b,$$

we could also have written $\dfrac{2A}{h} - b = a$.

NOTE TO STUDENT: *Fully worked-out solutions to all of the Practice Problems can be found at the back of the text starting at page SP-1*

▲ **Practice Problem 4** Solve for x.

$$H = \frac{3}{4}(2y + x) \qquad x = \frac{4H - 6y}{3}$$

Developing Your Study Skills

Keep Trying

You may be one of those students who have had much difficulty with mathematics in the past and who are sure that you cannot do well in this course. Perhaps you are thinking, "I have never been any good at mathematics," or "I have always hated mathematics," or "Math always scares me," or "I have not had any math for so long that I have forgotten it all." You may have even picked up the label "math anxiety" and attached it to yourself. That is most unfortunate, and it is time for you to reprogram your thinking. Replace those negative thoughts with more positive ones. You need to say things like, "I will give this math class my best shot," or "I can learn mathematics if I work at it," or "I will do better than I have done in previous math classes." You will be pleasantly surprised at the difference a positive attitude makes!

We live in a highly technical world, and you cannot afford to give up on the study of mathematics. Dropping mathematics may prevent you from entering certain career fields that you may find interesting. You may not have to take math courses as high-level as calculus, but such courses as intermediate algebra, finite math, college algebra, and trigonometry may be necessary. Learning mathematics can open new doors for you.

Learning mathematics is a process that takes time and effort. You will find that regular study and daily practice are necessary to strengthen your skills and to help you to grow academically. This process will lead you toward success in mathematics. Then, as you become more successful, your confidence in your ability to do mathematics will grow.

Verbal and Writing Skills

1. Temperature The formula for calculating the temperature in degrees Fahrenheit when you know the temperature in degrees Celsius is $F = \frac{9}{5}C + 32$. Explain in your own words how you would solve this equation for C.

Multiply each term by 5. Then subtract 160 from both sides. Then divide each side by 9. We would obtain $\frac{5F - 160}{9} = C$.

▲2. Geometry The formula for finding the area of a trapezoid with an altitude of 9 meters and bases of b meters and c meters is given by the equation $A = \frac{9}{2}(b + c)$. Explain in your own words how you would solve this equation for b.

Use the distributive property to obtain $A = \frac{9}{2}b + \frac{9}{2}c$.

Multiply each term by 2. Subtract $9c$ from both sides. Then divide each side by 9. We would obtain $\frac{2A - 9c}{9} = b$ or $\frac{2A}{9} - c = b$.

Applications

▲3. Geometry The formula for the area of a triangle is $A = \frac{1}{2}ab$, where b is the *base* of the triangle and a is the *altitude* of the triangle.

(a) Use this formula to find the base of a triangle that has an area of 60 square meters and an altitude of 12 meters.
$b = 10$ m

(b) Use this formula to find the altitude of a triangle that has an area of 88 square meters and a base of 11 meters.
$a = 16$ m

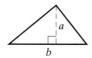

4. Simple Interest The formula for calculating simple interest is $I = Prt$, where P is the *principal* (amount of money invested), r is the *rate* at which the money is invested, and t is the *time*.

(a) Use this formula to find how long it would take to earn $720 in interest on an investment of $3000 at the rate of 6%.
$720 = 3000(0.06)t$
$t = 4$ yr

(b) Use this formula to find the rate of interest if $5000 earns $400 interest in 2 years.
$400 = 5000(2)r$
$r = 0.04$ or 4%

(c) Use this formula to find the amount of money invested if the interest earned was $120 and the rate of interest was 5% over 3 years.
$120 = 0.05(3)P$
$P = \$800$

5. The equation $4x + 3y = 18$ describes a line and is written in standard form.

(a) Solve for the variable y.
$y = \frac{18 - 4x}{3}$ or $y = -\frac{4}{3}x + 6$

(b) Use this result to find y with $x = -3$.
$y = 10$

6. The equation $-3x + 5y = 15$ describes a line and is written in standard form.

(a) Solve for the variable x.
$x = \frac{15 - 5y}{-3} = \frac{5y - 15}{3}$ or $x = \frac{5}{3}y - 5$

(b) Use this result to find x with $y = -6$.
$x = -15$

In each formula or equation, solve for the variable indicated.

Area of a triangle

▲7. $A = \frac{1}{2}bh$ Solve for b.
$b = \frac{2A}{h}$

▲8. $A = \frac{1}{2}bh$ Solve for h.
$h = \frac{2A}{b}$

Simple interest formula

9. $I = Prt$ Solve for P.
$P = \frac{I}{rt}$

10. $I = Prt$ Solve for t.
$t = \frac{I}{Pr}$

Slope–intercept form of a line

11. $y = mx + b$ Solve for m.

$m = \dfrac{y - b}{x}$

12. $y = mx + b$ Solve for b.

$b = y - mx$

Standard form of a line

13. $8x - 12y = 24$ Solve for y.

$y = \dfrac{2}{3}x - 2$

14. $9x - 15y = -12$ Solve for y.

$y = \dfrac{3}{5}x + \dfrac{4}{5}$

Slope–intercept form of a line

15. $y = -\dfrac{2}{3}x + 4$ Solve for x.

$x = -\dfrac{3}{2}y + 6$

16. $y = -\dfrac{3}{4}x + 9$ Solve for x.

$x = -\dfrac{4}{3}y + 12$

Standard form of a line

17. $ax + by = c$ Solve for y.

$y = \dfrac{c - ax}{b}$

18. $ax + by = c$ Solve for x.

$x = \dfrac{c - by}{a}$

Area of a circle

▲ **19.** $A = \pi r^2$ Solve for r^2.

$r^2 = \dfrac{A}{\pi}$

Surface area of a sphere

▲ **20.** $s = 4\pi r^2$ Solve for r^2.

$r^2 = \dfrac{s}{4\pi}$

Gravitational acceleration

21. $g = \dfrac{GM}{r^2}$ Solve for r^2.

$r^2 = \dfrac{GM}{g}$

Volume of sphere

▲ **22.** $V = \dfrac{4}{3}\pi r^3$ Solve for r^3.

$r^3 = \dfrac{3V}{4\pi}$

Simple interest formula

23. $A = P(1 + rt)$ Solve for t.

$t = \dfrac{A - P}{Pr}$

Area of a trapezoid

▲ **24.** $A = \dfrac{1}{2}a(b_1 + b_2)$ Solve for b_1.

$\dfrac{2A - ab_2}{a} = b_1$

Surface area of a right circular cylinder

▲ **25.** $S = 2\pi rh + 2\pi r^2$ Solve for h.

$h = \dfrac{S - 2\pi r^2}{2\pi r}$

26. $H = 5as + 10a^2$ Solve for s.

$\dfrac{H - 10a^2}{5a} = s$

Kinetic energy

27. $K = \dfrac{1}{2}mv^2$ Solve for m.

$m = \dfrac{2K}{v^2}$

28. $K = \dfrac{1}{2}mv^2$ Solve for v^2.

$v^2 = \dfrac{2K}{m}$

Volume of a rectangular prism

▲ **29.** $V = LWH$ Solve for L.

$L = \dfrac{V}{WH}$

▲ **30.** $V = LWH$ Solve for H.

$H = \dfrac{V}{LW}$

Volume of a cone

▲ **31.** $V = \dfrac{1}{3}\pi r^2 h$ Solve for r^2.

$r^2 = \dfrac{3V}{\pi h}$

▲ **32.** $V = \dfrac{1}{3}\pi r^2 h$ Solve for h.

$h = \dfrac{3V}{\pi r^2}$

Perimeter of a rectangle

▲ **33.** $P = 2L + 2W$ Solve for W.

$W = \dfrac{P - 2L}{2}$

Finding the nth term of an arithmetic sequence

34. $N = F + d(n - 1)$ Solve for d.

$d = \dfrac{N - F}{n - 1}$

Pythagorean Theorem

▲ **35.** $c^2 = a^2 + b^2$ Solve for a^2.

$a^2 = c^2 - b^2$

▲ **36.** $c^2 = a^2 + b^2$ Solve for b^2.

$b^2 = c^2 - a^2$

Temperature conversion formulas

37. $F = \dfrac{9}{5}C + 32$ Solve for C.

$C = \dfrac{5(F - 32)}{9}$ or $C = \dfrac{5F - 160}{9}$

38. $C = \dfrac{5}{9}(F - 32)$ Solve for F.

$F = \dfrac{9}{5}C + 32$ or $F = 1.8C + 32$

Ohm's law

39. $P = \dfrac{E^2}{R}$ Solve for R. $R = \dfrac{E^2}{P}$

Density formula (density, mass, and volume)

40. $d = \dfrac{m}{v}$ Solve for v. $v = \dfrac{m}{d}$

Area of a sector of a circle

▲ **41.** $A = \dfrac{\pi r^2 S}{360}$ Solve for S. $S = \dfrac{360A}{\pi r^2}$

▲ **42.** $A = \dfrac{\pi r^2 S}{360}$ Solve for r^2. $r^2 = \dfrac{360A}{\pi S}$

Applications

▲ **43.** *Geometry* Use the result you obtained in exercise 33 to solve the following problem. A farmer has a rectangular field with a perimeter of 5.8 miles and a length of 2.1 miles. Find the width of the field.
0.8 mi

44. Use the result you obtained in exercise 34 to solve the following problem. If you are given an arithmetic sequence where the first term (F) is 6, 3 is the value of n, and the third (nth) term is 24 (so N is 24), what is the difference (d)?
9

▲ **45.** *Geometry* Use the result you obtained in exercise 29 to solve the following problem. The foundation of a house is in the shape of a rectangular solid. The volume held by the foundation is 5940 cubic feet. The height of the foundation is 9 feet and the width is 22 feet. What is the length of the foundation?
30 ft

▲ **46.** *Geometry* Use the result you obtained in exercise 30 to solve the following problem. The fish tank at the Mandarin Danvers Restaurant is in the shape of a rectangular solid. The volume held by the tank is 3024 cubic inches. The length of the tank is 18 inches while the width of the tank is 14 inches. What is the height of the tank?
12 in.

47. *Tourism* The number of foreign visitors measured in thousands (V) admitted to the United States for a pleasure trip for any given year can be predicted by the equation $V = 1100x + 7050$, where x is the number of years since 1985. For example, if $x = 20$ (this would be the year 2005), the predicted number of visitors in thousands would be $1100(20) + 7050 = 29,050$. Thus we would predict that in 2005, a total of 29,050,000 visitors came to the United States for a pleasure trip. (*Source:* U.S. Immigration and Naturalization Service.)

(a) Solve this equation for x.
(b) Use the result of your answer in (a) to find the year in which the number of visitors will be predicted to be 36,750,000. (*Hint:* Let $V = 36,750$ in your answer for (a).)

(a) $\dfrac{V - 7050}{1100} = x$ **(b)** 2012

48. *Tourism* The number of foreign visitors from Europe measured in thousands (E) admitted to the United States from Europe for a pleasure trip for any given year can be predicted by the equation $E = 480x + 2400$, where x is the number of years since 1985. For example, if $x = 25$ (this would be the year 2010), the predicted number of visitors in thousands would be $480(25) + 2400 = 14,400$. Thus we would predict that in 2010, a total of 14,400,000 visitors from Europe will come to the United States for a pleasure trip. (*Source:* U.S. Immigration and Naturalization Service.)

(a) Solve this equation for x.
(b) Use the result of your answer in (a) to find the year in which the number of visitors from Europe will be predicted to be 13,920,000. (*Hint:* Let $E = 13,920$ in your answer for (a).)

(a) $\dfrac{E - 2400}{480} = x$ **(b)** 2009

To Think About

▲ **49.** In the formula $A = \frac{1}{2}ab$, if b doubles, what is the effect on A?   *A* doubles

▲ **50.** In the formula $A = \frac{1}{2}ab$, if both a and b double, what is the effect on A?   *A* increases four times

▲ **51.** In $A = \pi r^2$, if r doubles, what is the effect on A?   *A* increases four times

▲ **52.** In $A = \pi r^2$, if r is halved, what is the effect on A?   *A* becomes $\frac{1}{4}A$

Cumulative Review

53. **[0.5.4]** Find 20% of $80. $16

54. **[0.5.4]** What is 0.5% of 200?   1

55. **[0.3.1]** *Electronic Games* A very popular handheld electronic game requires $3\frac{1}{4}$ square feet of a certain type of durable plastic in the manufacturing process. How many square feet of durable plastic does this company need to make 12,000 handheld games?   39,000 ft^2

56. **[0.2.4]** *Spotlight Rental* The Superstar Lighting Company rents out giant spotlights that shine up in the sky to mark the location of special events, such as the opening of a movie, a major sports play-off game, or a huge sales event at an auto dealership. A giant spotlight was used for $4\frac{1}{3}$ hours on Saturday, $2\frac{3}{4}$ hours on Tuesday, and $3\frac{1}{2}$ hours on Wednesday. What was the total number of hours that the spotlight was in use? $10\frac{7}{12}$ hr

Quick Quiz 2.5

1. Solve for x.
$A = 3x + 2w$ $x = \dfrac{A - 2w}{3}$

2. Solve for a.
$A = \dfrac{1}{3}h(a + b)$ $a = \dfrac{3A - hb}{h}$

3. Solve for y.
$2ax(3 - y) = 2axy - 5$ $y = \dfrac{6ax + 5}{4ax}$

4. **Concept Check** Explain how you would solve for x in the equation $y = \frac{3}{8}x - 9$.   Answers may vary

Classroom Quiz 2.5   You may use these problems to quiz your students' mastery of Section 2.5.

1. Solve for y.
$A = 4x + 5y$ **Ans:** $y = \dfrac{A - 4x}{5}$

2. Solve for b.
$A = \dfrac{1}{4}h(a + b)$ **Ans:** $b = \dfrac{4A - ha}{h}$

3. Solve for y.
$4ax(5 - y) = 2axy - 3$
Ans: $y = \dfrac{20ax + 3}{6ax}$

① Interpreting Inequality Statements

We frequently speak of one value being greater than or less than another value. We say that "5 is less than 7" or "9 is greater than 4." These relationships are called **inequalities.** We can write inequalities in mathematics by using symbols. We use the symbol < to represent the words "**is less than.**" We use the symbol > to represent the words "**is greater than.**"

Statement in Words	Statement in Algebra
5 is less than 7.	$5 < 7$
9 is greater than 4.	$9 > 4$

Note: "5 is less than 7" and "7 is greater than 5" have the same meaning. Similarly, $5 < 7$ and $7 > 5$ have the same meaning. They represent two equivalent ways of describing the same relationship between the two numbers 5 and 7.

We can better understand the concept of inequality if we examine a number line.

We say that one number is greater than another if it is to the right of the other on the number line. Thus $7 > 5$, since 7 is to the right of 5.

What about negative numbers? We can say "-1 is greater than -3" and write it in symbols as $-1 > -3$ because we know that -1 lies to the right of -3 on the number line.

EXAMPLE 1 In each statement, replace the question mark with the symbol < or >.

(a) $3 \; ? \; -1$ **(b)** $-2 \; ? \; 1$ **(c)** $-3 \; ? \; -4$ **(d)** $0 \; ? \; 3$ **(e)** $-3 \; ? \; 0$

Solution

(a) $3 > -1$ Use >, since 3 is to the right of -1 on the number line.

(b) $-2 < 1$ Use <, since -2 is to the left of 1. (Or equivalently, we could say that 1 is to the right of -2.)

(c) $-3 > -4$ Note that -3 is to the right of -4.

(d) $0 < 3$

(e) $-3 < 0$

Practice Problem 1 In each statement, replace the question mark with the symbol < or >.

(a) $7 \; ? \; 2$ **(b)** $-2 \; ? \; -4$ **(c)** $-1 \; ? \; 2$ **(d)** $-8 \; ? \; -5$ **(e)** $0 \; ? \; -2$ **(f)** $5 \; ? \; -3$

   >   >   <   <   >   >

② Graphing an Inequality on a Number Line

Sometimes we will use an inequality to express the relationship between a variable and a number. $x > 3$ means that x could have the value of *any number* greater than 3.

Any number that makes an inequality true is called a **solution** of the inequality. The set of all numbers that make the inequality true is called the **solution set.** A picture that represents all of the solutions of an inequality is called a **graph** of the inequality. The inequality $x > 3$ can be graphed on a number line as follows:

Case 1

Student Learning Objectives

After studying this section, you will be able to:

① **Interpret inequality statements.**

② **Graph an inequality on a number line.**

③ **Translate English phrases into algebraic statements.**

Teaching Tip As a sidelight, you may want to point out that it is possible to use an inequality to express relationships between irrational and rational numbers. Draw attention to the decimal approximation of π. Although π is a nonterminating, nonrepeating decimal, we can say that $\pi > 3.14$ and show that this is true with a number line.

We will study irrational numbers extensively in Chapter 9. In this section we limit our study of inequalities to inequalities between rational numbers.

Teaching Example 1 In each statement, replace the question mark with the symbol < or >.

(a) $-5 \; ? \; 2$ **(b)** $4 \; ? \; -12$

(c) $-8 \; ? \; -7$ **(d)** $-4 \; ? \; 0$

(e) $0 \; ? \; 4$

Ans: (a) $-5 < 2$ **(b)** $4 > -12$

(c) $-8 < -7$ **(d)** $-4 < 0$

(e) $0 < 4$

Teaching Tip Although almost all elementary algebra books use this notation, it is not the only one in practice. There are several variations. In upper-level math courses the interval notations are usually employed. You may want to mention to your students that they may use an alternative notation if it is clear and mathematically acceptable. For example, $x \geq 2$ may be graphed as

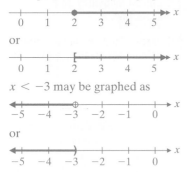

or

$x < -3$ may be graphed as

or

Teaching Example 2 State each mathematical relationship in words and then graph it.

(a) $x \geq -1$ (b) $x < 1$

(c) $x \leq 0$ (d) $\dfrac{5}{2} < x$

Ans:

(a) x is greater than or equal to -1

(b) x is less than 1

(c) x is less than or equal to 0

(d) $\dfrac{5}{2}$ is less than x *or* x is greater than $\dfrac{5}{2}$

Note that all of the points to the right of 3 are shaded. The open circle at 3 indicates that we do not include the point for the number 3.

Similarly, we can graph $x < -2$ as follows:

Case 2

Note that all of the points to the left of -2 are shaded.

Sometimes a variable will be either greater than or equal to a certain number. In the statement "x is greater than or equal to 3," we are implying that x could have the value of 3 or any number greater than 3. We write this as $x \geq 3$. We graph it as follows:

Case 3

Note that the closed circle at 3 indicates that we *do* include the point for the number 3.

Similarly, we can graph $x \leq -2$ as follows:

Case 4

Note: Be careful you do not confuse ∘———▸ with •———▸. It is important to decide if you need an open circle or a closed one. Case 1 and Case 2 use open circles. Case 3 and Case 4 use closed circles.

EXAMPLE 2 State each mathematical relationship in words and then graph it.

(a) $x < -2$ (b) $-3 < x$ (c) $x \geq -2$ (d) $x \leq -\dfrac{1}{2}$

Solution

(a) We state that "x is less than -2."

(b) We can state that "-3 is less than x" or, equivalently, that "x is greater than -3." Be sure you see that $-3 < x$ is equivalent to $x > -3$. Although both statements are correct, we *usually write the variable first* in a simple inequality containing a variable and a numerical value.

(c) We state that "x is greater than or equal to -2."

(d) We state that "x is less than or equal to $-\frac{1}{2}$."

Practice Problem 2 State each mathematical relationship in words and then graph it on the given number line.

(a) $x > 5$ x is greater than 5   (b) $x \leq -2$ x is less than or equal to -2

(c) $3 > x$ x is less than 3   (d) $x \geq -\dfrac{3}{2}$ x is greater than or equal to $-\dfrac{3}{2}$

Translating English Phrases into Algebraic Statements

We can translate many everyday situations into algebraic statements with an un-known value and an inequality symbol. This is the first step in solving word prob-lems using inequalities.

EXAMPLE 3 Translate each English statement into an algebraic statement.

(a) The police on the scene said that the car was traveling more than 80 miles per hour. (Use the variable s for speed.)

(b) The owner of the trucking company said that the payload of a truck must never exceed 4500 pounds. (Use the variable p for payload.)

Solution

(a) Since the speed must be greater than 80, we have $s > 80$.

(b) If the payload of the truck can never exceed 4500 pounds, then the payload must be always less than or equal to 4500 pounds. Thus we write $p \le 4500$.

Practice Problem 3 Translate each English statement into an algebraic statement.

(a) During the drying cycle, the temperature inside the clothes dryer must never exceed 180 degrees Fahrenheit. (Use the variable t for temperature.) $t \le 180$

(b) The bank loan officer said that the total consumer debt incurred by Wally and Mary must be less than $15,000 if they want to qualify for a mortgage to buy their first home. (Use the variable d for debt.) $d < 15,000$

Teaching Example 3 Translate each English statement into an algebraic statement.

(a) Only people who are more than 4 feet tall may ride the roller coaster. (Use h.)

(b) The total weight of the passengers of an elevator may not be more than 1200 pounds. (Use w.)

Ans: (a) $h > 4$ **(b)** $w \le 1200$

NOTE TO STUDENT: Fully worked-out solutions to all of the Practice Problems can be found at the back of the text starting at page SP-1

Developing Your Study Skills

Getting Organized for an Exam

Studying adequately for an exam requires careful preparation. Begin early so that you will be able to spread your review over several days. Even though you may still be learning new material at this time, you can be reviewing concepts previously learned in the chapter. Giving yourself plenty of time for review will take the pressure off. You need this time to process what you have learned and to tie concepts together.

Adequate preparation enables you to feel confident and to think clearly with less tension and anxiety.

2.6 EXERCISES

 MyMathLab®

 MathXL
PRACTICE

WATCH

DOWNLOAD

READ

REVIEW

Verbal and Writing Skills

1. Is the statement $5 > -6$ equivalent to the statement $-6 < 5$? Why?

Yes; both statements imply 5 is to the right of -6 on a number line

2. Is the statement $-8 < -3$ equivalent to the statement $-3 > -8$? Why?

Yes; both statements imply -3 is to the right of -8 on a number line

Replace the ? by $<$ or $>$.

3. $9 \: ? \: -3$ $>$

4. $-2 \: ? \: 5$ $<$

5. $-4 \: ? \: -2$ $<$

6. $-3 \: ? \: -6$ $>$

7. $\dfrac{3}{5} \: ? \: \dfrac{4}{7}$ $>$

8. $\dfrac{4}{6} \: ? \: \dfrac{7}{9}$ $<$

9. $-4.2 \: ? \: 3.5$ $<$

10. $-2.6 \: ? \: 7.5$ $<$

11. $-\dfrac{10}{3} \: ? \: -3$ $<$

12. $-6 \: ? \: -\dfrac{17}{3}$ $<$

13. $-\dfrac{5}{8} \: ? \: -\dfrac{3}{5}$ $<$

14. $-\dfrac{2}{3} \: ? \: -\dfrac{3}{4}$ $>$

Which is greater?

 15. $\dfrac{123}{4986}$ or 0.0247? 0.0247

16. $\dfrac{997}{6384}$ or 0.15613? $\dfrac{997}{6384}$

Graph each inequality on the number line.

17. $x > 7$

18. $x < 1$

19. $x \geq -6$

20. $x \leq -2$

21. $x < -\dfrac{1}{4}$

22. $x \leq -\dfrac{3}{2}$

23. $x \leq -5.3$

24. $x > -3.5$

25. $25 < x$

26. $35 \geq x$

Translate each graph to an inequality using the variable x.

27.

$x \geq -\dfrac{2}{3}$

28.

$x > -4.5$

29.

$x < -20$

30.

$x \leq \dfrac{5}{2}$

31.

3.2 3.3 3.4 3.5 3.6 3.7 3.8 3.9 4.0 4.1 4.2

$x \le 3.7$

32.

-25 -20 -15 -10 -5   0   5   10   15   20   25

$x > -10$

Translate each English statement into an inequality.

33. ***Full-Time Student*** At Normandale Community College the number of credits a student takes per semester must not be less than 12 to be considered full time. (Use the variable *c* for credits.) $c \ge 12$

34. ***BMI Category*** A person is considered underweight if his or her BMI (body mass index) measurement is smaller than 18.5. (Use the variable *B* for BMI.) $B < 18.5$

35. ***Height Limit*** In order for you to be allowed to ride the roller coaster at the theme park, your height must be at least 48 inches. (Use *h* for height.) $h \ge 48$

36. ***Boxing Category*** To box in the featherweight category, your weight must not exceed 126 pounds. (Use *w* for weight.) $w \le 126$

Cumulative Review

37. **[0.5.5]** ***Percent Accepted*** For the most coveted graduate study positions, only 16 out of 800 students are accepted. What percent are accepted? 2%

38. **[0.4.2]** Write the fraction $\frac{3}{8}$ as a percent. 37.5%

39. **[0.7.5]** ***Dry Cleaning*** The Apex Dry Cleaners raised the cost of dry cleaning a sports coat from $8 to $10. If the same percent of increase applies to all other dry cleaning also, what will the new price for dry cleaning a men's suit be if the old price was $12? $15

40. **[1.2.1]** ***Radar Picture*** A radar picture is taken of a portion of the Amazon rain forest. The radar detects that a cliff is 414 feet above sea level and that a giant tree trunk is directly below it in a body of water. The tree trunk is 81 feet below sea level. How far is the top of the cliff from the tree trunk? 495 ft

Quick Quiz 2.6 Graph each inequality on the given number line.

1. $x > 5$

0   1   2   3   4   5   6   7   8   9   10

2. $x \le -3.5$

-7 -6 -5 -4 -3 -2 -1   0   1   2   3

3. Translate the following graph to an inequality using the variable *x*. $x \ge -2$

-5 -4 -3 -2 -1   0   1   2   3   4   5

4. **Concept Check** Explain the difference between $12 < x$ and $x > 12$. Would the graph of these inequalities be the same or different? Answers may vary

Classroom Quiz 2.6 You may use these problems to quiz your students' mastery of Section 2.6.

Graph each inequality on a number line.

1. $x < 4$

Ans:

-5 -4 -3 -2 -1   0   1   2   3   4   5

2. $x \ge -2.5$

Ans:

-5 -4 -3 -2 -1   0   1   2   3   4   5

3. Translate the following graph to an inequality using the variable *x*. **Ans:** $x \le -6$

-9 -8 -7 -6 -5 -4 -3 -2 -1   0   1

After studying this section, you will be able to:

 1 Solve and graph inequalities on a number line.

Teaching Example 1 Perform the given operation and write a new inequality.

(a) $-5 < 2$ Add 7 to both sides.

(b) $4 > -12$ Subtract 1 from both sides.

(c) $6 > -12$ Multiply both sides by 4.

(d) $9 > 3$ Divide both sides by 3.

Ans: (a) $2 < 9$

(b) $3 > -13$

(c) $24 > -48$

(d) $3 > 1$

NOTE TO STUDENT: Fully worked-out solutions to all of the Practice Problems can be found at the back of the text starting at page SP-1

1 **Solving and Graphing Inequalities on a Number Line**

As stated in Section 2.6, the possible values that make an inequality true are called its **solutions.** Thus, when we **solve an inequality,** we are finding *all* the values that make it true. To solve an inequality, we simplify it to the point where we can clearly see all possible values for the variable. We've solved equations by adding, subtracting, multiplying by, and dividing by a particular value on both sides of the equation. Here we perform similar operations with inequalities with one important exception. We'll show some examples so that you can see how these operations can be used with inequalities just as with equations.

We will first examine the pattern that occurs when we perform these operations *with a positive value* on both sides of an inequality.

EXAMPLE 1

Notice that the inequality symbol remains the same when these operations are performed.

Practice Problem 1 Perform the given operation and write a new inequality.

(a) $9 > 6$ Add 4 to each side. $13 > 10$

(b) $-2 < 5$ Subtract 3 from both sides. $-5 < 2$

(c) $1 > -3$ Multiply both sides by 2. $2 > -6$

(d) $10 < 15$ Divide both sides by 5. $2 < 3$

Now let us examine what happens when we perform these operations *with a negative value.*

What happens to the inequality sign when we multiply both sides by a negative number? Since -8 is to the right of -12 on a number line, we know that the new inequality should be $-8 > -12$ if we want the statement to remain true. Notice how we reverse the direction of the inequality from $<$ (less than) to $>$ (greater than). Thus we have the following.

$$4 < 6 \longrightarrow \text{Multiply both sides by } -2. \longrightarrow -8 > -12$$

The same thing happens when we divide by a negative number. The inequality is reversed from $<$ to $>$. We know this, since -2 is to the right of -3 on a number line.

$$4 < 6 \longrightarrow \text{Divide both sides by } -2. \longrightarrow -2 > -3$$

EXAMPLE 2

Original Inequality			New Inequality
(a) $-2 < -1$	$\longrightarrow$ Multiply both sides by -3.	$\longrightarrow$	$6 > 3$
(b) $0 > -4$	$\longrightarrow$ Divide both sides by -2.	$\longrightarrow$	$0 < 2$
(c) $8 \geq 4$	$\longrightarrow$ Divide both sides by -4.	$\longrightarrow$	$-2 \leq -1$

Notice that we perform the arithmetic with signed numbers just as we always do. But the new inequality signs are reversed (from those of the original inequalities). *Whenever both sides of an inequality are multiplied or divided by a negative quantity, the direction of the inequality is reversed.*

Practice Problem 2

(a) $7 > 2$ Multiply each side by -2. $-14 < -4$

(b) $-3 < -1$ Multiply each side by -1. $3 > 1$

(c) $-10 \geq -20$ Divide each side by -10. $1 \leq 2$

(d) $-15 \leq -5$ Divide each side by -5. $3 \geq 1$

Teaching Example 2 Perform the given operation and write a new inequality.

(a) $-5 < 2$ Multiply both sides by -4.

(b) $4 > -12$ Divide both sides by -1.

(c) $9 > 3$ Divide both sides by -3.

Ans: (a) $20 > -8$

    (b) $-4 < 12$

    (c) $-3 < -1$

NOTE TO STUDENT: *Fully worked-out solutions to all of the Practice Problems can be found at the back of the text starting at page SP-1*

PROCEDURE FOR SOLVING INEQUALITIES

You may use the same procedures to solve inequalities that you did to solve equations *except* that the direction of an inequality is *reversed* if you *multiply* or *divide* both sides *by a negative number.*

It may be helpful to think over quickly what we have discussed here. The inequalities remain the same when we add a number to both sides or subtract a number from both sides of the equation. The inequalities remain the same when we multiply both sides by a positive number or divide both sides by a positive number.

RULES WHEN INEQUALITIES REMAIN THE SAME

For all real numbers a, b, and c:

1. If $a > b$, then $a + c > b + c$.

2. If $a > b$, then $a - c > b - c$.

3. If $a > b$ and c is a **positive number** $(c > 0)$, then $ac > bc$.

4. If $a > b$ and c is a **positive number** $(c > 0)$, then $\dfrac{a}{c} > \dfrac{b}{c}$.

If we multiply both sides of an inequality by a negative number or if we divide both sides of an inequality by a negative number, then the inequality is reversed.

RULES WHEN INEQUALITIES ARE REVERSED

For all real numbers a, b, and c:

1. If $a > b$, and c is a **negative number** $(c < 0)$, then $ac < bc$.

2. If $a > b$, and c is a **negative number** $(c < 0)$, then $\dfrac{a}{c} < \dfrac{b}{c}$.

The pattern is fairly simple. We could also make similar rule boxes for the cases where $a \geq b$, $a < b$, and $a \leq b$, but they are really not necessary. We simply remember that the inequality is reversed when we multiply or divide by a negative number. Otherwise the inequality remains unchanged.

EXAMPLE 3   Solve and graph $3x + 7 \geq 13$.

Solution

$$3x + 7 - 7 \geq 13 - 7 \qquad \text{Subtract 7 from both sides.}$$
$$3x \geq 6 \qquad \text{Simplify.}$$
$$\frac{3x}{3} \geq \frac{6}{3} \qquad \text{Divide both sides by 3.}$$
$$x \geq 2 \qquad \text{Simplify. Note that the direction of the inequality is not changed, since we have divided by a positive number.}$$

The graph is as follows:

Practice Problem 3   Solve and graph $8x - 2 < 3$.

   $x < \dfrac{5}{8}$

EXAMPLE 4   Solve and graph $5 - 3x > 7$.

Solution

$$5 - 5 - 3x > 7 - 5 \qquad \text{Subtract 5 from both sides.}$$
$$-3x > 2 \qquad \text{Simplify.}$$
$$\frac{-3x}{-3} < \frac{2}{-3} \qquad \text{Divide by } -3 \text{ and } \textbf{reverse the inequality} \text{ since we are dividing by a negative number.}$$
$$x < -\frac{2}{3} \qquad \text{Note the direction of the inequality.}$$

The graph is as follows:

Practice Problem 4   Solve and graph $4 - 5x > 7$.

   $x < -\dfrac{3}{5}$

Just like equations, some inequalities contain parentheses and fractions. The initial steps to solve these inequalities will be the same as those used to solve equations with parentheses and fractions. When the variable appears on both sides of the inequality, it is advisable to collect the x-terms on the left side of the inequality symbol.

EXAMPLE 5   Solve and graph $-\dfrac{13x}{2} \leq \dfrac{x}{2} - \dfrac{15}{8}$.

Solution

$$8\left(\frac{-13x}{2}\right) \leq 8\left(\frac{x}{2}\right) - 8\left(\frac{15}{8}\right) \qquad \text{Multiply all terms by LCD} = 8. \text{ We do } \textbf{not} \text{ reverse the direction of the inequality symbol since we are multiplying by a positive number.}$$
$$-52x \leq 4x - 15 \qquad \text{Simplify.}$$
$$-52x - 4x \leq 4x - 15 - 4x \qquad \text{Subtract } 4x \text{ from both sides.}$$
$$-56x \leq -15 \qquad \text{Combine like terms.}$$
$$\frac{-56x}{-56} \geq \frac{-15}{-56} \qquad \text{Divide both sides by } -56. \text{ We } \textbf{reverse} \text{ the direction of the inequality when we divide both sides by a negative number.}$$
$$x \geq \frac{15}{56}$$

The graph is as follows:

Practice Problem 5 Solve and graph

$$\frac{1}{2}x + 3 < \frac{2}{3}x. \quad x > 18$$

EXAMPLE 6 Solve and graph $\frac{1}{3}(3 - 2x) \le -4(x + 1)$.

Solution

$$1 - \frac{2x}{3} \le -4x - 4 \qquad \text{Remove parentheses.}$$

$$3(1) - 3\left(\frac{2x}{3}\right) \le 3(-4x) - 3(4) \qquad \text{Multiply all terms by LCD} = 3.$$

$$3 - 2x \le -12x - 12 \qquad \text{Simplify.}$$

$$3 - 2x + 12x \le -12x + 12x - 12 \qquad \text{Add } 12x \text{ to both sides.}$$

$$3 + 10x \le -12 \qquad \text{Combine like terms.}$$

$$3 - 3 + 10x \le -12 - 3 \qquad \text{Subtract 3 from both sides.}$$

$$10x \le -15 \qquad \text{Simplify.}$$

$$\frac{10x}{10} \le \frac{-15}{10} \qquad \begin{array}{l}\text{Divide both sides by 10. Since we are}\\\text{dividing by a \textbf{positive} number, the}\\\text{inequality is \textbf{not} reversed.}\end{array}$$

$$x \le -\frac{3}{2}$$

The graph is as follows:

Practice Problem 6 Solve and graph

$$\frac{1}{2}(3 - x) \le 2x + 5. \quad x \ge -\frac{7}{5}$$

CAUTION: The most common error students make when solving inequalities is forgetting to reverse the direction of the inequality symbol when multiplying or dividing both sides of the inequality by a negative number.

Normally when solving inequalities you solve for x by putting the variables on the left side. If you solve by placing the variables on the right side, you will end up with statements like $3 > x$. This is equivalent to $x < 3$. It is wise to express your answer with the variable on the left side.

EXAMPLE 7 A hospital director has determined that the costs of operating one floor of the hospital for an eight-hour shift must never exceed $2370. An expression for the cost of operating one floor of the hospital is $130n + 1200$, where n is the number of nurses. This expression is based on an estimate of $1200 in fixed costs and a cost of $130 per nurse for an eight-hour shift. Solve the inequality $130n + 1200 \le 2370$ to determine the number of nurses that may be on duty on this floor during an eight-hour shift if the director's cost control measure is to be followed.

Teaching Example 6 Solve and graph
$\frac{2}{5}(x - 4) > x - 1$.

Ans: $x < -1$

Teaching Example 7 Eleanor Abbot works for an insurance company. Her monthly income can be expressed as $1700 + 250n$, where n is the number of insurance policies she sells in that month. If Eleanor wants her monthly income never to be less than $4200, how many insurance policies does she need to sell each month?

Ans: $n \ge 10$

Solution

$$130n + 1200 \le 2370 \qquad \text{The inequality we must solve.}$$
$$130n + 1200 - 1200 \le 2370 - 1200 \qquad \text{Subtract 1200 from each side.}$$
$$130n \le 1170 \qquad \text{Simplify.}$$
$$\frac{130n}{130} \le \frac{1170}{130} \qquad \text{Divide each side by 130.}$$
$$n \le 9$$

The number of nurses on duty on this floor during an eight-hour shift must always be less than or equal to nine.

Practice Problem 7 The company president of Staywell, Inc., wants the monthly profits never to be less than $2,500,000. He has determined that an expression for monthly profit for the company is $2000n - 700,000$. In the expression, n is the number of exercise machines manufactured each month. The profit on each machine is $2000, and the $-$700,000 in the expression represents the fixed costs of running the manufacturing division.

Solve the inequality $2000n - 700,000 \ge 2,500,000$ to find how many machines must be made and sold each month to satisfy these financial goals.

$n \ge 1600$

PRACTICE   WATCH   DOWNLOAD   READ   REVIEW

Solve and graph the result.

1. $x + 7 \leq 4$ $\quad x \leq -3$

2. $x - 5 < -3$ $\quad x < 2$

3. $5x \leq 25$ $\quad x \leq 5$

4. $6x \geq -42$ $\quad x \geq -7$

5. $-2x < 18$ $\quad x > -9$

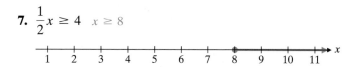

6. $-7x < 28$ $\quad x > -4$

7. $\frac{1}{2}x \geq 4$ $\quad x \geq 8$

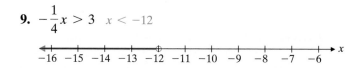

8. $\frac{1}{3}x \leq 2$ $\quad x \leq 6$

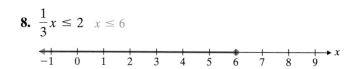

9. $-\frac{1}{4}x > 3$ $\quad x < -12$

10. $-\frac{1}{5}x < 10$ $\quad x > -50$

11. $2x - 3 < 4$ $\quad x < \frac{7}{2}$

12. $4x - 6 \geq -8$ $\quad x \geq -\frac{1}{2}$

13. $8 - 5x > 13$ $\quad x < -1$

14. $9 - 4x \leq 21$ $\quad x \geq -3$

15. $-4 + 5x < -3x + 8$ $\quad x < \frac{3}{2}$

16. $-6 - 4x < 1 - 6x$ $\quad x < \frac{7}{2}$

17. $\frac{5x}{6} - 5 > \frac{x}{6} - 9$ $\quad \begin{array}{l} 4x > -24 \\ x > -6 \end{array}$

18. $\frac{x}{4} - 2 < \frac{3x}{4} + 5$ $\quad \begin{array}{l} -2x < 28 \\ x > -14 \end{array}$

19. $2(3x + 4) > 3(x + 3)$ $\quad \begin{array}{l} 6x + 8 > 3x + 9 \\ 3x > 1 \\ x > \frac{1}{3} \end{array}$

20. $5(x - 3) \leq 2(x - 3)$ $\quad \begin{array}{l} 5x - 15 \leq 2x - 6 \\ 3x \leq 9 \\ x \leq 3 \end{array}$

Verbal and Writing Skills

21. Add -2 to both sides of the inequality $5 > 3$. What is the result? Why is the direction of the inequality not reversed? $\quad 3 > 1$; adding *any* number to both sides of an inequality does not reverse the direction.

22. Divide -3 into both sides of the inequality $-21 > -29$. What is the result? Why is the direction of the inequality reversed?

$7 < \frac{29}{3}$; whenever you divide both sides of an inequality by a negative number, the direction of the inequality is reversed.

179

Mixed Practice *Solve. Collect the variable terms on the left side of the inequality.*

23. $3x + 8 < 7x - 4$
$-4x < -12$
$x > 3$

24. $7x + 3 > 9x - 5$
$-2x > -8$
$x < 4$

25. $6x - 2 \geq 4x + 6$
$2x \geq 8$
$x \geq 4$

26. $5x - 5 \leq 2x + 10$
$3x \leq 15$
$x \leq 5$

27. $0.3(x - 1) < 0.1x - 0.5$
$0.2x < -0.2$
$2x < -2$
$x < -1$

28. $0.2(3 - x) + 0.1 > 0.1(x - 2)$
$-0.3x > -0.9$
$-3x > -9$
$x < 3$

29. $3 + 5(2 - x) \geq -3(x + 5)$
$3 + 10 - 5x \geq -3x - 15$
$-2x \geq -28$
$x \leq 14$

30. $9 - 3(2x - 1) \leq 4(x + 2)$
$9 - 6x + 3 \leq 4x + 8$
$-10x \leq -4$
$x \geq \dfrac{2}{5}$

31. $\dfrac{x + 6}{7} - \dfrac{6}{14} > \dfrac{x + 3}{2}$
$2x + 12 - 6 > 7x + 21$
$-5x > 15$
$x < -3$

32. $\dfrac{x + 5}{4} - \dfrac{3}{5} \geq \dfrac{x - 3}{20}$
$5x + 25 - 12 \geq x - 3$
$4x \geq -16$
$x \geq -4$

33. $\dfrac{1}{6} - \dfrac{1}{2}(3x + 2) < \dfrac{1}{3}\left(x - \dfrac{1}{2}\right)$
$1 - 9x - 6 < 2x - 1$ $x > -\dfrac{4}{11}$

34. $\dfrac{2}{3}(2x - 5) + 3 \geq \dfrac{1}{4}(3x + 1) - 5$
$16x - 40 + 36 \geq 9x + 3 - 60$ $x \geq -\dfrac{53}{7}$

Applications

35. *Course Average* To pass a course with a B grade, a student must have an average of 80 or greater. A student's grades on three tests are 75, 83, and 86. Solve the inequality $\dfrac{75 + 83 + 86 + x}{4} \geq 80$ to find what score the student must get on the next test to get a B average or better.   76 or greater

36. *Payment Options* Sharon sells very expensive European sports cars. She may choose to receive $10,000.00 or 8% of her sales as payment for her work. Solve the inequality $0.08x > 10,000$ to find how much she needs to sell to make the 8% offer a better deal.   More than $125,000 worth of cars

37. *Elephant Weight* The average African elephant weighs 268 pounds at birth. During the first three weeks of life the baby elephant will usually gain about 4 pounds per day. Assuming that growth rate, solve the inequality $268 + 4x \geq 300$ to find how many days it will be until a baby elephant weighs at least 300 pounds.   8 days or more

38. *Car Loan* Rennie is buying a used car that costs $4500. The deal called for a $600 down payment, and payments of $260 monthly. He wants to know whether he can pay off the car within a year. Solve the inequality $600 + 260x \geq 4500$ to find out the minimum number of months it will take to pay off the car.
It will take Rennie 15 months.

Cumulative Review

▲ **39.** **[1.8.2]** *Basketball Poster* Melinda is making a poster for her college basketball team. On the poster she is placing a life-sized picture of a basketball. When inflated properly, a basketball has a diameter of 9 inches. What is the area of the basketball on her poster? Use $\pi \approx 3.14$. Round your answer to the nearest tenth.
63.6 in.^{2}

▲ **40.** **[1.8.2]** *Seating Area* The seats of an outdoor amphitheater are arranged in the shape of a trapezoid. The altitude of the trapezoid is 60 feet. The bases of the trapezoid are 100 feet and 150 feet. What is the area of this seating area?   7500 ft^2

Quick Quiz 2.7 Solve and graph the result.

1. $4(x - 3) \leq 2x$ $x \leq 6$

2. $4 - 3(x + 2) + 5(x - 4) > 0$ $x > 11$

3. $\frac{1}{2}x + \frac{1}{4} \leq \frac{1}{8}(5x - 3)$ $x \geq 5$

4. Concept Check Explain the steps involved when solving the inequality $7 - 2(x - 4) < 7(x - 3)$.
Answers may vary

Classroom Quiz 2.7 You may use these problems to quiz your students' mastery of Section 2.7.

Solve and graph the result.

1. $5(x - 6) \leq 3x$ **Ans:** $x \leq 15$

2. $1 - 2(x + 2) + 5(x - 3) > 0$ **Ans:** $x > 6$

3. $\frac{1}{3}x + \frac{1}{5} \leq \frac{1}{15}(3x - 2)$ **Ans:** $x \leq -\frac{5}{2}$ or $x \leq -2\frac{1}{2}$

Putting Your Skills to Work: Use Math to Save Money

GAS PRICES

It's July 2008 in Stockton, California, and Sam needs to put gas in his car. He is on a street that has an ARCO gas station and a SHELL station. Sam will use his debit/credit card to pay for the gas. The ARCO station is charging $4.43 per gallon of gas while the SHELL station is charging $4.55 per gallon.

If Sam's goal is to save money it would seem obvious that he should go to ARCO, right? But Sam knows from experience it's not that simple. He knows

that ARCO will charge an extra $0.45 as an "ATM Transaction Fee" in addition to the gas he buys:

1. If Sam plans on buying just **one gallon** of gas, which gas station should he choose?
SHELL

2. If Sam plans on buying **three gallons** of gas, which gas station should he choose?
SHELL

3. If Sam plans on buying **four gallons** of gas, which gas station should he choose?
ARCO

4. If Sam plans on buying **ten gallons** of gas, which gas station should he choose?
ARCO

5. How many gallons of gas would Sam need to buy for the cost to be **exactly the same** at the two gas stations? Consider the results of Question 2 and Question 3 when formulating your answer.
3.75 gallons

6. Does the station where you normally get gas charge the same for cash or credit?
Answers may vary

7. Do you know if the station charges an "ATM transaction fee"?
Answers may vary

8. Has the increase in gas prices caused you to change your driving habits? If so, please explain.
Answers may vary

Topic	Procedure	Examples
Solving equations without parentheses or fractions, p. 146.	1. On both sides of the equation, combine like terms if possible. 2. Add or subtract terms on both sides of the equation in order to get all terms with the variable on one side of the equation. 3. Add or subtract a value on both sides of the equation to get all terms not containing the variable on the other side of the equation. 4. Divide both sides of the equation by the coefficient of the variable. 5. If possible, simplify the solution. 6. Check your solution by substituting the obtained value into the original equation.	Solve for x. $$5x + 2 + 2x = -10 + 4x + 3$$ $$7x + 2 = -7 + 4x$$ $$7x - 4x + 2 = -7 + 4x - 4x$$ $$3x + 2 = -7$$ $$3x + 2 - 2 = -7 - 2$$ $$3x = -9$$ $$\frac{3x}{3} = \frac{-9}{3}$$ $$x = -3$$ *Check:* Is -3 the solution of $$5x + 2 + 2x = -10 + 4x + 3?$$ $$5(-3) + 2 + 2(-3) \overset{?}{=} -10 + 4(-3) + 3$$ $$-15 + 2 - 6 \overset{?}{=} -10 + (-12) + 3$$ $$-13 - 6 \overset{?}{=} -22 + 3$$ $$-19 = -19 \checkmark$$
Solving equations with parentheses and/or fractions, pp. 148, 153.	1. Remove any parentheses. 2. Simplify, if possible. 3. If fractions exist, multiply all terms on both sides by the least common denominator of all the fractions. 4. Now follow the remaining steps for solving an equation without parentheses or fractions.	Solve for y. $$5(3y - 4) = \frac{1}{4}(6y + 4) - 48$$ $$15y - 20 = \frac{3}{2}y + 1 - 48$$ $$15y - 20 = \frac{3}{2}y - 47$$ $$2(15y) - 2(20) = 2\left(\frac{3}{2}y\right) - 2(47)$$ $$30y - 40 = 3y - 94$$ $$30y - 3y - 40 = 3y - 3y - 94$$ $$27y - 40 = -94$$ $$27y - 40 + 40 = -94 + 40$$ $$27y = -54$$ $$\frac{27y}{27} = \frac{-54}{27}$$ $$y = -2$$
Solving formulas for a specified variable, p. 162.	1. Remove any parentheses. 2. If fractions exist, multiply all terms on both sides by the LCD, which may be a variable. 3. Add or subtract terms on both sides of the equation in order to get all terms containing the *desired variable* on one side of the equation. 4. Add or subtract terms on both sides of the equation in order to get all other terms on the opposite side of the equation. 5. Divide both sides of the equation by the coefficient of the desired variable. This division may involve other variables. 6. Simplify, if possible. 7. (Optional) Check your solution by substituting the obtained expression into the original equation.	Solve for z. $B = \frac{1}{3}(hx + hz)$ First we remove parentheses. $$B = \frac{1}{3}hx + \frac{1}{3}hz$$ Now we multiply each term by 3. $$3(B) = 3\left(\frac{1}{3}hx\right) + 3\left(\frac{1}{3}hz\right)$$ $$3B = hx + hz$$ $$3B - hx = hx - hx + hz$$ $$3B - hx = hz$$ The coefficient of z is h, so we divide each side by h. $$\frac{3B - hx}{h} = z$$

Topic	Procedure	Examples
Solving inequalities, p. 175.	1. Follow the steps for solving an equation up until the multiplication or division step. 2. If you multiply or divide both sides of the inequality by a *positive number*, the direction of the inequality is not reversed. 3. If you multiply or divide both sides of the inequality by a *negative number*, the direction of the inequality is reversed.	Solve for x and graph your solution. $$\frac{1}{2}(3x - 2) \leq -5 + 5x - 3$$ First remove parentheses and simplify. $$\frac{3}{2}x - 1 \leq -8 + 5x$$ Now multiply each term by 2. $$2\left(\frac{3}{2}x\right) - 2(1) \leq 2(-8) + 2(5x)$$ $$3x - 2 \leq -16 + 10x$$ $$3x - 10x - 2 \leq -16 + 10x - 10x$$ $$-7x - 2 \leq -16$$ $$-7x - 2 + 2 \leq -16 + 2$$ $$-7x \leq -14$$ When we divide both sides by a negative number, the inequality is reversed. $$\frac{-7x}{-7} \geq \frac{-14}{-7}$$ $$x \geq 2$$ Graph of the solution:

Chapter 2 Review Problems

Sections 2.1–2.3

Solve for the variable. Noninteger answers may be left in fractional form or decimal form.

1. $5x = -35$ $x = -7$

2. $x - 19 = -22$ $x = -3$

3. $6 - 18x = 4 - 17x$ $x = 2$

4. $18 - 10x = 63 + 5x$ $x = -3$

5. $6x - 2(x + 3) = 5$ $x = \frac{11}{4}$ or 2.75

6. $1 - 2(6 - x) = 3x + 2$ $x = -13$

7. $x - (0.5x + 2.6) = 17.6$
$x = 40.4$

8. $-0.2(x + 1) = 0.3(x + 11)$
$x = -7$

9. $3(x - 2) = -4(5 + x)$
$x = -2$

10. $-\frac{2}{5}x = 16$ $x = -40$

11. $\frac{3}{4}x = 15$ $x = 20$

12. $4(2x + 3) = 5(x - 3)$ $x = -9$

13. $3(x - 3) = 13x + 21$
$x = -3$

14. $1.2x - 0.8 = 0.8x + 0.4$
$x = 3$

15. $3.2 - 0.6x = 0.4(x - 2)$
$x = 4$

16. $12 - x + 2 = 3x - 10 + 4x$
$x = 3$

17. $28 = 5x - (3x + 12)$
$x = 20$

18. $12 - 5x = -7x - 2$
$x = -7$

19. $2(3 - x) = 1 - (x - 2)$
$x = 3$

20. $4(x + 5) - 7 = 2(x + 3)$
$x = -\frac{7}{2}$ or -3.5

21. $0.9y + 3 = 0.4y + 1.5$
$y = -3$

22. $7y - 3.4 = 11.3$ $\;y = 2.1$

23. $3 = 2x + 5 - 3(x - 1)$ $\;x = 5$

24. $2(5x - 1) - 7 = 3(x - 1) + 5 - 4x$ $\;x = 1$

Section 2.4

Solve for the variable. Noninteger answers may be left in fractional form or decimal form.

25. $\dfrac{3}{4}x - 3 = \dfrac{1}{2}x + 2$ $\;x = 20$

26. $1 = \dfrac{5x}{6} + \dfrac{2x}{3}$ $\;x = \dfrac{2}{3}$

27. $\dfrac{7x}{5} = 5 + \dfrac{2x}{5}$ $\;x = 5$

28. $\dfrac{7x - 3}{2} - 4 = \dfrac{5x + 1}{3}$ $\;x = \dfrac{35}{11}$

29. $\dfrac{3x - 2}{2} + \dfrac{x}{4} = 2 + x$ $\;x = 4$

30. $\dfrac{-3}{2}(x + 5) = 1 - x$ $\;x = -17$

31. $\dfrac{-4}{3}(2x + 1) = -x - 2$ $\;x = \dfrac{2}{5}$ or 0.4

32. $\dfrac{1}{3}(x - 2) = \dfrac{x}{4} + 2$ $\;x = 32$

33. $\dfrac{1}{5}(x - 3) = 2 - \dfrac{x}{2}$ $\;x = \dfrac{26}{7}$

34. $\dfrac{3}{4} - \dfrac{2}{3}x = \dfrac{1}{3}x + \dfrac{3}{4}$ $\;x = 0$

35. $x + \dfrac{1}{8} + \dfrac{1}{2}x = \dfrac{1}{4}\left(-3x + \dfrac{1}{2}\right)$ $\;x = 0$

36. $\dfrac{3}{2}x - \dfrac{5}{6} + x = \dfrac{1}{2}x + \dfrac{2}{3}$ $\;x = \dfrac{3}{4}$

37. $-\dfrac{8}{3}x - 8 + 2x - 5 = -\dfrac{5}{3}$ $\;x = -17$

38. $3.5(2x + 3) = 2.4x - 1$ $\;x = -2.5$

39. $\dfrac{7}{12}(x - 3) = \dfrac{1}{3}x + 4$ $\;x = 23$

40. $\dfrac{1}{6} + \dfrac{1}{3}(x - 3) = \dfrac{1}{2}(x + 9)$ $\;x = -32$

41. $\dfrac{1}{7}(x + 5) - \dfrac{6}{14} = \dfrac{1}{2}(x + 3)$ $\;x = -\dfrac{17}{5}$ or -3.4

42. $\dfrac{1}{6}(8x + 3) = \dfrac{1}{2}(2x + 7)$ $\;x = 9$

43. $-\dfrac{2}{5}(3x + 1) = \dfrac{1}{3}(2 - x)$ $\;x = -\dfrac{16}{13}$

44. $\dfrac{7}{9}x + \dfrac{2}{3} = 5 + \dfrac{1}{3}x$ $\;x = 9.75$ or $\dfrac{39}{4}$

Section 2.5

Solve for the variable indicated.

45. Solve for y. $3x - y = 10$
$y = 3x - 10$

46. Solve for y. $5x + 2y + 7 = 0$
$y = \dfrac{-5x - 7}{2}$

47. Solve for r. $A = P(1 + rt)$
$r = \dfrac{A - P}{Pt}$

48. Solve for h. $A = 4\pi r^2 + 2\pi rh$
$h = \dfrac{A - 4\pi r^2}{2\pi r}$

49. Solve for p. $H = \dfrac{1}{3}(a + 2p + 3)$
$p = \dfrac{3H - a - 3}{2}$

50. Solve for y. $ax + by = c$
$y = \dfrac{c - ax}{b}$

51. **(a)** Solve for A. $x = \dfrac{ABC}{10}$ $\;A = \dfrac{10x}{BC}$
(b) Use your result to find A if $x = 6$, $B = -1$, and $C = 1.5$. $\;A = -40$

52. **(a)** Solve for w. $2l + 2w = P$ $\;w = \dfrac{P - 2l}{2}$
(b) Use your result to find w if $P = 34$ and $l = 10.5$.
$w = 6.5$

53. **(a)** Solve for h. $V = lwh$ $h = \dfrac{V}{lw}$

   (b) Use your result to find h when $V = 48, l = 2$, and $w = 4$. $h = 6$

Sections 2.6–2.7

Solve each inequality and graph the result.

54. $9 + 2x \le 6 - x$ $x \le -1$

55. $4x - 7 \ge -x + 8$ $x \ge 3$

56. $2x - 3 + x > 5(x + 1)$ $x < -4$

57. $-x + 4 < 3x + 16$ $x > -3$

58. $8 - \dfrac{1}{3}x \le x$ $x \ge 6$

59. $7 - \dfrac{3}{5}x > 4$ $x < 5$

60. $-4x - 14 < 4 - 2(3x - 1)$ $x < 10$

61. $3(x - 2) + 8 < 7x + 14$ $x > -3$

62. $\dfrac{1}{2}(2x + 3) > 10$ $x > \dfrac{17}{2}$

63. $5(1 - x) < 3(x - 1) - 2(3 - x)$ $x > \dfrac{7}{5}$

Use an inequality to solve.

64. Wages Julian earns $15 per hour as a plasterer's assistant. His employer determines that the current job allows him to pay $480 in wages to Julian. What are the maximum number of hours that Julian can work on this job? (*Hint:* Use $15h \le 480$.)

$15h \le 480$

$h \le 32 \text{ hr}$

65. Hiring a Substitute The cost of hiring a substitute elementary teacher for a day is $110. Let $n =$ the number of substitute teachers. Set up an inequality to determine how many times a substitute teacher may be hired if the school's budget for substitute teachers is $2420 per month. What is the maximum number of times a substitute teacher may be hired during a month? (*Hint:* Use $110n \le 2420$.)

$110n \le 2420$

$n \le 22$

Mixed Practice

Solve for the variable. Noninteger answers may be left in fractional form or decimal form.

66. $10(2x + 4) - 13 = 8(x + 7) - 3$

$x = \dfrac{13}{6}$

67. $-9x + 15 - 2x = 4 - 3x$

$x = \dfrac{11}{8}$

68. $-2(x - 3) = -4x + 3(3x + 2)$

$x = 0$

69. $\dfrac{1}{2} + \dfrac{5}{4}x = \dfrac{2}{5}x - \dfrac{1}{10} + 4$

$x = 4$

70. $\dfrac{1}{6}x - \dfrac{2}{3} = \dfrac{1}{3}(x - 4)$

$x = 4$

71. $\dfrac{1}{2}(x - 3) = \dfrac{1}{4}(3x - 1)$

$x = -5$

Solve for the variable indicated.

72. Solve for f. $10 + 3f = 5e - 2f$

$f = e - 2$

73. Solve for a. $W = \dfrac{2a + 4b}{3}$

$a = \dfrac{3W - 4b}{2}$

Solve each inequality and graph the result.

74. $5 - \dfrac{1}{2}x > 4$

$x < 2$

75. $2(x - 1) \ge 3(2 + x)$

$x \le -8$

76. $\dfrac{1}{3}(x + 2) \le \dfrac{1}{2}(3x - 5)$

$x \ge \dfrac{19}{7}$

77. $4(2 - x) - (-5x + 1) \ge -8$

$x \ge -15$

How Am I Doing? Chapter 2 Test

Remember to use your Chapter Test Prep Video CD to see the worked-out solutions to the test problems you want to review.

Note to Instructor: The Chapter 2 Test file in the TestGen program provides algorithms specifically matched to these problems so you can easily replicate this test for additional practice or assessment purposes.

Solve for the variable. Noninteger answers may be left in fractional form or decimal form.

1. $3x + 5.6 = 11.6$

2. $9x - 8 = -6x - 3$

3. $2(2y - 3) = 4(2y + 2)$

4. $\frac{1}{7}y + 3 = \frac{1}{2}y$

5. $4(7 - 4x) = 3(6 - 2x)$

6. $0.8x + 0.18 - 0.4x = 0.3(x + 0.2)$

7. $\frac{2y}{3} + \frac{1}{5} - \frac{3y}{5} + \frac{1}{3} = 1$

8. $3 - 2y = 2(3y - 2) - 5y$

9. $5(20 - x) + 10x = 165$

10. $5(x + 40) - 6x = 9x$

11. $-2(2 - 3x) = 76 - 2x$

12. $20 - (2x + 6) = 5(2 - x) + 2x$

In questions 13–17, solve for x.

13. $2x - 3 = 12 - 6x + 3(2x + 3)$

14. $\frac{1}{3}x - \frac{3}{4}x = \frac{1}{12}$

15. $\frac{3}{5}x + \frac{7}{10} = \frac{1}{3}x + \frac{3}{2}$

16. $\frac{15x - 2}{28} = \frac{5x - 3}{7}$

17. $\frac{2}{3}(x + 8) + \frac{3}{5} = \frac{1}{5}(11 - 6x)$

18. Solve for w. $A = 3w + 2P$

19. Solve for w. $\frac{2w}{3} = 4 - \frac{1}{2}(x + 6)$

20. Solve for a. $A = \frac{1}{2}h(a + b)$

21. Solve for y. $5ax(2 - y) = 3axy + 5$

22. Solve for B. $V = \frac{1}{3}Bh$

▲ **23.** Use your result from question 22 to find the area of the base (B) of a pyramid if the volume (V) is 140 cubic inches and the height (h) is 14 inches.

Solve and graph the inequality.

24. $3(x - 2) \geq 5x$

$\xleftarrow{\hspace{0.3cm}}{\underset{-5 \quad -4 \quad -3 \quad -2 \quad -1 \quad 0}{\longrightarrow}} x$

$x \leq -3$

25. $2 - 7(x + 1) - 5(x + 2) < 0$

$\xleftarrow{\hspace{0.3cm}}{\underset{-\frac{3}{2} \quad -\frac{5}{4} \quad -1 \quad -\frac{3}{4} \quad -\frac{1}{2} \quad -\frac{1}{4}}{\longrightarrow}} x$

$x > -\frac{5}{4}$

26. $5 + 8x - 4 < 2x + 13$

$\xleftarrow{\hspace{0.3cm}}{\underset{0 \quad 1 \quad 2 \quad 3 \quad 4 \quad 5}{\longrightarrow}} x$

$x < 2$

27. $\frac{1}{4}x + \frac{1}{16} \leq \frac{1}{8}(7x - 2)$

$\xleftarrow{\hspace{0.3cm}}{\underset{-1 \quad -\frac{1}{2} \quad 0 \quad \frac{1}{2} \quad 1 \quad \frac{3}{2}}{\longrightarrow}} x$

$x \geq \frac{1}{2}$

1.	$x = 2$
2.	$x = \frac{1}{3}$
3.	$y = -\frac{7}{2}$ or -3.5
4.	$y = 8.4$ or $\frac{42}{5}$
5.	$1 = x$
6.	$x = -1.2$
7.	$y = 7$
8.	$y = \frac{7}{3}$
9.	$x = 13$
10.	$20 = x$
11.	$x = 10$
12.	$x = -4$
13.	$x = 12$
14.	$x = -\frac{1}{5}$ or -0.2
15.	$x = 3$
16.	$x = 2$
17.	$x = -2$
18.	$w = \frac{A - 2P}{3}$
19.	$w = \frac{6 - 3x}{4}$
20.	$a = \frac{2A - hb}{h}$
21.	$y = \frac{10ax - 5}{8ax}$
22.	$B = \frac{3V}{h}$
23.	$B = 30$ in.^{2}
24.	See graph
25.	See graph
26.	See graph
27.	See graph

Approximately one-half of this test covers the content of Chapters 0 and 1. The remainder covers the content of Chapter 2. In questions 1–4, simplify.

1. $\dfrac{6}{7} - \dfrac{2}{3}$

2. $1\dfrac{3}{4} + 2\dfrac{1}{5}$

3. $4\dfrac{3}{8} \div \dfrac{1}{2}$

4. $(1.23)(0.56)$

5. Divide. $0.144 \div 1.2$

6. What is 4.3% of 830?

7. Multiply. $(-3)(-5)(-1)(2)(-1)$

8. Combine like terms. $3st - 8s^2t + 12st^2 + s^2t - 5st + 2st^2$

9. Simplify. $(5x)^2$

10. Simplify. $2\{3x - 4[5 - 3y(2 - x)]\}$

11. Solve for x. $-2(x + 5) = 4x - 15$

12. Solve for x. $\dfrac{1}{3}(x + 5) = 2x - 5$

13. Solve for y. $\dfrac{2y}{3} - \dfrac{1}{4} = \dfrac{1}{6} + \dfrac{y}{4}$

14. Solve for b. $3b - 4c = 9 - b$

15. Solve for b. $H = \dfrac{2}{3}(b + 4a)$

16. Solve for t. $I = Prt$

17. Solve for a. $A = \dfrac{ha}{2} + \dfrac{hb}{2}$

In problems 18–22, solve and graph the inequality.

18. $7x - 2 \le 3x + 10 - 2x$ $x \le 2$

19. $\dfrac{1}{2}(x - 5) \ge x - 4$ $x \le 3$

20. $4(2 - x) > 1 - 5x - 8$ $x > -15$

21. $x + \dfrac{5}{9} \le \dfrac{1}{3} + \dfrac{7}{9}x$ $x \le -1$

22. $7x - 13 \le 3(4 - 5x) - 3$ $x \le 1$

23. The football team will not let Chuck play unless he passes biology with a C (70 or better) average. There are five tests in the semester, and he has failed (0) the first one. However, he found a tutor and received an 82, an 89, and an 87 on the next three tests. Solve the inequality $\dfrac{0 + 82 + 89 + 87 + x}{5} \ge 70$ to find what his minimum score must be on the last test in order for him to pass the course and play football.

---

1. $\dfrac{4}{21}$

2. $\dfrac{79}{20}$ or $3\dfrac{19}{20}$

3. $\dfrac{35}{4}$ or $8\dfrac{3}{4}$

4. 0.6888

5. 0.12

6. 35.69

7. 30

8. $-2st - 7s^2t + 14st^2$

9. $25x^2$

10. $6x - 40 + 48y - 24xy$

11. $x = \dfrac{5}{6}$

12. $x = 4$

13. $y = 1$

14. $b = \dfrac{4c + 9}{4}$

15. $b = \dfrac{3H - 8a}{2}$

16. $t = \dfrac{I}{Pr}$

17. $a = \dfrac{2A}{h} - b$

18. See graph

19. See graph

20. See graph

21. See graph

22. See graph

23. 92

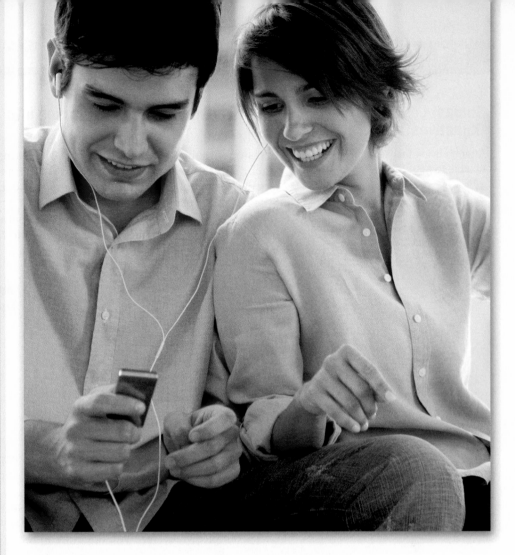

CHAPTER 3

How do people listen to music? LPs, eight tracks, and cassette tapes were once very popular. Today, most people listen to CDs or download music online. The way people purchase recorded music is constantly changing. The music recording industry uses the mathematics covered in this chapter daily to make predictions about music buying trends of the American public. Using these mathematical techniques, the recording industry can make wise business decisions and thus maximize profit.

Solving Applied Problems

Student Learning Objectives

After studying this section, you will be able to:

 Translate English phrases into algebraic expressions.

 Write an algebraic expression to compare two or more quantities.

 Translating English Phrases into Algebraic Expressions

One of the most useful applications of algebra is solving word problems. One of the first steps in solving a word problem is translating the conditions of the problem into algebra. In this section we show you how to translate common English phrases into algebraic symbols. This process is similar to translating between languages like Spanish and French.

Several English phrases describe the operation of addition. If we represent an unknown number by the variable x, all of the following phrases can be translated into algebra as $x + 3$.

English Phrases Describing Addition	Algebraic Expression	Diagram
Three *more than* a number		
The *sum of* a number and three		
A number *increased by* three	$x + 3$	
Three is *added to* a number.		
Three *greater than* a number		
A number *plus* three		

In a similar way we can use algebra to express English phrases that describe the operations of subtraction, multiplication, and division.

CAUTION: Since subtraction is not commutative, the order is essential. A number decreased by five is $x - 5$. It is not correct to say $5 - x$. Use extra care as you study each example. Make sure you understand the proper order.

English Phrases Describing Subtraction	Algebraic Expression	Diagram
A number *decreased by* four		
Four *less than* a number		
Four is *subtracted from* a number.		
Four *smaller than* a number	$x - 4$	
Four *fewer than* a number		
A number *diminished by* four		
A number *minus* 4		
The *difference between* a number and four		

English Phrases Describing Multiplication	Algebraic Expression	Diagram
Double a number		
Twice a number		
The *product* of two and a number	$2x$	
Two *of* a number		
Two *times* a number		

Since division is not commutative, the order is essential. A number divided by 3 is $\dfrac{x}{3}$. It is not correct to say $\dfrac{3}{x}$. Use extra care as you study each example.

English Phrases Describing Division	Algebraic Expression	Diagram

A number *divided by* five
One-*fifth* of a number
The *quotient* of a number and five

$\dfrac{x}{5}$

Teaching Tip Students who have had algebra before very often have very bad memories of "word problems." This is a time to tell them that they can learn to do word problems without anxiety if they master a few basic steps. Practicing translating English phrases into algebraic expressions will prove valuable in solving word problems.

Often other words are used in English instead of the word *number*. We can use a variable, such as *x*, here also.

EXAMPLE 1

English Phrase	Algebraic Expression
(a) A *quantity* is increased by five.	$x + 5$
(b) Double the *value*	$2x$
(c) One-third of the *weight*	$\dfrac{x}{3}$ or $\dfrac{1}{3}x$
(d) Twelve more than an *unknown number*	$x + 12$
(e) Seven less than a *number*	$x - 7$

Note that the algebraic expression for "seven less than a number" does not follow the order of the words in the English phrase. The variable or expression that follows the words *less than* always comes first.

seven less than *x*

$x - 7$

The variable or expression that follows the words *more than* technically comes before the plus sign. However, since addition is commutative, it also can be written after the plus sign.

Practice Problem 1 Write each English phrase as an algebraic expression.

(a) Four more than a number $x + 4$ **(b)** Triple a value $3x$
(c) Eight less than a number $x - 8$ **(d)** One-fourth of a height $\dfrac{1}{4}x$

More than one operation can be described in an English phrase. Sometimes parentheses must be used to make clear which operation is done first.

EXAMPLE 2

English Phrase	Algebraic Expression
(a) Seven more than double a number	$2x + 7$
	*Note that these are **not** the same.*
(b) The value of the number is increased by seven and then doubled.	$2(x + 7)$
	*Note that the word **then** tells us to add x and 7 before doubling.*
(c) One-half of the sum of a number and 3	$\dfrac{1}{2}(x + 3)$

Teaching Example 1 Write each English phrase as an algebraic expression.

(a) A quantity is decreased by seven.

(b) Quadruple a quantity.

(c) Thirteen greater than a value

(d) Seven-eighths of a number

(e) Six fewer than a number

Ans: (a) $x - 7$ (b) $4x$ (c) $x + 13$

(d) $\dfrac{7}{8}x$ (e) $x - 6$

Teaching Example 2 Write each English phrase as an algebraic expression.

(a) A number is decreased by four and then tripled.

(b) Five less than double a value

(c) Two-thirds of a value that has been increased by eight

Ans: (a) $3(x - 4)$ (b) $2x - 5$

(c) $\dfrac{2}{3}(x + 8)$

> **Practice Problem 2** Write each English phrase as an algebraic expression.
>
> **(a)** Eight more than triple a number $3x + 8$
>
> **(b)** A number is increased by eight and then it is tripled. $3(x + 8)$
>
> **(c)** One-third of the sum of a number and 4 $\frac{1}{3}(x + 4)$

② Writing an Algebraic Expression to Compare Two or More Quantities

Often in a word problem two or more quantities are described in terms of another. We will want to use a variable to represent one quantity and then write an algebraic expression using *the same variable* to represent the other quantity. Which quantity should we let the variable represent? We usually let the variable represent the quantity that is the basis of comparison: the quantity that the others are being *compared to*.

Teaching Example 3 Use a variable and an algebraic expression to describe the two quantities in the English sentence "Angelina is five years older than Maria."

Ans: $m =$ Maria's age

  $m + 5 =$ Angelina's age

EXAMPLE 3 Use a variable and an algebraic expression to describe the two quantities in the English sentence "Mike's salary is $2000 more than Fred's salary."

Solution   The two quantities that are being compared are Mike's salary and Fred's salary. Since Mike's salary is being *compared to* Fred's salary, we let the variable represent Fred's salary. The choice of the letter f helps us to remember that the variable represents Fred's salary.

$$\text{Let } f = \text{Fred's salary.}$$

Then $f + \$2000 =$ Mike's salary, since Mike's salary is $2000 *more than* Fred's.

> **Practice Problem 3** Use a variable and an algebraic expression to describe the two quantities in the English sentence "Marie works 17 hours per week less than Ann." $a =$ Ann's hours per week; $a - 17 =$ Marie's hours per week

Teaching Example 4 The length of a rectangle is 8 feet longer than half of the width. Use a variable and an algebraic expression to describe the length and the width. Draw a picture of the rectangle and label the length and width.

Ans: $w =$ the width

$\frac{1}{2}w + 8 =$ the length

$\frac{1}{2}w + 8$

w

EXAMPLE 4 The length of a rectangle is 3 meters shorter than twice the width. Use a variable and an algebraic expression to describe the length and the width. Draw a picture of the rectangle and label the length and width.

Solution   The length of the rectangle is being *compared to* the width. Use the letter w for width.

$$\text{Let } w = \text{the width.}$$

$$\underbrace{3 \text{ meters shorter than twice the width}}$$

$$\text{Then } 2w - 3 = \text{the length.}$$

A picture of the rectangle is shown.

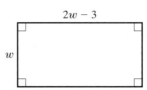

$2w - 3$

w

PRACTICE PROBLEM 4

$l = 2w + 5$

w

> **Practice Problem 4** The length of a rectangle is 5 meters longer than double the width. Use a variable and an algebraic expression to describe the length and the width. Draw a picture of the rectangle and label the length and width. $w =$ width; $2w + 5 =$ length

EXAMPLE 5 The first angle of a triangle is triple the second angle. The third angle of the triangle is 12° more than the second angle. Describe each angle algebraically. Draw a diagram of the triangle and label its parts.

Solution Since the first and third angles are described in terms of the second angle, we let the variable represent the number of degrees in the second angle.

Let s = the number of degrees in the second angle.

Then $3s$ = the number of degrees in the first angle.

And $s + 12$ = the number of degrees in the third angle.

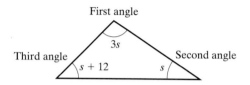

First angle

Third angle

$3s$

$s + 12$

Second angle

s

Practice Problem 5 The first angle of a triangle is 16° less than the second angle. The third angle is double the second angle. Describe each angle algebraically. Draw a diagram of the triangle and label its parts.

First angle = $s - 16$; Second angle = s; Third angle = $2s$

Some comparisons will involve fractions.

EXAMPLE 6 A theater manager was examining the records of attendance for last year. The number of people attending the theater in January was one-half of the number of people attending the theater in February. The number of people attending the theater in March was three-fifths of the number of people attending the theater in February. Use algebra to describe the attendance each month.

Solution What are we looking for? The *number of people* who attended the theater *each month*. The basis of comparison is February. That is where we begin.

Let f = the number of people who attended in February.

Then $\frac{1}{2}f$ = the number of people who attended in January.

And $\frac{3}{5}f$ = the number of people who attended in March.

Practice Problem 6 The college dean noticed that in the spring the number of students on campus was two-thirds of the number of students on campus in the fall. She also noticed that in the summer the number of students on campus was one-fifth the number of students on campus in the fall. Use algebra to describe the number of students on campus in each of these three time periods.

x = number of students in fall;

$\frac{2}{3}x$ = number of students in spring;

$\frac{1}{5}x$ = number of students in summer

Verbal and Writing Skills *Write an algebraic expression for each quantity. Let x represent the unknown value.*

1. a quantity increased by 5 $x + 5$

2. nine greater than a number $x + 9$

3. twelve less than a number $x - 12$

4. a value decreased by seven $x - 7$

5. one-eighth of a quantity $\frac{1}{8}x$ or $\frac{x}{8}$

6. one-half of a quantity $\frac{1}{2}x$ or $\frac{x}{2}$

7. twice a quantity $2x$

8. triple a number $3x$

9. three more than half of a number $3 + \frac{1}{2}x$

10. five more than one-third of a number $5 + \frac{1}{3}x$

11. double a quantity increased by nine $2x + 9$

12. ten times a number increased by 1 $10x + 1$

13. one-third of the sum of a number and seven $\frac{1}{3}(x + 7)$

14. one-fourth of the sum of a number and 5 $\frac{1}{4}(x + 5)$

15. one-third of a number reduced by twice the same number $\frac{1}{3}x - 2x$

16. one-fifth of a number reduced by double the same number $\frac{1}{5}x - 2x$

17. five times a quantity decreased by eleven $5x - 11$

18. four less than seven times a number $7x - 4$

Write an algebraic expression for each of the quantities being compared.

19. *Stock Value* The value of a share of IBM stock on that day was $74.50 more than the value of a share of AT&T stock.
x = value of a share of AT&T stock
$x + 74.50$ = value of a share of IBM stock

20. *Investments* One day in September 2007, the value of one share of Xerox stock was $49.10 less than the value of a share of Target stock.
x = value of a share of Target stock
$x - 49.10$ = value of a share of Xerox stock

▲ **21.** *Geometry* The length of the rectangle is 7 inches more than double the width.
$2w + 7$ = length of rectangle
w = width of rectangle

▲ **22.** *Geometry* The length of the rectangle is 3 meters more than triple the width.
$3w + 3$ = length of rectangle
w = width of rectangle

23. *Cookie Sales* The number of boxes of cookies sold by Sarah was 43 fewer than the number of boxes of cookies sold by Keiko. The number of boxes of cookies sold by Imelda was 53 more than the number sold by Keiko.
x = number of boxes of cookies sold by Keiko
$x - 43$ = number of boxes of cookies sold by Sarah
$x + 53$ = number of boxes of cookies sold by Imelda

24. *April Rainfall* The average April rainfall in Savannah, Georgia, is about 13 inches more than that of Burlington, Vermont. The average rainfall in Phoenix, Arizona, is about 28 inches less than that of Burlington.
x = average rainfall of Burlington
$x + 13$ = average rainfall of Savannah
$x - 28$ = average rainfall of Phoenix

▲ **25.** *Geometry* The first angle of a triangle is 16 degrees less than the second angle. The third angle of the triangle is double the second angle.
1st angle = $s - 16$; 2nd angle = s; 3rd angle = $2s$

▲ **26.** *Geometry* The first angle of a triangle is 19 degrees more than the third angle. The second angle is triple the third angle.
1st angle = $t + 19$; 2nd angle = $3t$; 3rd angle = t

27. *Exports* The value of the exports of Japan was twice the value of the exports of Canada.
v = value of exports of Canada
$2v$ = value of exports of Japan

28. *Olympic Medals* The total number of medals won by Canada in all Winter Olympic Games through 2004 was 2 more than Switzerland.
m = medals won by Switzerland
$m + 2$ = medals won by Canada

▲ **29.** *Geometry* The first angle of a triangle is triple the second angle. The third angle of the triangle is 14 degrees less than the second angle.
1st angle $= 3x$; 2nd angle $= x$; 3rd angle $= x - 14$

30. *Book Cost* The cost of Hiro's biology book was $13 more than the cost of his history book. The cost of his English book was $27 less than the cost of his history book.
$x =$ the cost of his history book in dollars
$x + 13 =$ the cost of his biology book in dollars
$x - 27 =$ the cost of his English book in dollars

To Think About

Kayak Rentals *The following bar graph depicts the number of people who rented sea kayaks at Essex Boat Rental during the summer of 2007. Use the bar graph to answer exercises 31 and 32.*

31. Write an expression for the number of men who rented sea kayaks at Essex Boat Rental in each age category. Start by using x for the number of men aged 16 to 24 who rented kayaks.
$x =$ number of men aged 16 to 24
$x + 82 =$ number of men aged 25 to 34
$x - 25 =$ number of men aged 35 to 44
$x - 110 =$ number of men aged 45 and above

32. Write an expression for the number of women who rented sea kayaks at Essex Boat Rental. Start by using y for the number of women aged 25 to 34 who rented kayaks.
$y =$ number of women aged 25 to 34
$y - 133 =$ number of women aged 16 to 24
$y - 17 =$ number of women aged 35 to 44
$y - 110 =$ number of women aged 45 and above

Rental of Sea Kayaks by Age Group

Cumulative Review *Solve for the variable.*

33. [2.4.1] $x + \dfrac{1}{2}(x - 3) = 9$ $x = 7$

34. [2.4.1] $\dfrac{3}{5}x - 3(x - 1) = 9$ $x = -\dfrac{5}{2}$ or $-2\dfrac{1}{2}$

Quick Quiz 3.1 Write an algebraic expression for each quantity. Let x represent the unknown value.

1. Ten greater than a number $x + 10$ or $10 + x$

2. Five less than double a number $2x - 5$

3. The first angle of a triangle is 15 degrees more than the second angle. The third angle of the triangle is double the second angle. Write an algebraic expression for the measure of each of the three angles.

First angle $= x + 15$; Second angle $= x$; Third angle $= 2x$

4. Concept Check Explain how you would decide whether to use $\frac{1}{3}(x + 7)$ or $\frac{1}{3}x + 7$ as the algebraic expression for the phrase "one-third of the sum of a number and seven." Answers may vary

Classroom Quiz 3.1 You may use these problems to quiz your students' mastery of Section 3.1.

Write an algebraic expression for each quantity. Let x represent the unknown value.

1. Seven less than a number **Ans:** $x - 7$

2. Eight more than triple a number. **Ans:** $8 + 3x$ or $3x + 8$

Write an algebraic expression for the measure of each of the three angles.

3. The first angle of a triangle is 20 degrees more than the second angle. The third angle of the triangle is double the second angle.

Ans: First angle $= x + 20$; Second angle $= x$; Third angle $= 2x$

In Section 0.7 we introduced a simple three-step procedure to solve applied problems. You have had an opportunity to use that approach to solve word problems in Exercises 0.7 and in the Applications exercises in Chapters 1 and 2. Now we are going to focus our attention on solving applied problems that require the use of variables, translating English phrases into algebraic expressions, and setting up equations. The process is a little more involved. Some students find the following outline a helpful way to keep organized while solving such problems.

1. *Understand the problem.*
   (a) Read the word problem carefully to get an overview.
   (b) Determine what information you will need to solve the problem.
   (c) Draw a sketch. Label it with the known information. Determine what needs to be found.
   (d) Choose a variable to represent one unknown quantity.
   (e) If necessary, represent other unknown quantities in terms of that very same variable.

2. *Write an equation.*
   (a) Look for key words to help you to translate the words into algebraic symbols and expressions.
   (b) Use a given relationship in the problem or an appropriate formula to write an equation.

3. *Solve and state the answer.*

4. *Check.*
   (a) Check the solution in the original equation. Is the answer reasonable?
   (b) Be sure the solution to the equation answers the question in the word problem. You may need to do some additional calculations if it does not.

1 Solving Number Problems

EXAMPLE 1 Two-thirds of a number is eighty-four. What is the number?

Solution

1. **Understand the problem.** Draw a sketch.

   Let $x =$ the unknown number.

2. **Write an equation.**

$$\underbrace{\text{Two-thirds of a number}}_{\frac{2}{3}x} \quad \underbrace{\text{is}}_{=} \quad \underbrace{\text{eighty-four.}}_{84}$$

3. Solve and state the answer.

$$\frac{2}{3}x = 84$$

$$3\left(\frac{2}{3}x\right) = 3(84) \quad \text{Multiply both sides of the equation by 3.}$$

$$2x = 252$$

$$\frac{2x}{2} = \frac{252}{2} \quad \text{Divide both sides by 2.}$$

$$x = 126$$

The number is 126.

4. Check. Is two-thirds of 126 eighty-four?

$$\frac{2}{3}(126) \stackrel{?}{=} 84$$

$$84 = 84 \quad \checkmark$$

Practice Problem 1 Three-fourths of a number is negative eighty-one. What is the number? -108

NOTE TO STUDENT: *Fully worked-out solutions to all of the Practice Problems can be found at the back of the text starting at page SP-1*

Learning to solve problems like Examples 1 and 2 is a very useful skill. You will find that learning the material in Chapter 3 will be much easier if you can master the procedure used in these two examples.

EXAMPLE 2 Five more than six times a quantity is three hundred five. Find the number.

Teaching Example 2 Eleven fewer than eight times a number is 117. Find the number.

Ans: 16

Solution

1. Understand the problem. Read the problem carefully. You may not need to draw a sketch.
Let $x =$ the unknown quantity.

2. Write an equation.

Five more than	six times a number	is	three hundred five.
5 +	6x	=	305

3. Solve and state the answer. You may want to rewrite the equation to make it easier to solve.

$$6x + 5 = 305$$

$$6x + 5 - 5 = 305 - 5 \quad \text{Subtract 5 from both sides.}$$

$$6x = 300$$

$$\frac{6x}{6} = \frac{300}{6} \quad \text{Divide both sides by 6.}$$

$$x = 50$$

The quantity, or number, is 50.

4. Check. Is five more than six times 50 three hundred five?

$$6(50) + 5 \stackrel{?}{=} 305$$

$$300 + 5 \stackrel{?}{=} 305$$

$$305 = 305 \quad \checkmark$$

Practice Problem 2 Two less than triple a number is forty-nine. Find the number. 17

EXAMPLE 3 The larger of two numbers is three more than twice the smaller. The sum of the numbers is thirty-nine. Find each number.

Solution

1. **Understand the problem.** Read the problem carefully. The problem refers to *two* numbers. We must write an algebraic expression for *each number* before writing the equation. The larger number is being compared to the smaller number. We want to use *one variable* to describe each number.

Let s = the smaller number.

Then $2s + 3$ = the larger number.

three more than twice the smaller number

2. **Write an equation.** The sum of the numbers is thirty-nine.

$$s + (2s + 3) \quad = \quad 39$$

3. **Solve.**
$$s + (2s + 3) = 39$$
$$3s + 3 = 39 \quad \text{Combine like terms.}$$
$$3s = 36 \quad \text{Subtract 3 from each side.}$$
$$s = 12 \quad \text{Divide both sides by 3.}$$

4. **Check.**
$$12 + [2(12) + 3] \overset{?}{=} 39$$
$$39 = 39 \quad ✓$$

The solution checks, but have we solved the word problem? We need to find *each* number. 12 is the smaller number. Substitute 12 into the expression $2s + 3$ to find the larger number.

$$2s + 3 = 2(12) + 3 = 27$$

The smaller number is 12. The larger number is 27.

Practice Problem 3 Consider two numbers. The second number is twelve less than triple the first number. The sum of the two numbers is twenty-four. Find each number. The first number is 9. The second number is 15.

② Using the Mathematics Blueprint to Solve Applied Word Problems

To facilitate understanding more involved word problems, we will use a Mathematics Blueprint similar to the one we used in Section 0.7. This format is a simple way to organize facts, determine what to set variables equal to, and select a method or approach that will assist you in finding the desired quantity. You will find using this form helpful, particularly in those cases when you read through a word problem and mentally say to yourself, "Now where do I begin?" You begin by responding to the headings of the blueprint. Soon a procedure for solving the problem will emerge.

Mathematics Blueprint for Problem Solving			
Gather the Facts	Assign the Variable	Basic Formula or Equation	Key Points to Remember

EXAMPLE 4 The mean annual snowfall in Juneau, Alaska, is 105.8 inches. This is 20.2 inches less than three times the annual snowfall in Boston. What is the annual snowfall in Boston?

Solution

Understand the problem and write an equation.

Mathematics Blueprint for Problem Solving

Gather the Facts	Assign the Variable	Basic Formula or Equation	Key Points to Remember
Snowfall in Juneau is 105.8 inches. This is 20.2 inches less than three times the snowfall in Boston.	We do not know the snowfall in Boston. Let b = annual snowfall in Boston. Then $3b - 20.2$ = annual snowfall in Juneau.	Set $3b - 20.2$ equal to 105.8, which is the snowfall in Juneau.	All measurements of snowfall are recorded in inches.

Juneau's snowfall is 20.2 less than three times Boston's snowfall.

$$105.8 = 3b - 20.2$$

Solve and state the answer. You may want to rewrite the equation to make it easier to solve.

$$3b - 20.2 = 105.8$$
$$3b = 126 \quad \text{Add 20.2 to both sides.}$$
$$b = 42 \quad \text{Divide both sides by 3.}$$

The annual snowfall in Boston is 42 inches.

Check. Reread the word problem. Work backward.

Three times 42 is 126.
126 less 20.2 is 105.8.

Is this the annual snowfall in Juneau? Yes. ✓

Teaching Example 4 The mean annual snowfall in Caribou, Maine, is 110 inches. This is 26.8 inches more than four times the annual snowfall in Baltimore, Maryland. What is the annual snowfall in Baltimore?

Ans: 20.8 in.

Practice Problem 4 The maximum recorded rainfall for a 24-hour period in the United States occurred in Alvin, Texas, on July 25–26, 1979. The maximum recorded rainfall for a 24-hour period in Canada occurred in Ucluelet Brynnor Mines, British Columbia, on October 6, 1977. Alvin, Texas, received 43 inches of rain in that period. The amount recorded in Texas was 14 inches less than three times the amount recorded in Canada. How much rainfall was recorded at Ucluelet Brynnor Mines in Canada? (*Source:* National Oceanic and Atmospheric Administration.) 19 in.

Some word problems require a simple translation of the facts. Others require a little more detective work. You will not always need to use the Mathematics Blueprint to solve every word problem. As you gain confidence in problem solving, you will no doubt leave out some of the steps. We suggest that you use the procedure when you find yourself on unfamiliar ground. It is a powerful organizational tool.

 Using Formulas to Solve Word Problems

Sometimes the relationship between two quantities is so well understood that we have developed a formula to describe that relationship. We have already done some work with formulas in Section 2.5. The following examples show how you can use a formula to solve a word problem.

EXAMPLE 5 Two people travel in separate cars. They each travel a distance of 330 miles on an interstate highway. To maximize fuel economy, Fred travels at exactly 50 mph. Sam travels at exactly 55 mph. How much time did the trip take each person? (Use the formula distance = rate · time or $d = rt$.)

Fred's speed is 50 mph
Sam's speed is 55 mph
330 miles

Solution

Mathematics Blueprint for Problem Solving

Gather the Facts	Assign the Variable	Basic Formula or Equation	Key Points to Remember
Each person drives 330 miles. Fred drives at 50 mph. Sam drives at 55 mph.	Time is the unknown quantity for each driver. Use subscripts to denote different values of t. t_f = Fred's time t_s = Sam's time	distance = (rate)(time) or $d = rt$	The time is expressed in hours.

Substitute the known values into the formula and solve for t.

$$d = rt \qquad\qquad d = rt$$
$$330 = 50t_f \qquad 330 = 55t_s$$
$$6.6 = t_f \qquad\qquad 6 = t_s$$

It took Fred 6.6 hours to drive 330 miles.   It took Sam 6 hours to drive 330 miles.

Check. Is this reasonable? Yes, you would expect Fred to take longer to drive the same distance because Fred is driving at a lower speed.

Note: You may wish to express 6.6 hours in hours and minutes. To change 0.6 hour to minutes, proceed as follows:

$$0.6 \ \text{hour} \cdot \frac{60 \ \text{minutes}}{1 \ \text{hour}} = (0.6)(60) \ \text{minutes} = 36 \ \text{minutes}$$

Thus, Fred drove for 6 hours and 36 minutes.

Practice Problem 5 Sarah left the city to visit her aunt and uncle, who live in a rural area north of the city. She traveled the 220-mile trip in 4 hours. On her way home she took a slightly longer route, which measured 225 miles on the car odometer. The return trip took 4.5 hours.

(a) What was her average speed on the trip leaving the city? 55 mph

(b) What was her average speed on the return trip? 50 mph

(c) On which trip did she travel faster and by how much?

(c) On the trip leaving the city by 5 mph

EXAMPLE 6 A teacher told Melinda that she had a course average of 78 based on her six math tests. When she got home, Melinda found five of her tests. She had scores of 87, 63, 79, 71, and 96 on the five tests. She could not find her sixth test. What score did she obtain on that test? (Use the formula that an average = the sum of scores ÷ the number of scores.)

Teaching Example 6 The average weekday high temperature last week was 83°. The high temperatures on Monday through Thursday were 75°, 78°, 84°, and 87°. What was the high temperature on Friday?

Ans: 91°

Solution

Mathematics Blueprint for Problem Solving

Gather the Facts	Assign the Variable	Basic Formula or Equation	Key Points to Remember
Her five known test scores are 87, 63, 79, 71, and 96. Her course average is 78.	We do not know the score Melinda received on her sixth test. Let x = the score on the sixth test.	average = $\dfrac{\text{sum of scores}}{\text{number of scores}}$	Since there are six test scores, we will need to divide the sum by 6.

When you average anything, you total up the sum of all the values and then divide it by the number of values.

We now write the equation for the average of six items. This involves adding the scores for all the tests and dividing by 6.

$$\frac{87 + 63 + 79 + 71 + 96 + x}{6} = 78$$

$$\frac{396 + x}{6} = 78 \qquad \text{Add the numbers in the numerator.}$$

$$6\left(\frac{396 + x}{6}\right) = 6(78) \qquad \begin{array}{l}\text{Multiply both sides of the equation by}\\ \text{6 to remove the fraction.}\end{array}$$

$$396 + x = 468 \qquad \text{Simplify.}$$

$$x = 72 \qquad \text{Subtract 396 from both sides to find } x.$$

Melinda's score on the sixth test was 72.

Check. To verify that this is correct, we check that the average of the 6 tests is 78.

$$\frac{87 + 63 + 79 + 71 + 96 + 72}{6} \stackrel{?}{=} 78$$

$$\frac{468}{6} \stackrel{?}{=} 78$$

$$78 = 78 \quad \checkmark$$

The problem checks. We know that the score on the sixth test was 72.

Practice Problem 6 Barbara's math course has four tests and one final exam. The final exam counts as much as two tests. Barbara has test scores of 78, 80, 100, and 96 on her four tests. What grade does she need on the final exam if she wants to have a 90 average for the course? 93

Solve. Check your solution.

1. What number minus 543 gives 718?
$x - 543 = 718$ $x = 1261$

2. What number added to 74 gives 265?
$x + 74 = 265$ $x = 191$

3. A number divided by eight is 296. What is the number?
$\frac{x}{8} = 296$ $x = 2368$

4. Eighteen less than a number is 23. What is the number?
$x - 18 = 23$ $x = 41$

5. Seventeen greater than a number is 199. Find the number. $17 + x = 199$ $x = 182$

6. Three times a number is one. What is the number?
$3x = 1$ $x = \frac{1}{3}$

7. A number is doubled and then increased by seven. The result is ninety-three. What is the original number?
$2x + 7 = 93$ $x = 43$

8. Four less than nine times a number is one hundred twenty-two. Find the original number.
$9x - 4 = 122$ $x = 14$

9. When eighteen is reduced by two-thirds of a number, the result is 12. Find the number.
$18 - \frac{2}{3}x = 12$ $x = 9$

10. Twice a number is increased by one-third the same number. The result is 42. Find the number.
$2x + \frac{1}{3}x = 42$ $x = 18$

11. Eight less than triple a number is the same as five times the same number. Find the number.
$3x - 8 = 5x$ $x = -4$

12. Ten less than double a number is the same as seven times the number. Find the number.
$2x - 10 = 7x$ $x = -2$

13. A number, half of that number, and one-third of that number are added. The result is 22. What is the original number?
$x + \frac{1}{2}x + \frac{1}{3}x = 22$ $x = 12$

14. A number, twice that number, and one-third of that number are added. The result is 20. What is the original number?
$x + 2x + \frac{1}{3}x = 20$ $x = 6$

Applications *Solve. Check to see if your answer is reasonable.*

15. *Motorcycle Inventory* A Harley Davidson motorcycle shop maintains an inventory of four times as many new bikes as used bikes. If there are 60 new bikes, how many used bikes are now in stock?
$4x = 60$ $x = 15$
There are 15 used bikes in stock.

16. *Shirt Sizes* The Enro Shirt Outlet store maintains an inventory of five times as many shirts in regular men's sizes (size XL and smaller) than it does in large men's sizes (size 2X and larger). There are 430 shirts in inventory in the store that are regular men's sizes. How many shirts in large men's sizes are in inventory?
$5x = 430$ $x = 86$
There are 86 shirts in large men's sizes (size 2X and larger) in the store.

17. *Wildfires* Between 1/1/06 and 9/25/06 there were 82,955 wildfires in the United States. This was 14,725 less than twice the number of wildfires between 1/1/03 and 9/25/03. How many wildfires were there during the period between 1/1/03 and 9/25/03? (*Source:* www.nifc.gov) $2x - 14,725 = 82,955$ $x = 48,840$
There were 48,840 wildfires between 1/1/03 and 9/25/03.

18. *On-Demand Movies* Joseph just changed his cable television plan to include on-demand movies for $2.95 each. The charge for the basic cable package is $24.95 per month. If his bill one month was $39.70, how many on-demand movies did he watch?
$24.95 + 2.95x = 39.70$ $x = 5$
He watched 5 movies.

19. *CD Purchase* The sale price of a new Panasonic compact disc player is $218 at a local discount store. At the store where this sale is going on, each new CD is on sale for $11 each. If Kyle purchases a player and some CDs for $284, how many CDs did he purchase?
$218 + 11x = 284$ $11x = 66$ He purchased 6 CDs.

20. *iPod Nano* Kerry just bought a new Apple iPod Nano for $129. On the same day, she bought several albums from iTunes for $11.99 each. If she spent $224.92 total, how many albums did she buy?
$129 + 11.99x = 224.92$ $x = 8$
She bought 8 albums.

21. One-Day Sale Raquelle went to Weller's Department Store's one-day sale. She bought two blouses for $38 each and a pair of shoes for $49. She also wanted to buy some jewelry. Each item of jewelry was bargain priced at $11.50 each. If she brought $171 with her, how many pieces of jewelry could she buy?

$(2)(38) + 49 + 11.50x = 171 \quad x = 4$

She could buy 4 items of jewelry.

22. Waiting Tables Brad is a waiter at the "Steaks-Are-Us" restaurant. He gets paid $5.75 per hour, and he can keep his tips. He knows that his tips average $8.80 per table. If he worked an eight-hour shift and took home $169.20, how many tables did he serve?

$(5.75)(8) + 8.80x = 169.20 \quad x = 14$

He served 14 tables.

23. TGV On May 18, 1990, the fastest speed of any national railroad was achieved by the French high-speed train *Train à Grande Vitesse* (TGV) as it traveled over a distance from Cortalain to Tours, France. A commentator said that this speed was so fast that if it continued at that rate, the train would travel 6404 miles in 20 hours. How fast did the train travel on that date? (*Source:* www.guinnessworldrecords.com)

$r = \dfrac{d}{t} \qquad r = \dfrac{6404}{20} = 320.2$

The train traveled at 320.2 mi/hr.

24. Submarine Speed In 1958, the nuclear-powered submarine *Nautilus* took 6 days, 12 hours to travel submerged 5068 km across the Atlantic Ocean from Portsmouth, England, to New York City. What was its average speed, in kilometers per hour, for this trip? (Round to the nearest whole number.)

$r = \dfrac{d}{t}$

$r = \dfrac{5068}{156} \approx 32$

Its average speed was about 32 km/hr.

25. Gravity It has been shown that the force of gravity on a planet varies with the mass of the planet. The force of gravity on Jupiter, for example, is about two and a half times that of Earth. Using this information, approximately how much would a 220-lb astronaut weigh on Jupiter?

$J = 2.5E \quad J = 2.5(220) = 550$

The astronaut would weigh 550 lb on Jupiter.

26. Sunday Comics Charles Schulz, the creator of *Peanuts,* once estimated that he had drawn close to 2600 Sunday comics over his career. At that time, about how many years had he been drawing *Peanuts?*

$52x = 2600$

$x = \dfrac{2600}{52} = 50$

He had been drawing *Peanuts* for 50 years.

27. In-Line Skating Two in-line skaters, Nell and Kristin, start from the same point and skate in the same direction. Nell skates at 12 miles per hour and Kristin skates at 14 miles per hour. If they can keep up that pace for 2.5 hours, how far apart will they be at the end of that time?

$(14)(2.5) - (12)(2.5) = x$

$x = 5$

They will be 5 miles apart.

28. Bus Travel Two Greyhound buses leave the Jackson City station at the same time. One travels north at 55 miles per hour. The other travels south at 60 miles per hour. In how many hours will the two buses be 460 hours apart?

$460 = 55t + 60t$

$460 = 115t$

$t = 4$

They will be 460 miles apart in 4 hours.

29. Travel Routes Nella drove from Albuquerque, New Mexico, to the Garden of the Gods rock formation in Colorado Springs. It took her six hours to travel 312 miles over the mountain road. She came home on the highway. On the highway she took five hours to travel 320 miles. How fast did she travel using the mountain route? How much faster (in miles per hour) did she travel using the highway route?

She traveled 52 miles per hour on the mountain road. It was 12 mph faster on the highway route.

30. Travel Speeds Allison drives 30 miles per hour through the city and 55 miles per hour on the New Jersey Turnpike. She drove 90 miles from Battery Park to the Jersey Shore. How much of the time was city driving if she spent 1.2 hours on the turnpike?

$(55)(1.2) + 30x = 90 \quad x = 0.8$

0.8 hours or 48 minutes was city driving.

31. Grading System Danielle's chemistry professor has a complicated grading system. Each lab counts once and each test counts as two labs. At the end of the semester there is a final lab that counts as three regular labs. Danielle received scores of 84, 87, and 93 on her labs, and scores of 89 and 94 on her tests. What score must she get on the final lab to receive an A (an A is 90)?

$$\frac{84 + 87 + 93 + 89 + 89 + 94 + 94 + x + x + x}{10} = 90$$

$x = 90$

She must score 90 on the final lab.

32. Politicians' Salaries In 2007, Speaker of the House Nancy Pelosi and Vice President Dick Cheney both earned the same annual salary. President George W. Bush earned $27,400 more than twice what Pelosi and Cheney each earned. The average of the three salaries was $257,533.33. What did each earn in 2007? Round your answers to the nearest whole dollar. (*Source:* www.govspot.com)

Let x = salary of Pelosi and Cheney.

$$\frac{x + x + (2x + 27,400)}{3} = 257,533.33 \quad x = 186,300$$

Pelosi and Cheney each earned $186,300. Bush earned $400,000.

To Think About

33. Cricket Chirps In warmer climates, approximate temperature predictions can be made by counting the number of chirps of a cricket during a minute. The Fahrenheit temperature decreased by forty is equivalent to one-fourth of the number of cricket chirps.

(a) Write an equation for this relationship.

(b) Approximately how many chirps per minute should be recorded if the temperature is 90°F?

(c) If a person recorded 148 cricket chirps in a minute, what would be the Fahrenheit temperature according to this formula?

(a) $F - 40 = \dfrac{x}{4}$ **(b)** $90 - 40 = \dfrac{x}{4}$ $x = 200$ 200 chirps

(c) $F - 40 = \dfrac{148}{4}$ $F = 77$ 77°F

Cumulative Review *Simplify.*

34. [1.6.1] $5x(2x^2 - 6x - 3)$
$10x^3 - 30x^2 - 15x$

35. [1.6.1] $-2a(ab - 3b + 5a)$
$-2a^2b + 6ab - 10a^2$

36. [1.7.2] $7x - 3y - 12x - 8y + 5y$
$-5x - 6y$

37. [1.7.2] $5x^2y - 7xy^2 - 8xy - 9x^2y$
$-4x^2y - 7xy^2 - 8xy$

Quick Quiz 3.2

1. A number is tripled and then decreased by 15. The result is 36. What is the original number?   17

2. The sum of one-half of a number, one-sixth of the number, and one-eighth of the number is thirty-eight. Find the original number.   48

3. James scored 84, 89, 73, and 80 on four tests. What must he score on the next test to have an 80 average in the course?   74

4. Concept Check Explain how you would set up an equation to solve the following problem.

Phil purchased two shirts for $23 each and then purchased several pairs of socks. The socks were priced at $0.75 per pair. How many pairs of socks did he purchase if the total cost was $60.25?

Answers may vary

Classroom Quiz 3.2 You may use these problems to quiz your students' mastery of Section 3.2.

1. A number is doubled and then increased by 19. The result is 55. What is the original number?   **Ans:** 18

2. The sum of one-half of a number, one-fourth of the number, and one-eighth of the number is twenty-eight. Find the original number.   **Ans:** 32

3. James scored 85, 89, 76, and 83 on four tests. What must he score on the next test to have an 85 average in the course?   **Ans:** 92

3.3 SOLVING WORD PROBLEMS: COMPARISONS

Solving Word Problems Involving Comparisons

Many real-life problems involve comparisons. We often compare quantities such as length, height, or income. Sometimes not all the information is known about the quantities that are being compared. You need to identify each quantity and write an algebraic expression that describes the situation in the word problem.

EXAMPLE 1 The Center City Animal Hospital treated a total of 18,360 dogs and cats last year. The hospital treated 1376 more dogs than cats. How many dogs were treated last year? How many cats were treated last year?

Solution

1. Understand the problem.

What information is given?   The combined number of dogs and cats is 18,360.

What is being compared?   There were 1376 more dogs than cats.

If you compare one quantity to another, usually the second quantity is represented by the variable. Since we are comparing the number of dogs to the number of cats, we start with the number of cats.

Let c = the number of cats treated at the hospital.

Then $c + 1376$ = the number of dogs treated at the hospital.

2. Write an equation.

The number of cats   plus   the number of dogs   is   18,360.

$\qquad c \qquad\qquad + \qquad (c + 1376) \qquad = \qquad 18{,}360$

3. Solve and state the answer.

$$c + c + 1376 = 18{,}360$$
$$2c + 1376 = 18{,}360 \qquad \text{Combine like terms.}$$
$$2c + 1376 - 1376 = 18{,}360 - 1376 \qquad \text{Subtract 1376 from both sides.}$$
$$2c = 16{,}984$$
$$\frac{2c}{2} = \frac{16{,}984}{2} \qquad \text{Divide both sides by 2.}$$
$$c = 8492 \qquad \text{The number of cats treated is 8492.}$$
$$c + 1376 = 8492 + 1376 = 9868 \qquad \text{The number of dogs treated is 9868.}$$

4. Check. The number of dogs treated plus the number of cats treated should total 18,360.

$$8492 + 9868 \overset{?}{=} 18{,}360$$
$$18{,}360 = 18{,}360 \checkmark$$

Practice Problem 1 A deck hand on a fishing boat is working with a rope that measures 89 feet. He needs to cut it into two pieces. The long piece must be 17 feet longer than the short piece. Find the length of each piece of rope.

Short piece = 36 ft; long piece = 53 ft

If the word problem contains three unknown quantities, determine the basis of comparison for two of the quantities.

EXAMPLE 2 An airport filed a report showing the number of plane departures from the airport during each month last year. The number of departures in March was 50 more than the number of departures in January. In July, the number

Student Learning Objective

After studying this section, you will be able to:

 Solve word problems involving comparisons.

Teaching Example 1 Bay City Auto sold 516 used vehicles last year. They sold 188 fewer used trucks than used cars. How many used trucks and used cars did they sell last year?

Ans: 164 used trucks and 352 used cars

NOTE TO STUDENT: Fully worked-out solutions to all of the Practice Problems can be found at the back of the text starting at page SP-1

of departures was 150 less than triple the number of departures in January. In those three months, the airport had 2250 departures. How many departures were recorded for each month?

Solution

1. ***Understand the problem.*** What is the basis of comparison?

   The number of departures in March is compared to the number in January.
   The number of departures in July is compared to the number in January.
   Express this algebraically. It may help to underline the key phrases.

   $$\text{Let } j = \text{the departures in January.}$$

   March had 50 more than January

   $$\text{Then } j + 50 = \text{the departures in March.}$$

   July had 150 less than triple January

   $$\text{So } 3j - 150 = \text{the departures in July.}$$

2. ***Write an equation.***

number of departures in January	+	number of departures in March	+	number of departures in July	=	three months' total departures
j	+	$(j + 50)$	+	$(3j - 150)$	=	2250

3. ***Solve and state the answer.***

   $$j + (j + 50) + (3j - 150) = 2250$$
   $$5j - 100 = 2250 \quad \text{Combine like terms.}$$
   $$5j = 2350 \quad \text{Add 100 to each side.}$$
   $$j = 470 \quad \text{Divide both sides by 5.}$$

   Now, if $j = 470$, then

   $$j + 50 = 470 + 50 = 520$$

   and

   $$3j - 150 = 3(470) - 150 = 1410 - 150 = 1260.$$

   The number of departures in January was 470; the number of departures in March was 520; the number of departures in July was 1260.

4. ***Check.*** Do these answers seem reasonable? Yes. Do these answers agree with all the statements in the word problem?

   Is the number of departures in March 50 more than those in January?

   $$520 \overset{?}{=} 50 + 470$$
   $$520 = 520 \quad ✓$$

   Is the number of departures in July 150 less than triple those in January?

   $$1260 \overset{?}{=} 3(470) - 150$$
   $$1260 \overset{?}{=} 1410 - 150$$
   $$1260 = 1260 \quad ✓$$

   Is the total number of departures in the three months equal to 2250?

   $$470 + 520 + 1260 \overset{?}{=} 2250$$
   $$2250 = 2250 \quad ✓$$

   Yes, all conditions are satisfied. The three answers are correct.

Practice Problem 2 A social services worker was comparing the cost incurred by three families in heating their homes for the year. The first family had an annual heating bill that was $360 more than that of the second family. The third family had a heating bill that was $200 less than double the heating bill of the second family. The total annual heating bill for the three families was $3960. What was the annual heating bill for each family?

First family: $1310; second family: $950; third family: $1700

NOTE TO STUDENT: Fully worked-out solutions to all of the Practice Problems can be found at the back of the text starting at page SP-1

▲ **EXAMPLE 3** A small plot of land is in the shape of a rectangle. The length is 7 meters longer than the width. The perimeter of the rectangle is 86 meters. Find the dimensions of the rectangle.

Solution

1. Understand the problem. Read the problem:

> What information is given?
>
> *The perimeter of the rectangle is 86 meters.*
>
> What is being compared?
>
> *The length is being compared to the width.*

Express this algebraically and draw a picture.

Let w = the width.
Then $w + 7$ = the length.

Reread the problem: What are you being asked to do?

> *Find the dimensions of the rectangle. The dimensions of the rectangle are the length and the width of the rectangle.*

2. Write an equation. The perimeter is the total distance around the rectangle.

$$w + (w + 7) + w + (w + 7) = 86$$

3. Solve and state the answer.

$$w + (w + 7) + w + (w + 7) = 86$$
$$4w + 14 = 86 \quad \text{Combine like terms.}$$
$$4w = 72 \quad \text{Subtract 14 from both sides.}$$
$$w = 18 \quad \text{Divide both sides by 4.}$$

The width of the rectangle is 18 meters. What is the length?

$$w + 7 = \text{the length}$$
$$18 + 7 = 25$$

The length of the rectangle is 25 meters.

4. Check. Put the actual dimensions in your drawing and add the lengths of the sides. (See the figure in the margin.) Is the sum 86 meters? ✓

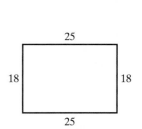

▲ **Practice Problem 3** A farmer purchased 720 meters of wire fencing to enclose a pasture. The pasture is in the shape of a triangle. The first side of the triangle is 30 meters less than the second side. The third side is one-half as long as the second side. Find the dimensions of the triangle.

First side: 270 m; second side: 300 m; third side: 150 m

Teaching Example 3 Lakeland College has a rectangular athletic field with a perimeter of 340 yards. The length of the field is 30 yards more than the width. Find the length and the width of the field.

Ans: The athletic field is 100 yards long and 70 yards wide.

Teaching Tip Remind students that when doing perimeter problems, they must take the time to determine what geometric figure is involved. Students often confuse the perimeter of a rectangle and the perimeter of a triangle.

Applications *Solve. Check to see if your answer is reasonable. Have you answered the question that was asked?*

1. **Parade Banner** For their homecoming parade, the students of Wheaton College have created a colorful banner, 47 meters in length, that is made of two pieces of parachute material. The short piece is 17 meters shorter than the long piece. Find the length of each piece.

   The long piece is 32 meters long.
   The short piece is 15 meters long.

2. **Copper Wire** A copper conducting wire measures 84 centimeters in length. It is cut into two pieces. The shorter piece is 5 centimeters shorter than the long piece. Find the length of each piece.

   The long piece is 44.5 centimeters long.
   The short piece is 39.5 centimeters long.

3. **Salaries** James and Lin got married and got new jobs in Chicago. Lin earns $3400 more per year than James. Together they earn $82,300. How much do each of them earn per year?

   James earns $39,450 per year.
   Lin earns $42,850 per year.

4. **Auto Insurance** Dick and Barbara moved to Ipswich and purchased a home. They paid $1856 per year to insure their two cars. Dick said it cost $358 more per year to insure his car than it did to insure Barbara's car. How much did it cost to insure Dick's car? How much did it cost to insure Barbara's car?

   Dick's car cost $1107 per year.
   Barbara's car cost $749 per year.

5. **Volunteering** Three siblings—David, Kate, and Sarah—committed themselves to volunteer 100 hours for Habitat for Humanity to build new housing. Sarah worked 15 hours more than Dave. Kate worked 5 hours fewer than Dave. How many hours did each of these siblings volunteer?

   David worked 30 hours.
   Sarah worked 45 hours.
   Kate worked 25 hours.

6. **Farmers' Market** Lyle was selling apples, plums, and peaches at the local farmers' market. He sold 18 more pounds of apples than pounds of plums. He sold 9 pounds less of peaches than pounds of plums. He sold a total of 69 pounds of fruit. How many pounds of each fruit did he sell?

   20 pounds of plums
   38 pounds of apples
   11 pounds of peaches

7. **Home States** Professor Fullerton was demonstrating the ease of making a bar graph using information about students' home states. He was amazed that his class of 32 represented only 3 states. There were 8 more students born in Texas than were born in Oklahoma. The number of students born in Arizona was 9 fewer than the number born in Oklahoma. How many were born in each state?

   11 students born in Oklahoma
   19 students born in Texas
   2 students born in Arizona

8. **Multiplex** An 800-seat multiplex is divided into three theaters. Theater One is the smallest. Theater Two has 180 more seats than Theater One. Theater Three has 60 seats fewer than twice the number in Theater One. How large is each theater?

   170 seats in Theater One
   350 seats in Theater Two
   280 seats in Theater Three

▲ 9. **Bridge Support** The length of a rectangular piece of steel in a bridge is 3 meters less than double the width. The perimeter of the piece of steel is 42 meters. Find the length of the piece of steel. Find the width of the piece of steel.

   The length is 13 meters. The width is 8 meters.

▲ 10. **State Park** A state park in Colorado has a perimeter of 92 miles. The park is in the shape of a rectangle. The length of the park is 30 miles less than triple the width of the park. Find the length of the park. Find the width of the park.

   The length is 27 miles. The width is 19 miles.

▲ **11.** *Painting Frame* Rashau made a rectangular frame for her latest oil painting. The length is 20 centimeters more than triple the width. The perimeter of the frame is 96 centimeters. Find the length and the width of the frame.

The length is 41 centimeters.
The width is 7 centimeters.

12. *Rug Dimensions* Marta purchased a rectangular rug for her dorm room. The perimeter of the rug is 288 inches. The length of the rug is 24 inches more than the width. Find the length and the width of the rug.

The width is 60 inches.
The length is 84 inches.

▲ **13.** *Jewelry Box* A solid-gold jewelry box was found in the underwater palace of Cleopatra just off the shore of Alexandria. The length of the rectangular box is 35 centimeters less than triple the width. The perimeter of the box is 190 centimeters. Find the length and width of Cleopatra's solid-gold jewelry box.

The width is 32.5 centimeters.
The length is 62.5 centimeters.

14. *Giant Chocolate Bar* The main attraction at the 1998 International Milka Chocolate Festival in Austria was a giant chocolate bar. The length of the bar was 3.4 feet more than twice the width and the perimeter was 45.8 feet. Find the length and width of the chocolate bar.

The width was 6.5 feet.
The length was 16.4 feet.

15. *Running Speed* The top running speed of a cheetah is double the top running speed of a jackal. The top running speed of an elk is 10 miles per hour faster than that of a jackal. If each of these three animals could run at top speed for an hour (which, of course, is not possible), they could run a combined distance of 150 miles. What is the top running speed of each of these three animals? (*Source:* American Museum of Natural History)

The cheetah can run at 70 miles per hour.
The jackal can run at 35 miles per hour.
The elk can run at 45 miles per hour.

16. *River Length* The Missouri River is 149 miles longer than double the length of the Snake River. The Potomac River is 796 miles shorter than the Snake River. The combined lengths of these three rivers is 3685 miles. What is the length of each of the three rivers? (*Source:* U.S. Department of the Interior)

The Missouri River is 2315 miles long.
The Snake River is 1083 miles long.
The Potomac River is 287 miles long.

▲ **17.** *Kite Design* Mr. Shen wished to enter the Orange County Kite Competition with an innovative kite of his own design. He did tests that showed an irregular quadrilateral design was best. (See diagram.) The longest side (A) was twice the length of the shortest side (C). Side B was one and one-half times the shortest side (C). Side D needed to be 3" (3 inches) longer than side C. If the competition required that the perimeter of the kite must be 58", what were the dimensions of Mr. Shen's kite?

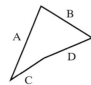

x = length C
$x + 2x + 1.5x + x + 3 = 58$
x = length C = 10"
$2x$ = length A = 20"
$1.5x$ = length B = 15"
$x + 3$ = length D = 13"

▲ **18.** *Pennant Design* The Tri-County schools are participating in a Renaissance Faire. The team from Appleton High School has designed a pennant to fly over its team tent. The pennant is in the shape of a triangle. The longest side is 5" (5 inches) less than twice the middle length side. The shortest side is 3" longer than half the middle length side. If the team captain bought 72" of satin edging for the banner, and there were 4" left over after the pennant was made, what are the dimensions of the pennant?

x = middle side length
$x + 2x - 5 + 0.5x + 3 = 68$
x = middle side length   = 20"
$2x - 5$ = longest side length   = 35"
$0.5x + 3$ = shortest side length   = 13"

To Think About

▲ **19.** *Geometry* A small square is constructed. Then a new square is made by increasing each side by 2 meters. The perimeter of the new square is 3 meters shorter than five times the length of one side of the original square. Find the dimensions of the original square.

The original square was 11 meters on each side.

▲ **20.** *Geometry* A rectangle is constructed. The length of this rectangle is double the width. Then a new rectangle is made by increasing each side by 3 meters. The perimeter of the new rectangle is 2 meters greater than four times the length of the old rectangle. Find the dimensions of the original rectangle.

The original rectangle measured 5 meters by 10 meters.

Cumulative Review *Simplify.*

21. **[1.6.1]** $-4x(2x^2 - 3x + 8)$
$-8x^3 + 12x^2 - 32x$

22. **[1.6.1]** $5a(ab + 6b - 2a)$
$5a^2b + 30ab - 10a^2$

23. **[1.7.2]** $-7x + 10y - 12x - 8y - 2$
$-19x + 2y - 2$

24. **[1.7.2]** $3x^2y - 6xy^2 + 7xy + 6x^2y$
$9x^2y - 6xy^2 + 7xy$

Quick Quiz 3.3

1. The length of a rectangular yard is 80 yards longer than double the width. The perimeter of the yard is 610 yards. Find the width and the length of the yard.
Width: 75 yd
Length: 230 yd

2. There are three times as many people in Jim's psychology class as in his art history class. There are 14 more people in Jim's algebra class than in his art history class. The total enrollment of the three classes is 79 students. How many students are enrolled in each class?
Psychology: 39 students
Art history: 13 students
Algebra: 27 students

3. A triangular support beam of the roof of a large warehouse has a perimeter of 73 meters. The first side is two-thirds of the second side. The third side is 15 meters shorter than the second side. What are the lengths of the three sides of the triangular support beam?
First side: 22 m
Second side: 33 m
Third side: 18 m

4. **Concept Check** Explain how you would set up an equation to solve the following problem.

When Don and Laurie got married, Laurie earned $2600 more per year than Don. Together they earned $71,200 per year. How much did each of them earn for the year?   Answers may vary

Classroom Quiz 3.3 You may use these problems to quiz your students' mastery of Section 3.3.

1. The length of a rectangular yard is 70 yards shorter than triple the width. The perimeter of the yard is 380 yards. Find the width and the length of the yard.
Ans: Width: 65 yd
      Length: 125 yd

2. There are four times as many people in Sally's sociology class as in her music appreciation class. There are 13 more people in Sally's algebra class than in her music appreciation class. The total enrollment of the three classes is 85 students. How many students are enrolled in each class?
Ans: Sociology: 48 students
      Music appreciation: 12 students
      Algebra: 25 students

3. A triangular support beam of the roof of an apartment building has a perimeter of 83 meters. The first side is three-fourths of the second side. The third side is 6 meters longer than the second side. What are the lengths of the three sides of the triangular support beam?
Ans: First side: 21 m
      Second side: 28 m
      Third side: 34 m

How are you doing with your homework assignments in Sections 3.1 to 3.3? Do you feel you have mastered the material so far? Do you understand the concepts you have covered? Before you go further in the textbook, take some time to do each of the following problems.

3.1

Write an algebraic expression. Use the variable x to represent the unknown number.

1. Three times a quantity is then decreased by 40.
2. One-half of a number is increased by 12.
3. First a number is multiplied by five. Then the result is divided by 12.
4. One-fourth of the sum of a number and 10

3.2

Translate the information into an equation. Then solve the equation in order to answer the question.

5. A number is doubled and then increased by five. The result is −47. What is the original number?
6. A number is divided by four, and eight is added to the result. A value of 25 is obtained. What is the original number?
7. At Lucas Ford, the monthly rental on a Ford Focus is $310. The cost of a new car is $12,400. In how many months will the rental cost equal the cost of purchasing a new car?
8. Bill Tupper makes $24 per hour for a 40-hour week. He earns time and a half for every hour over 40 hours during that one week. If Bill earned $1140 last week, how many overtime hours did he work?
9. The average salary of five workers at Verizon is $50,200. The first person earns $40,000 per year. A second person earns $35,000 per year. A third person earns $13,500 more per year than the first person. The fourth person earns an unknown annual salary. The fifth person earns $2500 per year more than the fourth person. How much is the annual salary of the fourth person? How much is the annual salary of the fifth person?

3.3

10. An electrician needs to cut a 10-foot wire into two pieces. He needs the longer piece to be 1.5 times the length of the shorter piece. How long is each piece?
11. Andy is doing his reading for his History of World Civilizations class. He must read a total of 260 pages in the next three days. On Monday he reads 25 pages less than twice the number of pages he read on Sunday. On Tuesday he reads 15 pages less than three times the number of pages he read on Sunday. If he reached his reading goal, how many pages did he read on each day?
12. In 2006, the third most expensive car in the world was the SSC Ultimate Aero. The second most expensive was the Pagani Zonda Roadster, priced at $12,000 more than the SSC. The most expensive car was the Bugatti Vegron, priced at $118,000 less than twice the price of the SSC. If the total price of the three cars was $2,514,000, how much did each of the cars cost? (*Source:* www.sportscarcup.com)

Now turn to page SA-7 for the answer to each of these problems. Each answer also includes a reference to the objective in which the problem is first taught. If you missed any of these problems, you should stop and review the Examples and Practice Problems in the referenced objective. A little review now will help you master the material in the upcoming sections of the text.

1. $3x - 40$

2. $\dfrac{1}{2}x + 12$

3. $\dfrac{5x}{12}$ or $\dfrac{5}{12}x$

4. $\dfrac{1}{4}(x + 10)$ or $\dfrac{x + 10}{4}$

5. -26

6. 68

7. 40 months

8. 5 hr

9. The fourth person earns $60,000 per year. The fifth person earns $62,500 per year.

10. The short piece is 4 feet long. The long piece is 6 feet long.

11. He read 50 pages on Sunday, 75 pages on Monday, and 135 pages on Tuesday.

12. SSC: $655,000; Pagani: $667,000; Bugatti: $1,192,000

3.4 SOLVING WORD PROBLEMS: THE VALUE OF MONEY AND PERCENTS

The problems we now present are frequently encountered in business. They deal with money: buying, selling, and renting items; earning and borrowing money; and the value of collections of stamps or coins. Many applications require an understanding of the use of percents and decimals. Review Sections 0.4 and 0.5 if you are weak in these skills.

 Solving Problems Involving Periodic Rate Charges

EXAMPLE 1 A business executive rented a car. The Supreme Car Rental Agency charged $39 per day and $0.28 per mile. The executive rented the car for two days and the total rental cost was computed to be $176. How many miles did the executive drive the rented car?

Solution

1. *Understand the problem.* How do you calculate the cost of renting a car?

total cost = per-day cost + mileage cost

What is known?

It cost $176 to rent the car for two days.

What do you need to find?

The number of miles the car was driven.

Choose a variable:

Let m = the number of miles driven in the rented car.

2. *Write an equation.* Use the relationship for calculating the total cost.

per-day cost	+	mileage cost	=	total cost
(39)(2)	+	(0.28)m	=	176

3. *Solve and state the answer.*

$$(39)(2) + (0.28)(m) = 176$$

$$78 + 0.28m = 176 \quad \text{Simplify the equation.}$$

$$0.28m = 98 \quad \text{Subtract 78 from both sides.}$$

$$\frac{0.28m}{0.28} = \frac{98}{0.28} \quad \text{Divide both sides by 0.28.}$$

$$m = 350 \quad \text{Simplify.}$$

The executive drove 350 miles.

4. *Check.* Does this seem reasonable? If he drove the car 350 miles in two days, would it cost $176?

$$(\text{cost of \$39 per day for 2 days}) + (\text{cost of \$0.28 per mile for 350 miles})$$
$$\overset{?}{=} \text{total cost of \$176}$$

$$(\$39)(2) + (350)(\$0.28) \overset{?}{=} \$176$$

$$\$78 + \$98 \overset{?}{=} \$176$$

$$\$176 = \$176 \quad ✓$$

Practice Problem 1 Alfredo wants to rent a truck to move to Florida. He has determined that the cheapest rental rates for a truck of the correct size are from a local company that will charge him $25 per day and $0.20 per mile. He has not yet completed an estimate of the mileage of the trip, but he knows that he will need the truck for three days. He has allowed $350 in his moving budget for the truck. How far can he travel for a rental cost of exactly $350? 1375 mi

NOTE TO STUDENT: Fully worked-out solutions to all of the Practice Problems can be found at the back of the text starting at page SP-1

 Solving Percent Problems

Many applied situations require finding a percent of an unknown number. If we want to find 23% of $400, we multiply 0.23 by 400: 0.23(400) = 92. If we want to find 23% of an unknown number, we can express this using algebra by writing 0.23*n*, where *n* represents the unknown number.

EXAMPLE 2 A sofa was marked with the following sign: "The price of this sofa has been reduced by 23%. You can save $138 if you buy now." What was the original price of the sofa?

Teaching Example 2 At Cragmore Valley College, tuition last semester was 96% of the tuition charged this semester. Last semester's tuition was $120 per credit. What is the tuition per credit this semester?

Ans: $125

Solution

1. *Understand the problem.*

Let *s* = the original price of the sofa.

Then 0.23*s* = the amount of the price reduction, which is $138.

2. *Write an equation and solve.*

$$0.23s = 138 \quad \text{Write the equation.}$$

$$\frac{0.23s}{0.23} = \frac{138}{0.23} \quad \text{Divide each side of the equation by 0.23.}$$

$$s = 600 \quad \text{Simplify.}$$

The original price of the sofa was $600.

3. *Check.* Is $600 a reasonable answer? ✓   Does 23% of $600 = $138? ✓

Practice Problem 2 John earns a commission of 38% of the cost of every set of encyclopedias that he sells. Last year he earned $4560 in commissions. What was the cost of the encyclopedias that he sold last year? $12,000

EXAMPLE 3 Hector received a pay raise this year. The raise was 6% of last year's salary. This year he will earn $15,900. What was his salary last year before the raise?

Teaching Example 3 The average cost of automobile insurance in Fremont County this year has increased 4% from last year's average cost. If the average cost of car insurance this year is $780, what was last year's average cost?

Ans: $750

Solution

1. *Understand the problem.* What do we need to find?

Hector's salary last year.

What do we know?

Hector received a 6% pay raise and now earns $15,900.

What does this mean?

Reword the problem: *This year's salary of* $15,900 *is* 6% *more than last year's salary.*

Choose a variable:

Let *x* = Hector's salary last year.

Then 0.06*x* = the amount of the raise.

2. Write an equation and solve.

Last year's salary	+	the amount of his raise	=	this year's salary	
x	+	$0.06x$	=	15,900	Write the equation.
$1.00x$	+	$0.06x$	=	15,900	Rewrite x as $1.00x$.
		$1.06x$	=	15,900	Combine like terms.
		$\dfrac{1.06x}{1.06}$	=	$\dfrac{15,900}{1.06}$	Divide by 1.06.
		x	=	15,000	Simplify.

Thus Hector's salary was $15,000 last year before the raise.

3. Check. Does it seem reasonable that Hector's salary last year was $15,000? The check is up to you.

Practice Problem 3 The price of Betsy's new car is 7% more than the price of a similar model last year. She paid $13,910 for her car this year. What would a similar model have cost last year? $13,000

Calculator

 Interest

You can use a calculator to find simple interest. Find the interest on $450 invested at an annual rate of 6.5% for 15 months. Notice that the time is in months. Since the interest formula $I = prt$ is in years, you need to change 15 months to years by dividing 15 by 12.

Enter: 15 ÷ 12 =

Display: 1.25

Leave this on the display and multiply as follows:

1.25 × 450 ×

6.5 % =

The display should read

36.5625

which would round to $36.56.

Try finding the simple interest in the following investments.

(a) $9516 invested at 12% for 30 months

(b) $593 borrowed at 8% for 5 months

Ans:

(a) $2854.80

(b) $19.77

3 Solving Investment Problems Involving Simple Interest

Interest is a charge for borrowing money or an income from investing money. Interest rates affect our lives. They affect the national economy and they affect a consumer's ability to borrow money for big purchases. For these reasons, a student of mathematics should be able to solve problems involving interest.

There are two basic types of interest: simple and compound. **Simple interest** is computed by multiplying the amount of money borrowed or invested (which is called the *principal*) times the rate of interest times the period of time over which it is borrowed or invested (usually measured in years unless otherwise stated).

$$\text{Interest} = \text{principal} \times \text{rate} \times \text{time}$$
$$I = prt$$

You often hear of banks offering a certain interest rate *compounded* quarterly, monthly, weekly, or daily. In **compound interest** the amount of interest is added to the amount of the original principal at the end of each time period, so future interest is based on the sum of both principal and previous interest. Most financial institutions use compound interest in their transactions.

Problems involving compound interest may be solved by:

1. Repeated calculations using the simple interest formula.
2. Using a compound interest table.
3. Using exponential functions, a topic that is usually covered in a higher-level college algebra course.

All examples and exercises in this chapter will involve **simple interest.**

EXAMPLE 4 Find the interest on $3000 borrowed at a simple interest rate of 18% for one year.

Solution

$$I = prt \qquad \text{The simple interest formula.}$$
$$I = (3000)(0.18)(1) \qquad \text{Substitute the values of the variables: principal} = 3000,$$
$$\text{the rate} = 18\% = 0.18, \text{the time} = \text{one year.}$$
$$I = 540$$

Thus the interest charge for borrowing $3000 for one year at a simple interest rate of 18% is $540.

Practice Problem 4 Find the interest on $7000 borrowed at a simple interest rate of 12% for one year. $840

Teaching Example 4 Find the interest on $6400 borrowed at a simple interest rate of 11% for one year.

Ans: $704

Now we apply this concept to a word problem about investments.

EXAMPLE 5 A woman invested an amount of money in two accounts for one year. She invested some at 8% simple interest and the rest at 6% simple interest. Her total amount invested was $1250. At the end of the year she had earned $86 in interest. How much money had she invested in each account?

Solution

Teaching Example 5 A man invested $2500 in two accounts for one year. He invested some of the money at 7% simple interest and the rest at 9% simple interest. At the end of one year, he had earned $205 in interest. How much money did he invest in each account?

Ans: $1000 at 7%, 1500 at 9%

Mathematics Blueprint for Problem Solving

Gather the Facts	Assign the Variable	Basic Formula or Equation	Key Points to Remember
$1250 is invested: part at 8% interest, part at 6% interest. The total interest for the year was $86.	x = amount invested at 8%. $1250 - x$ = amount invested at 6%. $0.08x$ = amount of interest for x dollars at 8%. $0.06(1250 - x)$ = amount of interest for $1250 - x$ dollars at 6%.	Interest earned at 8% + interest earned at 6% = total interest earned during the year, which is $86.	Be careful to write $1250 - x$ for the amount of money invested at 6%. The order of the amounts is $1250 - x$. Do not use $x - 1250$.

$$
\begin{array}{ccccc}
\text{interest} & & \text{interest} & & \text{total interest} \\
\text{earned} & + & \text{earned} & = & \text{earned during} \\
\text{at 8\%} & & \text{at 6\%} & & \text{the year} \\
0.08x & + & 0.06(1250 - x) & = & 86
\end{array}
$$

Note: Be sure you write $(1250 - x)$ for the amount of money invested at 6%. Students often write it backwards by mistake. It is *not* correct to use $(x - 1250)$ instead of $(1250 - x)$. The order of the terms is very important.

Solve and state the answer.

$0.08x + 75 - 0.06x = 86$	Remove parentheses.
$0.02x + 75 = 86$	Combine like terms.
$0.02x = 11$	Subtract 75 from both sides.
$\dfrac{0.02x}{0.02} = \dfrac{11}{0.02}$	Divide both sides by 0.02.
$x = 550$	The amount invested at 8% interest is $550.
$1250 - x = 1250 - 550 = 700$	The amount invested at 6% interest is $700.

Check. Are these values reasonable? Yes. Do the amounts equal $1250?

$$\$550 + \$700 \overset{?}{=} \$1250$$
$$\$1250 = \$1250 \checkmark$$

Would these amounts earn $86 interest in one year invested at the specified rates?

$$0.08(\$550) + 0.06(\$700) \overset{?}{=} \$86$$
$$\$44 + \$42 \overset{?}{=} \$86$$
$$\$86 = \$86 \checkmark$$

Practice Problem 5 A woman invested her savings of $8000 in two accounts that each calculate interest only once per year. She placed one amount in a special notice account that yields 9% annual interest. The remainder she placed in a tax-free All-Savers account that yields 7% annual interest. At the end of the year, she had earned $630 in interest from the two accounts together. How much had she invested in each account? $3500 at 9%, $4500 at 7%

 Solving Coin Problems

Coin problems provide an unmatched opportunity to use the concept of *value*. We must make a distinction between *how many coins* there are and the *value* of the coins.

Consider the next example. Here we know *the value* of some coins, but do not know *how many* we have.

EXAMPLE 6 When Bob got out of math class, he had to make a long-distance call. He had exactly enough dimes and quarters to make a phone call that would cost $2.55. He had one fewer quarter than he had dimes. How many coins of each type did he have?

Solution Let d = the number of dimes.

Then $d - 1$ = the number of quarters.

The total value of the coins was $2.55. How can we represent the value of the dimes and the value of the quarters? Think.

Each dime is worth $0.10.

5 dimes are worth $(5)(0.10) = 0.50.$

d dimes are worth $(d)(0.10) = 0.10d.$

Each quarter is worth $0.25.$

8 quarters are worth $(8)(0.25) = 2.00.$

$(d - 1)$ quarters are worth
$(d - 1)(0.25) = 0.25(d - 1).$

Now we can write an equation for the total value.

$$\text{(value of dimes)} + \text{(value of quarters)} = \$2.55$$

$$0.10d \quad + \quad 0.25(d-1) \quad = \quad 2.55$$

$0.10d + 0.25d - 0.25 = 2.55$	Remove parentheses.
$0.35d - 0.25 = 2.55$	Combine like terms.
$0.35d = 2.80$	Add 0.25 to both sides.
$\dfrac{0.35d}{0.35} = \dfrac{2.80}{0.35}$	Divide both sides by 0.35.
$d = 8$	Simplify.

$$d - 1 = 8 - 1 = 7$$

Thus Bob had eight dimes and seven quarters.

Check. Is this answer reasonable? Yes. Does Bob have one fewer quarter than he has dimes?

$$8 - 7 \stackrel{?}{=} 1$$
$$1 = 1 \quad \checkmark$$

Are eight dimes and seven quarters worth $2.55?

$$8(\$0.10) + 7(\$0.25) \stackrel{?}{=} \$2.55$$
$$\$0.80 + \$1.75 \stackrel{?}{=} \$2.55$$
$$\$2.55 = \$2.55 \quad \checkmark$$

Practice Problem 6 Ginger has five more quarters than dimes. She has $5.10 in change. If she has only quarters and dimes, how many coins of each type does she have?   11 dimes and 16 quarters

---

EXAMPLE 7 Michele and her two children returned from the grocery store with $2.80 in change. She had twice as many quarters as nickels. She had two more dimes than nickels. How many nickels, dimes, and quarters did she have?

Solution

Mathematics Blueprint for Problem Solving

Gather the Facts	Assign the Variable	Basic Formula or Equation	Key Points to Remember
Michele had $2.80 in change. She had twice as many quarters as nickels. She had two more dimes than nickels.	x = number of nickels. $2x$ = number of quarters. $x + 2$ = number of dimes. $0.05x$ = value of the nickels. $0.25(2x)$ = value of the quarters. $0.10(x + 2)$ = value of the dimes.	The value of the nickels + the value of the dimes + the value of the quarters = $2.80.	Don't add the number of coins to get $2.80. You must add the values of the coins!

(value of nickels) + (value of dimes) + (value of quarters) = \$2.80

$$0.05x + 0.10(x + 2) + 0.25(2x) = 2.80$$

Solve.

$0.05x + 0.10x + 0.20 + 0.50x = 2.80$	Remove parentheses.
$0.65x + 0.20 = 2.80$	Combine like terms.
$0.65x = 2.60$	Subtract 0.20 from both sides.
$\dfrac{0.65x}{0.65} = \dfrac{2.60}{0.65}$	Divide both sides by 0.65.
$x = 4$	Simplify. Michele had four nickels.
$2x = 2(4) = 8$	She had eight quarters.
$x + 2 = 4 + 2 = 6$	She had six dimes.

When Michele left the grocery store she had four nickels, eight quarters, and six dimes.

Check. Is the answer reasonable? Yes. Did Michele have twice as many quarters as nickels?

$$(4)(2) \stackrel{?}{=} 8 \qquad 8 = 8 \ \checkmark$$

Did she have two more dimes than nickels?

$$4 + 2 \stackrel{?}{=} 6 \qquad 6 = 6 \ \checkmark$$

Do four nickels, eight quarters, and six dimes have a value of \$2.80?

$$4(\$0.05) + 8(\$0.25) + 6(\$0.10) \stackrel{?}{=} \$2.80$$
$$\$0.20 + \$2.00 + \$0.60 \stackrel{?}{=} \$2.80$$
$$\$2.80 = \$2.80 \ \checkmark$$

NOTE TO STUDENT: *Fully worked-out solutions to all of the Practice Problems can be found at the back of the text starting at page SP-1*

Practice Problem 7 A young boy told his friend that he had twice as many nickels as dimes in his pocket. He also said that he had four more quarters than dimes. He said that he had \$2.35 in change in his pocket. Can you determine how many nickels, dimes, and quarters he had? 3 dimes, 6 nickels, and 7 quarters

Developing Your Study Skills

Applications or Word Problems

Applications or word problems are the very life of mathematics! They are the reason for doing mathematics because they teach you how to put into use the mathematical skills you have developed. Learning mathematics without ever doing word problems is similar to learning all the skills of a sport without ever playing a game or learning all the notes on an instrument without ever playing a song.

The key to success is practice. Make yourself do as many problems as you can. If you need help organizing your facts, use the Mathematics Blueprint. You may not be able to do all problems correctly at first, but keep trying. Do not give up whenever you reach a difficult one. If you cannot solve it, just try another one. Then come back and try it again later.

A misconception among students when they begin studying word problems is that each problem is different. At first the problems may seem this way, but as you practice more and more, you will begin to see the similarities, the different "types." You will see patterns in solving problems, which will enable you to solve them more easily.

Applications

Solve. All problems involving interest refer to simple interest.

1. *Raking Leaves* Paul has a job raking leaves for a neighbor. He makes $6.50 an hour, plus $0.75 for each bag he fills. Last Saturday he worked three hours and made a total of $27.75. How many bags of leaves did he fill?

$27.75 = 3(6.50) + 0.75x$

11 bags

2. *Waiting Tables* Isabel is considering a job as a waitress. She would be paid $5 per hour plus tips. The waitstaff have told her that an average tip at that restaurant is $4.50 per table. If Isabel works 25 hours per week, how many tables would she have to serve in order to make $296 for the week?

$5(25) + 4.50x = 296$

38 tables

3. *Overtime* Kate has a summer job as a swim instructor to earn her college tuition. She gets paid $8 per hour for the first 40 hours and $12 per hour for each hour in the week worked over 40 hours. Kate's goal is to earn at least $380 per week. How many hours of overtime per week will she need to achieve her goal?

$8(40) + 12x = 380$

5 hours

4. *Overtime* Pieter has a summer job as a construction assistant to earn his college tuition. He gets paid $7 per hour for the first 40 hours and $10.50 per hour for each hour in the week worked over 40 hours. Pieter's goal is to earn at least $385 per week. How many hours of overtime per week will he need to achieve his goal?

$7(40) + 10.50x = 385$

10 hours

5. *Layaway* Mrs. Peterson's triplets all want to attend the local private academy, where they are required to wear uniforms. Mrs. Peterson knows that the only way she can afford these uniforms is to shop early, put the uniforms on layaway, and pay a little every week. The total uniform cost for the three girls is $1817.75. If she put down a deposit of $600 and she could afford $105 per week, how long would it take her to pay off the uniforms? (Round to nearest whole number.)

$600.00 + 105.00x = 1817.75$ $x \approx 11.6$

12 weeks

6. *Party Costs* The Swedish Chef Catering Company charges $50.00 for setup, $85.00 for cleanup, and the special menu that Judy Carter ordered cost $23.50 per person. If Judy Carter's bill came to $558.00, how many people came to the party?

$50.00 + 85.00 + 23.50x = 558.00$

18 people

7. *Camera Sale* The camera Melissa wanted for her birthday is on sale at 28% off the usual price. The amount of the discount is $100.80. What was the original price of the camera?

$0.28x = 100.80$ $x = 360$

The original price was $360.

8. *Work Force* The number of women working full-time in Springfield has risen 12% this year. This means 216 more women have full-time jobs. What was the number of women working full-time last year?

$0.12x = 216$ $x = 1800$

1800 women

9. *Salary* The cost of living last year went up 3%. Fortunately, Alice Swanson got a 3% raise in her salary from last year. This year she is earning $22,660. How much did she make last year?

$1.03x = 22,660$

$22,000

10. *Stock Profit* A speculator bought stocks and later sold them for $5136, making a profit of 7%. How much did the stocks cost him?

$1.07x = 5136$

$4800

11. *Investments* Katerina Lubov inherited some money and invested it at 6% simple interest. At the end of the year, the total amount of her original principal and the interest was $12,720. How much did she originally invest?

$1.06x = 12,720$

$12,000

12. *Investments* Robert Campbell invested some money at 8% simple interest. At the end of the year, the total amount of his original principal and the interest was $7560. How much did he originally invest?

$1.08x = 7560$

$7000

13. **Trust Fund** Mr. and Mrs. Wright set up a trust fund for their children last year. Part of it is earning 7% simple interest per year while the rest of it is earning 5% simple interest per year. They placed $5000 in the trust fund. In one year the trust fund has earned $310. How much did they invest at each interest rate?

$0.07x + 0.05(5000 - x) = 310$
$3000 at 7%, $2000 at 5%

14. **Investments** Anne and Michael invested $8000 last year in tax-free bonds. Some of the bonds earned 8% simple interest while the rest earned 6% simple interest. At the end of the year, they had earned $580 in interest. How much did they invest at each interest rate?

$0.08x + 0.06(8000 - x) = 580$
$5000 at 8%, $3000 at 6%

15. **Investments** Plymouth Rock Bank invested $400,000 last year in mutual funds. The conservative fund earned 8% simple interest. The growth fund earned 12% simple interest. At the end of the year, the bank had earned $38,000 from these mutual funds. How much did it invest in each fund?

$0.08x + 0.12(400,000 - x) = 38,000$
$250,000 in the conservative fund,
$150,000 in the growth fund

16. **Investments** Millennium Securities last year invested $600,000 in mutual funds. The international fund earned 11% simple interest. The high-tech fund earned 7% simple interest. At the end of the year, the company had earned $50,000 in interest. How much did it invest in each fund?

$0.11x + 0.07(600,000 - x) = 50,000$
$200,000 in the international fund,
$400,000 in the high-tech fund

17. **Investments** Dave Horn invested half of his money at 5%, one-third of his money at 4%, and the rest of his money at 3.5%. If his total annual investment income was $530, how much had he invested?

x = amount of investment
$\frac{x}{2}(0.05) + \frac{x}{3}(0.04) + \frac{x}{6}(0.035) = 530$
$12,000

18. **Investments** When Karin Schumann won the lottery, she decided to invest half her winnings in a mutual fund paying 4% interest, one-third of it in a credit union paying 4.5% interest, and the rest in a bank CD paying 3% interest. If her annual investment income was $2400, how much money had she invested?

$\frac{x}{2}(0.04) + \frac{x}{3}(0.045) + \frac{x}{6}(0.03) = 2400$
$60,000

19. **Coin Bank** Little Melinda has nickels and quarters in her bank. She has four fewer nickels than quarters. She has $3.70 in the bank. How many coins of each type does she have?

$0.05(x - 4) + 0.25x = 3.70$ 13 quarters
$0.30x = 3.90$ 9 nickels
$x = 13$

20. **Coin Change** Reggie's younger brother had several coins when he returned from his paper route. He had a total of $5.35 in dimes and quarters. He had six more quarters than he had dimes. How many of each coin did he have?

$0.25(x + 6) + 0.10x = 5.35$ 11 dimes
$0.35x = 3.85$ 17 quarters
$x = 11$

21. **Coin Change** A newspaper carrier has $3.75 in change. He has three more quarters than dimes but twice as many nickels as quarters. How many coins of each type does he have?

$0.25(x + 3) + 0.10x + 0.05(2x + 6) = 3.75$ 18 nickels
$0.45x = 2.70$ 6 dimes
$x = 6$ 9 quarters

22. **Coin Change** Tim Whitman has $4.50 in change in his desk drawer to use on the vending machines downstairs. He has four more quarters than dimes. He has three times as many nickels as dimes. How many coins of each type does he have?

$0.25(x + 4) + 0.10x + 0.05(3x) = 4.50$
$x = 7$ He has 7 dimes, 11 quarters, and 21 nickels.

23. **Stamps** Ramon was office manager for a small law firm. He made the weekly postage purchase. He bought regular 37¢ stamps for most of the firm, but he also picked up some 40¢ breast cancer stamps at the request of some of the lawyers. He noticed that the number of breast cancer stamps was 8 fewer than the number of regular stamps and that he spent $19.90. How many of each stamp did he buy?

x = number of regular stamps
$x - 8$ = number of breast cancer stamps
$x(0.37) + (x - 8)(0.40) = 19.90$
$37x + 40x - 320 = 1990$
$77x = 2310$
$x = 30$
30 regular stamps; 22 breast cancer stamps

24. **Ticket Prices** Mario's Marionettes donated free tickets for their show to the local Boys & Girls Club. They claimed that the ticket value was $176.75. A child's ticket cost $5.50 and an adult's ticket cost $8.75. If there were three times as many children's tickets as adults' tickets, how many adults and children got to attend the show for free?

x = number of adults' tickets
$3x$ = number of children's tickets
$8.75x + 5.50(3x) = 176.75$
$875x + 1650x = 17675$
$2525x = 17675$
$x = 7$
7 adults; 21 children

25. *Paper Currency* Jim and Amy sold many pieces of furniture and made $1380 during their garage sale. They had sixteen more $10 bills than $50 bills. They had one more than three times as many $20 bills as $50 bills. How many of each denomination did they have?

$10(x + 16) + 20(3x + 1) + 50x = 1380$
Twenty-six $10 bills, thirty-one $20 bills, ten $50 bills

26. *Paper Currency* Roberta Burgess came home with $325 in tips from two nights on her job as a waitress. She had $20 bills, $10 bills, and $5 bills. She discovered that she had three times as many $5 bills as she had $10 bills. She also found that she had 4 fewer $20 bills than she had $10 bills. How many of each denomination did she have?

$5(3x) + 10x + 20(x - 4) = 325$
Nine $10 bills, twenty-seven $5 bills, five $20 bills

27. *Salary* Madelyn Logan is an office furniture dealer who earns an $18,000 base salary. She also earns a 4% commission on sales. How much must she sell to earn a total of $55,000?

$925,000 worth of furniture

28. *Animal Shelter* The local animal shelter accepts abandoned cats and dogs. They usually receive twice as many cats as dogs. They estimate that 80% of the cats and 60% of the dogs that come in need some kind of medical treatment. If they treated 286 animals last year, how many cats and dogs did they take in?

130 dogs; 260 cats

Cumulative Review *Perform the operations in the proper order.*

29. **[1.5.1]** $5(3) + 6 \div (-2)$
$15 - 3 = 12$

30. **[1.5.1]** $5(-3) - 2(12 - 15)^2 \div 9$
$-15 - 2(9) \div 9 = -17$

Evaluate for $a = -1$ and $b = 4$.

31. **[1.8.1]** $a^2 - 2ab + b^2$
$1 + 8 + 16 = 25$

32. **[1.8.1]** $a^3 + ab^2 - b - 5$
$-1 - 16 - 4 - 5 = -26$

Quick Quiz 3.4

1. Charles is renting a large copy machine for his company and has been given a budget of $1250. The machine he wants to rent will require a $224 installation fee and a $114-per-month rental contract. How many months will Charles be able to rent the copy machine with his current budget? 9 months

2. Last year the cost of Marcia's health plan went up 7%. The cost for coverage for Marcia and her family this year is $12,412. What did it cost last year before the increase went into effect? $11,600

3. Walter and Barbara invested $5000 in two CDs last year. One of the CDs paid 4% interest. The other one paid 5% interest. At the end of one year they earned $228 in simple interest. How much did they invest at each interest rate?
$2200 at 4%
$2800 at 5%

4. **Concept Check** Explain how you would set up an equation to solve the following problem.

Robert has $2.75 in change consisting of nickels, dimes, and quarters. He has twice as many dimes as quarters. He has one more nickel than he has quarters. How many of each coin does he have?
Answers may vary

Classroom Quiz 3.4 You may use these problems to quiz your students' mastery of Section 3.4.

1. Stephanie is leasing a car for her business. The car she has selected requires a down payment of $2500 and will cost $245 per month to lease. She has a car leasing budget of $6910. How many months will Stephanie be able to lease the car with her current budget? **Ans:** 18 months

2. Last year the cost of Phil's health plan went up 6%. The cost for coverage for Phil and his family this year is $12,932. What did it cost last year before the increase went into effect? **Ans:** $12,200

3. Steve and Bianca invested $6000 in two CDs last year. One of the CDs paid 4% interest. The other one paid 6% interest. At the end of one year they earned $314 in simple interest. How much did they invest at each interest rate?
Ans: $2300 at 4%
  $3700 at 6%

Student Learning Objectives

After studying this section, you will be able to:

1 Find the area and the perimeter of two-dimensional objects. Find the missing angle of a triangle.

2 Find the volume and the surface area of three-dimensional objects.

3 Solve more complicated geometric problems.

1 Finding Area, Perimeter, and Missing Angles

In Section 1.8 we reviewed a number of area and perimeter formulas. We repeat this list of formulas here for convenience.

AREA AND PERIMETER FORMULAS

A **parallelogram** is a four-sided figure with opposite sides parallel. In a parallelogram, opposite sides are equal and opposite angles are equal.

Perimeter = the sum of all four sides

Area = ab

A **rectangle** is a parallelogram with all interior angles measuring 90°.

Perimeter = $2l + 2w$

Area = lw

A **square** is a rectangle with all four sides equal.

Perimeter = $4s$

Area = s^2

A **trapezoid** is a four-sided figure with two sides parallel. The parallel sides are called the bases of the trapezoid.

Perimeter = the sum of all four sides

Area = $\frac{1}{2}a(b_1 + b_2)$

A **triangle** is a closed plane figure with three sides.

Perimeter = the sum of the three sides

Area = $\frac{1}{2}ab$

A **circle** is a plane curve consisting of all points at an equal distance from a given point called the center.

Circumference is the distance around a circle. A **radius** is a line segment from the center of the circle to a point on the circle. A **diameter** is a line segment across the circle that passes through the center with endpoints on the circle.

Circumference = $2\pi r$

Area = πr^2

π is a constant associated with circles. It is an irrational number that is approximately 3.141592654. We usually use 3.14 as a sufficiently accurate approximation. Thus we write $\pi \approx 3.14$ for most of our calculations involving π.

We frequently encounter triangles in word problems. There are four important facts about triangles, which we list for convenient reference.

TRIANGLE FACTS

1. The sum of the interior angles of any triangle is 180°. That is,

measure of $\angle A$ + measure of $\angle B$ + measure of $\angle C = 180°$.

2. An **equilateral** triangle is a triangle with three sides equal in length and three angles that measure 60° each.

Equilateral triangle

3. An **isosceles** triangle is a triangle with two equal sides. The two angles opposite the equal sides are also equal.

Isosceles triangle

measure $\angle A$ = measure $\angle B$

4. A **right** triangle is a triangle with one angle that measures 90°.

Right triangle

Teaching Example 1 Find the area of a triangular sail whose altitude is sixteen feet and whose base is eight feet.

Ans: 64 ft²

▲ **EXAMPLE 1**   Find the area of a triangular window whose base is 16 inches and whose altitude is 19 inches.

Solution   We can use the formula for the area of a triangle. Substitute the known values in the formula and solve for the unknown.

$A = \dfrac{1}{2}ab$ Write the formula for the area of a triangle.

$ = \dfrac{1}{2}(16\text{ in.})(19\text{ in.})$ Substitute the known values in the formula.

$ = \dfrac{1}{2}(16)(19)(\text{in.})(\text{in.})$ Simplify.

$A = 152\text{ in.}^2$ or   152 square inches

The area of the triangle is 152 square inches.

Practice Problem 1   Find the area of a triangle whose base is 14 inches and whose altitude is 20 inches.   140 in.²

19 in.

16 in.

NOTE TO STUDENT: Fully worked-out solutions to all of the Practice Problems can be found at the back of the text starting at page SP-1

EXAMPLE 2   The area of an NBA basketball court is 4700 square feet. Find the width of the court if the length is 94 feet.

Solution   Draw a diagram.

94 feet

4700 square feet w

Teaching Example 2 The area of a rectangular classroom is 616 square feet. The width of the room is 22 feet. Find the length of the classroom.

Ans: 28 ft

Write the formula for the area of a rectangle and solve for the unknown value.

$$A = lw$$

$$4700 \text{ (ft)}^2 = (94 \text{ ft})(w) \quad \text{Substitute the known values into the formula.}$$

$$\frac{4700 \text{ (ft)(ft)}}{94 \text{ ft}} = \frac{94 \text{ ft}}{94 \text{ ft}} w \quad \text{Divide both sides by 94 feet.}$$

$$50 \text{ ft} = w$$

The width of the basketball court is 50 feet.

▲ **Practice Problem 2**   The area of a rectangular field is 120 square yards. If the width of the field is 8 yards, what is the length?   15 yd

---

▲ **EXAMPLE 3**   The area of a trapezoid is 400 square inches. The altitude is 20 inches and one of the bases is 15 inches. Find the length of the other base.

Solution

$$A = \frac{1}{2}a\,(b_1 + b_2) \qquad \text{Write the formula for the area of a trapezoid.}$$

$$400(\text{in.})^2 = \frac{1}{2}(20 \text{ in.})(15 \text{ in.} + b_2) \qquad \text{Substitute the known values.}$$

$$400(\text{in.})(\text{in.}) = 10(\text{in.})(15 \text{ in.} + b_2) \qquad \text{Simplify.}$$

$$400(\text{in.})(\text{in.}) = 150(\text{in.})(\text{in.}) + (10 \text{ in.})b_2 \qquad \text{Remove parentheses.}$$

$$250(\text{in.})(\text{in.}) = (10 \text{ in.})b_2 \qquad \text{Subtract 150 in.}^2 \text{ from both sides.}$$

$$25 \text{ in.} = b_2 \qquad \text{Divide both sides by 10 in.}$$

The other base is 25 inches long.

▲ **Practice Problem 3**   The area of a trapezoid is 256 square feet. The bases are 12 feet and 20 feet. Find the altitude.   16 ft

---

▲ **EXAMPLE 4**   Find the area of a circular sign whose diameter is 14 inches. (Use 3.14 for π.) Round to the nearest square inch.

Solution $A = \pi r^2$ Write the formula for the area of a circle.

Note that the length of the diameter is given in the word problem. We need to know the length of the radius to use the formula. Since $d = 2r$, then 14 in. $= 2r$ and $r = 7$ in.

$$A = \pi(7 \text{ in.})(7 \text{ in.}) \qquad \text{Substitute known values into the formula.}$$

$$= (3.14)(7 \text{ in.})(7 \text{ in.})$$

$$= (3.14)(49 \text{ in.}^2)$$

$$= 153.86 \text{ in.}^2$$

Rounded to the nearest square inch, the area of the circle is approximately 154 square inches.

In this example, you were asked to find the area of the circle. Do not confuse the formula for area with the formula for circumference. $A = \pi r^2$, while $C = 2\pi r$. Remember that area involves square units, so it is only natural that in the area formula you would square the radius.

▲ **Practice Problem 4**   Find the circumference of a circle whose radius is 15 meters. (Use 3.14 for π.) Round your answer to the nearest meter.   94 m

▲ **EXAMPLE 5** Find the perimeter of a parallelogram whose longer sides are 4 feet and whose shorter sides are 2.6 feet.

Solution   Draw a picture.

4 ft

2.6 ft

The perimeter is the distance around a figure. To find the perimeter, add the lengths of the sides. Since the opposite sides of a parallelogram are equal, we can write the following.

$$P = 2(4 \text{ feet}) + 2(2.6 \text{ feet})$$
$$= 8 \text{ feet} + 5.2 \text{ feet}$$
$$= 13.2 \text{ feet}$$

The perimeter of the parallelogram is 13.2 feet.

▲ **Practice Problem 5** Find the perimeter of an equilateral triangle with a side that measures 15 centimeters.   45 cm

EXAMPLE 6   The smallest angle of an isosceles triangle measures 24°. The other two angles are larger. What are the measurements of the other two angles?

Solution   We know that in an isosceles triangle the measures of two angles are equal. We know that the sum of the measures of all three angles is 180°. Both of the larger angles must be different from 24°, therefore these two larger angles must be equal.
   Let x = the measure in degrees of each of the larger angles.

Then we can write $24° + x + x = 180°$
$$24° + 2x = 180° \quad \text{Add like terms.}$$
$$2x = 156° \quad \text{Subtract 24 from each side.}$$
$$x = 78° \quad \text{Divide each side by 2.}$$

Thus the measures of each of the other two angles of the triangle must be 78°.

Practice Problem 6   The largest angle on an isosceles triangle measures 132°. The other two angles are smaller. What are the measurements of the other two angles?   They each measure 24°.

24°

② Finding Volume and Surface Area

Now let's examine three-dimensional figures. **Surface area** is the total area of the faces of a figure. You can find surface area by calculating the area of each face and then finding the sum. **Volume** is the measure of the amount of space inside a figure. Some formulas for the surface area and the volume of regular figures can be found in the following table.

GEOMETRIC FORMULAS: THREE-DIMENSIONAL FIGURES

Rectangular prism   Note that all of the faces are rectangles.

h = height

Surface area = $2lw + 2wh + 2lh$

Volume = lwh

l = length w = width

GEOMETRIC FORMULAS: THREE-DIMENSIONAL FIGURE (continued)

Sphere

r = radius

Surface area = $4\pi r^2$

Volume = $\dfrac{4}{3}\pi r^3$

Right circular cylinder

r = radius
h = height

Surface area = $2\pi rh + 2\pi r^2$
Volume = $\pi r^2 h$

Teaching Example 7 Find the volume of a sphere with diameter 10 inches. (Use 3.14 for π.) Round to the nearest cubic inch.

Ans: 523 in.^{3}

▲ **EXAMPLE 7** Find the volume of a sphere with radius 4 centimeters. (Use 3.14 for π.) Round your answer to the nearest cubic centimeter.

Solution

$$V = \frac{4}{3}\pi r^3 \qquad \text{Write the formula for the volume of a sphere.}$$

$$= \frac{4}{3}(3.14)(4 \text{ cm})^3 \quad \text{Substitute the known values into the formula.}$$

$$= \frac{4}{3}(3.14)(64) \text{ cm}^3 \approx 267.946667 \text{ cm}^3$$

Rounded to the nearest whole number, the volume of the sphere is 268 cubic centimeters.

NOTE TO STUDENT: *Fully worked-out solutions to all of the Practice Problems can be found at the back of the text starting at page SP-1*

▲ **Practice Problem 7** Find the surface area of a sphere with radius 5 meters. Round your answer to the nearest square meter. 314 m^2

Teaching Example 8 Find the area of a paper label on a can of vegetables if the height of the can is 6 inches and the radius is 1.5 inches. (Use 3.14 for π.)

Ans: 56.52 in.^{2}

▲ **EXAMPLE 8** A can is made of aluminum. It has a flat top and a flat bottom. The height is 5 inches and the radius is 2 inches. How much aluminum is needed to make the can? How much aluminum is needed to make 10,000 cans?

Solution

Understand the problem. What do we need to find? Reword the problem.

We need to find the total surface area of a cylinder.

You may calculate the area of each piece and find the sum or you may use the formula.

We will use 3.14 to approximate π.

$$\text{Surface area} = 2\pi rh + 2\pi r^2$$
$$= 2(3.14)(2 \text{ in.})(5 \text{ in.}) + 2(3.14)(2 \text{ in.})^2$$
$$= 62.8 \text{ in.}^2 + 25.12 \text{ in.}^2 = 87.92 \text{ in.}^2$$

87.92 square inches of aluminum are needed to make each can.

Have we answered all the questions in the word problem? Reread the problem. How much aluminum is needed to make 10,000 cans?

(aluminum for 1 can)(10,000) = (87.92 in.^{2})(10,000) = 879,200 square inches

It would take 879,200 square inches of aluminum to make 10,000 cans.

▲ **Practice Problem 8** Sand is stored in a cylindrical drum that is 4 feet high and has a radius of 3 feet. How much sand can be stored in the drum? Round your answer to the nearest cubic foot. 113 ft^3

Teaching Example 9 The owners of a house that measures 40 feet by 20 feet want to spread mulch in a two-foot band around the house. The mulch will cost $1.75 for each square foot that needs to be covered. Find the cost of the mulch.

Ans: $448

③ Solving More Complicated Geometric Problems

▲ **EXAMPLE 9** A quarter-circle (of radius 1.5 yards) is connected to two rectangles with dimensions as labeled on the sketch in the margin. You need to lay a strip of carpet in your house according to this sketch. How many square yards of carpeting will be needed on the floor? (Use 3.14 for π. Round your final answer to the nearest tenth.)

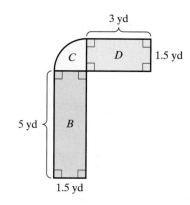

Solution The desired area is the sum of three areas, which we will call B, C, and D. Area B and area D are rectangular in shape. It is relatively easy to find the areas of these shapes.

$$A_B = (5 \text{ yd})(1.5 \text{ yd}) = 7.5 \text{ yd}^2 \qquad A_D = (3 \text{ yd})(1.5 \text{ yd}) = 4.5 \text{ yd}^2$$

Area C is one-fourth of a circle. The radius of the circle is 1.5 yd.

$$A_C = \frac{\pi r^2}{4} = \frac{(3.14)(1.5 \text{ yd})^2}{4} = \frac{(3.14)(2.25) \text{ yd}^2}{4} = 1.76625 \text{ yd}^2 \approx 1.8 \text{ yd}^2$$

The total area is $A_B + A_D + A_C \approx 7.5 \text{ yd}^2 + 4.5 \text{ yd}^2 + 1.8 \text{ yd}^2 \approx 13.8 \text{ yd}^2$. 13.8 square yards of carpeting will be needed to cover the floor.

▲ **Practice Problem 9** John has a swimming pool that measures 8 feet by 12 feet. He plans to make a concrete walkway around the pool that is 3 feet wide. What will be the total cost of the walkway at $12 per square foot? $1872

Teaching Tip Discuss the answer to Example 9. Point out that this is the amount of carpeting to be placed on the floor. There would undoubtedly be some waste when the carpeting is cut to these dimensions. Ask what would be the amount of waste if area C were cut from a square piece of carpeting 1.5 yd on a side.

Round all values to the nearest hundredth.

$$A_{\text{square}} = (1.5 \text{ yd})(1.5 \text{ yd})$$
$$= 2.25 \text{ yd}^2$$
$$A_{\text{square}} - A_C = 2.25 \text{ yd}^2 - 1.77 \text{ yd}^2$$
$$= 0.48 \text{ yd}^2$$

This would be 0.48 yd^2 of waste.

EXAMPLE 10 Find the weight of the water in a full cylinder containing water if the height of the cylinder is 5 feet and the radius is 2 feet. 1 cubic foot of water weighs 62.4 pounds. Round your answer to the nearest ten pounds. Use 3.14 for π.

Solution

Understand the problem.
Draw a diagram.

Write the formula for the volume of a cylinder.

$$V = \pi r^2 h$$
$$= (3.14)(2 \text{ ft})^2(5 \text{ ft})$$
$$= (3.14)(4 \text{ ft}^2)(5 \text{ ft})$$
$$= 62.8 \text{ ft}^3$$

The volume of the cylinder is 62.8 cubic feet.

Each cubic foot of water weighs 62.4 pounds. This can be written as 62.4 lb/ft^3. Since the volume of the cylinder is 62.8 ft^3, we multiply.

$$\text{Weight of the water} = (62.8 \text{ ft}^3)\left(\frac{62.4 \text{ lb}}{1 \text{ ft}^3}\right)$$
$$= (62.8)(62.4) \text{ lb} = 3918.72 \text{ lb}$$

The weight of the water in the cylinder is 3920 rounded to the nearest ten pounds.

Teaching Example 10 One tank at the city Aquarium is 6 feet long, 4 feet wide, and 5 feet high. What is the weight of the water in the tank when the tank is completely full? (1 cubic foot of water weighs 62.4 pounds.) Round your answer to the nearest ten pounds.

Ans: 7490 lb

Practice Problem 10 Find the weight of the water in a full rectangular container that is 5 feet wide, 6 feet long, and 8 feet high. Remember, 1 cubic foot of water weighs 62.4 pounds. Round your answer to the nearest 100 pounds.

15,000 lb

Verbal and Writing Skills

Geometry Fill in the blank to complete each sentence.

▲ **1.** Perimeter is the ___distance around___ a plane figure.

▲ **2.** ___Circumference___ is the distance around a circle.

▲ **3.** Area is a measure of the amount of ___surface___ in a region.

▲ **4.** A ___trapezoid___ is a four-sided figure with exactly two sides parallel.

▲ **5.** The sum of the interior angles of any triangle is ___180°___.

▲ **6.** An equilateral triangle is a triangle in which each angle measures ___60°___.

▲ **7.** *Geometry* Find the area of a triangle whose altitude is 24 inches and whose base is 13 inches.
$$A = \frac{1}{2}(24)(13) = 156 \text{ in.}^2$$

▲ **8.** *Geometry* A triangle has an altitude of 9 m and a base of 21 m. Find the area of the triangle.
$$A = \frac{1}{2}(9)(21) = 94.5 \text{ m}^2$$

▲ **9.** *Geometry* Find the area of a parallelogram whose altitude is 14 inches and whose base is 7 inches.
$$A = (14)(7) = 98 \text{ in.}^2$$

▲ **10.** *Geometry* Find the area of a parallelogram whose altitude is 15 meters and whose base is 10 meters.
$$A = (15)(10) = 150 \text{ m}^2$$

▲ **11.** *Geometry* The area of a triangle is 90 square feet. Find the altitude if the base is 30 feet.
$$90 = \frac{1}{2}(a)(30)$$
$$90 = 15a$$
$$a = 6 \text{ ft}$$

▲ **12.** *Geometry* The area of a triangle is 120 square inches. Find the altitude if the base is 24 inches.
$$120 = \frac{1}{2}(a)(24)$$
$$120 = 12a$$
$$a = 10 \text{ in.}$$

▲ **13.** *Geometry* The perimeter of a parallelogram is 46 inches. If the length of one side is 14 inches, what is the length of a side adjacent to it?
$$2(14) + 2x = 46$$
$$x = 9 \text{ in.}$$

▲ **14.** *Geometry* If one side of a parallelogram measures 18.5 cm and an adjacent side measures 17.5 cm, what is the perimeter of the parallelogram?
$$P = 2(18.5) + 2(17.5)$$
$$P = 72 \text{ cm}$$

▲ **15.** *Geometry* Find the area of a circular sign whose radius is 7.00 feet. (Use 3.14 to approximate π.) 153.86 ft^2

▲ **16.** *Flower Bed* Find the area of a circular flower bed whose diameter is 6 meters. (Use 3.14 to approximate π.) 28.26 m^2

▲ **17.** *Mercury* The diameter of the planet Mercury is approximately 3032 miles. Find the distance around its equator. (Use 3.14 to approximate π.)

$$C = \pi d$$
$$C = (3.14)(3032)$$
$$C = 9520.48 \text{ mi}$$

▲ **18.** *Olympic Medals* The medals made for the 2008 Beijing Olympics measure 70 mm, or about 2.76 inches, in diameter. Find the area of the medals in square inches. (Use 3.14 to approximate π.) Round your answer to the nearest hundredth.

$$A = \pi r^2$$
$$A = (3.14)(1.38)^2$$
$$A \approx 5.98 \text{ in.}^2$$

▲ **19.** *Geometry* The area of a trapezoid is 600 square inches and the bases are 20 inches and 30 inches. Find the altitude.
$$600 = \frac{1}{2}a(30 + 20) \quad a = 24 \text{ in.}$$

▲ **20.** *Geometry* The area of a trapezoid is 900 square inches. The bases are 40 inches and 50 inches. Find the altitude.
$$900 = \frac{1}{2}a(40 + 50)$$
$$a = 20 \text{ in.}$$

Applications

▲ **21.** ***Streets of Manhattan*** In midtown Manhattan, the street blocks have a uniform size of 80 meters north-south by 280 meters east-west. If a typical New York City neighborhood is two blocks east-west by three blocks north-south, how much area does it cover?

length = 3(80)
width = 2(280)
  area = (240)(560) = 134,400 m²

▲ **22.** ***Computer Cases*** A computer manufacturing plant makes its computer cases from plastic. Two sides of a triangular scrap of plastic measure 7 cm and 25 cm.

(a) If the perimeter of the plastic piece is 56 cm, how long is the third side?
  7 + 25 + x = 56
        x = 24   The third side is 24 cm.

(b) If each side of the triangular piece is decreased by 0.1 cm, what would the new perimeter be? 55.7 cm

▲ **23.** ***Geometry*** The circumference of a circle is 31.4 centimeters. Find the radius. (Use 3.14 to approximate π.)
31.4 = (3.14)(2r)
  r = 5 cm

▲ **24.** ***Geometry*** The perimeter of an equilateral triangle is 27 inches. Find the length of each side of the triangle.
27 = 3x
  x = 9 in.

▲ **25.** ***Tires*** An automobile tire has a diameter of 64 cm. What is the circumference of the tire? (Use 3.14 to approximate π.)
C = πd = 3.14(64) = 200.96 cm

▲ **26.** ***Golden Ratio*** The ancient Greeks discovered that rectangles where the length-to-width ratio was 8 to 5 (known as the golden ratio) were most pleasing to the eye. Kayla wants to create a painting that will be framed by some antique molding that she found. If she calculates that the molding can be used to make a frame of 104 inches in perimeter, what dimensions should her painting be so that it is visually appealing?

2(8x) + 2(5x) = 104
        x = 4
Length should be 8(4) = 32 in.
  Width should be 5(4) = 20 in.

▲ **27.** ***Geometry*** Two angles of a triangle measure 47 degrees and 59 degrees. What is the measure of the third angle?   x + 47 + 59 = 180
          x = 74°

▲ **28.** ***Geometry*** A right triangle has one angle that measures 77 degrees. What does the other acute angle measure?   x + 77 = 90
        x = 13°

▲ **29.** ***Geometry*** Each of the equal angles of an isosceles triangle is twice as large as the third angle. What is the measure of each angle?
Third angle = x; equal angles = 2x
2x + 2x + x = 180
      x = 36
Third angle = 36°; equal angles = 72°

▲ **30.** ***Geometry*** The measure of the first angle in a triangle is triple the measure of the second angle. The measure of the third angle is 10 degrees more than the second angle. What is the measure of each angle?
Second angle = x, first angle = 3x, third angle = x + 10
3x + x + (x + 10) = 180
        x = 34
Second angle = 34°, first angle = 102°, third angle = 44°

▲ **31.** ***Playhouse Construction*** Roger constructs children's playhouses. At the top of the houses he places an isosceles triangle to form the roof. The largest angle measures 120°. What are the measurements of the other two angles in this triangle?
They each measure 30°.

▲ **32.** ***TV Antenna*** The largest angle of an isosceles triangle used in holding a cable TV antenna measures 146°. The other two angles are smaller. What are the measurements of the other two angles in the triangular piece of the cable TV antenna?
They each measure 17°.

▲ **33.** *Parking Lot* A triangular region of the community college parking lot was measured. The measure of the first angle of this triangle is double the measure of the second angle. The measure of the third angle is 19° greater than double the measure of the first angle. What are the measurements of all three angles?

First angle is 46°;
second angle is 23°;
third angle is 111°

▲ **34.** *Lobster Pot* A lobster pot used in Gloucester, Massachusetts, has a small triangular piece of wood at the entry point where the lobster enters. The measure of the first angle of this triangle is triple the measure of the second angle. The third angle of the triangle is 30° less than double the measure of the first angle. What are the measurements of all the three angles?

First angle is 63°;
second angle is 21°;
third angle is 96°

▲ **35.** *Garden Border* Kim wants to put a small fence in the shape of a circle around a tree in her backyard. She wants the radius of the circle to be 3 feet, leaving room to plant flowers around the tree. How many feet of fencing does she need to buy if it is sold by the foot? (Use 3.14 to approximate π.)   19 ft

▲ **36.** *Rock Garden* Nate and Jenny want to order large rocks to use as a border around three rectangular gardens. Each garden measures 4 feet by 9 feet, and each rock will each measure about 6 inches across. How many rocks do they need to order?   156 rocks

▲ **37.** *Painting a Shed* The cost for paint to cover the exterior of the four sides of a shed was $40. A 5-gallon can of paint costs $20 and will cover 102 square feet. The shed consists of four sides, each shaped like a trapezoid. They each have one base of 8 feet and one base of 9 feet. What is the altitude of each trapezoid?

6 ft

▲ **38.** *Painting a Sign* The cost for paint to cover both sides of a large road-side sign (front side and back side) was $150. A 5-gallon can of paint costs $25 and will cover 250 square feet. The sign is constructed in the shape of a trapezoid. It has one base measuring 35 feet and one base measuring 40 feet. What is the altitude of this trapezoid?   20 ft

▲ **39.** *Geometry* What is the volume of a cylinder whose height is 8 inches and whose radius is 10 inches? (Use 3.14 to approximate π.)

$V = (3.14)(8)(10)(10) = 2512$ in.^{3}

▲ **40.** *Gas Cylinder* A cylinder holds propane gas. The volume of the cylinder is 235.5 cubic feet. Find the height if the radius is 5 feet. (Use 3.14 to approximate π.)

$235.5 = (3.14)(5)(5)x$
$x = 3$ ft

▲ **41.** *Weather Balloon* A spherical weather balloon needs to hold at least 175 cubic feet of helium to be buoyant enough to lift the meteorological instruments. Will a helium-filled balloon with a diameter of 8 feet stay aloft?

$V = \frac{4}{3}\pi r^3$

$V = \left(\frac{4}{3}\right)(3.14)(4)^3 = 267.95$ ft^3

Yes, it will stay aloft.

▲ **42.** *Storage Freezer* George is considering purchasing a new freezer. He has space for a freezer that measures 5 feet long and 2.5 feet high. He also needs to have 25 cubic feet of storage space. How wide can the freezer be?

$V = lwh$
$25 = (5)(x)(2.5)$
$25 = 12.5x$
$x = 2$ ft wide

▲ **43.** *Milk Container* A plastic cylinder made to hold milk is constructed with a solid top and bottom. The radius is 6 centimeters and the height is 4 centimeters. **(a)** Find the volume of the cylinder. **(b)** Find the total surface area of the cylinder. (Use 3.14 to approximate π.)

(a) $V = (3.14)(6)(6)(4) = 452.16$ cm^3

(b) $S = 2(3.14)(6)(4) + 2(3.14)(6)(6) = 376.8$ cm^2

▲ **44.** *Pyrex Sphere* A Pyrex glass sphere is made to hold liquids in a science lab. The radius of the sphere is 3 centimeters. **(a)** Find the volume of the sphere. **(b)** Find the total surface area of the sphere. (Use 3.14 to approximate π.)

(a) $V = \frac{4}{3}(3.14)(3)(3)(3) = 113.04$ cm^3

(b) $S = 4(3.14)(3)(3) = 113.04$ cm^2

▲ **45. Goat Tether** Dan Perkin's goat is attached to the corner of the 8-foot-by-4-foot pump house by an 8-foot length of rope. The goat eats the grass around the pump house as shown in the diagram. This pattern consists of three-fourths of a circle of radius 8 feet and one-fourth of a circle of radius 4 feet. What area of grass can Dan Perkin's goat eat? (Use 3.14 to approximate π.)

$A = 0.75(3.14)(64) + 0.25(3.14)(16)$
$A = 150.72 + 12.56 = 163.28 \text{ ft}^2$

▲ **46. Seven-Layer Torte** For a party, Julian is making a seven-layer torte, which consists of seven 9-inch-diameter cake layers with a cream topping on each. He knows that one recipe of cream topping will cover 65 square inches of cake. How many batches of the topping must he make to complete his torte? (Use 3.14 to approximate π.)

$A = 7(3.14)(4.5)^2 = 445.095 \text{ in.}^2$
$445.095/65 \approx 7 \text{ batches}$

▲ **47. Walkway** A cement walkway is to be poured. It consists of the two rectangles with dimensions as shown in the diagram and a quarter of a circle with radius 1.5 yards. (Use 3.14 to approximate π.) **(a)** How many square yards will the walkway be? **(b)** If a painter paints it for $2.50 per square yard, how much will the painting cost? **(a)** $A = 22.77 \text{ yd}^2$ **(b)** about $56.93

▲ **48. Aluminum Plate** An aluminum plate is made according to the following dimensions. The outer radius is 10 inches and the inner radius is 3 inches. (Use 3.14 to approximate π.) **(a)** How many square inches in area is the aluminum plate (the shaded region in the sketch)? **(b)** If the cost to construct the plate is $7.50 per square inch, what is the construction cost? **(a)** 285.74 in.^{2} **(b)** $2143.05

49. Toy Car Track New Toy Company is producing a racing car set, complete with plastic track. The track is a solid surface and has a straight edge with circular ends. See the following sketch for dimensions. What is the distance around the outside of track? (Use 3.14 to approximate π.)

$P = 2\pi r + 2x$
$P = 2(3.14)(7) + 2(15)$
$P = 73.96 \text{ in.}$

50. Uniform Design The Viceroy Soccer team has designed new uniforms. The front of their jerseys will have a blue rectangle, with a red V inside. (See the diagram.) The plain jersey shirt costs $19.00. The red color costs $0.05 per square inch and the blue color costs $0.02 per square inch.

(a) What is the area of the red V?

Area V $= \dfrac{1}{2}(10)(10) - \dfrac{1}{2}(4)(7) = 36 \text{ in.}^2$

(b) What is the area of the blue color?

Area blue $= (12)(10) - 36 = 84 \text{ in.}^2$

(c) What is the cost per jersey?

Cost $= 19.00 + 0.05(36) + 0.02(84) = \$22.48$

▲ **51. *Painting Frame*** It cost $120.93 to frame a painting with a flat aluminum strip. This was calculated at $1.39 per foot for materials and labor. Find the dimensions of the painting if its length is 3 feet less than four times its width.

$L = 4W - 3;$ $(2W + 2L)1.39 = 120.93$
$[2W + 2(4W - 3)]1.39 = 120.93$
$(10W - 6)1.39 = 120.93$
$W = 9.3 \text{ ft}; L = 32.4 \text{ ft}$

▲ **52. *Driveway*** A concrete driveway measures 18 yards by 4 yards. The driveway will be 0.5 yards thick. **(a)** How many cubic yards of concrete will be needed? **(b)** How much will it cost if concrete is $3.50 per cubic yard?
(a) 36 yd^3 **(b)** $126

To Think About

▲ **53. *Moon*** Assume that a very long rope supported by poles that are 3 feet tall is stretched around the moon. (Neglect gravitational pull and assume that the rope takes the shape of a large circle.) The radius of the moon is approximately 1080 miles. How much longer would you need to make the rope if you wanted the rope to be supported by poles that are 4 feet tall? (Use 3.14 to approximate π.)
$2(3.14)(1080 \cdot 5280 + 4) - 2(3.14)(1080 \cdot 5280 + 3) = 2(3.14) = 6.28 \text{ ft}$

Cumulative Review *Simplify.*

54. [1.9.1] $2(x - 6) - 3[4 - x(x + 2)]$
$3x^2 + 8x - 24$

55. [1.9.1] $-5(x + 3) - 2[4 + 3(x - 1)]$
$-11x - 17$

56. [0.7.1] In 2002, 203 billion pieces of mail were delivered in the United States. By 2006, that number had increased by 5%. How many pieces of mail were delivered in 2006? (*Source:* www.usps.com) Round your answer to the nearest billion.
213 billion

57. [0.7.1] During the first six months of 2005, 92,558 hybrid cars were sold in the United States. During the same period of 2004, 36,276 hybrids were sold. Find the percent increase in the number of hybrids sold between the first six months of 2004 and the first six months of 2005. (*Source:* www.allabouthybridcars.com) Round to the nearest percent.   155%

3.6 USING INEQUALITIES TO SOLVE WORD PROBLEMS

① Translating English Sentences into Inequalities

Real-life situations often involve inequalities. For example, if a student wishes to maintain a certain grade point average (GPA), you may hear him or her say, "I have to get a grade of at least an 87 on the final exam."

What does "at least an 87" mean? What grade does the student need to achieve? 87 would fit the condition. What about 88? The student would be happy with an 88. In fact the student would be happy with any grade that *is greater than or equal to* an 87. We can write this as

$$\text{grade} \geq 87.$$

Notice that the English phrase "at least an 87" translates mathematically as "greater than or equal to an 87."

Be careful when translating English phrases that involve inequalities into algebra. If you are not sure which symbol to use, try a few numbers to see if they fit the condition of the problem. Choose the mathematical symbol accordingly.

Here is a list that may help you.

Student Learning Objectives

After studying this section, you will be able to:

① Translate English sentences into inequalities.

② Solve applied problems using inequalities.

English Phrases Describing Inequality	Algebraic Expression
Greater than	
Is more than	
Is greater than	>
Less than	
Is less than	
Is smaller than	<
Greater than or equal to	
Is greater than or equal to	
Is not less than	≥
Is at least	
Cannot be less than	
Less than or equal to	
Is less than or equal to	
Is not greater than	≤
Is at most	
Is not more than	
Does not exceed	

EXAMPLE 1 Translate each sentence into an inequality.

(a) The perimeter of the rectangle must be no more than 70 inches.
(b) The average number of errors must be less than two mistakes per page.
(c) His income must be at least $950 per week.
(d) The new car will cost at most $12,070.

Solution

(a) Perimeter must be *no more than* 70.

Perimeter ≤ 70

Try 80. Does 80 satisfy this requirement? (That is, is 80 more than 70?) No, it does not. Do 60 and 50 satisfy this requirement? Yes. Use the "less than or equal to" symbol.

(b) Average number must be *less than* 2.

Average < 2

(c) Income must be *at least* $950.

Income ≥ 950

"At least" means "not less than." The income must be $950 or more. Use ≥.

Teaching Example 1 Translate each English statement into an inequality.

(a) The height of the fence must be at least 6 feet.

(b) The weight of the truck can be no more than 8 tons.

(c) The speed of the typist must be more than 30 words per minute.

(d) The area of the square must be less than 36 square inches.

Ans: (a) height ≥ 6   **(b)** weight ≤ 8
    (c) speed > 30   **(d)** area < 36

(d) The new car will cost *at most* $12,070.

$$\text{Car} \quad \leq \quad 12{,}070$$

"At most $12,070" means it is the top price for the car. The cost of the car cannot be more than $12,070. Use ≤.

NOTE TO STUDENT: *Fully worked-out solutions to all of the Practice Problems can be found at the back of the text starting at page SP-1*

 Practice Problem 1 Translate each sentence into an inequality.

(a) The height of the person must be no greater than 6 feet. height ≤ 6

(b) The speed of the car must have been greater than 65 miles per hour. speed > 65

(c) The area of the room must be greater than or equal to 560 square feet.

(d) The profit margin cannot be less than $50 per television set. profit margin ≥ 50

(c) area ≥ 560

2 Solving Applied Problems Using Inequalities

We can use the same problem-solving techniques we used previously to solve word problems involving inequalities.

Teaching Example 2 A restaurant owner has a goal that her restaurant will serve an average of at least 85 customers per day. On Sunday through Friday of this week, the restaurant has served 108, 60, 64, 68, 79, and 98 customers. How many customers must be served on Saturday in order for the owner to meet her goal?

Ans: 118 or more customers

EXAMPLE 2 A manufacturing company makes 60-watt light bulbs. For every 1000 bulbs manufactured, a random sample of five bulbs is selected and tested. To meet quality control standards, the average number of hours these five bulbs last must be at least 900 hours. The test engineer tested five bulbs. Four of them lasted 850 hours, 1050 hours, 1000 hours, and 950 hours, respectively. How many hours must the fifth bulb last in order for the batch to meet quality control standards?

Solution

1. *Understand the problem.*

Let n = the number of hours the fifth bulb must last

The average that the five bulbs last *must be at least* 900 hours. We could write

$$\text{Average} \geq 900 \text{ hours.}$$

2. *Write an inequality.*

$$\text{Average of five bulbs} \geq 900 \text{ hours.}$$

$$\frac{850 + 1050 + 1000 + 950 + n}{5} \geq 900$$

3. *Solve and state the answer.*

$$\frac{850 + 1050 + 1000 + 950 + n}{5} \geq 900$$

$$\frac{3850 + n}{5} \geq 900 \qquad \text{Simplify the numerator.}$$

$$3850 + n \geq 4500 \qquad \text{Multiply both sides by 5.}$$

$$n \geq 650 \qquad \text{Subtract 3850 from each side.}$$

If the fifth bulb burns for 650 hours or more, the quality control standards will be met.

4. *Check.* Any value of 650 or larger should satisfy the original inequality. Suppose we pick $n = 650$ and substitute it into the inequality.

$$\frac{850 + 1050 + 1000 + 950 + 650}{5} \overset{?}{\geq} 900$$

$$\frac{4500}{5} \overset{?}{\geq} 900$$

$$900 \geq 900 \quad \checkmark$$

If we try $n = 700$, the left side of the inequality will be 910. $910 \geq 900$. In fact, if we try any number greater than 650 for n, the left side of the inequality will be greater than 900. The inequality will be valid for $n = 650$ or any larger value.

Practice Problem 2   A manufacturing company makes coils of $\frac{1}{2}$-inch maritime rope. For every 300 coils of rope made, a random sample of four coils of rope is selected and tested. To meet quality control standards, the average number of pounds that each coil will hold must be at least 1100 pounds. The test engineer tested four coils of rope. The first three held 1050 pounds, 1250 pounds, and 950 pounds, respectively. How many pounds must the last coil hold if the sample is to meet quality control standards?   At least 1150 pounds

EXAMPLE 3   Juan is selling supplies for restaurants in the Southwest. He gets paid $700 per month plus 8% of the value of all the supplies he sells to restaurants. If he wants to earn more than $3100 per month, what value of products will he need to sell?

Teaching Example 3 Theresa's phone company charges her $28 per month plus $0.06 per minute that she spends talking on the phone. If Theresa wants her phone bill to be less than $40 per month, how many minutes can she talk on the phone?

Ans: Less than 200 minutes

Solution

1. ***Understand the problem.*** Juan earns a fixed salary and a salary that depends on how much he sells (commission salary) each month.

Let $x =$ the value of the supplies he sells each month.

Then $0.08x =$ the amount of the commission salary he earns each month.

He wants his total income to be more than $3100 per month; thus

$$\text{total income} > 3100.$$

2. ***Write the inequality.***

The fixed income added to the commission income must be greater than 3100.

| 700 | + | $0.08x$ | > | 3100 |

3. ***Solve and state the answer.***

$$700 + 0.08x > 3100$$
$$0.08x > 2400 \quad \text{Subtract 700 from each side.}$$
$$x > 30,000 \quad \text{Divide each side by 0.08.}$$

Juan must sell more than $30,000 worth of supplies each month.

4. ***Check.*** Any value greater than 30,000 should satisfy the original inequality.

Suppose we pick $x = 30,001$ and substitute it into the inequality.

$$700 + 0.08(30,001) \overset{?}{>} 3100$$
$$700 + 2400.08 \overset{?}{>} 3100$$
$$3100.08 > 3100 \ \checkmark$$

Practice Problem 3   Rita is selling commercial fire and theft alarm systems to local companies. She gets paid $1400 per month plus 2% of the cost of all the systems she sells in one month. She wants to earn more than $2200 per month. What value of products will she need to sell each month?

More than $40,000 worth of products

Verbal and Writing Skills *Translate to an inequality.*

1. The cost is greater than $67,000.
$x > 67,000$

2. Anna wants her time in a 5-kilometer race to be less than 21 minutes.
$x < 21$

3. The number of people on the subway car must not be more than 90.
$x \leq 90$

4. The tax increase is more than $7890.
$x > 7890$

5. The height of stratocumulus clouds is not more than 1500 feet.
$h \leq 1500$

6. The volume of water in the aquarium must not drop below 35 gallons.
$v \geq 35$

7. To earn an A in algebra, Ramon cannot get less than a 93 average.
$x \geq 93$

8. The camp-ins at the Museum of Science in Boston cannot accommodate more than 600 people.
$x \leq 600$

Applications *Solve using an inequality.*

▲ **9.** *Window Design* A triangular window on Ross Camp's new Searay boat has two sides that measure 87 centimeters and 64 centimeters, respectively. The perimeter of the triangle must not exceed 291 centimeters. What are the possible values for the length of the third side of the window?
$87 + 64 + x \leq 291$
It must be less than or equal to 140 cm.

▲ **10.** *Package Size* The U.S. Postal Service provides guidelines for package size. One of the guidelines specifies the dimensions of the package: The length plus the girth (circumference around the widest part) cannot exceed 130 inches. If you are preparing a package that is 33 inches wide and 28 inches high, how long is the package permitted to be?
$L + 2(33) + 2(28) \leq 130$
The length must be 8 inches or less.

28″

33″

L

11. *Flower Delivery* Flora's Flowers uses the Speedy Delivery Service. Flora estimates that if she averages at least 24 deliveries per week, she will save money by hiring a student to deliver her flowers after school. Her contract with Speedy Delivery is up for renewal in 6 weeks. Flora tracks her deliveries for 5 weeks: 18, 40, 21, 7, and 36. At least how many deliveries must she average in week 6 to convince her to drop the delivery service?
$\dfrac{18 + 40 + 21 + 7 + 36 + x}{6} \geq 24$
$x \geq 22$ deliveries
She must have 22 or more deliveries in week 6.

12. *Course Average* Sophia has scored 85, 77, and 68 on three tests in her biology course. She has one final test to take and wants to obtain an average of 80 or more in the course. What possible values can she obtain on her final test and still achieve her goal?
$\dfrac{85 + 77 + 68 + x}{4} \geq 80$ She must score 90 or more.

▲ 13. *Garden Design* Joel plans to plant a flower garden in the front of his house, where he can fit a garden that is 8 feet wide. He wishes to use the can of wildflower mix that he got for his birthday. The can of seeds claims that it will luxuriously carpet an area of no more than 60 square feet. What are the possible dimensions for the depth of Joel's garden, so that it will be covered with wildflowers?

$8d \le 60$
$d \le 7.5$
The depth can be no more than 7.5 feet.

14. *Class Size* The Lee County Elementary School District noted that there was an increase of 1785 pupils during the fall semester. How many new teachers must they add to keep their average class size under 28? (Assume one teacher for each class.)

$1785/T < 28$
$T > 1785/28$
$T > 63.75$
They must add at least 64 new teachers.

15. *Truck Rental* Abby and Frank decided that to save money, they would move themselves from their apartment to their new house. They had to rent a truck to do this. Because their budget was very strict, they could afford to spend no more than $100. The rental truck costs $29.95 per day plus $0.49 per mile. If they needed the truck for two days, what was the limit on the mileage they could put on the truck?

$2(29.95) + 0.49x \le 100$
$x \le 81.8$
They can drive no more than 81.8 miles.

16. *Salary* Robin just graduated from college and has taken her first job. She is working for a computer sales company, Compusell. While she hopes to eventually get into management, she is presently working as a salesperson. Her earnings consist of a monthly salary of $1300, plus 6% of all her sales. To meet her monthly obligations, Robin wants to earn at least $2800 per month. What is the amount of sales she must make monthly to meet her goal?

$1300 + 0.06x \ge 2800$
$x \ge 25,000$
Her sales must be at least $25,000.

17. *Coolant* The coolant in Tom's car is effective for radiator temperatures of less than 110 degrees Celsius. What would be the corresponding temperature range in degrees Fahrenheit? (Use $C = \frac{5}{9}(F - 32)$ in your inequality.)

$\frac{5}{9}(F - 32) < 110$
$F < 230$
The range is less than 230°F.

18. *IQ* A person's IQ is found by multiplying mental age (m), as indicated by standard tests, by 100 and dividing this result by the chronological age (c). This gives the formula IQ $= 100\frac{m}{c}$. If a group of 14-year-old children has an IQ of not less than 80, what is the mental age range of this group?

$\frac{100m}{14} \ge 80$
$m \ge 11.2$
The mental age range is at least 11.2 years.

19. *Federal Aid* The amount of federal aid in billions of dollars given to state and local governments can be approximated by the equation $A = 19.2x + 124.8$, where x is the number of years since 1990. Use this equation to determine those years for which the amount of federal aid to state and local governments will exceed $528 billion. (*Source:* www.census.gov)

$19.2x + 124.8 > 528$
$x > 21$
The amount of federal aid will exceed $528 billion after 2011.

20. *Medicare* The number of people in millions who are enrolled in Medicare can be approximated by the equation $N = 0.52x + 34.5$, where x is the number of years since 1990. Use this equation to determine those years for which the number of people enrolled in Medicare will exceed 47.5 million. (*Source:* www.census.gov)

$0.52x + 34.5 > 47.5$
$x > 25$
The number of people enrolled in Medicare will exceed 47.5 million after 2015.

21. *Commission* Frank makes $16,000 per year plus a 6% commission on all sales over $30,000. He wants to make more than $22,000 this year. How many dollars in sales must he achieve to make this salary goal?
More than $130,000 in sales

22. *Commission* Judy makes $18,000 per year plus a 7% commission on all sales over $40,000. She wants to make more than $26,400 this year. How many dollars in sales must she achieve to make this salary goal?
More than $160,000 in sales

To Think About

23. *CD Manufacturing* A compact disc recording company's weekly costs can be described by the equation cost = 5000 + 7n, where *n* is the number of compact discs manufactured in one week. The equation for the amount of income produced from selling these *n* discs is income = 18n. At least how many discs need to be manufactured and sold in one week for the income to be greater than the cost?

$18n > 5000 + 7n$
$n > 454.545$ discs
455 or more discs would need to be sold.

24. *DVD Player Manufacturing* A company that manufactures DVD players has weekly costs described by the equation cost = 15,000 + 120n, where *n* is the number of DVD players manufactured in one week. The equation for the amount of income produced from selling these *n* DVD players is income = 250n. At least how many DVD players need to be manufactured and sold in one week for the income to be greater than the cost?

$250n > 15,000 + 120n$
$n > 115.38$
116 or more DVD players need to be manufactured and sold.

Cumulative Review *Solve.*

25. **[2.7.1]** $10 - 3x > 14 - 2x$
$x < -4$

26. **[2.7.1]** $2x - 3 \geq 7x - 18$
$x \leq 3$

27. **[2.7.1]** $30 - 2(x + 1) \leq 4x$
$x \geq 4\frac{2}{3}$

28. **[2.7.1]** $2(x + 3) - 22 < 4(x - 2)$
$x > -4$

Quick Quiz 3.6

1. Jane wants to plant a rectangular garden that is 8 feet wide. She wants to have a garden that is no more than 79.2 square feet in area. In order to comply with that plan, what should be the length of Jane's garden?
No more than 9.9 feet

2. The student activities office has four students who work part-time on a work-study program. Three of the students work 15 hours per week, 17 hours per week, and 12 hours per week. The office policy is that the average number of hours students work per week in the student activities office is less than 15.5 hours. What is the restriction for the number of hours that the fourth student can work?
Less than 18 hours per week

3. Chris makes $17,000 per year plus a 4% commission on all sales over $10,000. He wants to make more than $28,000 this year. How many dollars in sales must he achieve to make his salary goal?
More than $285,000 in sales

4. **Concept Check** Explain how you would set up an inequality to solve the following problem.
Caleb wants to spend no more than $150 to rent a truck to move things for his dad's company. The rental company charges $37.50 per day to rent the truck, and 19.5 cents per mile that the truck is driven. If he rented the truck for two days, how many miles could he drive the truck?
Answers may vary

Classroom Quiz 3.6 You may use these problems to quiz your students' mastery of Section 3.6.

1. Russ and Norma Camp want to build a rectangular summer cottage that is 18 feet wide. According to the zoning restrictions of the town, the cottage cannot have a floor plan that exceeds 405 square feet. In order to comply with the town restriction, what should be the length of the summer cottage? **Ans:** No more than 22.5 feet

2. The fleet of four delivery trucks for the city of Springfield must conform to a requirement that the average miles per gallon rating of the trucks cannot be less than 20.5 miles per gallon. Three of the trucks have the following ratings: 24 mpg, 19 mpg, and 18 mpg. In order to conform to this requirement, what must be the mpg rating of the fourth truck? **Ans:** At least 21 miles per gallon

3. Grace makes $18,000 per year plus a 5% commission on all sales over $20,000. She wants to make more than $32,000 this year. How many dollars in sales must she achieve to make this salary goal? **Ans:** More than $300,000 in sales

Putting Your Skills to Work: Use Math to Save Money

CHOOSING A CELL PHONE PLAN

Everyone wants to comparison shop in order to save money. Finding the best cell phone plan is one place where comparison shopping can really help, especially when you're trying to work within a budget. Consider the story of Jake.

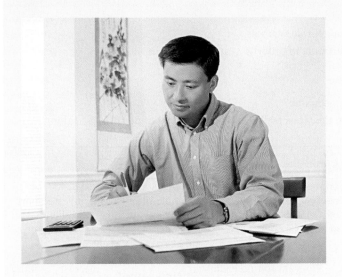

Jake is interested in purchasing a new cell phone plan. He would like to call his family, friends, and co-workers, as well as send/receive text and picture messages. He hopes he can afford to access the Internet, too, so he can check his e-mail with his phone. He wants to find a cell phone plan that would allow him to do these things for between $69 and $89 per month.

Analyzing the Options

He did some research, and he is considering these cell phone plans:

Calling Plan A: 300 minutes per month, $0.20 each additional minute: **$39.99 per month**

Calling Plan B: 450 minutes per month, $0.45 each additional minute: **$39.99 per month**

Calling Plan C: 600 minutes per month, $0.20 each additional minute: **$49.99 per month**

Calling Plan D: 900 minutes per month, $0.40 each additional minute: **$59.99 per month**

Which calling plan is the best choice if. . .

1. Jake plans to talk for 100 minutes per month?
   Either A or B

2. Jake plans to talk for 350 minutes per month? B

3. Jake plans to talk for 470 minutes per month? B

4. Jake plans to talk for 550 minutes per month? C

5. Jake plans to talk for 11 hours per month? D

6. Jake plans to talk for 16 hours, 40 minutes per month? D

Making the Best Choice to Save Money

Jake expects he will talk between 500 and 750 minutes per month, and is therefore considering Calling Plan C or Calling Plan D. Remember, Jake would like to send/receive text and picture messages. He'd also like to check his e-mail on the Internet with his phone. Jake's research also shows that "message bundles" are available at an additional charge to the calling plans.

Bundle A: 200 messages (text, picture, video, and IM) per month, $0.10 each additional message: **$5.00 per month**

Bundle B: 1500 messages (text, picture, video, and IM) per month, $0.05 each additional message: **$15.00 per month**

Bundle C: Unlimited text, picture, video, and IM messages per month: **$20.00 per month**

Bundle D: Unlimited web access and unlimited text, picture, video, and IM messages per month: **$35.00 per month**

Bundle D is the only bundle that would provide Jake with web access to check his e-mail in addition to messaging. If Jake wants to keep the cost of his cell phone service between $69 and $89 per month, determine the following:

7. The cost of Calling Plan C and Bundle D per month. $84.99

8. The cost of Calling Plan D and Bundle D per month $94.99

Jake is excited that Calling Plan C and Bundle D fit his budget. However, he is nervous about what will happen if he goes over his minutes with Calling Plan C.

9. Determine the total cost for one month if Jake talks for 750 minutes with Calling Plan C and Bundle D. $114.99

10. If you were Jake, which calling plan and bundle would you choose? Why? Answers may vary

Chapter 3 Organizer

Procedure for Solving Applied Problems

1. **Understand the problem.**
   (a) Read the word problem carefully to get an overview.
   (b) Determine what information you will need to solve the problem.
   (c) Draw a sketch. Label it with the known information. Determine what needs to be found.
   (d) Choose a variable to represent one unknown quantity.
   (e) If necessary, represent other unknown quantities in terms of that same variable.
2. **Write an equation.**
   (a) Look for key words to help you to translate the words into algebraic symbols.
   (b) Use a given relationship in the problem or an appropriate formula in order to write an equation.
3. **Solve and state the answer.**
4. **Check.**
   (a) Check the solution in the original equation. Is the answer reasonable?
   (b) Be sure the solution to the equation answers the question in the word problem. You may need to do some additional calculations if it does not.

EXAMPLE A

The perimeter of a rectangle is 126 meters. The length of the rectangle is 6 meters less than double the width. Find the dimensions of the rectangle.

1. **Understand the problem.**
   We want to find the length and the width of a rectangle whose perimeter is 126 meters.

   Let w = width.

   The length is *compared to* the width, so we start with the width.

   The length is 6 meters less than double the width.

   Then $l = 2w - 6$.

2. **Write an equation.**
   The perimeter of a rectangle is
   $P = 2w + 2l$.

   $126 = 2w + 2(2w - 6)$

3. **Solve.**
   $$126 = 2w + 4w - 12$$
   $$126 = 6w - 12$$
   $$138 = 6w$$
   $$23 = w \quad \text{The width is 23 meters.}$$
   $$2w - 6 = 2(23) - 6 =$$
   $$46 - 6 = 40$$

   The length is 40 meters.

4. **Check:**
   Is this reasonable? Yes. A rectangle 23 meters wide and 40 meters long seems to be about right for the perimeter to be 126 meters. Is the perimeter exactly 126 meters? Is the length exactly 6 meters less than double the width?

   $$2(23) + 2(40) \overset{?}{=} 126$$
   $$46 + 80 \overset{?}{=} 126$$
   $$126 = 126 \ \checkmark$$
   $$40 \overset{?}{=} 2(23) - 6$$
   $$40 \overset{?}{=} 46 - 6$$
   $$40 = 40 \ \checkmark$$

EXAMPLE B

Gina saved some money for college. She invested $2400 for one year and earned $225 in simple interest. She invested part of it at 12% and the rest of it at 9%. How much did she invest at each rate?

1. **Understand the problem.**
   We want to find each amount that was invested.

   $$\text{interest earned at 12\%} + \text{interest earned at 9\%} = \text{total interest of \$225}$$

   We let x represent one quantity of money.

   Let x = amount of money invested at 12%.

   We started with $2400. If we invest x at 12%, we still have $(2400 - x)$ left.

   Then $2400 - x$ = the amount of money invested at 9%.

   Interest = prt

   Interest at 12% $I_1 = 0.12x$

   Interest at 9% $I_2 = 0.09(2400 - x)$

2. **Write an equation.**
   $I_1 + I_2 = 225$.

   $0.12x + 0.09(2400 - x) = 225$

3. **Solve.**
   $$0.12x + 216 - 0.09x = 225$$
   $$0.03x + 216 = 225$$
   $$0.03x = 9$$
   $$x = \frac{9}{0.03}$$
   $$x = 300$$

   $300 was invested at 12%.
   $2400 - x = 2400 - 300 = 2100$
   $2100 was invested at 9%.

4. **Check:**
   Is this reasonable? Yes. Are the conditions of the problem satisfied? Does the total amount invested equal $2400? Will $2100 at 9% and $300 at 12% yield $225 in interest?

   $$2100 + 300 \overset{?}{=} 2400$$
   $$2400 = 2400 \ \checkmark$$

   $$0.09(2100) + 0.12(300) \overset{?}{=} 225$$
   $$189 + 36 \overset{?}{=} 225$$
   $$225 = 225 \ \checkmark$$

Topic	Procedure	Examples
Geometric formulas, p. 222.	Know the definitions of and formulas related to basic geometric shapes such as the triangle, rectangle, parallelogram, trapezoid, and circle. Review these formulas from the beginning of Section 3.5 as needed. Formulas for the volume and total surface area of three solids are also given there.	The volume of a right circular cylinder found in a four-cylinder car engine is determined by the formula $V = \pi r^2 h$. Find V if $r = 6$ centimeters and $h = 5$ centimeters. Use $\pi \approx 3.14$. $$V = 3.14(6 \text{ cm})^2(5 \text{ cm})$$ $$= 3.14(36)(5) \text{ cm}^3$$ $$= 565.2 \text{ cc (or cm}^3)$$ Four such cylinders would make this a 2260.8-cc engine.
Using inequalities to solve word problems, p. 233.	Be sure that the inequality you use fits the conditions of the word problem. **1.** *At least 87* means ≥ 87. **2.** *No less than $150* means $\geq \$150$. **3.** *At most $500* means $\leq \$500$. **4.** *No more than 70* means ≤ 70. **5.** *Less than 2* means < 2. **6.** *More than 10* means > 10.	Julie earns 2% on sales over $1000. How much must Julie make in sales to earn at least $300? Let x = amount of sales. $$(2\%)(x - \$1000) \geq \$300$$ $$0.02(x - 1000) \geq 300$$ $$x - 1000 \geq 15{,}000$$ $$x \geq 16{,}000$$ Julie must make sales of $16,000 or more (at least $16,000).

Chapter 3 Review Problems

Section 3.1

Write an algebraic expression. Use the variable x to represent the unknown value.

1. 19 more than a number $x + 19$

2. two-thirds of a number $\dfrac{2}{3}x$

3. half a number $\dfrac{1}{2}x$ or $\dfrac{x}{2}$

4. 18 less than a number $x - 18$

5. triple the sum of a number and 4 $3(x + 4)$

6. twice a number decreased by three $2x - 3$

Write an expression for each of the quantities compared. Use the letter specified.

7. **Workforce** The number of working people is four times the number of retired people. The number of unemployed people is one-half the number of retired people. (Use the letter r.)

r = the number of retired people
$4r$ = the number of working people
$0.5r$ = the number of unemployed people

▲ **8.** **Geometry** The length of the rectangle is 5 meters more than triple the width. (Use the letter w.)

$3w + 5$ = the length of the rectangle
w = the width of the rectangle

▲ **9.** **Geometry** A triangle has three angles A, B, and C. The number of degrees in angle A is double the number of degrees in angle B. The number of degrees in angle C is 17 degrees less than the number in angle B. (Use the letter b.)

b = the number of degrees in angle B
$2b$ = the number of degrees in angle A
$b - 17$ = the number of degrees in angle C

10. **Class Size** There are 29 more students in biology class than in algebra. There are one-half as many students in geology class as in algebra. (Use the letter a.)

a = the number of students in algebra
$a + 29$ = the number of students in biology
$0.5a$ = the number of students in geology

Section 3.2

Solve.

11. Fourteen is taken away from triple a number. The result is −5. What is the number?
    $x = 3$

12. When twice a number is reduced by seven, the result is −21. What is the number?
    $x = -7$

13. **Table Set** Six chairs and a table cost $450. If the table costs $210, how much does one chair cost?
    $40

14. **Age** Jon is twice as old as his cousin David. If Jon is 32, how old is David?
    16 years old

15. **Speed Rates** Two cars left San Francisco and drove to San Diego, 800 miles away. One car was driven at 60 miles an hour. The other car was driven at 65 miles an hour. How long did it take each car to make the trip? Round your answers to the nearest tenth.
    $t_1 \approx 13.3\,\text{hr}; t_2 \approx 12.3\,\text{hr}$

16. **Biology Score** Zach took four biology tests and got scores of 83, 86, 91, and 77. He needs to complete one more test. If he wants to get an average of 85 on all five tests, what score does he need on his last test?
    88

Section 3.3

▲ 17. **Geometry** An isosceles triangle has a perimeter of 85 feet. If one of the two equal sides measures 31 feet, what is the measure of the third side?
    23 ft

▲ 18. **Geometry** The measure of the second angle of a triangle is three times the measure of the first angle. The measure of the third angle is 12 degrees less than twice the measure of the first. Find the measure of each of the three angles.
    1st angle = 32°; 2nd angle = 96°; 3rd angle = 52°

19. **Rope** A piece of rope 50 yards long is cut into two pieces. One piece is three-fifths as long as the other. Find the length of each piece.
    31.25 yd and 18.75 yd

20. **Income** George and Heather have a combined income of $65,000. Heather earns $2000 more than half of what George earns. How much does each earn?
    George = $42,000
    Heather = $23,000

Section 3.4

21. **Electric Bill** The electric bill at Jane's house this month was $71.50. The charge is based on a flat rate of $25 per month plus a charge of $0.15 per kilowatt-hour of electricity used. How many kilowatt-hours of electricity were used?
    310 kilowatt-hours

22. **Car Rental** Abel rented a car from Sunshine Car Rentals. He was charged $39 per day and $0.25 per mile. He rented the car for three days and paid a rental cost of $187. How many miles did he drive the car? 280 mi

23. **Simple Interest** Boxell Associates has $7400 invested at 5.5% simple interest. After money was withdrawn, $242 was earned for one year on the remaining funds. How much was withdrawn?
    $3000

24. **CD Player** Jamie bought a new CD player for her car. When she bought it, she found that the price had been decreased by 18%. She was able to buy the CD player for $36 less than the original price. What was the original price?
    $200

25. *Investments* Peter and Shelly invested $9000 for one year. Part of it was invested at 12% and the remainder was invested at 8%. At the end of one year the couple had earned exactly $1000 in simple interest. How much did they invest at each rate?
$7000 at 12% and $2000 at 8%

26. *Investments* A man invests $5000 in a savings bank. He places part of it in a checking account that earns 4.5% and the rest in a regular savings account that earns 6%. His total annual income from this investment in simple interest is $270. How much was invested in each account?
$2000 at 4.5% and $3000 at 6%

27. *Coin Change* Mary has $3.75 in nickels, dimes, and quarters. She has three more quarters than dimes. She has twice as many nickels as quarters. How many of each coin does she have?
18 nickels; 6 dimes; 9 quarters

28. *Tip Jar* After the morning rush at the neighborhood coffee shop, the tip jar contained $9.80 in coins. There were two more quarters than nickels. There were three fewer dimes than nickels. How many coins of each type were there?
24 nickels; 21 dimes; 26 quarters

Section 3.5

Solve.

▲ **29. *Geometry*** Find the surface area of a tennis ball that has a diameter of 2.6 inches. (Use $\pi \approx 3.14$.) Round your answer to the nearest hundredth.
21.23 in.^{2}

▲ **30. *Trench Length*** Wilfredo and Tina bought a new screen house for camping. It was hexagonal (6 sides) in shape and 4 ft long on a side. If they wanted to dig a trench around their house so that rainwater would flow away, how long would the trench be? (You are looking for the perimeter of the screen house.)
24 ft

▲ **31. *Geometry*** Find the third angle of a triangle if one angle measures 62 degrees and the second angle measures 47 degrees. 71°

▲ **32. *Geometry*** Find the area of a triangle whose base is 8 miles and whose altitude is 10.5 miles. 42 mi^2

▲ **33. *Geometry*** Find the volume of a storage room that has a width of 6 feet, a length of 11 feet, and a height of 7 feet. 462 ft^3

▲ **34. *Geometry*** Find the volume of a basketball whose radius is 4.5 inches. (Use $\pi \approx 3.14$.) 381.51 in.^{3}

▲ **35. *Geometry*** The Clock Tower (also known as Big Ben) in London has four clock faces measuring 23 feet in diameter. Find the area of each clock face. (Use $\pi \approx 3.14$.) Round your answer to the nearest hundredth. 415.27 ft^2

▲ **36. *Geometry*** The perimeter of an isosceles triangle is 46 inches. The two equal sides are each 17 inches long. How long is the third side? 12 in.

▲ **37. *House Painting*** Jim wants to have the exterior siding on his house painted. Two sides of the house have a height of 18 feet and a length of 22 feet, and the other two sides of the house have a height of 18 feet and a length of 19 feet. The painter has quoted a price of $2.50 per square foot. The windows and doors of the house take up 100 square feet. How much will the painter charge to paint the siding? $3440

▲ **38. *Aluminum Cylinder*** Find the cost to build a cylinder out of aluminum if the radius is 5 meters, the height is 14 meters, and the cylinder will have a flat top and flat bottom. The cost factor has been determined to be $40.00 per square meter. (*Hint:* First find the total surface area.) (Use $\pi \approx 3.14$.)
$23,864

Section 3.6

Solve.

39. *Truck Rental* Brian rents a truck for $20 a day and $0.25 a mile. How far can he drive if he wants to rent the truck for three days and does not want to spend more than $100?
160 miles or less

40. *Income* Greta earns $25,000 per year in salary and 5% commission on her sales. How much were her sales if her annual income was more than $38,000?
More than $260,000

41. *Transcontinental Railroad* In 1862, two companies were granted the rights to build the transcontinental railroad from Omaha, Nebraska, to Sacramento, California. The Central Pacific railroad began in 1863 from Sacramento heading east. The Union Pacific began 24 months later, leaving from Omaha and heading west. The Central Pacific averaged 8.75 miles of track each month, while the Union Pacific averaged 20 miles of track per month. The two companies met at Promontory, Utah, after completing 1590 miles of track. How much track did each build?

$D = rt$
$1590 = 8.75t + 20(t - 24)$
$t = 72$ months
CP distance $= 8.75(72) = 630$ mi
UP distance $= 20(72 - 24) = 20(48) = 960$ mi

42. *Clothes Shopping* Michael plans to spend less than $70 on shirts and ties. He bought three shirts for $17.95 each. How much can he spend on ties?
Less than $16.15

Mixed Practice

Solve.

43. *TV Show* Wally timed a popular "30-minute" TV show. The show lasted 4 minutes longer than four times the amount of time that was devoted to commercials. How many minutes long was the actual show?

$24\frac{4}{5}$ min

44. *Balanced Weight* Faye conducted an experiment. To verify the weight of four identical steel balls, she placed them on a balance. She placed three of the balls on one side of the balance. On the other side she placed an 8-ounce weight and one of the balls. She discovered that these two sides exactly balanced on the scale. How much does each steel ball weigh? 4 oz

45. *Electricity Use* Fred and Nancy Sullivan are planning to purchase a new refrigerator. The most efficient model costs $870, but the electricity to operate it would cost only $41 per year. The less efficient model is the same size. It costs $744, but the electricity to operate it would cost $83 per year. How long will it be before the total cost is greater for the less efficient model? More than three years

46. *Buying vs. Renting* Custom Computer Works of Gloucester rents its office building and warehouse for $1800 per month. The owner is thinking of buying the building. He could purchase it with a down payment of $42,000. He would then make monthly mortgage payments of $1100. How many months would it take for his total costs to be less to purchase the building rather than rent it?
More than 60 months

47. *Bookcase Construction* Judy Carter is planning to build a bookcase for her home office. She has only 38 feet of lumber to make the unit. The bookcase is constructed of shelves as shown in the diagram. She wants the width of the bookcase to be 2.5 times the height. Find the width and the height of the bookcase she needs to construct.
Width $= 10$ ft;
height $= 4$ ft

48. *Retirement Plan* Rick Ponticelli is investing some money in a retirement plan. He plans to invest $100 a month the first year, $200 a month the second year, $300 a month the third year, and continue with that pattern. How many years will it be before he has invested $66,000 in his retirement plan? 10 yr

49. ***Basketball*** In a recent basketball game the Boston Celtics scored 99 points. They scored three times as many field goals as free throws. They also made 12 three-point baskets. How many field goals did they make? How many free throws did they make? (A field goal is worth two points and a free throw is worth one point.) 27 field goals; 9 free throws

▲ **50.** ***Room Dimensions*** The perimeter of a rectangular room in Sally's house is 76 feet. The length of the room is 10 feet less than double the width of the room. Find the dimensions of the room.
Width = 16 ft; length = 22 ft

51. ***Deck Building*** Ryan and Carl spent 28 hours building a deck. Ryan worked 8 less than twice as many hours as Carl. How many hours did each man put in?
Number of hours Carl worked = 12 hr
Number of hours Ryan worked = 16 hr

52. ***Investments*** Sarah Tanguay has $2000 more invested at 8.5% than she does at 9.75%. Both investments are at simple interest rates. If the annual return from each investment is the same, how much is invested at each rate?
Amount invested at 9.75% = $13,600
Amount invested at 8.5% = $15,600

▲ **53.** ***Geometry*** The first angle of a triangle measures 20° larger than the second angle of the triangle. The third angle measures twice as large as the second angle. Find the measure of each of these three angles.
First angle 60°; second angle 40°; third angle 80°

▲ **54.** ***Sign Construction*** A sign is made in the shape of a trapezoid. The altitude of the trapezoid is 12 feet. The two bases of the trapezoid are 15 feet and 19 feet. The sign is constructed of prime cedar that costs $6 per square foot. How much did the cedar that was used to make the sign cost? $1224

55. ***Final Exam Score*** Thelma has a math final exam that counts twice as much as a test. She had four tests with grades of 88, 77, 95, and 92. What score does she need to obtain on the final exam to get an average of 90 for the course? 94

56. ***Mileage Deduction*** Barbara drove her car 27,000 miles last year. She drove twice as much for pleasure as she did for business. She can deduct $0.33 per mile for all business miles. Last year she claimed a business mileage deduction of $2000. Was her deduction correct? Please explain your answer. If her deduction was not correct, please indicate the correct deduction.
No; she traveled 9000 miles for business; her correct deduction was $2970

57. ***Jet Travel*** A jet plane travels 32 miles in three minutes. How far can the plane travel in one hour if it continues at that speed? How many minutes would it take the jet to travel from Boston to Denver (approximately 1800 miles) at that same speed?
640 mi; approximately 169 min

58. ***Football*** A running back in a Wheaton College football game catches a pass and runs for the goal line 80 yards away. He can run a 50-yard dash in six seconds. If he runs at 80% of his speed in the 50-yard dash, how long will it take him to reach the goal line?
12 sec

59. ***Field Trip*** A Ford Explorer can hold a driver and four passengers. A Dodge Caravan can hold a driver and six passengers. The biology department is taking a total of 95 people on a field trip. A total of 20 of the 95 people are available to drive. Not all potential drivers will be needed. The college wants to use twice as many Caravans as Explorers because the Caravans get better gas mileage. How many of each kind of vehicle are needed? 5 Explorers, 10 Caravans

▲ **60.** ***Geometry*** The measure of the largest angle of a triangle is triple the measure of the smallest angle. The third angle is 12 degrees less than double the measure of the smallest angle. Find the measure of each angle.
Smallest angle = 32°
Largest angle = 96°
Third angle = 52°

61. ***Zorbing*** Zorbing is a sport, invented in New Zealand, where you are secured inside an inner capsule in a large transparent ball and rolled down a slope. The ball needs to be inflated with an inert gas before it can be used. An inflated Zorb has a diameter of 3 meters. If the Zorb itself weighs 80 kg, the zorber weighs 75 kg, and the gas used to inflate the ball weighs 2 kg per cubic meter, what is the total weight of the Zorb while in use? (Use 3.14 to approximate π.)
183.26 kg

Note to Instructor: The Chapter 3 Test file in the TestGen program provides algorithms specifically matched to these problems so you can easily replicate this test for additional practice or assessment purposes.

1. _35_

2. _36_

3. _−4_

4. First side = 20 m;
second side = 30 m;
third side = 16 m

5. Width = 20 m;
length = 47 m

6. 1st pollutant = 8 ppm;
2nd pollutant = 4 ppm;
3rd pollutant = 3 ppm

7. _15 months_

8. _$21,000_

9. $1400 at 14%
$2600 at 11%

10. 16 nickels, 7 dimes,
8 quarters

How Am I Doing? Chapter 3 Test

Remember to use your Chapter Test Prep Video CD to see the worked-out solutions to the test problems you want to review.

Solve.

1. A number is doubled and then decreased by 11. The result is 59. What is the original number?

2. The sum of one-half of a number, one-ninth of the number, and one-twelfth of the number is twenty-five. Find the original number.

3. Double the sum of a number and 5 is the same as fourteen more than triple the same number. Find the number.

▲ 4. A triangular region has a perimeter of 66 meters. The first side is two-thirds of the second side. The third side is 14 meters shorter than the second side. What are the lengths of the three sides of the triangular region?

▲ 5. A rectangle has a length 7 meters longer than double the width. The perimeter is 134 meters. Find the dimensions of the rectangle.

6. Three harmful pollutants were measured by a consumer group in the city. The total sample contained 15 parts per million (ppm) of three harmful pollutants. The amount of the first pollutant was double the second. The amount of the third pollutant was 75% of the second. How many parts per million of each pollutant were found?

7. Raymond has a budget of $1940 to rent a computer for his company office. The computer company he wants to rent from charges $200 for installation and service as a one-time fee. Then they charge $116 per month rental for the computer. How many months will Raymond be able to rent a computer with this budget?

8. Last year the yearly tuition at Westmont College went up 8%. This year's charge for tuition for the year is $22,680. What was it last year before the increase went into effect?

9. Franco invested $4000 in money market funds. Part was invested at 14% interest, the rest at 11% interest. At the end of each year the fund company pays interest. After one year he earned $482 in simple interest. How much was invested at each interest rate?

10. Mary has $3.50 in change. She has twice as many nickels as quarters. She has one fewer dime than she has quarters. How many of each coin does she have?

▲ **11.** Find the circumference of a circle with radius 34 inches. Use $\pi \approx 3.14$ as an approximation.

| 11. | 213.52 in. |

▲ **12.** Find the area of a trapezoid if the two bases are 10 inches and 14 inches and the altitude is 16 inches.

| 12. | 192 in.² |

▲ **13.** Find the volume of a sphere with radius 10 inches. Use $\pi \approx 3.14$ as an approximation and round your answer to the nearest cubic inch.

| 13. | 4187 in.³ |

▲ **14.** Find the area of a parallelogram with a base of 12 centimeters and an altitude of 8 centimeters.

| 14. | 96 cm² |

▲ **15.** How much would it cost to carpet the area shown in the figure if carpeting costs $12 per square yard?

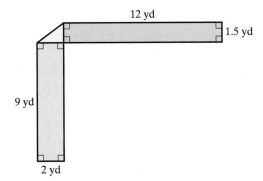

| 15. | $450 |

16. Cathy scored 76, 84, and 78 on three tests. How much must she score on the next test to have at least an 80 average?

| 16. | At least an 82 |

17. Carol earns $15,000 plus 5% commission on sales over $10,000. How much must she make in sales to earn more than $20,000?

| 17. | More than $110,000 |

▲ **18.** Andrew is putting down tile for a new gymnasium. He can lay 25 square feet of tile in an hour. How long will it take him to do a gymnasium 150 feet long and 100 feet wide?

| 18. | 600 hr |

Cumulative Test for Chapters 0–3

Approximately one-half of this test covers the content of Chapters 0–2. The remainder covers the content of Chapter 3.

1. Divide. $2.4\overline{)8.856}$

2. Add. $\dfrac{3}{8} + \dfrac{5}{12} + \dfrac{1}{2}$

3. Simplify. $4(4x - y + 5) - 3(6x - 2y)$

4. Evaluate. $12 - 3(2 - 4) + 12 \div 4$

5. Solve for b. $H = \dfrac{1}{2}(3a + 5b)$

6. Solve for x and graph your solution. $5x - 3 \le 2(4x + 1) + 4$

7. When a number is doubled and then increased by 15, the result is 1. Find the number.

8. The number of students in a literature class is 12 fewer than the number of students in sociology. The total enrollment for the two classes is 96 students. How many students are in each class?

▲ 9. A rectangle has a perimeter of 78 centimeters. The length of the rectangle is 11 centimeters longer than triple the width. Find the dimensions of the rectangle.

10. This year the sales for Mayflower Medical are 15% higher than last year. This year the sales were $265,000. What were last year's sales? Round your answer to the nearest dollar.

11. Hassan invested $7000 for one year. He invested part of it in a high-risk fund that pays 15% interest. He placed the rest in a safe, low-risk fund that pays 7% interest. At the end of one year he earned $730 in simple interest. How much did he invest in each of the two funds?

12. Linda has some dimes, nickels, and quarters in her purse. The total value of these coins is $2.55. She has three times as many dimes as quarters. She has three more nickels than quarters. How many coins of each type does she have?

▲ 13. Find the area of a triangle with an altitude of 13 meters and a base of 25 meters. What is the cost to construct such a triangle out of sheet metal that costs $4.50 per square meter?

▲ 14. Find the volume of a sphere that has a radius of 3.00 inches. How much will the contents of the sphere weigh if it is filled with a liquid that weighs 1.50 pounds per cubic inch? Use $\pi \approx 3.14$ as an approximation.

15. Melinda Tobey saw a glacier in Alaska that moves 65 feet per day. How long will it take the glacier to reach a cabin that is 1625 feet away?

---

1. 3.69

2. $\dfrac{31}{24}$ or $1\dfrac{7}{24}$

3. $-2x + 2y + 20$

4. 21

5. $\dfrac{2H - 3a}{5} = b$

6. $x \ge -3$

7. -7

8. Literature students = 42; sociology students = 54

9. Width = 7 cm; length = 32 cm

10. $230,435

11. $3000 at 15% $4000 at 7%

12. 7 nickels, 12 dimes, 4 quarters

13. $A = 162.5$ m²; $731.25

14. $V = 113.04$ in.³ 169.56 lb

15. 25 days

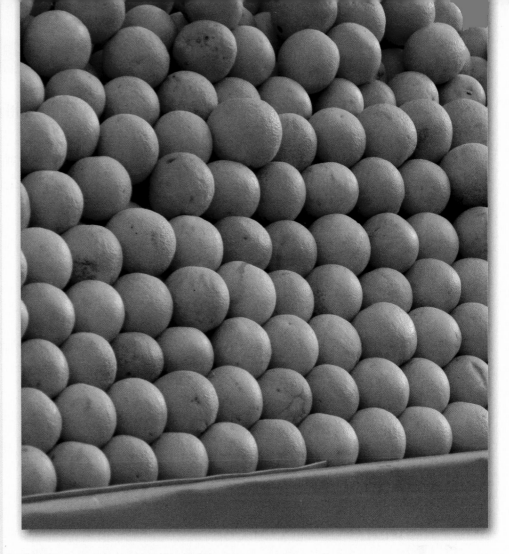

CHAPTER

4

In a fruit market, have you ever noticed a stack of oranges in a pyramid shape where each layer is square? Suppose you wanted to know how many oranges are in a pyramid with seven layers of oranges. After studying the mathematics in this chapter, you will have the knowledge to find out.

Exponents and Polynomials

1 Using the Product Rule to Multiply Exponential Expressions with Like Bases

Recall that x^2 means $x \cdot x$. That is, x appears as a factor two times. The 2 is called the **exponent.** The **base** is the variable x. The expression x^2 is called an **exponential expression.** What happens when we multiply $x^2 \cdot x^2$? Is there a pattern that will help us form a general rule?

$$(2^2)(2^3) = \overbrace{(2 \cdot 2)(2 \cdot 2 \cdot 2)}^{5 \text{ twos}} = 2^5$$

The exponent means 2 occurs 5 times as a factor.

$$(3^3)(3^4) = \overbrace{(3 \cdot 3 \cdot 3)(3 \cdot 3 \cdot 3 \cdot 3)}^{7 \text{ threes}} = 3^7$$

Notice that $3 + 4 = 7$.

$$(x^3)(x^5) = \overbrace{(x \cdot x \cdot x)(x \cdot x \cdot x \cdot x \cdot x)}^{8 \ x\text{'s}} = x^8$$

The sum of the exponents is $3 + 5 = 8$.

$$(y^4)(y^2) = \overbrace{(y \cdot y \cdot y \cdot y)(y \cdot y)}^{6 \ y\text{'s}} = y^6$$

The sum of the exponents is $4 + 2 = 6$.

We can state the pattern in words and then use variables.

> **THE PRODUCT RULE**
>
> To multiply two exponential expressions that have the same base, keep the base and *add the exponents.*
>
> $$x^a \cdot x^b = x^{a+b}$$

Be sure to notice that this rule applies only to expressions that have the *same base.* Here x represents the base, while the letters a and b represent the exponents that are added.

It is important that you apply this rule even when an exponent is 1. Every variable that does not have a written exponent is understood to have an exponent of 1. Thus $x^1 = x$, $y^1 = y$, and so on.

EXAMPLE 1 Multiply. (a) $x^3 \cdot x^6$ (b) $x \cdot x^5$

Solution

(a) $x^3 \cdot x^6 = x^{3+6} = x^9$

(b) $x \cdot x^5 = x^{1+5} = x^6$　　Note that the exponent of the first x is 1.

Practice Problem 1 Multiply. (a) $a^7 \cdot a^5 \quad a^{12}$ (b) $w^{10} \cdot w \quad w^{11}$

EXAMPLE 2 Simplify. (a) $y^5 \cdot y^{11}$ (b) $2^3 \cdot 2^5$ (c) $x^6 \cdot y^8$

Solution

(a) $y^5 \cdot y^{11} = y^{5+11} = y^{16}$

(b) $2^3 \cdot 2^5 = 2^{3+5} = 2^8$　　Note that the base does not change! Only the exponent changes.

(c) $x^6 \cdot y^8$　　The rule for multiplying exponential expressions does not apply since the bases are not the same. This cannot be simplified.

Practice Problem 2 Simplify, if possible.

(a) $x^3 \cdot x^9 \quad x^{12}$ (b) $3^7 \cdot 3^4 \quad 3^{11}$ (c) $a^3 \cdot b^2$ Cannot be simplified

We can now look at multiplying expressions such as $(2x^5)(3x^6)$.

The number 2 in $2x^5$ is called the **numerical coefficient.** Recall that a numerical coefficient is a number that is multiplied by a variable. When we multiply two expressions such as $2x^5$ and $3x^6$, we first multiply the numerical coefficients; we multiply the variables with exponents separately.

As you do the following problems, keep in mind the rule for multiplying expressions with exponents and the rules for multiplying signed numbers.

EXAMPLE 3 Multiply.

(a) $(2x^5)(3x^6)$ **(b)** $(5x^3)(x^6)$ **(c)** $(-6x)(-4x^5)$

Solution

(a) $(2x^5)(3x^6) = (2 \cdot 3)(x^5 \cdot x^6)$ Multiply the numerical coefficients.

$\qquad\qquad\quad = 6(x^5 \cdot x^6)$ Use the rule for multiplying expressions with exponents. Add the exponents.

$\qquad\qquad\quad = 6x^{11}$

(b) Every variable that does not have a visible numerical coefficient is understood to have a numerical coefficient of 1. Thus x^6 has a numerical coefficient of 1.

$$(5x^3)(x^6) = (5 \cdot 1)(x^3 \cdot x^6) = 5x^9$$

(c) $(-6x)(-4x^5) = (-6)(-4)(x^1 \cdot x^5) = 24x^6$ Remember that x has an exponent of 1.

Practice Problem 3 Multiply.

(a) $(-a^8)(a^4)$ $-a^{12}$ **(b)** $(3y^2)(-2y^3)$ $-6y^5$ **(c)** $(-4x^3)(-5x^2)$ $20x^5$

Teaching Example 3 Multiply.
(a) $(-5a^4)(2a^3)$
(b) $(y^8)(4y)$
(c) $(-m^2)(-9m^6)$
Ans: (a) $-10a^7$ **(b)** $4y^9$ **(c)** $9m^8$

Problems of this type may involve more than one variable or more than two factors.

EXAMPLE 4 Multiply. $(5ab)\left(-\frac{1}{3}a\right)(9b^2)$

Solution $\qquad (5ab)\left(-\frac{1}{3}a\right)(9b^2) = (5)\left(-\frac{1}{3}\right)(9)(a \cdot a)(b \cdot b^2)$

$\qquad\qquad\qquad\qquad\qquad\qquad = -15a^2b^3$

Practice Problem 4 Multiply.

$$(2xy)\left(-\frac{1}{4}x^2y\right)(6xy^3) -3x^4y^5$$

Teaching Example 4 Multiply.
$(2r^3s)(5s^4)\left(-\frac{3}{5}r\right)$
Ans: $-6r^4s^5$

② Using the Quotient Rule to Divide Exponential Expressions with Like Bases

Frequently, we must divide exponential expressions. Because division by zero is undefined, in all problems in this chapter we assume that the denominator of any variable expression is not zero. We'll look at division in three separate parts.

Suppose that we want to simplify $x^5 \div x^2$. We could do the division the long way.

$$\frac{x^5}{x^2} = \frac{(x)(x)(x)\cancel{(x)}\cancel{(x)}}{\cancel{(x)}\cancel{(x)}} = x^3$$

Here we are using the arithmetical property of reducing fractions (see Section 0.1). When the same factor appears in both numerator and denominator, that factor can be removed.

A simpler way is to *subtract the exponents*. Notice that the base remains the same.

THE QUOTIENT RULE (PART 1)

$\dfrac{x^a}{x^b} = x^{a-b}$ Use this form if the larger exponent is in the numerator and $x \neq 0$.

Teaching Example 5 Divide.

(a) $\dfrac{a^{13}}{a^5}$ (b) $\dfrac{7^{10}}{7^2}$ (c) $\dfrac{b^6}{b}$

Ans: (a) a^8 (b) 7^8 (c) b^5

EXAMPLE 5 Divide.

(a) $\dfrac{2^{16}}{2^{11}}$ (b) $\dfrac{x^5}{x^3}$ (c) $\dfrac{y^{16}}{y^7}$

Solution

(a) $\dfrac{2^{16}}{2^{11}} = 2^{16-11} = 2^5$ Note that the base does *not* change.

(b) $\dfrac{x^5}{x^3} = x^{5-3} = x^2$ (c) $\dfrac{y^{16}}{y^7} = y^{16-7} = y^9$

NOTE TO STUDENT: Fully worked-out solutions to all of the Practice Problems can be found at the back of the text starting at page SP-1

Practice Problem 5 Divide.

(a) $\dfrac{10^{13}}{10^7}$ 10^6 (b) $\dfrac{x^{11}}{x}$ x^{10} (c) $\dfrac{y^{18}}{y^8}$ y^{10}

Now we consider the situation where the larger exponent is in the denominator. Suppose that we want to simplify $x^2 \div x^5$.

$$\dfrac{x^2}{x^5} = \dfrac{\cancel{(x)}\cancel{(x)}}{\cancel{(x)}\cancel{(x)}(x)(x)(x)} = \dfrac{1}{x^3}$$

THE QUOTIENT RULE (PART 2)

$\dfrac{x^a}{x^b} = \dfrac{1}{x^{b-a}}$ Use this form if the larger exponent is in the denominator and $x \neq 0$.

Teaching Example 6 Divide.

(a) $\dfrac{m^2}{m^6}$ (b) $\dfrac{15^{12}}{15^{26}}$ (c) $\dfrac{r}{r^4}$

Ans: (a) $\dfrac{1}{m^4}$ (b) $\dfrac{1}{15^{14}}$ (c) $\dfrac{1}{r^3}$

EXAMPLE 6 Divide.

(a) $\dfrac{12^{17}}{12^{20}}$ (b) $\dfrac{b^7}{b^9}$ (c) $\dfrac{x^{20}}{x^{24}}$

Solution

(a) $\dfrac{12^{17}}{12^{20}} = \dfrac{1}{12^{20-17}} = \dfrac{1}{12^3}$ Note that the base does *not* change.

(b) $\dfrac{b^7}{b^9} = \dfrac{1}{b^{9-7}} = \dfrac{1}{b^2}$ (c) $\dfrac{x^{20}}{x^{24}} = \dfrac{1}{x^{24-20}} = \dfrac{1}{x^4}$

Practice Problem 6 Divide.

(a) $\dfrac{c^3}{c^4}$ $\dfrac{1}{c}$ (b) $\dfrac{10^{31}}{10^{56}}$ $\dfrac{1}{10^{25}}$ (c) $\dfrac{z^{15}}{z^{21}}$ $\dfrac{1}{z^6}$

When there are numerical coefficients, use the rules for dividing signed numbers to reduce fractions to lowest terms.

EXAMPLE 7 Divide.

(a) $\dfrac{5x^5}{25x^7}$

(b) $\dfrac{-12x^8}{4x^3}$

(c) $\dfrac{-16x^7}{-24x^8}$

Solution

(a) $\dfrac{5x^5}{25x^7} = \dfrac{1}{5x^{7-5}} = \dfrac{1}{5x^2}$

(b) $\dfrac{-12x^8}{4x^3} = -3x^{8-3} = -3x^5$

(c) $\dfrac{-16x^7}{-24x^8} = \dfrac{2}{3x^{8-7}} = \dfrac{2}{3x}$

Practice Problem 7 Divide.

(a) $\dfrac{-7x^7}{-21x^9}$ $\dfrac{1}{3x^2}$

(b) $\dfrac{15x^{11}}{-3x^4}$ $-5x^7$

(c) $\dfrac{23x^8}{46x^9}$ $\dfrac{1}{2x}$

You have to work very carefully if two or more variables are involved. Treat the coefficients and each variable separately.

EXAMPLE 8 Divide.

(a) $\dfrac{x^3y^2}{5xy^6}$

(b) $\dfrac{-3x^2y^5}{12x^6y^8}$

Solution

(a) $\dfrac{x^3y^2}{5xy^6} = \dfrac{x^2}{5y^4}$

(b) $\dfrac{-3x^2y^5}{12x^6y^8} = -\dfrac{1}{4x^4y^3}$

Practice Problem 8 Divide.

(a) $\dfrac{x^7y^9}{y^{10}}$ $\dfrac{x^7}{y}$

(b) $\dfrac{12x^5y^6}{-24x^3y^8}$ $-\dfrac{x^2}{2y^2}$

Suppose that a given base appears with the same exponent in the numerator and denominator of a fraction. In this case we can use the fact that *any nonzero number divided by itself is* 1.

EXAMPLE 9 Divide.

(a) $\dfrac{x^6}{x^6}$

(b) $\dfrac{3x^5}{x^5}$

Solution

(a) $\dfrac{x^6}{x^6} = 1$

(b) $\dfrac{3x^5}{x^5} = 3\left(\dfrac{x^5}{x^5}\right) = 3(1) = 3$

Practice Problem 9 Divide.

(a) $\dfrac{10^7}{10^7}$ 1

(b) $\dfrac{12a^4}{15a^4}$ $\dfrac{4}{5}$

Do you see that if we had subtracted exponents when simplifying $\dfrac{x^6}{x^6}$ we would have obtained x^0 in Example 9? So we can surmise that any number (except 0) to the 0 power equals 1. We can write this fact as a separate rule.

Teaching Example 7 Divide.

(a) $\dfrac{6y^7}{8y^6}$

(b) $\dfrac{7a^6}{-14a^{10}}$

(c) $\dfrac{27m^3}{33m}$

Ans: (a) $\dfrac{3y}{4}$ (b) $-\dfrac{1}{2a^4}$ (c) $\dfrac{9m^2}{11}$

Teaching Example 8 Divide.

(a) $\dfrac{15r^2s^7}{5r^6s}$

(b) $\dfrac{12a^2b}{18a^8b^4}$

Ans: (a) $\dfrac{3s^6}{r^4}$ (b) $\dfrac{2}{3a^6b^3}$

Teaching Example 9 Divide.

(a) $\dfrac{5^{10}}{5^{10}}$

(b) $\dfrac{-16y^7}{8y^7}$

Ans: (a) 1 (b) -2

THE QUOTIENT RULE (PART 3)

$$\frac{x^a}{x^a} = x^0 = 1 \qquad \text{if } x \neq 0 \qquad (0^0 \text{ remains undefined}).$$

TO THINK ABOUT: What Is 0 to the 0 Power? What about 0^0? Why is it undefined? $0^0 = 0^{1-1}$. If we use the quotient rule, $0^{1-1} = \frac{0}{0}$. Since division by zero is undefined, we must agree that 0^0 is undefined.

Teaching Example 10 Divide.

(a) $\dfrac{-r^0 s^3}{6^0 s}$ (b) $\dfrac{2m^6 b^0}{10m^6 b^4}$

Ans: (a) $-s^2$ (b) $\dfrac{1}{5b^4}$

EXAMPLE 10 Divide.

(a) $\dfrac{4x^0 y^2}{8^0 y^5 z^3}$ (b) $\dfrac{5x^2 y}{10x^2 y^3}$

Solution

(a) $\dfrac{4x^0 y^2}{8^0 y^5 z^3} = \dfrac{4(1)y^2}{(1)y^5 z^3} = \dfrac{4y^2}{y^5 z^3} = \dfrac{4}{y^3 z^3}$ (b) $\dfrac{5x^2 y}{10x^2 y^3} = \dfrac{1x^0}{2y^2} = \dfrac{(1)(1)}{2y^2} = \dfrac{1}{2y^2}$

Practice Problem 10 Divide.

(a) $\dfrac{-20a^3 b^8 c^4}{28a^3 b^7 c^5}$ $-\dfrac{5b}{7c}$ (b) $\dfrac{5x^0 y^6}{10x^4 y^8}$ $\dfrac{1}{2x^4 y^2}$

Teaching Tip After discussing Example 10, you may wish to point out to students that since $x^0 = 1$ whenever $x \neq 0$, it is also true that x could be an entire expression. Thus

$$\left(\frac{ab^2}{c}\right)^0 = 1 \text{ as long as } a \neq 0,$$
$$b \neq 0, c \neq 0$$

We can combine all three parts of the quotient rule we have developed.

THE QUOTIENT RULE

$\dfrac{x^a}{x^b} = x^{a-b}$ Use this form if the larger exponent is in the numerator and $x \neq 0$.

$\dfrac{x^a}{x^b} = \dfrac{1}{x^{b-a}}$ Use this form if the larger exponent is in the denominator and $x \neq 0$.

$\dfrac{x^a}{x^a} = x^0 = 1$ if $x \neq 0$.

We can combine the product rule and the quotient rule to simplify algebraic expressions that involve both multiplication and division.

Teaching Example 11 Simplify.

$\dfrac{(4ab)(-9a^2 b)}{-30a^6 b^4}$

Ans: $\dfrac{6}{5a^3 b^2}$

EXAMPLE 11 Simplify. $\dfrac{(8x^2 y)(-3x^3 y^2)}{-6x^4 y^3}$

Solution $\dfrac{(8x^2 y)(-3x^3 y^2)}{-6x^4 y^3} = \dfrac{-24x^5 y^3}{-6x^4 y^3} = 4x$

Practice Problem 11 Simplify. $\dfrac{(-6ab^5)(3a^2 b^4)}{16a^5 b^7}$ $-\dfrac{9b^2}{8a^2}$

Raising Exponential Expressions to a Power

How do we simplify an expression such as $(x^4)^3$? $(x^4)^3$ is x^4 raised to the third power. For this type of problem we say that we are raising a power to a power. A problem such as $(x^4)^3$ could be done by writing the following.

$$(x^4)^3 = x^4 \cdot x^4 \cdot x^4 \quad \text{By definition}$$
$$= x^{12} \qquad\qquad \text{By adding exponents}$$

Notice that when we add the exponents we get $4 + 4 + 4 = 12$. This is the same as multiplying 4 by 3. That is, $4 \cdot 3 = 12$. This process can be summarized by the following rule.

> **RAISING A POWER TO A POWER**
>
> To raise a power to a power, keep the same base and multiply the exponents.
>
> $$(x^a)^b = x^{ab}$$

Recall what happens when you raise a negative number to a power. $(-1)^2 = 1$. $(-1)^3 = -1$. In general,

$$(-1)^n = \begin{cases} +1 & \text{if } n \text{ is even} \\ -1 & \text{if } n \text{ is odd.} \end{cases}$$

EXAMPLE 12 Simplify.

(a) $(x^3)^5$ **(b)** $(2^7)^3$ **(c)** $(-1)^8$

Solution

(a) $(x^3)^5 = x^{3 \cdot 5} = x^{15}$ **(b)** $(2^7)^3 = 2^{7 \cdot 3} = 2^{21}$ **(c)** $(-1)^8 = +1$

Note that in both parts (a) and (b) the base does not change.

Practice Problem 12 Simplify.

(a) $(a^4)^3$ a^{12} **(b)** $(10^5)^2$ 10^{10} **(c)** $(-1)^{15}$ -1

Here are two rules involving products and quotients that are very useful: the product raised to a power rule and the quotient raised to a power rule. We'll illustrate each with an example.

If a product in parentheses is raised to a power, the parentheses indicate that *each factor* must be raised to that power.

$$(xy)^2 = x^2 y^2 \qquad (xy)^3 = x^3 y^3$$

> **PRODUCT RAISED TO A POWER**
>
> $$(xy)^a = x^a y^a$$

Teaching Tip Take a little extra time to make sure that students see where this rule comes from. They will remember it a lot longer if they can see some examples of how and why the rule works. Essentially, we want to reinforce the fact that raising to a positive integer power is repeated multiplication of the same factor. $(x^a)^b = x^a \cdot x^a \cdot x^a \cdots x^a$. The x^a occurs as a factor b times.

Teaching Example 12 Simplify.
(a) $(m^9)^5$ **(b)** $(9^2)^4$ **(c)** $(-1)^{16}$
Ans: (a) m^{45} **(b)** 9^8 **(c)** 1

Teaching Example 13 Simplify.

(a) $(mn)^3$ (b) $(4ab^3)^2$ (c) $(-r^6s)^4$

Ans: (a) m^3n^3 (b) $16a^2b^6$ (c) $r^{24}s^4$

EXAMPLE 13 Simplify.

(a) $(ab)^8$ (b) $(3x)^4$ (c) $(-2x^2)^3$

Solution

(a) $(ab)^8 = a^8b^8$ (b) $(3x)^4 = (3)^4x^4 = 81x^4$

(c) $(-2x^2)^3 = (-2)^3 \cdot (x^2)^3 = -8x^6$

Practice Problem 13 Simplify.

(a) $(3xy)^3$ $27x^3y^3$ (b) $(yz)^{37}$ $y^{37}z^{37}$ (c) $(-3x^3)^2$ $9x^6$

If a fractional expression within parentheses is raised to a power, the parentheses indicate that both numerator and denominator must be raised to that power.

$$\left(\frac{x}{y}\right)^5 = \frac{x^5}{y^5} \qquad \left(\frac{x}{y}\right)^2 = \frac{x^2}{y^2} \qquad \text{if } y \neq 0$$

QUOTIENT RAISED TO A POWER

$$\left(\frac{x}{y}\right)^a = \frac{x^a}{y^a} \qquad \text{if } y \neq 0$$

Teaching Example 14 Simplify.

(a) $\left(\frac{a}{b^2}\right)^7$ (b) $\left(\frac{-6y^7}{z}\right)^2$

Ans: (a) $\dfrac{a^7}{b^{14}}$ (b) $\dfrac{36y^{14}}{z^2}$

EXAMPLE 14 Simplify.

(a) $\left(\frac{x}{y}\right)^5$ (b) $\left(\frac{7}{w}\right)^4$

Solution

(a) $\left(\frac{x}{y}\right)^5 = \frac{x^5}{y^5}$ (b) $\left(\frac{7}{w}\right)^4 = \frac{7^4}{w^4} = \frac{2401}{w^4}$

Practice Problem 14 Simplify.

(a) $\left(\frac{x}{5}\right)^3$ $\dfrac{x^3}{125}$ (b) $\left(\frac{4a}{b}\right)^6$ $\dfrac{4096a^6}{b^6}$

Many expressions can be simplified by using the previous rules involving exponents. Be sure to take particular care to determine the correct sign, especially if there is a negative numerical coefficient.

Teaching Example 15 Simplify. $\left(\dfrac{3xy^6}{-x^3z^0}\right)^2$

Ans: $\dfrac{9y^{12}}{x^4}$

EXAMPLE 15 Simplify. $\left(\dfrac{-3x^2z^0}{y^3}\right)^4$

Solution $\left(\dfrac{-3x^2z^0}{y^3}\right)^4 = \left(\dfrac{-3x^2}{y^3}\right)^4$ Simplify inside the parentheses first. Note that $z^0 = 1$.

$= \dfrac{(-3)^4x^8}{y^{12}}$ Apply the rules for raising a power to a power. Notice that we wrote $(-3)^4$ and not -3^4. We are raising -3 to the fourth power.

$= \dfrac{81x^8}{y^{12}}$ Simplify the coefficient: $(-3)^4 = +81$.

NOTE TO STUDENT: Fully worked-out solutions to all of the Practice Problems can be found at the back of the text starting at page SP-1

Practice Problem 15 Simplify. $\left(\dfrac{-2x^3y^0z}{4xz^2}\right)^5$ $-\dfrac{x^{10}}{32z^5}$

We list here the rules of exponents we have discussed in Section 4.1.

$$x^a \cdot x^b = x^{a+b}$$

$$\frac{x^a}{x^b} = \begin{cases} x^{a-b} & \text{if} \quad a > b \\[2mm] \dfrac{1}{x^{b-a}} & \text{if} \quad b > a \\[2mm] x^0 = 1 & \text{if} \quad a = b \end{cases} \qquad x \neq 0$$

$$(x^a)^b = x^{ab}$$

$$(xy)^a = x^a y^a$$

$$\left(\frac{x}{y}\right)^a = \frac{x^a}{y^a} \qquad y \neq 0$$

Developing Your Study Skills

Why Is Review Necessary?

You master a course in mathematics by learning the concepts one step at a time. Thus understanding of mathematics is built step-by-step, with each step a supporting foundation for the next. The process is a carefully designed procedure, so no steps can be skipped. A student of mathematics needs to realize the importance of this building process to succeed.

Because new concepts depend on those previously learned, students often need to take time to review. The reviewing process will strengthen understanding and skills, which may be weak due to a lack of mastery or the passage of time. Review at the right time on the right concepts can strengthen previously learned skills and make progress possible.

Timely, periodic review of previously learned mathematical concepts is absolutely necessary for mastery of new concepts. You may have forgotten a concept or grown a bit rusty in applying it. Reviewing is the answer. Make use of the cumulative review problems in your textbook, whether they are assigned or not. Look back to previous chapters whenever you have forgotten how to do something. Review the chapter organizers from previous chapters. Study the examples and practice some exercises to refresh your understanding.

Be sure that you understand and can perform the computations of each new concept. This will enable you to move successfully on to the next one.

Remember, mathematics is a step-by-step building process. Learn each concept and reinforce and strengthen with review whenever necessary.

Verbal and Writing Skills

1. Write in your own words the product rule for exponents.

When you multiply exponential expressions with the same base, keep the base the same and add the exponents.

2. To be able to use the rules of exponents, what must be true of the bases?

The bases must be the same.

3. If the larger exponent is in the denominator, the quotient rule states that $\dfrac{x^a}{x^b} = \dfrac{1}{x^{b-a}}$. Provide an example to show why this is true.

A sample example is $\dfrac{2^2}{2^3} = \dfrac{\cancel{2} \cdot \cancel{2}}{\cancel{2} \cdot \cancel{2} \cdot 2} = \dfrac{1}{2} = \dfrac{1}{2^{3-2}}$

In exercises 4 and 5, identify the numerical coefficient, the base(s), and the exponent(s).

4. $-8x^5y^2$ -8; x, y; 5 and 2

5. $6x^{11}y$ 6; x, y; 11 and 1

6. Evaluate **(a)** $3x^0$ and **(b)** $(3x)^0$. **(c)** Why are the results different?

(a) $3x^0 = 3(1) = 3$ **(b)** $(3x)^0 = 1$ **(c)** In **(a)** the coefficient is multiplied by x^0 or 1. In **(b)** the coefficient is in the parentheses and is part of the base that is raised to the zero power, which is 1.

Write in simplest exponent form.

7. $2 \cdot 2 \cdot a \cdot a \cdot a \cdot b$ 2^2a^3b

8. $5 \cdot x \cdot x \cdot x \cdot y \cdot y$ $5x^3y^2$

9. $(-5)(x)(y)(z)(y)(x)(x)(z)$ $-5x^3y^2z^2$

10. $(-8)(a)(b)(a)(a)(c)(b)$ $-8a^3b^2c$

Multiply. Leave your answer in exponent form.

11. $(7^4)(7^6)$ 7^{10}

12. $(4^3)(4^5)$ 4^8

13. $(5^{10})(5^{16})$ 5^{26}

14. $(6^5)(6^8)$ 6^{13}

15. $x^4 \cdot x^8$ x^{12}

16. $a^8 \cdot a^{12}$ a^{20}

17. $t^{15} \cdot t$ t^{16}

18. $z^{16} \cdot z^{10}$ z^{26}

Multiply.

19. $-5x^4(4x^2)$ $-20x^6$

20. $6x^2(-9x^3)$ $-54x^5$

21. $(5x)(10x^2)$ $50x^3$

22. $(-4x^2)(-3x)$ $12x^3$

23. $(2xy^3)(9x^2y^5)$ $18x^3y^8$

24. $(4ab)(8ab^5)$ $32a^2b^6$

25. $\left(\dfrac{2}{5}xy^3\right)\left(\dfrac{1}{3}x^2y^2\right)$ $\dfrac{2}{15}x^3y^5$

26. $\left(\dfrac{4}{5}x^5y\right)\left(\dfrac{15}{16}x^2y^4\right)$ $\dfrac{3}{4}x^7y^5$

27. $(1.1x^2z)(-2.5xy)$

$-2.75x^3yz$

28. $(2.3wx^4)(-3.5xy^4)$

$-8.05wx^5y^4$

29. $(8a)(2a^3b)(0)$

0

30. $(5ab)(2a^2)(0)$

0

31. $(-16x^2y^4)(-5xy^3)$

$80x^3y^7$

32. $(-12x^4y)(-7x^5y^3)$

$84x^9y^4$

33. $(-8x^3y^2)(3xy^5)$

$-24x^4y^7$

34. $(9x^2y^6)(-11x^3y^3)$

$-99x^5y^9$

35. $(-2x^3y^2)(0)(-3x^4y)$

0

36. $(-4x^8y^2)(13y^3)(0)$

0

37. $(8a^4b^3)(-3x^2y^5)$

$-24a^4b^3x^2y^5$

38. $(5x^3y)(-2w^4z)$

$-10w^4x^3yz$

39. $(2x^2y)(-3y^3z^2)(5xz^4)$

$-30x^3y^4z^6$

40. $(3ab)(5a^2c)(-2b^2c^3)$

$-30a^3b^3c^4$

Divide. Leave your answer in exponent form. Assume that all variables in any denominator are nonzero.

41. $\dfrac{y^{12}}{y^5}$ y^7

42. $\dfrac{x^{13}}{x^3}$ x^{10}

43. $\dfrac{y^5}{y^8}$ $\dfrac{1}{y^3}$

44. $\dfrac{b^{20}}{b^{23}}$ $\dfrac{1}{b^3}$

45. $\dfrac{11^{18}}{11^{30}}$ $\dfrac{1}{11^{12}}$

46. $\dfrac{8^9}{8^{12}}$ $\dfrac{1}{8^3}$

47. $\dfrac{2^{17}}{2^{10}}$ 2^7

48. $\dfrac{7^{18}}{7^9}$ 7^9

49. $\dfrac{a^{13}}{4a^5}$ $\dfrac{a^8}{4}$

50. $\dfrac{b^{16}}{5b^{13}}$ $\dfrac{b^3}{5}$

51. $\dfrac{x^7}{y^9}$ $\dfrac{x^7}{y^9}$

52. $\dfrac{x^{20}}{y^3}$ $\dfrac{x^{20}}{y^3}$

53. $\dfrac{48x^5y^3}{24xy^3}$ $2x^4$

54. $\dfrac{45a^4b^3}{15a^4b^2}$ $3b$

55. $\dfrac{16x^5y}{-32x^2y^3}$ $-\dfrac{x^3}{2y^2}$

56. $\dfrac{-36x^3y^7}{72x^5y}$ $-\dfrac{y^6}{2x^2}$

57. $\dfrac{1.8f^4g^3}{54f^2g^8}$ $\dfrac{f^2}{30g^5}$

58. $\dfrac{3.1s^5t^3}{62s^8t}$ $\dfrac{t^2}{20s^3}$

59. $\dfrac{-85x^5y^{10}}{-5xy^7}$ $17x^4y^3$

60. $\dfrac{-60x^6y^3}{-4x^6y}$ $15y^2$

61. $\dfrac{8^0x^2y^3}{16x^5y}$ $\dfrac{y^2}{16x^3}$

62. $\dfrac{3^2x^3y^7}{3^0x^5y^2}$ $\dfrac{9y^5}{x^2}$

63. $\dfrac{18a^6b^3c^0}{24a^5b^3}$ $\dfrac{3a}{4}$

64. $\dfrac{12a^7b^8}{16a^3b^8c^0}$ $\dfrac{3a^4}{4}$

65. $\dfrac{12c^2}{42ab}$ $\dfrac{2c^2}{7ab}$

66. $\dfrac{70a^2c^3}{40b}$ $\dfrac{7a^2c^3}{4b}$

Simplify.

67. $(x^2)^6$ x^{12}

68. $(w^5)^8$ w^{40}

69. $(x^3y)^5$ $x^{15}y^5$

70. $(ab^4)^2$ a^2b^8

71. $(rs^2)^6$ r^6s^{12}

72. $(m^3n^2)^5$ $m^{15}n^{10}$

73. $(3a^3b^2c)^3$ $27a^9b^6c^3$

74. $(2x^4yz^3)^2$ $4x^8y^2z^6$

75. $(-3a^4)^2$ $9a^8$

76. $(-2a^5)^4$ $16a^{20}$

77. $\left(\dfrac{x}{2m^4}\right)^7$ $\dfrac{x^7}{128m^{28}}$

78. $\left(\dfrac{p^5}{6x}\right)^5$ $\dfrac{p^{25}}{7776x^5}$

79. $\left(\dfrac{5x}{7y^2}\right)^2$ $\dfrac{25x^2}{49y^4}$

80. $\left(\dfrac{2a^4}{3b^3}\right)^4$ $\dfrac{16a^{16}}{81b^{12}}$

81. $(-3a^2b^3c^0)^4$ $81a^8b^{12}$

82. $(-2a^5b^2c^0)^5$ $-32a^{25}b^{10}$

83. $(-2x^3y^0z)^3$ $-8x^9z^3$

84. $(-4xy^0z^4)^3$ $-64x^3z^{12}$

85. $\dfrac{(3x)^5}{(3x^2)^3}$ $\dfrac{9}{x}$

86. $\dfrac{(6y^3)^4}{(6y)^2}$ $36y^{10}$

Mixed Practice

87. $(-5a^2b^3)^2(ab)$ $25a^5b^7$

88. $(-3a^3b)^3(a^2b)$ $-27a^{11}b^4$

89. $\left(\dfrac{8}{y^5}\right)^2$ $\dfrac{64}{y^{10}}$

90. $\left(\dfrac{4}{x^6}\right)^3$ $\dfrac{64}{x^{18}}$

91. $\left(\dfrac{2x}{y^3}\right)^4$ $\dfrac{16x^4}{y^{12}}$

Cumulative Review *Simplify.*

92. **[1.2.1]** $-3 - 8$ -11

93. **[1.1.5]** $-17 + (-32) + (-24) + 27$ -46

94. **[1.3.1]** $\left(-\dfrac{3}{5}\right)\left(-\dfrac{2}{15}\right)$ $\dfrac{2}{25}$

95. **[1.3.3]** $-\dfrac{5}{4} \div \dfrac{5}{16}$ -4

Rain Forests *Brazil, the largest country in South America, boasts a land area of approximately 8,511,960 sq km. In 1970, the Amazon rain forest covered 4,100,000 sq km of Brazil's land area. By 2006, the rain forest in Brazil only covered 3,400,254 sq km. (Source: en.wikipedia.org.) Round your answers to the nearest tenth of a percent.*

96. **[0.7.1]** What was the percent decrease in the rain forest from 1970 to 2006?

$\dfrac{4,100,000 - 3,400,254}{4,100,000} = \dfrac{699,746}{4,100,000} \approx 0.1707$, so the rain forest decreased by approximately 17.1%

97. **[0.7.1]** If Brazil is said to lose 2.3% of its rain forest per year, how much rain forest was lost during 2007? $(3,400,254)(0.023) \approx 78,206$ km^2

Quick Quiz 4.1 Simplify the following.

1. $(2x^2y^3)(-5xy^4)$ $-10x^3y^7$

2. $\dfrac{-28x^6y^6}{35x^3y^8}$ $-\dfrac{4x^3}{5y^2}$

3. $(-3x^3y^5)^4$ $81x^{12}y^{20}$

4. **Concept Check** Explain the steps you would need to follow to simply the expression.

$$\dfrac{(4x^3)^2}{(2x^4)^3}$$

Answers may vary

Classroom Quiz 4.1 You may use these problems to quiz your students' mastery of Section 4.1.

Simplify the following.

1. $(6xy^8)(-3x^4y^4)$ **Ans:** $-18x^5y^{12}$

2. $\dfrac{-81x^4y^4}{-27xy^8}$ **Ans:** $\dfrac{3x^3}{y^4}$

3. $(-4x^2y^7)^3$ **Ans:** $-64x^6y^{21}$

1 Using Negative Exponents

If n is an integer and $x \neq 0$, then x^{-n} is defined as follows:

DEFINITION OF A NEGATIVE EXPONENT

$$x^{-n} = \frac{1}{x^n}, \quad x \neq 0$$

EXAMPLE 1 Write with positive exponents.

(a) y^{-3} (b) z^{-6} (c) w^{-1}

Solution

(a) $y^{-3} = \dfrac{1}{y^3}$ (b) $z^{-6} = \dfrac{1}{z^6}$ (c) $w^{-1} = \dfrac{1}{w^1} = \dfrac{1}{w}$

Practice Problem 1 Write with positive exponents.

(a) x^{-12} $\dfrac{1}{x^{12}}$ (b) w^{-5} $\dfrac{1}{w^5}$ (c) z^{-2} $\dfrac{1}{z^2}$

Teaching Example 1 Write with positive exponents.

(a) a^{-5} (b) m^{-4} (c) b^{-16}

Ans: (a) $\dfrac{1}{a^5}$ (b) $\dfrac{1}{m^4}$ (c) $\dfrac{1}{b^{16}}$

NOTE TO STUDENT: *Fully worked-out solutions to all of the Practice Problems can be found at the back of the text starting at page SP-1*

To evaluate a numerical expression with a negative exponent, first write the expression with a positive exponent. Then simplify.

EXAMPLE 2 Evaluate.

(a) 3^{-2} (b) 2^{-5}

Solution

(a) $3^{-2} = \dfrac{1}{3^2} = \dfrac{1}{9}$ (b) $2^{-5} = \dfrac{1}{2^5} = \dfrac{1}{32}$

Practice Problem 2 Evaluate.

(a) 4^{-3} $\dfrac{1}{64}$ (b) 2^{-4} $\dfrac{1}{16}$

Teaching Example 2 Evaluate.

(a) 5^{-2} (b) 6^{-3}

Ans: (a) $\dfrac{1}{25}$ (b) $\dfrac{1}{216}$

All the previously studied laws of exponents are true for any integer exponent. These laws are summarized in the following box. Assume that $x, y \neq 0$.

LAWS OF EXPONENTS

The Product Rule

$$x^a \cdot x^b = x^{a+b}$$

The Quotient Rule

$$\frac{x^a}{x^b} = x^{a-b} \quad \text{Use if } a > b. \qquad \frac{x^a}{x^b} = \frac{1}{x^{b-a}} \quad \text{Use if } a < b.$$

Power Rules

$$(xy)^a = x^a y^a, \qquad (x^a)^b = x^{ab}, \qquad \left(\frac{x}{y}\right)^a = \frac{x^a}{y^a}$$

By using the definition of a negative exponent and the properties of fractions, we can derive two more helpful properties of exponents. Assume that $x, y \neq 0$.

PROPERTIES OF NEGATIVE EXPONENTS

$$\frac{1}{x^{-n}} = x^n \qquad \frac{x^{-m}}{y^{-n}} = \frac{y^n}{x^m}$$

Teaching Example 3 Simplify. Write the expression with no negative exponents.

(a) $\dfrac{5}{y^{-7}}$ (b) $\dfrac{a^{-1}b^4}{c^{-2}}$ (c) $r^5 s^{-6}$

Ans: (a) $5y^7$ (b) $\dfrac{b^4 c^2}{a}$ (c) $\dfrac{r^5}{s^6}$

EXAMPLE 3 Simplify. Write the expression with no negative exponents.

(a) $\dfrac{1}{x^{-6}}$ (b) $\dfrac{x^{-3}y^{-2}}{z^{-4}}$ (c) $x^{-2}y^3$

Solution

(a) $\dfrac{1}{x^{-6}} = x^6$ (b) $\dfrac{x^{-3}y^{-2}}{z^{-4}} = \dfrac{z^4}{x^3 y^2}$ (c) $x^{-2}y^3 = \dfrac{y^3}{x^2}$

NOTE TO STUDENT: *Fully worked-out solutions to all of the Practice Problems can be found at the back of the text starting at page SP-1*

Practice Problem 3 Simplify. Write the expression with no negative exponents.

(a) $\dfrac{3}{w^{-4}}$ $3w^4$ (b) $\dfrac{x^{-6}y^4}{z^{-2}}$ $\dfrac{y^4 z^2}{x^6}$ (c) $x^{-6}y^{-5}$ $\dfrac{1}{x^6 y^5}$

Teaching Example 4 Simplify. Write the expression with no negative exponents.

(a) $(2a^{-1}b^3)^{-5}$ (b) $\dfrac{a^{-3}b^{-2}}{a^{-4}b^3}$

Ans: (a) $\dfrac{a^5}{32b^{15}}$ (b) $\dfrac{a}{b^5}$

EXAMPLE 4 Simplify. Write the expression with no negative exponents.

(a) $(3x^{-4}y^2)^{-3}$ (b) $\dfrac{x^2 y^{-4}}{x^{-5}y^3}$

Solution

(a) $(3x^{-4}y^2)^{-3} = 3^{-3}x^{12}y^{-6} = \dfrac{x^{12}}{3^3 y^6} = \dfrac{x^{12}}{27y^6}$

(b) $\dfrac{x^2 y^{-4}}{x^{-5}y^3} = \dfrac{x^2 x^5}{y^4 y^3} = \dfrac{x^7}{y^7}$ First rewrite the expression so that only positive exponents appear. Then simplify using the product rule.

Practice Problem 4 Simplify. Write the expression with no negative exponents.

(a) $(2x^4y^{-5})^{-2}$ $\dfrac{y^{10}}{4x^8}$ (b) $\dfrac{y^{-3}z^{-4}}{y^2 z^{-6}}$ $\dfrac{z^2}{y^5}$

---

② Using Scientific Notation

One common use of negative exponents is in writing numbers in scientific notation. Scientific notation is most useful in expressing very large and very small numbers.

SCIENTIFIC NOTATION

A positive number is written in **scientific notation** if it is in the form $a \times 10^n$, where $1 \le a < 10$ and n is an integer.

EXAMPLE 5   Write in scientific notation.

(a) 4567

(b) 157,000,000

Solution

(a) $4567 = 4.567 \times 1000$ To change 4567 to a number that is greater than 1 but less than 10, we move the decimal point three places to the left. We must then multiply the number by a power of 10 so that we do not change the value of the number. Use 1000.

$$= 4.567 \times 10^3$$

(b) $157,000,000 = 1.57000000 \times 100000000$

8 places   8 zeros

$$= 1.57 \times 10^8$$

Practice Problem 5   Write in scientific notation.

(a) 78,200 7.82×10^4

(b) 4,786,000 4.786×10^6

Numbers that are smaller than 1 will have a negative power of 10 if they are written in scientific notation.

EXAMPLE 6   Write in scientific notation.

(a) 0.061

(b) 0.000052

Solution

(a) We need to write 0.061 as a number that is greater than 1 but less than 10. In which direction do we move the decimal point?

$0.061 = 6.1 \times 10^{-2}$ Move the decimal point 2 places to the right.

(b) $0.000052 = 5.2 \times 10^{-5}$ Why?

Practice Problem 6   Write in scientific notation.

(a) 0.98 9.8×10^{-1}

(b) 0.000092 9.2×10^{-5}

The reverse procedure transforms scientific notation into ordinary decimal notation.

EXAMPLE 7   Write in decimal notation.   **(a)** 1.568×10^2 **(b)** 7.432×10^{-3}

Solution

(a) $1.568 \times 10^2 = 1.568 \times 100$

$$= 156.8$$

Alternative Method

$1.568 \times 10^2 = 156.8$ The exponent 2 tells us to move the decimal point 2 places to the right.

(b) $7.432 \times 10^{-3} = 7.432 \times \dfrac{1}{1000}$

$$= 0.007432$$

Alternative Method

$7.432 \times 10^{-3} = 0.007432$ The exponent -3 tells us to move the decimal point 3 places to the left.

Practice Problem 7   Write in decimal notation.

(a) 1.93×10^6 1,930,000

(b) 8.562×10^{-5} 0.00008562

Calculator

Scientific Notation

Most scientific calculators can display only eight digits at one time. Numbers with more than eight digits are usually shown in scientific notation. 1.12 E 08 or 1.12 8 means 1.12×10^8. You can use a calculator to compute with large numbers by entering the numbers using scientific notation. For example,

$$(7.48 \times 10^{24}) \times (3.5 \times 10^8)$$

is entered as follows.

7.48 [EXP] 24 [×]

3.5 [EXP] 8 [=]

Display: [2.618 E 33]

or [2.618   33]

Note: Some calculators have an [EE] key instead of [EXP].

The distance light travels in one year is called a *light-year*. A light-year is a convenient unit of measure to use when investigating the distances between stars.

EXAMPLE 8 A light-year is a distance of 9,460,000,000,000,000 meters. Write this in scientific notation.

Solution 9,460,000,000,000,000 meters $= 9.46 \times 10^{15}$ meters

Practice Problem 8 Astronomers measure distances to faraway galaxies in parsecs. A parsec is a distance of 30,900,000,000,000,000 meters. Write this in scientific notation. 3.09×10^{16} m

To perform a calculation involving very large or very small numbers, it is usually helpful to write the numbers in scientific notation and then use the laws of exponents to do the calculation.

EXAMPLE 9 Use scientific notation and the laws of exponents to find the following. Leave your answer in scientific notation.

(a) $(32,000,000)(1,500,000,000,000)$

(b) $\dfrac{0.00063}{0.021}$

Solution

(a) $(32,000,000)(1,500,000,000,000)$

$= (3.2 \times 10^7)(1.5 \times 10^{12})$ Write each number in scientific notation.

$= 3.2 \times 1.5 \times 10^7 \times 10^{12}$ Rearrange the order. Remember that multiplication is commutative.

$= 4.8 \times 10^{19}$ Multiply 3.2×1.5. Multiply $10^7 \times 10^{12}$.

(b) $\dfrac{0.00063}{0.021} = \dfrac{6.3 \times 10^{-4}}{2.1 \times 10^{-2}}$ Write each number in scientific notation.

$= \dfrac{6.3}{2.1} \times \dfrac{10^{-4}}{10^{-2}}$ Rearrange the order. We are actually using the definition of multiplication of fractions.

$= \dfrac{6.3}{2.1} \times \dfrac{10^2}{10^4}$ Rewrite with positive exponents.

$= 3.0 \times 10^{-2}$

Practice Problem 9 Use scientific notation and the laws of exponents to find the following. Leave your answer in scientific notation.

(a) $(56,000)(1,400,000,000)$ 7.84×10^{13} **(b)** $\dfrac{0.000111}{0.00000037}$ 3.0×10^2

When we use scientific notation, we are often writing approximate numbers. We must include some zeros so that the decimal point can be properly located. However, all other digits except for these zeros are considered **significant digits.** The number 34.56 has four significant digits. The number 0.0049 has two significant digits. The zeros are considered placeholders. The number 634,000 has three significant digits (unless we have specific knowledge to the contrary). The zeros are considered placeholders. We sometimes round numbers to a specific number of significant digits. For example, 0.08746 rounded to two significant digits is 0.087. When we round 1,348,593 to three significant digits, we obtain 1,350,000.

EXAMPLE 10 The approximate distance from Earth to the star Polaris is 208 parsecs. A parsec is a distance of approximately 3.09×10^{13} kilometers. How long would it take a space probe traveling at 40,000 kilometers per hour to reach the star? Round to three significant digits.

Solution

1. Understand the problem. Recall that the distance formula is

$$\text{distance} = \text{rate} \times \text{time}.$$

We are given the distance and the rate. We need to find the time.
Let's take a look at the distance. The distance is given in parsecs, but the rate is given in kilometers per hour. We need to change the distance to kilometers. We are told that a parsec is approximately 3.09×10^{13} kilometers. That is, there are 3.09×10^{13} kilometers per parsec. We use this information to change 208 parsecs to kilometers.

$$208 \text{ parsecs} = \frac{(208 \text{ parsecs})(3.09 \times 10^{13} \text{ kilometers})}{1 \text{ parsec}} = 642.72 \times 10^{13} \text{ kilometers}$$

Teaching Example 10 The distance from Earth to the closest star, Proxima Centauri, is approximately 1.29 parsecs. How many hours would it take a space probe traveling at 50,000 kilometers per hour to reach this star? Round to three significant digits.

Ans: 797 million hours

2. Write an equation. Use the distance formula.

$$d = r \times t$$

3. Solve the equation and state the answer. Substitute the known values into the formula and solve for the unknown, time.

$$642.72 \times 10^{13} \text{ km} = \frac{40,000 \text{ km}}{1 \text{ hr}} \times t$$

$$6.4272 \times 10^{15} \text{ km} = \frac{4 \times 10^4 \text{ km}}{1 \text{ hr}} \times t \qquad \text{Change the numbers to scientific notation.}$$

$$\frac{6.4272 \times 10^{15} \text{ km}}{\dfrac{4 \times 10^4 \text{ km}}{1 \text{ hr}}} = t \qquad \text{Divide both sides by } \frac{4 \times 10^4 \text{ km}}{1 \text{ hr}}.$$

$$\frac{(6.4272 \times 10^{15} \text{ k\hspace{-0.4em}m})(1 \text{ hr})}{4 \times 10^4 \text{ k\hspace{-0.4em}m}} = t$$

$$1.6068 \times 10^{11} \text{ hr} = t$$

1.6068×10^{11} is 160.68×10^9 or 160.68 billion hours. The space probe will take approximately 160.68 billion hours to reach the star.

Reread the problem. Are we finished? What is left to do? We need to round the answer to three significant digits. Rounding to three significant digits, we have

$$160.68 \times 10^9 \approx 161 \times 10^9.$$

This is approximately 161 billion hours or a little more than 18 million years.

4. Check. Unless you have had a great deal of experience working in astronomy, it would be difficult to determine whether this is a reasonable answer. You may wish to reread your analysis and redo your calculations as a check.

Practice Problem 10 The average distance from Earth to the distant star Betelgeuse is 159 parsecs. How many hours would it take a space probe to travel from Earth to Betelgeuse at a speed of 50,000 kilometers per hour? Round to three significant digits. 9.83×10^{10} hr

Simplify. Express your answer with positive exponents. Assume that all variables are nonzero.

1. x^{-4} $\dfrac{1}{x^4}$

2. y^{-5} $\dfrac{1}{y^5}$

3. 3^{-4} $\dfrac{1}{81}$

4. 5^{-3} $\dfrac{1}{125}$

5. $\dfrac{1}{y^{-8}}$ y^8

6. $\dfrac{1}{z^{-10}}$ z^{10}

7. $\dfrac{x^{-4}y^{-5}}{z^{-6}}$ $\dfrac{z^6}{x^4y^5}$

8. $\dfrac{x^{-6}y^{-2}}{z^{-5}}$ $\dfrac{z^5}{x^6y^2}$

9. a^3b^{-2} $\dfrac{a^3}{b^2}$

10. a^5b^{-8} $\dfrac{a^5}{b^8}$

11. $(2x^{-3})^{-3}$ $\dfrac{x^9}{8}$

12. $(4x^{-4})^{-2}$ $\dfrac{x^8}{16}$

13. $3x^{-2}$ $\dfrac{3}{x^2}$

14. $4y^{-4}$ $\dfrac{4}{y^4}$

15. $(3xy^2)^{-2}$ $\dfrac{1}{9x^2y^4}$

16. $(4x^2y^4)^{-3}$ $\dfrac{1}{64x^6y^{12}}$

Mixed Practice

17. $\dfrac{3xy^{-2}}{z^{-3}}$ $\dfrac{3xz^3}{y^2}$

18. $\dfrac{4x^{-2}y^{-3}}{y^4}$ $\dfrac{4}{x^2y^7}$

19. $\dfrac{(3x)^{-2}}{(3x)^{-3}}$ $3x$

20. $\dfrac{(2ab^2)^{-3}}{(2ab^2)^{-4}}$ $2ab^2$

21. $a^{-1}b^3c^{-4}d$ $\dfrac{b^3d}{ac^4}$

22. $x^{-5}y^{-2}z^3$ $\dfrac{z^3}{x^5y^2}$

23. $(8^{-2})(2^3)$ $\dfrac{1}{8}$

24. $(9^2)(3^{-3})$ 3

25. $\left(\dfrac{3x^0y^2}{z^4}\right)^{-2}$ $\dfrac{z^8}{9y^4}$

26. $\left(\dfrac{2a^3b^0}{c^2}\right)^{-3}$ $\dfrac{c^6}{8a^9}$

27. $\dfrac{x^{-2}y^{-3}}{x^4y^{-2}}$ $\dfrac{1}{x^6y}$

28. $\dfrac{a^{-6}b^3}{a^{-2}b^{-5}}$ $\dfrac{b^8}{a^4}$

Write in scientific notation.

29. 123,780
1.2378×10^5

30. 5,786,100
5.7861×10^6

31. 0.063
6.3×10^{-2}

32. 0.000742
7.42×10^{-4}

33. 889,610,000,000
8.8961×10^{11}

34. 7,652,000,000
7.652×10^9

35. 0.00000342
3.42×10^{-6}

36. 0.0000000058
5.8×10^{-9}

In exercises 37–42, write in decimal notation.

37. 3.02×10^5
302,000

38. 8.137×10^7
81,370,000

39. 4.7×10^{-4}
0.00047

40. 6.53×10^{-3}
0.00653

41. 9.83×10^5
983,000

42. 3.5×10^{-8}
0.000000035

43. *Bamboo Growth* The growth rate of some species of bamboo is 0.0000237 miles per hour. Write this in scientific notation. 2.37×10^{-5} mi/hr

44. *Neptune* Neptune is 2.793×10^9 miles from the sun. Write this in decimal notation. 2,793,000,000 mi

45. *Astronomical Unit* The astronomical unit (AU) is a unit of length approximately equal to 1.496×10^8 km. Write this in decimal notation. 149,600,000 km

46. *Gold Atom* The average volume of an atom of gold is 0.0000000000000000000001695 cubic centimeters. Write this in scientific notation. 1.695×10^{-23} cm^3

Evaluate by using scientific notation and the laws of exponents. Leave your answer in scientific notation.

47. $(42{,}000{,}000)(150{,}000{,}000)$ 6.3×10^{15}

48. $(55{,}000{,}000{,}000)(16{,}000{,}000)$ 8.8×10^{17}

49. $\dfrac{(5{,}000{,}000)(16{,}000)}{8{,}000{,}000{,}000}$ 1.0×10^1

50. $(0.0075)(0.0000002)(0.001)$ 1.5×10^{-12}

51. $(0.003)^4$ 8.1×10^{-11}

52. $(500{,}000)^4$ 6.25×10^{22}

53. $(150{,}000{,}000)(0.00005)(0.002)(30{,}000)$ 4.5×10^5

54. $\dfrac{(1{,}600{,}000)(0.00003)}{2400}$ 2×10^{-2}

Applications

National Debt *In July 2006, the national debt was about* 8.42×10^{12} *dollars.* (*Source:* www.factfinder.census.gov.)

55. The Census Bureau estimates that in July 2006, the population of the United States was 2.99×10^8 people. If the national debt were evenly divided among every person in the country, how much debt would be assigned to each individual? Round to three significant digits. 2.82×10^4 dollars

56. The Census Bureau estimates that in July 2006, the number of people in the United States who were 18 years or older was approximately 2.26×10^8. If the national debt were evenly divided among every person 18 years or older in the country, how much would be assigned to each individual? Round to three significant digits. 3.73×10^4 dollars

57. *Watch Hand* The tip of a $\frac{1}{3}$-inch-long hour hand on a watch travels at a speed of 0.00000275 miles per hour. How far has it traveled in a day? 6.6×10^{-5} mi

58. *Neutron Mass* The mass of a neutron is approximately 1.675×10^{-27} kilogram. Find the mass of 180,000 neutrons. 3.015×10^{-22} kg

59. *Mission to Pluto* In January 2006, a spacecraft called *New Horizons* began its journey to Pluto. The trip is 3.5×10^9 miles long and will take $9\frac{1}{2}$ years. How many miles will *New Horizons* travel per year? (*Source:* www.pluto.jhuapl.edu.) About 3.68×10^8 mi/yr

60. *Moles per Molecule* Avogadro's number says that there are approximately 6.02×10^{23} molecules/mole. How many molecules can one expect in 0.00483 mole? 2.908×10^{21} molecules

61. *Construction Cost* In 1990 the cost for construction of new private buildings was estimated at $\$3.61 \times 10^{11}$. By 2010 the estimated cost for construction of new private buildings will be $\$9.36 \times 10^{11}$. What is the percent of increase from 1990 to 2010? Round to the nearest tenth of a percent. (*Source:* U.S. Census Bureau.) 159.3%

62. *Construction Cost* In 1990 the cost for construction of new public buildings was estimated at $\$1.07 \times 10^{11}$. By 2010 the estimated cost for construction of new public buildings will be $\$3.06 \times 10^{11}$. What is the percent of increase from 1990 to 2010? Round to the nearest tenth of a percent. (*Source:* U.S. Census Bureau.) 186.0%

Cumulative Review *Simplify.*

63. **[1.2.1]** $-2.7 - (-1.9)$ -0.8

64. **[1.4.2]** $(-1)^{33}$ -1

65. **[1.1.4]** $-\dfrac{3}{4} + \dfrac{5}{7}$ $-\dfrac{1}{28}$

Quick Quiz 4.2 Simplify and write your answer with only positive exponents.

1. $3x^{-3}y^2z^{-4}$ $\dfrac{3y^2}{x^3z^4}$

2. $\dfrac{4a^3b^{-4}}{8a^{-5}b^{-3}}$ $\dfrac{a^8}{2b}$

3. Write in scientific notation. 0.00876 8.76×10^{-3}

4. **Concept Check** Explain how you would simplify a problem like the following so that your answer had only positive exponents.

$$(4x^{-3}y^4)^{-3} \quad \text{Answers may vary}$$

Classroom Quiz 4.2 You may use these problems to quiz your students' mastery of Section 4.2.

Simplify and write your answer with only positive exponents.

1. $4^{-2}x^4y^{-3}z^{-2}$

Ans: $\dfrac{x^4}{16y^3z^2}$

2. $\dfrac{10a^{-2}b^{-3}}{15a^4b^{-5}}$

Ans: $\dfrac{2b^2}{3a^6}$

3. Write in scientific notation. 0.0004387

Ans: 4.387×10^{-4}

1 Recognizing Polynomials and Determining Their Degrees

A **polynomial** in x is the sum of a finite number of terms of the form ax^n, where a is any real number and n is a whole number. Usually these polynomials are written in descending powers of the variable, as in

$$5x^3 + 3x^2 - 2x - 5 \quad \text{and} \quad 3.2x^2 - 1.4x + 5.6.$$

A **multivariable polynomial** is a polynomial with more than one variable. The following are multivariable polynomials:

$$5xy + 8, \quad 2x^2 - 7xy + 9y^2, \quad 17x^3y^9$$

The **degree of a term** is the sum of the exponents of all of the variables in the term. For example, the degree of $7x^3$ is three. The degree of $4xy$ is two. The degree of $10x^4y^2$ is six.

The **degree of a polynomial** is the highest degree of all of the terms in the polynomial. For example, the degree of $5x^3 + 8x^2 - 20x - 2$ is three. The degree of $6xy - 4x^2y + 2xy^3$ is four.

The polynomial 0 is said to have **no degree.** A polynomial consisting of a constant only is said to have degree 0.

There are special names for polynomials with one, two, or three terms.

A **monomial** has *one* term:

$$5a, \quad 3x^3yz^4, \quad 12xy$$

A **binomial** has *two* terms:

$$7x + 9y, \quad -6x - 4, \quad 5x^4 + 2xy^2$$

A **trinomial** has *three* terms:

$$8x^2 - 7x + 4, \quad 2ab^3 - 6ab^2 - 15ab, \quad 2 + 5y + y^4$$

EXAMPLE 1 State the degree of the polynomial, and whether it is a monomial, a binomial, or a trinomial.

(a) $5xy + 3x^3$ (b) $-7a^5b^2$ (c) $8x^4 - 9x - 15$

Solution

(a) This polynomial is of degree 3. It has two terms, so it is a binomial.

(b) The sum of the exponents is $5 + 2 = 7$. Therefore this polynomial is of degree 7. It has one term, so it is a monomial.

(c) This polynomial is of degree 4. It has three terms, so it is a trinomial.

Practice Problem 1 State the degree of the polynomial, and whether it is a monomial, a binomial, or a trinomial.

(a) $-7x^5 - 3xy$ (b) $22a^3b^4$ (c) $-3x^3 + 3x^2 - 6x$

(a) Degree 5; binomial   (b) Degree 7; monomial   (c) Degree 3; trinomial

2 Adding Polynomials

We usually write a polynomial in x so that the exponents on x decrease from left to right. For example, the polynomial

$$5x^2 - 6x + 2$$

is said to be written in **decreasing order** since each exponent is decreasing as we move from left to right.

You can add, subtract, multiply, and divide polynomials. Let us take a look at addition. To add two polynomials, we add their like terms.

EXAMPLE 2 Add. $(5x^2 - 6x - 12) + (-3x^2 - 9x + 5)$

Solution

$$
\begin{aligned}
(5x^2 - 6x - 12) + (-3x^2 - 9x + 5) &= [5x^2 + (-3x^2)] + [-6x + (-9x)] + [-12 + 5] \\
&= [(5 - 3)x^2] + [(-6 - 9)x] + [-12 + 5] \\
&= 2x^2 + (-15x) + (-7) \\
&= 2x^2 - 15x - 7
\end{aligned}
$$

Practice Problem 2 Add. $(-8x^3 + 3x^2 + 6) + (2x^3 - 7x^2 - 3)$

$$-6x^3 - 4x^2 + 3$$

The numerical coefficients of polynomials may be any real number. Thus polynomials may have numerical coefficients that are decimals or fractions.

EXAMPLE 3 Add. $\left(\frac{1}{2}x^2 - 6x + \frac{1}{3}\right) + \left(\frac{1}{5}x^2 - 2x - \frac{1}{2}\right)$

Solution

$$
\begin{aligned}
\left(\frac{1}{2}x^2 - 6x + \frac{1}{3}\right) + \left(\frac{1}{5}x^2 - 2x - \frac{1}{2}\right) &= \left[\frac{1}{2}x^2 + \frac{1}{5}x^2\right] + [-6x + (-2x)] + \left[\frac{1}{3} + \left(-\frac{1}{2}\right)\right] \\
&= \left[\left(\frac{1}{2} + \frac{1}{5}\right)x^2\right] + [(-6 - 2)x] + \left[\frac{1}{3} + \left(-\frac{1}{2}\right)\right] \\
&= \left[\left(\frac{5}{10} + \frac{2}{10}\right)x^2\right] + [-8x] + \left[\frac{2}{6} - \frac{3}{6}\right] \\
&= \frac{7}{10}x^2 - 8x - \frac{1}{6}
\end{aligned}
$$

Practice Problem 3 Add.

$$\left(-\frac{1}{3}x^2 - 6x - \frac{1}{12}\right) + \left(\frac{1}{4}x^2 + 5x - \frac{1}{3}\right) \qquad -\frac{1}{12}x^2 - x - \frac{5}{12}$$

EXAMPLE 4 Add. $(1.2x^3 - 5.6x^2 + 5) + (-3.4x^3 - 1.2x^2 + 4.5x - 7)$

Solution   Group like terms.

$$
\begin{aligned}
(1.2x^3 - 5.6x^2 + 5) + (-3.4x^3 - 1.2x^2 + 4.5x - 7) &= (1.2 - 3.4)x^3 + (-5.6 - 1.2)x^2 + 4.5x + (5 - 7) \\
&= -2.2x^3 - 6.8x^2 + 4.5x - 2
\end{aligned}
$$

Practice Problem 4 Add.

$$(3.5x^3 - 0.02x^2 + 1.56x - 3.5) + (-0.08x^2 - 1.98x + 4)$$

$$3.5x^3 - 0.1x^2 - 0.42x + 0.5$$

Teaching Example 2 Add.
$(-x^4 + 5x^2 - 17x) + (4x^4 + x^2 + 11x)$
Ans: $3x^4 + 6x^2 - 6x$

Teaching Example 3 Add.
$\left(-\frac{2}{3}x^3 + \frac{5}{8}x^2 - 3x + \frac{1}{6}\right) +$
$\left(\frac{1}{2}x^3 + \frac{1}{4}x^2 + \frac{5}{9}\right)$
Ans: $-\frac{1}{6}x^3 + \frac{7}{8}x^2 - 3x + \frac{13}{18}$

Teaching Example 4 Add. $(2.7x^3 - 3.5x + 8.6)$
$+(-0.4x^3 + 1.3x^2 - 6.4)$
Ans: $2.3x^3 + 1.3x^2 - 3.5x + 2.2$

Subtracting Polynomials

Recall that subtraction of real numbers can be defined as adding the opposite of the second number. Thus $a - b = a + (-b)$. That is, $3 - 5 = 3 + (-5)$. A similar method is used to subtract two polynomials.

To subtract two polynomials, change the sign of each term in the second polynomial and then add.

Teaching Example 5 Subtract.
$(11x^2 - 2x + 4) - (-5x^2 - 3x + 1)$

Ans: $16x^2 + x + 3$

EXAMPLE 5 Subtract. $(7x^2 - 6x + 3) - (5x^2 - 8x - 12)$

Solution We change the sign of each term in the second polynomial and then add.

$$(7x^2 - 6x + 3) - (5x^2 - 8x - 12) = (7x^2 - 6x + 3) + (-5x^2 + 8x + 12)$$
$$= (7 - 5)x^2 + (-6 + 8)x + (3 + 12)$$
$$= 2x^2 + 2x + 15$$

NOTE TO STUDENT: Fully worked-out solutions to all of the Practice Problems can be found at the back of the text starting at page SP-1

Practice Problem 5 Subtract.

$$(5x^3 - 15x^2 + 6x - 3) - (-4x^3 - 10x^2 + 5x + 13)$$

$9x^3 - 5x^2 + x - 16$

As mentioned previously, polynomials may involve more than one variable. When subtracting polynomials in two variables, you will need to use extra care in determining which terms are like terms. For example $6x^2y$ and $5x^2y$ are like terms. In a similar fashion, $3xy$ and $8xy$ are like terms. However, $7xy^2$ and $15x^2y^2$ are not like terms. Every exponent of every variable in the two terms must be the same if the terms are to be like terms. You will use this concept in Example 6.

Teaching Example 6 Subtract.
$(8a^2b - 14ab + 6b) - (-ab + 2b + b^2)$

Ans: $8a^2b - 13ab + 4b - b^2$

EXAMPLE 6 Subtract.

$$(-6x^2y - 3xy + 7xy^2) - (5x^2y - 8xy - 15x^2y^2)$$

Solution Change the sign of each term in the second polynomial and add. Look for like terms.

$$(-6x^2y - 3xy + 7xy^2) + (-5x^2y + 8xy + 15x^2y^2)$$
$$= (-6 - 5)x^2y + (-3 + 8)xy + 7xy^2 + 15x^2y^2$$
$$= -11x^2y + 5xy + 7xy^2 + 15x^2y^2$$

Nothing further can be done to combine these four terms.

Practice Problem 6 Subtract.

$$(x^3 - 7x^2y + 3xy^2 - 2y^3) - (2x^3 + 4xy - 6y^3)$$

$-x^3 - 7x^2y + 3xy^2 - 4xy + 4y^3$

Evaluating Polynomials to Predict a Value

Sometimes polynomials are used to predict values. In such cases we need to **evaluate** the polynomial. We do this by substituting a known value for the variable and determining the value of the polynomial.

EXAMPLE 7 Automobiles sold in the United States have become more fuel efficient over the years due to regulations from Congress. The number of miles per gallon obtained by the average automobile in the United States can be described by the polynomial

$$0.3x + 12.9,$$

where x is the number of years since 1970. (*Source:* U.S. Federal Highway Administration.) Use this polynomial to estimate the number of miles per gallon obtained by the average automobile in

(a) 1972   **(b)** 2012

Teaching Example 7 Town planners believe that the population of Bradenton can be estimated by the polynomial $0.5x^2 + 65x + 28{,}750$, where x is the number of years since 1985. Use this polynomial to estimate the population of the town in **(a)** 2000 and **(b)** 2015. Of course, you will want to round your answer to the nearest whole person.

Ans: (a) 29,838   **(b)** 31,150

Solution

(a) The year 1972 is two years later than 1970, so $x = 2$.

Thus the number of miles per gallon obtained by the average automobile in 1972 can be estimated by evaluating $0.3x + 12.9$ when $x = 2$.

$$0.3(2) + 12.9 = 0.6 + 12.9$$
$$= 13.5$$

We estimate that the average car in 1972 obtained 13.5 miles per gallon.

(b) The year 2012 will be 42 years after 1970, so $x = 42$.

Thus the estimated number of miles per gallon obtained by the average automobile in 2012 can be predicted by evaluating $0.3x + 12.9$ when $x = 42$.

$$0.3(42) + 12.9 = 12.6 + 12.9$$
$$= 25.5$$

We therefore predict that the average car in 2012 will obtain 25.5 miles per gallon.

Practice Problem 7 The number of miles per gallon obtained by the average truck in the United States can be described by the polynomial $0.03x + 5.4$, where x is the number of years since 1970. (*Source:* U.S. Federal Highway Administration.) Use this polynomial to estimate the number of miles per gallon obtained by the average truck in

(a) 1974   5.52 mi/gal   **(b)** 2011   6.63 mi/gal

Verbal and Writing Skills

1. State in your own words a definition for a polynomial in x and give an example.

A polynomial in x is the sum of a finite number of terms of the form ax^n, where a is any real number and n is a whole number. An example is $3x^2 - 5x - 9$.

2. State in your own words a definition for a multivariable polynomial and give an example.

A multivariable polynomial is a polynomial with more than one variable. An example is $5x^2y - 7xy + 8y$.

3. State in your own words how to determine the degree of a polynomial in x.

The degree of a polynomial in x is the largest exponent of x in any of the terms of the polynomial.

4. State in your own words how to determine the degree of a multivariable polynomial.

First find the degree of each term. This is the sum of the exponents of all the variables in each term. The degree of the polynomial is the highest degree of all the terms of the polynomial.

State the degree of the polynomial and whether it is a monomial, a binomial, or a trinomial.

5. $6x^3y$
Degree 4; monomial

6. $5xy^6$
Degree 7; monomial

7. $20x^5 + 6x^3 - 7x$
Degree 5; trinomial

8. $13x^4 - 12x + 20$
Degree 4; trinomial

9. $4x^2y^3 - 7x^3y^3$
Degree 6; binomial

10. $9x^4y^3 - 6xy^5$
Degree 7; binomial

Add.

11. $(-3x + 15) + (8x - 43)$
$5x - 28$

12. $(5x - 11) + (-7x + 34)$
$-2x + 23$

13. $(6x^2 + 5x - 6) + (-8x^2 - 3x + 5)$
$-2x^2 + 2x - 1$

14. $(3x^2 - 2x - 5) + (-7x^2 + 8x - 2)$
$-4x^2 + 6x - 7$

15. $\left(\frac{1}{2}x^2 + \frac{1}{3}x - 4\right) + \left(\frac{1}{3}x^2 + \frac{1}{6}x - 5\right)$
$\frac{5}{6}x^2 + \frac{1}{2}x - 9$

16. $\left(\frac{1}{4}x^2 - \frac{2}{3}x - 10\right) + \left(-\frac{1}{3}x^2 + \frac{1}{9}x + 2\right)$
$-\frac{1}{12}x^2 - \frac{5}{9}x - 8$

17. $(3.4x^3 - 7.1x + 3.4) + (2.2x^2 - 6.1x - 8.8)$
$3.4x^3 + 2.2x^2 - 13.2x - 5.4$

18. $(-4.6x^3 + 5.6x - 0.3) + (9.8x^2 + 4.5x - 1.7)$
$-4.6x^3 + 9.8x^2 + 10.1x - 2$

Subtract.

19. $(2x - 19) - (-3x + 5)$
$5x - 24$

20. $(5x - 5) - (6x - 3)$
$-x - 2$

21. $\left(\frac{2}{5}x^2 - \frac{1}{2}x + 5\right) - \left(\frac{1}{3}x^2 - \frac{3}{7}x - 6\right)$
$\frac{1}{15}x^2 - \frac{1}{14}x + 11$

22. $\left(\frac{3}{8}x^2 - \frac{2}{3}x - 7\right) - \left(\frac{2}{3}x^2 - \frac{1}{2}x + 2\right)$
$-\frac{7}{24}x^2 - \frac{1}{6}x - 9$

23. $(4x^3 + 3x) - (x^3 + x^2 - 5x)$
$3x^3 - x^2 + 8x$

24. $(6x^3 + x^2 - x + 1) - (2x^3 - x^2 + 5)$
$4x^3 + 2x^2 - x - 4$

25. $(0.5x^4 - 0.7x^2 + 8.3) - (5.2x^4 + 1.6x + 7.9)$
$-4.7x^4 - 0.7x^2 - 1.6x + 0.4$

26. $(1.3x^4 - 3.1x^3 + 6.3x) - (x^4 - 5.2x^2 + 6.5x)$
$0.3x^4 - 3.1x^3 + 5.2x^2 - 0.2x$

Perform the indicated operations.

27. $(8x + 2) + (x - 7) - (3x + 1)$
$6x - 6$

28. $(x - 5) - (3x + 8) - (5x - 2)$
$-7x - 11$

29. $(5x^2y - 6xy^2 + 2) + (-8x^2y + 12xy^2 - 6)$
$-3x^2y + 6xy^2 - 4$

30. $(7x^2y^2 - 6xy + 5) + (-15x^2y^2 - 6xy + 18)$
$-8x^2y^2 - 12xy + 23$

31. $(3x^4 - 4x^2 - 18) - (2x^4 + 3x^3 + 6)$
$x^4 - 3x^3 - 4x^2 - 24$

32. $(2b^3 + 3b - 5) - (-3b^3 + 5b^2 + 7b)$
$5b^3 - 5b^2 - 4b - 5$

Prisons *The number of prisoners held in federal and state prisons, measured in thousands, can be described by the polynomial* $-2.06x^2 + 77.82x + 743$. *The variable* x *represents the number of years since 1990. (Source: www.ojp.usdoj.gov.)*

33. Estimate the number of prisoners in 1990. 743,000

34. Estimate the number of prisoners in 2000. 1,315,200

35. According to the polynomial, by how much did the prison population increase from 2002 to 2007? 90,400

36. According to the polynomial, by how much will the prison population decrease from 2008 to 2012? 18,320

Applications

▲ **37.** **Geometry** The lengths and the widths of the following three rectangles are labeled. Create a polynomial that describes the sum of the *area* of these three rectangles. $3x^2 + 12x$

▲ **38.** **Geometry** The dimensions of the sides of the following figure are labeled. Create a polynomial that describes the *perimeter* of this figure. $7x + 67$

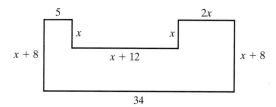

Cumulative Review

39. [2.5.1] Solve for x. $3y - 8x = 2$ $x = \dfrac{3y - 2}{8}$

40. [2.5.1] Solve for B. $3A = 2BCD$ $B = \dfrac{3A}{2CD}$

41. [2.5.1] Solve for b. $A = \dfrac{1}{2}h(b + c)$ $\dfrac{2A}{h} - c = b$

42. [2.5.1] Solve for d. $B = \dfrac{5xy}{d}$ $d = \dfrac{5xy}{B}$

Quick Quiz 4.3 Combine.

1. $(3x^2 - 5x + 8) + (-7x^2 - 6x - 3)$
$-4x^2 - 11x + 5$

2. $(2x^2 - 3x - 7) - (-4x^2 + 6x + 9)$
$6x^2 - 9x - 16$

3. $(5x - 3) - (2x - 4) + (-6x + 7)$
$-3x + 8$

4. **Concept Check** Explain how you would determine the degree of the following polynomial and how you would decide if it is a monomial, a binomial, or a trinomial.

$$2xy^2 - 5x^3y^4$$

Answers may vary

Classroom Quiz 4.3 You may use these problems to quiz your students' mastery of Section 4.3.

Combine.

1. $(-5x^2 + 6x - 8) + (8x^2 - 7x - 9)$
Ans: $3x^2 - x - 17$

2. $(4x^2 + 5x - 3) - (-2x^2 - 8x + 10)$
Ans: $6x^2 + 13x - 13$

3. $(7x - 2) + (-4x - 5) - (2x - 3)$
Ans: $x - 4$

1. $-24x^3y^5$

2. $-\dfrac{7y^3}{5x^7}$

3. $\dfrac{4x^5}{y^9}$

4. $81x^{20}y^4$

5. $\dfrac{x^6}{16y^8}$

6. $\dfrac{x^5}{3y^5}$

7. 5.874×10^4

8. 9.362×10^{-5}

9. $2.3x^2 - 0.4x - 5.4$

10. $4x^2 + 2x - 8$

11. $\dfrac{1}{6}x^3 + \dfrac{3}{8}x^2 + 3x$

12. $\dfrac{5}{16}x^2 - \dfrac{3}{10}x - \dfrac{3}{8}$

How are you doing with your homework assignments in Sections 4.1 to 4.3? Do you feel you have mastered the material so far? Do you understand the concepts you have covered? Before you go further in the textbook, take some time to do each of the following problems.

Simplify all answers. Reduce all fractions.

4.1

Multiply.

1. $(8x^2y^3)(-3xy^2)$

Divide. Assume that all variables are nonzero. Leave your answer with only positive exponents.

2. $-\dfrac{35xy^6}{25x^8y^3}$

3. $\dfrac{60x^7y^0}{15x^2y^9}$

Simplify.

4. $(-3x^5y)^4$

4.2

Simplify. Express your answer with positive exponents. Assume that all variables are nonzero.

5. $(4x^{-3}y^4)^{-2}$

6. $\dfrac{4x^4y^{-3}}{12x^{-1}y^2}$

7. Write in scientific notation. 58,740

8. Write in scientific notation. 0.00009362

4.3

Perform the indicated operations.

9. $(2x^2 + 0.5x - 2) + (0.3x^2 - 0.9x - 3.4)$

10. $(3x^2 + 7x - 10) - (-x^2 + 5x - 2)$

11. $\left(\dfrac{1}{2}x^3 + \dfrac{1}{4}x^2 - 2x\right) - \left(\dfrac{1}{3}x^3 - \dfrac{1}{8}x^2 - 5x\right)$

12. $\left(\dfrac{1}{16}x^2 + \dfrac{1}{8}\right) + \left(\dfrac{1}{4}x^2 - \dfrac{3}{10}x - \dfrac{1}{2}\right)$

Now turn to page SA-9 for the answer to each of these problems. Each answer also includes a reference to the objective in which the problem is first taught. If you missed any of these problems, you should stop and review the Examples and Practice Problems in the referenced objective. A little review now will help you master the material in the upcoming sections of the text.

① Multiplying a Monomial by a Polynomial

We use the distributive property to multiply a monomial by a polynomial. Remember, the distributive property states that for real numbers a, b, and c,

$$a(b + c) = ab + ac.$$

EXAMPLE 1 Multiply. $3x^2(5x - 2)$

Solution

$$
\begin{aligned}
3x^2(5x - 2) &= 3x^2(5x) + 3x^2(-2) && \text{Use the distributive property.} \\
&= (3 \cdot 5)(x^2 \cdot x) + (3)(-2)x^2 \\
&= 15x^3 - 6x^2
\end{aligned}
$$

Practice Problem 1 Multiply. $4x^3(-2x^2 + 3x)$ $-8x^5 + 12x^4$

Try to do as much of the multiplication as you can mentally.

EXAMPLE 2 Multiply.

(a) $2x(x^2 + 3x - 1)$ **(b)** $-2xy^2(x^2 - 2xy - 3y^2)$

Solution

(a) $2x(x^2 + 3x - 1) = 2x^3 + 6x^2 - 2x$

(b) $-2xy^2(x^2 - 2xy - 3y^2) = -2x^3y^2 + 4x^2y^3 + 6xy^4$

Notice in part (b) that you are multiplying each term by the negative expression $-2xy^2$. This will change the sign of each term in the product.

Practice Problem 2 Multiply.

(a) $-3x(x^2 + 2x - 4)$ **(b)** $6xy(x^3 + 2x^2y - y^2)$

(a) $-3x^3 - 6x^2 + 12x$ **(b)** $6x^4y + 12x^3y^2 - 6xy^3$

When we multiply by a monomial, the monomial may be on the right side.

EXAMPLE 3 Multiply. $(x^2 - 2x + 6)(-2xy)$

Solution $(x^2 - 2x + 6)(-2xy) = -2x^3y + 4x^2y - 12xy$

Practice Problem 3 Multiply.

$(-6x^3 + 4x^2 - 2x)(-3xy)$ $18x^4y - 12x^3y + 6x^2y$

② Multiplying Two Binomials

We can build on our knowledge of the distributive property and our experience with multiplying monomials to learn how to multiply two binomials. Let's suppose that we want to multiply $(x + 2)(3x + 1)$. We can use the distributive property. Since a can represent any quantity, let $a = x + 2$. Then let $b = 3x$ and $c = 1$. We now have the following.

$$
\begin{aligned}
a(b + c) &= ab + ac \\
(x + 2)(3x + 1) &= (x + 2)(3x) + (x + 2)(1) \\
&= 3x^2 + 6x + x + 2 \\
&= 3x^2 + 7x + 2
\end{aligned}
$$

Student Learning Objectives

After studying this section, you will be able to:

 Multiply a monomial by a polynomial.

 Multiply two binomials.

Teaching Example 1 Multiply.
$3a^4(-5a^2 + 2a)$

Ans: $-15a^6 + 6a^5$

Teaching Example 2 Multiply.

(a) $-7x(5x^2 - x - 3)$

(b) $3a^3b(a + 5ab - 2a^2b^4)$

Ans: (a) $-35x^3 + 7x^2 + 21x$

    (b) $3a^4b + 15a^4b^2 - 6a^5b^5$

Teaching Example 3 Multiply.
$(-3y^4 + 7y^3 + 2y - 8)(x^2y)$

Ans: $-3x^2y^5 + 7x^2y^4 + 2x^2y^2 - 8x^2y$

Let's take another look at the original problem, $(x + 2)(3x + 1)$. This time we will assign a letter to each term in the binomials. That is, let $a = x$, $b = 2$, $c = 3x$, and $d = 1$. Using substitution, we have the following.

$$
\begin{aligned}
(x + 2)(3x + 1) &= (a + b)(c + d) \\
&= (a + b)c + (a + b)d \\
&= ac + bc + ad + bd \\
&= (x)(3x) + (2)(3x) + (x)(1) + (2)(1) \quad \text{By substitution} \\
&= 3x^2 + 6x + x + 2 \\
&= 3x^2 + 7x + 2
\end{aligned}
$$

Teaching Tip Some students prefer to multiply the four products in a different order, particularly if they have a learning disability. This is OK and students should be given this freedom.

How does this compare with the preceding result?

The distributive property shows us *how* the problem can be done and *why* it can be done. In actual practice there is a memory device to help students remember the steps involved. It is often referred to as FOIL. The letters FOIL stand for the following.

> F multiply the *F*irst terms
>
> O multiply the *O*uter terms
>
> I multiply the *I*nner terms
>
> L multiply the *L*ast terms

The FOIL letters are simply a way to remember the four terms in the final product and how they are obtained. Let's return to our original problem.

$(x + 2)(3x + 1)$ F Multiply the *first* terms to obtain $3x^2$.

$(x + 2)(3x + 1)$ O Multiply the *outer* terms to obtain x.

$(x + 2)(3x + 1)$ I Multiply the *inner* terms to obtain $6x$.

$(x + 2)(3x + 1)$ L Multiply the *last* terms to obtain 2.

The result so far is $3x^2 + x + 6x + 2$. These four terms are the same four terms that we obtained when we multiplied using the distributive property. We can combine the like terms to obtain the final answer: $3x^2 + 7x + 2$. Now let's study the FOIL method in a few examples.

Teaching Example 4 Multiply. $(x + 2)(7x - 1)$

Ans: $7x^2 + 13x - 2$

EXAMPLE 4 Multiply. $(2x - 1)(3x + 2)$

Solution

$$
\begin{array}{c}
\text{First} \qquad\qquad \text{Last} \\
(2x - 1)(3x + 2) \\
\text{Inner} \\
\text{Outer}
\end{array}
\qquad
\begin{aligned}
&\text{First} + \text{Outer} + \text{Inner} + \text{Last} \\
&\quad F \qquad\quad O \qquad\quad I \qquad\quad L \\
&= 6x^2 + \quad 4x \quad - \quad 3x \quad - \quad 2 \\
&= 6x^2 + x - 2 \quad \text{Combine like terms.}
\end{aligned}
$$

NOTE TO STUDENT: *Fully worked-out solutions to all of the Practice Problems can be found at the back of the text starting at page SP-1*

Practice Problem 4 Multiply. $(5x - 1)(x - 2)$ $5x^2 - 11x + 2$

EXAMPLE 5 Multiply. $(3a - 2b)(4a - b)$

Solution

$$= 12a^2 - 3ab - 8ab + 2b^2$$
$$= 12a^2 - 11ab + 2b^2$$

Practice Problem 5 Multiply. $(8a - 5b)(3a - b)$ $24a^2 - 23ab + 5b^2$

After you have done several problems, you may be able to combine the outer and inner products mentally.

In some problems the inner and outer products cannot be combined.

EXAMPLE 6 Multiply. $(3x + 2y)(5x - 3z)$

Solution

$$(3x + 2y)(5x - 3z) \qquad = 15x^2 - 9xz + 10xy - 6yz$$

Since there are no like terms, we cannot combine any terms.

Practice Problem 6 Multiply. $(3a + 2b)(2a - 3c)$ $6a^2 - 9ac + 4ab - 6bc$

EXAMPLE 7 Multiply. $(7x - 2y)^2$

Solution

$(7x - 2y)(7x - 2y)$ When we square a binomial, it is the same as multiplying the binomial by itself.

$$= 49x^2 - 14xy - 14xy + 4y^2$$

$$= 49x^2 - 28xy + 4y^2$$

Practice Problem 7 Multiply. $(3x - 2y)^2$ $9x^2 - 12xy + 4y^2$

We can multiply binomials containing exponents that are greater than 1. That is, we can multiply binomials containing x^2 or y^3, and so on.

Teaching Example 8 Multiply.
$(m^4 - 3n^3)(7m^4 - n^3)$

Ans: $7m^8 - 22m^4n^3 + 3n^6$

EXAMPLE 8 Multiply. $(3x^2 + 4y^3)(2x^2 + 5y^3)$

Solution $(3x^2 + 4y^3)(2x^2 + 5y^3) = 6x^4 + 15x^2y^3 + 8x^2y^3 + 20y^6$
$$= 6x^4 + 23x^2y^3 + 20y^6$$

Practice Problem 8 Multiply. $(2x^2 + 3y^2)(5x^2 + 6y^2)$ $10x^4 + 27x^2y^2 + 18y^4$

Teaching Example 9 What is the area in square feet of a square room that has $(3x - 5)$ feet as the measure of all of its sides?

Ans: $(9x^2 - 30x + 25)\,\text{ft}^2$

▲ **EXAMPLE 9** The width of a living room is $(x + 4)$ feet. The length of the room is $(3x + 5)$ feet. What is the area of the room in square feet?

$x + 4$

$3x + 5$

Solution $A = (\text{length})(\text{width}) = (3x + 5)(x + 4)$
$$= 3x^2 + 12x + 5x + 20$$
$$= 3x^2 + 17x + 20$$

There are $(3x^2 + 17x + 20)$ square feet in the room.

NOTE TO STUDENT: Fully worked-out solutions to all of the Practice Problems can be found at the back of the text starting at page SP-1

▲ **Practice Problem 9** What is the area in square feet of a room that is $(2x - 1)$ feet wide and $(7x + 3)$ feet long? $(14x^2 - x - 3)\,\text{ft}^2$

Developing Your Study Skills

How to Review for an Exam

Reviewing adequately for an exam enables you to bring together the concepts you have learned over several sections. For your review, you will need to:

1. Reread your textbook. Make a list of any terms, rules, or formulas you need to know for the exam. Be sure you understand them all.
2. Reread your notes. Go over returned homework and quizzes. Redo the problems you missed.
3. Practice some of each type of problem covered in the chapter(s) you are to be tested on.
4. Use the end-of-chapter materials provided in your textbook. Read carefully through the chapter organizer. Do the review problems. Take the chapter test. When you are finished, check your answers. Redo any problems you missed.
5. Get help if any concepts give you difficulty.

4.4 EXERCISES

 MyMathLab

 Math XL
PRACTICE

WATCH

DOWNLOAD

READ

REVIEW

Multiply.

1. $-2x(6x^3 - x)$
$-12x^4 + 2x^2$

2. $5x(-3x^4 + 4x)$
$-15x^5 + 20x^2$

3. $3x^2(7x - 3)$
$21x^3 - 9x^2$

4. $2x^2(8x - 5)$
$16x^3 - 10x^2$

5. $2x^3(-2x^3 + 5x - 1)$
$-4x^6 + 10x^4 - 2x^3$

6. $5x^2(-3x^3 + 6x - 2)$
$-15x^5 + 30x^3 - 10x^2$

7. $\frac{1}{2}(2x + 3x^2 + 5x^3)$
$x + \frac{3}{2}x^2 + \frac{5}{2}x^3$

8. $\frac{2}{3}(4x + 6x^2 - 2x^3)$
$\frac{8}{3}x + 4x^2 - \frac{4}{3}x^3$

9. $(2x^3 - 4x^2 + 5x)(-x^2y)$
$-2x^5y + 4x^4y - 5x^3y$

10. $(3b^3 + 3b^2 - ab)(-5a^2)$
$-15a^2b^3 - 15a^2b^2 + 5a^3b$

11. $(3x^3 + x^2 - 8x)(3xy)$
$9x^4y + 3x^3y - 24x^2y$

12. $(2x^3 + x^2 - 6x)(2xy)$
$4x^4y + 2x^3y - 12x^2y$

13. $(x^3 - 3x^2 + 5x - 2)(3x)$
$3x^4 - 9x^3 + 15x^2 - 6x$

14. $(-2x^3 + 4x^2 - 7x + 3)(2x)$
$-4x^4 + 8x^3 - 14x^2 + 6x$

15. $(x^2y^2 - 6xy + 8)(-2xy)$
$-2x^3y^3 + 12x^2y^2 - 16xy$

16. $(x^2y^2 + 5xy - 9)(-3xy)$
$-3x^3y^3 - 15x^2y^2 + 27xy$

17. $(-7x^3 + 3x^2 + 2x - 1)(4x^2y)$
$-28x^5y + 12x^4y + 8x^3y - 4x^2y$

18. $(-5x^3 - 6x^2 + x - 1)(5xy^2)$
$-25x^4y^2 - 30x^3y^2 + 5x^2y^2 - 5xy^2$

19. $(3d^4 - 4d^2 + 6)(-2c^2d)$
$-6c^2d^5 + 8c^2d^3 - 12c^2d$

20. $(-4x^3 + 6x^2 - 5x)(-7xy^2)$
$28x^4y^2 - 42x^3y^2 + 35x^2y^2$

21. $6x^3(2x^4 - x^2 + 3x + 9)$
$12x^7 - 6x^5 + 18x^4 + 54x^3$

22. $8x^3(-2x^4 + 3x^2 - 5x - 14)$
$-16x^7 + 24x^5 - 40x^4 - 112x^3$

23. $-2x^3(8x^3 - 5x^2 + 6x)$
$-16x^6 + 10x^5 - 12x^4$

24. $-6x^4(-3x^3 + 8x^2 - 4x)$
$18x^7 - 48x^6 + 24x^5$

Multiply. Try to do most of the exercises mentally without writing down intermediate steps.

25. $(x + 5)(x + 7)$
$x^2 + 12x + 35$

26. $(x + 6)(x + 3)$
$x^2 + 9x + 18$

27. $(x + 6)(x + 2)$
$x^2 + 8x + 12$

28. $(x + 9)(x + 3)$
$x^2 + 12x + 27$

29. $(x - 8)(x + 2)$
$x^2 - 6x - 16$

30. $(x + 3)(x - 6)$
$x^2 - 3x - 18$

31. $(x - 5)(x - 4)$
$x^2 - 9x + 20$

32. $(x - 6)(x - 5)$
$x^2 - 11x + 30$

33. $(5x - 2)(-4x - 3)$
$-20x^2 - 7x + 6$

34. $(7x + 1)(-2x - 3)$
$-14x^2 - 23x - 3$

35. $(2x - 5)(x + 3y)$
$2x^2 + 6xy - 5x - 15y$

36. $(x - 6)(3x + 2y)$
$3x^2 + 2xy - 18x - 12y$

37. $(5x + 2)(3x - y)$
$15x^2 - 5xy + 6x - 2y$

38. $(3x + 4)(5x - y)$
$15x^2 - 3xy - 4y + 20x$

39. $(4y + 1)(5y - 3)$
$20y^2 - 7y - 3$

40. $(5y + 1)(6y - 5)$
$30y^2 - 19y - 5$

To Think About

41. What is wrong with this multiplication?
$(x - 2)(-3) = 3x - 6$
The signs are incorrect.
The result should be $-3x + 6$.

42. What is wrong with this answer?
$-(3x - 7) = -3x - 7$
The last term is incorrect.
The result should be $-3x + 7$.

43. What is the missing term?
$(5x + 2)(5x + 2) = 25x^2 + \underline{\quad 20x \quad} + 4$

44. Multiply the binomials and write a brief description of what is special about the result. $(5x - 1)(5x + 1)$
$25x^2 - 1$; it is the difference of two perfect squares

Mixed Practice *Multiply.*

45. $(4x - 3y)(5x - 2y)$
$20x^2 - 23xy + 6y^2$

46. $(3b - 5c)(2b - 7c)$
$6b^2 - 31bc + 35c^2$

47. $(7x - 2)^2$
$49x^2 - 28x + 4$

48. $(3x - 7)^2$
$9x^2 - 42x + 49$

49. $(4a + 2b)^2$
$16a^2 + 16ab + 4b^2$

50. $(5a + 3b)^2$
$25a^2 + 30ab + 9b^2$

51. $(0.2x + 3)(4x - 0.3)$
$0.8x^2 + 11.94x - 0.9$

52. $(0.5x - 2)(6x - 0.2)$
$3x^2 - 12.1x + 0.4$

53. $\left(\frac{1}{2}x + \frac{1}{3}\right)\left(\frac{1}{2}x - \frac{1}{4}\right)$
$\frac{1}{4}x^2 + \frac{1}{24}x - \frac{1}{12}$

54. $\left(\frac{1}{3}x + \frac{1}{5}\right)\left(\frac{1}{3}x - \frac{1}{2}\right)$
$\frac{1}{9}x^2 - \frac{1}{10}x - \frac{1}{10}$

55. $(2a - 3b)(x - 5b)$
$2ax - 10ab - 3bx + 15b^2$

56. $(4b - 5c)(a - 3b)$
$4ab - 12b^2 - 5ac + 15bc$

Find the area of the rectangle.

▲ **57.**

5x + 2

2x − 3

$10x^2 - 11x - 6$

▲ **58.**

7x + 3

4x − 6

$28x^2 - 30x - 18$

Cumulative Review

59. **[2.3.3]** Solve for x. $3(x - 6) = -2(x + 4) + 6x$
$x = -10$

60. **[2.3.3]** Solve for w. $3(w - 7) - (4 - w) = 11w$
$w = -\frac{25}{7}$ or $-3\frac{4}{7}$

61. **[3.4.4]** *Paper Currency* Heather returned from the bank with $375. She had one more $20 bill than she had $10 bills. The number of $5 bills she had was one less than triple the number of $10 bills. How many of each denomination did she have?   9 twenties; 8 tens; 23 fives

Social Security *Every year, the U.S. Department of the Treasury disburses billions of dollars in social security checks. The polynomial $17.7x + 240$ can be used to estimate the total value of social security checks sent in a year in billions of dollars, where x is the number of years since 1990. (Source: Statistical Abstract of the United States: 2007.)*

62. **[4.3.4]** Estimate the total value of social security checks sent in 1990. $240 billion

63. **[4.3.4]** Estimate the total value of social security checks sent in 2005. $505.5 billion

64. **[4.3.4]** Predict the total value of social security checks that will be sent in 2010.
$594 billion

65. **[4.3.4]** Predict the total value of social security checks that will be sent in 2014.
$664.8 billion

Quick Quiz 4.4 *Multiply.*

1. $(2x^2y^2 - 3xy + 4)(4xy^2)$ $8x^3y^4 - 12x^2y^3 + 16xy^2$

2. $(2x + 3)(3x - 5)$ $6x^2 - x - 15$

3. $(6a - 4b)(2a - 3b)$ $12a^2 - 26ab + 12b^2$

4. **Concept Check** Explain how you would multiply $(7x - 3)^2$.   Answers may vary

Classroom Quiz 4.4 *You may use these problems to quiz your students' mastery of Section 4.4.*

Multiply.

1. $(3xy^2 - 5xy - 2)(3x^2y)$ **Ans:** $9x^3y^3 - 15x^3y^2 - 6x^2y$

2. $(4x - 3)(5x - 2)$ **Ans:** $20x^2 - 23x + 6$

3. $(7a - 6b)(2a + 3b)$ **Ans:** $14a^2 + 9ab - 18b^2$

Multiplying Binomials of the Type $(a + b)(a - b)$

The case when you multiply $(x + y)(x - y)$ is interesting and deserves special consideration. Using the FOIL method, we find

$$(x + y)(x - y) = x^2 - xy + xy - y^2 = x^2 - y^2.$$

Notice that the sum of the inner product and the outer product is zero. We see that

$$(x + y)(x - y) = x^2 - y^2.$$

This works in all cases when the binomials are the sum and difference of the same two terms. That is, in one factor the terms are added, while in the other factor the same two terms are subtracted.

$$(5a + 2b)(5a - 2b) = 25a^2 - 10ab + 10ab - 4b^2$$
$$= 25a^2 - 4b^2$$

The product is the difference of the squares of the terms. That is, $(5a)^2 - (2b)^2$ or $25a^2 - 4b^2$.

Many students find it helpful to memorize this equation.

MULTIPLYING BINOMIALS: A SUM AND A DIFFERENCE

$$(a + b)(a - b) = a^2 - b^2$$

You may use this relationship to find the product quickly in cases where it applies. The terms must be the same and there must be a sum and a difference.

EXAMPLE 1 Multiply. $(7x + 2)(7x - 2)$

Solution

$$(7x + 2)(7x - 2) = (7x)^2 - (2)^2 = 49x^2 - 4$$

Practice Problem 1 Multiply. $(6x + 7)(6x - 7)$ $36x^2 - 49$

EXAMPLE 2 Multiply. $(5x - 8y)(5x + 8y)$

Solution

$$(5x - 8y)(5x + 8y) = (5x)^2 - (8y)^2 = 25x^2 - 64y^2$$

Practice Problem 2 Multiply. $(3x - 5y)(3x + 5y)$ $9x^2 - 25y^2$

Student Learning Objectives

After studying this section, you will be able to:

1 Multiply binomials of the type $(a + b)(a - b)$.

2 Multiply binomials of the type $(a + b)^2$ and $(a - b)^2$.

3 Multiply polynomials with more than two terms.

Teaching Tip The formula $(a + b)(a - b) = a^2 - b^2$ is useful now for multiplying and later in the course for factoring. Therefore, it is worthwhile to memorize it. Emphasize to your students the future value of the formula.

Teaching Example 1 Multiply. $(5m + 4)(5m - 4)$

Ans: $25m^2 - 16$

Teaching Example 2 Multiply. $(4r - 9s)(4r + 9s)$

Ans: $16r^2 - 81s^2$

2 **Multiplying Binomials of the Type $(a + b)^2$ and $(a - b)^2$**

A second case that is worth special consideration is a binomial that is squared. Consider the following problem.

$$(3x + 2)^2 = (3x + 2)(3x + 2)$$
$$= 9x^2 + 6x + 6x + 4$$
$$= 9x^2 + 12x + 4$$

If you complete enough problems of this type, you will notice a pattern. The answer always contains the square of the first term added to double the product of the first and last terms added to the square of the last term.

3x is the first term	2 is the last term	Square the first term: $(3x)^2$	Double the product of the first and last terms: $2(3x)(2)$	Square the last term: $(2)^2$
↓	↓	↓	↓	↓

$$(3x \quad + \quad 2)^2 \quad = \quad 9x^2 \quad + \quad 12x \quad + \quad 4$$

We can show the same steps using variables instead of words.

$$(a + b)^2 = a^2 + 2ab + b^2$$

There is a similar formula for the square of a difference:

$$(a - b)^2 = a^2 - 2ab + b^2$$

We can use this formula to simplify $(2x - 3)^2$.

$$(2x - 3)^2 = (2x)^2 - 2(2x)(3) + (3)^2$$
$$= 4x^2 - 12x + 9$$

You may wish to multiply this product using FOIL to verify.

These two types of products, the square of a sum and the square of a difference, can be summarized as follows.

A BINOMIAL SQUARED

$$(a + b)^2 = a^2 + 2ab + b^2$$
$$(a - b)^2 = a^2 - 2ab + b^2$$

Teaching Example 3 Multiply.

(a) $(4a + 7)^2$

(b) $(10m - 3n)^2$

Ans: (a) $16a^2 + 56a + 49$

(b) $100m^2 - 60mn + 9n^2$

EXAMPLE 3 Multiply.

(a) $(5y - 2)^2$ **(b)** $(8x + 9y)^2$

Solution

(a) $(5y - 2)^2 = (5y)^2 - (2)(5y)(2) + (2)^2$
$$= 25y^2 - 20y + 4$$

(b) $(8x + 9y)^2 = (8x)^2 + (2)(8x)(9y) + (9y)^2$
$$= 64x^2 + 144xy + 81y^2$$

Practice Problem 3 Multiply.

(a) $(4a - 9b)^2$ $16a^2 - 72ab + 81b^2$ **(b)** $(5x + 4)^2$ $25x^2 + 40x + 16$

NOTE TO STUDENT: Fully worked-out solutions to all of the Practice Problems can be found at the back of the text starting at page SP-1

Teaching Tip This is beyond a doubt the most common student error found in all of beginning algebra! Be sure to warn your students often to avoid this error!

CAUTION: $(a + b)^2 \neq a^2 + b^2$! The two sides are not equal! Squaring the sum $(a + b)$ does not give $a^2 + b^2$! Beginning algebra students often make this error. Make sure you remember that when you square a binomial, there is always a *middle term*.

$$(a + b)^2 = a^2 + 2ab + b^2$$

Sometimes a numerical example helps you to see this.

$$(3 + 4)^2 \neq 3^2 + 4^2$$
$$7^2 \neq 9 + 16$$
$$49 \neq 25$$

Notice that what is missing on the right is $2ab = 2 \cdot 3 \cdot 4 = 24$.

 Multiplying Polynomials with More Than Two Terms

We used the distributive property to multiply two binomials $(a + b)(c + d)$, and we obtained $ac + ad + bc + bd$. We could also use the distributive property to multiply the polynomials $(a + b)$ and $(c + d + e)$, and we would obtain $ac + ad + ae + bc + bd + be$. Let us see if we can find a direct way to multiply products such as $(3x - 2)(x^2 - 2x + 3)$. It can be done quickly using an approach similar to that used in arithmetic for multiplying whole numbers. Consider the following arithmetic problem.

$$
\begin{array}{r}
128 \\
\times\ \ 43 \\
\hline
384 \\
512\ \ \ \\
\hline
5504 \\
\end{array}
$$

$384 \leftarrow$ The product of 128 and 3

$512 \leftarrow$ The product of 128 and 4 moved one space to the left

$5504 \leftarrow$ The sum of the two partial products

Let us use a similar format to multiply the two polynomials. For example, multiply $(x^2 - 2x + 3)$ and $(3x - 2)$.

$$
\begin{array}{r}
x^2 - 2x + 3 \\
3x - 2 \\
\hline
-2x^2 + 4x - 6 \\
3x^3 - 6x^2 + 9x\ \ \ \ \ \ \\
\hline
3x^3 - 8x^2 + 13x - 6 \\
\end{array}
$$

This is often called **vertical multiplication.**

$-2x^2 + 4x - 6 \leftarrow$ The product $(x^2 - 2x + 3)(-2)$

$3x^3 - 6x^2 + 9x \leftarrow$ The product $(x^2 - 2x + 3)(3x)$ moved one space to the left so that like terms are underneath each other

$3x^3 - 8x^2 + 13x - 6 \leftarrow$ The sum of the two partial products

EXAMPLE 4 Multiply vertically. $(3x^3 + 2x^2 + x)(x^2 - 2x - 4)$

Solution

$$
\begin{array}{r}
3x^3 + 2x^2 + x \\
x^2 - 2x - 4 \\
\hline
-12x^3 - 8x^2 - 4x \\
-6x^4 - 4x^3 - 2x^2\ \ \ \ \ \ \ \\
3x^5 + 2x^4 + x^3\ \ \ \ \ \ \ \ \ \ \ \ \ \ \\
\hline
3x^5 - 4x^4 - 15x^3 - 10x^2 - 4x \\
\end{array}
$$

We place one polynomial over the other.

$-12x^3 - 8x^2 - 4x \leftarrow$ The product $(3x^3 + 2x^2 + x)(-4)$

$-6x^4 - 4x^3 - 2x^2 \leftarrow$ The product $(3x^3 + 2x^2 + x)(-2x)$

$3x^5 + 2x^4 + x^3 \leftarrow$ The product $(3x^3 + 2x^2 + x)(x^2)$

$3x^5 - 4x^4 - 15x^3 - 10x^2 - 4x \leftarrow$ The sum of the three partial products

Note that the answers for each partial product are placed so that like terms are underneath each other.

Practice Problem 4 Multiply vertically. $(4x^3 - 2x^2 + x)(x^2 + 3x - 2)$

$$4x^5 + 10x^4 - 13x^3 + 7x^2 - 2x$$

Teaching Example 4 Multiply vertically. $(x^3 - 5x^2 + 2x)(2x^2 + 3x - 1)$

Ans: $2x^5 - 7x^4 - 12x^3 + 11x^2 - 2x$

ALTERNATIVE METHOD: FOIL Horizontal Multiplication Some students prefer to do this type of multiplication using a horizontal format similar to the FOIL method. The following example illustrates this approach.

EXAMPLE 5 Multiply horizontally. $(x^2 + 3x + 5)(x^2 - 2x - 6)$

Solution We will use the distributive property repeatedly.

$$(x^2 + 3x + 5)(x^2 - 2x - 6) = x^2(x^2 - 2x - 6) + 3x(x^2 - 2x - 6) + 5(x^2 - 2x - 6)$$

$$= \boxed{x^4 - 2x^3 - 6x^2} + \boxed{3x^3 - 6x^2 - 18x} + \boxed{5x^2 - 10x - 30}$$

$$= x^4 + x^3 - 7x^2 - 28x - 30$$

Practice Problem 5 Multiply horizontally. $(2x^2 + 5x + 3)(x^2 - 3x - 4)$

$$2x^4 - x^3 - 20x^2 - 29x - 12$$

Some problems may need to be done in two or more separate steps.

EXAMPLE 6 Multiply. $(2x - 3)(x + 2)(x + 1)$

Solution We first need to multiply any two of the binomials. Let us select the first pair.

$$\underbrace{(2x - 3)(x + 2)}_{\text{Find this product first.}}(x + 1)$$

$$(2x - 3)(x + 2) = 2x^2 + 4x - 3x - 6$$
$$= 2x^2 + x - 6$$

Now we replace the first two factors with their resulting product.

$$\underbrace{(2x^2 + x - 6)}_{\text{First product}}(x + 1)$$

We then multiply again.

$$(2x^2 + x - 6)(x + 1) = (2x^2 + x - 6)x + (2x^2 + x - 6)1$$
$$= 2x^3 + x^2 - 6x + 2x^2 + x - 6$$
$$= 2x^3 + 3x^2 - 5x - 6$$

The vertical format of Example 4 is an alternative method for this type of problem.

$$
\begin{array}{r}
2x^2 + x - 6 \\
x + 1 \\
\hline
2x^2 + x - 6 \\
2x^3 + x^2 - 6x \\
\hline
2x^3 + 3x^2 - 5x - 6
\end{array}
$$

Be sure to use special care in writing the exponents correctly if problems have more than one variable.

Thus we have

$$(2x - 3)(x + 2)(x + 1) = 2x^3 + 3x^2 - 5x - 6.$$

Note that it does not matter which two binomials are multiplied first. For example, you could first multiply $(2x - 3)(x + 1)$ to obtain $2x^2 - x - 3$ and then multiply that product by $(x + 2)$ to obtain the same result.

Practice Problem 6 Multiply. $(3x - 2)(2x + 3)(3x + 2)$
(*Hint:* Rearrange the factors.) $18x^3 + 27x^2 - 8x - 12$

Sometimes we encounter a binomial raised to a third power. In such cases we would write out the binomial three times as a product and then multiply. So to evaluate $(3x + 4)^3$ we would first write $(3x + 4)(3x + 4)(3x + 4)$ and then follow the method of Example 6.

Verbal and Writing Skills

1. In the special case of $(a + b)(a - b)$, a binomial times a binomial is a _____binomial_____.

2. Identify which of the following could be the answer to a problem using the formula for $(a + b)(a - b)$. Why?
   - **(a)** $9x^2 - 16$
   - **(b)** $4x^2 + 25$
   - **(c)** $9x^2 + 12x + 4$
   - **(d)** $x^4 - 1$

   (a) and (d) (b) is eliminated since 25 is positive.
   (c) is eliminated because it has three terms and not two.

3. A student evaluated $(4x - 7)^2$ as $16x^2 + 49$. What is missing? State the correct answer.
   The middle term is missing. The answer should be
   $16x^2 - 56x + 49$.

4. The square of a binomial, $(a - b)^2$, always produces which of the following?
   - **(a)** binomial
   - **(b)** trinomial
   - **(c)** four-term polynomial   **(b)** trinomial

Use the formula $(a + b)(a - b) = a^2 - b^2$ to multiply.

5. $(y - 7)(y + 7)$
$y^2 - 49$

6. $(x + 5)(x - 5)$
$x^2 - 25$

7. $(x - 8)(x + 8)$
$x^2 - 64$

8. $(x + 10)(x - 10)$
$x^2 - 100$

9. $(6x - 5)(6x + 5)$
$36x^2 - 25$

10. $(4x - 9)(4x + 9)$
$16x^2 - 81$

11. $(2x - 7)(2x + 7)$
$4x^2 - 49$

12. $(3x - 10)(3x + 10)$
$9x^2 - 100$

13. $(5x - 3y)(5x + 3y)$
$25x^2 - 9y^2$

14. $(8a - 3b)(8a + 3b)$
$64a^2 - 9b^2$

15. $(0.6x + 3)(0.6x - 3)$
$0.36x^2 - 9$

16. $(5x - 0.2)(5x + 0.2)$
$25x^2 - 0.04$

Use the formula for a binomial squared to multiply.

17. $(2y + 5)^2$
$4y^2 + 20y + 25$

18. $(3x - 1)^2$
$9x^2 - 6x + 1$

19. $(5x - 4)^2$
$25x^2 - 40x + 16$

20. $(6x + 5)^2$
$36x^2 + 60x + 25$

21. $(7x + 3)^2$
$49x^2 + 42x + 9$

22. $(8x - 3)^2$
$64x^2 - 48x + 9$

23. $(3x - 7)^2$
$9x^2 - 42x + 49$

24. $(2x + 3y)^2$
$4x^2 + 12xy + 9y^2$

25. $\left(\dfrac{2}{3}x + \dfrac{1}{4}\right)^2$ $\dfrac{4}{9}x^2 + \dfrac{1}{3}x + \dfrac{1}{16}$

26. $\left(\dfrac{3}{4}x + \dfrac{1}{2}\right)^2$ $\dfrac{9}{16}x^2 + \dfrac{3}{4}x + \dfrac{1}{4}$

27. $(9xy + 4z)^2$
$81x^2y^2 + 72xyz + 16z^2$

28. $(7y - 3xz)^2$
$49y^2 - 42xyz + 9x^2z^2$

Mixed Practice   *Multiply. Use the special formula that applies.*

29. $(7x + 3y)(7x - 3y)$
$49x^2 - 9y^2$

30. $(12x - 5y)(12x + 5y)$
$144x^2 - 25y^2$

31. $(3c - 5d)^2$
$9c^2 - 30cd + 25d^2$

32. $(6c - d)^2$
$36c^2 - 12cd + d^2$

Multiply.

33. $(9a - 10b)(9a + 10b)$
$81a^2 - 100b^2$

34. $(11a + 6b)(11a - 6b)$
$121a^2 - 36b^2$

35. $(5x + 9y)^2$
$25x^2 + 90xy + 81y^2$

36. $(4x + 8y)^2$
$16x^2 + 64xy + 64y^2$

37. $(x^2 - x + 5)(x - 3)$
$x^3 - 4x^2 + 8x - 15$

38. $(x^2 + 2x - 6)(x + 7)$
$x^3 + 9x^2 + 8x - 42$

39. $(4x + 1)(x^3 - 2x^2 + x - 1)$
$4x^4 - 7x^3 + 2x^2 - 3x - 1$

40. $(3x - 1)(x^3 + x^2 - 4x - 2)$
$3x^4 + 2x^3 - 13x^2 - 2x + 2$

41. $(a^2 - 3a + 2)(a^2 + 4a - 3)$
$a^4 + a^3 - 13a^2 + 17a - 6$

42. $(x^2 + 4x - 5)(x^2 - 3x + 4)$
$x^4 + x^3 - 13x^2 + 31x - 20$

43. $(x + 3)(x - 1)(3x - 8)$
$3x^3 - 2x^2 - 25x + 24$

44. $(x - 7)(x + 4)(2x - 5)$
$2x^3 - 11x^2 - 41x + 140$

45. $(2x - 5)(x - 1)(x + 3)$
$2x^3 - x^2 - 16x + 15$

46. $(3x - 2)(x + 6)(x - 3)$
$3x^3 + 7x^2 - 60x + 36$

47. $(a - 5)(2a + 3)(a + 5)$
$2a^3 + 3a^2 - 50a - 75$

48. $(b + 7)(3b - 2)(b - 7)$
$3b^3 - 2b^2 - 147b + 98$

To Think About

▲ **49.** Find the volume of this object.
$24x^3 + 14x^2 - 11x - 6$

$(2x + 1)$

$(3x - 2)$ $(4x + 3)$

▲ **50.** *Geometry* The volume of a pyramid is $V = \frac{1}{3}Bh$, where B is the area of the base and h is the height. Find the volume of the following pyramid.

$\frac{1}{3}(x^3 - 5x^2 - 8x + 48)$

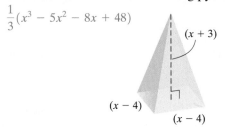

$(x + 3)$

$(x - 4)$

$(x - 4)$

Cumulative Review

51. **[3.4.3]** *Investments* An executive invested $18,000 in two accounts, one a cash-on-reserve account that yielded 7% simple interest, and the other a long-term certificate paying 11% simple interest. At the end of one year, the two accounts had earned her a total of $1540. How much had she invested in each account?
$11,000 at 7%; $7000 at 11%

▲ **52.** **[3.3.1]** *Room Dimensions* The perimeter of a rectangular room measures 34 meters. The width is 2 meters more than half the length. Find the dimensions of the room.
Width is 7 m; length is 10 m

4.6 DIVIDING POLYNOMIALS

1 Dividing a Polynomial by a Monomial

To divide a polynomial by a monomial, divide each term of the numerator by the denominator; then write the sum of the results. We are using the property of fractions that states that

$$\frac{a+b}{c} = \frac{a}{c} + \frac{b}{c}$$

DIVIDING A POLYNOMIAL BY A MONOMIAL

1. Divide each term of the polynomial by the monomial.

2. When dividing variables, use the property $\frac{x^a}{x^b} = x^{a-b}$.

EXAMPLE 1 Divide. $\dfrac{8y^6 - 8y^4 + 24y^2}{8y^2}$

Solution $\dfrac{8y^6 - 8y^4 + 24y^2}{8y^2} = \dfrac{8y^6}{8y^2} - \dfrac{8y^4}{8y^2} + \dfrac{24y^2}{8y^2} = y^4 - y^2 + 3$

Practice Problem 1 Divide. $\dfrac{15y^4 - 27y^3 - 21y^2}{3y^2}$ $5y^2 - 9y - 7$

Teaching Example 1 Divide.

$$\frac{12x^5 + 4x^3 - 20x^2}{4x^2}$$

Ans: $3x^3 + x - 5$

NOTE TO STUDENT: Fully worked-out solutions to all of the Practice Problems can be found at the back of the text starting at page SP-1

2 Dividing a Polynomial by a Binomial

Division of a polynomial by a binomial is similar to long division in arithmetic. Notice the similarity in the following division problems.

Division of a three-digit number by a two-digit number	Division of a polynomial by a binomial

$$
\begin{array}{r}
32 \\
21\overline{)672} \\
\underline{63} \\
42 \\
\underline{42} \\
0
\end{array}
$$

$$
\begin{array}{r}
3x + 2 \\
2x + 1\overline{)6x^2 + 7x + 2} \\
\underline{6x^2 + 3x} \\
4x + 2 \\
\underline{4x + 2} \\
0
\end{array}
$$

DIVIDING A POLYNOMIAL BY A BINOMIAL

1. Place the terms of the polynomial and binomial in descending order. Insert a 0 for any missing term.

2. Divide the first term of the polynomial by the first term of the binomial. The result is the first term of the answer.

3. Multiply the first term of the answer by the binomial and subtract the result from the first two terms of the polynomial. Bring down the next term to obtain a new polynomial.

4. Divide the new polynomial by the binomial using the process described in step 2.

5. Continue dividing, multiplying, and subtracting until the degree of the remainder is less than the degree of the binomial divisor.

6. Write the remainder as the numerator of a fraction that has the binomial divisor as its denominator.

EXAMPLE 2   Divide. $(x^3 + 5x^2 + 11x + 14) \div (x + 2)$

Solution

Step 1 The terms are arranged in descending order. No terms are missing.

Step 2 Divide the first term of the polynomial by the first term of the binomial. In this case, divide x^3 by x to get x^2.

$$x + 2 \overline{)x^3 + 5x^2 + 11x + 14} \quad \overset{x^2}{}$$

Step 3 Multiply x^2 by $x + 2$ and subtract the result from the first two terms of the polynomial, $x^3 + 5x^2$ in this case.

$$
\begin{array}{r}
x^2 \\
x + 2 \overline{)x^3 + 5x^2 + 11x + 14} \\
\underline{x^3 + 2x^2} \downarrow \\
3x^2 + 11x
\end{array}
$$

Bring down the next term.

Step 4 Continue to use the step 2 process. Divide $3x^2$ by x. Write the resulting $3x$ as the next term of the answer.

$$
\begin{array}{r}
x^2 + 3x \\
x + 2 \overline{)x^3 + 5x^2 + 11x + 14} \\
\underline{x^3 + 2x^2} \\
3x^2 + 11x
\end{array}
$$

Step 5 Continue multiplying, dividing, and subtracting until the degree of the remainder is less than the degree of the divisor. In this case, we stop when the remainder does not have an x.

$$
\begin{array}{r}
x^2 + 3x + 5 \\
x + 2 \overline{)x^3 + 5x^2 + 11x + 14} \\
\underline{x^3 + 2x^2} \\
3x^2 + 11x \\
\underline{3x^2 + 6x} \\
5x + 14 \\
\underline{5x + 10} \\
4
\end{array}
$$

← The remainder is 4.

Step 6 The answer is $x^2 + 3x + 5 + \dfrac{4}{x + 2}$.

To check the answer, we multiply $(x + 2)(x^2 + 3x + 5)$ and add the remainder, 4.

$$
\begin{aligned}
(x + 2)(x^2 + 3x + 5) + 4 &= x^3 + 5x^2 + 11x + 10 + 4 \\
&= x^3 + 5x^2 + 11x + 14
\end{aligned}
$$

This is the original polynomial. It checks.

Practice Problem 2   Divide.

$(x^3 + 10x^2 + 31x + 35) \div (x + 4)$ $x^2 + 6x + 7 + \dfrac{7}{x + 4}$

Take great care with the subtraction step when negative numbers are involved.

EXAMPLE 3 Divide. $(5x^3 - 24x^2 + 9) \div (5x + 1)$

Solution We must first insert $0x$ to represent the missing x-term. Then we divide $5x^3$ by $5x$.

$$
\begin{array}{r}
x^2 \phantom{{}+0x+9} \\
5x + 1 \overline{)\,5x^3 - 24x^2 + 0x + 9} \\
\underline{5x^3 + x^2} \\
-25x^2
\end{array}
$$

Note that we are subtracting:
$$-24x^2 - (+1x^2) = -24x^2 - 1x^2$$
$$= -25x^2$$

Next we divide $-25x^2$ by $5x$.

$$
\begin{array}{r}
x^2 - 5x \phantom{{}+9} \\
5x + 1 \overline{)\,5x^3 - 24x^2 + 0x + 9} \\
\underline{5x^3 + x^2} \\
-25x^2 + 0x \\
\underline{-25x^2 - 5x} \\
5x
\end{array}
$$

Note that we are subtracting:
$$0x - (-5x) = 0x + 5x = 5x$$

Finally, we divide $5x$ by $5x$.

$$
\begin{array}{r}
x^2 - 5x + 1 \\
5x + 1 \overline{)\,5x^3 - 24x^2 + 0x + 9} \\
\underline{5x^3 + x^2} \\
-25x^2 + 0x \\
\underline{-25x^2 - 5x} \\
5x + 9 \\
\underline{5x + 1} \\
8
\end{array}
$$

$\longleftarrow$ The remainder is 8.

The answer is $x^2 - 5x + 1 + \dfrac{8}{5x + 1}$.

To check, multiply $(5x + 1)(x^2 - 5x + 1)$ and add the remainder, 8.

$$(5x + 1)(x^2 - 5x + 1) + 8 = 5x^3 - 24x^2 + 1 + 8$$
$$= 5x^3 - 24x^2 + 9$$

This is the original polynomial. Our answer is correct.

Practice Problem 3 Divide.

$$(2x^3 - x^2 + 1) \div (x - 1) \quad 2x^2 + x + 1 + \frac{2}{x - 1}$$

Teaching Example 3 Divide.

$(2x^3 - 3x + 2) \div (x + 2)$

Ans: $2x^2 - 4x + 5 - \dfrac{8}{x + 2}$

NOTE TO STUDENT: Fully worked-out solutions to all of the Practice Problems can be found at the back of the text starting at page SP-1

Now we will perform the division by writing a minimum of steps. See if you can follow each step.

Teaching Example 4 Divide.
$(6x^3 + 5x^2 - 2x - 1) \div (3x + 1)$

Ans: $2x^2 + x - 1$

EXAMPLE 4 Divide and check. $(12x^3 - 11x^2 + 8x - 4) \div (3x - 2)$

Solution

$$
\begin{array}{r}
4x^2 - x + 2 \\
3x - 2 \overline{\smash{)}12x^3 - 11x^2 + 8x - 4} \\
\underline{12x^3 - 8x^2} \\
-3x^2 + 8x \\
\underline{-3x^2 + 2x} \\
6x - 4 \\
\underline{6x - 4} \\
0
\end{array}
$$

Check. $(3x - 2)(4x^2 - x + 2) = 12x^3 - 3x^2 + 6x - 8x^2 + 2x - 4$
$= 12x^3 - 11x^2 + 8x - 4$ Our answer is correct.

Practice Problem 4 Divide and check.

$(20x^3 - 11x^2 - 11x + 6) \div (4x - 3)$ $5x^2 + x - 2$

Developing Your Study Skills

Taking an Exam

Allow yourself plenty of time to get to your exam. You may even find it helpful to arrive a little early in order to collect your thoughts and ready yourself. This will help you feel more relaxed.

After you get your exam, you will find it helpful to do the following.

1. Take two or three moderately deep breaths. Inhale, then exhale slowly. You will feel your entire body begin to relax.

2. Write down on the back of the exam any formulas or ideas that you need to remember.

3. Look over the entire test quickly in order to pace yourself and use your time wisely. Notice how many points each question is worth. Spend more time on items of greater worth.

4. Read directions carefully and be sure to answer all questions clearly. Keep your work neat and easy to read.

5. Ask your instructor about anything that is not clear to you.

6. Answer the questions that are easiest for you first. Then come back to the more difficult ones.

7. Do not get bogged down on one question for too long because it may jeopardize your chances of finishing other problems. Leave a tough question and come back to it when you have time later.

8. Check your work. This will help you to catch minor errors.

9. Stay calm if others leave before you do. You are entitled to use the full amount of allotted time. You will do better on the exam if you take your time and work carefully.

Divide.

1. $\dfrac{25x^4 - 15x^2 + 20x}{5x}$

$5x^3 - 3x + 4$

2. $\dfrac{18b^5 - 12b^3 + 6b^2}{3b}$

$6b^4 - 4b^2 + 2b$

3. $\dfrac{8y^4 - 12y^3 - 4y^2}{4y^2}$

$2y^2 - 3y - 1$

4. $\dfrac{10y^4 - 35y^3 + 5y^2}{5y^2}$

$2y^2 - 7y + 1$

5. $\dfrac{81x^7 - 36x^5 - 63x^3}{9x^3}$

$9x^4 - 4x^2 - 7$

6. $\dfrac{49x^8 - 35x^6 - 56x^3}{7x^3}$

$7x^5 - 5x^3 - 8$

7. $(48x^7 - 54x^4 + 36x^3) \div 6x^3$

$8x^4 - 9x + 6$

8. $(72x^8 - 56x^5 - 40x^3) \div 8x^3$

$9x^5 - 7x^2 - 5$

Divide. Check your answers for exercises 9–16 by multiplication.

9. $\dfrac{6x^2 + 13x + 5}{2x + 1}$

$$
\begin{array}{r}
3x + 5 \\
2x+1\overline{)6x^2 + 13x + 5} \\
\underline{6x^2 + 3x} \\
10x + 5 \\
\underline{10x + 5} \\
0
\end{array}
$$

10. $\dfrac{12x^2 + 19x + 5}{3x + 1}$

$$
\begin{array}{r}
4x + 5 \\
3x+1\overline{)12x^2 + 19x + 5} \\
\underline{12x^2 + 4x} \\
15x + 5 \\
\underline{15x + 5} \\
0
\end{array}
$$

11. $\dfrac{x^2 - 8x - 17}{x - 5}$

$$
\begin{array}{r}
x - 3 \\
x-5\overline{)x^2 - 8x - 17} \\
\underline{x^2 - 5x} \\
-3x - 17 \\
\underline{-3x + 15} \\
-32
\end{array}
$$

12. $\dfrac{x^2 - 9x - 5}{x - 3}$

$$
\begin{array}{r}
x - 6 \\
x-3\overline{)x^2 - 9x - 5} \\
\underline{x^2 - 3x} \\
-6x - 5 \\
\underline{-6x + 18} \\
-23
\end{array}
$$

13. $\dfrac{3x^3 - x^2 + 4x - 2}{x + 1}$

$$
\begin{array}{r}
3x^2 - 4x + 8 \\
x+1\overline{)3x^3 - x^2 + 4x - 2} \\
\underline{3x^3 + 3x^2} \\
-4x^2 + 4x \\
\underline{-4x^2 - 4x} \\
8x - 2 \\
\underline{8x + 8} \\
-10
\end{array}
$$

14. $\dfrac{2x^3 - 3x^2 - 3x + 6}{x - 1}$

$$
\begin{array}{r}
2x^2 - x - 4 \\
x-1\overline{)2x^3 - 3x^2 - 3x + 6} \\
\underline{2x^3 - 2x^2} \\
-x^2 - 3x \\
\underline{-x^2 + x} \\
-4x + 6 \\
\underline{-4x + 4} \\
2
\end{array}
$$

15. $\dfrac{4x^3 + 4x^2 - 19x - 15}{2x + 5}$

$$
\begin{array}{r}
2x^2 - 3x - 2 \\
2x+5\overline{)4x^3 + 4x^2 - 19x - 15} \\
\underline{4x^3 + 10x^2} \\
-6x^2 - 19x \\
\underline{-6x^2 - 15x} \\
-4x - 15 \\
\underline{-4x - 10} \\
-5
\end{array}
$$

16. $\dfrac{6x^3 + 11x^2 - 8x + 5}{2x + 5}$

$$
\begin{array}{r}
3x^2 - 2x + 1 \\
2x+5\overline{)6x^3 + 11x^2 - 8x + 5} \\
\underline{6x^3 + 15x^2} \\
-4x^2 - 8x \\
\underline{-4x^2 - 10x} \\
2x + 5 \\
\underline{2x + 5} \\
0
\end{array}
$$

17. $\dfrac{10x^3 + 11x^2 - 11x + 2}{5x - 2}$

$$
\begin{array}{r}
2x^2 + 3x - 1 \\
5x-2\overline{)10x^3 + 11x^2 - 11x + 2} \\
\underline{10x^3 - 4x^2} \\
15x^2 - 11x \\
\underline{15x^2 - 6x} \\
-5x + 2 \\
\underline{-5x + 2} \\
0
\end{array}
$$

18. $\dfrac{5x^3 - 28x^2 - 20x + 21}{5x - 3}$

$$
\begin{array}{r}
x^2 - 5x - 7 \\
5x-3\overline{)5x^3 - 28x^2 - 20x + 21} \\
\underline{5x^3 - 3x^2} \\
-25x^2 - 20x \\
\underline{-25x^2 + 15x} \\
-35x + 21 \\
\underline{-35x + 21} \\
0
\end{array}
$$

19. $\dfrac{12y^3 - 22y^2 + 23y - 20}{3y - 4}$

$$
\begin{array}{r}
4y^2 - 2y + 5 \\
3y-4\overline{)12y^3 - 22y^2 + 23y - 20} \\
\underline{12y^3 - 16y^2} \\
-6y^2 + 23y \\
\underline{-6y^2 + 8y} \\
15y - 20 \\
\underline{15y - 20} \\
0
\end{array}
$$

20. $\dfrac{8y^3 + 10y^2 - 49y + 30}{4y - 5}$

$$
\begin{array}{r}
2y^2 + 5y - 6 \\
4y-5\overline{)8y^3 + 10y^2 - 49y + 30} \\
\underline{8y^3 - 10y^2} \\
20y^2 - 49y \\
\underline{20y^2 - 25y} \\
-24y + 30 \\
\underline{-24y + 30} \\
0
\end{array}
$$

21. $(y^3 - y^2 - 13y - 12) \div (y + 3)$

$$
\begin{array}{r}
y^2 - 4y - 1 \\
y + 3 \overline{)y^3 - y^2 - 13y - 12} \\
\underline{y^3 + 3y^2} \\
-4y^2 - 13y \\
\underline{-4y^2 - 12y} \\
-y - 12 \\
\underline{-y - 3} \\
-9
\end{array}
$$

22. $(y^3 - 2y^2 - 26y - 4) \div (y + 4)$

$$
\begin{array}{r}
y^2 - 6y - 2 \\
y + 4 \overline{)y^3 - 2y^2 - 26y - 4} \\
\underline{y^3 + 4y^2} \\
-6y^2 - 26y \\
\underline{-6y^2 - 24y} \\
-2y - 4 \\
\underline{-2y - 8} \\
4
\end{array}
$$

23. $(y^4 - 9y^2 - 5) \div (y - 2)$

$$
\begin{array}{r}
y^3 + 2y^2 - 5y - 10 \\
y - 2 \overline{)y^4 - 0y^3 - 9y^2 + 0y - 5} \\
\underline{y^4 - 2y^3} \\
2y^3 - 9y^2 \\
\underline{2y^3 - 4y^2} \\
-5y^2 + 0y \\
\underline{-5y^2 + 10y} \\
-10y - 5 \\
\underline{-10y + 20} \\
-25
\end{array}
$$

24. $(2y^4 + 3y^2 - 5) \div (y - 2)$

$$
\begin{array}{r}
2y^3 + 4y^2 + 11y + 22 \\
y - 2 \overline{)2y^4 + 0y^3 + 3y^2 + 0y - 5} \\
\underline{2y^4 - 4y^3} \\
4y^3 + 3y^2 \\
\underline{4y^3 - 8y^2} \\
11y^2 + 0y \\
\underline{11y^2 - 22y} \\
22y - 5 \\
\underline{22y - 44} \\
39
\end{array}
$$

Cumulative Review

25. **[0.7.1]** *Milk Prices* In 1995, the average price for a gallon of reduced-fat 2% milk was $2.36. By 2000, the price had increased by 20%. In 2005, the price increased 13% from the 2000 price. What was the average price of a gallon of reduced-fat 2% milk in 2000? What was the price in 2005? (*Source:* www.usda.gov.)

2000: $2.36 \times 1.20 \approx $2.83
2005: $2.83 \times 1.13 \approx $3.20

26. **[3.2.1]** *Page Numbers* Thomas was assigned to read a special two-page case study in his psychology book. He can't remember the page numbers, but does remember that the two numbers add to 341. What are the page numbers? 170 and 171

27. **[0.7.1]** *Hurricane Pattern* The National Hurricane Center has noticed an interesting pattern in the number of Atlantic hurricanes that begin during the period August 1 to October 31 each year. Examine the bar graph then answer the following questions. Round all answers to the nearest tenth.

(a) What was the mean number of hurricanes per year during the years 1999 to 2002? 6.5
(b) What was the mean number of hurricanes per year during the years 2003 to 2006? 7.5
(c) What was the percent of increase from the four-year period 1999 to 2002 to the four-year period 2003 to 2006? 15.4%
(d) If the same percent of increase for a four-year period continued during the years 2007 to 2010, what would be the mean number of hurricanes during the period 2007 to 2010? Round to the nearest tenth. 8.7

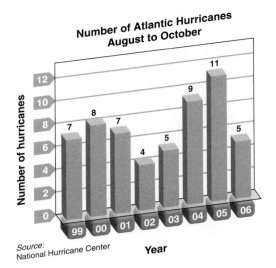

Number of Atlantic Hurricanes August to October

Number of hurricanes

Source: National Hurricane Center

Year

Quick Quiz 4.6 Divide.

1. $\dfrac{20x^5 - 64x^4 - 8x^3}{4x^2}$ $5x^3 - 16x^2 - 2x$

2. $(8x^3 + 2x^2 - 19x - 6) \div (2x + 3)$ $4x^2 - 5x - 2$

3. $(x^3 + 4x - 3) \div (x - 2)$ $x^2 + 2x + 8 + \dfrac{13}{x - 2}$

4. **Concept Check** Explain how you would check your answer to problem 3. Perform the check. Does your answer check? Answers may vary

Classroom Quiz 4.6 You may use these problems to quiz your students' mastery of Section 4.6.

Divide.

1. $\dfrac{21x^6 - 14x^5 + 49x^4}{7x^3}$
   Ans: $3x^3 - 2x^2 + 7x$

2. $(15x^3 + 14x^2 - 17x - 12) \div (3x + 4)$
   Ans: $5x^2 - 2x - 3$

3. $(x^3 + 2x^2 - 3) \div (x - 2)$
   Ans: $x^2 + 4x + 8 + \dfrac{13}{x - 2}$

Putting Your Skills to Work: Use Math to Save Money

TRIP TO THE GROCERY STORE

Is your dollar buying less each month at the grocery store? The rise in the cost of gas not only directly affects our budget, but also affects it indirectly by causing the price of groceries to increase. By using coupons, buying the store brand items, and purchasing staples in bulk, it is possible to get more for your money at the grocery store. Consider the story of Jenny.

Jenny is a single mom with two kids. Concerned with rising costs, she decided to track their grocery expenses for three months. In the first month they spent $450,

in the second month they spent $425, and in the third month they spent $460.

1. What are Jenny's average grocery expenses each month? $445

2. To help bring their monthly costs down, she and the kids started clipping coupons and cutting back on expensive treats. They saved 6% of their average cost by using the coupons and $30 by cutting out the treats. How much was their grocery bill the fourth month? $388.30

3. Jenny and the kids were pleased with the results, but knew they could do more. For the next month, she and the kids decided to choose store brand products when possible. They were happy to see that it saved them 20% off what they spent in the fourth month. How much was their grocery bill the fifth month? $310.64

4. Compare their original average costs and their costs in the fifth month and calculate the savings to the nearest tenth of a percent. 30.2%

5. Can you think of other ways that Jenny's family could lower their monthly grocery bill?
   Answers may vary

Topic	Procedure	Examples
Multiplying monomials, p. 250.	$$x^a \cdot x^b = x^{a+b}$$ 1. Multiply the numerical coefficients. 2. Add the exponents of a given base.	$$3^{12} \cdot 3^{15} = 3^{27}$$ $$(-3x^2)(6x^3) = -18x^5$$ $$(2ab)(4a^2b^3) = 8a^3b^4$$
Dividing monomials, p. 252.	$$\frac{x^a}{x^b} = \begin{cases} x^{a-b} & \text{Use if } a \text{ is greater than } b. \\ \dfrac{1}{x^{b-a}} & \text{Use if } b \text{ is greater than } a. \end{cases}$$ 1. Divide or reduce the fraction created by the quotient of the numerical coefficients. 2. Subtract the exponents of a given base.	$$\frac{16x^7}{8x^3} = 2x^4$$ $$\frac{5x^3}{25x^5} = \frac{1}{5x^2}$$ $$\frac{-12x^5y^7}{18x^3y^{10}} = -\frac{2x^2}{3y^3}$$
Exponent of zero, p. 254.	$$x^0 = 1 \quad \text{if } x \neq 0$$	$$5^0 = 1 \qquad \frac{x^6}{x^6} = 1$$ $$w^0 = 1 \qquad 3x^0y = 3y$$
Raising a power, product, or quotient to a power, pp. 255, 256.	$$(x^a)^b = x^{ab}$$ $$(xy)^a = x^a y^a$$ $$\left(\frac{x}{y}\right)^a = \frac{x^a}{y^a} \quad (y \neq 0)$$ 1. Raise the numerical coefficient to the power outside the parentheses. 2. Multiply the exponent outside the parentheses times the exponent inside the parentheses.	$$(x^9)^3 = x^{27}$$ $$(3x^2)^3 = 27x^6$$ $$\left(\frac{2x^2}{y^3}\right)^3 = \frac{8x^6}{y^9}$$ $$(-3x^4y^5)^4 = 81x^{16}y^{20}$$ $$(-5ab)^3 = -125a^3b^3$$
Negative exponents, pp. 261, 262.	If $x \neq 0$ and $y \neq 0$, then $$x^{-n} = \frac{1}{x^n}, \quad \frac{1}{x^{-n}} = x^n, \quad \frac{x^{-m}}{y^{-n}} = \frac{y^n}{x^m}$$	Write with positive exponents. $$x^{-6} = \frac{1}{x^6}, \quad \frac{1}{w^{-3}} = w^3, \quad \frac{w^{-12}}{z^{-5}} = \frac{z^5}{w^{12}}$$
Scientific notation, p. 262.	A positive number is written in scientific notation if it is in the form $a \times 10^n$, where $1 \leq a < 10$ and n is an integer.	$$2{,}568{,}000 = 2.568 \times 10^6$$ $$0.0000034 = 3.4 \times 10^{-6}$$
Add polynomials, p. 268.	To add two polynomials, we add their like terms.	$$(-7x^3 + 2x^2 + 5) + (x^3 + 3x^2 + x)$$ $$= -6x^3 + 5x^2 + x + 5$$
Subtracting polynomials, p. 270.	To subtract the polynomials, change all signs of the second polynomial and add the result to the first polynomial. $$a - b = a + (-b)$$	$$(5x^2 - 6) - (-3x^2 + 2) = (5x^2 - 6) + (3x^2 - 2)$$ $$= 8x^2 - 8$$
Multiplying a monomial by a polynomial, p. 275.	Use the distributive property. $$a(b + c) = ab + ac$$ $$(b + c)a = ba + ca$$	Multiply. $$-5x(2x^2 + 3x - 4) = -10x^3 - 15x^2 + 20x$$ $$(6x^3 - 5xy - 2y^2)(3xy) = 18x^4y - 15x^2y^2 - 6xy^3$$
Multiplying two binomials, pp. 275, 281, 282.	1. The product of the sum and difference of the same two terms: $$(a + b)(a - b) = a^2 - b^2$$ 2. The square of a binomial: $$(a + b)^2 = a^2 + 2ab + b^2$$ $$(a - b)^2 = a^2 - 2ab + b^2$$ 3. Use FOIL for other binomial multiplication. The middle terms can often be combined, giving a trinomial.	$$(3x + 7y)(3x - 7y) = 9x^2 - 49y^2$$ $$(3x + 7y)^2 = 9x^2 + 42xy + 49y^2$$ $$(3x - 7y)^2 = 9x^2 - 42xy + 49y^2$$ $$(3x - 5)(2x + 7) = 6x^2 + 21x - 10x - 35$$ $$= 6x^2 + 11x - 35$$

Topic	Procedure	Examples
Multiplying two polynomials, p. 283.	To multiply two polynomials, multiply each term of one by each term of the other. This method is similar to the multiplication of many-digit numbers.	Vertical method: $$\begin{array}{r} 3x^2 - 7x + 4 \\ \times \quad 3x - 1 \\ \hline -3x^2 + 7x - 4 \\ 9x^3 - 21x^2 + 12x \\ \hline 9x^3 - 24x^2 + 19x - 4 \end{array}$$ Horizontal method: $$(5x + 2)(2x^2 - x + 3)$$ $$= 10x^3 - 5x^2 + 15x + 4x^2 - 2x + 6$$ $$= 10x^3 - x^2 + 13x + 6$$
Multiplying three or more polynomials, p. 284.	1. Multiply any two polynomials. 2. Multiply the result by any remaining polynomials.	$$(2x + 1)(x - 3)(x + 4) = (2x^2 - 5x - 3)(x + 4)$$ $$= 2x^3 + 3x^2 - 23x - 12$$
Dividing a polynomial by a monomial, p. 287.	1. Divide each term of the polynomial by the monomial. 2. When dividing variables, use the property $$\frac{x^a}{x^b} = x^{a-b}.$$	Divide. $\quad (15x^3 + 20x^2 - 30x) \div (5x)$ $$= \frac{15x^3}{5x} + \frac{20x^2}{5x} - \frac{30x}{5x}$$ $$= 3x^2 + 4x - 6$$
Dividing a polynomial by a binomial, p. 287.	1. Place the terms of the polynomial and binomial in descending order. Insert a 0 for any missing term. 2. Divide the first term of the polynomial by the first term of the binomial. 3. Multiply the first term of the answer by the binomial, and subtract the result from the first two terms of the polynomial. Bring down the next term to obtain a new polynomial. 4. Divide the new polynomial by the binomial using the process described in step 2. 5. Continue dividing, multiplying, and subtracting until the degree of the remainder is less than the degree of the binomial divisor. 6. Write the remainder as the numerator of a fraction that has the binomial divisor as its denominator.	Divide. $$(8x^3 + 2x^2 - 13x + 7) \div (4x - 1)$$ $$\begin{array}{r} 2x^2 + x - 3 \\ 4x - 1 \overline{\smash{)}8x^3 + 2x^2 - 13x + 7} \\ \underline{8x^3 - 2x^2} \\ 4x^2 - 13x \\ \underline{4x^2 - x} \\ -12x + 7 \\ \underline{-12x + 3} \\ 4 \end{array}$$ The answer is $$2x^2 + x - 3 + \frac{4}{4x - 1}.$$

Chapter 4 Review Problems

Section 4.1

Simplify. In problems 1–12, leave your answer in exponent form.

1. $(-6a^2)(3a^5)$ $\quad -18a^7$

2. $(5^{10})(5^{13})$ $\quad 5^{23}$

3. $(3xy^2)(2x^3y^4)$ $\quad 6x^4y^6$

4. $(2x^3y^4)(-7xy^5)$ $\quad -14x^4y^9$

5. $\dfrac{7^{15}}{7^{27}}$ $\quad \dfrac{1}{7^{12}}$

6. $\dfrac{x^{12}}{x^{17}}$ $\quad \dfrac{1}{x^5}$

7. $\dfrac{y^{30}}{y^{16}}$ $\quad y^{14}$

8. $\dfrac{9^{13}}{9^{24}}$ $\quad \dfrac{1}{9^{11}}$

9. $\dfrac{-15xy^2}{25x^6y^6}$ $\quad -\dfrac{3}{5x^5y^4}$

10. $\dfrac{-12a^3b^6}{18a^2b^{12}}$ $\quad -\dfrac{2a}{3b^6}$

11. $(x^3)^8$ $\quad x^{24}$

12. $(b^5)^6$ $\quad b^{30}$

13. $(-3a^3b^2)^2$ $\quad 9a^6b^4$

14. $(3x^3y)^4$ $\quad 81x^{12}y^4$

15. $\left(\dfrac{5ab^2}{c^3}\right)^2$ $\quad \dfrac{25a^2b^4}{c^6}$

16. $\left(\dfrac{x^0y^3}{4w^5z^2}\right)^3$ $\quad \dfrac{y^9}{64w^{15}z^6}$

Section 4.2

Simplify. Write with positive exponents.

17. $a^{-3}b^5$ $\dfrac{b^5}{a^3}$

18. $m^8 p^{-5}$ $\dfrac{m^8}{p^5}$

19. $\dfrac{2x^{-6}}{y^{-3}}$ $\dfrac{2y^3}{x^6}$

20. $(2x^{-5}y)^{-3}$ $\dfrac{x^{15}}{8y^3}$

21. $(6a^4b^5)^{-2}$ $\dfrac{1}{36a^8b^{10}}$

22. $\dfrac{3x^{-3}}{y^{-2}}$ $\dfrac{3y^2}{x^3}$

23. $\dfrac{4x^{-5}y^{-6}}{w^{-2}z^8}$ $\dfrac{4w^2}{x^5y^6z^8}$

24. $\dfrac{3^{-3}a^{-2}b^5}{c^{-3}d^{-4}}$ $\dfrac{b^5c^3d^4}{27a^2}$

Write in scientific notation.

25. 156,340,200,000
   1.563402×10^{11}

26. 179,632
   1.79632×10^5

27. 0.00092
   9.2×10^{-4}

28. 0.00000174
   1.74×10^{-6}

Write in decimal notation.

29. 1.2×10^5
   120,000

30. 6.034×10^6
   6,034,000

31. 3×10^6
   3,000,000

32. 2.5×10^{-1}
   0.25

33. 4.32×10^{-5}
   0.0000432

34. 6×10^{-9}
   0.000000006

Perform the indicated calculation. Leave your answer in scientific notation.

35. $\dfrac{(28,000,000)(5,000,000,000)}{7000}$
   2.0×10^{13}

36. $(3.12 \times 10^5)(2.0 \times 10^6)(1.5 \times 10^8)$
   9.36×10^{19}

37. $(1.6 \times 10^{-3})(3.0 \times 10^{-5})(2.0 \times 10^{-2})$
   9.6×10^{-10}

38. $\dfrac{(0.00078)(0.000005)(0.00004)}{0.002}$
   7.8×10^{-11}

39. **Mission to Pluto** The *New Horizons* spacecraft began its 3.5×10^9-mile journey to Pluto in January 2006. If the cost of this mission is $0.20 per mile, find the total cost. (*Source:* www.pluto.jhuapl.edu.)
   7×10^8 dollars

40. **Atomic Clock** An atomic clock is based on the fact that cesium emits 9,192,631,770 cycles of radiation in one second. How many of these cycles occur in one day? Round to three significant digits.
   7.94×10^{14} cycles

41. **Computer Speed** Today's fastest modern computers can perform one operation in 1×10^{-8} second. How many operations can such a computer perform in 1 minute?
   6×10^9 operations

Section 4.3

Combine.

42. $(2.8x^2 - 1.5x + 3.4) + (2.7x^2 + 0.5x - 5.7)$
   $5.5x^2 - x - 2.3$

43. $(4.1x^2 + 1.8x - 3.3) + (1.5x^2 - 2.9x - 0.7)$
   $5.6x^2 - 1.1x - 4$

44. $(x^3 + x^2 - 6x + 2) - (2x^3 - x^2 - 5x - 6)$
   $-x^3 + 2x^2 - x + 8$

45. $(4x^3 - x^2 - x + 3) - (-3x^3 + 2x^2 + 5x - 1)$
   $7x^3 - 3x^2 - 6x + 4$

46. $\left(\dfrac{3}{5}x^2y - \dfrac{1}{3}x + \dfrac{3}{4}\right) - \left(\dfrac{1}{2}x^2y + \dfrac{2}{7}x + \dfrac{1}{3}\right)$
   $\dfrac{1}{10}x^2y - \dfrac{13}{21}x + \dfrac{5}{12}$

47. $\dfrac{1}{2}x^2 - \dfrac{3}{4}x + \dfrac{1}{5} - \left(\dfrac{1}{4}x^2 - \dfrac{1}{2}x + \dfrac{1}{10}\right)$
   $\dfrac{1}{4}x^2 - \dfrac{1}{4}x + \dfrac{1}{10}$

48. $(3x^2 - 5x) + (8x - 3) - (x^2 + 2)$
   $2x^2 + 3x - 5$

49. $(x^2 - 9) - (4x^2 + 5x) + (5x - 6)$
   $-3x^2 - 15$

Section 4.4

Multiply.

50. $(3x + 1)(5x - 1)$
$15x^2 + 2x - 1$

51. $(7x - 2)(4x - 3)$
$28x^2 - 29x + 6$

52. $(2x + 3)(10x + 9)$
$20x^2 + 48x + 27$

53. $5x(2x^2 - 6x + 3)$
$10x^3 - 30x^2 + 15x$

54. $(3x^3 + 4x^2 - 5x + 8)(-5x)$
$-15x^4 - 20x^3 + 25x^2 - 40x$

55. $(xy^2 + 5xy - 6)(-4xy^2)$
$-4x^2y^4 - 20x^2y^3 + 24xy^2$

56. $(5a + 7b)(a - 3b)$
$5a^2 - 8ab - 21b^2$

57. $(2x^2 - 3)(4x^2 - 5y)$
$8x^4 - 10x^2y - 12x^2 + 15y$

58. $-3x^2y(5x^4y + 3x^2 - 2)$
$-15x^6y^2 - 9x^4y + 6x^2y$

Section 4.5

Multiply.

59. $(4x + 3)^2$
$16x^2 + 24x + 9$

60. $(a + 5b)(a - 5b)$
$a^2 - 25b^2$

61. $(7x + 6y)(7x - 6y)$
$49x^2 - 36y^2$

62. $(5a - 2b)^2$
$25a^2 - 20ab + 4b^2$

63. $(8x + 9y)^2$
$64x^2 + 144xy + 81y^2$

64. $(x^2 + 7x + 3)(4x - 1)$
$4x^3 + 27x^2 + 5x - 3$

65. $(x - 6)(2x - 3)(x + 4)$
$2x^3 - 7x^2 - 42x + 72$

Section 4.6

Divide.

66. $(12y^3 + 18y^2 + 24y) \div (6y)$
$2y^2 + 3y + 4$

67. $(30x^5 + 35x^4 - 90x^3) \div (5x^2)$
$6x^3 + 7x^2 - 18x$

68. $(16x^3 - 24x^2 + 32x) \div (4x)$
$4x^2 - 6x + 8$

69. $(106x^6 - 24x^5 + 38x^4 + 26x^3) \div (2x^3)$
$53x^3 - 12x^2 + 19x + 13$

70. $(15x^2 + 11x - 14) \div (5x + 7)$
$3x - 2$

71. $(12x^2 - x - 63) \div (4x + 9)$
$3x - 7$

72. $(6x^3 + x^2 + 6x + 5) \div (2x - 1)$
$3x^2 + 2x + 4 + \dfrac{9}{2x - 1}$

73. $(2x^3 - x^2 + 3x - 1) \div (x + 2)$
$2x^2 - 5x + 13 - \dfrac{27}{x + 2}$

74. $(20x^2 + 3x - 2) \div (4x - 1)$
$5x + 2$

75. $(6x^2 + x - 9) \div (2x + 3)$
$3x - 4 + \dfrac{3}{2x + 3}$

76. $(x^3 - x - 24) \div (x - 3)$
$x^2 + 3x + 8$

77. $(2x^3 - 3x + 1) \div (x - 2)$
$2x^2 + 4x + 5 + \dfrac{11}{x - 2}$

Mixed Practice

Simplify. Leave your answer in exponent form.

78. $(4^6)(4^{23})$ 4^{29}

79. $\dfrac{6^{50}}{6^{35}}$ 6^{15}

Simplify. Write your answer with only positive exponents.

80. $(2x^3y^{-2})^{-2}$ $\dfrac{y^4}{4x^6}$

81. $\dfrac{-3ab^{-1}}{bc^{-2}}$ $\dfrac{-3ac^2}{b^2}$

82. $\left(\dfrac{3x^0y^{-2}}{z^3}\right)^{-2}$ $\dfrac{y^4z^6}{9}$

Write in scientific notation.

83. 0.000922 9.22×10^{-4}

Remove parentheses and simplify.

84. $(-6x^3 - 7x + 8) - (-2x^3 + 5x - 9)$
$-4x^3 - 12x + 17$

85. $-6x^2(2x^3 + 5x^2 - 2)$
$-12x^5 - 30x^4 + 12x^2$

86. Multiply. $(7x - 3)(2x - 1)(x + 2)$
$14x^3 + 15x^2 - 23x + 6$

87. Divide. $(6x^3 + 11x^2 - x - 2) \div (3x - 2)$
$2x^2 + 5x + 3 + \dfrac{4}{3x - 2}$

Applications

Solve. Express your answer in scientific notation.

88. *Fighting Disease* President Bush signed a foreign spending bill for fiscal year 2006 that included 2.8×10^9 dollars to fight HIV/AIDS, malaria, and tuberculosis worldwide. If the population of the United States in 2006 was 3×10^8 people, how much per person did the United States spend to fight these diseases? (*Source*: www.info.state.gov.) Write your answer in dollars and cents. About $9.33

89. *Population* The population of India in 2010 is projected to be 1.18×10^9. The projected population of Bangladesh in 2010 is 1.6×10^8. (*Source: Statistical Abstract of the United States: 2007.*) What is the total population of the two countries projected to be? (*Hint:* First write 1.6×10^8 as 0.16×10^9 before you do the calculation.)
1.34×10^9 people

90. *Electron Mass* The mass of an electron is approximately 9.11×10^{-28} gram. Find the mass of 30,000 electrons.
2.733×10^{-23} g

91. *Gray Whales* During feeding season, gray whales eat 3.4×10^5 pounds of food per day. If the feeding period lasts for 140 days, how many pounds of food total will a gray whale consume?
4.76×10^7 lb

To Think About

Find a polynomial that describes the shaded area.

▲ **92.**

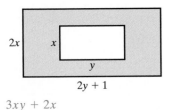

$3xy + 2x$

▲ **93.**

$2x^2 - 4y^2$

Note to Instructor: The Chapter 4 Test file in the TestGen program provides algorithms specifically matched to these problems so you can easily replicate this test for additional practice or assessment purposes.

Remember to use your Chapter Test Prep Video CD to see the worked-out solutions to the test problems you want to review.

Simplify. Leave your answer in exponent form.

1. $(3^{10})(3^{24})$

2. $\dfrac{25^{18}}{25^{34}}$

3. $(8^4)^6$

In questions 4–8, simplify.

4. $(-3xy^4)(-4x^3y^6)$

5. $\dfrac{-35x^8y^{10}}{25x^5y^{10}}$

6. $(-5xy^6)^3$

7. $\left(\dfrac{7a^7b^2}{3c^0}\right)^2$

8. $\dfrac{(3x^2)^3}{(6x)^2}$

9. Evaluate. 4^{-3}

In questions 10 and 11, simplify and write with only positive exponents.

10. $6a^{-4}b^{-3}c^5$

11. $\dfrac{3x^{-3}y^2}{x^{-4}y^{-5}}$

12. Write in scientific notation. 0.0005482

13. Write in decimal notation. 5.82×10^8

14. Multiply. Leave your answer in scientific notation.
$(4.0 \times 10^{-3})(3.0 \times 10^{-8})(2.0 \times 10^4)$

Combine.

15. $(2x^2 - 3x - 6) + (-4x^2 + 8x + 6)$

16. $(3x^3 - 4x^2 + 3) - (14x^3 - 7x + 11)$

Multiply.

17. $-7x^2(3x^3 - 4x^2 + 6x - 2)$

18. $(5x^2y^2 - 6xy + 2)(3x^2y)$

19. $(5a - 4b)(2a + 3b)$

20. $(3x + 2)(2x + 1)(x - 3)$

21. $(7x^2 + 2y^2)^2$

22. $(5s - 11t)(5s + 11t)$

23. $(3x - 2)(4x^3 - 2x^2 + 7x - 5)$

24. $(3x^2 - 5xy)(x^2 + 3xy)$

Divide.

25. $(15x^6 - 5x^4 + 25x^3) \div 5x^3$

26. $(8x^3 - 22x^2 - 5x + 12) \div (4x + 3)$

27. $(2x^3 - 6x - 36) \div (x - 3)$

Solve. Express your answer in scientific notation. Round to the nearest hundredth.

28. At the end of 2002, Saudi Arabia was estimated to have 2.618×10^{11} barrels of oil reserves. By one estimate, they have 86 years of reserves remaining. If they disbursed the oil in an equal amount each year, how many barrels of oil would they pump each year? (*Source:* United Nations Review of World Energy, 2003.)

29. A space probe is traveling from Earth to the planet Pluto at a speed of 2.49×10^4 miles per hour. How far would this space probe travel in one week?

1.	3^{34}
2.	$\dfrac{1}{25^{16}}$
3.	8^{24}
4.	$12x^4y^{10}$
5.	$-\dfrac{7x^3}{5}$
6.	$-125x^3y^{18}$
7.	$\dfrac{49a^{14}b^4}{9}$
8.	$\dfrac{3x^4}{4}$
9.	$\dfrac{1}{64}$
10.	$\dfrac{6c^5}{a^4b^3}$
11.	$3xy^7$
12.	5.482×10^{-4}
13.	$582,000,000$
14.	2.4×10^{-6}
15.	$-2x^2 + 5x$
16.	$-11x^3 - 4x^2 + 7x - 8$
17.	$-21x^5 + 28x^4 - 42x^3 + 14x^2$
18.	$15x^4y^3 - 18x^3y^2 + 6x^2y$
19.	$10a^2 + 7ab - 12b^2$
20.	$6x^3 - 11x^2 - 19x - 6$
21.	$49x^4 + 28x^2y^2 + 4y^4$
22.	$25s^2 - 121t^2$
23.	$12x^4 - 14x^3 + 25x^2 - 29x + 10$
24.	$3x^4 + 4x^3y - 15x^2y^2$
25.	$3x^3 - x + 5$
26.	$2x^2 - 7x + 4$
27.	$2x^2 + 6x + 12$
28.	3.04×10^9 barrels per year
29.	4.18×10^6 mi

1.	$\dfrac{1}{30}$
2.	-0.74
3.	$-\dfrac{6}{7}$
4.	$2x^2 - 13x$
5.	110.55
6.	66
7.	$x = -\dfrac{9}{2}$
8.	$x = 15$
9.	$x < -3$
10.	$d = \dfrac{3y - 8x}{11}$
11.	12,400 employees
12.	$199.20
13.	$20x^2 - 21x - 5$
14.	$9x^2 - 30x + 25$
15.	$6x^3 - 17x^2 - 26x - 8$
16.	$-20x^5y^8$
17.	$-\dfrac{2x^3}{3y^9}$
18.	$16x^{12}y^8$
19.	$\dfrac{9z^8}{w^2x^3y^4}$
20.	4.25×10^{10}
21.	5.6×10^{-4}
22.	4.0×10^{-35}
23.	$5x^3 + 7x^2 - 6x + 50$
24.	$-36x^3y^2 + 18x^2y^3 - 48xy^4$
25.	$6x^3 - 19x^2 + 18x - 5$
26.	$x + 5 + \dfrac{3}{x - 3}$

Approximately one-half of this test covers the content of Chapters 0 through 3. The remainder covers the content of Chapter 4.

In questions 1–4, simplify.

1. $\dfrac{11}{15} - \dfrac{7}{10}$

2. $(-3.7) \times (0.2)$

3. $\left(-4\dfrac{1}{2}\right) \div \left(5\dfrac{1}{4}\right)$

4. $7x(3x - 4) - 5x(2x - 3) - (3x)^2$

5. Find 6% of 1842.5

6. Evaluate $3a^2 + ab - 4b^2$ for $a = 5$ and $b = -1$.

In questions 7–9, solve.

7. $7x - 3(4 - 2x) = 14x - (3 - x)$

8. $\dfrac{2}{3}x + 6 = 4(x - 11)$

9. $2 - 5x > 17$

10. Solve for d: $8(x + d) = 3(y - d)$

11. A national walkout of 11,904 employees of the VBM Corp. occurred last month. This was 96% of the total number of employees. How many employees does VBM have?

12. How much interest would $3320 earn in one year if invested at 6% simple interest?

13. Multiply. $(4x - 5)(5x + 1)$

14. Multiply. $(3x - 5)^2$

15. Multiply. $(3x + 2)(2x + 1)(x - 4)$

In questions 16–18, simplify.

16. $(-4x^4y^5)(5xy^3)$

17. $\dfrac{14x^8y^3}{-21x^5y^{12}}$

18. $(-2x^3y^2z^0)^4$

19. Write with only positive exponents. $\dfrac{9x^{-3}y^{-4}}{w^2z^{-8}}$

20. Write in scientific notation. 42,500,000,000

21. Write in scientific notation. 0.00056

22. Calculate. Leave your answer in scientific notation. $\dfrac{(2.0 \times 10^{-12})(8.0 \times 10^{-20})}{4.0 \times 10^3}$

23. Subtract. $(x^3 - 3x^2 - 5x + 20) - (-4x^3 - 10x^2 + x - 30)$

Multiply.

24. $-6xy^2(6x^2 - 3xy + 8y^2)$

25. $(2x^2 - 3x + 1)(3x - 5)$

26. Divide. $(x^2 + 2x - 12) \div (x - 3)$

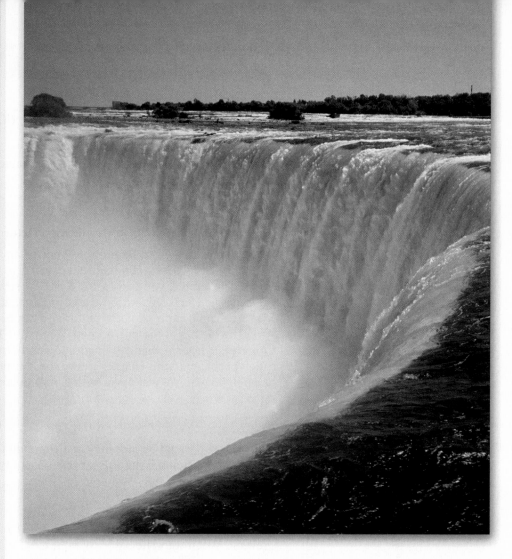

CHAPTER 5

Niagara Falls is an amazing wonder. Every year hundreds of thousands of tourists flock to see this spectacular waterfall. Besides its natural beauty, it also powers three hydroelectric stations that produce 4.4 gigawatts of power. Many of the equations related to hydroelectric power are quadratic equations similar to the ones studied in this chapter.

Factoring

Student Learning Objective

After studying this section, you will be able to:

 Factor polynomials whose terms contain a common factor.

1 Factoring Polynomials Whose Terms Contain a Common Factor

Recall that when two or more numbers, variables, or algebraic expressions are multiplied, each is called a **factor.**

$$3 \cdot 2 \qquad 3x^2 \cdot 5x^3 \qquad (2x - 3)(x + 4)$$

factor   factor   factor   factor   factor   factor

When you are asked **to factor** a number or an algebraic expression, you are being asked, "What factors, when multiplied, will give that number or expression?"

For example, you can factor 6 as $3 \cdot 2$ since $3 \cdot 2 = 6$. You can factor $15x^5$ as $3x^2 \cdot 5x^3$ since $3x^2 \cdot 5x^3 = 15x^5$. Factoring is simply the reverse of multiplying. 6 and $15x^5$ are simple expressions to factor and can be factored in different ways.

The factors of the polynomial $2x^2 + x - 12$ are not so easy to recognize. Factoring a polynomial changes an addition and/or subtraction problem into a multiplication problem. In this chapter we will be learning techniques for finding the factors of a polynomial. We will begin with **common factors.**

Teaching Example 1 Factor.

(a) $18a - 9b$ (b) $4ab + 3b$

Ans: (a) $9(2a - b)$ (b) $b(4a + 3)$

EXAMPLE 1   Factor.   **(a)** $3x - 6y$ **(b)** $9x + 2xy$

Solution   Begin by looking for a common factor, a factor that both terms have in common. Then rewrite the expression as a product.

(a) $3x - 6y = 3(x - 2y)$ This is true because $3(x - 2y) = 3x - 6y$.
(b) $9x + 2xy = x(9 + 2y)$ This is true because $x(9 + 2y) = 9x + 2xy$.

Some people find it helpful to think of factoring as the distributive property in reverse. When we write $3x - 6y = 3(x - 2y)$, we are doing the reverse of distributing the 3. This is the main point of this section. We are doing problems of the form $ca + cb = c(a + b)$. The common factor c becomes the first factor of our answer.

NOTE TO STUDENT: *Fully worked-out solutions to all of the Practice Problems can be found at the back of the text starting at page SP-1*

Practice Problem 1   Factor.

(a) $21a - 7b$ $7(3a - b)$ **(b)** $5xy + 8x$ $x(5y + 8)$

Teaching Tip You will want to stress that removing the largest common factor may involve removing a number, a variable, or both.

When we factor, we begin by looking for the **greatest common factor.** For example, in the polynomial $48x - 16y$, a common factor is 2. We could factor $48x - 16y$ as $2(24x - 8y)$. However, this is not complete. To factor $48x - 16y$ completely, we look for the greatest common factor of 48 and of 16.

$$48x - 16y = 16(3x - y)$$

Teaching Example 2 Factor.
$9a^3 - 36ab + 27a^2$

Ans: $9a(a^2 - 4b + 3a)$

EXAMPLE 2   Factor $24xy + 12x^2 + 36x^3$. Remember to remove the greatest common factor.

Solution   Find the greatest common factor of 24, 12, and 36. You may want to factor each number, or you may notice that 12 is a common factor. 12 is the greatest numerical common factor.

Notice also that x is a factor of each term. Thus, $12x$ is the greatest common factor.

$$24xy + 12x^2 + 36x^3 = 12x(2y + x + 3x^2)$$

Practice Problem 2   Factor $12a^2 + 16ab^2 - 12a^2b$. Be careful to remove the greatest common factor.

$4a(3a + 4b^2 - 3ab)$

FACTORING A POLYNOMIAL WITH COMMON FACTORS

1. Determine the greatest common numerical factor by asking, "What is the largest integer that will divide into the coefficients of all the terms?"

2. Determine the greatest common variable factor by first asking, "What variables are common to all the terms?" Then, for each variable that is common to all the terms ask, "What is the largest exponent of the variable that is common to all the terms?"

3. The common factors found in steps 1 and 2 are the first part of the answer.

4. After removing the common factors, what remains is placed in parentheses as the second factor.

Teaching Tip Explain to your students that to check that they have factored out the greatest common factor, they must do two things:

1. See if any further common factor exists within the terms in the parentheses. (If so, they did not correctly factor out the GREATEST common factor.)

2. Multiply the factors to check that the product exactly matches the original problem. That is, use
$$c(a + b) = ca + cb.$$

EXAMPLE 3   Factor.   **(a)** $12x^2 - 18y^2$ **(b)** $x^2y^2 + 3xy^2 + y^3$

Solution

(a) Note that the largest integer that is common to both terms is 6 (not 3 or 2).
$$12x^2 - 18y^2 = 6(2x^2 - 3y^2)$$

(b) Although y is common to all of the terms, we factor out y^2 since 2 is the largest exponent of y that is common to all terms. We do not factor out x, since x is not common to all of the terms.
$$x^2y^2 + 3xy^2 + y^3 = y^2(x^2 + 3x + y)$$

Teaching Example 3 Factor.

(a) $30a^2 + 18b$

(b) $2m^3n^5 - 5m^2n^4 + 2n^3$

Ans: (a) $6(5a^2 + 3b)$

(b) $n^3(2m^3n^2 - 5m^2n + 2)$

Practice Problem 3   Factor.

(a) $16a^3 - 24b^3$ $8(2a^3 - 3b^3)$ **(b)** $r^3s^2 - 4r^4s + 7r^5$ $r^3(s^2 - 4rs + 7r^2)$

Checking Is Very Important!   You can check any factoring problem by multiplying the factors you obtain. The result should be the same as the original polynomial.

EXAMPLE 4   Factor. $8x^3y + 16x^2y^2 - 24x^3y^3$

Solution   We see that 8 is the largest integer that will divide evenly into the three numerical coefficients. We can factor an x^2 out of each term. We can also factor y out of each term.
$$8x^3y + 16x^2y^2 - 24x^3y^3 = 8x^2y(x + 2y - 3xy^2)$$

Check.
$$8x^2y(x + 2y - 3xy^2) = 8x^3y + 16x^2y^2 - 24x^3y^3 \checkmark$$

Teaching Example 4 Factor.
$14a^3bc + 35a^2b^2c^2 - 21a^2b$

Ans: $7a^2b(2ac + 5bc^2 - 3)$

Practice Problem 4   Factor. $18a^3b^2c - 27ab^3c^2 - 45a^2b^2c^2$

$9ab^2c(2a^2 - 3bc - 5ac)$

EXAMPLE 5   Factor. $9a^3b^2 + 9a^2b^2$

Solution   We observe that both terms contain a common factor of 9. We can factor a^2 and b^2 out of each term. Thus the common factor is $9a^2b^2$.
$$9a^3b^2 + 9a^2b^2 = 9a^2b^2(a + 1)$$

Teaching Example 5 Factor.
$12x^4y^3 - 20x^3y^2 + 2x^2y$

Ans: $2x^2y(6x^2y^2 - 10xy + 1)$

CAUTION: Don't forget to include the 1 inside the parentheses in Example 5. The solution is wrong without it. You will see why if you try to check a result written without the 1.

Practice Problem 5   Factor and check. $30x^3y^2 - 24x^2y^2 + 6xy^2$

$6xy^2(5x^2 - 4x + 1)$

EXAMPLE 6 Factor. $3x(x - 4y) + 2(x - 4y)$

Solution Be sure you understand what are terms and what are factors of the polynomial in this example. There are two terms. The expression $3x(x - 4y)$ is one term. The expression $2(x - 4y)$ is the second term. Each term is made up of two factors. Observe that the binomial $(x - 4y)$ is a common factor of the terms. A common factor may be any type of polynomial. Thus we can factor out the common factor $(x - 4y)$.

$$3x(x - 4y) + 2(x - 4y) = (x - 4y)(3x + 2)$$

Practice Problem 6 Factor. $3(a + 5b) + x(a + 5b)$ $(a + 5b)(3 + x)$

EXAMPLE 7 Factor. $7x^2(2x - 3y) - (2x - 3y)$

Solution The common factor of the terms is $(2x - 3y)$. What happens when we factor out $(2x - 3y)$? What are we left with in the second term?
 Recall that $(2x - 3y) = 1(2x - 3y)$. Thus

$7x^2(2x - 3y) - (2x - 3y) = 7x^2(2x - 3y) - 1(2x - 3y)$ Rewrite the original expression.

$\qquad\qquad\qquad = (2x - 3y)(7x^2 - 1)$ Factor out $(2x - 3y)$.

NOTE TO STUDENT: Fully worked-out solutions to all of the Practice Problems can be found at the back of the text starting at page SP-1

Practice Problem 7 Factor. $8y(9y^2 - 2) - (9y^2 - 2)$ $(9y^2 - 2)(8y - 1)$

▲ **EXAMPLE 8** A computer programmer is writing a program to find the total area of 4 circles. She uses the formula $A = \pi r^2$. The radii of the circles are a, b, c, and d, respectively. She wants the final answer to be in factored form with the value of π occurring only once, in order to minimize the rounding error. Write the total area with a formula that has π occurring only once.

Solution

 For each circle, $A = \pi r^2$, where $r = a$, b, c, or d.
 The total area is $\pi a^2 + \pi b^2 + \pi c^2 + \pi d^2$.
 In factored form the total area $= \pi(a^2 + b^2 + c^2 + d^2)$.

▲ **Practice Problem 8** Use $A = \pi r^2$ to find the shaded area. The radius of the larger circle is b. The radius of the smaller circle is a. Write the shaded-area formula in factored form so that π appears only once. $\pi(b^2 - a^2)$

Verbal and Writing Skills *In exercises 1 and 2, write a word or words to complete each sentence.*

1. In the expression $3x^2 \cdot 5x^3$, $3x^2$ and $5x^3$ are called ____factors____ .

2. In the expression $3x^2 + 5x^3$, $3x^2$ and $5x^3$ are called ____terms____ .

3. We can factor $30a^4 + 15a^3 - 45a^2$ as $5a(6a^3 + 3a^2 - 9a)$. Is the factoring complete? Why or why not?
   No; $6a^3 + 3a^2 - 9a$ has a common factor of $3a$.

4. We can factor $4x^3 - 8x^2 + 20x$ as $4(x^3 - 2x^2 + 10x)$. Is the factoring complete?
   No; $x^3 - 2x^2 + 10x$ has a common factor of x.

Remove the largest possible common factor. Check your answers for exercises 5–28 by multiplication.

5. $3a^2 + 3a$
   $3a(a + 1)$

6. $2c^2 + 2c$
   $2c(c + 1)$

7. $21ab - 14ab^2$
   $7ab(3 - 2b)$

8. $18wz - 27w^2z$
   $9wz(2 - 3w)$

9. $2\pi rh + 2\pi r^2$
   $2\pi r(h + r)$

10. $9a^2b^2 - 36ab$
    $9ab(ab - 4)$

11. $5x^3 + 25x^2 - 15x$
    $5x(x^2 + 5x - 3)$

12. $8x^3 - 10x^2 - 14x$
    $2x(4x^2 - 5x - 7)$

13. $12ab - 28bc + 20ac$
    $4(3ab - 7bc + 5ac)$

14. $12xy - 18yz - 36xz$
    $6(2xy - 3yz - 6xz)$

15. $16x^5 + 24x^3 - 32x^2$
    $8x^2(2x^3 + 3x - 4)$

16. $36x^6 + 45x^4 - 18x^2$
    $9x^2(4x^4 + 5x^2 - 2)$

17. $14x^2y - 35xy - 63x$
    $7x(2xy - 5y - 9)$

18. $40a^2 - 16ab - 24a$
    $8a(5a - 2b - 3)$

19. $54x^2 - 45xy + 18x$
    $9x(6x - 5y + 2)$

20. $48xy - 24y^2 + 40y$
    $8y(6x - 3y + 5)$

21. $3xy^2 - 2ay + 5xy - 2y$
    $y(3xy - 2a + 5x - 2)$

22. $2ab^3 + 3xb^2 - 5b^4 + 2b^2$
    $b^2(2ab + 3x - 5b^2 + 2)$

23. $24x^2y - 40xy^2$
    $8xy(3x - 5y)$

24. $35abc^2 - 49ab^2c$
    $7abc(5c - 7b)$

25. $7x^3y^2 + 21x^2y^2$
    $7x^2y^2(x + 3)$

26. $8x^3y^2 + 32xy^2$
    $8xy^2(x^2 + 4)$

27. $8x^4y + 24x^2y^2 - 32x^2y$
    $8x^2y(x^2 + 3y - 4)$

28. $30x^2y^2 - 24xy^3 + 12x^2y$
    $6xy(5xy - 4y^2 + 2x)$

Hint: In exercises 29–42, refer to Examples 6 and 7.

29. $7a(x + 2y) - b(x + 2y)$
    $(x + 2y)(7a - b)$

30. $6(3a + b) - z(3a + b)$
    $(3a + b)(6 - z)$

31. $3x(x - 4) - 2(x - 4)$
    $(x - 4)(3x - 2)$

32. $5x(x - 7) + 3(x - 7)$
    $(x - 7)(5x + 3)$

33. $6b(2a - 3c) - 5d(2a - 3c)$
    $(2a - 3c)(6b - 5d)$

34. $7x(3y + 5z) - 6t(3y + 5z)$
    $(3y + 5z)(7x - 6t)$

35. $7c(b - a^2) - 5d(b - a^2) + 2f(b - a^2)$
    $(b - a^2)(7c - 5d + 2f)$

36. $2a(x^2 - y) - 5b(x^2 - y) + c(x^2 - y)$
    $(x^2 - y)(2a - 5b + c)$

37. $2a(xy - 3) - 4(xy - 3) - z(xy - 3)$
    $(xy - 3)(2a - 4 - z)$

38. $3c(bc - 3a) - 2(bc - 3a) - 6b(bc - 3a)$
    $(bc - 3a)(3c - 2 - 6b)$

39. $4a^3(a - 3b) + (a - 3b)$
    $(a - 3b)(4a^3 + 1)$

40. $(c + 4d) + 7c^2(c + 4d)$
    $(c + 4d)(1 + 7c^2)$

41. $(a + 2) - x(a + 2)$
    $(a + 2)(1 - x)$

42. $y(5x - 1) - (5x - 1)$
    $(5x - 1)(y - 1)$

Cumulative Review

43. **[0.7.1]** *Videotape Speed* A videotape plays 6.72 feet per minute for 2 hours at standard speed. It plays for 6 hours at extended long play speed. What is the rate of the tape at the slower speed?
2.24 ft/min

Coffee Production In 2006, the top six green (unroasted beans) coffee-producing countries of the world produced 5,350,000 metric tons of coffee beans. The percent of this 5.35 million metric tons produced by each of these countries is shown in the following graph. Use the graph to answer the following questions. (Source: en.wikipedia.org)

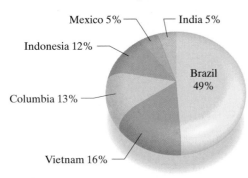

Percent of Coffee Produced by the Top Six Coffee-Producing Countries of the World in 2006

Mexico 5% — India 5%
Indonesia 12%
Brazil 49%
Columbia 13%
Vietnam 16%

44. **[0.5.4]** How many metric tons of coffee were produced in Vietnam?
856,000 metric tons

45. **[0.5.4]** How many metric tons of coffee were produced in Brazil?
2,621,500 metric tons

46. **[0.7.1]** The population of Vietnam in 2006 was approximately 85,000,000 people. How many pounds of coffee were produced in Vietnam for each person? (Round your answer to the nearest whole number. One metric ton is about 2205 pounds.)
About 22 lb/person

47. **[0.7.1]** The population of Brazil in 2006 was approximately 190,000,000 people. How many pounds of coffee were produced in Brazil for each person? (Round your answer to the nearest whole number. One metric ton is about 2205 pounds.)
About 30 lb/person

Quick Quiz 5.1 Remove the largest possible common factor.

1. $3x - 4x^2 + 2xy$
$x(3 - 4x + 2y)$

2. $20x^3 - 25x^2 - 5x$
$5x(4x^2 - 5x - 1)$

3. $8a(a + 3b) - 7b(a + 3b)$
$(a + 3b)(8a - 7b)$

4. Concept Check Explain how you would remove the greatest common factor from the following polynomial.

$$36a^3b^2 - 72a^2b^3$$

Answers may vary

Classroom Quiz 5.1 You may use these problems to quiz your students' mastery of Section 5.1.

Remove the largest possible common factor.

1. $8x - 16x^2 - 12xy$
Ans: $4x(2 - 4x - 3y)$

2. $39x^3 - 6x^2 - 3x$
Ans: $3x(13x^2 - 2x - 1)$

3. $7a(2a + 5b) - 9b(2a + 5b)$
Ans: $(2a + 5b)(7a - 9b)$

1 Factoring Expressions with Four Terms by Grouping

A common factor of a polynomial can be a number, a variable, or an algebraic expression. Sometimes the polynomial is written so that it is easy to recognize the common factor. This is especially true when the common factor is enclosed by parentheses.

EXAMPLE 1 Factor. $x(x - 3) + 2(x - 3)$

Solution Observe each term:

$$\underbrace{x(x - 3)}_{\substack{\text{first} \\ \text{term}}} + \underbrace{2(x - 3)}_{\substack{\text{second} \\ \text{term}}}$$

The common factor of the first and second terms is the quantity $(x - 3)$, so we have

$$x(x - 3) + 2(x - 3) = (x - 3)(x + 2).$$

Practice Problem 1 Factor. $3y(2x - 7) - 8(2x - 7)$ $(2x - 7)(3y - 8)$

Now let us face a new challenge. Think carefully. Try to follow this new idea. Suppose the polynomial in Example 1, $x(x - 3) + 2(x - 3)$, were written in the form $x^2 - 3x + 2x - 6$. (Note that this form is obtained by multiplying the factors of the first and second terms.) How would we factor a four-term polynomial like this?

In such cases we remove a common factor from the first two terms and a different common factor from the second two terms. That is, we would factor x from $x^2 - 3x$ and 2 from $2x - 6$.

$$x^2 - 3x + 2x - 6 = x(x - 3) + 2(x - 3)$$

Because the resulting terms have a common factor (the binomial enclosed by the parentheses), we would then proceed as we did in Example 1. This procedure for factoring is often called **factoring by grouping.**

EXAMPLE 2 Factor. $2x^2 + 3x + 6x + 9$

Solution

$$\underbrace{2x^2 + 3x}_{\substack{\text{Factor out a common} \\ \text{factor of } x \text{ from} \\ \text{the first two terms.}}} + \underbrace{6x + 9}_{\substack{\text{Factor out a common} \\ \text{factor of 3 from} \\ \text{the second two terms.}}}$$

$$\underbrace{x(2x + 3)} \qquad \underbrace{3(2x + 3)}$$

Note that the sets of parentheses in the two terms contain the same expression at this step.

The expression in parentheses is now a common factor of the terms. Now we finish the factoring.

$$2x^2 + 3x + 6x + 9 = x(2x + 3) + 3(2x + 3)$$
$$= (2x + 3)(x + 3)$$

Practice Problem 2 Factor. $6x^2 - 15x + 4x - 10$ $(2x - 5)(3x + 2)$

Student Learning Objective

After studying this section, you will be able to:

1 **Factor expressions with four terms by grouping.**

Teaching Example 1 Factor.
$3y(2y^2 + 7) - (2y^2 + 7)$

Ans: $(2y^2 + 7)(3y - 1)$

NOTE TO STUDENT: Fully worked-out solutions to all of the Practice Problems can be found at the back of the text starting at page SP-1

Teaching Example 2 Factor.
$3a^2 - 9a + 2a - 6$

Ans: $(a - 3)(3a + 2)$

EXAMPLE 3 Factor. $4x + 8y + ax + 2ay$

Solution   Factor out a common factor of 4 from the first two terms.

$$\overbrace{4x + 8y} + \underbrace{ax + 2ay} = \overbrace{4(x + 2y)} + \underbrace{a(x + 2y)}$$

Factor out a common factor of a from the second two terms.

$4(x + 2y) + a(x + 2y) = (x + 2y)(4 + a)$ The common factor of the terms is the expression in parentheses, $x + 2y$.

Practice Problem 3 Factor by grouping. $ax + 2a + 4bx + 8b$

$(x + 2)(a + 4b)$

NOTE TO STUDENT: *Fully worked-out solutions to all of the Practice Problems can be found at the back of the text starting at page SP-1*

In some problems the terms are out of order. In this case, we have to rearrange the order of the terms first so that the first two terms have a common factor.

EXAMPLE 4 Factor. $bx + 4y + 4b + xy$

Solution
$bx + 4y + 4b + xy = bx + 4b + xy + 4y$ Rearrange the terms so that the first terms have a common term.
$= b(x + 4) + y(x + 4)$ Factor out the common factor of b from the first two terms and the common factor of y from the second two terms.
$= (x + 4)(b + y)$

Practice Problem 4 Factor. $6a^2 + 5bc + 10ab + 3ac$ $(3a + 5b)(2a + c)$

Sometimes you will need to factor out a negative common factor from the second two terms to obtain two terms that contain the same parenthetical expression.

EXAMPLE 5 Factor. $2x^2 + 5x - 4x - 10$

Solution
$2x^2 + 5x - 4x - 10 = x(2x + 5) - 2(2x + 5)$ Factor out the common factor of x from the first two terms and the common factor of -2 from the second two terms.
$= (2x + 5)(x - 2)$

Notice that if you factored out a common factor of $+2$ in the first step, the two resulting terms would not contain the same parenthetical expression. If the expressions inside the two sets of parentheses are not exactly the same, you cannot express the polynomial as a product of two factors!

Practice Problem 5 Factor. $6xy + 14x - 15y - 35$ $(3y + 7)(2x - 5)$

EXAMPLE 6   Factor. $2ax - a - 2bx + b$

Solution

$2ax - a - 2bx + b = a(2x - 1) - b(2x - 1)$ Factor out the common factor of a from the first two terms. Factor out the common factor of $-b$ from the second two terms.

$\qquad\qquad\qquad = (2x - 1)(a - b)$ Since the two resulting terms contain the same parenthetical expression, we can complete the factoring.

Practice Problem 6   Factor. $3x + 6y - 5ax - 10ay$ $(x + 2y)(3 - 5a)$

Teaching Example 6 Factor.
$21mx - 3x - 7my + y$

Ans: $(7m - 1)(3x - y)$

CAUTION: Many students find that they make a factoring error in the first step of problems like Example 6. Watch that step very carefully. If you take your answer and multiply it by FOIL, you should obtain the original problem.

EXAMPLE 7   Factor and check your answer. $8ad + 21bc - 6bd - 28ac$

Solution   We observe that the first two terms do not have a common factor.

$8ad + 21bc - 6bd - 28ac = 8ad - 6bd - 28ac + 21bc$ Rearrange the order using the commutative property of addition.

$\qquad\qquad\qquad = 2d(4a - 3b) - 7c(4a - 3b)$ Factor out the common factor of $2d$ from the first two terms and the common factor of $-7c$ from the last two terms.

$\qquad\qquad\qquad = (4a - 3b)(2d - 7c)$ Factor out the common factor of $(4a - 3b)$.

To check, we multiply the two binomials using the FOIL procedure.

$(4a - 3b)(2d - 7c) = 8ad - 28ac - 6bd + 21bc$
$\qquad\qquad\qquad = 8ad + 21bc - 6bd - 28ac$ ✓   Rearrange the order of the terms. This is the original problem. Thus it checks.

Practice Problem 7   Factor and check your answer.

$\qquad 10ad + 27bc - 6bd - 45ac$ $(5a - 3b)(2d - 9c)$

Teaching Example 7 Factor.
$12mx + 5ny - 6my - 10nx$

Ans: $(2x - y)(6m - 5n)$

Teaching Tip When discussing the solution of Example 7, ask students what order they put their terms in before factoring. Usually you will get two or three different suggestions. By putting them all on the blackboard and factoring each way, you will be able to convince students that any order is OK as long as the first two terms have a common factor and as long as the binomial factor contained in each set of parentheses is the same.

Verbal and Writing Skills

1. To factor $3x^2 - 6xy + 5x - 10y$, we must first remove a common factor of $3x$ from the first two terms. What do we do with the last two terms? What should we get for the answer? We must remove a common factor of 5 from the last two terms. This will give us $3x(x - 2y) + 5(x - 2y)$. Then our final answer is $(x - 2y)(3x + 5)$.

2. To factor $5x^2 + 15xy - 2x - 6y$, we must first remove a common factor of 5 from the first two terms. What do we do with the last two terms? What should we get for the answer? We must remove a common factor of -2 from the last two terms. This will give us $5x(x + 3y) - 2(x + 3y)$. Then our final answer is $(x + 3y)(5x - 2)$.

Factor by grouping. Check your answers for exercises 3–26.

3. $ab - 3a + 4b - 12$ $(b - 3)(a + 4)$

4. $xy - x + 4y - 4$ $(y - 1)(x + 4)$

5. $x^3 - 4x^2 + 3x - 12$ $(x - 4)(x^2 + 3)$

6. $x^3 - 6x^2 + 2x - 12$ $(x - 6)(x^2 + 2)$

7. $2ax + 6bx - ay - 3by$ $(a + 3b)(2x - y)$

8. $4x + 8y - 3wx - 6wy$ $(x + 2y)(4 - 3w)$

9. $3ax + bx - 6a - 2b$ $(3a + b)(x - 2)$

10. $ad + 3a - d^2 - 3d$ $(d + 3)(a - d)$

11. $5a + 12bc + 10b + 6ac$ $(a + 2b)(5 + 6c)$

12. $4u^2 + v + 4uv + u$ $(u + v)(4u + 1)$

13. $6c - 12d + cx - 2dx$ $(c - 2d)(6 + x)$

14. $xy + 5x - 5y - 25$ $(y + 5)(x - 5)$

15. $y^2 - 2y - 3y + 6$ $(y - 2)(y - 3)$

16. $x^2 - 4x + 3x - 12$ $(x - 4)(x + 3)$

17. $54 - 6y + 9y - y^2$ $(9 - y)(6 + y)$

18. $ax - 3a - 2bx + 6b$ $(x - 3)(a - 2b)$

19. $6ax - y + 2ay - 3x$ $(3x + y)(2a - 1)$

20. $6tx + r - 3t - 2rx$ $(2x - 1)(3t - r)$

21. $2x^2 + 8x - 3x - 12$ $(x + 4)(2x - 3)$

22. $3y^2 - y + 9y - 3$ $(3y - 1)(y + 3)$

23. $t^3 - t^2 + t - 1$ $(t - 1)(t^2 + 1)$

24. $x^2 - 2x - xy + 2y$ $(x - 2)(x - y)$

25. $28x^2 + 8xy^2 + 21xw + 6y^2w$ $(7x + 2y^2)(4x + 3w)$

26. $15wx + 18x^2 + 12xy^2 + 10wy^2$ $(5w + 6x)(3x + 2y^2)$

To Think About

27. Although $6a^2 - 12bd - 8ad + 9ab = 6(a^2 - 2bd) - a(8d - 9b)$ is true, it is not the correct solution to the problem "Factor $6a^2 - 12bd - 8ad + 9ab$." Explain. Can this expression be factored?   We must rearrange the terms in a different order so that the expression in parentheses is the same in each case. We use the order $6a^2 - 8ad + 9ab - 12bd$ to factor $2a(3a - 4d) + 3b(3a - 4d) = (3a - 4d)(2a + 3b)$.

28. Tim was trying to factor $5x^2 - 3xy - 10x + 6y$. In his first step he wrote down $x(5x - 3y) + 2(-5x + 3y)$. Was he doing the problem correctly? What is the answer?   No. Each set of parentheses must contain exactly the same quantity. He should have factored out a -2 from the last two terms. So in his first step he should have written $x(5x - 3y) - 2(5x - 3y)$. The answer is $(5x - 3y)(x - 2)$.

Cumulative Review

29. [0.5.5] *Online Sales* In 2010, it is estimated that online retail sales in the United States will total \$328.6 billion. About \$119.1 billion of this will be from sales related to travel. What percent of the total will be travel sales? Round your answer to the nearest tenth of a percent. (*Source:* www.census.gov.)   36.2%

30. [3.3.1] *Professor Salaries* In 2006, the average salary for a professor at a public 2-year or 4-year institution was \$11,200 more than twice the average salary for an instructor. The two salaries total \$131,500. Find the average salaries of a professor and an instructor. (*Source:* www.census.gov.)   Instructor: \$40,100; professor: \$91,400

31. [0.7.1] *Wheat Production* During the years 2001–2002, 581 million metric tons of wheat were produced worldwide. By 2003–2004, total wheat production had decreased by 4.6%. During the years 2005–2006, worldwide wheat production had increased by 12.3% from the 2003–2004 amount. How many metric tons of wheat were produced in 2005–2006? (*Source:* www.census.gov.)   About 622,000,000 metric tons

Quick Quiz 5.2   Factor by grouping.

1. $7ax + 12a - 14x - 24$ $(7x + 12)(a - 2)$

2. $2xy^2 - 15 + 6x - 5y^2$ $(y^2 + 3)(2x - 5)$

3. $10xy - 3x + 40by - 12b$ $(10y - 3)(x + 4b)$

4. Concept Check Explain how you would factor the following polynomial.
$$10ax + b^2 + 2bx + 5ab$$
Answers may vary

Classroom Quiz 5.2   You may use these problems to quiz your students' mastery of Section 5.2.

Factor by grouping.

1. $8ax + 11a - 24x - 33$
  Ans: $(8x + 11)(a - 3)$

2. $3xy^2 - 14 + 6x - 7y^2$
  Ans: $(y^2 + 2)(3x - 7)$

3. $20b - 45by + 9xy - 4x$
  Ans: $(x - 5b)(9y - 4)$

Student Learning Objectives

After studying this section, you will be able to:

 Factor polynomials of the form $x^2 + bx + c$.

 Factor polynomials that have a common factor and a factor of the form $x^2 + bx + c$.

 Factoring Polynomials of the Form $x^2 + bx + c$

Suppose that you wanted to factor $x^2 + 5x + 6$. After some trial and error you *might* obtain $(x + 2)(x + 3)$, or you might get discouraged and not get an answer. If you did get these factors, you could check this answer by the FOIL method.

$$(x + 2)(x + 3) = x^2 + 3x + 2x + 6$$
$$= x^2 + 5x + 6$$

But trial and error can be a long process. There is another way. Let's look at the preceding equation again.

$$\begin{array}{cccc} \text{F} & \text{O} & \text{I} & \text{L} \\ (x + 2)(x + 3) = x^2 + & 3x + 2x & + 6 \\ = x^2 & + 5x & + 6 \end{array}$$

The first thing to notice is that the product of the first terms in the factors gives the first term of the polynomial. That is, $x \cdot x = x^2$.

The first term is the product of these terms.

$$x^2 + 5x + 6 \quad = \quad (x + 2)(x + 3)$$

The next thing to notice is that the sum of the products of the outer and inner terms in the factors produces the middle term of the polynomial. That is, $(x \cdot 3) + (2 \cdot x) = 3x + 2x = 5x$. Thus we see that the sum of the second terms in the factors, $2 + 3$, gives the coefficient of the middle term, 5.

Finally, note that the product of the last terms of the factors gives the last term of the polynomial. That is, $2 \cdot 3 = 6$.

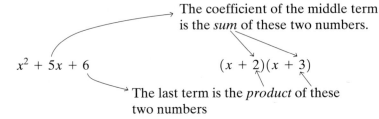

The coefficient of the middle term is the *sum* of these two numbers.

$$x^2 + 5x + 6 \qquad (x + 2)(x + 3)$$

The last term is the *product* of these two numbers

Let's summarize our observations in general terms and then try a few examples.

> **FACTORING TRINOMIALS OF THE FORM $x^2 + bx + c$**
>
> 1. The answer will be of the form $(x + m)(x + n)$.
>
> 2. m and n are numbers such that:
>   (a) When you multiply them, you get the last term, which is c.
>   (b) When you add them, you get the coefficient of the middle term, which is b.

Teaching Tip Students should be told that factoring trinomials of the form $x^2 + bx + c$ is one of the most common forms of factoring.

Teaching Example 1 Factor. $x^2 + 11x + 18$

Ans: $(x + 9)(x + 2)$

NOTE TO STUDENT: *Fully worked-out solutions to all of the Practice Problems can be found at the back of the text starting at page SP-1*

EXAMPLE 1 Factor. $x^2 + 7x + 12$

Solution The answer is of the form $(x + m)(x + n)$. We want to find the two numbers, m and n, that you can multiply to get 12 and add to get 7. The numbers are 3 and 4.

$$x^2 + 7x + 12 = (x + 3)(x + 4)$$

Practice Problem 1 Factor. $x^2 + 8x + 12$ $(x + 6)(x + 2)$

EXAMPLE 2   Factor. $x^2 + 12x + 20$

Solution   We want two numbers that have a product of 20 and a sum of 12. The numbers are 10 and 2.

$$x^2 + 12x + 20 = (x + \underline{10})(x + \underline{2})$$

Note: If you cannot think of the numbers in your head, write down the possible factors whose product is 20.

Product	Sum
$1 \cdot 20 = 20$	$1 + 20 = 21$
$2 \cdot 10 = 20$	$2 + 10 = 12$ ←
$4 \cdot 5 = 20$	$4 + 5 = 9$

Then select the pair whose sum is 12.   Select this pair.

Practice Problem 2   Factor. $x^2 + 17x + 30$ $(x + 2)(x + 15)$
You may find that it is helpful to list all the factors whose product is 30 first.

So far we have factored only trinomials of the form $x^2 + bx + c$, where b and c are positive numbers. The same procedure applies if b is a negative number and c is positive. Because m and n have a positive product and a negative sum, they must both be negative.

EXAMPLE 3   Factor. $x^2 - 8x + 15$

Solution   We want two numbers that have a product of $+15$ and a sum of -8. They must be negative numbers since the sign of the middle term is negative and the sign of the last term is positive.

the sum $-5 + (-3)$

$$x^2 - 8x + 15 = (x - 5)(x - 3)$$

Think: $(-5)(-3) = +15$
and $-5 + (-3) = -8.$

the product $(-5)(-3)$

Multiply using FOIL to check.

Practice Problem 3   Factor. $x^2 - 11x + 18$ $(x - 9)(x - 2)$

EXAMPLE 4   Factor. $x^2 - 9x + 14$

Solution   We want two numbers whose product is 14 and whose sum is -9. The numbers are -7 and -2. So

$$x^2 - 9x + 14 = (x - 7)(x - 2) \text{ or } (x - 2)(x - 7).$$

Practice Problem 4   Factor. $x^2 - 11x + 24$ $(x - 8)(x - 3)$

All the examples so far have had a positive last term. What happens when the last term is negative? If the last term is a negative number, one of the numbers m or n must be a positive number and the other must be a negative number. Why? The product of a positive number and a negative number is negative.

EXAMPLE 5 Factor. $x^2 - 3x - 10$

Solution We want two numbers whose product is -10 and whose sum is -3. The two numbers are -5 and $+2$.

$$x^2 - 3x - 10 = (x - 5)(x + 2)$$

Practice Problem 5 Factor. $x^2 - 5x - 24$ $(x - 8)(x + 3)$

What if we made a sign error and *incorrectly* factored the trinomial $x^2 - 3x - 10$ as $(x + 5)(x - 2)$? We could detect the error immediately since the sum of $+5$ and -2 is 3. We need a sum of -3!

EXAMPLE 6 Factor $y^2 + 10y - 24$. Check your answer.

Solution The two numbers whose product is -24 and whose sum is $+10$ are $+12$ and -2.

$$y^2 + 10y - 24 = (y + 12)(y - 2)$$

CAUTION: It is very easy to make a sign error in these problems. Make sure that you mentally multiply your answer by FOIL to obtain the original expression. Check each sign carefully.

Check. $(y + 12)(y - 2) = y^2 - 2y + 12y - 24 = y^2 + 10y - 24$ ✓

Practice Problem 6 Factor $y^2 + 17y - 60$. $(y + 20)(y - 3)$
Multiply your answer to check.

EXAMPLE 7 Factor. $x^2 - 16x - 36$

Solution We want two numbers whose product is -36 and whose sum is -16.
List all the possible factors of 36 (without regard to sign). Find the pair that has a difference of 16. We are looking for a difference because the signs of the factors are different.

Factors of 36	The Difference Between the Factors
36 and 1	35
18 and 2	16 ← This is the value we want.
12 and 3	9
9 and 4	5
6 and 6	0

Once we have picked the pair of numbers (18 and 2), it is not difficult to find the signs. For the coefficient of the middle term to be -16, we will have to add the numbers -18 and $+2$.

$$x^2 - 16x - 36 = (x - 18)(x + 2)$$

Practice Problem 7 Factor. $x^2 - 7x - 60$ $(x - 12)(x + 5)$
You may find it helpful to list the pairs of numbers whose product is 60.

At this point you should work several problems to develop your factoring skill. This is one section where you really need to drill by doing many problems.

Feel a little confused about the signs? If you do, you may find these facts helpful.

> **FACTS ABOUT FACTORING TRINOMIALS OF THE FORM $x^2 + bx + c$**
>
> The *two numbers m and n* will have the *same sign* if the last term of the polynomial is *positive*.
>
> $$x^2 + bx + c = (x \quad m)(x \quad n)$$
>
> 1. They will both be *positive* if the *coefficient* of the *middle* term is *positive*.
>
> $$x^2 + 5x + 6 = (x + 2)(x + 3)$$
>
> 2. They will both be *negative* if the *coefficient* of the *middle* term is *negative*.
>
> $$x^2 - 5x + 6 = (x - 2)(x - 3)$$
>
> The two numbers *m* and *n* will have *opposite signs* if the last term is *negative*.
>
> 1. The *larger* of the absolute values of the two numbers will be given a plus sign if the coefficient of the *middle term* is *positive*.
>
> $$x^2 + 6x - 7 = (x + 7)(x - 1)$$
>
> 2. The larger of the absolute values of the two numbers will be given a negative sign if the coefficient of the *middle term* is *negative*.
>
> $$x^2 - 6x - 7 = (x - 7)(x + 1)$$

Do not memorize these facts; rather, try to understand the pattern.

Sometimes the exponent of the first term of the polynomial will be greater than 2. If the exponent is an even power, it is a square. For example, $x^4 = (x^2)(x^2)$. Likewise, $x^6 = (x^3)(x^3)$.

EXAMPLE 8   Factor. $y^4 - 2y^2 - 35$

Solution   Think: $y^4 = (y^2)(y^2)$ This will be the first term of each set of parentheses.

$$(y^2 \quad)(y^2 \quad)$$

$(y^2 + \ )(y^2 - \ )$ The last term of the polynomial is negative. Thus the signs of *m* and *n* will be different.

$(y^2 + 5)(y^2 - 7)$ Now think of factors of 35 whose difference is 2.

Practice Problem 8   Factor. $a^4 + a^2 - 42$ $(a^2 + 7)(a^2 - 6)$

Teaching Example 8 Factor. $y^4 - y^2 - 6$

Ans: $(y^2 - 3)(y^2 + 2)$

Factoring Polynomials That Have a Common Factor and a Factor of the Form $x^2 + bx + c$

Some factoring problems require two steps. Often we must first factor out a common factor from each term of the polynomial. Once this is done, we may find that the other factor is a trinomial that can be factored using the methods previously discussed in this section.

EXAMPLE 9   Factor. $2x^2 + 36x + 160$

Solution

$2x^2 + 36x + 160 = 2(x^2 + 18x + 80)$ First factor out the common factor of 2 from each term of the polynomial.

$$= 2(x + 8)(x + 10)$$ Then factor the remaining polynomial.

The final answer is $2(x + 8)(x + 10)$. *Be sure to list all parts of the answer.*

Check. $2(x + 8)(x + 10) = 2(x^2 + 18x + 80) = 2x^2 + 36x + 160$ ✓

Thus we are sure that the answer is $2(x + 8)(x + 10)$.

Practice Problem 9   Factor. $3x^2 + 45x + 150$ $3(x + 5)(x + 10)$

Teaching Example 9 Factor. $3x^2 + 18x + 24$

Ans: $3(x + 4)(x + 2)$

EXAMPLE 10 Factor. $3x^2 + 9x - 162$

Solution

$$3x^2 + 9x - 162 = 3(x^2 + 3x - 54)$$ First factor out the common factor of 3 from each term of the polynomial.

$$= 3(x - 6)(x + 9)$$ Then factor the remaining polynomial.

The final answer is $3(x - 6)(x + 9)$. *Be sure you include the 3.*

Check. $3(x - 6)(x + 9) = 3(x^2 + 3x - 54) = 3x^2 + 9x - 162$ ✓

Thus we are sure that the answer is $3(x - 6)(x + 9)$.

Practice Problem 10 Factor. $4x^2 - 8x - 140$ $4(x + 5)(x - 7)$

CAUTION: Don't forget the common factor!

It is quite easy to forget to look for a greatest common factor as the first step of factoring a trinomial. Therefore, it is a good idea to examine your final answer in any factoring problem and ask yourself, "Can I factor out a common factor from any binomial contained inside a set of parentheses?" Often you will be able to see a common factor at that point if you missed it in the first step of the problem.

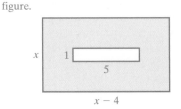
▲ **EXAMPLE 11** Find a polynomial in factored form for the shaded area in the figure.

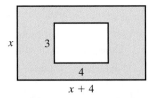

Solution To obtain the shaded area, we find the area of the larger rectangle and subtract from it the area of the smaller rectangle. Thus we have the following:

$$\text{shaded area} = x(x + 4) - (4)(3)$$
$$= x^2 + 4x - 12$$

Now we factor this polynomial to obtain the shaded area $= (x + 6)(x - 2)$.

▲ **Practice Problem 11** Find a polynomial in factored form for the shaded area in the figure. $(x + 5)(x - 4)$

Verbal and Writing Skills

Fill in the blanks.

1. To factor $x^2 + 5x + 6$, find two numbers whose ___product___ is 6 and whose ___sum___ is 5.

2. To factor $x^2 + 5x - 6$, find two numbers whose ___product___ is -6 and whose ___sum___ is 5.

Factor.

3. $x^2 + 2x + 1$
$(x + 1)(x + 1)$

4. $x^2 + 11x + 30$
$(x + 5)(x + 6)$

5. $x^2 + 12x + 35$
$(x + 5)(x + 7)$

6. $x^2 + 10x + 21$
$(x + 3)(x + 7)$

7. $x^2 - 4x + 3$
$(x - 3)(x - 1)$

8. $x^2 - 6x + 8$
$(x - 2)(x - 4)$

9. $x^2 - 11x + 28$
$(x - 7)(x - 4)$

10. $x^2 - 13x + 12$
$(x - 1)(x - 12)$

11. $x^2 + 5x - 24$
$(x + 8)(x - 3)$

12. $x^2 + 5x - 36$
$(x + 9)(x - 4)$

13. $x^2 - 13x - 14$
$(x - 14)(x + 1)$

14. $x^2 - 6x - 16$
$(x - 8)(x + 2)$

15. $x^2 + 2x - 35$
$(x + 7)(x - 5)$

16. $x^2 + 4x - 12$
$(x - 2)(x + 6)$

17. $x^2 - 2x - 24$
$(x - 6)(x + 4)$

18. $x^2 - 11x - 26$
$(x - 13)(x + 2)$

19. $x^2 + 15x + 36$
$(x + 12)(x + 3)$

20. $x^2 + 15x + 44$
$(x + 11)(x + 4)$

21. $x^2 - 10x + 24$
$(x - 6)(x - 4)$

22. $x^2 - 13x + 42$
$(x - 7)(x - 6)$

23. $x^2 + 13x + 30$
$(x + 3)(x + 10)$

24. $x^2 + 9x + 20$
$(x + 5)(x + 4)$

25. $x^2 - 6x + 5$
$(x - 5)(x - 1)$

26. $x^2 - 15x + 54$
$(x - 9)(x - 6)$

Mixed Practice *Look over your answers to exercises 3–26 carefully. Be sure that you are clear on your sign rules. Exercises 27–42 contain a mixture of all the types of problems in this section. Make sure you can do them all. Check your answers by multiplication.*

27. $a^2 + 6a - 16$
$(a + 8)(a - 2)$

28. $a^2 - 13a + 30$
$(a - 3)(a - 10)$

29. $x^2 - 12x + 32$
$(x - 4)(x - 8)$

30. $x^2 - 6x - 27$
$(x + 3)(x - 9)$

31. $x^2 + 4x - 21$
$(x + 7)(x - 3)$

32. $x^2 - 9x + 18$
$(x - 6)(x - 3)$

33. $x^2 + 15x + 56$
$(x + 7)(x + 8)$

34. $x^2 + 20x + 99$
$(x + 9)(x + 11)$

35. $y^2 + 4y - 45$
$(y + 9)(y - 5)$

36. $x^2 + 12x - 45$
$(x + 15)(x - 3)$

37. $x^2 + 9x - 36$
$(x + 12)(x - 3)$

38. $x^2 - 13x + 36$
$(x - 4)(x - 9)$

39. $x^2 - 2xy - 15y^2$
$(x + 3y)(x - 5y)$

40. $x^2 - 3xy - 28y^2$
$(x + 4y)(x - 7y)$

41. $x^2 - 16xy + 63y^2$
$(x - 7y)(x - 9y)$

42. $x^2 + 19xy + 48y^2$
$(x + 3y)(x + 16y)$

In exercises 43–54, first factor out the greatest common factor from each term. Then factor the remaining polynomial. Refer to Examples 9 and 10.

43. $4x^2 + 24x + 20$
$4(x + 5)(x + 1)$

44. $4x^2 + 28x + 40$
$4(x + 2)(x + 5)$

45. $6x^2 + 18x + 12$
$6(x + 1)(x + 2)$

46. $6x^2 + 24x + 18$
$6(x + 1)(x + 3)$

47. $5x^2 - 30x + 25$
$5(x - 1)(x - 5)$

48. $2x^2 - 20x + 32$
$2(x - 8)(x - 2)$

49. $3x^2 - 6x - 72$
$3(x + 4)(x - 6)$

50. $3x^2 - 12x - 63$
$3(x + 3)(x - 7)$

51. $7x^2 + 21x - 70$
$7(x + 5)(x - 2)$

52. $7x^2 + 7x - 84$
$7(x + 4)(x - 3)$

53. $3x^2 - 18x + 15$
$3(x - 1)(x - 5)$

54. $5x^2 - 35x + 60$
$5(x - 3)(x - 4)$

▲ **55.** *Geometry* Find a polynomial in factored form for the shaded area. Both figures are rectangles with dimensions as labeled. $(12 + x)(10 - x)$

▲ **56.** *Geometry* How much larger is the perimeter of the large rectangle than the perimeter of the small rectangle? $44 - (4x + 4) = 40 - 4x$

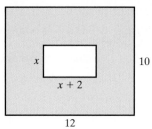

Cumulative Review

57. **[2.5.1]** Solve for P. $A = P + Prt$
(*Hint:* Factor out P on the right side of the equation as your first step.)
$$P = \frac{A}{1 + rt}$$

58. **[2.7.1]** Solve $2 - 3x \le 7$ for x. Graph the solution on the number line.
$$x \ge -\frac{5}{3}$$

59. **[3.2.3]** *Travel Speed* A new car that maintains a constant speed travels from Watch Hill, Rhode Island, to Greenwich, Connecticut, in 2 hours. A train, traveling 20 mph faster, makes the trip in $1\frac{1}{2}$ hours. How far is it from Watch Hill to Greenwich? (*Hint:* Let $c =$ the car's speed. First find the speed of the car and the speed of the train. Then be sure to answer the question.)
120 mi

60. **[3.6.2]** *Salary* Carla works in an electronics store. She is paid $600 per month plus 4% commission on her sales. She wants to earn $3800 or more this month. At least how much must she generate in sales? $80,000

Average Temperature The equation $T = 19 + 2M$ has been used by some meteorologists to predict the monthly average temperature for the small island of Menorca off the coast of Spain during the first 6 months of the year. The variable T represents the average monthly temperature measured in degrees Celsius. The variable M represents the number of months since January.

61. **[4.3.4]** What is the average temperature of Menorca during the month of April? 25°C

62. **[4.3.4]** During what month will the average temperature be 29°C? June

Using the Trial-and-Error Method

When the coefficient of the x^2-term in a trinomial of the form $ax^2 + bx + c$ is not 1, the trinomial is more difficult to factor. Several possibilities must be considered.

EXAMPLE 1 Factor. $2x^2 + 5x + 3$

Solution In order for the coefficient of the x^2-term of the polynomial to be 2, the coefficients of the x-terms in the factors must be 2 and 1. Thus $2x^2 + 5x + 3 = (2x\quad)(x\quad)$.

In order for the last term of the polynomial to be 3, the constants in the factors must be 3 and 1.

Since all signs in the polynomial are positive, we know that each factor in parentheses will contain only positive signs. However, we still have two possibilities. They are as follows:

$$(2x + 1)(x + 3)$$
$$(2x + 3)(x + 1)$$

We check them by multiplying by the FOIL method.

$$(2x + 1)(x + 3) = 2x^2 + 7x + 3 \quad \text{Wrong middle term}$$
$$(2x + 3)(x + 1) = 2x^2 + 5x + 3 \quad \text{Correct middle term}$$

Thus the correct answer is

$$(2x + 3)(x + 1) \quad \text{or} \quad (x + 1)(2x + 3).$$

Practice Problem 1 Factor. $2x^2 + 7x + 5$ $(2x + 5)(x + 1)$

Some problems have many more possibilities.

EXAMPLE 2 Factor. $4x^2 - 13x + 3$

Solution

The Different Factorizations of 4 Are:	The Factorization of 3 Is:
(2)(2)	(1)(3)
(1)(4)	

Let us list the possible factoring combinations and compute the middle term by the FOIL method. Note that the signs of the constants in both factors will be negative. Why?

Possible Factors	Middle Term	Correct?
$(2x - 3)(2x - 1)$	$-8x$	No
$(4x - 3)(x - 1)$	$-7x$	No
$(4x - 1)(x - 3)$	$-13x$	Yes

The correct answer is $(4x - 1)(x - 3)$ or $(x - 3)(4x - 1)$.
This method is called the **trial-and-error method.**

Practice Problem 2 Factor. $9x^2 - 64x + 7$ $(9x - 1)(x - 7)$

Teaching Example 1 Factor. $3x^2 + 8x + 5$

Ans: $(3x + 5)(x + 1)$

Teaching Tip Except for those students who remember quite well how to factor trinomials of the general form $ax^2 + bx + c$, most students greatly prefer to use the grouping method, which begins with Example 4. A few students will still prefer the trial-and-error method.

Teaching Example 2 Factor. $5x^2 - 13x + 6$

Ans: $(5x - 3)(x - 2)$

NOTE TO STUDENT: Fully worked-out solutions to all of the Practice Problems can be found at the back of the text starting at page SP-1

319

Teaching Example 3 Factor. $7y^2 + 11y - 6$

Ans: $(7y - 3)(y + 2)$

Teaching Tip Remind students that when factoring problems of the type $ax^2 + bx + c$, they should ALWAYS CHECK by multiplying the factors to see if the original trinomial is obtained.

EXAMPLE 3 Factor. $3x^2 - 2x - 8$

Solution

Factorization of 3	Factorizations of 8
$(3)(1)$	$(8)(1)$
	$(4)(2)$

Let us list only one-half of the possibilities. We'll let the constant in the first factor of each product be positive.

Possible Factors	Middle Term	Correct Factors?
$(x + 8)(3x - 1)$	$+23x$	No
$(x + 1)(3x - 8)$	$-5x$	No
$(x + 4)(3x - 2)$	$+10x$	No
$(x + 2)(3x - 4)$	$+2x$	No (but only because the sign is wrong)

So we just *reverse* the signs of the constants in the factors.

	Middle Term	Correct Factors?
$(x - 2)(3x + 4)$	$-2x$	Yes

The correct answer is

$$(x - 2)(3x + 4) \quad \text{or} \quad (3x + 4)(x - 2).$$

Practice Problem 3 Factor. $3x^2 - x - 14$ $(3x - 7)(x + 2)$

NOTE TO STUDENT: *Fully worked-out solutions to all of the Practice Problems can be found at the back of the text starting at page SP-1*

It takes a good deal of practice to readily factor problems of this type. The more problems you do, the more proficient you will become. The following method will help you factor more quickly.

 Using the Grouping Method

One way to factor a trinomial of the form $ax^2 + bx + c$ is to write it with four terms and factor by grouping, as we did in Section 5.2. For example, the trinomial $2x^2 + 13x + 20$ can be written as $2x^2 + 5x + 8x + 20$. Using the methods of Section 5.2, we factor it as follows.

$$2x^2 + 5x + 8x + 20 = x(2x + 5) + 4(2x + 5)$$
$$= (2x + 5)(x + 4)$$

We can factor all factorable trinomials of the form $ax^2 + bx + c$ in this way. We will use the following procedure.

> **GROUPING NUMBER METHOD FOR FACTORING TRINOMIALS OF THE FORM $ax^2 + bx + c$**
>
> 1. Obtain the grouping number ac.
> 2. Find the two numbers whose product is the grouping number and whose sum is b.
> 3. Use those numbers to write bx as the sum of two terms.
> 4. Factor by grouping.
> 5. Multiply to check.

Let's try the problem from Example 1.

EXAMPLE 4 Factor by grouping. $2x^2 + 5x + 3$

Solution

1. The grouping number is $(2)(3) = 6$.
2. The factors of 6 are $6 \cdot 1$ and $3 \cdot 2$. We choose the numbers 3 and 2 because their product is 6 and their sum is 5.
3. We write $5x$ as the sum $2x + 3x$.
4. Factor by grouping.

$$2x^2 + 5x + 3 = 2x^2 + 2x + 3x + 3$$
$$= 2x(x + 1) + 3(x + 1)$$
$$= (x + 1)(2x + 3)$$

5. Multiply to check.

$$(x + 1)(2x + 3) = 2x^2 + 3x + 2x + 3$$
$$= 2x^2 + 5x + 3 \ \checkmark$$

Practice Problem 4 Factor by grouping. $2x^2 + 7x + 5$ $(2x + 5)(x + 1)$

Teaching Example 4 Factor by grouping.
$3x^2 + 10x + 7$

Ans: $(3x + 7)(x + 1)$

EXAMPLE 5 Factor by grouping. $4x^2 - 13x + 3$

Solution

1. The grouping number is $(4)(3) = 12$.
2. The factors of 12 are $(12)(1)$ or $(4)(3)$ or $(6)(2)$. Note that the middle term of the polynomial is negative. Thus we choose the numbers -12 and -1 because their product is still 12 and their sum is -13.
3. We write $-13x$ as the sum $-12x + (-1x)$ or $-12x - 1x$.
4. Factor by grouping.

$$4x^2 - 13x + 3 = 4x^2 - 12x - 1x + 3$$
$$= 4x(x - 3) - 1(x - 3) \quad \text{Remember to factor out a } -1$$
                                                        from the last two terms so that
                                                        both sets of parentheses contain
                                                        the same expression.

$$= (x - 3)(4x - 1)$$

Practice Problem 5 Factor by grouping. $9x^2 - 64x + 7$ $(x - 7)(9x - 1)$

Teaching Example 5 Factor by grouping.
$6x^2 - 13x + 5$

Ans: $(3x - 5)(2x - 1)$

EXAMPLE 6 Factor by grouping. $3x^2 - 2x - 8$

Solution

1. The grouping number is $(3)(-8) = -24$.
2. We want two numbers whose product is -24 and whose sum is -2. They are -6 and 4.
3. We write $-2x$ as the sum $-6x + 4x$.
4. Factor by grouping.

$$3x^2 - 6x + 4x - 8 = 3x(x - 2) + 4(x - 2)$$
$$= (x - 2)(3x + 4)$$

Practice Problem 6 Factor by grouping. $3x^2 + 4x - 4$ $(x + 2)(3x - 2)$

Teaching Example 6 Factor by grouping.
$8x^2 - 2x - 3$

Ans: $(4x - 3)(2x + 1)$

To factor polynomials of the form $ax^2 + bx + c$, use the method, either trial-and-error or grouping, that works best for you.

③ Using the Common Factor Method

Some problems require first factoring out the greatest common factor and then factoring the trinomial by one of the two methods of this section.

Teaching Example 7 Factor. $4x^2 - 6x - 70$

Ans: $2(x - 5)(2x + 7)$

EXAMPLE 7 Factor. $9x^2 + 3x - 30$

Solution

$$9x^2 + 3x - 30 = 3(3x^2 + 1x - 10)$$ We first factor out the common factor of 3 from each term of the trinomial.

$$= 3(3x - 5)(x + 2)$$ We then factor the trinomial by the grouping method or by the trial-and-error method.

NOTE TO STUDENT: Fully worked-out solutions to all of the Practice Problems can be found at the back of the text starting at page SP-1

Practice Problem 7 Factor. $8x^2 + 8x - 6$ $2(2x - 1)(2x + 3)$

Be sure to remove the greatest common factor as the very first step.

Teaching Example 8 Factor. $30x^2 + 5x - 5$

Ans: $5(3x - 1)(2x + 1)$

EXAMPLE 8 Factor. $32x^2 - 40x + 12$

Solution

$$32x^2 - 40x + 12 = 4(8x^2 - 10x + 3)$$ We first factor out the greatest common factor of 4 from each term of the trinomial.

$$= 4(2x - 1)(4x - 3)$$ We then factor the trinomial by the grouping method or by the trial-and-error method.

Practice Problem 8 Factor. $24x^2 - 38x + 10$ $2(4x - 5)(3x - 1)$

Factor by the trial-and-error method. Check your answers using FOIL.

1. $4x^2 + 13x + 3$
$(4x + 1)(x + 3)$

2. $3x^2 + 11x + 10$
$(3x + 5)(x + 2)$

3. $5x^2 + 7x + 2$
$(5x + 2)(x + 1)$

4. $3x^2 + 4x + 1$
$(3x + 1)(x + 1)$

5. $4x^2 + 5x - 6$
$(4x - 3)(x + 2)$

6. $5x^2 + 13x - 6$
$(5x - 2)(x + 3)$

7. $2x^2 - 5x - 3$
$(2x + 1)(x - 3)$

8. $2x^2 - x - 6$
$(2x + 3)(x - 2)$

Factor by the grouping number method. Check your answers by using FOIL.

9. $9x^2 + 9x + 2$
$(3x + 1)(3x + 2)$

10. $4x^2 + 11x + 6$
$(4x + 3)(x + 2)$

11. $15x^2 - 34x + 15$
$(3x - 5)(5x - 3)$

12. $10x^2 - 29x + 10$
$(2x - 5)(5x - 2)$

13. $2x^2 + 3x - 20$
$(2x - 5)(x + 4)$

14. $6x^2 + 11x - 10$
$(3x - 2)(2x + 5)$

15. $8x^2 + 10x - 3$
$(4x - 1)(2x + 3)$

16. $5x^2 + 2x - 7$
$(5x + 7)(x - 1)$

Factor by any method.

17. $6x^2 - 5x - 6$
$(3x + 2)(2x - 3)$

18. $3x^2 - 13x - 10$
$(3x + 2)(x - 5)$

19. $10x^2 + 3x - 1$
$(5x - 1)(2x + 1)$

20. $6x^2 + x - 5$
$(6x - 5)(x + 1)$

21. $7x^2 - 5x - 18$
$(x - 2)(7x + 9)$

22. $9x^2 - 22x - 15$
$(9x + 5)(x - 3)$

23. $9y^2 - 13y + 4$
$(9y - 4)(y - 1)$

24. $5y^2 - 11y + 2$
$(5y - 1)(y - 2)$

25. $5a^2 - 13a - 6$
$(5a + 2)(a - 3)$

26. $3a^2 - 10a - 8$
$(3a + 2)(a - 4)$

27. $14x^2 + 17x - 6$
$(7x - 2)(2x + 3)$

28. $32x^2 + 36x - 5$
$(8x - 1)(4x + 5)$

29. $15x^2 + 4x - 4$
$(5x - 2)(3x + 2)$

30. $8x^2 - 11x + 3$
$(8x - 3)(x - 1)$

31. $12x^2 + 28x + 15$
$(6x + 5)(2x + 3)$

32. $24x^2 + 17x + 3$
$(8x + 3)(3x + 1)$

33. $12x^2 - 16x - 3$
$(6x + 1)(2x - 3)$

34. $12x^2 + x - 6$
$(4x + 3)(3x - 2)$

35. $3x^4 - 14x^2 - 5$
$(3x^2 + 1)(x^2 - 5)$

36. $4x^4 + 8x^2 - 5$
$(2x^2 - 1)(2x^2 + 5)$

37. $2x^2 + 11xy + 15y^2$
$(2x + 5y)(x + 3y)$

38. $15x^2 + 14xy + 3y^2$
$(5x + 3y)(3x + y)$

39. $5x^2 + 16xy - 16y^2$
$(5x - 4y)(x + 4y)$

40. $12x^2 + 11xy - 5y^2$
$(3x - y)(4x + 5y)$

Factor by first factoring out the greatest common factor. See Examples 7 and 8.

41. $10x^2 - 25x - 15$
$5(2x + 1)(x - 3)$

42. $20x^2 - 25x - 30$
$5(4x + 3)(x - 2)$

43. $6x^3 + 9x^2 - 60x$
$3x(2x - 5)(x + 4)$

44. $6x^3 - 16x^2 - 6x$
$2x(3x + 1)(x - 3)$

Mixed Practice *Factor.*

45. $5x^2 + 3x - 2$
$(5x - 2)(x + 1)$

46. $6x^2 + x - 2$
$(3x + 2)(2x - 1)$

47. $12x^2 - 38x + 20$
$2(3x - 2)(2x - 5)$

48. $20x^2 - 38x + 12$
$2(2x - 3)(5x - 2)$

49. $12x^3 - 20x^2 + 3x$
$x(6x - 1)(2x - 3)$

50. $6x^3 - 23x^2 + 20x$
$x(3x - 4)(2x - 5)$

51. $8x^2 + 16x - 10$
$2(2x - 1)(2x + 5)$

52. $16x^2 + 36x - 10$
$2(4x - 1)(2x + 5)$

Cumulative Review

53. [2.4.1] Solve. $7x - 3(4 - 2x) = 2(x - 3) - (5 - x)$

$x = \dfrac{1}{10}$

54. [2.5.1] Solve for x. $7x - 3y = 5$

$x = \dfrac{5 + 3y}{7}$

Overseas Travelers The double-bar graph shows the top states visited by foreign travelers (excluding travelers from Canada and Mexico) for the years 2000 and 2005. *(Source: www.census.gov.)*

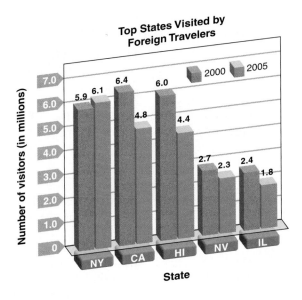

55. [0.7.1] (a) How many overseas travelers visited these five states in 2005? 19,400,000 travelers

(b) What percent of these travelers visited Hawaii (HI)?

About 22.7%

56. [0.7.1] Of the overseas visitors to these five states in 2000, what percent traveled to California (CA)?

About 27.4%

57. [0.7.1] What was the percent decrease in the number of visitors to California (CA) from 2000 to 2005?

25%

58. [0.7.1] What was the percent increase in the number of visitors to New York (NY) from 2000 to 2005?

About 3.4%

Quick Quiz 5.4 Factor by any method.

1. $12x^2 + 16x - 3$

$(2x + 3)(6x - 1)$

2. $10x^2 - 21x + 9$

$(5x - 3)(2x - 3)$

3. $6x^3 - 3x^2 - 30x$

$3x(2x - 5)(x + 2)$

4. Concept Check Explain how you would factor $10x^3 + 18x^2y - 4xy^2$.

Answers may vary

Classroom Quiz 5.4 You may use these problems to quiz your students' mastery of Section 5.4.

Factor by any method.

1. $30x^2 + 7x - 2$

Ans: $(5x + 2)(6x - 1)$

2. $20x^2 - 27x + 9$

Ans: $(5x - 3)(4x - 3)$

3. $6x^3 - 21x^2 - 45x$

Ans: $3x(x - 5)(2x + 3)$

How are you doing with your homework assignments in Sections 5.1 to 5.4? Do you feel you have mastered the material so far? Do you understand the concepts you have covered? Before you go further in the textbook, take some time to do each of the following problems.

Factor each of the following problems. Be sure to remove the greatest common factor in each case.

5.1

1. $6xy - 15z + 21$

2. $30x^2 - 45xy - 10x$

3. $7(4x - 5) - b(4x - 5)$

4. $2x(8y + 3z) - 5y(8y + 3z)$

5.2

5. $18 + 3x - 6y - xy$

6. $15x - 9xb + 20w - 12bw$

7. $x^3 - 5x^2 - 3x + 15$

8. $7a + 21b + 2ab + 6b^2$

5.3

9. $x^2 - 15x + 56$

10. $x^2 + 12x - 64$

11. $a^2 + 10ab + 21b^2$

12. $7x^2 - 14x - 245$
(*Hint:* First factor out the greatest common factor.)

5.4

13. $12x^2 + 17x - 5$

14. $3x^2 - 23x + 14$

15. $6x^2 + 17xy + 12y^2$

16. $14x^3 - 20x^2 - 16x$
(*Hint:* First factor out the greatest common factor.)

Now turn to page SA-11 for the answer to each of these problems. Each answer also includes a reference to the objective in which the problem is first taught. If you missed any of these problems, you should stop and review the Examples and Practice Problems in the referenced objective. A little review now will help you master the material in the upcoming sections of the text.

1. $3(2xy - 5z + 7)$

2. $5x(6x - 9y - 2)$

3. $(4x - 5)(7 - b)$

4. $(8y + 3z)(2x - 5y)$

5. $(6 + x)(3 - y)$

6. $(3x + 4w)(5 - 3b)$

7. $(x - 5)(x^2 - 3)$

8. $(a + 3b)(7 + 2b)$

9. $(x - 7)(x - 8)$

10. $(x + 16)(x - 4)$

11. $(a + 3b)(a + 7b)$

12. $7(x + 5)(x - 7)$

13. $(4x - 1)(3x + 5)$

14. $(x - 7)(3x - 2)$

15. $(2x + 3y)(3x + 4y)$

16. $2x(7x + 4)(x - 2)$

5.5 SPECIAL CASES OF FACTORING

As we proceed in this section you will be able to reduce the time it takes you to factor polynomials by quickly recognizing and factoring two special types of polynomials: the difference of two squares and perfect-square trinomials.

 Factoring the Difference of Two Squares

Recall the formula from Section 4.5:

$$(a + b)(a - b) = a^2 - b^2.$$

In reverse form we can use it for factoring.

> **DIFFERENCE OF TWO SQUARES**
>
> $$a^2 - b^2 = (a + b)(a - b)$$

We can state it in words in this way: "The difference of two squares can be factored into the sum and difference of those values that were squared."

Teaching Example 1 Factor. $36a^2 - 1$

Ans: $(6a + 1)(6a - 1)$

EXAMPLE 1 Factor. $9x^2 - 1$

Solution We see that the problem is in the form of the difference of two squares. $9x^2$ is a square and 1 is a square. So using the formula we can write the following.

$$9x^2 - 1 = (3x + 1)(3x - 1) \quad \text{Because } 9x^2 = (3x)^2 \text{ and } 1 = (1)^2$$

Practice Problem 1 Factor. $64x^2 - 1$ $(8x + 1)(8x - 1)$

NOTE TO STUDENT: Fully worked-out solutions to all of the Practice Problems can be found at the back of the text starting at page SP-1

Teaching Example 2 Factor. $81m^2 - 100$

Ans: $(9m + 10)(9m - 10)$

EXAMPLE 2 Factor. $25x^2 - 16$

Solution Again we use the formula for the difference of squares.

$$25x^2 - 16 = (5x + 4)(5x - 4) \quad \text{Because } 25x^2 = (5x)^2 \text{ and } 16 = (4)^2$$

Practice Problem 2 Factor. $36x^2 - 49$ $(6x + 7)(6x - 7)$

Sometimes the polynomial contains two variables.

Teaching Example 3 Factor. $64y^2 - 9a^2$

Ans: $(8y + 3a)(8y - 3a)$

EXAMPLE 3 Factor. $4x^2 - 49y^2$

Solution We see that

$$4x^2 - 49y^2 = (2x + 7y)(2x - 7y).$$

Practice Problem 3 Factor. $100x^2 - 81y^2$ $(10x + 9y)(10x - 9y)$

CAUTION: Please note that the difference-of-two-squares formula only works if the last term is negative. So if Example 3 had been to factor $4x^2 + 49y^2$, we would **not** have been able to factor the problem. We will examine this in more detail in Section 5.6.

Some problems may involve more than one step.

EXAMPLE 4   Factor. $81x^4 - 1$

Solution   We see that

$$81x^4 - 1 = (9x^2 + 1)(9x^2 - 1). \qquad \text{Because } 81x^4 = (9x^2)^2 \text{ and } 1 = (1)^2$$

Is the factoring complete? We can factor $9x^2 - 1$.

$$81x^4 - 1 = (9x^2 + 1)(3x + 1)(3x - 1) \qquad \text{Because } (9x^2 - 1) = (3x + 1)(3x - 1)$$

Practice Problem 4   Factor. $x^8 - 1$ $(x^4 + 1)(x^2 + 1)(x + 1)(x - 1)$

Teaching Example 4 Factor. $16a^4 - 1$

Ans: $(4a^2 + 1)(2a + 1)(2a - 1)$

② Factoring Perfect-Square Trinomials

There is a formula that will help us to very quickly factor certain trinomials, called **perfect-square trinomials.** Recall from Section 4.5 the formulas for binomials squared.

$$(a + b)^2 = a^2 + 2ab + b^2$$
$$(a - b)^2 = a^2 - 2ab + b^2$$

We can use these two equations in reverse form for factoring.

PERFECT-SQUARE TRINOMIALS

$$a^2 + 2ab + b^2 = (a + b)^2$$
$$a^2 - 2ab + b^2 = (a - b)^2$$

Teaching Tip The key to recognizing a perfect-square trinomial is determining if the first and last terms of the trinomial are perfect squares. Spend some time emphasizing to the class the definition of *perfect square* and the list of such values.

A perfect-square trinomial is a trinomial that is the result of squaring a binomial. How can we recognize a perfect-square trinomial?

1. The first and last terms are *perfect squares*.
2. The middle term is twice the product of the values whose squares are the first and last terms.

EXAMPLE 5   Factor. $x^2 + 6x + 9$

Solution   This is a perfect-square trinomial.

1. The first and last terms are perfect squares because $x^2 = (x)^2$ and $9 = (3)^2$.
2. The middle term, $6x$, is twice the product of x and 3.

Teaching Example 5 Factor. $y^2 + 4y + 4$

Ans: $(y + 2)^2$

Since $x^2 + 6x + 9$ is a perfect-square trinomial, we can use the formula

$$a^2 + 2ab + b^2 = (a + b)^2$$

with $a = x$ and $b = 3$. So we have

$$x^2 + 6x + 9 = (x + 3)^2.$$

Practice Problem 5 Factor. $x^2 + 10x + 25$ $(x + 5)^2$

Teaching Example 6 Factor.
$36m^2 - 84m + 49$

Ans: $(6m - 7)^2$

NOTE TO STUDENT: *Fully worked-out solutions to all of the Practice Problems can be found at the back of the text starting at page SP-1*

EXAMPLE 6 Factor. $4x^2 - 20x + 25$

Solution   This is a perfect-square trinomial. Note that $20x = 2(2x \cdot 5)$. Also note the negative sign. Thus we have the following.

$$4x^2 - 20x + 25 = (2x - 5)^2 \quad \text{Since } a^2 - 2ab + b^2 = (a - b)^2$$

Practice Problem 6 Factor. $25x^2 - 30x + 9$ $(5x - 3)^2$

A polynomial may have more than one variable and its exponents may be higher than 2. The same principles apply.

Teaching Example 7 Factor.

(a) $16y^2 + 40wy + 25w^2$

(b) $9m^6 - 6m^3 + 1$

Ans: (a) $(4y + 5w)^2$ **(b)** $(3m^3 - 1)^2$

EXAMPLE 7 Factor.

(a) $49x^2 + 42xy + 9y^2$ **(b)** $36x^4 - 12x^2 + 1$

Solution

(a) This is a perfect-square trinomial. Why?

$$49x^2 + 42xy + 9y^2 = (7x + 3y)^2 \quad \begin{array}{l}\text{Because } 49x^2 = (7x)^2, 9y^2 = (3y)^2, \\ \text{and } 42xy = 2(7x \cdot 3y)\end{array}$$

(b) This is a perfect-square trinomial. Why?

$$36x^4 - 12x^2 + 1 = (6x^2 - 1)^2 \quad \begin{array}{l}\text{Because } 36x^4 = (6x^2)^2, 1 = (1)^2, \\ \text{and } 12x^2 = 2(6x^2 \cdot 1)\end{array}$$

Practice Problem 7 Factor.

(a) $25x^2 + 60xy + 36y^2$ $(5x + 6y)^2$ **(b)** $64x^6 - 48x^3 + 9$ $(8x^3 - 3)^2$

Some polynomials appear to be perfect-square trinomials but are not. They were factored in other ways in Section 5.4.

EXAMPLE 8 Factor. $49x^2 + 35x + 4$

Solution This is *not* a perfect-square trinomial! Although the first and last terms are perfect squares since $(7x)^2 = 49x^2$ and $(2)^2 = 4$, the middle term, $35x$, is not double the product of 2 and $7x$! $35x \neq 28x$! So we must factor by trial and error or by grouping to obtain

$$49x^2 + 35x + 4 = (7x + 4)(7x + 1).$$

Practice Problem 8 Factor. $9x^2 + 15x + 4$ $(3x + 1)(3x + 4)$

Teaching Example 8 Factor.
$25x^2 - 30x - 16$

Ans: $(5x + 2)(5x - 8)$

3 Factoring Out a Common Factor Before Using a Special-Case Formula

For some polynomials, we first need to factor out the greatest common factor. Then we will find an opportunity to use the difference-of-two-squares formula or one of the perfect-square trinomial formulas.

Look carefully at Example 9. Can you identify the greatest common factor?

EXAMPLE 9 Factor. $12x^2 - 48$

Solution

$$12x^2 - 48 = 12(x^2 - 4)$$ First we factor out the greatest common factor, 12.

$$= 12(x + 2)(x - 2)$$ Then we use the difference-of-two-squares formula, $a^2 - b^2 = (a + b)(a - b)$.

Teaching Example 9 Factor. $7y^2 - 28a^2$

Ans: $7(y + 2a)(y - 2a)$

Practice Problem 9 Factor. $20x^2 - 45$ $5(2x + 3)(2x - 3)$

Look carefully at Example 10. Can you identify the greatest common factor?

EXAMPLE 10 Factor. $24x^2 - 72x + 54$

Solution

$$24x^2 - 72x + 54 = 6(4x^2 - 12x + 9)$$ First we factor out the greatest common factor, 6.

$$= 6(2x - 3)^2$$ Then we use the perfect-square trinomial formula, $a^2 - 2ab + b^2 = (a - b)^2$.

Teaching Example 10 Factor.
$45x^2 - 120x + 80$

Ans: $5(3x - 4)^2$

Practice Problem 10 Factor. $75x^2 - 60x + 12$ $3(5x - 2)^2$

Factor by using the difference-of-two-squares formula.

1. $9x^2 - 1$
$(3x + 1)(3x - 1)$

2. $25x^2 - 1$
$(5x + 1)(5x - 1)$

3. $81x^2 - 16$
$(9x + 4)(9x - 4)$

4. $100x^2 - 49$
$(10x + 7)(10x - 7)$

5. $x^2 - 49$
$(x + 7)(x - 7)$

6. $x^2 - 100$
$(x + 10)(x - 10)$

7. $25x^2 - 81$
$(5x + 9)(5x - 9)$

8. $49x^2 - 4$
$(7x + 2)(7x - 2)$

9. $x^2 - 25$
$(x + 5)(x - 5)$

10. $x^2 - 36$
$(x + 6)(x - 6)$

11. $1 - 16x^2$
$(1 + 4x)(1 - 4x)$

12. $1 - 64x^2$
$(1 + 8x)(1 - 8x)$

13. $16x^2 - 49y^2$
$(4x + 7y)(4x - 7y)$

14. $25x^2 - 81y^2$
$(5x + 9y)(5x - 9y)$

15. $36x^2 - 169y^2$
$(6x + 13y)(6x - 13y)$

16. $64x^2 - 81y^2$
$(8x + 9y)(8x - 9y)$

17. $100x^2 - 81$
$(10x + 9)(10x - 9)$

18. $16a^2 - 25$
$(4a + 5)(4a - 5)$

19. $25a^2 - 81b^2$
$(5a + 9b)(5a - 9b)$

20. $9x^2 - 49y^2$
$(3x + 7y)(3x - 7y)$

Factor by using the perfect-square trinomial formula.

21. $9x^2 + 6x + 1$
$(3x + 1)^2$

22. $25x^2 + 10x + 1$
$(5x + 1)^2$

23. $y^2 - 10y + 25$
$(y - 5)^2$

24. $y^2 - 12y + 36$
$(y - 6)^2$

25. $36x^2 - 60x + 25$
$(6x - 5)^2$

26. $16x^2 - 72x + 81$
$(4x - 9)^2$

27. $49x^2 + 28x + 4$
$(7x + 2)^2$

28. $25x^2 + 30x + 9$
$(5x + 3)^2$

29. $x^2 + 14x + 49$
$(x + 7)^2$

30. $x^2 + 8x + 16$
$(x + 4)^2$

31. $25x^2 - 40x + 16$
$(5x - 4)^2$

32. $64x^2 - 16x + 1$
$(8x - 1)^2$

33. $81x^2 + 36xy + 4y^2$
$(9x + 2y)^2$

34. $36x^2 + 60xy + 25y^2$
$(6x + 5y)^2$

35. $9x^2 - 30xy + 25y^2$
$(3x - 5y)^2$

36. $4x^2 - 28xy + 49y^2$
$(2x - 7y)^2$

Mixed Practice *Factor by using either the difference-of-two-squares or the perfect-square trinomial formula.*

37. $16a^2 + 72ab + 81b^2$
$(4a + 9b)^2$

38. $169a^2 + 26ab + b^2$
$(13a + b)^2$

39. $49x^2 - 42xy + 9y^2$
$(7x - 3y)^2$

40. $9x^2 - 30xy + 25y^2$
$(3x - 5y)^2$

41. $9x^2 + 42x + 49$
$(3x + 7)^2$

42. $25x^2 + 40x + 16$
$(5x + 4)^2$

43. $144x^2 - 1$
$(12x + 1)(12x - 1)$

44. $16x^2 - 121$
$(4x + 11)(4x - 11)$

45. $x^4 - 36$
$(x^2 + 6)(x^2 - 6)$

46. $x^4 - 49$
$(x^2 + 7)(x^2 - 7)$

47. $4x^4 - 20x^2 + 25$
$(2x^2 - 5)^2$

48. $36x^4 - 60x^2 + 25$
$(6x^2 - 5)^2$

To Think About

49. In Example 4, first we factored $81x^4 - 1$ as $(9x^2 + 1)$ $(9x^2 - 1)$, and then we factored $9x^2 - 1$ as $(3x + 1)$ $(3x - 1)$. Show why you cannot factor $9x^2 + 1$.
Because no matter what combination you try, you cannot multiply two binomials to equal $9x^2 + 1$.

50. What two numbers could replace the b in $25x^2 + bx + 16$ so that the resulting trinomial would be a perfect square? (*Hint:* One number is negative.)
± 40

51. What value could you give to c so that $16y^2 - 56y + c$ would become a perfect-square trinomial? Is there only one answer or more than one?
49; one answer

52. Jerome says that he can find two values of b so that $100x^2 + bx - 9$ will be a perfect square. Kesha says there is only one that fits, and Larry says there are none. Who is correct and why?
Larry: the last term must be positive in order to have a perfect square.

Factor by first looking for a greatest common factor. See Examples 9 and 10.

53. $16x^2 - 36$
$4(2x + 3)(2x - 3)$

54. $27x^2 - 75$
$3(3x + 5)(3x - 5)$

55. $147x^2 - 3y^2$
$3(7x + y)(7x - y)$

56. $16y^2 - 100x^2$
$4(2y + 5x)(2y - 5x)$

57. $16x^2 - 16x + 4$
$4(2x - 1)^2$

58. $45x^2 - 60x + 20$
$5(3x - 2)^2$

59. $98x^2 + 84x + 18$
$2(7x + 3)^2$

60. $128x^2 + 96x + 18$
$2(8x + 3)^2$

Mixed Practice *Factor. Be sure to look for common factors first.*

61. $x^2 + 16x + 63$
$(x + 9)(x + 7)$

62. $x^2 - 3x - 40$
$(x + 5)(x - 8)$

63. $2x^2 + 5x - 3$
$(2x - 1)(x + 3)$

64. $15x^2 - 11x + 2$
$(3x - 1)(5x - 2)$

65. $16x^2 - 121$
$(4x + 11)(4x - 11)$

66. $81x^2 - 36y^2$
$9(3x + 2y)(3x - 2y)$

67. $9x^2 + 42x + 49$
$(3x + 7)^2$

68. $9x^2 + 30x + 25$
$(3x + 5)^2$

69. $36x^2 - 36x + 9$
$9(2x - 1)^2$

70. $6x^2 + 60x + 150$
$6(x + 5)^2$

71. $2x^2 - 32x + 126$
$2(x - 9)(x - 7)$

72. $2x^2 - 32x + 110$
$2(x - 11)(x - 5)$

Cumulative Review

73. [4.6.2] Divide. $(x^3 + x^2 - 2x - 11) \div (x - 2)$
$x^2 + 3x + 4 + \dfrac{-3}{x - 2}$

74. [4.6.2] Divide. $(6x^3 + 11x^2 - 11x - 20) \div (3x + 4)$
$2x^2 + x - 5$

Iguana Diet *The green iguana can reach a length of 6 feet and weigh up to 18 pounds. Of the basic diet of the iguana, 40% should consist of greens such as lettuce, spinach, and parsley; 35% should consist of bulk vegetables such as broccoli, zucchini, and carrots; and 25% should consist of fruit.*

75. [0.7.1] If a certain iguana weighing 150 ounces has a daily diet equal to 2% of its body weight, compose a diet for it in ounces that will meet the iguana's one-day requirement for nutrition.
1.2 oz of greens, 1.05 oz of bulk vegetables, 0.75 oz of fruit

76. [0.7.1] If another iguana weighing 120 ounces has a daily diet equal to 3% of its body weight, compose a diet for it in ounces that will meet the iguana's one-day requirement for nutrition.
1.44 oz of greens, 1.26 oz of bulk vegetables, 0.9 oz of fruit

Quick Quiz 5.5 Factor the following.

1. $49x^2 - 81y^2$
$(7x + 9y)(7x - 9y)$

2. $9x^2 - 48x + 64$
$(3x - 8)^2$

3. $162x^2 - 200$
$2(9x + 10)(9x - 10)$

4. **Concept Check** Explain how to factor the polynomial $24x^2 + 120x + 150$.
Answers may vary

Classroom Quiz 5.5 You may use these problems to quiz your students' mastery of Section 5.5.

Factor the following.

1. $64x^2 - 121y^2$
Ans: $(8x + 11y)(8x - 11y)$

2. $25x^2 - 70x + 49$
Ans: $(5x - 7)^2$

3. $144x^2 - 100$
Ans: $4(6x + 5)(6x - 5)$

Student Learning Objectives

After studying this section, you will be able to:

 Identify and factor any polynomial that can be factored.

 Determine whether a polynomial is prime.

 Identifying and Factoring Polynomials

Often the various types of factoring problems are all mixed together. We need to be able to identify each type of polynomial quickly. The following table summarizes the information we have learned about factoring.

Many polynomials require more than one factoring method. When you are asked to factor a polynomial, it is expected that you will factor it completely. Usually, the first step is factoring out a common factor; then the next step will become apparent.

Carefully go through each example in the following Factoring Organizer. Be sure you understand each step that is involved.

Factoring Organizer

Number of Terms in the Polynomial	Identifying Name and/or Formula	Example
A. Any number of terms	**Common factor** The terms have a common factor consisting of a number, a variable, or both.	$2x^2 - 16x = 2x(x - 8)$ $3x^2 + 9y - 12 = 3(x^2 + 3y - 4)$ $4x^2y + 2xy^2 - wxy + xyz = xy(4x + 2y - w + z)$
B. Two terms	**Difference of two squares** First and last terms are perfect squares. $a^2 - b^2 = (a + b)(a - b)$	$16x^2 - 1 = (4x + 1)(4x - 1)$ $25y^2 - 9x^2 = (5y + 3x)(5y - 3x)$
C. Three terms	**Perfect-square trinomial** First and last terms are perfect squares. $a^2 + 2ab + b^2 = (a + b)^2$ $a^2 - 2ab + b^2 = (a - b)^2$	$25x^2 - 10x + 1 = (5x - 1)^2$ $16x^2 + 24x + 9 = (4x + 3)^2$
D. Three terms	**Trinomial of the form $x^2 + bx + c$** It starts with x^2. The constants of the two factors are numbers whose product is c and whose sum is b.	$x^2 - 7x + 12 = (x - 3)(x - 4)$ $x^2 + 11x - 26 = (x + 13)(x - 2)$ $x^2 - 8x - 20 = (x - 10)(x + 2)$
E. Three terms	**Trinomial of the form $ax^2 + bx + c$** It starts with ax^2, where a is any number but 1.	Use trial-and-error or the grouping number method to factor $12x^2 - 5x - 2$. 1. The grouping number is -24. 2. The two numbers whose product is -24 and whose sum is -5 are -8 and 3. 3. $12x^2 - 5x - 2 = 12x^2 + 3x - 8x - 2$ $\qquad = 3x(4x + 1) - 2(4x + 1)$ $\qquad = (4x + 1)(3x - 2)$
F. Four terms	**Factor by grouping** Rearrange the order if the first two terms do not have a common factor.	$wx - 6yz + 2wy - 3xz = wx + 2wy - 3xz - 6yz$ $\qquad = w(x + 2y) - 3z(x + 2y)$ $\qquad = (x + 2y)(w - 3z)$

Teaching Example 1 Factor.
(a) $5m^2 - 10m - 40$
(b) $abx - 3aby + 5bx - 15by$
(c) $32a^2 - 50b^2$
(d) $9x + 15 - 6x^2$

Ans: (a) $5(m - 4)(m + 2)$
(b) $b(x - 3y)(a + 5)$
(c) $2(4a + 5b)(4a - 5b)$
(d) $-3(2x - 5)(x + 1)$

EXAMPLE 1 Factor.

(a) $25x^3 - 10x^2 + x$

(b) $20x^2y^2 - 45y^2$

(c) $2ax + 4ay + 4x + 8y$

(d) $15x^2 - 3x^3 + 18x$

Solution

(a) $25x^3 - 10x^2 + x = x(25x^2 - 10x + 1)$ Factor out the common factor of x. The other factor is a perfect-square trinomial.

$\qquad\qquad\qquad\qquad = x(5x - 1)^2$

(b) $20x^2y^2 - 45y^2 = 5y^2(4x^2 - 9)$ Factor out the common factor of $5y^2$. The other factor is a difference of squares.

$$= 5y^2(2x + 3)(2x - 3)$$

(c) $2ax + 4ay + 4x + 8y = 2[ax + 2ay + 2x + 4y]$ Factor out the common factor of 2.

$$= 2[a(x + 2y) + 2(x + 2y)]$$ Factor the terms inside the bracket by the grouping method.

$$= 2(x + 2y)(a + 2)$$ Factor out the common factor of $(x + 2y)$.

(d) $15x^2 - 3x^3 + 18x = -3x^3 + 15x^2 + 18x$ Rearrange the terms in descending order of powers of x.

$$= -3x(x^2 - 5x - 6)$$ Factor out the common factor of $-3x$.

$$= -3x(x - 6)(x + 1)$$ Factor the trinomial.

Practice Problem 1   Factor. Be careful. These practice problems are mixed.

(a) $3x^2 - 36x + 108$ $3(x - 6)^2$ **(b)** $9x^4y^2 - 9y^2$ $9y^2(x^2 + 1)(x + 1)(x - 1)$
(c) $5x^3 - 15x^2y + 10x^2 - 30xy$ **(d)** $12x - 9 - 4x^2$

           (c) $5x(x - 3y)(x + 2)$ **(d)** $-(2x - 3)^2$

Teaching Tip After discussing Example 1, ask students to do the following problems as an additional class exercise.
Factor completely. (Be sure to factor out the common factor first!)
(a) $3ax + 9bx - 12aw - 36bw$
(b) $x^3 + 2x^2y - 15xy^2$
(c) $8 + 7x - x^2$

Ans:
(a) $3[ax + 3bx - 4aw - 12bw]$
     $= 3[x(a + 3b) - 4w(a + 3b)]$
     $= 3(a + 3b)(x - 4w)$
(b) $x(x^2 + 2xy - 15y^2)$
     $= x(x + 5y)(x - 3y)$
(c) $-1(x^2 - 7x - 8)$
     $= -1(x - 8)(x + 1)$ or the answer may be
     $(8 - x)(1 + x)$

NOTE TO STUDENT: Fully worked-out solutions to all of the Practice Problems can be found at the back of the text starting at page SP-1

2   Determining Whether a Polynomial Is Prime

Not all polynomials can be factored using the methods in this chapter. If we cannot factor a polynomial by elementary methods, we will identify it as a **prime** polynomial. If, after you have mastered the factoring techniques in this chapter, you encounter a polynomial that you cannot factor with these methods, you should feel comfortable enough to say, "The polynomial cannot be factored with the methods in this chapter, so it is prime," rather than "I can't do it—I give up!"

EXAMPLE 2   Factor, if possible. $x^2 + 6x + 12$

Solution   The factors of 12 are

$$(1)(12) \text{ or } (2)(6) \text{ or } (3)(4).$$

None of these pairs add up to 6, the coefficient of the middle term. Thus the problem cannot be factored by the methods of this chapter. It is prime.

Practice Problem 2   Factor. $x^2 - 9x - 8$ Prime

EXAMPLE 3   Factor, if possible. $25x^2 + 4$

Solution   We have a formula to factor the difference of two squares. There is no way to factor the sum of two squares. That is, $a^2 + b^2$ cannot be factored. Thus

$$25x^2 + 4 \text{ is prime.}$$

Practice Problem 3   Factor, if possible. $25x^2 + 82x + 4$ Prime

Teaching Example 2 Factor, if possible.
$x^2 - 6x - 8$
Ans: Prime

Teaching Example 3 Factor, if possible.
$4a^2 + 10a + 9$
Ans: Prime

Review the six basic types of factoring in the Factoring Organizer on page 332. Each of the six types is included in exercises 1–12. Be sure you can find two of each type.

Factor. Check your answer by multiplying.

1. $5a^2 - 3ab + 8a$ $a(5a - 3b + 8)$

2. $7x^2 + 9xy - 2x$ $x(7x + 9y - 2)$

3. $16x^2 - 25y^2$ $(4x + 5y)(4x - 5y)$

4. $64x^2 - 9y^2$ $(8x + 3y)(8x - 3y)$

5. $9x^2 - 12xy + 4y^2$ $(3x - 2y)^2$

6. $16x^2 + 24xy + 9y^2$ $(4x + 3y)^2$

7. $x^2 + 8x + 15$ $(x + 5)(x + 3)$

8. $x^2 + 15x + 54$ $(x + 9)(x + 6)$

9. $15x^2 + 7x - 2$ $(3x + 2)(5x - 1)$

10. $6x^2 + 13x - 5$ $(2x + 5)(3x - 1)$

11. $ax - 3cx + 3ay - 9cy$ $(a - 3c)(x + 3y)$

12. $bx - 2dx + 5by - 10dy$ $(b - 2d)(x + 5y)$

Mixed Practice *Factor, if possible. Be sure to factor completely. Always factor out the greatest common factor first, if one exists.*

13. $y^2 + 14y + 49$ $(y + 7)^2$

14. $y^2 + 16y + 64$ $(y + 8)^2$

15. $4x^2 - 12x + 9$ $(2x - 3)^2$

16. $16x^2 - 40x + 25$ $(4x - 5)^2$

17. $2x^2 - 11x + 12$ $(2x - 3)(x - 4)$

18. $3x^2 - 10x + 8$ $(3x - 4)(x - 2)$

19. $x^2 - 3xy - 70y^2$ $(x - 10y)(x + 7y)$

20. $x^2 - 6xy - 16y^2$ $(x - 8y)(x + 2y)$

21. $ax - 5a + 3x - 15$ $(a + 3)(x - 5)$

22. $by + 7b - 6y - 42$ $(b - 6)(y + 7)$

23. $16x - 4x^3$ $4x(4 - x^2)$
$4x(2 + x)(2 - x)$

24. $8y^2 + 10y - 12$ $2(4y^2 + 5y - 6)$
$2(4y - 3)(y + 2)$

25. $5x^3y^3 - 10x^2y^3 + 5xy^3$ $5xy^3(x^2 - 2x + 1)$
$5xy^3(x - 1)^2$

26. $2x^4y - 12x^3y + 18x^2y$ $2x^2y(x^2 - 6x + 9) = 2x^2y(x - 3)^2$

27. $3xyz^2 - 6xyz - 9xy$ $3xy(z^2 - 2z - 3)$
$3xy(z + 1)(z - 3)$

28. $-16x^2 - 2x - 32x^3$ $-2x(16x^2 + 8x + 1)$
$-2x(4x + 1)^2$

29. $3x^2 + 6x - 105$ $3(x^2 + 2x - 35)$
$3(x + 7)(x - 5)$

30. $4x^2 - 28x - 72$ $4(x^2 - 7x - 18)$
$4(x - 9)(x + 2)$

31. $3x^3 + 3x^2 - 36x$ $3x(x^2 + x - 12) = 3x(x + 4)(x - 3)$

32. $7x^2 + 3x - 2$ Prime

33. $7x^2 - 2x^4 + 4$ $-1(2x^4 - 7x^2 - 4)$
$-1(2x^2 + 1)(x + 2)(x - 2)$

34. $14x^2 - x^3 + 32x$ $-x(x^2 - 14x - 32) = -x(x - 16)(x + 2)$

35. $6x^2 - 3x + 2$ Prime

36. $4x^3 + 8x^2 - 60x$ $4x(x^2 + 2x - 15)$
$4x(x + 5)(x - 3)$

Remove the greatest common factor first. Then continue to factor, if possible.

37. $5x^2 + 10xy - 30y$
$5(x^2 + 2xy - 6y)$

38. $7a^2 + 21b - 42$
$7(a^2 + 3b - 6)$

39. $30x^3 + 3x^2y - 6xy^2$
$3x(10x^2 + xy - 2y^2)$
$3x(2x + y)(5x - 2y)$

40. $30x^3 - 25x^2y - 30xy^2$
$5x(6x^2 - 5xy - 6y^2) = 5x(2x - 3y)(3x + 2y)$

41. $24x^2 - 58x + 30$
$2(12x^2 - 29x + 15) = 2(3x - 5)(4x - 3)$

42. $12x^2 - 30x + 12$
$6(2x^2 - 5x + 2)$
$6(2x - 1)(x - 2)$

Cumulative Review

Carbon Dioxide Emissions *The bar graph below shows the amount of carbon dioxide emissions, in millions of metric tons, for the United States, China, and Europe. Use the graph to answer the following questions. (Source: www.eia.doe.gov.)*

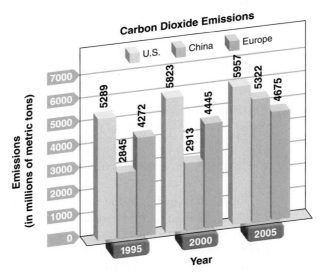

43. **[0.7.1]** Find the percent increase in the amount of carbon dioxide emissions by China from 2000 to 2005.
About 83%

44. **[0.7.1]** Does the United States or Europe have the larger percent increase in the amount of carbon dioxide emissions from 2000 to 2005?
Europe, with about a 5% increase

45. **[0.7.1]** If the same percent of increase in China that occurred between 2000 and 2005 continued from 2005 to 2010, how many tons of carbon dioxide will be emitted by China in 2010?
About 9739 million metric tons

46. **[0.7.1]** If the same percent of increase in Europe that occurred between 2000 and 2005 continued from 2005 to 2010, how many tons of carbon dioxide will be emitted by Europe in 2010?
About 4909 million metric tons

Quick Quiz 5.6

Completely factor the following, if possible.

1. $6x^2 - 17x + 12$ $(2x - 3)(3x - 4)$

2. $60x^2 - 9x - 6$ $3(4x + 1)(5x - 2)$

3. $25x^2 + 49$ Prime

4. **Concept Check** Explain how to completely factor $2x^2 + 6xw - 5x - 15w$. Answers may vary

Classroom Quiz 5.6 You may use these problems to quiz your students' mastery of Section 5.6.

Completely factor the following, if possible.

1. $15x^2 - 23x + 4$ **Ans:** $(5x - 1)(3x - 4)$ **2.** $30x^2 + 3x - 6$ **Ans:** $3(2x + 1)(5x - 2)$ **3.** $11x^2 + 3x + 1$ **Ans:** Prime

Student Learning Objectives

After studying this section, you will be able to:

 Solve quadratic equations by factoring.

 Use quadratic equations to solve applied problems.

 Solving Quadratic Equations by Factoring

In Chapter 2, we learned how to solve linear equations such as $3x + 5 = 0$ by finding the root (or value of x) that satisfied the equation. Now we turn to the question of how to solve equations like $3x^2 + 5x + 2 = 0$. Such equations are called **quadratic equations.** A quadratic equation is a polynomial equation in one variable that contains a variable term of degree 2 and no terms of higher degree.

> The *standard form* of a quadratic equation is $ax^2 + bx + c = 0$, where a, b, and c are real numbers and $a \neq 0$.

In this section, we will study quadratic equations in standard form, where a, b, and c are integers.

Many quadratic equations have two real number solutions (also called real **roots**). But how can we find them? The most direct approach is the factoring method. This method depends on a very powerful property.

> **ZERO FACTOR PROPERTY**
>
> If $a \cdot b = 0$, then $a = 0$ or $b = 0$.

Notice the word *or* in the zero factor property. When we make a statement in mathematics using this word, we intend it to mean *one or the other or both*. Therefore, the zero factor property states that if the product $a \cdot b$ is zero, then a can equal zero or b can equal zero or *both a and b* can equal zero. We can use this principle to solve quadratic equations. Before you start, make sure that the equation is in standard form.

> 1. Make sure the equation is set equal to zero.
> 2. Factor, if possible, the quadratic expression that equals zero.
> 3. Set each factor containing a variable equal to zero.
> 4. Solve the resulting equations to find each root.
> 5. Check each root.

Teaching Example 1 Solve the equation to find the two roots. $5x^2 + 7x - 6 = 0$

Ans: $\frac{3}{5}$ and -2

EXAMPLE 1 Solve the equation to find the two roots. $3x^2 + 5x + 2 = 0$

Solution

$$3x^2 + 5x + 2 = 0 \qquad \text{The equation is in standard form.}$$
$$(3x + 2)(x + 1) = 0 \qquad \text{Factor the quadratic expression.}$$
$$3x + 2 = 0 \qquad x + 1 = 0 \qquad \text{Set each factor equal to 0.}$$
$$3x = -2 \qquad x = -1 \qquad \text{Solve the equations to find the two roots.}$$
$$x = -\frac{2}{3}$$

The two roots (that is, solutions) are $-\frac{2}{3}$ and -1.

Check.   We can determine if the two numbers $-\frac{2}{3}$ and -1 are solutions to the equation. Substitute $-\frac{2}{3}$ for x in the *original equation*. If an identity results, $-\frac{2}{3}$ is a solution. Do the same for -1.

$$3x^2 + 5x + 2 = 0$$

$$3\left(-\frac{2}{3}\right)^2 + 5\left(-\frac{2}{3}\right) + 2 \stackrel{?}{=} 0$$

$$3\left(\frac{4}{9}\right) + 5\left(-\frac{2}{3}\right) + 2 \stackrel{?}{=} 0$$

$$\frac{4}{3} - \frac{10}{3} + 2 \stackrel{?}{=} 0$$

$$\frac{4}{3} - \frac{10}{3} + \frac{6}{3} \stackrel{?}{=} 0$$

$$0 = 0 \checkmark$$

$$3x^2 + 5x + 2 = 0$$

$$3(-1)^2 + 5(-1) + 2 \stackrel{?}{=} 0$$

$$3(1) + 5(-1) + 2 \stackrel{?}{=} 0$$

$$3 - 5 + 2 \stackrel{?}{=} 0$$

$$-2 + 2 \stackrel{?}{=} 0$$

$$0 = 0 \checkmark$$

Thus $-\frac{2}{3}$ and -1 are both roots of the equation $3x^2 + 5x + 2 = 0$.

Practice Problem 1   Solve the equation by factoring to find the two roots and check. $10x^2 - x - 2 = 0$

$$-\frac{2}{5} \text{ and } \frac{1}{2}$$

NOTE TO STUDENT: *Fully worked-out solutions to all of the Practice Problems can be found at the back of the text starting at page SP-1*

EXAMPLE 2   Solve the equation to find the two roots. $2x^2 + 13x - 7 = 0$

Solution

$$\begin{array}{ll} 2x^2 + 13x - 7 = 0 & \text{The equation is in standard form.} \\ (2x - 1)(x + 7) = 0 & \text{Factor.} \\ 2x - 1 = 0 \quad x + 7 = 0 & \text{Set each factor equal to 0.} \\ 2x = 1 \qquad x = -7 & \text{Solve the equations to find the two roots.} \\ x = \dfrac{1}{2} & \end{array}$$

The two roots are $\frac{1}{2}$ and -7.

Check.   If $x = \frac{1}{2}$, then we have the following.

$$2\left(\frac{1}{2}\right)^2 + 13\left(\frac{1}{2}\right) - 7 = 2\left(\frac{1}{4}\right) + 13\left(\frac{1}{2}\right) - 7$$

$$= \frac{1}{2} + \frac{13}{2} - \frac{14}{2} = 0 \checkmark$$

If $x = -7$, then we have the following.

$$2(-7)^2 + 13(-7) - 7 = 2(49) + 13(-7) - 7$$

$$= 98 - 91 - 7 = 0 \checkmark$$

Thus $\frac{1}{2}$ and -7 are both roots of the equation $2x^2 + 13x - 7 = 0$.

Practice Problem 2   Solve the equation to find the two roots.
$3x^2 + 11x - 4 = 0$

$$\frac{1}{3} \text{ and } -4$$

Teaching Example 2 Solve the equation to find the two roots. $2x^2 + 7x + 3 = 0$

Ans: $-\frac{1}{2}$ and -3

If the quadratic equation $ax^2 + bx + c = 0$ has no visible constant term, then $c = 0$. All such quadratic equations can be solved by factoring out a common factor and then using the zero factor property to obtain two solutions that are real numbers.

Teaching Example 3 Solve the equation to find the two roots. $4x^2 - x = 0$

Ans: 0 and $\dfrac{1}{4}$

EXAMPLE 3   Solve the equation to find the two roots. $7x^2 - 3x = 0$

Solution

$$
\begin{aligned}
7x^2 - 3x &= 0 &&\text{The equation is in standard form. Here } c = 0.\\
x(7x - 3) &= 0 &&\text{Factor out the common factor.}\\
x = 0 \quad 7x - 3 &= 0 &&\text{Set each factor equal to 0 by the zero factor property.}\\
7x &= 3 &&\text{Solve the equations to find the two roots.}\\
x &= \frac{3}{7}
\end{aligned}
$$

The two roots are 0 and $\frac{3}{7}$.

Check.   Verify that 0 and $\frac{3}{7}$ are the roots of $7x^2 - 3x = 0$.

NOTE TO STUDENT: Fully worked-out solutions to all of the Practice Problems can be found at the back of the text starting at page SP-1

Practice Problem 3   Solve the equation to find the two roots.
$7x^2 + 11x = 0$ 0 and $-\dfrac{11}{7}$

If the quadratic equation is not in standard form, we use the same basic algebraic methods we studied in Sections 2.1–2.3 to place the terms on one side and zero on the other so that we can use the zero factor property.

Teaching Example 4 Solve. $x - 5 + x^2 = 10 + 3x$

Ans: $x = -3,\ x = 5$

EXAMPLE 4   Solve. $x^2 = 12 - x$

Solution

$$
\begin{aligned}
x^2 &= 12 - x &&\text{The equation is not in standard form.}\\
x^2 + x - 12 &= 0 &&\text{Add } x \text{ and } -12 \text{ to both sides of the equation so that}\\
&&&\text{the left side is equal to zero; we can now factor.}\\
(x - 3)(x + 4) &= 0 &&\text{Factor.}\\
x - 3 = 0 \quad x + 4 &= 0 &&\text{Set each factor equal to 0 by the zero factor property.}\\
x = 3 \quad\quad x &= -4 &&\text{Solve the equations for } x.
\end{aligned}
$$

Check.   If $x = 3$: $(3)^2 \overset{?}{=} 12 - 3$ If $x = -4$: $(-4)^2 \overset{?}{=} 12 - (-4)$

$$9 \overset{?}{=} 12 - 3 \qquad\qquad\qquad\qquad 16 \overset{?}{=} 12 + 4$$
$$9 = 9 \ \checkmark \qquad\qquad\qquad\qquad\quad 16 = 16 \ \checkmark$$

Both roots check.

Practice Problem 4   Solve. $x^2 - 6x + 4 = -8 + x$ $x = 3,\ x = 4$

Teaching Example 5 Solve. $\dfrac{x^2 - 5x}{2} = 7$

Ans: $x = -2,\ x = 7$

EXAMPLE 5   Solve. $\dfrac{x^2 - x}{2} = 6$

Solution   We must first clear the fractions from the equation.

$$
\begin{aligned}
2\left(\frac{x^2 - x}{2}\right) &= 2(6) &&\text{Multiply each side by 2.}\\
x^2 - x &= 12 &&\text{Simplify.}\\
x^2 - x - 12 &= 0 &&\text{Place in standard form.}\\
(x - 4)(x + 3) &= 0 &&\text{Factor.}\\
x - 4 = 0 \quad x + 3 &= 0 &&\text{Set each factor equal to zero.}\\
x = 4 \quad\quad x &= -3 &&\text{Solve the equations for } x.
\end{aligned}
$$

The check is left to the student.

Practice Problem 5   Solve. $\dfrac{2x^2 - 7x}{3} = 5$ $x = -\dfrac{3}{2},\ x = 5$

② Using Quadratic Equations to Solve Applied Problems

Certain types of word problems—for example, some geometry applications—lead to quadratic equations. We'll show how to solve such word problems in this section.

It is particularly important to check the apparent solutions to the quadratic equation with conditions stated in the word problem. Often a particular solution to the quadratic equation will be eliminated by the conditions of the word problem.

▲ **EXAMPLE 6** Carlos lives in Mexico City. He has a rectangular brick walkway in front of his house. The length of the walkway is 3 meters longer than twice the width. The area of the walkway is 44 square meters. Find the length and width of the rectangular walkway.

Teaching Example 6 The length of a rectangle is 3 inches shorter than triple the width. The rectangle has an area of 60 square inches. Find the length and width of the rectangle.

Ans: Width = 5 in.; length = 12 in.

Solution

1. **Understand the problem.**
   Draw a picture.

   Let w = the width in meters.
   Then $2w + 3$ = the length in meters.

2. **Write an equation.**

$$\text{area} = (\text{width})(\text{length})$$
$$44 = w(2w + 3)$$

3. **Solve and state the answer.**

$44 = w(2w + 3)$	
$44 = 2w^2 + 3w$	Remove parentheses.
$0 = 2w^2 + 3w - 44$	Subtract 44 from both sides.
$0 = (2w + 11)(w - 4)$	Factor.
$2w + 11 = 0 \quad w - 4 = 0$	Set each factor equal to 0.
$2w = -11 \qquad w = 4$	Simplify and solve.
$w = -5\frac{1}{2}$	Although $-5\frac{1}{2}$ is a solution to the quadratic equation, it is not a valid solution to the word problem. It would not make sense to have a rectangle with a negative number as a width.

Since $w = 4$, the width of the walkway is 4 meters. The length is $2w + 3$, so we have $2(4) + 3 = 8 + 3 = 11$. Thus the length of the walkway is 11 meters.

4. **Check.** Is the length 3 meters more than twice the width?

$$11 \overset{?}{=} 3 + 2(4) \qquad 11 = 3 + 8 \quad ✓$$

Is the area of the rectangle 44 square meters?

$$4 \times 11 \overset{?}{=} 44 \qquad 44 = 44 \quad ✓$$

▲ **Practice Problem 6** The length of a rectangle is 2 meters longer than triple the width. The area of the rectangle is 85 square meters. Find the length and width of the rectangle. Width = 5 m; length = 17 m

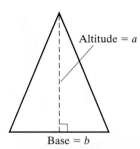

Altitude = a

Base = b

▲ **EXAMPLE 7** The top of a local cable television tower has several small triangular reflectors. The area of each triangle is 49 square centimeters. The altitude of each triangle is 7 centimeters longer than the base. Find the altitude and the base of one of the triangles.

Solution

Let b = the length of the base in centimeters
$b + 7$ = the length of the altitude in centimeters

To find the area of a triangle, we use

$$\text{area} = \frac{1}{2}(\text{altitude})(\text{base}) = \frac{1}{2}ab = \frac{ab}{2}.$$

$\dfrac{ab}{2} = 49$	Write an equation.
$\dfrac{(b + 7)(b)}{2} = 49$	Substitute the expressions for altitude and base.
$\dfrac{b^2 + 7b}{2} = 49$	Simplify.
$b^2 + 7b = 98$	Multiply each side of the equation by 2.
$b^2 + 7b - 98 = 0$	Place the quadratic equation in standard form.
$(b - 7)(b + 14) = 0$	Factor.
$b - 7 = 0 \qquad b + 14 = 0$	Set each factor equal to zero.
$b = 7 \qquad\quad b = -14$	Solve the equations for b.

We cannot have a base of -14 centimeters, so we reject the negative answer. The only possible solution is 7. So the base is 7 centimeters. The altitude is $b + 7 = 7 + 7 = 14$. The altitude is 14 centimeters. The triangular reflector has a base of 7 centimeters and an altitude of 14 centimeters.

Check. When you do the check, answer the following two questions.

1. Is the altitude 7 centimeters longer than the base?
2. Is the area of a triangle with a base of 7 centimeters and an altitude of 14 centimeters actually 49 square centimeters?

Teaching Example 7 A triangular window has an area of 30 square inches. The altitude of the triangle is 4 inches longer than the base. Find the altitude and the base of the window.

Ans: Base = 6 in.; altitude = 10 in.

NOTE TO STUDENT: Fully worked-out solutions to all of the Practice Problems can be found at the back of the text starting at page SP-1

▲ **Practice Problem 7** A triangle has an area of 35 square centimeters. The altitude of the triangle is 3 centimeters shorter than the base. Find the altitude and the base of the triangle. Altitude = 7 cm; base = 10 cm

Many problems in the sciences require the use of quadratic equations. You will study these in more detail if you take a course in physics or calculus in college. Often a quadratic equation is given as part of the problem.

When an object is thrown upward, its height (S) in meters is given, approximately, by the quadratic equation

$$S = -5t^2 + vt + h.$$

The letter h represents the initial height in meters. The letter v represents the initial velocity of the object thrown. The letter t represents the time in seconds starting from the time the object is thrown.

EXAMPLE 8   A tennis ball is thrown upward with an initial velocity of 8 meters/second. Suppose that the initial height above the ground is 4 meters. At what time t will the ball hit the ground?

4 m

Solution   In this case $S = 0$ since the ball will hit the ground. The initial upward velocity is $v = 8$ meters/second. The initial height is 4 meters, so $h = 4$.

$$S = -5t^2 + vt + h$$ Write an equation.

$$0 = -5t^2 + 8t + 4$$ Substitute all values into the equation.

$$5t^2 - 8t - 4 = 0$$ Isolate the terms on the left side. (Most students can factor more readily if the squared variable is positive.)

$$(5t + 2)(t - 2) = 0$$ Factor.

$$5t + 2 = 0 \qquad t - 2 = 0$$ Set each factor equal to 0.

$$5t = -2 \qquad t = 2$$ Solve the equations for t.

$$t = -\frac{2}{5}$$

Teaching Example 8   A baseball is thrown upward from an initial height of 3 meters. It has an initial velocity of 14 meters per second. At what time will the baseball hit the ground?

Ans: 3 sec

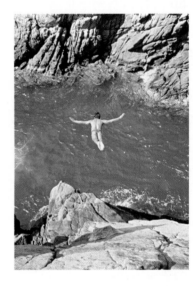

We want a positive time t in seconds; thus we do not use $t = -\frac{2}{5}$. Therefore, the ball will strike the ground 2 seconds after it is thrown.

Check.   Verify the solution.

Practice Problem 8   A Mexican cliff diver does a dive from a cliff 45 meters above the ocean. This constitutes free fall, so the initial velocity is $v = 0$, and since there is no upward velocity (no springboard), then $h = 45$ meters. How long will it be until he breaks the water's surface?   3 sec

Using the factoring method, solve for the roots of each quadratic equation. Be sure to place the equation in standard form before factoring. Check your answers.

1. $x^2 - 4x - 21 = 0$
$(x + 3)(x - 7) = 0$
$x = -3, x = 7$

2. $x^2 - x - 20 = 0$
$(x + 4)(x - 5) = 0$
$x = -4, x = 5$

3. $x^2 + 14x + 24 = 0$
$(x + 12)(x + 2) = 0$
$x = -12, x = -2$

4. $x^2 + 10x + 9 = 0$
$(x + 9)(x + 1) = 0$
$x = -9, x = -1$

5. $2x^2 - 7x + 6 = 0$
$(2x - 3)(x - 2) = 0$
$x = \dfrac{3}{2}, x = 2$

6. $2x^2 - 11x + 12 = 0$
$(2x - 3)(x - 4) = 0$
$x = \dfrac{3}{2}, x = 4$

7. $6x^2 - 13x = -6$
$(3x - 2)(2x - 3) = 0$
$x = \dfrac{2}{3}, x = \dfrac{3}{2}$

8. $10x^2 + 19x = 15$
$(5x - 3)(2x + 5) = 0$
$x = \dfrac{3}{5}, x = -\dfrac{5}{2}$

9. $x^2 + 13x = 0$
$x(x + 13) = 0$
$x = 0, x = -13$

10. $8x^2 - x = 0$
$x(8x - 1) = 0$
$x = 0, x = \dfrac{1}{8}$

11. $8x^2 = 72$
$8x^2 - 72 = 0$
$8(x + 3)(x - 3) = 0$
$x = 3, x = -3$

12. $9x^2 = 81$
$9x^2 - 81 = 0$
$9(x + 3)(x - 3) = 0$
$x = 3, x = -3$

13. $5x^2 + 3x = 8x$
$5(x^2 - x) = 0$
$5x(x - 1) = 0$
$x = 0, x = 1$

14. $6x^2 - 4x = 3x$
$x(6x - 7) = 0$
$x = 0, x = \dfrac{7}{6}$

15. $6x^2 = 16x - 8$
$6x^2 - 16x + 8 = 0$
$2(3x - 2)(x - 2) = 0$
$x = \dfrac{2}{3}, x = 2$

16. $24x^2 = -10x + 4$
$24x^2 + 10x - 4 = 0$
$2(4x - 1)(3x + 2) = 0$
$x = \dfrac{1}{4}, x = -\dfrac{2}{3}$

17. $(x - 5)(x + 2) = -4(x + 1)$
$x^2 + x - 6 = 0$
$(x + 3)(x - 2) = 0$
$x = -3, x = 2$

18. $(x - 5)(x + 4) = 2(x - 5)$
$x^2 - 3x - 10 = 0$
$(x + 2)(x - 5) = 0$
$x = -2, x = 5$

19. $4x^2 - 3x + 1 = -7x$
$(2x + 1)^2 = 0$
$x = -\dfrac{1}{2}$

20. $9x^2 - 2x + 4 = 10x$
$(3x - 2)^2 = 0$
$x = \dfrac{2}{3}$

21. $\dfrac{x^2}{3} - 5 + x = -5$
$x(x + 3) = 0$
$x = 0, x = -3$

22. $7 + \dfrac{x^2}{2} - x = 7$
$x(x - 2) = 0$
$x = 0, x = 2$

23. $\dfrac{x^2 + 10x}{8} = -2$
$x^2 + 10x + 16 = 0$
$(x + 8)(x + 2) = 0$
$x = -8, x = -2$

24. $\dfrac{x^2 + 7x}{5} = 6$
$x^2 + 7x - 30 = 0$
$(x + 10)(x - 3) = 0$
$x = -10, x = 3$

25. $\dfrac{10x^2 - 25x}{12} = 5$
$10x^2 - 25x - 60 = 0$
$5(2x + 3)(x - 4) = 0$
$x = -\dfrac{3}{2}, x = 4$

26. $\dfrac{12x^2 - 4x}{5} = 8$
$12x^2 - 4x - 40 = 0$
$4(3x + 5)(x - 2) = 0$
$x = -\dfrac{5}{3}, x = 2$

To Think About

27. Why can an equation in standard form with $c = 0$ (that is, an equation of the form $ax^2 + bx = 0$) always be solved?

You can always factor out x.

28. Martha solved $(x + 3)(x - 2) = 14$ as follows:

$$x + 3 = 14 \quad \text{or} \quad x - 2 = 14$$
$$x = 11 \quad \text{or} \quad x = 16$$

Josette said this had to be wrong because these values do not check. Explain what is wrong with Martha's method.

You must have the product of two or more factors equal to zero, not 14. Remember the zero factor property.

Applications

▲ **29.** *Geometry* The area of a rectangular garden is 140 square meters. The width is 3 meters longer than one-half of the length. Find the length and the width of the garden.

$$x\left(\frac{x}{2} + 3\right) = 140$$
$$x^2 + 6x - 280 = 0$$
$$(x + 20)(x - 14) = 0$$
$$\text{length} = 14 \text{ m}$$
$$\text{width} = 10 \text{ m}$$

▲ **30.** *Geometry* The area of a triangular sign is 33 square meters. The base of the triangle is 1 meter less than double the altitude. Find the altitude and the base of the sign.

$$\frac{1}{2}x(2x - 1) = 33$$
$$2x^2 - x - 66 = 0$$
$$(2x + 11)(x - 6) = 0$$
$$\text{altitude} = 6 \text{ m}$$
$$\text{base} = 11 \text{ m}$$

Forming Groups Suppose the number of students in a mathematics class is x. The teacher insists that each student participate in group work each class. The number of possible groups is:

$$G = \frac{x^2 - 3x + 2}{2}$$

31. The class has 13 students. How many possible groups are there?   66 groups

32. Four students withdraw from the class. How many fewer groups can be formed?   38 fewer groups

33. A teacher claims each student could be in 36 different groups. How many students are there?

10 students

34. The teacher wants each student to participate in a different group each day. There are 45 class days in the semester. How many students must be in the class?   11 students

Falling Object Use the following information for exercises 35 and 36. When an object is thrown upward, its height (S), in meters, is given (approximately) by the quadratic equation

$$S = -5t^2 + vt + h,$$

where v = the upward initial velocity in meters/second,

t = the time of flight in seconds, and

h = the height above level ground from which the object is thrown.

35. Johnny is standing on a platform 6 meters high and throws a ball straight up as high as he can at a velocity of 13 meters/second. At what time t will the ball hit the ground? How far from the ground is the ball after 2 seconds have elapsed from the time of the throw? (Assume that the ball is 6 meters from the ground when it leaves Johnny's hand.)

$-5t^2 + 13t + 6 = 0$
$(5t + 2)(t - 3) = 0$ It will hit the ground after 3 seconds.
$S = -5(2)^2 + 13(2) + 6$
   $= 12$ m above ground after 2 sec

36. You are standing on the edge of a cliff near Acapulco, overlooking the ocean. The place where you stand is 180 meters from the ocean. You drop a pebble into the water. ("Dropping" the pebble implies that there is no initial velocity, so $v = 0$.) How many seconds will it take to hit the water? How far has the pebble dropped after 3 seconds?

$0 = -5t^2 + 180$
$t^2 - 36 = 0$
   $t = 6$ It will hit the water in 6 seconds.
$S = -5(3)^2 + 180$
   $= 135$ m $180 - 135 = 45$ m
After 3 seconds it has dropped 45 meters.

Internal Phone Calls The technology and communication office of a local company has set up a new telephone system so that each employee has a separate telephone and extension number. They are studying the possible number of telephone calls that can be made from people in the office to other people in the office. They have discovered that the total number of possible telephone calls T is described by the equation $T = 0.5(x^2 - x)$, where x is the number of people in the office. Use this information to answer exercises 37–38.

37. If 70 people are presently employed at the office, how many possible telephone calls can be made between these 70 people?   2415 telephone calls

38. On the day after Thanksgiving, only a small number of employees were working at the office. It has been determined that on that day, a total of 105 different phone calls could have been made from people working in the office to other people working in the office. How many people worked on the day after Thanksgiving?   15 people

The same equation given above is used in the well-known "handshake problem." If there are n people at a party and each person shakes hands with every other person once, the number of handshakes that take place among the n people is $H = 0.5(n^2 - n)$.

39. Barry is hosting a holiday party and has invited 16 friends. If all his friends attend the party and everyone, including Barry, shakes hands with everyone else at the party, how many handshakes will take place?
136 handshakes

40. At a mathematics conference, 55 handshakes took place among the organizers during the opening meeting. How many organizers were there?
11 organizers

Cumulative Review   *Simplify.*

41. [4.1.1] $(2x^2y^3)(-5x^3y)$ $-10x^5y^4$

42. [4.1.1] $(3a^4b^5)(4a^6b^8)$ $12a^{10}b^{13}$

43. [4.1.2] $\dfrac{21a^5b^{10}}{-14ab^{12}}$ $-\dfrac{3a^4}{2b^2}$

44. [4.1.2] $\dfrac{18x^3y^6}{54x^8y^{10}}$ $\dfrac{1}{3x^5y^4}$

Quick Quiz 5.7   Solve.

1. $15x^2 - 8x + 1 = 0$ $x = \dfrac{1}{5}, x = \dfrac{1}{3}$

2. $4 + x(x - 2) = 7$ $x = 3, x = -1$

3. $4x^2 = 9x + 9$ $x = 3, x = -\dfrac{3}{4}$

4. Concept Check Explain how you would solve the following problem: A rectangle has an area of 65 square feet. The length of the rectangle is 3 feet longer than double the width. Find the length and the width of the rectangle.   Answers may vary

Classroom Quiz 5.7   You may use these problems to quiz your students' mastery of Section 5.7.

Solve.

1. $12x^2 + 5x - 2 = 0$
Ans: $x = -\dfrac{2}{3}, x = \dfrac{1}{4}$

2. $-24 + x(x - 4) = x$
Ans: $x = 8, x = -3$

3. $2x^2 = -5x + 7$
Ans: $x = -\dfrac{7}{2}, x = 1$

Putting Your Skills to Work: Use Math to Save Money

CAR REBATES/PAYMENTS

Basides a house, the biggest purchase many people make is a car. When shopping for automobile financing, customers need to take many factors into account, including the amount of the monthly loan payment and whether to choose a lower loan rate or a rebate. Consider the story of Megan.

Megan is interested in buying a $25,000 car and has found a loan that has a term of 5 years and a 9% interest rate. She wants to know what her monthly payments and total cost will be.

To calculate Megan's monthly car payment, we use the amortization formula

$$PMT = \frac{PV}{\left[1 - \dfrac{1}{(1 + I)^N}\right] \Big/ I}.$$

This formula uses the following variables:

- *PMT* is the monthly payment.
- *I* is the interest rate in decimal notation divided by 12, which is the amount of interest paid each month.
- *PV* is the present value or the amount being financed.
- *N* is 12 times the number of years of the loan, which is the number of payments needed to pay off the loan.

Here, $PV = \$25,000$, $I = \dfrac{0.09}{12} = 0.0075$, and

$N = 12 \times 5 = 60$.

1. Substitute these values into the amortization formula to find Megan's monthly payment (PMT), rounded to the nearest ten dollars. $520

2. What is the total amount that Megan will pay in car payments during the 5 years? About $31,200

3. What is the total amount of interest that Megan will pay for the car? About $6200

Suppose the car comes with the following options: (A) a $2000 rebate that Megan deducts from the price of the car before financing the remainder at 9%, or (B) no rebate and a 5-year, 0% interest rate.

4. What would Megan's monthly payment be using option A? Round your answer to the nearest ten dollars. $480

5. What is the total amount that Megan would pay in car payments using option A? About $28,800

6. What is the total amount that Megan would pay in car payments using option B? $25,000

7. Which option would save Megan the most money? Option B

Extension

8. Look in a newspaper and make the similar comparisons among finance rates and rebates. Answers may vary

Topic	Procedure	Examples
A. Common factor, p. 302.	Factor out the largest common factor from each term.	$2x^2 - 2x = 2x(x - 1)$ $3a^2 + 3ab - 12a = 3a(a + b - 4)$ $8x^4y - 24x^3 = 8x^3(xy - 3)$
Special cases **B. Difference of two squares,** p. 326. **C. Perfect-square trinomials,** p. 327.	If you recognize the special cases, you will be able to factor quickly. $a^2 - b^2 = (a + b)(a - b)$ $a^2 + 2ab + b^2 = (a + b)^2$ $a^2 - 2ab + b^2 = (a - b)^2$	$25x^2 - 36y^2 = (5x + 6y)(5x - 6y)$ $16x^4 - 1 = (4x^2 + 1)(2x + 1)(2x - 1)$ $25x^2 + 10x + 1 = (5x + 1)^2$ $49x^2 - 42xy + 9y^2 = (7x - 3y)^2$
D. Trinomials of the form $x^2 + bx + c$, p. 312.	Factor trinomials of the form $x^2 + bx + c$ by finding two numbers that have a product of c and a sum of b. If each term of the trinomial has a common factor, factor it out as the first step.	$x^2 - 18x + 77 = (x - 7)(x - 11)$ $x^2 + 7x - 18 = (x + 9)(x - 2)$ $5x^2 - 10x - 40 = 5(x^2 - 2x - 8)$ $\qquad\qquad\qquad = 5(x - 4)(x + 2)$
E. Trinomials of the form $ax^2 + bx + c$, where $a \neq 1$, p. 319.	Factor trinomials of the form $ax^2 + bx + c$ by the grouping number method or by the trial-and-error method.	$6x^2 + 11x - 10$ Grouping number $= -60$ Two numbers whose product is -60 and whose sum is $+11$ are $+15$ and -4. $6x^2 + 15x - 4x - 10 = 3x(2x + 5) - 2(2x + 5)$ $\qquad\qquad\qquad\qquad\quad = (2x + 5)(3x - 2)$
F. Four terms. Factor by grouping, p. 307.	Rearrange the terms if necessary so that the first two terms have a common factor. Then factor out the common factors. $ax + ay - bx - by = a(x + y) - b(x + y)$ $\qquad\qquad\qquad\qquad = (x + y)(a - b)$	$2ax^2 + 21 + 14x^2 + 3a$ $= 2ax^2 + 14x^2 + 3a + 21$ $= 2x^2(a + 7) + 3(a + 7)$ $= (a + 7)(2x^2 + 3)$
Prime polynomials, p. 333.	A polynomial that is not factorable is called prime.	$x^2 + y^2$ is prime. $x^2 + 5x + 7$ is prime.
Multistep factoring, p. 329.	Many problems require two or three steps of factoring. Always try to factor out the greatest common factor as the first step.	$3x^2 - 21x + 36 = 3(x^2 - 7x + 12)$ $\qquad\qquad\qquad\quad = 3(x - 4)(x - 3)$ $2x^3 - x^2 - 6x = x(2x^2 - x - 6)$ $\qquad\qquad\qquad = x(2x + 3)(x - 2)$ $25x^3 - 49x = x(25x^2 - 49)$ $\qquad\qquad\quad = x(5x + 7)(5x - 7)$ $8x^2 - 24x + 18 = 2(4x^2 - 12x + 9)$ $\qquad\qquad\qquad\quad = 2(2x - 3)^2$
Solving quadratic equations by factoring, p. 336.	1. Write as $ax^2 + bx + c = 0$. 2. Factor. 3. Set each factor equal to 0. 4. Solve the resulting equations.	Solve. $3x^2 + 5x = 2$ $3x^2 + 5x - 2 = 0$ $(3x - 1)(x + 2) = 0$ $3x - 1 = 0 \qquad x + 2 = 0$ $x = \dfrac{1}{3} \qquad\quad x = -2$

Topic	Procedure	Examples
Using quadratic equations to solve applied problems, p. 339.	Some word problems, like those involving the product of two numbers, area, and formulas with a squared variable, can be solved using the factoring methods we have shown.	The length of a rectangle is 4 less than three times the width. Find the length and width if the area is 55 square inches. Let w = width. Then $3w - 4$ = length. $$55 = w(3w - 4)$$ $$55 = 3w^2 - 4w$$ $$0 = 3w^2 - 4w - 55$$ $$0 = (3w + 11)(w - 5)$$ $$w = -\tfrac{11}{3} \quad \text{or} \quad w = 5$$ $-\tfrac{11}{3}$ is not a valid solution. Thus width = 5 inches and length = 11 inches.

Chapter 5 Review Problems

Section 5.1

Factor out the greatest common factor.

1. $12x^3 - 20x^2 y$ $4x^2(3x - 5y)$

2. $10x^3 - 35x^3 y$ $5x^3(2 - 7y)$

3. $24x^3 y - 8x^2 y^2 - 16x^3 y^3$ $8x^2 y(3x - y - 2xy^2)$

4. $45a^3 b^3 + 30a^2 b^3 - 15a^2 b^4$ $15a^2 b^3(3a + 2 - b)$

5. $3a^3 + 6a^2 - 9ab + 12a$ $3a(a^2 + 2a - 3b + 4)$

6. $2x - 4y + 6z + 12$ $2(x - 2y + 3z + 6)$

7. $2a(a + 3b) - 5(a + 3b)$ $(a + 3b)(2a - 5)$

8. $15x^3 y + 6xy^2 + 3xy$ $3xy(5x^2 + 2y + 1)$

Section 5.2

Factor by grouping.

9. $2ax + 5a - 8x - 20$ $(2x + 5)(a - 4)$

10. $a^2 - 4ab + 7a - 28b$ $(a - 4b)(a + 7)$

11. $x^2 y + 3y - 2x^2 - 6$ $(x^2 + 3)(y - 2)$

12. $30ax - 15ay + 42x - 21y$ $3(2x - y)(5a + 7)$

13. $15x^2 - 3x + 10x - 2$ $(5x - 1)(3x + 2)$

14. $30w^2 - 18w + 5wz - 3z$ $(5w - 3)(6w + z)$

Section 5.3

Factor completely. Be sure to factor out any common factors as your first step.

15. $x^2 + 6x - 27$ $(x + 9)(x - 3)$

16. $x^2 + 9x - 10$ $(x + 10)(x - 1)$

17. $x^2 + 14x + 48$ $(x + 6)(x + 8)$

18. $x^2 + 8xy + 15y^2$ $(x + 3y)(x + 5y)$

19. $x^4 + 13x^2 + 42$ $(x^2 + 7)(x^2 + 6)$

20. $x^4 - 2x^2 - 35$ $(x^2 - 7)(x^2 + 5)$

21. $6x^2 + 30x + 36$ $6(x + 2)(x + 3)$

22. $3x^2 + 39x + 36$ $3(x + 1)(x + 12)$

23. $2x^2 - 28x + 96$ $2(x - 6)(x - 8)$

24. $4x^2 - 44x + 120$ $4(x - 5)(x - 6)$

Section 5.4

Factor completely. Be sure to factor out any common factors as your first step.

25. $4x^2 + 7x - 15$ $(4x - 5)(x + 3)$

26. $12x^2 + 11x - 5$ $(3x - 1)(4x + 5)$

27. $15x^2 + 7x - 4$ $(5x + 4)(3x - 1)$

28. $6x^2 - 13x + 6$ $(3x - 2)(2x - 3)$

29. $2x^2 - x - 3$ $(2x - 3)(x + 1)$

30. $3x^2 + 2x - 8$ $(3x - 4)(x + 2)$

31. $20x^2 + 48x - 5$ $(10x - 1)(2x + 5)$

32. $20x^2 + 21x - 5$ $(5x - 1)(4x + 5)$

33. $6a^2 + 11a - 10$ $(3a - 2)(2a + 5)$

34. $6a^2 - 19a + 10$ $(3a - 2)(2a - 5)$

35. $6x^2 + 4x - 10$ $2(x - 1)(3x + 5)$

36. $6x^2 - 4x - 10$ $2(x + 1)(3x - 5)$

37. $4x^2 - 26x + 30$ $2(2x - 3)(x - 5)$

38. $4x^2 - 20x - 144$ $4(x - 9)(x + 4)$

39. $16x^2 + 16x - 60$ $4(2x - 3)(2x + 5)$

40. $24x^2 + 6x - 3$ $3(4x - 1)(2x + 1)$

41. $12x^2 + xy - 6y^2$ $(4x + 3y)(3x - 2y)$

42. $6x^2 + 5xy - 25y^2$ $(3x - 5y)(2x + 5y)$

Section 5.5

Factor these special cases. Be sure to factor out any common factors.

43. $49x^2 - y^2$ $(7x + y)(7x - y)$

44. $16x^2 - 36y^2$ $4(2x + 3y)(2x - 3y)$

45. $36x^2 - 60x + 25$ $(6x - 5)^2$

46. $16x^2 + 40x + 25$ $(4x + 5)^2$

47. $25x^2 - 36$ $(5x + 6)(5x - 6)$

48. $100x^2 - 9$ $(10x + 3)(10x - 3)$

49. $y^2 - 36x^2$ $(y + 6x)(y - 6x)$

50. $9y^2 - 25x^2$ $(3y + 5x)(3y - 5x)$

51. $36x^2 + 12x + 1$ $(6x + 1)^2$

52. $25x^2 - 20x + 4$ $(5x - 2)^2$

53. $16x^2 - 24xy + 9y^2$ $(4x - 3y)^2$

54. $49x^2 - 28xy + 4y^2$ $(7x - 2y)^2$

55. $2x^2 - 32$ $2(x + 4)(x - 4)$

56. $3x^2 - 27$ $3(x + 3)(x - 3)$

57. $28x^2 + 140x + 175$ $7(2x + 5)^2$

58. $72x^2 - 192x + 128$ $8(3x - 4)^2$

Section 5.6

If possible, factor each polynomial completely. If a polynomial cannot be factored, state that it is prime.

59. $4x^2 - 9y^2$ $(2x + 3y)(2x - 3y)$

60. $x^2 + 6x + 9$ $(x + 3)^2$

61. $x^2 - 9x + 18$ $(x - 3)(x - 6)$

62. $x^2 + 13x - 30$ $(x + 15)(x - 2)$

63. $5x^2 - 13x - 6$ $(x - 3)(5x + 2)$

64. $9x^2 - 9x - 4$ $(3x - 4)(3x + 1)$

65. $24x - 60$ $12(2x - 5)$

66. $8x^2y^2 - 4xy$ $4xy(2xy - 1)$

67. $50x^3y^2 + 30x^2y^2 - 10x^2y^2$ $10x^2y^2(5x + 2)$

68. $26a^3b - 13ab^3 + 52a^2b^4$ $13ab(2a^2 - b^2 + 4ab^3)$

69. $x^3 - 16x^2 + 64x$ $x(x - 8)^2$

70. $2x^2 + 40x + 200$ $2(x + 10)^2$

71. $3x^2 - 18x + 27$ $3(x - 3)^2$

72. $25x^3 - 60x^2 + 36x$ $x(5x - 6)^2$

73. $7x^2 - 9x - 10$ $(7x + 5)(x - 2)$

74. $4x^2 - 13x - 12$ $(4x + 3)(x - 4)$

75. $9x^3y - 4xy^3$ $xy(3x + 2y)(3x - 2y)$

76. $3x^3a^3 - 11x^4a^2 - 20x^5a$ $x^3a(3a + 4x)(a - 5x)$

77. $12a^2 + 14ab - 10b^2$ $2(3a + 5b)(2a - b)$

78. $121a^2 + 66ab + 9b^2$ $(11a + 3b)^2$

79. $7a - 7 - ab + b$ $(a - 1)(7 - b)$

Mixed Practice

If possible, factor each polynomial completely. If a polynomial cannot be factored, state that it is prime.

80. $18b - 42 + 3bc - 7c$ $(3b - 7)(6 + c)$

81. $10b + 16 - 24x - 15bx$ $(5b + 8)(2 - 3x)$

82. $5xb - 35x + 4by - 28y$ $(b - 7)(5x + 4y)$

83. $2a^2x - 15ax + 7x$ $x(2a - 1)(a - 7)$

84. $x^5 - 17x^3 + 16x$ $x(x + 4)(x - 4)(x + 1)(x - 1)$

85. $x^4 - 81y^{12}$ $(x^2 + 9y^6)(x + 3y^3)(x - 3y^3)$

86. $6x^4 - x^2 - 15$ $(3x^2 - 5)(2x^2 + 3)$

87. $28yz - 16xyz + x^2yz$ $yz(14 - x)(2 - x)$

88. $12x^3 + 17x^2 + 6x$ $x(3x + 2)(4x + 3)$

89. $16w^2 - 2w - 5$ $(2w + 1)(8w - 5)$

90. $12w^2 - 12w + 3$ $3(2w - 1)^2$

91. $4y^3 + 10y^2 - 6y$ $2y(2y - 1)(y + 3)$

92. $10y^2 + 33y - 7$ $(5y - 1)(2y + 7)$

93. $8y^{10} - 16y^8$ $8y^8(y^2 - 2)$

94. $9x^4 - 144$ $9(x^2 + 4)(x + 2)(x - 2)$

95. $x^2 - 6x + 12$ Prime

96. $8x^2 - 19x - 6$ Prime

97. $8y^5 + 4y^3 - 60y$ $4y(2y^2 - 5)(y^2 + 3)$

98. $9xy^2 + 3xy - 42x$ $3x(3y + 7)(y - 2)$

99. $16x^4y^2 - 56x^2y + 49$ $(4x^2y - 7)^2$

100. $128x^3y - 2xy$ $2xy(8x + 1)(8x - 1)$

101. $2ax + 5a - 10b - 4bx$ $(2x + 5)(a - 2b)$

102. $2x^3 - 9 + x^2 - 18x$ $(2x + 1)(x + 3)(x - 3)$

Section 5.7

Solve the following equations by factoring.

103. $x^2 + x - 20 = 0$ $x = -5, x = 4$

104. $2x^2 + 11x - 6 = 0$ $x = -6, x = \dfrac{1}{2}$

105. $7x^2 = 15x + x^2$ $x = 0, x = \dfrac{5}{2}$

106. $5x^2 - x = 4x^2 + 12$ $x = 4, x = -3$

107. $2x^2 + 9x - 5 = 0$ $x = -5, x = \dfrac{1}{2}$

108. $x^2 + 11x + 24 = 0$ $x = -8, x = -3$

109. $x^2 + 14x + 45 = 0$ $x = -5, x = -9$

110. $5x^2 = 7x + 6$ $x = -\dfrac{3}{5}, x = 2$

111. $3x^2 + 6x = 2x^2 - 9$ $x = -3$

112. $4x^2 + 9x - 9 = 0$ $x = -3, x = \dfrac{3}{4}$

113. $5x^2 - 11x + 2 = 0$ $x = \dfrac{1}{5}, x = 2$

Solve.

▲ **114.** *Geometry* The area of a triangle is 25 square inches. The altitude is 5 inches longer than the base. Find the length of the base and the altitude.
Base = 5 in., altitude = 10 in.

▲ **115.** *Geometry* The area of a rectangle is 30 square feet. The length of the rectangle is 4 feet shorter than double the width. Find the length and width of the rectangle.
Width = 5 ft, length = 6 ft

116. *Rocket Height* The height in feet that a model rocket attains is given by $h = -16t^2 + 80t + 96$, where t is the time measured in seconds. How many seconds will it take until the rocket finally reaches the ground? (*Hint:* At ground level $h = 0$.)   6 sec

117. *Output Power* An electronic technician is working with a 100-volt electric generator. The output power of the generator is given by the equation $p = -5x^2 + 100x$, where x is the amount of current measured in amperes and p is measured in watts. The technician wants to find the value for x when the power is 480 watts. Can you find the two answers?
8 amperes, 12 amperes

How Am I Doing? Chapter 5 Test

Remember to use your Chapter Test Prep Video CD to see the worked-out solutions to the test problems you want to review.

If possible, factor each polynomial completely. If a polynomial cannot be factored, state that it is prime.

1. $x^2 + 12x - 28$

2. $16x^2 - 81$

3. $10x^2 + 27x + 5$

4. $9a^2 - 30a + 25$

5. $7x - 9x^2 + 14xy$

6. $10xy + 15by - 8x - 12b$

7. $6x^3 - 20x^2 + 16x$

8. $5a^2c - 11ac + 2c$

9. $81x^2 - 100$

10. $9x^2 - 15x + 4$

11. $20x^2 - 45$

12. $36x^2 + 1$

13. $3x^3 + 11x^2 + 10x$

14. $60xy^2 - 20x^2y - 45y^3$

15. $81x^2 - 1$

16. $81y^4 - 1$

17. $2ax + 6a - 5x - 15$

18. $aw^2 - 8b + 2bw^2 - 4a$

19. $3x^2 - 3x - 90$

20. $2x^3 - x^2 - 15x$

Solve.

21. $x^2 + 14x + 45 = 0$

22. $14 + 3x(x + 2) = -7x$

23. $2x^2 + x - 10 = 0$

24. $x^2 - 3x - 28 = 0$

Solve using a quadratic equation.

▲ **25.** The park service is studying a rectangular piece of land that has an area of 91 square miles. The length of this piece of land is 1 mile shorter than double the width. Find the length and width of this rectangular piece of land.

Note to Instructor: The Chapter 5 Test file in the TestGen program provides algorithms specifically matched to these problems so you can easily replicate this test for additional practice or assessment purposes.

1.	$(x + 14)(x - 2)$
2.	$(4x + 9)(4x - 9)$
3.	$(5x + 1)(2x + 5)$
4.	$(3a - 5)^2$
5.	$x(7 - 9x + 14y)$
6.	$(2x + 3b)(5y - 4)$
7.	$2x(3x - 4)(x - 2)$
8.	$c(5a - 1)(a - 2)$
9.	$(9x + 10)(9x - 10)$
10.	$(3x - 1)(3x - 4)$
11.	$5(2x + 3)(2x - 3)$
12.	Prime
13.	$x(3x + 5)(x + 2)$
14.	$-5y(2x - 3y)^2$
15.	$(9x + 1)(9x - 1)$
16.	$(9y^2 + 1)(3y + 1)(3y - 1)$
17.	$(x + 3)(2a - 5)$
18.	$(a + 2b)(w + 2)(w - 2)$
19.	$3(x - 6)(x + 5)$
20.	$x(2x + 5)(x - 3)$
21.	$x = -5, x = -9$
22.	$x = -\dfrac{7}{3}, x = -2$
23.	$x = -\dfrac{5}{2}, x = 2$
24.	$x = 7, x = -4$
25.	Width = 7 mi, length = 13 mi

1.	20%
2.	0.54
3.	5.8
4.	$8x^4y^{10}$
5.	81
6.	$27x^2 + 6x - 8$
7.	$2x^3 - 12x^2 + 19x - 3$
8.	$x \geq -\dfrac{9}{2}$
9.	$x = 2$
10.	$x = -15$
11.	$q = \dfrac{p + 3}{5}$
12.	$(3x - 1)(2x - 1)$
13.	$(3x + 4)(2x - 1)$
14.	$(3x + 2)(3x - 1)$
15.	$(11x + 8y)(11x - 8y)$
16.	$-4(5x + 6)(4x - 5)$
17.	Prime
18.	$x(4x + 5)^2$
19.	$(4x^2 + b^2)(2x + b)(2x - b)$
20.	$(2x + 3)(a - 2b)$
21.	$3(5x - 3)(x + 1)$
22.	$x = -9, x = 4$
23.	$x = \dfrac{5}{3}, x = 2$
24.	Length = 20 ft, width = 15 ft

Approximately one-half of this test covers the content of Chapters 0–4. The remainder covers the content of Chapter 5.

1. What percent of 300 is 60?

2. What is 12% of 4.5?

Simplify.

3. $4.3 - 2(3.4 - 7.2) - 6.1$

4. $(-2x^3y^4)(-4xy^6)$

5. $(-3)^4$

6. $(9x - 4)(3x + 2)$

7. $(2x^2 - 6x + 1)(x - 3)$

Solve.

8. $5x - 6 \leq 7x + 3$

9. $3x - (7 - 5x) = 3(4x - 5)$

10. $\dfrac{1}{2}x - 3 = \dfrac{1}{4}(3x + 3)$

11. Solve for q. $3(q + 1) = 8q - p$

Factor each polynomial completely. If a polynomial cannot be factored, state that it is prime.

12. $6x^2 - 5x + 1$

13. $6x^2 + 5x - 4$

14. $9x^2 + 3x - 2$

15. $121x^2 - 64y^2$

16. $4x + 120 - 80x^2$

17. $x^2 + 5x + 9$

18. $16x^3 + 40x^2 + 25x$

19. $16x^4 - b^4$

20. $2ax - 4bx + 3a - 6b$

21. $15x^2 + 6x - 9$

Solve.

22. $x^2 + 5x - 36 = 0$

23. $3x^2 - 11x + 10 = 0$

Solve using a quadratic equation.

▲ **24.** Mr. Jerome's garden covered 300 square feet. If the width was 5 feet less than the length, find the measurements of the garden.

CHAPTER

6

Calculators are everywhere. The demand for calculators continues to increase throughout the country. Those who open stores specializing in calculators often find that their profits depend greatly on the amount of advertising they do. Could you use your mathematical skills to predict the profits of a calculator store? The mathematics you will learn in this chapter will help you predict the future profitability of a small business.

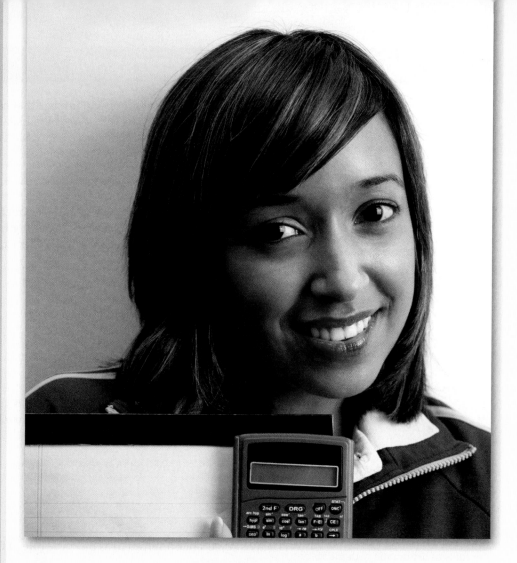

Rational Expressions and Equations

Student Learning Objective

After studying this section, you will be able to:

 Simplify rational expressions by factoring.

Recall that a rational number is a number that can be written as one integer divided by another integer, such as $3 \div 4$ or $\frac{3}{4}$. We usually use the word *fraction* to mean $\frac{3}{4}$. We can extend this idea to algebraic expressions. A **rational expression** is a polynomial divided by another polynomial, such as

$$(3x + 2) \div (x + 4) \quad \text{or} \quad \frac{3x + 2}{x + 4}.$$

The last fraction is sometimes also called a **fractional algebraic expression.** There is a special restriction for all fractions, including fractional algebraic expressions. The denominator of the fraction cannot be 0. For example, in the expression

$$\frac{3x + 2}{x + 4},$$

the denominator cannot be 0. Therefore, the value of x cannot be -4. The following important restriction will apply throughout this chapter. We state it here to avoid having to mention it repeatedly throughout this chapter.

> **RESTRICTION**
>
> The denominator of a rational expression cannot be zero. Any value of the variable that would make the denominator zero is not allowed.

We have discovered that fractions can be simplified (or reduced) in the following way.

$$\frac{15}{25} = \frac{3 \cdot \cancel{5}}{5 \cdot \cancel{5}} = \frac{3}{5}$$

This is sometimes referred to as the **basic rule of fractions** and can be stated as follows.

> **BASIC RULE OF FRACTIONS**
>
> For any rational expression $\frac{a}{b}$ and any polynomials a, b, and c (where $b \neq 0$ and $c \neq 0$),
>
> $$\frac{ac}{bc} = \frac{a}{b}.$$

We will examine several examples where a, b, and c are real numbers, as well as more involved examples where a, b, and c are polynomials. In either case we shall make extensive use of our factoring skills in this section.

One essential property is revealed by the basic rule of fractions: If the numerator and denominator of a given fraction are multiplied by the same nonzero quantity, an equivalent fraction is obtained. The rule can be used two ways. You can start with $\frac{ac}{bc}$ and end with the equivalent fraction $\frac{a}{b}$. Or, you can start with $\frac{a}{b}$ and end with the equivalent fraction $\frac{ac}{bc}$. In this section we focus on the first process.

Teaching Tip Taking the time to review simple rules of building and reducing fractions with numerical values will help to establish a clearer picture of how to simplify algebraic fractions. Taking the time to connect the rules of arithmetic with the rules of algebra is always profitable for the students taking a beginning algebra course.

Teaching Example 1 Reduce. $\frac{35}{42}$

Ans: $\frac{5}{6}$

NOTE TO STUDENT: Fully worked-out solutions to all of the Practice Problems can be found at the back of the text starting at page SP-1

EXAMPLE 1 Reduce. $\dfrac{21}{39}$

Solution $\dfrac{21}{39} = \dfrac{7 \cdot \cancel{3}}{13 \cdot \cancel{3}} = \dfrac{7}{13}$ Use the rule $\frac{ac}{bc} = \frac{a}{b}$. Let $c = 3$ because 3 is the greatest common factor of 21 and 39.

Practice Problem 1 Reduce. $\dfrac{28}{63} \quad \dfrac{4}{9}$

Simplifying Rational Expressions by Factoring

The process of reducing fractions shown in Example 1 is sometimes called *dividing out* common factors. When you do this, you are **simplifying the fraction.** Remember, only **factors** of both the numerator and the denominator can be divided out. To apply the basic rule of fractions, it is usually helpful if the numerator and denominator of the fraction are completely factored. You will need to use your factoring skills from Chapter 5 to accomplish this step.

EXAMPLE 2 Simplify. $\dfrac{4x + 12}{5x + 15}$

Solution

$$\dfrac{4x + 12}{5x + 15} = \dfrac{4(x + 3)}{5(x + 3)} \quad \begin{array}{l} \text{Factor 4 from the numerator.} \\ \text{Factor 5 from the denominator.} \end{array}$$

$$= \dfrac{4\,\cancel{(x + 3)}}{5\,\cancel{(x + 3)}} \quad \text{Apply the basic rule of fractions.}$$

$$= \dfrac{4}{5}$$

Practice Problem 2 Simplify. $\dfrac{12x - 6}{14x - 7} \quad \dfrac{6}{7}$

Teaching Example 2 Simplify. $\dfrac{3x - 15}{7x - 35}$

Ans: $\dfrac{3}{7}$

EXAMPLE 3 Simplify. $\dfrac{x^2 + 9x + 14}{x^2 - 4}$

Solution

$$= \dfrac{(x + 7)(x + 2)}{(x - 2)(x + 2)} \quad \begin{array}{l} \text{Factor the numerator.} \\ \text{Factor the denominator.} \end{array}$$

$$= \dfrac{(x + 7)\,\cancel{(x + 2)}}{(x - 2)\,\cancel{(x + 2)}} \quad \text{Apply the basic rule of fractions.}$$

$$= \dfrac{x + 7}{x - 2}$$

Teaching Example 3 Simplify. $\dfrac{x^2 - 3x + 2}{x^2 + 2x - 8}$

Ans: $\dfrac{x - 1}{x + 4}$

CAUTION: Do not try to remove terms that are added. In Example 3, do not try to remove the x^2 in the top and the x^2 in the bottom of the fraction. The basic rule of fractions applies only to quantities that are factors of both numerator and denominator.

Practice Problem 3 Simplify. $\dfrac{4x^2 - 9}{2x^2 - x - 3} \quad \dfrac{2x + 3}{x + 1}$

Some problems may involve more than one step of factoring. Always remember to factor out any common factors as the first step, if it is possible to do so.

EXAMPLE 4 Simplify. $\dfrac{x^3 - 9x}{x^3 + x^2 - 6x}$

Solution

$$= \dfrac{x(x^2 - 9)}{x(x^2 + x - 6)} \quad \begin{array}{l} \text{Factor out a common factor from the polynomials} \\ \text{in the numerator and the denominator.} \end{array}$$

$$= \dfrac{\cancel{x}\,\cancel{(x + 3)}(x - 3)}{\cancel{x}\,\cancel{(x + 3)}(x - 2)} \quad \begin{array}{l} \text{Factor each polynomial and apply the basic} \\ \text{rule of fractions.} \end{array}$$

$$= \dfrac{x - 3}{x - 2}$$

Teaching Example 4 Simplify.

$$\dfrac{x^3 - 4x^2 - 5x}{x^3 - 25x}$$

Ans: $\dfrac{x + 1}{x + 5}$

Practice Problem 4 Simplify. $\dfrac{x^3 - 16x}{x^3 - 2x^2 - 8x} \quad \dfrac{x + 4}{x + 2}$

When you are simplifying, be on the lookout for the special situation where *a factor in the denominator is the opposite of a factor in the numerator*. In such a case you should factor a negative number from one of the factors so that it becomes equivalent to the other factor and can be divided out. Look carefully at the following two examples.

Teaching Example 5 Simplify. $\dfrac{3x - 6}{10 - 5x}$

Ans: $-\dfrac{3}{5}$

EXAMPLE 5 Simplify. $\dfrac{5x - 15}{6 - 2x}$

Solution Notice that the variable term in the numerator, $5x$, and the variable term in the denominator, $-2x$, *are opposite in sign*. Likewise, the numerical terms -15 and 6 *are opposite in sign*. Factor out a negative number from the denominator.

$$\frac{5x - 15}{6 - 2x} = \frac{5(x - 3)}{-2(-3 + x)} \quad \begin{array}{l}\text{Factor 5 from the numerator.}\\ \text{Factor } -2 \text{ from the denominator.}\end{array}$$

Note that $(x - 3)$ and $(-3 + x)$ are equivalent since $(+x - 3) = (-3 + x)$.

$$= \frac{5\,\cancel{(x - 3)}}{-2\,\cancel{(-3 + x)}} \quad \text{Apply the basic rule of fractions.}$$

$$= -\frac{5}{2}$$

Note that $\dfrac{5}{-2}$ is not considered to be in simple form. We usually avoid leaving a negative number in the denominator. Therefore, to simplify, give the result as $-\dfrac{5}{2}$ or $\dfrac{-5}{2}$.

NOTE TO STUDENT: *Fully worked-out solutions to all of the Practice Problems can be found at the back of the text starting at page SP-1*

Practice Problem 5 Simplify. $\dfrac{8x - 20}{15 - 6x} \quad -\dfrac{4}{3}$

Teaching Example 6 Simplify.

$$\frac{1 - x^2}{3x^2 - 2x - 1}$$

Ans: $-\dfrac{1 + x}{3x + 1}$

EXAMPLE 6 Simplify. $\dfrac{2x^2 - 11x + 12}{16 - x^2}$

Solution

$$= \frac{(x - 4)(2x - 3)}{(4 - x)(4 + x)} \quad \begin{array}{l}\text{Factor the numerator and the denominator.}\\ \text{Observe that } (x - 4) \text{ and } (4 - x) \text{ are opposites.}\end{array}$$

$$= \frac{(x - 4)(2x - 3)}{-1(-4 + x)(4 + x)} \quad \text{Factor } -1 \text{ out of } (+4 - x) \text{ to obtain } -1(-4 + x).$$

$$= \frac{\cancel{(x - 4)}(2x - 3)}{-1\,\cancel{(-4 + x)}(4 + x)} \quad \begin{array}{l}\text{Apply the basic rule of fractions since } (x - 4)\\ \text{and } (-4 + x) \text{ are equivalent.}\end{array}$$

$$= \frac{2x - 3}{-1(4 + x)}$$

$$= -\frac{2x - 3}{4 + x}$$

Practice Problem 6 Simplify. $\dfrac{4x^2 + 3x - 10}{25 - 16x^2} \quad -\dfrac{x + 2}{5 + 4x}$

After doing Examples 5 and 6, you will notice a pattern. Whenever the factor in the numerator and the factor in the denominator are opposites, the value -1 results. We could actually make this a property.

For all monomials A and B where $A \neq B$, it is true that

$$\frac{A - B}{B - A} = -1.$$

You may use this property when reducing fractions if it is helpful to you. Otherwise, you may use the factoring method shown in Examples 5 and 6.

Some problems will involve two or more variables. In such cases, you will need to factor carefully and make sure that each set of parentheses contains the correct terms.

EXAMPLE 7   Simplify. $\dfrac{x^2 - 7xy + 12y^2}{2x^2 - 7xy - 4y^2}$

Solution

$$= \frac{(x - 4y)(x - 3y)}{(2x + y)(x - 4y)} \quad \text{Factor the numerator.}$$
$$ \quad\quad\quad\quad\quad\quad\quad\quad \text{Factor the denominator.}$$

$$= \frac{\cancel{(x - 4y)}(x - 3y)}{(2x + y)\cancel{(x - 4y)}} \quad \text{Apply the basic rule of fractions.}$$

$$= \frac{x - 3y}{2x + y}$$

Practice Problem 7   Simplify. $\dfrac{x^2 - 8xy + 15y^2}{2x^2 - 11xy + 5y^2}$ $\dfrac{x - 3y}{2x - y}$

Teaching Example 7 Simplify.

$$\frac{9x^2 - y^2}{6x^2 + 13xy - 5y^2}$$

Ans: $\dfrac{3x + y}{2x + 5y}$

EXAMPLE 8   Simplify. $\dfrac{6a^2 + ab - 7b^2}{36a^2 - 49b^2}$

Solution

$$= \frac{(6a + 7b)(a - b)}{(6a + 7b)(6a - 7b)} \quad \text{Factor the numerator.}$$
$$ \quad\quad\quad\quad\quad\quad\quad\quad\quad \text{Factor the denominator.}$$

$$= \frac{\cancel{(6a + 7b)}(a - b)}{\cancel{(6a + 7b)}(6a - 7b)} \quad \text{Apply the basic rule of fractions.}$$

$$= \frac{a - b}{6a - 7b}$$

Practice Problem 8   Simplify. $\dfrac{25a^2 - 16b^2}{10a^2 + 3ab - 4b^2}$ $\dfrac{5a - 4b}{2a - b}$

Teaching Example 8 Simplify.

$$\frac{16a^2 - 9b^2}{8a^2 - 2ab - 3b^2}$$

Ans: $\dfrac{4a + 3b}{2a + b}$

Simplify.

1. $\dfrac{3a - 9b}{a - 3b}$ $\dfrac{3(a - 3b)}{a - 3b} = 3$

2. $\dfrac{5x + 2y}{35x + 14y}$ $\dfrac{5x + 2y}{7(5x + 2y)} = \dfrac{1}{7}$

3. $\dfrac{6x + 18}{x^2 + 3x}$ $\dfrac{6(x + 3)}{x(x + 3)} = \dfrac{6}{x}$

4. $\dfrac{8x - 12}{2x^3 - 3x^2}$ $\dfrac{4(2x - 3)}{x^2(2x - 3)} = \dfrac{4}{x^2}$

5. $\dfrac{9x^2 + 6x + 1}{1 - 9x^2}$ $\dfrac{(3x + 1)(3x + 1)}{(1 - 3x)(1 + 3x)} = \dfrac{3x + 1}{1 - 3x}$

6. $\dfrac{4x^2 + 4x + 1}{1 - 4x^2}$ $\dfrac{(2x + 1)(2x + 1)}{(1 + 2x)(1 - 2x)} = \dfrac{2x + 1}{1 - 2x}$

7. $\dfrac{3a^2b(a - 2b)}{6ab^2}$ $\dfrac{a(a - 2b)}{2b}$

8. $\dfrac{9ab^2}{6a^2b^2(b + 3a)}$ $\dfrac{3}{2a(b + 3a)}$

9. $\dfrac{x^2 + x - 2}{x^2 - x}$ $\dfrac{(x + 2)(x - 1)}{x(x - 1)} = \dfrac{x + 2}{x}$

10. $\dfrac{x^2 + x - 12}{x^2 - 3x}$ $\dfrac{(x + 4)(x - 3)}{x(x - 3)} = \dfrac{x + 4}{x}$

11. $\dfrac{x^2 - 3x - 10}{3x^2 + 5x - 2}$ $\dfrac{(x - 5)(x + 2)}{(3x - 1)(x + 2)} = \dfrac{x - 5}{3x - 1}$

12. $\dfrac{4x^2 - 10x + 6}{2x^2 + x - 3}$ $\dfrac{2(2x - 3)(x - 1)}{(2x + 3)(x - 1)} = \dfrac{2(2x - 3)}{2x + 3}$

13. $\dfrac{x^2 + 4x - 21}{x^3 - 49x}$ $\dfrac{(x + 7)(x - 3)}{x(x + 7)(x - 7)} = \dfrac{x - 3}{x(x - 7)}$

14. $\dfrac{x^3 - 3x^2 - 40x}{x^2 + 10x + 25}$ $\dfrac{x(x - 8)(x + 5)}{(x + 5)(x + 5)} = \dfrac{x(x - 8)}{x + 5}$

15. $\dfrac{3x^2 - 11x - 4}{x^2 + x - 20}$ $\dfrac{(3x + 1)(x - 4)}{(x + 5)(x - 4)} = \dfrac{3x + 1}{x + 5}$

16. $\dfrac{x^2 - 9x + 18}{2x^2 - 9x + 9}$ $\dfrac{(x - 6)(x - 3)}{(2x - 3)(x - 3)} = \dfrac{x - 6}{2x - 3}$

17. $\dfrac{3x^2 - 8x + 5}{4x^2 - 5x + 1}$ $\dfrac{(3x - 5)(x - 1)}{(4x - 1)(x - 1)} = \dfrac{3x - 5}{4x - 1}$

18. $\dfrac{3y^2 + 10y + 3}{3y^2 - 14y - 5}$ $\dfrac{(3y + 1)(y + 3)}{(3y + 1)(y - 5)} = \dfrac{y + 3}{y - 5}$

19. $\dfrac{5x^2 - 27x + 10}{5x^2 + 3x - 2}$ $\dfrac{(5x - 2)(x - 5)}{(5x - 2)(x + 1)} = \dfrac{x - 5}{x + 1}$

20. $\dfrac{2x^2 + 2x - 12}{x^2 + 3x - 4}$ $\dfrac{2(x + 3)(x - 2)}{(x + 4)(x - 1)}$

Mixed Practice   *Take some time to review exercises 1–20 before you proceed with exercises 21–30.*

21. $\dfrac{12 - 3x}{5x^2 - 20x}$ $\dfrac{-3(-4 + x)}{5x(x - 4)} = -\dfrac{3}{5x}$

22. $\dfrac{20 - 4ab}{a^2b - 5a}$ $\dfrac{4(5 - ab)}{-a(-ab + 5)} = -\dfrac{4}{a}$

23. $\dfrac{2x^2 - 7x - 15}{25 - x^2}$ $\dfrac{(2x + 3)(x - 5)}{-1(5 + x)(-5 + x)} = \dfrac{-2x - 3}{5 + x}$

24. $\dfrac{49 - x^2}{2x^2 - 9x - 35}$ $\dfrac{-1(7 + x)(-7 + x)}{(2x + 5)(x - 7)} = \dfrac{-x - 7}{2x + 5}$

25. $\dfrac{(3x + 4)^2}{9x^2 + 9x - 4}$ $\dfrac{(3x + 4)(3x + 4)}{(3x + 4)(3x - 1)} = \dfrac{3x + 4}{3x - 1}$

26. $\dfrac{12x^2 - 11x - 5}{(4x - 5)^2}$ $\dfrac{(4x - 5)(3x + 1)}{(4x - 5)(4x - 5)} = \dfrac{3x + 1}{4x - 5}$

27. $\dfrac{2x^2 + 9x - 18}{30 - x - x^2}$ $\dfrac{(2x - 3)(x + 6)}{(6 + x)(5 - x)} = \dfrac{2x - 3}{5 - x}$

28. $\dfrac{4y^2 - y - 3}{8 - 7y - y^2}$ $\dfrac{(4y + 3)(y - 1)}{-1(8 + y)(-1 + y)} = \dfrac{4y + 3}{-y - 8}$

29. $\dfrac{a^2 + 2ab - 3b^2}{2a^2 + 5ab - 3b^2}$ $\dfrac{(a + 3b)(a - b)}{(2a - b)(a + 3b)} = \dfrac{a - b}{2a - b}$

30. $\dfrac{a^2 + 3ab - 10b^2}{3a^2 - 7ab + 2b^2}$ $\dfrac{(a + 5b)(a - 2b)}{(3a - b)(a - 2b)} = \dfrac{a + 5b}{3a - b}$

Cumulative Review *Multiply.*

31. [4.5.2] $(3x - 7)^2$
$9x^2 - 42x + 49$

32. [4.5.1] $(7x + 6y)(7x - 6y)$
$49x^2 - 36y^2$

33. [4.5.3] $(2x + 3)(x - 4)(x - 2)$
$2x^3 - 9x^2 - 2x + 24$

34. [4.5.3] $(x^2 - 3x - 5)(2x^2 + x + 3)$
$2x^4 - 5x^3 - 10x^2 - 14x - 15$

35. [0.3.2] ***Dividing Acreage*** Walter and Ann Perkins wish to divide $4\frac{7}{8}$ acres of farmland into three equal-sized house lots. What will be the acreage of each lot?
$1\dfrac{5}{8}$ acres

36. [0.3.1] ***Roasting a Turkey*** Ron and Mary Larson are planning to cook a $17\frac{1}{2}$-pound turkey. The directions suggest a cooking time of 22 minutes per pound for turkeys that weigh between 16 and 20 pounds. How many hours and minutes should they allow for an approximate cooking time?
6 hr, 25 min

Spanish Speakers *The number of people in Mexico who speak Spanish is 12,500,000 more than three times the number of Spanish speakers in the United States. The number of people who speak Spanish in the United States is 17,000,000 less than the number of Spanish speakers in Spain. The number of Spanish speakers in the United States is 22,500,000.*

37. [3.3.1] How many people speak Spanish in Spain?
39,500,000 people

38. [3.3.1] How many people speak Spanish in Mexico?
80,000,000 people

Quick Quiz 6.1 Simplify.

1. $\dfrac{x^3 + 3x^2}{x^3 - 2x^2 - 15x}$ $\dfrac{x}{x - 5}$

2. $\dfrac{6 - 2ab}{ab^2 - 3b}$ $-\dfrac{2}{b}$

3. $\dfrac{8x^2 + 6x - 5}{16x^2 + 40x + 25}$ $\dfrac{2x - 1}{4x + 5}$

4. Concept Check Explain why it is important to completely factor both the numerator and the denominator when simplifying
$$\dfrac{x^2y - y^3}{x^2y + xy^2 - 2y^3}.$$
Answers may vary

Classroom Quiz 6.1 You may use these problems to quiz your students' mastery of Section 6.1.
Simplify.

1. $\dfrac{x^3 + 5x^2}{x^3 - 6x^2 - 55x}$ **Ans:** $\dfrac{x}{x - 11}$

2. $\dfrac{10 - 2x}{4x^2 - 20x}$ **Ans:** $-\dfrac{1}{2x}$

3. $\dfrac{6x^2 - 13x - 8}{9x^2 - 48x + 64}$ **Ans:** $\dfrac{2x + 1}{3x - 8}$

Multiplying Rational Expressions

To multiply two rational expressions, we multiply the numerators and multiply the denominators. As before, the denominators cannot equal zero.

For any two rational expressions $\dfrac{a}{b}$ and $\dfrac{c}{d}$ where $b \neq 0$ and $d \neq 0$,

$$\frac{a}{b} \cdot \frac{c}{d} = \frac{ac}{bd}.$$

Simplifying or reducing fractions *prior to multiplying them* usually makes the computations easier to do. Leaving the reducing step until the end makes the simplifying process longer and increases the chance for error. This long approach should be avoided.

As an example, let's do the same problem two ways to see which one is easier. Let's simplify the following problem by multiplying first and then reducing the result.

$$\frac{5}{7} \times \frac{49}{125}$$

$$\frac{5}{7} \times \frac{49}{125} = \frac{245}{875} \qquad \text{Multiply the numerators and multiply the denominators.}$$

$$= \frac{7}{25} \qquad \text{Reduce the fraction. (\textit{Note:} It can take a bit of trial and error to discover how to reduce it.)}$$

Compare this with the following method, where we reduce the fractions prior to multiplying them.

$$\frac{5}{7} \times \frac{49}{125}$$

$$\frac{5}{7} \times \frac{7 \cdot 7}{5 \cdot 5 \cdot 5} \qquad \textbf{Step 1.} \text{ It is easier to factor first. We factor the numerator and denominator of the second fraction.}$$

$$= \frac{\cancel{5}}{\cancel{7}} \times \frac{\cancel{7} \cdot 7}{\cancel{5} \cdot 5 \cdot 5} = \frac{7}{25} \qquad \textbf{Step 2.} \text{ Then we apply the basic rule of fractions to divide out the common factors of 5 and 7 that appear in both a numerator and in a denominator.}$$

A similar approach can be used with the multiplication of rational expressions. We first factor the numerator and denominator of each fraction. Then we divide out any factor that is common to a numerator and a denominator. Finally, we multiply the remaining numerators and the remaining denominators.

Teaching Example 1 Multiply.

$$\frac{x^2 - 4x - 5}{2x^2 + 7x + 3} \cdot \frac{x^2 - 9}{x^2 - 8x + 15}$$

Ans: $\dfrac{x + 1}{2x + 1}$

> **EXAMPLE 1** Multiply. $\dfrac{x^2 - x - 12}{x^2 - 16} \cdot \dfrac{2x^2 + 7x - 4}{x^2 - 4x - 21}$
>
> **Solution**
>
> $$\frac{(x - 4)(x + 3)}{(x - 4)(x + 4)} \cdot \frac{(x + 4)(2x - 1)}{(x + 3)(x - 7)} \qquad \text{Factoring is always the first step.}$$
>
> $$= \frac{\cancel{(x - 4)}\cancel{(x + 3)}}{\cancel{(x - 4)}\cancel{(x + 4)}} \cdot \frac{\cancel{(x + 4)}(2x - 1)}{\cancel{(x + 3)}(x - 7)} \qquad \begin{array}{l}\text{Apply the basic rule of fractions.}\\ \text{(Three pairs of factors divide out.)}\end{array}$$
>
> $$= \frac{2x - 1}{x - 7} \qquad \text{The final answer.}$$

NOTE TO STUDENT: Fully worked-out solutions to all of the Practice Problems can be found at the back of the text starting at page SP-1

> **Practice Problem 1** Multiply. $\dfrac{6x^2 + 7x + 2}{x^2 - 7x + 10} \cdot \dfrac{x^2 + 3x - 10}{2x^2 + 11x + 5} \quad \dfrac{3x + 2}{x - 5}$

In some cases, it may take several steps to factor a given numerator or denominator. You should always check for a *common factor* as your first step.

EXAMPLE 2 Multiply. $\dfrac{x^4 - 16}{x^3 + 4x} \cdot \dfrac{2x^2 - 8x}{4x^2 + 2x - 12}$

Teaching Example 2 Multiply.

$$\frac{5x^2 - 10x - 15}{x^3 + x^2} \cdot \frac{x^3 - x^2}{x^2 - x - 6}$$

Ans: $\dfrac{5(x - 1)}{x + 2}$ or $\dfrac{5x - 5}{x + 2}$

Solution

$$= \frac{(x^2 + 4)(x^2 - 4)}{x(x^2 + 4)} \cdot \frac{2x(x - 4)}{2(2x^2 + x - 6)}$$

Factor each numerator and denominator. Factoring out the common factor first is very important.

$$= \frac{(x^2 + 4)(x + 2)(x - 2)}{x(x^2 + 4)} \cdot \frac{2x(x - 4)}{2(x + 2)(2x - 3)}$$

Factor again where possible.

$$= \frac{\cancel{(x^2 + 4)}\cancel{(x + 2)}(x - 2)}{\cancel{x}\cancel{(x^2 + 4)}} \cdot \frac{\cancel{2x}(x - 4)}{\cancel{2}\cancel{(x + 2)}(2x - 3)}$$

Divide out factors that appear in both a numerator and a denominator. (There are four such pairs of factors.)

$$= \frac{(x - 2)(x - 4)}{2x - 3} \quad \text{or} \quad \frac{x^2 - 6x + 8}{2x - 3}$$

Write the answer as one fraction. (Usually, if there is more than one factor in a numerator or denominator, the answer is left in factored form.)

Practice Problem 2 Multiply. $\dfrac{2y^2 - 6y - 8}{y^2 - y - 2} \cdot \dfrac{y^2 - 5y + 6}{2y^2 - 32} \cdot \dfrac{y - 3}{y + 4}$

Dividing Rational Expressions

For any two fractions $\frac{a}{b}$ and $\frac{c}{d}$, the operation of division can be performed by inverting the second fraction and multiplying it by the first fraction. When we invert a fraction, we are finding its *reciprocal*. Two numbers are **reciprocals** of each other if their product is 1. The reciprocal of $\frac{3}{5}$ is $\frac{5}{3}$. The reciprocal of 7 is $\frac{1}{7}$. The reciprocal of $\frac{a}{b}$ is $\frac{b}{a}$. Sometimes people state the rule for dividing fractions this way: "To divide two fractions, keep the first fraction unchanged and multiply by the reciprocal of the second fraction."

The definition for division of fractions is

$$\frac{a}{b} \div \frac{c}{d} = \frac{a}{b} \cdot \frac{d}{c}.$$

This property holds whether a, b, c, and d are polynomials or numerical values. (It is assumed, of course, that no denominator is zero.)

In the first step for dividing two rational expressions, invert the second fraction and rewrite the quotient as a product. Then follow the procedure for multiplying rational expressions.

EXAMPLE 3 Divide. $\dfrac{6x + 12y}{2x - 6y} \div \dfrac{9x^2 - 36y^2}{4x^2 - 36y^2}$

Teaching Example 3 Divide.

$$\frac{2x - 2y}{x^2 + 10xy + 21y^2} \div \frac{x^2 - xy}{x^3 - 9xy^2}$$

Ans: $\dfrac{2(x - 3y)}{x + 7y}$

Solution

$$= \frac{6x + 12y}{2x - 6y} \cdot \frac{4x^2 - 36y^2}{9x^2 - 36y^2}$$

Invert the second fraction and write the problem as the product of two fractions.

$$= \frac{6(x + 2y)}{2(x - 3y)} \cdot \frac{4(x^2 - 9y^2)}{9(x^2 - 4y^2)}$$

Factor each numerator and denominator.

$$= \frac{(3)(2)(x + 2y)}{2(x - 3y)} \cdot \frac{(2)(2)(x + 3y)(x - 3y)}{(3)(3)(x + 2y)(x - 2y)}$$

Factor again where possible.

$$= \frac{\cancel{(3)}\cancel{(2)}\cancel{(x + 2y)}}{\cancel{2}\cancel{(x - 3y)}} \cdot \frac{(2)(2)(x + 3y)\cancel{(x - 3y)}}{\cancel{(3)}(3)\cancel{(x + 2y)}(x - 2y)}$$

Divide out factors that appear in both a numerator and a denominator.

$$= \frac{(2)(2)(x + 3y)}{3(x - 2y)}$$

Write the result as one fraction.

$$= \frac{4(x + 3y)}{3(x - 2y)}$$

Simplify. (Usually, answers are left in this form.)

Although it is correct to write this answer as $\dfrac{4x + 12y}{3x - 6y}$, it is customary to leave the answer in factored form to ensure that the final answer is simplified.

NOTE TO STUDENT: *Fully worked-out solutions to all of the Practice Problems can be found at the back of the text starting at page SP-1*

Practice Problem 3   Divide. $\dfrac{x^2 + 5x + 6}{x^2 + 8x} \div \dfrac{2x^2 + 5x + 2}{2x^2 + x} \quad \dfrac{x + 3}{x + 8}$

A polynomial that is not in fraction form can be written as a fraction if you give it a denominator of 1.

Teaching Example 4 Divide.

$$\frac{x^2 - 1}{x + 3} \div (5 - 5x)$$

Ans: $-\dfrac{x + 1}{5(x + 3)}$

EXAMPLE 4   Divide. $\dfrac{15 - 3x}{x + 6} \div (x^2 - 9x + 20)$

Solution   Note that $x^2 - 9x + 20$ can be written as $\dfrac{x^2 - 9x + 20}{1}$.

$$= \frac{15 - 3x}{x + 6} \cdot \frac{1}{x^2 - 9x + 20}$$

Invert and multiply.

$$= \frac{-3(-5 + x)}{x + 6} \cdot \frac{1}{(x - 5)(x - 4)}$$

Factor where possible. Note that we had to factor -3 from the first numerator so that it would have a factor in common with the second denominator.

$$= \frac{-3\cancel{(-5 + x)}}{x + 6} \cdot \frac{1}{\cancel{(x - 5)}(x - 4)}$$

Divide out the common factor. $(-5 + x)$ is equivalent to $(x - 5)$.

$$= \frac{-3}{(x + 6)(x - 4)}$$

The final answer. Note that the answer can be written in several equivalent forms.

$$\text{or} \quad -\frac{3}{(x + 6)(x - 4)} \quad \text{or} \quad \frac{3}{(x + 6)(4 - x)}$$

Practice Problem 4   Divide. $\dfrac{x + 3}{x - 3} \div (9 - x^2) \quad \dfrac{1}{(x - 3)(3 - x)}$

CAUTION: It is logical to assume that the problems in Section 6.2 have at least one common factor that can be divided out. Therefore, if after factoring, you do not observe any common factors, you should be somewhat suspicious. In such cases, it would be wise to double-check your factoring steps for errors.

Verbal and Writing Skills

1. Before multiplying rational expressions, we should always first try to _factor the numerator and denominator completely and divide out any common factors._

2. Division of two rational expressions is done by keeping the first fraction unchanged and then _multiplying by the reciprocal of the second fraction_ .

Multiply.

3. $\dfrac{3x + 6}{x - 3} \cdot \dfrac{x^2 + 2x - 15}{x^2 + 4x + 4}$

$\dfrac{3(x + 2)}{x - 3} \cdot \dfrac{(x + 5)(x - 3)}{(x + 2)(x + 2)} = \dfrac{3(x + 5)}{x + 2}$

4. $\dfrac{6x - 6}{x + 3} \cdot \dfrac{x^2 - 3x - 18}{6x^2 - 6}$

$\dfrac{6(x - 1)}{x + 3} \cdot \dfrac{(x + 3)(x - 6)}{6(x + 1)(x - 1)} = \dfrac{x - 6}{x + 1}$

5. $\dfrac{24x^3}{4x^2 - 36} \cdot \dfrac{2x^2 + 6x}{16x^2}$

$\dfrac{24x^3}{4(x + 3)(x - 3)} \cdot \dfrac{2x(x + 3)}{16x^2} = \dfrac{3x^2}{4(x - 3)}$

6. $\dfrac{14x^2}{10x + 50} \cdot \dfrac{5x^3 - 125x}{35x^4}$

$\dfrac{14x^2}{10(x + 5)} \cdot \dfrac{5x(x + 5)(x - 5)}{35x^4} = \dfrac{x - 5}{5x}$

7. $\dfrac{x^2 + 3x - 10}{x^2 + x - 20} \cdot \dfrac{x^2 - 3x - 4}{x^2 + 4x + 3}$

$\dfrac{(x - 2)(x + 5)}{(x - 4)(x + 5)} \cdot \dfrac{(x - 4)(x + 1)}{(x + 3)(x + 1)} = \dfrac{x - 2}{x + 3}$

8. $\dfrac{x^2 - x - 20}{x^2 - 3x - 10} \cdot \dfrac{x^2 + 7x + 10}{x^2 + 4x - 5}$

$\dfrac{(x - 5)(x + 4)}{(x - 5)(x + 2)} \cdot \dfrac{(x + 2)(x + 5)}{(x + 5)(x - 1)} = \dfrac{x + 4}{x - 1}$

Divide.

9. $\dfrac{x + 6}{x - 8} \div \dfrac{x + 5}{x^2 - 6x - 16}$

$\dfrac{x + 6}{x - 8} \cdot \dfrac{(x - 8)(x + 2)}{x + 5} = \dfrac{(x + 6)(x + 2)}{x + 5}$

10. $\dfrac{x - 4}{x + 9} \div \dfrac{x - 7}{x^2 + 13x + 36}$

$\dfrac{x - 4}{x + 9} \cdot \dfrac{(x + 4)(x + 9)}{x - 7} = \dfrac{(x - 4)(x + 4)}{x - 7}$

11. $(5x + 4) \div \dfrac{25x^2 - 16}{5x^2 + 11x - 12}$

$\dfrac{5x + 4}{1} \cdot \dfrac{(x + 3)(5x - 4)}{(5x + 4)(5x - 4)} = x + 3$

12. $\dfrac{4x^2 - 25}{4x^2 - 20x + 25} \div (4x + 10)$

$\dfrac{(2x - 5)(2x + 5)}{(2x - 5)(2x - 5)} \cdot \dfrac{1}{2(2x + 5)} = \dfrac{1}{2(2x - 5)}$

13. $\dfrac{3x^2 + 12xy + 12y^2}{x^2 + 4xy + 3y^2} \div \dfrac{4x + 8y}{x + y}$

$\dfrac{3(x + 2y)(x + 2y)}{(x + 3y)(x + y)} \cdot \dfrac{(x + y)}{4(x + 2y)} = \dfrac{3(x + 2y)}{4(x + 3y)}$

14. $\dfrac{5x^2 + 10xy + 5y^2}{x^2 + 5xy + 6y^2} \div \dfrac{3x + 3y}{x + 2y}$

$\dfrac{5(x + y)(x + y)}{(x + 3y)(x + 2y)} \cdot \dfrac{(x + 2y)}{3(x + y)} = \dfrac{5(x + y)}{3(x + 3y)}$

Mixed Practice *Perform the operation indicated.*

15. $\dfrac{(x + 5)^2}{3x^2 - 7x + 2} \cdot \dfrac{x^2 - 4x + 4}{x + 5}$

$\dfrac{(x + 5)(x + 5)}{(3x - 1)(x - 2)} \cdot \dfrac{(x - 2)(x - 2)}{x + 5} = \dfrac{(x + 5)(x - 2)}{3x - 1}$

16. $\dfrac{3x^2 - 10x - 8}{(4x + 5)^2} \cdot \dfrac{4x + 5}{(x - 4)^2}$

$\dfrac{(3x + 2)(x - 4)}{(4x + 5)(4x + 5)} \cdot \dfrac{4x + 5}{(x - 4)(x - 4)} = \dfrac{3x + 2}{(4x + 5)(x - 4)}$

17. $\dfrac{x^2 + x - 30}{10 - 2x} \div \dfrac{x^2 + 4x - 12}{5x + 15}$

$\dfrac{(x + 6)(x - 5)}{-2(x - 5)} \cdot \dfrac{5(x + 3)}{(x + 6)(x - 2)}$

$= \dfrac{5(x + 3)}{-2(x - 2)}$ or $-\dfrac{5(x + 3)}{2(x - 2)}$

18. $\dfrac{x^2 + 3x - 28}{x^2 + 14x + 49} \div \dfrac{12 - 3x^2}{x^2 + 5x - 14}$

$\dfrac{(x + 7)(x - 4)}{(x + 7)(x + 7)} \cdot \dfrac{(x + 7)(x - 2)}{-3(x - 2)(x + 2)}$

$= \dfrac{x - 4}{-3(x + 2)}$ or $-\dfrac{x - 4}{3(x + 2)}$

19. $\dfrac{y^2 + 4y - 12}{y^2 + 2y - 24} \cdot \dfrac{y^2 - 16}{y^2 + 2y - 8}$

$\dfrac{(y + 6)(y - 2)}{(y + 6)(y - 4)} \cdot \dfrac{(y - 4)(y + 4)}{(y + 4)(y - 2)} = 1$

20. $\dfrac{4y^2 + 11y + 6}{8y^2 + 2y - 3} \cdot \dfrac{4y^2 - 8y + 3}{2y^2 + y - 6}$

$\dfrac{(4y + 3)(y + 2)}{(4y + 3)(2y - 1)} \cdot \dfrac{(2y - 1)(2y - 3)}{(2y - 3)(y + 2)} = 1$

21. $\dfrac{x^2 + 7x + 12}{2x^2 + 9x + 4} \div \dfrac{x^2 + 6x + 9}{2x^2 - x - 1}$

$\dfrac{(x + 4)(x + 3)}{(2x + 1)(x + 4)} \cdot \dfrac{(2x + 1)(x - 1)}{(x + 3)(x + 3)} = \dfrac{x - 1}{x + 3}$

22. $\dfrac{x^2 + 7x + 10}{2x^2 - 3x - 2} \cdot \dfrac{2x^2 - 11x - 6}{x^2 - x - 30}$

$\dfrac{(x + 5)(x + 2)}{(2x + 1)(x - 2)} \cdot \dfrac{(2x + 1)(x - 6)}{(x + 5)(x - 6)} = \dfrac{x + 2}{x - 2}$

Cumulative Review

23. **[2.3.2]** Solve. $6x^2 + 3x - 18 = 5x - 2 + 6x^2$

$x = -8$

24. **[4.5.3]** Multiply. $(7x^2 - x - 1)(x - 3)$

$7x^3 - 22x^2 + 2x + 3$

▲ **25.** **[3.1.2]** *Golden Gate Bridge* The Golden Gate Bridge has a total length (including approaches) of 8981 feet and a road width of 90 feet. The width of the sidewalk is 10.5 feet. (The sidewalk spans the entire length of the bridge.) Assume it would cost $x per square foot to resurface the road or the sidewalk. Write an expression for how much more it would cost to resurface the road than the sidewalk.

79.5(8981)(x) = $713,989.5x

▲ **26.** **[3.3.1]** *Garden Design* Harold Rafton planted a square garden bed. George Avis also planted a garden that was 2 feet less in width, but 3 feet longer in length than Harold's garden. If the area of George's garden was 36 square feet, find the dimensions of each garden.

Harold's garden is 6 ft by 6 ft.
George's garden is 9 ft by 4 ft.

Quick Quiz 6.2 Multiply.

1. $\dfrac{2x - 10}{x - 4} \cdot \dfrac{x^2 + 5x + 4}{x^2 - 4x - 5}$ $\quad \dfrac{2(x + 4)}{x - 4}$

2. $\dfrac{3x^2 - 13x - 10}{3x^2 + 2x} \cdot \dfrac{x^2 - 25x}{x^2 - 25}$ $\quad \dfrac{x - 25}{x + 5}$

Divide.

3. $\dfrac{2x^2 - 18}{3x^2 + 3x} \div \dfrac{x^2 + 6x + 9}{x^2 + 4x + 3}$ $\quad \dfrac{2(x - 3)}{3x}$

4. **Concept Check** Explain how you would divide $\dfrac{21x - 7}{9x^2 - 1} \div \dfrac{1}{3x + 1}$.

Answers may vary

Classroom Quiz 6.2 You may use these problems to quiz your students' mastery of Section 6.2.

Multiply.

1. $\dfrac{3x - 24}{x - 7} \cdot \dfrac{x^2 - 2x - 35}{x^2 - 6x - 16}$

Ans: $\dfrac{3(x + 5)}{x + 2}$

2. $\dfrac{4x^2 - 9}{4x^2 + 12x + 9} \cdot \dfrac{2}{6x - 9}$

Ans: $\dfrac{2}{3(2x + 3)}$

3. $\dfrac{2x^2 - 32}{x^2 - 5x + 6} \div \dfrac{2x^2 - 6x - 8}{x^2 - x - 2}$

Ans: $\dfrac{x + 4}{x - 3}$

Adding and Subtracting Rational Expressions with a Common Denominator

If rational expressions have the same denominator, they can be combined in the same way as arithmetic fractions. The numerators are added or subtracted and the denominator remains the same.

Student Learning Objectives

After studying this section, you will be able to:

1 Add and subtract rational expressions with a common denominator.

2 Determine the LCD for rational expressions with different denominators.

3 Add and subtract rational expressions with different denominators.

> **ADDING RATIONAL EXPRESSIONS**
>
> For any rational expressions $\dfrac{a}{b}$ and $\dfrac{c}{b}$,
>
> $$\frac{a}{b} + \frac{c}{b} = \frac{a + c}{b} \qquad \text{where } b \neq 0.$$

EXAMPLE 1 Add. $\dfrac{5a}{a + 2b} + \dfrac{6a}{a + 2b}$

Solution

$$\frac{5a}{a + 2b} + \frac{6a}{a + 2b} = \frac{5a + 6a}{a + 2b} = \frac{11a}{a + 2b}$$

Note that the denominators are the same. Only add the numerators. Keep the same denominator. Do not change the denominator.

Teaching Example 1 Add. $\dfrac{3x + 1}{2x - 5} + \dfrac{x - 4}{2x - 5}$

Ans: $\dfrac{4x - 3}{2x - 5}$

Practice Problem 1 Add. $\dfrac{2s + t}{2s - t} + \dfrac{s - t}{2s - t} \qquad \dfrac{3s}{2s - t}$

Teaching Tip A comprehensive example with both numerical and algebraic fractions helps students tie in previous knowledge of fractions and understand when a common denominator is not needed (multiplication and division) and when it is needed (addition and subtraction).

> **SUBTRACTING RATIONAL EXPRESSIONS**
>
> For any rational expressions $\dfrac{a}{b}$ and $\dfrac{c}{b}$,
>
> $$\frac{a}{b} - \frac{c}{b} = \frac{a - c}{b} \qquad \text{where } b \neq 0.$$

EXAMPLE 2 Subtract. $\dfrac{3x}{(x + y)(x - 2y)} - \dfrac{8x}{(x + y)(x - 2y)}$

Solution

$$\frac{3x}{(x + y)(x - 2y)} - \frac{8x}{(x + y)(x - 2y)} = \frac{3x - 8x}{(x + y)(x - 2y)}$$

Write as one fraction.

$$= \frac{-5x}{(x + y)(x - 2y)}$$

Simplify.

Teaching Example 2 Subtract.

$$\frac{5y + 1}{(y + 2)(2y + 3)} - \frac{3y}{(y + 2)(2y + 3)}$$

Ans: $\dfrac{2y + 1}{(y + 2)(2y + 3)}$

Practice Problem 2 Subtract.

$$\frac{b}{(a - 2b)(a + b)} - \frac{2b}{(a - 2b)(a + b)} \qquad \frac{-b}{(a - 2b)(a + b)}$$

Determining the LCD for Rational Expressions with Different Denominators

How do we add or subtract rational expressions when the denominators are not the same? First we must find the **least common denominator** (LCD). You need to be clear on how to find a least common denominator and how to add and subtract fractions from arithmetic before you attempt this section. Review Sections 0.1 and 0.2 if you have any questions about this topic.

HOW TO FIND THE LCD OF TWO OR MORE RATIONAL EXPRESSIONS

1. Factor each denominator completely.
2. The LCD is a product containing each *different factor*.
3. If a factor occurs more than once in any one denominator, the LCD will contain that factor repeated the greatest number of times that it occurs in any one denominator.

Teaching Example 3 Find the LCD.

$$\frac{8}{15x + 5}, \frac{3}{6x + 2}$$

Ans: $10(3x + 1)$

EXAMPLE 3 Find the LCD. $\dfrac{5}{2x - 4}, \dfrac{6}{3x - 6}$

Solution Factor each denominator.

$$2x - 4 = 2(x - 2) \qquad 3x - 6 = 3(x - 2)$$

The different factors are 2, 3, and $(x - 2)$. Since no factor appears more than once in any one denominator, the LCD is the product of these three factors.

$$\text{LCD} = (2)(3)(x - 2) = 6(x - 2)$$

Practice Problem 3 Find the LCD. $\dfrac{7}{6x + 21}, \dfrac{13}{10x + 35}$ $\quad 15(2x + 7)$

Teaching Example 4 Find the LCD.

(a) $\dfrac{7}{6x^3yz^2}, \dfrac{9}{4xy^4z^3}$

(b) $\dfrac{4}{x^2 + 6x + 8}, \dfrac{13}{x^2 - x - 20}$

Ans: (a) $12x^3y^4z^3$

(b) $(x + 2)(x + 4)(x - 5)$

EXAMPLE 4 Find the LCD.

(a) $\dfrac{5}{12ab^2c}, \dfrac{13}{18a^3bc^4}$

(b) $\dfrac{8}{x^2 - 5x + 4}, \dfrac{12}{x^2 + 2x - 3}$

Solution If a factor occurs more than once in any one denominator, the LCD will contain that factor repeated the greatest number of times that it occurs in any one denominator.

(a) $12ab^2c = 2 \cdot 2 \cdot 3 \cdot \quad a \cdot \quad b \cdot b \cdot c$

$18a^3bc^4 = \begin{vmatrix} 2 \cdot 3 \cdot 3 \cdot a \cdot a \cdot a \cdot b \cdot \end{vmatrix} c \cdot c \cdot c \cdot c$

$\qquad\qquad 2 \cdot 2 \cdot 3 \cdot 3 \cdot a \cdot a \cdot a \cdot b \cdot b \cdot c \cdot c \cdot c \cdot c$

$\quad \text{LCD} = 2^2 \cdot 3^2 \cdot a^3 \cdot b^2 \cdot c^4 = 36a^3b^2c^4$

(b) $x^2 - 5x + 4 = (x - 4)(x - 1)$

$\quad x^2 + 2x - 3 = \qquad (x - 1)(x + 3)$

$\qquad \text{LCD} = (x - 4)(x - 1)(x + 3)$

Practice Problem 4 Find the LCD.

(a) $\dfrac{3}{50xy^2z}, \dfrac{19}{40x^3yz}$

(b) $\dfrac{2}{x^2 + 5x + 6}, \dfrac{6}{3x^2 + 5x - 2}$

(a) $200x^3y^2z$ (b) $(x + 3)(x + 2)(3x - 1)$

Adding and Subtracting Rational Expressions with Different Denominators

If two rational expressions have different denominators, we first change them to equivalent rational expressions with the least common denominator. Then we add or subtract the numerators and keep the common denominator.

EXAMPLE 5   Add. $\dfrac{5}{xy} + \dfrac{2}{y}$

Solution   The denominators are different. We must find the LCD. The two factors are x and y. We observe that the LCD is xy.

$$\frac{5}{xy} + \frac{2}{y} = \frac{5}{xy} + \frac{2}{y} \cdot \frac{x}{x} \qquad \text{Multiply the second fraction by } \frac{x}{x}.$$

$$= \frac{5}{xy} + \frac{2x}{xy} \qquad \text{Now each fraction has a common denominator of } xy.$$

$$= \frac{5 + 2x}{xy} \qquad \text{Write the sum as one fraction.}$$

Practice Problem 5   Add. $\dfrac{7}{a} + \dfrac{3}{abc}$ $\dfrac{7bc + 3}{abc}$

Teaching Example 5 Add. $\dfrac{3}{a} + \dfrac{7}{ab}$

Ans: $\dfrac{3b + 7}{ab}$

EXAMPLE 6   Add. $\dfrac{3x}{x^2 - y^2} + \dfrac{5}{x + y}$

Solution   We factor the first denominator so that $x^2 - y^2 = (x + y)(x - y)$. Thus, the factors of the denominators are $(x + y)$ and $(x - y)$. We observe that the LCD $= (x + y)(x - y)$.

$$\frac{3x}{(x + y)(x - y)} + \frac{5}{(x + y)} \cdot \frac{x - y}{x - y} \qquad \text{Multiply the second fraction by } \frac{x - y}{x - y}.$$

$$= \frac{3x}{(x + y)(x - y)} + \frac{5x - 5y}{(x + y)(x - y)} \qquad \text{Now each fraction has a common denominator of } (x + y)(x - y).$$

$$= \frac{3x + 5x - 5y}{(x + y)(x - y)} \qquad \text{Write the sum of the numerators over the common denominator.}$$

$$= \frac{8x - 5y}{(x + y)(x - y)} \qquad \text{Combine like terms.}$$

Practice Problem 6   Add. $\dfrac{2a - b}{a^2 - 4b^2} + \dfrac{2}{a + 2b}$ $\dfrac{4a - 5b}{(a + 2b)(a - 2b)}$

Teaching Example 6 Add. $\dfrac{5}{x + 3} + \dfrac{2x + 1}{x^2 - 9}$

Ans: $\dfrac{7x - 14}{(x + 3)(x - 3)}$

It is important to remember that the LCD is the smallest algebraic expression into which each denominator can be divided. For rational expressions the LCD must contain *each factor* that appears in any denominator. If the factor is repeated, the LCD must contain that factor the greatest number of times that it appears in any one denominator.

In many cases, the denominators in an addition or subtraction problem are not in factored form. You must factor each denominator to determine the LCD. Combine like terms in the numerator; then determine whether that final numerator can be factored. If so, you may be able to simplify the fraction.

EXAMPLE 7 Add. $\dfrac{5}{x^2 - y^2} + \dfrac{3x}{x^3 + x^2y}$

Solution

$$\frac{5}{x^2 - y^2} + \frac{3x}{x^3 + x^2y}$$

$$= \frac{5}{(x + y)(x - y)} + \frac{3x}{x^2(x + y)}$$

Factor the two denominators. Observe that the LCD is $x^2(x + y)(x - y)$.

$$= \frac{5}{(x + y)(x - y)} \cdot \frac{x^2}{x^2} + \frac{3x}{x^2(x + y)} \cdot \frac{x - y}{x - y}$$

Multiply each fraction by the appropriate expression to obtain a common denominator of $x^2(x + y)(x - y)$.

$$= \frac{5x^2}{x^2(x + y)(x - y)} + \frac{3x^2 - 3xy}{x^2(x + y)(x - y)}$$

$$= \frac{5x^2 + 3x^2 - 3xy}{x^2(x + y)(x - y)}$$

Write the sum of the numerators over the common denominator.

$$= \frac{8x^2 - 3xy}{x^2(x + y)(x - y)}$$

Combine like terms.

$$= \frac{x(8x - 3y)}{x^2(x + y)(x - y)}$$

Divide out the common factor x in the numerator and denominator and simplify.

$$= \frac{8x - 3y}{x(x + y)(x - y)}$$

Practice Problem 7 Add.

$$\frac{7a}{a^2 + 2ab + b^2} + \frac{4}{a^2 + ab} \qquad \frac{7a^2 + 4a + 4b}{a(a + b)^2}$$

It is very easy to make a sign mistake when subtracting two fractions. You will find it helpful to place parentheses around the numerator of the second fraction so that you will not forget to subtract the entire numerator.

EXAMPLE 8 Subtract. $\dfrac{3x + 4}{x - 2} - \dfrac{x - 3}{2x - 4}$

Solution Factor the second denominator.

$$= \frac{3x + 4}{x - 2} - \frac{x - 3}{2(x - 2)}$$

Observe that the LCD is $2(x - 2)$.

$$= \frac{2}{2} \cdot \frac{3x + 4}{x - 2} - \frac{x - 3}{2(x - 2)}$$

Multiply the first fraction by $\frac{2}{2}$ so that the resulting fraction will have the common denominator.

$$= \frac{2(3x + 4) - (x - 3)}{2(x - 2)}$$

Write the indicated subtraction as one fraction. Note the parentheses around $x - 3$.

$$= \frac{6x + 8 - x + 3}{2(x - 2)}$$

Remove the parentheses in the numerator.

$$= \frac{5x + 11}{2(x - 2)}$$

Combine like terms.

Practice Problem 8 Subtract.

$$\frac{x + 7}{3x - 9} - \frac{x - 6}{x - 3} \qquad \frac{-2x + 25}{3(x - 3)}$$

To avoid making errors when subtracting two fractions, some students find it helpful to change subtraction to addition of the opposite of the second fraction. In other words, we use the property that $\dfrac{a}{b} - \dfrac{c}{b} = \dfrac{a}{b} + \dfrac{-c}{b}$.

EXAMPLE 9   Subtract. $\dfrac{8x}{x^2 - 16} - \dfrac{4}{x - 4}$

Solution

$\dfrac{8x}{x^2 - 16} - \dfrac{4}{x - 4}$

$= \dfrac{8x}{(x + 4)(x - 4)} + \dfrac{-4}{x - 4}$

Factor the first denominator. Use the property that $\dfrac{a}{b} - \dfrac{c}{b} = \dfrac{a}{b} + \dfrac{-c}{b}$.

$= \dfrac{8x}{(x + 4)(x - 4)} + \dfrac{-4}{x - 4} \cdot \dfrac{x + 4}{x + 4}$

Multiply the second fraction by $\dfrac{x + 4}{x + 4}$.

$= \dfrac{8x + (-4)(x + 4)}{(x + 4)(x - 4)}$

Write the sum of the numerators over the common denominator.

$= \dfrac{8x - 4x - 16}{(x + 4)(x - 4)}$

Remove parentheses.

$= \dfrac{4x - 16}{(x + 4)(x - 4)}$

Combine like terms. Note that the numerator can be factored.

$= \dfrac{4\cancel{(x - 4)}}{(x + 4)\cancel{(x - 4)}}$

Since $(x - 4)$ is a *factor* of the numerator *and* the denominator, we may divide out the common factor.

$= \dfrac{4}{x + 4}$

Practice Problem 9   Subtract and simplify.

$$\dfrac{x - 2}{x^2 - 4} - \dfrac{x + 1}{2x^2 + 4x} \qquad \dfrac{x - 1}{2x(x + 2)}$$

Teaching Example 9 Subtract. $\dfrac{5a + 2}{a^2 + 3a} - \dfrac{2}{a}$

Ans: $\dfrac{3a - 4}{a(a + 3)}$

Teaching Tip Students will often incorrectly divide out common factors before adding or subtracting fractions. In Example 9 and similar problems, students will be tempted to divide out the $(x + 4)$ quantities when they arrive at the expression

$$\dfrac{8x + (-4)(x + 4)}{(x + 4)(x - 4)}.$$

Be sure to point out that dividing out such quantities at this step would not be correct. The *terms* in the numerator must be combined first. Stress the idea that common factors should be divided out only at the very last step of the problem. Remind students that the value they divide out must be a *factor* of the entire numerator and the entire denominator. They cannot divide out terms that are separated by a plus or a minus sign.

Verbal and Writing Skills

1. Suppose two rational expressions have denominators of $(x + 3)(x + 5)$ and $(x + 3)^2$. Explain how you would determine the LCD.

The LCD would be a product that contains each factor. However, any repeated factor in any one denominator must be repeated the greatest number of times it occurs in any one denominator. So the LCD would be $(x + 5)(x + 3)^2$.

2. Suppose two rational expressions have denominators of $(x - 4)^2(x + 7)$ and $(x - 4)^3$. Explain how you would determine the LCD.

The LCD would be a product that contains each factor. However, any repeated factor in any one denominator must be repeated the greatest number of times it occurs in any one denominator. So the LCD would be $(x - 4)^3(x + 7)$.

Perform the operation indicated. Be sure to simplify.

3. $\dfrac{x}{x + 5} + \dfrac{2x + 1}{5 + x}$ $\dfrac{3x + 1}{x + 5}$

4. $\dfrac{8}{7 + 2x} + \dfrac{x + 3}{2x + 7}$ $\dfrac{x + 11}{2x + 7}$

5. $\dfrac{3x}{x + 3} - \dfrac{x + 5}{x + 3}$ $\dfrac{2x - 5}{x + 3}$

6. $\dfrac{2x + 4}{x - 5} - \dfrac{x + 5}{x - 5}$ $\dfrac{x - 1}{x - 5}$

7. $\dfrac{8x + 3}{5x + 7} - \dfrac{6x + 10}{5x + 7}$ $\dfrac{2x - 7}{5x + 7}$

8. $\dfrac{6x + 5}{4x + 1} - \dfrac{3x + 4}{4x + 1}$ $\dfrac{3x + 1}{4x + 1}$

Find the LCD. Do not combine fractions.

9. $\dfrac{10}{3a^2b^3}, \dfrac{8}{ab^2}$
LCD $= 3a^2b^3$

10. $\dfrac{12}{5a^2}, \dfrac{9}{a^3}$
LCD $= 5a^3$

11. $\dfrac{5}{18x^2y^5}, \dfrac{7}{30x^3y^3}$
LCD $= 90x^3y^5$

12. $\dfrac{9}{14xy^3}, \dfrac{14}{35x^4y^2}$
LCD $= 70x^4y^3$

13. $\dfrac{9}{2x - 6}, \dfrac{5}{9x - 27}$
LCD $= 18(x - 3)$

14. $\dfrac{12}{3x + 12}, \dfrac{5}{5x + 20}$
LCD $= 15(x + 4)$

15. $\dfrac{8}{x + 3}, \dfrac{15}{x^2 - 9}$
LCD $= x^2 - 9$

16. $\dfrac{13}{x^2 - 16}, \dfrac{7}{x - 4}$
LCD $= x^2 - 16$

17. $\dfrac{7}{3x^2 + 14x - 5}, \dfrac{4}{9x^2 - 6x + 1}$
LCD $= (x + 5)(3x - 1)^2$

18. $\dfrac{4}{2x^2 - 9x - 35}, \dfrac{3}{4x^2 + 20x + 25}$
LCD $= (x - 7)(2x + 5)^2$

Add.

19. $\dfrac{7}{ab} + \dfrac{3}{b}$ $\dfrac{7 + 3a}{ab}$

20. $\dfrac{8}{cd} + \dfrac{9}{d}$ $\dfrac{8 + 9c}{cd}$

21. $\dfrac{3}{x + 7} + \dfrac{8}{x^2 - 49}$

$\dfrac{3}{x + 7}\left(\dfrac{x - 7}{x - 7}\right) + \dfrac{8}{(x + 7)(x - 7)} = \dfrac{3x - 13}{(x + 7)(x - 7)}$

22. $\dfrac{5}{x^2 - 2x + 1} + \dfrac{3}{x - 1}$

$\dfrac{5}{(x - 1)(x - 1)} + \dfrac{3}{(x - 1)}\left(\dfrac{x - 1}{x - 1}\right) = \dfrac{3x + 2}{(x - 1)(x - 1)}$

23. $\dfrac{4y}{y + 1} + \dfrac{y}{y - 1}$

$\dfrac{4y}{y + 1}\left(\dfrac{y - 1}{y - 1}\right) + \dfrac{y}{y - 1}\left(\dfrac{y + 1}{y + 1}\right) = \dfrac{5y^2 - 3y}{(y + 1)(y - 1)}$

24. $\dfrac{5}{y - 3} + \dfrac{2}{y + 3}$

$\dfrac{5}{y - 3}\left(\dfrac{y + 3}{y + 3}\right) + \dfrac{2}{y + 3}\left(\dfrac{y - 3}{y - 3}\right) = \dfrac{7y + 9}{(y + 3)(y - 3)}$

25. $\dfrac{6}{5a} + \dfrac{5}{3a + 2}$

$\dfrac{6}{5a}\left(\dfrac{3a + 2}{3a + 2}\right) + \dfrac{5}{3a + 2}\left(\dfrac{5a}{5a}\right) = \dfrac{43a + 12}{5a(3a + 2)}$

26. $\dfrac{3}{2a + 5} + \dfrac{4}{7a}$

$\dfrac{3}{2a + 5}\left(\dfrac{7a}{7a}\right) + \dfrac{4}{7a}\left(\dfrac{2a + 5}{2a + 5}\right) = \dfrac{29a + 20}{7a(2a + 5)}$

27. $\dfrac{2}{3xy} + \dfrac{1}{6yz}$

$\dfrac{2}{3xy}\left(\dfrac{2z}{2z}\right) + \dfrac{1}{6yz}\left(\dfrac{x}{x}\right) = \dfrac{4z + x}{6xyz}$

28. $\dfrac{5}{4xy} + \dfrac{5}{12yz}$

$\dfrac{5}{4xy}\left(\dfrac{3z}{3z}\right) + \dfrac{5}{12yz}\left(\dfrac{x}{x}\right) = \dfrac{15z + 5x}{12xyz}$

Subtract.

29. $\dfrac{5x + 6}{x - 3} - \dfrac{x - 2}{2x - 6}$ $\dfrac{9x + 14}{2(x - 3)}$

30. $\dfrac{7x + 3}{x - 4} - \dfrac{x - 3}{2x - 8}$ $\dfrac{13x + 9}{2(x - 4)}$

31. $\dfrac{3x}{x^2 - 25} - \dfrac{2}{x + 5}$ $\dfrac{x + 10}{(x + 5)(x - 5)}$

32. $\dfrac{7x}{x^2 - 9} - \dfrac{6}{x + 3}$ $\dfrac{x + 18}{(x + 3)(x - 3)}$

33. $\dfrac{a + 3b}{2} - \dfrac{a - b}{5}$

$\dfrac{a + 3b}{2}\left(\dfrac{5}{5}\right) + \dfrac{-a + b}{5}\left(\dfrac{2}{2}\right) = \dfrac{3a + 17b}{10}$

34. $\dfrac{3b - a}{6} - \dfrac{a + b}{5}$

$\dfrac{3b - a}{6}\left(\dfrac{5}{5}\right) + \dfrac{-a - b}{5}\left(\dfrac{6}{6}\right) = \dfrac{9b - 11a}{30}$

35. $\dfrac{8}{2x - 3} - \dfrac{6}{x + 2}$

$\dfrac{8}{2x - 3}\left(\dfrac{x + 2}{x + 2}\right) + \dfrac{-6}{(x + 2)}\left(\dfrac{2x - 3}{2x - 3}\right)$

$= \dfrac{-4x + 34}{(2x - 3)(x + 2)}$

36. $\dfrac{6}{3x - 4} - \dfrac{5}{4x - 3}$

$\dfrac{6}{3x - 4}\left(\dfrac{4x - 3}{4x - 3}\right) + \dfrac{-5}{4x - 3}\left(\dfrac{3x - 4}{3x - 4}\right)$

$= \dfrac{9x + 2}{(3x - 4)(4x - 3)}$

37. $\dfrac{x}{x^2 + 2x - 3} - \dfrac{x}{x^2 - 5x + 4}$

$\dfrac{x}{(x + 3)(x - 1)}\left(\dfrac{x - 4}{x - 4}\right) + \dfrac{-x}{(x - 1)(x - 4)}\left(\dfrac{x + 3}{x + 3}\right)$

$= \dfrac{-7x}{(x + 3)(x - 1)(x - 4)}$

38. $\dfrac{1}{x^2 - 2x} - \dfrac{5}{x^2 - 4x + 4}$

$\dfrac{1}{x(x - 2)}\left(\dfrac{x - 2}{x - 2}\right) + \dfrac{-5}{(x - 2)(x - 2)}\left(\dfrac{x}{x}\right)$

$= \dfrac{-4x - 2}{x(x - 2)^2}$

39. $\dfrac{3}{x^2 + 9x + 20} + \dfrac{1}{x^2 + 10x + 24}$

$\dfrac{3}{(x + 4)(x + 5)}\left(\dfrac{x + 6}{x + 6}\right) + \dfrac{1}{(x + 4)(x + 6)}\left(\dfrac{x + 5}{x + 5}\right)$

$= \dfrac{4x + 23}{(x + 4)(x + 5)(x + 6)}$

40. $\dfrac{2}{x^2 - 3x - 10} + \dfrac{5}{x^2 - 2x - 15}$

$\dfrac{2}{(x + 2)(x - 5)}\left(\dfrac{x + 3}{x + 3}\right) + \dfrac{5}{(x + 3)(x - 5)}\left(\dfrac{x + 2}{x + 2}\right)$

$= \dfrac{7x + 16}{(x + 2)(x + 3)(x - 5)}$

41. $\dfrac{3x - 8}{x^2 - 5x + 6} + \dfrac{x + 2}{x^2 - 6x + 8}$

$\dfrac{3x - 8}{(x - 2)(x - 3)}\left(\dfrac{x - 4}{x - 4}\right) + \dfrac{x + 2}{(x - 2)(x - 4)}\left(\dfrac{x - 3}{x - 3}\right)$

$= \dfrac{4x^2 - 21x + 26}{(x - 2)(x - 4)(x - 3)} = \dfrac{4x - 13}{(x - 4)(x - 3)}$

42. $\dfrac{3x + 5}{x^2 + 4x + 3} + \dfrac{-x + 5}{x^2 + 2x - 3}$

$\dfrac{3x + 5}{(x + 1)(x + 3)}\left(\dfrac{x - 1}{x - 1}\right) + \dfrac{(-x + 5)}{(x + 3)(x - 1)} \cdot \left(\dfrac{x + 1}{x + 1}\right)$

$= \dfrac{2x}{(x + 1)(x - 1)}$

Mixed Practice *Add or subtract.*

43. $\dfrac{6x}{y - 2x} - \dfrac{5x}{2x - y}$

$\dfrac{6x}{y - 2x} - \dfrac{5x}{2x - y}\left(\dfrac{-1}{-1}\right) = \dfrac{11x}{y - 2x}$

44. $\dfrac{8b}{2b - a} - \dfrac{3b}{a - 2b}$

$\dfrac{8b}{2b - a} - \dfrac{3b}{a - 2b}\left(\dfrac{-1}{-1}\right) = \dfrac{11b}{2b - a}$

45. $\dfrac{3y}{8y^2 + 2y - 1} - \dfrac{5y}{2y^2 - 9y - 5}$

$\dfrac{3y}{(4y - 1)(2y + 1)}\left(\dfrac{y - 5}{y - 5}\right) + \dfrac{-5y}{(2y + 1)(y - 5)}\left(\dfrac{4y - 1}{4y - 1}\right)$

$= \dfrac{-17y^2 - 10y}{(4y - 1)(2y + 1)(y - 5)}$

46. $\dfrac{2x}{x^2 + 5x + 6} - \dfrac{x + 1}{x^2 + 2x - 3}$

$\dfrac{2x}{(x + 2)(x + 3)}\left(\dfrac{x - 1}{x - 1}\right) + \dfrac{(-x - 1)}{(x + 3)(x - 1)}\left(\dfrac{x + 2}{x + 2}\right)$

$= \dfrac{x^2 - 5x - 2}{(x + 2)(x + 3)(x - 1)}$

47. $\dfrac{4y}{y^2 + 4y + 3} + \dfrac{2}{y + 1}$

$\dfrac{4y}{(y + 1)(y + 3)} + \dfrac{2}{y + 1}\left(\dfrac{y + 3}{y + 3}\right) = \dfrac{6}{y + 3}$

48. $\dfrac{y - 23}{y^2 - y - 20} + \dfrac{2}{y - 5}$

$\dfrac{y - 23}{(y - 5)(y + 4)} + \dfrac{2}{y - 5}\left(\dfrac{y + 4}{y + 4}\right) = \dfrac{3}{y + 4}$

Cumulative Review

49. **[2.4.1]** Solve. $\frac{1}{3}(x - 2) + \frac{1}{2}(x + 3) = \frac{1}{4}(3x + 1)$

$4x - 8 + 6x + 18 = 9x + 3$

$x = -7$

50. **[2.5.1]** Solve for y. $5ax = 2(ay - 3bc)$

$y = \dfrac{5ax + 6bc}{2a}$

51. **[2.7.1]** Solve for x. $x - \frac{1}{5}x > \frac{1}{2} + \frac{1}{10}x$ $\quad x > \frac{5}{7}$

52. **[4.1.3]** Simplify. $(3x^3y^4)^4$

$81x^{12}y^{16}$

53. **[2.7.1]** *Commuting Costs* A subway token costs $1.50. A monthly unlimited-ride subway pass costs $50. How many days per month would you have to use the subway to go to work (assume one subway token to get to work and one subway token to get back home) in order for it to be cheaper to buy a monthly subway pass?

At least 17 days

54. **[0.7.1]** *Languages in Finland* In Finland, 91.5% of the population speaks Finnish. The other official language is Swedish, spoken by 5.5% of the population. One of the minority languages is Sámi, spoken by 0.03% of the population. (*Source:* www.stat.fi.) In 2006 there were 5,200,000 people in Finland. How many more people spoke Swedish than Sámi?

284,440 people

Quick Quiz 6.3 Perform the indicated operation. Simplify.

1. $\dfrac{3}{x^2 - 2x - 8} + \dfrac{2}{x - 4}$ $\quad \dfrac{2x + 7}{(x + 2)(x - 4)}$

2. $\dfrac{2x + y}{xy} - \dfrac{b - y}{by}$ $\quad \dfrac{bx + by + xy}{bxy}$

3. $\dfrac{2}{x^2 - 9} + \dfrac{3}{x^2 + 7x + 12}$

$\dfrac{5x - 1}{(x + 3)(x - 3)(x + 4)}$

4. **Concept Check** Explain how to find the LCD of the fractions $\dfrac{3}{10xy^2z}, \dfrac{6}{25x^2yz^3}$.

Answers may vary

Classroom Quiz 6.3 You may use these problems to quiz your students' mastery of Section 6.3.

Perform the indicated operation. Simplify.

1. $\dfrac{5}{x^2 - 4x - 21} + \dfrac{3}{x - 7}$

Ans: $\dfrac{3x + 14}{(x + 3)(x - 7)}$

2. $\dfrac{3a + 2b}{ab} - \dfrac{b - a}{bc}$

Ans: $\dfrac{3ac + 2bc - ab + a^2}{abc}$

3. $\dfrac{3}{x^2 - 16} + \dfrac{5}{x^2 + 6x + 8}$

Ans: $\dfrac{8x - 14}{(x + 4)(x - 4)(x + 2)}$

How are you doing with your homework assignments in Sections 6.1 to 6.3? Do you feel you have mastered the material so far? Do you understand the concepts you have covered? Before you go further in the textbook, take some time to do each of the following problems.

Simplify.

6.1

1. $\dfrac{8x - 48}{x^2 - 6x}$

2. $\dfrac{2x^2 - 7x - 15}{x^2 - 12x + 35}$

3. $\dfrac{y^2 + 2y + 1}{2x^2 - 2x^2y^2}$

4. $\dfrac{5x^2 - 23x + 12}{5x^2 + 7x - 6}$

6.2

5. $\dfrac{12a^2}{2x + 10} \cdot \dfrac{8x + 40}{16a^3}$

6. $\dfrac{x - 5}{x^2 + 5x - 14} \cdot \dfrac{x^2 + 12x + 35}{15 - 3x}$

7. $\dfrac{x^2 - 9}{2x + 6} \div \dfrac{2x^2 - 5x - 3}{4x^2 - 1}$

8. $\dfrac{3a^2 + 7a + 2}{4a^2 + 11a + 6} \div \dfrac{6a^2 - 13a - 5}{16a^2 - 9}$

6.3

9. $\dfrac{x - 3y}{xy} - \dfrac{4a - y}{ay}$

10. $\dfrac{7}{2x - 4} + \dfrac{-14}{x^2 - 4}$

11. $\dfrac{2x}{x^2 + 10x + 21} + \dfrac{x - 3}{x + 7}$

12. $\dfrac{4}{x^2 - 2x - 3} - \dfrac{5x}{x^2 + 5x + 4}$

1. $\dfrac{8}{x}$

2. $\dfrac{2x + 3}{x - 7}$

3. $\dfrac{y + 1}{2x^2(1 - y)}$

4. $\dfrac{x - 4}{x + 2}$

5. $\dfrac{3}{a}$

6. $\dfrac{x + 5}{-3(x - 2)}$ or $-\dfrac{x + 5}{3(x - 2)}$

7. $\dfrac{2x - 1}{2}$

8. $\dfrac{4a - 3}{2a - 5}$

9. $\dfrac{xy - 3ax - 3ay}{axy}$

10. $\dfrac{7}{2(x + 2)}$

11. $\dfrac{x^2 + 2x - 9}{(x + 3)(x + 7)}$

12. $\dfrac{-5x^2 + 19x + 16}{(x - 3)(x + 1)(x + 4)}$

Now turn to page SA-13 for the answer to each of these problems. Each answer also includes a reference to the objective in which the problem is first taught. If you missed any of these problems, you should stop and review the Examples and Practice Problems in the referenced objective. A little review now will help you master the material in the upcoming sections of the text.

Teaching Tip Before presenting Example 1 to the class, you might want to ask the class, "How would you simplify the expression

$$\frac{\dfrac{1}{7}}{\dfrac{2}{15} + \dfrac{1}{3}}?$$

" Usually some class members will suggest finding an LCD and adding the two fractions in the denominator. Continue to work out the problem.

$$\frac{\dfrac{1}{7}}{\dfrac{2}{15} + \dfrac{1}{3}} = \frac{\dfrac{1}{7}}{\dfrac{2}{15} + \dfrac{5}{15}} = \frac{\dfrac{1}{7}}{\dfrac{7}{15}}$$

$$= \frac{1}{7} \div \frac{7}{15} = \frac{1}{7} \cdot \frac{15}{7} = \frac{15}{49}$$

Then bridge the gap by saying, "In a similar fashion we can simplify a complex fraction that contains rational expressions." Then go on to show students how to do Example 1.

Teaching Example 1 Simplify. $\dfrac{\dfrac{1}{x} + \dfrac{2}{5}}{\dfrac{1}{y}}$

Ans: $\dfrac{y(5 + 2x)}{5x}$

NOTE TO STUDENT: Fully worked-out solutions to all of the Practice Problems can be found at the back of the text starting at page SP-1

 Simplifying Complex Rational Expressions by Adding or Subtracting in the Numerator and Denominator

A **complex rational expression** (also called a **complex fraction**) has a fraction in the numerator or in the denominator, or both.

$$\frac{3 + \dfrac{2}{x}}{\dfrac{x}{7} + 2} \qquad \frac{\dfrac{x}{y} + 1}{2} \qquad \frac{\dfrac{a + b}{3}}{\dfrac{x - 2y}{4}}$$

The bar in a complex rational expression is both a grouping symbol and a symbol for division.

$$\frac{\dfrac{a + b}{3}}{\dfrac{x - 2y}{4}} \quad \text{is equivalent to} \quad \left(\frac{a + b}{3}\right) \div \left(\frac{x - 2y}{4}\right)$$

We need a procedure for simplifying complex rational expressions.

> **PROCEDURE TO SIMPLIFY A COMPLEX RATIONAL EXPRESSION: ADDING AND SUBTRACTING**
>
> 1. Add or subtract so that you have a single fraction in the numerator and in the denominator.
> 2. Divide the fraction in the numerator by the fraction in the denominator. This is done by inverting the fraction in the denominator and multiplying it by the numerator.

EXAMPLE 1 Simplify. $\dfrac{\dfrac{1}{x}}{\dfrac{2}{y^2} + \dfrac{1}{y}}$

Solution

Step 1 Add the two fractions in the denominator.

$$\frac{\dfrac{1}{x}}{\dfrac{2}{y^2} + \dfrac{1}{y} \cdot \dfrac{y}{y}} = \frac{\dfrac{1}{x}}{\dfrac{2 + y}{y^2}}$$

Step 2 Divide the fraction in the numerator by the fraction in the denominator.

$$\frac{1}{x} \div \frac{2 + y}{y^2} = \frac{1}{x} \cdot \frac{y^2}{2 + y} = \frac{y^2}{x(2 + y)}$$

Practice Problem 1 Simplify. $\dfrac{\dfrac{1}{a} + \dfrac{1}{a^2}}{\dfrac{2}{b^2}}$ $\dfrac{b^2(a + 1)}{2a^2}$

A complex rational expression may contain two or more fractions in the numerator and the denominator.

EXAMPLE 2

Simplify. $\dfrac{\dfrac{1}{x} + \dfrac{1}{y}}{\dfrac{3}{a} - \dfrac{2}{b}}$

Teaching Example 2 Simplify. $\dfrac{\dfrac{5}{m} - \dfrac{6}{n}}{\dfrac{4}{a^2} + \dfrac{1}{a}}$

Ans: $\dfrac{a^2(5n - 6m)}{mn(4 + a)}$

Solution We observe that the LCD of the fractions in the numerator is xy. The LCD of the fractions in the denominator is ab.

$= \dfrac{\dfrac{1}{x} \cdot \dfrac{y}{y} + \dfrac{1}{y} \cdot \dfrac{x}{x}}{\dfrac{3}{a} \cdot \dfrac{b}{b} - \dfrac{2}{b} \cdot \dfrac{a}{a}}$ Multiply each fraction by the appropriate value to obtain common denominators.

$= \dfrac{\dfrac{y + x}{xy}}{\dfrac{3b - 2a}{ab}}$ Add the two fractions in the numerator.

Subtract the two fractions in the denominator.

$= \dfrac{y + x}{xy} \cdot \dfrac{ab}{3b - 2a}$ Invert the fraction in the denominator and multiply it by the numerator.

$= \dfrac{ab(y + x)}{xy(3b - 2a)}$ Write the answer as one fraction.

Practice Problem 2

Simplify. $\dfrac{\dfrac{1}{a} + \dfrac{1}{b}}{\dfrac{x}{2} - \dfrac{5}{y}}$ $\dfrac{2y(b + a)}{ab(xy - 10)}$

For some complex rational expressions, factoring may be necessary to determine the LCD and to combine fractions.

EXAMPLE 3 Simplify. $\dfrac{\dfrac{1}{x^2 - 1} + \dfrac{2}{x + 1}}{x}$

Teaching Example 3 Simplify.

$\dfrac{\dfrac{2x}{x - 3} - \dfrac{x^2}{x^2 - 2x - 3}}{7}$

Ans: $\dfrac{x(x + 2)}{7(x + 1)(x - 3)}$

Solution We need to factor $x^2 - 1$.

$= \dfrac{\dfrac{1}{(x + 1)(x - 1)} + \dfrac{2}{x + 1} \cdot \dfrac{x - 1}{x - 1}}{x}$ The LCD for the fractions in the numerator is $(x + 1)(x - 1)$.

$= \dfrac{\dfrac{1 + 2x - 2}{(x + 1)(x - 1)}}{x}$ Add the two fractions in the numerator.

$= \dfrac{2x - 1}{(x + 1)(x - 1)} \cdot \dfrac{1}{x}$ Simplify the numerator. Invert the fraction in the denominator and multiply.

$= \dfrac{2x - 1}{x(x + 1)(x - 1)}$ Write the answer as one fraction.

Practice Problem 3

Simplify. $\dfrac{\dfrac{x}{x^2 + 4x + 3} + \dfrac{2}{x + 1}}{x + 1}$ $\dfrac{3(x + 2)}{(x + 1)^2(x + 3)}$

When simplifying complex rational expressions, always check to see if the final fraction can be reduced or simplified.

Teaching Example 4 Simplify.

$$\frac{\dfrac{7}{x}}{\dfrac{4}{x-2}-\dfrac{3}{x^2-2x}}$$

Ans: $\dfrac{7(x-2)}{4x-3}$

EXAMPLE 4

Simplify. $\dfrac{\dfrac{3}{a+b}-\dfrac{3}{a-b}}{\dfrac{5}{a^2-b^2}}$

Solution The LCD of the two fractions in the numerator is $(a+b)(a-b)$.

$$=\frac{\dfrac{3}{a+b}\cdot\dfrac{a-b}{a-b}-\dfrac{3}{a-b}\cdot\dfrac{a+b}{a+b}}{\dfrac{5}{a^2-b^2}}$$

$$=\frac{\dfrac{3a-3b}{(a+b)(a-b)}-\dfrac{3a+3b}{(a+b)(a-b)}}{\dfrac{5}{a^2-b^2}}$$

Study carefully how we combine the two fractions in the numerator. Do you see how we obtain $-6b$?

$$=\frac{\dfrac{-6b}{(a+b)(a-b)}}{\dfrac{5}{(a+b)(a-b)}}$$ Factor a^2-b^2 as $(a+b)(a-b)$.

$$=\frac{-6b}{(a+b)(a-b)}\cdot\frac{(a+b)(a-b)}{5}$$ Since $(a+b)(a-b)$ are factors in both a numerator and a denominator, they may be divided out.

$$=\frac{-6b}{5}\quad\text{or}\quad -\frac{6b}{5}$$

Practice Problem 4

Simplify. $\dfrac{\dfrac{6}{x^2-y^2}}{\dfrac{1}{x-y}+\dfrac{3}{x+y}}\cdot\dfrac{3}{2x-y}$

NOTE TO STUDENT: Fully worked-out solutions to all of the Practice Problems can be found at the back of the text starting at page SP-1

② Simplifying Complex Rational Expressions Using the LCD

There is another way to simplify complex rational expressions: Multiply the numerator and denominator of the complex fraction by the least common denominator of all the denominators appearing in the complex fraction.

PROCEDURE TO SIMPLIFY A COMPLEX RATIONAL EXPRESSION: MULTIPLYING BY THE LCD

1. Determine the LCD of all individual denominators occurring in the numerator and denominator of the complex rational expression.
2. Multiply both the numerator and the denominator of the complex rational expression by the LCD.
3. Simplify, if possible.

EXAMPLE 5

Simplify by multiplying by the LCD. $\dfrac{\dfrac{5}{ab^2} - \dfrac{2}{ab}}{3 - \dfrac{5}{2a^2b}}$

Teaching Example 5 Simplify by multiplying by the LCD. $\dfrac{\dfrac{9}{m} - \dfrac{4}{m^2n}}{\dfrac{1}{mn} - 2}$

Ans: $\dfrac{9mn - 4}{m - 2m^2n}$

Solution   The LCD of all the denominators in the complex rational expression is $2a^2b^2$.

$$= \dfrac{2a^2b^2\left(\dfrac{5}{ab^2} - \dfrac{2}{ab}\right)}{2a^2b^2\left(3 - \dfrac{5}{2a^2b}\right)}$$

$$= \dfrac{2a^2b^2\left(\dfrac{5}{ab^2}\right) - 2a^2b^2\left(\dfrac{2}{ab}\right)}{2a^2b^2(3) - 2a^2b^2\left(\dfrac{5}{2a^2b}\right)} \qquad \text{Multiply each term by } 2a^2b^2.$$

$$= \dfrac{10a - 4ab}{6a^2b^2 - 5b} \qquad \text{Simplify.}$$

Practice Problem 5   Simplify by multiplying by the LCD.

$$\dfrac{\dfrac{2}{3x^2} - \dfrac{3}{y}}{\dfrac{5}{xy} - 4} \qquad \dfrac{2y - 9x^2}{15x - 12x^2y}$$

So that you can compare the two methods, we will redo Example 4 by multiplying by the LCD.

EXAMPLE 6

Simplify by multiplying by the LCD. $\dfrac{\dfrac{3}{a + b} - \dfrac{3}{a - b}}{\dfrac{5}{a^2 - b^2}}$

Teaching Example 6 Simplify by multiplying by the LCD. $\dfrac{\dfrac{7}{x}}{\dfrac{4}{x - 2} - \dfrac{3}{x^2 - 2x}}$

Ans: $\dfrac{7(x - 2)}{4x - 3}$

Solution   The LCD of all individual fractions contained in the complex fraction is $(a + b)(a - b)$.

$$= \dfrac{(a + b)(a - b)\left(\dfrac{3}{a + b}\right) - (a + b)(a - b)\left(\dfrac{3}{a - b}\right)}{(a + b)(a - b)\left(\dfrac{5}{(a + b)(a - b)}\right)} \qquad \begin{array}{l}\text{Multiply each term by} \\ \text{the LCD.}\end{array}$$

$$= \dfrac{3(a - b) - 3(a + b)}{5} \qquad \text{Simplify.}$$

$$= \dfrac{3a - 3b - 3a - 3b}{5} \qquad \text{Remove parentheses.}$$

$$= -\dfrac{6b}{5} \qquad \text{Simplify.}$$

Practice Problem 6   Simplify by multiplying by the LCD.

$$\dfrac{\dfrac{6}{x^2 - y^2}}{\dfrac{7}{x - y} + \dfrac{3}{x + y}} \qquad \dfrac{3}{5x + 2y}$$

Simplify.

1. $\dfrac{\dfrac{3}{x}}{\dfrac{2}{x^2}+\dfrac{5}{x}}$　$\dfrac{3x}{2+5x}$

2. $\dfrac{\dfrac{4}{x}}{\dfrac{7}{x}+\dfrac{2}{x^2}}$　$\dfrac{4x}{7x+2}$

3. $\dfrac{\dfrac{3}{a}+\dfrac{3}{b}}{\dfrac{3}{ab}}$　$a+b$

4. $\dfrac{\dfrac{1}{a}+\dfrac{2}{b}}{\dfrac{3}{ab}}$　$\dfrac{2a+b}{3}$

5. $\dfrac{\dfrac{x}{6}-\dfrac{1}{3}}{\dfrac{2}{3x}+\dfrac{5}{6}}$　$\dfrac{x^2-2x}{4+5x}$

6. $\dfrac{\dfrac{x}{8}-\dfrac{1}{2}}{\dfrac{3}{8x}+\dfrac{3}{4}}$　$\dfrac{x^2-4x}{3+6x}$

7. $\dfrac{\dfrac{7}{5x}-\dfrac{1}{x}}{\dfrac{3}{5}+\dfrac{2}{x}}$　$\dfrac{2}{3x+10}$

8. $\dfrac{\dfrac{8}{x}-\dfrac{2}{3x}}{\dfrac{2}{3}+\dfrac{5}{x}}$　$\dfrac{22}{2x+15}$

9. $\dfrac{\dfrac{5}{x}+\dfrac{3}{y}}{3x+5y}$　$\dfrac{1}{xy}$

10. $\dfrac{\dfrac{1}{x}+\dfrac{1}{y}}{x+y}$　$\dfrac{1}{xy}$

11. $\dfrac{4-\dfrac{1}{x^2}}{2+\dfrac{1}{x}}$　$\dfrac{2x-1}{x}$

12. $\dfrac{1-\dfrac{36}{x^2}}{1-\dfrac{6}{x}}$　$\dfrac{x+6}{x}$

13. $\dfrac{\dfrac{2}{x+6}}{\dfrac{2}{x-6}-\dfrac{2}{x^2-36}}$　$\dfrac{x-6}{x+5}$

14. $\dfrac{\dfrac{9}{x^2-1}}{\dfrac{4}{x-1}-\dfrac{2}{x+1}}$　$\dfrac{9}{2(x+3)}$

15. $\dfrac{a+\dfrac{3}{a}}{\dfrac{a^2+2}{3a}}$　$\dfrac{3a^2+9}{a^2+2}$

16. $\dfrac{x+\dfrac{4}{x}}{\dfrac{x^2+3}{4x}}$　$\dfrac{4x^2+16}{x^2+3}$

17. $\dfrac{\dfrac{3}{x-3}}{\dfrac{1}{x^2-9}+\dfrac{2}{x+3}}$　$\dfrac{3x+9}{2x-5}$

18. $\dfrac{\dfrac{5}{x+4}}{\dfrac{1}{x-4}-\dfrac{2}{x^2-16}}$　$\dfrac{5x-20}{x+2}$

19. $\dfrac{\dfrac{2}{y-1}+2}{\dfrac{2}{y+1}-2}$　$\dfrac{y+1}{-y+1}$

20. $\dfrac{\dfrac{y}{y+1}+1}{\dfrac{2y+1}{y-1}}$　$\dfrac{y-1}{y+1}$

To Think About

21. Consider the complex fraction $\dfrac{\dfrac{4}{x+3}}{\dfrac{5}{x}-1}$. What values are not allowable replacements for the variable x?
No expression in any denominator can be allowed to be zero, since division by zero is undefined. So -3, 5, and 0 are not allowable replacements for the variable x.

22. Consider the complex fraction $\dfrac{\dfrac{5}{x-2}}{\dfrac{6}{x}+1}$. What values are not allowable replacements for the variable x?
No expression in any denominator can be allowed to be zero, since division by zero is undefined. So 2, -6, and 0 are not allowable replacements for the variable x.

Cumulative Review

23. **[2.5.1]** Solve for w. $P = 2(l + w)$

$w = \dfrac{P - 2l}{2}$

24. **[2.7.1]** Solve and graph. $7 + x < 11 + 5x$

$x > -1$

25. **[3.2.1]** When nine is subtracted from double a number, the result is the same as one-half of the same number. What is the number?

6

26. **[0.7.1]** *Urban Sprawl* A neighborhood in Hopewell, New Jersey, is upset because a big office complex is being built in the neighborhood. Many of the citizens of Hopewell moved to their town to get away from urban sprawl. An investment firm will be building a complex planned to cover 3.5 million square feet. This initial estimate could grow to 5.5 million square feet. What percent of increase would this be?

Approximately 57%

Quick Quiz 6.4 Simplify.

1. $\dfrac{\dfrac{a}{4b} - \dfrac{1}{3}}{\dfrac{5}{4b} - \dfrac{4}{a}}$ $\dfrac{a(3a - 4b)}{3(5a - 16b)}$

2. $\dfrac{a + b}{\dfrac{1}{a} + \dfrac{1}{b}}$ ab

3. $\dfrac{\dfrac{10}{x^2 - 25}}{\dfrac{3}{x + 5} + \dfrac{2}{x - 5}}$ $\dfrac{2}{x - 1}$

4. Concept Check To simplify the following complex fraction, explain how you would add the two fractions in the numerator.

$$\dfrac{\dfrac{7}{x - 3} + \dfrac{15}{2x - 6}}{\dfrac{2}{x + 5}},$$

Answers may vary

Classroom Quiz 6.4 You may use these problems to quiz your students' mastery of Section 6.4.

Simplify.

1. $\dfrac{\dfrac{x}{5y} - \dfrac{1}{4}}{\dfrac{3}{4y} - \dfrac{2}{x}}$

Ans: $\dfrac{x(4x - 5y)}{5(3x - 8y)}$

2. $\dfrac{\dfrac{5}{x} + \dfrac{5}{y}}{10x + 10y}$

Ans: $\dfrac{1}{2xy}$

3. $\dfrac{\dfrac{8}{x^2 - 16}}{\dfrac{2}{x + 4} + \dfrac{3}{x - 4}}$

Ans: $\dfrac{8}{5x + 4}$

Teaching Tip You may want to show the connection more precisely between Section 2.4, which students have previously studied, and this present section. You might find this approach helpful. "In Section 2.4 you solved equations such as the following one."

$$\frac{x}{4} - 1 = \frac{x}{5}$$
$$LCD = (4)(5) = 20$$
$$20\left(\frac{x}{4}\right) - 20(1) = 20\left(\frac{x}{5}\right)$$
$$5x - 20 = 4x$$
$$5x - 4x = 20$$
$$x = 20$$

"In this section you will be doing very similar problems. This time, however, the denominators will contain variables."

Teaching Example 1 Solve for x and check your solution. $\dfrac{1}{x} - \dfrac{1}{2} = \dfrac{7}{2x}$

Ans: $x = -5$

NOTE TO STUDENT: *Fully worked-out solutions to all of the Practice Problems can be found at the back of the text starting at page SP-1*

Teaching Example 2 Solve and check.

$$\frac{5}{3} = \frac{x}{x-2}$$

Ans: $x = 5$

 Solving Equations Involving Rational Expressions That Have Solutions

In Section 2.4 we developed procedures to solve linear equations containing fractions whose denominators were numbers. In this section we use a similar approach to solve equations containing fractions whose denominators are polynomials. It would be wise for you to review Section 2.4 briefly *before you begin this section*. It will be especially helpful to carefully study Examples 1 and 2.

TO SOLVE AN EQUATION CONTAINING RATIONAL EXPRESSIONS

1. Determine the LCD of all the denominators.
2. Multiply each term of the equation by the LCD.
3. Solve the resulting equation.
4. Check your solution. Exclude from your solution any value that would make the LCD equal to zero. If such a value is obtained, there is *no solution*.

EXAMPLE 1 Solve for x and check your solution. $\dfrac{5}{x} + \dfrac{2}{3} = -\dfrac{3}{x}$

Solution

$$3x\left(\frac{5}{x}\right) + 3x\left(\frac{2}{3}\right) = 3x\left(-\frac{3}{x}\right) \quad \text{Observe that the LCD is } 3x. \text{ Multiply each term by } 3x.$$
$$15 + 2x = -9$$
$$2x = -9 - 15 \quad \text{Subtract 15 from both sides.}$$
$$2x = -24$$
$$x = -12 \quad \text{Divide both sides by 2.}$$

Check.
$$\frac{5}{-12} + \frac{2}{3} \stackrel{?}{=} -\frac{3}{-12} \quad \text{Replace each } x \text{ by } -12.$$
$$-\frac{5}{12} + \frac{8}{12} \stackrel{?}{=} \frac{3}{12}$$
$$\frac{3}{12} = \frac{3}{12} \quad \checkmark \quad \text{It checks.}$$

Practice Problem 1 Solve for x and check your solution.

$$\frac{3}{x} + \frac{4}{5} = -\frac{2}{x} \qquad x = -\frac{25}{4} \text{ or } -6\frac{1}{4} \text{ or } -6.25$$

EXAMPLE 2 Solve and check. $\dfrac{6}{x+3} = \dfrac{3}{x}$

Solution Observe that the LCD = $x(x+3)$.

$$x(x+3)\left(\frac{6}{x+3}\right) = x(x+3)\left(\frac{3}{x}\right) \quad \text{Multiply both sides by } x(x+3).$$
$$6x = 3(x+3) \quad \text{Simplify. Do you see how this is done?}$$
$$6x = 3x + 9 \quad \text{Remove parentheses.}$$
$$3x = 9 \quad \text{Subtract } 3x \text{ from both sides.}$$
$$x = 3 \quad \text{Divide both sides by 3.}$$

Check. $\dfrac{6}{3+3} \overset{?}{=} \dfrac{3}{3}$ Replace each x by 3.

$\dfrac{6}{6} = \dfrac{3}{3}$ ✓ It checks.

Practice Problem 2 Solve and check.

$$\frac{4}{2x+1} = \frac{6}{2x-1} \quad x = -\frac{5}{2} \text{ or } -2\frac{1}{2} \text{ or } -2.5$$

It is sometimes necessary to factor denominators before the correct LCD can be determined.

EXAMPLE 3 Solve and check. $\dfrac{3}{x+5} - 1 = \dfrac{4-x}{2x+10}$

Teaching Example 3 Solve and check.

$$\frac{2}{x+1} + \frac{5}{x^2-1} = \frac{1}{x-1}$$

Ans: $x = -2$

Solution

$\dfrac{3}{x+5} - 1 = \dfrac{4-x}{2(x+5)}$ Factor $2x+10$. We determine that the LCD is $2(x+5)$.

$2(x+5)\left(\dfrac{3}{x+5}\right) - 2(x+5)(1) = 2(x+5)\left[\dfrac{4-x}{2(x+5)}\right]$ Multiply each term by the LCD.

$2(3) - 2(x+5) = 4 - x$ Simplify.

$6 - 2x - 10 = 4 - x$ Remove parentheses.

$-2x - 4 = 4 - x$ Combine like terms.

$-4 = 4 + x$ Add $2x$ to both sides.

$-8 = x$ Subtract 4 from both sides.

Check. $\dfrac{3}{-8+5} - 1 \overset{?}{=} \dfrac{4-(-8)}{2(-8)+10}$ Replace each x in the original equation by -8.

$\dfrac{3}{-3} - 1 \overset{?}{=} \dfrac{4+8}{-16+10}$

$-1 - 1 \overset{?}{=} \dfrac{12}{-6}$

$-2 = -2$ ✓ It checks. The solution is -8.

Practice Problem 3 Solve and check.

$$\frac{x-1}{x^2-4} = \frac{2}{x+2} + \frac{4}{x-2} \quad x = -1$$

NOTE TO STUDENT: Fully worked-out solutions to all of the Practice Problems can be found at the back of the text starting at page SP-1

Determining Whether an Equation Involving Rational Expressions Has No Solution

Equations containing rational expressions sometimes appear to have solutions when in fact they do not. By this we mean that the "solutions" we get by using completely correct methods are, in actuality, not solutions.

In the case where a value makes a denominator in the equation equal to zero, we say it is not a solution to the equation. Such a value is called an **extraneous solution.** An extraneous solution is an apparent solution that does *not* satisfy the original equation. If all of the apparent solutions of an equation are extraneous solutions, we say that the equation has **no solution.** It is important that you check all apparent solutions in the original equation.

EXAMPLE 4 Solve and check. $\dfrac{y}{y - 2} - 4 = \dfrac{2}{y - 2}$

Solution Observe that the LCD is $y - 2$.

$$(y - 2)\left(\frac{y}{y - 2}\right) - (y - 2)(4) = (y - 2)\left(\frac{2}{y - 2}\right) \quad \text{Multiply each term by } (y - 2).$$

$y - 4(y - 2) = 2$	Simplify. Do you see how this is done?
$y - 4y + 8 = 2$	Remove parentheses.
$-3y + 8 = 2$	Combine like terms.
$-3y = -6$	Subtract 8 from both sides.
$\dfrac{-3y}{-3} = \dfrac{-6}{-3}$	Divide both sides by -3.
$y = 2$	2 is only an apparent solution.

This equation has no solution.

Why? We can see immediately that $y = 2$ is not a solution of the original equation. When we substitute 2 for y in a denominator, the denominator is equal to zero and the expression is undefined.

Check. $\dfrac{y}{y - 2} - 4 = \dfrac{2}{y - 2}$ Suppose that you try to check the apparent solution by substituting 2 for y.

$$\frac{2}{2 - 2} - 4 \stackrel{?}{=} \frac{2}{2 - 2}$$

$$\frac{2}{0} - 4 = \frac{2}{0} \quad \text{This does not check since you do not obtain a real number when you divide by zero.}$$

These expressions are not defined.

There is no such number as $2 \div 0$. We see that 2 does *not* check. This equation has **no solution.**

Practice Problem 4 Solve and check.

$$\frac{2x}{x + 1} = \frac{-2}{x + 1} + 1 \quad \text{No solution}$$

Solve and check problems 1–16.

1. $\dfrac{7}{x} + \dfrac{3}{4} = \dfrac{-2}{x}$

$x = -12$

2. $\dfrac{8}{x} + \dfrac{2}{5} = \dfrac{-2}{x}$

$x = -25$

3. $\dfrac{-1}{4x} + \dfrac{3}{2} = \dfrac{5}{x}$

$x = \dfrac{7}{2}$ or $3\dfrac{1}{2}$ or 3.5

4. $\dfrac{1}{3x} + \dfrac{5}{6} = \dfrac{2}{x}$

$x = 2$

5. $\dfrac{5x + 3}{3x} = \dfrac{7}{3} - \dfrac{9}{x}$

$5x + 3 = 7x - 27$

$x = 15$

6. $\dfrac{2x + 7}{5x} - \dfrac{1}{x} = \dfrac{8}{5}$

$2x + 7 - 5 = 8x$

$x = \dfrac{1}{3}$

7. $\dfrac{x + 5}{3x} = \dfrac{1}{2}$

$2x + 10 = 3x$

$x = 10$

8. $\dfrac{x - 6}{5x} = \dfrac{1}{3}$

$3x - 18 = 5x$

$x = -9$

9. $\dfrac{6}{3x - 5} = \dfrac{3}{2x}$

$12x = 9x - 15$

$x = -5$

10. $\dfrac{1}{x + 6} = \dfrac{4}{x}$

$x = 4x + 24$

$x = -8$

11. $\dfrac{2}{2x + 5} = \dfrac{4}{x - 4}$

$2x - 8 = 8x + 20$

$x = -\dfrac{14}{3}$ or $-4\dfrac{2}{3}$

12. $\dfrac{3}{x + 5} = \dfrac{3}{3x - 2}$

$9x - 6 = 3x + 15$

$x = \dfrac{7}{2}$ or $3\dfrac{1}{2}$ or 3.5

13. $\dfrac{2}{x} + \dfrac{x}{x + 1} = 1$

$2x + 2 + x^2 = x^2 + x$

$x = -2$

14. $\dfrac{5}{2} = 3 + \dfrac{2x + 7}{x + 6}$

$5x + 30 = 6x + 36 + 4x + 14$

$-4 = x$

15. $\dfrac{85 - 4x}{x} = 7 - \dfrac{3}{x}$

$85 - 4x = 7x - 3$

$x = 8$

16. $\dfrac{63 - 2x}{x} = 2 - \dfrac{5}{x}$

$63 - 2x = 2x - 5$

$x = 17$

Mixed Practice *Solve and check. If there is no solution, say so.*

17. $\dfrac{1}{x + 4} - 2 = \dfrac{3x - 2}{x + 4}$

$1 - 2x - 8 = 3x - 2$

$x = -1$

18. $\dfrac{2}{x + 5} - 1 = \dfrac{3x - 4}{x + 5}$

$2 - x - 5 = 3x - 4$

$x = \dfrac{1}{4}$ or 0.25

19. $\dfrac{2}{x - 6} - 5 = \dfrac{2(x - 5)}{x - 6}$

$2 - 5x + 30 = 2x - 10$

$6 = x$

No solution

20. $7 - \dfrac{x}{x + 5} = \dfrac{5}{5 + x}$

$35 + 7x - x = 5$

$-5 = x$

No solution

21. $\dfrac{2}{x + 1} - \dfrac{1}{x - 1} = \dfrac{2x}{x^2 - 1}$

$2x - 2 - x - 1 = 2x$

$-3 = x$

22. $\dfrac{8x}{4x^2 - 1} = \dfrac{3}{2x + 1} + \dfrac{3}{2x - 1}$

$8x = 6x - 3 + 6x + 3$

$x = 0$

23. $\dfrac{y + 1}{y^2 + 2y - 3} = \dfrac{1}{y + 3} - \dfrac{1}{y - 1}$

$y + 1 = y - 1 - y - 3$

$y = -5$

24. $\dfrac{6}{x - 5} + \dfrac{3x + 1}{x^2 - 2x - 15} = \dfrac{5}{x + 3}$

$6x + 18 + 3x + 1 = 5x - 25$

$x = -11$

25. $\dfrac{2x}{x + 4} - \dfrac{8}{x - 4} = \dfrac{2x^2 + 32}{x^2 - 16}$

$2x^2 - 8x - 32 - 8x = 2x^2 + 32$

$-4 = x$

No solution

26. $\dfrac{4x}{x + 3} - \dfrac{12}{x - 3} = \dfrac{4x^2 + 36}{x^2 - 9}$

$4x^2 - 12x - 12x - 36 = 4x^2 + 36$

$-3 = x$

No solution

27. $\dfrac{4}{x^2 - 1} + \dfrac{7}{x + 1} = \dfrac{5}{x - 1}$

$4 + 7x - 7 = 5x + 5$

$x = 4$

28. $\dfrac{7}{4x^2 - 1} + \dfrac{2}{2x + 1} = \dfrac{3}{2x - 1}$

$7 + 4x - 2 = 6x + 3$

$x = 1$

29. $\dfrac{x + 11}{x^2 - 5x + 4} + \dfrac{3}{x - 1} = \dfrac{5}{x - 4}$

$x + 11 + 3x - 12 = 5x - 5$

$x = 4$

No solution

30. $\dfrac{6}{x - 3} = \dfrac{-5}{x - 2} + \dfrac{-5}{x^2 - 5x + 6}$

$6(x - 2) = -5(x - 3) - 5$

$x = 2$

No solution

To Think About *In each of the following equations, what values are not allowable replacements for the variable x? Do not solve the equation.*

31. $\dfrac{3x}{x - 2} - \dfrac{4x}{x - 4} = \dfrac{3}{x^2 - 6x + 8}$

$x = 2, x = 4$

32. $\dfrac{3x}{x - 1} - \dfrac{5x}{2x + 3} = \dfrac{7}{2x^2 + 5x + 3}$

$x = 1, x = -\dfrac{3}{2}, x = -1$

Cumulative Review

33. **[5.4.1]** Factor. $8x^2 - 2x - 1$

$(4x + 1)(2x - 1)$

34. **[2.3.3]** Solve. $5(x - 2) = 8 - (3 + x)$

$x = \dfrac{5}{2}$ or $2\dfrac{1}{2}$ or 2.5

▲ **35.** **[3.3.1]** *Geometry* The perimeter of a rectangular computer monitor is 44 inches. The length is 8 inches less than twice the width. Find the dimensions of the monitor.

$2x + 2(2x - 8) = 44$ Width = 10 in.; length = 12 in.

$x = 10$

36. **[0.7.1]** *Mortgage Payments* Wally and Adele Panzas plan to purchase a new home and borrow $115,000. They plan to take out a 25-year mortgage at an annual interest rate of 8.75%. The bank will charge them $8.23 per month for each $1000 of mortgage. What will their monthly payments be?

$946.45

Quick Quiz 6.5 Solve. If there is no solution, say so.

1. $\dfrac{3}{4x} - \dfrac{5}{6x} = 2 - \dfrac{1}{2x}$ $x = \dfrac{5}{24}$

2. $\dfrac{x}{x - 1} - \dfrac{2}{x} = \dfrac{1}{x - 1}$ $x = 2$

3. $\dfrac{6}{x^2 - 2x - 8} + \dfrac{5}{x + 2} = \dfrac{1}{x - 4}$

No solution

4. **Concept Check** Explain how to find the LCD for the following equation. Do not solve the equation.

$$\dfrac{x}{x^2 - 9} + \dfrac{2}{3x - 9} = \dfrac{5}{2x + 6} + \dfrac{3}{2x^2 - 18}$$

Answers may vary

Classroom Quiz 6.5 You may use these problems to quiz your students' mastery of Section 6.5.

Solve. If there is no solution, say so.

1. $\dfrac{2}{5x} - \dfrac{1}{15x} = 2 - \dfrac{2}{3x}$

Ans: $x = \dfrac{1}{2}$ or 0.5

2. $3 - \dfrac{5x}{x + 2} = \dfrac{10}{x + 2}$

Ans: No solution

3. $\dfrac{3}{x} + \dfrac{5}{x + 2} = \dfrac{9x - 3}{x^2 + 2x}$

Ans: $x = 9$

6.6 RATIO, PROPORTION, AND OTHER APPLIED PROBLEMS

1 Solving Problems Involving Ratio and Proportion

A **ratio** is a comparison of two quantities. You may be familiar with ratios that compare miles to hours or miles to gallons. A ratio is often written as a quotient in the form of a fraction. For example, the ratio of 7 to 9 can be written as $\frac{7}{9}$.

A **proportion** is an equation that states that two ratios are equal. For example,

$$\frac{7}{9} = \frac{21}{27}, \quad \frac{2}{3} = \frac{10}{15}, \quad \text{and} \quad \frac{a}{b} = \frac{c}{d} \quad \text{are proportions.}$$

Let's take a closer look at the last proportion. We can see that the LCD of the fractional equation is bd.

$$(b\,d)\frac{a}{b} = (b\,d)\frac{c}{d} \quad \text{Multiply each side by the LCD.}$$
$$da = bc$$
$$ad = bc \qquad \text{Since multiplication is commutative, } da = ad.$$

Thus we have proved the following.

THE PROPORTION EQUATION

If

$$\frac{a}{b} = \frac{c}{d}, \quad \text{then} \quad ad = bc$$

for all real numbers a, b, c, and d, where $b \neq 0$ and $d \neq 0$.

This is sometimes called **cross multiplying.** It can be applied only if you have *one* fraction and nothing else on each side of the equation.

EXAMPLE 1 Michael took 5 hours to drive 245 miles on the turnpike. At the same rate, how many hours will it take him to drive a distance of 392 miles?

Solution

1. *Understand the problem.* Let x = the number of hours it will take to drive 392 miles. If 5 hours are needed to drive 245 miles, then x hours are needed to drive 392 miles.

2. *Write an equation.* We can write this as a proportion. Compare time to distance in each ratio.

$$\begin{array}{c} \text{Time} \longrightarrow \\ \text{Distance} \longrightarrow \end{array} \frac{5 \text{ hours}}{245 \text{ miles}} = \frac{x \text{ hours}}{392 \text{ miles}} \begin{array}{c} \longleftarrow \text{Time} \\ \longleftarrow \text{Distance} \end{array}$$

3. *Solve and state the answer.*

$$5(392) = 245x \quad \text{Cross-multiply.}$$
$$\frac{1960}{245} = x \qquad \text{Divide both sides by 245.}$$
$$8 = x$$

It will take Michael 8 hours to drive 392 miles.

4. *Check.* Is $\frac{5}{245} = \frac{8}{392}$? Do the computation and see.

Practice Problem 1 It took Brenda 8 hours to drive 420 miles. At the same rate, how long would it take her to drive 315 miles? 6 hr

Student Learning Objectives

After studying this section, you will be able to:

1. Solve problems involving ratio and proportion.

2. Solve problems involving similar triangles.

3. Solve distance problems involving rational expressions.

4. Solve work problems.

Teaching Example 1 Hector drove 452 miles in 8 hours. If he travels at the same rate, how far will he drive in 6 hours tomorrow?

Ans: 339 mi

NOTE TO STUDENT: *Fully worked-out solutions to all of the Practice Problems can be found at the back of the text starting at page SP-1*

Teaching Example 2 If $\frac{7}{8}$ inch on a map represents an actual distance of 42 miles, how long is the distance represented by $1\frac{1}{2}$ inches on the same map?

Ans: 72 mi

EXAMPLE 2 If $\frac{3}{4}$ inch on a map represents an actual distance of 20 miles, how long is the distance represented by $4\frac{1}{8}$ inches on the same map?

Solution Let x = the distance represented by $4\frac{1}{8}$ inches.

Initial measurement on map $\longrightarrow$ $\dfrac{3}{4}$ $\quad$ $4\frac{1}{8}$ $\longleftarrow$ Second measurement on the map

Initial distance $\longrightarrow$ $\dfrac{3}{4}{20} = \dfrac{4\frac{1}{8}}{x}$ $\longleftarrow$ Second distance

$$\left(\frac{3}{4}\right)(x) = (20)\left(4\frac{1}{8}\right) \quad \text{Cross-multiply.}$$

$$\left(\frac{3}{4}\right)(x) = (\overset{5}{\cancel{20}})\left(\frac{33}{\underset{2}{\cancel{8}}}\right) \quad \text{Write } 4\frac{1}{8} \text{ as } \frac{33}{8} \text{ and simplify.}$$

$$\frac{3x}{4} = \frac{165}{2} \quad \text{Multiply the fractions.}$$

$$4\left(\frac{3x}{4}\right) = \overset{2}{\cancel{4}}\left(\frac{165}{\cancel{2}}\right) \quad \text{Multiply each side by 4.}$$

$$3x = 330 \quad \text{Simplify.}$$

$$x = 110 \quad \text{Divide both sides by 3.}$$

$4\frac{1}{8}$ inches on the map represents an actual distance of 110 miles.

Practice Problem 2 If $\frac{5}{8}$ inch on a map represents an actual distance of 30 miles, how long is the distance represented by $2\frac{1}{2}$ inches on the same map?

120 mi

② Solving Problems Involving Similar Triangles

Similar triangles are triangles that have the same shape, but may be different sizes. For example, if you draw a triangle on a sheet of paper, place the paper in a photocopy machine, and make a copy that is reduced by 25%, you would create a triangle that is similar to the original triangle. The two triangles will have the same shape. The corresponding sides of the triangles will be proportional. The corresponding angles of the triangles will be equal.

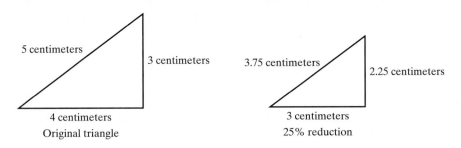

Original triangle

25% reduction

You can use the proportion equation to show that the corresponding sides of the triangles above are proportional. In fact, you can use the proportion equation to find an unknown length of a side of one of two similar triangles.

Teaching Example 3 A ramp is 50 feet long and rises up 35 feet. A ramp at the same angle rises up 28 feet. How long is the second ramp?

Ans: 40 ft

EXAMPLE 3 A ramp is 32 meters long and rises up 15 meters. A ramp at the same angle is 9 meters long. How high is the second ramp?

Solution To answer this question, we find the length of side x in the following two similar triangles.

Ramp A $\qquad$ Ramp B

Ramp A, longest side $\longrightarrow$ $\dfrac{32}{9} = \dfrac{15}{x}$ $\longleftarrow$ Shortest side, ramp A
Ramp B, longest side $\longrightarrow$ $\longleftarrow$ Shortest side, ramp B

$$32x = (9)(15) \quad \text{Cross-multiply.}$$
$$32x = 135 \quad \text{Multiply.}$$
$$x = \frac{135}{32} \quad \text{Divide both sides by 32.}$$
$$\text{or} \quad x = 4\frac{7}{32}$$

The ramp is $4\frac{7}{32}$ meters tall.

▲ **Practice Problem 3** Triangle C is similar to triangle D. Find the length of side x. Express your answer as a mixed number.

NOTE TO STUDENT: Fully worked-out solutions to all of the Practice Problems can be found at the back of the text starting at page SP-1

13 centimeters   16 centimeters x 18 centimeters $14\frac{5}{8}$ cm

Triangle C   Triangle D

We can also use similar triangles for indirect measurement—for instance, to find the height of an object that is too tall to measure using standard measuring devices. When the sun shines on two vertical objects at the same time, the shadows and the objects form similar triangles.

▲ **EXAMPLE 4** A woman who is 5 feet tall casts a shadow that is 8 feet long. At the same time of day, a building casts a shadow that is 72 feet long. How tall is the building?

Teaching Example 4 A woman who is 5.5 feet tall casts a shadow that is 4 feet long. At the same time of day, a tree casts a shadow that is 50 feet long. How tall is the tree?

Ans: 68.75 ft

Solution

1. *Understand the problem.* First we draw a sketch. We do not know the height of the building, so we call it x.

Woman: 5 feet
8-foot shadow

Building: x feet
72-foot shadow

2. *Write an equation and solve.*

Height of woman $\longrightarrow$ $\dfrac{5}{8} = \dfrac{x}{72}$ $\longleftarrow$ Height of building
Length of woman's shadow $\longrightarrow$ $\longleftarrow$ Length of building's shadow

$$(5)(72) = 8x \quad \text{Cross-multiply.}$$
$$360 = 8x$$
$$45 = x$$

The height of the building is 45 feet.

▲ **Practice Problem 4** A man who is 6 feet tall casts a shadow that is 7 feet long. At the same time of day, a large flagpole casts a shadow that is 38.5 feet long. How tall is the flagpole? 33 ft

In problems such as Example 4, we are assuming that the building and the person are standing exactly perpendicular to the ground. In other words, each triangle is assumed to be a right triangle. In other similar triangle problems, if the triangles are not right triangles you must be careful that the corresponding angles in the two triangles are the same.

3 **Solving Distance Problems Involving Rational Expressions**

Some distance problems are solved using equations with rational expressions. We will need the formula Distance = Rate × Time, $D = RT$, which we can write in the form $T = \dfrac{D}{R}$. In the United States, distances are usually measured in miles. In European countries, distances are usually measured in kilometers.

EXAMPLE 5 A French commuter airline flies from Paris to Avignon. Plane A flies at a speed that is 50 kilometers per hour faster than plane B. Plane A flies 500 kilometers in the amount of time that plane B flies 400 kilometers. Find the speed of each plane.

Solution

1. **Understand the problem.** Let s = the speed of plane B in kilometers per hour.

   Then $s + 50$ = the speed of plane A in kilometers per hour. Make a simple table for D, R, and T.

   Since $T = \dfrac{D}{R}$, for each plane we divide the expression for D by the expression for R and write it in the table in the column for time.

	D	R	$T = \dfrac{D}{R}$
Plane A	500	$s + 50$	?
Plane B	400	s	?

	D	R	$T = \dfrac{D}{R}$
Plane A	500	$s + 50$	$\dfrac{500}{s + 50}$
Plane B	400	s	$\dfrac{400}{s}$

2. **Write an equation and solve.** Each plane flies the same amount of time. That is, the time for plane A equals the time for plane B.

$$\frac{500}{s + 50} = \frac{400}{s}$$

You can solve this equation using the methods in Section 6.5 or you may cross-multiply. Here we will cross-multiply.

$$500s = (s + 50)(400) \quad \text{Cross-multiply.}$$
$$500s = 400s + 20{,}000 \quad \text{Remove parentheses.}$$
$$100s = 20{,}000 \quad \text{Subtract } 400s \text{ from each side.}$$
$$s = 200 \quad \text{Divide each side by 100.}$$

Plane B travels 200 kilometers per hour. Since

$$s + 50 = 200 + 50 = 250,$$

plane A travels 250 kilometers per hour.

Practice Problem 5 Two European freight trains traveled toward Paris for the same amount of time. Train A traveled 180 kilometers, while train B traveled 150 kilometers. Train A traveled 10 kilometers per hour faster than train B. What was the speed of each train? Train B traveled 50 km/hr. Train A traveled 60 km/hr.

Teaching Example 6 Theresa and Jeanette are hired to paint the main hallway on each floor of a three-story building. The hallways on each floor are the same size. Jeanette takes 6 hours to paint the hallway on the third floor. Theresa is more experienced and needs only 4 hours to paint the hallway on the second floor. If they work together to paint the first-floor hallway, how long will it take them? Express your answer in hours and minutes.

Ans: 2 hr, 24 min

4 Solving Work Problems

Some applied problems involve the length of time needed to do a job. These problems are often referred to as work problems.

EXAMPLE 6 Reynaldo can sort a huge stack of mail on an old sorting machine in 9 hours. His brother Carlos can sort the same amount of mail using a newer sorting machine in 8 hours. How long would it take them to do the job working together? Express your answer in hours and minutes. Round to the nearest minute.

Solution

1. **Understand the problem.** Let's do a little reasoning.

   If Reynaldo can do the job in 9 hours, then in *1 hour* he could do $\frac{1}{9}$ of the job.

   If Carlos can do the job in 8 hours, then in *1 hour* he could do $\frac{1}{8}$ of the job.

   Let x = the number of hours it takes Reynaldo and Carlos to do the job together. In *1 hour* together they could do $\frac{1}{x}$ of the job.

2. **Write an equation and solve.** The amount of work Reynaldo can do in 1 hour plus the amount of work Carlos can do in 1 hour must be equal to the amount of work they could do together in 1 hour.

Amount of work done by Reynaldo	+	Amount of work done by Carlos	=	Amount of work done together
$\frac{1}{9}$	+	$\frac{1}{8}$	=	$\frac{1}{x}$

Let us solve for x. We observe that the LCD is $72x$.

$$72x\left(\frac{1}{9}\right) + 72x\left(\frac{1}{8}\right) = 72x\left(\frac{1}{x}\right) \quad \text{Multiply each term by the LCD.}$$

$$8x + 9x = 72 \quad \text{Simplify.}$$
$$17x = 72 \quad \text{Combine like terms.}$$
$$x = \frac{72}{17} \quad \text{Divide each side by 17.}$$
$$x = 4\frac{4}{17}$$

To change $\frac{4}{17}$ of an hour to minutes we multiply.

$$\frac{4}{17} \text{ hour} \times \frac{60 \text{ minutes}}{1 \text{ hour}} = \frac{240}{17} \text{ minutes, which is approximately 14.118 minutes}$$

To the nearest minute this is 14 minutes. Thus doing the job together will take 4 hours and 14 minutes.

Practice Problem 6 John Tobey and Dave Wells obtained night custodian jobs at a local factory while going to college part-time. Using the buffer machine, John can buff all the floors in the building in 6 hours. Dave takes a little longer and can do all the floors in the building in 7 hours. Their supervisor bought another buffer machine. How long will it take John and Dave to do all the floors in the building working together, each with his own machine? Express your answer in hours and minutes. Round to the nearest minute. 3 hr, 14 min

Calculator

 Reciprocals

You can find $\frac{1}{x}$ for any value of x on a scientific calculator by using the key labeled $\boxed{x^{-1}}$ or the key labeled $\boxed{1/x}$. For example, to find $\frac{1}{9}$, we use $9\,\boxed{x^{-1}}$ or $9\,\boxed{1/x}$. The display will read 0.11111111. Therefore we can solve Example 6 as follows:

$$9\,\boxed{x^{-1}}\, \boxed{+}\, 8\,\boxed{x^{-1}}\, \boxed{=}$$

The display will read 0.2361111. Thus we have obtained the equation $0.2361111 = \frac{1}{x}$. Now this is equivalent to $x = \frac{1}{0.2361111}$. (Do you see why?) Thus we enter $0.2361111\,\boxed{x^{-1}}$, and the display reads 4.2352943. If we round to the nearest hundredth, we have $x \approx 4.24$ hours, which is approximately equal to our answer of $4\frac{4}{17}$ hours.

MyMathLab

PRACTICE   WATCH   DOWNLOAD   READ   REVIEW

Solve.

1. $\dfrac{4}{9} = \dfrac{8}{x}$ $x = 18$

2. $\dfrac{5}{12} = \dfrac{x}{8}$ $x = \dfrac{10}{3}$ or $3\dfrac{1}{3}$

3. $\dfrac{x}{17} = \dfrac{12}{5}$ $x = \dfrac{204}{5}$ or $40\dfrac{4}{5}$ or 40.8

4. $\dfrac{16}{x} = \dfrac{3}{4}$ $x = \dfrac{64}{3}$ or $21\dfrac{1}{3}$

5. $\dfrac{9.1}{8.4} = \dfrac{x}{6}$ $x = 6.5$

6. $\dfrac{5}{x} = \dfrac{17.5}{4.2}$ $x = 1.2$

7. $\dfrac{7}{x} = \dfrac{40}{130}$ $x = \dfrac{91}{4}$ or $22\dfrac{3}{4}$ or 22.75

8. $\dfrac{x}{18} = \dfrac{13}{2}$ $x = 117$

Applications   *Use a proportion to answer exercises 9–16.*

9. ***Map Scale*** The scale on the AAA map of Colorado is approximately $\frac{3}{4}$ inch to 15 miles. If the distance from Denver to Pueblo measures 5.5 inches on the map, how far apart are the two cities?   110 mi

10. ***Recipe Ratios*** Nella Coates's recipe for Shoofly Pie contains $\frac{3}{4}$ cup of unsulfured molasses. This recipe makes a small pie that serves 8 people. If she makes the larger pie that serves 12 people and the ratio of molasses to people remains the same, how much molasses will she need for the larger pie?

$1\dfrac{1}{8}$ cups of molasses

11. ***Exchange Rates*** Robyn spent two months traveling in New Zealand. The day she arrived, the exchange rate was 1.3 New Zealand dollars per U.S. dollar.

   (a) If she exchanged $500 U.S. dollars when she arrived, how many New Zealand dollars did she receive?   650 New Zealand dollars

   (b) Three days later the value of the New Zealand dollar dropped to 1.15 New Zealand dollars per U.S. dollar. How much less money would she have received had she waited three days to exchange her money?   75 New Zealand dollars less

12. ***Exchange Rates*** Sean spent a semester studying in Germany. On the day he arrived in Berlin, the exchange rate for the Euro was 0.69 Euro per U.S. dollar. Sean converted $350 to Euros that day.

   (a) How many Euros did Sean receive for his $350?   241.5 Euros

   (b) On his way home Sean decided to spend a week in London. He had €200 (200 Euros) that he wanted to change into British pounds. If the exchange rate was 0.72 British pounds per Euro, how many pounds did he receive?   144 British pounds

13. ***Speed Units*** Alfonse and Melinda are taking a drive in Mexico. They know that a speed of 100 kilometers per hour is approximately equal to 62 miles per hour. They are now driving on a Mexican road that has a speed limit of 90 kilometers per hour. How many miles per hour is the speed limit? Round to the nearest mile per hour.   56 mi/hr

14. ***Baggage Weight*** Dick and Anne took a trip to France. Their suitcases were weighed at the airport and the weight recorded was 39 kilograms. If 50 kilograms is equivalent to 110 pounds, how many pounds did their suitcases weigh? Round to the nearest pound.   86 lb

15. ***Map Scale*** On a map the distance between two mountains is $3\frac{1}{2}$ inches. The actual distance between the mountains is 136 miles. Russ is camped at a location that on the map is $\frac{3}{4}$ inch from the base of the mountain. How many miles is he from the base of the mountain? Round to the nearest mile.   29 mi

16. ***Construction Scales*** Maria is adding a porch that is 18 feet long to her house. On the drawing done by the carpenter the length is shown as 11 inches. The drawing shows that the width of the porch is 8 inches. How many feet wide will the porch be? Round to the nearest foot.   13 ft

Geometry *Triangles A and B are similar. Use them to answer exercises 17–20. Leave your answers as fractions.*

Triangle A   Triangle B

▲ **17.** If $x = 20$ in., $y = 29$ in., and $m = 13$ in., find the length of side n.

$n = \dfrac{377}{20}$ or $18\dfrac{17}{20}$ in.

▲ **18.** If $p = 14$ in., $m = 17$ in., and $z = 23$ in., find the length of side x.

$x = \dfrac{391}{14}$ or $27\dfrac{13}{14}$ in.

▲ **19.** If $x = 175$ meters, $n = 40$ meters, and $m = 35$ meters, find the length of side y.

$y = 200$ m

▲ **20.** If $z = 18$ cm, $y = 25$ cm, and $n = 9$ cm, find the length of side p.

$p = \dfrac{162}{25}$ or $6\dfrac{12}{25}$ cm

Geometry *Just as we have discussed similar triangles, other geometric shapes can be similar. Similar geometric shapes will have sides that are proportional. Quadrilaterals abcd and ghjk are similar. Use them to answer exercises 21–24. Leave your answers as fractions.*

▲ **21.** If $a = 5$ ft, $d = 8$ ft, and $g = 7$ ft, find the length of side k.

$k = \dfrac{56}{5}$ or $11\dfrac{1}{5}$ ft

▲ **22.** If $j = 10$ in., $k = 13$ in., and $c = 8$ in., find the length of side d.

$d = \dfrac{52}{5}$ or $10\dfrac{2}{5}$ in.

▲ **23.** If $b = 20$ m, $h = 24$ m, and $d = 32$ m, find the length of side k.

$k = \dfrac{768}{20}$ or $38\dfrac{2}{5}$ m

▲ **24.** If $a = 16$ cm, $d = 19$ cm, and $k = 23$ cm, find the length of side g.

$g = \dfrac{368}{19}$ or $19\dfrac{7}{19}$ cm

Use a proportion to solve.

▲ **25.** ***Geometry*** A rectangle whose width-to-length ratio is approximately 5 to 8 is called a **golden rectangle** and is said to be pleasing to the eye. Using this ratio, what should the length of a rectangular picture be if its width is to be 30 inches?

48 in.

▲ **26.** ***Shadows*** Samantha is 5.5 feet tall and notices that she casts a shadow of 9 feet. At the same time, the new sculpture at the local park casts a shadow of 20 feet. How tall is the sculpture? Round your answer to the nearest foot.

12 ft

▲ **27.** ***Floral Displays*** Floral designers often create arrangements where the flower height to container ratio is 5 to 3. The FIU Art Museum wishes to create a floral display for the opening of a new show. They know they want to use an antique Chinese vase from their collection that is 13 inches high. How tall will the entire flower arrangement be if they use this standard ratio? (Round your answer to the nearest inch.)   35 in.

▲ **28.** *Securing Wires* A wire line helps to secure a radio transmission tower. The wire measures 23 meters from the tower to the ground anchor pin. The wire is secured 14 meters up on the tower. If a second wire is secured 130 meters up on the tower and is extended from the tower at the same angle as the first wire, how long would the second wire need to be to reach an anchor pin on the ground? Round to the nearest meter.
214 m

130 meters

14 meters   23 meters

29. *Acceleration* Ben Hale is driving his new Toyota Camry on Interstate 90 at 45 miles per hour. He accelerates at the rate of 3 miles per hour every 2 seconds. How fast will he be traveling after accelerating for 11 seconds?
61.5 mi/hr

30. *Braking* Tim Newitt is driving a U-Haul truck to Chicago. He is driving at 55 miles per hour and has to hit the brakes because of heavy traffic. His truck slows at the rate of 2 miles per hour for every 3 seconds. How fast will he be traveling 10 seconds after he hits the brakes?
Approximately 48.3 mi/hr

31. *Flight Speeds* A Montreal commuter airliner travels 40 kilometers per hour faster than the television news helicopter over the city. The commuter airliner travels 1250 kilometers during the same time that the television news helicopter travels only 1050 kilometers. How fast does the commuter airliner fly? How fast does the television news helicopter fly?
Commuter airliner, 250 km/hr; helicopter, 210 km/hr

32. *Driving Speeds* Melissa drove to Dallas while Marcia drove to Houston in the same amount of time. Melissa drove 360 kilometers, while Marcia drove 280 kilometers. Melissa traveled 20 kilometers per hour faster than Marcia on her trip. What was the average speed in kilometers per hour for each woman?
Marcia, 70 km/hr; Melissa, 90 km/hr

33. *Fluff Containers* Marshmallow fluff comes in only two sizes, a $7\frac{1}{2}$-oz glass jar and a 16-oz plastic tub. At a local market, the 16-oz tub costs $1.49 and the $7\frac{1}{2}$-oz jar costs $0.79.

 (a) How much does marshmallow fluff in the glass jar cost per oz? (Round your answer to the nearest cent.) $0.11

 (b) How much does marshmallow fluff in the plastic tub cost per oz? (Round your answer to the nearest cent.) $0.09

 (c) If the marshmallow fluff company decided to add a third size, a 40-oz bucket, how much would the price be if it was at the same unit price as the 16-oz plastic tub? (*Hint:* Set up a proportion. Do *not* use your answer from part (b). Round your answer to the nearest cent.) $3.73

34. *Green Tea* Won Ling is a Chinese tea importer in Boston's Chinatown. He charges $12.25 for four sample packs of his famous green tea. The packs are in the following sizes: 25 grams, 40 grams, 50 grams, and 60 grams.

 (a) How much is Won Ling charging per gram for his green tea? $0.07

 (b) How much would you pay for a 60-gram pack if he were willing to sell that one by itself? $4.20

 (c) How much would you pay for an 800-gram package of green tea if it cost the same amount per gram? $56

35. ***Raking Leaves*** When all the leaves have fallen at Fred and Suzie's house in Concord, New Hampshire, Suzie can rake the entire yard in 6 hours. When Fred does it alone, it takes him 8 hours. How long would it take them to rake the yard together? Round to the nearest minute.

$3\frac{3}{7}$ hr or 3 hr, 26 min

36. ***Meal Preparation*** To celebrate Diwali, a major Indian festival, Deepak and Alpa host a party each year for all their friends. Deepak can decorate and prepare all the food in 6 hours. Alpa takes 5 hours to decorate and prepare the food. How long would it take if they decorated and made the food together? Round your answer to the nearest minute.

$2\frac{8}{11}$ hr or 2 hr, 44 min

Cumulative Review

37. [4.2.2] Write in scientific notation. 0.000892465
8.92465×10^{-4}

38. [4.2.2] Write in decimal notation. 5.82×10^8
582,000,000

39. [4.2.1] Write with positive exponents. $\dfrac{x^{-3}y^{-2}}{z^4w^{-8}}$

$\dfrac{w^8}{x^3y^2z^4}$

40. [4.2.1] Evaluate. $\left(\dfrac{2}{3}\right)^{-3}$

$\dfrac{27}{8}$ or $3\dfrac{3}{8}$

Quick Quiz 6.6 Solve for x.

1. $\dfrac{16.5}{2.1} = \dfrac{x}{7}$ $x = 55$

2. Last week Jeff Slater noted that 164 of the 205 flights to Chicago's O'Hare Airport flying out of Logan Airport departed on time. During the next week 215 flights left Logan Airport for Chicago's O'Hare Airport. If the same ratio holds, how many of those flights would he expect to depart on time?
172 flights

3. While hiking in the White Mountains, Phil, Melissa, Noah, and Olivia saw a tall tree that cast a shadow 34 feet long. They observed at the same time that a 6-foot-tall person cast a shadow that was 8.5 feet long. How tall is the tree?   24 ft

4. Concept Check Mike found that his car used 18 gallons of gas to travel 396 miles. He needs to take a trip of 450 miles and wants to know how many gallons of gas it will take. He set up the equation $\frac{18}{x} = \frac{450}{396}$. Explain what error he made and how he should correctly solve the problem.
Answers may vary

Classroom Quiz 6.6 You may use these problems to quiz your students' mastery of Section 6.6. Solve for x.

1. $\dfrac{2.1}{2.8} = \dfrac{x}{4.4}$ **Ans:** $x = 3.3$

2. Larry Coates observed that from 7 A.M. to 8 A.M. on the Revere Beach Parkway, 558 out of the 620 cars that drove by had only one occupant. If that same ratio held true for the 8 A.M. to 9 A.M. drive time, how many of the 930 cars that drove by would he expect to have only one occupant?   **Ans:** 837 cars

3. While hiking in the Colorado Rockies, Greg, Marcia, Lexi, and Caleb found a vertical cliff that cast a shadow 26 feet long on the level field below the cliff. At the same time, the 6-foot-tall park ranger cast a shadow 7.5 feet long. How tall is the cliff?   **Ans:** 20.8 ft

Putting Your Skills to Work: Use Math to Save Money

CREDIT CARD FOLLIES

Do you have a credit card? Do you make the minimum payment, or do you try to pay more? Almost all students in college have a major credit card, and most college students pay the monthly minimum payment. What are the implications of making the minimum payment? Consider the story of Adam.

By his junior year in college Adam owed $8000 on his credit card. Each month he made the monthly minimum payment. Some months he needed to charge new items. Adam's math teacher told him that college students often accumulate over $2000 in finance charges on their credit cards a year. Surprised by that amount, Adam decided to add up his monthly finance charges. They are listed below.

January	$138	July	$170
February	$142	August	$183
March	$157	September	$192
April	$158	October	$203
May	$160	November	$210
June	$165	December	$214

1. How much did he accumulate in finance charges during the year?
   $2092

2. What was his average monthly finance charge?
   $174.33

3. Discuss some ways that Adam can reduce the amount of interest he pays each year.
   Answers may vary

4. How much do you (or your parents) pay on your credit card each month in finance charges? How much do you (or your parents) pay in a year?
   Answers may vary

Topic	Procedure	Examples
Simplifying rational expressions, p. 355.	1. Factor the numerator and denominator. 2. Divide out any factor common to both the numerator and denominator.	$$\frac{36x^2 - 16y^2}{18x^2 + 24xy + 8y^2} = \frac{4(3x + 2y)(3x - 2y)}{2(3x + 2y)(3x + 2y)}$$ $$= \frac{2(3x - 2y)}{3x + 2y}$$
Multiplying rational expressions, p. 360.	1. Factor all numerators and denominators. 2. Simplify the resulting rational expression as described above.	$$\frac{x^2 - y^2}{x^2 + 2xy + y^2} \cdot \frac{x^2 + 4xy + 3y^2}{x^2 - 4xy + 3y^2}$$ $$= \frac{(x + y)(x - y)}{(x + y)(x + y)} \cdot \frac{(x + y)(x + 3y)}{(x - y)(x - 3y)}$$ $$= \frac{x + 3y}{x - 3y}$$
Dividing rational expressions, p. 361.	1. Invert the second fraction and rewrite the problem as a product. 2. Multiply the rational expressions.	$$\frac{14x^2 + 17x - 6}{x^2 - 25} \div \frac{4x^2 - 8x - 21}{x^2 + 10x + 25}$$ $$= \frac{(2x + 3)(7x - 2)}{(x + 5)(x - 5)} \cdot \frac{(x + 5)(x + 5)}{(2x - 7)(2x + 3)}$$ $$= \frac{(7x - 2)(x + 5)}{(x - 5)(2x - 7)}$$
Adding rational expressions, p. 365.	1. If the denominators differ, factor them and determine the least common denominator (LCD). 2. Use multiplication to change each fraction into an equivalent one with the LCD as the denominator. 3. Add the numerators; put the answer over the LCD. 4. Simplify as needed.	$$\frac{x - 1}{x^2 - 4} + \frac{x - 1}{3x + 6} = \frac{x - 1}{(x + 2)(x - 2)} + \frac{x - 1}{3(x + 2)}$$ $$\text{LCD} = 3(x + 2)(x - 2)$$ $$\frac{x - 1}{(x + 2)(x - 2)} + \frac{x - 1}{3(x + 2)}$$ $$= \frac{x - 1}{(x + 2)(x - 2)} \cdot \frac{3}{3} + \frac{x - 1}{3(x + 2)} \cdot \frac{x - 2}{x - 2}$$ $$= \frac{3x - 3 + x^2 - 3x + 2}{3(x + 2)(x - 2)}$$ $$= \frac{x^2 - 1}{3(x + 2)(x - 2)}$$ $$= \frac{(x + 1)(x - 1)}{3(x + 2)(x - 2)}$$
Subtracting rational expressions, p. 365.	Move the subtraction sign to the numerator of the second fraction. Add. Simplify if possible. $$\frac{a}{b} - \frac{c}{b} = \frac{a}{b} + \frac{-c}{b}$$	$$\frac{5x}{x - 2} - \frac{3x + 4}{x - 2} = \frac{5x}{x - 2} + \frac{-(3x + 4)}{x - 2}$$ $$= \frac{5x - 3x - 4}{x - 2}$$ $$= \frac{2x - 4}{x - 2}$$ $$= \frac{2(x - 2)}{x - 2} = 2$$

(Continued on next page)

Topic	Procedure	Examples
Simplifying complex rational expressions, p. 374.	1. Add the two fractions in the numerator. 2. Add the two fractions in the denominator. 3. Divide the fraction in the numerator by the fraction in the denominator. This is done by inverting the fraction in the denominator and multiplying by the numerator. 4. Simplify.	$$\dfrac{\dfrac{x}{x^2-4}+\dfrac{1}{x+2}}{\dfrac{3}{x+2}-\dfrac{4}{x-2}}$$ $$=\dfrac{\dfrac{x}{(x+2)(x-2)}+\dfrac{1}{x+2}\cdot\dfrac{x-2}{x-2}}{\dfrac{3}{x+2}\cdot\dfrac{x-2}{x-2}-\dfrac{4}{x-2}\cdot\dfrac{x+2}{x+2}}$$ $$=\dfrac{\dfrac{x+x-2}{(x+2)(x-2)}}{\dfrac{3x-6-4x-8}{(x+2)(x-2)}}$$ $$=\dfrac{2x-2}{(x+2)(x-2)}\div\dfrac{-x-14}{(x+2)(x-2)}$$ $$=\dfrac{2(x-1)}{(x+2)(x-2)}\cdot\dfrac{(x+2)(x-2)}{-x-14}$$ $$=\dfrac{2(x-1)}{-x-14}\text{ or }-\dfrac{2(x-1)}{x+14}\text{ or }\dfrac{-2(x-1)}{x+14}$$
Solving equations involving rational expressions, p. 380.	1. Determine the LCD of all denominators. 2. Note what values will make the LCD equal to 0. These are excluded from your solutions. 3. Multiply each side by the LCD, distributing as needed. 4. Solve the resulting polynomial equation. 5. Check. Be sure to exclude those values found in step 2.	$$\dfrac{3}{x-2}=\dfrac{4}{x+2}$$ $\text{LCD}=(x-2)(x+2).$ $$(x-2)(x+2)\dfrac{3}{x-2}=\dfrac{4}{x+2}(x-2)(x+2)$$ $$3(x+2)=4(x-2)$$ $$3x+6=4x-8$$ $$-x=-14$$ $$x=14$$ $\text{Check: }\dfrac{3}{14-2}\overset{?}{=}\dfrac{4}{14+2}$ $$\dfrac{3}{12}\overset{?}{=}\dfrac{4}{16}$$ $$\dfrac{1}{4}=\dfrac{1}{4}\checkmark$$
Solving applied problems with proportions, p. 385.	1. Organize the data. 2. Write a proportion equating the respective parts. Let x represent the value that is not known. 3. Solve the proportion.	Renee can make five cherry pies with 3 cups of flour. How many cups of flour does she need to make eight cherry pies? $$\dfrac{5\text{ cherry pies}}{3\text{ cups flour}}=\dfrac{8\text{ cherry pies}}{x\text{ cups flour}}$$ $$\dfrac{5}{3}=\dfrac{8}{x}$$ $$5x=24$$ $$x=\dfrac{24}{5}$$ $$x=4\dfrac{4}{5}$$ $4\dfrac{4}{5}$ cups of flour are needed for eight cherry pies.

Section 6.1

Simplify.

1. $\dfrac{bx}{bx - by}$ $\dfrac{x}{x - y}$

2. $\dfrac{4x - 4y}{5y - 5x}$ $-\dfrac{4}{5}$

3. $\dfrac{2x^2 + 5x - 3}{2x^2 - 9x + 4}$ $\dfrac{x + 3}{x - 4}$

4. $\dfrac{3x^2 + 7x + 2}{3x^2 + 13x + 4}$ $\dfrac{x + 2}{x + 4}$

5. $\dfrac{x^2 - 9}{x^2 - 10x + 21}$ $\dfrac{x + 3}{x - 7}$

6. $\dfrac{2x^2 + 18x + 40}{3x + 15}$ $\dfrac{2(x + 4)}{3}$

7. $\dfrac{x^3 - 4x^2}{x^3 - x^2 - 12x}$ $\dfrac{x}{x + 3}$

8. $\dfrac{2x^2 + 7x - 15}{25 - x^2}$ $\dfrac{2x - 3}{5 - x}$

9. $\dfrac{2x^2 - 2xy - 24y^2}{2x^2 + 5xy - 3y^2}$ $\dfrac{2(x - 4y)}{2x - y}$

10. $\dfrac{4 - y^2}{3y^2 + 5y - 2}$ $\dfrac{2 - y}{3y - 1}$

11. $\dfrac{5x^3 - 10x^2}{25x^4 + 5x^3 - 30x^2}$ $\dfrac{x - 2}{(5x + 6)(x - 1)}$

12. $\dfrac{16x^2 - 4y^2}{4x - 2y}$ $4x + 2y$

Section 6.2

Multiply or divide.

13. $\dfrac{2x^2 + 6x}{3x^2 - 27} \cdot \dfrac{x^2 + 3x - 18}{4x^2 - 4x}$ $\dfrac{x + 6}{6(x - 1)}$

14. $\dfrac{y^2 + 8y + 16}{5y^2 + 20y} \div \dfrac{y^2 + 7y + 12}{2y^2 + 5y - 3}$ $\dfrac{2y - 1}{5y}$

15. $\dfrac{2y^2 + 3y - 2}{2y^2 + y - 1} \div \dfrac{2y^2 + y - 1}{2y^2 - 3y - 2}$ $\dfrac{(y + 2)(2y + 1)(y - 2)}{(2y - 1)(y + 1)^2}$

16. $\dfrac{6y^2 + 13y - 5}{9y^2 + 3y} \div \dfrac{4y^2 + 20y + 25}{12y^2}$ $\dfrac{4y(3y - 1)}{(3y + 1)(2y + 5)}$

17. $\dfrac{3xy^2 + 12y^2}{2x^2 - 11x + 5} \div \dfrac{2xy + 8y}{8x^2 + 2x - 3}$ $\dfrac{3y(4x + 3)}{2(x - 5)}$

18. $\dfrac{11}{x - 2} \cdot \dfrac{2x^2 - 8}{44}$ $\dfrac{x + 2}{2}$

19. $\dfrac{x^2 - 5xy - 24y^2}{2x^2 - 2xy - 24y^2} \cdot \dfrac{4x^2 + 4xy - 24y^2}{x^2 - 10xy + 16y^2}$ $\dfrac{2(x + 3y)}{x - 4y}$

20. $\dfrac{2x^2 + 10x + 2}{8x - 8} \cdot \dfrac{3x - 3}{4x^2 + 20x + 4}$ $\dfrac{3}{16}$

Section 6.3

Add or subtract.

21. $\dfrac{6}{y+2} + \dfrac{2}{3y}$ $\dfrac{4(5y+1)}{3y(y+2)}$

22. $3 + \dfrac{2}{x+1} + \dfrac{1}{x}$ $\dfrac{3x^2+6x+1}{x(x+1)}$

23. $\dfrac{7}{x+2} + \dfrac{3}{x-4}$ $\dfrac{10x-22}{(x+2)(x-4)}$

24. $\dfrac{2}{x^2-9} + \dfrac{x}{x+3}$ $\dfrac{(x-1)(x-2)}{(x+3)(x-3)}$

25. $\dfrac{x}{y} + \dfrac{3}{2y} + \dfrac{1}{y+2}$ $\dfrac{2xy+4x+5y+6}{2y(y+2)}$

26. $\dfrac{4}{a} + \dfrac{2}{b} + \dfrac{3}{a+b}$ $\dfrac{(2a+b)(a+4b)}{ab(a+b)}$

27. $\dfrac{3x+1}{3x} - \dfrac{1}{x}$ $\dfrac{3x-2}{3x}$

28. $\dfrac{x+4}{x+2} - \dfrac{1}{2x}$ $\dfrac{2x^2+7x-2}{2x(x+2)}$

29. $\dfrac{27}{x^2-81} + \dfrac{3}{2(x+9)}$ $\dfrac{3}{2(x-9)}$

30. $\dfrac{1}{x^2+7x+10} - \dfrac{x}{x+5}$ $\dfrac{1-2x-x^2}{(x+5)(x+2)}$

Section 6.4

Simplify.

31. $\dfrac{\dfrac{4}{3y}-\dfrac{2}{y}}{\dfrac{1}{2y}+\dfrac{1}{y}}$ $-\dfrac{4}{9}$

32. $\dfrac{\dfrac{5}{x}+\dfrac{1}{2x}}{\dfrac{x}{4}+x}$ $\dfrac{22}{5x^2}$

33. $\dfrac{w-\dfrac{4}{w}}{1+\dfrac{2}{w}}$ $w-2$

34. $\dfrac{1-\dfrac{w}{w-1}}{1+\dfrac{w}{1-w}}$ 1

35. $\dfrac{1+\dfrac{1}{y^2-1}}{\dfrac{1}{y+1}-\dfrac{1}{y-1}}$ $-\dfrac{y^2}{2}$

36. $\dfrac{\dfrac{1}{y}+\dfrac{1}{x+y}}{1+\dfrac{2}{x+y}}$ $\dfrac{x+2y}{y(x+y+2)}$

37. $\dfrac{\dfrac{1}{a+b}-\dfrac{1}{a}}{b}$ $\dfrac{-1}{a(a+b)}$ or $-\dfrac{1}{a(a+b)}$

38. $\dfrac{\dfrac{2}{a+b}-\dfrac{3}{b}}{\dfrac{1}{a+b}}$ $\dfrac{-3a-b}{b}$ or $-\dfrac{3a+b}{b}$

39. $\dfrac{x + 5y}{x - 6y} \div \left(\dfrac{1}{5y} - \dfrac{1}{x + 5y} \right)$

$\dfrac{5y(x + 5y)^2}{x(x - 6y)}$

40. $\left(\dfrac{1}{x + 2y} - \dfrac{1}{x - y} \right) \div \dfrac{2x - 4y}{x^2 - 3xy + 2y^2}$

$\dfrac{-3y}{2(x + 2y)}$ or $-\dfrac{3y}{2(x + 2y)}$

Section 6.5

Solve for the variable. If there is no solution, say so.

41. $\dfrac{8a - 1}{6a + 8} = \dfrac{3}{4}$ $a = 2$

42. $\dfrac{8}{a - 3} = \dfrac{12}{a + 3}$ $a = 15$

43. $\dfrac{2x - 1}{x} - \dfrac{1}{2} = -2$ $x = \dfrac{2}{7}$

44. $\dfrac{5 - x}{x} - \dfrac{7}{x} = -\dfrac{3}{4}$ $x = -8$

45. $\dfrac{5}{2} - \dfrac{2y + 7}{y + 6} = 3$ $y = -4$

46. $\dfrac{5}{4} - \dfrac{1}{2x} = \dfrac{1}{x} + 2$ $x = -2$

47. $\dfrac{7}{8x} - \dfrac{3}{4} = \dfrac{1}{4x} + \dfrac{1}{2}$ $x = \dfrac{1}{2}$

48. $\dfrac{1}{3x} + 2 = \dfrac{5}{6x} - \dfrac{1}{2}$ $x = \dfrac{1}{5}$

49. $\dfrac{3}{y - 3} = \dfrac{3}{2} + \dfrac{y}{y - 3}$ No solution

50. $\dfrac{3x}{x^2 - 4} - \dfrac{2}{x + 2} = -\dfrac{4}{x - 2}$ $x = -\dfrac{12}{5}$ or $-2\dfrac{2}{5}$ or -2.4

51. $\dfrac{9}{2} - \dfrac{7y - 4}{y + 2} = -\dfrac{1}{4}$ $y = 6$

52. $\dfrac{3y - 1}{3y} - \dfrac{6}{5y} = \dfrac{1}{y} - \dfrac{4}{15}$ $y = 2$

53. $\dfrac{y + 18}{y^2 - 16} = \dfrac{y}{y + 4} - \dfrac{y}{y - 4}$ $y = -2$

54. $\dfrac{4}{x^2 - 1} = \dfrac{2}{x - 1} + \dfrac{2}{x + 1}$ No solution

55. $\dfrac{3y + 1}{y^2 - y} - \dfrac{3}{y - 1} = \dfrac{4}{y}$ $y = \dfrac{5}{4}$ or $1\dfrac{1}{4}$ or 1.25

56. $\dfrac{3}{y - 2} + \dfrac{4}{3y + 2} = \dfrac{1}{2 - y}$ $y = 0$

Section 6.6

Solve.

57. $\dfrac{x}{4} = \dfrac{7}{10}$ $x = \dfrac{14}{5}$ or $2\dfrac{4}{5}$ or 2.8

58. $\dfrac{8}{5} = \dfrac{2}{x}$ $x = \dfrac{5}{4}$ or $1\dfrac{1}{4}$ or 1.25

59. $\dfrac{33}{10} = \dfrac{x}{8}$ $x = \dfrac{132}{5}$ or $26\dfrac{2}{5}$ or 26.4

60. $\dfrac{16}{x} = \dfrac{24}{9}$ $x = 6$

61. $\dfrac{13.5}{0.6} = \dfrac{360}{x}$ $x = 16$

62. $\dfrac{2\frac{1}{2}}{3\frac{1}{4}} = \dfrac{7}{x}$ $x = \dfrac{91}{10}$ or $9\dfrac{1}{10}$ or 9.1

Use a proportion to answer each question.

63. **Paint Needs** A 5-gallon can of paint will cover 240 square feet. How many gallons of paint will be needed to cover 400 square feet? Round to the nearest tenth of a gallon. 8.3 gal

64. **Recipe Ratios** Aunt Lexie uses 3 pounds of sugar to make 100 cookies. How many cookies can she make with 5 pounds of sugar? Round to the nearest whole cookie. 167 cookies

65. **Gas Usage** Leila found that her car used 10 gallons of gas to travel 360 miles. She plans to drive 1400 miles from her home to New York City. How many gallons of gas will she use if her car continues to consume gas at the same rate? Round your answer to the nearest tenth of a gallon. 38.9 gal

66. **Map Scale** On a map of Texas, the distance between El Paso and Dallas is 4 inches. The actual distance between these cities is 640 miles. Houston and Dallas are 1.5 inches apart on the same map. How many miles apart are Houston and Dallas? 240 mi

67. **Travel Speeds** A train travels 180 miles in the same time that a car travels 120 miles. The speed of the train is 20 miles per hour faster than the speed of the car. Find the speed of the train and the speed of the car. Train, 60 mi/hr; car, 40 mi/hr

68. **Baking** Sam works at a popular bakery and can make the muffins and bagels for the morning rush in 3 hours. The new baker can do all of the baking in 5 hours. How long would it take Sam and the new baker to prepare for the morning rush if they worked together? $1\dfrac{7}{8}$ hr or about 1 hr and 53 min

▲ **69.** **Shadows** Mary takes a walk across a canyon in New Mexico. She stands 5.75 feet tall and her shadow is 3 feet long. At the same time, the shadow from the peak of the canyon wall casts a shadow that is 95 feet long. How tall is the peak of the canyon? Round to the nearest foot. 182 ft

▲ **70.** **Shadows** A flagpole that is 8 feet tall casts a shadow of 3 feet. At the same time of day, a tall office building in the city casts a shadow of 450 feet. How tall is the office building? 1200 ft

71. **Window Cleaning** As part of their spring cleaning routine, Tina and Mathias wash all of their windows, inside and out. Tina can do this job in 4 hours. When Mathias washes the windows, it takes him 6 hours. How long would it take them if they worked together? $2\dfrac{2}{5}$ hr or 2 hr and 24 min

72. **Plowing Fields** Sally runs the family farm in Boone, Iowa. She can plow the fields of the farm in 20 hours. Her daughter Brenda can plow the fields of the farm in 30 hours. If they have two identical tractors, how long would it take Brenda and Sally to plow the fields of the farm if they worked together? 12 hr

Mixed Practice

Perform the indicated operation. Simplify.

73. $\dfrac{a^2 + 2a - 8}{6a^2 - 3a^3} - \dfrac{a + 4}{3a^2}$

74. $\dfrac{4a^3 + 20a^2}{2a^2 + 13a + 15} \quad \dfrac{4a^2}{2a + 3}$

75. $\dfrac{x^2 - y^2}{x^2 + 4xy + 3y^2} \cdot \dfrac{x^2 + xy - 6y^2}{x^2 + xy - 2y^2} \quad \dfrac{x - 2y}{x + 2y}$

76. $\dfrac{x^2 - 6xy - 16y^2}{x^2 + 4xy - 21y^2} \cdot \dfrac{x^2 - 8xy + 15y^2}{x^2 - 3xy - 10y^2} \quad \dfrac{x - 8y}{x + 7y}$

77. $\dfrac{x}{x + 3} + \dfrac{9x + 18}{x^2 + 3x} \quad \dfrac{x + 6}{x}$

78. $\dfrac{x - 30}{x^2 - 5x} + \dfrac{x}{x - 5} \quad \dfrac{x + 6}{x}$

79. $\dfrac{a + b}{ax + ay} - \dfrac{a + b}{bx + by} \quad \dfrac{b^2 - a^2}{ab(x + y)}$

80. $\dfrac{\dfrac{5}{3x} + \dfrac{2}{9x}}{\dfrac{3}{x} + \dfrac{8}{3x}} \quad \dfrac{1}{3}$

81. $\dfrac{\dfrac{4}{5y} - \dfrac{8}{y}}{y + \dfrac{y}{5}} \quad \dfrac{-6}{y^2} \text{ or } -\dfrac{6}{y^2}$

82. $\dfrac{x - 3y}{x + 2y} \div \left(\dfrac{2}{y} - \dfrac{12}{x + 3y} \right) \quad \dfrac{y(x + 3y)}{2(x + 2y)}$

83. $\dfrac{7}{x + 2} = \dfrac{4}{x - 4} \quad x = 12$

84. $\dfrac{2x - 1}{3x - 8} = \dfrac{5}{8} \quad x = -32$

85. $2 + \dfrac{4}{b - 1} = \dfrac{4}{b^2 - b} \quad b = -2$

86. *Wire Weight* If 30 meters of wire weighs 8 kilograms, what will 40 meters of the same kind of wire weigh? Round your answer to the nearest tenth.   10.7 kg

87. *Driving Rates* Ming drove 270 miles in $4\dfrac{1}{2}$ hours. At that same rate, how far would she drive in 7 hours?
420 mi

Note to Instructor: The Chapter 6 Test file in the TestGen program provides algorithms specifically matched to these problems so you can easily replicate this test for additional practice or assessment purposes.

How Am I Doing? Chapter 6 Test

Remember to use your Chapter Test Prep Video CD to see the worked-out solutions to the test problems you want to review.

Perform the operation indicated. Simplify.

1. $\dfrac{2ac + 2ad}{3a^2c + 3a^2d}$

2. $\dfrac{8x^2 - 2x^2y^2}{y^2 + 4y + 4}$

3. $\dfrac{x^2 + 2x}{2x - 1} \cdot \dfrac{10x^2 - 5x}{12x^3 + 24x^2}$

4. $\dfrac{x + 2y}{12y^2} \cdot \dfrac{4y}{x^2 + xy - 2y^2}$

5. $\dfrac{2a^2 - 3a - 2}{a^2 + 5a + 6} \div \dfrac{a^2 - 5a + 6}{a^2 - 9}$

6. $\dfrac{1}{a^2 - a - 2} + \dfrac{3}{a - 2}$

7. $\dfrac{x - y}{xy} - \dfrac{a - y}{ay}$

8. $\dfrac{3x}{x^2 - 3x - 18} - \dfrac{x - 4}{x - 6}$

9. $\dfrac{\dfrac{x}{3y} - \dfrac{1}{2}}{\dfrac{4}{3y} - \dfrac{2}{x}}$

10. $\dfrac{\dfrac{6}{b} - 4}{\dfrac{5}{bx} - \dfrac{10}{3x}}$

11. $\dfrac{2x^2 + 3xy - 9y^2}{4x^2 + 13xy + 3y^2}$

12. $\dfrac{1}{x + 4} - \dfrac{2}{x^2 + 6x + 8}$

In questions 13–18, solve for x. Check your answers. If there is no solution, say so.

13. $\dfrac{4}{3x} - \dfrac{5}{2x} = 5 - \dfrac{1}{6x}$

14. $\dfrac{x - 3}{x - 2} = \dfrac{2x^2 - 15}{x^2 + x - 6} - \dfrac{x + 1}{x + 3}$

15. $3 - \dfrac{7}{x + 3} = \dfrac{x - 4}{x + 3}$

16. $\dfrac{3}{3x - 5} = \dfrac{7}{5x + 4}$

17. $\dfrac{9}{x} = \dfrac{13}{5}$

18. $\dfrac{9.3}{2.5} = \dfrac{x}{10}$

19. A random check of America West air flights last month showed that 113 of the 150 flights checked arrived on time. If the inspectors check 200 flights next month, how many can be expected to be on time? (Round to the nearest whole number.)

20. In northern Michigan the Gunderson family heats their home with firewood. They used $100 worth of wood in 25 days. Mr. Gunderson estimates that he needs to burn wood at that rate for about 92 days during the winter. If that is so, how much will the 92-day supply of wood cost?

▲21. A hiking club is trying to construct a rope bridge across a canyon. A 6-foot construction pole held upright casts a 7-foot shadow. At the same time of day, a tree at the edge of the canyon casts a shadow that exactly covers the distance that is needed for the rope bridge. The tree is exactly 87 feet tall. How long should the rope bridge be? Round to the nearest foot.

Answers:

1. $\dfrac{2}{3a}$

2. $\dfrac{2x^2(2 - y)}{(y + 2)}$

3. $\dfrac{5}{12}$

4. $\dfrac{1}{3y(x - y)}$

5. $\dfrac{2a + 1}{a + 2}$

6. $\dfrac{3a + 4}{(a + 1)(a - 2)}$

7. $\dfrac{x - a}{ax}$

8. $-\dfrac{x + 2}{x + 3}$

9. $\dfrac{x}{4}$

10. $\dfrac{6x}{5}$

11. $\dfrac{2x - 3y}{4x + y}$

12. $\dfrac{x}{(x + 2)(x + 4)}$

13. $x = -\dfrac{1}{5}$

14. $x = 4$

15. No solution

16. $x = \dfrac{47}{6}$

17. $x = \dfrac{45}{13}$

18. $x = 37.2$

19. 151 flights

20. $368

21. 102 ft

Cumulative Test for Chapters 0–6

Approximately one-half of this test covers the content of Chapters 0–5. The remainder covers the content of Chapter 6.

1. To the nearest thousandth, how much error can be tolerated in the length of a wire that is supposed to be 2.57 centimeters long if specifications allow an error of no more than 0.25%?

2. In 2005, the value of all U.S. agricultural exports was $62,958,000,000. The value of agricultural exports to Canada was $10,570,000,000. What percent of the total value of agricultural exports went to Canada? (*Source: Statistical Abstract of the United States: 2007.*) Round your answer to the nearest tenth of a percent.

3. Henry is thinking about buying a new car for $22,500. His state presently has a sales tax of 6.5%. In a few weeks the state will raise the sales tax to 7%. How much will he save in sales tax if he purchases the car before the sales tax rate is raised?

4. Solve. $5(x - 3) - 2(4 - 2x) = 7(x - 1) - (x - 2)$

5. Solve for h. $A = \pi r^2 h$

6. Solve and graph on a number line. $4(2 - x) < 3$

7. Solve for x. $\dfrac{3}{5}x - 2 \le \dfrac{1}{2}x + \dfrac{3}{5}$

8. Factor. $3ax + 3bx - 2ay - 2by$

9. Factor. $8a^3 - 38a^2b - 10ab^2$

10. Simplify. Write the answer with positive exponents. $\dfrac{-16a^3b^{-1}}{4a^{-2}b^5}$

11. Simplify. $\dfrac{4x^2 - 25}{2x^2 + 9x - 35}$

Perform the indicated operations.

12. $\dfrac{x^2 - 4}{x^2 - 25} \cdot \dfrac{3x^2 - 14x - 5}{3x^2 + 6x}$

13. $\dfrac{2x^2 - 9x + 9}{8x - 12} \div \dfrac{x^2 - 3x}{2x}$

14. $\dfrac{5}{2x + 4} + \dfrac{3}{x - 3}$

15. $\dfrac{8}{c^2 - 4} - \dfrac{2}{c^2 - 5c + 6}$

Solve for x.

16. $\dfrac{3x - 2}{3x + 2} = 2$

17. $\dfrac{x - 3}{x} = \dfrac{x + 2}{x + 3}$

In questions 18 and 19, simplify.

18. $\dfrac{\dfrac{1}{x - 3} + \dfrac{5}{x^2 - 9}}{\dfrac{6x}{x + 3}}$

19. $\dfrac{\dfrac{3}{a} + \dfrac{2}{b}}{\dfrac{5}{a^2} - \dfrac{2}{b^2}}$

20. Solve for x. $\dfrac{2x + 1}{3} = \dfrac{4}{5}$

21. Jane is looking at a road map. The distance between two cities is 130 miles. This distance is represented by $2\frac{1}{2}$ inches on the map. She sees that the distance she has to drive today is a totally straight interstate highway that is 4 inches long on the map. How many miles does she have to drive?

22. Roberto is working as a telemarketing salesperson for a large corporation. In the last 22 telephone calls he has made, he was able to make a sale five times. His goal is to make 110 sales this month. If his rate for making a sale continues as in the past, how many phone calls must he make?

---

1. 0.006 cm

2. 16.8%

3. $112.50

4. $x = 6$

5. $h = \dfrac{A}{\pi r^2}$

6. $x > 1.25$ (see graph)

7. $x \le 26$

8. $(a + b)(3x - 2y)$

9. $2a(4a + b)(a - 5b)$

10. $\dfrac{4a^5}{b^6}$

11. $\dfrac{2x + 5}{x + 7}$

12. $\dfrac{(x - 2)(3x + 1)}{3x(x + 5)}$

13. $\dfrac{1}{2}$

14. $\dfrac{11x - 3}{2(x + 2)(x - 3)}$

15. $\dfrac{6c - 28}{(c + 2)(c - 2)(c - 3)}$

16. $x = -2$

17. $x = -\dfrac{9}{2}$ or $-4\dfrac{1}{2}$ or -4.5

18. $\dfrac{x + 8}{6x(x - 3)}$

19. $\dfrac{3ab^2 + 2a^2b}{5b^2 - 2a^2}$

20. $x = \dfrac{7}{10}$ or 0.7

21. 208 mi

22. 484 phone calls

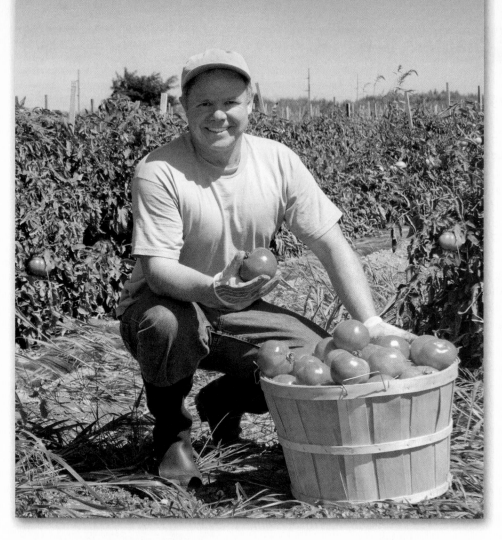

Graphing and Functions

7

Vegetable gardens both small and large are present in every state of the United States. Many Americans enjoy the produce from their gardens. Large vegetable farms produce millions of dollars worth of fresh vegetables that are exported daily to countries all around the world. The graphing techniques covered in this chapter can be used to predict how much a vegetable crop is worth.

① Plotting a Point, Given the Coordinates

Often we can better understand an idea if we see a picture. This is the case with many mathematical concepts, including those relating to algebra. We can illustrate algebraic relationships with drawings called **graphs.** Before we can draw a graph, however, we need a frame of reference.

In Chapter 1 we showed that any real number can be represented on a number line. Look at the following number line. The arrow indicates the positive direction.

To form a **rectangular coordinate system,** we draw a second number line vertically. We construct it so that the 0 point on each number line is exactly at the same place. We refer to this location as the **origin.** The horizontal number line is often called the **x-axis.** The vertical number line is often called the **y-axis.** Arrows show the positive direction for each axis.

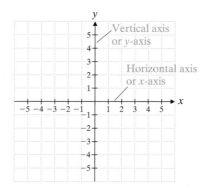

We can represent a point in this rectangular coordinate system by using an **ordered pair** of numbers. For example, $(5, 2)$ is an ordered pair that represents a point in the rectangular coordinate system. The numbers in an ordered pair are often referred to as the **coordinates** of the point. The first number is called the **x-coordinate** and it represents the distance from the origin measured along the horizontal or x-axis. If the x-coordinate is positive, we count the proper number of squares to the right (that is, in the positive direction). If the x-coordinate is negative, we count to the left. The second number in the pair is called the **y-coordinate** and it represents the distance from the origin measured along the y-axis. If the y-coordinate is positive, we count the proper number of squares upward (that is, in the positive direction). If the y-coordinate is negative, we count downward.

$$(5, 2)$$
$$x\text{-coordinate} \quad \text{ } \quad y\text{-coordinate}$$

Suppose the directory for the map on the right indicated that you would find a certain street in the region B5. To find the street you would first scan across the horizontal scale until you found section B; from there you would scan up the map until you hit section 5 along the vertical scale. As we will see in the next example, plotting a point in the rectangular coordinate system is much like finding a street on a map with grids.

EXAMPLE 1 Plot the point $(5, 2)$ on a rectangular coordinate system. Label this as point A.

Solution Since the x-coordinate is 5, we first count 5 units to the right on the x-axis. Then, because the y-coordinate is 2, we count 2 units up from the point

Student Learning Objectives

After studying this section, you will be able to:

① Plot a point, given the coordinates.

② Determine the coordinates of a plotted point.

③ Find ordered pairs for a given linear equation.

Teaching Example 1 Plot the point $(6, 1)$ on the coordinate system. Label this point as C.

Ans:

405

where we stopped on the *x*-axis. This locates the point corresponding to (5, 2). We mark this point with a dot and label it *A*.

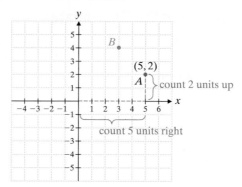

Practice Problem 1 Plot the point (3, 4) on the preceding coordinate system. Label this point as *B*.

It is important to remember that the first number in an ordered pair is the *x*-coordinate and the second number is the *y*-coordinate. The ordered pairs (5, 2) and (2, 5) represent different points.

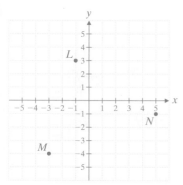

EXAMPLE 2 Plot each point on the following coordinate system. Label the points *F*, *G*, and *H*, respectively.

(a) (−5, 3)　　　　　　**(b)** (2, −6)　　　　　　**(c)** (−4, −5)

Solution

(a) Notice that the *x*-coordinate, −5, is negative. On the coordinate grid, negative *x*-values appear to the left of the origin. Thus, we will begin by counting 5 squares to the left, starting at the origin. Since the *y*-coordinate, 3, is positive, we will count 3 units up from the point where we stopped on the *x*-axis.

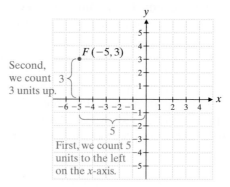

(b) The *x*-coordinate is positive. Begin by counting 2 squares to the right of the origin. Then count down because the *y*-coordinate is negative.

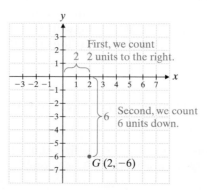

(c) The *x*-coordinate is negative. Begin by counting 4 squares to the left of the origin. Then count down because the *y*-coordinate is negative.

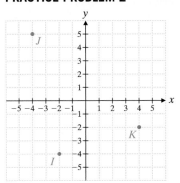

Practice Problem 2 Use the coordinate system in the margin to plot each point. Label the points *I*, *J*, and *K*, respectively.

(a) $(-2, -4)$ **(b)** $(-4, 5)$ **(c)** $(4, -2)$

EXAMPLE 3 Plot the following points.

$F: (0, 5)$ $G: \left(3, \frac{3}{2}\right)$ $H: (-6, 4)$ $I: (-3, -4)$

$J: (-4, 0)$ $K: (2, -3)$ $L: (6.5, -7.2)$

Solution These points are plotted in the figure.

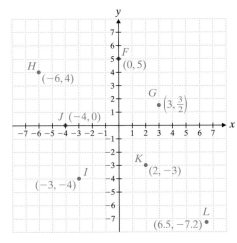

Note: When you are plotting decimal values like $(6.5, -7.2)$, plot the point halfway between 6 and 7 in the *x*-direction (for the 6.5) and at your best approximation of -7.2 in the *y*-direction.

Practice Problem 3 Plot the following points. Label each point with both the letter and the ordered pair. Use the coordinate system provided in the margin.

$A: (3, 7); B: (0, -6); C: (3, -4.5); D: \left(-\frac{7}{2}, 2\right)$

Teaching Example 3 Use the coordinate system below to plot each point. Label each point with both the letter and the ordered pair.

$M: (1, 0)$ $N: (-3, -2.5)$ $P: (-4.2, -6.5)$

$Q: \left(\frac{1}{2}, 4\right)$ $\cdot R: \left(-5, -\frac{3}{2}\right)$ $T: \left(\frac{5}{2}, 1\right)$

$S: (0, -5)$

Ans:

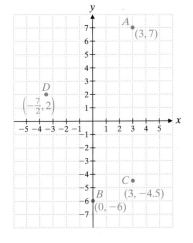

---

2 **Determining the Coordinates of a Plotted Point**

Sometimes we need to find the coordinates of a point that has been plotted. First, we count the units we need on the *x*-axis to get as close as possible to the point. Next we count the units up or down we need to go from the *x*-axis to reach the point.

Teaching Example 4 What ordered pair of numbers represents point C on the graph below?

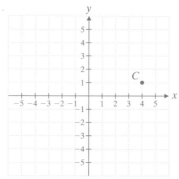

Ans: $(4, 1)$

EXAMPLE 4 What ordered pair of numbers represents point A on the graph below?

Solution If we move along the x-axis until we get as close as possible to A, we end up at the number 5. Thus we obtain 5 as the first number of the ordered pair. Then we count 4 units upward on a line parallel to the y-axis to reach A. So we obtain 4 as the second number of the ordered pair. Thus point A is represented by the ordered pair $(5, 4)$.

Practice Problem 4 What ordered pair of numbers represents point B on the graph below? $B = (2, 7)$

Teaching Example 5 Write the coordinates of each point plotted on the following graph.

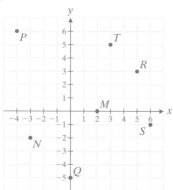

Ans: M: $(2, 0)$ N: $(-3, -2)$ P: $(-4, 6)$
Q: $(0, -5)$ R: $(5, 3)$ T: $(3, 5)$
S: $(6, -1)$

EXAMPLE 5 Write the coordinates of each point plotted on the following graph.

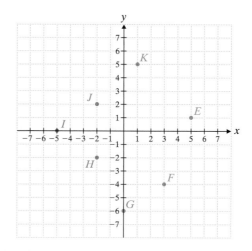

Solution The coordinates of each point are as follows.

$E = (5, 1)$ $I = (-5, 0)$

$F = (3, -4)$ $J = (-2, 2)$ Be very careful that you put the x-coordinate

$G = (0, -6)$ $K = (1, 5)$ first and the y-coordinate second. Be careful
                                       that each sign is correct.
$H = (-2, -2)$

PRACTICE PROBLEM 5

Practice Problem 5 Give the coordinates of each point plotted on the graph in the margin.

3 Finding Ordered Pairs for a Given Linear Equation

Equations such as $3x + 2y = 5$ and $6x + y = 3$ are called linear equations in two variables.

A **linear equation in two variables** is an equation that can be written in the form $Ax + By = C$, where A, B, and C are real numbers but A and B are not *both* zero.

Replacement values for x and y that make *true mathematical statements* of the equation are called *truth values,* and an ordered pair of these truth values is called a **solution.**

Consider the equation $3x + 2y = 5$. The ordered pair $(1, 1)$ is a solution to the equation because when we replace x by 1 and y by 1 in the equation, we obtain a true statement.

$$3(1) + 2(1) = 5 \quad \text{or} \quad 3 + 2 = 5$$

Likewise $(-1, 4)$, $(3, -2)$, $(5, -5)$, and $(7, -8)$ are also solutions to the equation. In fact, there are an infinite number of solutions to any given linear equation in two variables.

The linear equations that we work with are not always written in the form $Ax + By = C$, but are sometimes solved for y, as in $y = -6x + 3$. Consider the equation $y = -6x + 3$. The ordered pair $(2, -9)$ is a solution to the equation. When we replace x by 2 and y by -9 we obtain a true mathematical statement:

$$(-9) = -6(2) + 3 \quad \text{or} \quad -9 = -12 + 3.$$

If one value of an ordered-pair solution to a linear equation is known, the other can be quickly obtained. To do so, we replace the proper variable in the equation by the known value. Then using the methods learned in Chapter 2, we solve the resulting equation for the other variable.

EXAMPLE 6 Find the missing coordinate to complete the following ordered-pair solutions for the equation $2x + 3y = 15$.

(a) $(0, ?)$ **(b)** $(?, 1)$

Solution

(a) In the ordered pair $(0, ?)$ we know that $x = 0$. Replace x by 0 in the equation.

$$2x + 3y = 15$$
$$2(0) + 3y = 15$$
$$0 + 3y = 15$$
$$y = 5$$

Thus we have the ordered pair $(0, 5)$.

(b) In the ordered pair $(?, 1)$, we *do not know* the value of x. However, we do know that $y = 1$. So we start by replacing the variable y by 1. We will end up with an equation with one variable, x. We can then solve for x.

$$2x + 3y = 15$$
$$2x + 3(1) = 15$$
$$2x + 3 = 15$$
$$2x = 12$$
$$x = 6$$

Thus we have the ordered pair $(6, 1)$.

Practice Problem 6 Find the missing coordinate to complete the following ordered-pair solutions for the equation $3x - 4y = 12$.

(a) $(0, ?)$ $(0, -3)$ **(b)** $(?, 3)$ $(8, 3)$ **(c)** $(?, -6)$ $(-4, -6)$

Teaching Example 6 Find the missing coordinate to complete the following ordered-pair solutions for the equation $5x - 2y = 10$.

(a) $(?, 0)$ **(b)** $(-2, ?)$

Ans: (a) $(2, 0)$ **(b)** $(-2, -10)$

NOTE TO STUDENT: Fully worked-out solutions to all of the Practice Problems can be found at the back of the text starting at page SP-1

In examining data from real-world situations, we often find that plotting data points shows useful trends. In such cases, it is often necessary to use a different scale, one that displays only positive values.

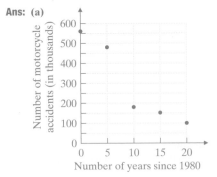
EXAMPLE 7 The number of motor vehicle accidents in millions is recorded in the following table for the years 1980 to 2000.

(a) Plot points that represent this data on the given coordinate system.

(b) What trends are apparent from the plotted data?

Number of Years Since 1980	Number of Motor Vehicle Accidents (in Millions)
0	18
5	19
10	12
15	11
20	15

Source: U.S. National Highway Traffic Safety Administration

Solution

(a)

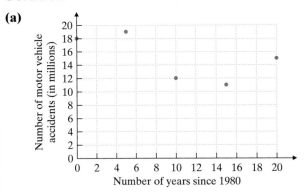

(b) From 1980 to 1985, there was a slight increase in the number of accidents. From 1985 to 1995, there was a significant decrease in the number of accidents. From 1995 to 2000, there was a moderate increase in the number of accidents.

Practice Problem 7 The number of motor vehicle deaths in thousands is recorded in the following table for the years 1980 to 2000.

(a) Plot points that represent this data on the given coordinate system.

(b) What trends are apparent from the plotted data?

Number of Years Since 1980	Number of Motor Vehicle Deaths (in Thousands)
0	51
5	44
10	45
15	42
20	43

Source: U.S. National Highway Traffic Safety Administration

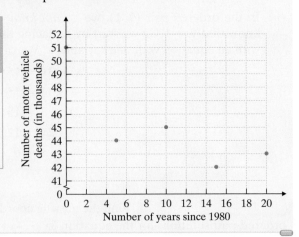

Motor vehicle deaths were significantly high in 1980. In 1985–2000 the number of motor vehicle deaths was relatively stable.

Verbal and Writing Skills

Unless otherwise indicated, assume each grid line represents one unit.

1. What is the *x*-coordinate of the origin? 0

2. What is the *y*-coordinate of the origin? 0

3. Explain why (5, 1) is referred to as an *ordered* pair of numbers.

The order in which you write the numbers matters. The graph of (5, 1) is not the same as the graph of (1, 5).

4. Explain how you would locate a point (4, 3) on graph paper.

Count from the origin 4 squares to the right. From that location count 3 squares up. Place a dot at this final location.

5. Plot the following points.

$J: (-4, 3.5)$ $K: (6, 0)$

$L: (5, -6)$ $M: (0, -4)$

6. Plot the following points.

$R: (-3, 0)$ $S: (3.5, 4)$

$T: (-2, -2.5)$ $V: (0, 5)$

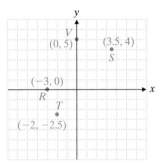

Consider the points plotted on the graph at right.

7. Give the coordinates for points *R, S, X,* and *Y*.

$R: (-3, -5)$

$S: \left(-4\frac{1}{2}, 0\right)$

$X: (3, -5)$

$Y: \left(2\frac{1}{2}, 6\right)$

8. Give the coordinates for points *T, V, W,* and *Z*.

$T: (-6, 4), V: \left(3\frac{1}{2}, -2\right), W: \left(6\frac{1}{2}, 2\right), Z: \left(0, -3\frac{1}{2}\right)$

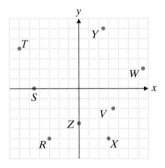

In exercises 9 and 10, six points are plotted in each figure. List all the ordered pairs needed to represent the points.

9.

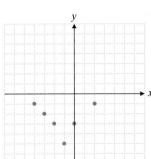

$(-4, -1)$
$(-3, -2)$
$(-2, -3)$
$(-1, -5)$
$(0, -3)$
$(2, -1)$

10.

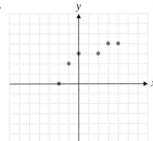

$(-2, 0)$ $(-1, 2)$ $(0, 3)$
$(2, 3)$ $(3, 4)$ $(4, 4)$

Find the missing coordinate to complete the ordered-pair solution to the given linear equation.

11. $y = 4x + 7$
(a) $(0,)$ $(0, 7)$
(b) $(2,)$ $(2, 15)$

12. $y = 6x + 5$
(a) $(0,)$ $(0, 5)$
(b) $(3,)$ $(3, 23)$

13. $y = -6x + 5$
(a) $(-1,)$ $(-1, 11)$
(b) $(3,)$ $(3, -13)$

14. $y = -4x + 9$
(a) $(-2,)$ $(-2, 17)$
(b) $(5,)$ $(5, -11)$

15. $3x - 4y = 11$
(a) $(-3,)$ $(-3, -5)$
(b) $(, 1)$ $(5, 1)$

16. $5x - 2y = 9$
(a) $(7,)$ $(7, 13)$
(b) $(, -7)$ $(-1, -7)$

17. $3x + 2y = -6$
(a) $(-2,)$ $(-2, 0)$
(b) $(, 3)$ $(-4, 3)$

18. $-4x + 5y = -20$
(a) $(10,)$ $(10, 4)$
(b) $(, -8)$ $(-5, -8)$

19. $x + \dfrac{1}{3}y = 5$
(a) $(, 18)$ $(-1, 18)$
(b) $\left(\dfrac{8}{3}, \right)$ $\left(\dfrac{8}{3}, 7\right)$

20. $3x + \dfrac{1}{4}y = 11$ (a) $(, 8)$ $(3, 8)$ (b) $\left(\dfrac{13}{4}, \right)$ $\left(\dfrac{13}{4}, 5\right)$

Using Road Maps *The preceding map shows a portion of New York, Connecticut, and Massachusetts. Like many maps used in driving or flying, it has horizontal and vertical grid markers for ease of use. For example, Newburgh, New York, is located in grid B3. Use the grid labels to indicate the locations of the following cities.*

21. Lynbrook, New York
B5

22. Hampton Bays, New York
D5

23. Athol, Massachusetts
E1

24. Pittsfield, Massachusetts
D1

25. Hartford, Connecticut
D3

26. Waterbury, Connecticut
C3

27. *CD Shipments* According to music CD manufacturers, the number of CDs shipped decreased significantly from 2001 to 2006. The number of CDs shipped during these years is recorded in the following table, and is measured in millions. For example, 882 in the second column means 882 million, or 882,000,000.

(a) Plot points that represent this data on the given rectangular coordinate system.

(b) What trends are apparent from the plotted data?

Number of Years Since 2001	Number of CDs Shipped (in Millions)
0	882
1	803
2	746
3	767
4	705
5	615

Source: www.riaa.com

(b) The number of CDs shipped decreased overall between 2001 and 2006, with a slight increase from 2003 to 2004.

28. *U.S. Oil Consumption* The number of barrels of oil used per day in the United States for selected years starting in 1995 is recorded in the following table. The number of barrels is measured in millions. For example, 18 in the second column means 18 million, or 18,000,000 barrels.

(a) Plot points that represent the data on the given rectangular coordinate system.

(b) What trends are apparent from the plotted data?

Number of Years Since 1995	Oil Consumption per Day (in Millions of Barrels)
0	18
5	20
10	22
15	23.5*
20	24.5*
25	26.5*

*Estimate
Source: www.energy.senate.gov

(b) From 1995 to 2005, the number of barrels consumed per day increased by the same amount within each 5-year period. It is estimated that from 2005 to 2020, consumption will continue to increase.

29. *Buying Books Online* The amount of money U.S. consumers spend buying books online continues to increase at a significant rate. The following chart records the amount spent each year from 2003 to 2006.

Year	Total Amount Spent (in Billions of Dollars)
2003	3.2
2004	3.6
2005	3.9
2006*	4.2

*Estimate
Source: Statistical Abstract of the United States: 2007.

(a) Plot points that represent the data on the given rectangular coordinate system.

(b) Based on your graph, estimate the amount consumers spent on buying books online in 2007.

An estimated $4.5 billion was spent buying books online in 2007.

30. *Buying Personal Computers Online* The amount of money U.S. consumers spend buying personal computers online continues to increase at a significant rate. The following chart records the amount spent each year from 2003 to 2006.

Year	Total Amount Spent (in Billions of Dollars)
2003	53.9
2004	67.3
2005	80.9
2006*	95.3

*Estimate
Source: Statistical Abstract of the United States: 2007

(a) Plot points that represent the data on the given rectangular coordinate system.

(b) Based on your graph, estimate the amount consumers spent buying personal computers online in 2007.

An estimated $110 billion was spent buying personal computers online in 2007.

Cumulative Review

31. **[6.3.1]** Simplify. $\dfrac{8x}{x+y} + \dfrac{8y}{x+y}$ $\quad \dfrac{8x+8y}{x+y} = 8$

32. **[6.2.1]** Simplify. $\dfrac{x^2 - y^2}{x^2 + 4xy + 3y^2} \cdot \dfrac{x^2 - 9y^2}{x^2 - 6xy + 9y^2}$

$\dfrac{(x+y)(x-y)}{(x+y)(x+3y)} \cdot \dfrac{(x+3y)(x-3y)}{(x-3y)(x-3y)} = \dfrac{x-y}{x-3y}$

Quick Quiz 7.1

1. Plot and label the following points.

$A: (3, -4)$
$B: (-6, -2)$
$C: (0, 5)$
$D: (-3, 6)$

Find the missing coordinate to complete the ordered-pair solution to the given linear equation.

2. $y = -5x - 7$
  (a) $(-2, \ )$ $\ (-2, 3)$
  (b) $(3, \ )$ $\ (3, -22)$
  (c) $(0, \ )$ $\ (0, -7)$

3. $4x - 3y = -12$
  (a) $(3, \ )$ $\ (3, 8)$
  (b) $(\ , -8)$ $\ (-9, -8)$
  (c) $(\ , 10)$ $\ (4.5, 10)$

4. **Concept Check** Explain how you would find the missing coordinate to complete the ordered-pair solution to the equation $2.5x + 3y = 12$ if the ordered pair was of the form $(\ , -6)$.
Answers may vary

Classroom Quiz 7.1 You may use these problems to quiz your students' mastery of Section 7.1.

1. Plot and label the following points.

$E: (3, 5)$
$F: (-4, 0)$
$G: (-2, -6)$
$H: (1, -3)$

Find the missing coordinate to complete the ordered-pair solution to the given linear equation.

2. $y = -3x - 9$
  (a) $(-3, \ )$ **Ans:** $(-3, 0)$
  (b) $(4, \ )$ **Ans:** $(4, -21)$
  (c) $(0, \ )$ **Ans:** $(0, -9)$

3. $-2x + 5y = -30$
  (a) $(-10, \ )$ **Ans:** $(-10, -10)$
  (b) $(\ , -3)$ **Ans:** $(7.5, -3)$
  (c) $(\ , 8)$ **Ans:** $(35, 8)$

Student Learning Objectives

After studying this section, you will be able to:

① Graphing a Linear Equation by Plotting Three Ordered Pairs

We have seen that a solution to a linear equation in two variables is an ordered pair. The graph of an ordered pair is a point. Thus we can graph an equation by graphing the points corresponding to its ordered-pair solutions.

A linear equation in two variables has an infinite number of ordered-pair solutions. We can see that this is true by noting that we can substitute any number for x in the equation and solve it to obtain a y-value. For example, if we substitute $x = 0, 1, 2, 3, \ldots$ into the equation $y = -x + 3$ and solve for y, we obtain the ordered-pair solutions $(0, 3), (1, 2), (2, 1), (3, 0), \ldots$. (If desired, substitute these values into the equation to convince yourself.) If we plot these points on a rectangular coordinate system, we notice that they fall on a straight line.

It turns out that all of the points corresponding to the ordered-pair solutions of $y = -x + 3$ lie on this line, and the line extends forever in both directions. A similar statement can be made about any linear equation in two variables.

① Graph a linear equation by plotting three ordered pairs.

② Graph a straight line by plotting its intercepts.

③ Graph horizontal and vertical lines.

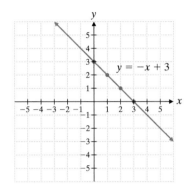

The graph of any linear equation in two variables is a straight line.

From geometry, we know that two points determine a line. Thus to graph a linear equation in two variables, we need only graph two ordered-pair solutions of the equation and then draw the line that passes through them. Having said this, we recommend that you use three points to graph a line. Two points will determine where the line is. The third point verifies that you have drawn the line correctly. For ease in plotting, it is better if the ordered pairs contain integers.

Teaching Tip Some students do not realize that when we draw the equation of a straight line, we draw arrow tips at each end. Other students will draw the arrow tips but not know what they mean. The purpose of the arrowhead is to indicate that the straight line continues forever in that direction. Later in this course students will graph some lines that do not continue forever, and those lines will not have arrowheads at each end.

TO GRAPH A LINEAR EQUATION

1. Look for three ordered pairs that are solutions to the equation.

2. Plot the points.

3. Draw a line through the points.

Teaching Example 1 Find three ordered pairs that satisfy $x + 2y = 6$. Then graph the resulting straight line.

Ans:

 Find three ordered pairs that satisfy $2x + y = 4$. Then graph the resulting straight line.

Solution Since we can choose any value for x, we choose numbers that are convenient. To organize the results, we will make a table of values. We will let $x = 0$, $x = 1$, and $x = 3$, respectively. We write these numbers under x in our table of values. For each of these x-values, we find the corresponding y-value in the equation $2x + y = 4$.

$2x + y = 4$	$2x + y = 4$	$2x + y = 4$
$2(0) + y = 4$	$2(1) + y = 4$	$2(3) + y = 4$
$0 + y = 4$	$2 + y = 4$	$6 + y = 4$
$y = 4$	$y = 2$	$y = -2$

Table of Values

x	y
0	4
1	2
3	-2

415

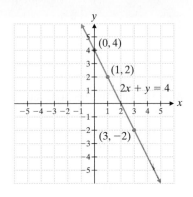

We record these results by placing each y-value in the table next to its corresponding x-value. Keep in mind that these values represent ordered pairs, each of which is a solution to the equation. To make calculating and graphing easier, we choose integer values whenever possible. If we plot these ordered pairs and connect the three points, we get a straight line that is the graph of the equation $2x + y = 4$. The graph of the equation is shown in the figure at the left.

Practice Problem 1   Find three ordered pairs that satisfy $x + y = 10$. Then graph the resulting straight line. Use the given coordinate system.

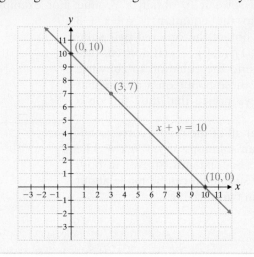

Teaching Example 2 Graph
$3x - 1 = 2y + 5$.

Ans:

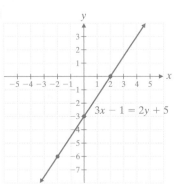

EXAMPLE 2   Graph $5x - 4y + 2 = 2$.

Solution   First, we simplify the equation $5x - 4y + 2 = 2$ by subtracting 2 from each side.

$$5x - 4y + 2 - 2 = 2 - 2$$
$$5x - 4y = 0$$

Since we are free to choose any value of x, $x = 0$ is a natural choice. Calculate the value of y when $x = 0$.

$$5(0) - 4y = 0$$
$$-4y = 0$$
$$y = 0 \quad \text{Remember: Any number times 0 is 0.}$$
$$\text{Since } -4y = 0, y \text{ must equal 0.}$$

Now let's see what happens when $x = 1$.

$$5(1) - 4y = 0$$
$$5 - 4y = 0$$
$$-4y = -5$$
$$y = \frac{-5}{-4} \quad \text{or} \quad \frac{5}{4} \quad \text{This is not an easy number to graph.}$$

A better choice for a replacement of x is a number that is divisible by 4. Let's see why. Let $x = 4$ and let $x = -4$.

$$5(4) - 4y = 0 \qquad\qquad 5(-4) - 4y = 0$$
$$20 - 4y = 0 \qquad\qquad -20 - 4y = 0$$
$$-4y = -20 \qquad\qquad -4y = 20$$
$$y = \frac{-20}{-4} \quad \text{or} \quad 5 \qquad y = \frac{20}{-4} \quad \text{or} \quad -5$$

Now we can put these numbers into our table of values and graph the line.

Table of Values

x	y
0	0
4	5
−4	−5

PRACTICE PROBLEM 2

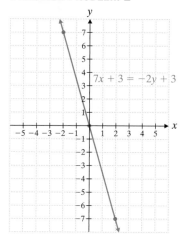

Practice Problem 2 Graph $7x + 3 = -2y + 3$ on the coordinate system in the margin.

NOTE TO STUDENT: *Fully worked-out solutions to all of the Practice Problems can be found at the back of the text starting at page SP-1*

TO THINK ABOUT: An Alternative Approach   In Example 2, we picked values of x and found the corresponding values for y. An alternative approach is to solve the equation for the variable y *first*.

$$5x - 4y = 0$$
$$-4y = -5x \qquad \text{Subtract } 5x \text{ from each side.}$$
$$\frac{-4y}{-4} = \frac{-5x}{-4} \qquad \text{Divide each side by } -4.$$
$$y = \frac{5}{4}x$$

Now let $x = -4$, $x = 0$, and $x = 4$, and find the corresponding values of y. Explain why you would choose multiples of 4 as replacements of x in this equation. Graph the equation and compare it to the graph in Example 2.

In the previous two examples we began by picking values for x. We could just as easily have chosen values for y.

Teaching Tip Some instructors prefer that students always solve a linear equation for y and then substitute in the x-values. If that is your own viewpoint, then before finding points for the equation in Example 2, first rewrite the equation in the form $y = \frac{5}{4}x$.

 Graphing a Straight Line by Plotting Its Intercepts

What values should we pick for x and y? Which points should we use for plotting? For many straight lines it is easiest to pick the two *intercepts*. Some lines have only one intercept. We will discuss these separately.

> The **x-intercept** of a line is the point where the line crosses the x-axis; it has the form $(a, 0)$. The **y-intercept** of a line is the point where the line crosses the y-axis; it has the form $(0, b)$.

INTERCEPT METHOD OF GRAPHING

To graph an equation using intercepts, we:

1. Find the x-intercept by letting $y = 0$ and solving for x.

2. Find the y-intercept by letting $x = 0$ and solving for y.

3. Find one additional ordered pair so that we have three points with which to plot the line.

EXAMPLE 3   Use the intercept method to graph $5y - 3x = 15$.

Solution

$$\text{Let } y = 0. \quad 5(0) - 3x = 15 \quad \text{Replace } y \text{ by } 0.$$
$$-3x = 15 \quad \text{Divide both sides by } -3.$$
$$x = -5$$

The ordered pair $(-5, 0)$ is the x-intercept.

$$\text{Let } x = 0. \quad 5y - 3(0) = 15 \quad \text{Replace } x \text{ by } 0.$$
$$5y = 15 \quad \text{Divide both sides by } 5.$$
$$y = 3$$

The ordered pair $(0, 3)$ is the y-intercept.
We find another ordered pair to have a third point.

$$\text{Let } y = 6. \quad 5(6) - 3x = 15 \quad \text{Replace } y \text{ by } 6.$$
$$30 - 3x = 15 \quad \text{Simplify.}$$
$$-3x = -15 \quad \text{Subtract 30 from both sides.}$$
$$x = \frac{-15}{-3} \quad \text{or} \quad 5$$

The ordered pair is $(5, 6)$.
Our table of values is

x	y
-5	0
0	3
5	6

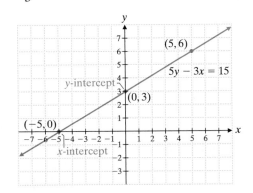

Practice Problem 3   Use the intercept method to graph $2y - x = 6$. Use the given coordinate system.

Teaching Tip You may wish to show the general condition that determines that the graph of an equation of the form $Ax + By = C$ goes through the origin. That is, point out that when $Ax + By = 0$, the graph of the equation goes through the origin. Ask students to find the x- and y-intercepts, respectively. Remind students that $A \neq 0$ or $B \neq 0$ or both.

x-intercept	*y-intercept*
Let $y = 0$	Let $x = 0$
$Ax + By = 0$	$Ax + By = 0$
$Ax + B(0) = 0$	$A(0) + By = 0$
$Ax = 0$	$By = 0$
$x = 0$	$y = 0$

Thus the x-intercept is $(0, 0)$ and the y-intercept is $(0, 0)$.

You may wish to provide a specific example such as $3x + 2y = 0$. Since the intercept is at the origin, $(0, 0)$, have students look for two other points to graph.

TO THINK ABOUT: Lines That Go Through the Origin   Can you draw all straight lines by the intercept method? Not really. Some straight lines may go through the origin and have only one intercept. If a line goes through the origin, it will have an equation of the form $Ax + By = 0$, where $A \neq 0$ or $B \neq 0$ or both. Examples are $3x + 4y = 0$ and $5x - 2y = 0$. In such cases you should plot two additional points besides the origin. Be sure to simplify each equation before attempting to graph it.

Graphing Horizontal and Vertical Lines

You will notice that the x-axis is a horizontal line. It is the line $y = 0$, since for any value of x, the value of y is 0. Try a few points. The points $(1, 0)$, $(3, 0)$, and $(-2, 0)$ all lie on the x-axis. Any horizontal line will be parallel to the x-axis. Lines such as $y = 5$ and $y = -2$ are horizontal lines. What does $y = 5$ mean? It means that for any value of x, y is 5. Likewise $y = -2$ means that for any value of x, $y = -2$.

How can we recognize the equation of a line that is horizontal, that is, parallel to the x-axis?

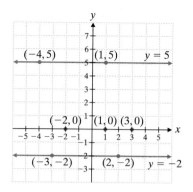

> If the graph of an equation is a straight line that is parallel to the x-axis (that is, a horizontal line), the equation will be of the form $y = b$, where b is some real number.

EXAMPLE 4 Graph $y = -3$.

Solution You could write the equation as $0x + y = -3$. Then it is clear that for any value of x that you substitute, you will always obtain $y = -3$. Thus, as shown in the figure, $(4, -3)$, $(0, -3)$, and $(-3, -3)$ are all ordered pairs that satisfy the equation $y = -3$. Since the y-coordinate of every point on this line is -3, it is easy to see that the horizontal line will be 3 units below the x-axis.

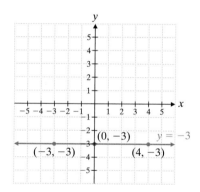

Teaching Example 4 Graph $y = 4$.

Ans:

Practice Problem 4 Graph $2y - 3 = 0$ on the given coordinate system.

Teaching Tip It is important for students to recognize and graph the equation of a horizontal line. Remind them that the straight line that is graphed in Example 4 can have the form $y = -3$, or the form $5y = -15$, or the form $4y + 11 = -1$. Both of the latter equations can be simplified to the form $y = -3$. This type of simplification should be done prior to graphing.

Notice that the y-axis is a vertical line. This is the line $x = 0$, since for any y, x is 0. Try a few points. The points $(0, 2)$, $(0, -3)$, and $\left(0, \frac{1}{2}\right)$ all lie on the y-axis. Any vertical line will be parallel to the y-axis. Lines such as $x = 2$ and $x = -3$ are vertical lines. Think of what $x = 2$ means. It means that for any value of y, x is 2. The graph of $x = 2$ is a vertical line two units to the right of the y-axis.

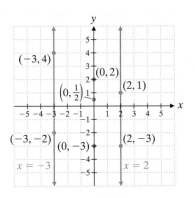

How can we recognize the equation of a line that is vertical, that is, parallel to the y-axis?

If the graph of an equation is a straight line that is parallel to the y-axis (that is, a vertical line), the equation will be of the form $x = a$, where a is some real number.

Teaching Example 5 Graph $x = -2$.

Ans:

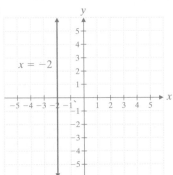

EXAMPLE 5 Graph $x = 5$.

Solution This can be done immediately by drawing a vertical line 5 units to the right of the origin. The x-coordinate of every point on this line is 5.

The equation $x - 5 = 0$ can be rewritten as $x = 5$ and graphed as shown.

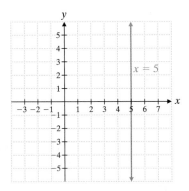

Practice Problem 5 Graph $x + 3 = 0$ on the following coordinate system.

Verbal and Writing Skills

Unless otherwise indicated, assume each grid line represents one unit.

1. Is the point $(-2, 5)$ a solution to the equation $2x + 5y = 0$? Why or why not?

No, replacing x by -2 and y by 5 in the equation does not result in a true statement.

2. The graph of a linear equation in two variables is a _____straight_____ _____line_____ .

3. The x-intercept of a line is the point where the line crosses the _____x-axis_____ .

4. The graph of the equation $y = b$ is a _____horizontal_____ line.

Complete the ordered pairs so that each is a solution of the given linear equation. Then plot each solution and graph the equation by connecting the points by a straight line.

5. $y = -2x + 1$
$(0, 1)$
$(-2, 5)$
$(1, -1)$

6. $y = -3x - 4$
$(-2, 2)$
$(-1, -1)$
$(0, -4)$

7. $y = x - 4$
$(0, -4)$
$(2, -2)$
$(4, 0)$

  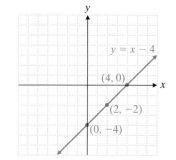

8. $y = 3x - 1$
$(0, -1)$
$(2, 5)$
$(-1, -4)$

9. $y = -2x + 3$
$(0, 3)$
$(2, -1)$
$(4, -5)$

10. $y = 2x - 5$
$(0, -5)$
$(2, -1)$
$(4, 3)$

11. $y = 3x + 2$
$(-1, -1)$
$(0, 2)$
$(1, 5)$

  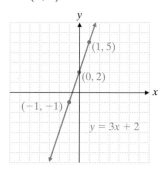

Graph each equation by plotting three points and connecting them.

12. $y = -3x + 2$

13. $3x - 2y = 0$

14. $2y - 5x = 0$

  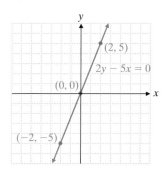

15. $y = -\dfrac{3}{4}x + 3$

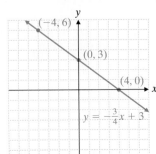

16. $y = \dfrac{2}{3}x + 2$

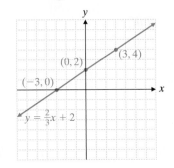

17. $4x + 3y = 12$

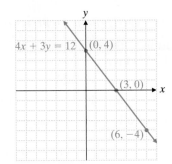

18. $-2x + 4y = 12$

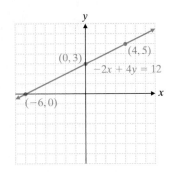

Graph each equation by plotting the intercepts and one other point.

19. $y = 6 - 2x$

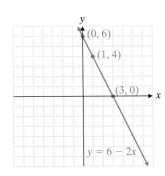

20. $y = 4 - 2x$

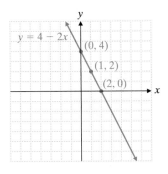

21. $x + 3 = 6y$

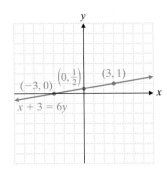

22. $x - 6 = 2y$

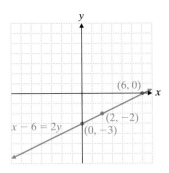

Mixed Practice *Graph the equation. Be sure to simplify the equation before graphing it.*

23. $y - 2 = 3y$

24. $3y + 1 = 7$

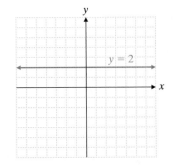

25. $2x + 9 = 5x$

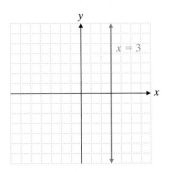

26. $3x - 4 = -13$

27. $2x + 5y - 2 = -12$

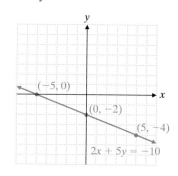

28. $3x - 4y - 5 = -17$

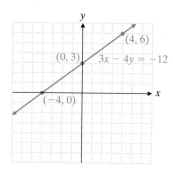

Applications

29. *Cross-Country Skiing* The number of calories burned by an average person while cross-country skiing is given by the equation $C = 8m$, where m is the number of minutes. (*Source:* National Center for Health Statistics.) Graph the equation for $m = 0, 15, 30, 45, 60,$ and 75.

30. *Calories Burned While Jogging* The number of calories burned by an average person while jogging is given by the equation $C = \dfrac{28}{3}m$, where m is the number of minutes. (*Source:* National Center for Health Statistics.) Graph the equation for $m = 0, 15, 30, 45, 60,$ and 75.

31. *Foreign Students in the United States* The number of foreign students enrolled in college in the United States can be approximated by the equation $S = 11t + 395$, where t stands for the number of years since 1990, and S is the number of foreign students (in thousands). (*Source:* www.opendoors.iienetwork.org.) Graph the equation for $t = 0, 4, 8, 16$.

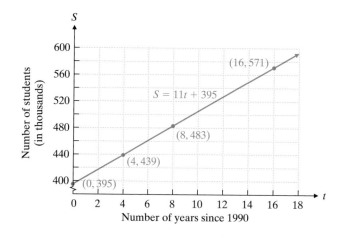

32. *Restaurant Sales* The amount of money U.S. consumers spend on food and beverages in restaurants can be approximated by the equation $S = 16.8t + 223$, where S is the sales in billions of dollars and t is the number of years since 1990. (*Source: Statistical Abstract of the United States: 2007.*) Graph the equation for $t = 0, 6, 10, 14, 18$.

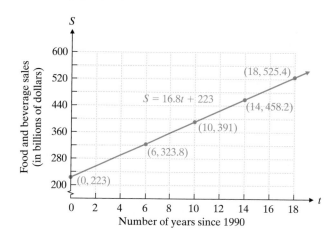

Cumulative Review

33. [6.5.1] Solve. $\dfrac{x}{x+4} + \dfrac{x-1}{x-4} = 2$

$x^2 - 4x + x^2 + 3x - 4 = 2x^2 - 32$

$x = 28$

34. [2.7.1] Solve and graph on a number line. $4 - 3x \le 18$

$x \ge -\dfrac{14}{3}$

Quick Quiz 7.2 Graph each equation.

1. $y = -2x + 4$

2. $y = \dfrac{2}{5}x - 3$

3. $4y + 1 = x + 9$

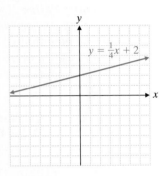

4. Concept Check In graphing the equation $3y - 7x = 0$, what is the most important ordered pair to obtain before drawing a graph of the line? Why is that ordered pair so essential to drawing the graph? Answers may vary

Classroom Quiz 7.2 You may use these problems to quiz your students' mastery of Section 7.2.

Graph each equation.

1. $y = 3x - 2$
 Ans:

2. $y = -\dfrac{4}{5}x + 3$
 Ans:

3. $3y + 2 = x - 4$
 Ans:

Finding the Slope of a Line Given Two Points on the Line

We often use the word *slope* to describe the incline (the steepness) of a hill. A carpenter or a builder will refer to the *pitch* or *slope* of a roof. The slope is the change in the vertical distance compared to the change in the horizontal distance as you go from one point to another point along the roof. If the change in the vertical distance is greater than the change in the horizontal distance, the slope will be steep. If the change in the horizontal distance is greater than the change in the vertical distance, the slope will be gentle.

Student Learning Objectives

After studying this section, you will be able to:

1 **Find the slope of a line given two points on the line.**

2 **Find the slope and y-intercept of a line given its equation.**

3 **Write the equation of a line given the slope and y-intercept.**

4 **Graph a line using the slope and y-intercept.**

5 **Find the slopes of parallel and perpendicular lines.**

In a coordinate plane, the **slope** of a straight line is defined by the change in y divided by the change in x.

$$\text{Slope} = \frac{\text{change in } y}{\text{change in } x}$$

Consider the line drawn through points A and B in the figure. If we measure the change from point A to point B in the x-direction and the y-direction, we will have an idea of the steepness (or the slope) of the line. From point A to point B, the change in y-values is from 2 to 4, a *change of* 2. From point A to point B, the change in x-values is from 1 to 5, a *change of* 4. Thus

$$\text{slope} = \frac{\text{change in } y}{\text{change in } x} = \frac{2}{4} = \frac{1}{2}.$$

Informally, we can describe this move as the rise over the run: $\text{slope} = \dfrac{\text{rise}}{\text{run}}$.

We now state a more formal (and more frequently used) definition.

DEFINITION OF SLOPE OF A LINE

The **slope** of any *nonvertical* straight line that contains the points with coordinates (x_1, y_1) and (x_2, y_2) is defined by the difference ratio

$$\text{slope} = m = \frac{y_2 - y_1}{x_2 - x_1} \qquad \text{where } x_2 \neq x_1.$$

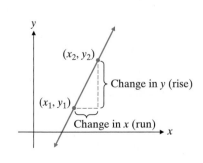

The use of subscripted terms such as x_1, x_2, and so on, is just a way of indicating that the first x-value is x_1 and the second x-value is x_2. Thus (x_1, y_1) are the coordinates of the first point and (x_2, y_2) are the coordinates of the second point. The letter m is commonly used for the slope.

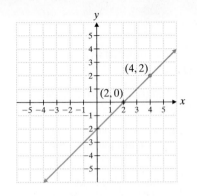

Teaching Example 1 Find the slope of the line that passes through $(-1, 1)$ and $(2, 3)$.

Ans: $m = \dfrac{2}{3}$

EXAMPLE 1 Find the slope of the line that passes through $(2, 0)$ and $(4, 2)$.

Solution Let $(2, 0)$ be the first point (x_1, y_1) and $(4, 2)$ be the second point (x_2, y_2).

$$\overset{\overbrace{\hspace{2em}2\hspace{2em}}}{\underset{\underbrace{\hspace{2em}2\hspace{2em}}}{(2, 0) \qquad\qquad (4, 2)}} \qquad m = \frac{2}{2} = 1$$

By the formula, $\quad \text{slope} = m = \dfrac{y_2 - y_1}{x_2 - x_1} = \dfrac{2 - 0}{4 - 2} = \dfrac{2}{2} = 1$.

A sketch of the line is shown in the figure at the left.

Note that the slope of the line will be the same if we let $(4, 2)$ be the first point (x_1, y_1) and $(2, 0)$ be the second point (x_2, y_2).

$$\overset{\overbrace{\hspace{2em}-2\hspace{2em}}}{\underset{\underbrace{\hspace{2em}-2\hspace{2em}}}{(4, 2) \qquad\qquad (2, 0)}}$$

$$m = \frac{y_2 - y_1}{x_2 - x_1} = \frac{0 - 2}{2 - 4} = \frac{-2}{-2} = 1$$

Thus, given two points, it does not matter which you call (x_1, y_1) and which you call (x_2, y_2).

CAUTION: Be careful, however, not to put the x's in one order and the y's in another order when finding the slope from two points on a line.

Practice Problem 1 Find the slope of the line that passes through $(6, 1)$ and $(-4, -1)$.

$$m = \frac{1}{5}$$

It is a good idea to have some concept of what different slopes mean. In downhill skiing, a very gentle slope used for teaching beginning skiers might drop one foot vertically for each 10 feet horizontally. The slope would be $\frac{1}{10}$. The speed of a skier on a hill with such a gentle slope would be only about 6 miles per hour.

A triple diamond slope for experts might drop 11 feet vertically for each 10 feet horizontally. The slope would be $\frac{11}{10}$. The speed of a skier on such an expert trail would be in the range of 60 miles per hour.

It is important to see how positive and negative slopes affect the graphs of lines.

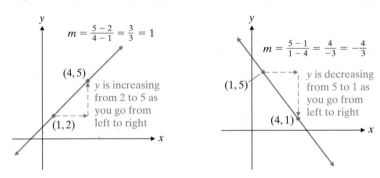

1. If the y-values increase as you go from left to right, the slope of the line is positive.

2. If the y-values decrease as you go from left to right, the slope of the line is negative.

EXAMPLE 2   Find the slope of the line that passes through $(-3, 2)$ and $(2, -4)$.

Solution   Let $(-3, 2)$ be (x_1, y_1) and $(2, -4)$ be (x_2, y_2).

$$m = \frac{y_2 - y_1}{x_2 - x_1} = \frac{-4 - 2}{2 - (-3)} = \frac{-4 - 2}{2 + 3} = \frac{-6}{5} = -\frac{6}{5}$$

The slope of this line is negative. We would expect this, since the y-value decreased from 2 to -4 as the x-value increased. What does the graph of this line look like? Plot the points and draw the line to verify.

Practice Problem 2   Find the slope of the line that passes through $(2, 0)$ and $(-1, 1)$.

$$m = -\frac{1}{3}$$

TO THINK ABOUT: Using the Slope to Describe a Line   Describe the line in Practice Problem 2 by looking at its slope. Then verify by drawing the graph.

EXAMPLE 3   Find the slope of the line that passes through the given points.

(a) $(0, 2)$ and $(5, 2)$ **(b)** $(-4, 0)$ and $(-4, -4)$

Solution

(a) Take a moment to look at the y-values. What do you notice? What does this tell you about the line? Now calculate the slope.

$$m = \frac{2 - 2}{5 - 0} = \frac{0}{5} = 0$$

Since any two points on a horizontal line will have the same y-value, the slope of a horizontal line is 0.

(b) Take a moment to look at the x-values. What do you notice? What does this tell you about the line? Now calculate the slope.

$$m = \frac{-4 - 0}{-4 - (-4)} = \frac{-4}{0}$$

Recall that division by 0 is undefined.

Example 3(b)

The slope of a vertical line is undefined. We say that a vertical line has **no slope.**
    Notice in our definition of slope that $x_2 \ne x_1$. Thus it is not appropriate to use the formula for slope for the points in (b). We did so to illustrate what would happen if $x_2 = x_1$. We get an impossible situation, $\frac{y_2 - y_1}{0}$. Now you can see why we include the restriction $x_2 \ne x_1$ in our definition.

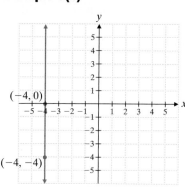

Practice Problem 3   Find the slope of the line that passes through the given points.

(a) $(-5, 6)$ and $(-5, 3)$ No slope   **(b)** $(-7, -11)$ and $(3, -11)$ $m = 0$

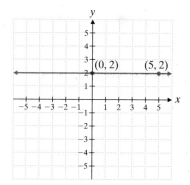

SLOPE OF A STRAIGHT LINE

Positive slope
Line goes upward to the right

1. *Lines with positive slopes go upward as x increases* (that is, as you go from left to right).

Negative slope
Line goes downward to the right

2. *Lines with negative slopes go downward as x increases* (that is, as you go from left to right).

Zero slope
Horizontal line

3. Horizontal lines have a slope of 0.

Undefined slope
Vertical line

4. A vertical line is said to have undefined slope. The slope of a vertical line is not defined. In other words, a vertical line has no slope.

Teaching Tip To show that three points lie on the same line, have students find the slope between pairs of points. For example, show that $(1, 3)$, $(3, 7)$, $(5, 11)$ are on the same line.

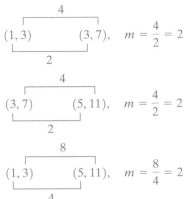

$(1, 3) \qquad (3, 7), \quad m = \dfrac{4}{2} = 2$

$(3, 7) \qquad (5, 11), \quad m = \dfrac{4}{2} = 2$

$(1, 3) \qquad (5, 11), \quad m = \dfrac{8}{4} = 2$

The points have a *common* difference ratio (slope). They must be on the *same* line.

② Finding the Slope and *y*-intercept of a Line Given Its Equation

Recall that the equation of a line is a linear equation in two variables. This equation can be written in several different ways. A very useful form of the equation of a straight line is the slope–intercept form. This form can be derived in the following way. Suppose that a straight line with slope m crosses the y-axis at a point $(0, b)$. Consider any other point on the line and label the point (x, y). Then we have the following.

$$\frac{y_2 - y_1}{x_2 - x_1} = m \qquad \text{Definition of slope.}$$

$$\frac{y - b}{x - 0} = m \qquad \text{Substitute } (0, b) \text{ for } (x_1, y_1) \text{ and } (x, y) \text{ for } (x_2, y_2).$$

$$\frac{y - b}{x} = m \qquad \text{Simplify.}$$

$$y - b = mx \qquad \text{Multiply both sides by } x.$$

$$y = mx + b \quad \text{Add } b \text{ to both sides.}$$

This form of a linear equation immediately reveals the slope of the line, m, and the y-coordinate of the point where the line intercepts (crosses) the y-axis, b.

> **SLOPE–INTERCEPT FORM OF A LINE**
>
> The slope–intercept form of the equation of the line that has slope m and y-intercept $(0, b)$ is given by
> $$y = mx + b.$$

By using algebraic operations, we can write any linear equation in slope–intercept form and use this form to identify the slope and the y-intercept of the line.

EXAMPLE 4   What is the slope and the y-intercept of the line $5x + 3y = 2$?

Solution   We want to solve for y and get the equation in the form $y = mx + b$. We need to isolate the y-variable.

$$5x + 3y = 2$$
$$3y = -5x + 2 \qquad \text{Subtract } 5x \text{ from both sides.}$$
$$y = \frac{-5x + 2}{3} \qquad \text{Divide both sides by 3.}$$
$$y = -\frac{5}{3}x + \frac{2}{3} \qquad \text{Using the property } \frac{a + b}{c} = \frac{a}{c} + \frac{b}{c}, \text{ write the right-hand side as two fractions.}$$

The *slope* is $-\dfrac{5}{3}$.   The *y-intercept* is $\left(0, \dfrac{2}{3}\right)$.

Practice Problem 4   What is the slope and the y-intercept of the line $4x - 2y = -5$?

Slope $= 2$; y-intercept $= \left(0, \dfrac{5}{2}\right)$

③ Writing the Equation of a Line Given the Slope and y-Intercept

If we know the slope of a line and the y-intercept, we can write the equation of the line, $y = mx + b$.

EXAMPLE 5   Find an equation of the line with slope $\frac{2}{5}$ and y-intercept $(0, -3)$.
(a) Write the equation in slope–intercept form, $y = mx + b$.
(b) Write the equation in the form $Ax + By = C$.

Solution

(a) We are given that $m = \frac{2}{5}$ and $b = -3$. Thus we have the following.
$$y = mx + b$$
$$y = \frac{2}{5}x + (-3)$$
$$y = \frac{2}{5}x - 3$$

(b) Recall, for the form $Ax + By = C$, that A, B, and C are integers. We first clear the equation of fractions. Then we move the x-term to the left side.

$$5y = 5\left(\frac{2x}{5}\right) - 5(3) \qquad \text{Multiply each term by 5.}$$
$$5y = 2x - 15 \qquad\qquad \text{Simplify.}$$
$$-2x + 5y = -15 \qquad\quad \text{Subtract } 2x \text{ from each side.}$$
$$2x - 5y = 15 \qquad\qquad \text{Multiply each term by } -1. \text{ The form } Ax + By = C \text{ is usually written with } A \text{ as a positive integer.}$$

Teaching Example 4 What is the slope and the y-intercept of the line $x - 3y = 2$?

Ans: Slope $= \dfrac{1}{3}$; y-intercept $= \left(0, -\dfrac{2}{3}\right)$

NOTE TO STUDENT: Fully worked-out solutions to all of the Practice Problems can be found at the back of the text starting at page SP-1

Teaching Example 5 Find an equation of the line with slope $\frac{3}{4}$ and y-intercept $(0, 1)$.
(a) Write the equation in slope–intercept form.
(b) Write the equation in the form $Ax + By = C$.

Ans: (a) $y = \dfrac{3}{4}x + 1$
(b) $3x - 4y = -4$

Teaching Example 6 Graph the line with slope $m = \frac{5}{3}$ and y-intercept $(0, -4)$.

Ans:

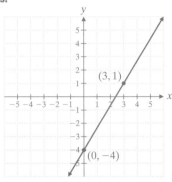

Practice Problem 5 Find an equation of the line with slope $-\frac{3}{7}$ and y-intercept $\left(0, \frac{2}{7}\right)$.

(a) Write the equation in slope–intercept form. $y = -\frac{3}{7}x + \frac{2}{7}$

(b) Write the equation in the form $Ax + By = C$. $3x + 7y = 2$

4 Graphing a Line Using the Slope and *y*-Intercept

If we know the slope of a line and the y-intercept, we can draw the graph of the line.

EXAMPLE 6 Graph the line with slope $m = \frac{2}{3}$ and y-intercept $(0, -3)$. Use the given coordinate system.

Solution Recall that the y-intercept is the point where the line crosses the y-axis. We need a point to start the graph with. So we plot the point $(0, -3)$ on the y-axis.

Recall that slope $= \dfrac{\text{rise}}{\text{run}}$. Since the slope of this line is $\frac{2}{3}$, we will go up (rise)

PRACTICE PROBLEM 6

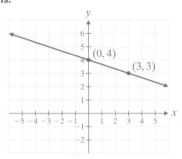

2 units and go over (run) to the right 3 units from the point $(0, -3)$. Look at the figure to the right. This is the point $(3, -1)$. Plot the point. Draw a line that connects the two points $(0, -3)$ and $(3, -1)$.

This is the graph of the line with slope $\frac{2}{3}$ and y-intercept $(0, -3)$.

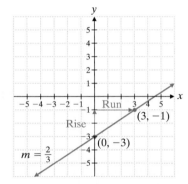

Teaching Example 7 Graph the equation $y = -\frac{1}{3}x + 4$.

Ans:

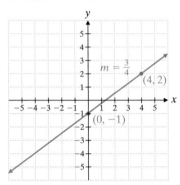

Practice Problem 6 Graph the line with slope $= \frac{3}{4}$ and y-intercept $(0, -1)$. Use the coordinate system in the margin.

EXAMPLE 7 Graph the equation $y = -\frac{1}{2}x + 4$. Use the following coordinate system.

Solution Begin with the y-intercept. Since $b = 4$, plot the point $(0, 4)$. Now look at the slope $-\frac{1}{2}$. This can be written as $\frac{-1}{2}$. Begin at $(0, 4)$ and go *down* 1 unit and to the right 2 units. This is the point $(2, 3)$. Plot the point. Draw the line that connects the points $(0, 4)$ and $(2, 3)$.

This is the graph of the equation $y = -\frac{1}{2}x + 4$.

PRACTICE PROBLEM 7

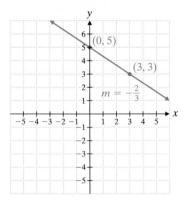

Practice Problem 7 Graph the equation $y = -\frac{2}{3}x + 5$. Use the coordinate system in the margin.

TO THINK ABOUT: Verifying Your Answer Explain why you count down 1 unit and move to the right 2 units to represent the slope $-\frac{1}{2}$. Could you have done this in another way? Try it. Verify that this is the same line.

 Finding the Slopes of Parallel and Perpendicular Lines

Parallel lines are two straight lines that never touch. Look at the parallel lines in the figure. Notice that the slope of line a is -3 and the slope of line b is also -3. Why do you think the slopes must be equal? What would happen if the slope of line b were -1? Graph it and see.

Graphing Calculator

 Graphing Lines

You can graph a line given in the form $y = mx + b$ using a graphing calculator. For example, to graph $y = 2x + 4$, enter the right-hand side of the equation in the $Y =$ editor of your calculator and graph. Choose an appropriate window to show all the intercepts. The following window is -10 to 10 by -10 to 10.
Display:

Try graphing other equations given in slope–intercept form.

PARALLEL LINES

Parallel lines are two straight lines that never touch.
Parallel lines have the same slope but different y-intercepts.

$$m_1 = m_2$$

Perpendicular lines are two lines that meet at a $90°$ angle. Look at the perpendicular lines in the figure at the right. The slope of line c is -3. The slope of line d is $\frac{1}{3}$. Notice that

$$(-3)\left(\frac{1}{3}\right) = \left(-\frac{3}{1}\right)\left(\frac{1}{3}\right) = -1.$$

You may wish to draw several pairs of perpendicular lines to determine whether the product of their slopes is always -1.

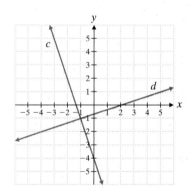

PERPENDICULAR LINES

Perpendicular lines are two lines that meet at a $90°$ angle.
Perpendicular lines have slopes whose product is -1. If m_1 and m_2 are slopes of perpendicular lines, then

$$m_1 m_2 = -1 \quad \text{or} \quad m_1 = -\frac{1}{m_2}.$$

Graphing Calculator

Graphing Parallel Lines

If two equations are in the form $y = mx + b$, then it will be obvious that they are parallel because the slope will be the same. On a graphing calculator, graph both of these equations:

$$y = -2x + 6$$
$$y = -2x - 4$$

Use a window of -10 to 10 for both x and y.
Display:

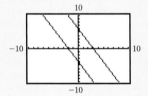

EXAMPLE 8 Line h has a slope of $-\frac{2}{3}$.

(a) If line f is parallel to line h, what is its slope?

(b) If line g is perpendicular to line h, what is its slope?

Solution

(a) Parallel lines have the same slope. Line f has a slope of $-\frac{2}{3}$.

(b) Perpendicular lines have slopes whose product is -1.

$$m_1 m_2 = -1$$

$$-\frac{2}{3} m_2 = -1 \qquad \text{Substitute } -\frac{2}{3} \text{ for } m_1.$$

$$\left(-\frac{3}{2}\right)\left(-\frac{2}{3}\right)m_2 = -1\left(-\frac{3}{2}\right) \qquad \text{Multiply both sides by } -\frac{3}{2}.$$

$$m_2 = \frac{3}{2}$$

Thus line g has a slope of $\frac{3}{2}$.

Practice Problem 8 Line h has a slope of $\frac{1}{4}$.

(a) If line j is parallel to line h, what is its slope? $\frac{1}{4}$

(b) If line k is perpendicular to line h, what is its slope? -4

EXAMPLE 9 The equation of line l is $y = -2x + 3$.

(a) What is the slope of a line that is parallel to line l?

(b) What is the slope of a line that is perpendicular to line l?

Solution

(a) Looking at the equation, we can see that the slope of line l is -2. The slope of a line that is parallel to line l is -2.

(b) Perpendicular lines have slopes whose product is -1.

$$m_1 m_2 = -1$$

$$(-2)m_2 = -1 \qquad \text{Substitute } -2 \text{ for } m_1.$$

$$m_2 = \frac{1}{2} \qquad \text{Because } (-2)\left(\frac{1}{2}\right) = -1.$$

The slope of a line that is perpendicular to line l is $\frac{1}{2}$.

Practice Problem 9 The equation of line n is $y = \frac{2}{3}x - 4$.

(a) What is the slope of a line that is parallel to line n? $\frac{2}{3}$

(b) What is the slope of a line that is perpendicular to line n? $-\frac{3}{2}$

PRACTICE   WATCH   DOWNLOAD   READ   REVIEW

Verbal and Writing Skills

1. Can you find the slope of the line passing through $(5, -12)$ and $(5, -6)$? Why or why not?
   No; division by zero is impossible, so the slope is undefined.

2. Can you find the slope of the line passing through $(6, -2)$ and $(-8, -2)$? Why or why not?
   Yes; the x-values are not equal, so we can use the definition. The slope is 0.

Find the slope of the straight line that passes through the given pair of points.

3. $(4, 1)$ and $(6, 7)$ 3

4. $(7, 5)$ and $(12, 10)$ 1

5. $(5, 10)$ and $(6, 5)$ -5

6. $(4, 7)$ and $(1, 13)$ -2

7. $(-2, 1)$ and $(3, 4)$ $\dfrac{3}{5}$

8. $(5, 6)$ and $(-3, 1)$ $\dfrac{5}{8}$

9. $(-6, -5)$ and $(2, -7)$ $-\dfrac{1}{4}$

10. $(-8, -3)$ and $(4, -9)$ $-\dfrac{1}{2}$

11. $(-3, 0)$ and $(0, -4)$ $-\dfrac{4}{3}$

12. $(0, 5)$ and $(5, 3)$ $-\dfrac{2}{5}$

13. $\left(\dfrac{3}{4}, -4\right)$ and $(2, -8)$ $-\dfrac{16}{5}$

14. $\left(\dfrac{5}{3}, -2\right)$ and $(3, 6)$ 6

Find the slope and the y-intercept.

15. $y = 8x + 9$ $m = 8; (0, 9)$

16. $y = 2x + 10$ $m = 2; (0, 10)$

17. $3x + y - 4 = 0$ $m = -3; (0, 4)$

18. $8x + y + 7 = 0$ $m = -8; (0, -7)$

19. $y = -\dfrac{8}{7}x + \dfrac{3}{4}$ $m = -\dfrac{8}{7}; \left(0, \dfrac{3}{4}\right)$

20. $y = \dfrac{5}{3}x - \dfrac{4}{5}$ $m = \dfrac{5}{3}; \left(0, -\dfrac{4}{5}\right)$

21. $y = -6x$ $m = -6; (0, 0)$

22. $y = -2$ $m = 0; (0, -2)$

23. $6x + y = \dfrac{4}{5}$ $m = -6; \left(0, \dfrac{4}{5}\right)$

24. $2x + y = -\dfrac{3}{4}$ $m = -2; \left(0, -\dfrac{3}{4}\right)$

25. $3x + 4y - 6 = 0$ $m = -\dfrac{3}{4}; \left(0, \dfrac{3}{2}\right)$

26. $5x + 3y - 1 = 0$ $m = -\dfrac{5}{3}; \left(0, \dfrac{1}{3}\right)$

27. $7x - 3y = 4$ $m = \dfrac{7}{3}; \left(0, -\dfrac{4}{3}\right)$

28. $9x - 4y = 18$ $m = \dfrac{9}{4}; \left(0, -\dfrac{9}{2}\right)$

Write the equation of the line in slope–intercept form.

29. $m = \dfrac{3}{5}$, y-intercept $(0, 3)$ $y = \dfrac{3}{5}x + 3$

30. $m = \dfrac{2}{3}$, y-intercept $(0, 5)$ $y = \dfrac{2}{3}x + 5$

31. $m = 4$, y-intercept $(0, -5)$ $y = 4x - 5$

32. $m = 2$, y-intercept $(0, -1)$ $y = 2x - 1$

33. $m = -\dfrac{5}{4}$, y-intercept $\left(0, -\dfrac{3}{4}\right)$ $y = -\dfrac{5}{4}x - \dfrac{3}{4}$

34. $m = -4$, y-intercept $\left(0, \dfrac{1}{2}\right)$ $y = -4x + \dfrac{1}{2}$

Graph the line $y = mx + b$ for the given values.

35. $m = \dfrac{3}{4}$, $b = -4$

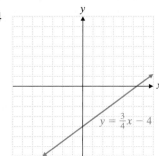

$y = \frac{3}{4}x - 4$

36. $m = \dfrac{1}{3}$, $b = -2$

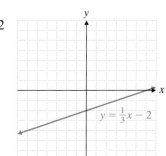

$y = \frac{1}{3}x - 2$

37. $m = -\dfrac{5}{3}, b = 2$

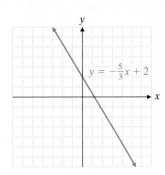

38. $m = -\dfrac{3}{2}, b = 4$

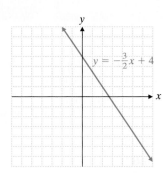

In exercises 39–44, graph the line.

39. $y = \dfrac{2}{3}x + 2$

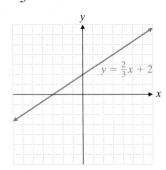

40. $y = \dfrac{3}{4}x + 1$

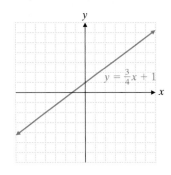

41. $y + 2x = 3$

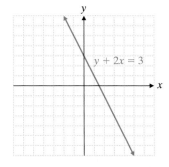

42. $y + 4x = 5$

43. $y = 2x$

44. $y = 3x$

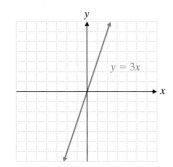

45. A line has a slope of $\dfrac{5}{6}$.

    (a) What is the slope of the line parallel to it? $\dfrac{5}{6}$

    (b) What is the slope of the line perpendicular to it? $-\dfrac{6}{5}$

46. A line has a slope of $\dfrac{11}{4}$.

    (a) What is the slope of the line parallel to it? $\dfrac{11}{4}$

    (b) What is the slope of the line perpendicular to it? $-\dfrac{4}{11}$

47. A line has a slope of 6.

    (a) What is the slope of the line parallel to it? 6

    (b) What is the slope of the line perpendicular to it? $-\dfrac{1}{6}$

48. A line has a slope of $-\dfrac{5}{8}$.

    (a) What is the slope of the line parallel to it? $-\dfrac{5}{8}$

    (b) What is the slope of the line perpendicular to it? $\dfrac{8}{5}$

49. The equation of a line is $y = \dfrac{2}{3}x + 6$.

    (a) What is the slope of a line parallel to it? $\dfrac{2}{3}$

    (b) What is the slope of a line perpendicular to it? $-\dfrac{3}{2}$

50. The equation of a line is $y = 3x - 8$.

    (a) What is the slope of a line parallel to it? 3

    (b) What is the slope of a line perpendicular to it? $-\dfrac{1}{3}$

To Think About

51. Do the points $(3, -4)$, $(18, 6)$, and $(9, 0)$ all lie on the same line? If so, what is the equation of the line?
Yes; $2x - 3y = 18$

52. Do the points $(2, 1)$, $(-3, -2)$, and $(7, 4)$ lie on the same line? If so, what is the equation of the line?
Yes; $3x - 5y = 1$

53. *Home Health Aides* It is projected that one of the fastest-growing occupations in the United States between 2004 and 2014 will be home health aides. During this 10-year period, the number of home health aides can be approximated by the equation $y = 5(7x + 125)$, where x is the number of years since 2004 and y is the number of employees in thousands. (*Source:* www.bls.gov.)

(a) Write the equation in slope–intercept form.
$y = 35x + 625$

(b) Find the slope and the y-intercept.
$m = 35$; $(0, 625)$

(c) In this specific equation, what is the meaning of the slope? What does it indicate?
The slope is the amount of increase in the number of home health aides in the United States per year.

54. *Physician Assistants* It is projected that one of the fastest-growing occupations in the United States between 2004 and 2014 will be physician assistants. During this 10-year period, the number of physician assistants can be approximated by the equation $y = \dfrac{1}{10}(31x + 620)$, where x is the number of years since 2004 and y is the number of employees in thousands. (*Source:* www.bls.gov.)

(a) Write the equation in slope–intercept form.
$y = \dfrac{31}{10}x + 62$

(b) Find the slope and the y-intercept.
$m = \dfrac{31}{10}$; y-intercept is $(0, 62)$

(c) In this specific equation, what is the meaning of the slope? What does it indicate?
The slope is the amount of increase in the number of physician assistants in the United States per year.

Cumulative Review *Solve for x and graph the solution.*

55. [2.7.1] $3x + 8 > 2x + 12$ $x > 4$

56. [2.7.1] $\dfrac{1}{4}x + 3 > \dfrac{2}{3}x + 2$
$3x + 36 > 8x + 24$
$12 > 5x$
$x < \dfrac{12}{5}$

57. [2.7.1] $\dfrac{1}{2}(x + 2) \le \dfrac{1}{3}x + 5$
$3x + 6 \le 2x + 30$
$x \le 24$

58. [2.7.1] $7x - 2(x + 3) \le 4(x - 7)$
$7x - 2x - 6 \le 4x - 28$
$x \le -22$

Quick Quiz 7.3

1. Find the slope of the straight line that passes through the points $(-2, 5)$ and $(-6, 3)$.
$\dfrac{1}{2}$

2. Find the slope and the y-intercept of the line $6x - 2y - 3 = 0$.
$m = 3$; $\left(0, -\dfrac{3}{2}\right)$

3. Write the equation of the line in slope–intercept form that passes through $(0, -5)$ and has a slope of $-\dfrac{5}{7}$.
$y = -\dfrac{5}{7}x - 5$

4. **Concept Check** The equation of a given line is $y = \dfrac{7}{11}x - \dfrac{3}{4}$. Explain how you would find the slope of a line that is perpendicular to that line.
Answers may vary

Classroom Quiz 7.3 You may use these problems to quiz your students' mastery of Section 7.3.

1. Find the slope of the straight line that passes through the points $(5, -7)$ and $(-2, -6)$. **Ans:** $-\dfrac{1}{7}$

2. Find the slope and the y-intercept of the line $3x + 8y - 5 = 0$. **Ans:** $m = -\dfrac{3}{8}$; $\left(0, \dfrac{5}{8}\right)$

3. Write the equation of the line in slope–intercept form that passes through $(0, -7)$ and has a slope of $\frac{3}{8}$. **Ans:** $y = \dfrac{3}{8}x - 7$

1. _____ See graph grid

$A\ (3, 9)$
$B\ (7, -7)$
$C\ (-7, -8)$
2. _____ $D\ (-4, 3)$

3. _____ See ordered pairs

4. _____ See table

5. _____ See graph

6. _____ See graph

How are you doing with your homework assignments in Sections 7.1 to 7.3? Do you feel you have mastered the material so far? Do you understand the concepts you have covered? Before you go further in the textbook, take some time to do each of the following problems.

7.1

1. Plot the following points.

$E:\ (-6, 3)$
$F:\ (5, -4)$
$G:\ (-2, -8)$
$H:\ (6.5, 3.5)$

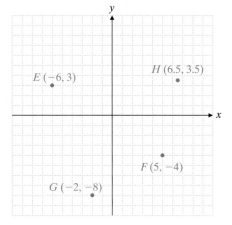

2. Give the coordinates for points A, B, C, D.

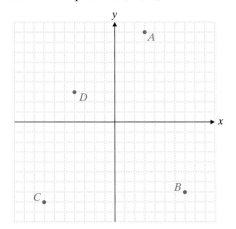

3. Complete the ordered pairs for the equation $y = -7x + 3$.

$(4, -25)$
$(0, 3)$
$(-2, 17)$

4. Complete the table for the equation $6x - 5y = -30$.

x	y
0	6
−5	0
−10	−6

7.2

5. Graph $y = -4x - 3$.

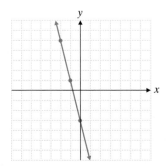

6. Graph $y = \dfrac{3}{4}x - 1$.

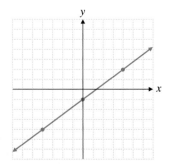

7. Graph $3x - 5y = 0$.

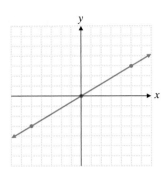

8. Graph $-5x + 2 = y + 3$.

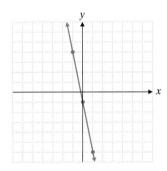

7.3

9. Find the slope of the straight line that passes through $(-2, 5)$ and $(0, 1)$.

10. Find the slope and y-intercept of $-12x + 6y + 8 = 0$.

11. A line has a slope of $\dfrac{3}{4}$. What is the slope of a line perpendicular to it?

12. An equation of a line is $3y - 5x + 10 = 0$. What is the slope of a line parallel to it?

Now turn to page SA-16 for the answer to each of these problems. Each answer also includes a reference to the objective in which the problem is first taught. If you missed any of these problems, you should stop and review the Examples and Practice Problems in the referenced objective. A little review now will help you master the material in the upcoming sections of the text.

7.	See graph
8.	See graph
9.	-2
10.	Slope $= 2$ y-intercept $= \left(0, -\dfrac{4}{3}\right)$
11.	$-\dfrac{4}{3}$
12.	$\dfrac{5}{3}$

1 Writing an Equation of a Line Given a Point and the Slope

If we know the slope of a line and the y-intercept, we can write the equation of the line in slope–intercept form. Sometimes we are given the slope and a point on the line. We use the information to find the y-intercept. Then we can write the equation of the line.

It may be helpful to summarize our approach.

> **TO FIND AN EQUATION OF A LINE GIVEN A POINT AND THE SLOPE**
>
> 1. Substitute the given values of x, y, and m into the equation $y = mx + b$.
>
> 2. Solve for b.
>
> 3. Use the values of b and m to write the equation in the form $y = mx + b$.

Teaching Example 1 Find an equation of the line that passes through $(2, -3)$ with slope $-\frac{1}{2}$.

Ans: $y = -\dfrac{1}{2}x - 2$

EXAMPLE 1 Find an equation of the line that passes through $(-3, 6)$ with slope $-\frac{2}{3}$.

Solution We are given the values $m = -\frac{2}{3}$, $x = -3$, and $y = 6$.

$$y = mx + b$$
$$6 = \left(-\frac{2}{3}\right)(-3) + b \quad \text{Substitute known values.}$$
$$6 = 2 + b$$
$$4 = b$$

An equation of the line is $y = -\frac{2}{3}x + 4$.

Practice Problem 1 Find an equation of the line that passes through $(-8, 12)$ with slope $-\frac{3}{4}$. $\quad y = -\dfrac{3}{4}x + 6$

2 Writing an Equation of a Line Given Two Points

Our procedure can be extended to the case for which two points are given.

Teaching Example 2 Find an equation of the line that passes through $(3, 2)$ and $(6, 1)$.

Ans: $y = -\dfrac{1}{3}x + 3$

EXAMPLE 2 Find an equation of the line that passes through $(2, 5)$ and $(6, 3)$.

Solution We first find the slope of the line. Then we proceed as in Example 1.

$$m = \frac{y_2 - y_1}{x_2 - x_1}$$
$$m = \frac{3 - 5}{6 - 2} \quad \begin{array}{l}\text{Substitute } (x_1, y_1) = (2, 5) \text{ and} \\ (x_2, y_2) = (6, 3) \text{ into the formula.}\end{array}$$
$$= \frac{-2}{4} = -\frac{1}{2}$$

Choose either point, say $(2, 5)$, to substitute into $y = mx + b$ as in Example 1.

$$5 = -\frac{1}{2}(2) + b$$

$$5 = -1 + b$$

$$6 = b$$

An equation of the line is $y = -\frac{1}{2}x + 6$.

Note: We could have substituted the slope and the other point, $(6, 3)$, into the slope–intercept form and arrived at the same answer. Try it.

Practice Problem 2 Find an equation of the line that passes through $(3, 5)$ and $(-1, 1)$. $y = x + 2$

NOTE TO STUDENT: *Fully worked-out solutions to all of the Practice Problems can be found at the back of the text starting at page SP-1*

Teaching Example 3 What is the equation of the line in the figure below?

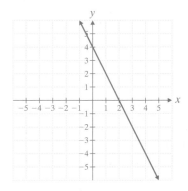

Ans: $y = -2x + 4$

Teaching Tip You may want to point out that we chose $(5, -2)$ because it is easy to determine the coordinates of this point since it lies on an intersection of the grid lines.

③ Writing an Equation of a Line Given a Graph of the Line

EXAMPLE 3 What is the equation of the line in the figure at right?

Solution First, look for the *y*-intercept. The line crosses the *y*-axis at $(0, 4)$. Thus $b = 4$.
Second, find the slope.

$$m = \frac{\text{change in } y}{\text{change in } x}$$

Look for another point on the line. We choose $(5, -2)$. Count the number of vertical units from 4 to -2 (rise). Count the number of horizontal units from 0 to 5 (run).

$$m = \frac{-6}{5}$$

Now, using $m = -\frac{6}{5}$ and $b = 4$, we can write an equation of the line.

$$y = mx + b$$

$$y = -\frac{6}{5}x + 4$$

PRACTICE PROBLEM 3

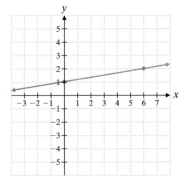

Practice Problem 3 What is the equation of the line in the figure at the right?

$$y = \frac{1}{6}x + 1$$

Find an equation of the line that has the given slope and passes through the given point.

1. $m = 4, (-3, 0)$
$y = 4x + 12$

2. $m = 3, (2, -2)$
$y = 3x - 8$

3. $m = -2, (3, 5)$
$y = -2x + 11$

4. $m = -1, (5, 0)$
$y = -x + 5$

5. $m = -3, \left(\frac{1}{2}, 2\right)$
$y = -3x + \frac{7}{2}$

6. $m = -2, \left(3, \frac{1}{3}\right)$
$y = -2x + \frac{19}{3}$

7. $m = \frac{1}{4}, (4, 5)$
$y = \frac{1}{4}x + 4$

8. $m = \frac{2}{3}, (3, -2)$
$y = \frac{2}{3}x - 4$

Write an equation of the line passing through the given points.

9. $(3, -12)$ and $(-4, 2)$
$y = -2x - 6$

10. $(-1, 8)$ and $(0, 5)$
$y = -3x + 5$

11. $(2, -6)$ and $(-1, 6)$
$y = -4x + 2$

12. $(-1, 1)$ and $(5, 7)$
$y = x + 2$

13. $(3, 5)$ and $(-1, -15)$
$y = 5x - 10$

14. $(-1, -19)$ and $(2, 2)$
$y = 7x - 12$

15. $\left(1, \frac{5}{6}\right)$ and $\left(3, \frac{3}{2}\right)$
$y = \frac{1}{3}x + \frac{1}{2}$

16. $(2, 0)$ and $\left(\frac{3}{2}, \frac{1}{2}\right)$
$y = -x + 2$

Mixed Practice

17. Find an equation of the line with a slope of -3 that passes through the point $(-1, 3)$.
$y = -3x$

18. Find an equation of the line with a slope of $-\frac{1}{2}$ that passes through the point $(6, -2)$.
$y = -\frac{1}{2}x + 1$

19. Find an equation of the line that passes through $(2, -3)$ and $(-1, 6)$. $y = -3x + 3$

20. Find an equation of the line that passes through $(1, -8)$ and $(2, -14)$. $y = -6x - 2$

Write an equation of each line.

21.

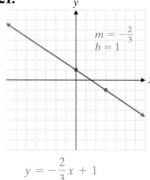

$m = -\frac{2}{3}$
$b = 1$

$y = -\frac{2}{3}x + 1$

22.

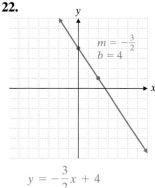

$m = -\frac{3}{2}$
$b = 4$

$y = -\frac{3}{2}x + 4$

23.

$m = \frac{2}{3}$
$b = -4$

$y = \frac{2}{3}x - 4$

24.

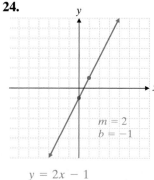

$m = 2$
$b = -1$

$y = 2x - 1$

25.

$y = -\frac{2}{3}x$

26.

$y = \frac{3}{2}x$

27.

$y = -2$

28.

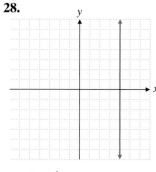

$x = 4$

To Think About *Find an equation of the line that fits each description.*

29. Passes through $(7, -2)$ and has zero slope $y = -2$

30. Passes through $(4, 6)$ and has undefined slope $x = 4$

31. Passes through $(4, -6)$ and is perpendicular to the x-axis $x = 4$

32. Passes through $(-3, 5)$ and is perpendicular to the y-axis $y = 5$

33. Passes through $(0, 5)$ and is parallel to $y = \frac{1}{3}x + 4$ $y = \frac{1}{3}x + 5$

34. Passes through $(0, 5)$ and is perpendicular to $y = \frac{1}{3}x + 4$ $y = -3x + 5$

35. Passes through $(2, 3)$ and is perpendicular to $y = 2x - 9$ $y = -\frac{1}{2}x + 4$

36. Passes through $(2, 9)$ and is parallel to $y = 5x - 3$ $y = 5x - 1$

37. *Population Growth* The growth of the population of the United States during the period from 1980 to 2008 can be approximated by an equation of the form $y = mx + b$, where x is the number of years since 1980 and y is the population measured in millions. (*Source:* U.S. Census Bureau.) Find the equation if two ordered pairs that satisfy it are (0, 227) and (10, 251). $y = 2.4x + 227$

38. *Home Equity Loans* The amount of debt outstanding on home equity loans in the United States during the period from 1993 to 2008 can be approximated by an equation of the form $y = mx + b$, where x is the number of years since 1993 and y is the debt measured in billions of dollars. (*Source:* Board of Governors of the Federal Reserve System.) Find the equation if two ordered pairs that satisfy it are (1, 280) and (6, 500). $y = 44x + 236$

Cumulative Review

39. **[6.5.1]** $\dfrac{3}{t} - \dfrac{2}{t-1} = \dfrac{4}{t}$ $3t - 3 - 2t = 4t - 4$ $t = \dfrac{1}{3}$

40. **[6.2.1]** Simplify. $\dfrac{4x^2 - 25}{4x^2 - 20x + 25} \cdot \dfrac{x^2 - 9}{2x^2 + x - 15}$ $\dfrac{(2x + 5)(x - 3)}{(2x - 5)^2}$

41. **[0.7.1]** *Basketball Sneakers* A pair of basketball sneakers sells for $80. The next week the sneakers go on sale for 15% off. The third week there is a coupon in the newspaper offering a 10% discount off the second week's price. How much would you have paid for the sneakers during the third week if you had used the coupon? $61.20

42. **[3.4.1]** *Cell Phone Costs* Dave and Jane Wells have a cell phone. The plan they subscribe to costs $50 per month. It includes 200 free minutes of calling time. Each minute after the 200 is charged at the rate of $0.21 per minute. Last month their cell phone bill was $68.90. How many total minutes did they use their cell phone during the month? 290 min

Quick Quiz 7.4

1. Write the equation in slope–intercept form of the line that passes through the point $(3, -5)$ and has a slope of $\frac{2}{3}$. $y = \frac{2}{3}x - 7$

2. Write the equation in slope–intercept form of the line that passes through the points $(-2, 7)$ and $(-4, -5)$. $y = 6x + 19$

3. Write the equation of the line that passes through $(4, 5)$ and $(4, -2)$. What is the slope of this line? $x = 4$; the slope is undefined

4. **Concept Check** How would you find the equation of the line that passes through $(-2, -3)$ and has zero slope? Answers may vary

Classroom Quiz 7.4 *You may use these problems to quiz your students' mastery of Section 7.4.*

1. Write the equation in slope–intercept form of the line that passes through the point $(-4, -5)$ and has a slope of $\frac{3}{4}$.

Ans: $y = \frac{3}{4}x - 2$

2. Write the equation in slope–intercept form of the line that passes through the points $(-3, 6)$ and $(-5, -2)$.

Ans: $y = 4x + 18$

3. Write the equation of the line that passes through $(3, -7)$ and $(3, 9)$. What is the slope of this line?

Ans: $x = 3$; the slope is undefined.

After studying this section, you will be able to:

 Graph linear inequalities in two variables.

In Section 2.7 we discussed inequalities in one variable. Look at the inequality $x < -2$ (x is less than -2). Some of the solutions to the inequality are -3, -5, and $-5\frac{1}{2}$. In fact all numbers to the left of -2 on a number line are solutions. The graph of the inequality is given in the following figure. Notice that the open circle at -2 indicates that -2 is *not* a solution.

$$x < -2$$

Now we will extend our discussion to consider linear inequalities in two variables.

Teaching Example 1 Graph $x - 2y \geq 2$.

Ans:

 Graphing Linear Inequalities in Two Variables

Consider the inequality $y \geq x$. The solution of the inequality is the set of all possible ordered pairs that when substituted into the inequality will yield a true statement. Which ordered pairs will make the statement $y \geq x$ true? Let's try some.

$(0, 6)$	$(-2, 1)$	$(1, -2)$	$(3, 5)$	$(4, 4)$
$6 \geq 0$, true	$1 \geq -2$, true	$-2 \geq 1$, false	$5 \geq 3$, true	$4 \geq 4$, true

$(0, 6)$, $(-2, 1)$, $(3, 5)$, and $(4, 4)$ are solutions to the inequality $y \geq x$. In fact, every point at which the y-coordinate is greater than or equal to the x-coordinate is a solution to the inequality. This is shown by the solid line and the shaded region in the graph at the left.

Is there an easier way to graph a linear inequality in two variables? It turns out that we can graph such an inequality by first graphing the associated linear equation and then testing one point that is not on that line. That is, we can change the inequality symbol to an equals sign and graph the equation. If the inequality symbol is $\geq$ or $\leq$, we use a solid line to indicate that the points on the line are included in the solution of the inequality. If the inequality symbol is $>$ or $<$, we use a dashed line to indicate that the points on the line are not included in the solution of the inequality. Then we test one point that is not on the line. If the point is a solution to the inequality, we shade the region on the side of the line that includes the point. If the point is not a solution, we shade the region on the other side of the line.

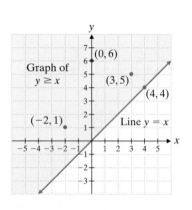

Graph of $y \geq x$

PRACTICE PROBLEM 1

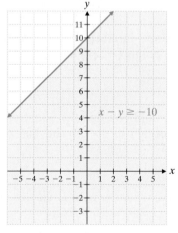

NOTE TO STUDENT: *Fully worked-out solutions to all of the Practice Problems can be found at the back of the text starting at page SP-1*

EXAMPLE 1 Graph $5x + 3y > 15$. Use the given coordinate system.

Solution We begin by graphing the line $5x + 3y = 15$. You may use any method discussed previously to graph the line. Since there is no equals sign in the inequality, we will draw a dashed line to indicate that the line is *not* part of the solution set.

Look for a test point. The easiest point to test is $(0, 0)$. Substitute $(0, 0)$ for (x, y) in the inequality.

$$5x + 3y > 15$$
$$5(0) + 3(0) > 15$$
$$0 > 15 \quad \text{false}$$

$(0, 0)$ is not a solution. Shade the region on the side of the line that does *not* include $(0, 0)$.

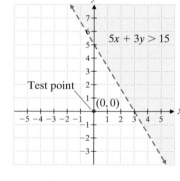

Practice Problem 1 Graph $x - y \geq -10$. Use the coordinate system in the margin.

GRAPHING LINEAR INEQUALITIES

1. Replace the inequality symbol by an equality symbol. Graph the line.
   (a) The line will be solid if the inequality is $\geq$ or $\leq$.
   (b) The line will be dashed if the inequality is $>$ or $<$.

2. Test the point $(0, 0)$ in the inequality if $(0, 0)$ does not lie on the graphed line in step 1.
   (a) If the inequality is true, shade the side of the line that includes $(0, 0)$.
   (b) If the inequality is false, shade the side of the line that does not include $(0, 0)$.

3. If the point $(0, 0)$ is a point on the line, choose another test point and proceed accordingly.

Teaching Example 2 Graph $y > 2x$.

Ans:

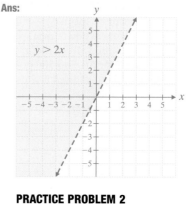

EXAMPLE 2 Graph $2y \leq -3x$.

Solution

Step 1 Graph $2y = -3x$. Since $\leq$ is used, the line should be a solid line.

Step 2 We see that the line passes through $(0, 0)$.

Step 3 Choose another test point. We will choose $(-3, -3)$.

$$2y \leq -3x$$
$$2(-3) \leq -3(-3)$$
$$-6 \leq 9 \quad \text{true}$$

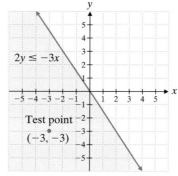

Shade the region that includes $(-3, -3)$, that is, the region below the line.

Practice Problem 2 Graph $2y > x$ on the coordinate system in the margin.

PRACTICE PROBLEM 2

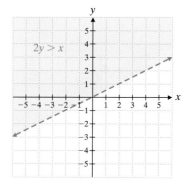

Teaching Example 3 Graph $y \leq 4$.

Ans:

If we are graphing the inequality $x < -2$ on the coordinate plane, the solution will be a region. Notice that this is very different from the solution of $x < -2$ on a number line discussed earlier.

EXAMPLE 3 Graph $x < -2$.

Solution

Step 1 Graph $x = -2$. Since $<$ is used, the line should be dashed.

Step 2 Test $(0, 0)$ in the inequality.

$$x < -2$$
$$0 < -2 \quad \text{false}$$

Shade the region that does not include $(0, 0)$, that is, the region to the left of the line $x = -2$. Observe that every point in the shaded region has an x-value that is less than -2.

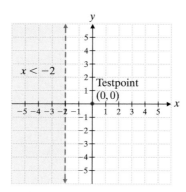

Practice Problem 3 Graph $y \geq -3$ on the coordinate system in the margin.

PRACTICE PROBLEM 3

Verbal and Writing Skills

1. Does it matter what point you use as your test point? Justify your response.

No, all points in one region will be solutions to the inequality, while all points in the other region will not be solutions. Thus testing any point will give the same result, as long as the test point is not on the boundary line.

2. Explain when to use a solid line or a dashed line when graphing linear inequalities in two variables.

Use a solid line for inequalities containing ≤ or ≥. Use a dashed line for inequalities containing < or >.

Graph the region described by the inequality.

3. $y < 2x - 4$

4. $y > 1 - 3x$

5. $2x - 3y < 6$

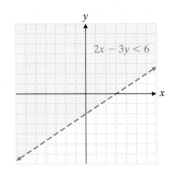

6. $3x + 2y < -6$

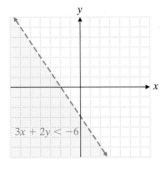

7. $2x - y \geq 3$

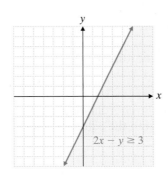

8. $3x - y \geq 4$

9. $y \geq 4x$

10. $y \leq -2x$

11. $y < -\dfrac{1}{2}x$

12. $y > \dfrac{1}{5}x$

13. $x \geq 2$

14. $y \leq -2$

15. $2x - 3y + 6 \geq 0$

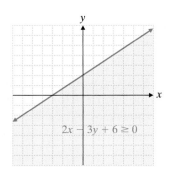

16. $3x + 4y - 8 \leq 0$

17. $2x > -3y$

18. $3x \leq -2y$

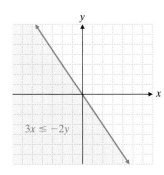

19. $2x > 3 - y$

20. $x > 4 - y$

21. $x > -2y$

22. $x < -3y$

Cumulative Review

23. **[5.7.1]** Solve for x: $3x^2 + 2x - 8 = 0$

$x = \dfrac{4}{3}, x = -2$

24. **[7.4.2]** What is the slope–intercept form of the equation for the line through $(5, -2)$ and $(10, 0)$?

$y = \dfrac{2}{5}x - 4$

Quick Quiz 7.5

1. When you graph the inequality $y > -3x + 1$, should you use a solid line for a boundary or a dashed line for a boundary? Why?

Use a dashed line. If the inequality has a $>$ or a $<$ symbol, the points on the line itself are not included. This is indicated by a dashed line.

2. Graph the region described by $3y \le -7x$.

3. Graph the region described by $-5x + 2y > -3$.

4. **Concept Check** Explain how you would determine if you should shade above the line or below the line if you were to graph the inequality $y > -3x + 4$ using $(0, 0)$ as a test point.

Answers may vary

Classroom Quiz 7.5 You may use these problems to quiz your students' mastery of Section 7.5.

1. When you graph the inequality $y \le -6x - 2$, should you use a solid line for a boundary or a dashed line for a boundary? Why?

Ans: Use a solid line. If the inequality has a $\le$ or a $\ge$ symbol, the points on the line itself are included. This is indicated by a solid line.

3. Graph the region described by $3x + 5y \le -2$.

Ans:

2. Graph the region described by $5y > 2x$.

Ans:

7.6 FUNCTIONS

1 Understanding the Meanings of a Relation and a Function

Thus far you have studied linear equations in two variables. You have seen that such an equation can be represented by a table of values, by the algebraic equation itself, and by a graph.

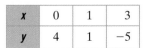

x	0	1	3
y	4	1	−5

$$y = -3x + 4$$

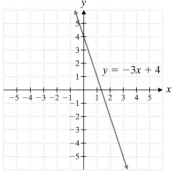

Student Learning Objectives

After studying this section, you will be able to:

1 Understand the meanings of a relation and a function.

2 Graph simple nonlinear equations.

3 Determine whether a graph represents a function.

4 Use function notation.

The solutions to the linear equation are all the ordered pairs that satisfy the equation (make the equation true). They are all the points that lie on the graph of the line. These ordered pairs can be represented in a table of values. Notice the relationship between the ordered pairs. We can choose any value for x. But once we have chosen a value for x, the value of y is determined. For the preceding equation, if x is 0, then y must be 4. We say that x is the **independent variable** and that y is the **dependent variable.**

Mathematicians call such a pairing of two values a *relation*.

DEFINITION OF A RELATION

A **relation** is any set of ordered pairs.

All the first coordinates in all of the ordered pairs of the relation make up the **domain** of the relation. All the second coordinates in all of the ordered pairs make up the **range** of the relation. Notice that the definition of a relation is very broad. Some relations cannot be described by an equation. These relations may simply be a set of discrete ordered pairs.

EXAMPLE 1 State the domain and range of the relation.
$$\{(5, 7), (9, 11), (10, 7), (12, 14)\}$$

Solution The domain consists of all the first coordinates in the ordered pairs.

Domain

$$\{(5, 7), (9, 11), (10, 7), (12, 14)\}$$

Range

The range consists of all the second coordinates in the ordered pairs. We usually list the values of a domain or range from smallest to largest.

The domain is $\{5, 9, 10, 12\}$.
The range is $\{7, 11, 14\}$.   We list 7 only once.

Practice Problem 1 State the domain and range of the relation.
$$\{(-3, -5), (3, 5), (0, -5), (20, 5)\}$$

The domain is $\{-3, 0, 3, 20\}$. The range is $\{-5, 5\}$.

Teaching Example 1 State the domain and range of the relation.
$$\{(1, 7), (6, 3), (-1, 7), (0, 5)\}$$

Ans: The domain is $\{-1, 0, 1, 6\}$.
     The range is $\{3, 5, 7\}$.

NOTE TO STUDENT: Fully worked-out solutions to all of the Practice Problems can be found at the back of the text starting at page SP-1

Some relations have the special property that no two different ordered pairs have the same first coordinate. Such relations are called **functions.** The relation $y = -3x + 4$ is a function. If we substitute a value for x, we get just one value for y. Thus no two ordered pairs will have the same x-coordinate and different y-coordinates.

DEFINITION OF A FUNCTION

A **function** is a relation in which no two different ordered pairs have the same first coordinate.

Teaching Example 2 Determine whether the relation is a function.

(a) $\{(1, 2), (5, 9), (8, 6), (5, 1)\}$

(b) $\{(-3, 7), (0, 4), (2, 2), (5, 7)\}$

Ans: (a) Not a function

    (b) Function

EXAMPLE 2   Determine whether the relation is a function.

(a) $\{(3, 9), (4, 16), (5, 9), (6, 36)\}$

(b) $\{(7, 8), (9, 10), (12, 13), (7, 14)\}$

Solution

(a) Look at the ordered pairs. No two ordered pairs have the same first coordinate. Thus this set of ordered pairs defines a function. Note that the ordered pairs $(3, 9)$ and $(5, 9)$ have the same second coordinate, but the relation is still a function. It is the first coordinates that cannot be the same.

(b) Look at the ordered pairs. Two different ordered pairs, $(7, 8)$ and $(7, 14)$, have the same first coordinate. Thus this relation is *not* a function.

Practice Problem 2   Determine whether the relation is a function.

(a) $\{(-5, -6), (9, 30), (-3, -3), (8, 30)\}$ Function

(b) $\{(60, 30), (40, 20), (20, 10), (60, 120)\}$ Not a function

A functional relationship is often what we find when we analyze two sets of data. Look at the following table of values, which compares Celsius temperature with Fahrenheit temperature. Is there a relationship between degrees Fahrenheit and degrees Celsius? Is the relation a function?

Temperature

°F	23	32	41	50
°C	−5	0	5	10

Since every Fahrenheit temperature produces a unique Celsius temperature, we would expect this to be a function. We can verify our assumption by looking at the formula $C = \frac{5}{9}(F - 32)$ and its graph. The formula is a linear equation, and its graph is a line with slope $\frac{5}{9}$ and y-intercept at about -17.8. The relation is a function. In the equation given here, notice that the *dependent variable* is C, since the value of C *depends* on the value of F. We say that F is the *independent variable*. The *domain* can be described as the set of possible values of the independent variable. The *range* is the set of corresponding values of the dependent variable. Scientists believe that the coldest temperature possible is approximately $-273°C$. They call this temperature **absolute zero.** Thus,

Domain = {all possible Fahrenheit temperatures from absolute zero to infinity}

Range = {all corresponding Celsius temperatures from $-273°C$ to infinity}.

| EXAMPLE 3 | Each of the following tables contains some data pertaining to a relation. Determine whether the relation suggested by the table is a function. If it is a function, identify the domain and range.

(a) Circle

Radius	1	2	3	4	5
Area	3.14	12.56	28.26	50.24	78.5

(b) $4000 Loan at 8% for a Minimum of One Year

Time (yr)	1	2	3	4	5
Interest	$320	$665.60	$1038.85	$1441.96	$1877.31

Solution

(a) Looking at the table, we see that no two different ordered pairs have the same first coordinate. The area of a circle is a function of the length of the radius.

Next we need to identify the independent variable to determine the domain. Sometimes it is easier to identify the dependent variable. Here we notice that the area of the circle depends on the length of the radius. Thus radius is the independent variable. Since a negative length does not make sense, the radius cannot be a negative number. Although only integer radius values are listed in the table, the radius of a circle can be any nonnegative real number.

$$\text{Domain} = \{\text{all nonnegative real numbers}\}$$
$$\text{Range} = \{\text{all nonnegative real numbers}\}$$

(b) No two different ordered pairs have the same first coordinate. Interest is a function of time.

Since the amount of interest paid on a loan depends on the number of years (term of the loan), interest is the dependent variable and time is the independent variable. Negative numbers do not apply in this situation. Although the table includes only integer values for the time, the length of a loan in years can be any real number that is greater than or equal to 1.

$$\text{Domain} = \{\text{all real numbers greater than or equal to 1}\}$$
$$\text{Range} = \{\text{all positive real numbers greater than or equal to } \$320\}$$

| Practice Problem 3 | Determine whether the relation suggested by the table is a function. If it is a function, identify the domain and the range.

(a) 28 Mpg at $4.16 per Gallon

Distance	0	28	42	56	70
Cost	$0	$4.16	$6.24	$8.32	$10.40

It is a function.
Domain = {all nonnegative real numbers};
Range = {all nonnegative real numbers}

(b) Store's Inventory of Shirts

Number of Shirts	5	10	5	2	8
Price of Shirt	$20	$25	$30	$45	$50

Not a function.

TO THINK ABOUT: Is It a Function? Look at the following bus schedule. Determine whether the relation is a function. Which is the independent variable? Explain your choice.

Bus Schedule

Bus Stop	Main St.	8th Ave.	42nd St.	Sunset Blvd.	Cedar Lane
Time	7:00	7:10	7:15	7:30	7:39

Teaching Example 3 Determine whether the relation suggested by the table is a function. If it is a function, identify the domain and the range.

(a)
Earnings at Pay Rate of $10 per Hour

Hours Worked	1	2	3	4
Earnings	$10	$20	$30	$40

(b)
Airline Ticket Prices

Number of Air Miles	185	250	385	250
Ticket Price	$240	$270	$405	$325

Ans:
(a) It is a function.
Domain = {all nonnegative real numbers};
Range = {all nonnegative real numbers}

(b) Not a function

NOTE TO STUDENT: Fully worked-out solutions to all of the Practice Problems can be found at the back of the text starting at page SP-1

Graphing Nonlinear Equations

You can graph nonlinear equations solved for y using a graphing calculator. For example, graph $y = x^2 - 2$ on a graphing calculator using an appropriate window. Display:

PRACTICE PROBLEM 4

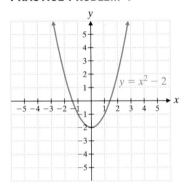

Teaching Example 4 Graph $y = x^2 + 3$.

Ans:

PRACTICE PROBLEM 5

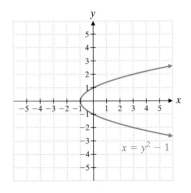

2 Graphing Simple Nonlinear Equations

Thus far in this chapter we have graphed linear equations in two variables. We now turn to graphing a few nonlinear equations. We will need to plot more than three points to get a good idea of what the graph of a nonlinear equation will look like.

EXAMPLE 4 Graph $y = x^2$.

Solution Begin by constructing a table of values. We select values for x and then determine by the equation the corresponding values of y. We will include negative values for x as well as positive values. We then plot the ordered pairs and connect the points with a smooth curve.

x	$y = x^2$	y
-2	$y = (-2)^2 = 4$	4
-1	$y = (-1)^2 = 1$	1
0	$y = (0)^2 = 0$	0
1	$y = (1)^2 = 1$	1
2	$y = (2)^2 = 4$	4

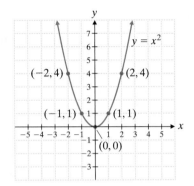

This type of curve is called a *parabola*. We will study this type of curve more extensively in Chapter 10.

Practice Problem 4 Graph $y = x^2 - 2$ on the coordinate system in the margin.

Some equations are solved for x. Usually, in those cases we pick values of y and then obtain the corresponding values of x from the equation.

EXAMPLE 5 Graph $x = y^2 + 2$.

Solution Since the equation is solved for x, we start by picking a value of y. We will find the value of x from the equation in each case. We will select $y = -2$ first. Then we substitute it into the equation to obtain x. For convenience in graphing, we will repeat the y column at the end so that it is easy to write the ordered pairs (x, y).

y	$x = y^2 + 2$	x	y
-2	$x = (-2)^2 + 2 = 4 + 2 = 6$	6	-2
-1	$x = (-1)^2 + 2 = 1 + 2 = 3$	3	-1
0	$x = (0)^2 + 2 = 0 + 2 = 2$	2	0
1	$x = (1)^2 + 2 = 1 + 2 = 3$	3	1
2	$x = (2)^2 + 2 = 4 + 2 = 6$	6	2

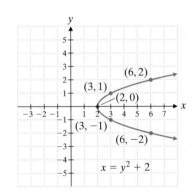

Practice Problem 5 Graph $x = y^2 - 1$ on the coordinate system in the margin.

If the equation involves fractions with variables in the denominator, we must use extra caution. Remember that you may never divide by zero.

EXAMPLE 6   Graph $y = \dfrac{4}{x}$.

Solution   It is important to note that x cannot be zero because division by zero is not defined. $y = \frac{4}{0}$ is not allowed! Observe that when we draw the graph we get two separate branches that do not touch.

x	$y = \dfrac{4}{x}$	y
-4	$y = \dfrac{4}{-4} = -1$	-1
-2	$y = \dfrac{4}{-2} = -2$	-2
-1	$y = \dfrac{4}{-1} = -4$	-4
0	We cannot divide by zero.	There is no value.
1	$y = \dfrac{4}{1} = 4$	4
2	$y = \dfrac{4}{2} = 2$	2
4	$y = \dfrac{4}{4} = 1$	1

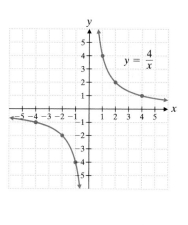

Practice Problem 6   Graph $y = \dfrac{6}{x}$ on the coordinate system in the margin.

③ Determining Whether a Graph Represents a Function

Can we tell whether a graph represents a function? Recall that a function cannot have two different ordered pairs with the same first coordinate. That is, each value of x must have a separate, unique value of y. Look at the graph of the function $y = x^2$ in Example 4. Each x-value has a unique y-value. Look at the graph of $x = y^2 + 2$ in Example 5. At $x = 3$ there are two y-values, 1 and -1. In fact, for every x-value greater than 2, there are two y-values. $x = y^2 + 2$ is not a function.

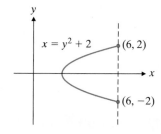

Observe that we can draw a vertical line through $(6, 2)$ and $(6, -2)$. Any graph that is not a function will have at least one region in which a vertical line will cross the graph more than once.

VERTICAL LINE TEST

If a vertical line can intersect the graph of a relation more than once, the relation is not a function. If no such line can be drawn, then the relation is a function.

Teaching Example 5   Graph $x = y^2 - 2$.

Ans:

PRACTICE PROBLEM 6

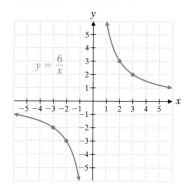

Teaching Example 6   Graph $y = -\dfrac{2}{x}$.

Ans:

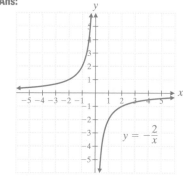

Teaching Example 7 Determine whether each of the following is the graph of a function.

(a)

(b)

(c)

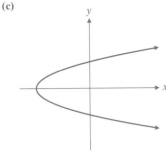

Ans:

(a) Function

(b) Function

(c) Not a function

EXAMPLE 7 Determine whether each of the following is the graph of a function.

(a)   (b)   (c)

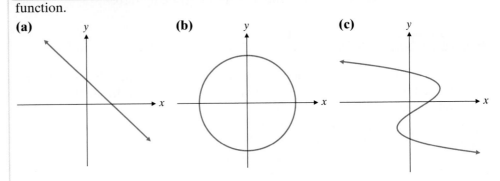

Solution

(a) The graph of the straight line is a function. Any vertical line will cross this straight line in only one location.

(b) and (c) Each of these graphs is not the graph of a function. In each case there exists a vertical line that will cross the curve in more than one place.

(a)   (b)   (c)

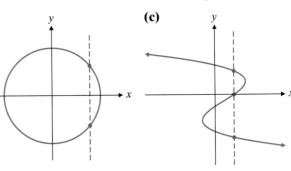

Practice Problem 7 Determine whether each of the following is the graph of a function.

(a)   (b)   (c)

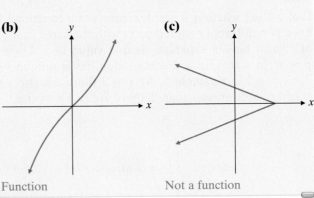

Not a function   Function   Not a function

4 Using Function Notation

We have seen that an equation like $y = 2x + 7$ is a function. For each value of x, the equation assigns a unique value to y. We could say "y is a function of x." If we name the function f, this statement can be symbolized by using the **function notation** $y = f(x)$. Many times we avoid using the y-variable completely and write the function as $f(x) = 2x + 7$.

CAUTION: Be careful. The notation $f(x)$ does not mean f multiplied by x.

EXAMPLE 8   If $f(x) = 3x^2 - 4x + 5$, find each of the following.

(a) $f(-2)$ **(b)** $f(4)$ **(c)** $f(0)$

Solution

(a) $f(-2) = 3(-2)^2 - 4(-2) + 5 = 3(4) - 4(-2) + 5 = 12 + 8 + 5 = 25$

(b) $f(4) = 3(4)^2 - 4(4) + 5 = 3(16) - 4(4) + 5 = 48 - 16 + 5 = 37$

(c) $f(0) = 3(0)^2 - 4(0) + 5 = 3(0) - 4(0) + 5 = 0 - 0 + 5 = 5$

Practice Problem 8   If $f(x) = -2x^2 + 3x - 8$, find each of the following.

(a) $f(2)$ -10 **(b)** $f(-3)$ -35 **(c)** $f(0)$ -8

When evaluating a function, it is helpful to place parentheses around the value that is being substituted for x. Taking the time to do this will minimize sign errors in your work.

Some functions are useful in medicine, anthropology, and forensic science. For example, the approximate length of a man's femur (thigh bone) is given by the function $f(x) = 0.53x - 17.03$, where x is the height of the man in inches. If a man is 68 inches tall, $f(68) = 0.53(68) - 17.03 = 19.01$. A man 68 inches tall would have a femur length of approximately 19.01 inches.

Source: www.nsbri.org

Verbal and Writing Skills

1. What are the three ways you can describe a function?
Using a table of values, an algebraic equation, or a graph

2. What is the difference between a function and a relation?
A relation is any set of ordered pairs. A function is a relation in which no two different ordered pairs have the same first coordinate.

3. The domain of a function is the set of ___possible___ ___values___ of the ___independent___ variable.

4. The range of a function is the set of ___corresponding___ ___values___ of the ___dependent___ variable.

5. How can you tell whether a graph is the graph of a function?
If a vertical line can intersect the graph more than once, the relation is not a function. If no such line exists, then the relation is a function.

6. Without drawing a graph, how could you tell if the equation $x = y^2$ is a function or not?
The ordered pairs $(4, 2)$ and $(4, -2)$ both satisfy the equation. They both have the same first coordinate but different second coordinates. Since we have two different pairs with the same first coordinate, the relation is *not* a function.

(a) Find the domain and range of the relation. (b) Determine whether the relation is a function.

7. $\left\{ \left(\frac{2}{5}, 4\right), \left(3, \frac{2}{5}\right), \left(-3, \frac{2}{5}\right), \left(\frac{2}{5}, -1\right) \right\}$

    (a) Domain $= \left\{ -3, \frac{2}{5}, 3 \right\}$;

        Range $= \left\{ -1, \frac{2}{5}, 4 \right\}$

    (b) Not a function

8. $\left\{ \left(\frac{3}{8}, -6\right), (-6, 5), \left(\frac{3}{8}, 2\right), (5, -6) \right\}$

    (a) Domain $= \left\{ -6, \frac{3}{8}, 5 \right\}$;

        Range $= \{-6, 2, 5\}$

    (b) Not a function

9. $\{(6, 2.5), (3, 1.5), (0, 0.5)\}$

    (a) Domain $= \{0, 3, 6\}$;

        Range $= \{0.5, 1.5, 2.5\}$

    (b) Function

10. $\{(7.3, 1), (0, 8), (2, 1)\}$

    (a) Domain $= \{0, 2, 7.3\}$;

        Range $= \{1, 8\}$

    (b) Function

11. $\{(12, 1), (14, 3), (1, 12), (9, 12)\}$

    (a) Domain $= \{1, 9, 12, 14\}$;

        Range $= \{1, 3, 12\}$

    (b) Function

12. $\{(5.6, 8), (5.8, 6), (6, 5.8), (5, 6)\}$

    (a) Domain $= \{5, 5.6, 5.8, 6\}$;

        Range $= \{5.8, 6, 8\}$

    (b) Function

13. $\{(3, 75), (5, 95), (3, 85), (7, 100)\}$

    (a) Domain $= \{3, 5, 7\}$;

        Range $= \{75, 85, 95, 100\}$

    (b) Not a function

14. $\{(85, 3), (95, 11), (110, 15), (110, 20)\}$

    (a) Domain $= \{85, 95, 110\}$;

        Range $= \{3, 11, 15, 20\}$

    (b) Not a function

Graph the equation.

15. $y = x^2 + 3$

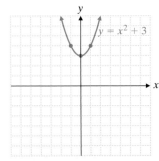

16. $y = x^2 - 1$

17. $y = 2x^2$

18. $y = \frac{1}{2}x^2$

19. $x = -2y^2$

20. $x = \frac{1}{2}y^2$

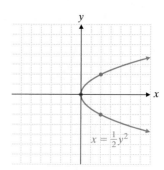

21. $x = y^2 - 4$

22. $x = 2y^2$

23. $y = \frac{2}{x}$

24. $y = -\frac{2}{x}$

25. $y = \frac{4}{x^2}$

26. $y = -\frac{6}{x^2}$

27. $x = (y + 1)^2$

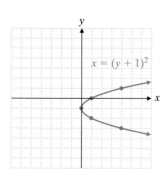

28. $y = (x - 3)^2$

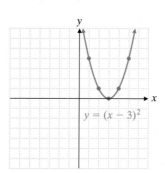

29. $y = \dfrac{4}{x - 2}$

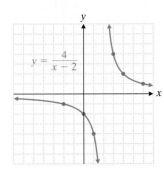

30. $x = \dfrac{2}{y + 1}$

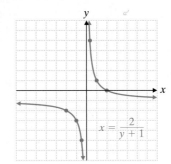

Determine whether each relation is a function.

31.

Function

32.

Function

33.

Not a function

34.

Not a function

35.

Function

36.

Not a function

37.

Not a function

38.

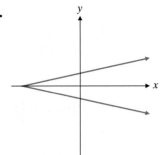

Not a function

Given the following functions, find the indicated values.

39. $f(x) = 2 - 3x$
   (a) $f(-8)$ 26   **(b)** $f(0)$ 2   **(c)** $f(2)$ −4

40. $f(x) = 7 - 4x$
   (a) $f(-2)$ 15   **(b)** $f(1)$ 3   **(c)** $f(3)$ −5

41. $f(x) = 2x^2 - x + 3$
   (a) $f(0)$ 3   **(b)** $f(-3)$ 24   **(c)** $f(2)$ 9

42. $f(x) = -x^2 - x + 5$
   (a) $f(0)$ 5   **(b)** $f(-1)$ 5   **(c)** $f(2)$ −1

Applications

43. *California Population* During a recent population growth period in California, from 1995 to 2005, the approximate population of the state measured in millions could be predicted by the function $f(x) = 0.02x^2 + 0.08x + 31.6$, where x is the number of years since 1995. (*Source:* www.census.gov.) Find $f(0), f(4),$ and $f(10)$. Graph the function. What pattern do you observe?

$f(0) = 31.6, f(4) = 32.24, f(10) = 34.4$
The curve slopes more steeply for larger values of x.
The growth rate is increasing as x gets larger.

44. *Kentucky Population* During a recent population growth period in Kentucky, from 1995 to 2005, the approximate population of the state measured in thousands could be predicted by the function $f(x) = -0.64x^2 + 30.2x + 3860$, where x is the number of years since 1995. (*Source:* www.census.gov.) Find $f(0), f(5),$ and $f(10)$. Use the function to predict the population in 2011 and 2015. Graph the function. What pattern do you observe?

$f(0) = 3860, f(5) = 3995, f(10) = 4098$
$f(16) \approx 4179, f(20) \approx 4208$
The curve slopes upward more steeply for smaller values of x. The growth rate is decreasing as x gets larger.

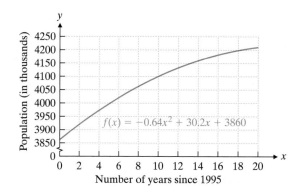

Cumulative Review *Simplify.*

45. **[4.3.2]** $(2x^2 - 2x + 1) + (-6x^2 - x + 7)$
$-4x^2 - 3x + 8$

46. **[4.3.3]** $(-3x^2 + 4x + 1) - (x^2 + 2x - 5)$
$-4x^2 + 2x + 6$

47. **[4.5.3]** $(4x^2 - 2x + 3)(x + 2)$
$4x^3 + 6x^2 - x + 6$

48. **[4.6.2]** $(6x^2 + 10x - 5) \div (3x - 1)$
$2x + 4 + \dfrac{-1}{3x - 1}$

Quick Quiz 7.6

1. Is this relation a function? $\{(5, 7), (7, 5), (5, 5)\}$ Why?
No. Two different ordered pairs have the same first coordinate.

2. For $f(x) = 3x^2 - 4x + 2$:
(a) Find $f(-3)$. 41   **(b)** Find $f(4)$. 34

3. For $g(x) = \dfrac{7}{x - 3}$:
(a) Find $g(2)$. -7 **(b)** Find $g(-5)$. $-\dfrac{7}{8}$

4. **Concept Check** In the relation $\{(3, 4), (5, 6), (3, 8), (2, 9)\}$, why is there a different number of elements in the domain than in the range? Answers may vary

Classroom Quiz 7.6 You may use these problems to quiz your students' mastery of Section 7.6.

1. Is this relation a function? $\{(-2, 8), (2, 3), (-3, -2)\}$ Why?
Ans: Yes. No two different ordered pairs have the same first coordinate.

2. For $f(x) = -2x^2 + 3x - 2$:
(a) Find $f(-2)$. **Ans:** -16 **(b)** Find $f(5)$. **Ans:** -37

3. For $g(x) = \dfrac{8}{2 - x}$:
(a) Find $g(3)$. **Ans:** -8 **(b)** Find $g(-4)$. **Ans:** $\dfrac{4}{3}$

Putting Your Skills to Work: Use Math to Save Money

RETIREMENT PLANNING

It is never too early to begin thinking about what you will be doing twenty, thirty, or more years from now. For example, how do you envision your retirement? One thing you will need when you retire is income. Social Security is projected to run out of funds, so we cannot be sure that we will have this cushion to fall back on. It is therefore important to consider investing money in a retirement account. Consider the story of Louis.

Louis accepts a job that pays $42,000 per year. He begins investing money in his retirement account at the age of 25 with a goal of retiring at the age of 65. His company has a retirement option that deducts approximately 5% of his salary and invests it in a retirement account that pays a yearly average of 8%.

To calculate Louis's projected earnings at retirement, we use the retirement formula

$$FV = PMT \times \frac{(1 + I)^N - 1}{I},$$

which has the following variables:

- FV is the future value of the retirement account.
- PMT is the monthly payment contributed to the retirement account.
- I is the interest rate in decimal notation divided by 12, which is the amount of interest being earned each month.
- N is 12 times the number of years of the account, which is the number of payments made into the account.

Here, FV, the future value, is the variable we will be solving for.

Louis's PMT is $175, since 5% of Louis's annual salary is $2100, and that amount divided by 12 equals $175.

Also, $I = \frac{0.08}{12} \approx 0.0067$, and $N = 12 \times 40 = 480$, since Louis will be working for 40 years.

1. Substitute these values into the retirement formula to find the future value of Louis's retirement account when he turns 65. Round your answer to the nearest ten thousand dollars. Approximately $620,000

2. How much money will Louis personally contribute into his retirement account? $84,000

3. Approximately how much interest is the account projected to earn? Approximately $536,000

Making It Personal for You

4. Look in the classified ads or contact the human resources department of a company whose jobs interest you. Find out about the starting salary of a job, as well as the contributions that are made to a retirement account. Using your age and the age at which you wish to retire, calculate the amount of money you could have in a retirement account when you retire. Answers may vary

5. How much money would you have contributed by the time you retire? Answers may vary

6. How much interest would you have earned by the time you retire? Answers may vary

Topic	Procedure	Examples
Graphing straight lines, p. 415.	An equation of the form $$Ax + By = C$$ has a graph that is a straight line. To graph such an equation, plot any three points; two give the line and the third checks it. (Where possible, use the *x*- and *y*-intercepts.)	Graph $3x + 2y = 6$. <table><tr><td>*x*</td><td>*y*</td></tr><tr><td>0</td><td>3</td></tr><tr><td>4</td><td>-3</td></tr><tr><td>2</td><td>0</td></tr></table>
Finding the slope given two points, p. 425.	Nonvertical lines passing through distinct points (x_1, y_1) and (x_2, y_2) have slope $$m = \frac{y_2 - y_1}{x_2 - x_1}.$$ The slope of a horizontal line is 0. The slope of a vertical line is undefined.	What is the slope of the line through $(2, 8)$ and $(5, 1)$? $$m = \frac{1 - 8}{5 - 2} = -\frac{7}{3}$$
Finding the slope and *y*-intercept of a line given the equation, p. 428.	**1.** Rewrite the equation in the form $y = mx + b$. **2.** The slope is *m*. **3.** The *y*-intercept is $(0, b)$.	Find the slope and *y*-intercept. $$3x - 4y = 8$$ $$-4y = -3x + 8$$ $$y = \tfrac{3}{4}x - 2$$ The slope is $\frac{3}{4}$. The *y*-intercept is $(0, -2)$.
Finding the equation of a line given the slope and *y*-intercept, p. 429.	The slope–intercept form of the equation of a line is $$y = mx + b.$$ The slope is *m* and the *y*-intercept is $(0, b)$.	Find an equation of the line with *y*-intercept $(0, 7)$ and with slope $m = 3$. $$y = 3x + 7$$
Graphing a line using slope and *y*-intercept, p. 430.	**1.** Plot the *y*-intercept. **2.** Starting from $(0, b)$, plot a second point using the slope. $$\text{slope} = \frac{\text{rise}}{\text{run}}$$ **3.** Draw a line that connects the two points.	Graph $y = -4x + 1$. First plot the *y*-intercept at $(0, 1)$. Slope $= -4$ or $\frac{-4}{1}$
Finding the slopes of parallel and perpendicular lines, p. 431.	Parallel lines have the same slope. Perpendicular lines have slopes whose product is -1.	Line *q* has a slope of 2. The slope of a line parallel to *q* is 2. The slope of a line perpendicular to *q* is $-\frac{1}{2}$.
Finding the equation of a line through a point with a given slope, p. 438.	**1.** Substitute the known values into the equation $y = mx + b$. **2.** Solve for *b*. **3.** Use the values of *m* and *b* to write the slope–intercept form of the equation.	Find an equation of the line through $(3, 2)$ with slope $m = \frac{4}{5}$. $$y = mx + b \qquad 2 = \frac{4}{5}(3) + b$$ $$2 = \frac{12}{5} + b$$ $$-\frac{2}{5} = b$$ An equation is $y = \frac{4}{5}x - \frac{2}{5}$.

(Continued on next page)

Topic	Procedure	Examples
Finding the equation of a line through two points, p. 438.	1. Find the slope. 2. Use the procedure when given a point and the slope.	Find the equation of the line through $(3, 2)$ and $(13, 10)$. $$m = \frac{y_2 - y_1}{x_2 - x_1} = \frac{10 - 2}{13 - 3} = \frac{8}{10} = \frac{4}{5}$$ We choose the point $(3, 2)$. $y = mx + b \qquad 2 = \frac{4}{5}(3) + b$ $2 = \frac{12}{5} + b$ $-\frac{2}{5} = b$ The equation is $y = \frac{4}{5}x - \frac{2}{5}$.
Graphing linear inequalities, p. 442.	1. Graph as if it were an equation. If the inequality symbol is $>$ or $<$, use a dashed line. If the inequality symbol is $\geq$ or $\leq$, use a solid line. 2. Look for a test point. The easiest test point is $(0, 0)$, unless the line passes through $(0, 0)$. In that case, choose another test point. 3. Substitute the coordinates of the test point into the inequality. 4. If it is a true statement, shade the side of the line containing the test point. If it is a false statement, shade the side of the line that does *not* contain the test point.	Graph $y \geq 3x + 2$. Graph the line $y = 3x + 2$. Use a solid line. Test $(0, 0)$. $\qquad 0 \geq 3(0) + 2$ $\qquad\qquad\qquad 0 \geq 2 \quad$ false Shade the side of the line that does not contain $(0, 0)$.
Determining whether a relation is a function, p. 448.	A function is a relation in which no two different ordered pairs have the same first coordinate.	Is this relation a function? $\{(5, 7), (3, 8), (5, 10)\}$ It is *not* a function since $(5, 7)$ and $(5, 10)$ are two different ordered pairs with the same x-coordinate, 5.
Determining whether a graph represents a function, p. 451.	If a vertical line can intersect the graph of a relation more than once, the relation is not a function. If no such line exists, the relation is a function.	Does this graph represent a function? Yes. Any vertical line will intersect it at most once.

Chapter 7 Review Problems

Section 7.1

1. Plot and label the following points.
A: $(2, -3)$ B: $(-1, 0)$ C: $(3, 2)$ D: $(-2, -3)$

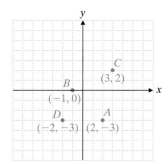

2. Give the coordinates of each point.

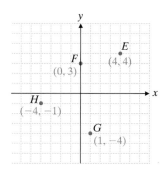

Complete the ordered pairs so that each is a solution to the given equation.

3. $y = 7 - 3x$
  (a) $(0, 7)$ **(b)** $(-1, 10)$

4. $2x + 5y = 12$
  (a) $(1, 2)$ **(b)** $(-4, 4)$

5. $x = 6$
  (a) $(6, -1)$ **(b)** $(6, 3)$

Section 7.2

6. Graph $3y = 2x + 6$.

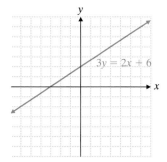

7. Graph $5y + x = -15$.

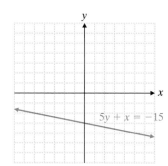

8. Graph $2y + 4x = -8 + 2y$.

Section 7.3

9. Find the slope of the line passing through $(5, -3)$ and $\left(2, -\frac{1}{2}\right)$.

$m = -\dfrac{5}{6}$

10. Find the slope and y-intercept of the line $9x - 11y + 15 = 0$.

$m = \dfrac{9}{11}; \left(0, \dfrac{15}{11}\right)$

11. Write an equation of the line with slope $-\frac{1}{2}$ and y-intercept $(0, 3)$.

$y = -\dfrac{1}{2}x + 3$

12. The equation of a line is $y = \frac{3}{5}x - 2$. What is the slope of a line perpendicular to that line?

$m = -\dfrac{5}{3}$

13. Graph $y = -\dfrac{1}{2}x + 3$.

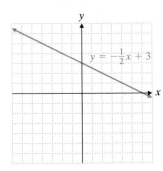

14. Graph $2x - 3y = -12$.

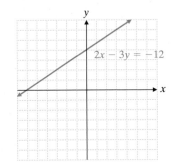

15. Graph $6x - 2y = 6 + 6x$.

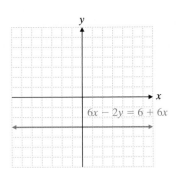

Section 7.4

16. Write an equation of the line passing through $(2, 5)$ having a slope of 1.
$y = x + 3$

17. Write an equation of the line passing through $(3, -4)$ having a slope of -6.
$y = -6x + 14$

18. Write an equation of the line passing through $(-1, 4)$ having a slope of $-\frac{1}{3}$.
$y = -\dfrac{1}{3}x + \dfrac{11}{3}$

19. Write an equation of the line passing through $(3, 7)$ and $(-6, 7)$.
$y = 7$

20. What is the slope of a line parallel to a line whose equation is $y = -\frac{2}{3}x + 4$?
$-\dfrac{2}{3}$

21. What is the slope of a line perpendicular to the line whose equation is $-2x + 6y = 3$?
-3

Write an equation of the graph.

22.

23.

24.

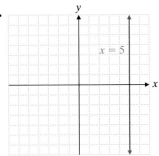

Section 7.5

25. Graph $y < \frac{1}{3}x + 2$.

26. Graph $3y + 2x \geq 12$.

27. Graph $x \leq 2$.

Section 7.6

Determine the domain and range of the relation. Determine whether the relation is a function.

28. $\{(5, -6), (-6, 5), (-5, 5), (-6, -6)\}$
   Domain: $\{-6, -5, 5\}$; Range: $\{-6, 5\}$; not a function

29. $\{(2, -3)(5, -3)(6, 4)(-2, 4)\}$
   Domain $= \{-2, 2, 5, 6\}$;
   Range $= \{-3, 4\}$;
   function

In exercises 30–32, determine whether the graphs represent a function.

30.

Function

31.

Not a function

32.

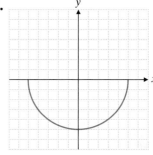

Function

33. Graph $y = x^2 - 5$.

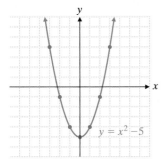

34. Graph $x = y^2 + 3$.

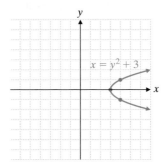

35. Graph $y = (x - 3)^2$.

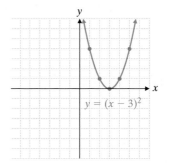

Given the following functions, find the indicated values.

36. $f(x) = 7 - 6x$ **(a)** $f(0)$ 7   **(b)** $f(-4)$ 31

37. $g(x) = -2x^2 + 3x + 4$ **(a)** $g(-1)$ −1   **(b)** $g(3)$ −5

38. $f(x) = 8x - 3$ **(a)** $f(-1)$ −11   **(b)** $f\left(\dfrac{1}{2}\right)$ 1

39. $f(x) = \dfrac{2}{x + 4}$ **(a)** $f(-2)$ 1   **(b)** $f(6)$ $\dfrac{1}{5}$

40. $f(x) = x^2 - 2x + \dfrac{3}{x}$ **(a)** $f(-1)$ 0   **(b)** $f(3)$ 4

Applications

Monthly Electric Bill *Russ and Norma Camp found that their monthly electric bill could be calculated by the equation* $y = 30 + 0.09x$. *In this equation y represents the amount of the monthly bill in dollars and x represents the number of kilowatt-hours used during the month.*

41. What would be their monthly bill if they used 2000 kilowatt-hours of electricity?

$210

42. What would be their monthly bill if they used 1600 kilowatt-hours of electricity?

$174

43. Write the equation in the form $y = mx + b$, and determine the numerical value of the y-intercept. What is the significance of this y-intercept? What does it tell us?

$y = 0.09x + 30$; (0, 30); it tells us that if Russ and Norma use no electricity, the minimum cost is $30.

44. If the equation is placed in the form $y = mx + b$, what is the numerical value of the slope? What is the significance of this slope? What does it tell us?

$m = 0.09$; the electric bill increases $0.09 for each kilowatt-hour of use.

45. If Russ and Norma have a monthly bill of $147, how many kilowatt-hours of electricity did they use?

1300 kilowatt-hours

46. If Russ and Norma have a monthly bill of $246, how many kilowatt-hours of electricity did they use?

2400 kilowatt-hours

Petroleum Consumption *Scientists have found that the amount of petroleum consumed worldwide for a given year can be predicted by the equation* $y = 1.15x + 65.6$, *where x is the number of years since 1990 and y is the number of barrels of petroleum consumed per day measured in millions. Use this equation to answer the following questions. (Source: www.eia.doe.gov.)*

47. How many barrels of petroleum were consumed in 2000?

77.1 million barrels

48. How many barrels of petroleum will be consumed in 2015?

94.35 million barrels

49. The given equation is in the form $y = mx + b$. What is the slope and what is its significance? What does it tell us?

The slope is 1.15. The slope tells us the rate of change per year. In this case it tells us that for each additional year, the world uses 1.15 million more barrels of petroleum.

50. The given equation is in the form $y = mx + b$. What is the y-intercept? What is the significance of this y-intercept? What does it tell us?

The y- intercept is (0, 65.6). This tells us that when $x = 0$, which is the year 1990, a total of 65.6 million barrels of petroleum were consumed worldwide.

51. In what year will the consumption of petroleum reach 96.65 million barrels?

2017

52. In what year will the consumption of petroleum reach 100.1 million barrels?

2020

Mixed Practice

53. Graph $5x + 3y = -15$.

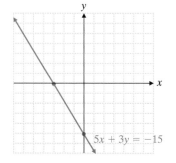

54. Graph $y = \frac{3}{4}x - 3$.

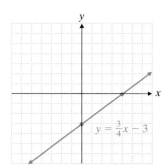

55. Find the slope of a line through $(2, -7)$ and $(-3, -5)$.

$-\dfrac{2}{5}$

56. What is the slope and the y-intercept of the line $7x + 6y - 10 = 0$?

$m = -\dfrac{7}{6}; \left(0, \dfrac{5}{3}\right)$

57. Write an equation of a line that passes through $(3, -5)$ and has a slope of $\frac{2}{3}$.

$y = \dfrac{2}{3}x - 7$

58. Graph $y < -2x + 1$.

Job Losses in Manufacturing Economists in the Labor Department are concerned about the continued job loss in the manufacturing industry. The number of people employed in manufacturing jobs in the United States can be predicted by the equation $y = -269x + 17{,}020$, where x is the number of years since 1994 and y is the number of employees in **thousands** in the manufacturing industry. Use this data to answer the following questions. (Source: www.bls.gov.)

59. How many people were employed in manufacturing in 1994? In 2000? In 2008?

17,020,000 people in 1994
15,406,000 people in 2000
13,254,000 people in 2008

60. Use your answers for 59 to draw a graph of the equation $y = -269x + 17{,}020$.

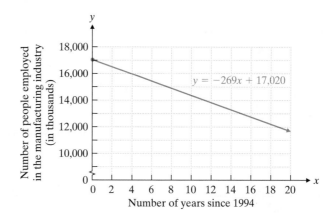

61. What is the slope of this equation? What is its significance?

The slope is −269. The slope tells us that the number of people employed in manufacturing decreases each year by 269 thousand. In other words, employment in manufacturing goes down 269,000 people each year.

62. What is the y-intercept of this equation? What is its significance?

The y-intercept is (0, 17,020). This tells us that in the year 1994, the number of manufacturing jobs was 17,020 thousands, which is 17,020,000.

63. Use the equation to predict in what year the number of manufacturing jobs will be 11,640,000.

2014

64. Use the equation to predict in what year the number of manufacturing jobs will be 10,295,000.

2019

How Am I Doing? Chapter 7 Test

Remember to use your Chapter Test Prep Video CD to see the worked-out solutions to the test problems you want to review.

1. See graph

2. See graph

3. See graph

4. See graph

5. $m = -\dfrac{3}{2}; \left(0, \dfrac{5}{2}\right)$

6. $m = 1$

7. $y = \dfrac{1}{2}x - 4$

8. $m = 0$

9. $y = -\dfrac{3}{2}x + \dfrac{7}{2}$

10. $x = 2$, undefined

1. Plot and label the following points.
B: $(6, 1)$ C: $(-4, -3)$
D: $(-3, 0)$ E: $(5, -2)$

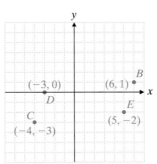

2. Graph the line $6x - 3 = 5x - 2y$.

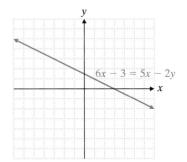

3. Graph the line $8x + 2y = -4$.

4. Graph $y = \frac{2}{3}x - 4$.

5. What is the slope and the y-intercept of the line $3x + 2y - 5 = 0$?

6. Find the slope of the line that passes through $(8, 6)$ and $(-3, -5)$.

7. Write an equation for the line that passes through $(4, -2)$ and has a slope of $\frac{1}{2}$.

8. Find the slope of the line through $(-3, 11)$ and $(6, 11)$.

9. Find an equation for the line passing through $(5, -4)$ and $(-3, 8)$.

10. Write an equation for the line through $(2, 7)$ and $(2, -2)$. What is the slope of this line?

11. Graph the region described by $4y \le 3x$.

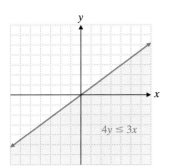

12. Graph the region described by $-3x - 2y > 10$.

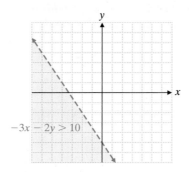

13. Is this relation a function? $\{(2, -8), (3, -7), (2, 5)\}$ Why?

14. Look at the relation graphed below. Is this relation a function? Why?

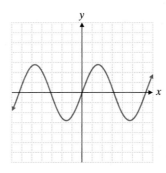

15. Graph $y = 2x^2 - 3$.

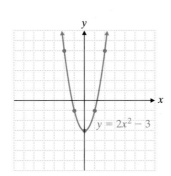

x	y
-2	5
-1	-1
0	-3
1	-1
2	5

16. For $f(x) = -x^2 - 2x - 3$:
   (a) Find $f(0)$.
   (b) Find $f(-2)$.

17. For $g(x) = \dfrac{3}{x - 4}$:
   (a) Find $g(3)$.
   (b) Find $g(-11)$.

11. _____ See graph _____

12. _____ See graph _____

13. No; two different ordered pairs have the same first coordinate

14. Yes; any vertical line passes through no more than one point on the graph

15. _____ See graph _____

16. (a) -3

   (b) -3

17. (a) -3

   (b) $-\dfrac{1}{5}$

Cumulative Test for Chapters 0–7

1. ___$-27x^9y^{12}z^3$___

2. ___$-\dfrac{2y^{10}}{3x^8}$___

3. ___$x = 7$___

4. ___$R = 40$___

5. ___$x = -4$___

6. ___$w = \dfrac{A - lh}{l}$ or $w = \dfrac{A}{l} - h$___

7. ___$2(5a + 7b)(5a - 7b)$___

8. ___$(3x + 1)(3x - 4)$___

9. ___$2(x - 1)$___

10. ___$y = 2x - 7$___

11. ___$x = 7$___

12. ___$y = -\dfrac{2}{3}x - 3$___

13. ___$m = 0$___

14. ___$m = \dfrac{4}{3}$___

Approximately one-half of this test covers the content of Chapters 0–6. The remainder covers the content of Chapter 7.

1. Simplify. $(-3x^3y^4z)^3$

2. Simplify; leave the answer with positive exponents only. $\dfrac{-16x^{-3}y^4}{24x^5y^{-6}}$

Solve.

3. $2 - 3(4 - x) = x - (3 - x)$

4. $\dfrac{600}{R} = \dfrac{600}{R + 50} + \dfrac{500(R + 20)}{R^2 + 50R}$

5. $x = 8 - 2^4 \div 8 - 10$

6. Solve $A = lw + lh$ for w.

7. Factor. $50a^2 - 98b^2$

8. Factor. $9x^2 - 9x - 4$

9. Simplify. $\dfrac{x - \dfrac{1}{x}}{\dfrac{1}{2} + \dfrac{1}{2x}}$

10. Find an equation of the line through $(5, 3)$ and $(2, -3)$.

In questions 11 and 12, give the equation of the line that fits each description.

11. Passes through $(7, -4)$ and is vertical

12. Passes through $(3, -5)$ with slope $-\dfrac{2}{3}$

13. Find the slope of the line through $(-8, -3)$ and $(11, -3)$.

14. What is the slope of the line $4x - 3y = 6$?

15. Graph $y = \frac{2}{3}x - 4$.

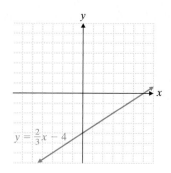

16. Graph $3x + 8 = 5x$.

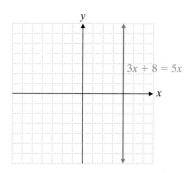

17. Graph the region $2x + 5y \leq -10$.

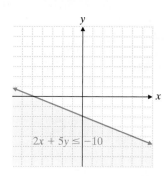

18. Graph $x = y^2 + 3$. Is this a function?

x	y
7	-2
4	-1
3	0
4	1
7	2

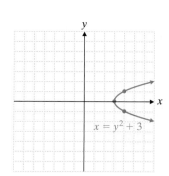

19. Is this relation a function? $\{(3, 10), (10, -3), (0, -3), (5, 10)\}$

20. For $f(x) = 3x^2 - 2x + \dfrac{4}{x}$:

(a) Find $f(-1)$.   (b) Find $f(2)$.

---

15. _See graph_

16. _See graph_

17. _See graph_

18. _See graph; no_

19. _Yes_

20. (a) _1_

(b) _10_

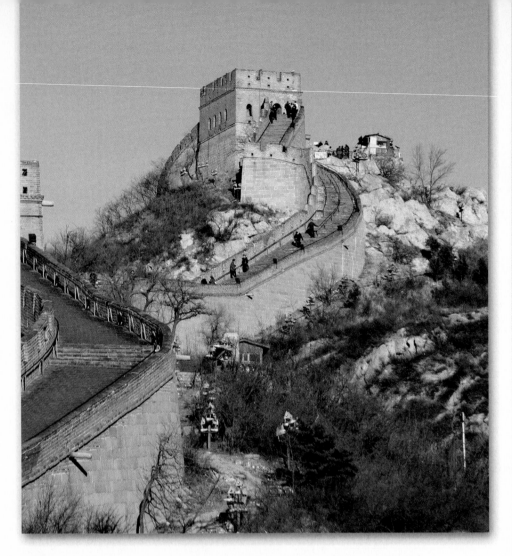

Systems of Equations

CHAPTER

8

The Great Wall of China is a sight that many people come to see firsthand. Tourist, travelers to the 2008 Beijing Olympics, business officials, and government leaders come each day to see the remains of this remarkable historic structure. But how many soldiers manned this ancient structure in historic times? How many tons of stone were used to construct this wall? These and many other historical questions are being answered today by mathematicians working with archaeologists and historians. They use the mathematics of this chapter to help answer these interesting questions.

In Chapter 7 we examined linear equations and inequalities in two variables. In this chapter we will examine systems of linear equations and inequalities. Two or more equations or inequalities in several variables that are considered simultaneously are called a **system of equations** or a **system of inequalities.** Recall that the graph of a linear equation is a straight line. If we have the graphs of two linear equations on one coordinate system, how might they relate to one another?

The lines may intersect.

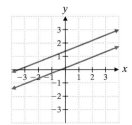

The lines may be parallel.

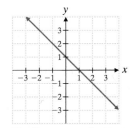

The lines may coincide.

The solutions to a system of linear equations are those points that both lines have in common. Note that parallel lines have no points in common. Thus the second system has no solution. Intersecting lines intersect at one point. Thus the first system has one solution. Lines that coincide have an infinite number of points in common. Thus the third system has an infinite number of solutions. You can determine the solution to a system of equations by graphing.

Using Graphing to Solve a System of Linear Equations with a Unique Solution

To solve a system of equations by graphing, graph both equations on the same coordinate system. If the lines intersect, the system has a unique solution. The ordered pair of coordinates of the point of intersection is the solution to the system of equations.

EXAMPLE 1 Solve by graphing.

$$-2x + 3y = 6$$
$$2x + 3y = 18$$

Solution Graph both equations on the same coordinate system and determine the point of intersection.

We will graph the system by first obtaining a table of values for each equation. First we solve the first equation for y.

$$-2x + 3y = 6$$
$$3y = 2x + 6$$

$$y = \frac{2}{3}x + 2$$

Let $x = 0.$ $y = \frac{2}{3}(0) + 2 = 0 + 2 = 2$

Let $x = -3.$ $y = \frac{2}{3}(-3) + 2 = -2 + 2 = 0$

Let $x = 6.$ $y = \frac{2}{3}(6) + 2 = 4 + 2 = 6$

x	y
0	2
-3	0
6	6

Student Learning Objectives

After studying this section, you will be able to:

1 Use graphing to solve a system of linear equations with a unique solution.

2 Graph a system of linear equations to determine the type of solution.

3 Solve applied problems by graphing.

Graphing Calculator

Graphing a System of Linear Equations

To solve the system

$$3x + 4y = -28$$
$$x - 4y = 12,$$

we first need to solve each equation for y. This will give us the equivalent system

$$y = -0.75x - 7$$
$$y = 0.25x - 3.$$

We graph this system with a scale of -10 to 10 for both x and y.
Display.

If we use the Zoom function, we will locate the intersection at $x = -4$, $y = -4$.

Teaching Example 1 Solve by graphing.
$$x + 2y = 6$$
$$x - y = 3$$

Ans:

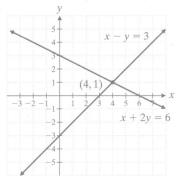

471

This allows us to use the three sets of ordered pairs to graph $-2x + 3y = 6$. Next we solve the second equation for y.

$$2x + 3y = 18$$
$$3y = -2x + 18$$

$$y = -\frac{2}{3}x + 6$$

x	y
0	6
-3	8
6	2

Let $x = 0$. $y = -\frac{2}{3}(0) + 6 = 0 + 6 = 6$

Let $x = -3$. $y = -\frac{2}{3}(-3) + 6 = 2 + 6 = 8$

Let $x = 6$. $y = -\frac{2}{3}(6) + 6 = -4 + 6 = 2$

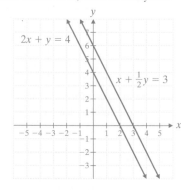

NOTE TO STUDENT: *Fully worked-out solutions to all of the Practice Problems can be found at the back of the text starting at page SP-1*

This allows us to use these three ordered pairs to graph $2x + 3y = 18$. (Of course, each equation could be graphed by any method we have discussed in the text. It is not necessary to use this method.) The graph of the system is shown in the figure at the right. The lines intersect at the point for which $x = 3$ and $y = 4$. Thus the unique solution to the system of equations is $(3, 4)$. Always check your answer.

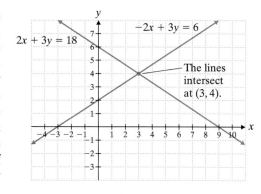

$-2(3) + 3(4) \stackrel{?}{=} 6$ $2(3) + 3(4) \stackrel{?}{=} 18$

$6 = 6$ $18 = 18$ ✓

A linear system of equations that has one solution is said to be **consistent.**

Practice Problem 1 Solve by graphing. $x + y = 12$
$$-x + y = 4$$

Teaching Example 2 Solve by graphing.

$$2x + y = 4$$
$$x + \frac{1}{2}y = 3$$

Ans: No solution; inconsistent system

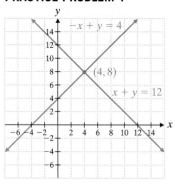

Graphing a System of Linear Equations to Determine the Type of Solution

Two lines in a plane may intersect, may never intersect (be parallel lines), or may be the same line. The corresponding system of equations will have one solution, no solution, or an infinite number of solutions, respectively. Thus far we have focused on systems of linear equations that have a unique solution. We now draw your attention to the other two cases.

EXAMPLE 2 Solve by graphing.

$$3x - y = 1$$
$$3x - y = -7$$

Solution We graph both equations on the same coordinate system. The graph is at the right. Notice that the lines are parallel. They do not intersect. Hence there is no solution to this system of equations. A system of linear equations that has no solution is called **inconsistent.**

PRACTICE PROBLEM 2

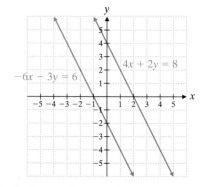

Practice Problem 2 Solve by graphing. $4x + 2y = 8$
$$-6x - 3y = 6$$

EXAMPLE 3 Solve by graphing.

$$x + y = 4$$
$$3x + 3y = 12$$

Solution We graph both equations on the same coordinate system at the right. Notice that the equations represent the same line (coincide). The coordinates of every point on the line will satisfy both equations. That is, every point on the line represents a solution to $x + y = 4$ and to $3x + 3y = 12$. Thus there is *an infinite number of solutions* to this system. Such equations are said to be **dependent.**

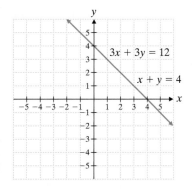

Practice Problem 3 Solve by graphing.

$$3x - 9y = 18$$
$$-4x + 12y = -24$$

PRACTICE PROBLEM 3

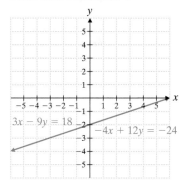

TO THINK ABOUT: Determining the Type of Solution Without Graphing

Without graphing, how could you tell that the system in Example 2 represented parallel lines? Without graphing, how could you tell that the equations in Example 3 are dependent? Without graphing, determine whether each of the following systems is inconsistent (has no solution) or consistent (has a solution). Are any of the systems of equations dependent?

(a) $2x - 3y = 6$
   $6y = 4x + 1$

(b) $3x + y = 2$
   $2x + y = 3$

(c) $x - 2y = 1$
   $3x - 6y = 3$

 Solving Applied Problems by Graphing

EXAMPLE 4 Walter and Barbara need some plumbing repairs done at their house. They called two plumbing companies for estimates of the work that needed to be done. Roberts Plumbing and Heating charges $40 for a house call and then $35 per hour for labor. Instant Plumbing Repairs charges $70 for a house call and then $25 per hour for labor.

(a) Create a cost equation for each company where y is the total cost of plumbing repairs and x is the number of hours of labor. Write a system of equations.

(b) Graph the two equations using the values $x = 0, 3,$ and 6.

(c) Determine from your graph how many hours of plumbing repairs would be required for the two companies to charge the same.

(d) Determine from your graph which company charges less if the estimated amount of time to complete the plumbing repairs is 4 hours.

Solution

(a) For each company we will obtain an equation of the form

$$y = (\text{cost of house call}) + (\text{cost per hour})x.$$

Roberts Plumbing and Heating charges $40 for a house call and $35 per hour. Thus we obtain the first equation, $y = 40 + 35x$.

Instant Plumbing Repairs charges $70 for a house call and $25 per hour. Thus we obtain the second equation, $y = 70 + 25x$.

This yields the following system of equations:

$$y = 40 + 35x$$
$$y = 70 + 25x.$$

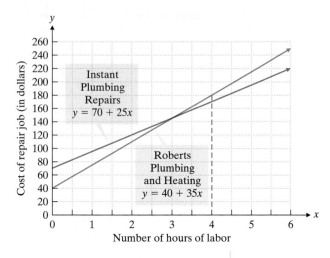

(b) We will graph the system by first obtaining a table of values for each equation.

Roberts Plumbing and Heating $y = 40 + 35x$

Let $x = 0$. $y = 40 + 35(0) = 40$

Let $x = 3$. $y = 40 + 35(3) = 145$

Let $x = 6$. $y = 40 + 35(6) = 250$

x	y
0	40
3	145
6	250

Instant Plumbing Repairs $y = 70 + 25x$

Let $x = 0$. $y = 70 + 25(0) = 70$

Let $x = 3$. $y = 70 + 25(3) = 145$

Let $x = 6$. $y = 70 + 25(6) = 220$

x	y
0	70
3	145
6	220

The graph of the system is to the left.

(c) We see that the graphs of the two lines intersect at (3, 145). Thus the two companies will charge the same if 3 hours of plumbing repairs are required.

(d) We draw a dashed line at $x = 4$. We see that the blue line representing Roberts Plumbing and Heating is higher than the red line representing Instant Plumbing Repairs after 3 hours. Thus the cost would be less if Walter and Barbara use Instant Plumbing Repairs for 4 hours of work.

Practice Problem 4   Jonathan and Joanne Wells need some electrical work done on their new house. They obtained estimates from two companies. Bill Tupper's Electrical Service charges $100 for a house call and $30 per hour for labor. Wire for Hire charges $50 for a house call and $40 per hour for labor.

(a) Create a cost equation for each company where y is the total cost of the electrical work and x is the number of hours of labor. Write a system of equations. $y = 100 + 30x; y = 50 + 40x$

(b) Graph the two equations using the values $x = 0, 4$, and 8.

(c) Determine from your graph how many hours of electrical repairs would be required for the two companies to charge the same.

(d) Determine from your graph which company charges less if the estimated amount of time to complete the electrical repairs to the house is 6 hours.

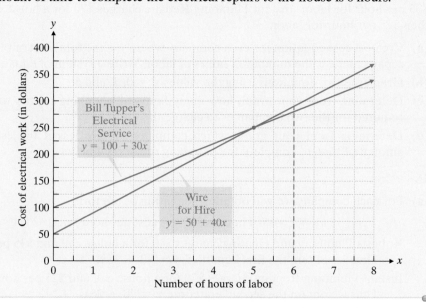

(c) The companies charge the same for 5 hours.
(d) At 6 hours the cost is less for Bill Tupper's Electrical Service.

PRACTICE PROBLEM 4(b)

x	y		x	y
0	100		0	50
4	220		4	210
8	340		8	370

Verbal and Writing Skills

1. In the system $y = 2x - 5$ and $y = 2x + 6$, the lines have the same slope but different y-intercepts. What does this tell you about the graph of this system?

The lines are parallel. One line intersects the y-axis at -5. The other line intersects the y-axis at 6. There is no solution. The system is inconsistent.

2. In the system $y = -3x + 4$ and $y = -3x + 4$, the lines have the same slope and the same y-intercepts. What does this tell you about the graph of the system?

The two lines coincide. They are the same line. There are an infinite number of solutions. The equations are dependent.

3. Before you graph a system of equations, if you notice that the lines have different slopes, you can conclude that they ___intersect___ and that the system has ___one___ solution.

4. If lines have the same slope, you cannot conclude that the lines are parallel. Why? How else might the lines relate to each other?

They may coincide.

5. Two lines have different slopes but the same y-intercept. What does this tell you about the graphs of the lines and about the solution of the system?

The lines intersect at the y-intercept. The y-intercept is the solution of the system.

6. In the system $y = -3x + 5$ and $y = -3x - 2$, what can you say about the number of solutions?

The lines have the same slope but different y-intercepts. Therefore, the lines are parallel. There is no solution to this system of equations.

Solve by graphing. If there isn't a unique solution to the system, state the reason.

7. $x - y = 3$
$x + y = 5$

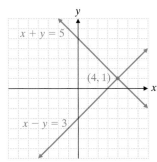

8. $x - y = 4$
$-x - y = 6$

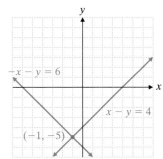

9. $y = -3x$
$y = 2x + 5$

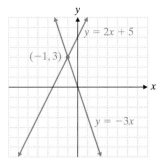

10. $x + 4y = 6$
$-x + 2y = 12$

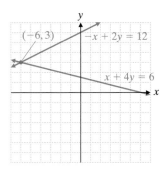

11. $4x + y = 5$
$3x - 2y = 12$

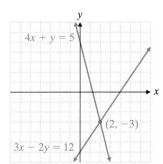

12. $x + 2y = 10$
$3x - 2y = 6$

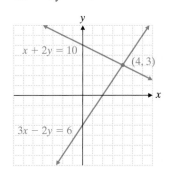

13. $-2x + y - 3 = 0$
$4x + y + 3 = 0$

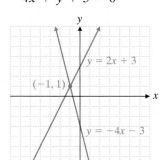

14. $x - 2y - 10 = 0$
$2x + 3y - 6 = 0$

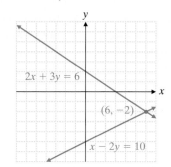

15. $3x - 2y = -18$
$2x + 3y = 14$

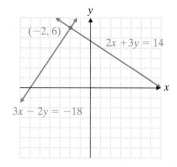

Mixed Practice *Solve by graphing. If there isn't a unique solution to the system, state the reason.*

16. $3x + 2y = -10$
$-2x + 3y = 24$

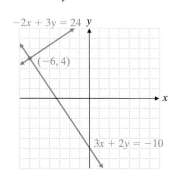

17. $y = \dfrac{3}{4}x + 7$
$y = -\dfrac{1}{2}x + 2$

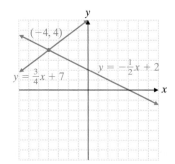

18. $y = \dfrac{5}{7}x - 2$
$y = \dfrac{1}{3}x + \dfrac{2}{3}$

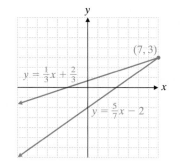

19. $3x - 2y = -4$
$-9x + 6y = -9$

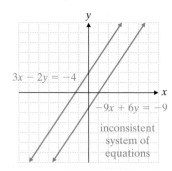

20. $4x - 6y = 8$
$-2x + 3y = -4$

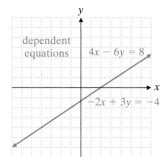

21. $y - 2x - 6 = 0$
$\dfrac{1}{2}y - 3 = x$

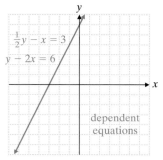

22. $2y + x - 6 = 0$
$y + \dfrac{1}{2}x = 4$

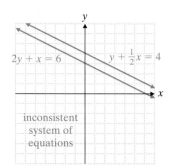

23. $y = \dfrac{1}{2}x - 2$
$y = \dfrac{2}{3}x - 1$

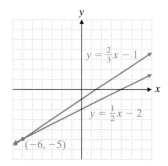

24. $2y + 6x - 8 = 0$
$y = \dfrac{3}{2}x + 4$

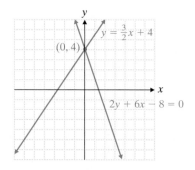

25. ***Telephone Charges*** The Newtown Telephone Company advertises that if you use their 10-10 service, you will always pay only 7¢ per minute, no matter where you call within the continental United States. Its major rival, the Tryus Telephone Company, charges only 3¢ per minute, but you must pay a 36¢ "connection fee" per call.

(a) Create a cost equation for each telephone company, where y is the total cost for each phone call and x is the number of minutes the phone call lasts. Write a system of equations.

(b) Graph the two equations using the values $x = 1$, 4, 8, and 12. Fill in a table for those values of x and y.

(c) Determine from your graph how many minutes long your phone call must be for the cost of using each telephone company to be the same. 9 min

(d) Determine from your graph which company would cost you less if you typically make phone calls of 10 minutes or more. Tryus Telephone Company

26. ***Landscaping Costs*** Fred and Amy want to have shrubs planted in front of their new house. They have obtained estimates from two landscaping companies. Camp Property Care charges $100 for an initial consultation and $60 an hour for labor. Manchester Landscape Designs charges $200 for an initial consultation and $40 an hour for labor.

(a) Create a cost equation for each company where y is the total cost of the landscaping work and x is the number of hours of labor. Write a system of equations.

(b) Graph the two equations using the values $x = 0, 5$, and 10. Fill in a table for those values of x and y.

(c) Determine from your graph how many hours of labor would be required for the cost of landscaping from each of the two companies to be the same. 5 hr

(d) Determine from your graph which company would charge less if 7 hours of landscaping are required. Manchester Landscape Designs

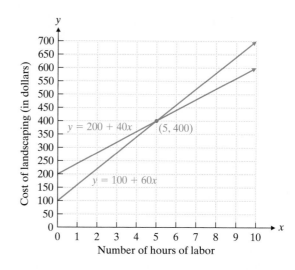

(a) Newtown Telephone Company $y = 7x$;
Tryus Telephone Company $y = 36 + 3x$

(b)

x	$y = 7x$
1	7
4	28
8	56
12	84

x	$y = 36 + 3x$
1	39
4	48
8	60
12	72

(a) Camp Property Care $y = 100 + 60x$
Manchester Landscape Designs $y = 200 + 40x$

(b)

x	$y = 100 + 60x$
0	100
5	400
10	700

x	$y = 200 + 40x$
0	200
5	400
10	600

If you have a graphing calculator, solve each system by graphing the equations and estimating the point of intersection to the nearest hundredth. You will need to adjust the window for the appropriate x- and y-values in order to see the region where the straight lines intersect.

27. $88x + 57y = 683.10$
$95x - 48y = 7460.64$
$(47.52, -61.38)$

28. $64x + 99y = -8975.52$
$58x - 73y = 3503.58$
$(-29.61, -71.52)$

Cumulative Review

Tourist Visits to Machu Picchu *In 2005, Machu Picchu, the 15th-century Incan citadel perched between the Andes and the Amazon jungle, drew a record 400,000 visitors. This was two and two-thirds times higher than in 2002. In 2001, the government set a limit of 2500 visitors daily, although the average is 2000 daily visitors. (Source: www .transitionsabroad.com.)*

29. **[3.3.1]** How many people visited Machu Picchu in 2002? 150,000 people

30. **[0.5.5]** What percent of the daily limit actually visits Machu Picchu daily? 80%

31. **[0.7.1]** According to the government, what is the maximum number of people per year that should be allowed to visit Machu Picchu? What is the actual average number of visitors per year?

912,500 visitors; 730,000 visitors

32. **[0.7.1]** Suppose a company wants to install and operate a tram that charges $120 roundtrip from the nearby city of Cuzco to Machu Picchu. If the company wants to earn at least $125,000 daily, how many visitors to Machu Picchu do they need to transport daily? What percent of the daily limit would this be?

At least 1042 visitors; about 41.7%

Quick Quiz 8.1

1. Solve the following system by graphing.

$$x - 3y = 6$$
$$4x + 3y = 9$$

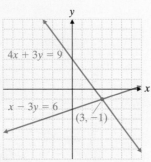

2. Find the slope and the y-intercept of each of the following lines.

$$y = \frac{3}{5}x + 2$$
$$5y - 3x = 35$$

What does this tell you if you want to find the solution to the system of equations by the graphing method?

The slope of each line is $\frac{3}{5}$. The y-intercept of the first line is $(0, 2)$. The y-intercept of the second line is $(0, 7)$. Since the lines have the same slopes but different y-intercepts, they are parallel lines. Therefore there is no solution. It is an inconsistent system of equations.

3. What will happen when you try to find the solution of the following system of equations by graphing?

$$y = 3x - 2$$
$$-6x + 2y = -4$$

When graphed, the two equations yield the same line. The equations are dependent. There is an infinite number of solutions.

4. **Concept Check** You are attempting to find the solution to this system of equations by graphing.

$$-3x + 4y = -16$$
$$0 = 6x - 8y + 16$$

What will you discover? Answers may vary

Classroom Quiz 8.1 You may use these problems to quiz your students' mastery of Section 8.1.

1. Solve the following system by graphing.

$$2x + 3y = 6$$
$$2x + y = -2$$

Ans:

2. Find the slope and the y-intercept of each of the following lines.

$$y = \frac{4}{3}x + 2$$
$$6y - 8x - 12 = 0$$

What does this tell you if you want to find the solution to the system of equations by the graphing method?

Ans: The slope of each line is $\frac{4}{3}$. The y-intercept of both lines is $(0, 2)$. If you graph them you will get the same line. They are dependent equations. There is an infinite number of solutions.

3. What will happen when you try to find the solution of the following system of equations by graphing?

$$4x + y + 4 = 0$$
$$\frac{3}{4}y = -3x + \frac{1}{4}$$

Ans: The two equations when graphed yield two parallel lines. They both have the slope of -4, but they have different y-intercepts. So it is an inconsistent system of equations. There is no solution.

8.2 SOLVING A SYSTEM OF EQUATIONS IN TWO VARIABLES BY THE SUBSTITUTION METHOD

1 Solving a System of Two Linear Equations with Integer Coefficients by the Substitution Method

Since the solution to a system of two linear equations may be a point where the two lines intersect, we would expect the solution to be an ordered pair. Since the point must lie on both lines, the ordered pair (x, y) must satisfy both equations. For example, the solution to the system $3x + 2y = 6$ and $x + y = 3$ is $(0, 3)$. Let's check.

$$3x + 2y = 6 \qquad\qquad x + y = 3$$
$$3(0) + 2(3) \overset{?}{=} 6 \qquad\qquad 0 + 3 \overset{?}{=} 3$$
$$6 = 6 \checkmark \qquad\qquad\qquad 3 = 3 \checkmark$$

Given a system of linear equations, we can develop algebraic methods for finding a solution. These methods reduce the system to one equation in one variable, which we already know how to solve. Once we have solved for this variable, we can substitute this known value into any one of the original equations and solve for the second variable.

The first method we will discuss is the **substitution method.**

PROCEDURE FOR SOLVING A SYSTEM OF EQUATIONS BY THE SUBSTITUTION METHOD

1. Solve one of the two equations for one variable. If possible, solve for a variable with a coefficient of 1 or −1.

2. Substitute the expression you obtain in step 1 for this variable in the *other* equation.

3. You now have one equation with one variable. Solve this equation to find the value for that one variable.

4. Substitute this value for the variable into one of the original equations to obtain a value for the other variable.

5. Check the solution in each equation to verify your results.

EXAMPLE 1 Find the solution.
$$x - 2y = 7$$
$$-5x + 4y = -5$$

Solution

Step 1 Solve one equation for one variable.

$x - 2y = 7$ The first equation is the easiest one in which to isolate a variable.
  $x = 7 + 2y$ Add $2y$ to both sides to solve for x.

Step 2 Substitute the expression into the other equation.

$ -5x + 4y = -5$ This is the original second equation.
$-5(\,7 + 2y\,) + 4y = -5$ Substitute the value $7 + 2y$ for x in this equation.

Step 3 Solve this equation.

$-35 - 10y + 4y = -5$ Remove the parentheses.
$ -35 - 6y = -5$ Combine like terms.
$-6y = 30$ Add 35 to both sides.
$y = -5$ Divide both sides by −6.

Teaching Tip It is important for students to understand the concept that a solution must be verified in *both* linear equations before we can be certain that it is a solution. After explaining the idea of a solution to a system of equations, present students with the following question.

"Here is a system of equations.
$$4x - 5y = 22$$
$$x + 2y = 1$$

Is $(3, -2)$ a solution to the system?"

The answer is *no*. The values $x = 3$ and $y = -2$ check in the first equation, but not in the second equation. Many students will check only the first equation and make an erroneous conclusion.

Teaching Example 1 Find the solution.
$$2x + y = 4$$
$$-x + 2y = -7$$

Ans: $(3, -2)$

Step 4 Obtain the value of the other variable.

We will now use the value $y = -5$ in one of the equations that contains both variables to find the value for x.

$$x - 2y = 7 \qquad \text{The easiest equation to use is the one that we solved for } x.$$
$$x - 2(-5) = 7 \qquad \text{Replace } y \text{ by } -5.$$
$$x + 10 = 7 \qquad \text{Simplify.}$$
$$x = -3 \qquad \text{Subtract 10 from both sides.}$$

The solution is $(-3, -5)$.

Step 5 *Check.* To be sure that we have the correct solution, we will need to check that the values obtained for x and y can be substituted into *both original* equations to obtain true mathematical statements.

$$x - 2y = 7 \qquad\qquad -5x + 4y = -5$$
$$-3 - 2(-5) \overset{?}{=} 7 \qquad\qquad -5(-3) + 4(-5) \overset{?}{=} -5$$
$$-3 + 10 \overset{?}{=} 7 \qquad\qquad 15 - 20 \overset{?}{=} -5$$
$$7 = 7 \ \checkmark \qquad\qquad\qquad -5 = -5 \ \checkmark$$

NOTE TO STUDENT: *Fully worked-out solutions to all of the Practice Problems can be found at the back of the text starting at page SP-1*

Practice Problem 1 Find the solution.

$$5x + 3y = 19$$
$$2x - y = 12 \quad (5, -2)$$

---

② Solving a System of Two Linear Equations with Fractional Coefficients by the Substitution Method

If a system of equations contains fractions, clear the equations of fractions *before* performing any other steps.

Teaching Example 2 Find the solution.

$$2x + 5y = 3$$
$$\frac{1}{3}x - \frac{2}{3}y = 2$$

Ans: $(4, -1)$

EXAMPLE 2 Find the solution.

$$\frac{3}{2}x + y = \frac{5}{2}$$
$$-y + 2x = -1$$

Solution We want to clear the first equation of fractions. We observe that the LCD of the fractions is 2.

$$2\left(\frac{3}{2}x\right) + 2\,(y) = 2\left(\frac{5}{2}\right) \qquad \text{Multiply each term of the equation by the LCD.}$$

$$3x + 2y = 5 \qquad \text{This is equivalent to } \frac{3}{2}x + y = \frac{5}{2}.$$

Now follow the five-step procedure.

Step 1 Solve for one variable. We will use the second equation.

$$-y + 2x = -1$$
$$-y = -1 - 2x \qquad \text{Add } -2x \text{ to both sides.}$$
$$y = 1 + 2x \qquad \text{Multiply each term by } -1.$$

Step 2 Substitute the resulting expression into the other equation.

$$3x + 2(1 + 2x) = 5$$

Step 3 Solve this equation for the variable.

$$3x + 2 + 4x = 5 \quad \text{Remove parentheses.}$$
$$7x + 2 = 5$$
$$7x = 3$$
$$x = \frac{3}{7}$$

Step 4 Find the value of the second variable. We will use the second equation.

$$-y + 2x = -1$$
$$-y + 2\left(\frac{3}{7}\right) = -1$$
$$-y + \frac{6}{7} = -1$$
$$y - \frac{6}{7} = 1$$
$$y = 1 + \frac{6}{7} = \frac{13}{7}$$

The solution to the system is $\left(\frac{3}{7}, \frac{13}{7}\right)$.

Step 5 *Check.*

$$\frac{3}{2}x + y = \frac{5}{2} \qquad\qquad -y + 2x = -1$$
$$\frac{3}{2}\left(\frac{3}{7}\right) + \left(\frac{13}{7}\right) \stackrel{?}{=} \frac{5}{2} \qquad -\left(\frac{13}{7}\right) + 2\left(\frac{3}{7}\right) \stackrel{?}{=} -1$$
$$\frac{9}{14} + \frac{13}{7} \stackrel{?}{=} \frac{5}{2} \qquad\qquad -\frac{13}{7} + \frac{6}{7} \stackrel{?}{=} -\frac{7}{7}$$
$$\frac{9}{14} + \frac{26}{14} \stackrel{?}{=} \frac{35}{14} \qquad\qquad -\frac{7}{7} = -\frac{7}{7} \checkmark$$
$$\frac{35}{14} = \frac{35}{14} \checkmark$$

Practice Problem 2 Find the solution. Be sure to check your answer.

$$\frac{1}{3}x - \frac{1}{2}y = 1$$
$$x + 4y = -8 \quad (0, -2)$$

Graphing Calculator

Finding an Approximate Solution to a System of Equations

We can solve systems of equations graphically by using a graphing calculator. For example, to solve the system of equations in Example 2, first rewrite each equation in slope–intercept form.

$$y = -\tfrac{3}{2}x + \tfrac{5}{2}$$
$$y = 2x + 1$$

Then graph $y_1 = -\frac{3}{2}x + \frac{5}{2}$ and $y_2 = 2x + 1$ on the same screen. Display:

Next you can use the Trace and Zoom features to find the intersection of the two lines. Some graphing calculators have a command to find and calculate the intersection. Display:

Rounded to four decimal places, the solution is (0.4286, 1.8571). Observe that

$$\frac{3}{7} = 0.4285714\ldots$$

and $\dfrac{13}{7} = 1.8571428\ldots,$

so the answer agrees with the answer found in Example 2.

Find the solution to each system of equations by the substitution method. Check your answers. Write your solution in the form (x, y), (a, b), (p, q), or (s, t).

1. $4x + 3y = 9$
$x = 3y + 6$
$(3, -1)$

2. $x = 4 - 4y$
$-x + 2y = 2$
$(0, 1)$

3. $2x + y = 4$
$2x - y = 0$
$(1, 2)$

4. $7x - 3y = -10$
$x + 3y = \quad 2$
$(-1, 1)$

5. $5x + 2y = 5$
$3x + \quad y = 4$
$(3, -5)$

6. $4x + 2y = 4$
$3x + \quad y = 4$
$(2, -2)$

7. $4x - 3y = -9$
$x - \quad y = -3$
$(0, 3)$

8. $3x - \quad y = 3$
$x + 2y = -6$
$(0, -3)$

9. $2a - 3b = \quad 0$
$3a + \quad b = 22$
$(6, 4)$

10. $a - 4b = 21$
$5a + \quad b = \quad 0$
$(1, -5)$

11. $3x - \quad y = \quad 3$
$x + 3y = 11$
$(2, 3)$

12. $4x - \quad y = \quad 9$
$3x - 5y = 11$
$(2, -1)$

13. $p + 2q - 4 = 0$
$7p - \quad q - 3 = 0$
$\left(\dfrac{2}{3}, \dfrac{5}{3}\right)$

14. $7s + 2t - 10 = 0$
$5s - \quad t + \quad 5 = 0$
$(0, 5)$

15. $3x - \quad y - 9 = 0$
$8x + 5y - 1 = 0$
$(2, -3)$

16. $8x + 2y - 7 = 0$
$-2x - \quad y + 2 = 0$
$\left(\dfrac{3}{4}, \dfrac{1}{2}\right)$

Mixed Practice *Find the solution to each system of equations. Write your answer in the form (x, y).*

17. $\dfrac{5}{3}x + \dfrac{1}{3}y = -3$
$-2x + 3y = 24$
$(-3, 6)$

18. $-x + \quad y = -4$
$\dfrac{3}{7}x + \dfrac{2}{3}y = \quad 5$
$(7, 3)$

19. $4x + 5y = \quad 2$
$\dfrac{1}{5}x + \quad y = -\dfrac{7}{5}$
$(3, -2)$

20. $x + 2y = -6$
$\dfrac{2}{3}x + \dfrac{1}{3}y = -3$
$(-4, -1)$

21. $\dfrac{4}{7}x + \dfrac{2}{7}y = \quad 2$
$3x + \quad y = 13$
$(6, -5)$

22. $x - 4y = 11$
$\dfrac{2}{3}x + \dfrac{1}{2}y = \quad 1$
$(3, -2)$

23. $3x + y = -1$
$-2x - y = -1$
$(-2, 5)$

24. $x - 2y = -4$
$5x - 4y = \quad 4$
$(4, 4)$

25. $3x - 4y = 36$
$y + 3 = 3(x + 1)$
$(-4, -12)$

26. $2x + 3y = -15$
$y + 15 = 2(x + 1)$
$(3, -7)$

27. $2x = 4(2y + 2)$
$3(x - 3y) + 2 = 17$
$(8, 1)$

28. $2(2x - y - 11) = 1 - y$
$3(3x + y) - 17 = 2(4x + y)$
$(8, 9)$

To Think About

29. How many equations do you think you would need to solve a system with three unknowns? With seven unknowns? Explain your reasoning.
3; 7; An extra equation is needed for each additional variable to reduce the system to one equation with one unknown.

30. The point where the graphs of two linear equations intersect is the ___solution___ to the system of linear equations.

31. The solution to a system of two equations must satisfy ___both___ equations.

32. How many solutions will a system of equations have if the graphs of the lines intersect? Justify your answer.
One. Since the solution must satisfy both equations, the point must lie on both lines. This is the point of intersection.

33. How many solution(s) will a system of equations have if the graphs of the equations are parallel lines? Why?
No solution; parallel lines have no points in common

Applications

34. *Construction Costs* Tim Martinez is involved in a large construction project in the city. He needs to rent a heavy construction crane for several weeks. He is considering renting from one of two companies. Boston Construction will rent a crane for an initial delivery charge of $1500 and a rental fee of $900 per week. North End Contractors will rent a crane for an initial delivery charge of $500 and a rental fee of $1000 per week.

(a) Create a cost equation for each company where y is the total cost of renting a crane and x is the number of weeks the crane is rented. Write a system of equations.

(b) Solve the system of equations by the substitution method to find out how many weeks would be required for the two companies to charge the same. What would the cost be for each company?
10 weeks; $10,500

(c) Tim remembers that last year he considered renting from the same two companies. For the number of weeks he needed the crane, the cost of renting from one company was $4,000 more than the other. How many weeks was the rental last year? Which company was less expensive for that period last year?   50 weeks; Boston Construction

(a) $y = 1500 + 900x$
      $y = 500 + 1000x$

35. *Snow Removal Costs* Fred Driscoll has just moved from the sunny Southwest to the wintry Northeast, outside Buffalo, New York. His farmhouse has a very long driveway and Fred has little experience with snow. He decides to hire a snow removal company for the winter. Lake Erie Plowing requires a $50 nonrefundable reservation fee at the start of the winter and charges $90 every time they come out to plow. Adirondack Plowing requires a reservation fee of $100 but only charges $80 each time they come out to plow.

(a) Create a cost equation for each plowing company where y is the total cost of plowing and x is the number of times Fred needs to be plowed out. Write a system of equations.

(b) Solve the system of equations by the substitution method to find out how many times Fred would have to be plowed out for the cost of the plowing companies to be the same. How many times would this be? What would be the cost for each company?   5 times plowing; $500

(c) Fred calculated that if he chose one company over the other, based on the average number of times that his neighbors claimed he would need to be plowed, he would save himself $80. How many times do his neighbors expect Fred to need plowing? Which company should he hire to save that money?   13 times; Adirondack Plowing

(a) $y = 90x + 50$
      $y = 80x + 100$

Cumulative Review

36. **[7.3.2]** Find the slope and y-intercept of $7x + 11y = 19$.

$$m = -\frac{7}{11}; \left(0, \frac{19}{11}\right)$$

37. **[7.5.1]** Graph $6x + 3y \geq 9$.

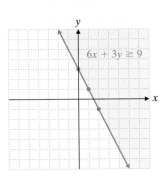

38. **[0.7.1]** *Gas Mileage* Darrin purchased a 2008 Infiniti G37 Coupe as soon as the car was available. It has a 20-gallon gasoline tank and gets 24 miles per gallon on the highway. The average price for premium gasoline (recommended for this car) was $4.20 per gallon in August 2008. Darrin travels 60 miles roundtrip on the highway each day to work and back home. How much does gasoline cost per day for Darrin to drive to work? $10.50

39. **[3.4.2]** *Holiday Spending* In 2007, the average consumer spent $64.82 on Halloween candy, decorations, and costumes. The average amount spent on costumes was $23.33. What percent of the total was spent on decorations and candy? If the average 18–24-year-old spent 10% more on Halloween expenses than the average consumer, how much did the average 18–24-year-old consumer spend on Halloween? About 64%; about $71.30

Quick Quiz 8.2 Find the solution to each system of equations by the substitution method. Write your solution in the form (x, y).

1. $10x + 3y = 8$
$\quad\ 2x + \ y = 2$ $\left(\frac{1}{2}, 1\right)$

2. $-3x - \ y = -9$
$\quad\ -x + 2y = -10$ $(4, -3)$

3. $-14x - 4y + 20 = 0$
$\quad\ 5x - \ y + \ 5 = 0$ $(0, 5)$

4. Concept Check Explain how you would solve the following system by the substitution method.

$$3x - \ y = -4$$
$$-2x + 4y = 36$$

Answers may vary

Classroom Quiz 8.2 You may use these problems to quiz your students' mastery of Section 8.2.

Find the solution to each system of equations by the substitution method. Write your solution in the form (x, y).

1. $\quad 8x + 5y = -2$
$\quad\ -6x + \ y = -8$
Ans: $(1, -2)$

2. $\ 4x - \ y = -26$
$\quad -x + 3y = \ 23$
Ans: $(-5, 6)$

3. $\quad x + 2y + \ 3 = 15$
$\quad 3x - 2y + 28 = \ 0$
Ans: $(-4, 8)$

① **Using the Addition Method to Solve a System of Two Linear Equations with Integer Coefficients**

The substitution method is useful for solving a system of equations when the coefficient of one variable is 1. In the following system, the variable x in the first equation has a coefficient of 1, and we can easily solve for x.

$$x - 2y = 7 \Rightarrow x = 7 + 2y$$
$$-5x + 4y = -5$$

This makes the substitution method a natural choice for solving this system. But we may not always have 1 as a coefficient of a variable. We need a method of solving systems of equations that will work for integer coefficients, fractional coefficients, and decimal coefficients. One such method is the **addition method.**

EXAMPLE 1 Solve by addition.

$$5x + 2y = 7 \quad \textbf{(1)}$$
$$3x - y = 13 \quad \textbf{(2)}$$

Solution We would like the coefficients of the y-terms to be opposites. One should be $+2y$ and the other $-2y$. Therefore, we multiply each term of equation **(2)** by 2.

$2(3x) - 2(y) = 2(13)$	Multiply each term of equation **(2)** by 2.
$5x + 2y = 7$	We now have an equivalent system of equations.
$\underline{6x - 2y = 26}$	Add the two equations. This will eliminate the y-variable.
$11x = 33$	
$x = 3$	Divide both sides by 11.
$5(3) + 2y = 7$	Substitute $x = 3$ into one of the original equations. Arbitrarily, we pick equation **(1)**.
$15 + 2y = 7$	Remove parentheses.
$2y = -8$	Subtract 15 from both sides.
$y = -4$	Divide both sides by 2.

The solution is $(3, -4)$.

Check. Replace x by 3 and y by -4 in *both* original equations.

$$5x + 2y = 7 \qquad\qquad 3x - y = 13$$
$$5(3) + 2(-4) \overset{?}{=} 7 \qquad 3(3) - (-4) \overset{?}{=} 13$$
$$15 - 8 \overset{?}{=} 7 \qquad\qquad 9 + 4 \overset{?}{=} 13$$
$$7 = 7 \checkmark \qquad\qquad\quad 13 = 13 \checkmark$$

Practice Problem 1 Solve by addition.

$$3x + y = 7$$
$$5x - 2y = 8 \quad (2, 1)$$

Student Learning Objectives

After studying this section, you will be able to:

① Use the addition method to solve a system of two linear equations with integer coefficients.

② Use the addition method to solve a system of two linear equations with fractional coefficients.

③ Use the addition method to solve a system of two linear equations with decimal coefficients.

Teaching Example 1 Solve by addition.
$$-x + 2y = 8$$
$$2x + 3y = 5$$

Ans: $(-2, 3)$

NOTE TO STUDENT: Fully worked-out solutions to all of the Practice Problems can be found at the back of the text starting at page SP-1

Notice that when we added the two equations together, one variable was eliminated. The addition method is therefore often called the **elimination method.**

For convenience, we will make a list of the steps we use to solve a system of equations by the addition method.

PROCEDURE FOR SOLVING A SYSTEM OF EQUATIONS BY THE ADDITION METHOD

1. Multiply each term of one or both equations by some nonzero integer so that the coefficients of one of the variables are opposites.

2. Add the equations of this new system so that one variable is eliminated.

3. Solve the resulting equation for the remaining variable.

4. Substitute the value found in step 3 into either of the original equations to find the value of the other variable.

5. Check your solution in both of the original equations.

Care should be used when the solution of a system contains fractions.

Teaching Example 2 Solve by addition.
$$5x + 2y = 4$$
$$3x + 5y = -1$$
Ans: $\left(\dfrac{22}{19}, -\dfrac{17}{19} \right)$

EXAMPLE 2 Solve by addition.

$$3x + 4y = 17 \quad \textbf{(1)}$$
$$2x + 7y = 19 \quad \textbf{(2)}$$

Solution Suppose we want the x-terms to have opposite coefficients. One equation could have $+6x$ and the other $-6x$. This would happen if we multiply equation **(1)** by $+2$ and equation **(2)** by -3.

$$(2)(3x) + (2)(4y) = (2)(17) \qquad \text{Multiply each term of equation (1) by 2.}$$
$$(-3)(2x) + (-3)(7y) = (-3)(19) \qquad \text{Multiply each term of equation (2) by } -3.$$

(Often this multiplication step can be done mentally.)

$$\begin{aligned} 6x + 8y &= 34 \\ -6x - 21y &= -57 \\ \hline -13y &= -23 \end{aligned}$$

We now have an equivalent system of equations.

The coefficients of the x-terms are opposites.

Add the two equations to eliminate the variable x.

$$y = \frac{-23}{-13} = \frac{23}{13} \qquad \text{Divide both sides by } -13.$$

Substitute $y = \dfrac{23}{13}$ into one of the original equations.

$$3x + 4\left(\frac{23}{13} \right) = 17$$

$$3x + \frac{92}{13} = 17 \qquad \text{Solve for } x.$$

$$13(3x) + 13\left(\frac{92}{13} \right) = 13(17) \qquad \text{Multiply both sides by 13 to clear the fraction.}$$

$$39x + 92 = 221 \qquad \text{Remove the parentheses.}$$

$$39x = 129 \qquad \text{Subtract 92 from both sides.}$$

$$x = \frac{43}{13} \qquad \text{Divide both sides by 39. The solution is } \left(\frac{43}{13}, \frac{23}{13} \right).$$

Teaching Tip After you have discussed Example 2 with students, present them with the following systems of equations as an additional class exercise. In each case ask them by what they should multiply the top equation and by what they should multiply the bottom equation to most easily eliminate the x-variable.

(a) $\quad 2x + 5y = \quad 3$
$\quad -5x + 3y = -23$

(b) $\quad 5x + 4y = \quad 5$
$\quad -7x + 6y = -36$

(c) $\quad 9x + 5y = \quad 14$
$\quad 4x - 7y = \quad -3$

Ans:

(a) Multiply top by 5, bottom by 2

(b) Multiply top by 7, bottom by 5

(c) Multiply top by 4, bottom by -9

Alternative solution: You could also have eliminated the y-variable. For example, if you multiply equation **(1)** by 7 and equation **(2)** by -4, you obtain the equivalent system shown here.

$$21x + 28y = 119$$
$$-8x - 28y = -76$$

You can add these equations to eliminate the y-variable. Since the numbers involved in this approach are somewhat larger, it is probably wiser to eliminate the x-variable in this example.

CAUTION: A common error occurs when students forget that we want to obtain two terms with *opposite signs* that when added will equal zero. If all the coefficients in the equations are positive, such as in Example 2, it will be necessary to multiply *one* of the equations by a negative number.

Practice Problem 2 Solve by addition.

$$4x + 5y = 17 \quad \left(\frac{59}{13}, -\frac{3}{13}\right)$$
$$3x + 7y = 12$$

NOTE TO STUDENT: Fully worked-out solutions to all of the Practice Problems can be found at the back of the text starting at page SP-1

Using the Addition Method to Solve a System of Two Linear Equations with Fractional Coefficients

If the system of equations has fractional coefficients, you should first clear each equation of the fractions. To do so, you will need to multiply each term in the equation by the LCD of the fractions.

EXAMPLE 3 Solve.

$$x - \frac{5}{2}y = \frac{5}{2} \quad \textbf{(1)}$$

$$\frac{4}{3}x + y = \frac{23}{3} \quad \textbf{(2)}$$

Teaching Example 3 Solve by addition.

$$x + \frac{2}{3}y = 5$$

$$\frac{2}{5}x + y = 2$$

Ans: $(5, 0)$

Solution

$$2(x) - 2\left(\frac{5}{2}y\right) = 2\left(\frac{5}{2}\right) \qquad \text{Multiply each term of equation (1) by 2.}$$

$$2x - 5y = 5$$

$$3\left(\frac{4}{3}x\right) + 3(y) = 3\left(\frac{23}{3}\right) \qquad \text{Multiply each term of equation (2) by 3.}$$

$$4x + 3y = 23$$

We now have an equivalent system of equations that does not contain fractions.

$$2x - 5y = 5 \quad \textbf{(3)}$$
$$4x + 3y = 23 \quad \textbf{(4)}$$

Let us eliminate the x-variable. We want the coefficients of x to be opposites.

$$(-2)(2x) - (-2)5y = (-2)(5) \quad \text{Multiply each term of equation (3) by } -2.$$
$$\begin{array}{rl} -4x + 10y = -10 & \text{We now have an equivalent system of equations.} \\ \underline{4x + 3y = 23} & \text{The coefficients of the } x\text{-terms are opposites.} \\ 13y = 13 & \text{Add the two equations to eliminate the variable } x. \\ y = 1 & \text{Divide both sides by 13.} \\ 4x + 3(1) = 23 & \text{Substitute } y = 1 \text{ into equation (4).} \\ 4x + 3 = 23 & \\ 4x = 20 & \\ x = 5 & \end{array}$$

The solution is $(5, 1)$.

Check. Check the solution in both of the *original* equations.

Practice Problem 3 Solve.

$$\frac{2}{3}x - \frac{3}{4}y = 3$$

$$-2x + y = 6 \quad (-9, -12)$$

Using the Addition Method to Solve a System of Two Linear Equations with Decimal Coefficients

Some linear equations will have decimal coefficients. It will be easier to work with the equations if we change the decimal coefficients to integer coefficients. To do so, we will multiply each term of the equation by a power of 10.

EXAMPLE 4 Solve.

$$0.12x + 0.05y = -0.02 \quad \textbf{(1)}$$
$$0.08x - 0.03y = -0.14 \quad \textbf{(2)}$$

Solution Since the decimals are hundredths, we will multiply each term of both equations by 100.

$$100(0.12x) + 100(0.05y) = 100(-0.02)$$
$$100(0.08x) + 100(-0.03y) = 100(-0.14)$$

$$12x + 5y = -2 \quad \textbf{(3)} \qquad \text{We now have an equivalent system of equations that has integer coefficients.}$$

$$8x - 3y = -14 \quad \textbf{(4)}$$

We will eliminate the variable y. We want the coefficients of the y-terms to be opposites.

$$\begin{array}{ll} 36x + 15y = -6 & \text{Multiply equation \textbf{(3)} by 3 and equation \textbf{(4)} by 5.} \\ \underline{40x - 15y = -70} & \\ 76x \qquad\quad = -76 & \text{Add the equations.} \\ x = -1 & \text{Solve for } x. \\ 12(-1) + 5y = -2 & \text{Replace } x \text{ by } -1 \text{ in equation \textbf{(3)}, and solve for } y. \\ -12 + 5y = -2 & \\ 5y = -2 + 12 & \\ 5y = 10 & \\ y = 2 & \end{array}$$

The solution is $(-1, 2)$.

Check. We substitute $x = -1$ and $y = 2$ into the original equations. Most students probably would rather not use equation **(1)** and equation **(2)**! However, *we must check the solutions in the original equations* if we want to be sure of our answers. It is possible to make an error going from equations **(1)** and **(2)** to equations **(3)** and **(4)**. The solution could satisfy equations **(3)** and **(4)**, but might not satisfy equations **(1)** and **(2)**.

$$\begin{array}{ll} 0.12x + 0.05y = -0.02 & 0.08x - 0.03y = -0.14 \\ 0.12(-1) + 0.05(2) \stackrel{?}{=} -0.02 & 0.08(-1) - 0.03(2) \stackrel{?}{=} -0.14 \\ -0.12 + 0.10 \stackrel{?}{=} -0.02 & -0.08 - 0.06 \stackrel{?}{=} -0.14 \\ -0.02 = -0.02 \ \checkmark & -0.14 = -0.14 \ \checkmark \end{array}$$

Practice Problem 4 Solve.

$$0.2x + 0.3y = -0.1$$
$$0.5x - 0.1y = -1.1 \quad (-2, 1)$$

CAUTION: A common error when solving a system is to find the value for x and then stop. You have not solved a system in x and y until you find both the x- and y-values that make both equations true.

Verbal and Writing Skills *Look at the following systems of equations. Decide which variable to eliminate in each system, and explain how you would eliminate that variable.*

1. $3x + 2y = 5$ Eliminate y by multiplying the second
   $5x - y = 3$ equation by 2, then adding the equations.

2. $2x - 9y = 1$ Eliminate x by multiplying the first equa-
   $2x + 3y = 2$ tion by -1, then adding the equations.

3. $4x - 3y = 10$ Eliminate y by multiplying the first equa-
   $5x + 4y = 0$ tion by 4, multiplying the second equation by 3, then adding the equations.

4. $7x + 6y = 13$ Eliminate x by multiplying the first equa-
   $-2x + 5y = 3$ tion by 2, multiplying the second equation by 7, then adding the equations.

Find the solution by the addition method. Check your answers.

5. $-x + y = -3$
   $-2x - y = 6$
   $(-1, -4)$

6. $-x + y = 2$
   $x + y = 4$
   $(1, 3)$

7. $2x + 3y = 1$
   $x - 2y = 4$
   $(2, -1)$

8. $2x + y = 4$
   $3x - 2y = -1$
   $(1, 2)$

9. $6x - y = -7$
   $6x + 2y = 5$
   $\left(-\frac{1}{2}, 4\right)$

10. $3x - 2y = -11$
   $3x + y = 1$
   $(-1, 4)$

11. $5x - 15y = 9$
   $-x + 10y = 1$
   $\left(3, \frac{2}{5}\right)$

12. $-4x + 5y = -16$
   $8x + y = -1$
   $\left(\frac{1}{4}, -3\right)$

13. $2x + 5y = 2$
   $3x + y = 3$
   $(1, 0)$

14. $5x - 3y = 14$
   $2x - y = 6$
   $(4, 2)$

15. $8x + 6y = -2$
   $10x - 9y = -8$
   $\left(-\frac{1}{2}, \frac{1}{3}\right)$

16. $4x + 9y = 0$
   $8x - 5y = -23$
   $\left(-\frac{9}{4}, 1\right)$

17. $2x + 3y = -8$
   $5x + 4y = -34$
   $(-10, 4)$

18. $3x - 5y = 11$
   $4x + 3y = 5$
   $(2, -1)$

19. $4x + 3y = 1$
   $2x - 5y = -19$
   $(-2, 3)$

20. $-4x - 5y = 2$
   $6x + 9y = -6$
   $(2, -2)$

21. $12x - 6y = -2$
   $-9x - 7y = -10$
   $\left(\frac{1}{3}, 1\right)$

22. $2x - 4y = -22$
   $-6x + 3y = 3$
   $(3, 7)$

Verbal and Writing Skills *Before using the addition method to solve each system, you will need to simplify the equation(s). Explain how you would change the fractional coefficients to integer coefficients in each system.*

23. $\frac{1}{4}x - \frac{3}{4}y = 2$

   $2x + 3y = 1$

   Multiply equation (1) by 4.

24. $5x + 2y = 9$

   $\frac{4}{9}x - \frac{2}{3}y = 4$

   Multiply equation (2) by 9.

25. $\frac{1}{2}x + \frac{2}{3}y = \frac{1}{3}$

   $\frac{3}{4}x - \frac{4}{5}y = 2$

   Multiply equation (1) by 6.
   Multiply equation (2) by 20.

26. $\frac{1}{4}x + \frac{1}{3}y = 15$

   $\frac{1}{5}x + \frac{3}{10}y = 13$

   Multiply equation (1) by 12.
   Multiply equation (2) by 10.

Find the solution. Check your answers.

27. $x + \dfrac{5}{4}y = \dfrac{9}{4}$

$\dfrac{2}{5}x - y = \dfrac{3}{5}$

$\left(2, \dfrac{1}{5}\right)$

28. $\dfrac{2}{3}x + y = 2$

$x + \dfrac{1}{2}y = 7$

$(9, -4)$

29. $\dfrac{x}{6} + \dfrac{y}{2} = -\dfrac{1}{2}$

$x - 9y = 21$

$(3, -2)$

30. $\dfrac{3}{2}x - \dfrac{y}{8} = -1$

$16x + 3y = -28$

$(-1, -4)$

31. $\dfrac{1}{3}x + 10y = -2$

$-3x + 10y = -12$

$\left(3, -\dfrac{3}{10}\right)$

32. $9x - 5y = 14$

$\dfrac{3}{10}x + \dfrac{5}{2}y = 1$

$\left(\dfrac{5}{3}, \dfrac{1}{5}\right)$

33. $\dfrac{2}{3}x + \dfrac{3}{5}y = -\dfrac{1}{5}$

$\dfrac{1}{4}x + \dfrac{1}{3}y = \dfrac{1}{4}$

$(-3, 3)$

34. $\dfrac{x}{5} + \dfrac{1}{2}y = \dfrac{9}{10}$

$\dfrac{x}{3} + \dfrac{3}{4}y = \dfrac{5}{4}$

$(-3, 3)$

Verbal and Writing Skills *Explain how you would change each system to an equivalent system with integer coefficients. Write the equivalent system.*

35. $0.5x - 0.3y = 0.1$

$5x + 3y = 6$

Multiply equation (1) by 10;
$5x - 3y = 1$
$5x + 3y = 6$

36. $0.08x + y = 0.05$

$2x - 0.1y = 3$

Multiply equation (1) by 100,
multiply equation (2) by 10;
$8x + 100y = 5$
$20x - y = 30$

37. $4x + 0.5y = 9$

$0.2x - 0.05y = 1$

Multiply equation (1) by 10,
multiply equation (2) by 100;
$40x + 5y = 90$
$20x - 5y = 100$

Solve for x and y using the addition method.

38. $0.5x - 0.2y = 0.4$

$-0.3x + 0.4y = 0.6$

$(2, 3)$

39. $0.2x + 0.3y = 0.4$

$0.5x + 0.4y = 0.3$

$(-1, 2)$

40. $0.02x - 0.04y = 0.26$

$0.07x - 0.09y = 0.66$

$(3, -5)$

41. $0.04x - 0.03y = 0.05$

$0.05x + 0.08y = -0.76$

$(-4, -7)$

42. $0.4x - 5y = -1.2$

$-0.03x + 0.5y = 0.14$

$\left(2, \dfrac{2}{5}\right)$

43. $-0.6x - 0.08y = -4$

$3x + 2y = 4$

$(8, -10)$

Mixed Practice *Solve for x and y using the addition method.*

44. $x + 6y = -1$

$-2x - 8y = 4$

$\left(-4, \dfrac{1}{2}\right)$

45. $3x - y = -8$

$-9x + 2y = 18$

$\left(-\dfrac{2}{3}, 6\right)$

46. $\dfrac{5}{4}x + y = 16$

$x - \dfrac{5}{3}y = -2$

$(8, 6)$

47. $\dfrac{1}{3}x - \dfrac{1}{4}y = 0$

$\dfrac{1}{6}x - \dfrac{1}{2}y = -6$

$(12, 16)$

48. $0.2x + 0.1y = 1.1$

$0.01x + 0.03y = 0.18$

$(3, 5)$

49. $0.05x - 0.04y = -0.08$

$0.2x + 0.3y = 0.6$

$(0, 2)$

To Think About *Find the solution.*

50. $4(x - 2y) = 5 - (y - 3x)$
$-5(x + 1) = y - 6x$
$(5, 0)$

51. $3(3x + 2) = y + 5x$
$4 = -2(x - y)$
$\left(-\dfrac{4}{3}, \dfrac{2}{3}\right)$

Cumulative Review

52. **[6.6.3]** *Race Car Speeds* Carlos and Bobby are professional race car drivers. They are in a 100-lap race on an oval track that is 4.75 miles long. Carlos makes 4 complete laps during the time that Bobby is out of the race for repairs to his car. If Carlos maintains an average speed of 95 miles per hour for the entire race, what average speed will Bobby have to achieve in order to beat Carlos? (Round to the nearest whole number.) 99 mi/hr or greater

53. **[6.6.1]** *Baseball* At one point during a recent season, the Boston Red Sox had won 45 baseball games and lost 55 games. Noah and Olivia were discussing the potential for the Red Sox to win 50 percent of all their games for the season. If the team had 62 games remaining at that point, how many of the future games would they need to win in order to achieve that percent? 36 games

54. **[5.6.1]** Factor $64x^2 - 4$ completely.
$4(4x + 1)(4x - 1)$

55. **[5.6.1]** Factor. $9x^2 - 15x + 4$
$(3x - 4)(3x - 1)$

Quick Quiz 8.3 Find the solution to the system by the addition method. Express your answer in the form (x, y).

1. $3x + 5y = -1$
$-5x + 4y = -23$
$(3, -2)$

2. $-7x + 3y = -31$
$4x + 6y = 10$
$(4, -1)$

3. $\dfrac{1}{3}x + y = \dfrac{8}{3}$
$\dfrac{4}{5}x - \dfrac{2}{5}y = \dfrac{18}{5}$
$(5, 1)$

4. Concept Check Explain how you would obtain an equivalent system that does not have decimals if you wanted to solve the following system.

$$0.5x + 0.4y = -3.4$$
$$0.02x + 0.03y = -0.08$$

Answers may vary

Classroom Quiz 8.3 You may use these problems to quiz your students' mastery of Section 8.3.

Find the solution to the system by the addition method. Express your answer in the form (x, y).

1. $5x + 3y = 27$
$-7x + 2y = -13$
Ans: $(3, 4)$

2. $3x + 2y = 10$
$10x - 8y = 48$
Ans: $(4, -1)$

3. $\dfrac{1}{2}x - \dfrac{1}{4}y = 1$
$\dfrac{4}{3}x + 4y = 12$
Ans: $(3, 2)$

1. $(-2, 4)$

2. $(-3, -2)$

3. $(6, -4)$

4. $(0, -2)$

5. $(-2, 3)$

6. $(1, -2)$

7. $(7, 1)$

8. $\left(-3, \dfrac{1}{2}\right)$

9. $(-9, -12)$

10. $(1, -3)$

How are you doing with your homework assignments in Sections 8.1 to 8.3? Do you feel you have mastered the material so far? Do you understand the concepts you have covered? Before you go further in the textbook, take some time to do each of the following problems.

8.1

Solve by graphing.

1.
$$2x - y = -8$$
$$-x + 2y = 10$$

2.
$$-2x + y = 4$$
$$x - 3y = 3$$

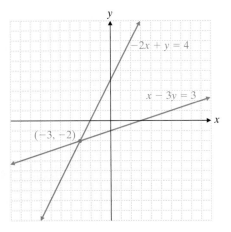

8.2

Solve by the substitution method.

3.
$$5x + y = 26$$
$$3x + 2y = 10$$

4.
$$3x - 5y = 10$$
$$x + 4y = -8$$

5.
$$2x + 3y = 5$$
$$\frac{1}{2}x + \frac{1}{6}y = -\frac{1}{2}$$

6.
$$7x - 5y - 17 = 0$$
$$x + 8y + 15 = 0$$

8.3

Solve by the addition method.

7.
$$3x - 2y = 19$$
$$x + y = 8$$

8.
$$3x + 10y = -4$$
$$2x - 8y = -10$$

9.
$$\frac{2}{3}x - \frac{3}{4}y = 3$$
$$\frac{1}{9}x + \frac{1}{12}y = -2$$

10.
$$0.6x - 0.5y = 2.1$$
$$0.4x - 0.3y = 1.3$$

Now turn to page SA-20 for the answer to each of these problems. Each answer also includes a reference to the objective in which the problem is first taught. If you missed any of these problems, you should stop and review the Examples and Practice Problems in the referenced objective. A little review now will help you master the material in the upcoming sections of the text.

Choosing an Appropriate Method to Solve a System of Equations Algebraically

At this point we will review the algebraic methods for solving systems of linear equations and discuss the advantages and disadvantages of each method.

Method	Advantage	Disadvantage
Substitution	Works well if one or more variables has a coefficient of 1 or −1.	Often becomes difficult to use if no variable has a coefficient of 1 or −1.
Addition	Works well if equations have fractional or decimal coefficients. Works well if no variable has a coefficient of 1 or −1.	None

EXAMPLE 1 Select a method and solve each system of equations.

(a) $x + y = 3080$
$2x + 3y = 8740$

(b) $5x - 2y = 19$
$-3x + 7y = 35$

Solution

(a) Since there are x- and y-values that have coefficients of 1, we will select the substitution method.

$$y = 3080 - x \qquad \text{Solve the first equation for } y.$$
$$2x + 3(3080 - x) = 8740 \qquad \text{Substitute the expression into the second equation.}$$
$$2x + 9240 - 3x = 8740 \qquad \text{Remove parentheses.}$$
$$-1x = -500 \qquad \text{Simplify.}$$
$$x = 500 \qquad \text{Divide each side by } -1.$$
$$y = 3080 - x$$
$$= 3080 - 500 \qquad \text{Substitute 500 for } x.$$
$$= 2580 \qquad \text{Simplify.}$$

The solution is $(500, 2580)$.

(b) Because none of the x- and y-variables has a coefficient of 1 or −1, we select the addition method. We choose to eliminate the y-variable. Thus, we would like the coefficients of y to be −14 and 14.

$$7(5x) - 7(2y) = 7(19) \qquad \text{Multiply each term of the first equation by 7.}$$
$$2(-3x) + 2(7y) = 2(35) \qquad \text{Multiply each term of the second equation by 2.}$$
$$35x - 14y = 133 \qquad \text{We now have an equivalent system of equations.}$$
$$\underline{-6x + 14y = 70}$$
$$29x = 203 \qquad \text{Add the two equations.}$$
$$x = 7 \qquad \text{Divide each side by 29.}$$

Substitute $x = 7$ into one of the original equations.

$$5(7) - 2y = 19$$
$$35 - 2y = 19 \qquad \text{Solve for } y.$$
$$-2y = -16$$
$$y = 8$$

The solution is $(7, 8)$.

Student Learning Objectives

After studying this section, you will be able to:

 Choose an appropriate method to solve a system of equations algebraically.

 Use algebraic methods to identify inconsistent and dependent systems.

Teaching Example 1 Select a method and solve each system of equations.

(a) $x - y = 61$
$5x + y = 137$

(b) $4x + 3y = 1$
$6x - 5y = 11$

Ans: **(a)** $(33, -28)$
(b) $(1, -1)$

NOTE TO STUDENT: Fully worked-out solutions to all of the Practice Problems can be found at the back of the text starting at page SP-1

Practice Problem 1 Select a method and solve each system of equations.

(a) $3x + 5y = 1485$

$\quad x + 2y = 564$ $(150, 207)$

(b) $7x + 6y = 45$

$\quad 6x - 5y = -2$ $(3, 4)$

2 Using Algebraic Methods to Identify Inconsistent and Dependent Systems

Recall that an inconsistent system of linear equations is a system of parallel lines. Since parallel lines never intersect, the system has no solution. Can we determine this algebraically?

Teaching Tip Note that when *all letters* vanish but nonzero numbers don't vanish and the result is *not true,* we have parallel lines and *no solution.*

EXAMPLE 2 Solve algebraically.

$$3x - y = -1$$
$$3x - y = -7$$

Solution Clearly, the addition method would be very convenient in this case.

$\quad 3x - y = -1$ Keep the first equation unchanged.

$\underline{-3x + y = +7}$ Multiply each term in the second equation by -1.

$\qquad\quad 0 = 6$ Add the two equations.

Teaching Example 2 Solve algebraically.

$$5x - 2y = 6$$
$$-10x + 4y = 1$$

Ans: No solution

Notice we have $0 = 6$, which we know is not true. The statement $0 = 6$ is inconsistent with known mathematical facts. No possible x- and y-values can make this equation true. Thus there is **no solution to this system of equations.**

This example shows us that, if we obtain a mathematical statement that is not true (inconsistent with known facts), we can identify the entire system as inconsistent. If graphed, these lines would be parallel.

Practice Problem 2 Solve algebraically.

$$4x + 2y = 2$$
$$-6x - 3y = 6$$ No solution

What happens if we try to solve a dependent system of equations algebraically? Recall that a dependent system consists of two lines that coincide (are the same line).

Teaching Tip Note that when *all letters* vanish and the result is a *true statement* (including $0 = 0$), the two equations represent the *same line* and have *an infinite number of solutions.*

EXAMPLE 3 Solve algebraically.

$$x + y = 4$$
$$3x + 3y = 12$$

Solution Let us use the substitution method.

$\qquad\qquad\qquad y = 4 - x$ Solve the first equation for y.

$\quad 3x + 3(4 - x) = 12$ Substitute $4 - x$ for y in the second equation.

$\quad 3x + 12 - 3x = 12$ Remove parentheses.

$\qquad\qquad\quad 12 = 12$

Teaching Example 3 Solve algebraically.

$$x - y = 2$$
$$-5x + 5y = -10$$

Ans: Infinite number of solutions

Notice we have $12 = 12$, which is always true. Thus we obtain an equation that is true for any value of x. An equation that is always true is called an **identity.** All the solutions of one equation of the system are also solutions of the other equation.

Thus the lines coincide. There is an **infinite number of solutions to this system of equations.**

This example shows us that, if we obtain a mathematical statement that is always true (an identity), we can identify the equations as dependent. There is an unlimited number of solutions to a system that has dependent equations.

Practice Problem 3   Solve algebraically.

$$3x - 9y = 18$$
$$-4x + 12y = -24 \quad \text{Infinite number of solutions}$$

TO THINK ABOUT: Two Linear Equations with Two Variables   Now is a good time to look back over what we have learned. When you graph a system of two linear equations, what possible kinds of graphs will you obtain?

What will happen when you try to solve a system of two linear equations using algebraic methods? How many solutions are possible in each case? The following chart may help you to organize your answers to these questions.

Graph	Number of Solutions	Algebraic Interpretation
Two lines intersect at one point	**One unique solution**	You obtain one value for x and one value for y. For example, $$x = 6, \quad y = -3.$$
Parallel lines	**No solution**	You obtain an equation that is inconsistent with known facts. For example, $$0 = 6.$$ The system of equations is inconsistent.
Lines coincide	**Infinite number of solutions**	You obtain an equation that is always true. For example, $$8 = 8.$$ The equations are dependent.

Think about each of the three possibilities in this chart. You need to understand each one. As you do each of exercises 5–28, you will need to identify which of the three possibilities is involved.

Teaching Tip After discussing Example 3, ask students to do the following as an additional class exercise.

Find the solution for each of the following systems if possible. If there is no solution, so state. If the system has an infinite number of solutions, so state.

(a) $-4x + 3y = 0$
$\quad\ \ 5x - 6y = 9$

(b) $\ \ 6x - 2y = \ \ 8$
$\quad -9x + 3y = -12$

(c) $-15x - 18y = 10$
$\quad\ \ \ 5x + \ \ 6y = -2$

Ans: (a) One solution: $(-3, -4)$
$\qquad$ **(b)** Infinite number of solutions
$\qquad$ **(c)** No solution

Verbal and Writing Skills

1. If there is no solution to a system of linear equations, the graphs of the equations are <u>parallel lines</u> . Solving the system algebraically, you will obtain an equation that is <u>inconsistent</u> with known facts.

2. If an algebraic attempt at solving a system of linear equations results in an identity, the system is said to be <u>dependent</u> . There are an <u>infinite</u> number of solutions, and the graphs of the lines <u>coincide</u> .

3. If there is exactly one solution, the graphs of the equations <u>intersect</u> . This system is said to be <u>independent</u> and <u>consistent</u> .

4. A student solves a system of equations and obtains the equation $-36x = 0$. He is unsure of what to do next. What should he do next? What should he conclude about the system of equations?
He should solve for x, $x = 0$, then substitute back into one of the original equations and solve for y. The system is independent and consistent.

If possible, solve by an algebraic method (substitution or addition method, without the use of graphing). Otherwise, state that the problem has no solution or an infinite number of solutions.

5. $-2x - 3y = 15$
$5x + 2y = 1$
$(3, -7)$

6. $3x + 4y = 8$
$-5x - 3y = 5$
$(-4, 5)$

7. $3x - 4y = 2$
$-5x + 6y = -12$
$(18, 13)$

8. $4x + 3y = 7$
$6x + 5y = 15$
$(-5, 9)$

9. $2x - 4y = 5$
$-4x + 8y = 9$
No solution

10. $3x + 6y = 12$
$x + 2y = 7$
No solution

11. $-5x + 2y = 2$
$15x - 6y = -6$
Infinite number of solutions

12. $6x - 4y = 8$
$-9x + 6y = -12$
Infinite number of solutions

13. $5x - 3y = 13$
$7x + 2y = 43$
$(5, 4)$

14. $-4x + 5y = 10$
$6x - 7y = 8$
$(55, 46)$

15. $3x - 2y = 70$
$0.6x + 0.5y = 50$
$(50, 40)$

16. $5x - 3y = 10$
$0.9x + 0.4y = 30$
$(20, 30)$

17. $0.2x - 0.3y = 0.1$
$-0.5x + 0.8y = 0$
$(8, 5)$

18. $0.05x + 0.04y = 0.12$
$0.03x - 0.08y = 0.28$
$(4, -2)$

19. $\frac{4}{3}x + \frac{1}{2}y = 1$
$\frac{1}{3}x - y = -\frac{1}{2}$ $\left(\frac{1}{2}, \frac{2}{3}\right)$

20. $\frac{3}{5}x - \frac{3}{4}y = \frac{1}{5}$
$2x + \frac{3}{2}y = 2$ $\left(\frac{3}{4}, \frac{1}{3}\right)$

21. $\frac{2}{3}x + \frac{1}{6}y = 2$
$\frac{1}{4}x - \frac{1}{2}y = -\frac{3}{4}$ $\left(\frac{7}{3}, \frac{8}{3}\right)$

22. $\frac{1}{2}x + \frac{2}{5}y = \frac{9}{10}$
$\frac{2}{3}x + \frac{2}{15}y = \frac{1}{2}$ $\left(\frac{2}{5}, \frac{7}{4}\right)$

Mixed Practice *Solve by any algebraic method. If there is not one solution to a system of equations, state the reason.*

23. $4x + 3y = -2$
$5x - 2y = 9$
$(1, -2)$

24. $3x = 4(y + 9)$
$y + 3 = 3(x + 1)$
$(-4, -12)$

25. $\frac{2}{3}x + y = 1$
$\frac{3}{4}x + \frac{1}{2}y = \frac{3}{4}$ $\left(\frac{3}{5}, \frac{3}{5}\right)$

26. $0.8x - 0.3y = 0.7$
$1.2x + 0.6y = 4.2$
$(2, 3)$

27. $x - 2y + 2 = 0$

$3(x - 2y + 1) = 15$

No solution. The system is inconsistent.

28. $3(x - 2y + 2) = -3$

$-\dfrac{1}{3}x + \dfrac{2}{3}y = 1$

Infinite number of solutions. The equations are dependent.

To Think About *Remove parentheses and solve the system for* (a, b).

29. $2(a + 3) = b + 1$

$3(a - b) = a + 1$

$(-4, -3)$

30. $3(a + b) - 2 = b + 2$

$a - 2(b - 1) = 3a - 4$

$(-2, 5)$

Cumulative Review *Solve for* x.

31. **[2.4.1]** $8x + 15\left(\dfrac{2}{3}x - \dfrac{2}{5}\right) = -10$ $x = -\dfrac{2}{9}$

32. **[2.3.3]** $0.2(x + 0.3) + x = 0.3(x + 0.5)$ $x = 0.1$

In 2007, federal income taxes for single filers could be estimated using the following chart.

If taxable income is over . . .	But not over . . .	Estimated income tax is . . .
$0	$7825	10% of amount over $0
$7825	$31,850	$782.50 + 15% of amount over $7825
$31,850	$77,100	$4386.25 + 25% of amount over $31,850
$77,100	$160,850	$15,698.75 + 28% of amount over $77,100

Source: www.irs.gov

33. **[0.7.1]** In 2007, Margot earned $27,600. Find the estimated federal income tax she had to pay.
$3748.75

34. **[0.7.1]** In 2007, Dean earned $32,900. During the year, a total of $5033 was deducted from his paychecks for federal income tax. Was this more or less than what he should have paid? Estimate the amount of his federal income tax refund.   More; $384.25

Quick Quiz 8.4 Solve if possible. Otherwise, state that the problem has no solution or has an infinite number of solutions.

1. $12 - x = 3(x + y)$

$3x + 14 = 4x + 5(x + y)$

$(9, -8)$

2. $-4x + 6y = 2$

$12y = 8x + 12$

No solution

3. $\dfrac{1}{2}x - \dfrac{1}{4}y = 1$

$\dfrac{4}{3}x + 4y = 12$

$(3, 2)$

4. **Concept Check** Explain how you would remove the fractions in order to solve the following system.

$\dfrac{3}{10}x + \dfrac{2}{5}y = \dfrac{1}{2}$

$\dfrac{1}{8}x - \dfrac{3}{16}y = -\dfrac{1}{2}$ Answers may vary

Classroom Quiz 8.4 You may use these problems to quiz your students' mastery of Section 8.4.

Solve if possible. Otherwise, state that the problem has no solution or has an infinite number of solutions.

1. $5(x + 1) = 4(1 - y)$

$7(x + 2) - 3 = 9 - 6y$ **Ans:** $\left(1, -\dfrac{3}{2}\right)$

2. $9x - 6y = 3$

$4y = 6x - 2$ **Ans:** infinite number of solutions

3. $\dfrac{1}{2}x + \dfrac{1}{3}y = \dfrac{2}{3}$

$\dfrac{2}{3}x + \dfrac{2}{5}y = \dfrac{14}{15}$ **Ans:** $(2, -1)$

Using a System of Equations to Solve Word Problems

Word problems can be solved in a variety of ways. In Chapter 3 and throughout the text, you have used one equation with one unknown to describe situations found in real-life applications. Some of these same problems can be solved using two equations in two unknowns. The method you use will depend on what you find works best for you. Become familiar with both methods before you decide.

EXAMPLE 1 A worker in a large post office is trying to verify the rate at which two electronic card-sorting machines operate. Yesterday the first machine sorted for 3 minutes and the second machine sorted for 4 minutes. The total workload both machines processed during that time period was 10,300 cards. Two days ago the first machine sorted for 2 minutes and the second machine for 3 minutes. The total workload both machines processed during that time period was 7400 cards. Can you determine the number of cards per minute sorted by each machine?

Solution

1. *Understand the problem.* The number of cards processed by two different machines on two different days provides the basis for two linear equations with two unknowns.

   The two equations will represent what occurred on two different days:

   (1) yesterday and

   (2) two days ago.

   The number of cards sorted by each machine is what we need to find.

   Let x = the number of cards per minute sorted by the first machine and

   y = the number of cards per minute sorted by the second machine.

2. *Write the equations.* Yesterday the first machine sorted for 3 minutes and the second machine sorted for 4 minutes, processing a total of 10,300 cards.

first machine	second machine	total no. of cards
3(no. of cards per minute)	4(no. of cards per minute)	
$3x$ +	$4y$ =	10,300 **(1)**

   Two days ago the first machine sorted for 2 minutes and the second machine sorted for 3 minutes, processing a total of 7400 cards.

first machine	second machine	total no. of cards
2(no. of cards per minute)	3(no. of cards per minute)	
$2x$ +	$3y$ =	7400 **(2)**

3. *Solve and state the answer.* We will use the addition method to solve this system.

$$3x + 4y = 10,300 \quad \textbf{(1)}$$
$$2x + 3y = 7400 \quad \textbf{(2)}$$

$6x + 8y = 20,600$	Multiply equation **(1)** by 2.
$\underline{-6x - 9y = -22,200}$	Multiply equation **(2)** by -3.
$-1y = -1600$	Add the two equations to eliminate x and solve for y.
$y = 1600$	
$2x + 3(1600) = 7400$	Substitute the value of y into equation **(2)**.
$2x + 4800 = 7400$	Simplify and solve for x.
$2x = 2600$	
$x = 1300$	

Thus, the first machine sorts 1300 cards per minute and the second machine sorts 1600 cards per minute.

4. Check. We can verify each statement in the problem.

Yesterday: Were 10,300 cards sorted?

$$3(1300) + 4(1600) \stackrel{?}{=} 10{,}300$$

$$3900 + 6400 \stackrel{?}{=} 10{,}300$$

$$10{,}300 = 10{,}300 \checkmark$$

Two days ago: Were 7400 cards sorted?

$$2(1300) + 3(1600) \stackrel{?}{=} 7400$$

$$2600 + 4800 \stackrel{?}{=} 7400$$

$$7400 = 7400 \checkmark$$

Practice Problem 1 After a recent disaster, the Red Cross brought in a pump to evacuate water from a basement. After 3 hours a larger pump was brought in and the two ran for an additional 5 hours, removing 49,000 gallons of water. The next day, after running both pumps for 3 hours, the larger one was taken to another site. The smaller pump finished the job after another 2 hours. If on the second day 30,000 gallons were pumped, how much water does each pump remove per hour?   Smaller pump, 3000 gal/hr;   larger pump, 5000 gal/hr

NOTE TO STUDENT: Fully worked-out solutions to all of the Practice Problems can be found at the back of the text starting at page SP-1

EXAMPLE 2 Fred is considering working for a company that sells encyclopedias. Fred found out that all starting representatives receive the same annual base salary and a standard commission of a certain percentage of the sales they make during the first year. He has been told that one representative sold $50,000 worth of encyclopedias her first year and that she earned $14,000. He was able to find out that another representative sold $80,000 worth of encyclopedias and that he earned $17,600. Determine the base salary and the commission rate of a beginning sales representative.

Teaching Example 2 While shopping together, Jan and Celia found a good sale on summer clothing for children. Jan bought 8 shirts and 5 pairs of shorts for her children. Celia bought 4 shirts and 7 pairs of shorts for her children. Jan paid $86 and Celia paid $70. How much did each shirt and each pair of shorts cost?

Ans: Each shirt cost $7 and each pair of shorts cost $6.

Solution   The earnings of the two different sales representatives will represent the two equations, each of the form

$$\text{base salary} + \text{commission} = \text{total earnings}.$$

What is unknown? The base salary and the commission rate are what we need to find.

Let b = the base salary and

c = the commission rate.

$b + 50{,}000c = 14{,}000$ **(1)**   Earnings for the first sales representative.

$b + 80{,}000c = 17{,}600$ **(2)**   Earnings for the second sales representative.

We will use the addition method to solve these equations.

$$-b - 50{,}000c = -14{,}000 \quad \text{Multiply equation \textbf{(1)} by } -1.$$

$$\underline{b + 80{,}000c = 17{,}600} \quad \text{Leave equation \textbf{(2)} unchanged.}$$

$$30{,}000c = 3600 \quad \text{Add the two equations and solve for } c.$$

$$c = \frac{3600}{30{,}000}$$

$$c = 0.12 \qquad \text{The commission rate is 12\%.}$$

$$b + (50{,}000)(0.12) = 14{,}000 \qquad \text{Substitute this value into equation \textbf{(1)}.}$$

$$b + 6000 = 14{,}000 \qquad \text{Simplify and solve for } b.$$

$$b = 8000 \qquad \text{The base salary is \$8000 per year.}$$

Thus the base salary is $8000 per year, and the commission rate is 12%.

NOTE TO STUDENT: Fully worked-out solutions to all of the Practice Problems can be found at the back of the text starting at page SP-1

Practice Problem 2 Rick has two cars, one using premium gas and the other using regular gas. Last week Rick bought 7 gallons of unleaded premium and 8 gallons of unleaded regular gasoline and paid $47.15. This week he purchased 8 gallons of unleaded premium and 4 gallons of unleaded regular and paid $38.20. He forgot to record how much each type of fuel cost per gallon. Can you determine these values? Premium, $3.25/gal; regular, $3.05/gal

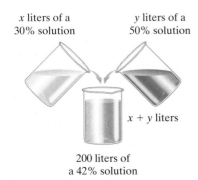

x liters of a
30% solution

y liters of a
50% solution

x + *y* liters

200 liters of
a 42% solution

Teaching Example 3 A store owner wants to mix 2 types of nuts together to obtain 30 pounds of nuts that will sell for $2.50 per pound. One type of nut sells for $1.50 per pound, and the other type sells for $3.00 per pound. How many pounds of each type of nut should she use in order to obtain the desired mix?

Ans: 10 lb of the nuts that cost $1.50/lb; 20 lb of the nuts that cost $3.00/lb

EXAMPLE 3 A lab technician is required to prepare 200 liters of a solution. The prepared solution must contain 42% fungicide. The technician wishes to combine a solution that contains 30% fungicide with a solution that contains 50% fungicide. How much of each solution should he use?

Solution *Understand the problem.* The technician needs to make one solution from two different solutions. We need to find the amount (number of liters) of each solution that will give us 200 liters of a 42% solution. The two unknowns are easy to identify.

Let x = the number of liters of the 30% solution needed and
y = the number of liters of the 50% solution needed.

Now, what are the two equations? We know that the 200 liters of the final solution will be made by combining x and y.

$$x + y = 200 \quad (1)$$

The second piece of information concerns the percent of fungicide in each solution. See the picture at left.

$$0.3x + 0.5y = 0.42(200) \quad (2)$$

Thus our system of equations is

$$x + y = 200 \quad (1)$$
$$0.3x + 0.5y = 84. \quad (2)$$

We will solve this system by the addition method.

$1.0x + 1.0y = 200$	Rewrite equation **(1)** in an equivalent form.
$\underline{-0.6x - 1.0y = -168}$	Multiply equation **(2)** by -2.
$0.4x = 32$	Solve for x.
$\dfrac{0.4x}{0.4} = \dfrac{32}{0.4}$	
$x = 80$	
$80 + y = 200$	Substitute the value $x = 80$ into one of the
$y = 120$	original equations and solve for y.

The technician should use 80 liters of the 30% fungicide and 120 liters of the 50% fungicide to obtain the required solution.

Check. Do the amounts total 200 liters?

$$80 + 120 = 200 \quad \checkmark$$

Do the amounts yield a 42% strength mixture?

$$0.30x + 0.50y = 0.42(200)$$
$$0.30(80) + 0.50(120) \overset{?}{=} 84$$
$$24 + 60 = 84 \quad \checkmark$$

Practice Problem 3 Dwayne Laboratories needs to ship 4000 liters of H_2SO_4 (sulfuric acid). Dwayne keeps in stock solutions that are 20% H_2SO_4 and 80% H_2SO_4. If the strength of the solution shipped is to be 65%, how should Dwayne's stock be mixed? 1000 L of 20% solution; 3000 L of 80% solution

EXAMPLE 4 Mike recently rode his boat on Lazy River. He took a 48-mile trip up the river traveling against the current in exactly 3 hours. He refueled and made the return trip in exactly 2 hours. What was the speed of his boat in still water and the speed of the current on the river?

Teaching Example 4 Carol took her boat on a 36-mile trip up the West Fork River. The trip one way, against the current, took 3 hours. The return trip took 1.5 hours. What was the speed of Carol's boat in still water, and what was the speed of the current?

Ans: Boat speed in still water was 18 mi/hr; current speed was 6 mi/hr

Solution Let x = the speed of the boat in still water in miles/hour and
y = the speed of the river current in miles/hour.

You may want to draw a picture to show the boat traveling in each direction.

|← Distance is 48 miles →|

Current

Against the current boat takes 3 hours

Current

With the current boat takes 2 hours

Traveling *against* the current, the boat took 3 hours to travel 48 miles. We can calculate the rate (speed) the boat traveled upstream. Use distance = rate · time or $d = r \cdot t$.

$$48 = 3 \cdot r \qquad r = 16 \text{ miles/hour}$$

The boat was traveling *against* the current. Thus the speed of the boat in still water minus the speed of the current is 16.

$$x - y = 16 \quad \textbf{(1)}$$

Traveling with the current, the boat took 2 hours to travel 48 miles. We can calculate the rate (speed) the boat traveled downstream. Use $d = r \cdot t$.

$$48 = 2 \cdot r \qquad r = 24 \text{ miles/hour}$$

The boat was traveling *with* the current. Thus the speed of the boat in still water plus the speed of the current is 24.

$$x + y = 24 \quad \textbf{(2)}$$

This system is solved most easily by the addition method.

$$
\begin{aligned}
x - y &= 16 \quad \textbf{(1)} \\
x + y &= 24 \quad \textbf{(2)} \\
\hline
2x &= 40 \qquad \text{Add the two equations.} \\
x &= 20 \qquad \text{Solve for } x \text{ by dividing both sides by 2.} \\
x + y &= 24 \quad \textbf{(2)} \qquad \text{Substitute the value of } x = 20 \text{ into one of the original} \\
& \qquad\qquad\qquad\quad \text{equations and solve for } y. \text{ We'll use equation } \textbf{(2)}. \\
20 + y &= 24 \\
y &= 4
\end{aligned}
$$

Thus the speed of Mike's boat in still water was 20 miles/hour and the speed of the current in Lazy River was 4 miles/hour.

Check. Can you verify these answers?

Practice Problem 4 Mr. Caminetti traveled downstream in his boat a distance of 72 miles in a time of 3 hours. His return trip up the river against the current took 4 hours. Find the speed of the boat in still water. Find the speed of the current in the river. Boat speed in still water is 21 mph; current speed is 3 mph

Applications *Solve using two equations with two variables.*

1. **Airline Travel** Sarah took the advertising department from her company on a round trip to Chicago to meet with a potential client. Including Sarah, a total of 14 people took the trip. She was able to purchase coach tickets for $250 and first-class tickets for $1150. She used her total budget for airfare for the trip, which was $10,700. How many first-class tickets did she buy? How many coach tickets did she buy?

$x + y = 14$
$250x + 1150y = 10,700$
8 first-class tickets
6 coach tickets

2. **Baseball** Walter had to purchase tickets for 11 people in his office to go to a Red Sox game. He paid $449 to buy the 11 tickets. Infield Grandstand seats cost $44. Right Field Roof seats cost $37. How many of each kind of ticket did he purchase?

$x + y = 11$
$37x + 44y = 449$
6 Infield Grandstand seats
5 Right Field Roof seats

3. **Art Dealer** An art dealer sold 12 etchings for $455. He sold some of them at $35 each and the rest at $40 each. How many etchings did he sell at each price?

$\left. \begin{array}{l} x + y = 12 \\ 35x + 40y = 455 \end{array} \right\}$ Five etchings at $35 each; seven etchings at $40 each

4. **High School Musical** El Segundo High School put on their annual musical. The students sold 650 tickets for a value of $4375. If orchestra seats cost $7.50 and balcony seats cost $3.50, how many of each kind of seat were sold?

$\left. \begin{array}{l} x + y = 650 \\ 7.5x + 3.5y = 4375 \end{array} \right\}$ 525 orchestra seats; 125 balcony seats

5. **Income of a Nurse** Sharyn worked as a Registered Nurse in London last summer. The first week she was there she worked a total of 23 hours at a city hospital. She was paid $13.50 an hour for day shift and $16.50 an hour for night shift work. She earned $352.50 for her first week of work. How many hours did she spend on the day shift? How many hours did she spend on the night shift?

$x + y = 23$
$13.50x + 16.50y = 352.50$
9 hr on day shift
14 hr on night shift

6. **Income of a Substitute Teacher** Tony worked as a substitute teacher last month. He substituted a total of 18 days. The first part of the month he worked in Tucson, Arizona, where he was paid $75 a day. Then he moved to Oregon, where he was paid $135 a day to substitute. During the month he earned $2010. How many days did he substitute in Tucson? How many days did he substitute in Oregon?

$x + y = 18$
$75x + 135y = 2010$
7 days in Tucson
11 days in Oregon

▲ 7. **Home Office Construction** The basement of your house is only partially finished. Currently, there is a room that you use for storage. You want to convert this space to a home office, but you want more space than is available in the room now. You decide to knock down two walls and enlarge the room. The existing room has a perimeter of 38 feet. If you extend the width of the room by 5 feet and make the new length 6 feet less than double the original length, the new perimeter will be 56 feet. What are the dimensions of the existing storage room? What are the dimensions of the new room?

$2L + 2W = 38$
$2(2L - 6) + 2(W + 5) = 56$
Original length = 10 ft
Original width = 9 ft
New length = 14 ft
New width = 14 ft

existing room

enlarged room

8. *Engineering Club* The engineering club at MCTC ran a coffee and donut sale to raise money for the club's activities. The first week, they made a profit of $200. The second week, they sold three times as many donuts, but half the number of coffees, and made a profit of $225. If the profit for each donut was $0.50 and the profit for each coffee was $0.75, how many donuts and coffees did they sell the first week?

$.50x + .75y = 200$
$.50(3x) + .75(.5y) = 225$
200 coffees
100 donuts

9. *Automobile Radiators* During her beginning mechanics course, Rose learned that her leaking radiator holds 16 quarts of water. She temporarily patched the leak when the radiator was partially full. At that time it was 50% antifreeze. After she filled the remaining space in the radiator with an 80% antifreeze solution and waited a little while, she found that it was 65% antifreeze. How many quarts of solution were in the radiator just before Rose added antifreeze? How many quarts of 80% antifreeze solution did she add?

$x + \quad y = 16$
$0.50x + 0.80y = 0.65(16)$
8 quarts of 50% antifreeze solution were in the radiator originally. She added 8 quarts of 80% antifreeze solution.

10. *Manufacturing Chocolate* A candy company wants to produce 20 kilograms of a special 45% fat content chocolate to conform with customer dietary demands. To obtain this, a 50% fat content Hawaiian chocolate is combined with a 30% fat content domestic chocolate in a special melting process. How many kilograms of the 50% fat content chocolate and how many kilograms of the 30% fat content chocolate are used to make the required 20 kilograms?

$x + \quad y = 20$
$0.50x + 0.30y = 0.45(20)$
15 kg of 50% fat content chocolate
5 kg of 30% fat content chocolate

11. *Hiking Club* The hiking club wants to sell 50 pounds of an energy snack food that will cost them only $1.80 per pound. The mixture will be made up of nuts that cost $2.00 per pound and raisins that cost $1.50 per pound. How much of each should they buy to obtain 50 pounds of the desired mixture?

$x + \quad y = 50$ } 30 lb of nuts
$2.00x + 1.50y = 50(1.80)$ } 20 lb of raisins

12. *Flower Costs* How many bunches of local daisies that cost $2.50 per bunch should be mixed with imported daisies that cost $3.50 per bunch to obtain 100 bunches of daisies that will cost $2.90 per bunch?

$x + \quad y = 100$ } 40 bunches of imported daisies
$2.50x + 3.50y = 2.90(100)$ } 60 bunches of local daisies

13. *Airline Travel* An airplane traveled between two cities that are 3000 kilometers apart. The trip against the wind took 6 hours. The return trip with the benefit of the wind was 5 hours. What is the wind speed in kilometers per hour? What is the speed of the plane in still air?

$\dfrac{3000}{6} = r - w$
$\dfrac{3000}{5} = r + w$ } The wind speed is 50 km/hr.
The plane's speed in still air is 550 km/hr.

14. *Airline Travel* An airplane flies between Boston and Cleveland. During a roundtrip flight, the plane flew a distance of 630 miles each way. The trip with a tailwind (the wind traveling the same direction as the plane) took 3 hours. The return trip traveling against the wind took 3.5 hours. Find the speed of the wind. Find the speed of the airplane in still air.

$3(p + w) = 630$ } The plane flies at 195 mph.
$3.5(p - w) = 630$ } The wind speed is 15 mph.

Mixed Practice *Solve the following applied problems using a system of equations.*

15. *Department Store Operations* Soledad Rivera's job was to ready the cash registers for the day's business at Foley's Department Store, in Houston. Each cash register began the day with the same amount of cash, $320. Each cash register received $25 in $1 bills and $55 in $5 bills. But from there each department had its own requirements. The Cosmetics department wanted two less than double the number of $10 bills and four fewer $20 bills than Jewelry wanted. How many $10 bills and $20 bills did these departments want?

$240 = 10x + 20y$
$340 = 20x + 20y$
Jewelry wanted 10 $10 bills and 7 $20 bills.
Cosmetics wanted 18 $10 bills and 3 $20 bills.

16. *Canoe Trip* The Outdoor Adventures Club sponsored a canoe trip. The participants paddled 4 miles upstream from the canoe rental headquarters to a popular swimming and picnicking spot. They ate lunch, rested, and headed back down to where they'd started. The trip upstream took 2 hours and the trip back took only 1 hour. How fast were they paddling on average? How fast was the current?

$4 = (p - c)2$
$4 = (p + c)1$
p = speed of paddling = 3 mph
c = speed of current = 1 mph

17. *Fruit Purchase* Hank purchased 7 pounds of oranges and 5 pounds of apples and it cost him $8.78. His wife went to the store the next day and purchased for the same prices per pound a total of 3 pounds of oranges and 8 pounds of apples, for a total of $8.39. What is the price per pound of oranges? What is the price per pound of apples?

Oranges are $0.69 per pound.
Apples are $0.79 per pound.

18. *Work Rates* Tom and Dave worked sorting discarded bottles for a recycle center. Tom worked for 3 hours and Dave worked for 2 hours on Saturday. They were able to sort through 2250 bottles. On Monday Tom worked for 2 hours and Dave worked for 3 hours. On Monday they sorted through 2650 bottles. How many bottles per hour can Tom sort? How many bottles per hour can Dave sort?

Tom can sort 290 bottles per hour.
Dave can sort 690 bottles per hour.

19. *Theater Tickets* The North Shore Community College theater department sold tickets for their annual production. General admission tickets were $8 and student admission tickets were $5.50. On opening night, 342 people attended the show. The receipts for the night totaled $2211. How many general admission tickets were sold? How many student admission tickets were sold?

132 general admission tickets;
210 student admission tickets

20. *Airplane Flight* Walter and Mary Jones flew from Boston to Denver to visit their daughter in college. The flight of the plane was a distance of 2000 miles. It took 5 hours to fly west against the jet stream. Coming back home with the jet stream the trip only took 4 hours. What was the speed of the jet stream? What was the speed of the plane in still air?

The speed of the jet stream was 50 mph. The speed of the plane was 450 mph.

21. *Population Growth* In 2005 the estimated population of Maryland was 5,467,000 and was projected to grow moderately. In the same year, the population of Missouri was 5,718,000 and was projected to grow at a slower rate. The relationship between the population of Maryland in thousands (y) and the number of years (x) since 2005 can be approximated by $40x - y = -5467$. A similar relationship for Missouri can be approximated by $53x - 2y = -11,447$. (*Source:* www.census.gov.) Solve the system of equations to determine the approximate year that the two states will have the same population. What will be the population of each state during that year? $x = 19$, which represents 2024; 6,227,000 people

22. *Population Growth* In 2005 the estimated population of Iowa was 3,040,000 and was projected to grow at a slow rate. In the same year, the population of Kansas was 3,108,000 and was projected to grow more rapidly. The relationship between the population of Iowa in thousands (y) and the number of years (x) since 2005 can be approximated by $50x - 10y = -29,410$. A similar relationship for Kansas can be approximated by $175x - 10y = -27,610$. (*Source:* www.census.gov.) Solve the system of equations to determine the approximate year that the two states will have the same population. What will be the population of each state during that year? $x = 14.4$, which corresponds to 2019; 3,013,000 people

Cumulative Review

23. **[6.3.3]** Add. $\dfrac{2}{x-1} + \dfrac{3}{x+1} + 1 \quad \dfrac{x^2 + 5x - 2}{x^2 - 1}$

24. **[1.8.1]** Evaluate when $x = -2$ and $y = 3$.
$2x^2 - 3y + 4y^2 \quad 35$

Quick Quiz 8.5

1. A college football game at Hampton University is played to a sell-out crowd of 18,500 people. Student tickets are $9 while general admission tickets are $15. A total of $217,500 was collected from ticket sales. How many people purchased student tickets? How many people purchased general admission tickets?

10,000 people purchased student tickets;
8500 people purchased general admission tickets

2. Last year the computer lab purchased 8 Dell PCs and 12 Apple iMacs. The cost of these computers was $20,800. This year the computer company said it would sell computers to the lab at the same cost per computer. This year the lab purchased 12 Dell PCs and 7 Apple iMacs for $18,000. How much did each Dell PC cost? How much did each Apple iMac cost?

Each Dell PC cost $800; each Apple iMac cost $1200

3. The LaBells chartered a jet from Denver to Richmond, Virginia. The flight plan required that the plane travel exactly 1500 miles. The trip was aided by the jet stream and took only 3 hours. The return trip of 1500 miles was traveling against the jet stream. That trip took 5 hours. What is the speed of the plane in still air? What is the speed of the jet stream?

Speed of plane in still air is 400 mi/hr; speed of jet stream is 100 mi/hr

4. **Concept Check** A new company is making a drink that is 45% pure fruit juice. They make a test batch of 500 gallons of the new drink. They are using some juice that is 50% pure fruit juice and some juice that is 30% pure fruit juice. They want to find out how many gallons of each of these two types they will need. Explain how you would set up a system of two equations using the variables x and y to solve this problem.

Answers may vary

Classroom Quiz 8.5 You may use these problems to quiz your students' mastery of Section 8.5.

1. The first week that *Titanic* was screened at the Liberty Tree Mall, a total of 22,000 adults attended the film. This was a record attendance for one film in one week at this theater. Ticket prices for that showing were $10 for regular showings and $8 for the matinee. A total of $204,000 was collected from ticket sales. How many people purchased tickets for a regular showing that week? How many people purchased tickets for a matinee that week?   **Ans:** 14,000 people; 8,000 people

2. Chris Smith drove to Nick's Roast Beef and brought the student activities office 7 super beef sandwiches and 5 orders of onion rings last week. The order came to $50. This week he brought the office 9 super beef sandwiches and 8 orders of onion rings. This week's order came to $69. How much does one super beef sandwich cost? How much does one order of onion rings cost?   **Ans:** $5; $3

3. An airplane traveled from Los Angeles to Boston aided by the jet stream. The flight plan required that the plane travel exactly 3000 miles. The trip took only 5 hours. The return trip of 3000 miles was traveling against the jet stream. That trip took 6 hours. What was the speed of the plane in still air? What was the speed of the jet stream?   **Ans:** 550 mph; 50 mph

Putting Your Skills to Work: Use Math to Save Money

ESTIMATING DISCOUNTS

There is always a sale somewhere claiming we can "save" 25%. The only true way to save money is not to spend it at all. However, when we need to purchase an item, looking for a bargain is a good idea. When we shop it helps if we can figure discounts quickly in our heads. Consider the story of Kate.

Kate needs to purchase two easy chairs. She has found the chairs at two different stores. Store A regularly sells each chair for $400, but now has the chairs on sale for 20% off. Store B sells each chair for $500, but if you buy one at full price, you get the second on sale for 20% off. In order to help her decide which store

was offering the best deal, Kate estimated the total price using the following values:

Approximate Values of Percents of Discounts

$10\% = \dfrac{1}{10}$	$20\% = \dfrac{1}{5}$	$25\% = \dfrac{1}{4}$	$30\% = \sim\dfrac{1}{3}$	$50\% = \dfrac{1}{2}$

1. Use the estimates to determine which store is offering the better buy.　Store A

2. Use the table to estimate how much you would save on a $28 shirt if it were reduced by each of the following percentages.
   (a) 10%　Approximately $3
   (b) 20%　Approximately $6
   (c) 30%　Approximately $9

3. Use the table to find approximately how much you would save on a $62 pair of pants if they are on sale at
   (a) 20% off.　Approximately $12
   (b) 25% off.　Approximately $16
   (c) 30% off.　Approximately $19

4. Amanda was shopping at a department store offering 20% off everything in the store. Use the table to estimate the amount Amanda would save if she purchased a lamp for $52, a rug for $36, and two pillows for $15 each.　Approximately $24

5. Henry was shopping for shoes at the same department store as Amanda. He found a pair that he liked for $95. He had found a coupon in the newspaper for an additional 10% off one item in the store. The cashier informs Henry that first 20% will be taken off the original price. Then, an additional 10% will be taken off at the cash register to get the final sale price. Estimate Henry's savings.　Approximately $27

6. Murray needs to buy six drapes for his living room. Store A sells the drapes for $40 each and is having a 25%-off sale. Store B sells the drapes for $30 and for every two drapes bought at full price, the third is 10% off. Which store is offering the better buy?　Store B

7. What shortcuts do you use to estimate without using a calculator?　Answers may vary

Topic	Procedure	Examples
Solving a system of equations by graphing, p. 471.	Graph both equations on the same coordinate system. One of the following will be true. **1.** The lines intersect. The point of intersection is the solution to the system. Such a system is consistent.	**1.**
	2. The lines are parallel. There is no point of intersection. The system has no solution. Such a system is inconsistent.	**2.**
	3. The lines coincide. Every point on the line represents a solution. There is an infinite number of solutions. Such a system has dependent equations.	**3.**
Solving a system of equations: substitution method, p. 479.	**1.** Solve for one variable in terms of the other variable in one of the two equations. **2.** Substitute the expression you obtain for this variable into the other equation. **3.** Solve this equation to find the value of the variable. **4.** Substitute this value for the variable into one of the equations to obtain a value for the other variable. **5.** Check the solution in both original equations.	Solve. $$2x + 3y = -6$$ $$3x - y = 13$$ Solve the second equation for y. $$-y = -3x + 13$$ $$y = 3x - 13$$ Substitute this expression into the first equation. $$2x + 3(3x - 13) = -6$$ $$2x + 9x - 39 = -6$$ $$11x - 39 = -6$$ $$11x = 33$$ $$x = 3$$ Substitute $x = 3$ into $y = 3x - 13$. $$y = 3(3) - 13 = 9 - 13 = -4$$ *Check.* $2(3) + 3(-4) \stackrel{?}{=} -6 \qquad 3(3) - (-4) \stackrel{?}{=} 13$ $6 - 12 \stackrel{?}{=} -6 \qquad 9 - (-4) \stackrel{?}{=} 13$ $-6 = -6 \checkmark \qquad 9 + 4 = 13 \checkmark$ The solution is $(3, -4)$.

(*Continued on next page*)

Topic	Procedure	Examples
Solving a system of equations: addition method, p. 485.	**1.** Multiply each term of one or both equations by some nonzero integer so that the coefficients of one of the variables are opposites. **2.** Add the equations of this new system so that one variable is eliminated. **3.** Solve the resulting equation for the remaining variable. **4.** Substitute the value found into either of the original equations to find the value of the other variable. **5.** Check your solution in both of the original equations.	Solve. $$3x - 2y = 0 \quad \textbf{(1)}$$ $$-4x + 3y = 1 \quad \textbf{(2)}$$ Eliminate the x's: a common multiple of 3 and 4 is 12. Multiply equation **(1)** by 4 and equation **(2)** by 3; then add. $$12x - 8y = 0$$ $$\underline{-12x + 9y = 3}$$ $$y = 3$$ Substitute $y = 3$ into either equation and solve for x. Let us pick $3x - 2y = 0$. $$3x - 2(3) = 0 \quad 3x - 6 = 0 \quad x = 2$$ *Check.* $$3(2) - 2(3) \overset{?}{=} 0 \qquad -4(2) + 3(3) \overset{?}{=} 1$$ $$6 - 6 \overset{?}{=} 0 \qquad\qquad -8 + 9 \overset{?}{=} 1$$ $$0 = 0 \ \checkmark \qquad\qquad\qquad 1 = 1 \ \checkmark$$ The solution is $(2, 3)$.
Choosing a method to solve a system of equations, p. 493.	**1.** Substitution works well if at least one variable has a coefficient of 1 or -1. **2.** Addition works well for integer, fractional, or decimal coefficients.	Use substitution. $$x + y = 8$$ $$2x - 3y = 9$$ Use addition. $$\frac{3}{5}x - \frac{1}{4}y = 4$$ $$\frac{1}{5}x + \frac{3}{4}y = 8$$
To identify inconsistent and dependent systems algebraically, p. 494.	**1.** If you obtain an equation that is inconsistent with known facts, such as $0 = 2$, the system is inconsistent. There is no solution. **2.** If you obtain an equation that is always true, such as $0 = 0$, the equations are dependent. There is an infinite number of solutions.	Solve. $$x + 2y = 1$$ $$3x + 6y = 12$$ Multiply the first equation by -3. $$-3x - 6y = -3$$ $$\underline{3x + 6y = 12}$$ $$0 = 9$$ 0 is not equal to 9, so there is *no solution*. The system is inconsistent. Solve. $$2x + y = 1$$ $$6x + 3y = 3$$ Multiply the first equation by -3. $$-6x - 3y = -3$$ $$\underline{6x + 3y = \ \ 3}$$ $$0 = 0$$ Since 0 is always equal to 0, the equations are dependent. There is an infinite number of solutions.

Topic	Procedure	Examples
Solving word problems using systems of equations, p. 498.	**1.** Understand the problem. Choose a variable to represent each unknown quantity. **2.** Write a system of equations in two variables. **3.** Solve the system of equations and state the answer. **4.** Check the answer.	Apples sell for \$0.35 per pound. Oranges sell for \$0.40 per pound. Nancy bought 12 pounds of apples and oranges for \$4.45. How many pounds of each fruit did she buy? Let x = number of pounds of apples and $\quad\quad y$ = number of pounds of oranges. 12 pounds in all were purchased. $x + y = 12$ The purchase cost \$4.45. $0.35x + 0.40y = 4.45$ Multiply the second equation by 100 to obtain the following system. $$x + \quad y = \quad 12$$ $$35x + 40y = 445$$ Multiply the top equation by -35 and add the equations. $$-35x - 35y = -420$$ $$\underline{35x + 40y = \quad 445}$$ $$5y = \quad 25$$ $$y = 5$$ Substitute $y = 5$ into $x + y = 12$. $$x + 5 = 12$$ $$x = 7$$ Nancy purchased 7 pounds of apples and 5 pounds of oranges. *Check.* Are there 12 pounds of apples and oranges? $$7 + 5 = 12 \;\checkmark$$ Would this purchase cost \$4.45? $$0.35(7) + 0.40(5) \stackrel{?}{=} 4.45$$ $$2.45 + 2.00 \stackrel{?}{=} 4.45$$ $$4.45 = 4.45 \;\checkmark$$

Chapter 8 Review Problems

Section 8.1

Solve by graphing the lines and finding the point of intersection.

1. $2x + 3y = 0$
$\quad -x + 3y = 9$

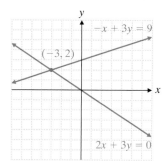

2. $-3x + y = -2$
$\quad -2x - y = -8$

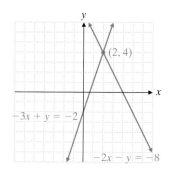

3. $2x - y = 6$
$6x + 3y = 6$

4. $2x - y = 1$
$3x + y = -6$

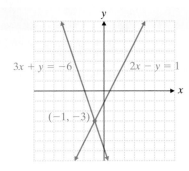

Section 8.2

Solve by the substitution method. Check your solution.

5. $x + y = 6$
$-2x + y = -3$
$(3, 3)$

6. $x + 3y = 18$
$2x + y = 11$
$(3, 5)$

7. $3x - 2y = 3$
$x - \dfrac{1}{3}y = 8$
$(15, 21)$

8. $0.5x + y = 16$
$4x - 2y = 8$
$(8, 12)$

Section 8.3

Solve by the addition method. Check your solution.

9. $6x - 2y = 10$
$2x + 3y = 7$
$(2, 1)$

10. $4x - 5y = -4$
$-3x + 2y = 3$
$(-1, 0)$

11. $5x - 4y = 2$
$-3x + 10y = -5$
$\left(0, -\dfrac{1}{2}\right)$

12. $9x + 2y = -5$
$12x - y = 8$
$\left(\dfrac{1}{3}, -4\right)$

Section 8.4

Solve by any appropriate method. If it is not possible to obtain one solution, state the reason. Write the solution in the form (x, y).

13. $6x - y = 33$
$6x + 7y = 9$
$(5, -3)$

14. $2x + 5y = 29$
$-3x + 10y = -26$
$(12, 1)$

15. $7x + 3y = 2$
$-8x - 7y = 2$
$\left(\dfrac{4}{5}, -\dfrac{6}{5}\right)$

16. $7x + 5y = 7$
$-3x - 4y = 1$
$\left(\dfrac{33}{13}, -\dfrac{28}{13}\right)$

17. $4x + 5y = 4$
$-5x - 8y = 2$
$(6, -4)$

18. $7x - 2y = 1$
$-5x + 3y = -7$
$(-1, -4)$

19. $2x - 4y = 6$
$-3x + 6y = 7$
No solution, inconsistent system

20. $4x - 7y = 8$
$5x + 9y = 81$
$(9, 4)$

21. $2x - 9y = 0$
$3x + 5 = 6y$
$\left(-3, -\dfrac{2}{3}\right)$

22. $2x + 10y = 1$
$-4x - 20y = -2$
Infinite number of solutions, dependent equations

23. $5x - 8y = 3x + 12$
$7x + y = 6y - 4$
$(-2, -2)$

24. $1 + x - y = y + 4$
$4(x - y) = 3 - x$
$(-1, -2)$

25. $3x + y = 9$
$x - 2y = 10$
$(4, -3)$

26. $x + 12y = 0$
$3x - 5y = -2$
$\left(-\dfrac{24}{41}, \dfrac{2}{41}\right)$

27. $2(x + 3) = y + 4$
$4x - 2y = -4$
Infinite number of solutions, dependent equations

28. $x + y = 3000$
$x - 2y = -120$
$(1960, 1040)$

Mixed Practice

Solve by any appropriate method. If it is not possible to obtain the solution, state the reason.

29. $5x + 4y + 3 = 23$
$8x - 3y - 4 = 75$
$(8, -5)$

30. $4x - 3y + 1 = 6$
$5x + 8y + 2 = -74$
$(-4, -7)$

31. $\dfrac{2x}{3} - \dfrac{3y}{4} = \dfrac{7}{12}$
$8x + 5y = 9$
$\left(\dfrac{29}{28}, \dfrac{1}{7}\right)$

32. $\dfrac{x}{2} + \dfrac{y}{5} = 4$
$\dfrac{x}{3} + \dfrac{y}{5} = \dfrac{10}{3}$
$(4, 10)$

33. $\dfrac{1}{5}a + \dfrac{1}{2}b = 6$
$\dfrac{3}{5}a - \dfrac{1}{2}b = 2$
$(10, 8)$

34. $\dfrac{2}{3}a + \dfrac{3}{5}b = -17$
$\dfrac{1}{2}a - \dfrac{1}{3}b = -1$
$(-12, -15)$

35. $4.8 + 0.6m = 0.9n$
$0.2m - 0.3n = 1.6$
No solution, inconsistent system

36. $8.4 - 0.8m = 0.4n$
$0.2m + 0.1n = 2.1$
Infinite number of solutions, dependent equations

37. $6s - 4t = 5$
$4s - 5t = -2$
$\left(\dfrac{33}{14}, \dfrac{16}{7}\right)$

38. $10s + 3t = 4$
$4s - 2t = -5$
$\left(-\dfrac{7}{32}, \dfrac{33}{16}\right)$

39. $3(x + 2) = -2 - (x + 3y)$
$3(x + y) = 3 - 2(y - 1)$
$(-5, 4)$

40. $13 - x = 3(x + y) + 1$
$14 + 2x = 5(x + y) + 3x$
$(9, -8)$

41. $0.2b = 1.4 - 0.3a$
$0.1b + 0.6 = 0.5a$
$(2, 4)$

42. $0.3a = 1.1 - 0.2b$
$0.3b = 0.4a - 0.9$
$(3, 1)$

43. $\dfrac{b}{5} = \dfrac{2}{5} - \dfrac{a - 3}{2}$
$4(a - b) = 3b - 2(a - 2)$
$(3, 2)$

44. $9(b + 4) = 2(2a + 5b)$
$\dfrac{b}{5} + \dfrac{a}{2} = \dfrac{18}{5}$
$(12, -12)$

Section 8.5

Solve.

45. *Museum of Modern Art* Admission to the Museum of Modern Art in New York City is $20 for adults and $12 for students. One Sunday, a total of 186 adults and students visited the museum. The receipts for that day totaled $3240. How many adults visited the museum that day? How many students?
126 adults; 60 students

46. *Fruit Packaging* A farmer near Napa has 252 pounds of apples to take to the farmer's market. He packages them into 2-lb and 5-lb bags to sell. If the number of 5-lb bags he used was double the number of 2-lb bags, how many of each type of bags did he prepare to bring to market?
21 2-lb bags; 42 5-lb bags

47. *Airplane Travel* A plane travels 1500 miles in 5 hours with the benefit of a tailwind. On the return trip it requires 6 hours to fly against the wind. Can you find the speed of the plane in still air and the wind speed?
275 mph in still air; 25 mph wind speed

48. *Chemical Solutions* A chemist has 20% and 30% acid solutions and needs 40 liters of a 25% acid solution. How much of each solution should he mix to get the 25% solution?
20 L of 20% solution; 20 L of 30% solution

49. *Scout Tree Sale* A boy scout troop is selling Christmas trees to raise money for a trip. On the first day they sold 79 trees. The balsam firs sold for $23 and the Norwegian pines were $28. The receipts for the day totaled $1942. How many trees of each type were sold?

54 balsam firs; 25 Norwegian pines

50. *Sanding Roads in Maine* Carl manages a highway department garage in Maine. A mixture of salt and sand is stored in the garage for use during the winter. Carl needs 24 tons of a salt/sand mixture that is 25% salt. He will combine shipments of 15% salt and shipments of 30% salt to achieve the desired 24 tons. How much of each type should he use?

8 tons of 15% salt; 16 tons of 30% salt

51. *Water Ski Boat* A water ski boat traveled 23 kilometers/hour going with the current. It went back in the opposite direction against the current and traveled only 15 kilometers/hour. What was the speed of the boat? What was the speed of the current?

Speed of the boat, 19 km/hr; speed of the current, 4 km/hr

52. *Manufacturing Costs* Fred is analyzing the cost of producing two different items at an electronics company. An electrical sensing device uses 5 grams of copper and requires 3 hours to assemble. A smaller sensing device made by the same company uses 4 grams of copper but requires 5 hours to assemble. The first device has a production cost of $27. The second device has a production cost of $32. How much does it cost the company for a gram of this type of copper? What is the hourly labor cost at this company? (Assume that production cost is obtained by adding the copper cost and the labor cost.)

$3 for a gram of copper; $4 hourly labor rate

Use the following information for exercises 53 and 54.

Toronto Blue Jays 2007 Season Ticket Prices

In-the-Action Seats	$195	Field Level Bases	$41
Premium Dugout Level	$59	Level Outfield	$23
Field Level Infield	$57	Upper Skydeck	$9

53. Connie and Dan Lacorazza took the youth group from their church to a Toronto Blue Jays game. They purchased 25 tickets. Some of the tickets were Level Outfield seats, while the rest were Upper Skydeck seats. The cost of all the tickets was $379. How many of each type of ticket did they buy?

11 Level Outfield seats; 14 Upper Skydeck seats

54. Russ and Norma Camp took a group of senior citizens to a Toronto Blue Jays game. They purchased 22 tickets. Some of the tickets were Field Level Bases seats, while the rest were Field Level Infield seats. The cost of all the tickets was $1062. How many of each type of ticket did they buy?

12 Field Level Bases seats; 10 Field Level Infield seats

Remember to use your Chapter Test Prep Video CD to see the worked-out solutions to the test problems you want to review.

Note to Instructor: The Chapter 8 Test file in the TestGen program provides algorithms specifically matched to these problems so you can easily replicate this test for additional practice or assessment purposes.

Solve by the method specified.

1. Substitution method

$$3x - y = -5$$
$$-2x + 5y = -14$$

2. Addition method

$$3x + 4y = 7$$
$$2x + 3y = 6$$

3. Graphing method

$$2x - y = 4$$
$$4x + y = 2$$

4. Any method

$$x + 3y = 12$$
$$2x - 4y = 4$$

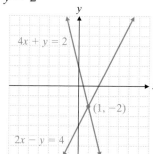

Solve by any method. If there is not one solution to a system, state why.

5.
$$2x - y = 5$$
$$-x + 3y = 5$$

6.
$$2x + 3y = 13$$
$$3x - 5y = 10$$

7.
$$\frac{2}{3}x - \frac{1}{5}y = 2$$
$$\frac{4}{3}x + 4y = 4$$

8.
$$3x - 6y = 5$$
$$-\frac{1}{2}x + y = \frac{7}{2}$$

9.
$$5x - 2 = y$$
$$10x = 4 + 2y$$

10.
$$0.3x + 0.2y = 0$$
$$1.0x + 0.5y = -0.5$$

11.
$$2(x + y) = 2(1 - y)$$
$$5(-x + y) = 2(23 - y)$$

12.
$$3(x - y) = 12 + 2y$$
$$8(y + 1) = 6x - 7$$

Solve.

13. Twice one number plus three times a second is one. Twice the second plus three times the first is nine. Find the numbers.

14. Both $8 and $12 tickets were sold for a basketball game. In all, 30,500 people paid for admission, resulting in a "gate" of $308,000. How many of each type of ticket were sold?

15. Five shirts and three pairs of slacks cost $172. Three of the same shirts and four pairs of the same slacks cost $156. How much does each shirt cost? How much is a pair of slacks?

16. You are producing a commemorative booklet for the 25th anniversary of your church. Ace Printers charges a $200 typesetting fee and $0.25 per booklet. Wong Printers charges a $250 typesetting fee but only $0.20 per booklet. How many booklets must you order for the cost to be the same? What is the cost?

17. A jet plane flew 2000 kilometers against the wind in a time of 5 hours. It refueled and flew back the same distance with the wind in 4 hours. Find the wind speed in kilometers per hour. Find the speed of the jet in still air in kilometers per hour.

1. $(-3, -4)$

2. $(-3, 4)$

3. $(1, -2)$; see graph

4. $(6, 2)$

5. $(4, 3)$

6. $(5, 1)$

7. $(3, 0)$

8. No solution, inconsistent system

9. Infinite number of solutions, dependent equations

10. $(-2, 3)$

11. $(-5, 3)$

12. $(-3.5, -4.5)$ or $\left(-\frac{7}{2}, -\frac{9}{2}\right)$

13. First number is 5; second number is -3

14. 16,000 of the $12 tickets; 14,500 of the $8 tickets

15. A shirt costs $20 and a pair of slacks costs $24.

16. 1000 booklets; $450 total cost

17. 50 km/hr speed of wind 450 km/hr speed of jet;

Cumulative Test for Chapters 0–8

Approximately one-half of this test covers the content of Chapters 0–7. The remainder covers the content of Chapter 8.

Simplify.

1. 15% of $42.75
(Round to the nearest cent.)

2. $\dfrac{-2^4 \cdot (2^2)}{2^3}$

3. $(3x^2 - 2x + 1)(5x - 3)$

4. $\dfrac{x^2 - 7x + 12}{x^2 - 9} \div (x^2 - 8x + 16)$

In questions 5 and 6, solve.

5. $-3(x + 2) < 4(x - 2)$

6. $6x^2 + x - 2 = 0$

7. 75% of what number is 30?

8. Solve for x. $\dfrac{1}{10}x + \dfrac{1}{3} = \dfrac{2}{5}x - \dfrac{1}{6}$

9. Factor $10t^2 - 25t^3 + 15t^4$.

10. Solve for x. $4x = 2(12 - 2x)$

Solve by any method. If there is not one solution to a system, state why.

11. $2x + y = 8$
$3x + 4y = -8$

12. $\dfrac{1}{3}x - \dfrac{1}{2}y = 4$

$\dfrac{3}{8}x - \dfrac{1}{4}y = 7$

13. $1.3x - 0.7y = 0.4$
$-3.9x + 2.1y = -1.2$

14. $10x - 5y = 45$
$3x - 8y = 7$

Solve.

15. The equations in a system are graphed and parallel lines are obtained. What is this system called? What does that mean?

16. How much water must be added to 60 kilograms of an 80% acid solution to produce a 50% solution?

17. A boat trip 6 miles upstream takes 3 hours. The return takes 90 minutes. How fast is the stream?

18. Two printers were used for a 5-hour job of printing 15,000 labels. The next day, a second run of the same labels took 7 hours because one printer broke down after 2 hours. How many labels can each printer process per hour?

CHAPTER

9

Sport parachuting has become popular throughout the world. Engaging in this sport requires accuracy and concentration because things happen very quickly. When a person jumps out of a plane and before the parachute is deployed, he or she will fall 2304 feet during the first 12 seconds. How fast could this person travel? How can the speed be calculated? These and other calculations involving sport parachuting require a knowledge of the mathematics covered in this chapter.

Radicals

Student Learning Objectives

After studying this section, you will be able to:

 1 Evaluate the square root of a perfect square.

2 Approximate the square root of a number that is not a perfect square.

Teaching Tip It is fairly important to take the time to clear up some possible confusion about negative signs and square roots.

(a) If someone asks what are the square roots of 121, the answers are +11 and −11.

(b) If someone asks us to evaluate $\sqrt{121}$, this is the principal or positive square root and thus the answer is +11.

(c) If someone asks us to evaluate $\sqrt{-121}$, at this point we would just state that this expression does not yield a real number.

(d) Finally, if a negative sign is in front of the radical, such as $-\sqrt{121}$, then the answer is −11.

Teaching Example 1 Find.

(a) $\sqrt{36}$ **(b)** $-\sqrt{81}$

Ans: (a) 6 **(b)** −9

Teaching Example 2 Find.

(a) $\sqrt{\dfrac{49}{64}}$ **(b)** $-\sqrt{\dfrac{1}{25}}$

Ans: (a) $\dfrac{7}{8}$ **(b)** $-\dfrac{1}{5}$

1 Evaluating the Square Root of a Perfect Square

How long is the side of a square whose area is 4? Recall the formula for the area of a square.

$$\text{area of a square} = s^2 \qquad \text{Our question then becomes, what number times itself is 4?}$$
$$s^2 = 4$$
$$s = 2 \qquad \text{because } (2)(2) = 4$$
$$s = -2 \qquad \text{because } (-2)(-2) = 4$$

We say that 2 is a **square root** of 4 because $(2)(2) = 4$. We can also say that −2 is a square root of 4 because $(-2)(-2) = 4$. Note that 4 is a **perfect square.** A square root of a perfect square is an integer.

The symbol $\sqrt{\ }$ is used in mathematics for the **principal square root** of a number, which is the nonnegative square root. The symbol itself $\sqrt{\ }$ is called the **radical sign.** If we want to find the negative square root of a number, we use the symbol $-\sqrt{\ }$.

DEFINITION OF PRINCIPAL SQUARE ROOT

For all nonnegative numbers N, the principal square root of N (written $\sqrt{N}$) is defined to be the nonnegative number a if and only if $a^2 = N$.

Notice in the definition that we did not use the words "for all *positive* numbers N" because we also want to include the square root of 0. $\sqrt{0} = 0$ since $0^2 = 0$.

Let us now examine the use of this definition with a few examples.

EXAMPLE 1 Find. **(a)** $\sqrt{144}$ **(b)** $-\sqrt{9}$

Solution

(a) $\sqrt{144} = 12$ since $(12)^2 = (12)(12) = 144$ and $12 \geq 0$.

(b) The symbol $-\sqrt{9}$ is read "the opposite of the square root of 9." Because $\sqrt{9} = 3$, we have $-\sqrt{9} = -3$.

Practice Problem 1 Find. **(a)** $\sqrt{64}$ 8 **(b)** $-\sqrt{121}$ −11

The number beneath the radical sign is called the **radicand.** The radicand will not always be an integer.

EXAMPLE 2 Find.

(a) $\sqrt{\dfrac{1}{4}}$ **(b)** $-\sqrt{\dfrac{4}{9}}$

Solution

(a) $\sqrt{\dfrac{1}{4}} = \dfrac{1}{2}$ since $\left(\dfrac{1}{2}\right)^2 = \left(\dfrac{1}{2}\right)\left(\dfrac{1}{2}\right) = \dfrac{1}{4}$. **(b)** $-\sqrt{\dfrac{4}{9}} = -\dfrac{2}{3}$ since $\sqrt{\dfrac{4}{9}} = \dfrac{2}{3}$.

Practice Problem 2 Find.

(a) $\sqrt{\dfrac{9}{25}}$ $\dfrac{3}{5}$ **(b)** $-\sqrt{\dfrac{121}{169}}$ $-\dfrac{11}{13}$

EXAMPLE 3   Find.   **(a)** $\sqrt{0.09}$ **(b)** $\sqrt{1600}$ **(c)** $\sqrt{225}$

Solution

(a) $\sqrt{0.09} = 0.3$ since $(0.3)(0.3) = 0.09$.

   Notice that $\sqrt{0.09}$ is *not* 0.03 because $(0.03)(0.03) = 0.0009$. Remember to count the decimal places when you multiply decimals. When finding the square root of a decimal, you should multiply to check your answer.

(b) $\sqrt{1600} = 40$ since $(40)(40) = 1600$.

(c) $\sqrt{225} = 15$

Teaching Example 3 Find.

(a) $\sqrt{400}$ (b) $\sqrt{0.0081}$ (c) $\sqrt{256}$

Ans: (a) 20   **(b)** 0.09   **(c)** 16

Practice Problem 3   Find.

(a) $-\sqrt{0.0036}$ -0.06 **(b)** $\sqrt{2500}$ 50   **(c)** $\sqrt{196}$ 14

2   Approximating the Square Root of a Number That Is Not a Perfect Square

Not all numbers are perfect squares. How can we find the square root of 2? That is, what number times itself equals 2? $\sqrt{2}$ is an irrational number. It cannot be written as $\frac{p}{q}$, where p and q are integers and $q \neq 0$. $\sqrt{2}$ is a nonterminating, nonrepeating decimal, and the best we can do is approximate its value. To do so, we can use the square root key on a calculator.

   To use a calculator, enter 2 and push the $\boxed{\sqrt{x}}$ or $\boxed{\sqrt{}}$ key.

$$\boxed{2} \;\; \boxed{\sqrt{x}} \;\; \boxed{1.4142136}$$

Remember, this is only an approximation. Most calculators are limited to an eight-digit display.

   To use the table of square roots, look under the x column until you find the number 2. Then look across the row to the $\sqrt{x}$ column. The value there is rounded to the nearest thousandth.

$$\sqrt{2} \approx 1.414$$

x	$\sqrt{x}$
1	1.000
2	1.414
3	1.732
4	2.000
5	2.236
6	2.449
7	2.646

EXAMPLE 4   Use a calculator or a table to approximate. Round to the nearest thousandth.

(a) $\sqrt{28}$ **(b)** $\sqrt{191}$

Teaching Example 4 Approximate. Round to the nearest thousandth.

(a) $\sqrt{11}$ (b) $\sqrt{161}$

Ans: (a) 3.317   **(b)** 12.689

Solution

(a) Using a calculator, we have 28 $\boxed{\sqrt{x}}$ 5.291502622. Thus $\sqrt{28} \approx 5.292$.

(b) Using a calculator, we have 191 $\boxed{\sqrt{x}}$ 13.82027496. Thus $\sqrt{191} \approx 13.820$.

Practice Problem 4   Approximate. Round to the nearest thousandth.

(a) $\sqrt{13}$ 3.606   **(b)** $\sqrt{35}$ 5.916   **(c)** $\sqrt{127}$ 11.269

Teaching Tip It is helpful to point out to students a reason for wanting to approximate square roots. For example, the distance between consecutive bases in a baseball diamond is exactly 90 feet. However, the distance from home plate to second base is exactly $\sqrt{16,200}$ feet. To approximate this distance to the nearest thousandth, we would use a calculator to obtain 127.279 feet, rounded to the nearest thousandth of a foot. If we wanted to check that a baseball diamond was accurately constructed, we might need that approximate value.

On a professional softball diamond, the distance from home plate to second base is exactly $\sqrt{7200}$ feet. Approximate this value using a calculator.

SIDELIGHT: A Topic for Further Thought

Let's look at the radicand carefully. Does the radicand need to be a nonnegative number? What if we write $\sqrt{-4}$? Is there a number that you can square to get -4?

Obviously, there is *no real number* that you can square to get -4. We know that $(2)^2 = 4$ and $(-2)^2 = 4$. Any real number that is squared will be nonnegative. We therefore conclude that $\sqrt{-4}$ does not represent a real number. Because we want to work with real numbers, our definition of $\sqrt{n}$ requires that n be nonnegative. Thus $\sqrt{-4}$ is *not a real number*.

You may encounter a more sophisticated number system called **complex numbers** in a higher-level mathematics course. In the complex number system, negative numbers have square roots.

Verbal and Writing Skills

1. Define a principal square root in your own words.
The principal square root of N, where $N \geq 0$, is a nonnegative number a that has the property $a^2 = N$.

2. Is $\sqrt{576} = 24$? Why or why not?
Yes; $(24)(24) = 576$

3. Is $\sqrt{0.9} = 0.3$? Why or why not?
No; $(0.3)(0.3) = 0.09$

4. How would you find $\sqrt{90}$?
Use a calculator or square root table.

Find the two square roots of each number.

5. 9 ± 3

6. 4 ± 2

7. 49 ± 7

8. 36 ± 6

Find the square root. Do not use a calculator or a table of square roots.

9. $\sqrt{16}$ 4

10. $\sqrt{100}$ 10

11. $\sqrt{81}$ 9

12. $\sqrt{25}$ 5

13. $-\sqrt{36}$ -6

14. $-\sqrt{25}$ -5

15. $\sqrt{0.81}$ 0.9

16. $\sqrt{0.64}$ 0.8

17. $\sqrt{\dfrac{36}{121}}$ $\dfrac{6}{11}$

18. $\sqrt{\dfrac{25}{81}}$ $\dfrac{5}{9}$

19. $\sqrt{\dfrac{49}{64}}$ $\dfrac{7}{8}$

20. $\sqrt{\dfrac{64}{169}}$ $\dfrac{8}{13}$

21. $\sqrt{625}$ 25

22. $\sqrt{1225}$ 35

23. $-\sqrt{10,000}$ -100

24. $-\sqrt{40,000}$ -200

Mixed Practice *Find the square root. Do not use a calculator or a table of square roots.*

25. $\sqrt{169}$ 13

26. $\sqrt{225}$ 15

27. $-\sqrt{\dfrac{1}{64}}$ $-\dfrac{1}{8}$

28. $\sqrt{\dfrac{81}{100}}$ $\dfrac{9}{10}$

29. $\sqrt{\dfrac{9}{16}}$ $\dfrac{3}{4}$

30. $-\sqrt{\dfrac{16}{49}}$ $-\dfrac{4}{7}$

31. $\sqrt{0.0036}$ 0.06

32. $\sqrt{0.0009}$ 0.03

33. $\sqrt{19,600}$ 140

34. $\sqrt{28,900}$ 170

35. $-\sqrt{289}$ -17

36. $-\sqrt{324}$ -18

Use a calculator or the square root table to approximate to the nearest thousandth.

37. $\sqrt{27}$ 5.196

38. $\sqrt{33}$ 5.745

39. $\sqrt{74}$ 8.602

40. $\sqrt{79}$ 8.888

41. $-\sqrt{183}$ -13.528

42. $-\sqrt{123}$ -11.091

43. $-\sqrt{195}$ -13.964

44. $-\sqrt{182}$ -13.491

Applications

▲ **45.** *Tree Care* A tree is 12 feet tall. A wire that stretches from a point that is 5 feet from the base of the tree to the top of the tree is $\sqrt{169}$ feet long. Exactly how many feet long is the wire?   13 ft

▲ **46.** *Extension Ladder* A garage is 15 feet tall. A ladder that stretches from a point that is 8 feet from the base of the garage to the top of the garage is $\sqrt{289}$ feet long. Exactly how many feet long is the ladder?   17 ft

Time for an Object to Fall If we ignore wind resistance, the time in seconds it takes for an object to fall s feet is given by $t = \frac{1}{4}\sqrt{s}$. Thus an object will fall 100 feet in $t = \frac{1}{4}\sqrt{100} = \frac{1}{4}(10) = 2.5$ seconds.

47. How long would it take an object dropped off a building to fall 900 feet?   7.5 sec

48. How long would it take an object dropped out of a tower to fall 1600 feet?   10 sec

To Think About: Finding Higher Roots *In addition to finding the square root of a number, we can also find the cube root of a number. To find the cube root of N (which is written as $\sqrt[3]{N}$), we find a number that, when cubed, will equal N.*

For example, $\sqrt[3]{8} = 2$ since $2^3 = 8$. We can also find the cube root of a negative number—for example, $\sqrt[3]{-8} = -2$ since $(-2)^3 = -8$. The cube root of a positive number is positive. The cube root of a negative number is negative. Therefore, each real number has only one real cube root.

The concept of roots can be extended to fourth and fifth roots. For example, $\sqrt[4]{16} = 2$ since $2^4 = 16$. Also, $\sqrt[5]{32} = 2$ since $2^5 = 32$.

Find each higher root.

49. $\sqrt[3]{27}$ 3

50. $\sqrt[3]{125}$ 5

51. $\sqrt[3]{-64}$ −4

52. $\sqrt[4]{256}$ 4

53. $\sqrt[4]{81}$ 3

54. $\sqrt[4]{625}$ 5

55. $\sqrt[5]{243}$ 3

56. $\sqrt[5]{1024}$ 4

57. Is there a real number that equals $\sqrt[4]{-16}$? Why?
No; a fourth power of a real number must be nonnegative.

58. Is there a real number that equals $\sqrt[4]{-81}$? Why?
No; a fourth power of a real number must be nonnegative.

Cumulative Review

59. **[8.4.1]** Solve. $3x + 2y = 8$
$7x - 3y = 11$
$9x + 6y = 24$
$14x - 6y = 22$ $x = 2; y = 1$

60. **[8.4.2]** Solve if possible. $2x - 3y = 1$
$-8x + 12y = 4$
No solution, inconsistent system of equations

61. **[0.7.1]** *Household Heating Costs* Among homes using natural gas, the average expenditure to heat homes during the winter of 2006–07 was $813. For the winter of 2007–08, the average expenditure was expected to be $900. What was the expected percent of increase? If the same percent of increase occurred the following winter, what would be the average household expenditure to heat a home using natural gas for the winter of 2008–09? (*Source:* www.eia.doe.gov.)
The percent of increase from 2006–07 to 2007–08 was 10.7%. The average household expenditure for 2008–09 would be $996.30.

62. **[0.7.1]** *Household Heating Costs* Among homes using heating oil, the average expenditure to heat homes during the winter of 2006–07 was $1466. For the winter of 2007–08, the average expenditure was expected to be $1841. What was the expected percent of increase? If the same percent of increase occurred the following winter, what would be the average household expenditure to heat a home using oil for the winter of 2008–09? (*Source:* www.eia.doe.gov.)
The percent of increase from 2006–07 to 2007–08 was 25.6%. The average household expenditure for 2008–09 would be $2312.30.

Quick Quiz 9.1 Find the square root. Do not use a calculator or a table of square roots.

1. $\sqrt{169}$ 13

2. $\sqrt{\dfrac{25}{64}}$ $\dfrac{5}{8}$

3. $\sqrt{0.49}$ 0.7

4. Concept Check Explain how you would go about determining $\sqrt{32,400}$ without using a calculator or a table of square roots. Answers may vary

Classroom Quiz 9.1 You may use these problems to quiz your students' mastery of Section 9.1.
Find the square root. Do not use a calculator or a table of square roots.

1. $\sqrt{196}$ **Ans:** 14

2. $\sqrt{\dfrac{81}{169}}$ **Ans:** $\dfrac{9}{13}$

3. $\sqrt{0.25}$ **Ans:** 0.5

Simplifying a Radical Expression with a Radicand That Is a Perfect Square

We know that $\sqrt{25} = \sqrt{5^2} = 5$, $\sqrt{9} = \sqrt{3^2} = 3$, and $\sqrt{36} = \sqrt{6^2} = 6$. If we write the radicand as a square of a nonnegative number, the square root is this nonnegative base. That is, $\sqrt{x^2} = x$ when x is nonnegative. We can use this idea to simplify radicals.

EXAMPLE 1 Find.

(a) $\sqrt{9^4}$ **(b)** $\sqrt{126^2}$ **(c)** $\sqrt{17^6}$

Solution

(a) Using the law of exponents, we rewrite 9^4 as a square.

$\sqrt{9^4} = \sqrt{(9^2)^2}$ Raise a power to a power: $(9^2)^2 = 9^4$.

$= 9^2$ To check, multiply: $(9^2)(9^2) = 9^4$.

(b) $\sqrt{126^2} = 126$

(c) $\sqrt{17^6} = \sqrt{(17^3)^2}$ Raise a power to a power: $(17^3)^2 = 17^6$.

$= 17^3$ To check, multiply: $(17^3)(17^3) = 17^6$.

Practice Problem 1 Find.

(a) $\sqrt{6^2}$ 6   **(b)** $\sqrt{13^4}$ 13^2 **(c)** $\sqrt{18^{12}}$ 18^6

This same concept can be used with variable expressions. We will restrict the values of each variable to be nonnegative if the variable is under the radical sign. Thus *all variable radicands in this chapter* are assumed to *represent positive numbers or zero*.

EXAMPLE 2 Find.

(a) $\sqrt{x^6}$ **(b)** $\sqrt{y^{24}}$

Solution

(a) $\sqrt{x^6} = \sqrt{(x^3)^2}$ By the law of exponents, $(x^3)^2 = x^6$.

$= x^3$

(b) $\sqrt{y^{24}} = y^{12}$

Practice Problem 2 Find.

(a) $\sqrt{y^{18}}$ y^9 **(b)** $\sqrt{x^{30}}$ x^{15}

In each of these examples we found the square root of a perfect square, and thus the radical sign disappeared. When you find the square root of a perfect square, do *not* leave a radical sign in your answer.

Often the coefficient of a variable radical will not be written in exponent form. You will want to be able to recognize the squares of numbers from 1 to 15, 20, and 25. If you can, you will be able to find many square roots faster mentally than by using a calculator or a table of square roots.

Before you begin, you need to know the multiplication rule for square roots.

Student Learning Objectives

After studying this section, you will be able to:

 1 Simplify a radical expression with a radicand that is a perfect square.

2 Simplify a radical expression with a radicand that is not a perfect square.

Teaching Example 1 Find.

(a) $\sqrt{4^{10}}$ **(b)** $\sqrt{13^8}$ **(c)** $\sqrt{23^2}$

Ans: **(a)** 4^5 **(b)** 13^4 **(c)** 23

Teaching Example 2 Find.

(a) $\sqrt{a^8}$ **(b)** $\sqrt{m^{14}}$

Ans: **(a)** a^4 **(b)** m^7

> **MULTIPLICATION RULE FOR SQUARE ROOTS**
>
> For all nonnegative numbers a and b,
>
> $$\sqrt{a} \cdot \sqrt{b} = \sqrt{ab} \quad \text{and} \quad \sqrt{ab} = \sqrt{a} \cdot \sqrt{b}.$$

Teaching Example 3 Find.

(a) $\sqrt{16a^6}$ (b) $\sqrt{49b^2c^{16}}$ (c) $\sqrt{m^{20}n^{12}}$

Ans: (a) $4a^3$ (b) $7bc^8$ (c) $m^{10}n^6$

EXAMPLE 3 Find.

(a) $\sqrt{225x^2}$ (b) $\sqrt{x^6y^{14}}$ (c) $\sqrt{169x^8y^{10}}$

Solution

(a) $\sqrt{225x^2} = 15x$ Using the multiplication rule for square roots,
$$\sqrt{225x^2} = \sqrt{225}\sqrt{x^2} = 15x.$$

(b) $\sqrt{x^6y^{14}} = x^3y^7$

(c) $\sqrt{169x^8y^{10}} = 13x^4y^5$

To check each answer, square it. The result should be the expression under the radical sign.

NOTE TO STUDENT: *Fully worked-out solutions to all of the Practice Problems can be found at the back of the text starting at page SP-1*

Practice Problem 3 Find.

(a) $\sqrt{625y^4}$ $25y^2$ (b) $\sqrt{x^{16}y^{22}}$ x^8y^{11} (c) $\sqrt{121x^{12}y^6}$ $11x^6y^3$

② Simplifying a Radical Expression with a Radicand That Is Not a Perfect Square

Most of the time when we encounter a square root, the radicand is not a perfect square. Thus, when we simplify the square root algebraically, we must retain the radical sign. Still, being able to simplify such radical expressions is a useful skill because it allows us to combine them, as we shall see in Section 9.3.

Teaching Example 4 Simplify.

(a) $\sqrt{18}$ (b) $\sqrt{24}$ (c) $\sqrt{45}$

Ans: (a) $3\sqrt{2}$ (b) $2\sqrt{6}$ (c) $3\sqrt{5}$

EXAMPLE 4 Simplify.

(a) $\sqrt{20}$ (b) $\sqrt{50}$ (c) $\sqrt{48}$

Solution   To begin, when you look at a radicand, look for perfect squares.

(a) $\sqrt{20} = \sqrt{4 \cdot 5}$ Note that 4 is a perfect square, and we can write 20 as $4 \cdot 5$.
$\quad\quad\,\, = \sqrt{4}\sqrt{5}$ Recall that $\sqrt{4} = 2$.
$\quad\quad\,\, = 2\sqrt{5}$

(b) $\sqrt{50} = \sqrt{25 \cdot 2}$
$\quad\quad\,\, = \sqrt{25}\sqrt{2}$
$\quad\quad\,\, = 5\sqrt{2}$

(c) $\sqrt{48} = \sqrt{16 \cdot 3}$
$\quad\quad\,\, = \sqrt{16}\sqrt{3}$
$\quad\quad\,\, = 4\sqrt{3}$

Practice Problem 4 Simplify.

(a) $\sqrt{98}$ $7\sqrt{2}$ (b) $\sqrt{12}$ $2\sqrt{3}$ (c) $\sqrt{75}$ $5\sqrt{3}$

The same procedure can be used with square root radical expressions containing variables. The key is to think of squares. That is, think, "How can I rewrite the expression so that it contains a perfect square?"

EXAMPLE 5 Simplify.

(a) $\sqrt{x^3}$

(b) $\sqrt{x^7y^9}$

Teaching Example 5 Simplify.

(a) $\sqrt{a^5}$ **(b)** $\sqrt{m^{11}n^3}$

Ans: **(a)** $a^2\sqrt{a}$ **(b)** $m^5n\sqrt{mn}$

Solution

(a) Recall that the law of exponents tells us to add the exponents when multiplying two expressions. Because we want one of these expressions to be a perfect square, we think of the exponent 3 as a sum of an even number and 1: $3 = 2 + 1$.

$$\sqrt{x^3} = \sqrt{x^2}\sqrt{x} \quad \text{Because } (x^2)(x) = x^3 \text{ by the law of exponents.}$$
$$= x\sqrt{x}$$

In general, to simplify a square root that has a variable with an exponent in the radicand, first write the square root as a product of the form $\sqrt{x^n}\sqrt{x}$, where n is the largest possible even exponent.

(b) $\sqrt{x^7y^9} = \sqrt{x^6}\sqrt{x}\sqrt{y^8}\sqrt{y}$
$$= x^3\sqrt{x}y^4\sqrt{y}$$
$$= x^3y^4\sqrt{xy}$$

In the final simplified form, we place the variable factors with exponents first and the radical factor second.

Practice Problem 5 Simplify.

(a) $\sqrt{x^{11}}$ $x^5\sqrt{x}$

(b) $\sqrt{x^5y^3}$ $x^2y\sqrt{xy}$

In summary, to simplify a square root radical, you factor each part of the expression under the radical sign and simplify the perfect squares.

EXAMPLE 6 Simplify.

(a) $\sqrt{12y^5}$

(b) $\sqrt{18x^3y^7w^{10}}$

Teaching Example 6 Simplify.

(a) $\sqrt{8a^7}$ **(b)** $\sqrt{50m^9n^2t}$

Ans: **(a)** $2a^3\sqrt{2a}$ **(b)** $5m^4n\sqrt{2mt}$

Solution

(a) $\sqrt{12y^5} = \sqrt{4 \cdot 3 \cdot y^4 \cdot y}$
$$= 2y^2\sqrt{3y} \quad \text{Remember, } y^4 = (y^2)^2 \text{ and } \sqrt{(y^2)^2} = y^2.$$

(b) $\sqrt{18x^3y^7w^{10}} = \sqrt{9 \cdot 2 \cdot x^2 \cdot x \cdot y^6 \cdot y \cdot w^{10}}$
$$= 3xy^3w^5\sqrt{2xy}$$

Practice Problem 6 Simplify.

(a) $\sqrt{48x^{11}}$ $4x^5\sqrt{3x}$

(b) $\sqrt{121x^6y^7z^8}$ $11x^3y^3z^4\sqrt{y}$

Throughout the world there has been a significant increase in the production and consumption of energy from alternative sources. These are primarily geothermal, solar, wind, and waste recycling energy production. For example, in 1990 1.7 quadrillion Btu (British thermal units) of energy from alternative sources were produced in the world. By 2000 this had increased to 3 quadrillion Btu (that is, 3,000,000,000,000,000 Btu). Some scientists predict that by 2010 this will increase to $\sqrt{192}$ quadrillion Btu. When simplified this can be written as $8\sqrt{3}$ quadrillion Btu. (*Source:* U.S. Energy Information Administration.)

Verbal and Writing Skills

1. Is $\sqrt{6^2} = \left(\sqrt{6}\right)^2$?

Yes. $\sqrt{6^2} = \sqrt{6 \cdot 6} = \sqrt{36} = 6$
$\left(\sqrt{6}\right)^2 = \left(\sqrt{6}\right)\left(\sqrt{6}\right) = \sqrt{6 \cdot 6} = \sqrt{36} = 6$

2. Why is it *not correct* to say
$\sqrt{30} = \sqrt{(-6)(-5)} = \sqrt{-6}\sqrt{-5}$?

The Rule $\sqrt{ab} = \sqrt{a} \cdot \sqrt{b}$ is only valid for *nonnegative* numbers a, b. In this case $a = -6$, $b = -5$, and they are negative numbers.

3. Is $\sqrt{(-9)^2} = -\sqrt{9^2}$?

No. $\sqrt{(-9)^2} = \sqrt{(-9)(-9)} = \sqrt{81} = 9$
but $-\sqrt{9^2} = -\sqrt{81} = -9$. The two expressions are not the same.

4. Is $\sqrt{2^4} = \left(\sqrt{2}\right)^4$?

Yes. $\sqrt{2^4} = \sqrt{2 \cdot 2 \cdot 2 \cdot 2} = \sqrt{16} = 4$
$\left(\sqrt{2}\right)^4 = \left(\sqrt{2}\right)\left(\sqrt{2}\right)\left(\sqrt{2}\right)\left(\sqrt{2}\right) = \sqrt{2 \cdot 2 \cdot 2 \cdot 2}$
$= \sqrt{16} = 4$

Simplify. Leave the answer in exponent form.

5. $\sqrt{8^2}$ 8

6. $\sqrt{2^2}$ 2

7. $\sqrt{18^4}$ 18^2

8. $\sqrt{13^4}$ 13^2

9. $\sqrt{9^6}$ 9^3

10. $\sqrt{5^{12}}$ 5^6

11. $\sqrt{33^8}$ 33^4

12. $\sqrt{51^{16}}$ 51^8

13. $\sqrt{5^{140}}$ 5^{70}

14. $\sqrt{7^{160}}$ 7^{80}

Simplify. Assume that all variables represent positive numbers.

15. $\sqrt{x^{12}}$ x^6

16. $\sqrt{x^{30}}$ x^{15}

17. $\sqrt{t^{18}}$ t^9

18. $\sqrt{w^{22}}$ w^{11}

19. $\sqrt{y^{26}}$ y^{13}

20. $\sqrt{w^{14}}$ w^7

21. $\sqrt{36x^8}$ $6x^4$

22. $\sqrt{64x^6}$ $8x^3$

23. $\sqrt{144x^2}$ $12x$

24. $\sqrt{196y^{10}}$ $14y^5$

25. $\sqrt{x^6y^4}$ x^3y^2

26. $\sqrt{x^2y^{18}}$ xy^9

27. $\sqrt{16x^2y^{20}}$ $4xy^{10}$

28. $\sqrt{49a^4b^2}$ $7a^2b$

29. $\sqrt{100a^{10}b^6}$ $10a^5b^3$

30. $\sqrt{81x^6y^{10}}$ $9x^3y^5$

31. $\sqrt{24}$ $2\sqrt{6}$

32. $\sqrt{27}$ $3\sqrt{3}$

33. $\sqrt{45}$ $3\sqrt{5}$

34. $\sqrt{20}$ $2\sqrt{5}$

35. $\sqrt{18}$ $3\sqrt{2}$

36. $\sqrt{32}$ $4\sqrt{2}$

37. $\sqrt{72}$ $6\sqrt{2}$

38. $\sqrt{60}$ $2\sqrt{15}$

39. $\sqrt{90}$ $3\sqrt{10}$

40. $\sqrt{125}$ $5\sqrt{5}$

41. $\sqrt{128}$ $8\sqrt{2}$

42. $\sqrt{96}$ $4\sqrt{6}$

43. $\sqrt{8x^3}$ $2x\sqrt{2x}$

44. $\sqrt{12y^5}$ $2y^2\sqrt{3y}$

45. $\sqrt{27w^5}$ $3w^2\sqrt{3w}$

46. $\sqrt{25x^5}$ $5x^2\sqrt{x}$

47. $\sqrt{32z^9}$ $4z^4\sqrt{2z}$

48. $\sqrt{24y^4}$ $2y^2\sqrt{6}$

49. $\sqrt{28x^5y^7}$ $2x^2y^3\sqrt{7xy}$

50. $\sqrt{45x^5y^4}$ $3x^2y^2\sqrt{5x}$

Mixed Practice   *Simplify. Assume all variables represent positive numbers.*

51. $\sqrt{48y^3w}$ $4y\sqrt{3yw}$

52. $\sqrt{12x^2y^3}$ $2xy\sqrt{3y}$

53. $\sqrt{75x^2y^3}$ $5xy\sqrt{3y}$

54. $\sqrt{50x^7y^6}$ $5x^3y^3\sqrt{2x}$

55. $\sqrt{64x^{10}}$ $8x^5$

56. $\sqrt{49x^6}$ $7x^3$

57. $\sqrt{y^7}$ $y^3\sqrt{y}$

58. $\sqrt{w^9}$ $w^4\sqrt{w}$

59. $\sqrt{x^5y^{11}}$ $x^2y^5\sqrt{xy}$

60. $\sqrt{x^3y^{13}}$ $xy^6\sqrt{xy}$

61. $\sqrt{135x^5y^7}$ $3x^2y^3\sqrt{15xy}$

62. $\sqrt{180a^5bc^2}$ $6a^2c\sqrt{5ab}$

63. $\sqrt{72a^2b^4c^5}$ $6ab^2c^2\sqrt{2c}$

64. $\sqrt{90ab^3c^6}$ $3bc^3\sqrt{10ab}$

65. $\sqrt{169a^3b^8c^5}$ $13ab^4c^2\sqrt{ac}$

66. $\sqrt{121a^8b^5c^{13}}$ $11a^4b^2c^6\sqrt{bc}$

To Think About *Simplify. Assume that all variables represent positive numbers.*

67. $\sqrt{(x+5)^2}$ $x+5$

68. $\sqrt{x^2+10x+25}$ $x+5$

69. $\sqrt{16x^2+8x+1}$ $4x+1$

70. $\sqrt{16x^2+24x+9}$ $4x+3$

71. $\sqrt{x^2y^2+14xy^2+49y^2}$ $y(x+7)$

72. $\sqrt{16x^2+16x+4}$ $2(2x+1)$

Cumulative Review

73. [8.2.1] Solve. $3x-2y=14$
$$y=x-5$$
$3x-2(x-5)=14 \Rightarrow x=4; y=-1$

74. [2.4.1] Solve. $\dfrac{1}{2}(x-4)=\dfrac{1}{3}x-5$
$x=-18$

75. [0.7.1] *Natural Gas Consumption* The chart shows how worldwide natural gas consumption is expected to increase. In 2010, it is expected that 115 trillion cubic feet of natural gas will be used worldwide. In 2020, worldwide use is expected to be 142 trillion cubic feet. Assuming the same annual rate of increase in natural gas consumption, estimate how much energy will be used in 2033.
177.1 trillion ft^3

World Natural Gas Consumption

Source: www.eia.doe.gov

76. [0.7.1] *Natural Gas Consumption* In 2010, Russia is expected to produce 26% of the natural gas consumed worldwide. By 2015, Russia is expected to produce 29% of all natural gas used worldwide. How many more trillion cubic feet of natural gas will Russia produce in 2015 than in 2010?
7.22 trillion ft^3

77. [0.7.1] *Natural Gas Consumption* What is the expected percent of increase in worldwide natural gas consumption from 2010 to 2020?
About 23.5%

Quick Quiz 9.2 *Simplify. Assume that all variables represent positive numbers.*

1. $\sqrt{120}$
$2\sqrt{30}$

2. $\sqrt{81x^3y^4}$
$9xy^2\sqrt{x}$

3. $\sqrt{135x^8y^5}$
$3x^4y^2\sqrt{15y}$

4. Concept Check Explain how you would simplify the following expression.
$$\sqrt{y^6z^7}$$
Answers may vary

Classroom Quiz 9.2 You may use these problems to quiz your students' mastery of Section 9.2.
Simplify. Assume that all variables represent positive numbers.

1. $\sqrt{96}$
   Ans: $4\sqrt{6}$

2. $\sqrt{121x^5y^5}$
   Ans: $11x^2y^2\sqrt{xy}$

3. $\sqrt{72x^{10}y^5}$
   Ans: $6x^5y^2\sqrt{2y}$

Teaching Example 1 Combine.

(a) $6\sqrt{5} - 2\sqrt{5}$

(b) $7\sqrt{y} + 4\sqrt{y} - \sqrt{y}$

Ans: (a) $4\sqrt{5}$ (b) $10\sqrt{y}$

Adding and Subtracting Radical Expressions with Like Radicals

Recall that when you simplify an algebraic expression, you combine like terms. $8b + 5b = 13b$ because both terms contain the same variable, b, and you can add the coefficients. This same idea is used when combining radicals. To add or subtract square root radicals, the numbers or the algebraic expressions under the radical signs must be the same. That is, the radicals must be **like radicals.**

EXAMPLE 1 Combine.

(a) $5\sqrt{2} - 8\sqrt{2}$

(b) $7\sqrt{a} + 3\sqrt{a} - 5\sqrt{a}$

Solution First check to see if the radicands are the same. If they are, you can combine the radicals.

(a) $5\sqrt{2} - 8\sqrt{2} = (5 - 8)\sqrt{2} = -3\sqrt{2}$

(b) $7\sqrt{a} + 3\sqrt{a} - 5\sqrt{a} = (7 + 3 - 5)\sqrt{a} = 5\sqrt{a}$

Practice Problem 1 Combine.

(a) $7\sqrt{11} + 4\sqrt{11}$ $11\sqrt{11}$

(b) $4\sqrt{t} - 7\sqrt{t} + 6\sqrt{t} - 2\sqrt{t}$ $\sqrt{t}$

Be careful. Not all the terms in an expression will have like radicals.

EXAMPLE 2 Combine. $5\sqrt{2a} + 3\sqrt{a} - 7\sqrt{2} + 3\sqrt{2a}$

Solution The only terms that have the same radicand are $5\sqrt{2a}$ and $3\sqrt{2a}$. These terms may be combined. All other terms stay the same.

$$5\sqrt{2a} + 3\sqrt{a} - 7\sqrt{2} + 3\sqrt{2a} = 8\sqrt{2a} + 3\sqrt{a} - 7\sqrt{2}$$

Teaching Example 2 Combine.

$$3\sqrt{ab} - 2\sqrt{a} + 5\sqrt{2a} - \sqrt{ab}$$

Ans: $2\sqrt{ab} - 2\sqrt{a} + 5\sqrt{2a}$

Practice Problem 2 Combine. $3\sqrt{x} - 2\sqrt{xy} - 5\sqrt{y} + 7\sqrt{xy}$

$3\sqrt{x} + 5\sqrt{xy} - 5\sqrt{y}$

Adding and Subtracting Radical Expressions That Must First Be Simplified

Sometimes it is necessary to simplify one or more radicals before their terms can be combined. Be sure to combine only like radicals.

Teaching Example 3 Combine.

(a) $\sqrt{8} - 6\sqrt{2}$

(b) $\sqrt{20} + \sqrt{18} - \sqrt{45}$

Ans: (a) $-4\sqrt{2}$

  (b) $-\sqrt{5} + 3\sqrt{2}$ or $3\sqrt{2} - \sqrt{5}$

EXAMPLE 3 Combine.

(a) $2\sqrt{3} + \sqrt{12}$

(b) $\sqrt{12} - \sqrt{27} + \sqrt{50}$

Solution

(a) $2\sqrt{3} + \sqrt{12} = 2\sqrt{3} + \sqrt{4\cdot3}$ Look for perfect square factors.

                $= 2\sqrt{3} + 2\sqrt{3}$ Simplify and combine like radicals.

                $= 4\sqrt{3}$

(b) $\sqrt{12} - \sqrt{27} + \sqrt{50} = \sqrt{4\cdot3} - \sqrt{9\cdot3} + \sqrt{25\cdot2}$

                  $= 2\sqrt{3} - 3\sqrt{3} + 5\sqrt{2}$ Only $2\sqrt{3}$ and $-3\sqrt{3}$

                  $= -\sqrt{3} + 5\sqrt{2}$ or $5\sqrt{2} - \sqrt{3}$ can be combined.

Practice Problem 3 Combine.

(a) $\sqrt{50} - \sqrt{18} + \sqrt{98}$ $9\sqrt{2}$

(b) $\sqrt{12} + \sqrt{18} - \sqrt{50} + \sqrt{27}$ $5\sqrt{3} - 2\sqrt{2}$

EXAMPLE 4 Combine. $\sqrt{2a} + \sqrt{8a} + \sqrt{27a}$

Solution

$$\sqrt{2a} + \sqrt{8a} + \sqrt{27a} = \sqrt{2a} + \sqrt{4 \cdot 2a} + \sqrt{9 \cdot 3a}$$

Look for perfect square factors.

$$= \sqrt{2a} + 2\sqrt{2a} + 3\sqrt{3a}$$

Be careful. $\sqrt{2a}$ and $\sqrt{3a}$ are **not** like radicals. The expressions under the radical signs must be the same.

$$= 3\sqrt{2a} + 3\sqrt{3a}$$

Practice Problem 4 Combine.

$$\sqrt{9x} + \sqrt{8x} - \sqrt{4x} + \sqrt{50x} \quad \sqrt{x} + 7\sqrt{2x}$$

Special care should be taken if the original radical has a numerical coefficient. If the radical can be simplified, the two resulting numerical coefficients should be multiplied.

EXAMPLE 5 Combine. $3\sqrt{20} + 4\sqrt{45} - 2\sqrt{80}$

Solution

$$3\sqrt{20} + 4\sqrt{45} - 2\sqrt{80} = 3 \cdot \sqrt{4} \cdot \sqrt{5} + 4 \cdot \sqrt{9} \cdot \sqrt{5} - 2 \cdot \sqrt{16} \cdot \sqrt{5}$$

$$= 3 \cdot 2 \cdot \sqrt{5} + 4 \cdot 3 \cdot \sqrt{5} - 2 \cdot 4 \cdot \sqrt{5}$$

$$= 6\sqrt{5} + 12\sqrt{5} - 8\sqrt{5}$$

$$= 10\sqrt{5}$$

Practice Problem 5 Combine.

$$\sqrt{27} - 4\sqrt{3} + 2\sqrt{75} \quad 9\sqrt{3}$$

EXAMPLE 6 Combine. $3a\sqrt{8a} + 2\sqrt{50a^3}$

Solution

$$3a\sqrt{8a} + 2\sqrt{50a^3} = 3a\sqrt{4}\sqrt{2a} + 2\sqrt{25a^2}\sqrt{2a}$$

$$= 3a \cdot 2 \cdot \sqrt{2a} + 2 \cdot 5a \cdot \sqrt{2a}$$

$$= 6a\sqrt{2a} + 10a\sqrt{2a}$$

$$= 16a\sqrt{2a}$$

(*Note:* If you are unsure of the last step, show the use of the distributive property in performing the addition.)

$$6a\sqrt{2a} + 10a\sqrt{2a} = (6 + 10)a\sqrt{2a}$$

$$= 16a\sqrt{2a}$$

Practice Problem 6 Combine.

$$2a\sqrt{12a} + 3\sqrt{27a^3} \quad 13a\sqrt{3a}$$

Verbal and Writing Skills

1. List the steps involved in simplifying an expression such as $\sqrt{75x} + \sqrt{48x} + \sqrt{16x^2}$.
1. Simplify each radical term.
2. Combine like radicals.

2. Mario said that, since $\sqrt{12}$ and $\sqrt{3}$ are not like radicals, $\sqrt{12} + \sqrt{3}$ cannot be simplified. What do you think?
Simplify $\sqrt{12}$ as $2\sqrt{3}$ and then combine. That is, $\sqrt{12} + \sqrt{3} = 2\sqrt{3} + \sqrt{3} = 3\sqrt{3}$.

Combine, if possible. Do not use a calculator or a table of square roots.

3. $3\sqrt{5} - \sqrt{5} + 4\sqrt{5}$
$6\sqrt{5}$

4. $\sqrt{7} + 2\sqrt{7} - 6\sqrt{7}$
$-3\sqrt{7}$

5. $\sqrt{2} + 8\sqrt{3} - 5\sqrt{3} + 4\sqrt{2}$
$5\sqrt{2} + 3\sqrt{3}$

6. $\sqrt{5} - \sqrt{6} + 3\sqrt{5} - 2\sqrt{6}$
$4\sqrt{5} - 3\sqrt{6}$

7. $\sqrt{5} + \sqrt{45}$
$\sqrt{5} + 3\sqrt{5} = 4\sqrt{5}$

8. $\sqrt{7} - \sqrt{63}$
$\sqrt{7} - 3\sqrt{7} = -2\sqrt{7}$

9. $\sqrt{80} + 4\sqrt{20}$
$4\sqrt{5} + 8\sqrt{5} = 12\sqrt{5}$

10. $\sqrt{80} + 2\sqrt{20}$
$4\sqrt{5} + 4\sqrt{5} = 8\sqrt{5}$

11. $3\sqrt{8} - 5\sqrt{2}$
$6\sqrt{2} - 5\sqrt{2} = \sqrt{2}$

12. $2\sqrt{27} - 4\sqrt{3}$
$6\sqrt{3} - 4\sqrt{3} = 2\sqrt{3}$

13. $-\sqrt{2} + \sqrt{18} + \sqrt{98}$
$-\sqrt{2} + 3\sqrt{2} + 7\sqrt{2} = 9\sqrt{2}$

14. $-\sqrt{54} + \sqrt{24} + 5\sqrt{6}$
$-3\sqrt{6} + 2\sqrt{6} + 5\sqrt{6} = 4\sqrt{6}$

15. $\sqrt{25} + \sqrt{72} + 3\sqrt{12}$
$5 + 6\sqrt{2} + 6\sqrt{3}$

16. $4\sqrt{12} + 2\sqrt{16} - \sqrt{44}$
$8\sqrt{3} + 8 - 2\sqrt{11}$

17. $3\sqrt{20} - \sqrt{50} + \sqrt{18}$
$6\sqrt{5} - 5\sqrt{2} + 3\sqrt{2} = 6\sqrt{5} - 2\sqrt{2}$

18. $\sqrt{50} + \sqrt{28} - \sqrt{9}$
$5\sqrt{2} + 2\sqrt{7} - 3$

19. $8\sqrt{2x} + \sqrt{50x}$
$8\sqrt{2x} + 5\sqrt{2x} = 13\sqrt{2x}$

20. $6\sqrt{5y} - \sqrt{20y}$
$6\sqrt{5y} - 2\sqrt{5y} = 4\sqrt{5y}$

21. $1.2\sqrt{3x} - 0.5\sqrt{12x}$
$1.2\sqrt{3x} - \sqrt{3x} = 0.2\sqrt{3x}$

22. $-1.5\sqrt{2a} + 0.2\sqrt{8a}$
$-1.5\sqrt{2a} + 0.4\sqrt{2a} = -1.1\sqrt{2a}$

23. $\sqrt{20y} + 2\sqrt{45y} - \sqrt{5y}$
$2\sqrt{5y} + 6\sqrt{5y} - \sqrt{5y} = 7\sqrt{5y}$

24. $\sqrt{72w} - 3\sqrt{2w} - \sqrt{50w}$
$6\sqrt{2w} - 3\sqrt{2w} - 5\sqrt{2w} = -2\sqrt{2w}$

Mixed Practice *Combine if possible. Do not use a calculator or a table of square roots.*

25. $\sqrt{50} - 3\sqrt{8}$
$-1\sqrt{2}$ or $-\sqrt{2}$

26. $\sqrt{75} - 5\sqrt{27}$
$-10\sqrt{3}$

27. $3\sqrt{28x} - 5x\sqrt{63x}$
$6\sqrt{7x} - 15x\sqrt{7x}$

28. $-2b\sqrt{45b} + 5\sqrt{80b}$
$-6b\sqrt{5b} + 20\sqrt{5b}$

29. $5\sqrt{8x^3} - 3x\sqrt{50x}$
$10x\sqrt{2x} - 15x\sqrt{2x} = -5x\sqrt{2x}$

30. $-2\sqrt{27y^3} + y\sqrt{12y}$
$-6y\sqrt{3y} + 2y\sqrt{3y} = -4y\sqrt{3y}$

31. $5x\sqrt{48} - 2x\sqrt{75}$
$10x\sqrt{3}$

32. $2x\sqrt{45} - 7x\sqrt{80}$
$-22x\sqrt{5}$

33. $2\sqrt{6y^3} - 2y\sqrt{54}$
$2y\sqrt{6y} - 6y\sqrt{6}$

34. $2x\sqrt{12x^3} - 6\sqrt{5x^2}$
$4x^2\sqrt{3x} - 6x\sqrt{5}$

35. $5x\sqrt{8x} - 24\sqrt{50x^3}$
$-110x\sqrt{2x}$

36. $6\sqrt{18x^3} - 5x\sqrt{72x}$
$-12x\sqrt{2x}$

Applications

▲ **37.** *Mars Mission* A planned unmanned mission to Mars may possibly require a small vehicle to make scientific measurements as it travels along the perimeter of a region like the one shown below. If the region is rectangular and the dimensions are as labeled, find the exact distance to be traveled by the vehicle in one trip around the perimeter.

$(8\sqrt{5} + 6\sqrt{3})$ mi

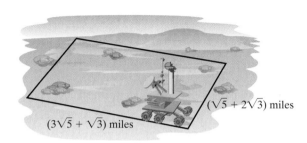

$(\sqrt{5} + 2\sqrt{3})$ miles

$(3\sqrt{5} + \sqrt{3})$ miles

▲ **38.** *Mars Mission* Suppose that on the second day of the mission to Mars, the vehicle must travel along the perimeter of a right triangle with dimensions as labeled below. Find the exact distance to be traveled by the vehicle in one trip around the perimeter.

$(4 + \sqrt{14})$ mi

$\sqrt{14}$ miles $(2 - \sqrt{3})$ miles

$(2 + \sqrt{3})$ miles

Cumulative Review

39. [8.5.1] *Toy Store Purchases* Melanie went holiday shopping for her nieces and nephews. She bought 12 games for $121.88. Some of the games were $8.99 and some were $12.49. How many of each did Melanie buy?

8 games at $8.99 and 4 games at $12.49

40. [3.4.3] *Personal Investments* William invested a $20,000 inheritance in two places. He invested in a certificate of deposit with a return of 8% simple interest, and he invested the rest of the money at 4% simple interest in a money market fund. At the end of the year he earned a total of $1340 from these two places. How much did he invest in each place?

$13,500 at 8% invested in the certificate of deposit; $6500 at 4% invested in the money market fund

Quick Quiz 9.3 Combine if possible. Do not use a calculator or a table of square roots.

1. $\sqrt{98} + 3\sqrt{8}$

$13\sqrt{2}$

2. $5\sqrt{3} + 2\sqrt{75} - 3\sqrt{48}$

$3\sqrt{3}$

3. $\sqrt{9a} + \sqrt{8a} + \sqrt{49a} + \sqrt{32a}$

$10\sqrt{a} + 6\sqrt{2a}$

4. Concept Check Explain how you would combine $3\sqrt{8x^3} + 5x\sqrt{98x}$.

Answers may vary

Classroom Quiz 9.3 You may use these problems to quiz your students' mastery of Section 9.3.

Combine if possible. Do not use a calculator or a table of square roots.

1. $\sqrt{27} - 4\sqrt{75}$

Ans: $-17\sqrt{3}$

2. $5\sqrt{5} - 3\sqrt{20} + 4\sqrt{45}$

Ans: $11\sqrt{5}$

3. $\sqrt{27a} + \sqrt{36a} + \sqrt{48a} + \sqrt{25a}$

Ans: $7\sqrt{3a} + 11\sqrt{a}$

Student Learning Objectives

After studying this section, you will be able to:

1. Multiply monomial radical expressions.

2. Multiply a monomial radical expression by a polynomial.

3. Multiply two polynomial radical expressions.

Teaching Example 1 Multiply. $\sqrt{10y}\sqrt{2}$

Ans: $2\sqrt{5y}$

Teaching Example 2 Multiply.

(a) $\left(5\sqrt{11}\right)\left(7\sqrt{5}\right)$

(b) $\left(4\sqrt{2y}\right)\left(3y\sqrt{6y}\right)$

Ans: (a) $35\sqrt{55}$ (b) $24y^2\sqrt{3}$

Teaching Example 3 Find the area of a rectangular state park that measures $\sqrt{85}$ miles long and $\sqrt{20}$ miles wide.

Ans: $10\sqrt{17}\ mi^2$

1 Multiplying Monomial Radical Expressions

Recall the basic rule for multiplication of square root radicals,

$$\sqrt{a}\sqrt{b} = \sqrt{ab}.$$

We will use this concept to multiply square root radical expressions. Note that the direction "multiply" means to express the product in simplest form.

EXAMPLE 1 Multiply. $\sqrt{7}\sqrt{14x}$

Solution $\sqrt{7}\sqrt{14x} = \sqrt{98x}$

We do *not* stop here, because the radical $\sqrt{98x}$ can be simplified.

$$\sqrt{98x} = \sqrt{49\cdot 2x} = \sqrt{49}\sqrt{2x} = 7\sqrt{2x}$$

Practice Problem 1 Multiply. $\sqrt{3a}\sqrt{6a}$ $3a\sqrt{2}$

EXAMPLE 2 Multiply.

(a) $\left(2\sqrt{3}\right)\left(5\sqrt{7}\right)$ 　　　　(b) $\left(2a\sqrt{3a}\right)\left(3\sqrt{6a}\right)$

Solution

(a) $\left(2\sqrt{3}\right)\left(5\sqrt{7}\right) = 10\sqrt{21}$ Multiply the coefficients: $(2)(5) = 10$.

　　　　　　　　　　　　　　　Multiply the radicals: $\sqrt{3}\sqrt{7} = \sqrt{21}$.

(b) $\left(2a\sqrt{3a}\right)\left(3\sqrt{6a}\right) = 6a\sqrt{18a^2}$

　　　　　　　　　　　　　$= 6a\sqrt{9}\sqrt{2}\sqrt{a^2}$ Simplify $\sqrt{18a^2}$.

　　　　　　　　　　　　　$= 18a^2\sqrt{2}$ 　　Multiply $(6a)(3)(a)$.

Practice Problem 2 Multiply.

(a) $\left(2\sqrt{3}\right)\left(5\sqrt{5}\right)$ $10\sqrt{15}$ 　　　(b) $\left(4\sqrt{3x}\right)\left(2x\sqrt{6x}\right)$ $24x^2\sqrt{2}$

EXAMPLE 3 Find the area of a farm field that measures $\sqrt{9400}$ feet long and $\sqrt{4800}$ feet wide. Express your answer as a simplified radical.

Solution   We multiply.

$$\left(\sqrt{9400}\right)\left(\sqrt{4800}\right) = \left(10\sqrt{94}\right)\left(40\sqrt{3}\right)$$

$$= 400\sqrt{282}\text{ square feet (ft}^2)$$

Practice Problem 3 Find the area of a computer chip that measures $\sqrt{180}$ millimeters long and $\sqrt{150}$ millimeters wide. $30\sqrt{30}\ mm^2$

Multiplying a Monomial Radical Expression by a Polynomial

Recall that a binomial consists of two terms. $\sqrt{2} + \sqrt{5}$ is a binomial. We use the distributive property when multiplying a binomial by another factor.

EXAMPLE 4 Multiply and simplify. $\sqrt{5}(\sqrt{2} + 3\sqrt{7})$

Solution
$$= \sqrt{5}\sqrt{2} + 3\sqrt{5}\sqrt{7}$$
$$= \sqrt{10} + 3\sqrt{35}$$

In similar fashion, we can multiply a trinomial by another factor.

Teaching Example 4 Multiply and simplify.
$$\sqrt{3}(6\sqrt{5} - 2\sqrt{6})$$
Ans: $6\sqrt{15} - 6\sqrt{2}$

Practice Problem 4 Multiply and simplify.
$$\sqrt{6}(\sqrt{3} + 2\sqrt{8}) \quad 3\sqrt{2} + 8\sqrt{3}$$

We discover that $\sqrt{5} \cdot \sqrt{5} = 5$ and $\sqrt{3} \cdot \sqrt{3} = 3$. These are specific cases of the multiplication rule for square root radicals.

> For any nonnegative real number a,
> $$\sqrt{a} \cdot \sqrt{a} = a.$$

EXAMPLE 5 Multiply. $\sqrt{a}(3\sqrt{a} - 2\sqrt{5})$

Solution
$$= 3\sqrt{a}\sqrt{a} - 2\sqrt{5}\sqrt{a} \quad \text{Use the distributive property.}$$
$$= 3a - 2\sqrt{5a}$$

Teaching Example 5 Multiply and simplify.
$$2\sqrt{y}(\sqrt{5y} - 6\sqrt{3})$$
Ans: $2y\sqrt{5} - 12\sqrt{3y}$

Practice Problem 5 Multiply.
$$2\sqrt{x}(4\sqrt{x} - x\sqrt{2}) \quad 8x - 2x\sqrt{2x}$$

Be sure to simplify all radicals after multiplying.

 Multiplying Two Polynomial Radical Expressions

Recall the FOIL method used to multiply two binomials. The same method can be used for radical expressions.

Algebraic Expressions
$$(2x + y)(x - 2y) = 2x^2 - 4xy + xy - 2y^2$$
$$= 2x^2 - 3xy - 2y^2$$

Radical Expressions
$$(2\sqrt{3} + \sqrt{5})(\sqrt{3} - 2\sqrt{5}) = 2\sqrt{9} - 4\sqrt{15} + \sqrt{15} - 2\sqrt{25}$$
$$= 2(3) - 3\sqrt{15} - 2(5)$$
$$= 6 - 3\sqrt{15} - 10$$
$$= -4 - 3\sqrt{15}$$

Let's look at this procedure more closely.

EXAMPLE 6 Multiply. $\left(\sqrt{2} + 5\right)\left(\sqrt{2} - 3\right)$

Solution

$$\left(\sqrt{2} + 5\right)\left(\sqrt{2} - 3\right)$$

$$= \sqrt{4} - 3\sqrt{2} + 5\sqrt{2} - 15 \quad \text{Multiply to obtain the four products.}$$
$$= 2 + 2\sqrt{2} - 15 \quad \text{Simplify } \sqrt{4}; \text{ combine the middle terms.}$$
$$= -13 + 2\sqrt{2}$$

Practice Problem 6 Multiply.

$$\left(\sqrt{2} + \sqrt{6}\right)\left(2\sqrt{2} - \sqrt{6}\right) \quad -2 + 2\sqrt{3}$$

Be especially watchful if there is a number outside the radical sign. You must multiply the numerical coefficients very carefully.

EXAMPLE 7 Multiply. $\left(\sqrt{2} - 3\sqrt{6}\right)\left(\sqrt{2} + \sqrt{6}\right)$

Solution

$$\left(\sqrt{2} - 3\sqrt{6}\right)\left(\sqrt{2} + \sqrt{6}\right)$$

$$= \sqrt{4} + \sqrt{12} - 3\sqrt{12} - 3\sqrt{36} \quad \text{Multiply to obtain the four products.}$$
$$= 2 - 2\sqrt{12} - 18 \quad \text{Simplify; combine the middle terms.}$$
$$= -16 - 4\sqrt{3} \quad \text{Simplify } -2\sqrt{12}.$$

Practice Problem 7 Multiply.

$$\left(\sqrt{6} + \sqrt{5}\right)\left(\sqrt{2} + 2\sqrt{5}\right) \quad 2\sqrt{3} + 2\sqrt{30} + \sqrt{10} + 10$$

If an expression with radicals is squared, write the expression as the product of two binomials and multiply.

EXAMPLE 8 Multiply. $\left(2\sqrt{3} - \sqrt{6}\right)^2$

Solution

$$= \left(2\sqrt{3} - \sqrt{6}\right)\left(2\sqrt{3} - \sqrt{6}\right)$$
$$= 4\sqrt{9} - 2\sqrt{18} - 2\sqrt{18} + \sqrt{36}$$
$$= 12 - 4\sqrt{18} + 6$$
$$= 18 - 12\sqrt{2}$$

Practice Problem 8 Multiply.

$$\left(3\sqrt{5} - \sqrt{10}\right)^2 \quad 55 - 30\sqrt{2}$$

Multiply. Be sure to simplify any radicals in your answer. Do not use a calculator or a table of square roots.

1. $\sqrt{3}\sqrt{10}$ $\sqrt{30}$

2. $\sqrt{5}\sqrt{7}$ $\sqrt{35}$

3. $\sqrt{2}\sqrt{22}$ $2\sqrt{11}$

4. $\sqrt{3}\sqrt{15}$ $3\sqrt{5}$

5. $\sqrt{18}\sqrt{3}$ $3\sqrt{6}$

6. $\sqrt{28}\sqrt{3}$ $2\sqrt{21}$

7. $(4\sqrt{5})(3\sqrt{2})$ $12\sqrt{10}$

8. $(2\sqrt{7})(3\sqrt{3})$ $6\sqrt{21}$

9. $\sqrt{5}\sqrt{10a}$ $5\sqrt{2a}$

10. $\sqrt{5b}\sqrt{20}$ $10\sqrt{b}$

11. $(3\sqrt{10x})(2\sqrt{5x})$ $30x\sqrt{2}$

12. $(3\sqrt{10x})(4\sqrt{5x})$ $60x\sqrt{2}$

13. $(-4\sqrt{8a})(-2\sqrt{3a})$ $16a\sqrt{6}$

14. $(4\sqrt{12a})(2\sqrt{6a})$ $48a\sqrt{2}$

15. $(-3\sqrt{ab})(2\sqrt{b})$ $-6b\sqrt{a}$

16. $(-2\sqrt{x})(5\sqrt{xy})$ $-10x\sqrt{y}$

17. $\sqrt{5}(\sqrt{2}+\sqrt{3})$ $\sqrt{10}+\sqrt{15}$

18. $\sqrt{10}(\sqrt{2}+\sqrt{3})$ $2\sqrt{5}+\sqrt{30}$

19. $\sqrt{3}(2\sqrt{6}+5\sqrt{15})$
$6\sqrt{2}+15\sqrt{5}$

20. $\sqrt{6}(5\sqrt{12}-4\sqrt{3})$
$30\sqrt{2}-12\sqrt{2}=18\sqrt{2}$

21. $2\sqrt{x}(\sqrt{x}-8\sqrt{5})$
$2x-16\sqrt{5x}$

22. $-3\sqrt{b}(2\sqrt{a}+3\sqrt{b})$
$-6\sqrt{ab}-9b$

23. $\sqrt{6}(\sqrt{2}-3\sqrt{6}+2\sqrt{10})$
$2\sqrt{3}-18+4\sqrt{15}$

24. $\sqrt{10}(\sqrt{5}-3\sqrt{10}+5\sqrt{2})$
$5\sqrt{2}-30+10\sqrt{5}$

25. $(2\sqrt{3}+\sqrt{6})(\sqrt{3}-2\sqrt{6})$
$6-3\sqrt{18}-12=-6-9\sqrt{2}$

26. $(3\sqrt{5}+\sqrt{3})(\sqrt{5}+\sqrt{3})$
$15+4\sqrt{15}+3=18+4\sqrt{15}$

27. $(5+3\sqrt{2})(3+\sqrt{2})$
$15+14\sqrt{2}+6=21+14\sqrt{2}$

28. $(7-2\sqrt{3})(4+\sqrt{3})$
$28-1\sqrt{3}-6=22-\sqrt{3}$

29. $(2\sqrt{7}-3\sqrt{3})(\sqrt{7}+\sqrt{3})$
$14-\sqrt{21}-9=5-\sqrt{21}$

30. $(5\sqrt{6}-\sqrt{2})(\sqrt{6}+3\sqrt{2})$
$30+14\sqrt{12}-6=24+28\sqrt{3}$

31. $(\sqrt{3}+2\sqrt{6})(2\sqrt{3}-\sqrt{6})$
$6+3\sqrt{18}-12=-6+9\sqrt{2}$

32. $(\sqrt{2}+3\sqrt{10})(2\sqrt{2}-\sqrt{10})$
$4+5\sqrt{20}-30=-26+10\sqrt{5}$

33. $(3\sqrt{7}-\sqrt{8})(\sqrt{8}+2\sqrt{7})$
$3\sqrt{56}+42-8-2\sqrt{56}=2\sqrt{14}+34$

34. $(\sqrt{12}-\sqrt{5})(\sqrt{5}+2\sqrt{12})$
$\sqrt{60}+24-5-2\sqrt{60}=-2\sqrt{15}+19$

35. $(2\sqrt{6}-1)^2$
$24-4\sqrt{6}+1=25-4\sqrt{6}$

36. $(3\sqrt{2}+2)^2$
$18+12\sqrt{2}+4=22+12\sqrt{2}$

37. $(\sqrt{3}+5\sqrt{2})^2$
$3+10\sqrt{6}+50=53+10\sqrt{6}$

38. $(\sqrt{3}-2\sqrt{7})^2$
$3-4\sqrt{21}+28=31-4\sqrt{21}$

39. $(3\sqrt{5}-\sqrt{3})^2$
$45-6\sqrt{15}+3=48-6\sqrt{15}$

40. $(2\sqrt{3}+5\sqrt{5})^2$
$12+20\sqrt{15}+125=137+20\sqrt{15}$

Mixed Practice *Multiply. Be sure to simplify any radicals in your answer.*

41. $(3\sqrt{5})(2\sqrt{8})$
$12\sqrt{10}$

42. $(5\sqrt{3})(\sqrt{24})$
$30\sqrt{2}$

43. $\sqrt{7}(\sqrt{2}+3\sqrt{14}-\sqrt{6})$
$\sqrt{14}+21\sqrt{2}-\sqrt{42}$

44. $\sqrt{6}(\sqrt{6}+5\sqrt{10}-\sqrt{3})$
$6+10\sqrt{15}-3\sqrt{2}$

45. $(5\sqrt{2}-3\sqrt{7})(3\sqrt{2}+\sqrt{7})$
$9-4\sqrt{14}$

46. $(6\sqrt{5}+7\sqrt{6})(2\sqrt{5}-\sqrt{6})$
$18+8\sqrt{30}$

47. $2\sqrt{a}(3\sqrt{b}+\sqrt{ab}-2\sqrt{a})$
$6\sqrt{ab}+2a\sqrt{b}-4a$

48. $3\sqrt{y}(2\sqrt{6}-\sqrt{x}+3\sqrt{y})$
$6\sqrt{6y}-3\sqrt{xy}+9y$

Multiply the following binomials. Be sure to simplify your answer. Do not use a calculator or a table of square roots.

49. $\left(3x\sqrt{y} + \sqrt{5}\right)\left(3x\sqrt{y} - \sqrt{5}\right)$ $9x^2y - 5$

50. $\left(4a\sqrt{5} - 2\sqrt{2}\right)\left(4a\sqrt{5} + 2\sqrt{2}\right)$ $80a^2 - 8$

 Applications *Ramp Construction   A wheelchair access ramp is constructed with the dimensions as listed on the diagram. The gray shaded triangle is a right triangle.*

▲ **51.** Find the area of the shaded portion of the wheelchair ramp.
12 ft^2

▲ **52.** Find the perimeter of the shaded portion of the wheelchair ramp. Leave your answer as a radical in simplified form.
$\left(13\sqrt{2} + \sqrt{290}\right)$ ft

$\sqrt{290}$ ft

$\sqrt{2}$ feet

$\sqrt{288}$ feet

Cumulative Review

53. [5.5.1] Factor. $64a^2 - 25b^2$
$(8a + 5b)(8a - 5b)$

54. [5.4.1] Factor. $4x^2 - 20x + 25$
$(2x - 5)^2$

55. [1.8.2] *Speed in Knots* Ships often measure their speed in knots (or nautical miles per hour). The regular mile (statute mile) that we are more familiar with is 5280 feet. A nautical mile is about 6076 feet. Write a formula of the form $m = ak$, where m is the number of statute miles, k is the number of nautical miles, and a is a constant number (in this case a fraction). Use the formula to find out how fast a Coast Guard Cutter is going in miles per hour if it is traveling at 35 knots.

$m = \left(\dfrac{1519}{1320}\right)k$ Approximately 40.3 mph

56. [0.7.1] *Car Insurance* Phil and Melissa LaBelle found that the collision and theft portion of their car insurance could be reduced by 15% if they purchased an auto security device. They purchased and had such a device installed for $350. The yearly collision and theft portion of their car insurance is $280. How many years will it take for the security device to pay for itself?

$8\dfrac{1}{3}$ yr

Quick Quiz 9.4   Multiply. Be sure to simplify any radicals in your answer.

1. $\left(2\sqrt{x}\right)\left(4\sqrt{y}\right)\left(3\sqrt{xy}\right)$ $24xy$

2. $\sqrt{3}\left(\sqrt{8} + 2\sqrt{3} - 4\sqrt{5}\right)$ $2\sqrt{6} + 6 - 4\sqrt{15}$

3. $\left(4\sqrt{2} - 5\sqrt{3}\right)\left(2\sqrt{2} + 3\sqrt{3}\right)$ $-29 + 2\sqrt{6}$

4. Concept Check Explain how to simplify $\left(3\sqrt{3} - 2\right)^2$.
Answers may vary

Classroom Quiz 9.4   You may use these problems to quiz your students' mastery of Section 9.4.

Multiply. Be sure to simplify any radicals in your answer.

1. $\left(3\sqrt{2x}\right)\left(2\sqrt{2y}\right)\left(4\sqrt{xy}\right)$ **Ans:** $48xy$

2. $\sqrt{2}\left(\sqrt{30} + 3\sqrt{2} - 4\sqrt{10}\right)$ **Ans:** $2\sqrt{15} + 6 - 8\sqrt{5}$

3. $\left(3\sqrt{5} - 2\sqrt{2}\right)\left(4\sqrt{5} + 2\sqrt{2}\right)$ **Ans:** $52 - 2\sqrt{10}$

How are you doing with your homework assignments in Sections 9.1 to 9.4? Do you feel you have mastered the material so far? Do you understand the concepts you have covered? Before you go further in the textbook, take some time to do each of the following problems.

Assume all variables are positive.

9.1

Find the square roots, if possible. (If it is impossible to obtain a real number for an answer, say so.) Approximate to the nearest thousandth.

1. $\sqrt{49}$　　　　　**2.** $\sqrt{\dfrac{25}{64}}$　　　　　**3.** $\sqrt{81}$

4. $-\sqrt{100}$　　　　　**5.** $\sqrt{5}$　　　　　**6.** $-\sqrt{121}$

9.2

Simplify.

7. $\sqrt{0.04}$　　　　　**8.** $\sqrt{0.36}$　　　　　**9.** $\sqrt{x^4}$

10. $\sqrt{25x^4y^6}$　　　　　**11.** $\sqrt{50}$　　　　　**12.** $\sqrt{x^5}$

13. $\sqrt{144a^2b^3}$　　　　　**14.** $\sqrt{32x^5}$

9.3

Combine, if possible.

15. $\sqrt{98} + \sqrt{128}$　　　　　**16.** $3\sqrt{2} - \sqrt{8} + \sqrt{18}$

17. $2\sqrt{28} + 3\sqrt{63} - \sqrt{49}$　　　　　**18.** $3\sqrt{8} + \sqrt{12} + \sqrt{50} - 4\sqrt{75}$

19. $5\sqrt{2x} - 7\sqrt{8x}$　　　　　**20.** $-4x\sqrt{75} - 2x\sqrt{27}$

9.4

Multiply. Simplify your answer.

21. $\left(5\sqrt{2}\right)\left(3\sqrt{15}\right)$　　　　　**22.** $\left(4x\sqrt{6}\right)\left(x\sqrt{10}\right)$

23. $\sqrt{2}\left(\sqrt{6} + 3\sqrt{2}\right)$　　　　　**24.** $\left(\sqrt{7} - 3\right)^2$

25. $\left(\sqrt{11} - \sqrt{10}\right)\left(\sqrt{11} + \sqrt{10}\right)$　　　　　**26.** $\left(2\sqrt{3} - \sqrt{5}\right)^2$

27. $\left(\sqrt{x} + 5\right)\left(\sqrt{x} - 2\right)$　　　　　**28.** $\left(3\sqrt{7} + 4\sqrt{3}\right)\left(5\sqrt{7} - 2\sqrt{3}\right)$

Now turn to page SA-22 for the answer to each of these problems. Each answer also includes a reference to the objective in which the problem is first taught. If you missed any of these problems, you should stop and review the Examples and Practice Problems in the referenced objective. A little review now will help you master the material in the upcoming sections of the text.

1.	7
2.	$\dfrac{5}{8}$
3.	9
4.	-10
5.	2.236
6.	-11
7.	0.2
8.	0.6
9.	x^2
10.	$5x^2y^3$
11.	$5\sqrt{2}$
12.	$x^2\sqrt{x}$
13.	$12ab\sqrt{b}$
14.	$4x^2\sqrt{2x}$
15.	$15\sqrt{2}$
16.	$4\sqrt{2}$
17.	$13\sqrt{7} - 7$
18.	$11\sqrt{2} - 18\sqrt{3}$
19.	$-9\sqrt{2x}$
20.	$-26x\sqrt{3}$
21.	$15\sqrt{30}$
22.	$8x^2\sqrt{15}$
23.	$2\sqrt{3} + 6$
24.	$16 - 6\sqrt{7}$
25.	1
26.	$17 - 4\sqrt{15}$
27.	$x + 3\sqrt{x} - 10$
28.	$81 + 14\sqrt{21}$

Student Learning Objectives

After studying this section, you will be able to:

1. Use the quotient rule for square roots to simplify a fraction involving radicals.

2. Rationalize the denominator of a fraction with a square root in the denominator.

3. Rationalize the denominator of a fraction with a binomial denominator containing at least one square root.

Teaching Example 1 Simplify.

(a) $\sqrt{\dfrac{1}{64}}$ (b) $\dfrac{\sqrt{32}}{\sqrt{2}}$

Ans: (a) $\dfrac{1}{8}$ (b) 4

Using the Quotient Rule for Square Roots to Simplify a Fraction Involving Radicals

Just as there is a multiplication rule for square roots, there is a quotient rule for square roots. The rules are similar.

> **QUOTIENT RULE FOR SQUARE ROOTS**
>
> For all positive numbers a and b,
> $$\frac{\sqrt{a}}{\sqrt{b}} = \sqrt{\frac{a}{b}} \quad \text{and} \quad \sqrt{\frac{a}{b}} = \frac{\sqrt{a}}{\sqrt{b}}.$$

This can be used to divide square root radicals or to simplify square root radical expressions involving division. We will be using both parts of the quotient rule.

EXAMPLE 1 Simplify.

(a) $\dfrac{\sqrt{75}}{\sqrt{3}}$ (b) $\sqrt{\dfrac{25}{36}}$

Solution

(a) We notice that 3 is a factor of 75. We use the quotient rule to rewrite the expression.

$$\frac{\sqrt{75}}{\sqrt{3}} = \sqrt{\frac{75}{3}}$$
$$= \sqrt{25} \quad \text{Divide.}$$
$$= 5 \quad \text{Simplify.}$$

(b) Since both 25 and 36 are perfect squares, we will rewrite this as the quotient of square roots.

$$\sqrt{\frac{25}{36}} = \frac{\sqrt{25}}{\sqrt{36}}$$
$$= \frac{5}{6}$$

Practice Problem 1 Simplify.

(a) $\dfrac{\sqrt{98}}{\sqrt{2}}$ 7 (b) $\sqrt{\dfrac{81}{100}}$ $\dfrac{9}{10}$

Teaching Example 2 Simplify. $\sqrt{\dfrac{w^8 x^3}{16}}$

Ans: $\dfrac{w^4 x\sqrt{x}}{4}$

EXAMPLE 2 Simplify. $\sqrt{\dfrac{20}{x^6}}$

Solution

$$\sqrt{\frac{20}{x^6}} = \frac{\sqrt{20}}{\sqrt{x^6}} = \frac{\sqrt{4}\sqrt{5}}{x^3} = \frac{2\sqrt{5}}{x^3} \quad \text{Don't forget to simplify } \sqrt{20} \text{ as } 2\sqrt{5}.$$

Practice Problem 2 Simplify.

$\sqrt{\dfrac{50}{a^4}}$ $\dfrac{5\sqrt{2}}{a^2}$

Rationalizing the Denominator of a Fraction with a Square Root in the Denominator

Sometimes, when calculating with fractions that contain radicals, it is advantageous to have an integer in the denominator. If a fraction has a radical in the denominator, we **rationalize the denominator.** That is, we multiply to change the fraction to an equivalent one that has an integer in the denominator. Remember, we do not want to change the value of the fraction. Thus we will multiply the numerator and the denominator by the same number.

EXAMPLE 3 Simplify. $\dfrac{3}{\sqrt{2}}$

Solution Think, "What times 2 will make a perfect square?"

$$\frac{3}{\sqrt{2}} = \frac{3}{\sqrt{2}} \times 1 = \frac{3}{\sqrt{2}} \times \frac{\sqrt{2}}{\sqrt{2}} = \frac{3\sqrt{2}}{\sqrt{4}} = \frac{3\sqrt{2}}{2}$$

Practice Problem 3 Simplify.

$$\frac{9}{\sqrt{7}} \qquad \frac{9\sqrt{7}}{7}$$

(Note that a fraction is not simplified unless the denominator is rationalized.)

Teaching Example 3 Simplify. $\dfrac{2}{\sqrt{11}}$

Ans: $\dfrac{2\sqrt{11}}{11}$

When rationalizing a denominator containing a square root radical, we will want to use the smallest possible radical that will yield the square root of a perfect square. Very often we will not use the radical that is in the denominator.

EXAMPLE 4 Simplify.

(a) $\dfrac{\sqrt{7}}{\sqrt{8}}$
(b) $\dfrac{3}{\sqrt{x^3}}$

Solution

(a) Think, "What times 8 will make a perfect square?"

$$\frac{\sqrt{7}}{\sqrt{8}} = \frac{\sqrt{7}}{\sqrt{8}} \times \frac{\sqrt{2}}{\sqrt{2}} = \frac{\sqrt{14}}{\sqrt{16}} = \frac{\sqrt{14}}{4}$$

(b) Think, "What times x^3 will give an even exponent?"

$$\frac{3}{\sqrt{x^3}} \times \frac{\sqrt{x}}{\sqrt{x}} = \frac{3\sqrt{x}}{\sqrt{x^4}} = \frac{3\sqrt{x}}{x^2}$$

We could have simplified $\sqrt{x^3}$ before we rationalized the denominator.

$$\frac{3}{\sqrt{x^3}} = \frac{3}{x\sqrt{x}} \times \frac{\sqrt{x}}{\sqrt{x}} = \frac{3\sqrt{x}}{x\sqrt{x^2}} = \frac{3\sqrt{x}}{(x)(x)} = \frac{3\sqrt{x}}{x^2}$$

Practice Problem 4 Simplify.

(a) $\dfrac{\sqrt{2}}{\sqrt{12}} \qquad \dfrac{\sqrt{6}}{6}$
(b) $\dfrac{6a}{\sqrt{a^7}} \qquad \dfrac{6\sqrt{a}}{a^3}$

Teaching Example 4 Simplify.

(a) $\dfrac{\sqrt{5}}{\sqrt{12}}$
(b) $\dfrac{9y}{\sqrt{y^5}}$

Ans: (a) $\dfrac{\sqrt{15}}{6}$
(b) $\dfrac{9\sqrt{y}}{y^2}$

Teaching Example 5 Simplify. $\dfrac{\sqrt{7}}{\sqrt{45y}}$

Ans: $\dfrac{\sqrt{35y}}{15y}$

EXAMPLE 5 Simplify. $\dfrac{\sqrt{2}}{\sqrt{27x}}$

Solution Since it is not apparent what we should multiply 27 by to obtain a perfect square, we will begin by simplifying the denominator.

$$\frac{\sqrt{2}}{\sqrt{27x}} = \frac{\sqrt{2}}{3\sqrt{3x}}$$

Now it is easy to see that we multiply by $\dfrac{\sqrt{3x}}{\sqrt{3x}}$ to rationalize the denominator.

$$= \frac{\sqrt{2}}{3\sqrt{3x}} \times \frac{\sqrt{3x}}{\sqrt{3x}}$$

$$= \frac{\sqrt{6x}}{3\sqrt{9x^2}}$$

$$= \frac{\sqrt{6x}}{9x}$$

NOTE TO STUDENT: Fully worked-out solutions to all of the Practice Problems can be found at the back of the text starting at page SP-1

Practice Problem 5 Simplify.

$$\frac{\sqrt{5x}}{\sqrt{8x}} \quad \frac{\sqrt{10}}{4}$$

Teaching Tip Take the time to explain what a conjugate is. Then quiz students in class by asking them to identify the conjugate of each expression.

(a) $7 + \sqrt{3}$
conjugate: $7 - \sqrt{3}$

(b) $2\sqrt{5} + 7\sqrt{11}$
conjugate: $2\sqrt{5} - 7\sqrt{11}$

(c) $\sqrt{x} - 5$
conjugate: $\sqrt{x} + 5$

(d) $4\sqrt{x} - 5\sqrt{2}$
conjugate: $4\sqrt{x} + 5\sqrt{2}$

③ Rationalizing the Denominator of a Fraction with a Binomial Denominator Containing at Least One Square Root

Sometimes the denominator of a fraction is a binomial with a square root radical term. $\dfrac{1}{5 - 3\sqrt{2}}$ is such a fraction. How can we eliminate the radical term in the denominator? Recall that when we multiply $(a + b)(a - b)$, we obtain the square of a minus the square of b: $a^2 - b^2$. For example, $(x + 3)(x - 3) = x^2 - 9$. The resulting terms are perfect squares. We can use this idea to eliminate the radical in $5 - 3\sqrt{2}$.

$$\left(5 - 3\sqrt{2}\right)\left(5 + 3\sqrt{2}\right) = 5^2 + 15\sqrt{2} - 15\sqrt{2} - \left(3\sqrt{2}\right)^2 = 5^2 - \left(3\sqrt{2}\right)^2 = 25 - 18 = 7$$

Expressions like $\left(5 - 3\sqrt{2}\right)$ and $\left(5 + 3\sqrt{2}\right)$ are called **conjugates.**

Teaching Example 6 Simplify.

(a) $\dfrac{3x}{\sqrt{5} + 2}$ **(b)** $\dfrac{\sqrt{2}}{\sqrt{7} - \sqrt{2}}$

Ans: **(a)** $3x\sqrt{5} - 6x$ **(b)** $\dfrac{\sqrt{14} + 2}{5}$

EXAMPLE 6 Simplify.

(a) $\dfrac{2}{\sqrt{3} - 4}$ **(b)** $\dfrac{\sqrt{x}}{\sqrt{5} + \sqrt{3}}$

Solution

(a) The conjugate of $\sqrt{3} - 4$ is $\sqrt{3} + 4$.

$$\frac{2}{\sqrt{3} - 4} \cdot \frac{\sqrt{3} + 4}{\sqrt{3} + 4} = \frac{2\sqrt{3} + 8}{\left(\sqrt{3}\right)^2 + 4\sqrt{3} - 4\sqrt{3} - 4^2}$$

$$= \frac{2\sqrt{3} + 8}{3 - 16}$$

$$= \frac{2\sqrt{3} + 8}{-13}$$

$$= -\frac{2\sqrt{3} + 8}{13}$$

(b) The conjugate of $\sqrt{5} + \sqrt{3}$ is $\sqrt{5} - \sqrt{3}$.

$$\frac{\sqrt{x}}{\sqrt{5} + \sqrt{3}} \cdot \frac{\sqrt{5} - \sqrt{3}}{\sqrt{5} - \sqrt{3}} = \frac{\sqrt{5x} - \sqrt{3x}}{\left(\sqrt{5}\right)^2 - \sqrt{15} + \sqrt{15} - \left(\sqrt{3}\right)^2}$$

$$= \frac{\sqrt{5x} - \sqrt{3x}}{5 - 3}$$

$$= \frac{\sqrt{5x} - \sqrt{3x}}{2}$$

Be careful not to combine $\sqrt{5x}$ and $\sqrt{3x}$ in the numerator. They are not like radicals.

Practice Problem 6 Simplify.

(a) $\dfrac{4}{\sqrt{3} + \sqrt{5}}$ $\quad -2\left(\sqrt{3} - \sqrt{5}\right)$ **(b)** $\dfrac{\sqrt{a}}{\sqrt{10} - 3}$ $\quad \sqrt{10a} + 3\sqrt{a}$

Note that in Example 7 the numerator is a binomial.

EXAMPLE 7 Rationalize the denominator. $\dfrac{\sqrt{3} + \sqrt{2}}{\sqrt{3} - \sqrt{2}}$

Solution The conjugate of $\sqrt{3} - \sqrt{2}$ is $\sqrt{3} + \sqrt{2}$.

$$\frac{\sqrt{3} + \sqrt{2}}{\sqrt{3} - \sqrt{2}} \cdot \frac{\sqrt{3} + \sqrt{2}}{\sqrt{3} + \sqrt{2}}$$

$$= \frac{\sqrt{9} + \sqrt{6} + \sqrt{6} + \sqrt{4}}{\left(\sqrt{3}\right)^2 - \left(\sqrt{2}\right)^2} \qquad \text{Multiply.}$$

$$= \frac{3 + 2\sqrt{6} + 2}{3 - 2} \qquad \text{Simplify and combine like terms.}$$

$$= \frac{5 + 2\sqrt{6}}{1}$$

$$= 5 + 2\sqrt{6}$$

Practice Problem 7 Rationalize the denominator.

$$\frac{\sqrt{3} + \sqrt{5}}{\sqrt{3} - \sqrt{5}} \qquad -4 - \sqrt{15}$$

Note that in Examples 6 and 7 we used a raised dot · to indicate multiplication instead of the ×. You need to be familiar with both forms of multiplication notation.

Teaching Example 7 Rationalize the denominator. $\dfrac{4\sqrt{3} + 3\sqrt{2}}{5\sqrt{3} - 6\sqrt{2}}$

Ans: $32 + 13\sqrt{6}$

Simplify. Be sure to rationalize all denominators. Do not use a calculator or a table of square roots.

1. $\dfrac{\sqrt{12}}{\sqrt{3}}$ 2

2. $\dfrac{\sqrt{3}}{\sqrt{27}}$ $\dfrac{1}{3}$

3. $\dfrac{\sqrt{7}}{\sqrt{63}}$ $\dfrac{1}{3}$

4. $\dfrac{\sqrt{52}}{\sqrt{13}}$ 2

5. $\dfrac{\sqrt{98}}{\sqrt{2}}$ 7

6. $\dfrac{\sqrt{5}}{\sqrt{80}}$ $\dfrac{1}{4}$

7. $\dfrac{\sqrt{6}}{\sqrt{x^4}}$ $\dfrac{\sqrt{6}}{x^2}$

8. $\dfrac{\sqrt{12}}{\sqrt{x^2}}$ $\dfrac{2\sqrt{3}}{x}$

9. $\dfrac{\sqrt{18}}{\sqrt{a^4}}$ $\dfrac{3\sqrt{2}}{a^2}$

10. $\dfrac{\sqrt{24}}{\sqrt{b^6}}$ $\dfrac{2\sqrt{6}}{b^3}$

11. $\dfrac{3}{\sqrt{7}}$ $\dfrac{3\sqrt{7}}{7}$

12. $\dfrac{2}{\sqrt{10}}$ $\dfrac{\sqrt{10}}{5}$

13. $\dfrac{8}{\sqrt{15}}$ $\dfrac{8\sqrt{15}}{15}$

14. $\dfrac{9}{\sqrt{14}}$ $\dfrac{9\sqrt{14}}{14}$

15. $\dfrac{x\sqrt{x}}{\sqrt{2}}$ $\dfrac{x\sqrt{2x}}{2}$

16. $\dfrac{\sqrt{3y}}{\sqrt{6}}$ $\dfrac{3\sqrt{2y}}{6} = \dfrac{\sqrt{2y}}{2}$

17. $\dfrac{\sqrt{8}}{\sqrt{x}}$ $\dfrac{2\sqrt{2x}}{x}$

18. $\dfrac{\sqrt{18}}{\sqrt{y}}$ $\dfrac{3\sqrt{2y}}{y}$

19. $\dfrac{6}{\sqrt{28}}$ $\dfrac{3\sqrt{7}}{7}$

20. $\dfrac{3}{\sqrt{27}}$ $\dfrac{\sqrt{3}}{3}$

21. $\dfrac{6}{\sqrt{a}}$ $\dfrac{6\sqrt{a}}{a}$

22. $\dfrac{8}{\sqrt{b}}$ $\dfrac{8\sqrt{b}}{b}$

23. $\dfrac{x}{\sqrt{2x^5}}$ $\dfrac{\sqrt{2x}}{2x^2}$

24. $\dfrac{y}{\sqrt{5x^3}}$ $\dfrac{y\sqrt{5x}}{5x^2}$

25. $\dfrac{\sqrt{18}}{\sqrt{2x^3}}$ $\dfrac{3\sqrt{x}}{x^2}$

26. $\dfrac{\sqrt{24}}{\sqrt{6x^7}}$ $\dfrac{2\sqrt{x}}{x^4}$

27. $\sqrt{\dfrac{3}{5}}$ $\dfrac{\sqrt{15}}{5}$

28. $\sqrt{\dfrac{1}{10}}$ $\dfrac{\sqrt{10}}{10}$

29. $\sqrt{\dfrac{10}{21}}$ $\dfrac{\sqrt{210}}{21}$

30. $\sqrt{\dfrac{15}{17}}$ $\dfrac{\sqrt{255}}{17}$

31. $\dfrac{9}{\sqrt{32x}}$ $\dfrac{9\sqrt{2x}}{8x}$

32. $\dfrac{3}{\sqrt{50x}}$ $\dfrac{3\sqrt{2x}}{10x}$

Rationalize the denominator. Simplify your answer. Do not use a calculator or a table of square roots.

33. $\dfrac{4}{\sqrt{3}-1}$ $2\sqrt{3}+2$

34. $\dfrac{2}{\sqrt{6}+1}$ $\dfrac{2\sqrt{6}-2}{5}$

35. $\dfrac{8}{\sqrt{10}+\sqrt{2}}$ $\sqrt{10}-\sqrt{2}$

36. $\dfrac{5}{\sqrt{11}-\sqrt{6}}$ $\sqrt{11}+\sqrt{6}$

37. $\dfrac{\sqrt{6}}{\sqrt{6}-\sqrt{3}}$ $2+\sqrt{2}$

38. $\dfrac{\sqrt{3}}{\sqrt{5}+\sqrt{3}}$ $\dfrac{\sqrt{15}-3}{2}$

39. $\dfrac{\sqrt{7}}{\sqrt{8}+\sqrt{7}}$ $2\sqrt{14}-7$

40. $\dfrac{\sqrt{5}}{\sqrt{5}+\sqrt{6}}$ $\sqrt{30}-5$

41. $\dfrac{3x}{2\sqrt{2}-\sqrt{5}}$ $x(2\sqrt{2}+\sqrt{5})$

42. $\dfrac{4x}{2\sqrt{7}+2\sqrt{6}}$ $x(2\sqrt{7}-2\sqrt{6})$

43. $\dfrac{\sqrt{x}}{\sqrt{6}+\sqrt{2}}$ $\dfrac{\sqrt{6x}-\sqrt{2x}}{4}$

44. $\dfrac{\sqrt{x}}{\sqrt{11}+\sqrt{5}}$ $\dfrac{\sqrt{11x}-\sqrt{5x}}{6}$

45. $\dfrac{\sqrt{10}-\sqrt{3}}{\sqrt{10}+\sqrt{3}}$ $\dfrac{13-2\sqrt{30}}{7}$

46. $\dfrac{\sqrt{7}+\sqrt{3}}{\sqrt{7}-\sqrt{3}}$ $\dfrac{5+\sqrt{21}}{2}$

47. $\dfrac{4\sqrt{7}+3}{\sqrt{5}-\sqrt{2}}$ $\dfrac{4\sqrt{35}+3\sqrt{5}+3\sqrt{2}+4\sqrt{14}}{3}$

48. $\dfrac{4\sqrt{3}+2\sqrt{5}}{\sqrt{5}-\sqrt{3}}$ $3\sqrt{15}+11$

49. $\dfrac{4\sqrt{3}+2}{\sqrt{8}-\sqrt{6}}$ $5\sqrt{6}+8\sqrt{2}$

50. $\dfrac{3\sqrt{5}+4}{\sqrt{15}-\sqrt{3}}$ $\dfrac{19\sqrt{3}+7\sqrt{15}}{12}$

51. $\dfrac{x-25}{\sqrt{x}+5}$ $\sqrt{x}-5$

52. $\dfrac{x-36}{\sqrt{x}-6}$ $\sqrt{x}+6$

Applications

▲ **53. Geometry** A pyramid has a volume V and a height h. The length of one side of the square base is $s = \sqrt{\dfrac{3V}{h}}$.

(a) Simplify this expression by rationalizing the denominator.
$s = \dfrac{1}{h}\sqrt{3Vh}$

(b) Find the length of the side of a pyramid with a volume of 36 cubic inches and a height of 12 inches.
$s = 3$ in.

▲ **54. Geometry** A cylinder has a volume V and a height h. The radius of the cylinder is $r = \sqrt{\dfrac{V}{\pi h}}$.

(a) Simplify this expression by rationalizing the denominator.
$r = \dfrac{1}{\pi h}\sqrt{V\pi h}$

(b) Find the radius of a cylindrical storage silo of volume 6250 cubic feet and height 10π feet.
$r = \dfrac{25}{\pi}$ ft

▲ **55. Relay Tower** A reflector sheet for a microwave relay tower is shaped like a rectangle. The area of the rectangle is 3 square meters. Ideally the length should be exactly $\left(\sqrt{5} + 2\right)$ meters long. What should be the width of the rectangle? Express the answer exactly using radical expressions.
$\left(3\sqrt{5} - 6\right)$ m

▲ **56. Relay Tower** An engineer needs the approximate dimensions of the rectangle in exercise 57. Approximate the length and the width of that rectangle to the nearest thousandth.
Length, 4.236 m; width, 0.708 m

Cumulative Review

57. [2.3.3] Solve for x. Round your answer to the nearest hundredth. $-2(x - 5) = 8x - 3(x + 1) + 18$
$x \approx -0.71$

58. [1.8.2] Evaluate $a^2 - b \div a + a^3 - b$ for $a = -1$ and $b = 4$.
0

Quick Quiz 9.5 Simplify. Be sure to rationalize all denominators.

1. $\dfrac{3}{\sqrt{10}}$ $\dfrac{3\sqrt{10}}{10}$

2. $\dfrac{5x}{3\sqrt{2} + \sqrt{5}}$ $\dfrac{15x\sqrt{2} - 5x\sqrt{5}}{13}$

3. $\dfrac{\sqrt{6} + \sqrt{5}}{\sqrt{6} - \sqrt{5}}$ $11 + 2\sqrt{30}$

4. Concept Check Explain how you would simplify $\dfrac{x - 4}{\sqrt{x} + 2}$.   Answers may vary

Classroom Quiz 9.5 You may use these problems to quiz your students' mastery of Section 9.5.

Simplify. Be sure to rationalize all denominators.

1. $\dfrac{5}{\sqrt{8}}$ **Ans:** $\dfrac{5\sqrt{2}}{4}$

2. $\dfrac{3x}{2\sqrt{5} + \sqrt{3}}$ **Ans:** $\dfrac{6x\sqrt{5} - 3x\sqrt{3}}{17}$

3. $\dfrac{\sqrt{5} + \sqrt{3}}{\sqrt{5} - \sqrt{3}}$ **Ans:** $4 + \sqrt{15}$

Student Learning Objectives

After studying this section, you will be able to:

 Use the Pythagorean Theorem to solve applied problems.

 Solve radical equations.

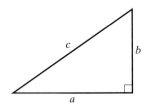

Using the Pythagorean Theorem to Solve Applied Problems

In ancient Greece, mathematicians studied a number of properties of right triangles. They proved that the square of the length of the longest side of a right triangle is equal to the sum of the squares of the lengths of the other two sides. This property is called the Pythagorean Theorem, in honor of the Greek mathematician Pythagoras (ca. 590 B.C.). The shorter sides, a and b, are referred to as the **legs** of the right triangle. The longest side, c, is called the **hypotenuse** of the right triangle. We state the theorem as follows.

> **PYTHAGOREAN THEOREM**
>
> In any right triangle, if c is the length of the hypotenuse and a and b are the lengths of the two legs, then
> $$c^2 = a^2 + b^2.$$

If we know any two sides of a right triangle, we can find the third side using this theorem.

Notice that the hypotenuse is always opposite the right angle. The smallest side is opposite the smallest angle.

Recall that in Section 9.1 we showed that if $s^2 = 4$, then $s = 2$ or -2. In general, if $x^2 = a$, then $x = \pm\sqrt{a}$. This is sometimes called "taking the square root of each side of an equation." The abbreviation $\pm\sqrt{a}$ means $+\sqrt{a}$ or $-\sqrt{a}$. It is read "plus or minus $\sqrt{a}$."

Teaching Example 1 Find the length of the hypotenuse of a right triangle with legs of 6 inches and 8 inches.

Ans: 10 in.

EXAMPLE 1 The ramp to Tony Pitkin's barn rises 5 feet over a horizontal distance of 12 feet. How long is the ramp? That is, find the length of the hypotenuse of a right triangle whose legs are 5 feet and 12 feet.

Solution

1. **Understand the problem.**
   Draw and label a diagram.

2. **Write an equation.** Write the Pythagorean Theorem. Substitute the known values into the equation.
$$c^2 = a^2 + b^2$$
$$c^2 = 5^2 + 12^2$$

3. **Solve and state the answer.**
$$c^2 = 25 + 144$$
$$c^2 = 169 \qquad \text{Take the square root of each side of the equation.}$$
$$c = \pm\sqrt{169}$$
$$= \pm 13$$

   The hypotenuse is 13 feet. We do not use -13 because length is not negative. Thus the length of the ramp is 13 feet.

4. **Check.** Substitute the values for a, b, and c into the Pythagorean Theorem and evaluate. Does $13^2 = 5^2 + 12^2$? Yes. The solution checks. ✓

NOTE TO STUDENT: Fully worked-out solutions to all of the Practice Problems can be found at the back of the text starting at page SP-1

Practice Problem 1 Find the length of the hypotenuse of a right triangle with legs of 9 centimeters and 12 centimeters. 15 cm

Sometimes you will need to find one of the legs of a right triangle given the hypotenuse and the other leg.

▲ **EXAMPLE 2**   Find the unknown leg of a right triangle that has a hypotenuse of 10 yards and one leg of $\sqrt{19}$ yards.

Solution   Draw a diagram. The diagram is in the margin on the right.

$$c^2 = a^2 + b^2 \qquad \text{Write the Pythagorean Theorem.}$$
$$10^2 = \left(\sqrt{19}\right)^2 + b^2 \quad \text{Substitute the known values into the equation.}$$
$$100 = 19 + b^2$$
$$81 = b^2$$
$$\pm 9 = b \qquad\qquad \text{Take the square root of each side of the equation.}$$

The leg is 9 yards long.

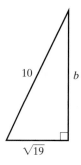

▲ **Practice Problem 2**   The hypotenuse of a right triangle is $\sqrt{17}$ meters. One leg is 1 meter long. Find the length of the other leg.   4 m

Teaching Example 2 Find the unknown leg of a right triangle that has a hypotenuse of $\sqrt{29}$ centimeters and one leg of 2 centimeters.

Ans: 5 cm

Sometimes the answer to a problem will be an irrational number such as $\sqrt{33}$. It is best to leave the answer in radical form unless you are asked for an approximate answer. To find the approximate value, use a calculator or a square root table. $\sqrt{33}$ is an exact answer.

▲ **EXAMPLE 3**   A 25-foot ladder is placed against a building. The foot of the ladder is 8 feet from the wall. At approximately what height does the top of the ladder touch the building? Round to the nearest tenth.

Solution

$$c^2 = a^2 + b^2$$
$$25^2 = 8^2 + b^2$$
$$625 = 64 + b^2$$
$$561 = b^2$$
$$\pm\sqrt{561} = b$$

Teaching Example 3 Topsfield is 15 miles north of Allentown. Bridgeton is 10 miles east of Allentown. How long is a road that travels in a straight line from Topsfield to Bridgeton? Round to the nearest tenth.

Ans: 18.0 mi

We want only the positive value for the distance, so $b = \sqrt{561}$. Using a calculator, we have 561 $\boxed{\sqrt{}}$ 23.68543856. Rounding, we obtain $b \approx 23.7$.

If you do not have a calculator, you will need to use the square root table in Appendix A. However, this square root table goes only as far as $\sqrt{200}$! To approximate larger radicals with the square root table, you will have to write the radical in a different fashion.

$b = \sqrt{3}\sqrt{187}$ Since $\sqrt{ab} = \sqrt{a}\sqrt{b}$ and $3 \cdot 187 = 561$.

$b \approx (1.732)(13.675)$ Replace each radical by the decimal approximation from the table.

$b \approx 23.7$ Multiply and round to the nearest tenth.

The ladder touches the building at a height of approximately 23.7 feet.

▲ **Practice Problem 3**   A support line is placed 3 meters away from the base of an 8-meter pole. If the support line is attached to the top of the pole and pulled tight (assume that it is a straight line), how long is the support line from the ground to the pole? Round to the nearest tenth.   8.5 m

Solving Radical Equations

A **radical equation** is an equation with a variable in one or more of the radicands. A square root radical equation can be simplified by squaring each side of the equation. The apparent solution to a radical equation *must* be checked by substitution into the *original* equation.

For some square root radical equations, you will need to *isolate the radical before* you square each side of the equation.

Teaching Example 4 Solve and check.

$$\sqrt{2x + 7} + 5 = 6$$

Ans: $x = -3$

> **EXAMPLE 4**   Solve and check. $1 + \sqrt{5x - 4} = 5$
>
> **Solution** $1 + \sqrt{5x - 4} = 5$ We want to isolate the radical first.
>
> $\qquad\qquad\qquad \sqrt{5x - 4} = 4$ Subtract 1 from each side to isolate the radical.
>
> $\qquad\qquad \left(\sqrt{5x - 4}\right)^2 = (4)^2$ Square each side.
>
> $\qquad\qquad\qquad\quad 5x - 4 = 16$ Simplify and solve for x.
>
> $\qquad\qquad\qquad\qquad 5x = 20$
>
> $\qquad\qquad\qquad\qquad\; x = 4$
>
> *Check.* $1 + \sqrt{5(4) - 4} \overset{?}{=} 5$
>
> $\qquad\qquad 1 + \sqrt{20 - 4} \overset{?}{=} 5$
>
> $\qquad\qquad\quad 1 + \sqrt{16} \overset{?}{=} 5$
>
> $\qquad\qquad\qquad 1 + 4 = 5$ ✓
>
> Thus 4 is the solution.
>
> **Practice Problem 4**   Solve and check. $\sqrt{3x - 2} - 7 = 0$ $x = 17$

NOTE TO STUDENT: *Fully worked-out solutions to all of the Practice Problems can be found at the back of the text starting at page SP-1*

Some problems will have an apparent solution, but that solution does not always check. An apparent solution that does not satisfy the original equation is called an **extraneous root.**

Teaching Example 5 Solve and check.

$$\sqrt{x - 5} + 2 = 0$$

Ans: No solution

> **EXAMPLE 5**   Solve and check. $\sqrt{x + 3} = -7$
>
> **Solution** $\qquad\quad \sqrt{x + 3} = -7$
>
> $\qquad\qquad \left(\sqrt{x + 3}\right)^2 = (-7)^2$ Square each side.
>
> $\qquad\qquad\qquad x + 3 = 49$ Simplify and solve for x.
>
> $\qquad\qquad\qquad\qquad x = 46$
>
> *Check.* $\qquad \sqrt{46 + 3} \overset{?}{=} -7$
>
> $\qquad\qquad\qquad \sqrt{49} \overset{?}{=} -7$
>
> $\qquad\qquad\qquad\quad 7 \neq -7!$ Does not check! The apparent solution is an extraneous root.
>
> There is no solution to Example 5.
>
> **Practice Problem 5**   Solve and check. $\sqrt{5x + 4} + 2 = 0$ No solution

Wait a minute! Is the original equation in Example 5 possible? No. $\sqrt{x + 3} = -7$ is an impossible statement. We defined the $\sqrt{}$ symbol to mean the positive square root of the radicand. Thus, it cannot be equal to a negative number. If we had noticed this in the first step, we could have written down "no solution" immediately.

When you square both sides of an equation you may obtain a quadratic equation. In such cases, you may obtain two apparent solutions, and you must check both of them.

EXAMPLE 6   Solve and check. $\sqrt{3x + 1} = x + 1.$

Solution

$$\sqrt{3x + 1} = x + 1$$
$$\left(\sqrt{3x + 1}\right)^2 = (x + 1)^2 \qquad \text{Square each side.}$$
$$3x + 1 = x^2 + 2x + 1 \quad \text{Simplify and set the equation equal to 0.}$$
$$0 = x^2 - x$$
$$0 = x(x - 1) \qquad \text{Factor the quadratic equation.}$$
$$x = 0 \quad \text{or} \quad x - 1 = 0 \qquad \text{Set each factor equal to 0 and solve.}$$
$$x = 1$$

Check.   If $x = 0$: $\sqrt{3(0) + 1} \overset{?}{=} 0 + 1$ If $x = 1$: $\sqrt{3(1) + 1} \overset{?}{=} 1 + 1$

$\sqrt{0 + 1} \overset{?}{=} 0 + 1$ $\sqrt{3 + 1} \overset{?}{=} 2$

$\sqrt{1} \overset{?}{=} 1$ $\sqrt{4} \overset{?}{=} 2$

$1 = 1 \ \checkmark$ $2 = 2 \ \checkmark$

Thus 0 and 1 are both solutions.

It is possible in these cases for both apparent solutions to check, for both to be extraneous, or for one to check and the other to be extraneous.

Practice Problem 6   Solve and check.

$$-2 + \sqrt{6x - 1} = 3x - 2 \quad x = \frac{1}{3}$$

EXAMPLE 7   Solve and check. $\sqrt{2x - 1} = x - 2$

Solution

$$\sqrt{2x - 1} = x - 2$$
$$\left(\sqrt{2x - 1}\right)^2 = (x - 2)^2 \qquad \text{Square each side.}$$
$$2x - 1 = x^2 - 4x + 4$$
$$0 = x^2 - 6x + 5 \qquad \text{Simplify and set the equation to 0.}$$
$$0 = (x - 5)(x - 1) \quad \text{Solve for } x.$$
$$x - 5 = 0 \quad \text{or} \quad x - 1 = 0$$
$$x = 5 \qquad\qquad x = 1$$

Teaching Example 6 Solve and check.
$$\sqrt{x - 1} + 1 = x$$
Ans: $x = 2, x = 1$

Teaching Example 7 Solve and check.
$$\sqrt{2x + 6} + 1 = x$$
Ans: $x = 5$

Check. If $x = 5$: $\sqrt{2(5) - 1} \overset{?}{=} 5 - 2$ If $x = 1$: $\sqrt{2(1) - 1} \overset{?}{=} 1 - 2$

$$\sqrt{10 - 1} \overset{?}{=} 3 \qquad\qquad \sqrt{2 - 1} \overset{?}{=} -1$$

$$\sqrt{9} \overset{?}{=} 3 \qquad\qquad \sqrt{1} \overset{?}{=} -1$$

$$3 = 3 \; \checkmark \qquad\qquad 1 \neq -1$$

It does not check. In this case 1 is an extraneous root.

Thus only 5 is a solution to this equation.

NOTE TO STUDENT: *Fully worked-out solutions to all of the Practice Problems can be found at the back of the text starting at page SP-1*

Practice Problem 7 Solve and check. $2 - x + \sqrt{x + 4} = 0$ $x = 5$

Example 8 is somewhat different in format. Take a minute to see how it compares to Examples 6 and 7.

Teaching Example 8 Solve and check.

$$\sqrt{5x - 4} = \sqrt{3x + 2}$$

Ans: $x = 3$

EXAMPLE 8 Solve and check. $\sqrt{3x + 3} = \sqrt{5x - 1}$

Solution Here there are two radicals. Each radical is already isolated.

$$\sqrt{3x + 3} = \sqrt{5x - 1}$$

$$\left(\sqrt{3x + 3}\right)^2 = \left(\sqrt{5x - 1}\right)^2 \quad \text{Square each side.}$$

$$3x + 3 = 5x - 1 \qquad \text{Simplify and solve for } x.$$

$$3 = 2x - 1$$

$$4 = 2x$$

$$2 = x$$

Check.

$$\sqrt{3(2) + 3} \overset{?}{=} \sqrt{5(2) - 1}$$

$$\sqrt{6 + 3} \overset{?}{=} \sqrt{10 - 1}$$

$$\sqrt{9} = \sqrt{9} \; \checkmark$$

Thus 2 is the solution.

Practice Problem 8 Solve and check. $\sqrt{2x + 1} = \sqrt{x - 10}$ No solution

Use the Pythagorean Theorem to find the length of the third side of each right triangle. Leave any irrational answers in radical form.

▲ **1.**

$$5^2 + 12^2 = c^2$$
$$169 = c^2$$
$$c = 13$$

▲ **2.**

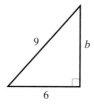

$$6^2 + b^2 = 9^2$$
$$b^2 = 45$$
$$b = 3\sqrt{5}$$

▲ **3.**

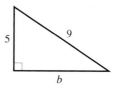

$$5^2 + b^2 = 9^2$$
$$b^2 = 56$$
$$b = 2\sqrt{14}$$

▲ **4.**

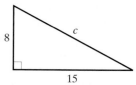

$$8^2 + 15^2 = c^2$$
$$64 + 225 = c^2$$
$$c = 17$$

Exercises 5–14 refer to a right triangle with legs a and b and hypotenuse c. Find the exact length of the missing side.

▲ **5.** $a = 7, b = 7$ Find c.
$$c^2 = 7^2 + 7^2 = 49 + 49 = 98$$
$$c = \sqrt{98} = 7\sqrt{2}$$

▲ **6.** $a = 8, b = 4$ Find c.
$$c^2 = 8^2 + 4^2 = 64 + 16 = 80$$
$$c = \sqrt{80} = 4\sqrt{5}$$

▲ **7.** $a = \sqrt{14}, b = 7$ Find c.
$$c^2 = \left(\sqrt{14}\right)^2 + 7^2 = 14 + 49 = 63$$
$$c = 3\sqrt{7}$$

▲ **8.** $a = \sqrt{21}, b = 5$ Find c.
$$c^2 = \left(\sqrt{21}\right)^2 + 5^2 = 21 + 25 = 46$$
$$c = \sqrt{46}$$

▲ **9.** $c = 15, b = 13$ Find a.
$$a^2 + 13^2 = 15^2$$
$$a^2 = 225 - 169 = 56$$
$$a = 2\sqrt{14}$$

▲ **10.** $c = 19, b = 16$ Find a.
$$a^2 + 16^2 = 19^2$$
$$a^2 = 381 - 256 = 125$$
$$a = 5\sqrt{5}$$

▲ **11.** $c = \sqrt{82}, a = 5$ Find b.
$$b^2 = \left(\sqrt{82}\right)^2 - 5^2 = 82 - 25 = 57$$
$$b = \sqrt{57}$$

▲ **12.** $a = \sqrt{5}, c = 7$ Find b.
$$b^2 = 7^2 - \left(\sqrt{5}\right)^2 = 49 - 5 = 44$$
$$b = 2\sqrt{11}$$

▲ **13.** $c = 13.8, b = 9.42$
Find a to the nearest hundredth.   10.08

▲ **14.** $a = 4.5, b = 6.34$
Find c to the nearest hundredth.   7.77

Applications *Draw a diagram and use the Pythagorean Theorem to solve. Round to the nearest tenth.*

▲ **15.** ***Tenting*** A guy rope is attached to the top of a tent pole. The tent pole is 8 feet tall. If the guy rope is pegged into the ground 5 feet from the tent, how long is the guy rope?
$$c^2 = 8^2 + 5^2 = 64 + 25 = 89$$
$$c = \sqrt{89} \approx 9.4 \text{ ft}$$

▲ **16.** ***Baseball*** A baseball diamond is a square. Each side of the square is 90 feet long. How far is it from home plate to second base? *Hint:* Draw the diagonal.
$$x^2 = 90^2 + 90^2$$
$$x^2 = 16,200$$
$$x \approx 127.3 \text{ ft}$$

▲ **17.** ***Flying a Kite*** A kite is flying on a string that is 100 feet long and is fastened to the ground at the other end. Assume that the string is a straight line. How high is the kite if it is flying above a point that is 20 feet away from where it is fastened to the ground?
$$x^2 + 20^2 = 100^2$$
$$x^2 = 10,000 - 400$$
$$x \approx 98.0 \text{ ft}$$

▲ **18.** ***Boat Mooring*** A boat must be moored 20 feet away from a dock. The boat will be 10 feet below the level of the dock at low tide. What is the minimum length of rope needed to go from the boat to the dock at low tide?
$$x^2 = 100 + 400$$
$$x = 10\sqrt{5} \approx 22.4 \text{ ft}$$

Radio Tower Construction To estimate the cost of erecting a radio tower, we need to know how many feet of support cable are needed.

▲ **19.** Ground anchor #2 is exactly 54 feet from the base of the tower (dashed line in diagram). The bottom ground support cable from level 6 to ground anchor #2 is attached to the tower at a height of 130 feet. The top ground support cable from level 6 to ground anchor #2 is attached to the tower at a height of 135 feet. How long is each support cable to level 6? (Round to the nearest hundredth of a foot.)

Top cable is 145.40 ft long; bottom cable is 140.77 ft long

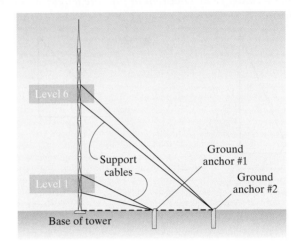

▲ **20.** Ground anchor #1 is exactly 28 feet from the base of the tower (dashed line in diagram). The bottom ground support cable from level 1 to ground anchor #1 is attached to the tower at a height of 42 feet. The top ground support cable from level 1 to ground anchor #1 is attached to the tower at a height of 47 feet. How long is each support cable to level 1? (Round to the nearest hundredth of a foot.)

Top cable is 54.71 ft long; bottom cable is 50.48 ft long

Solve for the variable. Check your solutions.

21. $\sqrt{x + 2} = 3$
$x + 2 = 9$
$x = 7$

22. $\sqrt{x - 8} = 4$
$x - 8 = 16$
$x = 24$

23. $\sqrt{2x + 7} = 5$
$2x + 7 = 25$
$x = 9$

24. $\sqrt{2x - 5} = 7$
$2x - 5 = 49$
$x = 27$

25. $\sqrt{2x + 2} = \sqrt{3x - 5}$
$2x + 2 = 3x - 5$
$7 = x$

26. $\sqrt{5x - 5} = \sqrt{4x + 1}$
$5x - 5 = 4x + 1$
$x = 6$

27. $\sqrt{2x} - 5 = 4$
$\sqrt{2x} = 9$
$2x = 81 \quad x = \dfrac{81}{2}$

28. $\sqrt{13x} - 2 = 5$
$13x = 49$
$x = \dfrac{49}{13}$

29. $\sqrt{3x + 10} = x$
$3x + 10 = x^2$
$0 = x^2 - 3x - 10$
$0 = (x - 5)(x + 2)$
$x = 5$ only

30. $\sqrt{6x - 5} = x$
$6x - 5 = x^2$
$0 = x^2 - 6x + 5$
$0 = (x - 5)(x - 1)$
$x = 5, x = 1$

31. $\sqrt{5y + 1} = y + 1$
$5y + 1 = y^2 + 2y + 1$
$y^2 - 3y = y(y - 3) = 0$
$y = 0, y = 3$

32. $\sqrt{2y + 9} = y + 3$
$2y + 9 = y^2 + 6y + 9$
$y^2 + 4y = 0$
$y(y + 4) = 0$
$y = 0$ only

33. $\sqrt{x + 3} = 3x - 1$
$x + 3 = 9x^2 - 6x + 1$
$9x^2 - 7x - 2 = 0$
$(9x + 2)(x - 1) = 0$
$x = 1$ only

34. $\sqrt{2x + 3} = 2x - 9$
$2x + 3 = 4x^2 - 36x + 81$
$4x^2 - 38x + 78 = 0$
$(2x - 13)(x - 3) = 0$
$x = \dfrac{13}{2}$ only

35. $\sqrt{3y + 1} - y = 1$
$3y + 1 = y^2 + 2y + 1$
$y^2 - y = 0$
$y = 0, y = 1$

36. $\sqrt{3y - 8} + 2 = y$
$3y - 8 = y^2 - 4y + 4$
$y^2 - 7y + 12 = 0$
$(y - 3)(y - 4) = 0$
$y = 3, y = 4$

37. $\sqrt{6y + 1} - 3y = y$
$6y + 1 = 16y^2$
$0 = 16y^2 - 6y - 1$
$0 = (8y + 1)(2y - 1)$
$y = \dfrac{1}{2}$ only

38. $\sqrt{2x + 3} - 2 = x$
$2x + 3 = x^2 + 4x + 4$
$0 = x^2 + 2x + 1$
$0 = (x + 1)^2$
$x = -1$

To Think About *Solve for x.*

39. $\sqrt{12x + 1} = 2\sqrt{6x} - 1$
$x = \dfrac{2}{3}$ only

40. $\sqrt{x - 2} + 3 = \sqrt{4x + 1}$
$x = 2, x = 6$

Cumulative Review

41. [6.5.1] Solve. $\dfrac{5x}{x - 4} = 5 + \dfrac{4x}{x - 4}$
$x = 5$

42. [4.4.2] Multiply mentally. $(7x - 2)^2$
$49x^2 - 28x + 4$

43. [7.6.4] $f(x) = 2x^2 - 3x + 6$
Find $f(-2)$. 20

44. [7.6.1] Is $x^2 + y^2 = 9$ a relation or a function?
Relation

Quick Quiz 9.6

1. In a right triangle with legs a and b and a hypotenuse of c, find the exact length of hypotenuse c if $a = \sqrt{3}$ and $b = 5$.
$c = 2\sqrt{7}$

Solve. Verify your solutions.

2. $8 - \sqrt{5x - 6} = 0$
$x = 14$

3. $\sqrt{6x + 1} = x - 1$
$x = 8$

4. Concept Check A student attempted to solve the equation $2 + \sqrt{1 - 8x} = x$ and found values of $x = -3$ and $x = -1$. Explain how you would check these values to see if they are solutions to the original equation.
Answers may vary

Classroom Quiz 9.6 You may use these problems to quiz your students' mastery of Section 9.6.

1. In a right triangle with legs a and b and a hypotenuse of c, find the exact length of leg a if $b = \sqrt{7}$ and $c = 7$. **Ans:** $a = \sqrt{42}$

Solve. Verify your solutions.

2. $4 - \sqrt{3x - 5} = 0$ **Ans:** $x = 7$

3. $\sqrt{2x + 13} = x + 7$ **Ans:** $x = -6$

Student Learning Objectives

After studying this section, you will be able to:

 Solve problems involving direct variation.

 Solve problems involving inverse variation.

 Identify the graphs of variation equations.

Teaching Example 1 When prescribing an antibiotic to children, the dosage varies directly as the weight of the child. If the dosage for a 32-pound child is 48 milligrams, what is the dosage for a 54-pound child?

Ans: 81 mg

NOTE TO STUDENT: Fully worked-out solutions to all of the Practice Problems can be found at the back of the text starting at page SP-1

Solving Problems Involving Direct Variation

Often in daily life there is a relationship between two measurable quantities. For example, we turn up the thermostat and the heating bill increases. We say that the heating bill varies directly as the temperature on the thermostat. If y **varies directly as** x, then $y = kx$, where k is a constant. The constant is often called the **constant of variation.** Consider the following example.

EXAMPLE 1 Cliff works part-time in a local supermarket while going to college. His salary varies directly as the number of hours worked. Last week he earned $33.60 for working 7 hours. This week he earned $52.80. How many hours did he work?

Solution Let $S =$ his salary,

$h =$ the number of hours he worked, and

$k =$ the constant of variation.

Since his salary varies directly as the number of hours he worked, we write

$$S = k \cdot h.$$

We can find the constant k by substituting the known values of $S = 33.60$ and $h = 7$.

$$33.60 = k \cdot 7 = 7k$$

$$\frac{33.60}{7} = \frac{7k}{7} \qquad \text{Solve for } k.$$

$$4.80 = k \qquad \text{The constant of variation is 4.80.}$$

$$S = 4.80h \qquad \text{Replace } k \text{ in the variation equation by 4.80.}$$

How many hours did he work to earn $52.80?

$$S = 4.80h \qquad \text{The direct variation equation with } k = 4.80.$$

$$52.80 = 4.80h \qquad \text{Substitute 52.80 for } S \text{ and solve for } h.$$

$$\frac{52.80}{4.80} = \frac{4.80h}{4.80}$$

$$11 = h$$

Cliff worked 11 hours this week.

Practice Problem 1 The *change* in temperature measured on the Celsius scale varies directly as the change measured on the Fahrenheit scale. The change in temperature from freezing water to boiling water is 100° on the Celsius scale and 180° on the Fahrenheit scale. If the Fahrenheit temperature drops 20°, what will be the change in temperature on the Celsius scale? Drop of $11\frac{1}{9}$°C

SOLVING A DIRECT VARIATION PROBLEM

1. Write the direct variation equation.

2. Solve for the constant k by substituting in known values.

3. Replace k in the direct variation equation by the value obtained in step 2.

4. Solve for the desired value.

EXAMPLE 2 If y varies directly as x, and $y = 20$ when $x = 3$, find the value of y when $x = 21$.

Solution

Step 1 $y = kx$

Step 2 To find k, we substitute $y = 20$ and $x = 3$.

$$20 = k(3)$$
$$20 = 3k$$
$$\frac{20}{3} = k$$

Step 3 We now write the variation equation with k replaced by $\frac{20}{3}$.

$$y = \frac{20}{3}x$$

Step 4 We replace x by 21 and find y.

$$y = \left(\frac{20}{\cancel{3}}\right)(\cancel{21}^{\,7}) = (20)(7) = 140$$

Thus $y = 140$ when $x = 21$.

Practice Problem 2 If y varies directly as x, and $y = 18$ when $x = 5$, find the value of y when $x = \frac{20}{23}$. $\quad y = \frac{72}{23}$

A number of real-life situations can be described by direct variation equations. The following table shows some of the more common forms of the direct variation equation. In each case, $k =$ the constant of variation.

SAMPLE DIRECT VARIATION SITUATIONS

Verbal Description	*Variation Equation*
y varies directly as x	$y = kx$
b varies directly as the square of c	$b = kc^2$
l varies directly as the cube of m	$l = km^3$
V varies directly as the square root of h	$V = k\sqrt{h}$

EXAMPLE 3 In a certain class of racing cars, the maximum speed varies directly as the square root of the horsepower of the engine. If a car with 225 horsepower can achieve a maximum speed of 120 mph, what speed could it achieve with 256 horsepower?

Solution Let $V =$ the maximum speed,
$\qquad\qquad h =$ the horsepower of the engine, and
$\qquad\qquad k =$ the constant of variation.

Step 1 Since the maximum speed (V) varies directly as the square root of the horsepower of the engine,
$$V = k\sqrt{h}.$$

Step 2 $120 = k\sqrt{225}$ $\quad$ Substitute known values of V and h.
$\qquad\;\;\; 120 = k \cdot 15$ $\qquad$ Solve for k.
$\qquad\qquad\; 8 = k$

Step 3 Now we can write the direct variation equation with the known value for k.
$$V = 8\sqrt{h}$$

Teaching Example 2 If y varies directly as x, and $y = 16$ when $x = 12$, find the value of y when $x = 36$.

Ans: $y = 48$

Teaching Example 3 The volume of water pouring from a hose varies directly as the square of the radius of the hose. Forty-eight gallons poured from a hose of radius 2 centimeters into a barrel. In the same period of time, how many gallons would pour from the hose if it had a radius of 4 centimeters?

Ans: 192 gal

Step 4 $V = 8\sqrt{256}$ Substitute the value of $h = 256$.
$\qquad = (8)(16)$ Solve for V.
$\qquad = 128$

Thus a car with 256 horsepower could achieve a maximum speed of 128 mph.

NOTE TO STUDENT: *Fully worked-out solutions to all of the Practice Problems can be found at the back of the text starting at page SP-1*

Practice Problem 3   A certain type of car has a stopping distance that varies directly as the square of its speed. If this car is traveling at a speed of 20 mph on an ice-covered road, it can stop in 60 feet. If the car is traveling at 40 mph on an icy road, what will its stopping distance be?   240 ft

2 Solving Problems Involving Inverse Variation

If one variable is a constant multiple of the reciprocal of another, the two variables are said to **vary inversely.** If y varies inversely as x, we express this by the equation $y = \dfrac{k}{x}$, where k is the constant of variation. Inverse variation problems can be solved by a four-step procedure similar to that used for direct variation problems.

> **SOLVING AN INVERSE VARIATION PROBLEM**
>
> 1. Write the inverse variation equation.
> 2. Solve for the constant k by substituting in known values.
> 3. Replace k in the inverse variation equation by the value obtained in step 2.
> 4. Solve for the desired value.

Teaching Example 4 If y varies inversely as x, and $y = 21$ when $x = 3$, find the value of y when $x = \frac{7}{3}$.

Ans: $y = 27$

EXAMPLE 4   If y varies inversely as x, and $y = 12$ when $x = 7$, find the value of y when $x = \dfrac{2}{3}$.

Solution

Step 1 $y = \dfrac{k}{x}$

Step 2 $12 = \dfrac{k}{7}$ Substitute known values of x and y to find k.

$\qquad 84 = k$

Step 3 $y = \dfrac{84}{x}$

Step 4   To find y when $x = \dfrac{2}{3}$, we substitute.

$$y = \frac{84}{\dfrac{2}{3}}$$

$$= \frac{84}{1} \div \frac{2}{3}$$

$$= \frac{\overset{42}{\cancel{84}}}{1} \cdot \frac{3}{\cancel{2}}$$

$$= 42 \cdot 3 = 126$$

Thus $y = 126$ when $x = \dfrac{2}{3}$.

Practice Problem 4   If y varies inversely as x, and $y = 8$ when $x = 15$, find the value of y when $x = \dfrac{3}{5}$. $y = 200$

EXAMPLE 5 A car manufacturer is thinking of reducing the size of the wheel used in a subcompact car. The number of times a car wheel must turn to cover a given distance varies inversely as the radius of the wheel. (Notice that this says that the smaller the wheel, the more times it must turn to cover a given distance.) A wheel with a radius of 0.35 meter must turn 400 times to cover a specified distance on a test track. How many times would it have to turn if the radius were reduced to 0.30 meter (see the sketch)?

Solution   Let n = the number of times the car wheel turns,
r = the radius of the wheel, and
k = the constant of variation.

Step 1   Since the number of turns varies inversely as the radius, we can write the following.

$$n = \frac{k}{r} \quad \text{Write the variation equation.}$$

Step 2 $400 = \dfrac{k}{0.35}$ Substitute known values of n and r to find k.

$140 = k$

Step 3 $n = \dfrac{140}{r}$ Use the variation equation where k is known.

How many times must the wheel turn if the radius is 0.30 meter?

Step 4 $n = \dfrac{140}{0.30}$ Substitute 0.30 for r.

$n = 466\dfrac{2}{3}$

The wheel would have to turn $466\dfrac{2}{3}$ times to cover the same distance if the radius were only 0.30 meter.

Practice Problem 5   Over the last three years, the market research division of a calculator company found that the volume of sales of scientific calculators varies inversely as the price of the calculator. One year 120,000 calculators were sold at $30 each. How many calculators were sold the next year when the price was $24 for each calculator?   150,000 calculators

The following table contains various forms of the inverse variation equation. In each case, k = the constant of variation.

SAMPLE INVERSE VARIATION SITUATIONS

Verbal Description	*Variation Equation*
y varies inversely as x	$y = \dfrac{k}{x}$
b varies inversely as the square of c	$b = \dfrac{k}{c^2}$
l varies inversely as the cube of m	$l = \dfrac{k}{m^3}$
d varies inversely as the square root of t	$d = \dfrac{k}{\sqrt{t}}$

EXAMPLE 6 The illumination of a light source varies inversely as the square of the distance from the source. The illumination measures 25 candlepower when a certain light is 4 meters away. Find the illumination when the light is 8 meters away (see figure).

4 m

8 m

Solution   Let I = the measurement of illumination,

d = the distance from the light source, and

k = the constant of variation.

Step 1   Since the illumination varies inversely as the square of the distance,

$$I = \frac{k}{d^2}.$$

Step 2   We evaluate the constant by substituting the given values.

$$25 = \frac{k}{4^2} \quad \text{Substitute } I = 25 \text{ and } d = 4.$$

$$25 = \frac{k}{16} \quad \text{Simplify and solve for } k.$$

$$400 = k$$

Step 3   We may now write the variation equation with the constant evaluated.

$$I = \frac{400}{d^2}$$

Step 4 $I = \dfrac{400}{8^2}$ Substitute a distance of 8 meters.

$= \dfrac{400}{64}$ Square 8.

$= \dfrac{25}{4} = 6.25$

The illumination is 6.25 candlepower when the light source is 8 meters away.

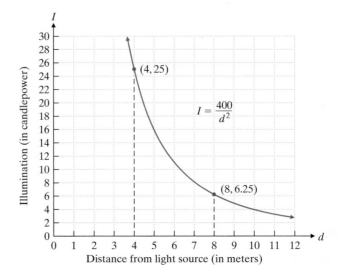

Practice Problem 6 If the amount of power in an electrical circuit is held constant, the resistance in the circuit varies inversely as the square of the amount of current. If the amount of current is 0.01 ampere, the resistance is 800 ohms. What is the resistance if the amount of current is 0.02 ampere?   200 ohms

NOTE TO STUDENT: Fully worked-out solutions to all of the Practice Problems can be found at the back of the text starting at page SP-1

 Identifying the Graphs of Variation Equations

The graphs of direct variation equations are shown in the following chart. Notice that as x increases, y also increases.

Variation Statement	Equation	Graph
1. y varies directly as x	$y = kx$	
2. y varies directly as x^2	$y = kx^2$	
3. y varies directly as x^3	$y = kx^3$	
4. y varies directly as the square root of x	$y = k\sqrt{x}$	

The graphs of inverse variation equations are shown in the following chart. Notice that as x increases, y decreases.

Variation Statement	Equation	Graph
5. y varies inversely as x	$y = \dfrac{k}{x}$	
6. y varies inversely as x^2	$y = \dfrac{k}{x^2}$	

1. If y varies directly as x, and $y = 9$ when $x = 2$, find y when $x = 16$.

$y = kx$
$9 = 2k$ $k = \dfrac{9}{2} \Rightarrow y = \dfrac{9}{2}x$ $y = \dfrac{9}{2}(16) = 72$

2. If y varies directly as x, and $y = 5$ when $x = 3$, find y when $x = 18$.

$y = kx$
$5 = 3k$ $k = \dfrac{5}{3} \Rightarrow y = \dfrac{5}{3}x$ $y = \dfrac{5}{3}(18) = 30$

3. If y varies directly as the cube of x, and $y = 12$ when $x = 2$, find y when $x = 7$.

$y = kx^3$
$12 = 8k$ $k = \dfrac{3}{2} \Rightarrow y = \dfrac{3}{2}x^3$ $y = \dfrac{1029}{2}$

4. If y varies directly as the square root of x, and $y = 7$ when $x = 4$, find y when $x = 25$.

$y = k\sqrt{x}$
$7 = k\sqrt{4}$ $k = \dfrac{7}{2} \Rightarrow y = \dfrac{7}{2}\sqrt{x}$ $y = \dfrac{35}{2}$

5. If y varies directly as the square of x, and $y = 900$ when $x = 25$, find y when $x = 30$.

$y = kx^2$
$900 = 625k$ $k = \dfrac{900}{625} \Rightarrow y = \dfrac{36}{25}x^2$ $y = \dfrac{36}{25}(900) = 1296$

6. If y varies directly as the square of x, and $y = 8000$ when $x = 16$, find y when $x = 5$.

$y = kx^2$
$8000 = 256k$ $k = \dfrac{8000}{256} \Rightarrow y = \dfrac{125}{4}x^2$ $y = \dfrac{125}{4}(25) = \dfrac{3125}{4}$

Applications

7. *Food Nutrition Facts* There are 100 calories in 20 grams of potato chips. The number of calories varies directly with the weight of the potato chips. How many calories will 45 grams of potato chips have?

$c = kw$ $100 = k20$ $k = 5$ $c = 5 \cdot 45$ $c = 225$ calories

8. *Stretching a Spring* Hooke's Law states that the force needed to stretch a spring varies directly with the amount the spring is stretched. If a force of 50 pounds is needed to stretch a spring 5 inches, how far will a force of 120 pounds stretch the same spring?

$F = kL$ $50 = k5$ $k = 10$ $120 = 10 \cdot L$ $L = 12$ in.

▲ **9.** *Storage Cube* The time it takes to fill a storage cube with sand varies directly as the cube of the side of the box. A storage cube that is 2.0 meters on each side (inside dimensions) can be filled in 7 minutes by a sand loader. How long will it take to fill a storage cube that is 4.0 meters on each side (inside dimensions)?

$T = ks^3$

$7 = k(2)^3$ $k = \dfrac{7}{8}$

$T = \dfrac{7}{8}(4)^3$

$T = 56$ min

2.0 m
2.0 m
2.0 m

10. *Weight on the Moon* The weight of an object on the surface of the moon varies directly as the weight of the object on the surface of Earth. An astronaut with his protective suit weighs 175 lb on Earth's surface; on the moon and wearing the same suit, he will weigh 29 lb. If his moon rover vehicle weighs 2000 pounds on Earth, how much will it weigh on the moon?

$M = kE \Rightarrow 29 = k(175) \Rightarrow M = \dfrac{29}{175} \cdot 2000 \approx 331.4$ lb

11. If y varies inversely as x, and $y = 18$ when $x = 5$, find y when $x = 8$.

$y = \dfrac{k}{x}$ $18 = \dfrac{k}{5}$ $k = 90$ $y = \dfrac{90}{8} = \dfrac{45}{4}$

12. If y varies inversely as x, and $y = 21$ when $x = 4$, find y when $x = 18$.

$y = \dfrac{k}{x}$ $21 = \dfrac{k}{4}$ $k = 84$ $y = \dfrac{84}{18} = \dfrac{42}{9}$

13. If y varies inversely as x, and $y = \dfrac{1}{3}$ when $x = 12$, find y when $x = 4$.

$y = \dfrac{k}{x}$ $\dfrac{1}{3} = \dfrac{k}{12}$ $k = 4 \Rightarrow y = \dfrac{4}{x}$ $y = 1$

14. If y varies inversely as x, and $y = \dfrac{1}{10}$ when $x = 30$, find y when $x = 1$.

$y = \dfrac{k}{x}$ $\dfrac{1}{10} = \dfrac{k}{30}$ $k = 3 \Rightarrow y = \dfrac{3}{x}$ $y = 3$

15. If y varies inversely as the square of x, and $y = 30$ when $x = 2$, find y when $x = 9$.

$y = \dfrac{k}{x^2}$ $k = x^2y$ $k = 120 \Rightarrow y = \dfrac{120}{81} = \dfrac{40}{27}$

16. If y varies inversely as the cube of x, and $y = \dfrac{1}{9}$ when $x = 3$, find y when $x = \dfrac{1}{2}$.

$y = \dfrac{k}{x^3}$ $k = x^3y$ $k = 3 \Rightarrow y = \dfrac{3}{\left(\dfrac{1}{2}\right)^3}$ $y = 24$

Applications

17. *Ice Cube Melting* The amount of time in minutes that it takes for an ice cube to melt varies inversely as the temperature of the water that the ice cube is placed in. When an ice cube is placed in 60°F water, it takes 2.3 minutes to melt. How long would it take for an ice cube of this size to melt if it were placed in 40°F water?

$$\text{time} = \frac{k}{\text{temperature}}$$

$$\text{time} = \frac{138}{\text{temperature}} \quad 3.45 \text{ min}$$

18. *Textile Mills* The textile mills in Lowell, Massachusetts, were operated with water power. The Merrimack River and its canals powered huge turbines that generated the electricity for the mills. In each mill room, a single drive-shaft supplied power to all machines by means of belts that ran to pulleys on the machines. The driveshaft turned at a constant speed, but the pulleys were different diameters, which provided for the different speeds of each machine. The speed of each machine varied inversely as the diameter of its pulley. If a machine with a 60-centimeter pulley turned at 1000 rpm, what was the speed of a machine with an 80-centimeter pulley?

$$y = \frac{k}{x}$$

$$k = 60\,(1000) = 60{,}000 \Rightarrow y = \frac{60{,}000}{80} \quad y = 750 \text{ rpm}$$

19. *Variation in Weight on Earth* The weight of an object near Earth's surface varies inversely as the square of its distance from the center of Earth. An object weighs 1000 pounds on Earth's surface. This is approximately 4000 miles from the center of Earth. How much will the object weigh 6000 miles from the center of Earth?

$$W = \frac{k}{D^2} \Rightarrow 1000 = \frac{k}{4000^2} \Rightarrow W = \frac{(1000)(4000)(4000)}{(6000)^2}$$

$$= \frac{4000}{9} \text{ or } 444\frac{4}{9} \text{ lb}$$

20. *Radio Waves* The wavelength, w, of a radio wave varies inversely as the frequency, f. A radio wave with a frequency of 1600 kilohertz has a length of 225 meters. What is the length of a radio wave with a frequency of 900 kilohertz?

$$w = \frac{k}{f} \quad 225 = \frac{k}{1600} \quad k = 360{,}000$$

$$w = \frac{360{,}000}{900} = 400 \text{ m}$$

21. *Distance Vision in a Boat* A person on the mast of a sailing ship 81 meters above the water can see approximately 12.3 kilometers out to sea. In general, the distance, d, you can see varies directly as the square root of your height, h, above the water. In this case, the equation is $d = \frac{15}{11}\sqrt{h}$, where d is measured in kilometers and h is measured in meters.

    (a) Find the distances you can see when $h = 121, 36,$ and 4. Round to the nearest tenth.
    15.0 km, 8.2 km, 2.7 km

    (b) Graph the equation using values of h of 121, 81, 36, and 4.

22. *Computer Files* Suppose a computer has a hard drive that can hold 2000 files if each file is 100,000 bytes long. The number of files, f, the hard drive can hold varies inversely as the size, s, of a file. In this case, the equation is $f = \dfrac{200{,}000{,}000}{s}$, where f is the number of files and s is the size of a file in bytes.

    (a) Find the number of files that can be held if $s = 400{,}000$, $s = 200{,}000$, and $s = 50{,}000$.
    500 files, 1000 files, 4000 files

    (b) Graph the equation using the values $s = 400{,}000$, $s = 200{,}000$, $s = 100{,}000$, and $s = 50{,}000$.

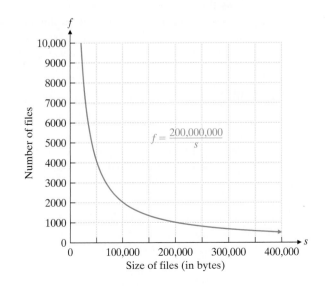

To Think About *Write an equation to describe the variation and solve.*

23. *Planetary Motion* Kepler's Third Law of Planetary Motion states that the square of the time it takes a planet to complete one revolution of the sun varies directly with the cube of the distance between the planet and the sun. Astronomers use 1 astronomical unit (au) as a measure of the distance from Earth to the sun.

   (a) Using this astronomical unit, find the constant k for the direct variation equation that represents Kepler's Third Law.

   (b) Pluto is 39.5 times as far from the sun as Earth is. Find the approximate length of the Plutonian year in terms of Earth years.

   (a) $t^2 = kd^3$
       Earth: $1^2 = k1^3$
       $k = 1$

   (b) Pluto: $t^2 = 1(39.5)^3$
          $t^2 = 61{,}629.875$
          $t = \sqrt{61{,}629.875}$
          ≈ 248.25 Earth years

24. *Sailboat Design* An important element of sailboat design is the calculation of the force of wind on the sail. The force varies directly with the square of the wind speed and directly with the area of the sail. For a certain class of racing boat, the force on 100 square feet of sail is 280 pounds when the wind speed is 6 mph. What would be the force on the sail if it were rigged for a storm (the amount of sail exposed to the wind is reduced by lashing part of the sail to the boom) so that only 60 square feet of sail were exposed and the wind speed were 40 mph?

$$F = kS^2 A$$
$$280 = k(6)^2(100)$$
$$F = \frac{(280)(40)(40)(60)}{3600}$$
$$\approx 7467 \text{ lb}$$

Cumulative Review

25. **[6.5.1]** Solve for a. $\dfrac{80{,}000}{320} = \dfrac{120{,}000}{320 + a}$

   $\text{LCD} = 320(320 + a)$
     $480 = 320 + a$
       $a = 160$

26. **[7.2.1]** Graph the equation $S = kh$, where $k = 4.80$. Use h for the horizontal axis and S for the vertical. Assume that h and S are nonnegative.

   $S = 4.80h$

Quick Quiz 9.7

1. The intensity of illumination from a light source varies inversely with the square of the distance from the light source. When a photoelectric cell is placed 2 inches from a light source the intensity is 45 lumens. Then it is moved so that it receives only 5 lumens. How far is it from the light source? 6 in.

2. The number of calories in a breakfast bar varies directly with the weight of the bar. There are 150 calories is a bar weighing 40 grams. How many calories are in a bar weighing 75 grams? 281.25 calories

3. The areas of similar triangle-shaped supports vary directly as the squares of their perimeters. A support that has a perimeter of 8 centimeters has an area of 6.4 square centimeters. A similar triangle-shaped support has a perimeter of 20 centimeters. What is its area? 40 cm^2

4. Concept Check The distance a body falls from rest varies directly as the square of the time it falls. If an object falls 64 feet in two seconds, explain how you would write an equation to describe this relationship. How would you find the constant k in this equation?
Answers may vary

Classroom Quiz 9.7 You may use these problems to quiz your students' mastery of Section 9.7.

1. The intensity of illumination from a light source varies inversely with the square of the distance from the light source. When a photoelectric cell is placed 3 inches from a light source the intensity is 32 lumens. Then it is moved so that it receives only 2 lumens. How far is it from the light source? **Ans:** 12 in.

2. The number of calories you work off while jogging varies directly with the distance that you jog. If a person works off 360 calories by jogging 30 minutes, how long would it take her to work off 600 calories? **Ans:** 50 min

3. The areas of similar rectangles vary directly as the square of the perimeters of the rectangles. A small rectangle with an area of 50 square inches has a perimeter of 30 inches. A larger rectangle that is similar has a perimeter of 60 inches. What is the area of the larger rectangle? **Ans:** 200 in.^{2}

Putting Your Skills to Work: Use Math to Save Money

FOOD AND RICE PRICES

If you've noticed an increase in food prices during recent months, you are not alone. Food prices are rising throughout the world. Reasons for this include bad weather that hampered crops, an increased demand for food, particularly in countries throughout Asia, and the global increase in fuel prices.

During the spring of 2008, food prices in the United States for staples such as bread, milk, eggs, and flour rose sharply. Milk prices jumped 26% from the previous year, while eggs jumped 50% from $1.45 per dozen to $2.18 per dozen (U.S. Labor Department). Overall food prices in the Dallas-Fort Worth area rose 7.1% between June of 2007 and June of 2008.

Rice is another food item that has seen a sharp increase in price. During the spring of 2008, two U.S. warehouse retail chains, Sam's Club and Costco, moved to limit how much rice customers could buy because of concerns about global supply and demand. The cost of rice in some locations increased by 50% from approximately $10 to $15 for a 25-pound bag.

Food Prices and You

Has the increase in food prices—or the increase in the cost of rice—had an impact on you and your family? How much do you spend on food each week? How about each month? How does that compare with what you were spending on food each month a year ago? Consider the story of Lucy.

Lucy and her family enjoy rice as part of their dinner four times a week. Lucy has a family of four that consumes approximately $\frac{2}{3}$ cup of rice with each meal.

1. How many cups of rice does Lucy prepare a week? $\frac{8}{3}$ or $2\frac{2}{3}$ cups

2. If a cup of raw rice weighs approximately 21 ounces, how many ounces of raw rice does Lucy's family consume each week? 56 oz

3. There are 16 ounces in a pound. Compute the number of pounds of raw rice Lucy's family eats each week. 3.5 lb

4. Last year she bought a 25-pound bag of rice for $9.25. She recently purchased a 25-pound bag of rice for $15.73.

   (a) What is the price increase per pound? $0.26/lb

   (b) What is the percent of increase? 70%

5. Approximately how much in today's prices is the cost per week of the rice consumed by Lucy's family? $2.20

6. How much rice do you eat per week? Has the rise in rice prices influenced your purchase of rice? Why or why not? Answers may vary

7. In what other food prices have you noticed an increase? What are some things you can do as a consumer to keep your costs low? Answers may vary

Topic	Procedure	Examples
Square roots, p. 516.	**1.** The square roots of perfect squares up to $\sqrt{225} = 15$ (or so) should be memorized. Approximate values of others can be found in tables or by calculator. **2.** Negative numbers do not have real number square roots.	$\sqrt{196} = 14 \qquad \sqrt{169} = 13$ $\sqrt{200} \approx 14.142 \qquad \sqrt{13} \approx 3.606$ $\sqrt{-5}$ This is not a real number.
Simplifying radicals, p. 521.	Assuming that a and b are nonnegative, $$\sqrt{ab} = \sqrt{a}\sqrt{b}$$ Square root radicals are simplified by taking the square roots of all factors that are perfect squares and leaving all other factors under the radical sign.	$\sqrt{48} = \sqrt{16}\sqrt{3} = 4\sqrt{3}$ $\sqrt{12x^2y^3z} = \sqrt{2^2x^2y^2}\sqrt{3yz}$ $\qquad = 2xy\sqrt{3yz}$ $\sqrt{a^{11}} = \sqrt{a^{10} \cdot a} = \sqrt{a^{10}}\sqrt{a} = a^5\sqrt{a}$
Adding and subtracting radicals, p. 526.	Simplify all radicals and then add or subtract like radicals.	$\sqrt{12} + \sqrt{18} + \sqrt{27} + \sqrt{50}$ $= \sqrt{4}\sqrt{3} + \sqrt{9}\sqrt{2} + \sqrt{9}\sqrt{3} + \sqrt{25}\sqrt{2}$ $= 2\sqrt{3} + 3\sqrt{2} + 3\sqrt{3} + 5\sqrt{2}$ $= 5\sqrt{3} + 8\sqrt{2}$
Multiplying radicals, p. 530.	Square root radicals are multiplied like polynomials. Assuming that a and b are nonnegative, apply the rule $\sqrt{a}\sqrt{b} = \sqrt{ab}$ and simplify all results.	$(2\sqrt{6} + \sqrt{3})(\sqrt{6} - 3\sqrt{3})$ $= 2\sqrt{36} - 6\sqrt{18} + \sqrt{18} - 3\sqrt{9}$ $= 2(6) - 5\sqrt{18} - 3(3)$ $= 12 - 5\sqrt{9}\sqrt{2} - 9$ $= 12 - 5(3)\sqrt{2} - 9$ $= 3 - 15\sqrt{2}$
Dividing radicals, p. 536.	For positive numbers a and b, $$\frac{\sqrt{a}}{\sqrt{b}} = \sqrt{\frac{a}{b}} \quad \text{and} \quad \sqrt{\frac{a}{b}} = \frac{\sqrt{a}}{\sqrt{b}}.$$	$\frac{\sqrt{80}}{\sqrt{5}} = \sqrt{\frac{80}{5}} = \sqrt{16} = 4$ $\sqrt{\frac{144}{25}} = \frac{\sqrt{144}}{\sqrt{25}} = \frac{12}{5}$
Conjugates, p. 538.	Note that the product of the sum and difference of two square root radicals is equal to the difference of their squares and thus contains no radicals. This sum and difference are called conjugates.	$(3\sqrt{6} + \sqrt{10})$ and $(3\sqrt{6} - \sqrt{10})$ are conjugates. $(3\sqrt{6} + \sqrt{10})(3\sqrt{6} - \sqrt{10})$ $= 9\sqrt{6^2} - \sqrt{10^2}$ $= 9 \cdot 6 - 10 = 54 - 10 = 44$
Rationalizing denominators, p. 537.	To simplify a fraction with a square root radical in the denominator, the radical (in the denominator) must be removed. **1.** Multiply the denominator by itself if it is a monomial. Multiply the numerator by the same quantity. **2.** Multiply the denominator by its conjugate if it is a binomial. Multiply the numerator by the same quantity.	$\frac{\sqrt{10}}{\sqrt{6}} = \frac{\sqrt{10} \cdot \sqrt{6}}{\sqrt{6} \cdot \sqrt{6}} = \frac{2\sqrt{15}}{6} = \frac{\sqrt{15}}{3}$ $\frac{\sqrt{3}}{\sqrt{3} - \sqrt{2}} = \frac{\sqrt{3}(\sqrt{3} + \sqrt{2})}{(\sqrt{3} - \sqrt{2})(\sqrt{3} + \sqrt{2})}$ $= \frac{\sqrt{9} + \sqrt{6}}{3 - 2} = 3 + \sqrt{6}$
Radicals with fractions, p. 537.	Note that $\sqrt{\frac{5}{3}}$ is the same as $\frac{\sqrt{5}}{\sqrt{3}}$ and so must be rationalized.	$\sqrt{\frac{5}{3}} = \frac{\sqrt{5}}{\sqrt{3}} \cdot \frac{\sqrt{3}}{\sqrt{3}} = \frac{\sqrt{15}}{3}$
Pythagorean Theorem, p. 542.	In any right triangle with hypotenuse of length c and legs of lengths a and b, $$c^2 = a^2 + b^2.$$	Find c to the nearest tenth if $a = 7$ and $b = 9$. $c^2 = 7^2 + 9^2 = 49 + 81 = 130$ $c = \sqrt{130} \approx 11.4$

Topic	Procedure	Examples
Radical equations, p. 544.	To solve an equation containing a square root radical: 1. Isolate the radical by itself on one side of the equation. 2. Square both sides. 3. Solve the resulting equation. 4. Check all apparent solutions. Extraneous roots may have been introduced in step 2.	Solve. $\sqrt{3x+2}-4=8$ $$\sqrt{3x+2}=12$$ $$(\sqrt{3x+2})^2=12^2$$ $$3x+2=144$$ $$3x=142$$ $$x=\frac{142}{3}$$ *Check:* $\sqrt{3\left(\frac{142}{3}\right)+2} \overset{?}{=} 12$ $$\sqrt{144}=12 \;\checkmark$$ So $x=\frac{142}{3}$.
Variation: direct, p. 550.	If y varies directly with x, there is a constant of variation k such that $y=kx$. Once k is determined, the value of y or x can easily be computed.	y varies directly with x. When $x=2$, $y=7$. $$y=kx$$ $$7=k(2) \quad \text{Substituting.}$$ $$k=\frac{7}{2} \quad\quad \text{Solving.}$$ $$y=\frac{7}{2}x$$ What is y when $x=8$? $$y=\frac{7}{2}x=\frac{7}{2}\cdot 8=28$$
Variation: inverse, p. 552.	If y varies inversely with x, there is a constant of variation k such that $$y=\frac{k}{x}.$$	y varies inversely with x. When x is 5, y is 12. What is y when x is 15? $$y=\frac{k}{x}$$ $$12=\frac{k}{5} \quad \text{Substituting.}$$ $$k=60 \quad \text{Solving.}$$ $$y=\frac{60}{x} \quad \text{Substituting.}$$ $$x=15 \quad y=\frac{60}{15}=4$$

Chapter 9 Review Problems

Section 9.1

Simplify or evaluate, if possible.

1. $\sqrt{36}$ 6

2. $\sqrt{25}$ 5

3. $\sqrt{169}$ 13

4. $\sqrt{196}$ 14

5. $-\sqrt{81}$ −9

6. $\sqrt{64}$ 8

7. $\sqrt{49}$ 7

8. $-\sqrt{121}$ −11

9. $-\sqrt{144}$ −12

10. $\sqrt{289}$ 17

11. $\sqrt{0.04}$ 0.2

12. $\sqrt{0.49}$ 0.7

13. $\sqrt{\dfrac{1}{100}}$ $\dfrac{1}{10}$

14. $\sqrt{\dfrac{64}{81}}$ $\dfrac{8}{9}$

Approximate, using the square root table in Appendix A or a calculator. Round to the nearest thousandth, if necessary.

15. $\sqrt{105}$ 10.247

16. $\sqrt{198}$ 14.071

17. $\sqrt{77}$ 8.775

18. $\sqrt{88}$ 9.381

Section 9.2

Simplify.

19. $\sqrt{27}$ $3\sqrt{3}$

20. $\sqrt{52}$ $2\sqrt{13}$

21. $\sqrt{28}$ $2\sqrt{7}$

22. $\sqrt{125}$ $5\sqrt{5}$

23. $\sqrt{40}$ $2\sqrt{10}$

24. $\sqrt{80}$ $4\sqrt{5}$

25. $\sqrt{x^8}$ x^4

26. $\sqrt{y^{10}}$ y^5

27. $\sqrt{x^5y^6}$ $x^2y^3\sqrt{x}$

28. $\sqrt{a^3b^4}$ $ab^2\sqrt{a}$

29. $\sqrt{16x^3y^5}$ $4xy^2\sqrt{xy}$

30. $\sqrt{98x^4y^6}$ $7x^2y^3\sqrt{2}$

31. $\sqrt{12x^5}$ $2x^2\sqrt{3x}$

32. $\sqrt{72x^9}$ $6x^4\sqrt{2x}$

33. $\sqrt{48x^8}$ $4x^4\sqrt{3}$

34. $\sqrt{125x^{12}}$ $5x^6\sqrt{5}$

35. $\sqrt{120a^3b^4c^5}$ $2ab^2c^2\sqrt{30ac}$

36. $\sqrt{121a^6b^4c}$ $11a^3b^2\sqrt{c}$

37. $\sqrt{56x^7y^9}$ $2x^3y^4\sqrt{14xy}$

38. $\sqrt{99x^{13}y^7}$ $3x^6y^3\sqrt{11xy}$

Section 9.3

Simplify.

39. $\sqrt{5} - \sqrt{20} + \sqrt{80}$ $3\sqrt{5}$

40. $5\sqrt{6} - \sqrt{24} + 2\sqrt{54}$ $9\sqrt{6}$

41. $x\sqrt{3} + 3x\sqrt{3} + \sqrt{27x^2}$ $7x\sqrt{3}$

42. $a\sqrt{2} + \sqrt{12a^2} + a\sqrt{98}$ $8a\sqrt{2} + 2a\sqrt{3}$

43. $5\sqrt{5} - 6\sqrt{20} + 2\sqrt{10}$ $-7\sqrt{5} + 2\sqrt{10}$

44. $2\sqrt{40} - 2\sqrt{90} + 3\sqrt{28}$ $-2\sqrt{10} + 6\sqrt{7}$

Section 9.4

Simplify.

45. $(2\sqrt{x})(3\sqrt{x^3})$ $6x^2$

46. $(-5\sqrt{a})(2\sqrt{ab})$ $-10a\sqrt{b}$

47. $(\sqrt{6a^2})(\sqrt{6b^5})$ $6ab^2\sqrt{b}$

48. $(-4x\sqrt{x})(2x\sqrt{x})$ $-8x^3$

49. $\sqrt{6}(\sqrt{24} - 5\sqrt{6})$ -18

50. $\sqrt{3}(\sqrt{48} - 8\sqrt{3})$ -12

51. $\sqrt{2}(\sqrt{5} - \sqrt{3} - 2\sqrt{2})$ $\sqrt{10} - \sqrt{6} - 4$

52. $\sqrt{5}(\sqrt{6} - 2\sqrt{5} + \sqrt{10})$ $\sqrt{30} - 10 + 5\sqrt{2}$

53. $\left(\sqrt{11} + 2\right)\left(2\sqrt{11} - 1\right)$ $20 + 3\sqrt{11}$

54. $\left(\sqrt{10} + 3\right)\left(3\sqrt{10} - 1\right)$ $27 + 8\sqrt{10}$

55. $\left(2 + 3\sqrt{6}\right)\left(4 - 2\sqrt{3}\right)$ $8 - 4\sqrt{3} + 12\sqrt{6} - 18\sqrt{2}$

56. $\left(5 - \sqrt{2}\right)\left(3 - \sqrt{12}\right)$ $15 - 3\sqrt{2} - 10\sqrt{3} + 2\sqrt{6}$

57. $\left(2\sqrt{3} + 3\sqrt{6}\right)^2$ $66 + 36\sqrt{2}$

58. $\left(5\sqrt{2} - 2\sqrt{6}\right)^2$ $74 - 40\sqrt{3}$

Section 9.5

Rationalize the denominator.

59. $\dfrac{1}{\sqrt{3x}}$ $\dfrac{\sqrt{3x}}{3x}$

60. $\dfrac{2y}{\sqrt{5}}$ $\dfrac{2y\sqrt{5}}{5}$

61. $\dfrac{x^2 y}{\sqrt{8}}$ $\dfrac{x^2 y \sqrt{2}}{4}$

62. $\dfrac{3ab}{\sqrt{2b}}$ $\dfrac{3a\sqrt{2b}}{2}$

63. $\sqrt{\dfrac{5}{6}}$ $\dfrac{\sqrt{30}}{6}$

64. $\sqrt{\dfrac{6}{11}}$ $\dfrac{\sqrt{66}}{11}$

65. $\dfrac{\sqrt{x^3}}{\sqrt{5x}}$ $\dfrac{x\sqrt{5}}{5}$

66. $\dfrac{\sqrt{a^5}}{\sqrt{6a}}$ $\dfrac{a^2\sqrt{6}}{6}$

67. $\dfrac{3}{\sqrt{5} + \sqrt{2}}$ $\sqrt{5} - \sqrt{2}$

68. $\dfrac{2}{\sqrt{6} - \sqrt{3}}$ $\dfrac{2\left(\sqrt{6} + \sqrt{3}\right)}{3}$

69. $\dfrac{1 - \sqrt{5}}{2 + \sqrt{5}}$ $-7 + 3\sqrt{5}$

70. $\dfrac{1 - \sqrt{3}}{3 + \sqrt{3}}$ $\dfrac{3 - 2\sqrt{3}}{3}$

Section 9.6

In exercises 71–77, use the Pythagorean Theorem to find the length of the missing side of a right triangle with legs a and b and hypotenuse c.

▲ **71.** $a = 5, b = 8$
  $c = \sqrt{89}$

▲ **72.** $c = \sqrt{11}, b = 3$
  $a = \sqrt{2}$

▲ **73.** $c = 5, a = 3.5$
  $b = \dfrac{\sqrt{51}}{2}$

▲ **74.** If $a = 2.400$ and $b = 2.000$, find c and round to the nearest thousandth.
  $c = 3.124$

▲ **75.** *Flagpole* A flagpole is 24 meters tall. A man stands 18 meters from the base of the pole. How far is it from the feet of the man to the top of the pole?
  30 m

▲ **76.** *Distances Between Cities* Kingman is a city in Arizona that is 143 miles west of Flagstaff. Fountain Hills is 120 miles south of Flagstaff. How far is it from Kingman to Fountain Hills? Round your answer to the nearest tenth of a mile.
  186.7 mi

▲ **77.** *Softball Diamond* The four bases in a fast-pitch soft-ball field form a square. The length of the diagonal (the distance from home plate to second base) is $\sqrt{7200}$ feet. How long is each side of the square?
  60 ft

Solve. Be sure to verify your answers.

78. $\sqrt{x - 4} = 5$ $x = 29$

79. $\sqrt{5x - 1} = 8$ $x = 13$

80. $\sqrt{1 - 5x} = \sqrt{9 - x}$ $x = -2$

81. $\sqrt{-5 + 2x} = \sqrt{1 + x}$ $x = 6$ **82.** $\sqrt{10x + 9} = -1 + 2x$ $x = 4$ **83.** $\sqrt{2x - 5} = 10 - x$ $x = 7$

84. $6 - \sqrt{5x - 1} = x + 1$ $x = 2$ **85.** $4x + \sqrt{x + 2} = 5x - 4$ $x = 7$

Section 9.7

86. If y varies directly with the square of x, and $y = 27$ when $x = 3$, find the value of y when $x = 4$.
$y = 48$

87. If y varies inversely with the square root of x, and $y = 2$ when $x = 25$, find the value of y when $x = 100$.
$y = 1$

88. If y varies inversely with the cube of x, and $y = 4$ when $x = 2$, find the value of y when $x = 4$.
$y = \dfrac{1}{2}$

89. *Illumination from a Light Source* The intensity of illumination from a light source varies inversely as the square of the distance from the light source. When an object is 4 meters from a light source, the intensity is 45 lumens. What is the intensity at a distance of 12 meters?   5 lumens

90. *Skid Marks of a Car* When an automobile driver slams on the brakes, the length of skid marks on the road varies directly with the square of the speed of the car. At 30 mph a certain car had skid marks 40 feet long. How long will the skid marks be if the car travels at 55 mph?
Approximately 134 ft

91. *Horsepower of a Boat* The horsepower that is needed to drive a racing boat through water varies directly with the cube of the speed of the boat. What will happen to the horsepower requirement if someone wants to double the maximum speed of a given boat?
It will be 8 times as much.

Mixed Practice   *Find the square root.*

92. $\sqrt{\dfrac{36}{121}}$ $\dfrac{6}{11}$

93. $\sqrt{0.0004}$ 0.02

Simplify.

94. $\sqrt{98x^6 y^3}$ $7x^3 y\sqrt{2y}$

Combine.

95. $3\sqrt{27} - 2\sqrt{75} + \sqrt{48}$ $3\sqrt{3}$

Multiply and simplify your answer.

96. $\left(\sqrt{3} - 2\sqrt{5}\right)\left(2\sqrt{5} + 3\sqrt{2}\right)$
$2\sqrt{15} + 3\sqrt{6} - 20 - 6\sqrt{10}$

97. $\left(4\sqrt{3} + 3\right)^2$
$57 + 24\sqrt{3}$

Rationalize the denominator.

98. $\dfrac{5}{\sqrt{12}}$ $\dfrac{5\sqrt{3}}{6}$

99. $\dfrac{\sqrt{3} + \sqrt{6}}{2\sqrt{3} + \sqrt{2}}$ $\dfrac{6 - \sqrt{6} + 6\sqrt{2} - 2\sqrt{3}}{10}$

Solve.

100. $\sqrt{2x - 3} = 9$
$x = 42$

101. $\sqrt{10x + 5} = 2x + 1$
$x = -\dfrac{1}{2}$ and $x = 2$

Remember to use your Chapter Test Prep Video CD to see the worked-out solutions to the test problems you want to review.

Note to Instructor: The Chapter 9 Test file in the TestGen program provides algorithms specifically matched to these problems so you can easily replicate this test for additional practice or assessment purposes.

Evaluate.

1. $\sqrt{121}$

2. $\sqrt{\dfrac{9}{100}}$

Simplify.

3. $\sqrt{48x^2y^7}$

4. $\sqrt{100x^3yz^4}$

Combine and simplify.

5. $8\sqrt{3} + 5\sqrt{27} - 5\sqrt{48}$

6. $\sqrt{4a} + \sqrt{8a} + \sqrt{36a} + \sqrt{18a}$

Multiply.

7. $\left(2\sqrt{a}\right)\left(3\sqrt{b}\right)\left(2\sqrt{ab}\right)$

8. $\sqrt{5}\left(\sqrt{10} + 2\sqrt{3} - 3\sqrt{5}\right)$

9. $\left(2\sqrt{3} + 5\right)^2$

10. $\left(4\sqrt{2} - \sqrt{5}\right)\left(3\sqrt{2} + \sqrt{5}\right)$

In questions 11–14, simplify.

11. $\sqrt{\dfrac{x}{5}}$

12. $\dfrac{3}{\sqrt{12}}$

13. $\dfrac{\sqrt{3} + 4}{5 + \sqrt{3}}$

14. $\dfrac{3a}{\sqrt{5} + \sqrt{2}}$

15. Use the table of square roots or a calculator to approximate $\sqrt{156}$. Round your answer to the nearest hundredth.

Find the missing side by using the Pythagorean Theorem.

▲ **16.**

▲ **17.**
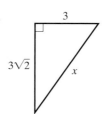

Solve. Verify your solutions.

18. $6 - \sqrt{2x + 1} = 0$

19. $x = 5 + \sqrt{x + 7}$

Solve.

20. The intensity of illumination from a light source varies inversely with the square of the distance from the light source. When a photoelectric cell is placed 8 inches from a light source, the intensity is 12 lumens. Then it is moved so that it receives only 3 lumens. How far is it from the light source?

21. Commission on the sale of office supplies varies directly with the amount of the sale. If a salesman earns $23.40 on sales of $780, what does he earn on sales of $2859?

22. The area of an equilateral triangle varies directly as the square of its perimeter. If the perimeter is 12 centimeters, the area is 6.93 square centimeters. What is the area of an equilateral triangle whose perimeter is 21 centimeters? Round to the nearest tenth.

1.	11
2.	$\dfrac{3}{10}$
3.	$4xy^3\sqrt{3y}$
4.	$10xz^2\sqrt{xy}$
5.	$3\sqrt{3}$
6.	$8\sqrt{a} + 5\sqrt{2a}$
7.	$12ab$
8.	$5\sqrt{2} + 2\sqrt{15} - 15$
9.	$37 + 20\sqrt{3}$
10.	$19 + \sqrt{10}$
11.	$\dfrac{\sqrt{5x}}{5}$
12.	$\dfrac{\sqrt{3}}{2}$
13.	$\dfrac{17 + \sqrt{3}}{22}$
14.	$a\left(\sqrt{5} - \sqrt{2}\right)$
15.	12.49
16.	$x = \sqrt{133}$
17.	$x = 3\sqrt{3}$
18.	$x = \dfrac{35}{2}$
19.	$x = 9$
20.	16 in.
21.	$85.77
22.	21.2 cm^2

Cumulative Test for Chapters 0–9

Approximately one-half of this test covers the content of Chapters 0–8. The remainder covers the content of Chapter 9.

In questions 1–3, simplify.

1. $\dfrac{28}{42}$

2. $5\dfrac{1}{3} + 1\dfrac{3}{4}$

3. $6\dfrac{1}{4} \div 6\dfrac{2}{3}$

4. Write 0.07% as a decimal.

Simplify.

5. $-11 - 16 + 8 + 4 - 13 + 31$

6. $(-3)^2 \cdot 4 - 8 \div 2 - (3 - 2)^3$

7. $4a^2bc^3 - 2abc^2 - 7a^2bc^3$

8. $(x - 1)(x - 4)(x + 4)$

9. $(4x - 5)^2$

10. $(3x - 11)(3x + 11)$

11. $\dfrac{4x}{x + 2} + \dfrac{4}{x + 2}$

12. $\dfrac{x^2 - 5x + 6}{x^2 - 5x - 6} \div \dfrac{x^2 - 4}{x^2 + 2x + 1}$

Simplify, if possible.

13. $\sqrt{98x^5y^6}$

14. $\sqrt{50} - \sqrt{98} + \sqrt{162}$

15. $\left(\sqrt{6} - \sqrt{3}\right)^2$

16. $\dfrac{\sqrt{3} + \sqrt{2}}{\sqrt{3} - \sqrt{2}}$

17. $\left(3\sqrt{6} - \sqrt{2}\right)\left(\sqrt{6} + 4\sqrt{2}\right)$

18. $\dfrac{2}{2 - \sqrt{5}}$

19. $-\sqrt{25}$

20. $\sqrt{-16}$

Use the Pythagorean Theorem to find the missing side of each triangle. Be sure to simplify your answer.

21. $a = 4, b = \sqrt{7}$

▲ 22. $a = 19, c = 21$

In questions 23 and 24, solve for the variable. Be sure to verify your answer(s).

23. $\sqrt{4x - 3} = x$

24. $\sqrt{6y + 13} - 1 = y$

25. If y varies inversely as x, and $y = 4$ when $x = 5$, find the value of y when $x = 2$.

26. The surface area of a balloon varies directly with the square of the cross-sectional radius. When the radius is 2 meters, the surface area is 1256 square meters. Find the surface area if the radius is 0.5 meter.

---

1. $\dfrac{2}{3}$

2. $7\dfrac{1}{12}$

3. $\dfrac{15}{16}$

4. 0.0007

5. 3

6. 31

7. $-3a^2bc^3 - 2abc^2$

8. $x^3 - x^2 - 16x + 16$

9. $16x^2 - 40x + 25$

10. $9x^2 - 121$

11. $\dfrac{4x + 4}{x + 2}$

12. $\dfrac{(x - 3)(x + 1)}{(x - 6)(x + 2)}$

13. $7x^2y^3\sqrt{2x}$

14. $7\sqrt{2}$

15. $9 - 6\sqrt{2}$

16. $5 + 2\sqrt{6}$

17. $10 + 22\sqrt{3}$

18. $-4 - 2\sqrt{5}$

19. -5

20. Not a real number

21. $c = \sqrt{23}$

22. $b = 4\sqrt{5}$

23. $x = 3, x = 1$

24. $y = 6$

25. $y = 10$

26. 78.5 m²

Firefighters need to use mathematics to determine water flow in fighting large fires. Calculations are made about the angle of the hose and the pressure needed to pump the water to the necessary height. Firefighters need to pass a mathematics exam in order to be appointed to city fire departments. Much of the mathematics that is needed to solve these problems is covered in this chapter.

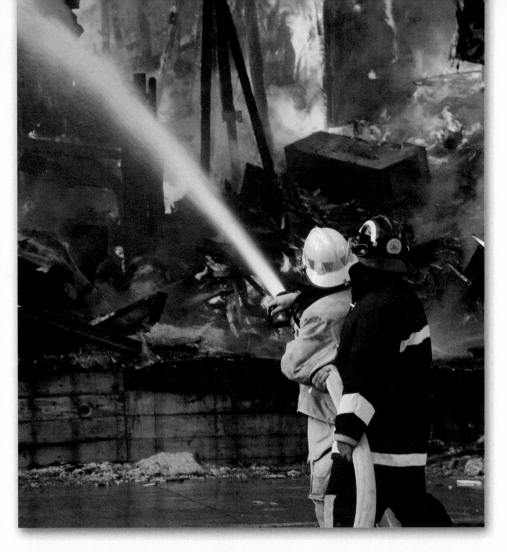

Quadratic Equations

1 Writing a Quadratic Equation in Standard Form

In Section 5.7 we introduced quadratic equations. A **quadratic equation** is a polynomial equation of degree two. For example,

$$3x^2 + 5x - 7 = 0, \quad \frac{1}{2}x^2 - 5x = 0, \quad 2x^2 = 5x + 9, \quad \text{and} \quad 3x^2 = 8$$

are all quadratic equations.

> The **standard form of a quadratic equation** is $ax^2 + bx + c = 0$, where a, b, and c are real numbers, and $a \neq 0$.

You will need to be able to recognize the real numbers that represent a, b, and c in specific equations. It will be easier to do so if these equations are placed in standard form. Note that it is usually easier to work with quadratic equations if a is positive, so in cases where a is negative, we recommend that you multiply the entire equation by -1. This is a suggestion and not a requirement.

EXAMPLE 1 Place each quadratic equation in standard form and identify the real numbers a, b, and c.

(a) $5x^2 - 6x + 3 = 0$ This equation is in standard form.

Solution $ax^2 + bx + c = 0$ Match each term to the standard form.

$a = 5, \quad b = -6, \quad c = 3$

(b) $2x^2 + 5x = 4$ The right-hand side is not zero. It is not in standard form.

Solution $2x^2 + 5x - 4 = 0$ Add -4 to each side of the equation.

$ax^2 + bx + c = 0$ Match each term to the standard form.

$a = 2, \quad b = 5, \quad c = -4$

(c) $-2x^2 + 15x + 4 = 0$ It is easier to work with quadratic equations if the first term is not negative.

Solution $2x^2 - 15x - 4 = 0$ Multiply each term on both sides of the equation by -1.

$ax^2 + bx + c = 0$

$a = 2, \quad b = -15, \quad c = -4$

(d) $7x^2 - 9x = 0$ This equation is in standard form.

Solution $ax^2 + bx + c = 0$ Note that the constant term is missing, so we know $c = 0$.

$a = 7, \quad b = -9, \quad c = 0$

Practice Problem 1 Place each quadratic equation in standard form and identify the real numbers a, b, and c. If $a < 0$, you can multiply each term of the equation by -1 to obtain an equivalent equation. This is not required, but most students find it helpful to have the first term be a positive number.

(a) $2x^2 + 12x - 9 = 0$ **(b)** $7x^2 = 6x - 8$

(c) $-x^2 - 6x + 3 = 0$ **(d)** $10x^2 - 12x = 0$

(a) $a = 2, b = 12, c = -9$ **(b)** $a = 7, b = -6, c = 8$ **(c)** $a = 1, b = 6, c = -3$ **(d)** $a = 10, b = -12, c = 0$

Solving Quadratic Equations of the Form
$ax^2 + bx = 0$ by Factoring

Notice that the terms in a quadratic equation of the form $ax^2 + bx = 0$ both have x as a factor. To solve such equations, begin by factoring out the x and any common numerical factor. Thus you may be able to factor out an x or an expression such as $5x$. Remember that you want to remove the *greatest common factor*. So if $5x$ is a common factor of each term, be sure to factor it out. Then use the zero factor property discussed in Section 5.7. This property states that if $a \cdot b = 0$, then either $a = 0$ or $b = 0$. Once each factor is set equal to 0, solve for x.

EXAMPLE 2   Solve. $7x^2 + 9x - 2 = -8x - 2$

Solution $7x^2 + 9x - 2 = -8x - 2$ The equation is not in standard form.

$7x^2 + 9x - 2 + 8x + 2 = 0$ Add $8x + 2$ to each side.

$7x^2 + 17x = 0$ Combine like terms.

$x(7x + 17) = 0$ Factor.

$x = 0$ $7x + 17 = 0$ Set each factor equal to zero.

$7x = -17$

$x = -\dfrac{17}{7}$

One of the solutions to the preceding equation is 0. This will always be true of equations of this form. One root will be zero. The other root will be a nonzero real number.

Practice Problem 2   Solve. $2x^2 - 7x - 6 = 4x - 6$

Teaching Example 2 Solve.

$6x^2 - 3x + 5 = x + 5$

Ans: $x = 0, x = \dfrac{2}{3}$

$x = 0, x = \dfrac{11}{2}$

 Solving Quadratic Equations of the Form
$ax^2 + bx + c = 0$ by Factoring

If an equation you are trying to solve contains fractions, clear the fractions by multiplying each term by the least common denominator of all the fractions in the equation. Sometimes an equation does not look like a quadratic equation. However, it takes the quadratic form once the fractions have been cleared. A note of caution: Always check a possible solution in the original equation. Any apparent solution that would make the denominator of any fraction in the original equation 0 is not a valid solution.

EXAMPLE 3   Solve and check. $8x - 6 + \dfrac{1}{x} = 0$

Solution

$8x - 6 + \dfrac{1}{x} = 0$ The equation has a fractional term.

$x(8x) - (x)(6) + x\left(\dfrac{1}{x}\right) = x(0)$ Multiply each term by the LCD, which is x.

$8x^2 - 6x + 1 = 0$ Simplify. The equation is now in standard form.

$(4x - 1)(2x - 1) = 0$ Factor.

$4x - 1 = 0$ $2x - 1 = 0$ Set each factor equal to 0.

$4x = 1$ $2x = 1$ Solve each equation for x.

$x = \dfrac{1}{4}$ $x = \dfrac{1}{2}$

Teaching Example 3 Solve and check.

$3x - 2 = \dfrac{1}{x}$

Ans: $x = 1, x = -\dfrac{1}{3}$

Check. Checking fractional roots is more difficult, but you should be able to do it if you work carefully.

$$\text{If } x = \frac{1}{4}: \quad 8\left(\frac{1}{4}\right) - 6 + \frac{1}{\frac{1}{4}} = 2 - 6 + 4 = 0 \quad \left(\text{since } 1 \div \frac{1}{4} = 1 \cdot \frac{4}{1} = 4\right)$$

$$0 = 0 \; \checkmark$$

$$\text{If } x = \frac{1}{2}: \quad 8\left(\frac{1}{2}\right) - 6 + \frac{1}{\frac{1}{2}} = 4 - 6 + 2 = 0 \quad \left(\text{since } 1 \div \frac{1}{2} = 1 \cdot \frac{2}{1} = 2\right)$$

$$0 = 0 \; \checkmark$$

Both roots check, so $\frac{1}{2}$ and $\frac{1}{4}$ are the two roots that satisfy the original equation

$$8x - 6 + \frac{1}{x} = 0.$$

It is also correct to write the answers in decimal form as 0.5 and 0.25. This will speed up the checking process, especially if you use a scientific calculator when performing the check.

NOTE TO STUDENT: Fully worked-out solutions to all of the Practice Problems can be found at the back of the text starting at page SP-1

Practice Problem 3 Solve and check.

$$3x + 10 - \frac{8}{x} = 0 \quad x = \frac{2}{3}, x = -4$$

If a quadratic equation is given to us in standard form, we examine it to see if it can be factored. If it is possible to factor the quadratic equation, then we use the zero factor property to find the solutions.

Teaching Example 4 Solve and check.

$$4x^2 + 5x - 6 = 0$$

Ans: $x = -2, x = \frac{3}{4}$

EXAMPLE 4 Solve and check. $6x^2 + 7x - 10 = 0$

Solution

$$(6x - 5)(x + 2) = 0 \qquad \text{Factor the quadratic equation.}$$
$$6x - 5 = 0 \qquad x + 2 = 0 \qquad \text{Set each factor equal to zero and solve for } x.$$
$$6x = 5 \qquad\qquad x = -2$$
$$x = \frac{5}{6}$$

Thus the solutions to the equation are $\frac{5}{6}$ and -2.

Check each solution in the original equation.

$$6\left(\frac{5}{6}\right)^2 + 7\left(\frac{5}{6}\right) - 10 \overset{?}{=} 0 \qquad 6(-2)^2 + 7(-2) - 10 \overset{?}{=} 0$$

$$6\left(\frac{25}{36}\right) + 7\left(\frac{5}{6}\right) - 10 \overset{?}{=} 0 \qquad 6(4) + 7(-2) - 10 \overset{?}{=} 0$$

$$\frac{25}{6} + \frac{35}{6} - 10 \overset{?}{=} 0 \qquad 24 + (-14) - 10 \overset{?}{=} 0$$

$$\frac{60}{6} - 10 \overset{?}{=} 0 \qquad\qquad 10 - 10 = 0 \; \checkmark$$

$$10 - 10 = 0 \; \checkmark$$

Practice Problem 4 Solve and check. $2x^2 + 9x - 18 = 0$

$$x = \frac{3}{2}, x = -6$$

It is important to place a quadratic equation in standard form before factoring. This will sometimes involve removing parentheses and combining like terms.

EXAMPLE 5 Solve. $x + (x - 6)(x - 2) = 2(x - 1)$

Solution

$$x + x^2 - 2x - 6x + 12 = 2x - 2 \qquad \text{Remove parentheses.}$$
$$x^2 - 7x + 12 = 2x - 2 \qquad \text{Combine like terms.}$$
$$x^2 - 9x + 14 = 0 \qquad \text{Add } -2x + 2 \text{ to each side.}$$
$$(x - 7)(x - 2) = 0 \qquad \text{Factor.}$$
$$x = 7, \quad x = 2 \qquad \text{Use the zero factor property.}$$

The solutions to the equation are 7 and 2. The check is left to the student.

Practice Problem 5 Solve. $-4x + (x - 5)(x + 1) = 5(1 - x)$

$$x = -2, x = 5$$

EXAMPLE 6 Solve. $\dfrac{8}{x + 1} - \dfrac{x}{x - 1} = \dfrac{2}{x^2 - 1}$

Solution The LCD $= (x + 1)(x - 1)$. We multiply each term by the LCD.

$$(x + 1)(x - 1)\left(\frac{8}{x + 1}\right) - (x + 1)(x - 1)\left(\frac{x}{x - 1}\right) = (x + 1)(x - 1)\left[\frac{2}{(x + 1)(x - 1)}\right]$$

$$8(x - 1) - x(x + 1) = 2 \qquad \text{Simplify.}$$
$$8x - 8 - x^2 - x = 2 \qquad \text{Remove parentheses.}$$
$$-x^2 + 7x - 8 = 2 \qquad \text{Combine like terms.}$$
$$-x^2 + 7x - 8 - 2 = 0 \qquad \text{Subtract 2 from each side.}$$
$$-x^2 + 7x - 10 = 0 \qquad \text{Simplify.}$$
$$x^2 - 7x + 10 = 0 \qquad \text{Multiply each term by } -1.$$
$$(x - 5)(x - 2) = 0 \qquad \text{Factor and solve for } x.$$
$$x = 5, \quad x = 2 \qquad \text{Use the zero factor property.}$$

Check. The check is left to the student.

Practice Problem 6 Solve.

$$\frac{10x + 18}{x^2 + x - 2} = \frac{3x}{x + 2} + \frac{2}{x - 1} \qquad x = \frac{14}{3}, x = -1$$

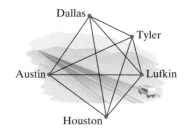

④ Solving Applied Problems Involving Quadratic Equations

A manager of a truck delivery service is determining the possible routes he may have to send trucks on during a given day. His delivery service must have the capability to send trucks between any two of the following Texas cities: Austin, Dallas, Tyler, Lufkin, and Houston. He made a rough map of the cities.

What is the maximum number of routes he may have to send trucks on in a given day, if there is a route between every two cities? One possible way to figure this out is to draw a straight line between every two cities. Let us assume that no three cities lie on a straight line. Therefore, the route between any two cities is distinct from all other routes. The figure to the right shows that we can draw exactly 10 lines, so the answer is that there are 10 truck routes. What if the manager has to service 18 cities? How many truck routes will there be then?

As the number of cities increases, it becomes quite difficult (and time consuming) to use a drawing to determine the number of routes. In a more advanced math

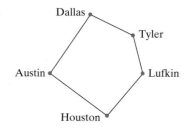

course, it can be proved that the maximum number of truck routes (t) can be found by using the number of cities (n) in the following equation.

$$t = \frac{n^2 - n}{2}$$

Thus, to find the number of truck routes for 18 cities, we merely substitute 18 for n.

$$t = \frac{(18)^2 - (18)}{2}$$
$$= \frac{324 - 18}{2}$$
$$= \frac{306}{2}$$
$$= 153$$

Thus 153 truck routes are needed to service 18 cities if no three of those cities lie on a straight line.

Teaching Example 7 How many cities can the company service with 45 truck routes?

Ans: 10 cities

EXAMPLE 7 A truck delivery company can handle a maximum of 36 truck routes in one day, and every two cities have a distinct truck route between them. How many separate cities can the truck company service?

Solution

$$t = \frac{n^2 - n}{2}$$

$$36 = \frac{n^2 - n}{2} \qquad \text{Substitute 36 for } t, \text{ the number of truck routes.}$$

$$72 = n^2 - n \qquad \text{Multiply both sides of the equation by 2.}$$

$$0 = n^2 - n - 72 \qquad \text{Subtract 72 from each side to obtain the standard form of a quadratic equation.}$$

$$0 = (n - 9)(n + 8) \qquad \text{Factor.}$$

$$n - 9 = 0 \qquad n + 8 = 0 \qquad \text{Set each factor equal to zero.}$$

$$n = 9 \qquad n = -8 \qquad \text{Solve for } n.$$

We reject -8 as a meaningless solution for this problem. We cannot have a negative number of cities. Thus our answer is that the company can service 9 cities.

Check. If the truck company services $n = 9$ cities, does that really result in a maximum number of $t = 36$ truck routes?

$$t = \frac{n^2 - n}{2} \qquad \text{The equation in which } t = \text{the number of truck routes and } n = \text{the number of cities to be connected.}$$

$$36 \stackrel{?}{=} \frac{(9)^2 - (9)}{2} \qquad \text{Substitute } t = 36 \text{ and } n = 9.$$

$$36 \stackrel{?}{=} \frac{81 - 9}{2}$$

$$36 \stackrel{?}{=} \frac{72}{2}$$

$$36 = 36 \ \checkmark$$

NOTE TO STUDENT: *Fully worked-out solutions to all of the Practice Problems can be found at the back of the text starting at page SP-1*

Practice Problem 7 How many cities can the company service with 28 truck routes?   8 cities

Write in standard form. Determine the values of a, b, and c.

1. $x^2 + 8x + 7 = 0$
$a = 1, b = 8, c = 7$

2. $x^2 + 6x - 5 = 0$
$a = 1, b = 6, c = -5$

3. $8x^2 - 11x = 0$
$a = 8, b = -11, c = 0$

4. $5x^2 + 15x = 0$
$a = 5, b = 15, c = 0$

5. $x^2 + 15x - 7 = 12x + 8$
$x^2 + 3x - 15 = 0$
$a = 1, b = 3, c = -15$

6. $4x^2 + 13x + 8 = x - 12$
$4x^2 + 12x + 20 = 0$
$a = 4, b = 12, c = 20$

Using the factoring method, solve for the roots of each quadratic equation. Be sure to place your equation in standard form before factoring.

7. $27x^2 - 9x = 0$
$9x(3x - 1) = 0$
$x = 0, x = \dfrac{1}{3}$

8. $6x + 18x^2 = 0$
$6x(1 + 3x) = 0$
$x = 0, x = -\dfrac{1}{3}$

9. $4x^2 - x = -9x$
$4x^2 + 8x = 0$
$4x(x + 2) = 0$
$x = 0, x = -2$

10. $3x^2 - x = 8x$
$3x^2 - 9x = 0$
$3x(x - 3) = 0$
$x = 0, x = 3$

11. $11x^2 - 13x = 8x - 3x^2$
$14x^2 - 21x = 0$
$7x(2x - 3) = 0$
$x = 0, x = \dfrac{3}{2}$

12. $60x^2 + 14x = 9x^2 - 20x$
$51x^2 + 34x = 0$
$17x(3x + 2) = 0$
$x = 0, x = -\dfrac{2}{3}$

13. $x^2 - 3x - 28 = 0$
$(x - 7)(x + 4) = 0$
$x = 7, x = -4$

14. $x^2 + 3x - 40 = 0$
$(x - 5)(x + 8) = 0$
$x = 5, x = -8$

15. $2x^2 + 11x - 6 = 0$
$(2x - 1)(x + 6) = 0$
$x = \dfrac{1}{2}, x = -6$

16. $2x^2 - 9x - 5 = 0$
$(2x + 1)(x - 5) = 0$
$x = -\dfrac{1}{2}, x = 5$

17. $3x^2 - 16x + 5 = 0$
$(3x - 1)(x - 5) = 0$
$x = \dfrac{1}{3}, x = 5$

18. $5x^2 - 21x + 4 = 0$
$(5x - 1)(x - 4) = 0$
$x = \dfrac{1}{5}, x = 4$

19. $x^2 = 5x + 14$
$x^2 - 5x - 14 = 0$
$(x - 7)(x + 2) = 0$
$x = 7, x = -2$

20. $x^2 = 5x + 24$
$x^2 - 5x - 24 = 0$
$(x - 8)(x + 3) = 0$
$x = 8, x = -3$

21. $12x^2 + 17x + 6 = 0$
$(4x + 3)(3x + 2) = 0$
$x = -\dfrac{3}{4}, x = -\dfrac{2}{3}$

22. $14x^2 + 29x + 12 = 0$
$(7x + 4)(2x + 3) = 0$
$x = -\dfrac{4}{7}, x = -\dfrac{3}{2}$

23. $a^2 + 9a + 39 = 7 - 3a$
$a^2 + 12a + 32 = 0$
$(a + 8)(a + 4) = 0$
$a = -8, a = -4$

24. $m^2 + 15m + 33 = 5 - m$
$m^2 + 16m + 28 = 0$
$(m + 14)(m + 2) = 0$
$m = -14, m = -2$

25. $6y^2 = -7y + 3$
$6y^2 + 7y - 3 = 0$
$(3y - 1)(2y + 3) = 0$
$y = \dfrac{1}{3}, y = -\dfrac{3}{2}$

26. $8y^2 = -2y + 3$
$8y^2 + 2y - 3 = 0$
$(4y + 3)(2y - 1) = 0$
$y = -\dfrac{3}{4}, y = \dfrac{1}{2}$

27. $25x^2 - 60x + 36 = 0$
$(5x - 6)(5x - 6) = 0$
$x = \dfrac{6}{5}$

Mixed Practice

28. $9x^2 - 24x + 16 = 0$
$(3x - 4)(3x - 4) = 0$
$x = \dfrac{4}{3}$

29. $(x - 5)(x + 2) = -10$
$x^2 - 3x = 0$
$x(x - 3) = 0$
$x = 0, x = 3$

30. $(x - 6)(x + 1) = 8$
$x^2 - 5x - 14 = 0$
$(x + 2)(x - 7) = 0$
$x = -2, x = 7$

31. $3x^2 + 8x - 10 = -6x - 10$
$3x^2 + 14x = 0$
$x(3x + 14) = 0$
$x = 0, x = -\dfrac{14}{3}$

32. $5x^2 - 7x + 12 = -8x + 12$
$5x^2 + x = 0$
$x(5x + 1) = 0$
$x = 0, x = -\dfrac{1}{5}$

33. $y(y - 7) = 2(5 - 2y)$
$y^2 - 3y - 10 = 0$
$(y + 2)(y - 5) = 0$
$y = -2, y = 5$

34. $x(x - 4) = 3(x - 2)$
$x^2 - 7x + 6 = 0$
$(x - 1)(x - 6) = 0$
$x = 1, x = 6$

35. $(x - 6)(x - 4) = 3$
$x^2 - 10x + 21 = 0$
$(x - 3)(x - 7) = 0$
$x = 3, x = 7$

36. $(y + 6)(y - 1) = 8$
$y^2 + 5y - 14 = 0$
$(y + 7)(y - 2) = 0$
$y = -7, y = 2$

37. $2x(x + 3) = (3x + 1)(x + 1)$
$x^2 - 2x + 1 = 0$
$(x - 1)(x - 1) = 0$
$x = 1$

38. $2(2x + 1)(x + 1) = x(7x + 1)$
$3x^2 - 5x - 2 = 0$
$(3x + 1)(x - 2) = 0$
$x = -\dfrac{1}{3}, x = 2$

Find the LCD and multiply each term by the LCD. Then solve. Check your solutions.

39. $x + \dfrac{8}{x} = 6$
$x^2 - 6x + 8 = 0$
$(x - 4)(x - 2) = 0$
$x = 4, x = 2$

40. $x - 1 = \dfrac{20}{x}$
$x^2 - x - 20 = 0$
$(x - 5)(x + 4) = 0$
$x = 5, x = -4$

41. $\dfrac{x}{7} - \dfrac{2}{x} = -\dfrac{5}{7}$
$x^2 + 5x - 14 = 0$
$(x + 7)(x - 2) = 0$
$x = -7, x = 2$

42. $\dfrac{x}{5} - \dfrac{7}{x} = \dfrac{2}{5}$
$x^2 - 2x - 35 = 0$
$(x + 5)(x - 7) = 0$
$x = -5, x = 7$

43. $\dfrac{3y + 5}{2} = \dfrac{-3}{y - 2}$
$3y^2 - y - 4 = 0$
$(3y - 4)(y + 1) = 0$
$y = \dfrac{4}{3}, y = -1$

44. $\dfrac{4x - 1}{3} = \dfrac{-1}{x - 2}$
$4x^2 - 9x + 5 = 0$
$(4x - 5)(x - 1) = 0$
$x = \dfrac{5}{4}, x = 1$

45. $\dfrac{4}{x} + \dfrac{3}{x + 5} = 2$
$2x^2 + 3x - 20 = 0$
$(2x - 5)(x + 4) = 0$
$x = \dfrac{5}{2}, x = -4$

46. $\dfrac{7}{x} + \dfrac{6}{x - 1} = 2$
$2x^2 - 15x + 7 = 0$
$(x - 7)(2x - 1) = 0$
$x = \dfrac{1}{2}, x = 7$

47. $\dfrac{24}{x^2 - 4} = 1 + \dfrac{2x - 6}{x - 2}$
$3x^2 - 2x - 40 = 0$
$(3x + 10)(x - 4) = 0$
$x = -\dfrac{10}{3}, x = 4$

48. $\dfrac{5x + 7}{x^2 - 9} = 1 + \dfrac{2x - 8}{x - 3}$
$3x^2 - 7x - 40 = 0$
$(3x + 8)(x - 5) = 0$
$x = -\dfrac{8}{3}, x = 5$

To Think About

49. Why can a quadratic equation in standard form with $c = 0$ always be solved? You can always factor out x.

50. Martha solved $(x + 3)(x - 2) = 14$ as follows.

$$x + 3 = 14 \qquad x - 2 = 14$$
$$x = 11 \qquad x = 16$$

Josette said this had to be wrong because these values do not check. Explain what is wrong with Martha's method.
You must have the product of two or more factors equal to 0, not 14. Remember the zero factor property.

51. The equation $ax^2 - 7x + c = 0$ has the roots $-\dfrac{3}{2}$ and 6. Find the values of a and c.
$a = \dfrac{14}{9}, c = -14$

52. The equation $2x^2 + bx - 15 = 0$ has the roots -3 and $\dfrac{5}{2}$. Find the value of b.
$b = 1$

Applications *Business Revenue The revenue for producing the label for the new Starblazer video game is given by the formula $R = -2.5n^2 + 80n - 280$, where R is the revenue (in dollars) and n is the number of labels produced (in hundreds).*

53. If the revenue for producing labels is \$320, what are the two values of the two different numbers of labels that may be produced?

$320 = -2.5n^2 + 80n - 280$
$0 = -2.5(n^2 - 32n + 240)$
$0 = -2.5(n - 12)(n - 20)$
$n = 1200, n = 2000$

54. If the revenue for producing labels is \$157.50, what are the two values of the different numbers of labels that may be produced?

$157.50 = -2.5n^2 + 80n - 280$
$0 = -2.5(n^2 - 32n + 175)$
$0 = -2.5(n - 25)(n - 7)$
$n = 2500, n = 700$

55. What is the average of the two answers you obtained in problem 53? Use the revenue equation to find the revenue for producing that number of labels. What is significant about that answer?

Average = 1600. Revenue for producing 1600 labels is \$360, which is more than producing either 1200 or 2000 labels.

56. What is the average of the two answers you obtained in problem 54? Use the revenue equation to find the revenue for producing that number of labels. What is significant about that answer?

Average = 1600. Revenue for producing 1600 labels is \$360, which is more than producing either 2500 or 700 labels.

57. Find the revenue for producing 1500 and 1700 labels. Compare the two revenues with the results from 53 and 55. What can you conclude?

Revenue for 1500 = \$357.50; revenue for 1700 = \$357.50. It appears the maximum revenue is at 1600 labels.

58. Find the revenue for producing 1400 and 1800 labels. Compare the two revenues with the results from 54 and 56. What can you conclude?

Revenue for 1400 = \$350; revenue for 1800 = \$350. It appears the maximum revenue is at 1600 labels.

Cumulative Review

59. **[6.4.1]** Simplify. $\dfrac{\dfrac{1}{x} + \dfrac{3}{x - 2}}{\dfrac{4}{x^2 - 4}}$

$\dfrac{(2x - 1)(x + 2)}{2x}$

60. **[6.3.3]** Add the fractions. $\dfrac{2x}{3x - 5} + \dfrac{2}{3x^2 - 11x + 10}$

$\dfrac{2x^2 - 4x + 2}{3x^2 - 11x + 10}$

Quick Quiz 10.1 Using the factoring method, solve for the roots of each quadratic equation.

1. $6x^2 - x = 15$

$x = \dfrac{5}{3}, x = -\dfrac{3}{2}$

2. $x^2 = 2(7x - 24)$

$x = 8, x = 6$

3. $x^2 = 2 - \dfrac{7}{3}x$

$x = -3, x = \dfrac{2}{3}$

4. **Concept Check** In the following problem, explain how you would place the quadratic equation in standard form.

$2 = \dfrac{5}{x + 1} + \dfrac{3}{x - 1}$ Answers may vary

Classroom Quiz 10.1 You may use these problems to quiz your students' mastery of Section 10.1.

Using the factoring method, solve for the roots of each quadratic equation.

1. $6x^2 - 5x = 4$

Ans: $x = \dfrac{4}{3}, x = -\dfrac{1}{2}$

2. $x(x - 7) = -10$

Ans: $x = 2, x = 5$

3. $x^2 = \dfrac{13}{2}x - 10$

Ans: $x = \dfrac{5}{2}, x = 4$

Solving Quadratic Equations Using the Square Root Property

Recall that the quadratic equation $x^2 = a$ has two possible solutions, $x = \sqrt{a}$ or $x = -\sqrt{a}$, where a is a nonnegative real number. That is, $x = \pm\sqrt{a}$. This basic idea is called the **square root property.**

> **SQUARE ROOT PROPERTY**
>
> If $x^2 = a$, then $x = \sqrt{a}$ or $x = -\sqrt{a}$, for all nonnegative real numbers a.

Teaching Example 1 Solve.

(a) $x^2 = 16$ (b) $x^2 = 24$

(c) $7x^2 - 56 = 0$

Ans: (a) $x = \pm 4$ (b) $x = \pm 2\sqrt{6}$

     (c) $x = \pm 2\sqrt{2}$

EXAMPLE 1 Solve.

(a) $x^2 = 49$ (b) $x^2 = 20$ (c) $5x^2 = 125$

Solution

(a) $x^2 = 49$ (b) $x^2 = 20$

$\quad x = \pm\sqrt{49}$ $\quad x = \pm\sqrt{20}$

$\quad x = \pm 7$ $\quad x = \pm 2\sqrt{5}$

(c) $5x^2 = 125$ Divide both sides by 5 before taking the square root.

$\quad x^2 = 25$

$\quad x = \pm\sqrt{25}$

$\quad x = \pm 5$

NOTE TO STUDENT: *Fully worked-out solutions to all of the Practice Problems can be found at the back of the text starting at page SP-1*

Practice Problem 1 Solve.

(a) $x^2 = 1$ (b) $x^2 - 50 = 0$ (c) $3x^2 = 81$

(a) $x = \pm 1$ (b) $x = \pm 5\sqrt{2}$ (c) $x = \pm 3\sqrt{3}$

Teaching Example 2 Solve.

$\quad 3x^2 - 7 = 13 - 2x^2$

Ans: $x = \pm 2$

EXAMPLE 2 Solve. $3x^2 + 5x = 18 + 5x + x^2$

Solution Simplify the equation by placing all the variable terms on the left and the constants on the right.

$$3x^2 + 5x = 18 + 5x + x^2$$
$$2x^2 = 18$$
$$x^2 = 9$$
$$x = \pm 3$$

Practice Problem 2 Solve. $4x^2 - 5 = 319$ $x = \pm 9$

The square root property can also be used if the term that is squared is a binomial.

Sometimes when we use the square root property we obtain an irrational number. This will occur if the number on the right side of the equation is not a perfect square.

EXAMPLE 3 Solve. $(3x + 1)^2 = 8$

Teaching Example 3 Solve. $(4x - 5)^2 = 28$

Ans: $x = \dfrac{5 \pm 2\sqrt{7}}{4}$

Solution $(3x + 1)^2 = 8$

$\qquad 3x + 1 = \pm\sqrt{8}$ Take the square root of both sides.

$\qquad 3x + 1 = \pm 2\sqrt{2}$ Simplify as much as possible.

Now we must solve the two equations expressed by the plus or minus statement.

$$3x + 1 = +2\sqrt{2} \qquad\qquad 3x + 1 = -2\sqrt{2}$$
$$3x = -1 + 2\sqrt{2} \qquad\qquad 3x = -1 - 2\sqrt{2}$$
$$x = \frac{-1 + 2\sqrt{2}}{3} \qquad\qquad x = \frac{-1 - 2\sqrt{2}}{3}$$

The roots of this quadratic equation are irrational numbers. They are

$$\frac{-1 + 2\sqrt{2}}{3} \quad \text{and} \quad \frac{-1 - 2\sqrt{2}}{3} \quad \text{or} \quad \frac{-1 \pm 2\sqrt{2}}{3}$$

We cannot simplify these roots further, so we leave them in this form.

Practice Problem 3 Solve.

$$(2x - 3)^2 = 12 \quad x = \frac{3 \pm 2\sqrt{3}}{2}$$

② Solving Quadratic Equations by Completing the Square

If a quadratic equation is not in a form where we can use the square root property, we can rewrite the equation by a method called **completing the square** to make it so. That is, we rewrite the equation with a perfect square on the left side so that we have an equation of the form $(ax + b)^2 = c$.

EXAMPLE 4 Solve. $x^2 + 12x = 4$

Teaching Example 4 Solve. $x^2 - 6x = 1$

Ans: $x = 3 \pm \sqrt{10}$

Solution Write out the equation $x^2 + 12x = 4$. Think of a way to determine what number to add to each side.

$$x^2 + 12x \qquad\quad = 4$$
$$(x + 6)(x + 6) = 4 + 36 \qquad \text{We need a +36 to complete the square.}$$

$$x^2 + 12x + 36 = 4 + 36 \qquad \text{Add +36 to both sides of the equation.}$$
$$(x + 6)^2 = 40 \qquad\qquad \text{Write the left side as a perfect square.}$$
$$x + 6 = \pm\sqrt{40} \qquad\quad \text{Solve for } x.$$
$$x + 6 = +2\sqrt{10} \qquad\qquad x + 6 = -2\sqrt{10}$$
$$x = -6 + 2\sqrt{10} \qquad\qquad x = -6 - 2\sqrt{10}$$

The two roots are $-6 + 2\sqrt{10}$ and $-6 - 2\sqrt{10}$ or $-6 \pm 2\sqrt{10}$.

Practice Problem 4 Solve. $x^2 + 10x = 3$ $x = -5 \pm 2\sqrt{7}$

Completing the square is a little more difficult when the coefficient of x is an odd number. Let's see what happens.

EXAMPLE 5   Solve. $x^2 - 3x - 1 = 0$

Solution   Write the equation so that all the x-terms are on the left and the constants are on the right.

$$x^2 - 3x \qquad = 1 \qquad \text{The missing number when added to itself must be } -3.$$

$$(x \qquad)(x \qquad) \qquad \text{Since } -\frac{3}{2} + \left(-\frac{3}{2}\right) = -\frac{6}{2} = -3, \text{ the missing number is } -\frac{3}{2}.$$

$$\left(x - \frac{3}{2}\right)\left(x - \frac{3}{2}\right) \qquad \text{Complete the pattern.}$$

$$x^2 - 3x + \frac{9}{4} = 1 + \frac{9}{4} \qquad \text{Complete the square by adding } \left(-\frac{3}{2}\right)\left(-\frac{3}{2}\right) = \frac{9}{4} \text{ to each side.}$$

$$\left(x - \frac{3}{2}\right)^2 = \frac{13}{4} \qquad \text{Solve for } x.$$

$$x - \frac{3}{2} = \pm\sqrt{\frac{13}{4}}$$

$$x - \frac{3}{2} = \pm\frac{\sqrt{13}}{2}$$

$$x - \frac{3}{2} = +\frac{\sqrt{13}}{2} \qquad\qquad x - \frac{3}{2} = -\frac{\sqrt{13}}{2}$$

$$x = \frac{3}{2} + \frac{\sqrt{13}}{2} \qquad\qquad x = \frac{3}{2} - \frac{\sqrt{13}}{2}$$

$$x = \frac{3 + \sqrt{13}}{2} \qquad\qquad x = \frac{3 - \sqrt{13}}{2}$$

The two roots are $\dfrac{3 \pm \sqrt{13}}{2}$.

Practice Problem 5   Solve.

$$x^2 - 5x = 7 \qquad x = \frac{5 \pm \sqrt{53}}{2}$$

NOTE TO STUDENT: *Fully worked-out solutions to all of the Practice Problems can be found at the back of the text starting at page SP-1*

If a, the coefficient of the squared variable, is not 1, we divide all terms of the equation by a so that the coefficient of the squared variable will be 1.

Let us summarize for future reference the steps we perform in order to solve a quadratic equation by completing the square.

COMPLETING THE SQUARE

1. Put the equation in the form $ax^2 + bx = c$. If $a \neq 1$, divide each term by a.

2. Square $\dfrac{b}{2}$ and add the result to both sides of the equation.

3. Factor the left side (a perfect square trinomial).

4. Use the square root property.

5. Solve the equations.

6. Check the solutions in the original equation.

EXAMPLE 6 Solve. $4x^2 + 4x - 3 = 0$

Teaching Example 6 Solve by completing the square. $3x^2 - 5x + 2 = 0$

Ans: $x = 1, x = \dfrac{2}{3}$

Solution

Step 1 $4x^2 + 4x = 3$ Place the constant on the right. This puts the equation in the form $ax^2 + bx = c$.

$x^2 + x = \dfrac{3}{4}$ Divide all terms by 4.

Step 2 $x^2 + 1x = \dfrac{3}{4}$ Take one-half the coefficient of x and square it. $\left(\dfrac{1}{2}\right)^2 = \dfrac{1}{4}$.

$x^2 + x + \dfrac{1}{4} = \dfrac{3}{4} + \dfrac{1}{4}$ Add $\dfrac{1}{4}$ to each side.

Step 3 $\left(x + \dfrac{1}{2}\right)^2 = 1$ Factor the left side.

Step 4 $x + \dfrac{1}{2} = \pm\sqrt{1}$ Use the square root property.

Step 5 $x + \dfrac{1}{2} = \pm 1$

$x + \dfrac{1}{2} = 1$ $x + \dfrac{1}{2} = -1$

$x = \dfrac{2}{2} - \dfrac{1}{2}$ $x = -\dfrac{2}{2} - \dfrac{1}{2}$

$x = \dfrac{1}{2}$ $x = -\dfrac{3}{2}$

Thus the two roots are $\dfrac{1}{2}$ and $-\dfrac{3}{2}$.

Step 6 $4\left(\dfrac{1}{2}\right)^2 + 4\left(\dfrac{1}{2}\right) - 3 \overset{?}{=} 0$ Check.

$4\left(\dfrac{1}{4}\right) + 4\left(\dfrac{1}{2}\right) - 3 \overset{?}{=} 0$

$1 + 2 - 3 = 0$ ✓

$4\left(-\dfrac{3}{2}\right)^2 + 4\left(-\dfrac{3}{2}\right) - 3 \overset{?}{=} 0$

$4\left(\dfrac{9}{4}\right) + 4\left(-\dfrac{3}{2}\right) - 3 \overset{?}{=} 0$

$9 - 6 - 3 = 0$ ✓

Both values check.

Practice Problem 6 Solve.

$8x^2 + 2x - 1 = 0$ $x = \dfrac{1}{4}, x = -\dfrac{1}{2}$

This method of completing the square will enable us to solve any quadratic equation that has real number roots. It is usually faster, however, to factor the quadratic equation if the polynomial is factorable.

Solve using the square root property.

1. $x^2 = 64$
$x = \pm 8$

2. $x^2 = 36$
$x = \pm 6$

3. $x^2 = 98$
$x = \pm 7\sqrt{2}$

4. $x^2 = 72$
$x = \pm 6\sqrt{2}$

5. $x^2 - 28 = 0$
$x = \pm 2\sqrt{7}$

6. $x^2 - 45 = 0$
$x = \pm 3\sqrt{5}$

7. $5x^2 = 45$
$x = \pm 3$

8. $3x^2 = 48$
$x = \pm 4$

9. $6x^2 = 120$
$x = \pm 2\sqrt{5}$

10. $3x^2 = 150$
$x = \pm 5\sqrt{2}$

11. $3x^2 - 375 = 0$
$x = \pm 5\sqrt{5}$

12. $4x^2 - 252 = 0$
$x = \pm 3\sqrt{7}$

13. $5x^2 + 13 = 73$
$5x^2 = 60$
$x = \pm 2\sqrt{3}$

14. $6x^2 + 5 = 53$
$x^2 = 8$
$x = \pm 2\sqrt{2}$

15. $13x^2 + 17 = 82$
$13x^2 = 65$
$x = \pm\sqrt{5}$

16. $6x^2 - 13 = 23$
$x = \pm\sqrt{6}$

17. $(x - 7)^2 = 16$
$x - 7 = \pm 4$
$x = 3, x = 11$

18. $(x - 6)^2 = 100$
$x - 6 = \pm 10$
$x = 16, x = -4$

19. $(x + 4)^2 = 6$
$x + 4 = \pm\sqrt{6}$
$x = -4 \pm \sqrt{6}$

20. $(x - 5)^2 = 36$
$x - 5 = \pm 6$
$x = 11, x = -1$

21. $(2x + 5)^2 = 2$
$2x + 5 = \pm\sqrt{2}$
$x = \dfrac{-5 \pm \sqrt{2}}{2}$

22. $(3x - 4)^2 = 6$
$3x - 4 = \pm\sqrt{6}$
$x = \dfrac{4 \pm \sqrt{6}}{3}$

23. $(3x - 1)^2 = 7$
$3x - 1 = \pm\sqrt{7}$
$x = \dfrac{1 \pm \sqrt{7}}{3}$

24. $(2x + 3)^2 = 10$
$2x + 3 = \pm\sqrt{10}$
$x = \dfrac{-3 \pm \sqrt{10}}{2}$

25. $(5x + 1)^2 = 18$
$5x + 1 = \pm 3\sqrt{2}$
$x = \dfrac{-1 \pm 3\sqrt{2}}{5}$

26. $(8x + 5)^2 = 20$
$8x + 5 = \pm 2\sqrt{5}$
$x = \dfrac{-5 \pm 2\sqrt{5}}{8}$

27. $(4x - 5)^2 = 54$
$4x - 5 = \pm 3\sqrt{6}$
$x = \dfrac{5 \pm 3\sqrt{6}}{4}$

28. $(5x - 3)^2 = 50$
$5x - 3 = \pm 5\sqrt{2}$
$x = \dfrac{3 \pm 5\sqrt{2}}{5}$

Solve by completing the square.

29. $x^2 - 6x = 11$
$(x - 3)^2 = 20$
$x = 3 \pm 2\sqrt{5}$

30. $x^2 - 2x = 4$
$(x - 1)^2 = 5$
$x = 1 \pm \sqrt{5}$

31. $x^2 + 8x - 20 = 0$
$x^2 + 8x + 16 = 20 + 16$
$(x + 4)^2 = 36$
$x = 2, x = -10$

32. $x^2 - 10x + 16 = 0$
$x^2 - 10x + 25 = -16 + 25$
$(x - 5)^2 = 9$
$x = 8, x = 2$

33. $x^2 - 12x - 5 = 0$
$x^2 - 12x + 36 = 5 + 36$
$(x - 6)^2 = 41$
$x = 6 \pm \sqrt{41}$

34. $x^2 + 16x + 30 = 0$
$x^2 + 16x + 64 = -30 + 64$
$(x + 8)^2 = 34$
$x = -8 \pm \sqrt{34}$

35. $x^2 - 7x = 0$
$x^2 - 7x + \dfrac{49}{4} = \dfrac{49}{4}$
$\left(x - \dfrac{7}{2}\right)^2 = \dfrac{49}{4}$
$x = 0, x = 7$

36. $x^2 - 9x = 0$
$x^2 - 9x + \dfrac{81}{4} = \dfrac{81}{4}$
$\left(x - \dfrac{9}{2}\right)^2 = \dfrac{81}{4}$
$x = 0, x = 9$

37. $5x^2 - 25x = 0$
$x^2 - 5x + \dfrac{25}{4} = \dfrac{25}{4}$
$\left(x - \dfrac{5}{2}\right)^2 = \dfrac{25}{4}$
$x = 0, x = 5$

38. $2x^2 - 3x = 20$
$x^2 - \dfrac{3}{2}x + \dfrac{9}{16} = 10 + \dfrac{9}{16}$
$\left(x - \dfrac{3}{4}\right)^2 = \dfrac{169}{16}$
$x = -\dfrac{5}{2}, x = 4$

39. $2x^2 - 7x = 9$
$x^2 - \dfrac{7}{2}x + \dfrac{49}{16} = \dfrac{49}{16} + \dfrac{9}{2}$
$\left(x - \dfrac{7}{4}\right)^2 = \dfrac{121}{16}$
$x = \dfrac{9}{2}, x = -1$

40. $2x^2 + 7x = 4$
$x^2 + \dfrac{7}{2}x + \dfrac{49}{16} = 2 + \dfrac{49}{16}$
$\left(x + \dfrac{7}{4}\right)^2 = \dfrac{81}{16}$
$x = \dfrac{1}{2}, x = -4$

Cumulative Review

41. **[8.3.1]** Solve. $3a - 5b = 8$
$\qquad 5a - 7b = 8$
$a = -4, b = -4$

42. **[6.5.1]** Solve. $\dfrac{x}{x+3} - \dfrac{2}{x} = \dfrac{x-2}{x^2 + 3x}$
$x = -1, x = 4$

43. **[3.5.1]** *Hiking Knife* While hiking in the Grand Canyon, Will had to use his hunting knife to clear a fallen branch from the path. Find the pressure on the blade of his knife if the knife is 8 in. long and sharpened to an edge 0.01 in. wide and he applies a force of 16 pounds. Use the formula $P = \frac{F}{A}$, where P is the pressure on the blade of the knife, F is the force in pounds placed on the knife, and A is the area exposed to an object by the sharpened edge of the knife. 200 lb/in.^{2}

44. **[0.7.1]** *Tire Defects* Darlene inspected a sample of 100 new steel-belted radial tires. Thirteen of the tires had defects in workmanship, while 9 of the tires had defects in materials. Of those defective tires, 6 had defects in both workmanship and materials. How many of the 100 tires had any kind of defect? (Assume that the only defects found were those in materials or workmanship.)   A total of 16 tires were defective. 6 had both kinds of defects, 7 had defects only in workmanship, and 3 had defects only in materials.

Quick Quiz 10.2   Solve using the square root property.

1. $4x^2 + 13 = 37$ $x = \pm\sqrt{6}$

2. $(3x - 2)^2 = 40$ $x = \dfrac{2 \pm 2\sqrt{10}}{3}$

Solve by completing the square.

3. $x^2 + 8x - 10 = 0$ $x = -4 \pm \sqrt{26}$

4. **Concept Check** Explain the first three steps of how you would solve the following problem by completing the square.
$$5x^2 - 10x + 2 = 0 \quad \text{Answers may vary}$$

Classroom Quiz 10.2   You may use these problems to quiz your students' mastery of Section 10.2.

Solve using the square root property.

1. $5x^2 + 19 = 54$ **Ans:** $x = \pm\sqrt{7}$

2. $(4x - 3)^2 = 50$ **Ans:** $x = \dfrac{3 \pm 5\sqrt{2}}{4}$

Solve by completing the square.

3. $x^2 + 10x - 20 = 0$ **Ans:** $x = -5 \pm 3\sqrt{5}$

10.3 USING THE QUADRATIC FORMULA TO FIND SOLUTIONS

Student Learning Objectives

After studying this section, you will be able to:

 Solve a quadratic equation using the quadratic formula.

 Find a decimal approximation for the real roots of a quadratic equation.

③ Determine whether a quadratic equation has no real solutions.

Teaching Tip Remind students to memorize the quadratic formula perfectly. Suggest that they memorize it over three separate days and write it down each day until they have it "absolutely perfect." A lot of students think they have it memorized but they will mix up one little detail—which of course makes every answer they obtain by their quadratic formula incorrect.

Teaching Example 1 Solve using the quadratic formula. $7x^2 + 2x - 5 = 0$

Ans: $x = -1, x = \dfrac{5}{7}$

① Solving a Quadratic Equation Using the Quadratic Formula

To find solutions of a quadratic equation, you can factor or you can complete the square when the trinomial is not factorable. If we use the method of completing the square on the general equation $ax^2 + bx + c = 0$, we can derive a general formula for the solution of any quadratic equation. This is called the **quadratic formula.**

> ### QUADRATIC FORMULA
>
> The roots of any quadratic equation of the form $ax^2 + bx + c = 0$, where $a, b,$ and c are real numbers and $a \neq 0$, are
> $$x = \frac{-b \pm \sqrt{b^2 - 4ac}}{2a}.$$

All you will need to know are the values of a, b, and c to solve a quadratic equation using the quadratic formula.

EXAMPLE 1 Solve using the quadratic formula. $3x^2 + 10x + 7 = 0$

Solution In our given equation, $3x^2 + 10x + 7 = 0$, we have

$a = 3$ (the coefficient of x^2), $b = 10$ (the coefficient of x), and
$c = 7$ (the constant term).

$$x = \frac{-b \pm \sqrt{b^2 - 4ac}}{2a} = \frac{-10 \pm \sqrt{(10)^2 - 4(3)(7)}}{2(3)}$$ Write the quadratic formula and substitute values for a, b, and c.

$$= \frac{-10 \pm \sqrt{100 - 84}}{6}$$ Simplify.

$$= \frac{-10 \pm \sqrt{16}}{6} = \frac{-10 \pm 4}{6}$$

$$x = \frac{-10 + 4}{6} = \frac{-6}{6} = -1$$ Using the positive sign.

$$x = \frac{-10 - 4}{6} = \frac{-14}{6} = -\frac{7}{3}$$ Using the negative sign.

Thus the two solutions are -1 and $-\dfrac{7}{3}$.

[*Note:* Here we obtain rational roots. We would obtain the same answer by factoring $3x^2 + 10x + 7 = 0$ as $(3x + 7)(x + 1) = 0$ and setting each factor $= 0$.]

Practice Problem 1 Solve using the quadratic formula. $x^2 - 7x + 6 = 0$

$x = 6, x = 1$

Often the roots will be irrational numbers. Sometimes these roots can be simplified. You should always leave your answer in simplest form.

Teaching Example 2 Solve.
$$3x^2 - 2x - 4 = 0$$
Ans: $x = \dfrac{1 \pm \sqrt{13}}{3}$

EXAMPLE 2 Solve. $2x^2 - 4x - 9 = 0$

Solution $a = 2,$ $b = -4,$ $c = -9$

$$x = \frac{-b \pm \sqrt{b^2 - 4ac}}{2a} = \frac{-(-4) \pm \sqrt{(-4)^2 - 4(2)(-9)}}{2(2)}$$

$$= \frac{4 \pm \sqrt{16 + 72}}{4} = \frac{4 \pm \sqrt{88}}{4}$$ Do not stop here!

Notice that we can simplify $\sqrt{88}$.

$$x = \frac{4 \pm 2\sqrt{22}}{4} = \frac{2(2 \pm \sqrt{22})}{4} = \frac{\overset{1}{\cancel{2}}(2 \pm \sqrt{22})}{\underset{2}{\cancel{4}}} = \frac{2 \pm \sqrt{22}}{2}$$

Be careful. Here we were able to divide the numerator and denominator by 2 because 2 was a factor of every term.

Practice Problem 2   Solve. $3x^2 - 8x + 3 = 0$

$$x = \frac{4 \pm \sqrt{7}}{3}$$

A quadratic equation *must* be written in the standard form $ax^2 + bx + c = 0$ *before* the quadratic formula can be used. Several algebraic steps may be needed to accomplish this objective. Also, since it is much easier to use the formula if a, b, and c are integers, we will avoid using fractional values.

EXAMPLE 3   Solve. $x^2 = 5 - \dfrac{3}{4}x$

Solution   First we obtain an equivalent equation that does not have fractions.

$$x^2 = 5 - \frac{3}{4}x$$

$$4(x^2) = 4(5) - 4\left(\frac{3}{4}x\right) \quad \text{Multiply each term by the LCD of 4.}$$

$$4x^2 = 20 - 3x \qquad \text{Simplify.}$$
$$4x^2 + 3x - 20 = 0 \qquad \text{Add } 3x - 20 \text{ to each side.}$$
$$a = 4, \quad b = 3, \quad c = -20 \qquad \text{Substitute } a = 4, b = 3, \text{ and } c = -20 \text{ into the quadratic formula.}$$

$$x = \frac{-3 \pm \sqrt{(3)^2 - 4(4)(-20)}}{2(4)}$$

$$x = \frac{-3 \pm \sqrt{9 + 320}}{8}$$

$$x = \frac{-3 \pm \sqrt{329}}{8}$$

Practice Problem 3   Solve.

$$x^2 = 7 - \frac{3}{5}x \quad x = \frac{-3 \pm \sqrt{709}}{10}$$

Teaching Example 3 Solve. $\dfrac{x}{2} + 1 = \dfrac{5}{4}x^2$

Ans: $x = \dfrac{1 \pm \sqrt{21}}{5}$

Teaching Tip It is possible to use the quadratic formula if a, b, and c are fractions, that is, rational numbers. However, the resulting calculations are usually more difficult. Therefore, strongly stress that students should always multiply each term by the LCD to eliminate the fractions.

② Finding a Decimal Approximation for the Real Roots of a Quadratic Equation

EXAMPLE 4   Find the roots of $3x^2 - 5x = 7$. Approximate to the nearest thousandth.

Solution
$$3x^2 - 5x = 7$$
$$3x^2 - 5x - 7 = 0 \quad \text{Place the equation in standard form.}$$
$$a = 3, \quad b = -5, \quad c = -7$$

$$x = \frac{-(-5) \pm \sqrt{(-5)^2 - 4(3)(-7)}}{2(3)} = \frac{5 \pm \sqrt{25 + 84}}{6}$$

$$= \frac{5 \pm \sqrt{109}}{6}$$

Teaching Example 4 Find the roots of $2x^2 = 5 - x$. Approximate to the nearest thousandth.

Ans: $x \approx 1.351$, $x \approx -1.851$

Using most scientific calculators, we can find $\dfrac{5 + \sqrt{109}}{6}$ with these keystrokes.

$$5 \boxed{+} \; 109 \boxed{\sqrt{}} \boxed{=} \boxed{\div} \; 6 \boxed{=} \quad 2.5733844$$

Rounding to the nearest thousandth, we have $x \approx 2.573$.

To find $\dfrac{5 - \sqrt{109}}{6}$, we use the following keystrokes.

$$5 \boxed{-} \; 109 \boxed{\sqrt{}} \boxed{=} \boxed{\div} \; 6 \boxed{=} \quad -0.9067178$$

Rounding to the nearest thousandth, we have $x \approx -0.907$.
If you do not have a calculator with a square root key, look up $\sqrt{109}$ in the square root table. $\sqrt{109} \approx 10.440$. Using this result, we have

$$x \approx \frac{5 + 10.440}{6} = \frac{15.440}{6} \approx 2.573 \quad \text{(rounded to the nearest thousandth)}$$

and

$$x \approx \frac{5 - 10.440}{6} = \frac{-5.440}{6} \approx -0.907 \quad \text{(rounded to the nearest thousandth).}$$

If you have a graphing calculator, there is an alternate method for finding a decimal approximation of the roots of a quadratic equation. In Section 10.4, you will cover graphing quadratic equations. You will discover that finding the roots of a quadratic equation of the form $ax^2 + bx + c = 0$ is equivalent to finding the values of x on the graph of $y = ax^2 + bx + c$, where $y = 0$. These may be approximated quite quickly with a graphing calculator.

NOTE TO STUDENT: *Fully worked-out solutions to all of the Practice Problems can be found at the back of the text starting at page SP-1*

Practice Problem 4 Find the roots of $2x^2 = 13x + 5$. Approximate to the nearest thousandth. $x \approx 6.864, x \approx -0.364$

③ Determining Whether a Quadratic Equation Has No Real Solutions

Teaching Example 5 Solve. $2x^2 + x = -1$

Ans: No real solution

EXAMPLE 5 Solve. $2x^2 + 5 = -3x$

Solution The equation is not in standard form.

$$2x^2 + 5 = -3x \qquad \text{Add } 3x \text{ to both sides.}$$

$$2x^2 + 3x + 5 = 0 \qquad \text{We can now find } a, b, \text{ and } c.$$

$$a = 2, \quad b = 3, \quad c = 5 \quad \text{Substitute these values into the quadratic formula.}$$

$$x = \frac{-3 \pm \sqrt{(3)^2 - 4(2)(5)}}{2(2)} = \frac{-3 \pm \sqrt{9 - 40}}{4} = \frac{-3 \pm \sqrt{-31}}{4}$$

There is no real number that is $\sqrt{-31}$. Since we are using only real numbers in this book, we say there is no solution to the problem. (*Note:* In more advanced math courses, these types of numbers, called complex numbers, will be studied.)

Practice Problem 5 Solve. $5x^2 + 2x = -3$ No real solution

We can tell whether the roots of any given quadratic equation are real numbers. Look at the quadratic formula.

$$\frac{-b \pm \sqrt{b^2 - 4ac}}{2a}$$

The expression under the radical sign is called the **discriminant.**

$$b^2 - 4ac \text{ is the discriminant.}$$

If the discriminant is a negative number, the roots are not real numbers, and there is no real number solution (no real roots) to the equation.

If the discriminant is a positive number, the roots are real numbers, and there are two real number solutions (two real roots) to the equation. In addition, if $b^2 - 4ac$ is a perfect square, the two real number solutions are rational numbers. If $b^2 - 4ac$ is not a perfect square, the two real number solutions are irrational. If the discriminant is equal to 0, there is one real number solution (one root) to the equation.

EXAMPLE 6 Determine whether $3x^2 = 5x - 4$ has real number solution(s).

Solution   First, we place the equation in standard form. Then we need only check the discriminant.

$$3x^2 = 5x - 4$$
$$3x^2 - 5x + 4 = 0$$
$$a = 3, \quad b = -5, \quad c = 4 \quad \text{Substitute the values for } a, b, \text{ and } c \text{ into the discriminant and evaluate.}$$

$$b^2 - 4ac = (-5)^2 - 4(3)(4) = 25 - 48 = -23$$

The discriminant is negative. Thus $3x^2 = 5x - 4$ has no real number solution(s).

Practice Problem 6 Determine whether each equation has real number solution(s).

(a) $5x^2 = 3x + 2$ **(b)** $2x^2 + 5 = 4x$

(a) Two real number solutions   **(b)** No real number solution

Teaching Example 6 Determine whether each equation has real number solution(s).

(a) $x^2 + 2 = 4x$ (b) $3x^2 = 2x - 3$

Ans: (a) Two real number solutions
      (b) No real number solutions

Verbal and Writing Skills

1. In the equation $3x^2 + 4x - 7 = 0$, what can we learn from the discriminant?
$\sqrt{b^2 - 4ac} = \sqrt{100} = 10$ Therefore, there are two rational roots.

2. In the equation $7x^2 - 6x + 2 = 0$, what can we learn from the discriminant?
$\sqrt{b^2 - 4ac} = \sqrt{-20}$ This is not a real number. There are no real solutions to this quadratic equation.

3. Is the equation $4x^2 = -5x + 6$ in standard form? If not, place it in standard form and find a, b, and c.
No. Standard form is $4x^2 + 5x - 6 = 0$. Then $a = 4$, $b = 5$, and $c = -6$.

4. Is the equation $8x = -9x^2 + 2$ in standard form? If not, place it in standard form and find a, b, and c.
No. Standard form is $9x^2 + 8x - 2 = 0$. Then $a = 9$, $b = 8$, and $c = -2$.

Solve using the quadratic formula. If there are no real roots, say so.

5. $x^2 + 3x - 10 = 0$
$x = \dfrac{-3 \pm \sqrt{9 + 40}}{2}$
$x = 2, x = -5$

6. $x^2 - 5x + 4 = 0$
$x = \dfrac{5 \pm \sqrt{25 - 16}}{2}$
$x = 1, x = 4$

7. $x^2 - 3x - 8 = 0$
$x = \dfrac{3 \pm \sqrt{9 + 32}}{2} = \dfrac{3 \pm \sqrt{41}}{2}$

8. $x^2 - x - 8 = 0$
$x = \dfrac{1 \pm \sqrt{1 + 32}}{2} = \dfrac{1 \pm \sqrt{33}}{2}$

9. $4x^2 + 7x - 2 = 0$
$x = \dfrac{-7 \pm \sqrt{49 + 32}}{8}$
$x = \dfrac{1}{4}, x = -2$

10. $3x^2 + 7x - 6 = 0$
$x = \dfrac{-7 \pm \sqrt{49 + 72}}{6}$
$x = \dfrac{2}{3}, x = -3$

11. $2x^2 = 3x + 20$
$x = \dfrac{3 \pm \sqrt{9 + 160}}{4}$
$x = 4, x = -\dfrac{5}{2}$

12. $9x - 2 = 3x^2$
$x = \dfrac{9 \pm \sqrt{81 - 24}}{6}$
$x = \dfrac{9 \pm \sqrt{57}}{6}$

13. $6x^2 - 3x = 1$
$x = \dfrac{3 \pm \sqrt{9 + 24}}{12}$
$x = \dfrac{3 \pm \sqrt{33}}{12}$

14. $3 = 7x - 4x^2$
$x = \dfrac{7 \pm \sqrt{49 - 48}}{8}$
$x = 1, x = \dfrac{3}{4}$

15. $x + \dfrac{3}{2} = 3x^2$
$0 = 6x^2 - 2x - 3$
$x = \dfrac{2 \pm \sqrt{4 + 72}}{12}$
$x = \dfrac{1 \pm \sqrt{19}}{6}$

16. $\dfrac{2}{3}x^2 + x = \dfrac{1}{2}$
$4x^2 + 6x - 3 = 0$
$x = \dfrac{-6 \pm \sqrt{36 + 48}}{8}$
$x = \dfrac{-3 \pm \sqrt{21}}{4}$

17. $\dfrac{x}{2} + \dfrac{5}{x} = \dfrac{7}{2}$
$x^2 - 7x + 10 = 0$
$x = \dfrac{7 \pm \sqrt{49 - 40}}{2}$
$x = 5, x = 2$

18. $\dfrac{1}{4} + \dfrac{5}{x} = \dfrac{x}{4}$
$x^2 - x - 20 = 0$
$x = \dfrac{1 \pm \sqrt{1 + 80}}{2}$
$x = 5, x = -4$

19. $5x^2 + 6x + 2 = 0$
$x = \dfrac{-6 \pm \sqrt{36 - 40}}{10}$
No real solution

20. $3x^2 - 4x + 2 = 0$
$x = \dfrac{4 \pm \sqrt{16 - 24}}{6}$
No real solution

21. $4y^2 = 10y - 3$
$4y^2 - 10y + 3 = 0$
$y = \dfrac{10 \pm \sqrt{100 - 48}}{8}$
$y = \dfrac{10 \pm 2\sqrt{13}}{8} = \dfrac{5 \pm \sqrt{13}}{4}$

22. $8y^2 = 8y - 1$
$8y^2 - 8y + 1 = 0$
$y = \dfrac{8 \pm \sqrt{64 - 32}}{16}$
$y = \dfrac{8 \pm 4\sqrt{2}}{16} = \dfrac{2 \pm \sqrt{2}}{4}$

23. $\dfrac{d^2}{2} + \dfrac{5d}{6} - 2 = 0$

$3d^2 + 5d - 12 = 0$

$d = \dfrac{-5 \pm \sqrt{25 + 144}}{6}$

$d = \dfrac{4}{3},\ d = -3$

24. $\dfrac{k^2}{4} - \dfrac{5k}{6} = \dfrac{2}{3}$

$3k^2 - 10k - 8 = 0$

$k = \dfrac{10 \pm \sqrt{100 + 96}}{6}$

$k = -\dfrac{2}{3},\ k = 4$

25. $4x^2 + 8x + 11 = 0$

No real solution

26. $5x^2 + 7x + 9 = 0$

No real solution

27. $x^2 + 12x + 36 = 0$

$x = -6$

28. $16x^2 - 8x + 1 = 0$

$x = \dfrac{1}{4}$

Use the quadratic formula to find the roots. Find a decimal approximation to the nearest thousandth.

29. $x^2 + 5x - 2 = 0$

$x = \dfrac{-5 \pm \sqrt{25 + 8}}{2}$

$x \approx 0.372,\ x \approx -5.372$

30. $x^2 + 8x - 8 = 0$

$x = \dfrac{-8 \pm \sqrt{64 + 32}}{2}$

$x \approx 0.899,\ x \approx -8.899$

31. $2x^2 - 7x - 5 = 0$

$x = \dfrac{7 \pm \sqrt{49 + 40}}{4}$

$x \approx 4.108,\ x \approx -0.608$

32. $3x^2 + 4x - 6 = 0$

$x = \dfrac{-4 \pm \sqrt{16 + 72}}{6}$

$x \approx 0.897,\ x \approx -2.230$

33. $5x^2 + 10x + 1 = 0$

$x = \dfrac{-5 \pm 2\sqrt{5}}{5}$

$x \approx -1.894,\ x \approx -0.106$

34. $2x^2 + 9x + 2 = 0$

$x = \dfrac{-9 \pm \sqrt{65}}{4}$

$x \approx -4.266,\ x \approx -0.234$

Mixed Practice *Solve for the variable. Choose the method you feel will work best.*

35. $6x^2 - 13x + 6 = 0$

$x = \dfrac{3}{2},\ x = \dfrac{2}{3}$

36. $6x^2 - 7x - 3 = 0$

$x = \dfrac{3}{2},\ x = -\dfrac{1}{3}$

37. $3(x^2 + 1) = 10x$

$3x^2 - 10x + 3 = 0$

$x = 3,\ x = \dfrac{1}{3}$

38. $3 + x(x + 2) = 18$

$x^2 + 2x - 15 = 0$

$x = 3,\ x = -5$

39. $(t + 5)(t - 3) = 7$

$t^2 + 2t - 22 = 0$

$t = \dfrac{-2 \pm \sqrt{4 + 88}}{2}$

$t = -1 \pm \sqrt{23}$

40. $(b + 4)(b - 7) = 4$

$b^2 - 3b - 32 = 0$

$b = \dfrac{3 \pm \sqrt{9 + 128}}{2}$

$b = \dfrac{3 \pm \sqrt{137}}{2}$

41. $y^2 - \dfrac{2}{5}y = 2$

$5y^2 - 2y - 10 = 0$

$y = \dfrac{1 \pm \sqrt{51}}{5}$

42. $y^2 - 3 = \dfrac{8}{3}y$

$3y^2 - 8y - 9 = 0$

$y = \dfrac{4 \pm \sqrt{43}}{3}$

43. $3x^2 - 13 = 0$

$x = \dfrac{\pm\sqrt{39}}{3}$

44. $5x^2 + 9x = 0$

$x = 0,\ x = -\dfrac{9}{5}$

45. $x(x - 2) = 7$

$x = 1 \pm 2\sqrt{2}$

46. $(4x - 1)^2 = 24$

$x = \dfrac{1 \pm 2\sqrt{6}}{4}$

To Think About

▲ **47.** *Swimming Pool Design* John and Chris Maney have designed a new in-ground swimming pool for their backyard. The pool is rectangular and measures 30 feet by 20 feet. They have also designed a tile border to go around the pool. In the diagram below, this border is *x* feet wide. The tile border covers 216 square feet in area. What is the width *x* of this tile border? (*Hint:* Find an expression for the area of the large rectangle and subtract an expression for the area of the small rectangle.)   2 ft

▲ **48.** *Lynn Campus Entrance* The Alumni Association of North Shore Community College has designed a new garden for the front entrance of the Lynn campus. It is rectangular and measures 40 feet by 30 feet. The association has designed a cement walkway to go around this rectangular garden. In the diagram below, this cement walkway is *x* feet wide. The entire cement walkway covers 456 square feet in area. What is the width *x* of this cement walkway? (*Hint:* Find an expression for the area of the large rectangle and subtract an expression for the area of the small rectangle.)   3 ft

Cumulative Review

49. **[4.6.2]** Divide. $(x^3 + 8x^2 + 17x + 6) \div (x + 3)$

$x^2 + 5x + 2$

50. **[2.5.1]** Solve for g. $2p = 3gf - 7p + 20$

$$\frac{9p - 20}{3f} = g$$

Quick Quiz 10.3 Solve by using the quadratic formula.

1. $9x^2 + 6x - 1 = 0$

$x = \dfrac{-1 \pm \sqrt{2}}{3}$

2. $4x^2 - 4x = 19$

$x = \dfrac{1 \pm 2\sqrt{5}}{2}$

3. $2 + x = 3x(x + 1)$

$x = \dfrac{-1 \pm \sqrt{7}}{3}$

4. Concept Check Explain how you would determine if the quadratic equation $5x^2 - 8x + 9 = 0$ has real solutions or has no real solutions.

Answers may vary

Classroom Quiz 10.3 You may use these problems to quiz your students' mastery of Section 10.3.

Solve by using the quadratic formula.

1. $3x^2 - 5x + 1 = 0$

Ans: $x = \dfrac{5 \pm \sqrt{13}}{6}$

2. $2x^2 - 6x = -1$

Ans: $x = \dfrac{3 \pm \sqrt{7}}{2}$

3. $2x(x + 4) = -11 - 3x$

Ans: $x = \dfrac{-11 \pm \sqrt{33}}{4}$

How are you doing with your homework assignments in Sections 10.1 to 10.3? Do you feel you have mastered the material so far? Do you understand the concepts you have covered? Before you go further in the textbook, take some time to do each of the following problems.

10.1

Solve by factoring.

1. $x^2 - 13x - 48 = 0$

2. $5x^2 + 7x = 14x$

3. $5x^2 = 22x - 8$

4. $-x + 1 = 6x^2$

5. $x(x + 9) = 4(x + 6)$

6. $\dfrac{5}{x + 2} = \dfrac{2x - 1}{5}$

Simplify all answers in problems 7–14.

10.2

Solve by taking the square root of each side of the equation.

7. $x^2 - 18 = 0$

8. $3x^2 + 1 = 76$

Solve by completing the square.

9. $x^2 + 6x + 1 = 0$

10. $2x^2 + 3x - 7 = 0$

10.3

Solve if possible using the quadratic formula.

11. $2x^2 + 4x - 5 = 0$

12. $4x^2 = x + 5$

13. $3x^2 + 8x + 1 = 0$

14. $5x^2 + 3 = 4x$

Now turn to page SA-24 for the answer to each of these problems. Each answer also includes a reference to the objective in which the problem is first taught. If you missed any of these problems, you should stop and review the Examples and Practice Problems in the referenced objective. A little review now will help you master the material in the upcoming sections of the text.

---

1. $x = 16, x = -3$

2. $x = 0, x = \dfrac{7}{5}$

3. $x = \dfrac{2}{5}, x = 4$

4. $x = \dfrac{1}{3}, x = -\dfrac{1}{2}$

5. $x = -8, x = 3$

6. $x = 3, x = -\dfrac{9}{2}$

7. $x = \pm 3\sqrt{2}$

8. $x = \pm 5$

9. $x = -3 \pm 2\sqrt{2}$

10. $x = \dfrac{-3 \pm \sqrt{65}}{4}$

11. $x = \dfrac{-2 \pm \sqrt{14}}{2}$

12. $x = \dfrac{5}{4}, x = -1$

13. $x = \dfrac{-4 \pm \sqrt{13}}{3}$

14. No real solution

Teaching Example 1 Graph.

(a) $y = \dfrac{1}{2}x^2$ (b) $y = -\dfrac{1}{2}x^2$

Ans:

(a)

x	y
-3	$\dfrac{9}{2}$
-2	2
-1	$\dfrac{1}{2}$
0	0
1	$\dfrac{1}{2}$
2	2
3	$\dfrac{9}{2}$

(b)

x	y
-3	$-\dfrac{9}{2}$
-2	-2
-1	$-\dfrac{1}{2}$
0	0
1	$-\dfrac{1}{2}$
2	-2
3	$-\dfrac{9}{2}$

(a)

(b)

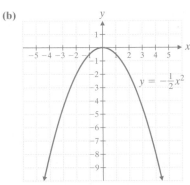

① Graphing Equations of the Form $y = ax^2 + bx + c$ Using Ordered Pairs

Recall that the graph of a linear equation is always a straight line. What is the graph of a quadratic equation? What does the graph look like if one of the variable terms is squared? We will use point plotting to look for a pattern. We begin with a table of values.

EXAMPLE 1 Graph.   (a) $y = x^2$ (b) $y = -x^2$

Solution

(a) $y = x^2$

x	y
-3	9
-2	4
-1	1
0	0
1	1
2	4
3	9

(b) $y = -x^2$

x	y
-3	-9
-2	-4
-1	-1
0	0
1	-1
2	-4
3	-9

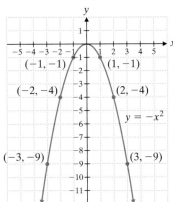

Practice Problem 1 Graph.   (a) $y = 2x^2$ (b) $y = -2x^2$

(a) $y = 2x^2$

x	y
-2	8
-1	2
0	0
1	2
2	8

(b) $y = -2x^2$

x	y
-2	-8
-1	-2
0	0
1	-2
2	-8

TO THINK ABOUT: When the Coefficient of x^2 Is Negative What happens to the graph when the coefficient of x^2 is negative? What happens when the coefficient of x^2 is an integer greater than 1 or less than -1? Graph $y = 3x^2$ and $y = -3x^2$. What happens when $-1 < a < 1$? Graph $y = \frac{1}{3}x^2$ and $y = -\frac{1}{3}x^2$.

The curves that you have been graphing are called **parabolas.** What happens to the graph of a parabola as the quadratic equation that describes it changes?

In the preceding example, the equations we worked with were relatively simple. They provided us with an understanding of the general shape of the graph of a quadratic equation. We have seen how changing the equation affects the graph of that equation. Let's look at an equation that is slightly more difficult to graph.

The **vertex** is the lowest point on a parabola that opens upward, or the highest point on a parabola that opens downward.

EXAMPLE 2 Graph $y = x^2 - 2x$. Identify the coordinates of the vertex.

Solution

x	y
-2	8
-1	3
0	0
1	-1	vertex
2	0
3	3
4	8

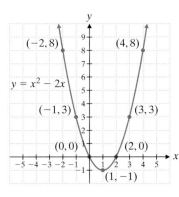

The vertex is at $(1, -1)$.

Notice that the x-intercepts are at $(0, 0)$ and $(2, 0)$. The x-coordinates are the solutions to the equation $x^2 - 2x = 0$.

$$x^2 - 2x = 0$$
$$x(x - 2) = 0$$
$$x = 0, \qquad x = 2$$

Practice Problem 2 Graph. For each table of values, use x-values from -4 to 2. Identify the coordinates of the vertex.

(a) $y = x^2 + 2x$

(b) $y = -x^2 - 2x$

(a)

x	y
-4	8
-3	3
-2	0
-1	-1
0	0
1	3
2	8

(b)

x	y
-4	-8
-3	-3
-2	0
-1	1
0	0
1	-3
2	-8

Graphing Calculator

Quadratic Equations

Graph $y = x^2 - 4x$.

Use the Trace and Zoom features. Use the feature that calculates the minimum point to find the vertex.

Thus the vertex is at $(2, -4)$.

Teaching Example 2 Graph $y = x^2 - 4x$. Use x-values from -1 to 5. Identify the coordinates of the vertex.

Ans: Vertex: $(2, -4)$

x	y
-1	5
0	0
1	-3
2	-4
3	-3
4	0
5	5

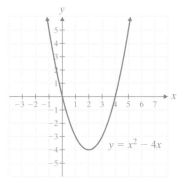

Graphing Equations of the Form $y = ax^2 + bx + c$ Using the Vertex Formula

Finding the vertex is the key to graphing a parabola. Let's take a look at some equations and identify the vertex of each parabola.

Equation: $y = x^2$ $y = x^2 - 3$ $y = (x - 1)^2$ $y = (x + 2)^2$ $y = x^2 - 2x$

                                          $y = x^2 - 2x + 1$ $y = x^2 + 4x + 4$

Vertex: $(0, 0)$ $(0, -3)$ $(1, 0)$ $(-2, 0)$ $(1, -1)$

Notice that the x-coordinate of the vertex of the parabola for each of the first two equations is 0. You may also notice that these two equations do not have middle terms. That is, for $y = ax^2 + bx + c$, $b = 0$. This is not true of the last three equations. As you may have already guessed, there is a relationship between the quadratic equation and the x-coordinate of the vertex of the parabola it describes.

> **VERTEX OF A PARABOLA**
>
> The x-coordinate of the vertex of the parabola described by $y = ax^2 + bx + c$ is
>
> $$x = \frac{-b}{2a}.$$

Once we have the x-coordinate, we can find the y-coordinate using the equation. We can plot the point and draw a general sketch of the curve. The sign of the squared term will tell us whether the graph opens upward or downward.

Teaching Example 3 $y = x^2 - 6x + 8$
Determine the vertex and the x-intercepts. Then sketch the graph.

Ans:

Vertex: $(3, -1)$ x-intercepts: $(2, 0), (4, 0)$

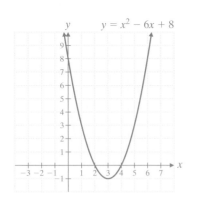

EXAMPLE 3 $y = -x^2 - 2x + 3$. Determine the vertex and the x-intercepts. Then sketch the graph.

Solution   We will first determine the coordinates of the vertex. We begin by finding the x-coordinate.

$$x = \frac{-b}{2a} = \frac{-(-2)}{2(-1)} = \frac{2}{-2} = -1$$

$y = -(-1)^2 - 2(-1) + 3$ Determine y when x is -1.

$ = 4$ The point $(-1, 4)$ is the vertex of the parabola.

Now we will determine the x-intercepts, those points where the graph crosses the x-axis. This occurs when $y = 0$. These are the solutions to the equation $-x^2 - 2x + 3 = 0$. Note that this equation is factorable.

$-x^2 - 2x + 3 = 0$ Find the solutions to the equation.

$-(x^2 + 2x - 3) = 0$

$-(x - 1)(x + 3) = 0$ Factor.

$x - 1 = 0$ $x + 3 = 0$ Solve for x.

$ x = 1$ $x = -3$ The x-intercepts are at $(1, 0)$ and at $(-3, 0)$.

We now have enough information to sketch the graph. Since $a = -1$, the parabola opens downward and the vertex $(-1, 4)$ is the highest point. The graph crosses the x-axis at $(-3, 0)$ and at $(1, 0)$.

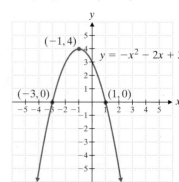

Practice Problem 3 $y = x^2 - 4x - 5$. Determine the vertex and the x-intercepts. Then sketch the graph.

NOTE TO STUDENT: *Fully worked-out solutions to all of the Practice Problems can be found at the back of the text starting at page SP-1*

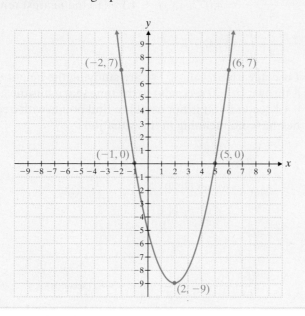

EXAMPLE 4   A person throws a ball into the air and its distance from the ground in feet is given by the equation $h = -16t^2 + 64t + 5$, where t is measured in seconds.

(a) Graph the equation $h = -16t^2 + 64t + 5$.

(b) What physical significance does the vertex of the graph have?

(c) After how many seconds will the ball hit the ground?

Solution

(a) Since the equation is a quadratic equation, the graph is a parabola. a is negative. Thus the parabola opens downward and the vertex is the highest point. We begin by finding the coordinates of the vertex. Our equation has ordered pairs (t, h) instead of (x, y). The t-coordinate of the vertex is

$$t = \frac{-b}{2a} = \frac{-64}{2(-16)} = \frac{-64}{-32} = 2.$$

Substitute $t = 2$ into the equation to find the h-coordinate of the vertex.

$$h = -16(2)^2 + 64(2) + 5 = -64 + 128 + 5 = 69$$

The vertex (t, h) is $(2, 69)$.

To sketch this graph, it would be helpful to know the h-intercept, that is, the point where $t = 0$.

$$h = -16(0)^2 + 64(0) + 5$$
$$= 5$$

The h-intercept is at $(0, 5)$.
The t-intercept is where $h = 0$. We will use the quadratic formula to find t when h is 0, that is, to find the roots of the quadratic equation $-16t^2 + 64t + 5 = 0$.

$$-16t^2 + 64t + 5 = 0$$

$$t = \frac{-64 \pm \sqrt{(64)^2 - 4(-16)(5)}}{2(-16)}$$

$$t = -\frac{64 \pm \sqrt{4096 + 320}}{-32} = -\frac{64 \pm \sqrt{4416}}{-32} = \frac{8 \pm \sqrt{69}}{4}$$

Teaching Example 4 A toy rocket is launched and its distance from the ground in feet is given by the equation $h = -16t^2 + 48t$, where t is measured in seconds.

(a) Graph the equation $h = -16t^2 + 48t$.

(b) What is the greatest height of the rocket?

(c) After how many seconds will the rocket hit the ground?

Ans: (a)

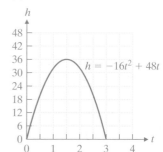

(b) 36 ft

(c) 3 sec

Using a calculator or a square root table, we have the following.

$$t = \frac{8 + 8.307}{4} \approx 4.077, \qquad t \approx 4.1 \quad \text{(to the nearest tenth)}$$

$$t = \frac{8 - 8.307}{4} \approx -0.077$$

We will not use negative t-values. We are studying the height of the ball from the time it is thrown ($t = 0$ seconds) to the time that it hits the ground (approximately $t = 4.1$ seconds). It is not useful to find points with negative values of t because they have no meaning in this situation.

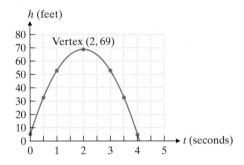

(b) The vertex represents the greatest height of the ball. The ball is 69 feet above the ground 2 seconds after it is thrown.

(c) The ball will hit the ground approximately 4.1 seconds after it is thrown. This is the point at which the graph intersects the t-axis, the positive root of the equation.

NOTE TO STUDENT: Fully worked-out solutions to all of the Practice Problems can be found at the back of the text starting at page SP-1

Practice Problem 4   A ball is thrown upward with a speed of 32 feet per second from a distance 10 feet above the ground. The height of the ball in feet is given by the equation $h = -16t^2 + 32t + 10$, where t is measured in seconds.

(a) Graph the equation $h = -16t^2 + 32t + 10$.

(b) What is the greatest height of the ball?   26 ft

(c) After how many seconds will the ball hit the ground?   about 2.3 sec

Verbal and Writing Skills

1. How many x-intercepts can a parabola have?
It may have 2, 1, or none.

2. How many vertices can a parabola have?
It will always have one vertex.

3. For a parabola in the form $y = ax^2 + bx + c$, what happens to the equation $x = \dfrac{-b}{2a}$ if $b = 0$?

If $b = 0$, then the parabola always has a vertex at $(0, c)$, that is, on the y-axis.

4. For a parabola in the form $y = ax^2 + bx + c$, what can be said about the parabola if $a < 0$?
It opens downward if $a < 0$.

Graph each equation.

5. $y = x^2 + 2$

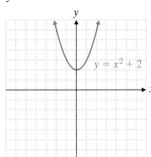

6. $y = x^2 - 4$

7. $y = -\frac{1}{3}x^2$

8. $y = -\frac{1}{2}x^2$

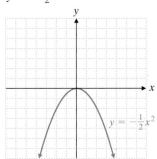

9. $y = 3x^2 - 1$

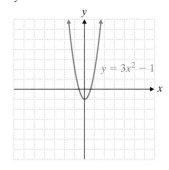

10. $y = 2x^2 + 3$

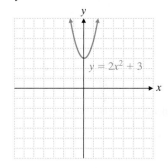

11. $y = (x - 2)^2$

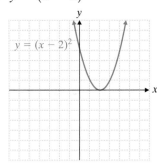

12. $y = (x + 1)^2$

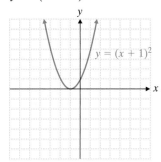

13. $y = -\frac{1}{2}x^2 + 4$

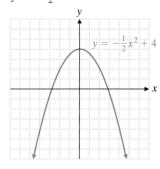

14. $y = -\frac{1}{3}x^2 + 6$

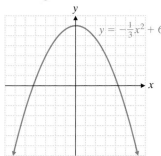

15. $y = \frac{1}{2}(x - 3)^2$

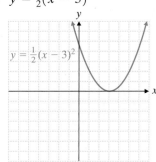

16. $y = \frac{1}{2}(x + 2)^2$

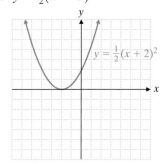

Graph each equation. Identify the coordinates of the vertex.

17. $y = x^2 + 4x$

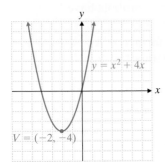

18. $y = -x^2 + 4x$

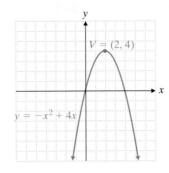

19. $y = -x^2 - 4x$

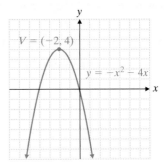

20. $y = x^2 - 4x$

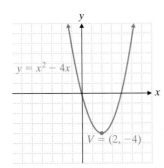

21. $y = -2x^2 + 8x$

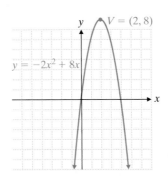

22. $y = x^2 + 6x$

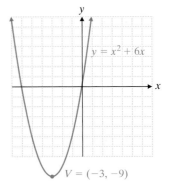

Determine the vertex and the x-intercepts. Then sketch the graph.

23. $y = x^2 + 2x - 3$

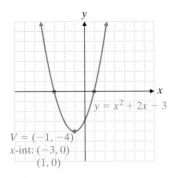

24. $y = x^2 - 4x + 3$

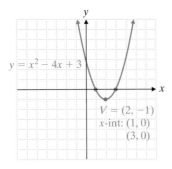

25. $y = -x^2 - 6x - 5$

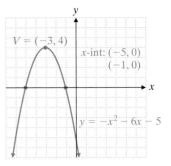

26. $y = -x^2 + 8x - 15$

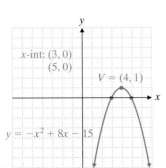

27. $y = x^2 + 6x + 9$

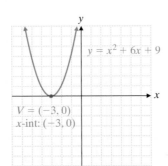

28. $y = -x^2 + 2x - 1$

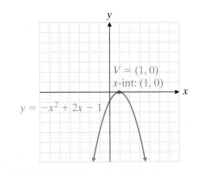

Applications

29. *Miniature Rockets* A miniature rocket is launched so that its height h in meters after t seconds is given by the equation $h = -4.9t^2 + 39.2t + 4$.

    (a) Draw a graph of the equation.

    (b) How high is the rocket after 2 seconds? 62.8 m

    (c) What is the maximum height the rocket will attain? 82.4 m

    (d) How long after the launch will the rocket hit the ground? About 8.1 sec

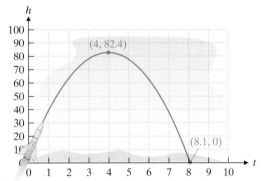

30. *Miniature Rockets* A miniature rocket is launched so that its height h in meters after t seconds is given by the equation $h = -4.9t^2 + 19.6t + 10$.

    (a) Draw a graph of the equation.

    (b) How high is the rocket after 3 seconds? 24.7 m

    (c) What is the maximum height the rocket will attain? 29.6 m

    (d) How long after the launch will the rocket hit the ground? About 4.46 sec

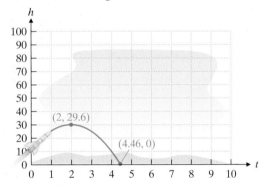

31. *Mosquito Population* The number of mosquitoes in some areas varies with the amount of rain. The number, N, of mosquitoes measured in millions in Hamilton, Massachusetts, in May can be predicted by the equation $N = 9x - x^2$, where x is the number of inches of rain during that month. (*Source*: Massachusetts Environmental Protection Agency.)

    (a) Graph the equation.

    (b) How many inches of rain produce the maximum number of mosquitoes in May? 4.5 in.

    (c) What is the maximum number of mosquitoes? 20.25 million

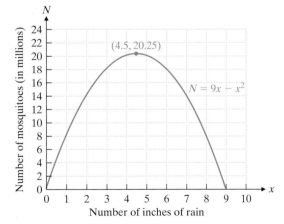

Use your graphing calculator to find the vertex, the y-intercept, and the x-intercepts of the parabolas described by the following equations. Round to the nearest thousandth when necessary.

 32. $y = -156x^2 - 289x + 1133$
Vertex $(-0.926, 1266.848)$
y-intercept $(0, 1133)$
x-intercepts $(1.923, 0)$ and $(-3.776, 0)$

 33. $y = -301x^2 - 167x + 1528$
Vertex $(-0.277, 1551.164)$
y-intercept $(0, 1528)$
x-intercepts $(1.993, 0)$ and $(-2.548, 0)$

Cumulative Review

34. **[9.7.2]** If y varies inversely as x, and $y = \dfrac{1}{3}$ when $x = 21$, find y when $x = 7$.

$y = \dfrac{7}{x}$ $y = 1$

35. **[9.7.1]** If y varies directly as x^2, and $y = 12$ when $x = 2$, find y when $x = 5$.

$y = 3x^2$ $y = 75$

36. **[6.6.3]** *Football* A football player picks up a ball from the field and runs 60 yards. He runs the first half of this distance at 20 feet per second. He runs the second half at 24 feet per second. How long does it take him to run the 60 yards?

8.25 sec

Quick Quiz 10.4 Consider the equation $y = x^2 - 6x + 5$.

1. What is the vertex of this curve? $(3, -4)$

2. What are the x-intercepts of this curve?
$(1, 0)$ and $(5, 0)$

3. Graph the equation.

4. **Concept Check** Explain how you would find the vertex of the equation $y = 4x^2 + 16x - 2$.
Answers may vary

Classroom Quiz 10.4 You may use these problems to quiz your students' mastery of Section 10.4.

Consider the equation $y = -x^2 - 8x - 12$.

1. What is the vertex of this curve?
Ans: $(-4, 4)$

2. What are the x-intercepts of this curve?
Ans: $(-6, 0)$ and $(-2, 0)$

3. Graph the equation.
Ans:

10.5 FORMULAS AND APPLIED PROBLEMS

1 Solving Applied Problems Using Quadratic Equations

We can use what we have learned about quadratic equations to solve word problems. Some word problems will involve geometry.

Recall that the Pythagorean Theorem states that in a right triangle, $a^2 + b^2 = c^2$, where a and b are the lengths of the legs and c is the length of the hypotenuse. We will use this theorem to solve word problems involving right triangles.

Student Learning Objective

After studying this section, you will be able to:

1 Solve applied problems using quadratic equations.

▲ **EXAMPLE 1** The hypotenuse of a right triangle is 25 meters in length. One leg is 17 meters longer than the other. Find the length of each leg.

Teaching Example 1 The hypotenuse of a right triangle is 10 inches in length. One leg is 2 inches shorter than the other. Find the length of each leg.

Ans: 8 in., 6 in.

Solution

1. **Understand the problem.** Draw a picture.
2. **Write an equation.** The problem involves a right triangle. Use the Pythagorean Theorem.

$$a^2 + b^2 = c^2$$
$$x^2 + (x + 17)^2 = 25^2$$

3. **Solve and state the answer.**

$$x^2 + (x + 17)^2 = 25^2$$
$$x^2 + x^2 + 34x + 289 = 625$$
$$2x^2 + 34x - 336 = 0$$
$$x^2 + 17x - 168 = 0$$
$$(x + 24)(x - 7) = 0$$
$$x + 24 = 0 \qquad x - 7 = 0$$
$$x = -24 \qquad x = 7$$

(Note that $x = -24$ is not a valid solution for this particular word problem.)

One leg is 7 meters in length. The other leg is $x + 17 = 7 + 17 = 24$ meters in length.

4. **Check.** Check the conditions in the original problem.
Do the two legs differ by 17 meters?

$$24 - 7 \stackrel{?}{=} 17 \qquad 17 = 17 \checkmark$$

Do the sides of the triangle satisfy the Pythagorean Theorem?

$$7^2 + 24^2 \stackrel{?}{=} 25^2$$
$$49 + 576 \stackrel{?}{=} 625$$
$$625 = 625 \checkmark$$

▲ **Practice Problem 1** The hypotenuse of a right triangle is 30 meters in length. One leg is 6 meters shorter than the other leg. Find the length of each leg.

24 m, 18 m

NOTE TO STUDENT: Fully worked-out solutions to all of the Practice Problems can be found at the back of the text starting at page SP-1

Sometimes it may help to use two variables in the initial part of the problem. Then one variable can be eliminated by substitution.

EXAMPLE 2   The ski club is renting a bus to travel to Mount Snow. The members agreed to share the cost of $180 equally. On the day of the ski trip, three members were sick with the flu and could not go. This increased the share of each person going on the trip by $10. How many people originally planned to attend?

Solution

1. *Understand the problem.*

   Let s = the number of students in the ski club

   c = the cost for each student in the original group

2. *Write an equation(s).*

   $$\text{number of students} \times \text{cost per student} = \text{total cost}$$
   $$s \cdot c = 180 \qquad \textbf{(1)}$$

   If three people are sick, the number of students drops by three, but the cost for each increases by $10. The total is still $180. Therefore, we have the following.

   $$(s - 3)(c + 10) = 180 \qquad \textbf{(2)}$$

3. *Solve and state the answer.*

   $$c = \frac{180}{s} \qquad \text{Solve equation \textbf{(1)} for } c.$$

   $$(s - 3)\left(\frac{180}{s} + 10\right) = 180 \qquad \text{Substitute } \frac{180}{s} \text{ for } c \text{ in equation \textbf{(2)}.}$$

   $$(s - 3)\left(\frac{180 + 10s}{s}\right) = 180 \qquad \text{Add } \frac{180}{s} + 10. \text{ Use the LCD, which is } s.$$

   $$(s - 3)(180 + 10s) = 180s \qquad \text{Multiply both sides of the equation by } s.$$

   $$10s^2 + 150s - 540 = 180s \qquad \text{Multiply the binomials.}$$

   $$10s^2 - 30s - 540 = 0$$

   $$s^2 - 3s - 54 = 0$$

   $$(s - 9)(s + 6) = 0$$

   $$s = 9 \qquad s = -6$$

   The number of students originally going on the ski trip was 9.

4. *Check.* Would the cost increase by $10 if the number of students dropped from nine to six? Nine people in the club would mean that each would pay $20 $(180 \div 9 = 20)$.

   If the number of students was reduced by three, there were six people who took the trip. Their cost was $30 each $(180 \div 6 = 30)$. The increase is $10.

   $$20 + 10 \overset{?}{=} 30$$
   $$30 = 30 \quad \checkmark$$

Practice Problem 2   Several friends from the University of New Mexico decided to charter a sport fishing boat for the afternoon while on a vacation at the coast. They agreed to share the rental cost of $240 equally. At the last minute, two of them could not go on the trip. This raised the cost for each person in the group by $4. How many people originally planned to charter the boat?   12 people

▲ **EXAMPLE 3**   Minette is fencing a garden that borders the back of a large barn. She has 120 feet of fencing. She would like a rectangular garden that measures 1350 square feet in area. She wants to use the back of the barn, so she needs to use fencing on three sides only. What dimensions should she use for her garden?

Solution

1. **Understand the problem.** Draw a picture.

   Let x = the width of the garden in feet and

   y = the length of the garden in feet.

2. **Write an equation(s).** Look at the drawing. The 120 feet of fencing will be needed for the width twice (two sides) and the length once (one side).

$$120 = 2x + y \qquad \textbf{(1)}$$

   The area formula is A = (width)(length).

$$1350 = xy \qquad \textbf{(2)}$$

3. **Solve and state the answer.**

$$y = 120 - 2x \qquad \text{Solve for } y \text{ in equation } \textbf{(1)}.$$
$$1350 = x(120 - 2x) \qquad \text{Substitute for } y \text{ in equation } \textbf{(2)}.$$
$$1350 = 120x - 2x^2 \qquad \text{Solve for } x.$$

$$2x^2 - 120x + 1350 = 0$$
$$x^2 - 60x + 675 = 0$$
$$(x - 15)(x - 45) = 0$$
$$x = 15 \qquad\qquad x = 45$$

First Solution: If the width is 15 feet, the length is 90 feet.

Check: $15 + 15 + 90 = 120$ ✓
                $(15)(90) = 1350$ ✓

Second Solution: If the width is 45 feet, the length is 30 feet.

Check: $45 + 45 + 30 = 120$ ✓
                $(45)(30) = 1350$ ✓

Thus we see that Minette has two choices that satisfy the given requirements (see the figures). She would probably choose the shape that is the most practical for her. (If her barn is shorter than 90 feet in length, she could not use the first solution!)

   Remark: Is there another way to do this problem? Yes. Use only the variable x to represent the length and width. If you let x = the width and $120 - 2x$ = the length, you can immediately write equation **(3)**, which is

$$1350 = x(120 - 2x). \qquad \textbf{(3)}$$

However, many students find that this is difficult to do, and they would prefer to use two variables. Work these problems in the way that is easiest for you.

▲ **Practice Problem 3**   An Arizona rancher has a small fenced-in area against a canyon wall. Therefore, he only needs three sides of fencing to make a rectangular holding area for cattle. He has 160 feet of fencing available. He wants the rectangular area to measure 3150 square feet. What dimensions should he use for the rectangular holding area?

If the width is 45 feet, the length is 70 feet. If the width is 35 feet, the length is 90 feet.

Teaching Example 3 Joe would like to use 50 feet of fencing to enclose a vegetable garden that he is planning for the spring. Joe's property is fenced in and he'll use the existing fence on one side of his garden. Joe wants to enclose 300 square feet. What dimensions should he use for his garden?

Ans: Joe can either make 2 sides of his garden 15 feet and the other side 20 feet, or he can make 2 sides 10 feet and the other side 30 feet.

First solution

Second solution

Applications *Solve.*

▲ **1. Geometry** The base of a triangle is 5 centimeters shorter than the altitude. The area of the triangle is 88 square centimeters. Find the lengths of the base and altitude.

$$\frac{1}{2}(x)(x - 5) = 88 \qquad \text{Altitude 16 cm,}$$
$$\text{base 11 cm}$$
$$x^2 - 5x - 176 = 0$$
$$(x - 16)(x + 11) = 0$$

▲ **2. Geometry** The area of a rectangle is 96 square meters. If the length is 4 more than the width, what are the dimensions of the rectangle?

$$\text{Width} = x = 8\,\text{m}$$
$$\text{Length} = x + 4 = 12\,\text{m}$$
$$x(x + 4) = 96 \qquad x = 8\,\text{m}$$
$$x^2 + 4x - 96 = 0$$
$$(x + 12)(x - 8) = 0$$

▲ **3. Garden Expansion** Emily wishes to expand the size of her front garden, which is square. Emily measures her front yard and decides she can increase the width by 4 feet and the length by 9 feet. If the area of the newly expanded garden is 266 square feet, what were the dimensions of the old garden?

$$(x + 4)(x + 9) = 266$$
$$x^2 + 13x - 230 = 0$$
$$(x - 10)(x + 23) = 0$$
10 ft by 10 ft

▲ **4. Addition to a House** Mike and Norma have a home with a very small rectangular room on the first floor that they use for a living room. This room is 2 feet longer than it is wide. They have decided to put on an addition to the house that will give them a larger rectangular room to meet the need for an adequate living room for their family. The length of the living room will increase by 10 feet. The width of the living room will increase by 8 feet. The area of the new, expanded living room will be 252 square feet. What are the dimensions of the present living room before the addition is built?

$$(x + 8)(x + 2 + 10) = 252 \qquad \text{So the present living}$$
$$(x + 8)(x + 12) = 252 \qquad \text{room is 6 feet wide. The}$$
$$x^2 + 20x - 156 = 0 \qquad \text{length is 2 feet longer.}$$
$$(x - 6)(x + 26) = 0 \qquad \text{The present living room}$$
$$x = 6 \qquad \text{is 8 feet long.}$$

5. Gift Expense To help out Nina and Tom with their new twins, the employees at Tom's office decided to give the new parents a gift certificate worth $420 to furnish the babies' room. The employees planned to share the cost equally. At the last minute, seven people from the executive staff chipped in, lowering the cost per person by $5. How many people originally planned to give the gift?   21 people

6. Septic Systems Everyone in the School Street Townhouse Association agreed to equally share the cost of a new $6000 septic system. However, before the system was installed, five additional units were built and the five new families were asked to contribute their share. This lowered the contribution of each family by $200. How many families originally agreed to share the cost of the new septic system?   10 families

7. Music Cost A fraternity was sponsoring a spring fling, for which the $400 DJ cost was to be split evenly among those members attending. If 20 more members than were expected came, reducing the cost per person by $1, find the number of fraternity members that attended.

$$xy = 400$$
$$(x + 20)(y - 1) = 400$$
$$x^2 + 20x - 8000 = 0$$
$$(x - 80)(x + 100) = 0$$
80 members were expected.
100 members attended.

8. Beach House Rental The Cornell University *Wild Roses* Ultimate team took a spring break trip to Savannah, Georgia, where they all agreed to chip in and rent a beach house for the week. The cost of the rental for the week was $1500, and each girl was assessed an equal amount. Just before they left New York, three more girls decided to go. The captain said that each girl would now pay $25 less. How many girls originally planned to go on the trip?

$$xy = 1500$$
$$(x + 3)(y - 25) = 1500$$
$$(x + 3)\left(\frac{1500}{x} - 25\right) = 1500$$
$$(x - 12)(x + 15) = 0$$
$$x = 12 \text{ girls originally planned to go}$$

▲ **9.** *Fenway Park* A security fence encloses a rectangular area on one side of Fenway Park in Boston. Three sides of fencing are used, since the fourth side of the area is formed by a building. The enclosed area measures 1250 square feet. Exactly 100 feet of fencing is used to fence in three sides of this rectangle. What are the possible dimensions that could have been used to construct this area? The width is 25 ft. The length along the building is 50 ft.

▲ **10.** *Garden Fencing* Henry Redbone needs to fence in his garden, to protect it from the rabbits who love to feast on his lettuce. He plans on planting the garden next to the barn, so that he only needs to fence in three sides of the garden. If Henry has 80 feet of fencing available, and he wants the garden to have an area of 750 square feet, what are the possible dimensions for the garden? Choices are 15 ft by 50 ft or 25 ft by 30 ft.

11. *Experimental Jet* A pilot is testing a new experimental craft. The cruising speed is classified information. In a test the jet traveled 2400 miles. The pilot revealed that if he had increased his speed 200 mph, the total trip would have taken 1 hour less. Determine the cruising speed of the jet. 600 mph

▲ **12.** *Gymnasium Floor Replacement* A gymnasium floor is being covered by square shock-absorbing tiles. The old gym floor required 864 square tiles. The new tiles are 2 inches larger in both length and width than the old tiles. The new flooring will require only 600 tiles. What is the length of a side of one of the new shock-absorbing tiles? (*Hint*: Since the area is the same in each case, write an expression for area with old tiles and one with new tiles. Set the two expressions equal to each other.) 12 in.

Use the following information to solve exercises 13–18.

Business Expansion *Two young college graduates opened a chain of Quick Print stores in southern California. The chain expanded during the late 1990s and early 2000s and many new stores were opened. The number of Quick Print Centers in operation during any year from 1998 to 2008 is given by the equation* $y = -2.5x^2 + 22.5x + 50$, *where x is the number of years since 1998.*

13. How many Quick Print Centers were in operation in 1998? 50 stores

14. How many Quick Print Centers were in operation in 2001? 95 stores

15. How many more Quick Print Centers were in operation in 2002 than in 2000? 15 stores

16. How many fewer Quick Print Centers were in operation in 2007 than in 2005? 35 stores

17. During what two years were there 100 Quick Print Centers? 2002 and 2003

18. During what two years were there 70 Quick Print Centers? 1999 and 2006

Cannon Shell Path Big Bertha, *an infamous World War I cannon, fired shells over an amazingly long distance. The path of a shell from this monstrous cannon can be modeled by the formula* $y = -0.018x^2 + 1.336x$, *where x is the horizontal distance the cannon shell traveled in miles and y is the height in miles the shell would reach on its flight.*

19. How far could *Big Bertha* send a shell? Round to the nearest whole number.
$0 = -0.018x^2 + 1.336x$
$0 = x(-0.018x + 1.336)$
$x = 0 \quad x = 74.2$ or about 74 mi

20. **(a)** How far into its journey would the shell achieve maximum height? Round to the nearest whole number. 37 mi
(b) What was the maximum height in miles? Round to the nearest whole number. 25 mi

Cumulative Review

21. **[2.7.1]** Graph on a number line the solution to $3x + 11 \geq 9x - 4$.

22. **[9.5.3]** Rationalize the denominator. $\dfrac{\sqrt{3}}{\sqrt{5} + 2}$

$\sqrt{15} - 2\sqrt{3}$

▲ **23.** **[3.3.1]** *Recycling Bin* A recycling bin that has a length of 3 feet, a width of 2 feet, and a height of 2 feet has three separate sections inside for sorting. Section 1 is for paper, section 2 is for glass and plastic, and section 3 is for metal. Sections 1 and 2 have identical measurements, but section 3 has $\frac{2}{3}$ the volume of section 1 or 2. If all three sections have the same width and height, how long is section 3?   0.75 ft

Quick Quiz 10.5

1. The hypotenuse of a right triangle is 13 yards long. One leg is 7 yards longer than the other leg. Find the length of each leg.   12 yd, 5 yd

When an object is thrown upward, its height (S) in meters is given approximately by the quadratic equation $S = -5t^2 + vt + h$

where v = the initial upward velocity in meters per second,

   t = the time of flight in seconds, and

   h = the height above level ground from which the object is thrown.

Suppose a ball is thrown upward with a velocity of 25 meters per second at a height of 30 meters above the ground.

2. How long after it is thrown will the ball hit the ground?   6 sec

3. What is the maximum height of the ball?   61.25 m

4. **Concept Check** The hiking club needs to raise $720 through dues paid by each member. However, 4 people dropped out of the club so the dues went up by $6 for each member. Explain how you could set up an equation to find out how many members were in the club prior to the dropout.   Answers may vary

Classroom Quiz 10.5   You may use these problems to quiz your students' mastery of Section 10.5.

1. The hypotenuse of a right triangle is 20 meters long. One leg is 4 meters shorter than the other leg. Find the length of each leg.   **Ans:** 12 m, 16 m

When an object is thrown upward, its height (S) in meters is given approximately by the quadratic equation $S = -5t^2 + vt + h$

where v = the initial upward velocity in meters per second,

   t = the time of flight in seconds, and

   h = the height above level ground from which the object is thrown.

Suppose a ball is thrown upward with a velocity of 5 meters per second at a height of 30 meters above the ground.

2. How long after it is thrown will the ball hit the ground?   **Ans:** 3 sec

3. What is the maximum height of the ball?   **Ans:** 50 m

Putting Your Skills to Work: Use Math to Save Money

ADJUST THE THERMOSTAT

Some Helpful Information

According to the U.S. Department of Energy, 45% of a typical home utility bill is for heating and/or cooling. We may not have any control over how fuel efficient our home was constructed, but we can control the setting on the thermostat. Consumer Reports suggests that every 1° change in our thermostat setting has a 2% impact on our utility bill. Consider the story of Maria and Josef.

A Specific Example

Maria and Josef use electricity to heat and cool their home, and their electricity bill averages $205 per month. They generally set their thermostat at 72° in the colder months and 78° in the warmer months.

Quick Calculations and Facts

1. If Maria and Josef set their thermostat at 68° in the colder months, how much per month could they expect to save?
   $16.40

2. If they adjust their thermostat to 84° in the warmer months, how much could they expect to save per month?
   $24.60

3. On the average, if Maria and Josef adjusted their thermostat setting 5° (either upward in the warmer months or downward in the colder

months), how much could they expect to save on their utilities per year?
About $246

Making Personal Applications to Your Own Life

4. Where do you have your thermostat set?
   Answers may vary

5. How much could you save on your utility bill by adjusting the settings either up or down?
   Answers may vary

HOME HEATING OIL PRICES

In parts of the United States where the climate turns cold during the winter, many homes are heated by oil. Mark lives just outside of Boston in a house heated by oil. During the winter of 2007, Mark paid $2.95 per gallon for home heating oil. The price of home heating oil has since increased to $4.45 per gallon. Looking ahead to the winter of 2008, Mark wants to budget enough money for his oil bills.

6. The oil company will only deliver 100 gallons of heating oil or more. Determine the cost of 100 gallons of home heating oil in December 2007 and November 2008. What is the difference in cost?
   $150

7. Mark knows he will need a 100-gallon oil delivery every month during the winter (November and December 2008, and January, February, and March 2009). What can Mark expect to pay in home heating oil costs for the five months of winter from November 2008 through March 2009?
   $2225

8. Mark typically keeps his thermostat set at 72° during the winter months. How much money would Mark save on his heating oil costs if he lowers his thermostat to 68° for the entire winter?
   4° lower = 8% savings, $2225 · 0.08 = $178

9. How much money would Mark save if he sets his thermostat to 64° during the winter?
   8° lower = 16% savings, $2225 · 0.16 = $356

10. How low would Mark need to set his thermostat for the winter in order to save one month's worth of heating costs ($445)?

10. $445 = 20% of $2225, 20% in savings = 10° lower, 72° − 10° = 62°

Topic	Procedure	Examples
Solving quadratic equations by factoring, p. 569.	1. Clear the equation of fractions, if necessary. 2. Write as $ax^2 + bx + c = 0$. 3. Factor. 4. Set each factor equal to 0. 5. Solve the resulting equations.	Solve. $x + \dfrac{1}{2} = \dfrac{3}{2x}$ Multiply each term by LCD $= 2x$. $$2x(x) + 2x\left(\dfrac{1}{2}\right) = 2x\left(\dfrac{3}{2x}\right)$$ $$2x^2 + x = 3$$ $$2x^2 + x - 3 = 0$$ $$(2x + 3)(x - 1) = 0$$ $$2x + 3 = 0 \qquad x - 1 = 0$$ $$2x = -3 \qquad x = 1$$ $$x = -\dfrac{3}{2}$$
Solving quadratic equations: taking the square root of each side of the equation, p. 576.	Begin solving quadratic equations of the form $ax^2 - c = 0$ by solving for x^2. Then use the property that if $x^2 = a$, $x = \pm\sqrt{a}$. This amounts to taking the square root of each side of the equation.	Solve. $2x^2 + 1 = 99$ $$2x^2 = 99 - 1$$ $$2x^2 = 98$$ $$\dfrac{2x^2}{2} = \dfrac{98}{2}$$ $$x^2 = 49$$ $$x = \pm\sqrt{49} = \pm 7$$
Solving quadratic equations: completing the square, p. 577.	1. Put the equation in the form $ax^2 + bx = c$. If $a \neq 1$, divide each term by a. 2. Square $\dfrac{b}{2}$ and add the result to both sides of the equation. 3. Factor the left side (a perfect square trinomial). 4. Use the square root property. 5. Solve the equations. 6. Check the solutions in the original equation.	Solve. $5x^2 - 3x - 7 = 0$ $$5x^2 - 3x = 7$$ $$x^2 - \dfrac{3}{5}x = \dfrac{7}{5}$$ $$x^2 - \dfrac{3}{5}x + \dfrac{9}{100} = \dfrac{7}{5} + \dfrac{9}{100}$$ $$\left(x - \dfrac{3}{10}\right)^2 = \dfrac{149}{100}$$ $$x - \dfrac{3}{10} = \pm\dfrac{\sqrt{149}}{10}$$ $$x = \dfrac{3}{10} \pm \dfrac{\sqrt{149}}{10}$$ $$x = \dfrac{3 \pm \sqrt{149}}{10}$$
Solving quadratic equations: quadratic formula, p. 582.	1. Put the equation in standard form: $ax^2 + bx + c = 0$. 2. Carefully determine the values of a, b, and c. 3. Substitute these into the formula $$x = \dfrac{-b \pm \sqrt{b^2 - 4ac}}{2a}.$$ 4. Simplify.	Solve. $3x^2 - 4x - 8 = 0$ $$a = 3, \ b = -4, \ c = -8$$ $$x = \dfrac{-(-4) \pm \sqrt{(-4)^2 - 4(3)(-8)}}{2(3)}$$ $$= \dfrac{4 \pm \sqrt{16 + 96}}{6}$$ $$= \dfrac{4 \pm \sqrt{112}}{6} = \dfrac{4 \pm \sqrt{16 \cdot 7}}{6}$$ $$= \dfrac{4 \pm 4\sqrt{7}}{6} = \dfrac{\cancel{2}\left(2 \pm 2\sqrt{7}\right)}{\cancel{2} \cdot 3} = \dfrac{2 \pm 2\sqrt{7}}{3}$$

Topic	Procedure	Examples
The discriminant, **p. 585.**	The *discriminant* in the quadratic formula is $b^2 - 4ac$. If it is negative, the quadratic equation has no real solutions (since it is not possible to take the square root of a negative number and get a real number).	What kinds of solutions does this equation have? $$3x^2 + 2x + 1 = 0$$ $$b^2 - 4ac = 2^2 - 4 \cdot 3 \cdot 1$$ $$= 4 - 12 = -8$$ This equation has no real solutions.
Properties of **parabolas, p. 591.**	**1.** The graph of $y = ax^2 + bx + c$ is a parabola. It opens up if $a > 0$ and down if $a < 0$. **2.** The vertex of a parabola is its lowest (if $a > 0$) or highest (if $a < 0$) point. Its x-coordinate is $x = \dfrac{-b}{2a}$.	The vertex of $y = 3x^2 + 4x - 11$ has x-coordinate $$x = \frac{-b}{2a} = \frac{-4}{6} = -\frac{2}{3}.$$ The parabola opens upward.
Graphing parabolas, **p. 592.**	Graph the vertex of a parabola and several other points to get a good idea of its graph. Usually, we find the x-intercepts if any exist. The following procedure is helpful. **1.** Determine if $a < 0$ or $a > 0$. **2.** Find the vertex. **3.** Find the x-intercepts. **4.** Plot one or two extra points if necessary. **5.** Draw the graph.	Graph. $y = x^2 - 2x - 8$ $a = 1, b = -2, c = -8; a > 0$, so the parabola opens upward. The x-coordinate of the vertex is $$x = \frac{-b}{2a} = \frac{-(-2)}{2} = 1.$$ If $x = 1, y = (1)^2 - 2(1) - 8 = -9$. The vertex is at $(1, -9)$. To find the x-intercepts, let $y = 0$. $$0 = x^2 - 2x - 8$$ $$0 = (x - 4)(x + 2)$$ $$x = 4, \quad x = -2$$ The x-intercepts are $(4, 0)$ and $(-2, 0)$.
Formulas and applied **problems, p. 599.**	Some word problems require the use of a quadratic equation. You may want to follow this four-step procedure for problem solving. **1.** Understand the problem. **2.** Write an equation. **3.** Solve and state the answer. **4.** Check.	The area of a rectangle is 48 square inches. The length is three times the width. Find the dimensions of the rectangle. **1.** Draw a picture. **2.** $(3w)w = 48$ **3.** $3w^2 = 48$ $\quad w^2 = 16$ $\quad\quad w = 4$ inches, $l = 12$ inches **4.** $(4)(12) = 48$

Chapter 10 Review Problems

Section 10.1 *Write in standard form. Determine the values of a, b, and c. Do not solve.*

1. $6x^2 = 5x - 8$
$6x^2 - 5x + 8 = 0$
$a = 6,\ b = -5,\ c = 8$

2. $9x^2 + 3x = -5x^2 + 16$
$14x^2 + 3x - 16 = 0$
$a = 14,\ b = 3,\ c = -16$

3. $(x + 3)(2x - 3) = x^2 + 5$
$x^2 + 3x - 14 = 0$
$a = 1, b = 3, c = -14$

4. $3x(4x + 1) = -x(5 - x)$
$12x^2 + 3x = -5x + x^2$
$11x^2 + 8x = 0$
$a = 11, b = 8, c = 0$

5. $\dfrac{3}{x^2} - \dfrac{6}{x} - 2 = 0$
$3 - 6x - 2x^2 = 0$
$-2x^2 - 6x + 3 = 0$
$a = -2, b = -6, c = 3$

6. $\dfrac{x}{x + 2} - \dfrac{3}{x - 2} = 5$
$4x^2 + 5x - 14 = 0$
$a = 4,\ b = 5,\ c = -14$

Place in standard form. Solve by factoring.

7. $x^2 + 26x + 25 = 0$
$(x + 25)(x + 1) = 0$
$x = -25,\ x = -1$

8. $x^2 + 16x + 64 = 0$
$(x + 8)(x + 8) = 0$
$x = -8$

9. $x^2 + 6x - 55 = 0$
$(x - 5)(x + 11) = 0$
$x = 5,\ x = -11$

10. $4x^2 - 16x + 15 = 0$
$(2x - 3)(2x - 5) = 0$
$x = \dfrac{3}{2}, x = \dfrac{5}{2}$

11. $9x^2 - 24x + 16 = 0$
$(3x - 4)(3x - 4) = 0$
$x = \dfrac{4}{3}$

12. $15x^2 = 26x - 8$
$15x^2 - 26x + 8 = 0$
$(5x - 2)(3x - 4) = 0$
$x = \dfrac{2}{5}, x = \dfrac{4}{3}$

13. $6x^2 = 13x - 5$
$6x^2 - 13x + 5 = 0$
$(3x - 5)(2x - 1) = 0$
$x = \dfrac{5}{3}, x = \dfrac{1}{2}$

14. $4x^2 = -5x + 9$
$4x^2 + 5x - 9 = 0$
$(x - 1)(4x + 9) = 0$
$x = 1, x = -\dfrac{9}{4}$

15. $x^2 + \dfrac{1}{6}x - 2 = 0$
$6x^2 + 1x - 12 = 0$
$(2x + 3)(3x - 4) = 0$
$x = -\dfrac{3}{2}, x = \dfrac{4}{3}$

16. $\dfrac{1}{2}x^2 = \dfrac{3}{4}x - \dfrac{1}{4}$
$2x^2 - 3x + 1 = 0$
$(2x - 1)(x - 1) = 0$
$x = \dfrac{1}{2}, x = 1$

17. $x^2 + \dfrac{2}{15}x = \dfrac{1}{15}$
$15x^2 + 2x - 1 = 0$
$(5x - 1)(3x + 1) = 0$
$x = \dfrac{1}{5}, x = -\dfrac{1}{3}$

18. $1 + \dfrac{13}{12x} - \dfrac{1}{3x^2} = 0$
$12x^2 + 13x - 4 = 0$
$(4x - 1)(3x + 4) = 0$
$x = \dfrac{1}{4}, x = -\dfrac{4}{3}$

19. $x^2 + 9x + 5 = x + 5$
$x^2 + 8x = 0$
$x(x + 8) = 0$
$x = 0, x = -8$

20. $x^2 - 2x + 7 = 7 + x$
$x^2 - 3x = 0$
$x(x - 3) = 0$
$x = 0, x = 3$

21. $1 + \dfrac{2}{3x + 4} = \dfrac{3}{3x + 2}$
$9x^2 + 15x = 0$
$3x(3x + 5) = 0$
$x = 0,\quad x = -\dfrac{5}{3}$

22. $2 - \dfrac{5}{x + 1} = \dfrac{3}{x - 1}$
$2x^2 - 8x = 0$
$2x(x - 4) = 0$
$x = 0,\ x = 4$

23. $5 + \dfrac{24}{2 - x} = \dfrac{24}{2 + x}$
$5x^2 - 48x - 20 = 0$
$(5x + 2)(x - 10) = 0$
$x = -\dfrac{2}{5}, x = 10$

24. $\dfrac{4}{9} = \dfrac{5x}{3} - x^2$
$9x^2 - 15x + 4 = 0$
$(3x - 1)(3x - 4) = 0$
$x = \dfrac{1}{3}, x = \dfrac{4}{3}$

Section 10.2 *Solve by taking the square root of each side.*

25. $x^2 - 8 = 41$
$x^2 = 49$
$x = \pm 7$

26. $x^2 + 11 = 92$
$x^2 = 81$
$x = \pm 9$

27. $x^2 - 5 = 17$
$x^2 = 22$
$x = \pm\sqrt{22}$

28. $x^2 + 11 = 50$
$x^2 = 39$
$x = \pm\sqrt{39}$

29. $2x^2 - 1 = 15$
$x^2 = 8$
$x = \pm 2\sqrt{2}$

30. $3x^2 + 4 = 154$
$x^2 = 50$
$x = \pm 5\sqrt{2}$

31. $3x^2 + 6 = 60$
$x^2 = 18$
$x = \pm 3\sqrt{2}$

32. $2x^2 - 5 = 43$
$x^2 = 24$
$x = \pm 2\sqrt{6}$

33. $(x - 4)^2 = 7$
$x - 4 = \pm\sqrt{7}$
$x = 4 \pm \sqrt{7}$

34. $(x - 2)^2 = 3$
$x - 2 = \pm\sqrt{3}$
$x = 2 \pm \sqrt{3}$

35. $(3x + 2)^2 = 28$
$3x + 2 = \pm 2\sqrt{7}$
$x = \dfrac{-2 \pm 2\sqrt{7}}{3}$

36. $(2x - 1)^2 = 32$
$2x - 1 = \pm 4\sqrt{2}$
$x = \dfrac{1 \pm 4\sqrt{2}}{2}$

Solve by completing the square.

37. $x^2 + 8x + 7 = 0$
$x^2 + 8x + 16 = 9$
$(x + 4)^2 = 9$
$x = -1, x = -7$

38. $x^2 + 14x + 33 = 0$
$x^2 + 14x + 49 = 16$
$(x + 7)^2 = 16$
$x = -3, x = -11$

39. $2x^2 - 8x - 90 = 0$
$x^2 - 4x + 4 = 49$
$(x - 2)^2 = 49$
$x = 9, x = -5$

40. $-5x^2 + 30x - 35 = 0$
$x^2 - 6x + 9 = -7 + 9$
$(x - 3)^2 = 2$
$x = 3 \pm \sqrt{2}$

41. $3x^2 + 6x - 6 = 0$
$x^2 + 2x + 1 = 3$
$(x + 1)^2 = 3$
$x = -1 \pm \sqrt{3}$

42. $2x^2 + 10x - 3 = 0$
$x^2 + 5x + \dfrac{25}{4} = \dfrac{31}{4}$
$\left(x + \dfrac{5}{2}\right)^2 = \dfrac{31}{4}$
$x = \dfrac{-5 \pm \sqrt{31}}{2}$

Section 10.3 *Solve using the quadratic formula.*

43. $x^2 + 4x - 6 = 0$
$x = \dfrac{-4 \pm \sqrt{16 + 24}}{2}$
$x = \dfrac{-4 \pm \sqrt{40}}{2}$
$x = -2 \pm \sqrt{10}$

44. $x^2 + 4x - 8 = 0$
$x = \dfrac{-4 \pm \sqrt{16 + 32}}{2}$
$x = -2 \pm 2\sqrt{3}$

45. $2x^2 - 7x + 4 = 0$
$x = \dfrac{7 \pm \sqrt{49 - 32}}{4}$
$x = \dfrac{7 \pm \sqrt{17}}{4}$

46. $2x^2 + 5x - 6 = 0$
$x = \dfrac{-5 \pm \sqrt{25 + 48}}{4}$
$x = \dfrac{-5 \pm \sqrt{73}}{4}$

47. $3x^2 - 8x - 4 = 0$
$x = \dfrac{8 \pm \sqrt{64 + 48}}{6}$
$x = \dfrac{4 \pm 2\sqrt{7}}{3}$

48. $4x^2 - 2x - 11 = 0$
$x = \dfrac{2 \pm \sqrt{4 + 176}}{8}$
$x = \dfrac{1 \pm 3\sqrt{5}}{4}$

49. $5x^2 - x = 2$
$x = \dfrac{1 \pm \sqrt{1 + 40}}{10}$
$x = \dfrac{1 \pm \sqrt{41}}{10}$

50. $3x^2 + 3x = 1$
$x = \dfrac{-3 \pm \sqrt{9 + 12}}{6}$
$x = \dfrac{-3 \pm \sqrt{21}}{6}$

Mixed Practice *Solve by any method. If there is no real number solution, say so.*

51. $2x^2 - 9x + 10 = 0$

$x = \dfrac{5}{2}, x = 2$

52. $4x^2 - 4x - 3 = 0$

$x = \dfrac{3}{2}, x = -\dfrac{1}{2}$

53. $25x^2 + 10x + 1 = 0$

$x = -\dfrac{1}{5}$

54. $2x^2 - 11x + 12 = 0$

$x = 4, x = \dfrac{3}{2}$

55. $3x^2 - 6x + 2 = 0$

$x = \dfrac{3 \pm \sqrt{3}}{3}$

56. $5x^2 - 7x = 8$

$x = \dfrac{7 \pm \sqrt{209}}{10}$

57. $4x^2 + 4x = x^2 + 5$

$x = \dfrac{-2 \pm \sqrt{19}}{3}$

58. $5x^2 + 7x + 1 = 0$

$x = \dfrac{-7 \pm \sqrt{29}}{10}$

59. $x^2 = 9x + 3$

$x = \dfrac{9 \pm \sqrt{93}}{2}$

60. $2x^2 = 9x + 5$

$x = -\dfrac{1}{2}, x = 5$

61. $-20x^2 - 15x = -8x^2 - 18$

$x = \dfrac{3}{4}, x = -2$

62. $2x^2 - 1 = 35$

$x = \pm 3\sqrt{2}$

63. $\dfrac{(y - 2)^2}{20} + 3 + y = 0$

$y = -8$ only

64. $\dfrac{(y + 2)^2}{5} + 2y = -9$

$y = -7$ only

65. $3x^2 + 1 = 6 - 8x$

$x = \dfrac{-4 \pm \sqrt{31}}{3}$

66. $4x^2 + 11x = 5x + 1$

$x = \dfrac{-3 \pm \sqrt{13}}{4}$

67. $8y(y + 1) = 7y + 9$

$y = -\dfrac{9}{8}, y = 1$

68. $\dfrac{3y - 2}{4} = \dfrac{y^2 - 2}{y}$

$y = -4, y = 2$

69. $\dfrac{y^2 + 5}{2y} = \dfrac{2y - 1}{3}$

$y = -3, y = 5$

70. $\dfrac{5x^2}{2} = x - \dfrac{7x^2}{2}$

$x = 0, x = \dfrac{1}{6}$

Section 10.4 *Graph each quadratic equation. Label the vertex.*

71. $y = 2x^2$

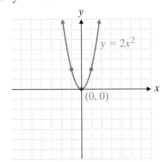

72. $y = x^2 + 4$

73. $y = x^2 - 3$

74. $y = -\dfrac{1}{2}x^2$

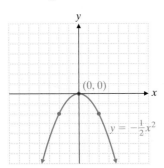

75. $y = x^2 - 3x - 4$

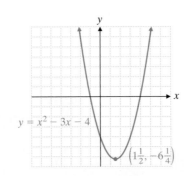

76. $y = \dfrac{1}{2}x^2 - 2$

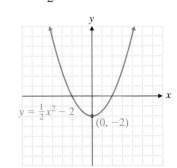

77. $y = -2x^2 + 12x - 17$

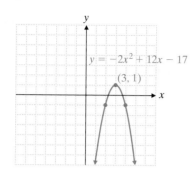

78. $y = -3x^2 - 2x + 4$

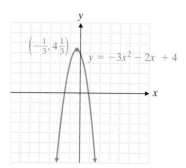

Section 10.5 *Solve.*

▲ **79.** *Dog Pen Fencing* Ken is building a dog pen against his house. He has 15 feet of fencing and wants to pen in an area of 28 square feet. What are the possible dimensions of the pen?

7 ft by 4 ft or 8 ft by 3.5 ft

▲ **80.** *Geometry* The hypotenuse of a right triangle is 20 centimeters. One leg of the right triangle is 4 centimeters shorter than the other. What is the length of each leg of the right triangle?

16 cm, 12 cm

▲ **81.** *TV Screen Dimensions* A television designer believes that the dimensions (length and width) of a TV screen are as important to sales as the size (measured by the diagonal) of the TV. If the diagonal of a newly designed TV is 30 inches and the length is 6 inches more than the width, what are the dimensions of the TV?

$(x)^2 + (x + 6)^2 = 30^2$ Width $= 18$ in.
$x^2 + 6x - 432 = 0$ Length $= 24$ in.
$(x + 24)(x - 18) = 0$
$x = 18$

82. *Space Probe* Suppose an advanced alien civilization sends a space ship to orbit around our planet. They send a probe to Earth to do research. The equation for the height of the probe is $h = -225t^2 + 202,500$, where h is the distance in feet from the surface of Earth t seconds after the probe leaves the mother ship. How long will it take the probe to reach Earth from the mother ship?

$0 = -225t^2 + 202,500$
$0 = -225(t^2 - 900)$
$0 = -225(t + 30)(t - 30)$
$t = 30$ sec

83. *Golf Club* Last year the golf club on campus raised $720 through dues assigned equally to each member. This year there were four fewer members. As a result, the dues for each member went up by $6. How many members were in the golf club last year?

24 members

84. *Automobile Trip* Alice drove 90 miles to visit her cousin. Her average speed on the trip home was 15 mph faster than her average speed on the trip there. Her total travel time for both trips was 3.5 hours. What was her average rate of speed on each trip?

Going, 45 mph; returning, 60 mph

Use the following information to solve exercises 85–88.

Lacorazza Pizza Chain The Lacorazza family opened a chain of pizza stores in New England in 1980. The number of pizza stores in operation during any year is given by the equation $y = -0.05x^2 + 2.5x + 60$, where x is the number of years since 1980.

85. According to the equation, how many pizza stores will still be in operation in 2030?

60 pizza stores

86. How many fewer pizza stores will still be in operation in 2040 than in 2030?

30 fewer pizza stores

87. During what two years will there be exactly 80 pizza stores in operation?

1990 and 2020

88. During what two years will there be exactly 90 pizza stores in operation?

2000 and 2010

Note to Instructor: The Chapter 10 Test file in the TestGen program provides algorithms specifically matched to these problems so you can easily replicate this test for additional practice or assessment purposes.

How Am I Doing? Chapter 10 Test

Remember to use your Chapter Test Prep Video CD to see the worked-out solutions to the test problems you want to review.

Solve by any desired method. If there is no real number solution, say so.

1. $5x^2 + 7x = 4$

2. $3x^2 + 13x = 10$

3. $2x^2 = 2x - 5$

4. $4x^2 - 19x = 4x - 15$

5. $12x^2 + 11x = 5$

6. $18x^2 + 32 = 48x$

7. $2x^2 - 11x + 3 = 5x + 3$

8. $5x^2 + 7 = 52$

9. $2x(x - 6) = 6 - x$

10. $x^2 - x = \dfrac{3}{4}$

Graph each quadratic equation. Locate the vertex.

11. $y = 3x^2 - 6x$

12. $y = -x^2 + 8x - 12$

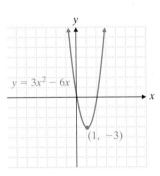

$y = 3x^2 - 6x$

$(1, -3)$

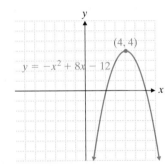

$(4, 4)$

$y = -x^2 + 8x - 12$

Solve.

▲**13.** The hypotenuse of a right triangle is 15 meters in length. One leg is 3 meters longer than the other leg. Find the length of each leg.

14. When an object is thrown upward, its height (S) in meters is given (approximately) by the quadratic equation

$$S = -5t^2 + vt + h,$$

where v = the initial upward velocity in meters per second,

t = the time of flight in seconds, and

h = the height above level ground from which the object is thrown.

Suppose that a ball is thrown upward with a velocity of 33 meters/second at a height of 14 meters above the ground. How long after it is thrown will it strike the ground?

Answers:

1. $x = \dfrac{-7 \pm \sqrt{129}}{10}$

2. $x = -5, \; x = \dfrac{2}{3}$

3. No real solution

4. $x = \dfrac{3}{4}, x = 5$

5. $x = -\dfrac{5}{4}, \; x = \dfrac{1}{3}$

6. $x = \dfrac{4}{3}$

7. $x = 0, x = 8$

8. $x = \pm 3$

9. $x = 6, \; x = -\dfrac{1}{2}$

10. $x = \dfrac{3}{2}, \; x = -\dfrac{1}{2}$

11. See graph

12. See graph

13. 9 m, 12 m

14. 7 sec

Practice Final Examination

The following questions cover the content of Chapters 0–10. Follow the directions for each question and simplify your answers.

1. Simplify.
$-2x + 3y\{7 - 2[x - (4x + y)]\}$

2. Evaluate if $x = -2$ and $y = 3$.
$2x^2 - 3xy - 4y$

3. Simplify. $(-3x^2y)(-6x^3y^4)$

4. Combine like terms.
$5x^2y - 6xy + 8xy - 3x^2y - 10xy$

5. Solve for x.
$\frac{1}{2}(x + 4) - \frac{2}{3}(x - 7) = 4x$

6. Solve for b. $p = 2a + 2b$

7. Solve for x and graph the resulting inequality on a number line.
$5x + 3 - (4x - 2) \le 6x - 8$

8. Multiply. $(2x + y)(x^2 - 3xy + 2y^2)$

9. Factor completely. $4x^2 - 18x - 10$

10. Factor completely. $3x^3 - 9x^2 - 30x$

11. Combine. $\dfrac{2}{x - 3} - \dfrac{3}{x^2 - x - 6} + \dfrac{4}{x + 2}$

12. Simplify. $\dfrac{\dfrac{3}{x} + \dfrac{5}{2x}}{1 + \dfrac{2}{x + 2}}$

13. Solve for x. $\dfrac{2}{x + 2} = \dfrac{4}{x - 2} + \dfrac{3x}{x^2 - 4}$

14. Find the slope of the line and then graph the line. $5x - 2y - 3 = 0$

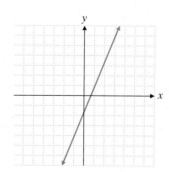

15. Find the equation of the line that has a slope of $-\dfrac{3}{4}$ and passes through the point $(-2, 5)$.

1. $-2x + 21y + 18xy + 6y^2$

2. 14

3. $18x^5y^5$

4. $2x^2y - 8xy$

5. $x = \dfrac{8}{5}$

6. $\dfrac{p - 2a}{2} = b$

7. $x \ge 2.6$; see graph

8. $2x^3 - 5x^2y + xy^2 + 2y^3$

9. $2(2x + 1)(x - 5)$

10. $3x(x - 5)(x + 2)$

11. $\dfrac{6x - 11}{(x + 2)(x - 3)}$

12. $\dfrac{11(x + 2)}{2x(x + 4)}$

13. $x = -\dfrac{12}{5}$

14.   Slope $= \dfrac{5}{2}$; see graph

15. $3x + 4y = 14$
or
$y = -\dfrac{3}{4}x + \dfrac{7}{2}$

16.   38.13 in.²

17. $x = -5, y = 2$

18. $a = -2, b = -3$

19. $x\sqrt{5x}$

20. $6\sqrt{3} - 12 + 12\sqrt{2}$

21. $-\dfrac{\sqrt{15} + \sqrt{35} + \sqrt{21} + 7}{2}$

22. $x = \dfrac{2}{3},\ x = -\dfrac{1}{4}$

23. $y = \dfrac{3 \pm \sqrt{7}}{2}$

24. $x = \pm 2$

25. $\sqrt{39}$

26.   5

27.   Width is 7 m, length is 12 m

28. $3000 at 10%, $4000 at 14%

29.   200 reserved-seat tickets. 160 general admission tickets

30.   Base is 8 m, altitude is 17 m

▲ **16.** Two quarter-circles are attached to a rectangle. The dimensions are indicated in the figure. Find the area of the region. Use 3.14 as an approximation for π.

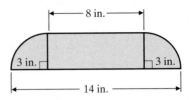

17. Solve for x and y.

$$2x + 7y = 4$$
$$-3x - 5y = 5$$

18. Solve for a and b.

$$a - \frac{3}{4}b = \frac{1}{4}$$
$$\frac{3}{2}a + \frac{1}{2}b = -\frac{9}{2}$$

19. Simplify and combine.
$$\sqrt{45x^3} + 2x\sqrt{20x} - 6\sqrt{5x^3}$$

20. Multiply and simplify.
$$\sqrt{6}\left(3\sqrt{2} - 2\sqrt{6} + 4\sqrt{3}\right)$$

21. Simplify by rationalizing the denominator. $\dfrac{\sqrt{3} + \sqrt{7}}{\sqrt{5} - \sqrt{7}}$

22. Solve for x. $12x^2 - 5x - 2 = 0$

23. Solve for y. $2y^2 = 6y - 1$ **24.** Solve for x. $4x^2 + 3 = 19$

▲ **25.** In the right triangle shown, sides b and c are given. Find the length of side a.

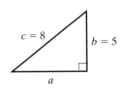

26. A number is tripled and then increased by 6. The result is 21. What is the original number?

▲ **27.** A rectangular region has a perimeter of 38 meters. The length of the rectangle is 2 meters shorter than double the width. What are the dimensions of the rectangle?

28. Margaret invested $7000 for one year. She placed part of the money in a tax-free bond earning 10% simple interest. She placed the rest in a money market account earning 14% simple interest. At the end of one year she had earned $860 in interest. How much did she invest at each interest rate?

29. A benefit concert was held on campus to raise scholarship money. A total of 360 tickets were sold. Admission prices were $5 for reserved seats and $3 for general admission. Total receipts were $1480. How many reserved-seat tickets were sold? How many general admission tickets were sold?

▲ **30.** The area of a triangle is 68 square meters. The altitude of the triangle is one meter longer than double the base of the triangle. Find the altitude and base of the triangle.

Appendix A Table of Square Roots

x	$\sqrt{x}$	x	$\sqrt{x}$	x	$\sqrt{x}$	x	$\sqrt{x}$	x	$\sqrt{x}$
1	1.000	41	6.403	81	9.000	121	11.000	161	12.689
2	1.414	42	6.481	82	9.055	122	11.045	162	12.728
3	1.732	43	6.557	83	9.110	123	11.091	163	12.767
4	2.000	44	6.633	84	9.165	124	11.136	164	12.806
5	2.236	45	6.708	85	9.220	125	11.180	165	12.845
6	2.449	46	6.782	86	9.274	126	11.225	166	12.884
7	2.646	47	6.856	87	9.327	127	11.269	167	12.923
8	2.828	48	6.928	88	9.381	128	11.314	168	12.961
9	3.000	49	7.000	89	9.434	129	11.358	169	13.000
10	3.162	50	7.071	90	9.487	130	11.402	170	13.038
11	3.317	51	7.141	91	9.539	131	11.446	171	13.077
12	3.464	52	7.211	92	9.592	132	11.489	172	13.115
13	3.606	53	7.280	93	9.644	133	11.533	173	13.153
14	3.742	54	7.348	94	9.695	134	11.576	174	13.191
15	3.873	55	7.416	95	9.747	135	11.619	175	13.229
16	4.000	56	7.483	96	9.798	136	11.662	176	13.266
17	4.123	57	7.550	97	9.849	137	11.705	177	13.304
18	4.243	58	7.616	98	9.899	138	11.747	178	13.342
19	4.359	59	7.681	99	9.950	139	11.790	179	13.379
20	4.472	60	7.746	100	10.000	140	11.832	180	13.416
21	4.583	61	7.810	101	10.050	141	11.874	181	13.454
22	4.690	62	7.874	102	10.100	142	11.916	182	13.491
23	4.796	63	7.937	103	10.149	143	11.958	183	13.528
24	4.899	64	8.000	104	10.198	144	12.000	184	13.565
25	5.000	65	8.062	105	10.247	145	12.042	185	13.601
26	5.099	66	8.124	106	10.296	146	12.083	186	13.638
27	5.196	67	8.185	107	10.344	147	12.124	187	13.675
28	5.292	68	8.246	108	10.392	148	12.166	188	13.711
29	5.385	69	8.307	109	10.440	149	12.207	189	13.748
30	5.477	70	8.367	110	10.488	150	12.247	190	13.784
31	5.568	71	8.426	111	10.536	151	12.288	191	13.820
32	5.657	72	8.485	112	10.583	152	12.329	192	13.856
33	5.745	73	8.544	113	10.630	153	12.369	193	13.892
34	5.831	74	8.602	114	10.677	154	12.410	194	13.928
35	5.916	75	8.660	115	10.724	155	12.450	195	13.964
36	6.000	76	8.718	116	10.770	156	12.490	196	14.000
37	6.083	77	8.775	117	10.817	157	12.530	197	14.036
38	6.164	78	8.832	118	10.863	158	12.570	198	14.071
39	6.245	79	8.888	119	10.909	159	12.610	199	14.107
40	6.325	80	8.944	120	10.954	160	12.649	200	14.142

Unless the value of $\sqrt{x}$ ends in 000, all values are rounded to the nearest thousandth.

Appendix B Metric Measurement and Conversion of Units

Student Learning Objectives

After studying this section, you will be able to:

1. Convert from one metric unit of measurement to another.

2. Convert between metric and U.S. units of measure.

3. Convert from one type of unit of measure to another.

1 Converting from One Metric Unit of Measurement to Another

Metric measurements are becoming more common in the United States. The metric system is used in many parts of the world and in the sciences. The basic unit of length in the metric system is the meter. For smaller measurements, centimeters are commonly used. There are 100 centimeters in 1 meter. For larger measurements, kilometers are commonly used. There are 1000 meters in 1 kilometer. The following tables give the metric units of measurement for length.

METRIC LENGTH

*1 kilometer (km)	=	1000 meters
1 hectometer (hm)	=	100 meters
1 dekameter (dam)	=	10 meters
*1 meter (m)	=	1 meter
1 decimeter (dm)	=	0.1 meter
*1 centimeter (cm)	=	0.01 meter
*1 millimeter (mm)	=	0.001 meter

The four most common units of metric measurement are indicated by an asterisk.

METRIC WEIGHT

1 kilogram (kg)	=	1000 grams
1 gram (g)	=	1 gram
1 milligram (mg)	=	0.001 gram

METRIC VOLUME

1 kiloliter (kL)	=	1000 liters
1 liter (L)	=	1 liter
1 milliliter (mL)	=	0.001 liter

One way to convert from one metric unit to another is to multiply by 1. We know mathematically that multiplying by 1 will yield an equivalent expression since $(a)(1) = a$. In our calculations when solving word problems, we treat dimension symbols much as we treat variables in algebra.

EXAMPLE 1 Fred drove a distance of 5.4 kilometers. How many meters is that?

Solution
$$5.4 \text{ km} \cdot 1 = 5.4 \text{ km} \cdot \frac{1000 \text{ m}}{1 \text{ km}} = \frac{(5.4)(1000)}{1} \cdot \frac{\text{km}}{\text{km}} \cdot \text{m}$$
$$= 5400 \text{ m}$$

The distance is 5400 meters.

Practice Problem 1   June's room is 3.4 meters wide. How many centimeters is that?

NOTE TO STUDENT: Fully worked-out solutions to all of the Practice Problems can be found at the back of the text starting at page SP-1

EXAMPLE 2   A chemist measured 67 milliliters of a solution. How many liters is that?

Solution
$$67 \text{ mL} \cdot \frac{0.001 \text{ L}}{1 \text{ mL}} = 0.067 \text{ L}$$

The chemist measured 0.067 liter of the solution.

Practice Problem 2   The container has 125 liters of water. How many kiloliters is that?

Converting between Metric and U.S. Units of Measure

The most common relationships between the U.S. and metric systems are listed in the following table. Most of these values are approximate.

METRIC CONVERSION RATIOS

LENGTH:

    1 inch = 2.54 centimeters
    39.37 inches = 1 meter
    1 mile = 1.61 kilometers
    0.62 mile = 1 kilometer

WEIGHT:

    1 pound = 454 grams
    2.20 pounds = 1 kilogram
    1 ounce = 28.35 grams
    0.0353 ounce = 1 gram

LIQUID CAPACITY:

    1 quart = 946 milliliters
    1.06 quarts = 1 liter
    1 gallon = 3.785 liters

EXAMPLE 3   A box weighs 190 grams. How many ounces is that? Round to the nearest hundredth.

Solution $190 \text{ g} \cdot \dfrac{0.0353 \text{ oz}}{1 \text{ g}} = 6.707 \text{ oz}$

$$\approx 6.71 \text{ oz} \quad \text{(rounded to the nearest hundredth)}$$

The box weighs approximately 6.71 ounces.

Practice Problem 3   A bag of groceries weighs 5.72 pounds. How many kilograms is that?

EXAMPLE 4 Juanita drives 23 kilometers to work each day. How many miles is the trip? Round to the nearest mile.

Solution $23 \text{ km} \cdot \dfrac{0.62 \text{ mi}}{1 \text{ km}} = 14.26 \text{ mi} \approx 14 \text{ mi}$ (rounded to the nearest mile)

Juanita drives approximately 14 miles to work each day.

Practice Problem 4 Carlos installed an electrical connection that is 8.00 centimeters long. How many inches long is the connection? Round to the nearest hundredth.

NOTE TO STUDENT: Fully worked-out solutions to all of the Practice Problems can be found at the back of the text starting at page SP-1

EXAMPLE 5 Anita purchased 42.0 gallons of gasoline for her car last month. How many liters did she purchase? Round to the nearest liter.

Solution $42.0 \text{ gal} \cdot \dfrac{3.785 \text{ L}}{1 \text{ gal}} = 158.97 \text{ L} \approx 159 \text{ L}$ (rounded to the nearest liter)

Anita purchased approximately 159 liters of gasoline last month.

Practice Problem 5 Warren purchased a 3-liter bottle of Coca-Cola. How many quarts of Coca-Cola is that? Round to the nearest hundredth of a quart.

3 Converting from One Type of Unit of Measure to Another

Sometimes you will need to convert from one type of unit of measure to another. For example, you may need to convert days to minutes or miles per hour to feet per second. Recall the U.S. units of measure.

Length	Time
12 inches = 1 foot	60 seconds = 1 minute
3 feet = 1 yard	60 minutes = 1 hour
5280 feet = 1 mile	24 hours = 1 day
1760 yards = 1 mile	7 days = 1 week
Weight	**Volume**
16 ounces = 1 pound	2 cups = 1 pint
2000 pounds = 1 ton	2 pints = 1 quart
	4 quarts = 1 gallon

EXAMPLE 6 A car was traveling at 50.0 miles per hour. How many feet per second was the car traveling? Round to the nearest tenth of a foot per second.

Solution $\dfrac{50 \text{ mi}}{\text{hr}} \cdot \dfrac{5280 \text{ ft}}{1 \text{ mi}} \cdot \dfrac{1 \text{ hr}}{60 \text{ min}} \cdot \dfrac{1 \text{ min}}{60 \text{ sec}} = \dfrac{(50)(5280) \text{ ft}}{(60)(60) \text{ sec}} = \dfrac{73.333\ldots \text{ ft}}{\text{sec}}$

The car was traveling at 73.3 feet per second, rounded to the nearest tenth of a foot per second.

Practice Problem 6 A speeding car was traveling at 70.0 miles per hour. How many feet per second was the car traveling? Round to the nearest tenth of a foot per second.

1. How many meters are in 34 km? 34,000 m

2. How many meters are in 128 km? 128,000 m

3. How many centimeters are in 57 m? 5700 cm

4. How many centimeters are in 46 m? 4600 cm

5. How many millimeters are in 25 cm? 250 mm

6. How many millimeters are in 63 cm? 630 mm

7. How many meters are in 563 mm? 0.563 m

8. How many meters are in 831 mm? 0.831 m

9. How many milligrams are in 29.4 g? 29,400 mg

10. How many milligrams are in 75.2 g? 75,200 mg

11. How many kilograms are in 98.4 g? 0.0984 kg

12. How many kilograms are in 62.7 g? 0.0627 kg

13. How many milliliters are in 7 L? 7000 mL

14. How many milliliters are in 12 L? 12,000 mL

15. How many kiloliters are in 4 mL? 0.000004 kL

16. How many kiloliters are in 3 mL? 0.000003 kL

Use the table of metric conversion ratios to find each of the following. Round all answers to the nearest tenth.

17. How many inches are in 4.2 cm?

$$4.2 \text{ cm} \times \frac{1 \text{ in.}}{2.54 \text{ cm}} \approx 1.7 \text{ in.}$$

18. How many inches are in 3.8 cm?

$$3.8 \text{ cm} \times \frac{1 \text{ in.}}{2.54 \text{ cm}} \approx 1.5 \text{ in.}$$

19. How many kilometers are in 14 mi?

$$14 \text{ mi} \times \frac{1.61 \text{ km}}{1 \text{ mi}} \approx 22.5 \text{ km}$$

20. How many kilometers are in 13 mi?

$$13 \text{ mi} \times \frac{1.61 \text{ km}}{1 \text{ mi}} \approx 20.9 \text{ km}$$

21. How many meters are in 110 in.?

$$110 \text{ in.} \times \frac{2.54 \text{ cm}}{1 \text{ in.}} = 279.4 \text{ cm} \approx 2.8 \text{ m}$$

22. How many meters are in 150 in.?

$$150 \text{ in.} \times \frac{2.54 \text{ cm}}{1 \text{ in.}} = 381 \text{ cm} \approx 3.8 \text{ m}$$

23. How many centimeters are in 7 in.?

$$7 \text{ in.} \times \frac{2.54 \text{ cm}}{1 \text{ in.}} \approx 17.8 \text{ cm}$$

24. How many centimeters are in 9 in.?

$$9 \text{ in.} \times \frac{2.54 \text{ cm}}{1 \text{ in.}} \approx 22.9 \text{ cm}$$

25. A box weighing 2.4 lb would weigh how many grams?

$$2.4 \text{ lb} \times \frac{454 \text{ g}}{1 \text{ lb}} = 1089.6 \text{ g}$$

26. A box weighing 1.6 lb would weigh how many grams?

$$1.6 \text{ lb} \times \frac{454 \text{ g}}{1 \text{ lb}} = 726.4 \text{ g}$$

27. A man weighs 78 kg. How many pounds does he weigh?

$$78 \text{ kg} \times \frac{2.2 \text{ lb}}{1 \text{ kg}} = 171.6 \text{ lb}$$

28. A woman weighs 52 kg. How many pounds does she weigh?

$$52 \text{ kg} \times \frac{2.2 \text{ lb}}{1 \text{ kg}} = 114.4 \text{ lb}$$

29. Ferrante purchased 3 qt of milk. How many liters is that?

$$3 \text{ qt} \times \frac{1 \text{ L}}{1.06 \text{ qt}} \approx 2.8 \text{ L}$$

30. Wong Tin purchased 1 gal of milk. How many liters is that?

$$1 \text{ gal} \times \frac{3.785 \text{ L}}{1 \text{ gal}} \approx 3.8 \text{ L}$$

Answer the following questions. Round all answers to the nearest hundredth.

31. How many inches are in 3050 miles?

$$3050 \text{ mi} \times \frac{5280 \text{ ft}}{1 \text{ mi}} \times \frac{12 \text{ in.}}{1 \text{ ft}} = 193,248,000 \text{ in.}$$

32. How many inches are in 4500 miles?

$$4500 \text{ mi} \times \frac{5280 \text{ ft}}{1 \text{ mi}} \times \frac{12 \text{ in.}}{1 \text{ ft}} = 285,120,000 \text{ in.}$$

33. A truck traveled at a speed of 40 feet per second. How many miles per hour is that?

$$\frac{40 \text{ ft}}{1 \text{ sec}} \times \frac{1 \text{ mi}}{5280 \text{ ft}} \times \frac{3600 \text{ sec}}{1 \text{ hr}} \approx 27.27 \text{ mph}$$

34. A car traveled at a speed of 55 feet per second. How many miles per hour is that?

$$\frac{55 \text{ ft}}{1 \text{ sec}} \times \frac{1 \text{ mi}}{5280 \text{ ft}} \times \frac{3600 \text{ sec}}{1 \text{ hr}} \approx 37.50 \text{ mph}$$

35. How many years are in 3,500,000 seconds? (Use the approximate value that 365 days = 1 year.)

$$3,500,000 \text{ sec} \times \frac{1 \text{ min}}{60 \text{ sec}} \times \frac{1 \text{ hr}}{60 \text{ min}} \times \frac{1 \text{ da}}{24 \text{ hr}} \times \frac{1 \text{ yr}}{365 \text{ da}} \approx 0.11 \text{ yr}$$

36. How many years are in 2,800,000 seconds? (Use the approximate value that 365 days = 1 year.)

$$2,800,000 \text{ sec} \times \frac{1 \text{ min}}{60 \text{ sec}} \times \frac{1 \text{ hr}}{60 \text{ min}} \times \frac{1 \text{ da}}{24 \text{ hr}} \times \frac{1 \text{ yr}}{365 \text{ da}} \approx 0.09 \text{ yr}$$

Appendix C Interpreting Data from Tables, Charts, and Graphs

1 Tables

A table is a device used to organize information into categories. Using it you can readily find details about each category.

EXAMPLE 1

Table of Nutritive Values of Certain Popular "Fast Foods"

Type of Sandwich	Calories	Protein (g)	Fat (g)	Cholesterol (g)	Sodium (mg)
Burger King Whopper	630	27	38	90	880
McDonald's Big Mac	500	25	26	100	890
Wendy's Bacon Cheeseburger	440	22	25	65	870
Burger King BK Broiler (chicken)	280	20	10	50	770
McDonald's McChicken	415	19	20	50	830
Wendy's Grilled Chicken	290	24	7	60	670

Source: U.S. Government Agencies and Food Manufacturers

(a) Which food item has the least amount of fat per serving?

(b) Which beef item has the least amount of calories per serving?

(c) How much more protein does a Burger King Whopper have than a McDonald's Big Mac?

Solution

(a) The least amount of fat, 7 grams, is in the Wendy's Grilled Chicken Sandwich.

(b) The least amount of calories, 440, for a beef sandwich is the Wendy's Bacon Cheeseburger.

(c) The Burger King Whopper has 2 grams of protein more than the McDonald's Big Mac.

Practice Problem 1

(a) Which sandwich has the lowest level of sodium?

(b) Which sandwich has the highest level of cholesterol?

NOTE TO STUDENT: Fully worked-out solutions to all of the Practice Problems can be found at the back of the text starting at page SP-1

Charts

A chart is a device used to organize information in which not every category is the same. Example 2 illustrates a chart containing different types of data.

EXAMPLE 2 The following chart shows how people in Topsfield indicated they spent their free time.

Survey of Use of Leisure Time

Category	Activity	Hours spent per week
Single men	Gym	6
	Outdoor sports	4
	Dating	7
	Watching pro sports	12
	Reading & TV	3
Single women	Gym	4
	Outdoor sports	2
	Dating	7
	Time with friends	10
	Reading & TV	9
Couples	Time with family	21
	Time as a couple	8
	Time with friends	4
	Reading & TV	9
Children	Watching TV	28
	Playing outside	8
	Reading	1

Use the chart to answer the following questions about people in Topsfield.

(a) What is the average amount of time a couple spends with family during the week?

(b) How much more time do children spend watching TV than playing outside?

(c) What activity do single men spend most of their time doing?

Solution

(a) The average amount of time a couple spends with family is 21 hours per week.

(b) Children spend 20 more hours per week watching TV than playing outside.

(c) Single men spend more time per week watching pro sports (12 hr) than on any other activity.

Practice Problem 2

(a) What two categories do single women spend the most time doing?

(b) What do couples spend most of their time doing?

(c) What is the most significant numerical difference in terms of the number of hours per week spent by single women versus single men?

3 Pictographs

A pictograph uses a visually appropriate symbol to represent an amount of items. A pictograph is used in Example 3.

EXAMPLE 3   Consider the following pictograph.

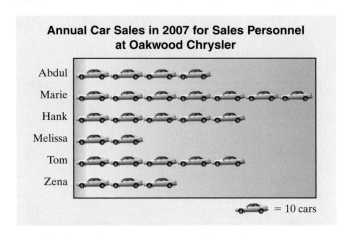

Annual Car Sales in 2007 for Sales Personnel at Oakwood Chrysler

= 10 cars

(a) How many cars did Melissa sell in 2007?

(b) Who sold the greatest number of cars?

(c) How many more cars did Tom sell than Zena?

Solution

(a) Melissa sold $2 \times 10 = 20$ cars.

(b) Marie sold the greatest number of cars.

(c) Tom sold $5 \times 10 = 50$ cars. Zena sold $3 \times 10 = 30$ cars. Now, $50 - 30 = 20$. Therefore, Tom sold 20 cars more than Zena.

NOTE TO STUDENT: Fully worked-out solutions to all of the Practice Problems can be found at the back of the text starting at page SP-1

Practice Problem 3

Approximate Number of Chain Pharmacy Stores in the United States in 2007

Source: Federal Food & Drug Administration   = 500 stores

(a) Approximately how many stores did Walgreens have?

(b) Approximately how many more stores did Rite Aid have than Medicine Shoppe?

(c) What was the combined number of Longs Drug stores and CVS stores?

Bar Graphs

A bar graph is helpful for making comparisons and noting changes or trends. A scale is provided so that the height of the bar graph indicates a specific number. A bar graph is displayed in Example 4. A bar graph may be represented horizontally or vertically. In either case the basic concepts of interpreting a bar graph are the same.

EXAMPLE 4 The approximate population of California by year is given in the following bar graph.

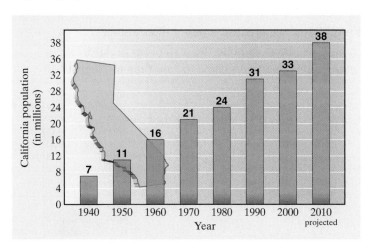

(a) What is the projected population of California in 2010?

(b) How much greater was the population of California in 1980 than in 1970?

Solution

(a) The projected population of California in 2010 is about 38 million.

(b) In 1980 it was 24 million. In 1970 it was 21 million. The population of California was approximately 3 million more people in 1980 than in 1970.

Practice Problem 4 The following bar graph depicts the number of accidents for U.S. air carriers on scheduled flight service.

Source: National Transportation Safety Board

(a) What two-year period had the greatest number of accidents?

(b) What was the increase in the number of accidents from the 1995–1996 period to the 1997–1998 period?

(c) What was the decrease in the number of accidents from the 1999–2000 period to the 2001–2002 period?

Line Graphs

A line graph is often used to display data when significant changes or trends are present. In a line graph, only a few points are actually plotted from measured values. The points are then connected by straight lines in order to show a trend. The intervening values between points may not exactly lie on the line. A line graph is displayed in Example 5.

EXAMPLE 5   The following line graph shows the number of customers per month coming to a restaurant in a tourist vacation community.

(a) What month had the greatest number of customers?

(b) How many customers came to the restaurant in April?

(c) How many more customers came in July than in August?

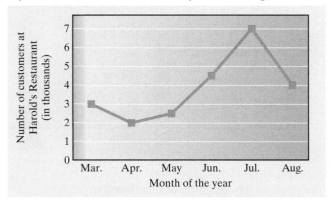

Solution

(a) More customers came during the month of July.

(b) Approximately 2000 customers came in April.

(c) In July there were 7000 customers, while in August there were 4000 customers. Thus, there were 3000 more customers in July than in August.

NOTE TO STUDENT: Fully worked-out solutions to all of the Practice Problems can be found at the back of the text starting at page SP-1

Practice Problem 5   The quality of the air is measured by the Air Quality Index (AQI). Air is tested daily in metropolitan areas and is reported as one of the following categories: good, moderate, unhealthy for sensitive groups, or unhealthy. The following line graph indicates the number of days that the air quality was considered unhealthy in the city of Baltimore during a 10-year period.

(a) What was the number of days in which the air was considered unhealthy in Baltimore in 2001?

(b) In what year did Baltimore have the most days in which the air was considered unhealthy?

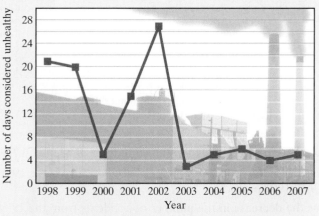

Source: U.S. Environmental Protection Agency

Pie Graphs and Circle Graphs

A pie graph or a circle graph indicates how a whole quantity is divided into parts. These graphs help you to visualize the size of the relative proportions of parts. Each piece of the pie or circle is called a sector. Example 6 uses a pie graph.

EXAMPLE 6 Together, the Great Lakes form the largest body of fresh water in the world. The total area of these five lakes is about 290,000 square miles, almost all of which is suitable for boating. The percentage of this total area taken up by each of the Great Lakes is shown in the pie graph.

(a) What percentage of the area is taken up by Lake Michigan?

(b) What lake takes up the largest percentage of area?

(c) How many square miles are taken up by Lake Huron and Lake Michigan together?

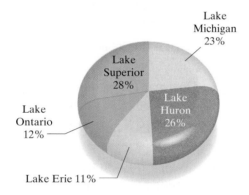

Solution

(a) Lake Michigan takes up 23% of the area.

(b) Lake Superior takes up the largest percentage.

(c) If we add 26 + 23, we get 49. Thus Lake Huron and Lake Michigan together take up 49% of the total area.

49% of 290,000 = (0.49)(290,000) = 142,100 square miles.

Practice Problem 6 Seattle receives on average about 37 inches of rain per year. However, the amount of rainfall per month varies significantly. The percent of rainfall that occurs during each quarter of the year is shown by the circle graph.

(a) What percent of the rain in Seattle falls between April and June?

(b) Forty percent of the rainfall occurs in what three-month period?

(c) How many inches of rain fall in Seattle from January to March?

In each case, study carefully the appropriate visual display, then answer the questions. Consider the following table in answering exercises 1–10.

Table of Facts of the Rocky Mountain States

State	Area in Square Miles	Date Admitted to the Union	Estimated 2006 Population	Number of Representatives in U.S. Congress	Popular Name
Colorado	104,247	1876	4,753,377	7	Centennial State
Idaho	83,557	1890	1,466,465	2	Gem State
Montana	147,138	1889	944,632	1	Treasure State
Nevada	110,540	1864	2,495,529	3	Silver State
Utah	84,916	1896	2,550,063	3	Beehive State
Wyoming	97,914	1890	515,004	1	Equality State

1. What is the area of Utah in square miles?
84,916 mi^2

2. What is the area of Montana in square miles?
147,138 mi^2

3. What is the 2006 estimated population of Colorado?
4,753,377 people

4. What is the 2006 estimated population of Nevada?
2,495,529 people

5. How many representatives in the U.S. Congress come from Idaho? 2 representatives

6. How many representatives in the U.S. Congress come from Wyoming? 1 representative

7. What is the popular name for Montana?
Treasure State

8. What is the popular name for Utah?
Beehive State

9. Which of these six states was the first one to be admitted to the Union? Nevada

10. In what year did two of these six states both get admitted to the Union? 1890

Use this pictograph to answer exercises 11–16.

11. How many homes were built in Tarrant County in the year 2007? 2000 homes

12. How many homes were built in Essex County in the year 2007? 1000 homes

13. In what county were the most homes built?
DuPage County

14. How many more homes were built in Tarrant County than Waverly County? 1400 homes

15. How many homes were built in Essex County and Northface County combined? 2800 homes

16. How many homes were built in DuPage County and Waverly County combined? 3000 homes

Use this pictograph to answer exercises 17–20.

17. How many apartment units are rented for under $250 per month? 4,000,000 apartment units

18. How many apartment units are rented for $800–$1249 per month? 3,200,000 apartment units

19. How many more apartment units are available in the $500–$799 range than in the $800–$1249 range?
7,200,000 apartment units

20. How many more apartment units are available in the $800–$1249 range than in the $1250 and up range?
2,400,000 apartment units

Number of New Homes Built in 2007 in Each of Five Counties

Approximate Number of Apartments in U.S. in 2004

Source: U.S. Bureau of the Census

Use this bar graph to answer exercises 21–26.

21. What was the population in Texas in 1950?
About 8 million people

22. What is the projected population in Texas in 2010?
About 25 million people

23. Between what two decades on this graph is the increase in population the greatest?
Between 2000 and 2010

24. Between what two decades was the increase in population the smallest?
Between 1960 and 1970; it increased by only one million

25. How many more people lived in Texas in 1970 than in 1950?   3 million more people

26. How many more people lived in Texas in 1980 than in 1960?   4 million people

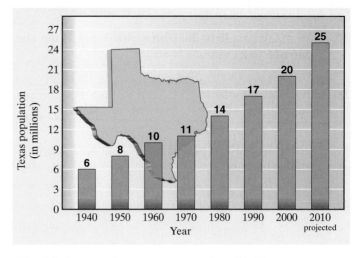

Use this bar graph to answer exercises 27–30.

27. According to the bar graph, how many people watched a sports event at least once in the last 12 months?
37 million people

28. According to the bar graph, how many people were involved in gardening at least once in the last 12 months?
55 million people

29. What two activities were the most common?
Movies and an exercise program

30. What two activities were the least common?
Watching sports and outdoor activities

Use this line graph to answer exercises 31–36.

Number of People Engaged in Various Activities at Least Once in the Last Twelve Months

Source: U.S. National Endowment for the Arts

31. What was the profit in 2004? $5.6 million

32. What was the profit in 2003? $3.5 million

33. How much greater was the profit in 2007 than 2006?
$2.5 million greater

34. In what year did the smallest profit occur?   2005

35. Between what two years did the profit decrease the most?   Between 2004 and 2005

36. Between what two years did the profit increase the most?   Between 2006 and 2007

Profit of Wentworth Construction Company

Use this line graph to answer exercises 37–40.

37. How many people in the United States are in the age group of 5 to 24 years?   74 million people

38. How many people in the United States are in the age group of 25 to 44 years?   83 million people

39. Thirty-three million people are in what age group?
65 years and above

40. Fifty-one million people are in what age group?
Age 45–64 years

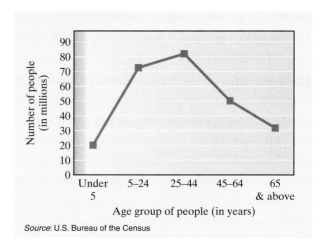

Source: U.S. Bureau of the Census

Use this circle graph to answer exercises 41–46.

41. What percent of the world's population is either Hindu or Buddhist?   19%

42. What percent of the world's population is either Muslim or nonreligious?   31%

43. What percent of the world's population is *not* Muslim?   81%

44. What percent of the world's population is *not* Christian?
64%

45. If there were approximately 6.5 billion people in the world in 2006, how many of them would we expect to be Hindu?   845 million people

46. If there are approximately 6.9 billion people in the world in 2010, how many of them would we expect to be Muslim?   1.311 billion people

Religious Faith Distribution in the World

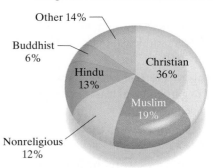

Source: United Nations Statistical Division

Use this circle graph to answer exercises 47–50.

47. What percent of the amount spent per child is spent on transportation and clothing?   19.5%

48. What percent of the amount spent per child is spent on food, health care, and housing?   61.3%

49. If the average two-child American family spends $10,600 per child each year, how much is spent on food? $2120

50. If the average two-child American family spends $10,600 per child each year, how much is spent on child care and education? $879.80

Distribution of Spending Per Child by "Average" Two-Child American Family in 2006

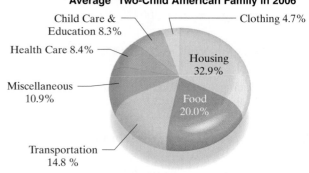

Source: United Nations Statistical Division

Appendix D Inductive and Deductive Reasoning

Using Inductive Reasoning to Reach a Conclusion

When we reach a conclusion based on specific observations, we are using **inductive reasoning.** Much of our early learning is based on simple cases of inductive reasoning. If a child touches a hot stove or other appliance several times and each time he gets burned, he is likely to conclude, "If I touch something that is hot, I will get burned." This is inductive reasoning. The child has thought about several actions and their outcomes and has made a conclusion or generalization.

The next few examples show how inductive reasoning can be used in mathematics.

Student Learning Objectives

After studying this section, you will be able to:

1 Use inductive reasoning to reach a conclusion.

2 Use deductive reasoning to reach a conclusion.

> **EXAMPLE 1** Find the next number in the sequence 10, 13, 16, 19, 22, 25, 28, ...

Solution We observe the pattern that each number is 3 more than the preceding number: $10 + 3 = 13$; $13 + 3 = 16$; and so on. Therefore, if we add 3 to 28, we conclude that the next number in the sequence is 31.

> **Practice Problem 1** Find the next number in the sequence.
> $$24, 31, 38, 45, 52, 59, 66, \ldots$$

NOTE TO STUDENT: Fully worked-out solutions to all of the Practice Problems can be found at the back of the text starting at page SP-1

> **EXAMPLE 2** Find the next number in the sequence 1, 8, 27, 64, 125, ...

Solution The sequence can be written as $1^3, 2^3, 3^3, 4^3, 5^3, \ldots$. Each successive integer is cubed. The next number would be 6^3 or 216.

> **Practice Problem 2** Find the next number in the sequence.
> $$3, 8, 15, 24, 35, 48, 63, 80, \ldots$$

> **EXAMPLE 3** Guess the next seven digits in the following irrational number:
> $$5.636336333633336 \ldots$$

Solution Between 6's there are the digits 3, 33, 333, 3333, and so on. The pattern is that the number of 3's keeps increasing by 1 each time. Thus the next seven digits are 3333363.

> **Practice Problem 3** Guess the next seven digits in the following irrational number:
> $$6.1213314441 \ldots$$

> **EXAMPLE 4** Find the next two figures that would appear in the sequence.

Solution We notice an alternating pattern: square, square, circle, circle, square. . . . We would next expect a square followed by a circle.

We notice a shading pattern of horizontal, vertical, horizontal, vertical, horizontal. . . . We would next expect vertical, then horizontal. Thus, the next two figures are

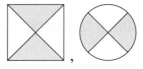

,

Practice Problem 4 Find the next two figures that would appear in the sequence.

How accurate is inductive reasoning? Do we always come to the right conclusion? Conclusions arrived at by inductive reasoning are always tentative. They may require further investigation. When we use inductive reasoning we are using specific data to reach a general conclusion. However, we may have reached the wrong conclusion. To illustrate:

Take the sequence of numbers 20, 10, 5. What is the next number? You might say 2.5. In the sequence 20, 10, 5, each number appears to be one-half of the preceding number. Thus we would predict 20, 10, 5, 2.5, ...

But wait! There is another possibility. Maybe the sequence is 20, 10, 5, −5, −10, −20, −25, −35, −40, ...

To know for sure which answer is correct, we would need more information, such as more numbers in the sequence to verify the pattern. **You should always treat inductive reasoning conclusions as tentative, requiring further verification.**

② Using Deductive Reasoning to Reach a Conclusion

Deductive reasoning requires us to take general facts, postulates, or accepted truths and use them to reach a specific conclusion. Suppose we know the following rules or "facts" of algebra:

> For all numbers a, b, and c, and for all numbers $d \neq 0$:
>
> 1. *Addition principle of equations:*   If $a = b$, then $a + c = b + c$.
>
> 2. *Division principle of equations:*   If $a = b$, then $\dfrac{a}{d} = \dfrac{b}{d}$.
>
> 3. *Multiplication principle of equations:*   If $a = b$, then $ac = bc$.
>
> 4. *Distributive property:* $a(b + c) = ab + ac$.

EXAMPLE 5 Use deductive reasoning and the four properties listed in the preceding box to justify each step in solving the equation.

$$2(7x - 2) = 38$$

Solution

Statement	Reason
1. $14x - 4 = 38$	**1.** Distributive property: $a(b + c) = ab + ac$ Here we distributed the 2.
2. $14x = 42$	**2.** Addition principle of equations: If $a = b$, then $a + c = b + c$ Here we added 4 to each side of the equation.

NOTE TO STUDENT: Fully worked-out solutions to all of the Practice Problems can be found at the back of the text starting at page SP-1

3. $\dfrac{14x}{14} = \dfrac{42}{14}$

$x = 3$

3. Division principle of equations:

If $a = b$, then $\dfrac{a}{d} = \dfrac{b}{d}$

Here we divided each side by 14.

Practice Problem 5   Use deductive reasoning and the four properties listed in the preceding box to justify each step in solving the equation.

$$\frac{1}{6}x = \frac{1}{3}x + 4$$

Sometimes we need to make conclusions about angles and lines in geometry. The following properties are useful. We will refer to angles 1, 2, 3, 4.

1. If two lines intersect, the opposite angles are equal.
  Here $\angle 1 = \angle 2$ and $\angle 4 = \angle 3$.

2. If two lines intersect, the adjacent angles are supplementary (they add up to 180°). Here, $\angle 4 + \angle 1 = 180°$, and $\angle 1 + \angle 3 = 180°$; also $\angle 3 + \angle 2 = 180°$, and $\angle 4 + \angle 2 = 180°$.

3. If a transversal (intersecting line) crosses two parallel lines, the alternate interior angles are equal. In the figure at right, line P is a transversal. If line M is parallel to line N, then $\angle 3 = \angle 4$ and $\angle 1 = \angle 2$.

4. If two alternate interior angles on each side of a transversal cutting two straight lines are equal, the two straight lines are parallel. If $\angle 3 = \angle 4$, then line M is parallel to line N.

5. Supplements of equal angles are equal. In the figure at right, if $\angle 1$ and $\angle 2$ are supplementary, and $\angle 3$ and $\angle 4$ are supplementary, then if $\angle 1 = \angle 4$, then $\angle 2 = \angle 3$.

6. Transitive property of equality:

If $a = b$ and $b = c$, then $a = c$.

We will now use these facts to prove some geometric conclusions.

EXAMPLE 6   Prove that line M is parallel to line N if $\angle 1 = \angle 3$.

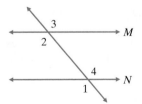

Solution

Statement	Reason
1. $\angle 3 = \angle 2$ and $\angle 4 = \angle 1$	**1.** If two lines intersect, the opposite angles are equal.
2. $\angle 2 = \angle 4$	**2.** Transitive property of equality.

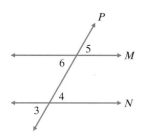

3. Therefore, line *M* is parallel to line *N*. (We write this as *M* ∥ *N*.)

3. If two alternate interior angles on each side of a transversal cutting two straight lines are equal, the straight lines are parallel.

Practice Problem 6   Refer to the diagram in the margin. Line *M* is parallel to line *N*. ∠5 = 26°. Prove ∠4 = 26°.

Now let us see if we can use deductive reasoning to solve the following problem.

EXAMPLE 7   For next semester, four professors from the psychology department have expressed the following desires for freshman courses. Each professor will teach only *one* freshman course next semester. The four freshman courses are General Psychology, Social Psychology, Psychology of Adjustment, and Educational Psychology.

1. Professors *A* and *B* don't want to teach General Psychology.

2. Professor *C* wants to teach Social Psychology.

3. Professor *D* will be happy to teach any course.

4. Professor *B* wants to teach Psychology of Adjustment.

Which professor will teach Educational Psychology, if all professors are given the courses they desire?

Solution   Let us organize the facts by listing the four freshman courses and each professor: *A*, *B*, *C*, and *D*.

General Psychology	Social Psychology	Psychology of Adjustment	Educational Psychology
A	*A*	*A*	*A*
B	*B*	*B*	*B*
C	*C*	*C*	*C*
D	*D*	*D*	*D*

Step	Reason
1. We cross off Professors *A* and *B* from the General Psychology list.	**1.** Professors *A* and *B* don't want to teach General Psychology.

General Psychology			
A̶	*A*	*A*	*A*
B̶	*B*	*B*	*B*
C	*C*	*C*	*C*
D	*D*	*D*	*D*

Step	Reason
2. We cross Professor *C* off every list (except Social Psychology) and mark that he will teach it.	**2.** Professor *C* wants to teach Social Psychology.

	Social Psychology		
A̶			
B̶			
C̶	☐ *C*	C̶	C̶
D			

Step	Reason

3. Professor *D* is thus the only person who can teach General Psychology. We cross him off every other list and mark that he will teach General Psychology.

3. Professor *D* is happy to teach any course.

General Psychology	
A̶	
B̶	
C̶	C̶ C̶ C̶
D	D̶ D̶ D̶

4. We cross out all courses for Professor *B* except Psychology of Adjustment.

4. Professor *B* wants to teach Psychology of Adjustment.

		Psychology of Adjustment	
A̶	A̶	A	A
B̶	B̶	B	B
C̶	C	C̶	C̶
D	D̶	D̶	D̶

5. Professor *A* will teach Educational Psychology.

5. He is the only professor left. All others are assigned.

Educational Psychology
A

Practice Problem 7 A Honda, Toyota, Mustang, and Camaro are parked side by side, but not in that order.

1. The Camaro is parked on the right end.

2. The Mustang is between the Honda and the Toyota.

3. The Honda is not next to the Camaro.

Which car is parked on the left end?

NOTE TO STUDENT: *Fully worked-out solutions to all of the Practice Problems can be found at the back of the text starting at page SP-1*

Find the next number in the sequence.

1. 2, 4, 6, 8, 10, 12, ...   14

2. 0, 5, 10, 15, 20, 25, ...   30

3. 7, 16, 25, 34, 43, ...   52

4. 12, 25, 38, 51, 64, ...   77

5. 1, 16, 81, 256, 625, ...   1296

6. 1, 6, 13, 22, 33, 46, ...   61

7. −7, 3, −6, 4, −5, 5, −4, 6, ...   −3

8. 2, −4, 8, −16, 32, −64, 128, ...   −256

9. $5x, 6x − 1, 7x − 2, 8x − 3, 9x − 4, ...$ $10x − 5$

10. $60x, 30x, 15x, 7.5x, 3.75x, 1.875x, ...$ $0.9375x$

In exercises 11–16, guess the next seven digits in each irrational number.

11. 8.181181118...   1111811

12. 3.043004300043...   0000430

13. 12.98987987698765...   9876549

14. 7.6574839201102938...   4756657

15. 2.14916253649...   6481100

16. 6.112223333...   4444455

17. Find the next row in this triangular pattern.

```
                1
             1     1
          1     2     1
       1     3     3     1
    1     4     6     4     1
 1     5    10    10     5     1
1   6   15   20   15   6   1
```

18. Find the next two figures that would appear in the sequence.

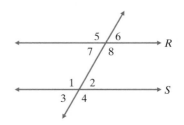

Use deductive reasoning and the four properties discussed in Example 5 to justify each step in solving each equation.

19. $12x − 30 = 6$
$x = 3$
Answers may vary

20. $3x + 7 = 4$
$x = −1$
Answers may vary

21. $4x − 3 = 3x − 5$
$x = −2$
Answers may vary

22. $2x − 9 = 4x + 5$
$x = −7$
Answers may vary

23. $8x − 3(x − 5) = 30$
$x = 3$
Answers may vary

24. $3x − 5(x − 1) = −5$
$x = 5$
Answers may vary

25. $\frac{1}{2}x + 6 = \frac{3}{2} + 6x$
$x = \frac{9}{11}$
Answers may vary

26. $\frac{4}{5}x − 3 = \frac{1}{10}x + \frac{3}{5}$
$x = \frac{36}{7}$
Answers may vary

In exercises 27–28, use deductive reasoning and the properties of geometry discussed in Example 6 to prove each statement.

27. If $\angle 1 = \angle 5$, prove that line P is parallel to line S.
Answers may vary

28. If line R is parallel to line S, prove that $\angle 6 = \angle 3$.
Answers may vary

29. Charlie, Brent, William, and Dave competed in the Olympics. These four divers placed first, second, third, and fourth in the competition.

1. Charlie ranked between Brent and Dave.
2. William did better than Brent.
3. Brent did better than Dave.

Who finished in each of the four places?
William was first, Brent was second, Charlie was third, and Dave was fourth.

30. The four floors of Hotel Royale are color-coded.

1. The blue floor is directly below the green floor.
2. The red floor is next to the yellow floor.
3. The green floor is above the red floor.
4. The blue floor is above the yellow floor.
5. There is no floor below the red floor.

What is the order of the colors from top to bottom of Hotel Royale?
Green, blue, yellow, red

31. There are four people named Peter, Michael, Linda, and Judy. Each of them has one occupation. They are a teacher, a butcher, a baker, and a candlestick maker.

1. Linda is the baker.
2. The teacher plans to marry the baker in October.
3. Judy and Peter were married last year.

Who is the teacher?
Michael

32. A president of a company, a lawyer, a salesman, and a doctor are seated at the head table at a banquet.

1. The lawyer prosecuted the doctor in a malpractice lawsuit and they should not sit together.
2. The salesman should be at one end of the table.
3. The lawyer should be at the far right.

From left to right how should they be seated?
From left to right the correct order is salesman, doctor, president, lawyer.

33. John is considering purchasing a new car when he starts his new full-time job. The car with the best gas mileage is a Honda Civic, the car with the most conveniently located dealer is a Toyota Corolla, the car with the lowest price is the Ford Focus, and the car that has the best handling is the Dodge Neon. His next-door neighbor has a Honda and he does not want to copy his neighbor. If he does not buy a Toyota then he will definitely not go to graduate school. He told his the best friend that he probably will not purchase a car based on its handling. He has definitely decided to go to graduate school. What car is he most likely to purchase?
Toyota Corolla

34. Detective Smith recently found a stolen Corvette abandoned near Skull Rocks. A small, empty boat was anchored 1 mile offshore with no person in the boat. Divers later found the body of a 35-year-old man $\frac{1}{2}$ mile farther out to sea than the boat. If the man entered the water after 10:00 A.M., then the current of the outgoing tide was strong enough to take a dead body a distance of $\frac{1}{2}$ mile farther out to sea. The coroner determined when the body was found at 4:00 P.M. that the man had been dead for seven hours. The detective determined that the man had died instantly when his head hit the rocks directly beneath the anchored boat. Could the man have acted alone based on the facts revealed so far?   No

35. Dr. Lowe is deciding whether he should do a root canal in order to save Fred's tooth or just remove the tooth and put in a false tooth. If Dr. Lowe works alone on the tooth, the procedure will take 2 hours and 15 minutes. Fred wants to save the tooth but he does not want the dental bill for Tuesday's work to be more than $200. Dr. Lowe charges $100 per hour for dental procedures. If the bill is under $150 it is because Dr. Lowe's efficient dental assistant is helping with the oral surgery. The dental procedure is scheduled for Tuesday, when Dr. Lowe does not have any dental assistants available. Will Fred's tooth be saved or will he get a false tooth?
Unless Fred is willing to pay more than $200, he will probably get a false tooth.

36. If Marcia passes the Bilingual Teacher's exam, she will stay in the Chicago area. If she devotes at least 50 hours of study time to practice her Spanish, she is confident that she can pass the exam. Either Marcia will stay in the Chicago area or she will move back to Massachusetts. If she did not pass the Bilingual Teacher's exam, it is because she needed 12 hours of study time per month. Marcia started studying for the exam in April. She will be moving to Massachusetts. When was the Bilingual Teacher's exam given, if we know that Marcia took the exam?
The exam took place in August or earlier in the same year.

Solutions to Practice Problems

Chapter 0

0.1 Practice Problems

1. (a) $\dfrac{10}{16} = \dfrac{2 \times 5}{2 \times 2 \times 2 \times 2} = \dfrac{5}{2 \times 2 \times 2} = \dfrac{5}{8}$

(b) $\dfrac{24}{36} = \dfrac{2 \times 2 \times 2 \times 3}{2 \times 2 \times 3 \times 3} = \dfrac{2}{3}$

(c) $\dfrac{36}{42} = \dfrac{2 \times 3 \times 2 \times 3}{2 \times 3 \times 7} = \dfrac{2 \times 3}{7} = \dfrac{6}{7}$

2. (a) $\dfrac{4}{12} = \dfrac{2 \times 2 \times 1}{2 \times 2 \times 3} = \dfrac{1}{3}$

(b) $\dfrac{25}{125} = \dfrac{5 \times 5 \times 1}{5 \times 5 \times 5} = \dfrac{1}{5}$

(c) $\dfrac{73}{146} = \dfrac{73 \times 1}{73 \times 2} = \dfrac{1}{2}$

3. (a) $\dfrac{18}{6} = \dfrac{2 \times 3 \times 3}{2 \times 3 \times 1} = 3$

(b) $\dfrac{146}{73} = \dfrac{73 \times 2}{73 \times 1} = 2$

(c) $\dfrac{28}{7} = \dfrac{2 \times 2 \times 7}{7} = 2 \times 2 = 4$

4. 56 out of 154 $= \dfrac{56}{154} = \dfrac{2 \times 7 \times 4}{2 \times 7 \times 11} = \dfrac{4}{11}$

5. (a) $\dfrac{12}{7} = 12 \div 7$ $7\overline{)12} \atop \;\;\underline{7} \atop \;\;5\;\text{Remainder}$

$\dfrac{12}{7} = 1\dfrac{5}{7}$

(b) $\dfrac{20}{5} = 20 \div 5$ $5\overline{)20} \atop \;\;\underline{20} \atop \;\;0\;\text{Remainder}$

$\dfrac{20}{5} = 4$

6. (a) $3\dfrac{2}{5} = \dfrac{(3 \times 5) + 2}{5} = \dfrac{15 + 2}{5} = \dfrac{17}{5}$

(b) $1\dfrac{3}{7} = \dfrac{(1 \times 7) + 3}{7} = \dfrac{7 + 3}{7} = \dfrac{10}{7}$

(c) $2\dfrac{6}{11} = \dfrac{(2 \times 11) + 6}{11} = \dfrac{22 + 6}{11} = \dfrac{28}{11}$

(d) $4\dfrac{2}{3} = \dfrac{(4 \times 3) + 2}{3} = \dfrac{12 + 2}{3} = \dfrac{14}{3}$

7. (a) $\dfrac{3}{8} = \dfrac{?}{24}$ $8 \times 3 = 24$ **(b)** $\dfrac{5}{6} = \dfrac{?}{30}$ $6 \times 5 = 30$

$\dfrac{3 \times 3}{8 \times 3} = \dfrac{9}{24}$ $\dfrac{5 \times 5}{6 \times 5} = \dfrac{25}{30}$

(c) $\dfrac{2}{7} = \dfrac{?}{56}$ $7 \times 8 = 56$

$\dfrac{2 \times 8}{7 \times 8} = \dfrac{16}{56}$

0.2 Practice Problems

1. (a) $\dfrac{3}{6} + \dfrac{2}{6} = \dfrac{3 + 2}{6} = \dfrac{5}{6}$

(b) $\dfrac{3}{11} + \dfrac{8}{11} = \dfrac{3 + 8}{11} = \dfrac{11}{11} = 1$

(c) $\dfrac{1}{8} + \dfrac{2}{8} + \dfrac{1}{8} = \dfrac{1 + 2 + 1}{8} = \dfrac{4}{8} = \dfrac{1}{2}$

(d) $\dfrac{5}{9} + \dfrac{8}{9} = \dfrac{5 + 8}{9} = \dfrac{13}{9}$ or $1\dfrac{4}{9}$

2. (a) $\dfrac{11}{13} - \dfrac{6}{13} = \dfrac{11 - 6}{13} = \dfrac{5}{13}$

(b) $\dfrac{8}{9} - \dfrac{2}{9} = \dfrac{8 - 2}{9} = \dfrac{6}{9} = \dfrac{2}{3}$

3. The LCD is 24, since 24 is exactly divisible by 8 and 12. There is no smaller number that is exactly divisible by 8 and 12.

4. Find the LCD using prime factors.

$\dfrac{8}{35}$ and $\dfrac{6}{15}$

$35 = 7 \cdot 5$
$15 = \;5 \cdot 3$
$\text{LCD} = 7 \cdot 5 \cdot 3 = 105$

5. Find the LCD of $\dfrac{5}{12}$ and $\dfrac{7}{30}$.

$12 = 3 \cdot 2 \cdot 2$
$30 = 3 2 \cdot 5$
$\text{LCD} = 3 \cdot 2 \cdot 2 \cdot 5 = 60$

6. Find the LCD of $\dfrac{2}{27}, \dfrac{1}{18},$ and $\dfrac{5}{12}$.

$27 = 3 \cdot 3 \cdot 3$
$18 = 3 \cdot 3 \cdot 2$
$12 = 3 \cdot 2 \cdot 2$
$\text{LCD} = 3 \cdot 3 \cdot 3 \cdot 2 \cdot 2 = 108$

7. From Practice Problem 3, the LCD is 24.

$\dfrac{5}{12} = \dfrac{5 \times 2}{12 \times 2} = \dfrac{10}{24}$ $\dfrac{1}{8} = \dfrac{1 \times 3}{8 \times 3} = \dfrac{3}{24}$

$\dfrac{5}{12} + \dfrac{1}{8} = \dfrac{10}{24} + \dfrac{3}{24} = \dfrac{13}{24}$

$\dfrac{13}{24}$ of the farm fields were planted in corn or soybeans.

8. We can see by inspection that both 5 and 25 divide exactly into 50.

$\dfrac{3}{5} = \dfrac{3 \times 10}{5 \times 10} = \dfrac{30}{50}$ $\dfrac{4}{25} = \dfrac{4 \times 2}{25 \times 2} = \dfrac{8}{50}$

$\dfrac{3}{5} + \dfrac{4}{25} + \dfrac{1}{50} = \dfrac{30}{50} + \dfrac{8}{50} + \dfrac{1}{50} = \dfrac{30 + 8 + 1}{50} = \dfrac{39}{50}$

9. Add $\dfrac{1}{49} + \dfrac{3}{14}$

First find the LCD.

$49 = 7 \cdot 7$
$14 = 7 \cdot 2$
$\text{LCD} = 7 \cdot 7 \cdot 2 = 98$

Then change to equivalent fractions and add.

$\dfrac{1}{49} = \dfrac{1 \times 2}{49 \times 2} = \dfrac{2}{98}$ $\dfrac{3}{14} = \dfrac{3 \times 7}{14 \times 7} = \dfrac{21}{98}$

$\dfrac{1}{49} + \dfrac{3}{14} = \dfrac{2}{98} + \dfrac{21}{98} = \dfrac{23}{98}$

10. $\dfrac{1}{12} - \dfrac{1}{30}$

First find the LCD.

$$12 = 2 \cdot 2 \cdot 3$$
$$30 = 2 \cdot 3 \cdot 5$$
$$\text{LCD} = 2 \cdot 2 \cdot 3 \cdot 5 = 60$$

Then change to equivalent fractions and subtract.

$$\dfrac{1}{12} = \dfrac{1 \times 5}{12 \times 5} = \dfrac{5}{60} \qquad \dfrac{1}{30} = \dfrac{1 \times 2}{30 \times 2} = \dfrac{2}{60}$$

$$\dfrac{1}{12} - \dfrac{1}{30} = \dfrac{5}{60} - \dfrac{2}{60} = \dfrac{3}{60} = \dfrac{1}{20}$$

11. $\dfrac{2}{3} + \dfrac{3}{4} - \dfrac{3}{8}$

First find the LCD.

$$3 = 3$$
$$4 = 2 \cdot 2$$
$$8 = 2 \cdot 2 \cdot 2$$
$$\text{LCD} = 3 \cdot 2 \cdot 2 \cdot 2 = 3 \cdot 8 = 24$$

$$\dfrac{2}{3} = \dfrac{2 \times 8}{3 \times 8} = \dfrac{16}{24} \qquad \dfrac{3}{4} = \dfrac{3 \times 6}{4 \times 6} = \dfrac{18}{24} \qquad \dfrac{3}{8} = \dfrac{3 \times 3}{8 \times 3} = \dfrac{9}{24}$$

Now combine the fractions.

$$\dfrac{2}{3} + \dfrac{3}{4} - \dfrac{3}{8} = \dfrac{16}{24} + \dfrac{18}{24} - \dfrac{9}{24} = \dfrac{25}{24} \text{ or } 1\dfrac{1}{24}$$

12. (a) The LCD of 3 and 5 is 15.

$$1\dfrac{2}{3} = \dfrac{5}{3} = \dfrac{5 \times 5}{3 \times 5} = \dfrac{25}{15} \qquad 2\dfrac{4}{5} = \dfrac{14}{5} = \dfrac{14 \times 3}{5 \times 3} = \dfrac{42}{15}$$

$$1\dfrac{2}{3} + 2\dfrac{4}{5} = \dfrac{25}{15} + \dfrac{42}{15} = \dfrac{67}{15} = 4\dfrac{7}{15}$$

(b) The LCD of 4 and 3 is 12.

$$5\dfrac{1}{4} = \dfrac{21}{4} = \dfrac{21 \times 3}{4 \times 3} = \dfrac{63}{12} \qquad 2\dfrac{2}{3} = \dfrac{8}{3} = \dfrac{8 \times 4}{3 \times 4} = \dfrac{32}{12}$$

$$5\dfrac{1}{4} - 2\dfrac{2}{3} = \dfrac{63}{12} - \dfrac{32}{12} = \dfrac{31}{12} = 2\dfrac{7}{12}$$

13.

$$4\dfrac{1}{5} + 6\dfrac{1}{2} + 4\dfrac{1}{5} + 6\dfrac{1}{2}$$
$$= \dfrac{21}{5} + \dfrac{13}{2} + \dfrac{21}{5} + \dfrac{13}{2}$$
$$= \dfrac{42}{10} + \dfrac{65}{10} + \dfrac{42}{10} + \dfrac{65}{10}$$
$$= \dfrac{214}{10} = \dfrac{107}{5} = 21\dfrac{2}{5}$$

The perimeter is $21\dfrac{2}{5}$ cm.

0.3 Practice Problems

1. (a) $\dfrac{2}{7} \times \dfrac{5}{11} = \dfrac{2 \cdot 5}{7 \cdot 11} = \dfrac{10}{77}$

(b) $\dfrac{1}{5} \times \dfrac{7}{10} = \dfrac{1 \cdot 7}{5 \cdot 10} = \dfrac{7}{50}$

(c) $\dfrac{9}{5} \times \dfrac{1}{4} = \dfrac{9 \cdot 1}{5 \cdot 4} = \dfrac{9}{20}$

(d) $\dfrac{8}{9} \times \dfrac{3}{10} = \dfrac{8 \cdot 3}{9 \cdot 10} = \dfrac{24}{90} = \dfrac{4}{15}$

2. (a) $\dfrac{3}{5} \times \dfrac{4}{3} = \dfrac{3 \cdot 4}{5 \cdot 3} = \dfrac{4 \cdot 3}{5 \cdot 3} = \dfrac{4}{5}$

(b) $\dfrac{9}{10} \times \dfrac{5}{12} = \dfrac{3 \cdot 3}{2 \cdot 5} \times \dfrac{5}{2 \cdot 2 \cdot 3} = \dfrac{3}{8}$

3. (a) $4 \times \dfrac{2}{7} = \dfrac{4}{1} \times \dfrac{2}{7} = \dfrac{4 \cdot 2}{1 \cdot 7} = \dfrac{8}{7} \text{ or } 1\dfrac{1}{7}$

(b) $12 \times \dfrac{3}{4} = \dfrac{12}{1} \times \dfrac{3}{4} = \dfrac{4 \cdot 3}{1} \times \dfrac{3}{4} = \dfrac{9}{1} = 9$

4. Multiply. $5\dfrac{3}{5}$ times $3\dfrac{3}{4}$

$$5\dfrac{3}{5} \times 3\dfrac{3}{4} = \dfrac{28}{5} \times \dfrac{15}{4} = \dfrac{4 \cdot 7}{5} \times \dfrac{3 \cdot 5}{4} = \dfrac{21}{1} = 21$$

The area of the field is 21 square miles.

5. $3\dfrac{1}{2} \times \dfrac{1}{14} \times 4 = \dfrac{7}{2} \times \dfrac{1}{14} \times \dfrac{4}{1} = \dfrac{7}{2} \times \dfrac{1}{2 \cdot 7} \times \dfrac{2 \cdot 2}{1} = 1$

6. (a) $\dfrac{2}{5} \div \dfrac{1}{3} = \dfrac{2}{5} \times \dfrac{3}{1} = \dfrac{6}{5} \text{ or } 1\dfrac{1}{5}$

(b) $\dfrac{12}{13} \div \dfrac{4}{3} = \dfrac{12}{13} \times \dfrac{3}{4} = \dfrac{4 \cdot 3}{13} \times \dfrac{3}{4} = \dfrac{9}{13}$

7. (a) $\dfrac{3}{7} \div 6 = \dfrac{3}{7} \div \dfrac{6}{1} = \dfrac{3}{7} \times \dfrac{1}{6} = \dfrac{3}{7} \times \dfrac{1}{2 \cdot 3} = \dfrac{1}{14}$

(b) $8 \div \dfrac{2}{3} = \dfrac{8}{1} \div \dfrac{2}{3} = \dfrac{8}{1} \times \dfrac{3}{2} = \dfrac{2 \cdot 4}{1} \times \dfrac{3}{2} = \dfrac{12}{1} = 12$

8. (a) $\dfrac{\frac{3}{11}}{\frac{5}{7}} = \dfrac{3}{11} \div \dfrac{5}{7} = \dfrac{3}{11} \times \dfrac{7}{5} = \dfrac{21}{55}$

(b) $\dfrac{\frac{12}{5}}{\frac{8}{15}} = \dfrac{12}{5} \div \dfrac{8}{15} = \dfrac{12}{5} \times \dfrac{15}{8} = \dfrac{3 \cdot 4}{5} \times \dfrac{3 \cdot 5}{2 \cdot 4} = \dfrac{9}{2} \text{ or } 4\dfrac{1}{2}$

9. (a) $1\dfrac{2}{5} \div 2\dfrac{1}{3} = \dfrac{7}{5} \div \dfrac{7}{3} = \dfrac{7}{5} \times \dfrac{3}{7} = \dfrac{3}{5}$

(b) $4\dfrac{2}{3} \div 7 = \dfrac{14}{3} \div \dfrac{7}{1} = \dfrac{14}{3} \times \dfrac{1}{7} = \dfrac{2 \cdot 7}{3} \times \dfrac{1}{7} = \dfrac{2}{3}$

(c) $\dfrac{1\frac{1}{5}}{1\frac{2}{7}} = 1\dfrac{1}{5} \div 1\dfrac{2}{7} = \dfrac{6}{5} \div \dfrac{9}{7} = \dfrac{6}{5} \times \dfrac{7}{9} = \dfrac{2 \cdot 3}{5} \times \dfrac{7}{3 \cdot 3} = \dfrac{14}{15}$

10. $64 \div 5\dfrac{1}{3} = \dfrac{64}{1} \div \dfrac{16}{3} = \dfrac{64}{1} \times \dfrac{3}{16} = \dfrac{4 \cdot 16}{1} \times \dfrac{3}{16} = \dfrac{12}{1} = 12$

He can fill 12 jars.

11. $126 \div 5\dfrac{1}{4} = \dfrac{126}{1} \div \dfrac{21}{4} = \dfrac{126}{1} \times \dfrac{4}{21} = \dfrac{6 \cdot 21}{1} \times \dfrac{4}{21} = 24$

The car got 24 miles per gallon.

0.4 Practice Problems

1. (a) $0.9 = \dfrac{9}{10} = $ nine tenths

(b) $0.09 = \dfrac{9}{100} = $ nine hundredths.

(c) $0.731 = \dfrac{731}{1000} = $ seven hundred thirty-one thousandths

(d) $1.371 = 1\dfrac{371}{1000} = $ one and three hundred seventy-one thousandths

(e) $0.0005 = \dfrac{5}{10,000} = $ five ten-thousandths

2. (a) $\dfrac{3}{8} = 0.375$

$$\begin{array}{r} 0.375 \\ 8\overline{)3.000} \\ \underline{2\ 4} \\ 60 \\ \underline{56} \\ 40 \\ \underline{40} \\ 0 \end{array}$$

(b) $\dfrac{7}{200} = 0.035$

$$\begin{array}{r} 0.035 \\ 200\overline{)7.000} \\ \underline{6\ 00} \\ 1\ 000 \\ \underline{1\ 000} \\ 0 \end{array}$$

(c) $\dfrac{33}{20} = 1.65$

$$
\begin{array}{r}
1.65 \\
20\overline{)33.00} \\
\underline{20} \\
13\,0 \\
\underline{12\,0} \\
1\,00 \\
\underline{1\,00} \\
0
\end{array}
$$

3. (a) $\dfrac{1}{6} = 0.1666\ldots$ or $0.1\overline{6}$

$$
\begin{array}{r}
0.166 \\
6\overline{)1.000} \\
\underline{6} \\
40 \\
\underline{36} \\
40 \\
\underline{36} \\
4
\end{array}
$$

(b) $\dfrac{5}{11} = 0.454545\ldots$ or $0.\overline{45}$

$$
\begin{array}{r}
0.4545 \\
11\overline{)5.0000} \\
\underline{4\,4} \\
60 \\
\underline{55} \\
50 \\
\underline{44} \\
60 \\
\underline{55} \\
5
\end{array}
$$

4. (a) $0.8 = \dfrac{8}{10} = \dfrac{4}{5}$ **(b)** $0.88 = \dfrac{88}{100} = \dfrac{22}{25}$

(c) $0.45 = \dfrac{45}{100} = \dfrac{9}{20}$ **(d)** $0.148 = \dfrac{148}{1000} = \dfrac{37}{250}$

(e) $0.612 = \dfrac{612}{1000} = \dfrac{153}{250}$ **(f)** $0.016 = \dfrac{16}{1000} = \dfrac{2}{125}$

5. (a)
$$
\begin{array}{r}
3.12 \\
5.08 \\
+\,1.42 \\
\hline
9.62
\end{array}
$$
(b)
$$
\begin{array}{r}
152.003 \\
-\,136.118 \\
\hline
15.885
\end{array}
$$
(c)
$$
\begin{array}{r}
1.1 \\
3.16 \\
+\,5.123 \\
\hline
9.383
\end{array}
$$
(d)
$$
\begin{array}{r}
1.0052 \\
-\,0.1234 \\
\hline
0.8818
\end{array}
$$

6. (a)
$$
\begin{array}{r}
0.0610 \\
5.0008 \\
+\,1.3000 \\
\hline
6.3618
\end{array}
$$
(b)
$$
\begin{array}{r}
18.000 \\
-\,0.126 \\
\hline
17.874
\end{array}
$$

7.
$$
\begin{array}{r}
0.5 \\
\times\,0.3 \\
\hline
0.15
\end{array}
$$
(one decimal place)
(one decimal place)
(two decimal places)

8.
$$
\begin{array}{r}
0.12 \\
\times\,\,\,0.4 \\
\hline
0.048
\end{array}
$$
(two decimal places)
(one decimal place)
(three decimal places)

9. (a)
$$
\begin{array}{r}
1.23 \\
\times\,0.005 \\
\hline
0.00615
\end{array}
$$
(two decimal places)
(three decimal places)
(five decimal places)

(b)
$$
\begin{array}{r}
0.003 \\
\times\,\,\,0.00002 \\
\hline
0.00000006
\end{array}
$$
(three decimal places)
(five decimal places)
(eight decimal places)

10.
$$
\begin{array}{r}
5.26 \\
6\overline{)31.56} \\
\underline{30} \\
1\,5 \\
\underline{1\,2} \\
36 \\
\underline{36} \\
0
\end{array}
$$
Each box of paper costs $5.26.

11
$$
\begin{array}{r}
300\,00. \\
.06\overline{)1800.00} \\
\underline{18} \\
00000
\end{array}
$$
Thus, $1800 \div 0.06 = 30{,}000$.

12.
$$
\begin{array}{r}
0.0036 \\
4.9\overline{)0.0\,1764} \\
\underline{147} \\
294 \\
\underline{294} \\
0
\end{array}
$$
Thus, $0.01764 \div 4.9 = 0.0036$.

13. (a) $0.0016 \times 100 = 0.16$
Move decimal point 2 places to the right.

(b) $2.34 \times 1000 = 2340$
Move decimal point 3 places to the right.

(c) $56.75 \times 10{,}000 = 567{,}500$
Move decimal point 4 places to the right.

14. (a) $\dfrac{5.82}{10} = 0.582$ (Move decimal point 1 place to the left.)

(b) $\dfrac{123.4}{1000} = 0.1234$ (Move decimal point 3 places to the left.)

(c) $\dfrac{0.00614}{10{,}000} = 0.000000614$ (Move decimal point 4 places to the left.)

0.5 Practice Problems

1. (a) $0.92 = 92\%$
(b) $0.418 = 41.8\%$
(c) $0.7 = 0.70 = 70\%$

2. (a) $0.0019 = 0.19\%$
(b) $0.0736 = 7.36\%$
(c) $0.0003 = 0.03\%$

3. (a) $3.04 = 304\%$
(b) $5.186 = 518.6\%$
(c) $2.1 = 2.10 = 210\%$

4. (a) $7\% = 0.07$
(b) $9.3\% = 0.093$
(c) $0.2\% = 0.002$

5. (a) $131\% = 1.31$
(b) $301.6\% = 3.016$
(c) $0.04\% = 0.0004$

6. Change the percents to decimals and multiply.
(a) 18% of 50 $= 0.18 \times 50 = 9$
(b) 4% of 64 $= 0.04 \times 64 = 2.56$
(c) 156% of 35 $= 1.56 \times 35 = 54.6$

7. (a) 4.2% of 38,000 $= 0.042 \times 38{,}000 = 1596$
His raise is $1596.
(b) $38{,}000 + 1596 = 39{,}596$
His new salary is $39,596.

8. $\dfrac{37}{148}$ reduces to $\dfrac{37 \cdot 1}{37 \cdot 4} = \dfrac{1}{4} = 0.25 = 25\%$

9. (a) $\dfrac{24}{48} = \dfrac{1}{2} = 0.5 = 50\%$ **(b)** $\dfrac{4}{25} = 0.16 = 16\%$

10. $\dfrac{430}{1256} = \dfrac{215}{628} \approx 0.342 \approx 34\%$
He scored approximately 34% of the time.

0.6 Practice Problems

1. Round 128,621 to 100,000. Round 378 to 400.
$100{,}000 \times 400 = 40{,}000{,}000$

2. Round $12\dfrac{1}{2}$ feet to 10 feet. Round $9\dfrac{3}{4}$ feet to 10 feet.
First room: $10 \times 10 = 100$ square feet
Round $11\dfrac{1}{4}$ feet to 10 feet. Round $18\dfrac{1}{2}$ feet to 20 feet.
Second room: $10 \times 20 = 200$ square feet
$100 + 200 = 300$
The estimate is 300 square feet.

3. (a) Round 422.8 miles to 400 miles. Round 19.3 gallons to 20 gallons.

$$\frac{20}{20\overline{)400}}$$

Roberta's truck gets about 20 miles per gallon.

(b) Round 3862 miles to 4000 miles.

$$\frac{200}{20\overline{)4000}}$$

She will use about 200 gallons of gas for her trip.

Round $3.69\frac{9}{10}$ to $4.00.

$200 \times \$4 = \$800$

The estimated cost is $800.

4. (a) Round 43,300 miles per hour to 40,000 miles per hour. Round 3,580,000,000 miles to 4,000,000,000 miles.

$$\frac{100,000}{40,000\overline{)4,000,000,000}}$$

It will take about 100,000 hours.

(b) Round 24 hours per day to 20 hours per day.

$$\frac{5000}{20\overline{)100,000}}$$

It will take about 5000 days.

5. $56.93\% = 0.5693$

Round 0.5693 to 0.6. Round $293,567.12 to $300,000.

$0.6 \times \$300,000 = \$180,000$

An estimate for the answer is $180,000.

0.7 Practice Problems

1.

Mathematics Blueprint for Problem Solving			
Gather the Facts	What Am I Solving for?	What Must I Calculate?	Key Points to Remember
Living Room measures $16\frac{1}{2}$ ft by $10\frac{1}{2}$ ft.	Area of room in square feet.	Multiply $16\frac{1}{2}$ ft by $10\frac{1}{2}$ ft to get the area in square feet.	9 sq feet = 1 sq yard
The carpet costs $20.00 per square yard.	Area of room in square yards. Cost of the carpet.	Divide the number of square feet by 9 to get the number of square yards. Multiply the number of square yards by $20.00.	

$16\frac{1}{2}$ ft $\times 10\frac{1}{2}$ ft $= \frac{33}{2}$ ft $\times \frac{21}{2}$ ft $= \frac{693}{4}$ or $173\frac{1}{4}$ square feet

$173\frac{1}{4} \div 9 = \frac{693}{4} \div \frac{9}{1} = \frac{693}{4} \times \frac{1}{9} = \frac{77}{4} = 19\frac{1}{4}$

$19\frac{1}{4}$ square yards of carpet are needed.

$19\frac{1}{4} \times 20 = \frac{77}{4} \times \frac{20}{1} = 385$

It will cost a minimum of $385.

Check:

Estimate area of room: $20 \times 10 = 200$ square feet
Estimate area in square yards: $200 \div 10 = 20$
Estimate the cost: $20 \times 20 = \$400.00$

This is close to our answer of $385.00. Our answer seems reasonable.

To Think About: 6.5 sq ft; 52.5 sq ft; $468

2. (a) $\frac{10}{55} \approx 0.182 \approx 18\%$

(b) $\frac{3,660,000}{13,240,000} \approx 0.276 \approx 28\%$

(c) $\frac{15}{55} \approx 0.273 \approx 27\%$

(d) $\frac{3,720,000}{13,240,000} \approx 0.281 \approx 28\%$

(e) We notice that 18% of the company's sales force is located in the Northwest, and they were responsible for 28% of the sales volume. The percent of sales compared to the percent of sales force is about 150%. 27% of the company's sales force is located in the Southwest, and they were responsible for 28% of the sales volume. The percent of sales compared to the percent of sales force is approximately 100%. It would appear that the Northwest sales force is more effective.

To Think About: 88; 35

Chapter 1 1.1 Practice Problems

1.

	Number	Integer	Rational Number	Irrational Number	Real Number
(a)	$-\frac{2}{5}$		X		X
(b)	1.515151...		X		X
(c)	-8	X	X		X
(d)	π			X	X

2. (a) Population growth of 1259 is +1259.

(b) Depreciation of $763 is −763.00.

(c) Wind-chill factor of minus 10 is −10.

3. (a) The additive inverse of $\frac{2}{5}$ is $-\frac{2}{5}$.

(b) The additive inverse of −1.92 is +1.92.

(c) The opposite of a loss of 12 yards on a football play is a gain of 12 yards on the play.

4. (a) $|-7.34| = 7.34$

(b) $\left|\frac{5}{8}\right| = \frac{5}{8}$

(c) $\left|\frac{0}{2}\right| = \frac{0}{2} = 0$

5. (a) $37 + 19$

$37 + 19 = 56$

$37 + 19 = +56$

(b) $-23 + (-35)$

$23 + 35 = 58$

$-23 + (-35) = -58$

6. $-\dfrac{3}{5} + \left(-\dfrac{4}{7}\right)$

$-\dfrac{21}{35} + \left(-\dfrac{20}{35}\right)$

$\dfrac{21}{35} + \dfrac{20}{35} = \dfrac{41}{35}$ or $1\dfrac{6}{35}$

$-\dfrac{21}{35} + \left(-\dfrac{20}{35}\right) = -\dfrac{41}{35}$ or $-1\dfrac{6}{35}$

7. $-12.7 + (-9.38)$

$12.7 + 9.38 = 22.08$

$-12.7 + (-9.38) = -22.08$

8. $-7 + (-11) + (-33)$

$= -18 + (-33)$

$= -51$

9. $-9 + 15$

$15 - 9 = 6$

$-9 + 15 = 6$

10. $-\dfrac{5}{12} + \dfrac{7}{12} + \left(-\dfrac{11}{12}\right)$

$= \dfrac{2}{12} + \left(-\dfrac{11}{12}\right) = -\dfrac{9}{12} = -\dfrac{3}{4}$

11. $-6.3 + (-8.0) + 3.5$

$= -14.3 + 3.5$

$= -10.8$

12.

-6	$+5$
-7	$+5$
-2	$+3$
-15	13

$-6 + 5 + (-7) + (-2) + 5 + 3 = -15 + 13 = -2$

13. (a) $-2.9 + (-5.7) = -8.6$

(b) $\dfrac{2}{3} + \left(-\dfrac{1}{4}\right)$

$= \dfrac{8}{12} + \left(-\dfrac{3}{12}\right) = \dfrac{5}{12}$

1.2 Practice Problems

1. $9 - (-3) = 9 + (+3) = 12$

2. $-12 - (-5) = -12 + (+5) = -7$

3. (a) $\dfrac{5}{9} - \dfrac{7}{9} = \dfrac{5}{9} + \left(-\dfrac{7}{9}\right) = -\dfrac{2}{9}$

(b) $-\dfrac{5}{21} - \left(-\dfrac{3}{7}\right) = -\dfrac{5}{21} + \dfrac{3}{7} = -\dfrac{5}{21} + \dfrac{9}{21} = \dfrac{4}{21}$

4. $-17.3 - (-17.3) = -17.3 + 17.3 = 0$

5. (a) $-21 - 9$

$= -21 + (-9)$

$= -30$

(b) $17 - 36$

$= 17 + (-36)$

$= -19$

(c) $12 - (-15)$

$= 12 + 15$

$= 27$

(d) $\dfrac{3}{5} - 2$

$= \dfrac{3}{5} + (-2)$

$= \dfrac{3}{5} + \left(-\dfrac{10}{5}\right) = -\dfrac{7}{5}$ or $-1\dfrac{2}{5}$

6. $350 - (-186) = 350 + 186 = 536$

The helicopter is 536 feet from the sunken vessel.

1.3 Practice Problems

1. (a) $(-6)(-2) = 12$

(b) $(7)(9) = 63$

(c) $\left(-\dfrac{3}{5}\right)\left(\dfrac{2}{7}\right) = -\dfrac{6}{35}$

(d) $\left(\dfrac{5}{6}\right)(-7) = \left(\dfrac{5}{6}\right)\left(-\dfrac{7}{1}\right) = -\dfrac{35}{6}$ or $-5\dfrac{5}{6}$

2. $(-5)(-2)(-6) = (+10)(-6) = -60$

3. (a) Positive; $-2(-3) = 6$

(b) Negative; $(-1)(-3)(-2) = 3(-2) = -6$

(c) Positive;

$-4\left(-\dfrac{1}{4}\right)(-2)(-6) = 1(-2)(-6) = -2(-6) = +12$ or 12

4. (a) $-36 \div (-2) = 18$

(b) $-49 \div 7 = -7$

(c) $\dfrac{50}{-10} = -5$

(d) $\dfrac{-39}{13} = -3$

5. (a) The numbers have the same sign, so the result will be positive. Divide the absolute values.

$$1.8\overset{\;\;\;7.}{\overline{)12.6}}$$

$-12.6 \div (-1.8) = 7$

(b) The numbers have different signs, so the result will be negative. Divide the absolute values.

$$0.9\overset{\;\;\;0.5}{\overline{)0.4\,5}}$$

$0.45 \div (-0.9) = -0.5$

6. $-\dfrac{5}{16} \div \left(-\dfrac{10}{13}\right) = \left(-\dfrac{5}{16}\right)\left(-\dfrac{13}{10}\right) = \left(-\dfrac{\overset{1}{\cancel{5}}}{16}\right)\left(-\dfrac{13}{\underset{2}{\cancel{10}}}\right) = \dfrac{13}{32}$

7. (a) $\dfrac{-12}{-\dfrac{4}{5}} = -\dfrac{12}{1} \div \left(-\dfrac{4}{5}\right) = -\dfrac{\cancel{12}}{1}\left(-\dfrac{5}{\cancel{4}}\right) = 15$

(b) $\dfrac{-\dfrac{2}{9}}{\dfrac{8}{13}} = -\dfrac{2}{9} \div \dfrac{8}{13} = -\dfrac{\cancel{2}}{9}\left(\dfrac{13}{\cancel{8}}\right) = -\dfrac{13}{36}$

8. (a) $6(-10) = -60$

The team lost approximately 60 yards with plays that were considered medium losses.

(b) $7(15) = 105$

The team gained approximately 105 yards with plays that were considered medium gains.

(c) $-60 + 105 = 45$

A total of 45 yards were gained during plays that were medium losses or medium gains.

1.4 Practice Problems

1. (a) $6(6)(6)(6) = 6^4$

(b) $-2(-2)(-2)(-2)(-2) = (-2)^5$

(c) $108(108)(108) = 108^3$

(d) $-11(-11)(-11)(-11)(-11)(-11) = (-11)^6$

(e) $(w)(w)(w) = w^3$

(f) $(z)(z)(z)(z) = z^4$

2. (a) $3^5 = (3)(3)(3)(3)(3) = 243$

(b) $2^2 = (2)(2) = 4$

$3^3 = (3)(3)(3) = 27$

$2^2 + 3^3 = 4 + 27 = 31$

3. (a) $(-3)^3 = -27$

(b) $(-2)^6 = 64$

(c) $-2^4 = -(2^4) = -16$

(d) $-(3^6) = -729$

4. (a) $\left(\frac{1}{3}\right)^3 = \left(\frac{1}{3}\right)\left(\frac{1}{3}\right)\left(\frac{1}{3}\right) = \frac{1}{27}$

(b) $(0.3)^4 = (0.3)(0.3)(0.3)(0.3) = 0.0081$

(c) $\left(\frac{3}{2}\right)^4 = \left(\frac{3}{2}\right)\left(\frac{3}{2}\right)\left(\frac{3}{2}\right)\left(\frac{3}{2}\right) = \frac{81}{16}$

(d) $3^4 = (3)(3)(3)(3) = 81$

$4^2 = (4)(4) = 16$

$(3)^4(4)^2 = (81)(16) = 1296$

(e) $4^2 - 2^4 = 16 - 16 = 0$

1.5 Practice Problems

1. $25 \div 5 \cdot 6 + 2^3$

$= 25 \div 5 \cdot 6 + 8$

$= 5 \cdot 6 + 8$

$= 30 + 8$

$= 38$

2. $(-4)^3 - 2^6 = -64 - 64 = -128$

3. $6 - (8 - 12)^2 + 8 \div 2$

$= 6 - (-4)^2 + 8 \div 2$

$= 6 - (16) + 8 \div 2$

$= 6 - 16 + 4$

$= -10 + 4$

$= -6$

4. $\left(-\frac{1}{7}\right)\left(-\frac{14}{5}\right) + \left(-\frac{1}{2}\right) \div \frac{3}{4}$

$= \left(-\frac{1}{7}\right)\left(-\frac{14}{5}\right) + \left(-\frac{1}{2}\right)\frac{4}{3}$

$= \frac{2}{5} + \left(-\frac{2}{3}\right)$

$= \frac{2 \cdot 3}{5 \cdot 3} + \left(-\frac{2 \cdot 5}{3 \cdot 5}\right)$

$= \frac{6}{15} + \left(-\frac{10}{15}\right)$

$= -\frac{4}{15}$

1.6 Practice Problems

1. (a) $3(x + 2y) = 3x + 3(2y) = 3x + 6y$

(b) $-2(a - 3b) = -2(a) + (-2)(-3b) = -2a + 6b$

2. (a) $-(-3x + y) = (-1)(-3x + y)$

$= (-1)(-3x) + (-1)(y)$

$= 3x - y$

3. (a) $\frac{3}{5}(a^2 - 5a + 25) = \left(\frac{3}{5}\right)(a^2) + \left(\frac{3}{5}\right)(-5a) + \left(\frac{3}{5}\right)(25)$

$= \frac{3}{5}a^2 - 3a + 15$

(b) $2.5(x^2 - 3.5x + 1.2)$

$= (2.5)(x^2) + (2.5)(-3.5x) + (2.5)(1.2)$

$= 2.5x^2 - 8.75x + 3$

4. $-4x(x - 2y + 3) = (-4)(x)(x) + (-4)(x)(-2)(y) + (-4)(x)(3)$

$= -4x^2 + 8xy - 12x$

5. $(3x^2 - 2x)(-4) = (3x^2)(-4) + (-2x)(-4) = -12x^2 + 8x$

6. $400(6x + 9y) = 400(6x) + 400(9y)$

$= 2400x + 3600y$

The area of the field in square feet is $2400x + 3600y$.

1.7 Practice Problems

1. (a) $5a$ and $8a$ are like terms.

$2b$ and $-4b$ are like terms.

(b) y^2 and $-7y^2$ are like terms. These are the only like terms.

2. (a) $16y^3 + 9y^3 = (16 + 9)y^3 = 25y^3$

(b) $5a + 7a + 4a = (5 + 7 + 4)a = 16a$

3. $-8y^2 - 9y^2 + 4y^2 = (-8 - 9 + 4)y^2 = -13y^2$

4. (a) $1.3x + 3a - 9.6x + 2a = -8.3x + 5a$

(b) $5ab - 2ab^2 - 3a^2b + 6ab = 11ab - 2ab^2 - 3a^2b$

(c) There are no like terms in the expression $7x^2y - 2xy^2 - 3x^2y^2 - 4xy$, so no terms can be combined.

5. $5xy - 2x^2y + 6xy^2 - xy - 3xy^2 - 7x^2y$

$= 5xy - xy - 2x^2y - 7x^2y + 6xy^2 - 3xy^2$

$= 4xy - 9x^2y + 3xy^2$

6. $\frac{1}{7}a^2 + 2a^2 = \frac{1}{7}a^2 + \frac{2}{1}a^2 = \frac{1}{7}a^2 + \frac{2 \cdot 7}{1 \cdot 7}a^2$

$= \frac{1}{7}a^2 + \frac{14}{7}a^2 = \frac{15}{7}a^2$

$-\frac{5}{12}b - \frac{1}{3}b = -\frac{5}{12}b - \frac{1 \cdot 4}{3 \cdot 4}b = -\frac{5}{12}b - \frac{4}{12}b$

$= -\frac{9}{12}b = -\frac{3}{4}b$

Thus, our solution is $\frac{15}{7}a^2 - \frac{3}{4}b$.

7. $5a(2 - 3b) - 4(6a + 2ab) = 10a - 15ab - 24a - 8ab$

$= -14a - 23ab$

1.8 Practice Problems

1. $4 - \frac{1}{2}x = 4 - \frac{1}{2}(-8)$

$= 4 + 4$

$= 8$

2. (a) $4x^2 = 4(-3)^2 = 4(9) = 36$

(b) $(4x)^2 = [4(-3)]^2 = [-12]^2 = 144$

3. $2x^2 - 3x = 2(-2)^2 - 3(-2)$

$= 2(4) - 3(-2)$

$= 8 + 6$

$= 14$

4. Area of a triangle is $A = \frac{1}{2}ab$

altitude $= 3$ meters (m)

base $= 7$ meters (m)

$A = \frac{1}{2}(3 \text{ m})(7 \text{ m})$

$= \frac{1}{2}(3)(7)(\text{m})(\text{m})$

$= \left(\frac{3}{2}\right)(7)(\text{m})^2$

$= \frac{21}{2}(\text{m})^2$

$= 10.5$ square meters

5. Area of a circle is

$A = \pi r^2$

$r = 3$ meters

$A \approx 3.14(3 \text{ m})^2$

$= 3.14(9)(\text{m})^2$

$= 28.26$ square meters

6. Formula $C = \frac{5}{9}(F - 32)$

$= \frac{5}{9}(68 - 32)$

$= \frac{5}{9}(36)$

$= 5(4)$

$= 20°$ Celsius

7. Use the formula.

$k \approx 1.61(r)$

$= 1.61(35)$ Replace r by 35.

$= 56.35$

The truck is traveling at approximately 56.35 kilometers per hour. It is violating the minimum speed limit.

1.9 Practice Problems

1. $5[4x - 3(y - 2)]$
$= 5[4x - 3y + 6]$
$= 20x - 15y + 30$

2. $-3[2a - (3b - c) + 4d]$
$= -3[2a - 3b + c + 4d]$
$= -6a + 9b - 3c - 12d$

3. $3[4x - 2(1 - x)] - [3x + (x - 2)]$
$= 3[4x - 2 + 2x] - [3x + x - 2]$
$= 3[6x - 2] - [4x - 2]$
$= 18x - 6 - 4x + 2$
$= 14x - 4$

4. $-2\{5x - 3[2x - (3 - 4x)]\}$
$= -2\{5x - 3[2x - 3 + 4x]\}$
$= -2\{5x - 3[6x - 3]\}$
$= -2\{5x - 18x + 9\}$
$= -2\{-13x + 9\}$
$= 26x - 18$

Chapter 2 2.1 Practice Problems

1. $x + 14 = 23$
$x + 14 + (-14) = 23 + (-14)$
$x + 0 = 9$
$x = 9$

Check. $x + 14 = 23$
$9 + 14 \overset{?}{=} 23$
$23 = 23$ ✓

2. $17 = x - 5$
$17 + 5 = x - 5 + 5$
$22 = x + 0$
$22 = x$

Check. $17 = x - 5$
$17 \overset{?}{=} 22 - 5$
$17 = 17$ ✓

3. $0.5 - 1.2 = x - 0.3$
$-0.7 = x - 0.3$
$-0.7 + 0.3 = x - 0.3 + 0.3$
$-0.4 = x$

Check. $0.5 - 1.2 = x - 0.3$
$0.5 - 1.2 \overset{?}{=} -0.4 - 0.3$
$-0.7 = -0.7$ ✓

4. $x + 8 = -22 + 6$
$-2 + 8 \overset{?}{=} -22 + 6$
$6 \neq -16$ This is not true.
Thus $x = -2$ is not a solution. Solve to find the solution.
$x + 8 = -22 + 6$
$x + 8 = -16$
$x + 8 - 8 = -16 - 8$
$x = -24$

5. $\dfrac{1}{20} - \dfrac{1}{2} = x + \dfrac{3}{5}$
$\dfrac{1}{20} - \dfrac{1 \cdot 10}{2 \cdot 10} = x + \dfrac{3 \cdot 4}{5 \cdot 4}$
$\dfrac{1}{20} - \dfrac{10}{20} = x + \dfrac{12}{20}$
$-\dfrac{9}{20} = x + \dfrac{12}{20}$
$-\dfrac{9}{20} + \left(-\dfrac{12}{20}\right) = x + \dfrac{12}{20} + \left(-\dfrac{12}{20}\right)$
$-\dfrac{21}{20} = x$

$x = -\dfrac{21}{20}$ or $-1\dfrac{1}{20}$

Check.
$\dfrac{1}{20} - \dfrac{1}{2} = x + \dfrac{3}{5}$
$\dfrac{1}{20} - \dfrac{1}{2} \overset{?}{=} -\dfrac{21}{20} + \dfrac{3}{5}$
$\dfrac{1}{20} - \dfrac{10}{20} \overset{?}{=} -\dfrac{21}{20} + \dfrac{12}{20}$
$-\dfrac{9}{20} = -\dfrac{9}{20}$ ✓

2.2 Practice Problems

1. $\dfrac{1}{8}x = -2$
$8\left(\dfrac{1}{8}x\right) = 8(-2)$
$8\left(\dfrac{1}{8}\right)x = -16$
$x = -16$

2. $9x = 72$
$\dfrac{9x}{9} = \dfrac{72}{9}$
$x = 8$

3. $6x = 50$
$\dfrac{6x}{6} = \dfrac{50}{6}$
$x = \dfrac{25}{3}$

4. $-27x = 54$
$\dfrac{-27x}{-27} = \dfrac{54}{-27}$
$x = -2$

5. $-x = 36$
$-1x = 36$
$\dfrac{-1x}{-1} = \dfrac{36}{-1}$
$x = -36$

6. $-51 = -6x$
$\dfrac{-51}{-6} = \dfrac{-6x}{-6}$
$\dfrac{17}{2} = x$

7. $16.2 = 5.2x - 3.4x$
$16.2 = 1.8x$
$\dfrac{16.2}{1.8} = \dfrac{1.8x}{1.8}$
$9 = x$

2.3 Practice Problems

1. $9x + 2 = 38$
$9x + 2 + (-2) = 38 + (-2)$
$9x = 36$
$\dfrac{9x}{9} = \dfrac{36}{9}$
$x = 4$

Check. $9(4) + 2 \overset{?}{=} 38$
$36 + 2 \overset{?}{=} 38$
$38 = 38$ ✓

2. $13x = 2x - 66$
$13x + (-2x) = 2x + (-2x) - 66$
$11x = -66$
$\dfrac{11x}{11} = \dfrac{-66}{11}$
$x = -6$

3. $3x + 2 = 5x + 2$
$3x + (-3x) + 2 = 5x + (-3x) + 2$
$2 = 2x + 2$
$2 + (-2) = 2x + 2 + (-2)$
$0 = 2x$
$\dfrac{0}{2} = \dfrac{2x}{2}$
$0 = x$

Check. $3(0) + 2 \overset{?}{=} 5(0) + 2$
$2 = 2$ ✓

4. $-z + 8 - z = 3z + 10 - 3$
$-2z + 8 = 3z + 7$
$-2z + 2z + 8 = 3z + 2z + 7$
$8 = 5z + 7$
$8 + (-7) = 5z + 7 + (-7)$
$1 = 5z$
$\dfrac{1}{5} = \dfrac{5z}{5}$
$\dfrac{1}{5} = z$

5. $4x - (x + 3) = 12 - 3(x - 2)$
$4x - x - 3 = 12 - 3x + 6$
$3x - 3 = -3x + 18$
$3x + 3x - 3 = -3x + 3x + 18$
$6x - 3 = 18$
$6x - 3 + 3 = 18 + 3$
$6x = 21$
$\dfrac{6x}{6} = \dfrac{21}{6}$
$x = \dfrac{7}{2}$

Check. $4\left(\dfrac{7}{2}\right) - \left(\dfrac{7}{2} + 3\right) \overset{?}{=} 12 - 3\left(\dfrac{7}{2} - 2\right)$
$14 - \dfrac{13}{2} \overset{?}{=} 12 - 3\left(\dfrac{3}{2}\right)$
$\dfrac{28}{2} - \dfrac{13}{2} \overset{?}{=} \dfrac{24}{2} - \dfrac{9}{2}$
$\dfrac{15}{2} = \dfrac{15}{2}$ ✓

6.
$$4(-2x - 3) = -5(x - 2) + 2$$
$$-8x - 12 = -5x + 10 + 2$$
$$-8x - 12 = -5x + 12$$
$$-8x + 8x - 12 = -5x + 8x + 12$$
$$-12 = 3x + 12$$
$$-12 - 12 = 3x + 12 - 12$$
$$-24 = 3x$$
$$\frac{-24}{3} = \frac{3x}{3}$$
$$-8 = x$$

7.
$$0.3x - 2(x + 0.1) = 0.4(x - 3) - 1.1$$
$$0.3x - 2x - 0.2 = 0.4x - 1.2 - 1.1$$
$$-1.7x - 0.2 = 0.4x - 2.3$$
$$-1.7x + 1.7x - 0.2 = 0.4x + 1.7x - 2.3$$
$$-0.2 = 2.1x - 2.3$$
$$-0.2 + 2.3 = 2.1x - 2.3 + 2.3$$
$$2.1 = 2.1x$$
$$\frac{2.1}{2.1} = \frac{2.1x}{2.1}$$
$$1 = x$$

8.
$$5(2z - 1) + 7 = 7z - 4(z + 3)$$
$$10z - 5 + 7 = 7z - 4z - 12$$
$$10z + 2 = 3z - 12$$
$$10z - 3z + 2 = 3z - 3z - 12$$
$$7z + 2 = -12$$
$$7z + 2 - 2 = -12 - 2$$
$$7z = -14$$
$$\frac{7z}{7} = \frac{-14}{7}$$
$$z = -2$$

Check.
$$5[2(-2) - 1] + 7 \overset{?}{=} 7(-2) - 4[-2 + 3]$$
$$5[-4 - 1] + 7 \overset{?}{=} 7(-2) - 4[1]$$
$$5(-5) + 7 \overset{?}{=} -14 - 4$$
$$-25 + 7 \overset{?}{=} -18$$
$$-18 = -18 \;\checkmark$$

2.4 Practice Problems

1.
$$\frac{3}{8}x - \frac{3}{2} = \frac{1}{4}x$$
$$8\left(\frac{3}{8}x - \frac{3}{2}\right) = 8\left(\frac{1}{4}x\right)$$
$$\left(\frac{8}{1}\right)\left(\frac{3}{8}\right)(x) - \left(\frac{8}{1}\right)\left(\frac{3}{2}\right) = \left(\frac{8}{1}\right)\left(\frac{1}{4}\right)(x)$$
$$3x - 12 = 2x$$
$$3x + (-3x) - 12 = 2x + (-3x)$$
$$-12 = -x$$
$$12 = x$$

2.
$$\frac{5x}{4} - 1 = \frac{3x}{4} + \frac{1}{2}$$
$$4\left(\frac{5x}{4}\right) - 4(1) = 4\left(\frac{3x}{4}\right) + 4\left(\frac{1}{2}\right)$$
$$5x - 4 = 3x + 2$$
$$5x - 3x - 4 = 3x - 3x + 2$$
$$2x - 4 = 2$$
$$2x - 4 + 4 = 2 + 4$$
$$2x = 6$$
$$\frac{2x}{2} = \frac{6}{2}$$
$$x = 3$$

Check.
$$\frac{5(3)}{4} - 1 \overset{?}{=} \frac{3(3)}{4} + \frac{1}{2}$$
$$\frac{15}{4} - 1 \overset{?}{=} \frac{9}{4} + \frac{1}{2}$$
$$\frac{15}{4} - \frac{4}{4} \overset{?}{=} \frac{9}{4} + \frac{2}{4}$$
$$\frac{11}{4} = \frac{11}{4} \;\checkmark$$

3.
$$\frac{x + 6}{9} = \frac{x}{6} + \frac{1}{2}$$
$$\frac{x}{9} + \frac{6}{9} = \frac{x}{6} + \frac{1}{2}$$
$$18\left(\frac{x}{9}\right) + 18\left(\frac{6}{9}\right) = 18\left(\frac{x}{6}\right) + 18\left(\frac{1}{2}\right)$$
$$2x + 12 = 3x + 9$$
$$2x - 2x + 12 = 3x - 2x + 9$$
$$12 = x + 9$$
$$12 - 9 = x + 9 - 9$$
$$3 = x$$

4.
$$\frac{1}{2}(x + 5) = \frac{1}{5}(x - 2) + \frac{1}{2}$$
$$\frac{x}{2} - \frac{5}{2} = \frac{x}{5} - \frac{2}{5} + \frac{1}{2}$$
$$10\left(\frac{x}{2}\right) - 10\left(\frac{5}{2}\right) = 10\left(\frac{x}{5}\right) - 10\left(\frac{2}{5}\right) + 10\left(\frac{1}{2}\right)$$
$$5x + 25 = 2x - 4 + 5$$
$$5x + 25 = 2x + 1$$
$$5x - 2x + 25 = 2x - 2x + 1$$
$$3x + 25 = 1$$
$$3x + 25 - 25 = 1 - 25$$
$$3x = -24$$
$$\frac{3x}{3} = \frac{-24}{3}$$
$$x = -8$$

Check.
$$\frac{1}{2}(-8 + 5) = \frac{1}{5}(-8 - 2) + \frac{1}{2}$$
$$\frac{1}{2}(-3) = \frac{1}{5}(-10) + \frac{1}{2}$$
$$-\frac{3}{2} = -2 + \frac{1}{2}$$
$$-\frac{3}{2} = -\frac{4}{2} + \frac{1}{2}$$
$$-\frac{3}{2} = -\frac{3}{2} \;\checkmark$$

5.
$$2.8 = 0.3(x - 2) + 2(0.1x - 0.3)$$
$$2.8 = 0.3x - 0.6 + 0.2x - 0.6$$
$$10(2.8) = 10(0.3x) - 10(0.6) + 10(0.2x) - 10(0.6)$$
$$28 = 3x - 6 + 2x - 6$$
$$28 = 5x - 12$$
$$28 + 12 = 5x - 12 + 12$$
$$40 = 5x$$
$$\frac{40}{5} = \frac{5x}{5}$$
$$8 = x$$

2.5 Practice Problems

1.
$$d = rt$$
$$\frac{4565}{8.3} = \frac{8.3r}{8.3}$$
$$550 = r$$
The average rate of speed was 550 miles per hour.

2. Solve for m.
$$E = mc^2$$
$$\frac{E}{c^2} = \frac{mc^2}{c^2}$$
$$\frac{E}{c^2} = m$$

3.
$$8 - 2y + 3x = 0$$
$$8 - 2y = -3x$$
$$-2y = -3x - 8$$
$$\frac{-2y}{-2} = \frac{-3x - 8}{-2}$$
$$y = \frac{-3x}{-2} + \frac{-8}{-2}$$
$$y = \frac{3}{2}x + 4$$

4.
$$H = \frac{3}{4}(2y + x)$$
$$H = \frac{6y}{4} + \frac{3x}{4}$$
$$4(H) = 4\left(\frac{6y}{4}\right) + 4\left(\frac{3x}{4}\right)$$
$$4H = 6y + 3x$$
$$4H - 6y = 3x$$
$$\frac{4H - 6y}{3} = \frac{3x}{3}$$
$$\frac{4H - 6y}{3} = x$$

2.6 Practice Problems

1. (a) $7 > 2$ **(b)** $-2 > -4$ **(c)** $-1 < 2$
(d) $-8 < -5$ **(e)** $0 > -2$ **(f)** $5 > -3$

2. (a) x is greater than 5

(b) x is less than or equal to -2

(c) x is less than 3 (or 3 is greater than x)

(d) x is greater than or equal to $-\dfrac{3}{2}$

3. (a) Since the temperature can never exceed 180 degrees, then the temperature must always be less than or equal to 180 degrees. Thus, $t \le 180$.
(b) Since the debt must be less than 15,000, we have $d < 15,000$.

2.7 Practice Problems

1. (a) $9 > 6$ **(b)** $-2 < 5$ **(c)** $1 > -3$ **(d)** $10 < 15$
$13 > 10$ $-5 < 2$ $2 > -6$ $2 < 3$

2. (a) $7 > 2$ **(b)** $-3 < -1$
$-14 < -4$ $3 > 1$
(c) $-10 \ge -20$ **(d)** $-15 \le -5$
$1 \le 2$ $3 \ge 1$

3.
$$8x - 2 < 3$$
$$8x - 2 + 2 < 3 + 2$$
$$8x < 5$$
$$\frac{8x}{8} < \frac{5}{8}$$
$$x < \frac{5}{8}$$

4.
$$4 - 5x > 7$$
$$4 - 4 - 5x > 7 - 4$$
$$-5x > 3$$
$$\frac{-5x}{-5} < \frac{3}{-5}$$
$$x < -\frac{3}{5}$$

5.
$$\frac{1}{2}x + 3 < \frac{2}{3}x$$
$$6\left(\frac{1}{2}x\right) + 6(3) < 6\left(\frac{2}{3}x\right)$$
$$3x + 18 < 4x$$
$$3x - 4x + 18 < 4x - 4x$$
$$-x + 18 < 0$$
$$-x + 18 - 18 < 0 - 18$$
$$-x < -18$$
$$\frac{-x}{-1} > \frac{-18}{-1}$$
$$x > 18$$

6.
$$\frac{1}{2}(3 - x) \le 2x + 5$$
$$\frac{3}{2} - \frac{1}{2}x \le 2x + 5$$
$$2\left(\frac{3}{2}\right) - 2\left(\frac{1}{2}x\right) \le 2(2x) + 2(5)$$
$$3 - x \le 4x + 10$$
$$3 - x - 4x \le 4x - 4x + 10$$
$$3 - 5x \le 10$$
$$3 - 3 - 5x \le 10 - 3$$
$$-5x \le 7$$
$$\frac{-5x}{-5} \ge \frac{7}{-5}$$
$$x \ge -\frac{7}{5}$$

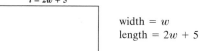

7.
$$2000n - 700,000 \ge 2,500,000$$
$$2000n - 700,000 + 700,000 \ge 2,500,000 + 700,000$$
$$2000n \ge 3,200,000$$
$$\frac{2000n}{2000} \ge \frac{3,200,000}{2000}$$
$$n \ge 1600$$

Chapter 3 3.1 Practice Problems

1. (a) $x + 4$ **(b)** $3x$ **(c)** $x - 8$ **(d)** $\dfrac{1}{4}x$

2. (a) $3x + 8$ **(b)** $3(x + 8)$ **(c)** $\dfrac{1}{3}(x + 4)$

3. Let a = Ann's hours per week.
Then $a - 17$ = Marie's hours per week.

4.

width = w
length = $2w + 5$

5. 1st angle = $s - 16$
2nd angle = s
3rd angle = $2s$

6. Let x = the number of students in the fall.
$\dfrac{2}{3}x$ = the number of students in the spring.
$\dfrac{1}{5}x$ = the number of students in the summer.

3.2 Practice Problems

1. Let x = the unknown number.
$$\frac{3}{4}x = -81$$
$$4\left(\frac{3}{4}x\right) = 4(-81)$$
$$3x = -324$$
$$\frac{3x}{3} = \frac{-324}{3}$$
$$x = -108$$
The number is -108.

2. Let x = the unknown number.
$$3x - 2 = 49$$
$$3x - 2 + 2 = 49 + 2$$
$$3x = 51$$
$$\frac{3x}{3} = \frac{51}{3}$$
$$x = 17$$
The number is 17.

3. Let x = the first number.
$$x + (3x - 12) = 24$$
$$4x - 12 = 24$$
$$4x = 36$$
$$x = 9$$
First number is 9. Second number = $3(9) - 12 = 15$.

4. Let rainfall in Canada = x
Then rainfall in Texas = $3x - 14$
$$3x - 14 = 43$$
$$3x = 57$$
$$x = 19$$
19 inches of rain was recorded in Canada.

5. (a) $d = rt$
$$220 = 4r$$
$$55 = r$$
Leaving the city, her average speed was 55 mph.

(b) $d = rt$
$$225 = 4.5r$$
$$50 = r$$
On the return trip, her average speed was 50 mph.

(c) She traveled 5 mph faster on the trip leaving the city.

6. Let x = her final exam score. Since the final counts as two tests, divide her total score by 6.
$$\frac{78 + 80 + 100 + 96 + x + x}{6} = 90$$
$$\frac{354 + 2x}{6} = 90$$
$$6\left(\frac{354 + 2x}{6}\right) = 6(90)$$
$$354 + 2x = 540$$
$$2x = 186$$
$$x = 93$$
She needs a 93 on the final exam.

3.3 Practice Problems

1. Let x = the length of the short piece.
Then $x + 17$ = the length of the long piece.
$$x + (x + 17) = 89$$
$$2x + 17 = 89$$
$$2x + 17 - 17 = 89 - 17$$
$$2x = 72$$
$$\frac{2x}{2} = \frac{72}{2}$$
$$x = 36$$
$$x + 17 = 36 + 17 = 53$$
The length of the short piece is 36 feet and the length of the long piece is 53 feet.

2. Let x = the second family's bill.
Then $x + 360$ = the first family's bill and
$2x - 200$ = the third family's bill.
$$x + (x + 360) + (2x - 200) = 3960$$
$$4x + 160 = 3960$$
$$4x = 3800$$
$$x = 950$$
$$x + 360 = 950 + 360 = 1310$$
$$2x - 200 = 2(950) - 200 = 1900 - 200 = 1700$$
The first family's bill was $1310, the second family's bill was $950, and the third family's bill was $1700.

3. Let x = the length of the second side.
Then $x - 30$ = the length of the first side and
$\frac{1}{2}x$ = the length of the third side.
$$x + (x - 30) + \frac{1}{2}x = 720$$
$$\frac{5}{2}x - 30 = 720$$
$$5x - 60 = 1440$$
$$5x = 1500$$
$$x = 300$$

$$x - 30 = 300 - 30 = 270$$
$$\frac{1}{2}x = \frac{1}{2}(300) = 150$$
The first side is 270 meters, the second side is 300 meters, and the third side is 150 meters.

3.4 Practice Problems

1. Let m = the number of miles.
$$3(25) + (0.20)m = 350$$
$$75 + 0.20m = 350$$
$$0.20m = 275$$
$$\frac{0.20m}{0.20} = \frac{275}{0.20}$$
$$m = 1375$$
He can travel 1375 miles for $350.

2. Let x = the cost of the encyclopedias he sold.
$$0.38x = 4560$$
$$\frac{0.38x}{0.38} = \frac{4560}{0.38}$$
$$x = 12,000$$
He sold $12,000 worth of encyclopedias last year.

3. Let x = the price last year.
$$x + 0.07x = 13,910$$
$$1.07x = 13,910$$
$$\frac{1.07x}{1.07} = \frac{13,910}{1.07}$$
$$x = 13,000$$
A similar model would have cost $13,000 last year.

4. $I = prt$
$$I = (7000)(0.12)(1)$$
$$I = 840$$
The interest charge on borrowing $7000 for one year at a simple interest rate of 12% is $840.

5. Let x = the amount invested at 9%.
Then $8000 - x$ was invested at 7%.
$$0.09x + 0.07(8000 - x) = 630$$
$$0.09x + 560 - 0.07x = 630$$
$$0.02x + 560 = 630$$
$$0.02x = 70$$
$$\frac{0.02x}{0.02} = \frac{70}{0.02}$$
$$x = 3500$$
$$8000 - x = 8000 - 3500 = 4500$$
Therefore, she invested $3500 at 9% and $4500 at 7%.

6. Let x = the number of dimes.
Then $x + 5$ = the number of quarters.
$$0.10x + 0.25(x + 5) = 5.10$$
$$0.10x + 0.25x + 1.25 = 5.10$$
$$0.35x + 1.25 = 5.10$$
$$0.35x = 3.85$$
$$\frac{0.35x}{0.35} = \frac{3.85}{0.35}$$
$$x = 11$$
$$x + 5 = 11 + 5 = 16$$
Therefore, she has 11 dimes and 16 quarters.

7. Let x = the number of dimes.
Then $2x$ = the number of nickels and
$x + 4$ = the number of quarters.
$$0.05(2x) + 0.10x + 0.25(x + 4) = 2.35$$
$$0.1x + 0.10x + 0.25x + 1 = 2.35$$
$$0.45x + 1 = 2.35$$
$$0.45x = 1.35$$
$$\frac{0.45x}{0.45} = \frac{1.35}{0.45}$$
$$x = 3$$
$$2x = 2(3) = 6$$
$$x + 4 = 3 + 4 = 7$$
Therefore, the boy has 3 dimes, 6 nickels, and 7 quarters.

3.5 Practice Problems

1. $A = \dfrac{1}{2}ab$

$= \dfrac{1}{2}(20 \text{ in.})(14 \text{ in.})$

$= (10)(14)(\text{in.})(\text{in.})$

$= 140$ square inches

2.
$$A = lw$$
$$120 \text{ (yd)}^2 = (l)(8 \text{ yd})$$
$$\dfrac{120 \text{ (yd)(yd)}}{8 \text{ yd}} = \dfrac{8 \text{ yd}}{8 \text{ yd}}l$$
$$15 \text{ yd} = l$$

3.
$$A = \dfrac{1}{2}a(b_1 + b_2)$$
$$256 \text{ (ft)}^2 = \dfrac{1}{2}a(12 \text{ ft} + 20 \text{ ft})$$
$$256 \text{ (ft)(ft)} = \dfrac{1}{2}(a)(32 \text{ ft})$$
$$256 \text{ (ft)(ft)} = (16 \text{ ft})a$$
$$16 \text{ ft} = a$$
The altitude is 16 feet.

4. $C = 2\pi r$

$C = 2(3.14)(15)$

$C = 94.2$

Rounded to the nearest meter, the circumference is approximately 94 meters.

5. $P = a + b + c$

$P = 15 \text{ cm} + 15 \text{ cm} + 15 \text{ cm}$

$= 45 \text{ cm}$

The perimeter of the triangle is 45 cm.

6.

$$132° + x + x = 180°$$
$$132° + 2x = 180°$$
$$2x = 48°$$
$$x = 24°$$
Both angles measure 24°.

7. Surface Area $= 4\pi r^2$

$= 4(3.14)(5 \text{ m})^2$

$= 314 \text{ m}^2$

Rounded to the nearest square meter, the surface area is 314 square meters.

8. $V = \pi r^2 h$

$= (3.14)(3 \text{ ft})^2(4 \text{ ft})$

$= 113.04 \text{ ft}^3$

Rounded to the nearest cubic foot, 113 cubic feet of sand can be stored in the drum.

9. Calculate the area of the pool.

$A = lw$

$A = (12 \text{ ft})(8 \text{ ft}) = 96 \text{ ft}^2$.

Now add 6 feet to the length and 6 feet to the width of the pool and calculate the area.

$A = lw$

$A = (18 \text{ ft})(14 \text{ ft}) = 252 \text{ ft}^2$.

Now subtract the areas.

$252 \text{ ft}^2 - 96 \text{ ft}^2 = 156 \text{ ft}^2$ at \$12 per square foot.

$156 \times 12 = 1872$

The cost would be \$1872.

10. $V = lwh$

$V = (6 \text{ ft})(5 \text{ ft})(8 \text{ ft}) = 240 \text{ ft}^3$

Weight $= 240 \text{ ft}^3 \times \dfrac{62.4 \text{ lb}}{1 \text{ ft}^3} = 14{,}976 \text{ lb} \approx 15{,}000 \text{ lb}$

3.6 Practice Problems

1. (a) height ≤ 6 **(b)** speed > 65

(c) area ≥ 560 **(d)** profit margin ≥ 50

2. Let $x =$ the number of pounds the fourth rope holds.
$$\dfrac{1050 + 1250 + 950 + x}{4} \geq 1100$$
$$\dfrac{3250 + x}{4} \geq 1100$$
$$x + 3250 \geq 4400$$
$$x \geq 1150$$
The rope must hold at least 1150 pounds.

3. Let $x =$ the value of the systems that she sells.
$$1400 + 0.02x > 2200$$
$$0.02x > 800$$
$$x > 40{,}000$$
Rita must sell more than \$40,000 worth of products each month.

Chapter 4 4.1 Practice Problems

1. (a) $a^7 \cdot a^5 = a^{7+5} = a^{12}$ **(b)** $w^{10} \cdot w = w^{10+1} = w^{11}$

2. (a) $x^3 \cdot x^9 = x^{3+9} = x^{12}$ **(b)** $3^7 \cdot 3^4 = 3^{7+4} = 3^{11}$

(c) $a^3 \cdot b^2 = a^3 \cdot b^2$ (cannot be simplified)

3. (a) $(-a^8)(a^4) = (-1 \cdot 1)(a^8 \cdot a^4)$

$= -1(a^8 \cdot a^4)$

$= -1a^{12}$

$= -a^{12}$

(b) $(3y^2)(-2y^3) = (3)(-2)(y^2 \cdot y^3) = -6y^5$

(c) $(-4x^3)(-5x^2) = (-4)(-5)(x^3 \cdot x^2) = 20x^5$

4. $(2xy)\left(-\dfrac{1}{4}x^2y\right)(6xy^3) = (2)\left(-\dfrac{1}{4}\right)(6)(x \cdot x^2 \cdot x)(y \cdot y \cdot y^3)$

$= -3x^4y^5$

5. (a) $\dfrac{10^{13}}{10^7} = 10^{13-7} = 10^6$ **(b)** $\dfrac{x^{11}}{x} = x^{11-1} = x^{10}$

(c) $\dfrac{y^{18}}{y^8} = y^{18-8} = y^{10}$

6. (a) $\dfrac{c^3}{c^4} = \dfrac{1}{c^{4-3}} = \dfrac{1}{c^1} = \dfrac{1}{c}$ **(b)** $\dfrac{10^{31}}{10^{56}} = \dfrac{1}{10^{56-31}} = \dfrac{1}{10^{25}}$

(c) $\dfrac{z^{15}}{z^{21}} = \dfrac{1}{z^{21-15}} = \dfrac{1}{z^6}$

7. (a) $\dfrac{-7x^7}{-21x^9} = \dfrac{1}{3x^{9-7}} = \dfrac{1}{3x^2}$ **(b)** $\dfrac{15x^{11}}{-3x^4} = -5x^{11-4} = -5x^7$

(c) $\dfrac{23x^8}{46x^9} = \dfrac{1}{2x^{9-8}} = \dfrac{1}{2x}$

8. (a) $\dfrac{x^7y^9}{y^{10}} = \dfrac{x^7}{y}$ **(b)** $\dfrac{12x^5y^6}{-24x^3y^8} = -\dfrac{x^2}{2y^2}$

9. (a) $\dfrac{10^7}{10^7} = 1$ **(b)** $\dfrac{12a^4}{15a^4} = \dfrac{4}{5}\left(\dfrac{a^4}{a^4}\right) = \dfrac{4}{5}(1) = \dfrac{4}{5}$

10. (a) $\dfrac{-20a^3b^8c^4}{28a^3b^7c^5} = -\dfrac{5a^0b}{7c} = -\dfrac{5(1)b}{7c} = -\dfrac{5b}{7c}$

(b) $\dfrac{5x^0y^6}{10x^4y^8} = \dfrac{5(1)y^6}{10x^4y^8} = \dfrac{1}{2x^4y^2}$

11. $\dfrac{(-6ab^5)(3a^2b^4)}{16a^5b^7} = \dfrac{-18a^3b^9}{16a^5b^7} = -\dfrac{9b^2}{8a^2}$

12. (a) $(a^4)^3 = a^{4 \cdot 3} = a^{12}$ **(b)** $(10^5)^2 = 10^{5 \cdot 2} = 10^{10}$

(c) $(-1)^{15} = -1$

13. (a) $(3xy)^3 = (3)^3x^3y^3 = 27x^3y^3$

(b) $(yz)^{37} = y^{37}z^{37}$

(c) $(-3x^3)^2 = (-3)^2(x^3)^2 = 9x^6$

14. (a) $\left(\dfrac{x}{5}\right)^3 = \dfrac{x^3}{5^3} = \dfrac{x^3}{125}$ **(b)** $\left(\dfrac{4a}{b}\right)^6 = \dfrac{(4a)^6}{b^6} = \dfrac{4^6a^6}{b^6} = \dfrac{4096a^6}{b^6}$

15. $\left(\dfrac{-2x^3y^0z}{4xz^2}\right)^5 = \left(\dfrac{-x^2}{2z}\right)^5 = \dfrac{(-1)^5(x^2)^5}{2^5z^5} = -\dfrac{x^{10}}{32z^5}$

4.2 Practice Problems

1. (a) $x^{-12} = \dfrac{1}{x^{12}}$ **(b)** $w^{-5} = \dfrac{1}{w^5}$ **(c)** $z^{-2} = \dfrac{1}{z^2}$

2. (a) $4^{-3} = \dfrac{1}{4^3} = \dfrac{1}{64}$ **(b)** $2^{-4} = \dfrac{1}{2^4} = \dfrac{1}{16}$

3. (a) $\dfrac{3}{w^{-4}} = 3w^4$ **(b)** $\dfrac{x^{-6}y^4}{z^{-2}} = \dfrac{y^4 z^2}{x^6}$ **(c)** $x^{-6}y^{-5} = \dfrac{1}{x^6 y^5}$

4. (a) $(2x^4 y^{-5})^{-2} = 2^{-2}x^{-8}y^{10} = \dfrac{y^{10}}{2^2 x^8} = \dfrac{y^{10}}{4x^8}$

(b) $\dfrac{y^{-3}z^{-4}}{y^2 z^{-6}} = \dfrac{z^6}{y^2 y^3 z^4} = \dfrac{z^6}{y^5 z^4} = \dfrac{z^2}{y^5}$

5. (a) $78{,}200 = 7.82 \times 10{,}000 = 7.82 \times 10^4$

    (b) $4{,}786{,}000 = 4.786 \times 1{,}000{,}000 = 4.786 \times 10^6$

6. (a) $0.98 = 9.8 \times 10^{-1}$

    (b) $0.000092 = 9.2 \times 10^{-5}$

7. (a) $1.93 \times 10^6 = 1.93 \times 1{,}000{,}000 = 1{,}930{,}000$

    (b) $8.562 \times 10^{-5} = 8.562 \times \dfrac{1}{100{,}000} = 0.00008562$

8. $30{,}900{,}000{,}000{,}000{,}000 \text{ meters} = 3.09 \times 10^{16} \text{ meters}$

9. (a) $(56{,}000)(1{,}400{,}000{,}000) = (5.6 \times 10^4)(1.4 \times 10^9)$

$$= (5.6)(1.4)(10^4)(10^9)$$
$$= 7.84 \times 10^{13}$$

(b) $\dfrac{0.000111}{0.00000037} = \dfrac{1.11 \times 10^{-4}}{3.7 \times 10^{-7}}$

$$= \dfrac{1.11}{3.7} \times \dfrac{10^{-4}}{10^{-7}}$$
$$= \dfrac{1.11}{3.7} \times \dfrac{10^7}{10^4}$$
$$= 0.3 \times 10^3$$
$$= 3.0 \times 10^2$$

10. $159 \text{ parsecs} = (159 \text{ parsecs})\dfrac{(3.09 \times 10^{13} \text{ kilometers})}{1 \text{ parsec}}$

$$= 491.31 \times 10^{13} \text{ kilometers}$$

$d = r \times t$

$$491.31 \times 10^{13} \text{ km} = \dfrac{50{,}000 \text{ km}}{1 \text{ hr}} \times t$$

$$4.9131 \times 10^{15} \text{ km} = \dfrac{5 \times 10^4 \text{ km}}{1 \text{ hr}} \times t$$

$$\dfrac{4.9131 \times 10^{15} \text{ km}}{\dfrac{5 \times 10^4 \text{ km}}{1 \text{ hr}}} = t$$

$$\dfrac{4.9131 \times 10^{15} \text{ km} \,(1 \text{ hr})}{5.0 \times 10^4 \text{ km}} = t$$

$$0.98262 \times 10^{11} \text{ hr} = t$$

It would take the probe about 9.83×10^{10} hours.

4.3 Practice Problems

1. (a) This polynomial is of degree 5. It has two terms, so it is a binomial.

    (b) This polynomial is of degree 7, since the sum of the exponents is $3 + 4 = 7$. It has one term, so it is a monomial.

    (c) This polynomial is of degree 3. It has three terms, so it is a trinomial.

2. $(-8x^3 + 3x^2 + 6) + (2x^3 - 7x^2 - 3)$

$$= [-8x^3 + 2x^3] + [3x^2 + (-7x^2)] + [6 + (-3)]$$
$$= [(-8 + 2)x^3] + [(3 - 7)x^2] + [6 - 3]$$
$$= -6x^3 + (-4x^2) + 3$$
$$= -6x^3 - 4x^2 + 3$$

3. $\left(-\dfrac{1}{3}x^2 - 6x - \dfrac{1}{12}\right) + \left(\dfrac{1}{4}x^2 + 5x - \dfrac{1}{3}\right)$

$$= \left[-\dfrac{1}{3}x^2 + \dfrac{1}{4}x^2\right] + [-6x + 5x] + \left[-\dfrac{1}{12} + \left(-\dfrac{1}{3}\right)\right]$$
$$= \left[\left(-\dfrac{1}{3} + \dfrac{1}{4}\right)x^2\right] + [(-6 + 5)x] + \left[-\dfrac{1}{12} + \left(-\dfrac{1}{3}\right)\right]$$
$$= \left[\left(-\dfrac{4}{12} + \dfrac{3}{12}\right)x^2\right] + [-x] + \left[-\dfrac{1}{12} - \dfrac{4}{12}\right]$$
$$= -\dfrac{1}{12}x^2 - x - \dfrac{5}{12}$$

4. $(3.5x^3 - 0.02x^2 + 1.56x - 3.5) + (-0.08x^2 - 1.98x + 4)$

$$= 3.5x^3 + (-0.02 - 0.08)x^2 + (1.56 - 1.98)x + (-3.5 + 4)$$
$$= 3.5x^3 - 0.1x^2 - 0.42x + 0.5$$

5. $(5x^3 - 15x^2 + 6x - 3) - (-4x^3 - 10x^2 + 5x + 13)$

$$= (5x^3 - 15x^2 + 6x - 3) + (4x^3 + 10x^2 - 5x - 13)$$
$$= (5 + 4)x^3 + (-15 + 10)x^2 + (6 - 5)x + (-3 - 13)$$
$$= 9x^3 - 5x^2 + x - 16$$

6. $(x^3 - 7x^2 y + 3xy^2 - 2y^3) - (2x^3 + 4xy - 6y^3)$

$$= (x^3 - 7x^2 y + 3xy^2 - 2y^3) + (-2x^3 - 4xy + 6y^3)$$
$$= (1 - 2)x^3 - 7x^2 y + 3xy^2 - 4xy + (-2 + 6)y^3$$
$$= -x^3 - 7x^2 y + 3xy^2 - 4xy + 4y^3$$

7. (a) 1974 is 4 years later than 1970, so $x = 4$.

$0.03(4) + 5.4 = 0.12 + 5.4 = 5.52$

We estimate that the average truck in 1974 obtained 5.52 miles per gallon.

    (b) 2011 is 41 years later than 1970, so $x = 41$.

$0.03(41) + 5.4 = 1.23 + 5.4 = 6.63$

We predict that the average truck in 2011 will obtain 6.63 miles per gallon.

4.4 Practice Problems

1. $4x^3(-2x^2 + 3x) = 4x^3(-2x^2) + 4x^3(3x)$

$$= 4(-2)(x^3 \cdot x^2) + (4 \cdot 3)(x^3 \cdot x)$$
$$= -8x^5 + 12x^4$$

2. (a) $-3x(x^2 + 2x - 4) = -3x^3 - 6x^2 + 12x$

    (b) $6xy(x^3 + 2x^2 y - y^2) = 6x^4 y + 12x^3 y^2 - 6xy^3$

3. $(-6x^3 + 4x^2 - 2x)(-3xy) = 18x^4 y - 12x^3 y + 6x^2 y$

4. $(5x - 1)(x - 2) = 5x^2 - 10x - x + 2 = 5x^2 - 11x + 2$

5. $(8a - 5b)(3a - b) = 24a^2 - 8ab - 15ab + 5b^2$

$$= 24a^2 - 23ab + 5b^2$$

6. $(3a + 2b)(2a - 3c) = 6a^2 - 9ac + 4ab - 6bc$

7. $(3x - 2y)(3x - 2y) = 9x^2 - 6xy - 6xy + 4y^2$

$$= 9x^2 - 12xy + 4y^2$$

8. $(2x^2 + 3y^2)(5x^2 + 6y^2) = 10x^4 + 12x^2 y^2 + 15x^2 y^2 + 18y^4$

$$= 10x^4 + 27x^2 y^2 + 18y^4$$

9. $A = (\text{length})(\text{width}) = (7x + 3)(2x - 1)$

$$= 14x^2 - 7x + 6x - 3$$
$$= 14x^2 - x - 3$$

There are $(14x^2 - x - 3)$ square feet in the room.

4.5 Practice Problems

1. $(6x + 7)(6x - 7) = (6x)^2 - (7)^2 = 36x^2 - 49$

2. $(3x - 5y)(3x + 5y) = (3x)^2 - (5y)^2 = 9x^2 - 25y^2$

3. (a) $(4a - 9b)^2 = (4a)^2 - 2(4a)(9b) + (9b)^2$

$$= 16a^2 - 72ab + 81b^2$$

    (b) $(5x + 4)^2 = (5x)^2 + 2(5x)(4) + (4)^2 = 25x^2 + 40x + 16$

4.
$$\begin{array}{r} 4x^3 - 2x^2 + x \\ x^2 + 3x - 2 \\ \hline -8x^3 + 4x^2 - 2x \\ 12x^4 - 6x^3 + 3x^2 \\ 4x^5 - 2x^4 + x^3 \\ \hline 4x^5 + 10x^4 - 13x^3 + 7x^2 - 2x \end{array}$$

5. $(2x^2 + 5x + 3)(x^2 - 3x - 4)$
$= 2x^2(x^2 - 3x - 4) + 5x(x^2 - 3x - 4) + 3(x^2 - 3x - 4)$
$= 2x^4 - 6x^3 - 8x^2 + 5x^3 - 15x^2 - 20x + 3x^2 - 9x - 12$
$= 2x^4 - x^3 - 20x^2 - 29x - 12$

6. $(3x - 2)(2x + 3)(3x + 2) = (3x - 2)(3x + 2)(2x + 3)$
$= [(3x)^2 - 2^2](2x + 3)$
$= (9x^2 - 4)(2x + 3)$

$$\begin{array}{r} 9x^2 - 4 \\ 2x + 3 \\ \hline 27x^2 + 0x - 12 \\ 18x^3 + 0x^2 - 8x \\ \hline 18x^3 + 27x^2 - 8x - 12 \end{array}$$

Thus we have
$(3x - 2)(2x + 3)(3x + 2) = 18x^3 + 27x^2 - 8x - 12.$

4.6 Practice Problems

1. $\dfrac{15y^4 - 27y^3 - 21y^2}{3y^2} = \dfrac{15y^4}{3y^2} - \dfrac{27y^3}{3y^2} - \dfrac{21y^2}{3y^2} = 5y^2 - 9y - 7$

2.
$$\begin{array}{r} x^2 + 6x + 7 \\ x + 4 \overline{\smash{\big)}\, x^3 + 10x^2 + 31x + 35} \\ \underline{x^3 + 4x^2} \\ 6x^2 + 31x \\ \underline{6x^2 + 24x} \\ 7x + 35 \\ \underline{7x + 28} \\ 7 \end{array}$$
Ans: $x^2 + 6x + 7 + \dfrac{7}{x + 4}$

3.
$$\begin{array}{r} 2x^2 + x + 1 \\ x - 1 \overline{\smash{\big)}\, 2x^3 - x^2 + 0x + 1} \\ \underline{2x^3 - 2x^2} \\ x^2 + 0x \\ \underline{x^2 - x} \\ x + 1 \\ \underline{x - 1} \\ 2 \end{array}$$
Ans: $2x^2 + x + 1 + \dfrac{2}{x - 1}$

4.
$$\begin{array}{r} 5x^2 + x - 2 \\ 4x - 3 \overline{\smash{\big)}\, 20x^3 - 11x^2 - 11x + 6} \\ \underline{20x^3 - 15x^2} \\ 4x^2 - 11x \\ \underline{4x^2 - 3x} \\ -8x + 6 \\ \underline{-8x + 6} \\ 0 \end{array}$$
Ans: $5x^2 + x - 2$

Check. $(4x - 3)(5x^2 + x - 2)$
$= 20x^3 + 4x^2 - 8x - 15x^2 - 3x + 6$
$= 20x^3 - 11x^2 - 11x + 6$

Chapter 5   5.1 Practice Problems

1. (a) $21a - 7b = 7(3a - b)$ because $7(3a - b) = 21a - 7b$.
   (b) $5xy + 8x = x(5y + 8)$ because $x(5y + 8) = 5xy + 8x$.

2. 4 is the greatest numerical common factor and a is a factor of each term. Thus, $4a$ is the greatest common factor.
$12a^2 + 16ab^2 - 12a^2b = 4a(3a + 4b^2 - 3ab)$

3. (a) The largest integer common to both terms is 8.
   $16a^3 - 24b^3 = 8(2a^3 - 3b^3)$

   (b) r^3 is common to all the terms.
   $r^3s^2 - 4r^4s + 7r^5 = r^3(s^2 - 4rs + 7r^2)$

4. We can factor $9, a, b^2$, and c out of each term.
$18a^3b^2c - 27ab^3c^2 - 45a^2b^2c^2 = 9ab^2c(2a^2 - 3bc - 5ac)$

5. We can factor $6, x$, and y^2 out of each term.
$30x^3y^2 - 24x^2y^2 + 6xy^2 = 6xy^2(5x^2 - 4x + 1)$
  Check: $6xy^2(5x^2 - 4x + 1) = 30x^3y^2 - 24x^2y^2 + 6xy^2$

6. $3(a + 5b) + x(a + 5b) = (a + 5b)(3 + x)$

7. $8y(9y^2 - 2) - (9y^2 - 2) = 8y(9y^2 - 2) - 1(9y^2 - 2)$
$= (9y^2 - 2)(8y - 1)$

8. $\pi b^2 - \pi a^2 = \pi(b^2 - a^2)$

5.2 Practice Problems

1. $3y(2x - 7) - 8(2x - 7) = (2x - 7)(3y - 8)$

2. $6x^2 - 15x + 4x - 10 = 3x(2x - 5) + 2(2x - 5)$
$= (2x - 5)(3x + 2)$

3. $ax + 2a + 4bx + 8b = a(x + 2) + 4b(x + 2)$
$= (x + 2)(a + 4b)$

4. $6a^2 + 5bc + 10ab + 3ac = 6a^2 + 10ab + 3ac + 5bc$
$= 2a(3a + 5b) + c(3a + 5b)$
$= (3a + 5b)(2a + c)$

5. $6xy + 14x - 15y - 35 = 2x(3y + 7) - 5(3y + 7)$
$= (3y + 7)(2x - 5)$

6. $3x + 6y - 5ax - 10ay = 3(x + 2y) - 5a(x + 2y)$
$= (x + 2y)(3 - 5a)$

7. $10ad + 27bc - 6bd - 45ac = 10ad - 6bd - 45ac + 27bc$
$= 2d(5a - 3b) - 9c(5a - 3b)$
$= (5a - 3b)(2d - 9c)$
  Check: $(5a - 3b)(2d - 9c) = 10ad - 45ac - 6bd + 27bc$
$= 10ad + 27bc - 6bd - 45ac$

5.3 Practice Problems

1. The two numbers that you can multiply to get 12 and add to get 8 are 6 and 2. $x^2 + 8x + 12 = (x + 6)(x + 2)$

2. The two numbers that have a product of 30 and a sum of 17 are 2 and 15. $x^2 + 17x + 30 = (x + 2)(x + 15)$

3. The two numbers that have a product of $+18$ and a sum of -11 must both be negative. The numbers are -9 and -2.
$x^2 - 11x + 18 = (x - 9)(x - 2)$

4. The two numbers whose product is 24 and whose sum is -11 are -8 and -3. $x^2 - 11x + 24 = (x - 8)(x - 3)$ or $(x - 3)(x - 8)$

5. The two numbers whose product is -24 and whose sum is -5 are -8 and $+3$. $x^2 - 5x - 24 = (x - 8)(x + 3)$ or $(x + 3)(x - 8)$

6. The two numbers whose product is -60 and whose sum is $+17$ are $+20$ and -3. $y^2 + 17y - 60 = (y + 20)(y - 3)$
  Check: $(y + 20)(y - 3) = y^2 - 3y + 20y - 60 = y^2 + 17y - 60$

7. List the possible factors of 60.

Factors of 60	Difference	
60 and 1	59	
30 and 2	28	
20 and 3	17	
15 and 4	11	
12 and 5	7	← Desired Value
10 and 6	4	

For the coefficient of the middle term to be -7, we must add -12 and $+5$. $x^2 - 7x - 60 = (x - 12)(x + 5)$

8. $a^4 = (a^2)(a^2)$
The two numbers whose product is -42 and whose sum is 1 are 7 and -6. $a^4 + a^2 - 42 = (a^2 + 7)(a^2 - 6)$

9. $3x^2 + 45x + 150 = 3(x^2 + 15x + 50)$
$= 3(x + 5)(x + 10)$

10. $4x^2 - 8x - 140 = 4(x^2 - 2x - 35)$
$= 4(x + 5)(x - 7)$

11. $x(x + 1) - 4(5) = x^2 + x - 20$
$= (x + 5)(x - 4)$

5.4 Practice Problems

1. To get a first term of $2x^2$, the coefficients of x in the factors must be 2 and 1. To get a last term of 5, the constants in the factors must be 1 and 5. Possibilities:

$(2x + 1)(x + 5) = 2x^2 + 11x + 5$

$(2x + 5)(x + 1) = 2x^2 + 7x + 5$

Thus $2x^2 + 7x + 5 = (2x + 5)(x + 1)$ or $(x + 1)(2x + 5)$.

2. The different factorizations of 9 are (3)(3) and (1)(9). The only factorization of 7 is (1)(7).

Possible Factors	Middle Term	Correct?
$(3x - 7)(3x - 1)$	$-24x$	No
$(9x - 7)(x - 1)$	$-16x$	No
$(9x - 1)(x - 7)$	$-64x$	Yes

The correct answer is $(9x - 1)(x - 7)$ or $(x - 7)(9x - 1)$.

3. The only factorization of 3 is (3)(1). The different factorizations of 14 are (14)(1) and (7)(2).

Possible Factors	Middle Term	Correct Factors?
$(3x + 14)(x - 1)$	$+11x$	No
$(3x + 1)(x - 14)$	$-41x$	No
$(3x + 7)(x - 2)$	$+x$	No (wrong sign)
$(3x + 2)(x - 7)$	$-19x$	No

To get the correct sign on the middle term, reverse the signs of the constants in the factors.

The correct answer is $(3x - 7)(x + 2)$ or $(x + 2)(3x - 7)$.

4. The grouping number is $2 \cdot 5 = 10$. The two numbers whose product is 10 and whose sum is 7 are 5 and 2. Write $7x$ as the sum $5x + 2x$.

$2x^2 + 7x + 5 = 2x^2 + 5x + 2x + 5$

$\qquad = x(2x + 5) + 1(2x + 5)$

$\qquad = (2x + 5)(x + 1)$

5. The grouping number is $9 \cdot 7 = 63$. The two numbers with a product of 63 and a sum of -64 are -63 and -1. Write $-64x$ as the sum $-63x - 1x$.

$9x^2 - 64x + 7 = 9x^2 - 63x - x + 7$

$\qquad = 9x(x - 7) - 1(x - 7)$

$\qquad = (x - 7)(9x - 1)$

6. The grouping number is $3(-4) = -12$. The two numbers with a product of -12 and a sum of 4 are 6 and -2. Write $4x$ as the sum $6x - 2x$.

$3x^2 + 4x - 4 = 3x^2 + 6x - 2x - 4$

$\qquad = 3x(x + 2) - 2(x + 2)$

$\qquad = (x + 2)(3x - 2)$

7. $8x^2 + 8x - 6 = 2(4x^2 + 4x - 3)$

$\qquad = 2(2x - 1)(2x + 3)$

8. $24x^2 - 38x + 10 = 2(12x^2 - 19x + 5)$

$\qquad = 2(4x - 5)(3x - 1)$

5.5 Practice Problems

1. $64x^2 - 1 = (8x + 1)(8x - 1)$ because $64x^2 = (8x)^2$ and $1 = (1)^2$.

2. $36x^2 - 49 = (6x + 7)(6x - 7)$ because $36x^2 = (6x)^2$ and $49 = (7)^2$.

3. $100x^2 - 81y^2 = (10x + 9y)(10x - 9y)$

4. $x^8 - 1 = (x^4 + 1)(x^4 - 1)$

$\qquad = (x^4 + 1)(x^2 + 1)(x^2 - 1)$

$\qquad = (x^4 + 1)(x^2 + 1)(x + 1)(x - 1)$

5. The first and last terms are perfect squares: $x^2 = (x)^2$ and $25 = (5)^2$. The middle term, $10x$, is twice the product of x and 5.

$x^2 + 10x + 25 = (x + 5)^2$

6. $30x = 2(5x \cdot 3)$

Also note the negative sign. $25x^2 - 30x + 9 = (5x - 3)^2$

7. (a) $25x^2 = (5x)^2$, $36y^2 = (6y)^2$, and $60xy = 2(5x \cdot 6y)$.

$25x^2 + 60xy + 36y^2 = (5x + 6y)^2$

(b) $64x^6 = (8x^3)^2$, $9 = (3)^2$, and $48x^3 = 2(8x^3 \cdot 3)$.

$64x^6 - 48x^3 + 9 = (8x^3 - 3)^2$

8. $9x^2 = (3x)^2$ and $4 = (2)^2$, but $15x \neq 2(3x \cdot 2) = 12x$.

$9x^2 + 15x + 4 = (3x + 1)(3x + 4)$

9. $20x^2 - 45 = 5(4x^2 - 9)$

$\qquad = 5(2x + 3)(2x - 3)$

10. $75x^2 - 60x + 12 = 3(25x^2 - 20x + 4)$

$\qquad = 3(5x - 2)^2$

5.6 Practice Problems

1. (a) $3x^2 - 36x + 108$

$\qquad = 3(x^2 - 12x + 36)$

$\qquad = 3(x - 6)^2$

(b) $9x^4y^2 - 9y^2$

$\qquad = 9y^2(x^4 - 1)$

$\qquad = 9y^2(x^2 + 1)(x^2 - 1)$

$\qquad = 9y^2(x^2 + 1)(x + 1)(x - 1)$

(c) $5x^3 - 15x^2y + 10x^2 - 30xy = 5x(x^2 - 3xy + 2x - 6y)$

$\qquad = 5x[x(x - 3y) + 2(x - 3y)]$

$\qquad = 5x(x - 3y)(x + 2)$

(d) $12x - 9 - 4x^2 = -4x^2 + 12x - 9$

$\qquad = -(4x^2 - 12x + 9)$

$\qquad = -(2x - 3)^2$

2. $x^2 - 9x - 8$

The factorizations of -8 are $(-2)(4)$, $(2)(-4)$, $(-8)(1)$, and $(-1)(8)$.

None of these pairs will add up to be the coefficient of the middle term. Thus the polynomial cannot be factored. It is prime.

3. $25x^2 + 82x + 4$

Check to see if this is a perfect square trinomial.

$2[(5)(2)] = 2(10) = 20$

This is not the coefficient of the middle term. The grouping number is 100. No factors add to 82. It is prime.

5.7 Practice Problems

1. $10x^2 - x - 2 = 0$

$(5x + 2)(2x - 1) = 0$

$5x + 2 = 0 \qquad 2x - 1 = 0$

$5x = -2 \qquad\quad 2x = 1$

$x = -\dfrac{2}{5} \qquad\quad x = \dfrac{1}{2}$

Check: $10\left(-\dfrac{2}{5}\right)^2 - \left(-\dfrac{2}{5}\right) - 2 \stackrel{?}{=} 0 \qquad 10\left(\dfrac{1}{2}\right)^2 - \dfrac{1}{2} - 2 \stackrel{?}{=} 0$

$10\left(\dfrac{4}{25}\right) + \dfrac{2}{5} - 2 \stackrel{?}{=} 0 \qquad 10\left(\dfrac{1}{4}\right) - \dfrac{1}{2} - 2 \stackrel{?}{=} 0$

$\dfrac{8}{5} + \dfrac{2}{5} - 2 \stackrel{?}{=} 0 \qquad\qquad \dfrac{5}{2} - \dfrac{1}{2} - 2 \stackrel{?}{=} 0$

$\dfrac{10}{5} - \dfrac{10}{5} \stackrel{?}{=} 0 \qquad\qquad\quad \dfrac{4}{2} - 2 \stackrel{?}{=} 0$

$0 = 0 \qquad\qquad\qquad\quad 2 - 2 \stackrel{?}{=} 2$

$\qquad\qquad\qquad\qquad\qquad\qquad 0 = 0$

Thus $-\dfrac{2}{5}$ and $\dfrac{1}{2}$ are both roots of the equation.

2. $3x^2 + 11x - 4 = 0$

$(3x - 1)(x + 4) = 0$

$3x - 1 = 0 \qquad x + 4 = 0$

$3x = 1 \qquad\qquad x = -4$

$x = \dfrac{1}{3}$

Thus $\dfrac{1}{3}$ and -4 are both roots of the equation.

3. $7x^2 + 11x = 0$
$x(7x + 11) = 0$
$x = 0 \qquad 7x + 11 = 0$
$\qquad\qquad 7x = -11$
$\qquad\qquad x = \dfrac{-11}{7}$

Thus 0 and $-\frac{11}{7}$ are both roots of the equation.

4. $x^2 - 6x + 4 = -8 + x$
$x^2 - 7x + 12 = 0$
$(x - 3)(x - 4) = 0$
$x - 3 = 0 \qquad x - 4 = 0$
$x = 3 \qquad\quad x = 4$
The roots are 3 and 4.

5. $\dfrac{2x^2 - 7x}{3} = 5$

$3\left(\dfrac{2x^2 - 7x}{3}\right) = 3(5)$

$2x^2 - 7x = 15$
$2x^2 - 7x - 15 = 0$
$(2x + 3)(x - 5) = 0$
$2x + 3 = 0 \qquad x - 5 = 0$
$x = -\dfrac{3}{2} \qquad\quad x = 5$

The roots are $-\frac{3}{2}$ and 5.

6. Let w = width, then
$3w + 2$ = length.
$(3w + 2)w = 85$
$3w^2 + 2w = 85$
$3w^2 + 2w - 85 = 0$
$(3w + 17)(w - 5) = 0$
$3w + 17 = 0 \qquad w - 5 = 0$
$w = -\dfrac{17}{3} \qquad\quad w = 5$

The only valid answer is width = 5 meters
$\qquad\qquad$ length = $3(5) + 2 = 17$ meters

7. Let b = base.
$b - 3$ = altitude.
$\dfrac{b(b - 3)}{2} = 35$

$\dfrac{b^2 - 3b}{2} = 35$

$b^2 - 3b = 70$
$b^2 - 3b - 70 = 0$
$(b + 7)(b - 10) = 0$
$b + 7 = 0$
$\qquad b = -7$
This is not a valid answer.
$b - 10 = 0$
$\qquad b = 10$
Thus the base = 10 centimeters
altitude = $10 - 3 = 7$ centimeters

8. $-5t^2 + 45 = 0$
$-5(t^2 - 9) = 0$
$t^2 - 9 = 0$
$(t + 3)(t - 3) = 0$
$t + 3 = 0 \qquad t - 3 = 0$
$t = -3 \qquad\quad t = 3$
$t = -3$ is not a valid answer.
Thus it will be 3 seconds before he breaks the water's surface.

Chapter 6   6.1 Practice Problems

1. $\dfrac{28}{63} = \dfrac{7 \cdot 4}{7 \cdot 9} = \dfrac{4}{9}$

2. $\dfrac{12x - 6}{14x - 7} = \dfrac{6(2x - 1)}{7(2x - 1)} = \dfrac{6}{7}$

3. $\dfrac{4x^2 - 9}{2x^2 - x - 3} = \dfrac{(2x - 3)(2x + 3)}{(2x - 3)(x + 1)} = \dfrac{2x + 3}{x + 1}$

4. $\dfrac{x^3 - 16x}{x^3 - 2x^2 - 8x} = \dfrac{x(x^2 - 16)}{x(x^2 - 2x - 8)} = \dfrac{x(x + 4)(x - 4)}{x(x + 2)(x - 4)} = \dfrac{x + 4}{x + 2}$

5. $\dfrac{8x - 20}{15 - 6x} = \dfrac{4(2x - 5)}{-3(-5 + 2x)} = \dfrac{4}{-3} = -\dfrac{4}{3}$

6. $\dfrac{4x^2 + 3x - 10}{25 - 16x^2} = \dfrac{(4x - 5)(x + 2)}{(5 + 4x)(5 - 4x)} = \dfrac{(4x - 5)(x + 2)}{-1(-5 + 4x)(5 + 4x)}$

$\qquad = \dfrac{x + 2}{-1(5 + 4x)} = -\dfrac{x + 2}{5 + 4x}$

7. $\dfrac{x^2 - 8xy + 15y^2}{2x^2 - 11xy + 5y^2} = \dfrac{(x - 3y)(x - 5y)}{(x - 5y)(2x - y)} = \dfrac{x - 3y}{2x - y}$

8. $\dfrac{25a^2 - 16b^2}{10a^2 + 3ab - 4b^2} = \dfrac{(5a + 4b)(5a - 4b)}{(5a + 4b)(2a - b)} = \dfrac{5a - 4b}{2a - b}$

6.2 Practice Problems

1. $\dfrac{6x^2 + 7x + 2}{x^2 - 7x + 10} \cdot \dfrac{x^2 + 3x - 10}{2x^2 + 11x + 5}$

$\quad = \dfrac{(2x + 1)(3x + 2)}{(x - 5)(x - 2)} \cdot \dfrac{(x + 5)(x - 2)}{(2x + 1)(x + 5)} = \dfrac{3x + 2}{x - 5}$

2. $\dfrac{2y^2 - 6y - 8}{y^2 - y - 2} \cdot \dfrac{y^2 - 5y + 6}{2y^2 - 32}$

$\quad = \dfrac{2(y^2 - 3y - 4)}{(y - 2)(y + 1)} \cdot \dfrac{(y - 3)(y - 2)}{2(y^2 - 16)}$

$\quad = \dfrac{2(y - 4)(y + 1)}{(y - 2)(y + 1)} \cdot \dfrac{(y - 3)(y - 2)}{2(y + 4)(y - 4)} = \dfrac{y - 3}{y + 4}$

3. $\dfrac{x^2 + 5x + 6}{x^2 + 8x} \div \dfrac{2x^2 + 5x + 2}{2x^2 + x}$

$\quad = \dfrac{x^2 + 5x + 6}{x^2 + 8x} \cdot \dfrac{2x^2 + x}{2x^2 + 5x + 2}$

$\quad = \dfrac{(x + 2)(x + 3)}{x(x + 8)} \cdot \dfrac{x(2x + 1)}{(2x + 1)(x + 2)} = \dfrac{x + 3}{x + 8}$

4. $\dfrac{x + 3}{x - 3} \div (9 - x^2) = \dfrac{x + 3}{x - 3} \div \dfrac{9 - x^2}{1} = \dfrac{x + 3}{x - 3} \cdot \dfrac{1}{9 - x^2}$

$\quad = \dfrac{x + 3}{x - 3} \cdot \dfrac{1}{(3 + x)(3 - x)}$

$\quad = \dfrac{1}{(x - 3)(3 - x)}$

6.3 Practice Problems

1. $\dfrac{2s + t}{2s - t} + \dfrac{s - t}{2s - t} = \dfrac{2s + t + s - t}{2s - t} = \dfrac{3s}{2s - t}$

2. $\dfrac{b}{(a - 2b)(a + b)} - \dfrac{2b}{(a - 2b)(a + b)}$

$\quad = \dfrac{b - 2b}{(a - 2b)(a + b)}$

$\quad = \dfrac{-b}{(a - 2b)(a + b)}$

3. $\dfrac{7}{6x + 21}, \dfrac{13}{10x + 35}$

$6x + 21 = 3(2x + 7)$
$10x + 35 = 5(2x + 7)$
$\quad$ LCD $= 3 \cdot 5 \cdot (2x + 7) = 15(2x + 7)$

4. (a) $\dfrac{3}{50xy^2z}, \dfrac{19}{40x^3yz}$

$50xy^2z = 2 \cdot \qquad 5 \cdot 5 \cdot x \cdot \qquad y \cdot y \cdot z$

$40x^3yz = 2 \cdot 2 \cdot 2 \cdot 5 \cdot \;\Big| \;x \cdot x \cdot x \cdot y \cdot \;\Big| \;z$

$\downarrow\downarrow\downarrow\downarrow\downarrow\downarrow\quad\downarrow\downarrow\downarrow\downarrow\downarrow$

$2 \cdot 2 \cdot 2 \cdot 5 \cdot 5 \cdot x \cdot x \cdot x \cdot y \cdot y \cdot z$

$\text{LCD} = 2^3 \cdot 5^2 \cdot x^3 \cdot y^2 \cdot z = 200x^3y^2z$

(b) $\dfrac{2}{x^2+5x+6}, \dfrac{6}{3x^2+5x-2}$

$x^2 + 5x + 6 = (x+3)(x+2)$

$3x^2 + 5x - 2 = \qquad (x+2)(3x-1)$

$\quad\downarrow\qquad\quad\downarrow\qquad\quad\downarrow$

$\text{LCD} = (x+3)(x+2)(3x-1)$

5. LCD $= abc$

$\dfrac{7}{a} + \dfrac{3}{abc} = \dfrac{7}{a} \cdot \dfrac{bc}{bc} + \dfrac{3}{abc} = \dfrac{7bc}{abc} + \dfrac{3}{abc} = \dfrac{7bc+3}{abc}$

6. $a^2 - 4b^2 = (a+2b)(a-2b)$

LCD $= (a+2b)(a-2b)$

$\dfrac{2a-b}{a^2-4b^2} + \dfrac{2}{a+2b}$

$= \dfrac{2a-b}{(a+2b)(a-2b)} + \dfrac{2}{(a+2b)} \cdot \dfrac{a-2b}{a-2b}$

$= \dfrac{2a-b}{(a+2b)(a-2b)} + \dfrac{2a-4b}{(a+2b)(a-2b)}$

$= \dfrac{2a-b+2a-4b}{(a+2b)(a-2b)}$

$= \dfrac{4a-5b}{(a+2b)(a-2b)}$

7. $\dfrac{7a}{a^2+2ab+b^2} + \dfrac{4}{a^2+ab}$

$= \dfrac{7a}{(a+b)^2} + \dfrac{4}{a(a+b)} \qquad \text{LCD} = a(a+b)^2$

$= \dfrac{7a^2}{a(a+b)^2} + \dfrac{4(a+b)}{a(a+b)^2} = \dfrac{7a^2+4a+4b}{a(a+b)^2}$

8. $\dfrac{x+7}{3x-9} - \dfrac{x-6}{x-3} = \dfrac{x+7}{3(x-3)} - \dfrac{x-6}{x-3}$

$= \dfrac{x+7}{3(x-3)} - \dfrac{x-6}{x-3} \cdot \dfrac{3}{3} = \dfrac{x+7-3(x-6)}{3(x-3)}$

$= \dfrac{x+7-3x+18}{3(x-3)} = \dfrac{-2x+25}{3(x-3)}$

9. $\dfrac{x-2}{x^2-4} - \dfrac{x+1}{2x^2+4x} = \dfrac{x-2}{(x+2)(x-2)} - \dfrac{x+1}{2x(x+2)}$

$= \dfrac{x-2}{(x+2)(x-2)} \cdot \dfrac{2x}{2x} - \dfrac{x+1}{2x(x+2)} \cdot \dfrac{x-2}{x-2}$

$= \dfrac{2x(x-2)-(x+1)(x-2)}{2x(x+2)(x-2)} = \dfrac{2x^2-4x-x^2+x+2}{2x(x+2)(x-2)}$

$= \dfrac{x^2-3x+2}{2x(x+2)(x-2)} = \dfrac{(x-1)(x-2)}{2x(x+2)(x-2)} = \dfrac{x-1}{2x(x+2)}$

6.4 Practice Problems

1. $\dfrac{\frac{1}{a}+\frac{1}{a^2}}{\frac{2}{b^2}} = \dfrac{\frac{1}{a}\cdot\frac{a}{a}+\frac{1}{a^2}}{\frac{2}{b^2}} = \dfrac{\frac{a+1}{a^2}}{\frac{2}{b^2}} = \dfrac{a+1}{a^2} \div \dfrac{2}{b^2}$

$= \dfrac{a+1}{a^2} \cdot \dfrac{b^2}{2} = \dfrac{b^2(a+1)}{2a^2}$

2. $\dfrac{\frac{1}{a}+\frac{1}{b}}{\frac{x}{2}-\frac{5}{y}} = \dfrac{\frac{1}{a}\cdot\frac{b}{b}+\frac{1}{b}\cdot\frac{a}{a}}{\frac{x}{2}\cdot\frac{y}{y}-\frac{5}{y}\cdot\frac{2}{2}} = \dfrac{\frac{b+a}{ab}}{\frac{xy-10}{2y}} = \dfrac{b+a}{ab} \cdot \dfrac{2y}{xy-10} = \dfrac{2y(b+a)}{ab(xy-10)}$

3. $\dfrac{\frac{x}{x^2+4x+3}+\frac{2}{x+1}}{x+1} = \dfrac{\frac{x}{(x+1)(x+3)}+\frac{2}{x+1}\cdot\frac{x+3}{x+3}}{x+1}$

$= \dfrac{\frac{x+2x+6}{(x+1)(x+3)}}{x+1} = \dfrac{3x+6}{(x+1)(x+3)} \cdot \dfrac{1}{(x+1)}$

$= \dfrac{3(x+2)}{(x+1)^2(x+3)}$

4. $\dfrac{\frac{6}{x^2-y^2}}{\frac{1}{x-y}+\frac{3}{x+y}} = \dfrac{\frac{6}{x^2-y^2}}{\frac{1}{x-y}\cdot\frac{x+y}{x+y}+\frac{3}{x+y}\cdot\frac{x-y}{x-y}}$

$= \dfrac{\frac{6}{x^2-y^2}}{\frac{x+y}{(x+y)(x-y)}+\frac{3x-3y}{(x+y)(x-y)}} = \dfrac{\frac{6}{(x+y)(x-y)}}{\frac{4x-2y}{(x+y)(x-y)}}$

$= \dfrac{6}{(x+y)(x-y)} \cdot \dfrac{(x+y)(x-y)}{2(2x-y)} = \dfrac{2\cdot3}{2(2x-y)} = \dfrac{3}{2x-y}$

5. The LCD of all the denominators is $3x^2y$.

$\dfrac{\frac{2}{3x^2}-\frac{3}{y}}{\frac{5}{xy}-4} = \dfrac{3x^2y\left(\frac{2}{3x^2}-\frac{3}{y}\right)}{3x^2y\left(\frac{5}{xy}-4\right)} = \dfrac{3x^2y\left(\frac{2}{3x^2}\right)-3x^2y\left(\frac{3}{y}\right)}{3x^2y\left(\frac{5}{xy}\right)-3x^2y(4)} = \dfrac{2y-9x^2}{15x-12x^2y}$

6. The LCD of all the denominators is $(x+y)(x-y)$.

$\dfrac{\frac{6}{x^2-y^2}}{\frac{7}{x-y}+\frac{3}{x+y}}$

$= \dfrac{(x+y)(x-y)\left(\frac{6}{(x+y)(x-y)}\right)}{(x+y)(x-y)\left(\frac{7}{x-y}\right)+(x+y)(x-y)\left(\frac{3}{x+y}\right)}$

$= \dfrac{6}{7(x+y)+3(x-y)}$

$= \dfrac{6}{7x+7y+3x-3y}$

$= \dfrac{6}{10x+4y}$

$= \dfrac{2\cdot3}{2(5x+2y)}$

$= \dfrac{3}{5x+2y}$

6.5 Practice Problems

1. $\dfrac{3}{x}+\dfrac{4}{5}=-\dfrac{2}{x}$ \qquad LCD $= 5x$

$5x\left(\dfrac{3}{x}\right)+5x\left(\dfrac{4}{5}\right)=5x\left(-\dfrac{2}{x}\right)$

$15+4x=-10$

$4x=-10-15$

$4x=-25$

$x=-\dfrac{25}{4}$

Check.

$\dfrac{3}{-\frac{25}{4}}+\dfrac{4}{5}\stackrel{?}{=}-\dfrac{2}{-\frac{25}{4}}$

$-\dfrac{12}{25}+\dfrac{4}{5}\stackrel{?}{=}\dfrac{8}{25}$

$-\dfrac{12}{25}+\dfrac{20}{25}\stackrel{?}{=}\dfrac{8}{25}$

$\dfrac{8}{25}=\dfrac{8}{25}$ ✓

2. LCD $= (2x + 1)(2x - 1)$

$$\frac{4}{2x + 1} = \frac{6}{2x - 1}$$

$$(2x + 1)(2x - 1)\left(\frac{4}{2x + 1}\right) = (2x + 1)(2x - 1)\left(\frac{6}{2x - 1}\right)$$

$$4(2x - 1) = 6(2x + 1)$$
$$8x - 4 = 12x + 6$$
$$-4x - 4 = 6$$
$$-4x = 10$$
$$x = -\frac{5}{2}$$

Check.

$$\frac{4}{2\left(-\frac{5}{2}\right) + 1} \overset{?}{=} \frac{6}{2\left(-\frac{5}{2}\right) - 1}$$

$$\frac{4}{-5 + 1} \overset{?}{=} \frac{6}{-5 - 1}$$

$$\frac{4}{-4} \overset{?}{=} \frac{6}{-6}$$

$$-1 = -1 \checkmark$$

3.

$$\frac{x - 1}{x^2 - 4} = \frac{2}{x + 2} + \frac{4}{x - 2}$$

$$\frac{x - 1}{(x + 2)(x - 2)} = \frac{2}{x + 2} + \frac{4}{x - 2}$$

$$(x + 2)(x - 2)\left[\frac{x - 1}{(x + 2)(x - 2)}\right]$$

$$= (x + 2)(x - 2)\left(\frac{2}{x + 2}\right) + (x + 2)(x - 2)\left(\frac{4}{x - 2}\right)$$

$$x - 1 = 2(x - 2) + 4(x + 2)$$
$$x - 1 = 2x - 4 + 4x + 8$$
$$x - 1 = 6x + 4$$
$$-5x - 1 = 4$$
$$-5x = 5$$
$$x = -1$$

Check.

$$\frac{-1 - 1}{(-1)^2 - 4} \overset{?}{=} \frac{2}{-1 + 2} + \frac{4}{-1 - 2}$$

$$\frac{-2}{-3} \overset{?}{=} \frac{2}{1} + \frac{4}{-3}$$

$$\frac{2}{3} \overset{?}{=} \frac{6}{3} - \frac{4}{3}$$

$$\frac{2}{3} = \frac{2}{3} \checkmark$$

4. The LCD is $x + 1$.

$$\frac{2x}{x + 1} = \frac{-2}{x + 1} + 1$$

$$(x + 1)\left(\frac{2x}{x + 1}\right) = (x + 1)\left(\frac{-2}{x + 1}\right) + (x + 1)(1)$$

$$2x = -2 + x + 1$$
$$2x = x - 1$$
$$x = -1 \text{ (but see the check)}$$

Check.

$$\frac{2(-1)}{-1 + 1} \overset{?}{=} \frac{-2}{-1 + 1} + 1$$

$$\frac{-2}{0} \overset{?}{=} \frac{-2}{0} + 1$$

These expressions are not defined; therefore, there is no solution to this equation.

6.6 Practice Problems

1. Let $x = $ the number of hours it will take to drive 315 miles.

$$\frac{8}{420} = \frac{x}{315}$$
$$8(315) = 420x$$
$$\frac{2520}{420} = x$$
$$6 = x$$

It would take Brenda 6 hours to drive 315 miles.

2. Let $x = $ the distance represented by $2\frac{1}{2}$ inches.

$$\frac{\frac{5}{8}}{30} = \frac{2\frac{1}{2}}{x}$$

$$\frac{5}{8}x = 30\left(2\frac{1}{2}\right)$$

$$\frac{5}{8}x = 30\left(\frac{5}{2}\right)$$

$$\frac{5}{8}x = 75$$

$$8\left(\frac{5}{8}x\right) = 8(75)$$

$$5x = 600$$
$$x = 120$$

Therefore $2\frac{1}{2}$ inches represents 120 miles.

3.

$$\frac{13}{x} = \frac{16}{18}$$
$$13(18) = 16x$$
$$234 = 16x$$
$$\frac{234}{16} = x$$

Side x has length

$$\frac{117}{8} = 14\frac{5}{8} \text{ cm.}$$

4.

$$\frac{6}{7} = \frac{x}{38.5}$$
$$6(38.5) = 7x$$
$$231 = 7x$$
$$x = 33 \text{ feet}$$

5. Let $x = $ the speed of train B. Then

$$\text{train A time} = \frac{180}{x + 10} \text{ and train B time} = \frac{150}{x}.$$

$$\frac{180}{x + 10} = \frac{150}{x}$$
$$180x = 150(x + 10)$$
$$180x = 150x + 1500$$
$$30x = 1500$$
$$x = 50$$

Train B traveled 50 kilometers per hour. Train A traveled $50 + 10 = 60$ kilometers per hour.

6.

	Number of Hours	Part of the Job Done in One Hour
John	6 hours	$\frac{1}{6}$
Dave	7 hours	$\frac{1}{7}$
John & Dave Together	x	$\frac{1}{x}$

LCD $= 42x$

$$\frac{1}{6} + \frac{1}{7} = \frac{1}{x}$$

$$42x\left(\frac{1}{6}\right) + 42x\left(\frac{1}{7}\right) = 42x\left(\frac{1}{x}\right)$$

$$7x + 6x = 42$$
$$13x = 42$$
$$x = 3\frac{3}{13}$$

$$\frac{3}{13} \text{ hour} \times \frac{60 \text{ min}}{1 \text{ hour}} = \frac{180}{13} \text{ min} \approx 13.846 \text{ min}$$

Thus, doing the job together will take 3 hours and 14 minutes.

Chapter 7 7.1 Practice Problems

1. Point B is 3 units to the right on the x-axis and 4 units up from the point where we stopped on the x-axis.

2. (a) Begin by counting 2 squares to the left, starting at the origin. Since the y-coordinate is negative, count 4 units down from the point where we stopped on the x-axis. Label the point I.

   (b) Begin by counting 4 squares to the left of the origin. Then count 5 units up because the y-coordinate is positive. Label the point J.

   (c) Begin by counting 4 units to the right of the origin. Then count 2 units down because the y-coordinate is negative. Label the point K.

3. The points are plotted in the figure.

4. Move along the x-axis to get as close as possible to B. We end up at 2. Thus the first number of the ordered pair is 2. Then count 7 units upward on a line parallel to the y-axis to reach B. So the second number of the ordered pair is 7. Thus, point B is represented by $(2, 7)$.

5. $A = (-2, -1)$; $B = (-1, 3)$; $C = (0, 0)$; $D = (2, -1)$; $E = (3, 1)$

6. (a) Replace x by 0 in the equation.
$$3x - 4y = 12$$
$$3(0) - 4y = 12$$
$$0 - 4y = 12$$
$$y = -3$$
The ordered pair is $(0, -3)$.

   (b) Replace the variable y by 3.
$$3x - 4y = 12$$
$$3x - 4(3) = 12$$
$$3x - 12 = 12$$
$$3x = 24$$
$$x = 8$$
The ordered pair is $(8, 3)$.

   (c) Replace the variable y by -6.
$$3x - 4y = 12$$
$$3x - 4(-6) = 12$$
$$3x + 24 = 12$$
$$3x = -12$$
$$x = -4$$
The ordered pair is $(-4, -6)$.

7. (a)

   (b) Motor vehicle deaths were significantly high in 1980. During 1985–2000, the number of motor vehicle deaths was relatively stable.

7.2 Practice Problems

1. Graph $x + y = 10$.
Let $x = 0$.
$$x + y = 10$$
$$0 + y = 10$$
$$y = 10$$

Let $x = 3$.
$$x + y = 10$$
$$3 + y = 10$$
$$y = 7$$

Let $x = 10$.
$$x + y = 10$$
$$10 + y = 10$$
$$y = 0$$
Plot the ordered pairs $(0, 10)$, $(3, 7)$, and $(10, 0)$.

2.
$$7x + 3 = -2y + 3$$
$$7x + 3 - 3 = -2y + 3 - 3$$
$$7x = -2y$$
$$7x + 2y = -2y + 2y$$
$$7x + 2y = 0$$

Let $x = 0$.
$$7(0) + 2y = 0$$
$$2y = 0$$
$$y = 0$$

Let $x = -2$.
$$7(-2) + 2y = 0$$
$$-14 + 2y = 0$$
$$2y = 14$$
$$y = 7$$

Let $x = 2$.
$$7(2) + 2y = 0$$
$$14 + 2y = 0$$
$$2y = -14$$
$$y = -7$$
Graph the ordered pairs $(0, 0)$, $(-2, 7)$, and $(2, -7)$.

3. $2y - x = 6$
Find the two intercepts.

Let $y = 0$.   Let $x = 0$.
$$2(0) - x = 6 \qquad 2y - 0 = 6$$
$$-x = 6 \qquad\qquad 2y = 6$$
$$x = -6 \qquad\qquad\ y = 3$$
The x-intercept is $(-6, 0)$, the y-intercept is $(0, 3)$.
Find a third point.
Let $y = 1$.
$$2(1) - x = 6$$
$$2 - x = 6$$
$$-x = 4$$
$$x = -4$$
Graph the ordered pairs $(-6, 0)$, $(0, 3)$, and $(-4, 1)$.

4. $2y - 3 = 0$
Solve for y.
$$2y = 3$$
$$y = \frac{3}{2}$$
This line is parallel to the x-axis. It is a horizontal line $1\frac{1}{2}$ units above the x-axis.

5. $x + 3 = 0$
Solve for x.
$$x = -3$$
This line is parallel to the y-axis. It is a vertical line 3 units to the left of the y-axis.

7.3 Practice Problems

1. $m = \dfrac{y_2 - y_1}{x_2 - x_1} = \dfrac{-1 - 1}{-4 - 6} = \dfrac{-2}{-10} = \dfrac{1}{5}$

2. $m = \dfrac{y_2 - y_1}{x_2 - x_1} = \dfrac{1 - 0}{-1 - 2} = \dfrac{1}{-3} = -\dfrac{1}{3}$

3. (a) $m = \dfrac{3 - 6}{-5 - (-5)} = \dfrac{-3}{0}$

$\dfrac{-3}{0}$ is undefined. Therefore there is no slope and the line is a vertical line through $x = -5$.

(b) $m = \dfrac{-11 - (-11)}{3 - (-7)} = \dfrac{0}{10} = 0$

$m = 0$. The line is a horizontal line through $y = -11$.

4. Solve for y.

$4x - 2y = -5$

$-2y = -4x - 5$

$y = \dfrac{-4x - 5}{-2}$

$y = 2x + \dfrac{5}{2}$ Slope $= 2$ y-intercept $= \left(0, \dfrac{5}{2}\right)$

5. (a) $y = mx + b$

$m = -\dfrac{3}{7}$ y-intercept $= \left(0, \dfrac{2}{7}\right), b = \dfrac{2}{7}$

$y = -\dfrac{3}{7}x + \dfrac{2}{7}$

(b) $y = -\dfrac{3}{7}x + \dfrac{2}{7}$

$7(y) = 7\left(-\dfrac{3}{7}x\right) + 7\left(\dfrac{2}{7}\right)$

$7y = -3x + 2$

$3x + 7y = 2$

6. y-intercept $= (0, -1)$. Thus the coordinates of the y-intercept for this line are $(0, -1)$. Plot the point. Slope is $\dfrac{\text{rise}}{\text{run}}$. Since the slope for this line is $\dfrac{3}{4}$, we will go up (rise) 3 units and go over (run) 4 units to the right from the point $(0, -1)$. This is the point $(4, 2)$.

7. $y = -\dfrac{2}{3}x + 5$

The y-intercept is $(0, 5)$ since $b = 5$. Plot the point $(0, 5)$. The slope is $-\dfrac{2}{3} = \dfrac{-2}{3}$. Begin at $(0, 5)$, go down 2 units and to the right 3 units. This is the point $(3, 3)$. Draw the line that connects the points $(0, 5)$ and $(3, 3)$.

8. (a) Parallel lines have the same slope. Line j has a slope of $\dfrac{1}{4}$.

(b) Perpendicular lines have slopes whose product is -1.

$m_1 m_2 = -1$

$\dfrac{1}{4}m_2 = -1$

$4\left(\dfrac{1}{4}\right)m_2 = -1(4)$

$m_2 = -4$

Thus line k has a slope of -4.

9. (a) The slope of line n is $\dfrac{2}{3}$. The slope of a line that is parallel to line n is $\dfrac{2}{3}$.

(b) $m_1 m_2 = -1$

$\dfrac{2}{3}m_2 = -1$

$m_2 = -\dfrac{3}{2}$

The slope of a line that is perpendicular to line n is $-\dfrac{3}{2}$.

7.4 Practice Problems

1. $y = mx + b$

$12 = -\dfrac{3}{4}(-8) + b$

$12 = 6 + b$

$6 = b$

An equation of the line is $y = -\dfrac{3}{4}x + 6$.

2. Find the slope.

$m = \dfrac{y_2 - y_1}{x_2 - x_1} = \dfrac{1 - 5}{-1 - 3} = \dfrac{-4}{-4} = 1$

Using either of the two points given, substitute x and y values into the equation $y = mx + b$.

$m = 1$ $x = 3$ and $y = 5$.

$y = mx + b$

$5 = 1(3) + b$

$5 = 3 + b$

$2 = b$

An equation of the line is $y = x + 2$.

3. The y-intercept is $(0, 1)$. Thus $b = 1$. Look for another point on the line. We choose $(6, 2)$. Count the number of vertical units from 1 to 2 (rise). Count the number of horizontal units from 0 to 6 (run). $m = \dfrac{1}{6}$

Now we can write an equation of the line.

$y = mx + b$

$y = \dfrac{1}{6}x + 1$

7.5 Practice Problems

1. Graph $x - y \geq -10$.

Begin by graphing the line $x - y = -10$. Use any method discussed previously. Since there is an equals sign in the inequality, draw a solid line to indicate that the line is part of the solution set. The easiest test point is $(0, 0)$. Substitute $x = 0, y = 0$ in the inequality.

$x - y \geq -10$

$0 - 0 \geq -10$

$0 \geq -10$ true

Therefore, shade the side of the line that includes the point $(0, 0)$.

2. Step 1   Graph $2y = x$. Since $>$ is used, the line should be a dashed line.
    Step 2   The line passes through $(0, 0)$.
    Step 3   Choose another test point, say $(-1, 1)$.
$$2y > x$$
$$2(1) > -1$$
$$2 > -1 \quad \text{true}$$
Shade the region that includes $(-1, 1)$, that is, the region above the line.

3. Step 1   Graph $y = -3$. Since $\geq$ is used, the line should be solid.
    Step 2   Test $(0, 0)$ in the inequality.
$$y \geq -3$$
$$0 \geq -3 \quad \text{true}$$
Shade the region that includes $(0, 0)$, that is, the region above the line $y = -3$.

7.6 Practice Problems

1. The domain is $\{-3, 0, 3, 20\}$. The range is $\{-5, 5\}$.
2. (a) Look at the ordered pairs. No two ordered pairs have the same first coordinate. Thus this set of ordered pairs defines a function.
   (b) Look at the ordered pairs. Two different ordered pairs, $(60, 30)$ and $(60, 120)$, have the same first coordinate. Thus this relation is not a function.
3. (a) Looking at the table, we see that no two different ordered pairs have the same first coordinate. The cost of gasoline is a function of the distance traveled.
    Note that cost depends on distance. Thus distance is the independent variable. Since a negative distance does not make sense, the domain is {all nonnegative real numbers}.
    The range is {all nonnegative real numbers}.
   (b) Looking at the table, we see two ordered pairs, $(5, 20)$ and $(5, 30)$, have the same first coordinate. Thus this relation is not a function.
4. Construct a table, plot the ordered pairs, and connect the points.

x	$y = x^2 - 2$	y
-2	$y = (-2)^2 - 2 = 2$	2
-1	$y = (-1)^2 - 2 = -1$	-1
0	$y = (0)^2 - 2 = -2$	-2
1	$y = (1)^2 - 2 = -1$	-1
2	$y = (2)^2 - 2 = 2$	2

5. Select values of y and then substitute them into the equation to obtain values of x.

y	$x = y^2 - 1$	x	y
-2	$x = (-2)^2 - 1 = 3$	3	-2
-1	$x = (-1)^2 - 1 = 0$	0	-1
0	$x = (0)^2 - 1 = -1$	-1	0
1	$x = (1)^2 - 1 = 0$	0	1
2	$x = (2)^2 - 1 = 3$	3	2

6. $y = \dfrac{6}{x}$

x	$y = \dfrac{6}{x}$	y
-3	$y = \dfrac{6}{-3} = -2$	-2
-2	$y = \dfrac{6}{-2} = -3$	-3
-1	$y = \dfrac{6}{-1} = -6$	-6
0	We cannot divide by 0.	
1	$y = \dfrac{6}{1} = 6$	6
2	$y = \dfrac{6}{2} = 3$	3
3	$y = \dfrac{6}{3} = 2$	2

7. (a) The graph of a vertical line is not a function.
   (b) This curve is a function. Any vertical line will cross the curve in only one location.
   (c) This curve is not the graph of a function. There exist vertical lines that will cross the curve in more than one place.
8. $f(x) = -2x^2 + 3x - 8$
   (a) $f(2) = -2(2)^2 + 3(2) - 8$
$$= -2(4) + 3(2) - 8$$
$$= -8 + 6 - 8$$
$$= -10$$
   (b) $f(-3) = -2(-3)^2 + 3(-3) - 8$
$$= -2(9) + 3(-3) - 8$$
$$= -18 - 9 - 8$$
$$= -35$$
   (c) $f(0) = -2(0)^2 + 3(0) - 8$
$$= -2(0) + 3(0) - 8$$
$$= 0 + 0 - 8$$
$$= -8$$

Chapter 8   8.1 Practice Problems

1. $x + y = 12$
     $-x + y = \ \ 4$
Graph the equations on the same coordinate system.
The lines intersect at the point $x = 4$, $y = 8$. Thus the solution to the system of equations is $(4, 8)$.

2. $4x + 2y = 8$
     $-6x - 3y = 6$
Graph both equations on the same coordinate system.
These lines are parallel. They do not intersect. Hence there is no solution to this system of equations.

3. $3x - 9y = 18$
$-4x + 12y = -24$
Graph both equations on the same coordinate system.
Notice both equations represent the same line. Thus there is an
infinite number of solutions to this system.

4. (a) Bill Tupper's Electrical Service charges $100 for a house call and
$30 per hour. Thus we obtain the first equation, $y = 100 + 30x$.
Wire for Hire charges $50 for a house call and $40 per hour. Thus
we obtain the second equation, $y = 50 + 40x$.

(b) Bill Tupper's Electrical Service $y = 100 + 30x$

Let $x = 0$. $y = 100 + 30(0) = 100$
Let $x = 4$. $y = 100 + 30(4) = 220$
Let $x = 8$. $y = 100 + 30(8) = 340$

x	y
0	100
4	220
8	340

Wire for Hire $y = 50 + 40x$

Let $x = 0$. $y = 50 + 40(0) = 50$
Let $x = 4$. $y = 50 + 40(4) = 210$
Let $x = 8$. $y = 50 + 40(8) = 370$

x	y
0	50
4	210
8	370

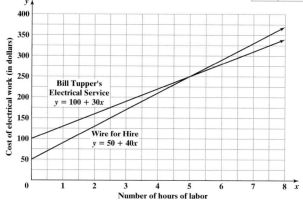

(c) We see that the graphs of the two lines intersect at $(5, 250)$. Thus
the two companies will charge the same if 5 hours of electrical
repairs are required.
(d) We draw a dashed line at $x = 6$. We see the line representing
Wire for Hire is higher than the line representing Bill Tupper's
Electrical Service after 5 hours. Thus the cost would be less if
we use Bill Tupper's Electrical Service for 6 hours of work.

8.2 Practice Problems

1. $5x + 3y = 19$
$2x - y = 12$

Step 1 Solve the second equation for y.
$2x - y = 12$
$-y = -2x + 12$
$y = 2x - 12$

Step 2 Now substitute $2x - 12$ for y in the first equation.
$5x + 3y = 19$
$5x + 3(2x - 12) = 19$

Step 3 Solve this equation.
$5x + 6x - 36 = 19$
$11x - 36 = 19$
$11x = 55$
$x = 5$

Step 4 Now obtain the value for the second variable.
$2x - y = 12$
$2(5) - y = 12$
$10 - y = 12$
$-y = 2$
$y = -2$
The solution is $(5, -2)$.

Step 5 Check.
$5x + 3y = 19$ $2x - y = 12$
$5(5) + 3(-2) \stackrel{?}{=} 19$ $2(5) - (-2) \stackrel{?}{=} 12$
$25 - 6 \stackrel{?}{=} 19$ $10 + 2 \stackrel{?}{=} 12$
$19 = 19$ ✓ $12 = 12$ ✓

2. $\frac{1}{3}x - \frac{1}{2}y = 1$
$x + 4y = -8$
Clear the first equation of fractions. We observe that the LCD is 6.
$$6\left(\frac{1}{3}x\right) - 6\left(\frac{1}{2}y\right) = 6(1)$$
$$2x - 3y = 6$$

Step 1 Solve the second equation for x.
$x + 4y = -8$
$x = -4y - 8$

Step 2 Substitute $-4y - 8$ for x in the equation $2x - 3y = 6$.
$2(-4y - 8) - 3y = 6$

Step 3 Solve this equation.
$-8y - 16 - 3y = 6$
$-11y - 16 = 6$
$-11y = 22$
$y = -2$

Step 4 Obtain the value of the second variable.
$x + 4y = -8$
$x + 4(-2) = -8$
$x - 8 = -8$
$x = 0$
The solution is $(0, -2)$.

Step 5 Check.
$\frac{1}{3}x - \frac{1}{2}y = 1$ $x + 4y = -8$
$\frac{1}{3}(0) - \frac{1}{2}(-2) \stackrel{?}{=} 1$ $0 + 4(-2) \stackrel{?}{=} -8$
$0 + 1 \stackrel{?}{=} 1$ $-8 = -8$ ✓
$1 = 1$ ✓

8.3 Practice Problems

1. $3x + y = 7$
$5x - 2y = 8$
Multiply the first equation by 2.
$2(3x) + 2(y) = 2(7)$
$6x + 2y = 14$
$\underline{5x - 2y = 8}$
$11x = 22$
$x = 2$
$5(2) - 2y = 8$
$10 - 2y = 8$
$-2y = -2$
$y = 1$
The solution is $(2, 1)$.

2. $4x + 5y = 17$
$3x + 7y = 12$
Multiply the first equation by -3 and the second equation by 4.
$-3(4x) + (-3)(5y) = (-3)(17)$
$4(3x) + 4(7y) = 4(12)$
$-12x - 15y = -51$
$\underline{12x + 28y = 48}$
$13y = -3$
$y = -\dfrac{3}{13}$

Substitute $y = -\dfrac{3}{13}$ into one of the original equations.

$4x + 5\left(-\dfrac{3}{13}\right) = 17$

$4x - \dfrac{15}{13} = 17$

$13(4x) - 13\left(\dfrac{15}{13}\right) = 13(17)$

$52x - 15 = 221$

$52x = 236$

$x = \dfrac{59}{13}$

The solution is $\left(\dfrac{59}{13}, -\dfrac{3}{13}\right)$.

3. $\dfrac{2}{3}x - \dfrac{3}{4}y = 3$
$-2x + y = 6$

Multiply the first equation by 12.

$12\left(\dfrac{2}{3}x\right) - 12\left(\dfrac{3}{4}y\right) = 12(3)$

$8x - 9y = 36$

We now have an equivalent system without fractions.

$8x - 9y = 36$
$-2x + y = 6$

Multiply the second equation by 4 to eliminate the x-variable.

$4(-2x) + 4(y) = 4(6)$
$8x - 9y = 36$
$\underline{-8x + 4y = 24}$
$-5y = 60$
$y = -12$

Substitute $y = -12$ into one of the original equations.

$-2x + (-12) = 6$
$-2x - 12 = 6$
$-2x = 18$
$x = -9$

The solution is $(-9, -12)$.

4. Multiply each term of each equation by 10.

$10(0.2x) + 10(0.3y) = 10(-0.1)$
$10(0.5x) - 10(0.1y) = 10(-1.1)$
$2x + 3y = -1$
$5x - y = -11$

Multiply the second equation by 3 to eliminate the y-variable.

$2x + 3y = -1$
$\underline{15x - 3y = -33}$
$17x = -34$
$x = -2$

Substitute $x = -2$ into one of the equations.

$2(-2) + 3y = -1$
$-4 + 3y = -1$
$3y = 3$
$y = 1$

The solution is $(-2, 1)$.

8.4 Practice Problems

1. (a) $3x + 5y = 1485$
$x + 2y = 564$
Solve for x in the second equation and solve using the substitution method.
$x = -2y + 564$
$3(-2y + 564) + 5y = 1485$
$-6y + 1692 + 5y = 1485$
$-y = -207$
$y = 207$
Substitute $y = 207$ into the second equation and solve for x.
$x + 2(207) = 564$
$x + 414 = 564$
$x = 150$
The solution is $(150, 207)$.

(b) $7x + 6y = 45$
$6x - 5y = -2$
Using the addition method, multiply the first equation by -6 and the second equation by 7.
$-6(7x) + (-6)(6y) = (-6)(45)$
$7(6x) - 7(5y) = 7(-2)$
$-42x - 36y = -270$
$\underline{42x - 35y = -14}$
$-71y = -284$
$y = 4$
Substitute $y = 4$ into one of the original equations and solve for x.
$7x + 6(4) = 45$
$7x + 24 = 45$
$7x = 21$
$x = 3$
The solution is $(3, 4)$.

2. $4x + 2y = 2$
$-6x - 3y = 6$
Use the addition method. Multiply the first equation by 3 and the second equation by 2.
$12x + 6y = 6$
$\underline{-12x - 6y = 12}$
$0 = 18$
The statement $0 = 18$ is not true. There is no solution to this system of equations.

3. $3x - 9y = 18$
$-4x + 12y = -24$
Use the addition method. Multiply the first equation by 4 and the second equation by 3.
$12x - 36y = 72$
$\underline{-12x + 36y = -72}$
$0 = 0$
The statement $0 = 0$ is true and can identify the equations as dependent. There is an infinite number of solutions to the system.

8.5 Practice Problems

1. Let x = number of gallons/hour pumped by the smaller pump and y = number of gallons/hour pumped by the larger pump.
1st day $8x + 5y = 49,000$
2nd day $5x + 3y = 30,000$
Use the addition method.
Multiply the first equation by 5 and the second equation by -8.
$40x + 25y = 245,000$
$\underline{-40x - 24y = -240,000}$
$y = 5000$
Substitute $y = 5000$ into one of the original equations and solve for x.
$8x + 5(5000) = 49,000$
$8x + 25,000 = 49,000$
$8x = 24,000$
$x = 3000$
The smaller pump removes 3000 gallons per hour.
The larger pump removes 5000 gallons per hour.

2. Let x = cost/gallon of unleaded premium gasoline and
y = cost/gallon of unleaded regular gasoline.
Last week's purchase $7x + 8y = 47.15$
This week's purchase $8x + 4y = 38.20$
Use the addition method. Multiply the first equation by 8 and the
second equation by -7.

$$56x + 64y = 377.20$$
$$-56x - 28y = -267.40$$
$$36y = 109.8$$
$$y = 3.05$$

Substitute $y = 3.05$ into one of the original equations and solve for x.

$$7x + 8(3.05) = 47.15$$
$$7x + 24.40 = 47.15$$
$$7x = 22.75$$
$$x = 3.25$$

Unleaded premium gasoline costs $3.25/gallon.
Unleaded regular gasoline costs $3.05/gallon.

3. Let x = the amount of 20% solution and y = the amount of 80%
solution.
Total amount of 65% solution
$x + y = 4000$
Amount of acid in 65% solution
$0.2x + 0.8y = 0.65(4000)$
Use the substitution method. Solve for x in the first equation.
$x = 4000 - y$
Substitute $x = 4000 - y$ into the second equation.

$$0.2(4000 - y) + 0.8y = 0.65(4000)$$
$$800 - 0.2y + 0.8y = 2600$$
$$800 + 0.6y = 2600$$
$$0.6y = 1800$$
$$y = 3000$$

Now substitute $y = 3000$ into the first equation.

$$x + y = 4000$$
$$x + 3000 = 4000$$
$$x = 1000$$

Therefore he will need 1000 liters of 20% solution and 3000 liters of
80% solution.

4. To go downstream with the current, we have as the rate of travel
$$r = \frac{d}{t} = \frac{72}{3} = 24 \text{ miles/hour}$$
To go upstream against the current, we have as the rate of travel
$$r = \frac{d}{t} = \frac{72}{4} = 18 \text{ miles/hour}$$
Let x = the speed of the boat in miles/hour and y = the speed of the
current in miles/hour.
Thus we have the equations $x + y = 24$ and $x - y = 18$. Use the
addition method to solve.

$$x + y = 24$$
$$x - y = 18$$
$$2x = 42$$
$$x = 21$$

Substitute $x = 21$ into one of the original equations and solve for y.

$$21 + y = 24$$
$$y = 3$$

Thus the speed of the boat in still water is 21 miles/hour and the
speed of the current is 3 miles/hour.

Chapter 9 9.1 Practice Problems

1. (a) $\sqrt{64} = 8$ because $8^2 = 64$.
(b) $-\sqrt{121} = -11$ since $\sqrt{121} = 11$.

2. (a) $\sqrt{\dfrac{9}{25}} = \dfrac{3}{5}$ since $\left(\dfrac{3}{5}\right)^2 = \left(\dfrac{3}{5}\right)\left(\dfrac{3}{5}\right) = \dfrac{9}{25}$.

(b) $-\sqrt{\dfrac{121}{169}} = -\dfrac{11}{13}$ since $\sqrt{\dfrac{121}{169}} = \dfrac{11}{13}$.

3. (a) $-\sqrt{0.0036} = -0.06$ since $(0.06)(0.06) = 0.0036$.
(b) $\sqrt{2500} = 50$ since $(50)(50) = 2500$.
(c) $\sqrt{196} = 14$

4. (a) $\sqrt{13} \approx 3.606$
(b) $\sqrt{35} \approx 5.916$
(c) $\sqrt{127} \approx 11.269$

9.2 Practice Problems

1. (a) $\sqrt{6^2} = 6$
(b) $\sqrt{13^4} = \sqrt{(13^2)^2} = 13^2$
(c) $\sqrt{18^{12}} = \sqrt{(18^6)^2} = 18^6$

2. (a) $\sqrt{y^{18}} = \sqrt{(y^9)^2} = y^9$
(b) $\sqrt{x^{30}} = \sqrt{(x^{15})^2} = x^{15}$

3. (a) $\sqrt{625y^4} = \sqrt{625}\sqrt{y^4} = 25y^2$
(b) $\sqrt{x^{16}y^{22}} = x^8y^{11}$
(c) $\sqrt{121x^{12}y^6} = 11x^6y^3$

4. (a) $\sqrt{98} = \sqrt{49 \cdot 2} = \sqrt{49}\sqrt{2} = 7\sqrt{2}$
(b) $\sqrt{12} = \sqrt{4 \cdot 3} = \sqrt{4}\sqrt{3} = 2\sqrt{3}$
(c) $\sqrt{75} = \sqrt{25 \cdot 3} = \sqrt{25}\sqrt{3} = 5\sqrt{3}$

5. (a) $\sqrt{x^{11}} = \sqrt{x^{10}}\sqrt{x} = x^5\sqrt{x}$
(b) $\sqrt{x^5y^3} = \sqrt{x^4}\sqrt{x}\sqrt{y^2}\sqrt{y} = x^2\sqrt{x}y\sqrt{y} = x^2y\sqrt{xy}$

6. (a) $\sqrt{48x^{11}} = \sqrt{16 \cdot 3 \cdot x^{10} \cdot x} = 4x^5\sqrt{3x}$
(b) $\sqrt{121x^6y^7z^8} = \sqrt{121 \cdot x^6 \cdot y^6 \cdot y \cdot z^8} = 11x^3y^3z^4\sqrt{y}$

9.3 Practice Problems

1. (a) $7\sqrt{11} + 4\sqrt{11} = (7 + 4)\sqrt{11} = 11\sqrt{11}$
(b) $4\sqrt{t} - 7\sqrt{t} + 6\sqrt{t} - 2\sqrt{t}$
$= (4 - 7 + 6 - 2)\sqrt{t}$
$= 1\sqrt{t}$
$= \sqrt{t}$

2. $3\sqrt{x} - 2\sqrt{xy} - 5\sqrt{y} + 7\sqrt{xy}$
$= 3\sqrt{x} + 5\sqrt{xy} - 5\sqrt{y}$

3. (a) $\sqrt{50} - \sqrt{18} + \sqrt{98}$
$= \sqrt{25 \cdot 2} - \sqrt{9 \cdot 2} + \sqrt{49 \cdot 2}$
$= 5\sqrt{2} - 3\sqrt{2} + 7\sqrt{2} = 9\sqrt{2}$
(b) $\sqrt{12} + \sqrt{18} - \sqrt{50} + \sqrt{27}$
$= \sqrt{4 \cdot 3} + \sqrt{9 \cdot 2} - \sqrt{25 \cdot 2} + \sqrt{9 \cdot 3}$
$= 2\sqrt{3} + 3\sqrt{2} - 5\sqrt{2} + 3\sqrt{3} = 5\sqrt{3} - 2\sqrt{2}$

4. $\sqrt{9x} + \sqrt{8x} - \sqrt{4x} + \sqrt{50x}$
$= \sqrt{9 \cdot x} + \sqrt{4 \cdot 2x} - \sqrt{4 \cdot x} + \sqrt{25 \cdot 2x}$
$= 3\sqrt{x} + 2\sqrt{2x} - 2\sqrt{x} + 5\sqrt{2x}$
$= \sqrt{x} + 7\sqrt{2x}$

5. $\sqrt{27} - 4\sqrt{3} + 2\sqrt{75}$
$= \sqrt{9} \cdot \sqrt{3} - 4 \cdot \sqrt{3} + 2\sqrt{25}\sqrt{3}$
$= 3 \cdot \sqrt{3} - 4\sqrt{3} + 2 \cdot 5 \cdot \sqrt{3}$
$= 3\sqrt{3} - 4\sqrt{3} + 10\sqrt{3} = 9\sqrt{3}$

6. $2a\sqrt{12a} + 3\sqrt{27a^3}$
$= 2a\sqrt{4}\sqrt{3a} + 3\sqrt{9a^2}\sqrt{3a}$
$= 2a \cdot 2 \cdot \sqrt{3a} + 3 \cdot 3a \cdot \sqrt{3a}$
$= 4a\sqrt{3a} + 9a\sqrt{3a}$
$= 13a\sqrt{3a}$

9.4 Practice Problems

1. $\sqrt{3a}\sqrt{6a} = \sqrt{18a^2} = \sqrt{9a^2 \cdot 2} = 3a\sqrt{2}$

2. (a) $(2\sqrt{3})(5\sqrt{5}) = 10\sqrt{15}$

(b) $(4\sqrt{3x})(2x\sqrt{6x}) = 8x\sqrt{18x^2}$

$$= 8x\sqrt{9}\sqrt{2}\sqrt{x^2} = 24x^2\sqrt{2}$$

3. $(\sqrt{180})(\sqrt{150}) = (6\sqrt{5})(5\sqrt{6})$

$$= 30\sqrt{30}\text{ mm}^2$$

4. $\sqrt{6}(\sqrt{3} + 2\sqrt{8})$

$$= \sqrt{6}\sqrt{3} + \sqrt{6}2\sqrt{8} = \sqrt{18} + 2\sqrt{48}$$

$$= \sqrt{9}\sqrt{2} + 2\sqrt{16}\sqrt{3}$$

$$= 3\sqrt{2} + 8\sqrt{3}$$

5. $2\sqrt{x}(4\sqrt{x} - x\sqrt{2})$

$$= 8\sqrt{x}\sqrt{x} - 2x\sqrt{x}\sqrt{2}$$

$$= 8x - 2x\sqrt{2x}$$

6. $(\sqrt{2} + \sqrt{6})(2\sqrt{2} - \sqrt{6})$

$$= 2\sqrt{4} - \sqrt{12} + 2\sqrt{12} - \sqrt{36}$$

$$= 2(2) + \sqrt{12} - 6$$

$$= 4 + 2\sqrt{3} - 6$$

$$= -2 + 2\sqrt{3}$$

7. $(\sqrt{6} + \sqrt{5})(\sqrt{2} + 2\sqrt{5})$

$$= \sqrt{12} + 2\sqrt{30} + \sqrt{10} + 2\sqrt{25}$$

$$= 2\sqrt{3} + 2\sqrt{30} + \sqrt{10} + 10$$

8. $(3\sqrt{5} - \sqrt{10})^2$

$$= (3\sqrt{5} - \sqrt{10})(3\sqrt{5} - \sqrt{10})$$

$$= 9\sqrt{25} - 3\sqrt{50} - 3\sqrt{50} + \sqrt{100}$$

$$= 45 - 6\sqrt{50} + 10$$

$$= 55 - 30\sqrt{2}$$

9.5 Practice Problems

1. (a) $\dfrac{\sqrt{98}}{\sqrt{2}} = \sqrt{\dfrac{98}{2}} = \sqrt{49} = 7$

(b) $\sqrt{\dfrac{81}{100}} = \dfrac{\sqrt{81}}{\sqrt{100}} = \dfrac{9}{10}$

2. $\sqrt{\dfrac{50}{a^4}} = \dfrac{\sqrt{50}}{\sqrt{a^4}} = \dfrac{\sqrt{25}\sqrt{2}}{a^2} = \dfrac{5\sqrt{2}}{a^2}$

3. $\dfrac{9}{\sqrt{7}} = \dfrac{9}{\sqrt{7}} \cdot \dfrac{\sqrt{7}}{\sqrt{7}} = \dfrac{9\sqrt{7}}{\sqrt{49}} = \dfrac{9\sqrt{7}}{7}$

4. (a) $\dfrac{\sqrt{2}}{\sqrt{12}} = \dfrac{\sqrt{2}}{\sqrt{12}} \cdot \dfrac{\sqrt{3}}{\sqrt{3}} = \dfrac{\sqrt{6}}{\sqrt{36}} = \dfrac{\sqrt{6}}{6}$

(b) $\dfrac{6a}{\sqrt{a^7}} = \dfrac{6a}{\sqrt{a^7}} \cdot \dfrac{\sqrt{a}}{\sqrt{a}} = \dfrac{6a\sqrt{a}}{\sqrt{a^8}} = \dfrac{6a\sqrt{a}}{a^4} = \dfrac{6\sqrt{a}}{a^3}$

5. $\dfrac{\sqrt{5x}}{\sqrt{8x}} = \dfrac{\sqrt{5x}}{2\sqrt{2x}} = \dfrac{\sqrt{5x}}{2\sqrt{2x}} \cdot \dfrac{\sqrt{2x}}{\sqrt{2x}} = \dfrac{\sqrt{10x^2}}{2\sqrt{4x^2}} = \dfrac{x\sqrt{10}}{4x} = \dfrac{\sqrt{10}}{4}$

6. (a) $\dfrac{4}{\sqrt{3} + \sqrt{5}} = \dfrac{4}{\sqrt{3} + \sqrt{5}} \cdot \dfrac{\sqrt{3} - \sqrt{5}}{\sqrt{3} - \sqrt{5}}$

$$= \dfrac{4\sqrt{3} - 4\sqrt{5}}{(\sqrt{3})^2 - \sqrt{15} + \sqrt{15} - (\sqrt{5})^2}$$

$$= \dfrac{4\sqrt{3} - 4\sqrt{5}}{3 - 5} = \dfrac{4\sqrt{3} - 4\sqrt{5}}{-2}$$

$$= \dfrac{4(\sqrt{3} - \sqrt{5})}{-2} = -2(\sqrt{3} - \sqrt{5})$$

(b) $\dfrac{\sqrt{a}}{\sqrt{10} - 3} = \dfrac{\sqrt{a}}{\sqrt{10} - 3} \cdot \dfrac{\sqrt{10} + 3}{\sqrt{10} + 3}$

$$= \dfrac{\sqrt{10a} + 3\sqrt{a}}{(\sqrt{10})^2 + 3\sqrt{10} - 3\sqrt{10} - 3^2}$$

$$= \dfrac{\sqrt{10a} + 3\sqrt{a}}{10 - 9} = \dfrac{\sqrt{10a} + 3\sqrt{a}}{1} = \sqrt{10a} + 3\sqrt{a}$$

7. $\dfrac{\sqrt{3} + \sqrt{5}}{\sqrt{3} - \sqrt{5}} = \dfrac{\sqrt{3} + \sqrt{5}}{\sqrt{3} - \sqrt{5}} \cdot \dfrac{\sqrt{3} + \sqrt{5}}{\sqrt{3} + \sqrt{5}}$

$$= \dfrac{\sqrt{9} + \sqrt{15} + \sqrt{15} + \sqrt{25}}{(\sqrt{3})^2 - (\sqrt{5})^2}$$

$$= \dfrac{3 + 2\sqrt{15} + 5}{3 - 5} = \dfrac{8 + 2\sqrt{15}}{-2}$$

$$= \dfrac{-2(-4 - \sqrt{15})}{-2} = -4 - \sqrt{15}$$

9.6 Practice Problems

1. $c^2 = a^2 + b^2$

$c^2 = 9^2 + 12^2$

$c^2 = 81 + 144$

$c^2 = 225$

$c = \pm\sqrt{225}$

$c = \pm 15$

The hypotenuse is 15 centimeters because length is not negative.

2. $c^2 = a^2 + b^2$

$(\sqrt{17})^2 = (1)^2 + b^2$

$17 = 1 + b^2$

$16 = b^2$

$\pm 4 = b$

The leg is 4 meters long.

3. $c^2 = 3^2 + 8^2$

$c^2 = 9 + 64$

$c^2 = 73$

$c = \pm\sqrt{73}$

The support line is $\sqrt{73} \approx 8.5$ meters long.

4. $\sqrt{3x - 2} - 7 = 0$

$\sqrt{3x - 2} = 7$

$(\sqrt{3x - 2})^2 = (7)^2$

$3x - 2 = 49$

$3x = 51$

$x = 17$

The solution is 17.

Check.

$\sqrt{3(17) - 2} - 7 \overset{?}{=} 0$

$\sqrt{51 - 2} - 7 \overset{?}{=} 0$

$\sqrt{49} - 7 \overset{?}{=} 0$

$7 - 7 \overset{?}{=} 0$

$0 = 0$ ✓

5. $\sqrt{5x + 4} + 2 = 0$

$\sqrt{5x + 4} = -2$

$(\sqrt{5x + 4})^2 = (-2)^2$

$5x + 4 = 4$

$5x = 0$

$x = 0$

Check.

$\sqrt{5(0) + 4} + 2 \overset{?}{=} 0$

$\sqrt{4} + 2 \overset{?}{=} 0$

$2 + 2 \overset{?}{=} 0$

$4 \neq 0$

No, this does not check. There is no solution.

6. $-2 + \sqrt{6x - 1} = 3x - 2$

$\sqrt{6x - 1} = 3x$

$(\sqrt{6x - 1})^2 = (3x)^2$

$6x - 1 = 9x^2$

$0 = 9x^2 - 6x + 1$

$0 = (3x - 1)^2$

$0 = 3x - 1$

$x = \dfrac{1}{3}$

Check.

$-2 + \sqrt{6\left(\dfrac{1}{3}\right) - 1} \overset{?}{=} 3\left(\dfrac{1}{3}\right) - 2$

$-2 + \sqrt{2 - 1} \overset{?}{=} 1 - 2$

$-2 + \sqrt{1} \overset{?}{=} -1$

$-2 + 1 \overset{?}{=} -1$

$-1 = -1$ ✓

The solution is $\dfrac{1}{3}$.

7. $2 - x + \sqrt{x + 4} = 0$

$\sqrt{x + 4} = x - 2$

$(\sqrt{x + 4})^2 = (x - 2)^2$

$x + 4 = x^2 - 4x + 4$

$0 = x^2 - 5x$

$0 = x(x - 5)$

$x = 0 \text{ or } x - 5 = 0$

$x = 5$

Check.

$2 - 0 + \sqrt{0 + 4} \overset{?}{=} 0$

$2 + \sqrt{4} \overset{?}{=} 0$

$2 + 2 \overset{?}{=} 0$

$4 \neq 0$

$x = 0$ does not check.

$2 - 5 + \sqrt{5 + 4} \overset{?}{=} 0$

$-3 + \sqrt{9} \overset{?}{=} 0$

$-3 + 3 \overset{?}{=} 0$

$0 = 0$ ✓

The only solution is 5.

8. $\sqrt{2x + 1} = \sqrt{x - 10}$

$(\sqrt{2x + 1})^2 = (\sqrt{x - 10})^2$

$2x + 1 = x - 10$

$x = -11$

Check.

$\sqrt{2(-11) + 1} \overset{?}{=} \sqrt{-11 - 10}$

$\sqrt{-22 + 1} \overset{?}{=} \sqrt{-21}$

$\sqrt{-21}$ is not a real number.

Thus, $x = -11$ is extraneous. There is no solution.

9.7 Practice Problems

1. Change of $100°C$ = change of $180°F$.

Let C = the change on the Celsius scale

f = the change on the Fahrenheit scale, and

k = the constant of variation.

$C = kf$

$100 = k \cdot 180$

$\dfrac{5}{9} = k$

Thus, $C = \dfrac{5}{9}f$.

Let $f = -20$.

$C = \dfrac{5}{9}(-20)$

$C = -11\dfrac{1}{9}$

The temperature drops $11\dfrac{1}{9}$ degrees Celsius.

2. $y = kx$

To find k we substitute $y = 18$ and $x = 5$.

$18 = k(5)$

$\dfrac{18}{5} = k$

We now write the variation equation with k replaced by $\dfrac{18}{5}$.

$y = \dfrac{18}{5}x$

Replace x by $\dfrac{20}{23}$ and solve for y.

$y = \dfrac{18}{5} \cdot \dfrac{20}{23}$

$y = \dfrac{72}{23}$

Thus $y = \dfrac{72}{23}$ when $x = \dfrac{20}{23}$.

3. Let d = the distance to stop the car,

s = the speed of the car, and

k = the constant of variation.

Since the distance varies directly as the square of the speed, we have

$d = ks^2$.

To evaluate k we substitute the known distance and speed.

$60 = k(20)^2$

$60 = k(400)$

$\dfrac{3}{20} = k$

Now write the variation equation with the known value for k.

$d = \dfrac{3}{20}s^2$

$d = \dfrac{3}{20}(40)^2$

$d = \dfrac{3}{20}(1600)$

$d = 240$

The distance to stop the car going 40 mph on an ice-covered road is 240 feet.

4. $y = \dfrac{k}{x}$

$8 = \dfrac{k}{15}$ Substitute known values of x and y to find k.

$120 = k$

$y = \dfrac{120}{x}$ Write the variation equation with k replaced by 120.

$y = \dfrac{120}{\dfrac{3}{5}} = \dfrac{120}{1} \cdot \dfrac{5}{3} = 200$ Find y when $x = \dfrac{3}{5}$.

Thus $y = 200$ when $x = \dfrac{3}{5}$.

5. Let V = the volume of sales,

p = the price of the calculator, and

k = the constant of variation.

Since volume varies inversely as the price, then

$V = \dfrac{k}{p}$.

$120,000 = \dfrac{k}{30}$ Evaluate k by substituting known values for V and p.

$3,600,000 = k$

$V = \dfrac{3,600,000}{24}$ Write the variation equation with the evaluated constant and substitute $p = 24$.

$V = 150,000$

Thus the volume increased to 150,000 calculators sold per year when the price was reduced to $24 per calculator.

6. Let R = the resistance in the circuit,

 i = the amount of current, and

 k = the constant of variation.

Since the resistance in the circuit varies inversely as the square of the amount of current,

$$R = \frac{k}{i^2}.$$

To evaluate k, substitute the known values for R and i.

$$800 = \frac{k}{(0.01)^2}$$

$$0.08 = k$$

$$R = \frac{0.08}{(0.02)^2}$$ Write the variation equation with the evaluated constant of $k = 0.08$ and substitute $i = 0.02$.

$$R = \frac{0.08}{0.0004}$$

$$R = 200$$

Thus the resistance is 200 ohms if the amount of current is increased to 0.02 ampere.

Chapter 10 10.1 Practice Problems

1. (a) $2x^2 + 12x - 9 = 0$

 $a = 2, b = 12, c = -9$

(b) $7x^2 = 6x - 8$

 $7x^2 - 6x + 8 = 0$

 $a = 7, b = -6, c = 8$

(c) $-x^2 - 6x + 3 = 0$

 $x^2 + 6x - 3 = 0$

 $a = 1, b = 6, c = -3$

(d) $10x^2 - 12x = 0$

 $a = 10, b = -12, c = 0$

2. $\qquad 2x^2 - 7x - 6 = 4x - 6$

 $2x^2 - 7x - 6 - 4x + 6 = 0$

 $\qquad\qquad 2x^2 - 11x = 0$

 $\qquad\qquad x(2x - 11) = 0$

 $x = 0 \qquad 2x - 11 = 0$

 $\qquad\qquad\qquad 2x = 11$

 $\qquad\qquad\qquad x = \frac{11}{2}$

The two roots are 0 and $\frac{11}{2}$.

3. $\qquad 3x + 10 - \frac{8}{x} = 0$

 $x(3x) + x(10) - x\left(\frac{8}{x}\right) = x(0)$

 $\qquad 3x^2 + 10x - 8 = 0$

 $\qquad (3x - 2)(x + 4) = 0$

 $\qquad 3x - 2 = 0 \qquad x + 4 = 0$

 $\qquad\qquad 3x = 2$

 $\qquad\qquad x = \frac{2}{3} \qquad x = -4$

Check.

$$3\left(\frac{2}{3}\right) + 10 - \frac{8}{\frac{2}{3}} \stackrel{?}{=} 0 \qquad 3(-4) + 10 - \frac{8}{-4} \stackrel{?}{=} 0$$

$$2 + 10 - 12 \stackrel{?}{=} 0 \qquad\qquad -12 + 10 + 2 \stackrel{?}{=} 0$$

$$0 = 0 \checkmark \qquad\qquad\qquad 0 = 0 \checkmark$$

Both roots check, so $x = \frac{2}{3}$ and $x = -4$ are the two roots that satisfy the equation $3x + 10 - \frac{8}{x} = 0$.

4. $\qquad 2x^2 + 9x - 18 = 0$

 $(2x - 3)(x + 6) = 0$

 $2x - 3 = 0 \qquad x + 6 = 0$

 $\qquad 2x = 3$

 $\qquad x = \frac{3}{2} \qquad x = -6$

Check.

$$2\left(\frac{3}{2}\right)^2 + 9\left(\frac{3}{2}\right) - 18 \stackrel{?}{=} 0 \qquad 2(-6)^2 + 9(-6) - 18 \stackrel{?}{=} 0$$

$$2\left(\frac{9}{4}\right) + \frac{27}{2} - 18 \stackrel{?}{=} 0 \qquad 2(36) - 54 - 18 \stackrel{?}{=} 0$$

$$\frac{9}{2} + \frac{27}{2} - 18 \stackrel{?}{=} 0 \qquad 72 - 54 - 18 \stackrel{?}{=} 0$$

$$\frac{36}{2} - 18 \stackrel{?}{=} 0 \qquad\qquad 18 - 18 \stackrel{?}{=} 0$$

$$18 - 18 \stackrel{?}{=} 0 \qquad\qquad 0 = 0 \checkmark$$

$$0 = 0 \checkmark$$

Both roots check, so $x = \frac{3}{2}$ and $x = -6$ are the two roots that satisfy the equation $2x^2 + 9x - 18 = 0$.

5. $-4x + (x - 5)(x + 1) = 5(1 - x)$

 $-4x + x^2 - 4x - 5 = 5 - 5x$

 $\qquad x^2 - 8x - 5 = 5 - 5x$

 $\qquad x^2 - 3x - 10 = 0$

 $\qquad (x + 2)(x - 5) = 0$

 $x + 2 = 0 \qquad x - 5 = 0$

 $x = -2 \qquad x = 5$

6. $\qquad \dfrac{10x + 18}{x^2 + x - 2} = \dfrac{3x}{x + 2} + \dfrac{2}{x - 1}$

 $\dfrac{10x + 18}{(x - 1)(x + 2)} = \dfrac{3x}{x + 2} + \dfrac{2}{x - 1}$ Multiply by the LCD.

 $10x + 18 = 3x(x - 1) + 2(x + 2)$

 $10x + 18 = 3x^2 - 3x + 2x + 4$

 $10x + 18 = 3x^2 - x + 4$

 $0 = 3x^2 - 11x - 14$

 $0 = (3x - 14)(x + 1)$

 $3x - 14 = 0 \qquad x + 1 = 0$

 $x = \dfrac{14}{3} \qquad x = -1$

7. t = the number of truck routes

 n = the number of cities they can service

 $t = \dfrac{n^2 - n}{2}$

 $28 = \dfrac{n^2 - n}{2}$

 $56 = n^2 - n$

 $0 = n^2 - n - 56$

 $0 = (n + 7)(n - 8)$

 $n + 7 = 0 \qquad n - 8 = 0$

 $n = -7 \qquad n = 8$

We reject $n = -7$. There cannot be a negative number of cities. Thus our answer is 8 cities.

10.2 Practice Problems

1. (a) $x^2 = 1$

 $x = \pm\sqrt{1}$

 $x = \pm 1$

(b) $x^2 - 50 = 0$

 $x^2 = 50$

 $x = \pm\sqrt{50}$

 $x = \pm 5\sqrt{2}$

(c) $3x^2 = 81$

 $x^2 = 27$

 $x = \pm\sqrt{27}$

 $x = \pm 3\sqrt{3}$

2. $4x^2 - 5 = 319$

 $4x^2 = 324$

 $x^2 = 81$

 $x = \pm 9$

3. $(2x - 3)^2 = 12$

$$2x - 3 = \pm\sqrt{12}$$
$$2x - 3 = \pm 2\sqrt{3}$$

$$2x - 3 = +2\sqrt{3} \qquad\qquad 2x - 3 = -2\sqrt{3}$$
$$2x = 3 + 2\sqrt{3} \qquad\qquad 2x = 3 - 2\sqrt{3}$$
$$x = \frac{3 + 2\sqrt{3}}{2} \qquad\qquad x = \frac{3 - 2\sqrt{3}}{2}$$

The roots are $\dfrac{3 \pm 2\sqrt{3}}{2}$.

4.
$$x^2 + 10x = 3$$
$$x^2 + 10x + (5)^2 = 3 + 25$$
$$(x + 5)^2 = 28$$
$$x + 5 = \pm\sqrt{28}$$
$$x + 5 = +2\sqrt{7} \qquad\qquad x + 5 = -2\sqrt{7}$$
$$x = -5 + 2\sqrt{7} \qquad\qquad x = -5 - 2\sqrt{7}$$

The roots are $-5 \pm 2\sqrt{7}$.

5.
$$x^2 - 5x = 7$$
$$x^2 - 5x + \left(\frac{5}{2}\right)^2 = 7 + \left(\frac{5}{2}\right)^2$$
$$\left(x - \frac{5}{2}\right)^2 = 7 + \frac{25}{4}$$
$$\left(x - \frac{5}{2}\right)^2 = \frac{53}{4}$$
$$x - \frac{5}{2} = \pm\sqrt{\frac{53}{4}}$$
$$x - \frac{5}{2} = \pm\frac{\sqrt{53}}{2}$$

$$x - \frac{5}{2} = +\frac{\sqrt{53}}{2} \qquad\qquad x - \frac{5}{2} = -\frac{\sqrt{53}}{2}$$
$$x = \frac{5}{2} + \frac{\sqrt{53}}{2} \qquad\qquad x = \frac{5}{2} - \frac{\sqrt{53}}{2}$$
$$x = \frac{5 + \sqrt{53}}{2} \qquad\qquad x = \frac{5 - \sqrt{53}}{2}$$

The roots are $\dfrac{5 \pm \sqrt{53}}{2}$.

6. $8x^2 + 2x - 1 = 0$
$$8x^2 + 2x = 1$$
$$x^2 + \frac{1}{4}x = \frac{1}{8}$$
$$x^2 + \frac{1}{4}x + \frac{1}{64} = \frac{1}{8} + \frac{1}{64}$$
$$\left(x + \frac{1}{8}\right)^2 = \frac{9}{64}$$
$$x + \frac{1}{8} = \pm\sqrt{\frac{9}{64}}$$
$$x + \frac{1}{8} = \pm\frac{3}{8}$$

$$x + \frac{1}{8} = \frac{3}{8} \qquad\qquad x + \frac{1}{8} = -\frac{3}{8}$$
$$x = \frac{3}{8} - \frac{1}{8} \qquad\qquad x = -\frac{3}{8} - \frac{1}{8}$$
$$x = \frac{2}{8} \qquad\qquad x = -\frac{4}{8}$$
$$x = \frac{1}{4} \qquad\qquad x = -\frac{1}{2}$$

The roots are $\dfrac{1}{4}$ and $-\dfrac{1}{2}$.

10.3 Practice Problems

1. $x^2 - 7x + 6 = 0$
$a = 1, b = -7, c = 6$

$$x = \frac{-b \pm \sqrt{b^2 - 4ac}}{2a} = \frac{-(-7) \pm \sqrt{(-7)^2 - 4(1)(6)}}{2(1)}$$
$$= \frac{7 \pm \sqrt{49 - 24}}{2}$$
$$= \frac{7 \pm \sqrt{25}}{2}$$
$$= \frac{7 \pm 5}{2}$$
$$x = \frac{7 + 5}{2} = \frac{12}{2} = 6 \qquad x = \frac{7 - 5}{2} = \frac{2}{2} = 1$$

The solutions are 6 and 1.

2. $3x^2 - 8x + 3 = 0$
$a = 3, b = -8, c = 3$

$$x = \frac{-b \pm \sqrt{b^2 - 4ac}}{2a} = \frac{-(-8) \pm \sqrt{(-8)^2 - 4(3)(3)}}{2(3)}$$
$$= \frac{8 \pm \sqrt{64 - 36}}{6}$$
$$= \frac{8 \pm \sqrt{28}}{6}$$
$$= \frac{8 \pm 2\sqrt{7}}{6}$$
$$x = \frac{2\left(4 \pm \sqrt{7}\right)}{6} = \frac{4 \pm \sqrt{7}}{3}$$

3.
$$x^2 = 7 - \frac{3}{5}x$$
$$5x^2 = 35 - 3x$$
$$5x^2 + 3x - 35 = 0$$
$$a = 5, b = 3, c = -35$$
$$x = \frac{-3 \pm \sqrt{(3)^2 - 4(5)(-35)}}{2(5)}$$
$$x = \frac{-3 \pm \sqrt{9 + 700}}{10}$$
$$x = \frac{-3 \pm \sqrt{709}}{10}$$

4.
$$2x^2 = 13x + 5$$
$$2x^2 - 13x - 5 = 0$$
$$a = 2, b = -13, c = -5$$
$$x = \frac{-(-13) \pm \sqrt{(-13)^2 - 4(2)(-5)}}{2(2)}$$
$$= \frac{13 \pm \sqrt{169 + 40}}{4} = \frac{13 \pm \sqrt{209}}{4}$$
$$x = \frac{13 + \sqrt{209}}{4} \qquad x = \frac{13 - \sqrt{209}}{4}$$
$$x \approx 6.864 \quad \text{and} \quad x \approx -0.364$$

5.
$$5x^2 + 2x = -3$$
$$5x^2 + 2x + 3 = 0$$
$$a = 5, b = 2, c = 3$$
$$x = \frac{-2 \pm \sqrt{(2)^2 - 4(5)(3)}}{2(5)}$$
$$= \frac{-2 \pm \sqrt{4 - 60}}{10} = \frac{-2 \pm \sqrt{-56}}{10}$$

There is no real number that is $\sqrt{-56}$.
There is no solution to this problem.

6. (a)
$$5x^2 = 3x + 2$$
$$5x^2 - 3x - 2 = 0$$
$$b^2 - 4ac = (-3)^2 - 4(5)(-2)$$
$$= 9 + 40 = 49$$

Because $49 > 0$ there will be two real roots that are rational numbers.

(b)
$$2x^2 + 5 = 4x$$
$$2x^2 - 4x + 5 = 0$$
$$a = 2, b = -4, c = 5$$
$$b^2 - 4ac = (-4)^2 - 4(2)(5)$$
$$= 16 - 40$$
$$= -24$$

Because $-24 < 0$, the equation $2x^2 + 5 = 4x$ has no real number solutions(s).

10.4 Practice Problems

1. (a) $y = 2x^2$

x	y
-2	8
-1	2
0	0
1	2
2	8

(b) $y = -2x^2$

x	y
-2	-8
-1	-2
0	0
1	-2
2	-8

2. (a) $y = x^2 + 2x$

x	y	
-4	8	
-3	3	
-2	0	
-1	-1	vertex
0	0	
1	3	
2	8	

(b) $y = -x^2 - 2x$

x	y	
-4	-8	
-3	-3	
-2	0	
-1	1	vertex
0	0	
1	-3	
2	-8	

3. $y = x^2 - 4x - 5$

$$x = \frac{-b}{2a} = \frac{-(-4)}{2(1)} = \frac{4}{2} = 2$$

If $x = 2$, then
$$y = (2)^2 - 4(2) - 5 = -9$$
The point $(2, -9)$ is the vertex.
x-intercepts:
$$x^2 - 4x - 5 = 0$$
$$(x + 1)(x - 5) = 0$$
$$x + 1 = 0 \qquad x - 5 = 0$$
$$x = -1 \qquad x = 5$$

The x-intercepts are the points $(-1, 0)$ and $(5, 0)$.

$$y = x^2 - 4x - 5$$

4. $h = -16t^2 + 32t + 10$

(a) Since the equation is a quadratic equation, the graph is a parabola. a is negative, therefore the parabola opens downward and the vertex is the highest point.

$$t = \frac{-b}{2a} = \frac{-32}{2(-16)} = \frac{-32}{-32} = 1$$

If $t = 1$, then $h = -16(1)^2 + 32(1) + 10$
$$h = -16 + 32 + 10$$
$$h = 26$$

The vertex (t, h) is $(1, 26)$.
Now find the h-intercept, the point where $t = 0$.
$$h = -16t^2 + 32t + 10$$
$$h = 10$$

The h-intercept is at $(0, 10)$.

Table Values	
t	**h**
0	10
0.5	22
1	26
1.5	22
2	10

(b) The ball is 26 feet above the ground one second after it is thrown. This is the greatest height of the ball.

(c) The ball hits the ground when $h = 0$.
$$0 = -16t^2 + 32t + 10$$
$$t = \frac{-32 \pm \sqrt{(32)^2 - 4(-16)(10)}}{2(-16)}$$
$$t = \frac{-32 \pm \sqrt{1024 + 640}}{-32}$$
$$t = \frac{-32 \pm \sqrt{1664}}{-32}$$
$$t = \frac{4 \pm \sqrt{26}}{4}$$
$$t \approx \frac{4 - 5.099}{4} \quad \text{and} \quad t \approx \frac{4 + 5.099}{4}$$
$$t \approx -0.275 \qquad\qquad t \approx 2.275$$

In this situation it is not useful to find points for negative values of t. Therefore $t \approx 2.3$ seconds. The ball will hit the ground approximately 2.3 seconds after it is thrown.

10.5 Practice Problems

1. $a^2 + b^2 = c^2$ Let $a = x$, then $b = x - 6$.
$$x^2 + (x - 6)^2 = 30^2$$
$$x^2 + x^2 - 12x + 36 = 900$$
$$2x^2 - 12x - 864 = 0$$
$$x^2 - 6x - 432 = 0$$
$$(x + 18)(x - 24) = 0$$
$$x + 18 = 0 \qquad\qquad x - 24 = 0$$
$$x = -18 \qquad\qquad x = 24$$

One leg is 24 meters, the other leg is $x - 6$ or $24 - 6 = 18$ meters.

2. Let c = cost for each person in the original group
s = number of people
Number of people × cost per person = total cost
$$s \cdot c = 240$$
If 2 people cannot go, then the number of people drops by 2 but the cost for each increases by \$4.
$$(s - 2)(c + 4) = 240$$
If $s \cdot c = 240$, then
$$c = \frac{240}{s}$$
$$(s - 2)\left(\frac{240}{s} + 4\right) = 240$$
$$(s - 2)\left(\frac{240 + 4s}{s}\right) = 240$$
$$(s - 2)(240 + 4s) = 240s$$
$$4s^2 + 232s - 480 = 240s$$
$$4s^2 - 8s - 480 = 0$$
$$s^2 - 2s - 120 = 0$$
$$(s - 12)(s + 10) = 0$$
$$s = 12 \qquad s = -10$$
The number of people originally going on the trip was 12.
Check. If 12 people originally planned the trip then
$$12 \times c = 240.$$
$$c = \$20 \text{ per person}$$
If 2 people dropped out, does this mean they must increase the cost of the trip by \$4?
$$10 \times c = 240$$
$$c = \$24.00 \text{ Yes.}$$

3. Let x = width
y = length
$$160 = 2x + y$$
The area of a rectangle is (width)(length).
$$A = xy \qquad \text{Substitute } y = 160 - 2x.$$
$$3150 = x(160 - 2x)$$
$$3150 = 160x - 2x^2$$
$$2x^2 - 160x + 3150 = 0$$
$$x^2 - 80x + 1575 = 0$$
$$(x - 45)(x - 35) = 0$$
$$x = 45 \quad \text{and} \quad x = 35$$
First solution:
If the width is 45 feet then the length is
$$y = 160 - 2(45)$$
$$= 160 - 90$$
$$= 70 \text{ feet}$$
Check. $\qquad 45 + 45 + 70 = 160$ ✓
$\qquad\qquad\qquad (45)(70) = 3150$ ✓

Second solution:
If the width is 35 feet then the length is
$$y = 160 - 2(35) = 160 - 70 = 90 \text{ feet}$$
Check. $\qquad 35 + 35 + 90 = 160$ ✓
$\qquad\qquad\qquad (35)(90) = 3150$ ✓

Appendix B
Practice Problems

1. $3.4 \text{ m} \cdot 1 = 3.4 \text{ m} \cdot \dfrac{100 \text{ cm}}{1 \text{ m}} = \dfrac{(3.4)(100)}{1} \cdot \dfrac{m}{m} \cdot \text{cm} = 340 \text{ cm}$

2. $125 \text{ L} \cdot \dfrac{0.001 \text{ kL}}{1 \text{ L}} = \dfrac{(125)(0.001)}{1} \cdot \dfrac{L}{L} \cdot \text{kL} = 0.125 \text{ kL}$

3. $5.72 \text{ lb} \cdot \dfrac{1 \text{ kg}}{2.20 \text{ lb}} = 2.6 \text{ kg}$

4. $8 \text{ cm} \cdot \dfrac{1 \text{ in.}}{2.54 \text{ cm}} \approx 3.15 \text{ in.}$

5. $3 \text{ L} \cdot \dfrac{1.06 \text{ qt}}{1 \text{ L}} = 3.18 \text{ qt}$

6. $\dfrac{70 \text{ mi}}{\text{hr}} \cdot \dfrac{5280 \text{ ft}}{1 \text{ mi}} \cdot \dfrac{1 \text{ hr}}{60 \text{ min}} \cdot \dfrac{1 \text{ min}}{60 \text{ sec}} = \dfrac{(70)(5280) \text{ ft}}{(60)(60) \text{ sec}} \approx 102.7 \text{ ft/sec}$

Appendix C
Practice Problems

1. (a) The lowest level of sodium, 670 mg, is in the Wendy's Grilled Chicken Sandwich.
(b) The highest level of cholesterol, 100 g, is in the McDonald's Big Mac.

2. (a) The two activities single women spend the most time doing are spending time with friends, and reading and watching TV.
(b) Couples spend more time per week with their family (21 hr) than any other activity.
(c) The greatest difference between single men and single women is the time spent reading and watching TV (6 hr).

3. (a) Walgreens had $12 \times 500 = 6000$ stores.
(b) Rite Aid had $10 \times 500 = 5000$ stores. Medicine Shoppe had $2 \times 500 = 1000$ stores. Since $5000 - 1000 = 4000$, Rite Aid had 4000 more stores than Medicine Shoppe.
(c) Longs Drug had $1 \times 500 = 500$ stores, and CVS had $13 \times 500 = 6500$ stores. There were $500 + 6500 = 7000$ total Longs Drug and CVS stores.

4. (a) The greatest number of accidents, 89, occurred during the 1999–2000 period.
(b) During the 1995–1996 period there were 61 accidents. During 1997–1998 there were 84 accidents. The increase in the number of accidents from 1995–1996 to 1997–1998 was 23.
(c) During the 1999–2000 period there were 89 accidents. During 2001–2002 there were 75 accidents. The decrease in the number of accidents from 1999–2000 to 2001–2002 was 14.

5. (a) There were 15 days in 2001 in which the air was considered unhealthy.
(b) The year with the most days in which the air was considered unhealthy was 2002.

6. (a) 15% of Seattle's rainfall falls between April and June.
(b) 40% of Seattle's rainfall occurs from October to December.
(c) 35% of the total rainfall occurs from January to March. This is $35\% \times 37 = 0.35 \times 37 = 12.95$ inches.

Appendix D
Practice Problems

1. Each number is 7 more than the preceding number. The next number in the sequence is $66 + 7 = 73$.

2. The sequence could be written $2^2 - 1, 3^2 - 1, 4^2 - 1, 5^2 - 1, \ldots$. The next number would be $10^2 - 1 = 99$.

3. Between 1's there are one 2, two 3's, and three 4's. Next would come four 5's, a 1, and then five 6's. The next seven digits are 5555166.

4. In the first figure, the bottom rectangle is shaded and the figure is rotated 90° clockwise to get the second figure. Then, in the third figure, the middle rectangle is shaded. This is rotated 90° clockwise to get the fourth figure. Then, the top rectangle is shaded and the figure is rotated. In the last figure, the bottom rectangle is shaded again and the sequence is repeated. The next two figures are

5.

Statement	Reason
1. $x = 2x + 24$	**1.** Multiplication principle: If $a = b$, then $ac = bc$. Here we multiplied by 6.
2. $-x = 24$	**2.** Addition principle: If $a = b$, then $a + c = b + c$. Here we added $-2x$ to each side of the equation.
3. $x = -24$	**3.** Multiplication principle: If $a = b$, then $ac = bc$. Here we multiplied each side by -1.

6.

Statement	Reason
1. $\angle 6 = \angle 5 = 26°$	**1.** If two lines intersect, the opposite angles are equal.
2. $\angle 4 = \angle 6 = 26°$	**2.** If a transversal crosses two parallel lines, the alternate interior angles are equal.

7. Let us organize the facts by listing the cars and each possible position: 1, 2, 3, 4, in which position 1 means far left and position 4 means far right.

Honda	Toyota	Mustang	Camaro
1	1	1	1
2	2	2	2
3	3	3	3
4	4	4	4

Step	Reason
1. We indicate that the Camaro is in position 4 and cross out the other 4's since no other car can be in position 4. We also cross out positions 1, 2, and 3 for the Camaro.	**1.** The Camaro is parked on the right end.

Honda	Toyota	Mustang	Camaro
1	1	1	~~1~~
2	2	2	~~2~~
3	3	3	~~3~~
~~4~~	~~4~~	~~4~~	**4**

Step	Reason
2. The Mustang cannot be in position 1 since it's between two cars. It also cannot be in position 3 as that would place it next to the Camaro. Therefore, the Mustang must be in position 2. Cross off the 2's in all other columns.	**2.** The Mustang is between the Honda and the Toyota.

Honda	Toyota	Mustang	Camaro
1	1	~~1~~	~~1~~
~~2~~	~~2~~	**2**	~~2~~
3	3	~~3~~	~~3~~
~~4~~	~~4~~	~~4~~	**4**

Step	Reason
3. The Honda must be in position 1 since it cannot be in position 3, next to the Camaro. This leaves position 3 for the Toyota.	**3.** The Honda is not next to the Camaro.

Honda	Toyota	Mustang	Camaro
1	~~1~~	~~1~~	~~1~~
~~2~~	~~2~~	**2**	~~2~~
~~3~~	**3**	~~3~~	~~3~~
~~4~~	~~4~~	~~4~~	**4**

The Honda, in position 1, is the car parked on the left end.

Answers to Selected Exercises

Chapter 0 **0.1 Exercises** **1.** 12 **3.** When two or more numbers are multiplied, each number that is multiplied is called a factor. In 2×3, 2 and 3 are factors. **5.** **7.** $\frac{3}{4}$ **9.** $\frac{1}{3}$ **11.** 5 **13.** $2\frac{5}{6}$ **15.** $12\frac{1}{3}$ **17.** $5\frac{3}{7}$ **19.** $\frac{16}{5}$ **21.** $\frac{33}{5}$ **23.** 8 **25.** 24 **27.** 21 **29.** 12 **31.** $15\frac{3}{4}$ **33.** $\frac{33}{160}$ **35.** $\frac{1}{4}$ **37.** $\frac{1}{2}$ **39.** $\frac{4}{9}$ **41.** $\frac{13}{16}$

Quick Quiz 0.1 **1.** $\frac{21}{23}$ **2.** $\frac{75}{11}$ **3.** $4\frac{19}{21}$ **4.** See Student Solutions Manual

0.2 Exercises **1.** Answers may vary. A sample answer is: 8 is exactly divisible by 4. **3.** 80 **5.** 20 **7.** 54 **9.** 90 **11.** $\frac{5}{8}$ **13.** $\frac{2}{7}$ **15.** $\frac{29}{24}$ or $1\frac{5}{24}$ **17.** $\frac{31}{63}$ **19.** $\frac{11}{15}$ **21.** $\frac{35}{36}$ **23.** $\frac{3}{2}$ or $1\frac{1}{2}$ **25.** $\frac{1}{2}$ **27.** $7\frac{11}{15}$ **29.** $1\frac{35}{72}$ **31.** $4\frac{11}{12}$ **33.** $6\frac{13}{28}$ **35.** $5\frac{19}{24}$ **37.** $4\frac{3}{7}$ **39.** 9 **41.** $\frac{23}{24}$ **43.** $7\frac{9}{16}$ **45.** $\frac{10}{21}$ **47.** $2\frac{7}{10}$ **49.** $24\frac{2}{3}$ mi **51.** $4\frac{1}{12}$ hr **53.** $A = 12$ in.; $B = 15\frac{7}{8}$ in. **55.** $\frac{9}{11}$ **56.** $\frac{133}{5}$

Quick Quiz 0.2 **1.** $\frac{5}{3}$ or $1\frac{2}{3}$ **2.** $7\frac{8}{15}$ **3.** $2\frac{5}{18}$ **4.** See Student Solutions Manual

0.3 Exercises **1.** First, change each number to an improper fraction. Look for a common factor in the numerator and denominator to divide by, and, if one is found, peform the division. Multiply the numerators. Multiply the denominators. **3.** $\frac{20}{7}$ or $2\frac{6}{7}$ **5.** $\frac{17}{30}$ **7.** $\frac{6}{25}$ **9.** $\frac{12}{5}$ or $2\frac{2}{5}$ **11.** $\frac{18}{5}$ or $3\frac{3}{5}$ **13.** $\frac{3}{5}$ **15.** $\frac{1}{7}$ **17.** 14 **19.** $\frac{7}{6}$ or $1\frac{1}{6}$ **21.** $\frac{15}{14}$ or $1\frac{1}{14}$ **23.** $\frac{8}{35}$ **25.** $1\frac{1}{3}$ **27.** $6\frac{1}{4}$ **29.** $\frac{8}{15}$ **31.** 6 **33.** $\frac{5}{4}$ or $1\frac{1}{4}$ **35.** $54\frac{3}{4}$ **37.** 40 **39.** 28 **41.** $\frac{3}{16}$ **43.** $71\frac{1}{2}$ yd **45.** $2\frac{4}{5}$ mi **47.** 55 **48.** 49

Quick Quiz 0.3 **1.** $\frac{5}{6}$ **2.** $14\frac{5}{8}$ **3.** $1\frac{8}{25}$ **4.** See Student Solutions Manual

0.4 Exercises **1.** 10, 100, 1000, 10,000, and so on **3.** 3, left **5.** 0.625 **7.** 0.2 **9.** $0.\overline{63}$ **11.** $\frac{4}{5}$ **13.** $\frac{5}{8}$ **15.** $\frac{17}{5}$ or $3\frac{2}{5}$ **17.** 2.09 **19.** 10.82 **21.** 261.208 **23.** 131.79 **25.** 30.282 **27.** 0.10575 **29.** 87.3 **31.** 2.64 **33.** 261.5 **35.** 0.508 **37.** 3450 **39.** 0.0076 **41.** 73,600 **43.** 0.73892 **45.** 8.22 **47.** 16.378 **49.** 2.12 **51.** 768.3 **53.** 2.6026 L **55.** No, there were 0.3 milligrams of copper in excess. **57.** $\frac{2}{3}$ **58.** $\frac{1}{6}$ **59.** $\frac{93}{100}$ **60.** $\frac{11}{10}$ or $1\frac{1}{10}$

Quick Quiz 0.4 **1.** 5.7078 **2.** 3.522 **3.** 28.8 **4.** See Student Solutions Manual

How Am I Doing? Sections 0.1–0.4 **1.** $\frac{3}{11}$ (obj. 0.1.2) **2.** $\frac{2}{5}$ (obj. 0.1.2) **3.** $3\frac{3}{4}$ (obj. 0.1.3) **4.** $\frac{33}{7}$ (obj. 0.1.3) **5.** 24 (obj. 0.1.4) **6.** 45 (obj. 0.1.4) **7.** LCD = 120 (obj. 0.2.2) **8.** $\frac{5}{7}$ (obj. 0.2.1) **9.** $\frac{19}{42}$ (obj. 0.2.3) **10.** $8\frac{5}{12}$ (obj. 0.2.4) **11.** $\frac{7}{18}$ (obj. 0.2.3) **12.** $\frac{4}{21}$ (obj. 0.2.3) **13.** $1\frac{33}{40}$ (obj. 0.2.4) **14.** $\frac{10}{9}$ or $1\frac{1}{9}$ (obj. 0.3.1) **15.** 42 (obj. 0.3.1) **16.** $\frac{7}{2}$ or $3\frac{1}{2}$ (obj. 0.3.2) **17.** $\frac{28}{39}$ (obj. 0.3.2) **18.** 0.375 (obj. 0.4.2) **19.** $0.8\overline{3}$ (obj. 0.4.2) **20.** 0.015 (obj. 0.4.2) **21.** 19.651 (obj. 0.4.4) **22.** 2.0664 (obj. 0.4.5) **23.** 0.45 (obj. 0.4.6) **24.** 2.246 (obj. 0.4.4)

0.5 Exercises **1.** Answers may vary. Sample answers follow. 19% means 19 out of 100 parts. Percent means per 100. 19% is really a fraction with a denominator of 100. In this case it would be $\frac{19}{100}$. **3.** 28% **5.** 56.8% **7.** 7.6% **9.** 239% **11.** 0.03 **13.** 0.004 **15.** 2.5 **17.** 0.0375 **19.** 5.2 **21.** 13 **23.** 72.8 **25.** 150% **27.** 5% **29.** 20% **31.** 80% **33.** $4.92 tip; $37.72 total bill **35.** 21% **37.** 540 gifts **39.** $1 an hr raise; $13.50/hr **41. (a)** $36,800 **(b)** $44,000 **43.** $612.99 **44.** $6220.50 **45.** 26.5 mi/gal **46.** 4.0 in.

Quick Quiz 0.5 **1.** 11.7 **2.** 15% **3.** 96.9 **4.** See Student Solutions Manual

0.6 Exercises **1.** 210,000 **3.** 18,000,000 **5.** 240 **7.** 20,000 **9.** 0.1 **11.** $4000.00 **13.** $400 **15.** $6400 **17.** 25 mi/gal **19.** $250,000 **21.** $60,000,000,000 **23.** $15,000 **25.** 10% **27.** 2.1 **28.** 60% **29.** $6.73 **30.** 270 tires

Quick Quiz 0.6 **1.** 18,000 **2.** 400,000 **3.** 20 **4.** See Student Solutions Manual

0.7 Exercises
1. $305.50 **3. (a)** $82\frac{1}{3}$ ft **(b)** 90 ft; $17.90 **5.** Jog $2\frac{2}{3}$ mi; walk $3\frac{1}{9}$ mi; rest $4\frac{4}{9}$ min; walk $1\frac{7}{9}$ mi

7. Betty; Melinda increases each activity by $\frac{2}{3}$ by day 3 but Betty increases each activity by $\frac{7}{9}$ by day 3. **9.** $4\frac{1}{2}$ mi **11.** $98,969.00

13. (a) $45,000 **(b)** $13,950 **15.** 18% **17.** 69%

Quick Quiz 0.7
1. 350 stones **2.** 46% **3.** $1215 **4.** See Student Solutions Manual

Chapter 0 Putting Your Skills to Work
1. $100, $300, $2000, $8000, $8000, $8000, $12,000 **2.** $365 **3.** 7 months

Chapter 0 Review Problems
1. $\frac{3}{4}$ **2.** $\frac{3}{10}$ **3.** $\frac{18}{41}$ **4.** $\frac{3}{5}$ **5.** $\frac{57}{8}$ **6.** $6\frac{4}{5}$ **7.** $26\frac{2}{3}$ **8.** 15 **9.** 5 **10.** 45 **11.** 22

12. $\frac{17}{20}$ **13.** $\frac{29}{24}$ or $1\frac{5}{24}$ **14.** $\frac{4}{15}$ **15.** $\frac{13}{30}$ **16.** $5\frac{23}{30}$ **17.** $6\frac{9}{20}$ **18.** $2\frac{29}{36}$ **19.** $1\frac{11}{12}$ **20.** $\frac{30}{11}$ or $2\frac{8}{11}$ **21.** $10\frac{1}{2}$ **22.** 50

23. $\frac{20}{7}$ or $2\frac{6}{7}$ **24.** $\frac{1}{16}$ **25.** $\frac{24}{5}$ or $4\frac{4}{5}$ **26.** $\frac{3}{20}$ **27.** 6 **28.** 7.201 **29.** 7.737 **30.** 29.561 **31.** 4.436 **32.** 0.03745

33. 362,341 **34.** 0.07956 **35.** 125.5 **36.** 0.00186 **37.** 0.07132 **38.** 1.3075 **39.** 90 **40.** 1.82 **41.** 0.5 **42.** 0.375

43. $\frac{9}{25}$ **44.** 0.014 **45.** 0.361 **46.** 0.0002 **47.** 1.253 **48.** 120 **49.** 3.96 **50.** 95% **51.** 12.5% **52.** 60%

53. 17,839,233 people **54.** 75% **55.** 400,000,000,000 **56.** 2500 **57.** 300,000 **58.** 10 **59.** $12,000 **60.** 25 **61.** $320

62. $800 **63.** $349.07 **64.** 100% **65.** 1840 mi; 1472 mi **66.** 1500 mi; 1050 mi **67.** $7\frac{1}{2}$ in.^{3} **68.** 200 boxes **69.** $462.80
70. 46 packages **71.** $1585.50 **72.** $401.25

How Am I Doing? Chapter 0 Test
1. $\frac{8}{9}$ (obj. 0.1.2) **2.** $\frac{4}{3}$ (obj. 0.1.2) **3.** $\frac{45}{7}$ (obj. 0.1.3) **4.** $11\frac{2}{3}$ (obj. 0.1.3)

5. $\frac{15}{8}$ or $1\frac{7}{8}$ (obj. 0.2.3) **6.** $4\frac{7}{8}$ (obj. 0.2.4) **7.** $\frac{5}{6}$ (obj. 0.2.4) **8.** $\frac{4}{3}$ or $1\frac{1}{3}$ (obj. 0.3.1) **9.** $\frac{5}{24}$ (obj. 0.3.1) **10.** $\frac{7}{2}$ or $3\frac{1}{2}$ (obj. 0.3.2)

11. $8\frac{1}{8}$ (obj. 0.3.1) **12.** $1\frac{21}{22}$ (obj. 0.3.2) **13.** 14.64 (obj. 0.4.4) **14.** 3.9897 (obj. 0.4.4) **15.** 1.312 (obj. 0.4.5) **16.** 73.85 (obj. 0.4.7)
17. 230 (obj. 0.4.6) **18.** 263.259 (obj. 0.4.7) **19.** 7.3% (obj. 0.5.2) **20.** 1.965 (obj. 0.5.3) **21.** 6.3 (obj. 0.5.4) **22.** 0.336 (obj. 0.5.4)
23. 6% (obj. 0.5.5) **24.** 30% (obj. 0.5.5) **25.** 18 computer chips (obj. 0.3.2) **26.** 100 (obj. 0.6.1) **27.** 700 (obj. 0.6.1)
28. 65% (obj. 0.7.1) **29.** 60 tiles (obj. 0.7.1)

Chapter 1
1.1 Exercises
1. Whole number, rational number, real number **3.** Irrational number, real number **5.** Rational number, real number **7.** Rational number, real number **9.** Irrational number, real number **11.** $-20,000$ **13.** $-37\frac{1}{2}$ **15.** $+7$ **17.** -8

19. 2.73 **21.** 1.3 **23.** $\frac{5}{6}$ **25.** -11 **27.** -50 **29.** $\frac{3}{10}$ **31.** $-\frac{7}{13}$ **33.** $\frac{1}{35}$ **35.** -19.2 **37.** 0.4 **39.** -14.16 **41.** -6

43. 0 **45.** $\frac{9}{20}$ **47.** -8 **49.** -3 **51.** 59 **53.** $\frac{7}{18}$ **55.** $\frac{2}{5}$ **57.** -2.21 **59.** 12 **61.** 0 **63.** 15.94 **65.** $167 profit

67. $-$3800 **69.** 9-yd gain **71.** 3500 **73.** $32,000,000 **75.** 18 **77.** $\frac{19}{16}$ or $1\frac{3}{16}$ **78.** $\frac{2}{3}$ **79.** $\frac{1}{12}$ **80.** $\frac{25}{34}$ **81.** 1.52
82. 0.65 **83.** 1.141 **84.** 0.26

Quick Quiz 1.1
1. -34 **2.** 0.5 **3.** $-\frac{13}{24}$ **4.** See Student Solutions Manual

1.2 Exercises
1. First change subtracting -3 to adding a positive three. Then use the rules for addition of two real numbers with different signs. Thus, $-8 - (-3) = -8 + 3 = -5$. **3.** -17 **5.** -4 **7.** -11 **9.** 8 **11.** 5 **13.** 0 **15.** 3 **17.** $-\frac{2}{5}$ **19.** $\frac{27}{20}$ or $1\frac{7}{20}$

21. $-\frac{19}{12}$ or $-1\frac{7}{12}$ **23.** -0.9 **25.** 4.47 **27.** $-\frac{17}{5}$ or $-3\frac{2}{5}$ **29.** $-\frac{14}{3}$ or $-4\frac{2}{3}$ **31.** -53 **33.** -73 **35.** 7.1 **37.** $\frac{35}{4}$ or $8\frac{3}{4}$

39. $-\frac{37}{6}$ or $-6\frac{1}{6}$ **41.** $-\frac{38}{35}$ or $-1\frac{3}{35}$ **43.** -8.5 **45.** $-\frac{29}{5}$ or $-5\frac{4}{5}$ **47.** 6.06 **49.** -5.047 **51.** 7 **53.** -48 **55.** -2 **57.** 0

59. -62 **61.** 1.6 **63.** $149 **65.** -21 **66.** -51 **67.** -19 **68.** 15°F **69.** $6\frac{2}{3}$ mi

Quick Quiz 1.2
1. 7 **2.** -1.9 **3.** $\frac{51}{56}$ **4.** See Student Solutions Manual

1.3 Exercises
1. To multiply two real numbers, multiply the absolute values. The sign of the result is positive if both numbers have the same sign, but negative if the two numbers have opposite signs. **3.** -20 **5.** 0 **7.** 49 **9.** 0.264 **11.** -4.5 **13.** $-\frac{3}{2}$ or $-1\frac{1}{2}$ **15.** $\frac{9}{11}$

17. $-\frac{5}{26}$ **19.** 4 **21.** 6 **23.** 15 **25.** -12 **27.** -130 **29.** -0.6 **31.** -0.9 **33.** $-\frac{3}{10}$ **35.** $\frac{20}{3}$ or $6\frac{2}{3}$ **37.** $\frac{7}{10}$ **39.** 14

41. $-\frac{5}{4}$ or $-1\frac{1}{4}$ **43.** $-\frac{2}{3}$ **45.** -24 **47.** -72 **49.** -16 **51.** -0.3 **53.** $-\frac{8}{35}$ **55.** $\frac{2}{27}$ **57.** 9 **59.** 4 **61.** 17 **63.** -72

65. -1 **67.** He gave $4.40 to each person and to himself. **69.** $235.60 **71.** Approximately 20 yd **73.** Approximately 70 yd

75. A gain of 20 yd **76.** -6.69 **77.** $-\frac{11}{6}$ or $-1\frac{5}{6}$ **78.** -15 **79.** -88

Quick Quiz 1.3 **1.** $-\frac{15}{8}$ or $-1\frac{7}{8}$ **2.** -120 **3.** 4 **4.** See Student Solutions Manual

1.4 Exercises **1.** The base is 4 and the exponent is 4. Thus you multiply $(4)(4)(4)(4) = 256$. **3.** The answer is negative. When you raise a negative number to an odd power the result is always negative. **5.** If you have parentheses surrounding the -2, then the base is -2 and the exponent is 4. The result is 16. If you do not have parentheses, then the base is 2. You evaluate to obtain 16 and then take the opposite of 16, which is -16. Thus $(-2)^4 = 16$ but $-2^4 = -16$. **7.** 5^7 **9.** w^2 **11.** p^4 **13.** $(3q)^3$ or 3^3q^3 **15.** 27 **17.** 81 **19.** 216 **21.** -27 **23.** 16

25. -25 **27.** $\frac{1}{16}$ **29.** $\frac{8}{125}$ **31.** 4.41 **33.** 0.0016 **35.** 256 **37.** -256 **39.** 161 **41.** -21 **43.** -128 **45.** 23 **47.** -576

48. -19 **49.** $-\frac{5}{3}$ or $-1\frac{2}{3}$ **50.** -8 **51.** 2.52

Quick Quiz 1.4 **1.** 256 **2.** 3.24 **3.** $\frac{27}{64}$ **4.** See Student Solutions Manual

1.5 Exercises **1.** $3(4) + 6(5)$ **3.** (a) 90 (b) 42 **5.** 12 **7.** 5 **9.** -29 **11.** 24 **13.** 21 **15.** 13 **17.** -6 **19.** 42

21. $\frac{9}{4}$ or $2\frac{1}{4}$ **23.** 0.848 **25.** $-\frac{7}{16}$ **27.** 5 **29.** $-\frac{23}{2}$ or $-11\frac{1}{2}$ **31.** 7.56 **33.** $\frac{1}{4}$ **35.** $2(-2) + 4(-1) + 10(0) + 2(+1)$

37. 2 above par **39.** 0.125 **40.** $-\frac{19}{12}$ or $-1\frac{7}{12}$ **41.** -1 **42.** $\frac{72}{125}$

Quick Quiz 1.5 **1.** -77 **2.** 16.96 **3.** 27 **4.** See Student Solutions Manual

How Am I Doing? Sections 1.1–1.5 **1.** -9 (obj. 1.1.4) **2.** $-\frac{41}{24}$ or $-1\frac{17}{24}$ (obj. 1.1.3) **3.** 1.24 (obj. 1.1.3) **4.** 4.8 (obj. 1.1.4)

5. 11 (obj. 1.2.1) **6.** $-\frac{29}{30}$ (obj. 1.2.1) **7.** 12.3 (obj. 1.2.1) **8.** 10 (obj. 1.2.1) **9.** -96 (obj. 1.3.1) **10.** $\frac{10}{11}$ (obj. 1.3.1)

11. -0.9 (obj. 1.3.3) **12.** $-\frac{10}{17}$ (obj. 1.3.3) **13.** 0.343 (obj. 1.4.2) **14.** 256 (obj. 1.4.2) **15.** -256 (obj. 1.4.2) **16.** $\frac{8}{27}$ (obj. 1.4.2)

17. 54 (obj. 1.4.2) **18.** 24 (obj. 1.5.1) **19.** 10 (obj. 1.5.1) **20.** 11 (obj. 1.5.1) **21.** -3.3 (obj. 1.5.1) **22.** $-\frac{9}{20}$ (obj. 1.5.1)

1.6 Exercises **1.** variable **3.** Here we are multiplying 4 by x by x. Since we know from the definition of exponents that x multiplied by x is x^2, this gives us an answer of $4x^2$.
5. Yes, $a(b - c)$ can be written as $a[b + (-c)]$.
$$3(10 - 2) = (3 \times 10) - (3 \times 2)$$
$$3 \times 8 = 30 - 6$$
$$24 = 24$$
7. $3x - 6y$ **9.** $-8a + 6b$ **11.** $9x + 3y$ **13.** $-8m - 24n$ **15.** $-x + 3y$ **17.** $-81x + 45y - 72$ **19.** $-10x + 2y - 12$

21. $10x^2 - 20x + 15$ **23.** $\frac{x^2}{5} + 2xy - \frac{4x}{5}$ **25.** $5x^2 + 10xy + 5xz$ **27.** $13.5x - 15$ **29.** $18x^2 + 3xy - 3x$ **31.** $-3x^2y - 2xy^2 + xy$

33. $-5a^2b - 10ab^2 + 20ab$ **35.** $2a^2 - 4a + \frac{8}{3}$ **37.** $0.36x^3 + 0.09x^2 - 0.15x$ **39.** $-1.32q^3 - 0.28qr - 4q$

41. $800(5x + 14y) = 4000x + 11{,}200y$ ft^2 **43.** $4x(3000 - 2y) = 12{,}000 - 8xy$ ft^2 **44.** -16 **45.** 64 **46.** 14 **47.** 4 **48.** 10

Quick Quiz 1.6 **1.** $-15a - 35b$ **2.** $-2x^2 + 8xy - 16x$ **3.** $-12a^2b + 15ab^2 + 27ab$ **4.** See Student Solutions Manual

1.7 Exercises **1.** A term is a number, a variable, or a product of numbers and variables. **3.** The two terms $5x$ and $-8x$ are like terms because they both have the variable x with the exponent of one. **5.** The only like terms are $7xy$ and $-14xy$ because the other two have different exponents even though they have the same variables. **7.** $-25b^2$ **9.** $6a^3 - 7a^2$ **11.** $-5x - 5y$ **13.** $7.1x - 3.5y$ **15.** $-2x - 8.7y$

17. $5p + q - 18$ **19.** $5bc - 6ac$ **21.** $x^2 - 10x + 3$ **23.** $-10y^2 - 16y + 12$ **25.** $-\frac{1}{15}x - \frac{2}{21}y$ **27.** $\frac{11}{20}a^2 - \frac{5}{6}b$ **29.** $-2rs + 2r$

31. $28a - 20b$ **33.** $-27ab - 11b^2$ **35.** $-2c - 10d^2$ **37.** $32x + 23$ **39.** $7a + 9b$ **41.** $14x - 8$ ft **43.** $-\frac{13}{12}$ or $-1\frac{1}{12}$ **44.** $-\frac{3}{8}$

45. $\frac{23}{50}$ **46.** $-\frac{15}{98}$

Quick Quiz 1.7 **1.** $\frac{13}{6}xy + \frac{5}{3}x^2y$ **2.** $0.6a^2b - 4.4ab^2$ **3.** $20x - 2y$ **4.** See Student Solutions Manual

1.8 Exercises **1.** -5 **3.** -12 **5.** $\frac{25}{2}$ or $12\frac{1}{2}$ **7.** -26 **9.** -1.3 **11.** $\frac{25}{4}$ or $6\frac{1}{4}$ **13.** 10 **15.** 5 **17.** -24 **19.** -20

21. 9 **23.** 39 **25.** -2 **27.** 44 **29.** -2 **31.** -9 **33.** 29 **35.** 32 **37.** 32 **39.** $-\frac{1}{2}$ **41.** 352 ft^2 **43.** 1.24 cm^2

45. 32 in.^2 **47.** $56{,}000 \text{ ft}^2$ **49.** 28.26 ft^2 **51.** $-78.5°C$ **53.** $\$2340.00$ **55.** $122°F; -67°F$ **57.** 16 **58.** $-x^2 + 2x - 4y$

Quick Quiz 1.8 **1.** 2 **2.** $\frac{9}{2}$ or $4\frac{1}{2}$ **3.** 33 **4.** See Student Solutions Manual

1.9 Exercises **1.** $-(3x + 2y)$ **3.** distributive   **5.** $3x + 6y$ **7.** $2c - 16d$ **9.** $x - 7y$ **11.** $8x^2 - 4x + 12$ **13.** $-2x + 26y$
15. $2x + 11y + 10$ **17.** $15a - 60ab$ **19.** $12a^2 - 21a - 18$ **21.** $3a^2 + 16b + 12b^2$ **23.** $-13a + 9b - 1$ **25.** $9b^2 + 36b - 12$
27. $12a^2 - 8b$ **29.** $1947.52°F$ **30.** $453{,}416 \text{ ft}^2$ **31.** 264.6 to 330.75 kg **32.** 19.845 to 30.87 kg

Quick Quiz 1.9 **1.** $-14x - 4y$ **2.** $-6x + 15y - 36$ **3.** $-8a - 24ab + 8b$ **4.** See Student Solutions Manual

Putting Your Skills to Work **1.** $\$1500$ **2.** $\$3450$ **3.** $\$6900$ **4.** About 55 months, or 4 years and 7 months
5. About $\$288$ **6.** $\$250$; about 28 months, or 2 years and 4 months   **7.** 2 months   **8.** $\$25{,}500$ **9.** Answers may vary
10. Answers may vary   **11.** Answers may vary

Chapter 1 Review Problems **1.** -8 **2.** -4.2 **3.** -9 **4.** 1.9 **5.** $-\frac{1}{3}$ **6.** $-\frac{7}{22}$ **7.** $\frac{1}{6}$ **8.** $-\frac{1}{2}$ **9.** 8 **10.** 13

11. -33 **12.** 9.2 **13.** $-\frac{13}{8}$ or $-1\frac{5}{8}$ **14.** $\frac{11}{24}$ **15.** -22.7 **16.** -88 **17.** -3 **18.** 13 **19.** 32 **20.** $\frac{5}{6}$ **21.** $-\frac{25}{7}$ or $-3\frac{4}{7}$

22. -72 **23.** 60 **24.** -30 **25.** -4 **26.** 16 **27.** -29 **28.** 1 **29.** $-\frac{1}{2}$ **30.** $-\frac{4}{7}$ **31.** -30 **32.** -5 **33.** -9.1

34. 0.9 **35.** 10.1 **36.** -1.2 **37.** 1.9 **38.** -1.3 **39.** 24 yd **40.** $-22°F$ **41.** $14{,}776 \text{ ft}$ **42.** Stock B   **43.** -243 **44.** 64

45. 625 **46.** $-\frac{8}{27}$ **47.** -81 **48.** 0.36 **49.** $\frac{25}{36}$ **50.** $\frac{27}{64}$ **51.** -44 **52.** 30 **53.** 3 **54.** $-21x + 7y$ **55.** $18x - 3x^2 + 9xy$

56. $-7x^2 + 3x - 11$ **57.** $-6xy^3 - 3xy^2 + 3y^3$ **58.** $-5a^2b + 3bc$ **59.** $-3x - 4y$ **60.** $-5x^2 - 35x - 9$ **61.** $10x^2 - 8x - \frac{1}{2}$

62. -55 **63.** 1 **64.** -4 **65.** -3 **66.** 0 **67.** 17 **68.** 15 **69.** $\$810$ **70.** $68°F$ to $77°F$ **71.** $\$75.36$ **72.** $\$8580.00$
73. $100{,}000 \text{ ft}^2$; $\$200{,}000$ **74.** 10.45 ft^2; $\$689.70$ **75.** $-2x + 42$ **76.** $-17x - 18$ **77.** $-2 + 10x$ **78.** $-12x^2 + 63x$
79. $5xy^3 - 6x^3y - 13x^2y^2 - 6x^2y$ **80.** $x - 10y + 35 - 15xy$ **81.** $-31a + 2b$ **82.** $15a - 15b - 10ab$ **83.** $-3x - 9xy + 18y^2$

84. $10x + 8xy - 32y$ **85.** -2.3 **86.** 8 **87.** $-\frac{22}{15}$ or $-1\frac{7}{15}$ **88.** $-\frac{1}{8}$ **89.** -1 **90.** -0.5 **91.** $\frac{81}{40}$ or $2\frac{1}{40}$ **92.** -8

93. 240 **94.** -25.42 **95.** $\$600$ **96.** 0.0081 **97.** -0.0625 **98.** 10 **99.** $-4.9x + 4.1y$ **100.** $-\frac{1}{9}$ **101.** $-\frac{2}{3}$

102. No; the dog's temperature is below the normal temperature of $101.48°F$.   **103.** $3y^2 + 12y - 7x - 28$ **104.** $-12x + 6y + 12xy$

How Am I Doing? Chapter 1 Test **1.** -0.3 (obj. 1.1.4)   **2.** 2 (obj. 1.2.1)   **3.** $-\frac{14}{3}$ or $-4\frac{2}{3}$ (obj. 1.3.1)   **4.** -70 (obj. 1.3.1)

5. 4 (obj. 1.3.3)   **6.** -3 (obj. 1.3.3)   **7.** -64 (obj. 1.4.2)   **8.** 2.56 (obj. 1.4.2)   **9.** $\frac{16}{81}$ (obj. 1.4.2)   **10.** 6.8 (obj. 1.5.1)   **11.** -25 (obj. 1.5.1)

12. $-5x^2 - 10xy + 35x$ (obj. 1.6.1)   **13.** $6a^2b^2 + 4ab^3 - 14a^2b^3$ (obj. 1.6.1)   **14.** $2a^2b + \frac{15}{2}ab$ (obj. 1.7.2)   **15.** $-1.8x^2y - 4.7xy^2$ (obj. 1.7.2)

16. $5a + 30$ (obj. 1.7.2)   **17.** $14x - 16y$ (obj. 1.7.2)   **18.** 122 (obj. 1.8.1)   **19.** 37 (obj. 1.8.1)   **20.** $\frac{13}{6}$ or $2\frac{1}{6}$ (obj. 1.8.1)

21. 96.6 km/hr (obj. 1.8.2)   **22.** $22{,}800 \text{ ft}^2$ (obj. 1.8.2)   **23.** $\$23.12$ (obj. 1.8.2)   **24.** 3 cans (obj. 1.8.2)   **25.** $3x - 6xy - 21y^2$ (obj. 1.9.1)
26. $-3a - 9ab + 3b^2 - 3ab^2$ (obj. 1.9.1)

Chapter 2 **2.1 Exercises** **1.** equals, equal   **3.** solution   **5.** Answers vary   **7.** $x = 4$ **9.** $x = 11$ **11.** $x = 17$ **13.** $x = -5$
15. $x = -13$ **17.** $x = 62$ **19.** $x = 15$ **21.** $x = 21$ **23.** $x = 0$ **25.** $x = 0$ **27.** $x = 21$ **29.** No; $x = 9$ **31.** No; $x = -14$

33. Yes   **35.** Yes   **37.** $x = -1.8$ **39.** $x = 0.6$ **41.** $x = 1$ **43.** $x = -\frac{1}{4}$ **45.** $x = -7$ **47.** $x = \frac{17}{6}$ or $2\frac{5}{6}$ **49.** $x = \frac{13}{12}$ or $1\frac{1}{12}$
51. $x = 5.2$ **53.** $x = 20.2$ **55.** $-2x - 4y$ **56.** $-2y^2 - 4y + 4$

Quick Quiz 2.1 **1.** $x = 14.3$ **2.** $x = -3.5$ **3.** $x = -8$ **4.** See Student Solutions Manual

2.2 Exercises **1.** 6 **3.** 7 **5.** $x = 36$ **7.** $x = -30$ **9.** $x = 80$ **11.** $x = -15$ **13.** $x = 4$ **15.** $x = 8$ **17.** $x = -\frac{8}{3}$

19. $x = 50$ **21.** $x = 15$ **23.** $x = -7$ **25.** $x = 0.2$ or $\frac{1}{5}$ **27.** $x = 4$ **29.** No; $x = -7$ **31.** Yes   **33.** $y = -0.03$ **35.** $t = \frac{8}{3}$

37. $y = -0.7$ **39.** $x = 3$ **41.** $x = -4$ **43.** $x = \frac{5}{12}$ **45.** $x = 1$ **47.** $m = 2$ **49.** $x = 27$ **51.** $x = -10.5$ **53.** $48 - 6 = 42$

54. $-27 - 10 = -37$ **55.** $5 + 16 = 21$ **56.** 22.5 tons   **57.** 27 earthquakes

Quick Quiz 2.2 1. $x = -38$ **2.** $x = 14$ **3.** $x = -12$ **4.** See Student Solutions Manual

2.3 Exercises 1. $x = 2$ **3.** $x = 6$ **5.** $x = -4$ **7.** $x = 13$ **9.** $x = 3.1$ **11.** $x = 28$ **13.** $x = -27$ **15.** $x = 8$
17. $x = 3$ **19.** $x = \dfrac{11}{2}$ **21.** $x = -9$ **23.** Yes   **25.** No; $x = -11$ **27.** $x = -1$ **29.** $x = 7$ **31.** $y = -1$ **33.** $x = 16$
35. $y = 3$ **37.** $x = 4$ **39.** $x = \dfrac{1}{4}$ **41.** $x = 2.5$ or $\dfrac{5}{2}$ **43.** $x = 6.5$ or $\dfrac{13}{2}$ **45.** $a = 0$ **47.** $x = -3.2$ or $-\dfrac{16}{5}$ **49.** $y = 2$
51. $x = -4$ **53.** $z = 5$ **55.** $a = -6.5$ **57.** $x = -\dfrac{2}{3}$ **59.** $x = -0.25$ or $-\dfrac{1}{4}$ **61.** $x = 8$ **63.** $x = -4.5$ or $-\dfrac{9}{2}$ **65.** $9xy - 8y^2$
66. $x - 9$ **67.** \$1023.94   **68. (a)** \$629.30   **(b)** \$647.28

Quick Quiz 2.3 1. $x = -\dfrac{4}{11}$ **2.** $x = 4$ **3.** $x = -\dfrac{8}{7}$ **4.** See Student Solutions Manual

2.4 Exercises 1. $x = -1$ **3.** $x = 1$ **5.** $x = 1$ **7.** $x = 12$ **9.** $y = 20$ **11.** $x = 3$ **13.** $x = \dfrac{7}{3}$ **15.** $x = -\dfrac{7}{2}$ or -3.5
17. Yes   **19.** No   **21.** $x = 1$ **23.** $x = 8$ **25.** $x = 2$ **27.** $x = -5$ **29.** $y = 4$ **31.** $x = -22$ **33.** $x = 2$ **35.** $x = -12$
37. $x = -\dfrac{5}{3}$ **39.** $x = -6$ **41.** No solution   **43.** Infinite number of solutions   **45.** $x = 0$ **47.** No solution   **49.** $-\dfrac{52}{3}$ or $-17\dfrac{1}{3}$
50. $\dfrac{22}{5}$ or $4\dfrac{2}{5}$ **51.** $572 - 975$ g   **52.** 3173 seats

Quick Quiz 2.4 1. $x = -\dfrac{7}{5}$ **2.** $x = \dfrac{19}{31}$ **3.** $x = \dfrac{11}{26}$ **4.** See Student Solutions Manual

How Am I Doing? Sections 2.1–2.4 1. $x = -9$ (obj. 2.1.1)   **2.** $x = \dfrac{15}{2}$ or 7.5 (obj. 2.1.1)   **3.** $x = 9$ (obj. 2.2.2)
4. $x = -8$ (obj. 2.2.2)   **5.** $x = \dfrac{2}{3}$ (obj. 2.3.1)   **6.** $x = \dfrac{1}{8}$ (obj. 2.3.2)   **7.** $x = \dfrac{3}{5}$ (obj. 2.3.3)   **8.** $x = \dfrac{15}{4}$ or 3.75 (obj. 2.3.3)
9. $x = -1.6$ or $-\dfrac{8}{5}$ (obj. 2.3.3)   **10.** $x = \dfrac{10}{7}$ (obj. 2.4.1)   **11.** $x = -3$ (obj. 2.4.1)   **12.** $x = \dfrac{9}{11}$ (obj. 2.4.1)   **13.** $x = 5$ (obj. 2.4.1)
14. $x = -\dfrac{1}{3}$ (obj. 2.4.1)

2.5 Exercises 1. Multiply each term by 5. Then subtract 160 from both sides. Then divide each side by 9. We would obtain $\dfrac{5F - 160}{9} = C$.

3. (a) 10 m **(b)** 16 m   **5. (a)** $y = -\dfrac{4}{3}x + 6$ or $y = \dfrac{18 - 4x}{3}$ **(b)** $y = 10$ **7.** $b = \dfrac{2A}{h}$ **9.** $P = \dfrac{I}{rt}$ **11.** $m = \dfrac{y - b}{x}$
13. $y = \dfrac{2}{3}x - 2$ **15.** $x = -\dfrac{3}{2}y + 6$ **17.** $y = \dfrac{c - ax}{b}$ **19.** $r^2 = \dfrac{A}{\pi}$ **21.** $r^2 = \dfrac{GM}{g}$ **23.** $t = \dfrac{A - P}{Pr}$ **25.** $h = \dfrac{S - 2\pi r^2}{2\pi r}$
27. $m = \dfrac{2K}{v^2}$ **29.** $L = \dfrac{V}{WH}$ **31.** $r^2 = \dfrac{3V}{\pi h}$ **33.** $w = \dfrac{P - 2L}{2}$ **35.** $a^2 = c^2 - b^2$ **37.** $C = \dfrac{5F - 160}{9}$ **39.** $R = \dfrac{E^2}{P}$
41. $S = \dfrac{360A}{\pi r^2}$ **43.** 0.8 mi   **45.** 30 ft   **47. (a)** $x = \dfrac{V - 7050}{1100}$ **(b)** 2012   **49.** A doubles   **51.** A increases 4 times
53. \$16   **54.** 1   **55.** 39,000 ft^2 **56.** $10\dfrac{7}{12}$ hr

Quick Quiz 2.5 1. $x = \dfrac{A - 2w}{3}$ **2.** $a = \dfrac{3A - bh}{h}$ **3.** $y = \dfrac{6ax + 5}{4ax}$ **4.** See Student Solutions Manual

2.6 Exercises 1. Yes, both statements imply 5 is to the right of -6 on a number line.   **3.** $>$ **5.** $<$ **7.** $>$ **9.** $<$ **11.** $<$ **13.** $<$
15. 0.0247   **17.**
19.
21.

23.
25.
27. $x \geq -\dfrac{2}{3}$ **29.** $x < -20$ **31.** $x \leq 3.7$ **33.** $c \geq 12$

35. $h \geq 48$ **37.** 2%   **38.** 37.5%   **39.** \$15   **40.** 495 ft

Quick Quiz 2.6 1.
2.
3. $x \geq -2$
4. See Student Solutions Manual

2.7 Exercises

1. $x \le -3$ **3.** $x \le 5$

5. $x > -9$ **7.** $x \ge 8$

9. $x < -12$ **11.** $x < \dfrac{7}{2}$

13. $x < -1$ **15.** $x < \dfrac{3}{2}$

17. $x > -6$ **19.** $x > \dfrac{1}{3}$

21. $3 > 1$ Adding any number to both sides of an inequality doesn't reverse the direction.   **23.** $x > 3$ **25.** $x \ge 4$ **27.** $x < -1$

29. $x \le 14$ **31.** $x < -3$ **33.** $x > -\dfrac{4}{11}$ **35.** 76 or greater   **37.** 8 days or more   **39.** 63.6 in.^{2} **40.** 7500 ft^2

Quick Quiz 2.7

1. $x \le 6$ **2.** $x > 11$

3. $x \ge 5$ **4.** See Student Solutions Manual

Putting Your Skills to Work

1. SHELL   **2.** SHELL   **3.** ARCO   **4.** ARCO   **5.** 3.75 gallons   **6.** Answers may vary   **7.** Answers may vary   **8.** Answers may vary

Chapter 2 Review Problems

1. $x = -7$ **2.** $x = -3$ **3.** $x = 2$ **4.** $x = -3$ **5.** $x = \dfrac{11}{4}$ or 2.75   **6.** $x = -13$ **7.** $x = 40.4$

8. $x = -7$ **9.** $x = -2$ **10.** $x = -40$ **11.** $x = 20$ **12.** $x = -9$ **13.** $x = -3$ **14.** $x = 3$ **15.** $x = 4$ **16.** $x = 3$

17. $x = 20$ **18.** $x = -7$ **19.** $x = 3$ **20.** $x = -\dfrac{7}{2}$ or -3.5 **21.** $y = -3$ **22.** $y = 2.1$ **23.** $x = 5$ **24.** $x = 1$ **25.** $x = 20$

26. $x = \dfrac{2}{3}$ **27.** $x = 5$ **28.** $x = \dfrac{35}{11}$ **29.** $x = 4$ **30.** $x = -17$ **31.** $x = \dfrac{2}{5}$ or 0.4   **32.** $x = 32$ **33.** $x = \dfrac{26}{7}$ **34.** $x = 0$

35. $x = 0$ **36.** $x = \dfrac{3}{4}$ **37.** $x = -17$ **38.** $x = -2.5$ **39.** $x = 23$ **40.** $x = -32$ **41.** $x = -\dfrac{17}{5}$ or -3.4 **42.** $x = 9$

43. $x = -\dfrac{16}{13}$ **44.** $x = 9.75$ or $\dfrac{39}{4}$ **45.** $y = 3x - 10$ **46.** $y = \dfrac{-5x - 7}{2}$ **47.** $r = \dfrac{A - P}{Pt}$ **48.** $h = \dfrac{A - 4\pi r^2}{2\pi r}$

49. $p = \dfrac{3H - a - 3}{2}$ **50.** $y = \dfrac{c - ax}{b}$ **51. (a)** $A = \dfrac{10x}{BC}$ **(b)** $A = -40$ **52. (a)** $w = \dfrac{P - 2l}{2}$ **(b)** $w = 6.5$

53. (a) $h = \dfrac{V}{lw}$ **(b)** $h = 6$ **54.** $x \le -1$ **55.** $x \ge 3$

56. $x < -4$ **57.** $x > -3$

58. $x \ge 6$ **59.** $x < 5$

60. $x < 10$ **61.** $x > -3$

62. $x > \dfrac{17}{2}$ **63.** $x > \dfrac{7}{5}$ **64.** $h \le 32$ hr   **65.** $n \le 22$

66. $x = \dfrac{13}{6}$ **67.** $x = \dfrac{11}{8}$ **68.** $x = 0$ **69.** $x = 4$ **70.** $x = 4$ **71.** $x = -5$ **72.** $f = e - 2$ **73.** $a = \dfrac{3W - 4b}{2}$

74. $x < 2$ **75.** $x \le -8$

76. $x \ge \dfrac{19}{7}$ **77.** $x \ge -15$

How Am I Doing? Chapter 2 Test

1. $x = 2$ (obj. 2.3.1)   **2.** $x = \dfrac{1}{3}$ (obj. 2.3.2)   **3.** $y = -\dfrac{7}{2}$ or -3.5 (obj. 2.3.3)

4. $y = 8.4$ or $\dfrac{42}{5}$ (obj. 2.4.1)   **5.** $x = 1$ (obj. 2.3.3)   **6.** $x = -1.2$ (obj. 2.3.3)   **7.** $y = 7$ (obj. 2.4.1)   **8.** $y = \dfrac{7}{3}$ or $2\dfrac{1}{3}$ (obj. 2.3.3)

9. $x = 13$ (obj. 2.3.3)   **10.** $x = 20$ (obj. 2.3.3)   **11.** $x = 10$ (obj. 2.3.3)   **12.** $x = -4$ (obj. 2.3.3)   **13.** $x = 12$ (obj. 2.3.3)

14. $x = -\dfrac{1}{5}$ (obj. 2.4.1)   **15.** $x = 3$ (obj. 2.4.1)   **16.** $x = 2$ (obj. 2.4.1)   **17.** $x = -2$ (obj. 2.4.1)   **18.** $w = \dfrac{A - 2P}{3}$ (obj. 2.5.1)

19. $w = \dfrac{6 - 3x}{4}$ (obj. 2.5.1)　　**20.** $a = \dfrac{2A - hb}{h}$ (obj. 2.5.1)　　**21.** $y = \dfrac{10ax - 5}{8ax}$ (obj. 2.5.1)　　**22.** $B = \dfrac{3V}{h}$ (obj. 2.5.1)

23. $B = 30$ in.^{2} (obj. 2.5.1)　　**24.** $x \le -3$ (obj. 2.7.1)

25. $x > -\dfrac{5}{4}$ (obj. 2.7.1)　　**26.** $x < 2$ (obj. 2.7.1)

27. $x \ge \dfrac{1}{2}$ (obj. 2.7.1)

Cumulative Test for Chapters 0–2 **1.** $\dfrac{4}{21}$　**2.** $\dfrac{79}{20}$ or $3\dfrac{19}{20}$　**3.** $\dfrac{35}{4}$ or $8\dfrac{3}{4}$　**4.** 0.6888　**5.** 0.12　**6.** 35.69　**7.** 30

8. $-2st - 7s^2t + 14st^2$　**9.** $25x^2$　**10.** $6x - 40 + 48y - 24xy$　**11.** $x = \dfrac{5}{6}$　**12.** $x = 4$　**13.** $y = 1$　**14.** $b = \dfrac{4c + 9}{4}$

15. $b = \dfrac{3H - 8a}{2}$　**16.** $t = \dfrac{I}{Pr}$　**17.** $a = \dfrac{2A}{h} - b$　**18.** $x \le 2$

19. $x \le 3$　　**20.** $x > -15$

21. $x \le -1$　　**22.** $x \le 1$　　**23.** 92

Chapter 3 3.1 Exercises
1. $x + 5$　**3.** $x - 12$　**5.** $\dfrac{1}{8}x$ or $\dfrac{x}{8}$　**7.** $2x$　**9.** $3 + \dfrac{1}{2}x$　**11.** $2x + 9$　**13.** $\dfrac{1}{3}(x + 7)$

15. $\dfrac{1}{3}x - 2x$　**17.** $5x - 11$　**19.** $x = $ value of a share of AT&T stock　　**21.** $2w + 7 = $ length
$\qquad\qquad\qquad\qquad\qquad\qquad x + 74.50 = $ value of a share of IBM stock　　　　　$w = $ width

23.　$x = $ number of boxes sold by Keiko　**25.** 1st angle $= s - 16$　**27.** $v = $ value of exports of Canada　**29.** 1st angle $= 3x$
$x - 43 = $ number of boxes sold by Sarah　　2nd angle $= s$　　　　$2v = $ value of exports of Japan　　2nd angle $= x$
$x + 53 = $ number of boxes sold by Imelda　　3rd angle $= 2s$　　　　　　　　　　　　　　　3rd angle $= x - 14$

31.　$x = $ men aged 16–24　　**33.** $x = 7$　**34.** $x = -\dfrac{5}{2}$ or $-2\dfrac{1}{2}$
$x + \ \ 82 = $ men aged 25–34
$x - \ \ 25 = $ men aged 35–44
$x - 110 = $ men aged 45 and above

Quick Quiz 3.1 **1.** $x + 10$ or $10 + x$　**2.** $2x - 5$　**3.** First angle: $x + 15$　**4.** See Student Solutions Manual
$\qquad\qquad\qquad\qquad\qquad\qquad\qquad\qquad\qquad\qquad\qquad$ Second angle: x
$\qquad\qquad\qquad\qquad\qquad\qquad\qquad\qquad\qquad\qquad\qquad$ Third angle: $2x$

3.2 Exercises **1.** 1261　**3.** 2368　**5.** 182　**7.** 43　**9.** 9　**11.** -4　**13.** 12　**15.** 15 used bikes　**17.** 48,840 wildfires
19. 6 CDs　**21.** 4 items　**23.** 320.2 mi/hr　**25.** 550 lb　**27.** 5 mi　**29.** She traveled 52 mph on the mountain road. It was 12 mph
faster on the highway route.　**31.** 90　**33. (a)** $F - 40 = \dfrac{x}{4}$ **(b)** 200 chirps **(c)** 77°F　**34.** $10x^3 - 30x^2 - 15x$
35. $-2a^2b + 6ab - 10a^2$　**36.** $-5x - 6y$　**37.** $-4x^2y - 7xy^2 - 8xy$

Quick Quiz 3.2 **1.** 17　**2.** 48　**3.** 74　**4.** See Student Solutions Manual

3.3 Exercises
1. The long piece is 32 meters long.　**3.** James earns $39,450 per year.　**5.** David worked 30 hours.　**7.** 11 students born in Oklahoma.
The short piece is 15 meters long.　　Lin earns $42,850 per year.　　Sarah worked 45 hours.　　19 students born in Texas.
$\qquad\qquad\qquad\qquad\qquad\qquad\qquad\qquad\qquad\qquad\qquad\qquad\qquad\qquad\qquad\qquad$ Kate worked 25 hours.　　2 students born in Arizona.

9. Length is 13 meters.　**11.** Length is 41 centimeters.　**13.** Width is 32.5 centimeters.　**15.** The cheetah can run 70 mph.　**17.** Length A $= 20$ in.
Width is 8 meters.　　Width is 7 centimeters.　　Length is 62.5 centimeters.　　The jackal can run 35 mph.　　Length B $= 15$ in.
$\qquad\qquad\qquad\qquad\qquad\qquad\qquad\qquad\qquad\qquad\qquad\qquad\qquad\qquad\qquad\qquad\qquad$ The elk can run 45 mph.　　Length C $= 10$ in.
$\qquad$ Length D $= 13$ in.

19. Original square was 11 m × 11 m　**21.** $-8x^3 + 12x^2 - 32x$　**22.** $5a^2b + 30ab - 10a^2$　**23.** $-19x + 2y - 2$　**24.** $9x^2y - 6xy^2 + 7xy$

Quick Quiz 3.3
1. Width is 75 yd　**2.** Psychology: 39 students　**3.** First side: 22 m　**4.** See Student Solutions Manual
Length is 230 yd　　Art history: 13 students　　Second side: 33 m
$\qquad\qquad\qquad\qquad\qquad$ Algebra: 27 students　　Third side: 18 m

How Am I Doing? Sections 3.1–3.3 **1.** $3x - 40$ (obj. 3.1.1)　**2.** $\dfrac{1}{2}x + 12$ (obj. 3.1.1)　**3.** $\dfrac{5x}{12}$ or $\dfrac{5}{12}x$ (obj. 3.1.1)

4. $\dfrac{1}{4}(x + 10)$ or $\dfrac{x + 10}{4}$ (obj. 3.1.1)　**5.** -26 (obj. 3.2.1)　**6.** 68 (obj. 3.2.1)　**7.** 40 months (obj. 3.2.2)　**8.** 5 hr (obj. 3.2.2)

9. Fourth person: $60,000
Fifth person: $62,500 (obj. 3.2.2)
10. Short piece: 4 feet
Long piece: 6 feet (obj. 3.3.1)
11. Sunday = 50 pages
Monday = 75 pages
Tuesday = 135 pages (obj. 3.3.1)
12. SSC: $655,000
Pagani: $667,000
Bugatti: $1,192,000 (obj. 3.3.1)

3.4 Exercises
1. 11 bags   **3.** 5 hr   **5.** 12 weeks   **7.** $360   **9.** $22,000   **11.** $12,000   **13.** $3000 at 7%
$2000 at 5%
15. Conservative fund = $250,000
Growth fund = $150,000
17. $12,000   **19.** 13 quarters
9 nickels
21. 18 nickels
6 dimes
9 quarters
23. 30 regular stamps
22 breast cancer stamps
25. $10 bills = 26
$20 bills = 31
$50 bills = 10
27. $925,000 worth of furniture   **29.** 12   **30.** −17   **31.** 25   **32.** −26

Quick Quiz 3.4
1. 9 months   **2.** $11,600   **3.** $2200 at 4%
$2800 at 5%
4. See Student Solutions Manual

3.5 Exercises
1. distance around   **3.** surface   **5.** 180°   **7.** 156 in.^{2} **9.** 98 in.^{2} **11.** 6 ft   **13.** 9 in.   **15.** 153.86 ft^2
17. 9520.48 mi   **19.** 24 in.   **21.** 134,400 m^2 **23.** 5 cm   **25.** 200.96 cm   **27.** 74°   **29.** Equal angles = 72°   **31.** 30° each
3rd angle = 36°

33. 1st angle = 46°
2nd angle = 23°
3rd angle = 111°
35. 19 ft   **37.** 6 ft   **39.** 2512 in.^{3} **41.** Yes   **43. (a)** V = 452.16 cm^3 **45.** 163.28 ft^2 **47. (a)** 22.77 yd^2
(b) S = 376.8 cm^2

(b) $56.93   **49.** 73.96 in.   **51.** 9.3 feet wide and 34.2 feet long   **53.** 6.28 ft   **54.** $3x^2 + 8x - 24$
55. $-11x - 17$ **56.** 213 billion   **57.** 155%

Quick Quiz 3.5
1. 116 ft^2 **2.** 113 in.^{3} **3.** $2280   **4.** See Student Solutions Manual

3.6 Exercises
1. $x > 67,000$ **3.** $x \le 90$ **5.** $h \le 1500$ **7.** $x \ge 93$ **9.** It must be less than or equal to 140 cm.   **11.** 22 or more
deliveries   **13.** Depth can be no more than 7.5 feet.   **15.** They can drive no more than 81.8 miles.   **17.** The range is less than 230 degrees
Fahrenheit.   **19.** The amount of federal aid will exceed $528 billion after 2011.   **21.** More than $130,000 in sales   **23.** 455 or more discs
would need to be sold.   **25.** $x < -4$ **26.** $x \le 3$ **27.** $x \ge 4\frac{2}{3}$ **28.** $x > -4$

Quick Quiz 3.6
1. No more than 9.9 feet   **2.** Less than 18 hours per week   **3.** More than $285,000 in sales
4. See Student Solutions Manual

Putting Your Skills to Work
1. Either A or B   **2.** B   **3.** B   **4.** C   **5.** D   **6.** D   **7.** $84.99   **8.** $94.99
9. $114.99   **10.** Answers may vary

Chapter 3 Review Problems
1. $x + 19$ **2.** $\frac{2}{3}x$ **3.** $\frac{1}{2}x$ or $\frac{x}{2}$ **4.** $x - 18$ **5.** $3(x + 4)$ **6.** $2x - 3$ **7.** r = the number of
retired people; $4r$ = the number of working people; $0.5r$ = the number of unemployed people   **8.** $3w + 5$ = the length; w = the width
9. b = the number of degrees in angle B; $2b$ = the number of degrees in angle A; $b - 17$ = the number of degrees in angle C
10. a = the number of students in algebra; $a + 29$ = the number of students in biology; $0.5a$ = the number of students in geology
11. $x = 3$ **12.** $x = -7$ **13.** $40   **14.** 16 years old   **15.** 13.3 hr; 12.3 hr   **16.** 88   **17.** 23 ft   **18.** 1st angle = 32°; 2nd angle = 96°;
3rd angle = 52°   **19.** 31.25 yd; 18.75 yd   **20.** George = $42,000; Heather = $23,000   **21.** 310 kilowatt-hours   **22.** 280 mi   **23.** $3000
24. $200   **25.** $7000 at 12%; $2000 at 8%   **26.** $2000 at 4.5%; $3000 at 6%   **27.** 18 nickels; 6 dimes; 9 quarters   **28.** 24 nickels; 21 dimes;
26 quarters   **29.** 21.23 in.^{2} **30.** $P = 24$ ft   **31.** 71°   **32.** 42 mi^2 **33.** 462 ft^3 **34.** 381.51 in.^{3} **35.** 415.27 ft^2 **36.** 12 in.
37. $3440   **38.** $23,864   **39.** 160 miles or less   **40.** More than $260,000   **41.** CP distance = 630 mi   **42.** Less than $16.15   **43.** $24\frac{4}{5}$ min
UP distance = 960 mi
44. 4 oz   **45.** More than 3 years   **46.** More than 60 months   **47.** The height is 4 feet. The width is 10 feet.   **48.** It will take 10 years.
49. They made 27 field goals and 9 free throws.   **50.** The width is 16 feet and the length is 22 feet.   **51.** Carl = 12 hr
Ryan = 16 hr
52. $13,600 at 9.75%   **53.** First angle measures 60°; second angle measures 40°; third angle measures 80°   **54.** $1224   **55.** 94
$15,600 at 8.5%
56. No, she traveled 9000 miles for business. Her correct deduction was $2970.   **57.** It can travel 640 miles in one hour. It would take approximately
169 minutes.   **58.** 12 sec   **59.** 5 Explorers and 10 Caravans   **60.** Smallest angle = 32°
largest angle = 96°
third angle = 52°
61. 183.26 kg

How Am I Doing? Chapter 3 Test
1. 35 (obj. 3.2.1)   **2.** 36 (obj. 3.2.1)   **3.** −4 (obj. 3.2.1)   **4.** First side = 20 m; second side = 30 m;
third side = 16 m (obj. 3.3.1)   **5.** Width = 20 m; length = 47 m (obj. 3.3.1)   **6.** 1st pollutant = 8 ppm; 2nd pollutant = 4 ppm; 3rd
pollutant = 3 ppm (obj. 3.3.1)   **7.** 15 months (obj. 3.4.1)   **8.** $21,000 (obj. 3.4.2)   **9.** $1400 at 14%; $2600 at 11% (obj. 3.4.3)
10. 16 nickels; 7 dimes; 8 quarters (obj. 3.4.4)   **11.** 213.52 in. (obj. 3.5.1)   **12.** 192 in.^{2} (obj. 3.5.1)   **13.** 4187 in.^{3} (obj. 3.5.2)
14. 96 cm^2 (obj. 3.5.1)   **15.** $450 (obj. 3.5.3)   **16.** At least an 82 (obj. 3.6.2)   **17.** More than $110,000 (obj. 3.6.2)   **18.** 600 hr (obj. 3.6.2)

Cumulative Test for Chapters 0–3

1. 3.69   **2.** $\frac{31}{24}$ or $1\frac{7}{24}$ **3.** $-2x + 2y + 20$ **4.** 21   **5.** $\frac{2H - 3a}{5} = b$

6. $x \geq -3$;

7. -7 **8.** Literature students = 42; sociology students = 54

9. Width = 7 cm; length = 32 cm   **10.** \$230,435   **11.** \$3000 at 15%; \$4000 at 7%   **12.** 7 nickels; 12 dimes; 4 quarters

13. $A = 162.5$ m²; \$731.25   **14.** $V = 113.04$ in.³; 169.56 lb   **15.** 25 days

Chapter 4

4.1 Exercises

1. When multiplying exponential expressions with the same base, keep the base the same and add the exponents.

3. $\frac{2^2}{2^3} = \frac{2 \cdot 2}{2 \cdot 2 \cdot 2} = \frac{1}{2} = \frac{1}{2^{3-2}}$ **5.** 6; x and y; 11 and 1   **7.** $2^2 a^3 b$ **9.** $-5x^3 y^2 z^2$ **11.** 7^{10} **13.** 5^{26} **15.** x^{12} **17.** t^{16} **19.** $-20x^6$

21. $50x^3$ **23.** $18x^3 y^8$ **25.** $\frac{2}{15} x^3 y^5$ **27.** $-2.75x^3 yz$ **29.** 0   **31.** $80x^3 y^7$ **33.** $-24x^4 y^7$ **35.** 0   **37.** $-24a^4 b^3 x^2 y^5$ **39.** $-30x^3 y^4 z^6$

41. y^7 **43.** $\frac{1}{y^3}$ **45.** $\frac{1}{11^{12}}$ **47.** 2^7 **49.** $\frac{a^8}{4}$ **51.** $\frac{x^7}{y^9}$ **53.** $2x^4$ **55.** $-\frac{x^3}{2y^2}$ **57.** $\frac{f^2}{30g^5}$ **59.** $17x^4 y^3$ **61.** $\frac{y^2}{16x^3}$ **63.** $\frac{3a}{4}$

65. $\frac{2c^2}{7ab}$ **67.** x^{12} **69.** $x^{15} y^5$ **71.** $r^6 s^{12}$ **73.** $27a^9 b^6 c^3$ **75.** $9a^8$ **77.** $\frac{x^7}{128m^{28}}$ **79.** $\frac{25x^2}{49y^4}$ **81.** $81a^8 b^{12}$ **83.** $-8x^9 z^3$

85. $\frac{9}{x}$ **87.** $25a^5 b^7$ **89.** $\frac{64}{y^{10}}$ **91.** $\frac{16x^4}{y^{12}}$ **92.** -11 **93.** -46 **94.** $\frac{2}{25}$ **95.** -4 **96.** The rain forest decreased by approximately 17.1%.   **97.** $\approx 78{,}206$ km² of rain forest lost

Quick Quiz 4.1

1. $-10x^3 y^7$ **2.** $-\frac{4x^3}{5y^2}$ **3.** $81x^{12} y^{20}$ **4.** See Student Solutions Manual

4.2 Exercises

1. $\frac{1}{x^4}$ **3.** $\frac{1}{81}$ **5.** y^8 **7.** $\frac{z^6}{x^4 y^5}$ **9.** $\frac{a^3}{b^2}$ **11.** $\frac{x^9}{8}$ **13.** $\frac{3}{x^2}$ **15.** $\frac{1}{9x^2 y^4}$ **17.** $\frac{3xz^3}{y^2}$ **19.** $3x$ **21.** $\frac{b^3 d}{ac^4}$ **23.** $\frac{1}{8}$

25. $\frac{z^8}{9y^4}$ **27.** $\frac{1}{x^6 y}$ **29.** 1.2378×10^5 **31.** 6.3×10^{-2} **33.** 8.8961×10^{11} **35.** 3.42×10^{-6} **37.** 302,000   **39.** 0.00047

41. 983,000   **43.** 2.37×10^{-5} mph   **45.** 149,600,000 km   **47.** 6.3×10^{15} **49.** 1.0×10^1 **51.** 8.1×10^{-11} **53.** 4.5×10^5

55. 2.82×10^4 dollars   **57.** 6.6×10^{-5} mi   **59.** About 3.68×10^8 mi/yr   **61.** 159.3%   **63.** -0.8 **64.** -1 **65.** $-\frac{1}{28}$

Quick Quiz 4.2

1. $\frac{3y^2}{x^3 z^4}$ **2.** $\frac{a^8}{2b}$ **3.** 8.76×10^{-3} **4.** See Student Solutions Manual

4.3 Exercises

1. A polynomial in x is the sum of a finite number of terms of the form ax^n, where a is any real number and n is a whole number. An example is $3x^2 - 5x - 9$.   **3.** The degree of a polynomial in x is the largest exponent of x in any of the terms of the polynomial.

5. Degree 4; monomial   **7.** Degree 5; trinomial   **9.** Degree 6; binomial   **11.** $5x - 28$ **13.** $-2x^2 + 2x - 1$ **15.** $\frac{5}{6} x^2 + \frac{1}{2} x - 9$

17. $3.4x^3 + 2.2x^2 - 13.2x - 5.4$ **19.** $5x - 24$ **21.** $\frac{1}{15} x^2 - \frac{1}{14} x + 11$ **23.** $3x^3 - x^2 + 8x$ **25.** $-4.7x^4 - 0.7x^2 - 1.6x + 0.4$

27. $6x - 6$ **29.** $-3x^2 y + 6xy^2 - 4$ **31.** $x^4 - 3x^3 - 4x^2 - 24$ **33.** 743,000   **35.** 90,400   **37.** $3x^2 + 12x$ **39.** $x = \frac{3y - 2}{8}$

40. $B = \frac{3A}{2CD}$ **41.** $\frac{2A}{h} - c = b$ **42.** $d = \frac{5xy}{B}$

Quick Quiz 4.3

1. $-4x^2 - 11x + 5$ **2.** $6x^2 - 9x - 16$ **3.** $-3x + 8$ **4.** See Student Solutions Manual

How Am I Doing? Sections 4.1–4.3

1. $-24x^3 y^5$ (obj. 4.1.1)   **2.** $-\frac{7y^3}{5x^7}$ (obj. 4.1.2)   **3.** $\frac{4x^5}{y^9}$ (obj. 4.1.2)   **4.** $81x^{20} y^4$ (obj. 4.1.3)

5. $\frac{x^6}{16y^8}$ (obj. 4.2.1)   **6.** $\frac{x^5}{3y^5}$ (obj. 4.2.1)   **7.** 5.874×10^4 (obj. 4.2.2)   **8.** 9.362×10^{-5} (obj. 4.2.2)   **9.** $2.3x^2 - 0.4x - 5.4$ (obj. 4.3.2)

10. $4x^2 + 2x - 8$ (obj. 4.3.3)   **11.** $\frac{1}{6} x^3 + \frac{3}{8} x^2 + 3x$ (obj. 4.3.3)   **12.** $\frac{5}{16} x^2 - \frac{3}{10} x - \frac{3}{8}$ (obj. 4.3.2)

4.4 Exercises

1. $-12x^4 + 2x^2$ **3.** $21x^3 - 9x^2$ **5.** $-4x^6 + 10x^4 - 2x^3$ **7.** $x + \frac{3}{2} x^2 + \frac{5}{2} x^3$ **9.** $-2x^5 y + 4x^4 y - 5x^3 y$

11. $9x^4 y + 3x^3 y - 24x^2 y$ **13.** $3x^4 - 9x^3 + 15x^2 - 6x$ **15.** $-2x^3 y^3 + 12x^2 y^3 - 16xy$ **17.** $-28x^5 y + 12x^4 y + 8x^3 y - 4x^2 y$

19. $-6c^2 d^5 + 8c^2 d^3 - 12c^2 d$ **21.** $12x^7 - 6x^5 + 18x^4 + 54x^3$ **23.** $-16x^6 + 10x^5 - 12x^4$ **25.** $x^2 + 12x + 35$ **27.** $x^2 + 8x + 12$

29. $x^2 - 6x - 16$ **31.** $x^2 - 9x + 20$ **33.** $-20x^2 - 7x + 6$ **35.** $2x^2 + 6xy - 5x - 15y$ **37.** $15x^2 - 5xy + 6x - 2y$ **39.** $20y^2 - 7y - 3$

41. The signs are incorrect. The result is $-3x + 6$.   **43.** $20x$ **45.** $20x^2 - 23xy + 6y^2$ **47.** $49x^2 - 28x + 4$ **49.** $16a^2 + 16ab + 4b^2$

51. $0.8x^2 + 11.94x - 0.9$ **53.** $\frac{1}{4}x^2 + \frac{1}{24}x - \frac{1}{12}$ **55.** $2ax - 10ab - 3bx + 15b^2$ **57.** $10x^2 - 11x - 6$ **59.** $x = -10$

60. $w = -\frac{25}{7}$ or $-3\frac{4}{7}$ **61.** 9 twenties; 8 tens; 23 fives   **62.** $240 billion   **63.** $505.5 billion   **64.** $594 billion   **65.** $664.8 billion

Quick Quiz 4.4 **1.** $8x^3y^4 - 12x^2y^3 + 16xy^2$ **2.** $6x^2 - x - 15$ **3.** $12a^2 - 26ab + 12b^2$ **4.** See Student Solutions Manual

4.5 Exercises **1.** binomial   **3.** The middle term is missing. The answer should be $16x^2 - 56x + 49$.   **5.** $y^2 - 49$ **7.** $x^2 - 64$

9. $36x^2 - 25$ **11.** $4x^2 - 49$ **13.** $25x^2 - 9y^2$ **15.** $0.36x^2 - 9$ **17.** $4y^2 + 20y + 25$ **19.** $25x^2 - 40x + 16$ **21.** $49x^2 + 42x + 9$

23. $9x^2 - 42x + 49$ **25.** $\frac{4}{9}x^2 + \frac{1}{3}x + \frac{1}{16}$ **27.** $81x^2y^2 + 72xyz + 16z^2$ **29.** $49x^2 - 9y^2$ **31.** $9c^2 - 30cd + 25d^2$ **33.** $81a^2 - 100b^2$

35. $25x^2 + 90xy + 81y^2$ **37.** $x^3 - 4x^2 + 8x - 15$ **39.** $4x^4 - 7x^3 + 2x^2 - 3x - 1$ **41.** $a^4 + a^3 - 13a^2 + 17a - 6$

43. $3x^3 - 2x^2 - 25x + 24$ **45.** $2x^3 - x^2 - 16x + 15$ **47.** $2a^3 + 3a^2 - 50a - 75$ **49.** $24x^3 + 14x^2 - 11x - 6$ **51.** $11,000 at 7%;

$7000 at 11%   **52.** Width $= 7$ m; length $= 10$ m

Quick Quiz 4.5 **1.** $49x^2 - 144y^2$ **2.** $6x^3 - x^2 - 19x - 6$ **3.** $15x^4 - 16x^3 - 8x^2 + 17x - 6$ **4.** See Student Solutions Manual

4.6 Exercises **1.** $5x^3 - 3x + 4$ **3.** $2y^2 - 3y - 1$ **5.** $9x^4 - 4x^2 - 7$ **7.** $8x^4 - 9x + 6$ **9.** $3x + 5$ **11.** $x - 3 - \frac{32}{x - 5}$

13. $3x^2 - 4x + 8 - \frac{10}{x + 1}$ **15.** $2x^2 - 3x - 2 - \frac{5}{2x + 5}$ **17.** $2x^2 + 3x - 1$ **19.** $4y^2 - 2y + 5$ **21.** $y^2 - 4y - 1 - \frac{9}{y + 3}$

23. $y^3 + 2y^2 - 5y - 10 - \frac{25}{y - 2}$ **25.** 2000: $2.83; 2005: $3.20   **26.** 170 and 171   **27. (a)** 6.5 **(b)** 7.5 **(c)** 15.4% **(d)** 8.7

Quick Quiz 4.6 **1.** $5x^3 - 16x^2 - 2x$ **2.** $4x^2 - 5x - 2$ **3.** $x^2 + 2x + 8 + \frac{13}{x - 2}$ **4.** See Student Solutions Manual

Putting Your Skills to Work **1.** $445   **2.** $388.30   **3.** $310.64   **4.** 30.2%   **5.** Answers may vary

Chapter 4 Review Problems **1.** $-18a^7$ **2.** 5^{23} **3.** $6x^4y^6$ **4.** $-14x^4y^9$ **5.** $\frac{1}{7^{12}}$ **6.** $\frac{1}{x^5}$ **7.** y^{14} **8.** $\frac{1}{9^{11}}$ **9.** $-\frac{3}{5x^5y^4}$

10. $-\frac{2a}{3b^6}$ **11.** x^{24} **12.** b^{30} **13.** $9a^6b^4$ **14.** $81x^{12}y^4$ **15.** $\frac{25a^2b^4}{c^6}$ **16.** $\frac{y^9}{64w^{15}z^6}$ **17.** $\frac{b^5}{a^3}$ **18.** $\frac{m^8}{p^5}$ **19.** $\frac{2y^3}{x^6}$ **20.** $\frac{x^{15}}{8y^3}$

21. $\frac{1}{36a^8b^{10}}$ **22.** $\frac{3y^2}{x^3}$ **23.** $\frac{4w^2}{x^5y^6z^8}$ **24.** $\frac{b^5c^3d^4}{27a^2}$ **25.** 1.563402×10^{11} **26.** 1.79632×10^5 **27.** 9.2×10^{-4} **28.** 1.74×10^{-6}

29. 120,000   **30.** 6,034,000   **31.** 3,000,000   **32.** 0.25   **33.** 0.0000432   **34.** 0.000000006   **35.** 2.0×10^{13} **36.** 9.36×10^{19}

37. 9.6×10^{-10} **38.** 7.8×10^{-11} **39.** 7×10^8 dollars   **40.** 7.94×10^{14} cycles   **41.** 6×10^9 operations   **42.** $5.5x^2 - x - 2.3$

43. $5.6x^2 - 1.1x - 4$ **44.** $-x^3 + 2x^2 - x + 8$ **45.** $7x^3 - 3x^2 - 6x + 4$ **46.** $\frac{1}{10}x^2y - \frac{13}{21}x + \frac{5}{12}$ **47.** $\frac{1}{4}x^2 - \frac{1}{4}x + \frac{1}{10}$

48. $2x^2 + 3x - 5$ **49.** $-3x^2 - 15$ **50.** $15x^2 + 2x - 1$ **51.** $28x^2 - 29x + 6$ **52.** $20x^2 + 48x + 27$ **53.** $10x^3 - 30x^2 + 15x$

54. $-15x^4 - 20x^3 + 25x^2 - 40x$ **55.** $-4x^2y^4 - 20x^2y^3 + 24xy^2$ **56.** $5a^2 - 8ab - 21b^2$ **57.** $8x^4 - 10x^2y - 12x^2 + 15y$

58. $-15x^6y^2 - 9x^4y + 6x^2y$ **59.** $16x^2 + 24x + 9$ **60.** $a^2 - 25b^2$ **61.** $49x^2 - 36y^2$ **62.** $25a^2 - 20ab + 4b^2$ **63.** $64x^2 + 144xy + 81y^2$

64. $4x^3 + 27x^2 + 5x - 3$ **65.** $2x^3 - 7x^2 - 42x + 72$ **66.** $2y^2 + 3y + 4$ **67.** $6x^3 + 7x^2 - 18x$ **68.** $4x^2 - 6x + 8$

69. $53x^3 - 12x^2 + 19x + 13$ **70.** $3x - 2$ **71.** $3x - 7$ **72.** $3x^2 + 2x + 4 + \frac{9}{2x - 1}$ **73.** $2x^2 - 5x + 13 + \frac{27}{x + 2}$ **74.** $5x + 2$

75. $3x - 4 + \frac{3}{2x + 3}$ **76.** $x^2 + 3x + 8$ **77.** $2x^2 + 4x + 5 + \frac{11}{x - 2}$ **78.** 4^{29} **79.** 6^{15} **80.** $\frac{y^4}{4x^6}$ **81.** $\frac{-3ac^2}{b^2}$ **82.** $\frac{y^4z^6}{9}$

83. 9.22×10^{-4} **84.** $-4x^3 - 12x + 17$ **85.** $-12x^5 - 30x^4 + 12x^2$ **86.** $14x^3 + 15x^2 - 23x + 6$ **87.** $2x^2 + 5x + 3 + \frac{4}{3x - 2}$

88. About $9.33   **89.** 1.34×10^9 people   **90.** 2.733×10^{-23} g   **91.** 4.76×10^7 lb   **92.** $3xy + 2x$ **93.** $2x^2 - 4y^2$

How Am I Doing? Chapter 4 Test **1.** 3^{34} (obj. 4.1.1)   **2.** $\frac{1}{25^{16}}$ (obj. 4.1.2)   **3.** 8^{24} (obj. 4.1.3)   **4.** $12x^4y^{10}$ (obj. 4.1.1)

5. $-\frac{7x^3}{5}$ (obj. 4.1.2)   **6.** $-125x^3y^{18}$ (obj. 4.1.3)   **7.** $\frac{49a^{14}b^4}{9}$ (obj. 4.1.3)   **8.** $\frac{3x^4}{4}$ (obj. 4.1.3)   **9.** $\frac{1}{64}$ (obj. 4.2.1)   **10.** $\frac{6c^5}{a^4b^3}$ (obj. 4.2.1)

11. $3xy^7$ (obj. 4.2.1)   **12.** 5.482×10^{-4} (obj. 4.2.2)   **13.** 582,000,000 (obj. 4.2.2)   **14.** 2.4×10^{-6} (obj. 4.2.2)   **15.** $-2x^2 + 5x$ (obj. 4.3.2)

16. $-11x^3 - 4x^2 + 7x - 8$ (obj. 4.3.3)   **17.** $-21x^5 + 28x^4 - 42x^3 + 14x^2$ (obj. 4.4.1)   **18.** $15x^4y^3 - 18x^3y^2 + 6x^2y$ (obj. 4.4.1)

19. $10a^2 + 7ab - 12b^2$ (obj. 4.4.2)   **20.** $6x^3 - 11x^2 - 19x - 6$ (obj. 4.5.3)   **21.** $49x^4 + 28x^2y^2 + 4y^4$ (obj. 4.5.2)   **22.** $25s^2 - 121t^2$ (obj. 4.5.1)

23. $12x^4 - 14x^3 + 25x^2 - 29x + 10$ (obj. 4.5.3)   **24.** $3x^4 + 4x^3y - 15x^2y^2$ (obj. 4.4.2)   **25.** $3x^3 - x + 5$ (obj. 4.6.1)

26. $2x^2 - 7x + 4$ (obj. 4.6.2)   **27.** $2x^2 + 6x + 12$ (obj. 4.6.2)   **28.** 3.04×10^9 barrels per year (obj. 4.2.2)   **29.** 4.18×10^6 mi (obj. 4.2.2)

Cumulative Test for Chapters 0–4 1. $\dfrac{1}{30}$ **2.** -0.74 **3.** $-\dfrac{6}{7}$ **4.** $2x^2 - 13x$ **5.** 110.55 **6.** 66 **7.** $x = -\dfrac{9}{2}$

8. $x = 15$ **9.** $x < -3$ **10.** $d = \dfrac{3y - 8x}{11}$ **11.** $12{,}400$ employees   **12.** $\$199.20$ **13.** $20x^2 - 21x - 5$ **14.** $9x^2 - 30x + 25$

15. $6x^3 - 17x^2 - 26x - 8$ **16.** $-20x^5y^8$ **17.** $-\dfrac{2x^3}{3y^9}$ **18.** $16x^{12}y^8$ **19.** $\dfrac{9z^8}{w^2x^3y^4}$ **20.** 4.25×10^{10} **21.** 5.6×10^{-4}

22. 4.0×10^{-35} **23.** $5x^3 + 7x^2 - 6x + 50$ **24.** $-36x^3y^2 + 18x^2y^3 - 48xy^4$ **25.** $6x^3 - 19x^2 + 18x - 5$ **26.** $x + 5 + \dfrac{3}{x - 3}$

Chapter 5   **5.1 Exercises 1.** factors   **3.** No; $6a^3 + 3a^2 - 9a$ has a common factor of $3a$.   **5.** $3a(a + 1)$ **7.** $7ab(3 - 2b)$
9. $2\pi r(h + r)$ **11.** $5x(x^2 + 5x - 3)$ **13.** $4(3ab - 7bc + 5ac)$ **15.** $8x^2(2x^3 + 3x - 4)$ **17.** $7x(2xy - 5y - 9)$
19. $9x(6x - 5y + 2)$ **21.** $y(3xy - 2a + 5x - 2)$ **23.** $8xy(3x - 5y)$ **25.** $7x^2y^2(x + 3)$ **27.** $8x^2y(x^2 + 3y - 4)$
29. $(x + 2y)(7a - b)$ **31.** $(x - 4)(3x - 2)$ **33.** $(2a - 3c)(6b - 5d)$ **35.** $(b - a^2)(7c - 5d + 2f)$ **37.** $(xy - 3)(2a - 4 - z)$
39. $(a - 3b)(4a^3 + 1)$ **41.** $(a + 2)(1 - x)$ **43.** 2.24 ft/min   **44.** $856{,}000$ metric tons   **45.** $2{,}621{,}500$ metric tons
46. About 22 lb/person   **47.** About 30 lb/person

Quick Quiz 5.1 1. $x(3 - 4x + 2y)$ **2.** $5x(4x^2 - 5x - 1)$ **3.** $(a + 3b)(8a - 7b)$ **4.** See Student Solutions Manual

5.2 Exercises 1. We must remove a common factor of 5 from the last two terms. This will give us $3x(x - 2y) + 5(x - 2y)$. Then our final answer
is $(x - 2y)(3x + 5)$.   **3.** $(b - 3)(a + 4)$ **5.** $(x - 4)(x^2 + 3)$ **7.** $(a + 3b)(2x - y)$ **9.** $(3a + b)(x - 2)$ **11.** $(a + 2b)(5 + 6c)$
13. $(c - 2d)(6 + x)$ **15.** $(y - 2)(y - 3)$ **17.** $(9 - y)(6 + y)$ **19.** $(3x + y)(2a - 1)$ **21.** $(x + 4)(2x - 3)$ **23.** $(t - 1)(t^2 + 1)$
25. $(7x + 2y^2)(4x + 3w)$ **27.** We must rearrange the terms in a different order so that the expression in parentheses is the same in each case.
We use the order $6a^2 - 8ad + 9ab - 12bd$ to factor $2a(3a - 4d) + 3b(3a - 4d) = (3a - 4d)(2a + 3b)$.   **29.** 36.2% **30.** Instructor: $\$40{,}100$;
professor: $\$91{,}400$ **31.** About $622{,}000{,}000$ metric tons

Quick Quiz 5.2 1. $(7x + 12)(a - 2)$ **2.** $(y^2 + 3)(2x - 5)$ **3.** $(10y - 3)(x + 4b)$ **4.** See Student Solutions Manual

5.3 Exercises 1. product, sum   **3.** $(x + 1)(x + 1)$ **5.** $(x + 5)(x + 7)$ **7.** $(x - 3)(x - 1)$ **9.** $(x - 7)(x - 4)$
11. $(x + 8)(x - 3)$ **13.** $(x - 14)(x + 1)$ **15.** $(x + 7)(x - 5)$ **17.** $(x - 6)(x + 4)$ **19.** $(x + 12)(x + 3)$ **21.** $(x - 6)(x - 4)$
23. $(x + 3)(x + 10)$ **25.** $(x - 5)(x - 1)$ **27.** $(a + 8)(a - 2)$ **29.** $(x - 4)(x - 8)$ **31.** $(x + 7)(x - 3)$ **33.** $(x + 7)(x + 8)$
35. $(y + 9)(y - 5)$ **37.** $(x + 12)(x - 3)$ **39.** $(x + 3y)(x - 5y)$ **41.** $(x - 7y)(x - 9y)$ **43.** $4(x + 5)(x + 1)$ **45.** $6(x + 1)(x + 2)$
47. $5(x - 1)(x - 5)$ **49.** $3(x + 4)(x - 6)$ **51.** $7(x + 5)(x - 2)$ **53.** $3(x - 1)(x - 5)$ **55.** $(12 + x)(10 - x)$ **57.** $P = \dfrac{A}{1 + rt}$

58. $x \geq -\dfrac{5}{3}$

59. 120 mi   **60.** $\$80{,}000$ **61.** $25°C$ **62.** June

Quick Quiz 5.3 1. $(x + 7)(x + 10)$ **2.** $(x - 6)(x - 8)$ **3.** $2(x + 6)(x - 8)$ **4.** See Student Solutions Manual

5.4 Exercises 1. $(4x + 1)(x + 3)$ **3.** $(5x + 2)(x + 1)$ **5.** $(4x - 3)(x + 2)$ **7.** $(2x + 1)(x - 3)$ **9.** $(3x + 1)(3x + 2)$
11. $(3x - 5)(5x - 3)$ **13.** $(2x - 5)(x + 4)$ **15.** $(4x - 1)(2x + 3)$ **17.** $(3x + 2)(2x - 3)$ **19.** $(5x - 1)(2x + 1)$
21. $(x - 2)(7x + 9)$ **23.** $(9y - 4)(y - 1)$ **25.** $(5a + 2)(a - 3)$ **27.** $(7x - 2)(2x + 3)$ **29.** $(5x - 2)(3x + 2)$
31. $(6x + 5)(2x + 3)$ **33.** $(6x + 1)(2x - 3)$ **35.** $(3x^2 + 1)(x^2 - 5)$ **37.** $(2x + 5y)(x + 3y)$ **39.** $(5x - 4y)(x + 4y)$
41. $5(2x + 1)(x - 3)$ **43.** $3x(2x - 5)(x + 4)$ **45.** $(5x - 2)(x + 1)$ **47.** $2(3x - 2)(2x - 5)$ **49.** $x(6x - 1)(2x - 3)$
51. $2(2x - 1)(2x + 5)$ **53.** $x = \dfrac{1}{10}$ **54.** $x = \dfrac{5 + 3y}{7}$ **55. (a)** $19{,}400{,}000$ travelers **(b)** About 22.7% **56.** About 27.4%
57. 25% **58.** About 3.4%

Quick Quiz 5.4 1. $(2x + 3)(6x - 1)$ **2.** $(5x - 3)(2x - 3)$ **3.** $3x(2x - 5)(x + 2)$ **4.** See Student Solutions Manual

How Am I Doing? Sections 5.1–5.4 1. $3(2xy - 5z + 7)$ (obj. 5.1.1)   **2.** $5x(6x - 9y - 2)$ (obj. 5.1.1)   **3.** $(4x - 5)(7 - b)$ (obj. 5.1.1)
4. $(8y + 3z)(2x - 5y)$ (obj. 5.1.1)   **5.** $(6 + x)(3 - y)$ (obj. 5.2.1)   **6.** $(3x + 4w)(5 - 3b)$ (obj. 5.2.1)   **7.** $(x - 5)(x^2 - 3)$ (obj. 5.2.1)
8. $(a + 3b)(7 + 2b)$ (obj. 5.2.1)   **9.** $(x - 7)(x - 8)$ (obj. 5.3.1)   **10.** $(x + 16)(x - 4)$ (obj. 5.3.1)   **11.** $(a + 3b)(a + 7b)$ (obj. 5.3.1)
12. $7(x + 5)(x - 7)$ (obj. 5.3.2)   **13.** $(4x - 1)(3x + 5)$ (obj. 5.4.1)   **14.** $(x - 7)(3x - 2)$ (obj. 5.4.1)   **15.** $(2x + 3y)(3x + 4y)$ (obj. 5.4.2)
16. $2x(7x + 4)(x - 2)$ (obj. 5.4.3)

5.5 Exercises 1. $(3x + 1)(3x - 1)$ **3.** $(9x + 4)(9x - 4)$ **5.** $(x + 7)(x - 7)$ **7.** $(5x + 9)(5x - 9)$ **9.** $(x + 5)(x - 5)$
11. $(1 + 4x)(1 - 4x)$ **13.** $(4x + 7y)(4x - 7y)$ **15.** $(6x + 13y)(6x - 13y)$ **17.** $(10x + 9)(10x - 9)$ **19.** $(5a + 9b)(5a - 9b)$
21. $(3x + 1)^2$ **23.** $(y - 5)^2$ **25.** $(6x - 5)^2$ **27.** $(7x + 2)^2$ **29.** $(x + 7)^2$ **31.** $(5x - 4)^2$ **33.** $(9x + 2y)^2$ **35.** $(3x - 5y)^2$
37. $(4a + 9b)^2$ **39.** $(7x - 3y)^2$ **41.** $(3x + 7)^2$ **43.** $(12x + 1)(12x - 1)$ **45.** $(x^2 + 6)(x^2 - 6)$ **47.** $(2x^2 - 5)^2$
49. No two binomials can be multiplied to obtain $9x^2 + 1$.   **51.** 49; one answer   **53.** $4(2x + 3)(2x - 3)$ **55.** $3(7x + y)(7x - y)$
57. $4(2x - 1)^2$ **59.** $2(7x + 3)^2$ **61.** $(x + 9)(x + 7)$ **63.** $(2x - 1)(x + 3)$ **65.** $(4x + 11)(4x - 11)$ **67.** $(3x + 7)^2$

69. $9(2x - 1)^2$ **71.** $2(x - 9)(x - 7)$ **73.** $x^2 + 3x + 4 + \dfrac{-3}{x - 2}$ **74.** $2x^2 + x - 5$ **75.** 1.2 oz of greens; 1.05 oz of bulk vegetables;

0.75 oz of fruit   **76.** 1.44 oz of greens; 1.26 oz of bulk vegetables; 0.9 oz of fruit

Quick Quiz 5.5   **1.** $(7x + 9y)(7x - 9y)$ **2.** $(3x - 8)^2$ **3.** $2(9x + 10)(9x - 10)$ **4.** See Student Solutions Manual

5.6 Exercises   **1.** $a(5a - 3b + 8)$ **3.** $(4x + 5y)(4x - 5y)$ **5.** $(3x - 2y)^2$ **7.** $(x + 5)(x + 3)$ **9.** $(3x + 2)(5x - 1)$
11. $(a - 3c)(x + 3y)$ **13.** $(y + 7)^2$ **15.** $(2x - 3)^2$ **17.** $(2x - 3)(x - 4)$ **19.** $(x - 10y)(x + 7y)$ **21.** $(a + 3)(x - 5)$
23. $4x(2 + x)(2 - x)$ **25.** $5xy^3(x - 1)^2$ **27.** $3xy(z + 1)(z - 3)$ **29.** $3(x + 7)(x - 5)$ **31.** $3x(x + 4)(x - 3)$
33. $-1(2x^2 + 1)(x + 2)(x - 2)$ **35.** Prime   **37.** $5(x^2 + 2xy - 6y)$ **39.** $3x(2x + y)(5x - 2y)$ **41.** $2(3x - 5)(4x - 3)$
43. About 83%   **44.** Europe, with an increase of about 5%   **45.** 9739.26 million metric tons   **46.** 4908.75 million metric tons

Quick Quiz 5.6   **1.** $(2x - 3)(3x - 4)$ **2.** $3(4x + 1)(5x - 2)$ **3.** Prime   **4.** See Student Solutions Manual

5.7 Exercises   **1.** $x = -3, x = 7$ **3.** $x = -12, x = -2$ **5.** $x = \dfrac{3}{2}, x = 2$ **7.** $x = \dfrac{2}{3}, x = \dfrac{3}{2}$ **9.** $x = 0, x = -13$ **11.** $x = 3, x = -3$

13. $x = 0, x = 1$ **15.** $x = \dfrac{2}{3}, x = 2$ **17.** $x = -3, x = 2$ **19.** $x = -\dfrac{1}{2}$ **21.** $x = 0, x = -3$ **23.** $x = -8, x = -2$

25. $x = -\dfrac{3}{2}, x = 4$ **27.** You can always factor out x.   **29.** $L = 14$ m; $W = 10$ m   **31.** 66 groups   **33.** 10 students

35. 12 m above ground after 2 sec   **37.** 2415 telephone calls   **39.** 136 handshakes   **41.** $-10x^5y^4$ **42.** $12a^{10}b^{13}$ **43.** $-\dfrac{3a^4}{2b^2}$ **44.** $\dfrac{1}{3x^5y^4}$

Quick Quiz 5.7   **1.** $x = \dfrac{1}{5}, x = \dfrac{1}{3}$ **2.** $x = 3, x = -1$ **3.** $x = 3, x = -\dfrac{3}{4}$ **4.** See Student Solutions Manual

Putting Your Skills to Work   **1.** $520   **2.** About $31,200   **3.** About $6200   **4.** $480   **5.** About $28,800   **6.** $25,000
7. Option B   **8.** Answers may vary

Chapter 5 Review Problems   **1.** $4x^2(3x - 5y)$ **2.** $5x^3(2 - 7y)$ **3.** $8x^2y(3x - y - 2xy^2)$ **4.** $15a^2b^3(3a + 2 - b)$
5. $3a(a^2 + 2a - 3b + 4)$ **6.** $2(x - 2y + 3z + 6)$ **7.** $(a + 3b)(2a - 5)$ **8.** $3xy(5x^2 + 2y + 1)$ **9.** $(2x + 5)(a - 4)$
10. $(a - 4b)(a + 7)$ **11.** $(x^2 + 3)(y - 2)$ **12.** $3(2x - y)(5a + 7)$ **13.** $(5x - 1)(3x + 2)$ **14.** $(5w - 3)(6w + z)$
15. $(x + 9)(x - 3)$ **16.** $(x + 10)(x - 1)$ **17.** $(x + 6)(x + 8)$ **18.** $(x + 3y)(x + 5y)$ **19.** $(x^2 + 7)(x^2 + 6)$ **20.** $(x^2 - 7)(x^2 + 5)$
21. $6(x + 2)(x + 3)$ **22.** $3(x + 1)(x + 12)$ **23.** $2(x - 6)(x - 8)$ **24.** $4(x - 5)(x - 6)$ **25.** $(4x - 5)(x + 3)$ **26.** $(3x - 1)(4x + 5)$
27. $(5x + 4)(3x - 1)$ **28.** $(3x - 2)(2x - 3)$ **29.** $(2x - 3)(x + 1)$ **30.** $(3x - 4)(x + 2)$ **31.** $(10x - 1)(2x + 5)$
32. $(5x - 1)(4x + 5)$ **33.** $(3a - 2)(2a + 5)$ **34.** $(3a - 2)(2a - 5)$ **35.** $2(x - 1)(3x + 5)$ **36.** $2(x + 1)(3x - 5)$
37. $2(2x - 3)(x - 5)$ **38.** $4(x - 9)(x + 4)$ **39.** $4(2x - 3)(2x + 5)$ **40.** $3(4x - 1)(2x + 1)$ **41.** $(4x + 3y)(3x - 2y)$
42. $(3x - 5y)(2x + 5y)$ **43.** $(7x + y)(7x - y)$ **44.** $4(2x + 3y)(2x - 3y)$ **45.** $(6x - 5)^2$ **46.** $(4x + 5)^2$ **47.** $(5x + 6)(5x - 6)$
48. $(10x + 3)(10x - 3)$ **49.** $(y + 6x)(y - 6x)$ **50.** $(3y + 5x)(3y - 5x)$ **51.** $(6x + 1)^2$ **52.** $(5x - 2)^2$ **53.** $(4x - 3y)^2$
54. $(7x - 2y)^2$ **55.** $2(x + 4)(x - 4)$ **56.** $3(x + 3)(x - 3)$ **57.** $7(2x + 5)^2$ **58.** $8(3x - 4)^2$ **59.** $(2x + 3y)(2x - 3y)$
60. $(x + 3)^2$ **61.** $(x - 3)(x - 6)$ **62.** $(x + 15)(x - 2)$ **63.** $(x - 3)(5x + 2)$ **64.** $(3x - 4)(3x + 1)$ **65.** $12(2x - 5)$
66. $4xy(2xy - 1)$ **67.** $10x^2y^2(5x + 2)$ **68.** $13ab(2a^2 - b^2 + 4ab^3)$ **69.** $x(x - 8)^2$ **70.** $2(x + 10)^2$ **71.** $3(x - 3)^2$
72. $x(5x - 6)^2$ **73.** $(7x + 5)(x - 2)$ **74.** $(4x + 3)(x - 4)$ **75.** $xy(3x + 2y)(3x - 2y)$ **76.** $x^3a(3a + 4x)(a - 5x)$
77. $2(3a + 5b)(2a - b)$ **78.** $(11a + 3b)^2$ **79.** $(a - 1)(7 - b)$ **80.** $(3b - 7)(6 + c)$ **81.** $(5b + 8)(2 - 3x)$ **82.** $(b - 7)(5x + 4y)$
83. $x(2a - 1)(a - 7)$ **84.** $x(x + 4)(x - 4)(x + 1)(x - 1)$ **85.** $(x^2 + 9y^6)(x + 3y^3)(x - 3y^3)$ **86.** $(3x^2 - 5)(2x^2 + 3)$
87. $yz(14 - x)(2 - x)$ **88.** $x(3x + 2)(4x + 3)$ **89.** $(2w + 1)(8w - 5)$ **90.** $3(2w - 1)^2$ **91.** $2y(2y - 1)(y + 3)$
92. $(5y - 1)(2y + 7)$ **93.** $8y^8(y^2 - 2)$ **94.** $9(x^2 + 4)(x - 2)(x + 2)$ **95.** Prime   **96.** Prime   **97.** $4y(2y^2 - 5)(y^2 + 3)$
98. $3x(3y + 7)(y - 2)$ **99.** $(4x^2y - 7)^2$ **100.** $2xy(8x + 1)(8x - 1)$ **101.** $(2x + 5)(a - 2b)$ **102.** $(2x + 1)(x + 3)(x - 3)$
103. $x = -5, x = 4$ **104.** $x = -6, x = \dfrac{1}{2}$ **105.** $x = 0, x = \dfrac{5}{2}$ **106.** $x = 4, x = -3$ **107.** $x = -5, x = \dfrac{1}{2}$ **108.** $x = -8, x = -3$
109. $x = -5, x = -9$ **110.** $x = -\dfrac{3}{5}, x = 2$ **111.** $x = -3$ **112.** $x = -3, x = \dfrac{3}{4}$ **113.** $x = \dfrac{1}{5}, x = 2$ **114.** Base = 5 in., altitude = 10 in.
115. Width = 5 ft, length = 6 ft   **116.** 6 sec   **117.** 8 amperes, 12 amperes

How Am I Doing? Chapter 5 Test   **1.** $(x + 14)(x - 2)$ (obj. 5.3.1)   **2.** $(4x + 9)(4x - 9)$ (obj. 5.5.1)   **3.** $(5x + 1)(2x + 5)$ (obj. 5.4.2)
4. $(3a - 5)^2$ (obj. 5.5.2)   **5.** $x(7 - 9x + 14y)$ (obj. 5.1.1)   **6.** $(2x + 3b)(5y - 4)$ (obj. 5.4.2)   **7.** $2x(3x - 4)(x - 2)$ (obj. 5.6.1)
8. $c(5a - 1)(a - 2)$ (obj. 5.6.1)   **9.** $(9x + 10)(9x - 10)$ (obj. 5.5.1)   **10.** $(3x - 1)(3x - 4)$ (obj. 5.6.1)   **11.** $5(2x + 3)(2x - 3)$ (obj. 5.6.1)
12. Prime (obj. 5.6.2)   **13.** $x(3x + 5)(x + 2)$ (obj. 5.6.1)   **14.** $-5y(2x - 3y)^2$ (obj. 5.6.1)   **15.** $(9x + 1)(9x - 1)$ (obj. 5.5.1)
16. $(9y^2 + 1)(3y + 1)(3y - 1)$ (obj. 5.5.1)   **17.** $(x + 3)(2a - 5)$ (obj. 5.6.1)   **18.** $(a + 2b)(w + 2)(w - 2)$ (obj. 5.6.1)
19. $3(x - 6)(x + 5)$ (obj. 5.6.1)   **20.** $x(2x + 5)(x - 3)$ (obj. 5.6.1)   **21.** $x = -5, x = -9$ (obj. 5.7.1)   **22.** $x = -\dfrac{7}{3}, x = -2$ (obj. 5.7.1)
23. $x = -\dfrac{5}{2}, x = 2$ (obj. 5.7.1)   **24.** $x = 7, x = -4$ (obj. 5.7.1)   **25.** Width = 7 mi, length = 13 mi (obj. 5.7.2)

Cumulative Test for Chapters 0–5   **1.** 20%   **2.** 0.54   **3.** 5.8   **4.** $8x^4y^{10}$ **5.** 81   **6.** $27x^2 + 6x - 8$ **7.** $2x^3 - 12x^2 + 19x - 3$
8. $x \geq -\dfrac{9}{2}$ **9.** $x = 2$ **10.** $x = -15$ **11.** $q = \dfrac{p + 3}{5}$ **12.** $(3x - 1)(2x - 1)$ **13.** $(3x + 4)(2x - 1)$ **14.** $(3x + 2)(3x - 1)$
15. $(11x + 8y)(11x - 8y)$ **16.** $-4(5x + 6)(4x - 5)$ **17.** Prime   **18.** $x(4x + 5)^2$ **19.** $(4x^2 + b^2)(2x + b)(2x - b)$
20. $(2x + 3)(a - 2b)$ **21.** $3(5x - 3)(x + 1)$ **22.** $x = -9, x = 4$ **23.** $x = \dfrac{5}{3}, x = 2$ **24.** Length = 20 ft, width = 15 ft

Chapter 6 6.1 Exercises 1. 3   3. $\dfrac{6}{x}$ 5. $\dfrac{3x+1}{1-3x}$ 7. $\dfrac{a(a-2b)}{2b}$ 9. $\dfrac{x+2}{x}$ 11. $\dfrac{x-5}{3x-1}$ 13. $\dfrac{x-3}{x(x-7)}$ 15. $\dfrac{3x+1}{x+5}$

17. $\dfrac{3x-5}{4x-1}$ 19. $\dfrac{x-5}{x+1}$ 21. $-\dfrac{3}{5x}$ 23. $\dfrac{-2x-3}{5+x}$ 25. $\dfrac{3x+4}{3x-1}$ 27. $\dfrac{2x-3}{5-x}$ 29. $\dfrac{a-b}{2a-b}$ 31. $9x^2-42x+49$

32. $49x^2-36y^2$ 33. $2x^3-9x^2-2x+24$ 34. $2x^4-5x^3-10x^2-14x-15$ 35. $1\dfrac{5}{8}$ acres   36. 6 hr, 25 min   37. 39,500,000 people
38. 80,000,000 people

Quick Quiz 6.1 1. $\dfrac{x}{x-5}$ **2.** $-\dfrac{2}{b}$ **3.** $\dfrac{2x-1}{4x+5}$ **4.** See Student Solutions Manual

6.2 Exercises 1. factor the numerator and denominator completely and divide out common factors   3. $\dfrac{3(x+5)}{x+2}$ 5. $\dfrac{3x^2}{4(x-3)}$ 7. $\dfrac{x-2}{x+3}$

9. $\dfrac{(x+6)(x+2)}{x+5}$ 11. $x+3$ 13. $\dfrac{3(x+2y)}{4(x+3y)}$ 15. $\dfrac{(x+5)(x-2)}{3x-1}$ 17. $\dfrac{-5(x+3)}{2(x-2)}$ or $\dfrac{5(x+3)}{-2(x-2)}$ or $-\dfrac{5(x+3)}{2(x-2)}$ 19. 1

21. $\dfrac{x-1}{x+3}$ 23. $x=-8$ 24. $7x^3-22x^2+2x+3$ 25. $79.5(8981)x=\$713,989.5x$ 26. Harold's is 6 ft by 6 ft. George's is 9 ft by 4 ft.

Quick Quiz 6.2 1. $\dfrac{2(x+4)}{x-4}$ **2.** $\dfrac{x-25}{x+5}$ **3.** $\dfrac{2(x-3)}{3x}$ **4.** See Student Solutions Manual

6.3 Exercises 1. The LCD would be a product that contains each factor. However, any repeated factor in any one denominator must be
repeated the greatest number of times it occurs in any one denominator. So the LCD would be $(x+5)(x+3)^2$.   3. $\dfrac{3x+1}{x+5}$ 5. $\dfrac{2x-5}{x+3}$

7. $\dfrac{2x-7}{5x+7}$ 9. $3a^2b^3$ 11. $90x^3y^5$ 13. $18(x-3)$ 15. x^2-9 17. $(x+5)(3x-1)^2$ 19. $\dfrac{7+3a}{ab}$ 21. $\dfrac{3x-13}{(x+7)(x-7)}$

23. $\dfrac{5y^2-3y}{(y+1)(y-1)}$ 25. $\dfrac{43a+12}{5a(3a+2)}$ 27. $\dfrac{4z+x}{6xyz}$ 29. $\dfrac{9x+14}{2(x-3)}$ 31. $\dfrac{x+10}{(x+5)(x-5)}$ 33. $\dfrac{3a+17b}{10}$ 35. $\dfrac{-4x+34}{(2x-3)(x+2)}$

37. $\dfrac{-7x}{(x+3)(x-1)(x-4)}$ 39. $\dfrac{4x+23}{(x+4)(x+5)(x+6)}$ 41. $\dfrac{4x-13}{(x-4)(x-3)}$ 43. $\dfrac{11x}{y-2x}$ 45. $\dfrac{-17y^2-10y}{(4y-1)(2y+1)(y-5)}$

47. $\dfrac{6}{y+3}$ 49. $x=-7$ 50. $y=\dfrac{5ax+6bc}{2a}$ 51. $x>\dfrac{5}{7}$ 52. $81x^{12}y^{16}$ 53. At least 17 days   54. 284,440 people

Quick Quiz 6.3 1. $\dfrac{2x+7}{(x+2)(x-4)}$ **2.** $\dfrac{bx+by+xy}{bxy}$ **3.** $\dfrac{5x-1}{(x+3)(x-3)(x+4)}$ **4.** See Student Solutions Manual

How Am I Doing? Sections 6.1–6.3 1. $\dfrac{8}{x}$ (obj. 6.1.1)   2. $\dfrac{2x+3}{x-7}$ (obj. 6.1.1)   3. $\dfrac{y+1}{2x^2(1-y)}$ (obj. 6.1.1)

4. $\dfrac{x-4}{x+2}$ (obj. 6.1.1)   5. $\dfrac{3}{a}$ (obj. 6.2.1)   6. $\dfrac{x+5}{-3(x-2)}$ or $-\dfrac{x+5}{3(x-2)}$ (obj. 6.2.1)   7. $\dfrac{2x-1}{2}$ (obj. 6.2.2)   8. $\dfrac{4a-3}{2a-5}$ (obj. 6.2.2)

9. $\dfrac{xy-3ax-3ay}{axy}$ (obj. 6.3.3)   10. $\dfrac{7}{2(x+2)}$ (obj. 6.3.3)   11. $\dfrac{x^2+2x-9}{(x+3)(x+7)}$ (obj. 6.3.3)   12. $\dfrac{-5x^2+19x+16}{(x-3)(x+1)(x+4)}$ (obj. 6.3.3)

6.4 Exercises 1. $\dfrac{3x}{2+5x}$ 3. $a+b$ 5. $\dfrac{x^2-2x}{4+5x}$ 7. $\dfrac{2}{3x+10}$ 9. $\dfrac{1}{xy}$ 11. $\dfrac{2x-1}{x}$ 13. $\dfrac{x-6}{x+5}$ 15. $\dfrac{3a^2+9}{a^2+2}$ 17. $\dfrac{3x+9}{2x-5}$

19. $\dfrac{y+1}{-y+1}$ 21. No expression in any denominator can be zero because division by zero is undefined. So $-3, 5,$ and 0 are not allowed.

23. $w=\dfrac{P-2l}{2}$ 24. $x>-1$ 25. 6   26. Approximately 57%

Quick Quiz 6.4 1. $\dfrac{a(3a-4b)}{3(5a-16b)}$ **2.** ab **3.** $\dfrac{2}{x-1}$ **4.** See Student Solutions Manual

6.5 Exercises 1. $x=-12$ 3. $x=\dfrac{7}{2}$ or $3\dfrac{1}{2}$ or 3.5   5. $x=15$ 7. $x=10$ 9. $x=-5$ 11. $x=-\dfrac{14}{3}$ or $-4\dfrac{2}{3}$ 13. $x=-2$

15. $x=8$ 17. $x=-1$ 19. No solution   21. $x=-3$ 23. $y=-5$ 25. No solution   27. $x=4$ 29. No solution

31. $x=2, x=4$ 33. $(4x+1)(2x-1)$ 34. $x=\dfrac{5}{2}$ or $2\dfrac{1}{2}$ or 2.5   35. Width = 10 in.; length = 12 in.   36. $946.45

Quick Quiz 6.5 1. $x=\dfrac{5}{24}$ **2.** $x=2$ **3.** No solution   **4.** See Student Solutions Manual

6.6 Exercises 1. $x=18$ 3. $x=\dfrac{204}{5}$ or $40\dfrac{4}{5}$ or 40.8   5. $x=6.5$ 7. $x=\dfrac{91}{4}$ or $22\dfrac{3}{4}$ or 22.75   9. 110 mi

11. (a) 650 New Zealand dollars   (b) 75 New Zealand dollars less   13. 56 mph   15. 29 mi   17. $\dfrac{377}{20}$ or $18\dfrac{17}{20}$ in.   19. 200 m

21. $k = \dfrac{56}{5}$ or $11\dfrac{1}{5}$ ft   **23.** $\dfrac{768}{20}$ or $38\dfrac{2}{5}$ m   **25.** 48 in.   **27.** 35 in.   **29.** 61.5 mph   **31.** Commuter = 250 km/hr   **33. (a)** $0.11
helicopter = 210 km/hr

(b) $0.09   **(c)** $3.73   **35.** $3\dfrac{3}{7}$ hr or 3 hr, 26 min   **37.** 8.92465×10^{-4} **38.** 582,000,000   **39.** $\dfrac{w^8}{x^3 y^2 z^4}$ **40.** $\dfrac{27}{8}$ or $3\dfrac{3}{8}$

Quick Quiz 6.6 **1.** $x = 55$ **2.** 172 flights   **3.** 24 ft   **4.** See Student Solutions Manual

Putting Your Skills to Work **1.** $2092   **2.** $174.33   **3.** Answers may vary   **4.** Answers may vary

Chapter 6 Review Problems **1.** $\dfrac{x}{x-y}$ **2.** $-\dfrac{4}{5}$ **3.** $\dfrac{x+3}{x-4}$ **4.** $\dfrac{x+2}{x+4}$ **5.** $\dfrac{x+3}{x-7}$ **6.** $\dfrac{2(x+4)}{3}$ **7.** $\dfrac{x}{x+3}$ **8.** $\dfrac{2x-3}{5-x}$

9. $\dfrac{2(x-4y)}{2x-y}$ **10.** $\dfrac{2-y}{3y-1}$ **11.** $\dfrac{x-2}{(5x+6)(x-1)}$ **12.** $4x+2y$ **13.** $\dfrac{x+6}{6(x-1)}$ **14.** $\dfrac{2y-1}{5y}$ **15.** $\dfrac{(y+2)(2y+1)(y-2)}{(2y-1)(y+1)^2}$

16. $\dfrac{4y(3y-1)}{(3y+1)(2y+5)}$ **17.** $\dfrac{3y(4x+3)}{2(x-5)}$ **18.** $\dfrac{x+2}{2}$ **19.** $\dfrac{2(x+3y)}{x-4y}$ **20.** $\dfrac{3}{16}$ **21.** $\dfrac{4(5y+1)}{3y(y+2)}$ **22.** $\dfrac{3x^2+6x+1}{x(x+1)}$

23. $\dfrac{10x-22}{(x+2)(x-4)}$ **24.** $\dfrac{(x-1)(x-2)}{(x+3)(x-3)}$ **25.** $\dfrac{2xy+4x+5y+6}{2y(y+2)}$ **26.** $\dfrac{(2a+b)(a+4b)}{ab(a+b)}$ **27.** $\dfrac{3x-2}{3x}$ **28.** $\dfrac{2x^2+7x-2}{2x(x+2)}$

29. $\dfrac{3}{2(x-9)}$ **30.** $\dfrac{1-2x-x^2}{(x+5)(x+2)}$ **31.** $-\dfrac{4}{9}$ **32.** $\dfrac{22}{5x^2}$ **33.** $w-2$ **34.** 1   **35.** $-\dfrac{y^2}{2}$ **36.** $\dfrac{x+2y}{y(x+y+2)}$

37. $\dfrac{-1}{a(a+b)}$ or $-\dfrac{1}{a(a+b)}$ **38.** $\dfrac{-3a-b}{b}$ or $-\dfrac{3a+b}{b}$ **39.** $\dfrac{5y(x+5y)^2}{x(x-6y)}$ **40.** $\dfrac{-3y}{2(x+2y)}$ or $-\dfrac{3y}{2(x+2y)}$ **41.** $a=2$ **42.** $a=15$

43. $x=\dfrac{2}{7}$ **44.** $x=-8$ **45.** $y=-4$ **46.** $x=-2$ **47.** $x=\dfrac{1}{2}$ **48.** $x=\dfrac{1}{5}$ **49.** No solution   **50.** $x=-\dfrac{12}{5}$ or $-2\dfrac{2}{5}$ or -2.4

51. $y=6$ **52.** $y=2$ **53.** $y=-2$ **54.** No solution   **55.** $y=\dfrac{5}{4}$ or $1\dfrac{1}{4}$ or 1.25   **56.** $y=0$ **57.** $x=\dfrac{14}{5}$ or $2\dfrac{4}{5}$ or 2.8

58. $x=\dfrac{5}{4}$ or $1\dfrac{1}{4}$ or 1.25   **59.** $x=\dfrac{132}{5}$ or $26\dfrac{2}{5}$ or 26.4   **60.** $x=6$ **61.** $x=16$ **62.** $x=\dfrac{91}{10}$ or $9\dfrac{1}{10}$ or 9.1   **63.** 8.3 gal

64. 167 cookies   **65.** 38.9 gal   **66.** 240 mi   **67.** Train = 60 mph   **68.** $1\dfrac{7}{8}$ hr or about 1 hr, 53 min   **69.** 182 ft   **70.** 1200 ft
car = 40 mph

71. $2\dfrac{2}{5}$ hr or 2 hr, 24 min   **72.** 12 hr   **73.** $-\dfrac{a+4}{3a^2}$ **74.** $\dfrac{4a^2}{2a+3}$ **75.** $\dfrac{x-2y}{x+2y}$ **76.** $\dfrac{x-8y}{x+7y}$ **77.** $\dfrac{x+6}{x}$ **78.** $\dfrac{x+6}{x}$

79. $\dfrac{b^2-a^2}{ab(x+y)}$ **80.** $\dfrac{1}{3}$ **81.** $-\dfrac{6}{y^2}$ **82.** $\dfrac{y(x+3y)}{2(x+2y)}$ **83.** $x=12$ **84.** $x=-32$ **85.** $b=-2$ **86.** 10.7 kg   **87.** 420 mi

How Am I Doing? Chapter 6 Test **1.** $\dfrac{2}{3a}$ (obj. 6.1.1)   **2.** $\dfrac{2x^2(2-y)}{(y+2)}$ (obj. 6.1.1)   **3.** $\dfrac{5}{12}$ (obj. 6.2.1)   **4.** $\dfrac{1}{3y(x-y)}$ (obj. 6.2.1)

5. $\dfrac{2a+1}{a+2}$ (obj. 6.2.2)   **6.** $\dfrac{3a+4}{(a+1)(a-2)}$ (obj. 6.3.3)   **7.** $\dfrac{x-a}{ax}$ (obj. 6.3.3)   **8.** $-\dfrac{x+2}{x+3}$ (obj. 6.3.3)   **9.** $\dfrac{x}{4}$ (obj. 6.4.2)

10. $\dfrac{6x}{5}$ (obj. 6.4.1)   **11.** $\dfrac{2x-3y}{4x+y}$ (obj. 6.1.1)   **12.** $\dfrac{x}{(x+2)(x+4)}$ (obj. 6.3.3)   **13.** $x=-\dfrac{1}{5}$ (obj. 6.5.1)   **14.** $x=4$ (obj. 6.5.1)

15. No solution (obj. 6.5.2)   **16.** $x=\dfrac{47}{6}$ (obj. 6.5.1)   **17.** $x=\dfrac{45}{13}$ (obj. 6.6.1)   **18.** $x=37.2$ (obj. 6.6.1)   **19.** 151 flights (obj. 6.6.1)

20. $368 (obj. 6.6.1)   **21.** 102 ft (obj. 6.6.2)

Cumulative Test for Chapters 0–6 **1.** 0.006 cm   **2.** 16.8%   **3.** $112.50   **4.** $x=6$ **5.** $h=\dfrac{A}{\pi r^2}$

6. $x > 1.25$ **7.** $x \le 26$ **8.** $(a+b)(3x-2y)$ **9.** $2a(4a+b)(a-5b)$ **10.** $-\dfrac{4a^5}{b^6}$ **11.** $\dfrac{2x+5}{x+7}$

12. $\dfrac{(x-2)(3x+1)}{3x(x+5)}$ **13.** $\dfrac{1}{2}$ **14.** $\dfrac{11x-3}{2(x+2)(x-3)}$ **15.** $\dfrac{6c-28}{(c+2)(c-2)(c-3)}$ **16.** $x=-2$ **17.** $x=-\dfrac{9}{2}$ or $-4\dfrac{1}{2}$ or -4.5

18. $\dfrac{x+8}{6x(x-3)}$ **19.** $\dfrac{3ab^2+2a^2b}{5b^2-2a^2}$ **20.** $x=\dfrac{7}{10}$ or 0.7   **21.** 208 mi   **22.** 484 phone calls

Chapter 7 7.1 Exercises **1.** 0   **3.** The order in which you write the numbers matters. The graph of $(5, 1)$ is not the same as
the graph of $(1, 5)$.   **5.**   **7.** $R: (-3, -5)$ **9.** $(-4, -1)$

$S: \left(-4\dfrac{1}{2}, 0\right)$ $(-3, -2)$

$X: (3, -5)$ $(-2, -3)$

$Y: \left(2\dfrac{1}{2}, 6\right)$ $(-1, -5)$

$(0, -3)$

$(2, -1)$

11. (a) $(0, 7)$ **(b)** $(2, 15)$ **13. (a)** $(-1, 11)$ **(b)** $(3, -13)$ **15. (a)** $(-3, -5)$ **(b)** $(5, 1)$ **17. (a)** $(-2, 0)$ **(b)** $(-4, 3)$

19. (a) $(-1, 18)$ **(b)** $\left(\dfrac{8}{3}, 7\right)$ **21.** B5   **23.** E1   **25.** D3

27. (a)

(b) The number of CDs shipped decreased overall between 2001 and 2006, with a slight increase from 2003 to 2004.

29. (a)

(b) An estimated \$4.5 billion was spent buying books online in 2007.

31. 8   **32.** $\dfrac{x - y}{x - 3y}$

Quick Quiz 7.1   1.

2. (a) $(-2, 3)$ **(b)** $(3, -22)$ **(c)** $(0, -7)$

3. (a) $(3, 8)$ **(b)** $(-9, -8)$ **(c)** $(4.5, 10)$

4. See Student Solutions Manual

7.2 Exercises   1. No, replacing x by -2 and y by 5 in the equation does not result in a true statement.   **3.** x-axis

5. $y = -2x + 1$
$(0, 1)$
$(-2, 5)$
$(1, -1)$

7. $y = x - 4$
$(0, -4)$
$(2, -2)$
$(4, 0)$

9. $y = -2x + 3$
$(0, 3)$
$(2, -1)$
$(4, -5)$

11. $y = 3x + 2$
$(-1, -1)$
$(0, 2)$
$(1, 5)$

13. $3x - 2y = 0$

15. $y = -\frac{3}{4}x + 3$

17. $4x + 3y = 12$

19. $y = 6 - 2x$

21. $x + 3 = 6y$

23. $y - 2 = 3y$

25. $2x + 9 = 5x$

27. $2x + 5y - 2 = -12$

29.

31.

33. 28   **34.** $x \geq -\dfrac{14}{3}$

Quick Quiz 7.2 **1.** **2.** **3.** **4.** See Student Solutions Manual

$y = -2x + 4$ $y = \frac{2}{5}x - 3$ $y = \frac{1}{4}x + 2$

7.3 Exercises **1.** No, division by zero is impossible, so the slope is undefined.   **3.** 3   **5.** −5   **7.** $\frac{3}{5}$ **9.** $-\frac{1}{4}$ **11.** $-\frac{4}{3}$ **13.** $-\frac{16}{5}$

15. $m = 8; (0, 9)$ **17.** $m = -3; (0, 4)$ **19.** $m = -\frac{8}{7}; \left(0, \frac{3}{4}\right)$ **21.** $m = -6; (0, 0)$ **23.** $m = -6; \left(0, \frac{4}{5}\right)$ **25.** $m = -\frac{3}{4}; \left(0, \frac{3}{2}\right)$

27. $m = \frac{7}{3}; \left(0, -\frac{4}{3}\right)$ **29.** $y = \frac{3}{5}x + 3$ **31.** $y = 4x - 5$ **33.** $y = -\frac{5}{4}x - \frac{3}{4}$

35.   **37.**   **39.**   **41.**   **43.**

$y = \frac{3}{4}x - 4$ $y = -\frac{5}{3}x + 2$ $y = \frac{2}{3}x + 2$ $y + 2x = 3$ $y = 2x$

45. (a) $\frac{5}{6}$ (b) $-\frac{6}{5}$ **47.** (a) 6 (b) $-\frac{1}{6}$ **49.** (a) $\frac{2}{3}$ (b) $-\frac{3}{2}$ **51.** Yes; $2x - 3y = 18$

53. (a) $y = 35x + 625$ (b) $m = 35; (0, 625)$ (c) The amount of increase in the number of home health aides in the U.S. per year.

55. $x > 4$ **56.** $x < \frac{12}{5}$

57. $x \le 24$ **58.** $x \le -22$

Quick Quiz 7.3 **1.** $\frac{1}{2}$ **2.** $m = 3; \left(0, -\frac{3}{2}\right)$ **3.** $y = -\frac{5}{7}x - 5$ **4.** See Student Solutions Manual

How Am I Doing? Sections 7.1–7.3 **1.** (obj. 7.1.1)   **2.** $A (3, 9)$
$B (7, -7)$
$C (-7, -8)$ (obj. 7.1.2)
$D (-4, 3)$

$E (-6, 3)$ $H (6.5, 3.5)$

$G (-2, -8)$ $F (5, -4)$

3. $(4, -25); (0, 3); (-2, 17)$ (obj. 7.1.3)   **4.** $(0, 6); (-5, 0); (-10, -6)$ (obj. 7.1.3)

5. (obj. 7.2.1)   **6.** (obj. 7.2.1)   **7.** (obj. 7.2.2)   **8.** (obj. 7.2.2)

$y = -4x - 3$ $y = \frac{3}{4}x - 1$ $3x - 5y = 0$ $y = -5x - 1$

9. -2 (obj. 7.3.1)   **10.** Slope = 2   **11.** $-\frac{4}{3}$ (obj. 7.3.5)   **12.** $\frac{5}{3}$ (obj. 7.3.5)

y-intercept $= \left(0, -\frac{4}{3}\right)$

(obj. 7.3.2)

7.4 Exercises **1.** $y = 4x + 12$ **3.** $y = -2x + 11$ **5.** $y = -3x + \frac{7}{2}$ **7.** $y = \frac{1}{4}x + 4$ **9.** $y = -2x - 6$ **11.** $y = -4x + 2$

13. $y = 5x - 10$ **15.** $y = \frac{1}{3}x + \frac{1}{2}$ **17.** $y = -3x$ **19.** $y = -3x + 3$ **21.** $y = -\frac{2}{3}x + 1$ **23.** $y = \frac{2}{3}x - 4$ **25.** $y = -\frac{2}{3}x$

27. $y = -2$ **29.** $y = -2$ **31.** $x = 4$ **33.** $y = \frac{1}{3}x + 5$ **35.** $y = -\frac{1}{2}x + 4$ **37.** $y = 2.4x + 227$ **39.** $\frac{1}{3}$ **40.** $\frac{(2x + 5)(x - 3)}{(2x - 5)^2}$

41. $61.20   **42.** 290 minutes

Quick Quiz 7.4 **1.** $y = \frac{2}{3}x - 7$ **2.** $y = 6x + 19$ **3.** $x = 4$; the slope is undefined   **4.** See Student Solutions Manual

7.5 Exercises **1.** No, all points in one region will be solutions to the inequality while all points in the other region will not be solutions. Thus testing any point will give the same result, as long as the point is not on the boundary line.

3.

5.

7.

9.

11.

13.

15.

17.

19.

21.

23. $x = \dfrac{4}{3}, x = -2$ **24.** $y = \dfrac{2}{5}x - 4$

Quick Quiz 7.5   **1.** Use a dashed line. If the inequality has a $<$ or a $>$ symbol, the points on the line itself are not included. This is indicated by a dashed line.   **2.**   **3.**   **4.** See Student Solutions Manual

7.6 Exercises   **1.** Using a table of values, an algebraic equation, or a graph   **3.** Possible values of the independent variable
5. If a vertical line can intersect the graph more than once, the relation is not a function. If no such line exists, then the relation is a function.

7. **(a)** Domain $= \left\{-3, \dfrac{2}{5}, 3\right\}$ **(b)** Not a function   **9.** **(a)** Domain $= \{0, 3, 6\}$ **(b)** Function
    Range $= \left\{-1, \dfrac{2}{5}, 4\right\}$ Range $= \{0.5, 1.5, 2.5\}$

11. **(a)** Domain $= \{1, 9, 12, 14\}$ **(b)** Function   **13.** **(a)** Domain $= \{3, 5, 7\}$ **(b)** Not a function
    Range $= \{1, 3, 12\}$ Range $= \{75, 85, 95, 100\}$

15.

$y = x^2 + 3$

17.

$y = 2x^2$

19.

$x = -2y^2$

21.

$x = y^2 - 4$

23.

$y = \dfrac{2}{x}$

25.

$y = \dfrac{4}{x^2}$

27.

$x = (y + 1)^2$

29.

$y = \dfrac{4}{x - 2}$

31. Function   **33.** Not a function   **35.** Function
37. Not a function   **39.** **(a)** 26 **(b)** 2 **(c)** -4
41. **(a)** 3 **(b)** 24 **(c)** 9
43. $f(0) = 31.6, f(4) = 32.24, f(10) = 34.4$
The curve slopes more steeply for larger values of x. The growth rate is increasing as x gets larger.

45. $-4x^2 - 3x + 8$ **46.** $-4x^2 + 2x + 6$ **47.** $4x^3 + 6x^2 - x + 6$ **48.** $2x + 4 + \dfrac{-1}{3x - 1}$

Quick Quiz 7.6   **1.** No. Two different ordered pairs have the same first coordinate.   **2.** **(a)** 41 **(b)** 34   **3.** **(a)** -7 **(b)** $-\dfrac{7}{8}$
4. See Student Solutions Manual

Putting Your Skills to Work   **1.** Approximately \$620,000   **2.** \$84,000   **3.** Approximately \$536,000   **4.** Answers may vary
5. Answers may vary   **6.** Answers may vary

Chapter 7 Review Problems   **1.**   **2.** [graph]   **3.** $(0, 7); (-1, 10)$ **4.** **(a)** $(1, 2)$ **(b)** $(-4, 4)$
5. **(a)** $(6, -1)$ **(b)** $(6, 3)$

6.

$3y = 2x + 6$

7.

$5y + x = -15$

8.

$2y + 4x = -8 + 2y$

9. $m = -\dfrac{5}{6}$ **10.** $m = \dfrac{9}{11}; \left(0, \dfrac{5}{11}\right)$ **11.** $y = -\dfrac{1}{2}x + 3$

12. $m = -\dfrac{5}{3}$ **13.**

$y = -\dfrac{1}{2}x + 3$

14.

$2x - 3y = -12$

15.

$6x - 2y = 6 + 6x$

16. $y = x + 3$ **17.** $y = -6x + 14$

18. $y = -\dfrac{1}{3}x + \dfrac{11}{3}$ **19.** $y = 7$ **20.** $-\dfrac{2}{3}$ **21.** -3

22. $y = \dfrac{2}{3}x - 3$ **23.** $y = -3x + 1$ **24.** $x = 5$

25.

$y < \dfrac{1}{3}x + 2$

26.

$3y + 2x \geq 12$

27.

$x \leq 2$

28. Domain: $\{-6, -5, 5\}$
Range: $\{-6, 5\}$
not a function

29. Domain: $\{-2, 2, 5, 6\}$
Range: $\{-3, 4\}$
function

30. Function   **31.** Not a function

32. Function

33.

$y = x^2 - 5$

34.

$x = y^2 + 3$

35.

$y = (x - 3)^2$

36. (a) 7 **(b)** 31   **37. (a)** -1 **(b)** -5 **38. (a)** -11 **(b)** 1

39. (a) 1 **(b)** $\dfrac{1}{5}$ **40. (a)** 0 **(b)** 4   **41.** $210   **42.** $174

43. $y = 0.09x + 30; (0, 30)$ tells us that if Russ and Norma use no electricity, the minimum cost will be $30.   **44.** $m = 0.09$; the electric bill increases $0.09 for each kilowatt-hour of use.   **45.** 1300 kilowatt-hours

46. 2400 kilowatt-hours   **47.** 77.1 million barrels   **48.** 94.35 million barrels   **49.** $m = 1.15$. This means that for each additional year, the world uses 1.15 million more barrels of petroleum.   **50.** y-intercept $= (0, 65.6)$. This means that when $x = 0$, which is the year 1990, a total of 65.6 million barrels of petroleum were consumed worldwide.   **51.** 2017   **52.** 2020

53.

$5x + 3y = -15$

54.

$y = \dfrac{3}{4}x - 3$

55. $-\dfrac{2}{5}$ **56.** $m = -\dfrac{7}{6}$; y-intercept $= \left(0, \dfrac{5}{3}\right)$ **57.** $y = \dfrac{2}{3}x - 7$

58.

$y < -2x + 1$

59. 17,020,000 people in 1994
15,406,000 people in 2000
13,254,000 people in 2008

60.

$y = -269x + 17,020$

Number of years since 1994

61. The slope is -269. The slope tells us that the number of people employed in manufacturing decreases each year by 269 thousand. In other words, employment in manufacturing goes down 269,000 people each year.   **62.** The y-intercept is $(0, 17,020)$. This tells us that in the year 1994, the number of manufacturing jobs was 17,020 thousand, which is 17,020,000.   **63.** 2014   **64.** 2019

How Am I Doing? Chapter 7 Test

1. (obj. 7.1.1)

2. (obj. 7.2.1)

$6x - 3 = 5x - 2y$

3. 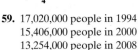 (obj. 7.2.1)

$8x + 2y = -4$

4. (obj. 7.2.1)

$y = \dfrac{2}{3}x - 4$

5. $m = -\dfrac{3}{2}; \left(0, \dfrac{5}{2}\right)$ (obj. 7.3.2)   **6.** $m = 1$ (obj. 7.3.1)   **7.** $y = \dfrac{1}{2}x - 4$ (obj. 7.3.3)   **8.** $m = 0$ (obj. 7.3.1)   **9.** $y = -\dfrac{3}{2}x + \dfrac{7}{2}$ (obj. 7.4.2)

10. $x = 2$; undefined (obj. 7.4.2)   **11.** (obj. 7.5.1)

$4y \leq 3x$

12. (obj. 7.5.1)

$-3x - 2y > 10$

13. No, two different ordered pairs have the same first coordinate. (obj. 7.6.1)   **14.** Yes, any vertical line passes through no more than one point on the graph. (obj. 7.6.3)

15.   (obj. 7.6.2)   **16. (a)** -3 **(b)** -3 (obj. 7.6.4)   **17. (a)** -3 **(b)** $-\dfrac{1}{5}$ (obj. 7.6.4)

$y = 2x^2 - 3$

Cumulative Test for Chapters 0–7
1. $-27x^9y^{12}z^3$ **2.** $-\dfrac{2y^{10}}{3x^8}$ **3.** $x = 7$ **4.** $R = 40$ **5.** $x = -4$

6. $w = \dfrac{A - lh}{l}$ or $w = \dfrac{A}{l} - h$ **7.** $2(5a + 7b)(5a - 7b)$ **8.** $(3x + 1)(3x - 4)$ **9.** $2(x - 1)$ **10.** $y = 2x - 7$

11. $x = 7$ **12.** $y = -\dfrac{2}{3}x - 3$ **13.** $m = 0$ **14.** $m = \dfrac{4}{3}$

15.   **16.**   **17.**   **18.** no   **19.** Yes   **20. (a)** 1 **(b)** 10

$(0, -4)$ $(3, -2)$

$y = \dfrac{2}{3}x - 4$ $3x + 8 = 5x$ $2x + 5y \le -10$ $x = y^2 + 3$

Chapter 8   8.1 Exercises
1. The lines are parallel. One line intersects the y-axis at -5. The other line intersects the y-axis at 6. There is no solution. The system is inconsistent.   **3.** intersect, one   **5.** The lines intersect at the y-intercept. The y-intercept is the solution of the system.

7.   **9.**   **11.**   **13.**   **15.**

$x + y = 5$ $(4, 1)$ $x - y = 3$ $y = 2x + 5$ $y = -3x$ $(-1, 3)$ $3x - 2y = 12$ $(2, -3)$ $4x + y = 5$ $y = 2x + 3$ $(-1, 1)$ $y = -4x - 3$ $2x + 3y = 14$ $(-2, 6)$ $3x - 2y = -18$

17.   **19.** inconsistent system of equations   **21.** $\dfrac{1}{2}y - x = 3$ $y - 2x = 6$ dependent equations   **23.**

$y = \dfrac{3}{4}x + 7$ $(-4, 4)$ $y = -\dfrac{1}{2}x + 2$ $-9x + 6y = -9$ $3x - 2y = -4$ $(-6, -5)$ $y = \dfrac{2}{3}x - 1$ $y = \dfrac{1}{2}x - 2$

25. (a) Newtown: $y = 7x$
Tryus: $y = 36 + 3x$ **(b)**

Tryus $y = 36 + 3x$ (9, 63)   Newtown $y = 7x$

x	1	4	8	12
$y = 7x$	7	28	56	84

x	1	4	8	12
$y = 36 + 3x$	39	48	60	72

(c) 9 min **(d)** Tryus   **27.** $(47.52, -61.38)$ **29.** 150,000 people   **30.** 80%   **31.** 912,500 visitors; 730,000 visitors
32. At least 1042 visitors; about 41.7%

Quick Quiz 8.1   **1.**
    2. The slope of each line is $\dfrac{3}{5}$. The y-intercept of the first line is $(0, 2)$. The y-intercept of the second line is $(0, 7)$. Since the lines have the same slopes but different y-intercepts, they are parallel lines. Therefore there is no solution. It is an inconsistent system of equations.

$4x + 3y = 9$ $(3, -1)$ $x - 3y = 6$

3. When graphed, the two equations yield the same line. The equations are dependent. There is an infinite number of solutions.
4. See Student Solutions Manual

8.2 Exercises   **1.** $(3, -1)$ **3.** $(1, 2)$ **5.** $(3, -5)$ **7.** $(0, 3)$ **9.** $(6, 4)$ **11.** $(2, 3)$ **13.** $\left(\dfrac{2}{3}, \dfrac{5}{3}\right)$ **15.** $(2, -3)$

17. $(-3, 6)$ **19.** $(3, -2)$ **21.** $(6, -5)$ **23.** $(-2, 5)$ **25.** $(-4, -12)$ **27.** $(8, 1)$ **29.** $3, 7$ **31.** both

33. No solution, parallel lines have no points in common   **35. (a)** $y = 90x + 50$ **(b)** 5 times plowing; $500   **(c)** 13 times; Adirondack Plowing
$y = 80x + 100$

36. $m = -\dfrac{7}{11}; \left(0, \dfrac{19}{11}\right)$ **37.**   **38.** $10.50   **39.** About 64%; about $71.30

Quick Quiz 8.2 **1.** $\left(\dfrac{1}{2}, 1\right)$ **2.** $(4, -3)$ **3.** $(0, 5)$ **4.** See Student Solutions Manual

8.3 Exercises **1.** Eliminate y by multiplying the second equation by 2, then adding the equations.
3. Eliminate y by multiplying the first equation by 4, multiplying the second equation by 3, then adding the equations.   **5.** $(-1, -4)$ **7.** $(2, -1)$
9. $\left(-\dfrac{1}{2}, 4\right)$ **11.** $\left(3, \dfrac{2}{5}\right)$ **13.** $(1, 0)$ **15.** $\left(-\dfrac{1}{2}, \dfrac{1}{3}\right)$ **17.** $(-10, 4)$ **19.** $(-2, 3)$ **21.** $\left(\dfrac{1}{3}, 1\right)$ **23.** Multiply equation (1) by 4.
25. Multiply equation (1) by 6.   **27.** $\left(2, \dfrac{1}{5}\right)$ **29.** $(3, -2)$ **31.** $\left(3, -\dfrac{3}{10}\right)$ **33.** $(-3, 3)$
Multiply equation (2) by 20.
35. Multiply equation (1) by 10.   **37.** Multiply equation (1) by 10.   **39.** $(-1, 2)$ **41.** $(-4, -7)$ **43.** $(8, -10)$ **45.** $\left(-\dfrac{2}{3}, 6\right)$
$5x - 3y = 1$ Multiply equation (2) by 100.
$5x + 3y = 6$ $40x + 5y = 90$
                     $20x - 5y = 100$
47. $(12, 16)$ **49.** $(0, 2)$ **51.** $\left(-\dfrac{4}{3}, \dfrac{2}{3}\right)$ **52.** 99 mph or greater   **53.** 36 games   **54.** $4(4x + 1)(4x - 1)$ **55.** $(3x - 4)(3x - 1)$

Quick Quiz 8.3 **1.** $(3, -2)$ **2.** $(4, -1)$ **3.** $(5, 1)$ **4.** See Student Solutions Manual

How Am I Doing? Sections 8.1–8.3 **1.**   (obj. 8.1.1)   **2.**   (obj. 8.1.1)   **3.** $(6, -4)$ (obj. 8.2.1)

4. $(0, -2)$ (obj. 8.2.1)   **5.** $(-2, 3)$ (obj. 8.2.2)   **6.** $(1, -2)$ (obj. 8.2.1)   **7.** $(7, 1)$ (obj. 8.3.1)   **8.** $\left(-3, \dfrac{1}{2}\right)$ (obj. 8.3.1)
9. $(-9, -12)$ (obj. 8.3.2)   **10.** $(1, -3)$ (obj. 8.3.3)

8.4 Exercises **1.** parallel lines, inconsistent   **3.** intersect, independent, consistent   **5.** $(3, -7)$ **7.** $(18, 13)$ **9.** No solution
11. Infinite number of solutions   **13.** $(5, 4)$ **15.** $(50, 40)$ **17.** $(8, 5)$ **19.** $\left(\dfrac{1}{2}, \dfrac{2}{3}\right)$ **21.** $\left(\dfrac{7}{3}, \dfrac{8}{3}\right)$ **23.** $(1, -2)$ **25.** $\left(\dfrac{3}{5}, \dfrac{3}{5}\right)$
27. No solution. The system is inconsistent.   **29.** $(-4, -3)$ **31.** $x = -\dfrac{2}{9}$ **32.** $x = 0.1$ **33.** $3748.75   **34.** More; $384.25

Quick Quiz 8.4 **1.** $(9, -8)$ **2.** No solution   **3.** $(3, 2)$ **4.** See Student Solutions Manual

8.5 Exercises **1.** 8 1st class   **3.** 5 @ $35   **5.** 9 hrs day shift   **7.** Original length $= 10$ ft   **9.** 8 qts of 50% antifreeze originally.
6 coach class   7 @ $40   14 hrs night shift   Original width $= 9$ ft   She added 8 quarts of 80% solution.
                                                            New length $= 14$ ft
                                                            New width $= 14$ ft
11. 30 lb nuts   **13.** 50 km/hr $=$ wind speed   **15.** Jewelry wanted 10 $10 bills and 7 $20 bills.   **17.** $0.69 per lb $=$ oranges
20 lb raisins   550 km/hr $=$ plane's speed   Cosmetics wanted 18 $10 bills and 3 $20 bills. $0.79 per lb $=$ apples
19. 132 general admission   **21.** $x = 19$, which represents 2024; 6,227,000 people   **23.** $\dfrac{x^2 + 5x - 2}{x^2 - 1}$ **24.** 35
210 student admission

Quick Quiz 8.5 **1.** 10,000 people purchased student tickets; 8500 people purchased general admission tickets   **2.** each Dell PC cost $800; each
Apple iMac cost $1200   **3.** 400 mi/hr $=$ speed of plane; 100 mi/hr $=$ speed of jetstream   **4.** See Student Solutions Manual

Putting Your Skills to Work **1.** Store A   **2. (a)** Approximately $3   **(b)** Approximately $6   **(c)** Approximately $9
3. (a) Approximately $12   **(b)** Approximately $16   **(c)** Approximately $19   **4.** Approximately $24   **5.** Approximately $27   **6.** Store B
7. Answers may vary

Chapter 8 Review Problems **1.**   **2.**   **3.**   **4.**

5. $(3, 3)$ **6.** $(3, 5)$ **7.** $(15, 21)$ **8.** $(8, 12)$ **9.** $(2, 1)$ **10.** $(-1, 0)$ **11.** $\left(0, -\dfrac{1}{2}\right)$ **12.** $\left(\dfrac{1}{3}, -4\right)$ **13.** $(5, -3)$

14. $(12, 1)$ **15.** $\left(\dfrac{4}{5}, -\dfrac{6}{5}\right)$ **16.** $\left(\dfrac{33}{13}, -\dfrac{28}{13}\right)$ **17.** $(6, -4)$ **18.** $(-1, -4)$ **19.** No solution, inconsistent system   **20.** $(9, 4)$

21. $\left(-3, -\dfrac{2}{3}\right)$ **22.** Infinite number of solutions, dependent equations   **23.** $(-2, -2)$ **24.** $(-1, -2)$ **25.** $(4, -3)$ **26.** $\left(-\dfrac{24}{41}, \dfrac{2}{41}\right)$

27. Infinite number of solutions, dependent equations   **28.** $(1960, 1040)$ **29.** $(8, -5)$ **30.** $(-4, -7)$ **31.** $\left(\dfrac{29}{28}, \dfrac{1}{7}\right)$ **32.** $(4, 10)$

33. $(10, 8)$ **34.** $(-12, -15)$ **35.** No solution, inconsistent system   **36.** Infinite number of solutions, dependent equations   **37.** $\left(\dfrac{33}{14}, \dfrac{16}{7}\right)$

38. $\left(-\dfrac{7}{32}, \dfrac{33}{16}\right)$ **39.** $(-5, 4)$ **40.** $(9, -8)$ **41.** $(2, 4)$ **42.** $(3, 1)$ **43.** $(3, 2)$ **44.** $(12, -12)$ **45.** 126 adults; 60 students

46. 21 2-lb bags; 42 5-lb bags   **47.** 275 mph in still air; 25 mph wind speed   **48.** 20 L of 20% solution; 20 L of 30% solution
49. 54 balsam firs; 25 Norwegian pines   **50.** 8 tons of 15% salt; 16 tons of 30% salt   **51.** Speed of the boat, 19 km/hr; speed of the current, 4 km/hr
52. $3 for a gram of copper; $4 hourly labor rate   **53.** 11 Level Outfield seats; 14 Upper Skydeck seats
54. 12 Field Level Bases seats; 10 Field Level Infield seats

How Am I Doing? Chapter 8 Test 1. $(-3, -4)$ (obj. 8.2.1)   **2.** $(-3, 4)$ (obj. 8.3.1)   **3.**   (obj. 8.1.1)   **4.** $(6, 2)$ (obj. 8.4.1)

5. $(4, 3)$ (obj. 8.4.1)   **6.** $(5, 1)$ (obj. 8.4.1)   **7.** $(3, 0)$ (obj. 8.4.1)   **8.** No solution, inconsistent system (obj. 8.4.1, 8.4.2)
9. Infinite number of solutions, dependent equations (obj. 8.4.1, 8.4.2)   **10.** $(-2, 3)$ (obj. 8.4.1)   **11.** $(-5, 3)$ (obj. 8.4.1)
12. $(-3.5, -4.5)$ (obj. 8.4.1)   **13.** $5, -3$ (obj. 8.5.1)   **14.** 16,000 of the $12 tickets; 14,500 of the $8 tickets (obj. 8.5.1)
15. A shirt costs $20 and a pair of slacks costs $24. (obj. 8.5.1)   **16.** 1000 booklets; $450 total cost (obj. 8.5.1)   **17.** 50 km/hr; 450 km/hr (obj. 8.5.1)

Cumulative Test for Chapters 0–8 1. $6.41   **2.** -8 **3.** $15x^3 - 19x^2 + 11x - 3$ **4.** $\dfrac{1}{(x + 3)(x - 4)}$ **5.** $x > \dfrac{2}{7}$ **6.** $x = \dfrac{1}{2}, -\dfrac{2}{3}$

7. 40   **8.** $x = \dfrac{5}{3}$ **9.** $5t^2(3t - 2)(t - 1)$ **10.** $x = 3$ **11.** $(8, -8)$ **12.** $(24, 8)$ **13.** Infinite number of solutions; dependent equations

14. $(5, 1)$ **15.** Inconsistent system; no solution   **16.** 36 kg of water   **17.** 1 mi/hr   **18.** The printer that broke prints 1200 labels/hour; the other printer prints 1800 labels/hour.

Chapter 9   **9.1 Exercises 1.** The principal square root of N, where $N \ge 0$, is a nonnegative number a that has the property $a^2 = N$.

3. No; $(0.3)(0.3) = 0.09$ **5.** ± 3 **7.** ± 7 **9.** 4   **11.** 9   **13.** -6 **15.** 0.9   **17.** $\dfrac{6}{11}$ **19.** $\dfrac{7}{8}$ **21.** 25   **23.** -100 **25.** 13

27. $-\dfrac{1}{8}$ **29.** $\dfrac{3}{4}$ **31.** 0.06   **33.** 140   **35.** -17 **37.** 5.196   **39.** 8.602   **41.** -13.528 **43.** -13.964 **45.** 13 ft   **47.** 7.5 sec
49. 3   **51.** -4 **53.** 3   **55.** 3   **57.** No, a 4th power must be nonnegative.   **59.** $(2, 1)$ **60.** No solution, inconsistent system
61. % of increase $= 10.7\%$; expenditure for 2008–2009 $= 996.30 **62.** % of increase $= 25.6\%$; expenditure for 2008–2009 $= 2312.30

Quick Quiz 9.1 1. 13   **2.** $\dfrac{5}{8}$ **3.** 0.7   **4.** See Student Solutions Manual

9.2 Exercises 1. Yes, $\sqrt{6^2} = \sqrt{6 \cdot 6} = \sqrt{36} = 6$ **3.** No, $\sqrt{(-9)^2} = \sqrt{(-9)(-9)} = \sqrt{81} = 9$ **5.** 8   **7.** 18^2 **9.** 9^3 **11.** 33^4
$\left(\sqrt{6}\right)^2 = \left(\sqrt{6}\right)\left(\sqrt{6}\right) = \sqrt{36} = 6$ but $-\sqrt{9^2} = -\sqrt{9 \cdot 9} = -\sqrt{81} = -9$

13. 5^{70} **15.** x^6 **17.** t^9 **19.** y^{13} **21.** $6x^4$ **23.** $12x$ **25.** $x^3 y^2$ **27.** $4xy^{10}$ **29.** $10a^5 b^3$ **31.** $2\sqrt{6}$ **33.** $3\sqrt{5}$ **35.** $3\sqrt{2}$

37. $6\sqrt{2}$ **39.** $3\sqrt{10}$ **41.** $8\sqrt{2}$ **43.** $2x\sqrt{2x}$ **45.** $3w^2\sqrt{3w}$ **47.** $4z^4\sqrt{2z}$ **49.** $2x^2 y^3 \sqrt{7xy}$ **51.** $4y\sqrt{3yw}$ **53.** $5xy\sqrt{3y}$

55. $8x^5$ **57.** $y^3\sqrt{y}$ **59.** $x^2 y^5 \sqrt{xy}$ **61.** $3x^2 y^3 \sqrt{15xy}$ **63.** $6ab^2 c^2 \sqrt{2c}$ **65.** $13ab^4 c^2 \sqrt{ac}$ **67.** $x + 5$ **69.** $4x + 1$

71. $y(x + 7)$ **73.** $(4, -1)$ **74.** $x = -18$ **75.** 177.1 trillion ft^3 **76.** 7.22 trillion ft^3 **77.** About 23.5%

Quick Quiz 9.2 1. $2\sqrt{30}$ **2.** $9xy^2\sqrt{x}$ **3.** $3x^4 y^2 \sqrt{15y}$ **4.** See Student Solutions Manual

9.3 Exercises 1. 1. Simplify each radical term.   **3.** $6\sqrt{5}$ **5.** $5\sqrt{2} + 3\sqrt{3}$ **7.** $4\sqrt{5}$ **9.** $12\sqrt{5}$ **11.** $\sqrt{2}$ **13.** $9\sqrt{2}$
2. Combine like radicals.

15. $5 + 6\sqrt{2} + 6\sqrt{3}$　**17.** $6\sqrt{5} - 2\sqrt{2}$　**19.** $13\sqrt{2x}$　**21.** $0.2\sqrt{3x}$　**23.** $7\sqrt{5y}$　**25.** $-\sqrt{2}$　**27.** $6\sqrt{7x} - 15x\sqrt{7x}$

29. $-5x\sqrt{2x}$　**31.** $10x\sqrt{3}$　**33.** $2y\sqrt{6y} - 6y\sqrt{6}$　**35.** $-110x\sqrt{2x}$　**37.** $\left(8\sqrt{5} + 6\sqrt{3}\right)$ mi　**39.** 8 games at \$8.99

40. \$13,500 in CD
　\$6500 in money market

4 games at \$12.49

Quick Quiz 9.3　1. $13\sqrt{2}$　**2.** $3\sqrt{3}$　**3.** $10\sqrt{a} + 6\sqrt{2a}$　**4.** See Student Solutions Manual

9.4 Exercises　1. $\sqrt{30}$　**3.** $2\sqrt{11}$　**5.** $3\sqrt{6}$　**7.** $12\sqrt{10}$　**9.** $5\sqrt{2a}$　**11.** $30x\sqrt{2}$　**13.** $16a\sqrt{6}$　**15.** $-6b\sqrt{a}$
17. $\sqrt{10} + \sqrt{15}$　**19.** $6\sqrt{2} + 15\sqrt{5}$　**21.** $2x - 16\sqrt{5x}$　**23.** $2\sqrt{3} - 18 + 4\sqrt{15}$　**25.** $-6 - 9\sqrt{2}$　**27.** $21 + 14\sqrt{2}$
29. $5 - \sqrt{21}$　**31.** $-6 + 9\sqrt{2}$　**33.** $2\sqrt{14} + 34$　**35.** $25 - 4\sqrt{6}$　**37.** $53 + 10\sqrt{6}$　**39.** $48 - 6\sqrt{15}$　**41.** $12\sqrt{10}$
43. $\sqrt{14} + 21\sqrt{2} - \sqrt{42}$　**45.** $9 - 4\sqrt{14}$　**47.** $6\sqrt{ab} + 2a\sqrt{b} - 4a$　**49.** $9x^2y - 5$　**51.** 12 ft^2　**53.** $(8a + 5b)(8a - 5b)$
54. $(2x - 5)^2$　**55.** $m = \left(\dfrac{1519}{1320}\right)k$; approx. 40.3 mph　**56.** $8\dfrac{1}{3}$ yr

Quick Quiz 9.4　1. $24xy$　**2.** $2\sqrt{6} + 6 - 4\sqrt{15}$　**3.** $-29 + 2\sqrt{6}$　**4.** See Student Solutions Manual

How Am I Doing? Sections 9.1–9.4　1. 7 (obj. 9.1.1)　**2.** $\dfrac{5}{8}$ (obj. 9.1.1)　**3.** 9 (obj. 9.1.1)　**4.** -10 (obj. 9.1.1)
5. 2.236 (obj. 9.1.2)　**6.** -11 (obj. 9.1.2)　**7.** 0.2 (obj. 9.2.1)　**8.** 0.6 (obj. 9.2.1)　**9.** x^2 (obj. 9.2.1)　**10.** $5x^2y^3$ (obj. 9.2.1)
11. $5\sqrt{2}$ (obj. 9.2.2)　**12.** $x^2\sqrt{x}$ (obj. 9.2.2)　**13.** $12ab\sqrt{b}$ (obj. 9.2.2)　**14.** $4x^2\sqrt{2x}$ (obj. 9.2.2)　**15.** $15\sqrt{2}$ (obj. 9.3.2)
16. $4\sqrt{2}$ (obj. 9.3.2)　**17.** $13\sqrt{7} - 7$ (obj. 9.3.2)　**18.** $11\sqrt{2} - 18\sqrt{3}$ (obj. 9.3.2)　**19.** $-9\sqrt{2x}$ (obj. 9.3.2)　**20.** $-26x\sqrt{3}$ (obj. 9.3.2)
21. $15\sqrt{30}$ (obj. 9.4.1)　**22.** $8x^2\sqrt{15}$ (obj. 9.4.1)　**23.** $2\sqrt{3} + 6$ (obj. 9.4.2)　**24.** $16 - 6\sqrt{7}$ (obj. 9.4.3)　**25.** 1 (obj. 9.4.3)
26. $17 - 4\sqrt{15}$ (obj. 9.4.3)　**27.** $x + 3\sqrt{x} - 10$ (obj. 9.4.3)　**28.** $81 + 14\sqrt{21}$ (obj. 9.4.3)

9.5 Exercises　1. 2　**3.** $\dfrac{1}{3}$　**5.** 7　**7.** $\dfrac{\sqrt{6}}{x^2}$　**9.** $\dfrac{3\sqrt{2}}{a^2}$　**11.** $\dfrac{3\sqrt{7}}{7}$　**13.** $\dfrac{8\sqrt{15}}{15}$　**15.** $\dfrac{x\sqrt{2x}}{2}$　**17.** $\dfrac{2\sqrt{2x}}{x}$　**19.** $\dfrac{3\sqrt{7}}{7}$　**21.** $\dfrac{6\sqrt{a}}{a}$

23. $\dfrac{\sqrt{2x}}{2x^2}$　**25.** $\dfrac{3\sqrt{x}}{x^2}$　**27.** $\dfrac{\sqrt{15}}{5}$　**29.** $\dfrac{\sqrt{210}}{21}$　**31.** $\dfrac{9\sqrt{2x}}{8x}$　**33.** $2\sqrt{3} + 2$　**35.** $\sqrt{10} - \sqrt{2}$　**37.** $2 + \sqrt{2}$　**39.** $2\sqrt{14} - 7$

41. $x\left(2\sqrt{2} + \sqrt{5}\right)$　**43.** $\dfrac{\sqrt{6x} - \sqrt{2x}}{4}$　**45.** $\dfrac{13 - 2\sqrt{30}}{7}$　**47.** $\dfrac{4\sqrt{35} + 3\sqrt{5} + 3\sqrt{2} + 4\sqrt{14}}{3}$　**49.** $5\sqrt{6} + 8\sqrt{2}$　**51.** $\sqrt{x} - 5$

53. (a) $s = \dfrac{1}{h}\sqrt{3Vh}$　**(b)** $s = 3$ in.　**55.** $\left(3\sqrt{5} - 6\right)$ m　**57.** $x \approx -0.71$　**58.** 0

Quick Quiz 9.5　1. $\dfrac{3\sqrt{10}}{10}$　**2.** $\dfrac{15x\sqrt{2} - 5x\sqrt{5}}{13}$　**3.** $11 + 2\sqrt{30}$　**4.** See Student Solutions Manual

9.6 Exercises　1. 13　**3.** $2\sqrt{14}$　**5.** $7\sqrt{2}$　**7.** $3\sqrt{7}$　**9.** $2\sqrt{14}$　**11.** $\sqrt{57}$　**13.** 10.08　**15.** 9.4 ft　**17.** 98.0 ft
19. Top cable $= 145.40$ ft; bottom cable $= 140.77$ ft　**21.** $x = 7$　**23.** $x = 9$　**25.** $x = 7$　**27.** $x = \dfrac{81}{2}$　**29.** $x = 5$ only　**31.** $y = 0, 3$
33. $x = 1$ only　**35.** $y = 0, 1$　**37.** $y = \dfrac{1}{2}$ only　**39.** $x = \dfrac{2}{3}$ only　**41.** $x = 5$　**42.** $49x^2 - 28x + 4$　**43.** 20　**44.** Relation

Quick Quiz 9.6　1. $2\sqrt{7}$　**2.** $x = 14$　**3.** $x = 8$　**4.** See Student Solutions Manual

9.7 Exercises　1. $y = 72$　**3.** $y = \dfrac{1029}{2}$　**5.** $y = 1296$　**7.** 225 calories　**9.** 56 min　**11.** $y = \dfrac{45}{4}$　**13.** $y = 1$　**15.** $y = \dfrac{40}{27}$
17. 3.45 min　**19.** $444\dfrac{4}{9}$ lb　**21. (a)** 15.0 km, 8.2 km, and 2.7 km　**23. (a)** $k = 1$　**(b)** 248.25 Earth years　**25.** 160
(b)　　**26.** $S = 4.80h$

Quick Quiz 9.7　1. 6 in.　**2.** 281.25 calories　**3.** 40 cm^2　**4.** See Student Solutions Manual

Putting Your Skills to Work　1. $\dfrac{8}{3}$ or $2\dfrac{2}{3}$ cups　**2.** 56 oz　**3.** 3.5 lb　**4. (a)** \$0.26/lb　**(b)** 70%　**5.** \$2.20
6. Answers may vary　**7.** Answers may vary

Chapter 9 Review Problems　1. 6　**2.** 5　**3.** 13　**4.** 14　**5.** -9　**6.** 8　**7.** 7　**8.** -11　**9.** -12　**10.** 17　**11.** 0.2
12. 0.7　**13.** $\dfrac{1}{10}$　**14.** $\dfrac{8}{9}$　**15.** 10.247　**16.** 14.071　**17.** 8.775　**18.** 9.381　**19.** $3\sqrt{3}$　**20.** $2\sqrt{13}$　**21.** $2\sqrt{7}$　**22.** $5\sqrt{5}$

23. $2\sqrt{10}$ **24.** $4\sqrt{5}$ **25.** x^4 **26.** y^5 **27.** $x^2y^3\sqrt{x}$ **28.** $ab^2\sqrt{a}$ **29.** $4xy^2\sqrt{xy}$ **30.** $7x^2y^3\sqrt{2}$ **31.** $2x^2\sqrt{3x}$ **32.** $6x^4\sqrt{2x}$
33. $4x^4\sqrt{3}$ **34.** $5x^6\sqrt{5}$ **35.** $2ab^2c^2\sqrt{30ac}$ **36.** $11a^3b^2\sqrt{c}$ **37.** $2x^3y^4\sqrt{14xy}$ **38.** $3x^6y^3\sqrt{11xy}$ **39.** $3\sqrt{5}$ **40.** $9\sqrt{6}$ **41.** $7x\sqrt{3}$
42. $8a\sqrt{2}+2a\sqrt{3}$ **43.** $-7\sqrt{5}+2\sqrt{10}$ **44.** $-2\sqrt{10}+6\sqrt{7}$ **45.** $6x^2$ **46.** $-10a\sqrt{b}$ **47.** $6ab^2\sqrt{b}$ **48.** $-8x^3$ **49.** -18
50. -12 **51.** $\sqrt{10}-\sqrt{6}-4$ **52.** $\sqrt{30}-10+5\sqrt{2}$ **53.** $20+3\sqrt{11}$ **54.** $27+8\sqrt{10}$ **55.** $8-4\sqrt{3}+12\sqrt{6}-18\sqrt{2}$
56. $15-3\sqrt{2}-10\sqrt{3}+2\sqrt{6}$ **57.** $66+36\sqrt{2}$ **58.** $74-40\sqrt{3}$ **59.** $\dfrac{\sqrt{3x}}{3x}$ **60.** $\dfrac{2y\sqrt{5}}{5}$ **61.** $\dfrac{x^2y\sqrt{2}}{4}$ **62.** $\dfrac{3a\sqrt{2b}}{2}$
63. $\dfrac{\sqrt{30}}{6}$ **64.** $\dfrac{\sqrt{66}}{11}$ **65.** $\dfrac{x\sqrt{5}}{5}$ **66.** $\dfrac{a^2\sqrt{6}}{6}$ **67.** $\sqrt{5}-\sqrt{2}$ **68.** $\dfrac{2(\sqrt{6}+\sqrt{3})}{3}$ **69.** $-7+3\sqrt{5}$ **70.** $\dfrac{3-2\sqrt{3}}{3}$
71. $c=\sqrt{89}$ **72.** $a=\sqrt{2}$ **73.** $b=\dfrac{\sqrt{51}}{2}$ **74.** $c=3.124$ **75.** 30 m   **76.** 186.7 mi   **77.** 60 ft   **78.** $x=29$ **79.** $x=13$
80. $x=-2$ **81.** $x=6$ **82.** $x=4$ **83.** $x=7$ **84.** $x=2$ **85.** $x=7$ **86.** $y=48$ **87.** $y=1$ **88.** $y=\dfrac{1}{2}$
89. 5 lumens   **90.** Approximately 134 ft   **91.** It will be 8 times as much.   **92.** $\dfrac{6}{11}$ **93.** 0.02   **94.** $7x^3y\sqrt{2y}$ **95.** $3\sqrt{3}$
96. $2\sqrt{15}+3\sqrt{6}-20-6\sqrt{10}$ **97.** $57+24\sqrt{3}$ **98.** $\dfrac{5\sqrt{3}}{6}$ **99.** $\dfrac{6-\sqrt{6}+6\sqrt{2}-2\sqrt{3}}{10}$ **100.** $x=42$ **101.** $x=-\dfrac{1}{2}, x=2$

How Am I Doing? Chapter 9 Test   **1.** 11 (obj. 9.1.1)   **2.** $\dfrac{3}{10}$ (obj. 9.1.1)   **3.** $4xy^3\sqrt{3y}$ (obj. 9.2.2)   **4.** $10xz^2\sqrt{xy}$ (obj. 9.2.2)
5. $3\sqrt{3}$ (obj. 9.3.2)   **6.** $8\sqrt{a}+5\sqrt{2a}$ (obj. 9.3.2)   **7.** $12ab$ (obj. 9.4.1)   **8.** $5\sqrt{2}+2\sqrt{15}-15$ (obj. 9.4.2)   **9.** $37+20\sqrt{3}$ (obj. 9.4.3)
10. $19+\sqrt{10}$ (obj. 9.4.3)   **11.** $\dfrac{\sqrt{5x}}{5}$ (obj. 9.5.2)   **12.** $\dfrac{\sqrt{3}}{2}$ (obj. 9.5.2)   **13.** $\dfrac{17+\sqrt{3}}{22}$ (obj. 9.5.3)   **14.** $a(\sqrt{5}-\sqrt{2})$ (obj. 9.5.3)
15. 12.49 (obj. 9.1.2)   **16.** $x=\sqrt{133}$ (obj. 9.6.1)   **17.** $x=3\sqrt{3}$ (obj. 9.6.1)   **18.** $x=\dfrac{35}{2}$ (obj. 9.6.2)   **19.** $x=9$ (obj. 9.6.2)
20. 16 in. (obj. 9.7.2)   **21.** \$85.77 (obj. 9.7.1)   **22.** 21.2 cm^2 (obj. 9.7.1)

Cumulative Test for Chapters 0–9   **1.** $\dfrac{2}{3}$ **2.** $7\dfrac{1}{12}$ **3.** $\dfrac{15}{16}$ **4.** 0.0007   **5.** 3   **6.** 31   **7.** $-3a^2bc^3-2abc^2$
8. $x^3-x^2-16x+16$ **9.** $16x^2-40x+25$ **10.** $9x^2-121$ **11.** $\dfrac{4x+4}{x+2}$ **12.** $\dfrac{(x-3)(x+1)}{(x-6)(x+2)}$ **13.** $7x^2y^3\sqrt{2x}$ **14.** $7\sqrt{2}$
15. $9-6\sqrt{2}$ **16.** $5+2\sqrt{6}$ **17.** $10+22\sqrt{3}$ **18.** $-4-2\sqrt{5}$ **19.** -5 **20.** Not a real number   **21.** $c=\sqrt{23}$
22. $b=4\sqrt{5}$ **23.** $x=1,3$ **24.** $y=6$ **25.** $y=10$ **26.** 78.5 m^2

Chapter 10   10.1 Exercises

1. $a=1$, $b=8$, $c=7$ **3.** $a=8$, $b=-11$, $c=0$ **5.** $a=1$, $b=3$, $c=-15$ **7.** $x=0$, $x=\dfrac{1}{3}$ **9.** $x=0$, $x=-2$ **11.** $x=0$, $x=\dfrac{3}{2}$ **13.** $x=7$, $x=-4$ **15.** $x=\dfrac{1}{2}$, $x=-6$ **17.** $x=\dfrac{1}{3}$, $x=5$
19. $x=7$, $x=-2$ **21.** $x=-\dfrac{3}{4}$, $x=-\dfrac{2}{3}$ **23.** $a=-8$, $a=-4$ **25.** $y=\dfrac{1}{3}$, $y=-\dfrac{3}{2}$ **27.** $x=\dfrac{6}{5}$ **29.** $x=0$, $x=3$ **31.** $x=0$, $x=-\dfrac{14}{3}$ **33.** $y=-2$, $y=5$
35. $x=3$, $x=7$ **37.** $x=1$ **39.** $x=4$, $x=2$ **41.** $x=-7$, $x=2$ **43.** $y=\dfrac{4}{3}$, $y=-1$ **45.** $x=\dfrac{5}{2}$, $x=-4$ **47.** $x=-\dfrac{10}{3}$, $x=4$ **49.** You can always factor out x.
51. $a=\dfrac{14}{9}$, $c=-14$ **53.** $n=1200$, $n=2000$ **55.** Average $=1600$, Revenue for 1600 = \$360, which is more than producing 1200 or 2000 labels   **57.** Revenue for 1500 = \$357.50, Revenue for 1700 = \$357.50, It appears the maximum revenue is at 1600 labels.
59. $\dfrac{(2x-1)(x+2)}{2x}$ **60.** $\dfrac{2x^2-4x+2}{3x^2-11x+10}$

Quick Quiz 10.1   **1.** $x=\dfrac{5}{3}$, $x=-\dfrac{3}{2}$ **2.** $x=8$, $x=6$ **3.** $x=-3$, $x=\dfrac{2}{3}$ **4.** See Student Solutions Manual

10.2 Exercises   **1.** $x=\pm8$ **3.** $x=\pm7\sqrt{2}$ **5.** $x=\pm2\sqrt{7}$ **7.** $x=\pm3$ **9.** $x=\pm2\sqrt{5}$ **11.** $x=\pm5\sqrt{5}$ **13.** $x=\pm2\sqrt{3}$
15. $x=\pm\sqrt{5}$ **17.** $x=3, x=11$ **19.** $x=-4\pm\sqrt{6}$ **21.** $x=\dfrac{-5\pm\sqrt{2}}{2}$ **23.** $x=\dfrac{1\pm\sqrt{7}}{3}$ **25.** $x=\dfrac{-1\pm3\sqrt{2}}{5}$
27. $x=\dfrac{5\pm3\sqrt{6}}{4}$ **29.** $x=3\pm2\sqrt{5}$ **31.** $x=2, x=-10$ **33.** $x=6\pm\sqrt{41}$ **35.** $x=0, x=7$ **37.** $x=0, x=5$

39. $x = \dfrac{9}{2}, x = -1$ **41.** $a = -4, b = -4$ **42.** $x = -1, x = 4$ **43.** 200 lb/in.^{2}

44. A total of 16 tires were defective. 6 had both kinds of defects. 7 had defects only in workmanship. 3 had defects only in materials.

Quick Quiz 10.2 **1.** $x = \pm\sqrt{6}$ **2.** $x = \dfrac{2 \pm 2\sqrt{10}}{3}$ **3.** $x = -4 \pm \sqrt{26}$ **4.** See Student Solutions Manual

10.3 Exercises

1. $\sqrt{b^2 - 4ac} = \sqrt{100} = 10$ **3.** No. Standard form is $4x^2 + 5x - 6 = 0$.
Therefore, there are two rational roots. Then $a = 4, b = 5$, and $c = -6$.

5. $x = 2, x = -5$ **7.** $x = \dfrac{3 \pm \sqrt{41}}{2}$ **9.** $x = \dfrac{1}{4}, x = -2$ **11.** $x = 4, x = -\dfrac{5}{2}$ **13.** $x = \dfrac{3 \pm \sqrt{33}}{12}$

15. $x = \dfrac{1 \pm \sqrt{19}}{6}$ **17.** $x = 5, x = 2$ **19.** No real solution **21.** $y = \dfrac{5 \pm \sqrt{13}}{4}$ **23.** $d = \dfrac{4}{3}, d = -3$ **25.** No real solution

27. $x = -6$ **29.** $x \approx 0.372, x \approx -5.372$ **31.** $x \approx 4.108, x \approx -0.608$ **33.** $x \approx -1.894, x \approx -0.106$ **35.** $x = \dfrac{3}{2}, x = \dfrac{2}{3}$

37. $x = 3, x = \dfrac{1}{3}$ **39.** $t = -1 \pm \sqrt{23}$ **41.** $y = \dfrac{1 \pm \sqrt{51}}{5}$ **43.** $x = \dfrac{\pm\sqrt{39}}{3}$ **45.** $x = 1 \pm 2\sqrt{2}$

47. 2 ft **49.** $x^2 + 5x + 2$ **50.** $\dfrac{9p - 20}{3f} = g$

Quick Quiz 10.3 **1.** $x = \dfrac{-1 \pm \sqrt{2}}{3}$ **2.** $x = \dfrac{1 \pm 2\sqrt{5}}{2}$ **3.** $x = \dfrac{-1 \pm \sqrt{7}}{3}$ **4.** See Student Solutions Manual

How Am I Doing? Sections 10.1–10.3 **1.** $x = 16, x = -3$ (obj. 10.1.2) **2.** $x = 0, x = \dfrac{7}{5}$ (obj. 10.1.3) **3.** $x = \dfrac{2}{5}, x = 4$ (obj. 10.1.3)

4. $x = \dfrac{1}{3}, x = -\dfrac{1}{2}$ (obj. 10.1.3) **5.** $x = -8, x = 3$ (obj. 10.1.3) **6.** $x = 3, x = -\dfrac{9}{2}$ (obj. 10.1.3) **7.** $x = \pm3\sqrt{2}$ (obj. 10.2.1)

8. $x = \pm5$ (obj. 10.2.1) **9.** $x = -3 \pm 2\sqrt{2}$ (obj. 10.2.2) **10.** $x = \dfrac{-3 \pm \sqrt{65}}{4}$ (obj. 10.2.2) **11.** $x = \dfrac{-2 \pm \sqrt{14}}{2}$ (obj. 10.3.1)

12. $x = \dfrac{5}{4}, x = -1$ (obj. 10.3.1) **13.** $x = \dfrac{-4 \pm \sqrt{13}}{3}$ (obj. 10.3.1) **14.** No real solution (obj. 10.3.3)

10.4 Exercises

1. It may have 2, 1, or none. **3.** If $b = 0$, then the parabola always has a vertex at $(0, c)$ which is on the y-axis.

5.
$y = x^2 + 2$

7.
$y = -\dfrac{1}{3}x^2$

9.
$y = 3x^2 - 1$

11.
$y = (x - 2)^2$

13.
$y = -\dfrac{1}{2}x^2 + 4$

15.
$y = \dfrac{1}{2}(x - 3)^2$

17.
$V = (-2, -4)$
$y = x^2 + 4x$

19.
$V = (-2, 4)$
$y = -x^2 - 4x$

21.
$V = (2, 8)$
$y = -2x^2 + 8x$

23.
$V = (-1, -4)$
x-int: $(-3, 0)$
$(1, 0)$ $y = x^2 + 2x - 3$

25.
$V = (-3, 4)$ x-int: $(-5, 0)$
$(-1, 0)$
$y = -x^2 - 6x - 5$

27.
$V = (-3, 0)$
x-int: $(-3, 0)$
$y = x^2 + 6x + 9$

29. (a)
(4, 82.4)
(8.1, 0)

(b) 62.8 m
(c) 82.4 m
(d) About 8.1 sec

31. (a)
(4.5, 20.25)
$N = 9x - x^2$
Number of mosquitoes in millions
Number of inches of rain

(b) 4.5 in.
(c) 20.25 million

33. Vertex: $(-0.277, 1551.164)$
y-intercept: $(0, 1528)$
x-intercepts: $(1.993, 0)$ and $(-2.548, 0)$

34. $y = 1$ **35.** $y = 75$ **36.** 8.25 sec

Quick Quiz 10.4 **1.** $(3, -4)$ **2.** $(1, 0)$ and $(5, 0)$ **3.**

$y = x^2 - 6x + 5$

4. See Student Solutions Manual

10.5 Exercises **1.** Base $= 11$ cm altitude $= 16$ cm   **3.** 10 ft by 10 ft   **5.** 21 people   **7.** 80 members were expected. 100 members attended.   **9.** Width $= 25$ ft length $= 50$ ft

11. 600 mph   **13.** 50 stores   **15.** 15 stores   **17.** 2002 and 2003   **19.** About 74 mi   **21.**

22. $\sqrt{15} - 2\sqrt{3}$ **23.** 0.75 ft

Quick Quiz 10.5 **1.** 12 yd, 5 yd   **2.** 6 sec   **3.** 61.25 m   **4.** See Student Solutions Manual

Putting Your Skills to Work **1.** \$16.40   **2.** \$24.60   **3.** About \$246   **4.** Answers may vary   **5.** Answers may vary
6. \$150   **7.** \$2225   **8.** \$178   **9.** \$356   **10.** $62°$

Chapter 10 Review Problems **1.** $a = 6, b = -5, c = 8$ **2.** $a = 14, b = 3, c = -16$ **3.** $a = 1, b = 3, c = -14$
4. $a = 11, b = 8, c = 0$ **5.** $a = -2, b = -6, c = 3$ **6.** $a = 4, b = 5, c = -14$ **7.** $x = -25, x = -1$ **8.** $x = -8$ **9.** $x = 5, x = -11$
10. $x = \dfrac{3}{2}, x = \dfrac{5}{2}$ **11.** $x = \dfrac{4}{3}$ **12.** $x = \dfrac{2}{5}, x = \dfrac{4}{3}$ **13.** $x = \dfrac{5}{3}, x = \dfrac{1}{2}$ **14.** $x = 1, x = -\dfrac{9}{4}$ **15.** $x = -\dfrac{3}{2}, x = \dfrac{4}{3}$ **16.** $x = \dfrac{1}{2}, x = 1$
17. $x = \dfrac{1}{5}, x = -\dfrac{1}{3}$ **18.** $x = \dfrac{1}{4}, x = -\dfrac{4}{3}$ **19.** $x = 0, x = -8$ **20.** $x = 0, x = 3$ **21.** $x = 0, x = -\dfrac{5}{3}$ **22.** $x = 0, x = 4$
23. $x = -\dfrac{2}{5}, x = 10$ **24.** $x = \dfrac{1}{3}, x = \dfrac{4}{3}$ **25.** $x = \pm 7$ **26.** $x = \pm 9$ **27.** $x = \pm\sqrt{22}$ **28.** $x = \pm\sqrt{39}$ **29.** $x = \pm 2\sqrt{2}$
30. $x = \pm 5\sqrt{2}$ **31.** $x = \pm 3\sqrt{2}$ **32.** $x = \pm 2\sqrt{6}$ **33.** $x = 4 \pm \sqrt{7}$ **34.** $x = 2 \pm \sqrt{3}$ **35.** $x = \dfrac{-2 \pm 2\sqrt{7}}{3}$ **36.** $x = \dfrac{1 \pm 4\sqrt{2}}{2}$
37. $x = -1, x = -7$ **38.** $x = -3, x = -11$ **39.** $x = 9, x = -5$ **40.** $x = 3 \pm \sqrt{2}$ **41.** $x = -1 \pm \sqrt{3}$ **42.** $x = \dfrac{-5 \pm \sqrt{31}}{2}$
43. $x = -2 \pm \sqrt{10}$ **44.** $x = -2 \pm 2\sqrt{3}$ **45.** $x = \dfrac{7 \pm \sqrt{17}}{4}$ **46.** $x = \dfrac{-5 \pm \sqrt{73}}{4}$ **47.** $x = \dfrac{4 \pm 2\sqrt{7}}{3}$ **48.** $x = \dfrac{1 \pm 3\sqrt{5}}{4}$
49. $x = \dfrac{1 \pm \sqrt{41}}{10}$ **50.** $x = \dfrac{-3 \pm \sqrt{21}}{6}$ **51.** $x = \dfrac{5}{2}, x = 2$ **52.** $x = \dfrac{3}{2}, x = -\dfrac{1}{2}$ **53.** $x = -\dfrac{1}{5}$ **54.** $x = 4, x = \dfrac{3}{2}$
55. $x = \dfrac{3 \pm \sqrt{3}}{3}$ **56.** $x = \dfrac{7 \pm \sqrt{209}}{10}$ **57.** $x = \dfrac{-2 \pm \sqrt{19}}{3}$ **58.** $x = \dfrac{-7 \pm \sqrt{29}}{10}$ **59.** $x = \dfrac{9 \pm \sqrt{93}}{2}$ **60.** $x = -\dfrac{1}{2}, x = 5$
61. $x = \dfrac{3}{4}, x = -2$ **62.** $x = \pm 3\sqrt{2}$ **63.** $y = -8$ only   **64.** $y = -7$ only   **65.** $x = \dfrac{-4 \pm \sqrt{31}}{3}$ **66.** $x = \dfrac{-3 \pm \sqrt{13}}{4}$
67. $y = -\dfrac{9}{8}, y = 1$ **68.** $y = -4, y = 2$ **69.** $y = -3, y = 5$ **70.** $x = 0, x = \dfrac{1}{6}$

71.

$y = 2x^2$

72.

$(0, 4)$
$y = x^2 + 4$

73.

$(0, -3)$
$y = x^2 - 3$

74.

$(0, 0)$
$y = -\dfrac{1}{2}x^2$

75.
$\left(1\dfrac{1}{2}, -6\dfrac{1}{4}\right)$
$y = x^2 - 3x - 4$

76.

$(0, -2)$
$y = \dfrac{1}{2}x^2 - 2$

77.

$(3, 1)$
$y = -2x^2 + 12x - 17$

78.

$\left(-\dfrac{1}{3}, 4\dfrac{1}{3}\right)$
$y = -3x^2 - 2x + 4$

79. 7 ft by 4 ft or 8 ft by 3.5 ft   **80.** 16 cm, 12 cm

81. Width $= 18$ in. length $= 24$ in.   **82.** $t = 30$ sec   **83.** 24 members   **84.** Going, 45 mph; returning, 60 mph   **85.** 60 pizza stores
86. 30 fewer pizza stores   **87.** 1990 and 2020   **88.** 2000 and 2010

How Am I Doing? Chapter 10 Test **1.** $x = \dfrac{-7 \pm \sqrt{129}}{10}$ (obj. 10.3.1)   **2.** $x = -5, x = \dfrac{2}{3}$ (obj. 10.1.3)

3. No real solution (obj. 10.3.3)   **4.** $x = \dfrac{3}{4}, x = 5$ (obj. 10.2.2)   **5.** $x = -\dfrac{5}{4}, x = \dfrac{1}{3}$ (obj. 10.2.1)   **6.** $x = \dfrac{4}{3}$ (obj. 10.2.2)

7. $x = 0, x = 8$ (obj. 10.3.1)   **8.** $x = \pm 3$ (obj. 10.3.1)   **9.** $x = -\dfrac{1}{2}, x = 6$ (obj. 10.3.1)   **10.** $x = \dfrac{3}{2}, x = -\dfrac{1}{2}$ (obj. 10.3.1)

11.   (obj. 10.4.1)   **12.**   (obj. 10.4.1)   **13.** 9 m, 12 m (obj. 10.5.1)   **14.** 7 sec (obj. 10.5.1)

$y = 3x^2 - 6x$

$y = -x^2 + 8x - 12$

Practice Final Examination **1.** $-2x + 21y + 18xy + 6y^2$ **2.** 14   **3.** $18x^5 y^5$ **4.** $2x^2 y - 8xy$ **5.** $x = \dfrac{8}{5}$ **6.** $b = \dfrac{p - 2a}{2}$

7. $x \geq 2.6$ **8.** $2x^3 - 5x^2 y + xy^2 + 2y^3$ **9.** $2(2x + 1)(x - 5)$ **10.** $3x(x - 5)(x + 2)$

11. $\dfrac{6x - 11}{(x + 2)(x - 3)}$ **12.** $\dfrac{11(x + 2)}{2x(x + 4)}$ **13.** $-\dfrac{12}{5}$ **14.** slope $= \dfrac{5}{2}$ **15.** $3x + 4y = 14$ or $y = -\dfrac{3}{4}x + \dfrac{7}{2}$

$5x - 2y - 3 = 0$

16. 38.13 in.^{2} **17.** $x = -5, y = 2$ **18.** $a = -2, b = -3$ **19.** $x\sqrt{5x}$ **20.** $6\sqrt{3} - 12 + 12\sqrt{2}$ **21.** $-\dfrac{\sqrt{15} + \sqrt{35} + \sqrt{21} + 7}{2}$

22. $x = \dfrac{2}{3}, x = -\dfrac{1}{4}$ **23.** $y = \dfrac{3 \pm \sqrt{7}}{2}$ **24.** $x = \pm 2$ **25.** $\sqrt{39}$ **26.** 5   **27.** Width 7 m, length 12 m   **28.** $3000 at 10%, $4000 at 14%

29. 200 reserved-seat tickets, 160 general admission tickets   **30.** Base 8 m, altitude 17 m

Appendix B Exercises

1. 34,000 m   **3.** 5700 cm   **5.** 250 mm   **7.** 0.563 m   **9.** 29,400 mg   **11.** 0.0984 kg   **13.** 7000 mL   **15.** 0.000004 kL   **17.** 1.7 in.
19. 22.5 km   **21.** 2.8 m   **23.** 17.8 cm   **25.** 1089.6 g   **27.** 171.6 lb   **29.** 2.8 L   **31.** 193,248,000 in.   **33.** 27.27 mph   **35.** 0.11 yr

Appendix C Exercises

1. 84,916 mi^2 **3.** 4,753,377 people   **5.** 2 representatives   **7.** Treasure State   **9.** Nevada   **11.** 2000 homes   **13.** DuPage County
15. 2800 homes   **17.** 4,000,000 apartment units   **19.** 7,200,000 more apartment units   **21.** 8,000,000 people   **23.** Between 2000 and 2010
25. 3 million more people   **27.** 37 million people   **29.** Movies and an exercise program   **31.** $5.6 million   **33.** $2.5 million greater
35. Between 2004 and 2005   **37.** 74 million people   **39.** 65 years and above   **41.** 19%   **43.** 81%   **45.** 845 million people   **47.** 19.5%
49. $2120

Appendix D Exercises

1. 14   **3.** 52   **5.** 1296   **7.** -3 **9.** $10x - 5$ **11.** 1111811   **13.** 9876549   **15.** 6481100   **17.** 1 6 15 20 15 6 1

19. $x = 3$; Answers may vary   **21.** $x = -2$; Answers may vary   **23.** $x = 3$; Answers may vary   **25.** $x = \dfrac{9}{11}$; Answers may vary

27. Answers may vary   **29.** William was first, Brent was second, Charlie was third, Dave was fourth   **31.** Michael
33. Toyota Corolla   **35.** Unless Fred is willing to pay over $200, he will probably get a false tooth.

Absolute value of a number (1.1)   The absolute value of a number x is the distance between 0 and the number x on the number line. It is written as $|x|$. $|x| = x$ if $x \geq 0$, but $|x| = -x$ if $x < 0$.

Altitude of a geometric figure (1.8)   The height of the geometric figure. In the three figures shown, the altitude is labeled a.

Altitude of a trapezoid   Altitude of a parallelogram

Altitude of a rhombus

Altitude of a triangle (1.8)   The height of any given triangle. In the three triangles shown, the altitude is labeled a.

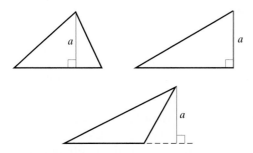

Associative property of addition (1.1)   If a, b, and c are real numbers, then

$$a + (b + c) = (a + b) + c.$$

This property states that if three numbers are added, it does not matter *which two numbers* are added first; the result will be the same.

Associative property of multiplication (1.3)   If a, b, and c are real numbers, then

$$a \times (b \times c) = (a \times b) \times c.$$

This property states that if three numbers are multiplied, it does not matter *which two numbers* are multiplied first; the result will be the same.

Base (1.4)   The number or variable that is raised to a power. In the expression 2^6, the number 2 is the base.

Base of a triangle (1.8)   The side of a triangle that is perpendicular to the altitude.

Binomial (4.3)   A polynomial of two terms. The expressions $a + 2b$, $6x^3 + 1$, and $5a^3b^2 + 6ab$ are all binomials.

Circumference of a circle (1.8)   The distance around a circle. The circumference of a circle is given by the formula $C = \pi d$ or $C = 2\pi r$, where d is the diameter of the circle and r is the radius of the circle.

Coefficient (4.1)   A coefficient is a factor or a group of factors in a product. In the term $4xy$ the coefficient of y is $4x$, but the coefficient of xy is 4. In the term $-5x^3y$ the coefficient of x^3y is -5.

Commutative property for addition (1.1)   If a and b are any real numbers, then $a + b = b + a$.

Commutative property for multiplication (1.3)   If a and b are any real numbers, then $ab = ba$.

Complex fraction (6.4)   A fraction that contains at least one fraction in the numerator or in the denominator or both. These three fractions are complex fractions:

$$\frac{7 + \dfrac{1}{x}}{x^2 + 2}, \qquad \frac{1 + \dfrac{1}{5}}{2 - \dfrac{1}{7}}, \qquad \text{and} \qquad \frac{\dfrac{1}{3}}{4}.$$

Constant (2.3)   Symbol or letter that is used to represent exactly one single quantity during a particular problem or discussion.

Coordinates of a point (7.1)   An ordered pair of numbers (x, y) that specifies the location of a point in a rectangular coordinate system.

Degree of a polynomial (4.3)   The degree of the highest-degree term of a polynomial. The degree of the polynomial $5x^3 + 2x^2 - 6x + 8$ is 3. The degree of the polynomial $5x^2y^2 + 3xy + 8$ is 4.

Degree of a term of a polynomial (4.3)   The sum of the exponents of the variables in the term. The degree of $3x^3$ is 3. The degree of $4x^5y^2$ is 7.

Denominator (0.1) and (6.1)   The bottom number or algebraic expression in a fraction. The denominator of

$$\frac{3x - 2}{x + 4}$$

is $x + 4$. The denominator of $\dfrac{3}{7}$ is 7. The denominator of a fraction may not be zero.

Dependent equations (8.1)   Two equations are dependent if every value that satisfies one equation satisfies the other. A system of two dependent equations in two variables will not have a unique solution.

Diagonal of a four-sided figure (1.8) A line connecting two nonadjacent corners of the figure. In each of the figures shown, line AC is a diagonal.

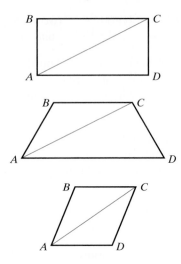

Difference (3.1) The result of subtracting one number or expression from another. The mathematical expression $x - 6$ can be written in words as the difference between x and 6.

Difference-of-two-squares polynomial (5.5) A polynomial of the form $a^2 - b^2$ that may be factored by using the formula

$$a^2 - b^2 = (a + b)(a - b).$$

Distributive property (1.6) For all real numbers a, b, and c, $a(b + c) = ab + ac$.

Dividend (0.4) The number that is to be divided by another. In the problem $30 \div 5 = 6$, the three parts are as follows:

Divisor (0.4) The number you divide into another.

Domain of a relation (7.6) In any relation, the set of values that can be used for the independent variable is called its domain. This is the set of all the first coordinates of the ordered pairs that define the relation.

Equilateral triangle (1.8) A triangle with three sides equal in length and three angles that each measure $60°$. Triangle ABC is an equilateral triangle.

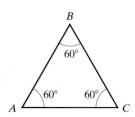

Even integers (1.3) Integers that are exactly divisible by 2, such as $\ldots, -4, -2, 0, 2, 4, 6, \ldots$.

Exponent (1.4) The number that indicates the power of a base. If the number is a positive integer, it indicates how many times the base is multiplied. In the expression 2^6, the exponent is 6.

Expression (4.3) A mathematical expression is any quantity using numbers and variables. Therefore, $2x$, $7x + 3$, and $5x^2 + 6x$ are all mathematical expressions.

Extraneous solution (6.5) and (9.6) An obtained solution to an equation that when substituted back into the original equation, does *not* yield an identity. $x = 2$ is an extraneous solution to the equation

$$\frac{x}{x - 2} - 4 = \frac{2}{x - 2}.$$

An extraneous solution is also called an extraneous root.

Factor (0.1) and (5.1) When two or more numbers, variables, or algebraic expressions are multiplied, each is called a factor. If we write $3 \cdot 5 \cdot 2$, the factors are 3, 5, and 2. If we write $2xy$, the factors are 2, x, and y. In the expression $(x - 6)(x + 2)$, the factors are $(x - 6)$ and $(x + 2)$.

Fractions

Algebraic fractions (6.1) The indicated quotient of two algebraic expressions.

$$\frac{x^2 + 3x + 2}{x - 4} \quad \text{and} \quad \frac{y - 6}{y + 8}$$

are algebraic fractions. In these fractions the value of the denominator cannot be zero.

Numerical fractions (0.1) A set of numbers used to describe parts of whole quantities. A numerical fraction can be represented by the quotient of two integers for which the denominator is not zero. The numbers $\frac{1}{5}, -\frac{2}{3}, \frac{8}{2}, -\frac{4}{31}, \frac{8}{1}$, and $-\frac{12}{1}$ are all numerical fractions. The set of rational numbers can be represented by numerical fractions.

Function (7.6) A relation in which no two different ordered pairs have the same first coordinate.

Hypotenuse of a right triangle (9.6) The side opposite the right angle in any right triangle. The hypotenuse is always the longest side of a right triangle. Side AB is the hypotenuse of the triangle below.

Identity (2.1) A statement that is always true. The equations $5 = 5$, $7 + 4 = 7 + 4$, and $x + 8 = x + 8$ are examples of identities.

Imaginary number (9.1) A complex number that is not a real number. An imaginary number can be created by taking the square root of a negative real number. The numbers $\sqrt{-9}$, $\sqrt{-7}$, and $\sqrt{-12}$ are all imaginary numbers.

Improper fraction (0.1) A numerical fraction whose numerator is larger than or equal to its denominator. $\frac{8}{3}, \frac{5}{2}$, and $\frac{7}{7}$ are improper fractions.

Inconsistent system of equations (8.1) A system of equations that does not have a solution.

Independent equations (8.1) Two equations that are not dependent are said to be independent.

Inequality (2.6), (2.7), and (7.5) A mathematical relationship between quantities that are not equal. $x \le -3$, $w > 5$, and $x < 2y + 1$ are mathematical inequalities.

Integers (1.1) The set of numbers . . . , $-5, -4, -3, -2, -1, 0, 1, 2, 3, 4, 5, \ldots$.

Intercepts of an equation (7.2) The point or points where the graph of the equation crosses the x-axis or the y-axis or both. (*See* x-intercept, y-intercept.)

Irrational number (9.1) A real number that cannot be expressed in the form $\frac{a}{b}$, where a and b are integers and $b \ne 0$. $\sqrt{2}, \pi, 5 + 3\sqrt{2}$, and $-4\sqrt{7}$ are irrational numbers.

Isosceles triangle (1.8) A triangle with two equal sides and two equal angles. Triangle ABC is an isosceles triangle. Angle BAC is equal to angle ACB. Side AB is equal in length to side BC.

Least common denominator of numerical fractions (0.2) The smallest whole number that is exactly divisible by all denominators of a group of fractions. The least common denominator (LCD) of $\frac{1}{6}, \frac{2}{3}$, and $\frac{3}{5}$ is 30. The least common denominator is also called the lowest common denominator.

Leg of a right triangle (9.6) One of the two shorter sides of a right triangle. Side AC and side BC are legs of triangle ACB.

Like terms (1.7) Terms that have identical variables and exponents. In the expression $5x^3 + 2xy^2 + 6x^2 - 3xy^2$, the term $2xy^2$ and the term $-3xy^2$ are like terms.

Linear equation in two variables (7.2) An equation of the form $Ax + By = C$, where A, B, and C are real numbers. The graph of a linear equation in two variables is a straight line.

Line of symmetry of a parabola (10.4) A line that can be drawn through a parabola such that, if the graph were folded on this line, the two halves of the curve would correspond exactly. The line of symmetry through the parabola formed by $y = ax^2 + bx + c$ is given by $x = \frac{-b}{2a}$. The line of symmetry of a parabola always passes through the vertex of the parabola.

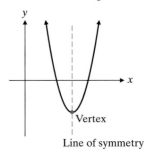

Mixed number (0.1) A number that consists of an integer written next to a proper fraction. $2\frac{1}{3}, 4\frac{6}{7}$, and $3\frac{3}{8}$ are all mixed numbers. Mixed numbers are sometimes called mixed fractions or mixed numerals.

Monomial (4.3) A polynomial of one term. The expressions $3xy$, $5a^2b^3cd$, and -6 are all monomials.

Natural numbers (0.1) The set of numbers 1, 2, 3, 4, 5, ... This set is also called the set of counting numbers.

Numeral (0.1) The symbol used to describe a number.

Numerator (0.1) The top number or algebraic expression in a fraction. The numerator of

$$\frac{x + 3}{5x - 2}$$

is $x + 3$. The numerator of $\frac{12}{13}$ is 12.

Numerical coefficient (4.1) The number that is multiplied by a variable or a group of variables. The numerical coefficient in $5x^3y^2$ is 5. The numerical coefficient in $-6abc$ is -6. The numerical coefficient in x^2y is 1. A numerical coefficient of 1 is not usually written.

Odd integers (1.3) Integers that are not exactly divisible by 2, such as $-3, -1, 1, 3, 5, 7, 9, \ldots$.

Opposite of a number (1.1) Two numbers that are the same distance from zero on the number line but lie on different sides of it are considered opposites. The opposite of -6 is 6. The opposite of $\frac{22}{7}$ is $-\frac{22}{7}$.

Ordered pair (7.1)   A pair of numbers presented in a specified order. An ordered pair is often used to specify a location on a graph. Every point in a rectangular coordinate system can be represented by an ordered pair (x, y).

Origin (7.1)   The point $(0, 0)$ in a rectangular coordinate system.

Parabola (10.4)   A curve created by graphing a quadratic equation. The curves shown are all parabolas. The graph of the equation $y = ax^2 + bx + c$, where $a \neq 0$, will always be a parabola.

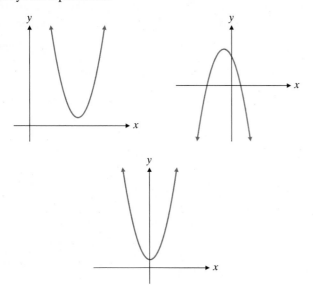

Parallel lines (7.3) and (8.1)   Two straight lines that never intersect. The graph of an inconsistent system of two linear equations in two variables will result in parallel lines.

Parallelogram (1.8)   A four-sided figure with opposite sides parallel. Figure $ABCD$ is a parallelogram.

Percent (0.5)   Hundredths or "per one hundred"; indicated by the % symbol. Thirty-seven hundredths $\left(\dfrac{37}{100}\right) = 37\%$ (thirty-seven percent).

Perfect square number (5.5)   A number that is the square of an integer. The numbers 1, 4, 9, 16, 25, 36, 49, 64, 81, 100, 121, 144, . . . are perfect square numbers.

Perfect square trinomial (5.5)   A polynomial of the form $a^2 + 2ab + b^2$ or $a^2 - 2ab + b^2$ that may be factored using one of the following formulas:

$$a^2 + 2ab + b^2 = (a + b)^2$$

or

$$a^2 - 2ab + b^2 = (a - b)^2.$$

Perimeter (1.8)   The distance around any plane figure. The perimeter of this triangle is 13. The perimeter of this rectangle is 20.

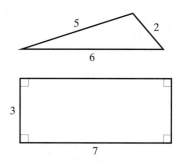

Pi (1.8)   An irrational number, denoted by the symbol π, that is approximately equal to 3.141592654. In most cases 3.14 can be used as a sufficiently accurate approximation for π.

Polynomial (4.3)   An expression that contains terms with nonnegative integer exponents. The expressions $5ab + 6$, $x^3 + 6x^2 + 3$, -12, and $x + 3y - 2$ *are all polynomials*. The expressions $x^{-2} + 2x^{-1}, 2\sqrt{x} + 6$, and $\dfrac{5}{x} + 2x^2$ *are not polynomials*.

Prime number (0.1)   Any natural number greater than 1 whose only natural number factors are 1 and itself. The first eight prime numbers are 2, 3, 5, 7, 11, 13, 17, and 19.

Prime polynomial (5.6)   A prime polynomial is a polynomial that cannot be factored by the methods of elementary algebra. $x^2 + x + 1$ is a prime polynomial.

Principal (3.4)   In monetary problems, the principal is the original amount of money invested or borrowed.

Principal square root (9.1)   For any given nonnegative number N, the principal square root of N (written $\sqrt{N}$) is the nonnegative number a if and only if $a^2 = N$. The principal square root of 25 $\left(\sqrt{25}\right)$ is 5.

Proper fraction (0.1)   A numerical fraction whose numerator is less than its denominator; $\dfrac{3}{7}, \dfrac{2}{5}$, and $\dfrac{8}{9}$ are proper fractions.

Proportion (6.6)   A proportion is an equation stating that two ratios are equal.

$$\frac{a}{b} = \frac{c}{d} \quad \text{where } b, d \neq 0$$

is a proportion.

Pythagorean Theorem (9.6)   In any right triangle, if c is the length of the hypotenuse and a and b are the lengths of the two legs, then $c^2 = a^2 + b^2$.

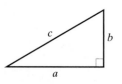

Quadratic equation (5.7) A quadratic equation is a polynomial equation with one variable that contains at least one term with the variable squared, but no term with the variable raised to a higher power. $5x^2 + 6x - 3 = 0$, $x^2 = 7$, and $5x^2 = 2x$ are all quadratic equations.

Quadratic formula (10.3) If $ax^2 + bx + c = 0$ and $a \neq 0$, then the solutions to the quadratic equation are given by

$$x = \frac{-b \pm \sqrt{b^2 - 4ac}}{2a}.$$

Quotient (0.4) The result of dividing one number or expression by another. In the problem $12 \div 4 = 3$, the quotient is 3.

Radical (9.1) An expression composed of a radical sign and a radicand. The expressions $\sqrt{5x}$, $\sqrt{\frac{3}{5}}$, $\sqrt{5x + b}$, and $\sqrt{10}$ are called radicals.

Radical equation (9.6) An equation that contains one or more radicals. $\sqrt{x + 4} = 12$, $\sqrt{x} = 3$, and $\sqrt{3x + 4} = x + 2$ are all radical equations.

Radical sign (9.1) The symbol $\sqrt{}$ used to indicate the square root of a number.

Radicand (9.1) The expression beneath the radical sign. The radicand in $\sqrt{5ab}$ is $5ab$.

Range of a relation (7.6) In any relation, the set of values that represents the dependent variable is called its range. This is the set of all the second coordinates of the ordered pairs that define the relation.

Ratio (6.6) The ratio of one number a to another number b is the quotient $a \div b$ or $\dfrac{a}{b}$.

Rational numbers (1.1) and (9.1) A number that can be expressed in the form $\dfrac{a}{b}$, where a and b are integers and $b \neq 0$. $\dfrac{7}{3}$, $-\dfrac{2}{5}$, $\dfrac{7}{-8}$, $\dfrac{5}{1}$, 1.62, and 2.7156 are rational numbers.

Rationalizing the denominator (9.5) The process of transforming a fraction that contains a radical in the denominator to an equivalent fraction that does not contain a radical in the denominator.

Original Fraction	Equivalent Fraction with Rationalized Denominator
$\dfrac{3}{\sqrt{2}}$	$\dfrac{3\sqrt{2}}{2}$

Rationalizing the expression (9.5) The process of transforming a radical that contains a fraction to an equivalent expression that does not contain a fraction inside a radical.

Original Radical	Equivalent Expression after It Has Been Rationalized
$\sqrt{\dfrac{3}{5}}$	$\dfrac{\sqrt{15}}{5}$

Real number (9.1) Any number that is rational or irrational. 2, 7, $\sqrt{5}$, $\dfrac{3}{8}$, π, $-\dfrac{7}{5}$, and $-3\sqrt{5}$ are all real numbers.

Rectangle (1.8) A four-sided figure with opposite sides parallel and each interior angle measuring 90°. The opposite sides of a rectangle are equal.

Relation (7.6) A relation is any set of ordered pairs.

Rhombus (1.8) A parallelogram with four equal sides. Figure $ABCD$ is a rhombus.

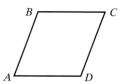

Right angle (1.8) An angle that measures 90°. Right angles are usually labeled in a sketch by using a small square to indicate that it is a right angle. Here, angle ABC is a right angle.

Right triangle (1.8) A triangle that contains a right angle.

Root of an equation (2.1) and (5.7) A value of the variable that makes an equation into a true statement. The root of an equation is also called the solution of an equation.

Scientific notation (4.2) A positive number is written in scientific notation if it is in the form $a \times 10^n$, where $1 \leq a < 10$ and n is an integer.

Slope–intercept form (7.3) The equation of a line that has slope m and the y-intercept at $(0, b)$ is given by $y = mx + b$.

Slope of a line (7.3) The ratio of change in y over the change in x for any two different points on a nonvertical line. The slope m is determined by

$$m = \frac{y_2 - y_1}{x_2 - x_1},$$

where $x_2 \neq x_1$ for any two points (x_1, y_1) and (x_2, y_2) on a nonvertical line.

Solution of a linear inequality (2.7)   The possible values that make a linear inequality true.

Solution of an equation (2.1)   A number that, when substituted into a given equation, yields an identity. The solution of an equation is also called the root of an equation.

Solution of an inequality in two variables (7.5)   The set of all possible ordered pairs that, when substituted into the inequality, will yield a true statement.

Solution to a system of two equations in two variables (8.2)   An ordered pair that can be substituted into each equation to obtain an identity in each case.

Square (1.8)   A rectangle with four equal sides.

Square root (9.1)   For any given nonnegative number N, the square root of N is the number a if $a^2 = N$. One square root of 16 is 4 since $(4)^2 = 16$. Another square root of 16 is -4 since $(-4)^2 = 16$. When we write $\sqrt{16}$, we want only the positive root, which is 4.

Standard form of a quadratic equation (5.7)   A quadratic equation that is in the form $ax^2 + bx + c = 0$.

Subscript of a variable (3.2)   A small number or letter written slightly below and to the right of a variable. In the expression $5 = 2(x - x_0)$, the subscript of x is 0. In the expression $t_f = 5(t_a - b)$, the subscript of the first t is f. The subscript of the second t is a. A subscript is used to indicate a different value of the variable.

System of equations (8.1)   A set of two or more equations that must be considered together.

Term (1.7)   A number, a variable, or a product of numbers and variables. For example, in the expression $a^3 - 3a^2b + 4ab^2 + 6b^3 + 8$, there are five terms. They are a^3, $-3a^2b$, $4ab^2$, $6b^3$, and 8. The terms of a polynomial are separated by plus or minus signs.

Trapezoid (1.8)   A four-sided figure with two sides parallel. The parallel sides are called the bases of the trapezoid. Figure $ABCD$ is a trapezoid.

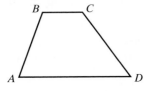

Trinomial (4.3)   A polynomial of three terms. The expressions $x^2 + 6x - 8$ and $a + 2b - 3c$ are trinomials.

Variable (1.4)   A letter that is used to represent a number or a set of numbers.

Variation (9.7)   An equation relating values of one variable to those of other variables. An equation of the form $y = kx$, where k is a constant, indicates *direct variation*. An equation of the form $y = \dfrac{k}{x}$, where k is a constant, indicates *inverse variation*. In both cases, k is called the *constant of variation*.

Vertex of a parabola (10.4)   The lowest point on a parabola opening upward or the highest point on a parabola opening downward. The x-coordinate of the vertex of the parabola formed by the equation $y = ax^2 + bx + c$ is given by $= \dfrac{-b}{2a}$.

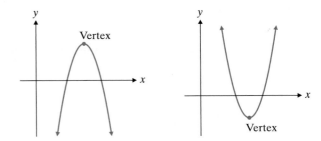

Vertical line test (7.6)   If a vertical line can intersect the graph of a relation more than once, the relation is not a function.

Whole numbers (0.1)   The set of numbers 0, 1, 2, 3, 4, 5, ...

x-intercept (7.2)   The ordered pair $(a, 0)$ is the x-intercept of a line if the line crosses the x-axis at $(a, 0)$. The x-intercept of line l on the following graph is $(4, 0)$.

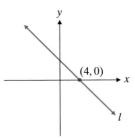

y-intercept (7.2)   The ordered pair $(0, b)$ is the y-intercept of a line if the line crosses the y-axis at $(0, b)$. The y-intercept of line p on the following graph is $(0, 3)$.

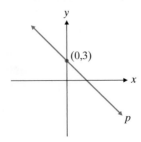

Beginning Algebra Subject Index

Beginning Algebra Applications Index

I-5

Photo Credits

CHAPTER 0   CO Comstock Complete   **p. 32** Getty Images—Stockbyte, Royalty Free   **p. 41** Adam Jones/Photo Researchers, Inc.
p. 60 Comstock Complete

CHAPTER 1   CO Comstock Complete   **p. 69** Peter Skinner/Photo Researchers, Inc.   **p. 102** AP Wide World Photos   **p. 117** John Coletti/Stock Boston   **p. 125** Jeanne White/Photo Researchers, Inc.

CHAPTER 2   CO Courtesy of www.istockphoto.com   **p. 145** Courtesy of www.istockphoto.com   **p. 145** Shutterstock   **p. 178** Photos.com.
p. 180 Photos.com.   **p. 181** Javier Larrea/Pixtal/Superstock Royalty Free

CHAPTER 3   CO Comstock Complete   **p. 203** Courtesy of www.istockphoto.com   **p. 204** Townsend P. Dickinson/The Image Works
p. 206 Courtesy of www.istockphoto.com   **p. 228** Courtesy of www.istockphoto.com   **p. 228** © Adrian Bradshaw/epa/CORBIS All Rights Reserved
p. 232 NASA/Johnson Space Center   **p. 239** Comstock Complete

CHAPTER 4   CO Courtesty of www.istockphoto.com   **p. 265** A. Riess (STSci)/NASA/John F. Kennedy Space Center
p. 293 Courtesy of www.istockphoto.com

CHAPTER 5   CO Courtesy of www.istockphoto.com   **p. 306** Courtesy of www.istockphoto.com   **p. 331** Courtesy of www.istockphoto.com
p. 341 Philip H. Coblentz/World Travel Images, Inc. Royalty Free

CHAPTER 6   CO Comstock Complete   **p. 394** Comstock Complete

CHAPTER 7   CO Courtesy of www.istockphoto.com

CHAPTER 8   CO Wong Tsu Shi/Shutterstock   **p. 506** Marcos Welsh/Art Life Images

CHAPTER 9   CO Alexander Gordeyev/Shutterstock   **p. 518** David Madison/Getty Images Inc.—Stone Allstock   **p. 550** Comstock Complete
p. 559 Photos.com.

CHAPTER 10   CO Comstock Complete   **p. 581** Courtesy of istockphoto.com   **p. 605** Tetra Images/Superstock Royalty Free

PROPERTIES OF REAL NUMBERS

	Addition	**Multiplication**
Commutative Properties	$a + b = b + a$	$ab = ba$
Associative Properties	$(a + b) + c = a + (b + c)$	$(ab)c = a(bc)$
Identity Properties	$a + 0 = 0 + a = a$	$a \cdot 1 = 1 \cdot a = a$
Inverse Properties	$a + (-a) = -a + a = 0$	$a \cdot \dfrac{1}{a} = \dfrac{1}{a} \cdot a = 1\ (a \neq 0)$
Distributive Property		$a(b + c) = ab + ac$

PROPERTIES OF EXPONENTS

If $x, y \neq 0$, then

$$x^a \cdot x^b = x^{a+b}$$

$$\frac{x^a}{x^b} = x^{a-b} \text{ if } a \geq b$$

$$\frac{x^a}{x^b} = \frac{1}{x^{b-a}} \text{ if } a < b$$

$$x^0 = 1$$

$$(x^a)^b = x^{ab}$$

$$(xy)^a = x^a y^a$$

$$\left(\frac{x}{y}\right)^a = \frac{x^a}{y^a}$$

ABSOLUTE VALUE

Definition

$$|x| = \begin{cases} x \text{ if } x \geq 0 \\ -x \text{ if } x < 0 \end{cases}$$

INEQUALITIES

If $a < b$, then for all values of c, $a + c < b + c$ and $a - c < b - c$.

If $a < b$ when c is a **positive number** $(c > 0)$, then $ac < bc$ and $\dfrac{a}{c} < \dfrac{b}{c}$.

If $a < b$ when c is a **negative number** $(c < 0)$, then $ac > bc$ and $\dfrac{a}{c} > \dfrac{b}{c}$.

FACTORING AND MULTIPLYING FORMULAS

Perfect-Square Trinomials	$a^2 + 2ab + b^2 = (a + b)^2$ $a^2 - 2ab + b^2 = (a - b)^2$
Difference of Two Squares	$a^2 - b^2 = (a + b)(a - b)$
Sum of Two Squares	$a^2 + b^2$ *cannot be factored*

PROPERTIES OF LINES AND SLOPES

The **slope** of any nonvertical line passing through (x_1, y_1) and (x_2, y_2) is $m = \dfrac{y_2 - y_1}{x_2 - x_1}\ (x_1 \neq x_2)$.

The **standard form** of the equation of a straight line is $Ax + By = C$.

The **slope–intercept form** of the equation of a straight line with slope m and y-intercept $(0, b)$ is $y = mx + b$.

A **horizontal line** has a slope of zero. The equation of a horizontal line can be written as $y = b$.

A **vertical line** has no slope (the slope is not defined for a vertical line).

The equation of a vertical line can be written as $x = a$.